Eco-dynamics and Environmental Perturbations in Mangrove Ecosystems

Susanta Kumar Chakraborty
Sankar Kumar Acharya

Eco-dynamics and Environmental Perturbations in Mangrove Ecosystems

Resilience, Adaptation, and Mitigation for Sustainable Conservation of the Indian Sundarbans

Springer

Susanta Kumar Chakraborty
Department of Zoology
Vidyasagar University
Midnapore, West Bengal, India

Sankar Kumar Acharya
Department of Agriculture Extension
Bidhan Chandra Krishi Viswavidyalaya
Mohanpur, West Bengal, India

ISBN 978-3-032-08709-6 ISBN 978-3-032-08710-2 (eBook)
https://doi.org/10.1007/978-3-032-08710-2

This Springer imprint is published by the registered company Springer Nature Switzerland AG
The registered company address is: Gewerbestrasse 11, 6330 Cham, Switzerland

Dedication

This book is dedicated to the committed environmental activists and down to earth people of Sundarbans on the Indian subcontinent who have been seamlessly and selflessly struggling towards the restoration and conservation of one of the world's most biologically productive, economically viable and ecologically unique ecosystems alongside acknowledging so many brilliant, belligerent and beautiful minds who selflessly delved into the details of an internationally acclaimed mangrove-estuarine complex through their commitment to theoretical and action-based research efforts.

Preface

Mangrove ecosystems are amalgamations of two subsystems, mangrove forests and adjoining brackish-water estuaries, which are in continuous interaction mainly through diurnal tidal processes that exchange necessary materials between the two for the sustenance and proper functioning of both subsystems. The mangrove forest subsystem is endowed with the unique power of its adaptability against the intense action of periodical tidal waves during both high and low tides, such that it thrives in the fluctuating salinity and other ecological factors operating on them through the different seasons of a year. Moreover, this unique ecological development in a tropical coastal environment, often being the counterpart of temperate salt marshes, provides many ecosystem services for the benefit of humans in the form of plant biomass (timber, fodder, thatching materials, shrimp, honey, fishery resources, etc.), as the protector of coastal belts by resisting soil erosion and acting as a buffer against natural hazards like cyclones and floods, as a bio-remediator by accumulating persistent pollutants and as a source of aesthetics, mainly because of its biodiversity, wealth, which supports the development of ecotourism.

Researchers have a moral responsibility to refrain from ravaging sensitive and fragile ecosystems like mangrove ecosystems and all the organisms living there, by not projecting local information but instead highlighting innovations that will ensure the protection and preservation of the diversity of species and their habitats by mitigating environmental harm. Part of that ensures the sustainable use of natural resources while adhering to the basic scientific principles of ecology and involving local people in such a way as not to deprive future generation from benefiting from the gifts of nature.

This book is an attempt to highlight the ecological uniqueness, biodiversity, threats, resilience, adaptation and mitigation efforts towards sustainable conservation following an integrated, comprehensive and holistic approach to unravelling the mysteries of the interrelationships among the ecological parameters, processes and resilience of the mangrove ecosystem of Sundarbans, India, emphasizing the society-driven adaptation and mitigation measures against environmental perturbations and pollution and towards achieving the goals of sustainable conservation for the entire mangrove ecosystem in general and its biodiversity in particular.

This book explores the prospects and role of proper environmental management to document the ecomanagement strategy of the mangrove ecosystem of Sundarbans, India, in view of ecological changes due to global warming–mediated climate change. It introduces the basic scientific principles of this ecosystem, traditional resource management practices counteracting modern ones, the ongoing stresses from pollution and the adaptation of the local people adhering to integrative and multidisciplinary approaches that aim to assess the risks, hazards and impacts not only on the structural components of the Sundarbans mangrove ecosystem but also on the basic needs of the people residing in and near this unique ecoregion and on the innovative adaptive and conservation strategies which are now being used to overcome such problems while achieving the goal of sustainable development.

The Sundarbans mangrove ecosystem at the saline waterbody–land interface as a transitional ecozone is endowed with a network of large estuaries and numerous creeks of different dimensions, and the mangrove forest, with its floral and faunal diversity, renders considerable ecological goods and services. These ecological services have been under human-mediated environmental threats that have tended to jeopardize natural biogeochemical cycles and disrupt the beneficial contributions of this ecosystem to humans leading to the deterioration of the ecological conditions of said ecosystem, which in turn ruins its regional, national and global biodiversity. The uniqueness of mangrove ecosystems has prompted ecologists, environmental planners and administrators to undertake proper ecorestoration measures by adhering to basic ecological principles to restore the ecodegraded mangrove ecosystem while adopting appropriate conservation strategies for sustainable development through a judicious understanding of resource availability and modes of resource use. The gathering of information through systematic research both by formal researchers and through community outreach programmes can help learn about the ecological potential of the concerned ecosystem and find prospects for the ecological and economic development of the environment.

Moreover, the multidimensional aspects of mangrove research revolves around the ecodynamics of the Sundarbans mangrove ecosystem on recognizing it as one the most precious resources in the world. The taxonomic diversity, biological adaptability and ecological interactions of this ecosystem render ecosystem services – despite environmental pollution and ecological perturbations – providing benefits to local people with mangrove forest–derived ecological goods and natural products, thereby paving the way for livelihood generation for local people. This ecosystem is subjected to multiple stress factors, both from anthropogenic and natural sources; ecological chaos and butterfly models; ecological and social resilience; social entropy; and various protection, preservation and conservation strategies. Against the backdrop of such research discourses, the prime challenge for the ecomanagement of this mangrove ecosystem depends not only on conventional research activities but also on developing innovative research ideas and methodological tools, including through community mobilization. Methodological innovations like ecological surveying for harvesting ground truth research information, participatory rural appraisal (PRA), applications of management information systems (MISs), decision support systems (DSSs), the blending of the human brain and artificial intelligence (AI), disaster forecasting and responses, the use of the global positioning system (GPS), adaptive comanagement by evolving collective and integrated resource management policies, integrated policy development issues pertaining to community mobilization and stewardship with their implications for mangrove ecosystem restoration and revitalization are parts of sustainable conservation.

The biodiversity of the mangrove ecosystem, catering to the basic needs of humans in terms of food, medicine, clothing, shelter materials, etc., offers multifarious ecosystem services, which have prompted resource managers to attach more importance to the incorporation of socioeconomics in evaluating the roles of biodiversity in a given ecosystem over the past century. However, biodiversity loss has since emerged as one of the major environmental issues at regional and global scales, which has necessitated developing methods for the assessment, monitoring and holistic management of such biodiversity in line with the goals of sustainable development. Therefore, ecomanagement for conserving biodiversity is possible only by integrating the scientific and socioeconomic components that can safeguard the biodiversity of natural habitats such as mangroves and pave the way for sustainable conservation. The momentum of urbanization and industrialization over the past centuries has brought about many changes in contemporary human lifestyles, which have disrupted the balance of cause and effect beyond the assimilation capacity of nature. Counteracting all the undesirable changes in the natural environment due to modern technology–driven industrial society gathered information on the ongoing deterioration of the environment in the recent past are required to be used.

The pace of such developments in India and other developing countries has had considerable harmful effects on the functioning of mangrove ecosystems. The Sundarbans mangrove ecosystem includes a vast expanse of green mangrove vegetation, a network of several major estuaries and hundreds of creeks and tributaries, though this area has suffered an unprecedented loss of resources over several decades according to statistics. The green coverage of this mangrove environment has been considerably reduced through the harvesting of plant resources, acidification due to increases in temperature, increasing turbidity and water hardness because of the accumulation of eroded products, salinity invasion, the overenrichment of water with nutrients (leading to eutrophication due to leakage and the mixing of water with sewage from cities), wastewater from industries and agricultural activities with the deposition and loading of gases and particulates from the exhaust of cars, furnaces and factories. All such problems can be tackled by better understanding the ongoing problems, emphasizing the prospects of using the Sundarbans mangrove ecosystem to more efficiently adapt to dumped waste, accumulated pollutants with toxic substances and the prospects of drafting mitigation strategies that integrate political will with appropriate economic means.

Information pertaining to ecological potential, economic use, disruptions to natural ecodynamics, increasing pollution and ecodegradation, leading to a drastic decline in the extent and biodiversity of the Sundarbans mangrove ecosystem is needed more than ever to face regional, national and global challenges through research. Sustainable environmental management renders a wide range of economic, environmental and social benefits. Attributes promoting water-quality monitoring, waste recycling and reuse, biodiversity conservation, pollution abatement, etc. tend to improve ecosystem services by requiring conserving ecological reserves which are supposed to have synergistic impacts on the respective eco-zones and bring holistic benefits by arresting environmental qualities degradation alongside improving environmental by way of the development of needed infrastructure. Appropriate ecomanagement processes certainly reduce economic losses and create green jobs in livelihood generation sectors such as ecotourism and sustainable fisheries in ecoregions under conservation processes. Conservation strategies adhering to basic ecological principles tend to protect and conserve the diversity of habitats and their species and ultimately help achieving sustainability through natural resource management, which is expected to open up options for keeping the natural resources within the carrying capacity of the ecosystem so that future generations may not be deprived of benefiting from the ecoregion's wealth of ecosystem services. Ecomanagement through proper ecological monitoring and governance is important for the conservation of mangrove ecosystems, where multiple pressures are making this sensitive and fragile ecosystem a threatened landscape. Such pro-environmental efforts together render holistic benefits to the humans by reducing economic losses due to environmental degradation alongside creating green jobs such as sustainable fisheries and ecotourism.

Readers are expected to be inspired and enriched by the multidimensional research inputs pertaining to mangrove ecosystems in general and the eco-biological uniqueness of the structural components of this dynamic ecosystem in particular. The publication of this book aims to attract, persuade and inspire readers by enriching them through discussions on the multidimensional research pertaining to the ecodynamics of mangrove ecosystems in general and the eco-biological uniqueness of the structural components of this unique ecosystem in particular. The major objectives of this book are to pool some salient and relevant ecological concepts and principles pertaining to mangrove ecosystems, highlighting interactions among the structural components of this ecosystem that make it among the most productive, economically viable and aesthetically celebrated ecosystems in the world. This book also takes into consideration all kinds of ongoing environmental perturbations imposing multidimensional stresses, including contributing factors, ecological governance, management and conservation against the backdrop of global ecological perspectives and local environmental conditions and of encouraging the active involvement of end users and governments.

The chapters of this book incorporate existing knowledge bases and the research outcomes of the research activities of authors on biodiversity, environmental history, ecology, pollution, eco-monitoring and conservation, mostly based on people's perceptions of mangrove ecosystems The target readers of this book include many people in academia (students, researchers and teachers), environmental planners, administrators, decision makers, local people and deep thinkers.

The prime objective of this book is to attract the attention of the international community not only to acquaint them with the diversity and aesthetic appeal of mangrove ecosystems but also to project salient ideas and underlying scientific principles to protect the functional components of this sensitive, fragile but ecologically productive mangrove ecosystem, highlighting their present ecological status, ongoing perturbations, contributing stress factors and modes of management, governance and conservation. This kind of endeavor will fall under various environmental and managerial categories with the active involvement of end users, the government, the private sector, civil society regional and international, beneficiaries, stakeholders and local people.

The book, titled "Ecodynamics and Environmental Perturbations in Mangrove Ecosystems: Resilience, Adaptation and Mitigation for the Sustainable Conservation of the Indian Sundarbans (a World Heritage Site)," was robustly conceived and became well organized into 10 chapters, of which the first six (Chapters: 1, 2, 3, 4, 5 and 6) explain the basic facts of the ecology, ecodynamics, environmental history, physiography, biodiversity and ecosystem services of the mangrove ecosystem in India. Chapter 7 explains the facts, figures and interpretations of environmental perturbations and pollution in the Sundarbans mangrove ecosystem alongwith their prospective remedies. Chapters 8 and 9 include an in-depth discussion on prospects of resource management and the methods used to address the problems of resource management, ecological chaos, ecological resilience, social entropy, adaptive comanagement, community mobilization and method-related innovations in protecting this ecologically viable, sensitive, biologically productive and aesthetically celebrated ecosystem through sustainable conservation. All of these 9 chapters are followed by an extensive conclusion as Chapter 10.

This book appears to fill the existing lacunae in mangrove research, going beyond current knowledge bases and thereby opening up new avenues in the understanding of the intricacy of the ecodynamics and ecological potential of the mangrove ecosystems of the world in general and the Sundarbans mangrove ecosystem in India in particular. The publication of this book is also expected to incorporate academic research information on sustainable ecological management into an interdisciplinary approach that aims to formulate frameworks, policies and perceptions for successful conservation. No single book can include in-depth analyses of every detail of the science of ecology and the methods of conservation for mangrove ecosystems especially considering the rapidly changing perspectives on how to analyse ecosystem functioning in different cultural contexts and for different policy priorities.

This book can be an ideal platform for students, researchers, social workers and politicians, naturalists, environmental planners and administrators around the globe to share their perspectives on a unique ecosystem, the mangrove ecosystem. This book is the product of interdisciplinary efforts and has tried to bring together research communities from different disciplines and practitioners at different levels to share experiences and research outcomes for the formulation of proper frameworks, ecomanagement policies and strategies for the successful conservation of this unique ecosystem.

This book is being considered unmatched when compared with other available books on the same subject, namely the Sundarbans mangrove ecosystem, a World Heritage Site, or any kind of research on it around the world. Readers across the globe may be waiting for a robust book which covers ecology, economy and sociology of this complex mangrove biosphere, which governs the dynamics of a huge geographical area, consisting of the ocean, seas, estuaries, coastal zones, mangroves and deltaic islands with various human activities, including agriculture on century-old alluvial deltas with high-intensity cropping sequences utilizing traditional knowledge bases of local people.

The existence, growth and regeneration of any robust and complex ecosystem has to confront, comply and reconcile with three sub-ecosystems experiencing interconnected ecological dynamics: core, periphery and far periphery. The Sundarbans mangrove ecosystem is no exception. The existence, growth and regenerative behaviours, dimensions and trajectories of any ecosystem depend on the kinetic interplay between these ecological factors.

The healthy existence, expanse and ecodynamics of the Sundarbans mangrove ecosystem render valuable ecosystem services that protect neighbouring cities and semi-urbanized areas against cyclones, which would lash off the life process of a city such as Kolkata, and they support local people living in the core and periphery sub-ecosystems. Therefore, the urban settlements nearer Sundarbans are the recipients of migrant populations, both humans and other animals, forest ecosystem.

The transformation of forest lands into agricultural lands may seem beneficial early on creating a backup for sustenance in the form of food, livelihood and the rural economy, but in the long run, there would be serious problems and threats contributed by pollutants, heavy metals, pesticides, microplastics, sewage, the destruction of natural flora and fauna, the ecodegradation of land, etc., and these tend to generate both ecological chaos and social entropy. The seamless destruction of mangroves, the extraction of shrimp and fin fish like hilsa, the mindless aggression of new settlements that ignore the subtle ecological fabric of the ecoregion will lead to ecological chaos and disruption. Increasing incidents of human–wildlife conflicts, the intensification of human trafficking (including in- and outmigration), the engulfing of croplands by a mantle of salts (leading to crop failures) and an increasing number of cyclones every year are adding to the ecological entropy in this area.

Readers are also expected to get benefit by developing their own understandings of the underlying scientific principles of the ecodynamics of the Sundarbans mangrove ecosystem and by building their confidence in working out potential applications in the field by not only going through the text strengthened by photographic documentation but also following up relevant references. This single book ambitiously aims to impart knowledge bases on the multidimensional aspects of ecology and the practice of conservation in relation to mangrove ecosystems, especially in view of rapidly changing and emerging perspectives on the ecology, biodiversity and methodological innovations on conservation practices of the threatened landscape mangrove ecosystem, by following interdisciplinary and holistic research approaches emphasizing on changing–cultural, economic and political perspectives.

Midnapore, West Bengal, India — Susanta Kumar Chakraborty
Mohanpur, West Bengal, India — Sankar Kumar Acharya
20th May, 2026

The Book and Its Rationale

Sundarbans features the splendour of the biodiversity and euphony of ecology on this planet. Perhaps being the world's most dynamic and evolving ecosystem, it spans over 10,000 km^2 across India and Bangladesh, of which India occupies around 40% of its geospatial distribution. This active deltaic region is among the largest in the world, measuring about 40,000 km^2 and generates all the components of an epic biosphere. Over centuries, it has drawn global attention from those who want to witness the grandeur of the Bengal tiger and mystic skylines of deep, dense, delightful mangroves. The estuarine crocodile, gangetic dolphin, olive ridley turtles, the delicious shrimp and crabs, and everything beyond the traditional community, struggling against and overcoming odds, are the phenomenal attractants of Sundarbans.

Sundarbans is majestic and replete with floral and faunal supremacy by count and diversity, yet the existence of Sundarbans has been under serious threat, especially over the past couple of decades. It has become vulnerable along its terrain and trajectories due to sea-level rise, an abrupt increase in the intensity of cyclonic storms, a sharp decline in the density of mangrove plants, the erosion of muddy embankments and the scaling up of social entropy and ecological chaos. Unplanned settlements following mass migration have led to the conversion of cropland into *bheris* for shrimp and prawn cultivation and poor periphery management of forest, which are erasing the prospects of this heritage site and eventually driving it towards doomsday. Perhaps needs and certainly enormous greed are identified as to be inextricably responsible for this decline, destruction of the region's rich biodiversity and mellifluent ecodynamism.

There has been no dearth of books, chapters and research articles being published almost every day on ecology, environment, pollution and conservation, of mangroves and coastal zones but what the world needs a comprehensive book that reflects how human motivations on mangroves are orchestrated and interwoven. This book has uniquely combined people's actions towards ecological stewardship to inquire into real and deeper meanings of human life with an emphasis on, studying the pedagogy of ecology through their struggle for existence. There is a dynamic interplay between the core, periphery and far periphery sub-ecosystems in the expanse and continuum of the Sundarbans mangrove ecosystem. Natural ecodynamisms of Sundarbans can be retrieved and saved by the application of participatory learning and action (PLA) techniques, not by insipid prescriptions made by self-proclaimed experts, and this book aims to incite people-based pro-environmental movement and to equip them with real-time data, value-added information and actionable dicta.

The people of the world are urgently in need of global interventions to save and restore both of the two gigantic ecosystems of the world, Amazon and Sundarbans, simply because any sort of ecological disruption to a holon of an ecosystem spans and engulfs other ecosystems on this planet, according to Lorenz's butterfly effect theory. The book is to be considered serendipitously exceptional in its global vision for ecorestoration and ecological stewardship approaches.

Numerous illustrations, graphs, photos, models and community incubation data from empirical studies over several decades have been splendidly furnished in this book, with citations and references to attract a global audience. In addition to its structural elegance, this book will provoke millions of minds by having them reflect on the crisis and encouraging them to engage in the restoration and exuberance of the beautiful Sundarbans of the Asian Sub-Continent, shared jointly by two neighboring countries India and Bangladesh.

Acknowledgements

Mangrove ecosystems in the tropical and subtropical regions of the worlds were developed by the continuous interactions of estuarine and mangrove forest subsystems and represent an ecologically sensitive, economically viable and aesthetically celebrated ecoregion. Located at the mangrove forest and estuarine interphase, characterized by tidal interplays and oscillating ecological settings, cover an area of around 20 million hectares across the world, of which Sundarbans mangrove ecosystem, shares two third portions with is the largest expanse of mangroves in the world, jointly shared by both India and Bangladesh.

This textbook dealing with the multidimensional aspects of mangrove ecosystems of the world in general and the Sundarbans mangroves in particular aims to generate proper conservation strategies for the protection and preservation of this fragile mangrove ecosystem in West Bengal, India, a World Heritage Site which is considered as a precious gift of nature with unique ecological characteristics, such as the only tidal forest with Bengal tigers, *Panthera tigris tigris*. This book had the privilege of having the support and assistance of several people in the process of writing it. The authors thank both the authorities of Vidyasagar University, at Midnapore (West), West Bengal, India, and Bidhan Chandra Krishi Viswavidyalay at Nadia, West Bengal for their support in facilitating the completion of the writing of this book. The authors gratefully recognize and thank all senior advisers, junior researchers and well-wishers for their valuable support throughout the entire journey of writing, from its conception to its subsequent crystallization and its gaining the approval of the publishing house Springer Nature, U.S.A.

The first author is privileged to acknowledge the research-sponsoring agencies: the University Grant Commission (UGC); the Department of Environment and Forest; the National Agriculture Technology Project (NATP) of the Indian Council of Agriculture Research (ICAR); the West Bengal Pollution Control Board; under the Government of West Bengal. I must gratefully acknowledge all the extraordinary people associated with Springer, especially Aaron, Clement and Meenahkumary, who by virtue of their sincere and considerate support acted as the driving force for writing this very ambitious book, covering multidimensional aspects of the mangrove ecosystem in India in general and Sundarbans mangrove ecosystem in particular. We would fail in our duty if we did not name some of the research students of the first author, Professor Susanta Kumar Chakraborty, such as the late Dr Sunirmal Giri, Dr Manoj Patra, Dr Ranjit Khalua, Dr Anupam Chandra, Dr Siddhartha Mishra, Dr Prasenjit Pradhan, Dr Mrinal De, Dr Diptiman Ghosh, Dr Tridip Datta, Dr Manjistha Bhattacharya, Dr Gurudas Chakraborty, Dr Samaresh Samanta, Dr Sudipto Ghorai, Dr Tapas Maity, Dr Debdas Jana, Dr Sumit Giri, Dr Subhasish Chatterjee, Dr Kishalay Paria, Dr Sujoy Middya, Dr Ram Ganguli, Dr Tilak Das Dr Arundhati Ganguli, Dr Munmun Ghosh, Dr Sankarson Roy, Dr. Nirmalaya Shee, Md Dr Abdullah Al-Helal (Khokon), Santu Paria, Anindita Das, Jayanti Roy, Joydev Shyamal and one nonteaching staff, Jagadish Mahata, for their active support and assistance throughout the entire process of preparing this book. The manifold support from the research students of the second author, such as Dr Arnab Banerjee, Dr Monirul Haque, Dr Amrit Sarkar, Dr Riti Chatterjee, Dr Swagata Patra, Dr Kabita Mondal and Dr N K Sharma deserves special mention.

The idea of writing this book on the Sundarbans mangrove ecosystem was conceived as this sensitive, fragile, ecologically viable, biologically productive and aesthetically celebrated ecosystem that has been under tremendous anthropogenic and natural stresses over the past century. Global climate change–mediated environmental perturbations like sea-level rise, land subsidence, salinity invasion and soil erosion have pushed many indigenous and endemic species of flora and fauna to the brink of extinction and cause the ecodegradation of the entire ecosystem. This book encompasses interpretations on the decadal ecological changes in the structural components and ecological parameters of this ecosystem, highlighting the basic fabric of ecosystem functioning, modern approaches to conservation strategies (resting on the stewardship of community involvement) and adaptive capacity while adhering to ecomanagement strategies or a hybrid concept featuring people-centric conservation efforts. All the chapters of the book are based on the existing knowledge base of the subject from international perspectives which have been blended with the authors' research experience in this field, citing several case studies and using nontechnical jargon to reach a wider audience. The target readers of this book are expected to include students of both undergraduate and postgraduate courses in almost all the disciplines in the physical, chemical, biological, environmental and social sciences; environmental planners; technicians; administrators; decision makers; and deep thinkers.

The authors are grateful for the help of Dr Ratnadeep Ray, Assistant Professor of Remote Sensing and Geospatial Information System (GIS) at Guru Nanak University, Kolkata, who prepared some relevant maps with remote-sensing and satellite images and who elaborately wrote the trend in decadal changes in ecological settings with those maps. Authors are greatful to Dr Dhriti Banerjee, the Director of Zoological Survey of India (Z.S.I.) for providing several useful publications of that century old prestigious research institute and also lending support for the proper identification of faunal materials. Authors are also thankful to Dr Gurupada Mondal, Joint Director, Zoological Survey of India for providing some useful information of new discoveries of biodiversity components in Sundarbans, India. Thanks are due to Dr GourGopal Maity, Retired Professor of Botany,

Kalyani University and Dr Ram Kumar Bhakat, Retired Associate Professor of Botany, Vidyasagar University for their valuable help identifying the plant samples. Active support rendered by Dr Sabyasachi Majumder, research scientist of marine biology, especially on foraminifera is thankfully acknowledged. Some unique photographs, especially of tigers in Sundarbans, were provided by Dr Anirban Ash, Associate Professor of Zoology, Burdwan University in Burdwan District, West Bengal, India. Mrs. Jhuma Chakraborty, Associate Professor of Philosophy, Tamralipta Mahavidyalay, Tamluk, in Midnapore (East) District, West Bengal, India, helped a lot in arranging numerous photographs and preparing some flowcharts for this book. Thanks are due to Dr. Asish Kumar Paul, Retired Professor of Geography, Vidyasagar University, Dr Santosh Kumar Sarker, Retired Professor of Calcutta University and Dr Phani Bhusan Ghosh, Former, Researcher of Wetland Research Institute, Kolkata for providing references and photos.

Although an author writes a book, if that book is a textbook, then it does not or should not belong to the author alone. Thousands of pieces of information generated by a multitude of research activities have been used in one form or another to enrich the subject of each and every chapter of this book. Therefore, writing a research-based textbook is not as easy as it appears to be, because it needs not only to highlight the current research on the relevant subjects but also to present a comparative view that includes earlier research outcomes undertaken by different researchers from different corners of the globe. In the end, the authors are thankful for and indebted to their family members, who have withstood all kinds of problems because of the authors' inadvertent absence and dissociation from daily affairs, without making complaints. Thanks are due to Mrs. Jhuma Bhattacharrya, Associate Professor of Tamralipta Mahavidyalaya for generouly helping arranging different photographs and in preparing several schematic diagrams.

Competing Interests The authors have no competing interests to declare that are relevant to the content of this manuscript.

Contents

About the Authors

Susanta Kumar Chakraborty is a former Professor of Zoology, Vidyasagar University, has been in active research since 1982 after obtaining his M.Sc in Zoology followed by M.Phil in Environmental Science in 1986 and Ph.D. in Marine Science in 1992 (all are from University of Calcutta). His research interests centred around on ecology and environmental management of coasts, mangroves, freshwater wetlands, rivers and estuaries, with an emphasis on biodiversity, rural biotechnology, ecosystem dynamics and environmental management. His active field based research activities on Sundarbans during last four decades in collaboration with several National and Regional Scientific agencies have several scientic documents including a comprehensive manuals on research methodologies mannuals. So far, he has successfully supervised the PhD thesis of 45 researchers, published more than 200 research papers in well-reputed international and national journals, achieving citations of more than 5000, h-index of 32 and i-10 index of 128, got international patents, written a number of books (including the text book *Riverine Ecology* (volumes I and II) and *Wetlands Ecology* published by Springer Nature, USA), completed 15 research projects, sanctioned by National and International sponsoring agencies, delivered more than a hundred talks as key note speaker, inaugurator, plenary speaker etc. across the world, received a number of prestigious awards, and shouldered a number of administrative responsibilities (as the Vice chancellor and Dean of the Faculty of Sciences, the Head of a department, a member of the

Executive Council, a member of the Finance Committee, as President of governing bodies, etc. and as Expert member at other universities and several government departments, like University Grant Commission (UGC), Indian Council of Agricultural Research (ICAR)). He has received a number of awards in recognition his research commitments.

Sankar Kumar Acharya, former Professor, Head of the department, Director and Dean of Agricultural Extension at Bidhan Chandra Krishi Viswavidyalaya (BCKV), Mohanpur, WB. His career was started as an Assistant Professor at BCKV in 1988 and has been in teaching, research and extension for over 38 years. An erudite teacher as well as an elegant speaker, he is internationally acclaimed for his unique research domain of social entropy and energy metabolism, social ecology and environmental sociology, enterprise ecology frameworks and conservation stewardship. He has published 253 research papers in national and international journals, supervised 26 PhD theses, earned several prestigious awards and authored 130 books. He served as an expert member for the Agriculture Commission and its Marketing and Extension Subcommittee for the Government of West Bengal; in World Wild Life Fund (WWF) projects for the Buxa Tiger Reserve Project; and at the University of Manchester on hunger, poverty and silence. He visited countries such as Italy, France, Germany, China, Sri Lanka and Bangladesh in connection with his academic commitments. He is a fellow of the West Bengal Academy of Science and Technology (WAST) and indian Society of Extension education (ISEE), New Delhi. He has a rich profile of editorial and reviewer contributions for national and international journals.

Chapter 1
Introduction

Abstract There is no definite proof when and why the term **'Mangroves'** first came into use in the scientific literature. It is probably derived from the Portuguese word **'Mangue'** meaning mangrove trees, or mangrove shrubs and the English **'Grove'** was added. The assemblage of mangrove plants along with other associated biotic components is called **'Mangal.'** With regard to the naming of a great landscape of the **'Sundarbans.'** several opinions have been put forward, the most acceptable of which is one that this name has been derived from the local name **'Sundari,'** for trees, scientifically designated *Heritiera fomes*, which was found in the Sundarbans mangrove forest in higher abundance around couple of centuries back. The other opinion is that as this mangrove forest looks very beautiful (which locally means **'Sundar'** in both Hindi and Bengali) and forest in the local language is **'Bon.'** Another proposition is that this mangrove forest is very close to the sea, locally named **'Samudra**.'

Most of the coastal zones are seen to be fringed with mangrove swamps (characteristics of tropical countries) and salt marshes (grown in the cooler environment of temperate countries), dotted with dense or scattered patches of halophytic vegetation. The Sundarbans mangrove ecosystem of India, being biologically productive, having ecological potential, and being aesthetically solemnized, and the only tidal forest with tiger, was declared a Tiger Reserve under Project Tiger in the 1973, National Park in the 1978, Biosphere Reserve in 1989, and the status of **'World's Heritage Site'** in 1987. The entire eco-region, characterized by the vast expanse of mangrove forest and intertidal mudflats among 104 deltaic islands, offers suitable habitats and an ecological niche for so many forms of organisms such as planktons (phytoplankton and zooplankton), benthos, nektons, benthic algae, and microbes, and is crisscrossed by a complex network of tidal estuaries and creeks. Through very complicated ecological processes interlinking two sub-systems composing the mangrove ecosystem, the forest subsystem and brackish water subsystems, several forms of natural resources such as food (fishery resources, honey), fodders, timbers, and leaves for constructing human shelters, and medicinal products are generated and render profuse benefits of human beings in a profuse manner. Owing to the development and action of so many human-mediated as well as natural ecological perturbations and environmental pollutions, this very ecosystem along with its rich

S. K. Chakraborty, S. K. Acharya, *Eco-dynamics and Environmental Perturbations in Mangrove Ecosystems*,
https://doi.org/10.1007/978-3-032-08710-2_1

biodiversity is now under the threat of eco-degradation, warranting for taking up proper conservation measures by formulating an appropriate framework for conservation strategies in tune with the guidelines of the *Convention on Wetlands of International Importance* (The Ramsar Convention), which was adopted in the city of Ramsar in Iran in 1971, which came into force in 1975. The mangroves are important for the local people, who, by developing their adaptability in the face of ongoing environmental odds, have been playing pivotal roles in safeguarding this gift of nature for their own sustenance on the one hand and for the overall maintenance of the entire environment on the other.

Keywords Sundarbans · Mangroves · Mangals · Ramsar Site · 'World's Heritage Site' · Project Tiger · Conservation

1.1 Basic Idea of the Coastal Zone and Sundarbans, India

The coastal zone, a junction of sea with land, extends in both seaward and landward directions where it includes the part of the sea that is influenced by the adjoining land mass and the part of the terrestrial environment that experiences direct or indirect impacts from the nearby sea. The coastal zone is characterized by having many unique geomorphological structures such as mangroves, corals, estuaries, creeks, intertidal rocky, mud, or sand flats, varied forms of brackish water, and backwater impoundments and dunes. All such geo-physical components have been undergoing tremendous changes over millennia because of anthropogenic activity. Being a transitional eco-zone at the interface of sea and land, it represents a dynamic natural system where the lithosphere, hydrosphere, and atmosphere always interact after being regulated by climatic and geological processes.

The ecological status of a specific coastal landform depends on the energy inputs into the coastal zone, utilizing which different inorganic and organic materials are formed (Pethic, 1992). The existing wealth of the coastal environment is thereby generated by the continuous interactions and adjustments of all of its structural components with the ongoing environmental changes over the geological and historical periods in the context of regional physiography, which ensure ecological stability, maintaining a dynamic equilibrium on temporal and spatial scales (Macnae, 1968; Kathiresan, 2001; Kathiresen and Bingham, 2001; Odum and Heald, 1975; Titus, 1988; Woodroffe et al., 1993).

The Sundarbans, located in the mangrove estuarine complex in the coastal belts of West Bengal, India, endowed with vast expanse of green mangrove forest is structured by perennial halophytic plants and the entire landscape is crisscrossed by innumerable tidal creeks and channels. The most common belief of the origin of Sundarbans involves a series of tectonic activities causing elevations and depressions of the gneissic hills that were present in the Gondwana belt around 200 million years ago. This resulted in the association of the gradual deposition of alluvial silt brought down from the folded Himalayas in the Northern parts of India throughout the centuries, which ultimately resulted the in the formation of these lower Gangetic deltaic islands.

The Sundarbans mangrove ecosystems, being one of the largest mangrove ecosystems in the world, has a complicated estuarine network encircling 104 deltaic islands, 54 of which have been reclaimed and inhabited by a highly dense human population. The area and extent of coverage of the Indian Sundarbans deltaic system shows that it is bounded by the Ichamati–Raimangal River in the east, by the Hooghly River in the West, by the Bay of Bengal in the south, and by the Dampier-1 Hodges line drawn in 1829–1830 in the north separating two mangrove-based ecosystems in two neighboring countries, Bangladesh and India. The total area of the Indian Sundarbans region is about 9630 km^2, of which the Reserve Forest occupies nearly 4260 km^2.

The entire region spreads over two administrative districts, namely South 24-Parganas (13 blocks) and North 24-Parganas (6 blocks), whereas Midnapore (East) District, the coastal tract of which was earlier covered by luxuriant mangrove vegetation and presently represents a contiguous part of the main Sundarbans. The Hooghly–Matla estuarine complex consists of ten major estuaries such as the Hooghly (Ganges and Bhagirathi), Mooriganga, Saptamukhi, Thakuran, Raimangal, Bidyadhari, Thakuran, Haribhanga, Matla, and Ajmalmari.

In 1973, a large portion of the Indian Sundarbans was declared the Sundarbans Tiger Reserve and in 1984, the status of a National Park was given. In 1987, UNESCO inscribed the Indian Sundarbans in the World Heritage List. In 2001, the entire Indian Sundarbans area was recognized by UNESCO as a Global Biosphere Reserve. Since 1977, four different wildlife sanctuaries were declared within the Indian Sundarbans. This deltaic tidal mangrove ecosystem is the only tide-dominated swampy area in the world inhabited by the threatened Royal Bengal tiger. Sundarbans mangrove forests, a contiguous landmass of both India and Bangladesh, have been considered one of the seven most important wetlands globally, and declared Ramsar Site (Paul, 2009; Chakraborty, 2021a, 2021b; Chakraborty et al., 2023).

1.2 Uniqueness of the Mangrove Ecosystem of the world in General and the Sundarbans Mangroves in Particular

The mangrove ecosystem, being an unique ecosystem with greater biological productivity, is very sensitive and fragile, located in the sea–land interphase, and is formed by the conglomeration of varieties of flora, fauna, and microorganisms that possess the adaptability and power of acclimatization under the unstable ecological conditions orchestrated by the oscillating ecological parameters of a tropical intertidal zone. The mangrove ecosystem of Sundarbans, India, lying between 88°85′ and 89°00′ north and between 21°32′ and 22°40′ east, represents one of the rarest tidal wetlands of the world in respect of its green expanse within more than hundred deltaic islands, endowed with a wide range of biodiversity components, distinctive eco-dynamics involving interactions among all structural components of two sub-systems, mangrove forest and adjoining estuarine, and the tremendous power of acting as a buffer system against different kinds of natural disasters. This tidal dominated deltaic–estuarine–mangrove complex lying on the deltas of the Ganges, Brahmaputra, and Meghna, is the largest of its kind, covering an area of 1,33,010 ha (10,000 km^2) of both land and water, the major part (6517 km^2) of which remains in Bangladesh whereas the other part (3483 km ^{2}) lies within India. The mangrove ecosystem in West Bengal, India, is traversed by a number of estuaries such as Mooriganga, Saptamukhi, Thakuran, Raimangal, Bidyadhari, Gosaba, Haribhanga, Matla, and Hooghly, all of which open into the Bay of Bengal after enjoying water flow from a north to south direction. The continuous deposition of sand, silt, and clay on the comparatively slower moving or stagnant water eventually leads to the creation of a mosaic of 104 deltaic islands of which 54 have already been reclaimed for developing human habitation. On entering and mixing of the river water loaded with silt and clay with the sea that has sand particles, a kind of change occurs in the flow patterns of all the water bodies, resulting churning movements of water that direct the vertical deposition of all the textural components of the sediments, which, with the passage of time, result in the formation of triangular deltas, A number of geomorphological and resultant hydrological changes governed by the balance of supply, redistribution, sequestering, and export of sediments, facilitating structuring and restructuring of this mangrove–estuarine complex, have converted it a dynamic, vulnerable but productive ecosystem.

1.3 Geological Past and Estuarine Networks of Sundarbans, India

The most widely acceptable definition of an estuary was proposed by Pritchard (1967): *"an estuary is a semi-enclosed coastal body of water which has a free connection with the sea and within which sea water is measurably diluted with fresh water due to land drainage,"* and estuaries have complex systems that receive chemical inputs from a variety of different sources.

In the process of tracing the geological past of mangroves, it was found that plenty of residual parts of mangroves were there even in the mega city, erstwhile Calcutta but presently Kolkata, indicating the presence of a rich diversity of mangrove plants in 9000–8000 years BP because of their southward migration during the time span of last several centuries (Umitsu, 1997). The mangroves of Sundarbans were presumed to settle, grow, and flourish over the deltaic plain, which was developed around 10,000 years BP by the accretion of sediments carried by the Ganga–Brahmaputra–Meghna river systems from the Himalayas. Quaternary sediments interpolated with the silt, clay, and sand were found to have been derived from the marine environment, as revealed from the analysis of the surface geology of the upper portion. The characteristic of tectonic genesis and geologically related deltaic Sundarbans in the Bengal basin could be derived from the Pre-Cambrian Indian shield in the west, the Indo-Barman ranges of the Cenozoic era in the east, and the Shillong plateau in the north east (Allison, 1998). Different species of mangrove vegetation were shifted from the north to south as a result of deforestation caused by the human settlements around 11,000 years BP after the last glacial period at very low temperatures (Tomlinson, 1986). The changing climatic conditions over the Sundarbans play determining roles in the distribution of flora and the dynamics of their ecological succession. Therefore, information pertaining to the bioclimatology plays a significant role in tracing the geological history of the tidal tropical forest, Sundarbans, which demonstrates the parts played by the mangrove assemblages of the Sundarbans as indicators of climate changes.

The origin of the Sundarbans mangrove–estuarine complex is comparatively recent, as deltaic formations of the Ganga–Brahmaputra system was supposed to have been initiated at the end of the Miocene and continued to result in the present state of development around 10,000 years ago (Pleistocene to recent years). Many geomorphological changes have taken place since the Tertiary period because of tectonic movements, mostly in and around northwestern parts of the state of Punjab with the simultaneous deposition of sediments from the southeastern flows of the River Ganges in the Bengal Basin, which consequently resulted in the development of the Deltas of the Sundarbans (Wadia, 1961; Gopal & Chauhan, 2006). In addition, the sediments carried by the riverine networks of two mighty rivers, Ganges and Brahmaputra, during the post-Pleistocene period was thought to bypass the deltaic networks to a great extent, which subsequently facilitated the rapid formation and maturation of the Bengal deep-sea fan (Biswas, 1991). The neo-tectonic movements within the Bengal Basin between the twelfth and sixteenth centuries forced the River Ganges to change its course eastward, which enabled its joining with the River Brahmaputra and that subsequently caused the shifting of the deltaic portion of the Bengal Basin toward the east (Morgan and McIntire, 1959). All those geological changes, especially the continuity of altered tectonic movements, had a considerable impact on the hydrology of the Bengal Basin because of the shifting of the freshwater flow eastward toward Bangladesh, coupled with an altered sedimentation pattern, which tended to impose salinity invasion into the Hooghly–Matla estuarine complexes of the Indian Sundarbans (Hazra et al., 2002; Gopal & Chauhan, 2006).

The coastal zone is the interface of the terrestrial landscape and sea where the shallower part of the sea is influenced by discharge from the adjoining land areas and the terrestrial zones are influenced by the action of the nearby sea, mainly by virtue of physical processes such as sedimental deposition, tidal actions, etc. The three sub-zones within the boundaries of the coastal zone play a significant role in the acceptance of various energy inputs and thereby reflects the presence of various geomorphic forms such as coastal dunes, intertidal belts, ripples, backwater, swamp, marshy land, and lagoon. The coastal interface occupies a large space in West Bengal and parts of the Northern Odissa coast for its macro-tidal environment, enormous water courses, and the flatness and lowland nature of the tidal flats through the direct interacting zones between materials and energy inputs. This subzone is again subdivided into a nearshore zone and a backshore zone on the basis of inundation limits. On the sea face, the nearshore zone includes the swash–backwash zone, the surf zone, and the breaker zone of wind waves under the high-water and low-water levels. On the sandy coast, the backshore zone, lying behind the landward limit of the nearshore zone, is affected by strong onshore winds and is reflected by geomorphic forms such as sand dunes, beach ridges, and extensive sand flats. However, on the inland tidal basins, estuarine sections, and on the tidal flats, particularly in the macro-tidal environment, the extension of the nearshore zone includes the intertidal zone and the sub-tidal zone under the rise and fall of tidal levels twice daily. On the muddy coast, the backshore zone lies above the average high-tide level or above the mean high-tide limit as the supra-tidal zone. The supra-tidal zone is characterized by the topographic forms of natural leaves, higher tidal flats, and mature tidal floodplains, which are influenced by the higher annual rise of tidal levels and storm water along this coast. The intertidal zone represents the part of the coastal belt that is restricted by the extreme highest point of the highest high-tide level and the extreme lowest marking of the lowest low-tide level. This intertidal zone forms an important part of the continental shelf, which occupies the shallower part of the landward direction of the coastal zone having an upper intertidal and a lower sub-tidal zone. Both the intertidal and sub-tidal zones representing the shallower portion of the coastal belt are included under the category of littoral zones, where the sub-tidal zone, unlike the intertidal zone, always remains under tidal water. The intertidal zone is again divided into supra-littoral (highest high-tide level), mid-littoral (mid-tide level), and infra-littoral zones (low-tide level) depending on the duration of tidal exposure and inundation (Paul, 2002a, b, 2009; Chakraborty, 2021a, b, 2023).

1.4 Estuarine Networks of the Sundarbans Mangrove Ecosystem

The estuarine network of the Indian Sundarbans, experiencing tropical climatic conditions, enjoys an important position on the global map and can be categorized as a River–Delta estuary, as advocated by Odum (1971) in addition to Pritchard's (1967) four categories of estuarine system.

Out of several major estuaries, five (Mooriganga, Saptamukhi, Thakuran, Matla, and Ajmalmari) have lost their connection with the main flow of the River Ganges because of the continuous deposition of sediments and they have been maintaining their estuarine properties only because of the monsoonal runoff (Cole and Vaidyaraman, 1966). This can be cited as a possible explanation of the cause of continuous cessation of the delta-building process in the west in contrast to the accelerated pace of development of deltas in the eastern part (Gopal & Chauhan, 2006). Larger amounts of sediment are transported due to tidal amplitudes ranging from 5.00 m to 5.50 m creating a dynamic depositional surface, with an influx of floods from the connected rivers in and around the Sundarbans coastal belts and landfall from tropical cyclones, to which mangroves become adjusted with variable environmental gradients (Paul, 2002a).

1.5 Settlements, Zonation, and Successional Establishment of Mangroves in Different Geomorphological Set-Ups

Three different geomorphic zones with distinctive ecological characteristics are used by different species of mangroves:

1. Marginal levees occupied by an elongated stretch of mangroves, the transverse extension of which is often curtailed by the embankment, which acts as a built barrier for the protection of adjoining cultivated land. The geomorphic character of this kind mangrove settlement is very like that of one described by Woodroffe.
2. At the confluence of two estuaries lenticular patches of mangroves occur at the crescentic point bar deposits.
3. The mudflats of the mid-channel bar are often occupied by sparse mangrove vegetation.

However, different species of mangroves often overlap each other.

A cluster of different species of mangroves occurring in different parts of the intertidal flats along the longitudinal profile of the estuaries such as *Excoecaria–Avicennia–Phoenix* combination in the upper stretch of the estuaries with higher salinity gradients; ii) *Avicennia–Aegialitis–Rhizophora* combination in the middle portion of the estuaries; and iii) *Avicennia–Ceriops–Rhizophora* combination downstream with greater freshwater influence.

Sequential patterns of settlements, growth, and propagation of mangroves in the confluence of rivers–estuaries with the sea. In the different sectors of the Sundarbans different forms of floral assemblages are observed, which are determined by different ecological parameters such as salinity, texture of the sediments, and sloping gradients of the intertidal belt. Two distinct types of zonation are seen: overlapping and non-overlapping types. The major plant species in the intertidal belts with modest sloping and clayey sediments are represented by *Avicennia* spp., *Excoecaria* spp., *Acanthus agallocha*, *Aegialitis* spp., *Rhizophora* spp., and *Ceriops* spp. In the upper parts of the intertidal belt with dense forest areas have mangroves belonging to different species overlapping with each other. However, some species exhibit a distinct non-overlapping cluster with other mangrove plants at the lowest intertidal zone. With the passage of time, there has been a gradual filling up of the space by the appearance of other species of mangroves, forming a cluster of mangrove species (Das & Bhattacharya, 1991b). The mangrove swamp of Sundarbans displays a unique heterogeneous combination of different species. The vegetation succession plays a significant role, as some species have been gradually diminishing.

1.6 Global Distribution of Mangroves

Enjoying worldwide circumtropical distribution across the globe and dominating almost one fourth of the world's tropical coastline, most of the mangroves are concentrated in the Indo-Pacific region. The total mangrove area covered by mangroves in around 30 tropical countries including various islands amounts to about 1,00,000 km^2 (Annon, 2003).

Global distribution of mangroves is determined by an array of interlinked and interdependent ecological attributes including several edaphic factors (texture of soils, porosity, transmissivity and permeability, proportionate distribution of different components of inorganic and organic nutrients, microbial load), geomorphological factors (rate of sedimentation in the process of deltaic island formation, stabilization of deposited sediments by flora, bioturbatory activities of fauna, gradients of sloping of intertidal belts), physical factors (amplitude and intensity of tides, micro- and macro-tidal cycles, current, salinity, and other water parameters), climatological conditions (rainfall, humidity, temperature, etc.), and biological factors (flora, fauna, and microbes). So far, 69 vascular plant species of mangroves that have several specific adaptations to thrive well in the rigorous and dynamic coastal marine–estuarine environment have been documented (Blasco, 1975; Chapman, 1976; Kathiresen and Bingham, 2001). These mangroves and their associated plant species help diversified forms of other faunal components (crustaceans, mollusks, polychaetes, finfishes, etc.), algal forms, both benthic and planktons, and microbial diversities (Chaudhuri & Choudhury, 1994; Annon, 2003; Chakraborty, 2011, 2013). Another 80 species of herbaceous and woody plants have been recorded as mangrove associates but do not have special halophytic adaptations for living in the intertidal habitats (Cannolly & Lee, 2007). The largest chunk of the world's mangroves (about two thirds of which occur jointly in Bangladesh and India in the Ganga–Brahmaputra–Meghna deltaic system of the Sundarbans, India. Other countries with mangroves include Indonesia, Thailand, Singapore, Brazil, Australia, Nigeria, and the USA (Spalding et al., 1997).

1.7 The Extent of Mangroves in India

Settlement, growth, proliferation, expansion, and distribution of mangroves vary among different coastal environments of India because of the differential impacts of a multitude of ecological, geological, and climatic factors such as water and soil quality, sloping gradients of intertidal belts, topography, water waves, direction and intensity, tidal amplitudes, rainfall, and temperature. The East Coast of India facilitates extensive formation of mangroves on the hundreds of newly formed deltas because of the continuous deposition of silts and clays on the gradual slopes of intertidal belts. The western coastline that have narrow intertidal belts experience intermittent precipitation deposition of more sand and comparatively higher salinity supports the fringing mangroves. Out of several water-quality parameters, salinity plays a determining role in limiting the distribution and zonation of mangroves. Maximum growth of mangroves is found to take place in polyhaline and mesohaline zones, which support a maximum number of species. Flora such as *Nypa frutigans*, *Kandelia candel*, *Sonneratia caseolaris*, and *Heritiera fomes* prefer to inhabit low-salinity zones upstream (0.05–5%). Landsat data of 1987 showed that India has 420,000 ha of mangroves including non-vegetated mudflats. Recent estimates show mangrove cover of approximately 3150 km^2, almost 80% of which occurs along the northeastern coast (Odisha and West Bengal) and in the Andaman and Nicobar group of islands. Approximately, 267,000 ha of the total Indian mangrove cover is in the deltaic regions, of which 78% is in the Ganges (Sundarbans) delta alone. Other deltaic regions, except Mahanadi, show relatively poor mangrove cover. Most degraded mangroves are found in the Gulf of Kutch, Gulf of Khambhat, and on the Kerala coast.

In the 1960s, the total expanse of mangroves in India covered an area of 6,81,976 ha, about 45% of which was shared by the Sundarbans mangrove ecosystem in its different deltaic islands of the Bay of Bengal (Blasco, 1975, 1977). It was also reported that one sixth of the mangroves in the country were accommodated on the Andaman and Nicobar Islands (Chakraborty & Naskar, 1988). Later, Saengar et al. (1983) estimated the total mangrove area in India at 3,56,500 ha and a survey in 1992 pointed out that the total area under mangroves in India including the mangrove area of the Andaman and Nicobar islands was 4,37,400 ha. The rapid pace of deforestation coupled with the overexploitation of mangroves have led to the development of open marshy land of approximately 1,00,000 ha, which, after being impacted by profuse pollution and increasing population pressure, became an ecologically degraded area (Blasco, 1975, 1977).

1.8 Seasons and Climate

The seasons in and around the Sundarbans mangrove ecosystem are well defined, each of 4 months' duration: pre-monsoon (March to June) with the highest temperature of around 42 °C and almost no rainfall; monsoon (July to October) with high rainfall (2000 ml) and a moderate temperature; and post-monsoon, which has the lowest temperature of around 10 °C with occasional rainfall. Winds blow with a higher intensity during May to September from south-easterly to south-westerly directions. Early pre-monsoon (February to April) experiences blowing of winds with highest intensity from the south and south-west. The post-monsoon (late October to January) is mainly a calm period with winds flowing from north-west (Annon, 2003). However, the prevailing climatic conditions have been affected by the global climate change-mediated occurrence of several natural disasters during the last couple of decades.

1.9 Excellence from the Point of View of Biodiversity

Mangroves in India occur an area of 4,991.68 km^2, more than 80% of which are located along the east coast of India and also in the Andaman and Nicobar islands. The Indian mangrove comprises about 50 species and some of the major mangrove species are *Avicennia marina*, *A. officinalis*, *A. alba*, *Rhizophora apiculata*,

R. mucronata, *Aegiceras*, and *Bruguiera*. The mangroves of Andaman and Nicobar, as well as Kerala, are dominated by *Rhizophora stylosa*, followed by *R. mucronata* and *Bruguiera gymnorhiza*. These mangrove-dominated environments harbor a rich biota of both terrestrial and marine origin. Sundarbans is considered a biodiversity-rich eco-region mainly because of the occurrence of a galaxy of floral and faunal components. The diversity of flora includes true halophytic mangrove plant species, mangrove-associated plants, algae, and phytoplankton. The faunal diversity ranges from the protozoa to mammals including benthic fauna, nekton, periphytons, insects, birds, reptiles, and amphibians. The unique faunal diversity of the deltaic mangrove islands of Sundarbans has drawn the attention of many researchers and scientists across the globe since the first publication of Stoliczka (1869) on the malacofauna of the brackish water systems of lower Bengal.

The wealth of biodiversity in Indian Sundarbans is formed by an assemblage of 150 species of algae, 163 species of fungi, 32 species of lichen, 34 species of true mangrove plants, and 40 species of their associated plant species, 250 species of fishes, 7 species of amphibians, 59 species of reptiles, around 200 species of birds, 39 species of mammals, besides numerous species of phytoplankton, zooplankton, ichthyoplankton, benthos, soil-inhabiting micro arthropods, and mangrove plant-dependent insects. All of the biotic structural components of this mangrove ecosystem display fluctuation in respect of species composition, population density, distributional pattern, biomass, and community structure spatially and temporally under the influence of the prevailing fluctuating ecological parameters such as temperature, precipitation, tidal amplitudes, pH, dissolved oxygen, salinity, nutrients, turbidity, etc., from one eco-zone to others during different periods of the year (Chakraborty, 2014).

1.10 Ecosystem Services of the Mangrove Ecosystem

Coastal ecosystems play important roles in the protection of the landmass, as a reservoir of bio-resources, the generation of meteorological information, the production of energy, the supplying of fossil fuels, the regulation of climatic oscillation, and forming avenues for recreation across the globe. The mangroves, considered the **"green wall of the sea,"** provide protection to the landmass from the waves of the sea, as Kolkata metropolitan area has been saved several times from cyclones because the mangroves act as a buffer to protect the Sundarbans. Therefore, the very presence of the dense mangrove forest represents as an assurance for the safety of the urban metropolis in the event of any cyclonic or storm surge.

Both ecological and economic benefits of the Sundarbans mangrove wetland are derived from its luxuriant resources, which include products from agriculture, aquaculture, fishery, forestry, wildlife, birds, and a valuable type of honey. Initiation of honey collection coincides with the full blossoming of several mangrove plant species during the period March to April, which continues for the next 2–3 months. Together with a leader, the honey collectors, locally known as '*Mouley*', set off into the forest on a prefixed date. The entire operation of honey collection from the entry to the forest to its end is popularly known as *Mahal*. The leader takes up the responsibility to drive the entire team moving in a definite pattern, usually in rows, toward the beehives tracking the course of the flying honeybees. The rock bees, *Apis dorsata*, start arriving in Sundarbans in the month of April from Eastern Himalaya. Fishing throughout the year plays an important role in the pattern of livelihood of the people of the Sundarbans, where the rivers, creeks, and channels vibrate with the presence of many species of fishes and shell fishes.

The mangrove ecosystem comprising intertidal woody halophytic vegetation and experiencing influences of diurnal tidal rhythms with both high and low tides, twice a day, develops very ecologically conducive conditions for the proliferation of biodiversity components on the one hand and the maintenance of ecological balance on the other (Aller, 1982, 1988, 2001; Chakraborty, 1995, 2017). This very specialized ecosystem renders so many benefits to humans in the form of ecosystem services and maintains the rural economy by providing food (fish, honey, etc.), timber, and leaves for constructing shelters, fuel wood, fresh oxygen-rich air, plenty of water with a wide range of salinity profiles, arresting soil erosion and ensuring the protection of the coast, acting as buffer against natural disasters such as floods, storms, cyclones, carbon sequestration, abatement of pollution through phytoremediation, and recharging of groundwater by maintaining the estuarine flow.

This detritus-based, highly productive coastal ecosystem derives its power for rendering ecosystem services because of the combined contribution of three primary producers (mangrove plants, benthic algae, and phytoplankton) and thereby promotes fisheries and aquaculture development much better than any other aquatic ecosystem all over the globe, mainly for two reasons. First, a huge quantity of energy, in the form of mangrove detritus-derived nutrients, are supplied from the mangrove forest subsystem to the estuarine subsystem (Odum & Heald, 1975; Alongi et al, 1989; Ashton 2002; Chakraborty, 2017) and as a strong positive correlation exists between the total expanse coupled with the biomass of the mangroves and the total fisheries' yield from adjacent water, different microhabitats of water and forest are used as a nursery for juveniles of both fin-fish and shellfish, as suitable shelters for crabs, oysters, clams, estuarine crocodiles, and snakes, as attractive resting and breeding grounds for turtles, and local residential and migratory birds (Macnae, 1974; Lee, 1995), Second, based on the natural availability of bioresources, profitable regional and international markets for bio-products of higher quality are developed with the aim of obtaining better profitability. Decomposition of mangrove plant litters produces nutrient-enriched detritus, which, after being exported from the mangrove forest subsystem to the estuarine subsystem, promotes biodiversity development and ensures higher biological productivity. Production and export of mangrove-based organic matter are found to have been influenced by several ecological factors such as temperature, pH, textural components of the sediments, disturbance through bioturbation, and other physical processes such as wind flow, precipitation, but other dominating regulatory factors appear to ensure the quality, quantity, and reactivity of organic matter. It has been established that there are characteristic differences between carbon and nitrogen concentration, carbon to nitrogen ratio (C:N), pH, etc., in intertidal belts, either with mangrove dominance or devoid of mangroves (Alongi et al., 1999). In addition, mangrove vegetation, by virtue of the root systems (pneumatophore and trap roots) resist soil erosion, absorb the surges of storms, cyclones, tsunamis, and even encounter the problems emanating from sea-level rise (SLR), especially in small islands and low-lying coastal zones (IPCC, 2007a, b).

1.11 Different Commodities Extracted from the Mangroves

The wooden stems and branches of mangrove plants as timbers and stakes are used for building shelters for local people. Dried wooden branches are also used as fuel. The grasses and some leaves of mangroves (*Avicennia* spp.) are also used as fodder. Fruits of *Sonneratia* spp. are also used for human consumption. Tannins are extracted from the barks of a number of mangrove plants such as *Rhizophora* and *Bruguiera* spp. The humus-enriched mangrove litter contributes detritus collected from the silty-clay-loaded bottom of mangrove forests mud and is used as manure. Extracts from the leaves and roots of *Derris heterophylla* are used for stupefying fish and applied as narcotics. Extracts of the leaves of *Acanthus* spp. are used for the treatments for rheumatic disorders and those of *Bruguiera* spp. for high blood pressure. Extracts from the barks and the leaves of *Excoecaria agallocha* also have medicinal value as those ingredients are used for various ailments.

The waterlogged depressed lands with tidal water in the vicinity of mangrove environments are commonly used for fisheries, prawn farming, and salt production. Bioactive compounds from mangrove-associated faunal components such as glycolproteins, polyphenols, peptides, and essential fatty acids, and all of those compounds that have rheological properties are used in food industries, and for antioxidant, anti-inflammatory, anti-hypertensive, and cardio-protective purposes.

1.12 Environmental Threats and Eco-restoration

Degradation of the mangrove ecosystem has been a continuous process at an alarming rate, resulting in severe ecological imbalance in the coastal environment. This mangrove ecosystem of the deltaic islands of Sundarbans, with its eco-dynamism, has been facing various threats: deforestation, increased salinity, soil erosion, pollution from industries, cyclonic destruction, an indiscriminate collection of the seeds of shrimps, unscientific aquaculture, and man–animal conflict.

Among several environmental threats, such as global warming, heavy metals, plastics and microplastics, agricultural waste, sewage, petroleum, and crude oils, biotic pressure appears to be the most challenging as it has been constantly increasing and commands for immediate protection and conservation of the affected ecosystem. In addition, the biodiversity and basic modes of ecosystem functioning are disrupted for several other reasons such as the reclamation of deltaic islands for human use, injudicious exploitation of fishes, floral, and other faunal components, deforestation, erosion, and unwanted accretion, salinity invasion, ecotourism, bio-invasion, and several forms of air, water, and soil pollution. Natural destruction is caused by intensive grazing, cyclones, attacks by pests, disruption of food chains and food web dynamics, floods, storms, cyclones, and occasional tsunamis.

1.13 Natural Threats to the Coastal Sundarbans and Eco-restoration Strategies

Fringing forest dominated by mangroves in India can be restored partially by afforestation along with taking up measures for improving the tidal drainage systems by artificially constructing canals, channels, or ditches fed by high-tide water to enable the brackish water to enter and reach into the inner parts of mangrove swamps. In the tropical environment of Sundarbans, India, mangrove forests are ecodegraded and lost, mostly to harvesting natural resources for human consumption and land use conversion for aquaculture and agriculture, also for human benefit (Richards, 1993; Blasco, 1984; Banerjee et al., 1989; Naskar & Guha Bakshi, 1987; Paul, 2002a, b). However, it becomes very difficult to restore the naturally degraded fringe forests of the mangrove–estuarine complex within the Sundarbans, India, because of the continuous impacts of so many marine-forcing factors such as coastal morpho-dynamics under the impacts of hydrological stress, tectonics-related land subsidence, and uplift by faulty tectonic movements, enunciation of hydrological stress factors from the drainage catchment areas of deltaic islands, deltas, and sediment starvation, hyper salinity, eutrophication, shifting river banks in catastrophic deltaic storms, over-wash of sand fan lobe encroachment, shoreline retreat due to erosion at the sea face, the active barrier rollover process, and the reduction in the freshwater supply, less precipitation, and declined discharges of freshwater into the coastal zones.

Eco-restoration of eco-degraded mangrove habitats is achieved through eco-rehabilitation, planting suitable species with the prime goal of increasing the functional capacity of coastal vegetation in order to enhance the protective and carbon-rich forest coverage, which has become instrumental in resisting the current of tidal water, and also to absorb future climatic shocks.

The afforestation of mangroves holds great promise in finding a solution for the restoration of eco-degraded habitats for ecological, environmental, and socioeconomic reasons. This kind of venture for the conservation and management of mangrove ecosystems are imperative, not only to prevent further deterioration of the overall environment but also to ensure the preservation of mangrove genetic resources.

In addition, innovative planning is required to develop a co-benefit regime for community-based adaptation through integrated land use for livelihood generation of the local inhabitants. This new innovative model integrating fishery, aquaculture, afforestation programs, silviculture, agriculture, etc., are implemented to restore fallow coastal lands into community-based livelihood adaptation practices, alongside reducing the intensity of inundation and problems associated with the enhancement of salinity. In addition, freshwater scarcity in the cultivation of agricultural crops and fish can be tackled. This kind of rational land use tends to improve household adaptation capacity through both short-term and long-term income generation by harnessing the potential scope of ecosystem-based adaptation in and around rehabilitated mangrove areas.

1.14 Pollution and Eco-degradation

The estuarine networks of Sundarbans, India, receive industrial and urban pollutants from point sources and large amounts of pesticides, fertilizers, and other organic pollutants from nonpoint sources. Escalating rates of urbanization and industrialization in the catchment areas of the estuarine networks of Sundarbans, India, are considered main sources of heavy metals, trace elements, petroleum hydrocarbons, pesticides, and microplastics. Anthropogenic pressures exploiting both biotic and abiotic structural components of coastal -mangrove ecosystem are leading to the disappearance, fragmentation, and outright destruction of habitats (Das & Bhattacharya, 1999a, b). Loss or decline of individual populations or the diversity of species results in decreasing ecological richness, alteration not only in the species composition but also in the ecosystem functioning, thereby leading to the degradation of communities and ecosystems. Such consequences relating to loss of biodiversity or a decrease in ecosystem goods and services beyond certain limits tend to disrupt the natural functioning of ecosystems by reducing ecological resilience and ultimately lead to ecosystem collapse (Chen & Twilley 1998; Conely et al, 2007). Such nonfunctioning of the ecosystem, after being affected by anthropogenic activities, fails to withstand the harmful effects of combined or isolated impacts of different categories of pollution. Population explosion and associated demands for the needs of livelihood and shelter put pressure on the natural resource bases and ultimately hamper the functioning of the ecosystem. In addition, increased emission of greenhouse gases (GHGs) and the resultant global warming lead to sea level rise (SLR) form the major threats for marine habitats in the intertidal and shallow regions. Anthropogenic emissions of GHGs and aerosols leading to cause global warming (Nakicenovic & Swart, 2000) which enhance globally averaged surface air temperature accompanied by increases and decreases in precipitation.to rise between 1.4 and 5.8 °C by 2100 relative to 1990 and the globally averaged sea level to rise between 0.09 and 0.88 m by 2100. But these projections vary from region to region in terms of the frequency and intensity of some extreme climate phenomena and will have impacts on natural and human systems.

Marine–coastal habitats of the Indian subcontinent and mangroves from the low-lying region are very susceptible to the inundation threats from SLR, which is the outcome of increased temperature and CO_2 levels, and alteration in the hydrological regime. However, mangroves are highly adaptable to stressful surroundings and can keep pace with the SLR in comparison with other biotic components. A supply of a good amount of allochthonous sediments at the bottom of the mangrove forest play vital roles in determining the survival of mangroves during relative SLR. However, the mangrove–estuarine regions are under continuous threat of habitat destruction from reclamation, especially for agriculture, aquaculture, urbanization, and industrialization.

Sea-level rise (SLR), being a serious global threat with overwhelming evidence, occurs because of climate change, and will continue for centuries even if the people of the world can stabilize the problems of GHG concentrations in the coming days (IPCC, 2007a). As coastal wetlands, including mangroves, experience the inundation and exposure rhythm, they generally remain just over the tidal range above mean sea level. The tidal range is an important determinant of the vulnerabilities of such eco-zones to SLR, and denotes the difference in elevation between the mean high tide and the mean low tide (Titus, 1988). Mangrove vegetation requires periodic flooding for the inundation and exposure for their own biological needs and therefore SLR tends to disrupt this kind of wetland in three significant ways: inundation, erosion, and saltwater intrusion. The common impact of SLR on the mangrove–estuarine ecosystem is that it compels the biota to migrate to areas in the upper landward direction, converting net change in the area of total wetlands after being determined by the slopes of the intertidal belts (Titus, 1988). A thorough study of the prospective impact of SLR on coastal–estuarine–mangrove complexes is supposed to enable factors to be identified on which the ecological balance of the wetland ecosystem is dependent and that can also show the history of sea levels in respect of the prevailing gradients of the sloping of the intertidal belts and bottom of the adjoining water bodies, geomorphological and sediment logical characteristics, different oceanographic processes such as the tidal amplitudes, patterns of currents and waves of water, and formation of tidal creeks, ripples, and dunes.

1.15 Causes of Eco-degradation of Mangrove Habitats: Assessment of Ecosystem Degradation, and Driver on Degree of Impact Trend

A number of ecological parameters, attributes, and processes have been identified as being the driving forces behind the eco-degradation of mangrove habitats.

The direct drivers also designated as proximate causes of degradation and loss of coastal mangrove ecosystems include the following:

- Land-use changes and habitat alteration leading to fragmentation, degradation, and loss of habitats.
- Nonjudicious exploitation of natural resources for commercial purposes and livelihood generation.
- Different forms of pollution, mostly by agricultural waste (chemical fertilizers and pesticides), sewage and eutrophication, hazardous chemicals such as heavy metals, petroleum and crude oils, and plastics and microplastics.

- Bioinvasion with the introduction of alien invasive species, which is also known as biological pollution.
- Anthropogenic climate-change-mediated global warming.

Indirect drivers, also designated as ultimate or root causes that underlie the proximate causes:

- Explosion of population necessitates increased demands for natural resources.
- Inequalities in society and uneven distribution of wealth causing poor people to be deprived of obtaining their minimum share for survival, whereas the wealthy people are responsible for the harmful effects of the overexploitation of natural resources, causing constraints toward sustainability.

1.16 Adaptation Against the Changing Stressed Environment

Adaptation in terms of both natural and social systems to the changing coastal environment is very necessary to combat the problems associated with the decline of diverse coastal resources such as mangroves, coral reefs, and water bodies packed with fishery resources, which tend to render a range of services, both ecological and economic, for the benefit of human well-being (Klein et al., 1998; Moberg & Ronnback, 2003; Nicholls et al., 2007; Nandy & Ahammad, 2012). Coastal adaptation also helps in understanding the responses of natural ecosystems and the degree of societal vulnerability to coastal environmental issues such as climate-change-induced SLR, inundation of low-lying deltaic plains of the Ganges–Brahmaputra estuarine networks, development of the ecological capacity to adapt to new and diverse socioeconomic situations (Nandy & Islam, 2010).

The Earth's climate system, being complex and nonlinear, exhibits adaptability and demonstrates a slow response to any external force such as GHG emissions with the dramatic reduction of the rates of GHG emissions in the near future tending to stabilize the atmospheric concentrations, resulting in an increase in surface air temperature by a few tenths of a degree per century, imposing threats to natural and human systems. This new situation can only be tackled by adopting appropriate adaptive measures alongwith continuous ecomonitoring (chakraborty et al, 2014).

1.17 Global Climate Change and Adaptation to It

The coastal ecosystems as a whole are affected by climate change and vulnerability which is found to be greater at an individual level owing to a lack of inequality in the distribution of coastal resources distribution, greater sensitivity of bio-resources to the impacts of climate change, inadequate information on land-use patterns, disparity of land ownership, lack of long-term economic empowerment of vulnerable people, and poor adaptive capacity, all of which make the survival of affected people very difficult. This has prompted proper eco-management strategies to be adopted, alongside undertaking adaptive measures mainly by eco-restoring the landscape and protecting natural resources to maintain ecosystem services rendered by the coastal ecosystem.

Adaptability by the individual, household, and community to climate change is facilitated by the extent of understanding of the local people of the problem, on the basis of which they are liable to change their behavior and life styles in order to do what is necessary to 'actively manage risks' (Bennett et al., 2014). Adaptation pertaining to climate change can be tackled by getting necessary broader-scale 'institutional support' from both national and international organizations. Individuals and community used to experience climate change in their daily walk of life and all of their experiences are reflected through a survey as follows: (i) encountering more frequent and powerful storms; (ii) unpredictability of the mode of occurrences of changing rainfall patterns; (iii) occurrence of flooding and increased freshwater in mangrove areas; (iv) SLR; (v) erosion on the embankment; (vi) increased intrusion of saltwater; (vii) intense coral bleaching; (viii) negative impact on tourism; (ix) impact on human health; (x) decrease in agricultural productivity.

However, a general lack of awareness and proper knowledge because of variabilities of human perception and moral values, economic conditions, and situational factors, tend to lead all together the hampering of 'actively managing risks' or hindering proactive planning for combating the nuisances of climate change (Ommer, 2007; Tuler et al., 2008).

1.18 Conservation, Protection, and Management of the Mangrove Ecosystem

Conservation, being the core component of environmental science, highlights the ways and means of maintaining and protecting the diversity of species and habitats by the sustainable use of natural resources, with options for providing future generations with optimal resource bases after being driven by moral imperative not to harm nature with its wealth of both living and nonliving components. Taking a sustainable conservation approach to protecting and conserving the mangrove ecosystem, which is being considered the threatened landscape of the world because of multidimensional threats enunciated for both natural and anthropogenic reasons.

The prime measure to ensure conservation and restoration of the threatened mangrove wetlands of the Sundarbans mangrove ecosystem is to build up artificial drainage ditches to revive the frequency of inundation and afforestation in the existing tidal mud flats along the shoreline and construction of embankments. Mangrove habitats are protected by the rules of the coastal regulation zone and committees are formed by governments at national level with the objective of coordinating an awareness-building process, in the form of workshops by imparting training, undertaking research, and performing management activities relevant to these habitats.

When considering emphasizing the ecological and socioeconomic values and threats, both natural conservation (waves, long-shore currents, SLR, erosion, subsidence, storms and cyclones, bioinvasion, alteration of hydrological cycles, mosquitoes) and anthropogenic (drainage for agriculture and forestry, stream channelization by dredging for navigation, various flood protection measures, construction of roads and embankments, conversion of aquatic bodies for aquaculture or mariculture, construction for storm protection and for arranging water supplies, irrigation, the large-scale application of chemical fertilizers and pesticides, the non-judicious disposal of solid waste, sediment diversion by artificially deep channels, mining of the bottom soil, profuse abstraction of groundwater for irrigation) to mangroves, a number of habitats rich in mangroves have been protected by assigning them the status of national parks, sanctuaries, and biosphere reserves. Environmental impact assessment is considered mandatory while considering any developmental projects that might have harmful impacts on the existing mangroves. Simultaneous large-scale mangrove afforestation for the rehabilitation of protected open mud flats is also taken care of.

Mangrove ecosystems are complex in nature, with the hydrologically active delta with meandering estuaries that continuously reshape the land, through both erosion and sedimentation, and a multidimensional long-term approach to monitoring the impact of multidimensional threats is a prerequisite. Remote-sensing data have proved to be useful for the evaluation of a reliable estimate of area coverage by mangrove vegetation. The mangrove regions under conservation and protection efforts require constant monitoring of the hydrological, ecological, and biological changes. Various goods and services provided by the vulnerable coastal zones appear to be vital from the point of view of enhancing the present and future adaptation capacity of a large human population residing in and around coastal areas by developing the proficiencies in multiple resource uses in the face of ongoing environmental perturbations and changing climatic threats from extreme natural disasters.

The following points need to be taken care of:

(i) Regular monitoring of mangrove ecosystems with remote-sensing data.
(ii) Existing luxuriant mangrove areas under private ownership or revenue departments should be protected and transferred to the Forest Department by amendments to existing legislation.
(iii) Luxuriant mangroves on virgin deltaic regions should be reserved as protected areas in the form of sanctuaries, national parks, genetic resource habitats, biosphere reserves, etc., prohibiting any kind of human activities such as modification, alteration, captive, and capture fisheries.
(iv) Captive and capture fisheries and aquaculture can only be permitted without damaging the existing vegetation.
(v) Local people should be involved in the plantation program and in the development of mangrove nurseries in the barren but inundated coastal wetlands.
(vi) Peoples are also required to be made aware of protecting the mangrove areas from illegal felling and poaching of wildlife.

(vii) Harvesting of natural resources in the form of honey, fish, fodder, fruits, timber, fuel, and charcoal is permitted, obtaining permission from the Forest Department in consultation with experts.
(viii) Undertaking any venture in the fishery sectors requires proper legislation clearance to avoid damage to the existing mangroves and pollution.
(ix) Tourism should be allowed in a restricted manner, not causing any harm to wildlife or mangroves.
(x) Awareness-building process should be taken up with the help of booklets, video, documentaries, lectures, exhibitions, posters sessions, and by post.

Therefore, the mainstream coastal development agenda must include conserving and restoring mangrove areas as a strategy for adapting to climate change. In addition, the abilities of the mangrove ecosystem to filter chemical pollutants and carry out natural bioconversion of organic compounds from both the water and the soil keep the environment cleaner by virtue of being marketable ecosystem services.

1.19 Recovery of Mangroves on Eco-restoration: Ecosystem Service, Production and Ecosystem Transformation

When the goal becomes using ecological engineering to meet human needs, it is necessarily based within the context of how best to deliver particular ecosystem services desired by local or regional communities. The resulting product may be the same as what would come from restoration but it need not be. As the focus of this perspective is on engineering designs to provide one or more ecosystem services instead of restoring ecosystems that provide the full suite of services for that ecosystem type, the product may be a novel/no-analog ecosystem or a technological solution that happens to use natural materials. For instance, when technological solutions are implemented in damaged streams to enhance desired functions, the aquatic ecosystem can change dramatically from what it was historically and the production of the desired ecosystem service(s) may or may not be sustainable from an energy or environmental standpoint. An example is an emerging form of urban stream **"restoration"** using principles of ecological engineering to deliver specific ecosystem services of the urban stream.

Mangrove ecosystems throughout the world suffer eco-degradation from changes in the land-use patterns, coastal development, logging, overflowing of pesticides, releasing of industrial effluents loaded with a lot of toxic substances, etc. The continued loss of mangrove forests across the globe worldwide underscores the necessity of undertaking restoration measures of forest structure and functions.

A case study has shown that mangroves require more than 30 years to recover from the impacts of a severe oil spill. Recovery indicators in the process of the successful restoration of mangroves include adequate tidal exchange, planting of healthy mangrove propagules, and marsh plants to provide a habitat that can be colonized more easily by mangrove trees, the maintenance of the density of trees, monitoring hydrology, and elevation gradients of the intertidal belts. All of these together in combined ways facilitate natural recovery processes to reestablish mangrove habitat over the long term. The recovery process in the mangrove restoration venture requires 20–30 years of counteracting the ongoing environmental stresses.

1.20 Conservation, Stewardship, Adaptive Co-management, and Conservation Policy

Alongside the monitoring and eco-auditing of interactive processes among land, water, and biodiversity in a mangrove ecosystem, the aim of tracking conservation and stewardship functions is required to undertake a proper conservation policy. Adaptive co-management is considered an important measure for evolving stewardship as collective and integrated resource management policies. Community resilience and ecosystem-based management appear to be important prerequisite to cope up with the ongoing environmental challenges. Conservation and stewardship go hand in hand in protecting the coastal ecosystem with the all-round involvement of local people.

Conservation remains futile unless stewardship is coupled with it. Hunger, malnutrition, and migration invade and alter ecosystem services; conservation of resources alone cannot be effective. We need to make the existence and functioning of the forest, for example, keep leveraging toward ensuring benefit for the human and livestock populace depending on their survival and growth. Saving water in a seepage tank can be more meaningful when there will be summer rice cultivation out of the water saved already. Similarly, in the restoration of mangroves, there must be a visible impact on livelihood generation and reduction of thrust of cyclonic storms. Exactly here we need stewardship and community mobilization. The process of eco-restoration should be reflected in the generation of wages and livelihood for the community of people. Effective stewardship requires an effective management information system (MIS). It is extremely important in managing the brunt of climate change in coastal villages of Sundarbans. It involves adaptive co-management as well. The following flow diagram offers a splendid opportunity for the smart management of climate change data in Namkhana Block of Sundarbans in which down to earth people's perception data on climate-change pattern, cyclonic storms, occasional droughts, tidal waves, and floods, etc. The community management information system (MIS), Critical Climate Service Center, Village Kiosk, Capacity building, and Communication Center which are being the developed and available infrastructural facilities operate together for constant monitoring, tracking, auditing, and micro-intervention planning for the robust yet resilient participatory resource management strategies for the coastal lines of Sundarbans. The proposed paradigm for micro data management ensures five basic functions, viz.:

(i) Collection of data: such as mangrove extraction, volume of shrimp collection, retention of residue tidal water, wage pattern;
(ii) Storing of data collected; analysis of data;
(iii) To delineate change patterns such as endangered or threatened species;
(iv) Retrieval of data as and when required;
(v) Delivery of data to village-level climate managers.

The monitoring data can undergo a log frame analysis to elucidate the present status, deliverable, and future prediction (Acharya 2022).

Adaptive co-management ensures a community approach in receiving, reacting, and responding to ecological signals and change outputs, including ecological perturbation. A systems perspective provides a logical framework for managing changes in social–ecological systems. To summarize briefly the previous sections, the dynamic interactions of ecological and social processes that characterize most of today's urgent problems necessitate a social–ecological framework for planning and stewardship. Any sustainable solution to a resource issue must be compatible with current social and ecological conditions and their likely future changes. A resource policy that is not ecologically, economically, and culturally sustainable is unlikely to be successful. Sustainable resource stewardship must therefore be multifaceted, recognizing the interactions among ecological, economic, and cultural variables and the important roles that past history and future events play in determining outcomes in specific situations (Berkes & Folke, 1998).

References

Acharya, A. (2022). Seasonal quantification of carbonate dissolution and CO_2 emission dynamics in the Indian Sundarbans estuaries regional studies in marine. *Science, 53*.

Aller, R. C. (1982). The effects of macrobenthos on chemical properties of marine sediment and overlying water. In P. L. McCall & M. J. S. Tevez (Eds.), *Animal–sediment relations*. Plenum Press, New York.

Aller, R. C. (1988). Benthic fauna and biogeochemical processes in marine sediments: the role of burrow structures. In T. H. Blackburn & J. Sorensen (Eds.), *Nitrogen cycling in coastal marine environments*. Wiley.

Aller, R. C. (2001). Transport and reactions in the bioirrigated zone. In B. P. Boudreau & B. B. Jørgensen (Eds.), *The benthic boundary layer: transport processes and biogeochemistry*. Oxford University Press.

Allison, M. A. (1998). Historical changes in the Ganges-Brahmaputra Delta front. *Journal of Coastal Research, 14*(4), 1269–1275.

Alongi, D. M., Boto, K. G., & Tirendi, F. (1989). Effect of exported mangrove litter on bacterial productivity and dissolved organic carbon fluxes in adjacent tropical nearshore sediments. *Mar Ecol. Prog. Set, 56*, 133–144.

Alongi, D. M., Tirendi, F., Dixon, P., Trott, L. A., & Brunskill, G. J. (1999). Mineralization of organic matter in intertidal sediments of a tropical semi-enclosed delta. *Estuarine, Coastal and Shelf Science, 48*, 451–467. https://doi.org/10.1006/ecss.1998.0465

Annon. (2003). *Mangrove ecosystem: Biodiversity and its influence on the natural recruitment of selected commercially important finfish and shellfish species in fisheries*. National Agricultural Technology Project (NATP). Indian Council of Agriculture Research (ICAR). Principal Investigator: George, J.P.; Co-PI: Chakraborty, S.K. and Damroy, S.N. (pp. 1–514).

Ashton, E. C. (2002). Mangrove sesarmid crab feeding experiments in peninsular Malaysia. *Journal of Experimental Marine Biology and Ecology, 273*, 97–119.

Banerjee, L. K., Sastry, A. R. K., & Nayar, M. P. (1989). *Mangroves in India: Identification manual* (p. 113). Botanical Survey of India.

Bennett, N., Dearden, P., & Peredo, A. M. (2014). Vulnerability to multiple stressors in coastal communities: A study of the Andaman Coast of Thailand. *Climate and Development*. https://doi.org/10.1080/17565529.2014.886993

Berkes, F., & Folke, C. (1998). *Linking social and ecological systems: Management practices and social mechanisms for building resilience*. Cambridge University Press.

Biswas, S. K. (1991). Classification of Indian sedimentary basins in the frame work of platetectonics. In S. K. Biswas et al. (Eds.), *Proceedings of second of second seminar on Petroliferous Basins of India* (Vol. I, pp. 1–46). Indian Petroleum Publishers.

Blasco, F. (1975). The mangroves of India. *Institut Français De Pondichéry. Travaux De La Section Scientifique Et Technique, 14*, 1–175.

Blasco, F. (1977). Outlines of ecology, botany and forestry of the mangals of the Indian subcontinent. In V. J. Chapman (Ed.), *Wet coastal ecosystems, ecosystems of the world –1* (pp. 241–260). Elsevier Scientific.

Blasco, F. (1984). Climate factors and the biology of mangrove plants. In S. C. Snedaker & J. G. Snedaker (Eds.), *The mangrove ecosystem: research methods* (pp. 18–35). UNESCO/SCOR.

Blythe, J. (2013). *Dynamics of fishers' responses to social-ecological change in coastal Mozambique: A resilience perspective.* Dissertation. University of Victoria, Victoria, British Columbia, Canada. https://dspace.library.uvic.ca:8443//handle/1828/4860

Cannolly, R. M., & Lee, S. Y. (2007). Mangroves and Saltmarh. In S. D. Connel & B. M. Gillanders (Eds.), *Marine Ecology* (pp. 485–512). Oxford University Press.

Chakraborty, S. K. (1995). Aquaculture potential of mangrove ecosystem of Sundarbans, India. In *Proceedings of Seminar on Fisheries—A multimillion dollar industry* (pp. 72–83).

Chakraborty, S. K. (2011). Mangrove ecosystem of Sundarbans, India: Biodiversity, ecology, threats and conservation. In J. N. Metras (Ed.), *Mangroves: Ecology, biology and taxonomy* (pp. 83–112). NOVA Publishers.

Chakraborty, S. K. (2013). Interactions of environmental variables determining the biodiversity of coastal mangrove ecosystem of West Bengal, India. *The Ecoscan Special Issue, 3*, 251–265.

Chakraborty, S. K. (2017). Ecological services on intertidal benthic fauna and the sustenance of coastal wetlands along the Midnapore(east) coast, West Bengal, India. In *Coastal wetlands: Alteration and remediation* (pp. 777–886). Springer.

Chakraborty, S. K. (2021a). *Riverine ecology (Vol. 1): Ecofunctionality of the physical environment of rivers resolution*. Springer Nature. 917 pages.

Chakraborty, S. K. (2021b). *Riverine ecology (Vol. 2): Biodiversity conservation, conflicts and resolution*. Springer Nature. 566 pages.

Chakraborty, S. K. (2023). Ocean ecosystem and its multidimensional eco-functionality and significance. In R. Brinkmann (Ed.), *The Palgrave handbook of global sustainability* (pp. 537–581). Northern Illinois University.

Chakraborty, P. K., & Naskar, K. R. (1988). Role of mangroves in estuarine fisheries development. In A. G. Jhingran & V. V. Sugunan (Eds.), *Conservation and source management of Inland Capture Fisheries Resources of India* (pp. 229–233). I.F.S.I.

Chakraborty, S. K., Maiti Dutta, S., Ghosh, P. B., Ray, R., & Paul, A. K. (2014). Impact of global warming on Sundarbans mangrove ecosystem, India: role of different assessment tools from ecosystem monitoring to molecular markers. In *Proceedings of the international conference on green India: strategic knowledge for combating climate change: prospects and challenges* (pp. 181–200).

Chakraborty, S. K., Sanyal, P., & Ray, R (2023). *Wetlands ecology: Eco-biological uniqueness of a Ramsar site (East Kolkata Wetlands, India)*. 709 pages.

Chapman, V. J. (1976). *Mangrove vegetation*. J. Cramer, Vaduz (Oxford Academic). pp 1–447.

Chaudhuri, A. B., & Choudhury, A. (1994). *Mangroves of the Sundarbans, India* (Vol. 1, pp. 1–247). IUCN.

Chen, R., & Twilley, R. R. (1998). A gap dynamic model of mangrove forest development along gradients of soil salinity and nutrient resources. *J. Ecol., 86*, 37–52.

Cole, C. V., & Vaidyaraman. (1966). Salinity distribution and effect of fresh water flows in the Hoogly river. In *Proceedings of Tenth Conference on Coastal Engineering, Takyo* (pp. 1312–1434). American Society of Civil Engineers.

Conely, D. J., Carstensen, J., Aertebjerg, G., Christensen, P. B., Dalsgaard, T., Flansen, J. L. S., & Josefson, A. B. (2007). Long-term changes and impacts of hypoxia in Danish coastal waters. *Ecological Applications, 17*, S165–S184.

Das, G. K., & Bhattacharya, A. (1999a). Destruction of certain encountered shellfish and finfish from the coastal waters of Sundarbans and its impact on the coastal ecosystem. *Journal of the Indian Society of Coastal Agricultural Research, 17*(1&2), 216–218.

Das, G. K., & Bhattacharya, A. (1999b). Rampant exploitation of post-larval penaeid prawn (Penaeus monodon) from the salt-water course of the deltaic Sundarbans, West Bengal—A threat to the balance of nature. *Journal of the Indian Society of Coastal Agricultural Research, 17*(1&2), 219–222.

Das, G. K. (2009a). Beach processes of Tidal Islands of Hugli Estuary, West Bengal. *Geographical Review of India, 71*(3), 240–248.

Das, G. K. (2009b). Grain size analysis of some beach sands from the Indian coasts. *Geographical Review of India, 71*(1), 10–18.

Das, G. K. (2010). Biodiversity in Sundarbans. In A. K. Dwivedi & S. C. Tripathi (Eds.), *Environment and biodiversity, environmental pollution and risk to biodiversity* (pp. 1–34). Lambert Academic Publishing AG & Saarbrucken.

Das, G. K. (2011). Impact of salinity and nutrients on the changing mangrove floristic—A case study from the River Flood Plains of Sundarbans, India. In *Biotic potential and the abiotic stress* (pp. 119–129). Lambert Academic Publishing AG & Saarbrucken.

Gopal, B., & Chauhan, M. (2006). Biodiversity and its conservation in the Sundarbans mangrove ecosystem. *Aquatic Sciences, 68*, 338–354.

Hazra, S., Ghosh, T., Dasgupta, R., & Sen, G. (2002). Sea level and associated changes in the Sundarbans. *Science and Culture., 68*(9–12), 309–321.

IPCC. (2007a). Climate change—2007. Impacts, adaptations and vulnerability. In *Contribution of Working Group II to the fourth assessment report of the Intergovernmental Panel on Climate Change*. Cambridge University Press.

IPCC. (2007b). Climate change 2007: Synthesis report. Contribution of Working Groups I, II and III to the fourth assessment report of the Intergovernmental Panel on Climate Change (R.K. Pachauri and A. Reisinger), pp. 104.

Kathiresen, K., & Bingham, B. L. (2001). Biology of mangroves and mangrove ecosystems. *Advances in Marine Biology, 40*, 81–251.

Klein, R. J. T., & Bateman, I. J. (1998). The recreational value of Cley Marshes nature reserve: An argument against managed retreat ? *Water and Environment, 12*, 280–285.

Lee, S. Y. (1995). Mangrove outwelling: a review. *Hydrobiologia, 295*, 203–212.

Macnae, W. (1968). A general account of the fauna and flora of mangrove swamps and forests in the Indo-West-Pacific region. *Advances in Marine Biology, 6*, 73–720.

Macnae, W. (1974). *Mangrove forest and fisheries* (p. 351). Indian Ocean Programme Publication Number–34. Indian Ocean Fishery Commission. OFC/DEV/74/34.

Moberg, F., & Ronnback, P. (2003). Ecosystem services of the tropical seascape: Interactions, substitutions and restoration. *Ocean & Coastal Management, 46*(1–2), 27–46.

Morgan, J. P., & MacIntire, W. G. (1959). Quaternary geology of the Bengal Basin, East Pakistan and India. *Bulletin of the Geological Society of America, 70*, 319–342.

Nakicenovic, N., & Swart, R. (2000). *Special report on emissions scenarios: A special report of working group III of the Intergovernmental Panel on Climate Change* (pp. 1–599). Cambridge University Press.

Nandy, P., & Ahammad, R. (2012). Navigating mangrove resilience through the ecosystem based adaptation: Lessons from Bangladesh. In D. J. Macintosh, R. Mahindapala, & M. Markopoulus (Eds.), *Sharing lessons on mangrove restoration*.

Nandy, P., & Islam, M. A. (2010). Climate resilient coastal zone development in Bangladesh: Participatory Governafor common resource management. In A. I. Ramanathan, P. Bhattacharya, T. Dittmar, M. B. K. Prasad, & B. R. Neupane (Eds.), *Management and sustainable development of coastal zone environments* (pp. 58–72). Capital Publishing Company.

Naskar, K. R., & Guhabakshi, D. N. (1987). *Mangrove swamps of the Sundarbans—An ecological perspective*. Naya Prokash. 263 pp.

Nicholls, R. J., Cooper, N., & Townend, I. H. (2007). The management of coastal flooding and erosion. In C. R. Thorne et al. (Eds.), *Future food and coastal erosion risks* (pp. 392–413). Thomas Telford.

Odum, E. P. (1971). *Fundamentals of ecology*. W.B. Saunders.

Odum, E. P., & Heald, E. J. (1975). Mangrove forest and aquatic productivity. In A. D. Haster (Ed.), *Coupling of land and water system, ecological studies* (Vol. 10, pp. 129–136). Springer.

Ommer, R. E. (2007). *Coasts under stress: Restructuring and social-ecological health*. McGill-Queens University Press.

Paul, A. K. (2002a). *Coastal geomorphology and environment* (pp. 1–582). ACB Publication.

Paul, A. K. (2002b). *Tsunami the disaster in Indian Ocean nations* (p. 125). ACB Publications.

Paul, A. K. (2009). Man-Environment interactions in the single largest mangrove forest along the shoreline of the Bay of Bengal: A case study of Sundarbans. In H. S. Sharma & V. S. Kole (Eds.), *Geomorphology in India* (pp. 263–286).

Pethic, J. (1992). *Natural change in coastal zone and management* (pp. 49–63). Thomas Telford.

Pritchard, D. W. (1967). What is an estuary: Physical viewpoint. In G. H. Lauf (Ed.), *Estuaries* (pp. 3–5). American Association of Advancement of Science, Science Publication No. 83.

Richards, J. F. (1993). Agricultural impacts in tropical wetlands: Rice paddies for mangroves in south and Southeast Asia. In M. Williams (Ed.), *Wetlands: A threatened landscape* (pp. 217–233). Blackwell.

Saengar, P., Heger, E. J., & Davie, J. D. S., (1983) *Global status of mangrove ecosystem*. Commission on ecology papers No#. International Union for conservation of nature and Natural Resources. The environmentalist Supp.

Spalding, et al. (1997). The global distribution and status of mangrove ecosystems. *International News Letter of Coastal. Management-Intercoast Network, 1*, 20–21.

Stoliczka, F. (1869). The malacology of Lower Bengal and the adjoining provinces. *Journal of Asiatic Society of Bengal, 38*, 86–111.

Titus, J. G. (1986). Greenhouse effect, sea level rise and coastal zone management. *Coastal zone Management Journal, 14*, 3.

Titus, J. G. 1988 Chapter 1. In Sealevel rise and wetland loss: An overview. Chapter 1, pp 1–35

Titus, J. G. (1991). Greenhouse effect and coastal wetland policy: How Americans could abandon an area the size of Massachusetts at minimum cost. *Environmental Management, 15*(1), 39–58.

Tomlinson, P. B. (1986). *The botany of mangroves*. Cambridge University Press. 433 pp.

Tuler, S., Agyeman, J., da Silva, P. P., LoRusso, K. R., & Kay, R. (2008). Assessing vulnerabilities: integrating information about driving forces that affect risks and resilience in fishing communities. *Human Ecology Review, 15*(2), 171–184. [online] URL: http://www.humanecologyreview.org/152.htm

Wadia. (1961). *Geology of India*. M.C. Muller & Co. Ltd.

Woodroffe, C. D., Mulrennan, M. E., & Chappel, J. (1993). Estuarine infill and coastal progradation, Southern van Diemen Gulf, Northern Australia. Nature 317, 711–713.

Umitsu, M. (1997). Landforms and flood in the Ganges delta and coastal lowland of Bangladesh. *Marine Geology, 20*(1), 76–87.

Chapter 2
Physiography, Geological Origin, and Historical Background of Sundarbans, India

Abstract The origin of the Sundarbans is very baffling as biologically complicated and ecologically mysterious mangroves are grown up at crisscrossed estuarine networks. Multidimensional ecological interaction processes command an interdisciplinary research approach dealing with different hypotheses which have been put forward to explain and interpret the origins of Sundarbans. The most acceptable proposition is that a series of tectonic activities resulting in elevations and depressions of the gneissic hills located on the Gondwana belt occurred around 200 million years ago. This kind of geological event coupled with the gradual deposition of alluvial silt brought from the folded Himalayas of Northern India throughout several centuries, helped the development of the deltaic islands of Sundarbans. Sundarbans, being the largest chunk of mangroves of the globe, encompass so many structural components in the form of biotic (flora, fauna, and microbes) and abiotic components (water and soils) within a complicated but unique aquatic system experiencing tidal exposure and inundation twice a day and embraces hundreds of rivers, rivulets, creeks, canals, and estuaries. The mangrove ecosystem of the Sundarbans, is recognized as the most productive ecosystem of the world and represents the only tidal habitat for tigers (*Panthera tigris tigris*) in the world. It has achieved international recognition as a 'World Heritage Site' for its diverse flora and fauna, and to protect this ecosystem, along with the diversified biodiversity, a number of conservation strategies in the form of the creation of a Biosphere Reserve, a World Heritage site, National Parks, tiger projects, and sanctuaries have been made. After its geological origin, the Sundarbans along with its all structural components and local people have experienced chequered historical paths in respect of ecology, administrative control over resource utilization, peoples' interactive association, biodiversity enrichment and loss etc. Most of the people of this region depend on seasonal mono-cropping of agriculture in the reclaimed zones, but a good number of people are forced to engage in other forms of livelihood generation such as honey collection, aquaculture, and capture fisheries, prawn-seed collection, timbering, rearing of domestic animals, and tourism.

S. K. Chakraborty, S. K. Acharya, *Eco-dynamics and Environmental Perturbations in Mangrove Ecosystems*,
https://doi.org/10.1007/978-3-032-08710-2_2

This estuarine–mangrove ecosystem, with distinctive deltaic islands of different shapes and sizes, has been originated through very intricate and dynamic geomorphological, physiographical, and biological evolutionary processes during the Plio–Pleistocene inter-glacial period. The complex network of rivers, tidal creeks, and inlets have been under continuous erosion and accretion because of the dynamic flow pattern, which in turn structures the complex geomorphic set-up of the area with the natural creation of several geomorphological structures such as dunes, mudflats, sand-flats, beaches, inlets, estuaries, creeks, and mangrove swamps. However, Sundarbans, as the only mangrove tiger habitat of the globe, is now under threat of severe coastal erosion, undesirable changes in the land-use patterns, global climate-change-mediated sea-level rise, and other kinds of pollution.

Keywords Sundarbans · Historical perspectives · Conservation strategies · Livelihood generation · Geomorphological evolution · Administrative history · Mangroves of India

2.1 Coastal Environment of West Bengal with Special Reference to Sundarbans

The coastal zone, along with continental shelves, share about 15% of the total oceanic area and contribute one-third of the ocean's total biological productivity supporting diversified biodiversity components and ecosystems such as mangroves, salt marshes, coral reefs, sea-grasses, lagoons, kelp forests, etc. The coastal zones of West Bengal, India, with the mighty mangrove ecosystem of Sundarbans, is characterized by the continuous interaction of estuarine networks with tide-dominated interconnected rivers, tidal creeks, channels or inlets, and a vast expanse of mangrove forests along with the associated faunal components. The morpho-dynamic changes in this mangrove–estuarine complex are determined significantly by the macrotidal coastline with specific tides and waves coupled with other geological and tectonic processes. An extensive blanket of flood plain deposits, low ridges of natural levees, and channel bars associated with various environmentally significant sedimentary structures characterize coastal West Bengal, India. The estuarine coastal zones under Sundarbans, India are endowed with mudflats, sand flats, coastal dunes, beaches, a network of tidal creeks with mangroves, back swamps, salt marshes, etc. However, construction of embankments for the protection of reclaimed landforms from flooding and erosion with a lot of human activities such as agriculture, aquaculture, and silviculture, etc., in Sundarbans impose a series of adverse impacts on the fragile ecology of the mangrove ecosystem.

The Sundarbans mangrove ecosystem, being a biologically productive, ecologically sensitive, and aesthetically celebrated landscape, attracts the attention of people across the world for myriad reasons, which has made it a unique place not only for scientific inquisitiveness but also tracing the history of this region, integrating the geography, literature, and socio-cultural perspectives of West Bengal, India, and Bangladesh (Blasco, 1984; Didar et al., 2018). The largest block of lush green tidal halophytic plants dominating the mangrove ecosystem of the world, named Sundarbans, is located at the largest fluvial–estuarine networks of the Ganges–Brahmaputra deltaic plain at the north-eastern coastal tracts of the Bay of Bengal and is endowed with its unique biodiversity components, flora, fauna, and microbes. Lying at the southern tip of West Bengal, India, the vibrant ecological condition of this ecosystem supporting an endless diversity of habitats along with their biodiversity components has converted this place not only to offer unique aesthetically loaded beauty but also to make an abode for the world's strongest tiger species, locally named Royal Bengal Tiger (*Panthera tigris tigris*, endangered); fishing cat (*Prionailurus viverrinus*; endangered); deer (*Axis axis*); monkeys (*Macaca mulatta*); pigs (*Sus scrofa*); otters (*Lutra lutra*); Gangetic dolphin (*Platanista gangetica*; endangered); Irrawaddy Dolphin (*Orcaella brevirostris*; vulnerable); estuarine crocodile (*Crocodylus porosus*); a number of very poisonous snakes including king cobra (*Ophiophagus hannah*; vulnerable); other reptiles (*Varanus bengalensis*, vulnerable); river terrapin (*Batagur baska*; critically endangered); red-crowned roofed turtle (*Batagur kachuga*; critically endangered); leatherback sea turtle (*Dermochelys coriacea*; critically endangered); olive ridley sea turtle (*Lepidochelys olivacea*; vulnerable); hawksbill sea turtle (*Eretmochelys imbricata*; critically endangered); and a hoard of both residential and migratory birds such as greater spotted eagle (*Aquila clanga*; vulnerable); darter (*Anhinga melanogaster*; near threatened); greater adjutant stork (*Leptoptilos dubius*; endangered); lesser adjutant stork (*Leptoptilos javanicus*; vulnerable); black-headed ibis (*Threskiornis melanocephalus*; near threatened), and all of which have made a blend of nature's harsh beauty with terror representing the entire region an attractive place, and challenging the very existence of the local peoples for their own sustenance through developing agriculture, fisheries, collecting honey, resisting floods and other natural disasters.

In view of the above, the present chapter discusses geomorphic evolution of different geomorphological settings with special reference to the geological past of different river courses of Eastern India, mainly, the river Ganges and Brahmaputra along with their tributaries leading to the deltaic formations of Sundarbans, and also focuses on different research information pertaining to the changing trend of people's perception of the mangrove ecosystem of Sundarbans on the Indian subcontinent during last couple of centuries.

2.2 Mangrove Ecosystem of Sundarbans and Its Ecological Uniqueness

Mangroves, being an critical biologically productive ecosystem, restricts their distribution only to tropical and also to some extent to sub-tropical zones of the world, with their greatest expanses in the Indian and Pacific Ocean regions. However, estimates have pointed out that mangroves in the Asia Pacific region have been disappearing by 1–2%/year because of factors such as drastic changes in the land use patterns through the clearance of mangroves for agriculture and for aquaculture, settlements of humans in the reclaimed areas, climate change, and sea-level rise in the deltaic systems of the Ganges–Brahmaputra (e.g., the Sundarbans of the Ganges–Brahmaputra experiencing dramatic change (Duke et al., 2007; Syvitski and Saito, 2007; Coleman, Huh, and Saito et al., 2007).

This estuarine coast of West Bengal, having the Sundarbans mangrove ecosystem, is endowed with a network of several major estuaries and creeks, has been under a process of continuous change because of the steady interaction of different natural (asymmetric tidal flow regime, waves, intensity, and direction of currents, river fluxes, sea-level change, erosion, salinity invasion, natural disasters such as cyclones, tsunami, storms, etc.), and manmade (pollution, deforestation, nonjudicious exploitation of natural resources, construction of fishing jetties and aquaculture farms, eco-tourism, introduction of exotic species, etc.) reasons in time and space, making the entire geomorphic set-up of the entire ecosystem a complex one. Continuous erosion followed by deposition of eroded materials, resulting in accretion, as observed in the banks of the estuarine flows, result in the modification of the slopping gradients of intertidal belts, development of point bars and mid-channel bars, alteration in the sizes and shapes of the sand dunes, as sandy deposits are eroded because of re-suspension followed by transportation by seaward tides with associated waves, whereas scouring and transportation of mud take place by suspension. The prevailing geotectonic activity on the Bengal Basin is influenced by the erosion and accretion process over the Sundarbans' mangrove estuarine deltaic system.

2.3 Physiographical Evolutionary History of Indian Mangroves and Counteracting Impacts from Anthropogenic Developmental Issues

2.3.1 Indian Mangroves and Their Distribution

In India, the Sundarbans region possesses the largest extent of mangrove cover, encompassing 43% of the total, followed by Gujarat, holding the second-largest share with 23% of the entire cover in India. Despite facing adverse environmental conditions characterized by high-energy tidal coasts and encountering ecological conditions, both of these ecoregions encompass a mangrove coverage of two thirds of the entire mangrove coverage in India. For instance, a humid and wet climate prevails in the Sundarbans, with higher biodiversity in contrast to the mangrove forests of Gujarat, existing in arid and dry conditions with a lower level of biodiversity. The Andaman and Nicobar Islands located in a low-energy tidal coast have very rich biodiversity and possess the third-largest mangrove forest in India, with a mangrove coverage of 13% of the total mangrove expanse in India.

The Sundarbans, a world heritage site, declared by the UNESCO in the year 1987., locating within the Gangetic Delta, is the largest mangrove wetland experiencing the highest sedimentation on a global scale. When considering both India and Bangladesh, the expanse of mangroves of the Indian Sundarbans jointly with the Bangladesh Sundarbans forms the largest mangrove forest, supporting innumerable flora and fauna, including the Royal Bengal Tiger (*Panthera tigris tigris*), Gangetic dolphins, estuarine crocodiles, horse shoe crabs, water monitor lizards, river terrapins, and fishing cats, and all of them have the power of adaptation to overcome and survive within tide-dominated tidal forests that are uniquely inhabited by them (Kathiresan et al., 2001, Kathiresan et al., 2005; Chakraborty, 2011, 2017).

Other states and provinces of India possessing mangroves are Odisha, Tamilnadu, Maharastra, Andhra Pradesh, Karnataka, Kerala, Maharastra, and Goa. India is ranked as the third most mangrove-based biodiversity-enriched country, following Indonesia and Australia (Ragavan et al., 2016). Bhitarkanika mangrove forest, located in the state of Odisha, is endowed with renowned mangrove genetic diversity of global concern, supporting the lives of so many birds and crocodiles, including the rare albino crocodile. Other coastal biodiversity components having global importance that are found within India's mangroves include: (i) the largest nesting site for olive ridley turtles along the Gahirmatha coast of Odisha; (ii) flamingos in the Run of Kachchh in Gujrat; (iii) sea-grass meadows inhabited by the sea cow (Dugong); (iv) coral reefs adorned with exquisite ornamental fish; and (v) intertidal mudflats bustling with a diverse array of migratory and resident bird species (Kathiresan et al., 2005).

2.3.2 Mangroves in Different Parts of India

Mangroves being a unique creation of nature settle, grow and flourish at the estuarine–land interphase, sustaining lot of natural incidents such as daily tidal inundation and exposure, threats of soil erosion, over-flooding, and thrusts from catastrophic cyclones, and overcoming all those odds with their inherent adaptive characteristics are extensively distributed along the sheltered areas of tidal creeks and channels, strait fringes, shallow bays, and embayed shores of the different deltaic islands. The open shore platforms experiencing direct wave attacks do not support the settlements and colonization of mangroves as they do not provide favorable environmental gradients. The prevailing environmental gradients control the stability of the sand spits and barrier bars, which promote the growth and distribution of mangroves over the tidal flats.

Mangrove forests, as an integral part of the mangrove ecosystem, act as a buffer against many natural disasters such as catastrophic storms, tsunamis, tidal waves, littoral currents, erosion, vertical accretion within the coastal fringes, and intertidal belts of different physiographic settings on the East Coast and West Coast of India, which are characterized by estuarine deltaic islands with swampy zones, clay-dominated swale and surface sediments, carbonate platforms of the coral fringed coasts, etc.

Mangroves cover an area of 356,500 km^2 along the coastline of India, which stretches for over 5700 km^2, representing 85% of the deltaic islands of Sundarbans in India. A few isolated patches of mangroves remain in Kerala. The isolated pockets of mangroves found at the Mahanadi, Godavari, Krishna, Narmada, and Tapti confluences are very vulnerable. Small populations of individual species are particularly vulnerable and in immediate danger of extinction. The Bombay region primarily has mangrove species such as *Acanthus ilicifolius* and stunted *Avicennia marina*, whereas degraded areas in Kutch in Gujrat support *A. marina*. Although a total of 58 species of halophytes have been recorded from all over India, overexploitation has led to the gradual elimination of species from many areas. The drastic decrease in mangrove forests across the country is apparent. The continued conservation and mangrove management of the Sundarbans, the largest tract of mangroves in India, therefore plays a significant role in the state of mangroves in the country.

Estuaries, being semi-enclosed coastal bodies of water, act as geological sinks for downstream supply of sediments and encroachment of tidal sediments into the embayment from the seaward side. The bidirectional water flow patterns within estuaries loaded with sediments are deposited on the wide shore-fringed tidal flats of the growing deltaic islands and then become suitable sites for the settlements and growth of mangroves and salt marshes. In India, the mouths of so many great estuaries near the confluence with the sea, such as Hooghly, Subarnarekha, Brahmani-Baitarani, Mahanadi, Godavari, Cauvery, Krishna, Mandovi, etc., possess these kinds of physiographic settings. Barrier bar lagoons and coastal embayments which also support the mangrove habitats depending on the fresh-water supply from riverine flows in the adjoining areas and influxes of saline water from the sea. Mixing of sea water with riverine fresh water promotes the deposition of fine sediments and influences formation of salinity gradients, as salinity tends to increase in the lagoons or coastal embayments after the disconnection of the landward distributary channels such as the barrier back tidal lagoon of the Subarnarekha estuary, Chilika lagoons, Hukitola bay, Muthupet of Cauvery delta, Coringa bay, Pitchavaram backwater system, etc., where mangrove vegetation settles and grows. Coastal belts in the states of Karnataka and Kerala exhibit similar kinds of physiographic settings on the side of the Arabian Sea where environmental gradients developed by the stability of sand spits and barrier bars govern the process of the growth and distribution of mangroves.

Along with the mangroves of Sundarbans, West Bengal, other states of India that possess mangroves along their maritime and estuarine belts are Odisha, Andhra Pradesh, Telangana, Tamil Nadu, Gujrat, Kerala, and Indian territories such as Andaman and Nicobar Islands.

2.3.2.1 Mangrove Belts of the East Coast of India

Mangroves of eastern coastal tracts of India flourish in the river deltaic systems with beach-ridge cheniers and tidal ranges, barrier lagoons characterized by a mixed wave–tide environment, carbonate platforms and **'ria'**-type coastal intertidal belts influenced by micro tidal coastal environment along the East Coast of India fringed with the Bay of Bengal and Andaman Sea.

West Bengal Coast in Sundarbans: Sundarbans of India located in the southern part of the river delta of the Ganges is characterized as having a large expanse of mangrove forest, covering an area of over 4264 km^2 on the tide-dominated deltaic depositional surface crisscrossed by tidal creeks or channels. Out of 104 deltaic islands within this mangrove estuarine complex, 54 have already been reclaimed, whereas the remaining 50 are still virgin, with diversified forms of flora and fauna suitable for surviving in this very rigorous and fragile ecological condition (Paul, 2002a; Chakraborty, 2011). The landscapes of deltaic islands of Sundarbans, India, exhibit different pictorial views in the different phases of diurnal and seasonal tidal cycles along with their species composition of mangrove plants, an ecological succession of biotic components, gradients in the elevation of intertidal belts, and the networks of estuaries with their connecting tidal creeks (Figs. 2.1, 2.2, 2.3, 2.4, 2.5, and 2.6).

Mangroves on Odisha coast: A vast expanse of mangrove forest with the highest diversity of mangrove floral species is found in the state of Odisha, India, especially in sheltered flats of Bhitarkanika Biosphere Reserve, where deposition of finer alluviums help the development of tidal forests with a meso-tidal range (2–2.5 m) on Dhamara and Baitarani rivers forming the curved shoreline of the Bay of Bengal. Hukitola Bay mangroves also are found within the coastal embayment of Mahanadi delta, where a number of sand spits and barrier bars protect the waves by acting as sand-ridge barriers along the seafront. Mangroves and salt marshes also occur on the estuary-fringed tidal flats of Mahanadi River and Devi River. In this state of Odisha, a major portion of the sheltered tidal flat of the shore fringe occurs at the islands of Brahmani River mouth, named Kalahandia, in Bhadrak District, and the maximum part is located at Bhitarkanika tidal flats separated by the river systems of Baitarani-Brahmani Delta at Kendrapara District (Figs. 2.7, 2.8, and 2.9). All of these eco-zones offer a favorable habitat for mangrove and salt marsh vegetation to grow and flourish in the sheltered shore-fringed protective barrier spit, endowed with sand dunes all along the low lying deltaic flat. Moreover, the state of Odisha possesses an extensive mangrove-dominated coastal belt particularly at the sheltered flats toward the north of Dhamara River. The sea Bay of Bengal, with its curved shoreline, provides a sheltered environment for deposition of finer alluviums as supporting materials needed for the growth of mangroves as tidal forests with a meso-tidal range (2–2.5 m). The presence of a number of sand spits and barrier bars at Hukitola Bay mangroves within the coastal embayment of Mahanadi delta protect the mangroves from the waves and swell of the open marine environment by acting as sand-ridge barriers along the seafront. This type of physiographic setting in turn provides sheltered water bodies behind, with a wide opening. Both the mangrove forest and salt marshes are also observed to occur along the estuary-fringed tidal flats of the Devi River mouth and the Mahanadi River mouth of this area. In the Chilika lagoon, which receives sea water from the Bay of Bengal and fresh water from two main rivers, Bhargavi and Daya, both the branches of the Mahanadi river provide suitable habitats for the growth of mangroves because of the deposition silts out of the mixing process of sediment-loaded sea water and fresh water (Fig. 2.10).

Mangroves on the Andaman Coast: In the group of islands of Andaman and Nicobar Islands, the very presence of carbonate banks on shore fringes and inner

Fig. 2.1 Landscapes of Sundarbans (01). (**A**) Vast stretch of productive estuarine water with less turbidity and more transparency, but the main flow is bifurcated by the development of a deltaic islands supported by lush green healthy mangroves at the core of Sundarbans Biosphere Reserve, away from any kind of direct human intervention; (**B**) Ecological successional stages in the mangrove ecosystem of Sundarbans, India depicting water and mudflat with newly settled *Avicennia alba* and then luxuriant growth of *Sonneratia apetala*.(**C**, **D**, **E** and **F**) Intermingling of mangrove forest composed of different species of mangrove plants on the different estuaties of Sundarbans, India, under the influence of different ecological gradients mainly developed by salinity, availability of sunlight, light penetration ability, turbidity, temperature, intensity of water current, tidal amplitudes, and the export and import of detritus contributed nutrients in between the mangrove forests and estuarine water

Fig. 2.2 Landscapes of Sundarbans (02). (**A**, **B**, **C**, **D**, **E** and **F**) The entire Sundarbans biosphere reserve is criss-crossed by thousands of tidal creeks, the banks of which are supported by an array of mangrove plant species displaying different species assemblages developed in tune with distinct ecological succession patterns. All of these creeks enjoying meandering pathways take the tidal water from the main stretch of estuary and also help to exchange materials (semi-decomposed and decomposed plant litters) in between the forest subsystem and the estuarine subsystem of the entire mangrove ecosystem of Sundarbans, on the Indian subcontinent

Fig. 2.3 Landscapes of Sundarbans (03). (**A**) Tidal creek with stable banks with hard compact clayey sediment supported by mangrove thorny shrubs, *Acanthus ilicifolius*, connecting the sea, Bay of Bengal with different aquaculture farms on the Midnapore (East) coast of West Bengal, India. (**B**) Tidal large creek with luxuriant growth of mangrove plants, *Aegiceras corniculatum*, providing the passage of fish shoal from the estuaries to mangrove forests and vice versa. (**C**, **D**, **E** and **F**) Different types of tidal creeks in the Sundarbans mangrove ecosystem during low tide showing their meandering pathways

Fig. 2.4 Landscapes of Sundarbans (04). (**A**, **B**, **C**, **D**, **E** and **F**) Estuaries of Sundarbans in India showing their different patterns of geomorphological configuration and vegetational assemblages in the adjoining mangrove forests. Bend of the main estuarine flow by the development and extension of mudflats stabilized by the settlement and growth of mangrove saline grass, *Porteresia coarctata*. (**B** and **F**) Vast expansion of estuary flanked by the deltaic islands reclaimed by humans and therefore growth of different mesophytic plant species along with true mangrove vegetation. (**C**, **D** and **F**) Exhibit luxuriant growth of mangroves on the bank of the estuaries such as *Bruguiera cylindrica, Aegiceras corniculatum*, and *Sonneratia apetala* at the time of high tide in Indian Sundarbans

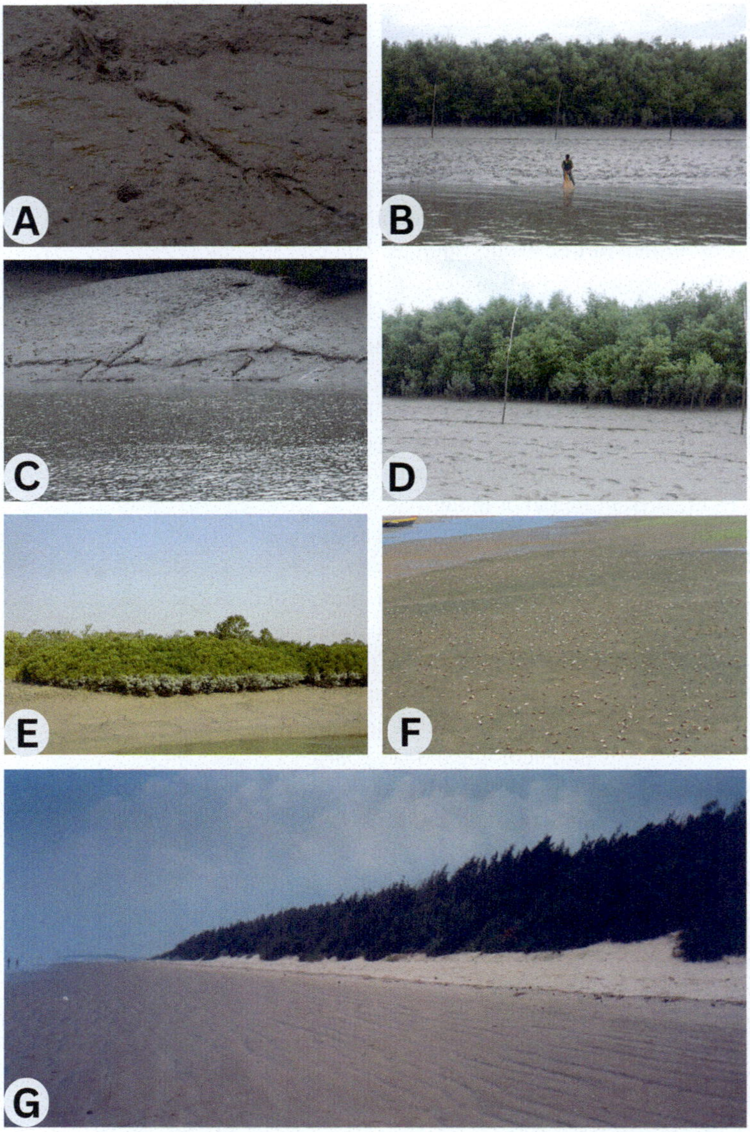

Fig. 2.5 Landscapes of Sundarbans (05) (**A**, **B**, **C**, **D** and **E**) Different intertidal mudflats in different parts of Sundarbans bridging the gap in between a dense mangrove ecosystem and shallower subtidal zones of the large estuarine networks and acting as the reservoir of a huge amount of mangrove leaf litter contributing detritus as organic nutrients promoting the settlement, growth, and proliferation of a galaxy of benthic fauna (macro-benthos, meiobenthos, and micro-benthos) in different parts of the intertidal zone (supra-littoral, mid-littoral, and infra-littoral). Images (**A** and **C**) also depict the initial phases of the development of creeks because of the differential patterns of tidal currents coupled with destabilization of surface sediment due to erosion processes. (**F**) Intertidal belt with loamy sediment with a higher percentage of sand particles, more than 80% of sands in the sediment support the lives of innumerable benthic fauna such as fiddler crabs, *Uca annulipes annulipes*, as depicted in this photo. (**G**) Sand flats with more than 95% sand particles and continued with dunes supported specific types of vegetation such as *Casuarina equisetifolia*

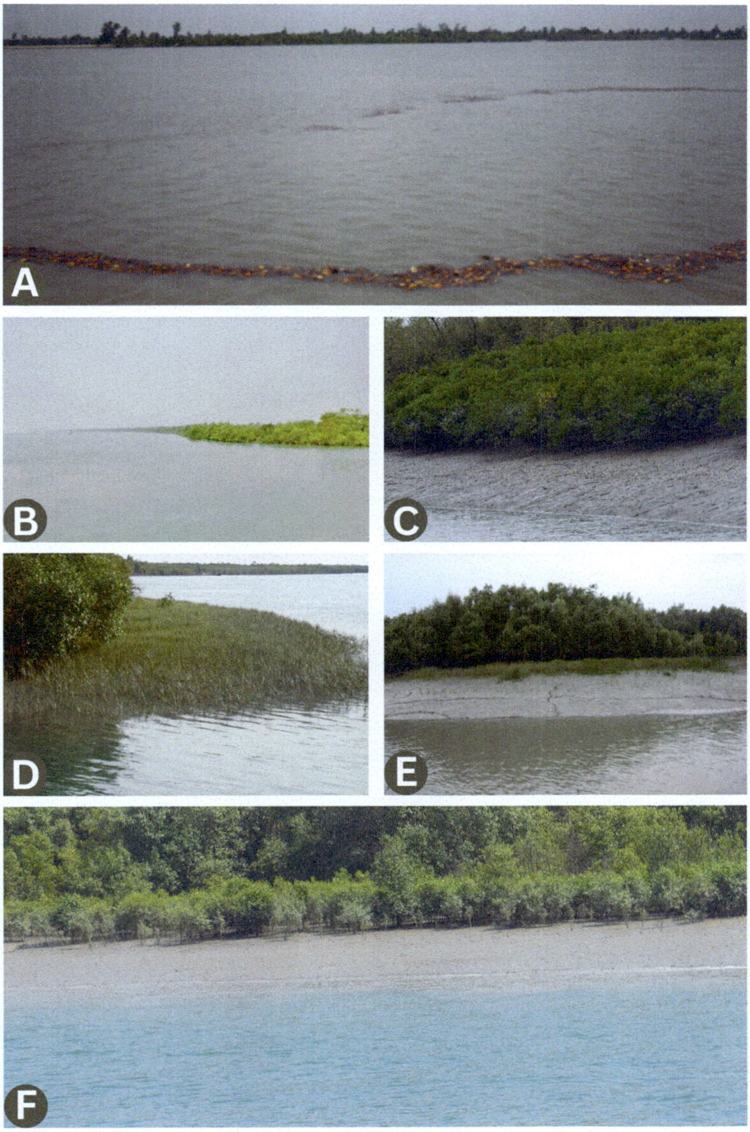

Fig. 2.6 Landscapes of Sundarbans (06). (**A**) Initial phase of low tide when lots of mangrove plant litter is carried away and redistributed in different parts of Sundarbans mangrove forest for further microbial decomposition to detritus-based nutrients. (**B**) Vast expanse of estuarine water flanked by luxuriant growth of the mangrove forest, which is inundated twice a day and supplies a huge amount of mangrove-contributed detritus as nutrients to the adjoining estuarine water. (**C**) Extensive expanse of mudflat supported by a specific type of mangrove vegetation, *Aegiceras corniculatum*. (**D**) Mudflat in the bend of estuarine flow and development of the grass belt, *Porteresia coarctata*, which acts as strong sediment binder and thereby resists the washing away of the sediments from this critical juncture of estuary–mangrove forest interaction zone. (**E** and **F**) Extensive mudflats exhibiting successive stages of development of mangrove vegetation above them

Fig. 2.7 Mangrove landscapes in Bhitarkanika National park, Odisha, India (1) (**A**) Jetty for the conveyance of people by water vessels. (**B**) Grooves of mangroves of the species *Kandelia candel* over a creek within this biosphere reserve. (**C**) Creek through the groves of *Xylocarpus mekongensis.* (**D**) Creek supported by mixed vegetation of mangroves on its bank including saline grass. (**E** and **F**) Mangroves represented mainly by the species *Aegiceras corniculatum.* (**f**) Setting sun on the mystic tract of the creek within Bhitarkanika National Park; (**G**) Nypa fruticans(Thunb) Wurmb on the supratidal zone surrounded by saline grass, Porteresia coarctata Takeoka. (**H**) Common creek guarded by dense mangroves when the rays of setting sun tinged the entire area

Fig. 2.8 Faunal diversity within Bhitarkanika National Park, India (2). (**A** and **B**) Tourist entry point of Bhitarkanika National Park, and one of the Ramsar Sites in India that has a vast expanse of mangroves with associated networks of tidal creeks. (**C**) Flock of birds, lesser sand plover at the interphase of mangrove forests and creek at Bhitarkanika National Park. (**D**) The rhesus monkey, *Macaca mulatta*, a gregarious form of the old-world monkey on the grass belt of this eco-region. (**D**, **E**, **F** and **G**) The Bhitarkanika National Park is famous for the higher abundance of estuarine crocodiles, *Crocodylus porosus,* the abundance of which has ensured great protection of the National Park, which has the highest diversity of mangrove plant species in India

Fig. 2.9 Mangrove Conservation efforts in Bhitarkanika National Parks, Odisha, India (3) (**A**, **B**, **C**, **D** and **E**) All those images depict different conservation strategies for the conservation of mangroves through an afforestation program: birds, crocodiles, and olive ridley sea turtles, *Lepidochelys olivacea*, which used to pay a visit each year to the Gahirmatha coast, close to the Bhitarkanika National Park, in huge numbers from the Pacific Ocean where the females lay their eggs and the hatched out juveniles again migrate to their habitats in the deep ocean. These conservation efforts in the laboratory, museum, and open fields promote the development of awareness of the conservation of the entire ecosystem. (**F** and **G**) Nesting ground of olive ridley sea turtle on the vast stretch of sandy intertidal belts which are protected

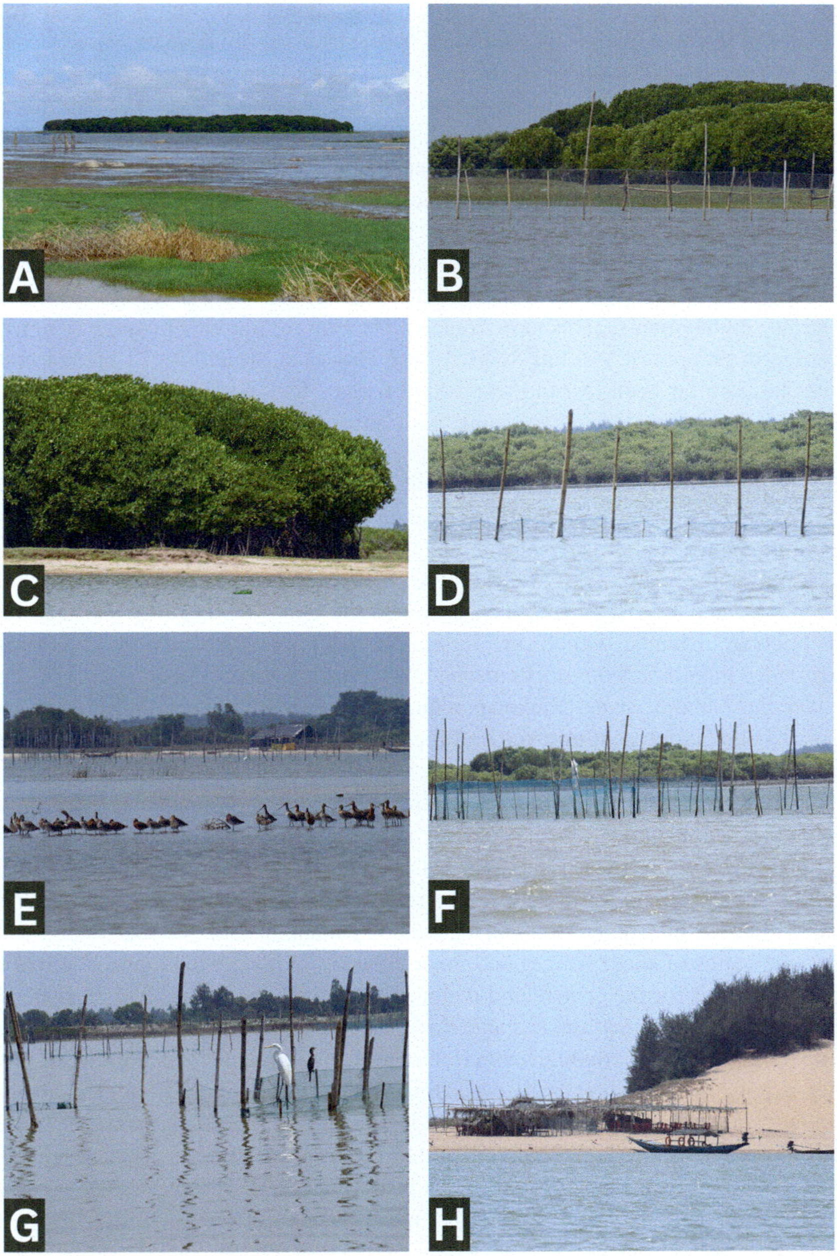

Fig. 2.10 Mangrove landscapes and around in Chilika lagoon of Odisha, India (4) (**A**, **B**, **C** and **D**) Healthy growth and expanse of mangroves at Chilika lagoon in the state of Odisha. (**E**) Profuse fishing activities in the biologically productive aquatic bodies fringed by mangroves, reflected by the abundant presence of several aquatic birds. (**F** and **G**) Fishing activities of local fishermen through indigenous methods (**H**) Sand flat continuing to large sand dunes supported by *Casuarina equisetifolia* on the parallel beach ridges as wind breakers and also to protect tidal flats

creeks support a wide range of environmental gradients that control the evergreen character of the mangroves. An additional fresh water supply is taken place because of the seepage flow of water or freshwater breaches into the creek margins and shore fringes, and the mixing of salt water with fresh water promotes the settlement, growth, and dispersal of the mangroves. Sediments with bio-encrusted surface on the carbonate banks enriched with bioclasts, chemogenic cements, micrites, coralline debris, and a low amount of siliciclastic materials have provided suitable ecological conditions for mangrove forests to flourish in some sheltered zones of Andaman and Nicobar Islands. In the Andaman and Nicobar Islands, nearing the shore fringes of the Bay of Bengal and the Andaman Sea, different, estuaries, channel straits, embayments, and shore-fringed platforms are found to possess mangrove coverage of an area of 966 km^2 (FSI, 1999). However, around 23.77 km^2 of mangroves have so far been lost in these island fringes for several geo-ecological reasons such as the emergence of land, land subsidence, and tsunami-like natural disasters (MOEF, 2012).

In the Andaman and Nicobar Islands, the luxuriant growth of mangroves with about 20 species is observed in sheltered flats of the inner parts of different estuaries, embayments, sheltered bays, and tidal inlets with back waters fed by river flows, larger along the shore fringes of the Arabian Sea on the west coast of India (Figs. 2.11 and 2.12). Eight species of mangroves exclusively occur in the Andaman and Nicobar Islands, which are *Lumnitzera littorea*, *Rhizophora lamarckii*, *Rhizophora mohanii*, *Sonneratia urama*, *Sonneratia gulngai*, *Sonneratia griffithii*, *S. lanceolata*, and *S. ovate* (Ragavan et al., 2016). Mangroves of the Andaman and Nicobar Islands represent **one-sixth** of the mangroves of the country (Chakraborty and Naskar, 1988).

Mangroves in the physiographic settings of this eco-region have been killed mostly by the incidence of earthquakes, tsunami waves, cyclone breaks, and by sediment drifts at high wave-energy levels. Higher and abrupt cementation of carbonate sediments considerably affect the mangrove regeneration process (Banerjee et al., 1989). In different groups of islands of Andaman and Nicobar Islands, the carbonate banks of shore fringes and inner creeks provide a wide range of environmental gradients for the growth and maintenance of the evergreen characteristics of mangroves. The breaches and seepage flow of fresh water into the shore fringes and creek margins develop conducive environmental conditions with an additional fresh water supply, which mixes with the saline water of the sea and thereby develop brackish water conditions to help formation of mangrove ecosystems under fragile conditions. Carbonate banks are constituted by coralline debris, micrites, chemogenic cements, bioclasts, and other siliciclastic materials. The sediments enriched with calcium carbonate with a bio-encrusted surface on the relatively sheltered shores of the Andaman and Nicobar Islands have provided suitable ecological support to the development of mangrove forests.

The physiographic setting at drowned bedrock valleys on the Andaman and Nicobar Islands with mangrove vegetation is tectonically not suitable and prone to the prospective impacts of earthquakes and tsunamis. This type of shore is controlled by encroached swelling sea water because of high tides and waves with

Fig. 2.11 Mangrove landscapes in the Andaman and Nicobar Islands (1) (**A**) Heterogeneous association of mangroves with other large mesophytic species. (**B**) Sampling of water and research survey of mangroves by a team of researchers. (**C**) Dense mangrove patch in the core areas of the mangrove forest in the Andaman Islands with estuarine water with higher transparency and less turbidity. (**D**) Mangrove fringe area with narrow sand flat but with less turbid water. (**E** and **F**) Shallow-water mangrove zone in the Andaman Islands acting as a suitable site for the breeding and nurturing of juvenile finfish and shell fish; (**G** and **H**) Mangrove belts in Andaman Islands under the process of eco-degradation due to overdeposition of sands

Fig. 2.12 Mangrove landscapes in the Andaman and Nicobar Islands (2). (**A**) An ideal shallow-water mangrove area with sand-flat and vast sheet of transparent water. (**B**) Brackish water impoundment near the mangrove belt used for the culture and fattening of edible crab, *Scylla serrata*. (**C**, **D** and **E**) Research survey at different eco-zones in the core area of the Andaman Islands. (**F**) Harvesting of fruit from the mangrove plant, *Xylocarpus granatum*. (**G** and **H**) Eco-degradation of mangrove areas because of different developmental activities near Port Blair of the Andaman Islands

higher intensities shift marine sediments to fill the shallow basins and water-accumulated valleys developed by the process of subsidence in the coastal belts. Continuous processes of uplift and subsidence events result in shifting of inter-tidal, sub-tidal and supra-tidal zones of the unstable coasts with both emerged and submerged forests. The hydraulic head of saline water pushes landward during high tides, transmitting of pore-water salinity through tidal pumping to facilitate the growth of mangroves on the carbonate sediments. However, from the landward side, the fresh-water breaches into the modern platform during low tide reduce the pore-water salinity of the shore fringes along the islands and therefore some less saline-tolerant mangrove species are observed to inhabit the upper-middle parts of the carbonate shore platform in the Andaman and Nicobar Islands.

Mangroves in the Andaman and Nicobar Islands are extensively distributed along the sheltered areas of shallow bays, channel strait fringes, tidal creeks, and embayed shores, which are under direct wave attacks and abrupt cementation of carbonate sediments in many places, making the settlements, colonization, and regeneration of mangroves difficult. Different natural disasters such as the incidence of tsunami waves, earthquakes, cyclone breaks, drifting of sediment onto the shore platforms at high wave-energy levels result in the mass mortality of mangroves. About 20 mangrove plant species occur in the mangrove forests of the Andaman and Nicobar Islands (Banerjee et al., 1989).

Mangroves in the coastal belts in the state of Telangana: In the state of Telangana, mangroves are developed on the shore-fringe forests in Coringa bay on Godavari delta along the low-lying delta-plain tidal flats fed by distributary-channel systems. The species diversity in the tidal flats of Mahanadi delta is found to be greater than in the Godavari delta. The mangrove forests and related wetlands distributed over 156 km^2 in the low-lying deltaic tracts on fringes and islands on the estuary in Krishna Delta (FSI, 1999). A large area of island mangroves in the deltaic islands is found to have been degraded owing to erosion in the Masulipatnam Bay resulting in shoreline changes of the Krishna River. From an overall perspective, the mangroves are degraded and a total of 59.91 km^2 mangrove-based areas in the Godavari and Krishna Deltas have been lost (MOEF, 2012). Mangroves in Godavari delta occur in the shore-fringe forests of Coringa bay behind the Godavari spit, and along the low-lying delta plain tidal flats fed by distributary channel systems in the Godavari delta of Telangana Province. The mangroves and mangrove associates in the Coringa Reserve Forest extend over the deltaic depositional tidal flats within the embayment settings and protected by Godavari spit from an open marine environment. The number of mangrove species is relatively poor in comparison with the tidal flats of Mahanadi delta. The mangrove forests in the Krishna Delta have been developed in the low-lying parts of the deltaic islands, back barriers, and estuary fringes of the Krishna Delta. Mangrove areas were distributed over an area of about 156 km^2 in deltaic islands (FSI1999) but in a large area, measuring of 59.91 km^2 mangroves are found to have been degraded owing to the erosion of the island platform in Masulipatnam Bay, by the development of salt pans and also by the alteration of shoreline towards the delta front areas of Krishna River (MOEF, 2012, in: Coastal Zones of India). The deltaic shores are found to be very much prone to wave-induced erosion and occasional occurrence of cyclone, with a significant height of storm surges (NCRMP, 2016). The number of mangrove species in the delta ranges from 32 to 35.

Kerala Coast: Mangroves occur in backwater fringes of the state of Kerala covering an extended area of 7.00 km^2 of the tidal flats along the shoreline of the Arabian Sea. The mangrove assemblage of Kerala comprises nine species of true mangroves and a few nonmangrove halophytes and other mangrove associates. Eco-restoration alongside promoting eco-tourism are being undertaken by the environmental planners and administrators of the state of Kerala in the face of continuous environmental threats (Fig. 2.13).

Mangroves in the coastal belts of the state of Tamil Nadu: In the coastal belt of Tamil Nadu, the barrier back tidal basin fed by the Coleroon River. Pichavaram is the area of luxuriant growth of mangroves (Fig. 2.14). Naturally controlled freshwater discharge through the Coleroon River into this coastal zone has created the conducive environmental gradient for the survival of salt-tolerant mangrove plant species (Selvam, 2003). The Muthupet area of the Cauvery delta, located behind the barrier, is fringed with a tidal basin experiencing a low tidal range (0.50–1.00 m), waves with a height of over 1.5 m, salt water influences and supports the development of the mangrove forest covering an area of 12.00 km^2, as observed in 1999, which increased to 32.55 km^2 in 2016 because of the undertaking of restoration programs. It has 20 different mangrove species (Venkatesan 1966). The Cauvery delta at Muthupet area in the state of Tamilnadu is surrounded by the districts of Tiruvarur, Thanjavur, Mayiladuthurai, and Nagapattinam, and is formed by the confluence of two rivers, Kaveri (the third largest river after Godavari and Krishna in Southern India) and Kollidam, forming **'Rangam Island,'** fringed with a tidal basin.

2.3.2.2 Mangrove Belts of the West Coast of India

Mangroves occur in several backwaters fed by river flows in the larger embayments, sheltered flats of the inner parts of estuaries, and sheltered bays along the shore fringes of the Arabian Sea on the west coast of India. Dense coverage of mangroves is found only in the shore fringes of the Gujarat coast, the Maharashtra coast, and the Goa coast, but the coverage becomes open and fragmented on the Karnataka coast and the Kerala coast. The composition of species compared with river deltas of the east coast is not diverse on the coast fringed by the Arabian Sea.

Gujarat coast: In the state of Gujarat, the Gulf of Kutch, the Gulf of Cambay, and the Gulf of Khambhat are the coastal areas where good patches of mangroves comprising eight to ten true mangrove species and seven to eight nonmangrove halophytes occur (Banerjee et al., 1989). The total coastal area of the Gujrat state with mangroves is about 1026 km^2 (FSI, 1999). This total mangrove-covered area encompassing saline flats and degraded scrub land is divisible into dense mangroves (732.20 km^2), very dense mangroves (248.41 km^2), degraded mangroves (29.789 km^2), and dense saltmarsh vegetation (18.25 km^2) (MOEF, 2012). The freshwater flows of Luni River, Sabarmati River, and Mahi River hardly reach the coastal zones owing to the aridity of the inland areas, but the coastal belt receives

Fig. 2.13 Ecological status of mangroves in different parts of the state Kerala, India. Present ecological status of mangroves in Kerala: (**A**) Vast expanse of mangrove on backwater intermingling with coconut trees, used for both fishing and water boating at Kattampally Kayaking, Kannur, Kerala; (**B**) Healthy growth and vast expanse of mangroves on back water, mainly represented by Rhizophora apiculata (**C**), (**D**) *Excoecaria agallocha*, (**D**) *Acanthus ilicifolius* at the back water of Kattampally Kayaking , Kannur, Kerala; (**E**) Stretches of mangroves on the bank of Perumba river Udumbunthala backwater at Kannur, Kerala reflects sound ecological health through proper eco-management of the region; (**D**) Assemblage of aquatic birds, Little cormorants, (Microcarba niger) in the fringe of mangroves at Udumbunthala backwater at Kannur, Kerala; (**E**) and (**H**): Mangroves represented by Aviccenia alba, (**A**) Marina , Ceriops decandra, etc.on the brackish water Perumba river, near Valapattanam bridge, Pappinisseri, Kerala used for fishing by local fishermen; (**F**) (**G**) and (**I**) Laxuriant growth of mangroves mainly dominated Rhizophora spp, *Excoecaria agallocha*. Avicennia spp in the Vembanadu back water in Kochi, Kerala and (**J**) Assemblage of aquatic birds, Little cormorants, (Microcarba niger) in the fringe of mangroves at Udumbunthala backwater at Kannur, Kerala

Fig. 2.13 (continued) **Ecological and economic prospects of mangroves in the heart of Kochi city , Kerala highlighting the environmental threats perspectives.** The famous traditional mode of fishing named as Chinese Mode of Fishing at the Vembanadu back water of Kochi, Kerala with Chinese Fishing Nets (about 10 meters high structures, operated by about 6 fishermen using counterweights); (**B**) Practicing of cage culture for harvesting fishes in controlled ways from the nutrients enriched mangrove based water bodies at Vembanadu back water, Kochi, Kerala; (**C**) Tourists from different parts of the world experience the wilderness of natural settings of mangroves at Vembanadu back water in Kochi, Kerala; (**D**) and (**E**): Natural ecological settings of mangrove forest in the Vembanadu back water, of Kochi, Kerala; (**F**) Deteriorating ecological condition through accumulation organic debris within mangroves at Vembanadu back water in Kochi, Kerala; (**G**) and (**H**) threats of bioinvasion of mesophytic plant species like Gliricidia sepium of Family Fabaceae, Thesapesia populnes, of family Malvaceae, Sesuvium spp, etc.; (**I**) Mangroves near shipping ward of Kochi, Kerala are well adapted sustaining the exposure of industrial activities and (**J**) Mangrove patches near shipping ward of Kochi, Kerala under the protection for eco-reclamation.

Fig. 2.14 Pichavaram Mangroves of Tamilnadu, India (**A**, **B**, **C**, **D**, **E**, **F**, **G**, and **H**) Ecologically healthy mangrove forest with highly productive less turbid water in the Pichavaram area of Tamilnadu where images (**D**, **F**) depict the shallow-water zone encircled by mangrove roots and branches to make this area suitable habitats for juvenile fish

some fresh water that has been discharged by the Narmada River and the Tapi River at the coast.

Maharastra coast: In the state of Maharastra, mangrove forests with about 11 mangrove plant species occur in the estuary fringes of Vaitarna River, Ulhas River, and Mashala creek, and adjoining the shore fringes of Raigad District to the south of the Konkan coast. The estimated coastal area in Maharashtra with mangrove vegetation was about 300 km^2, which was consequently reduced to 212.60 km^2 (MOEF, 2012).An additional area of 67.05 km^2 of extensive saltpans with more mangrove patches has also been found. Around 31.39 km^2 of the mangroves on the Maharashtra coast have been degraded to date. The fringes of the Ulhas estuary and Vaitarna estuary toward the north and the tidal Mashala creek in and around shore fringes to the south of Konkan coast of Raigad District support the development of mangroves and saltmarsh communities.

Karnatak coast: Karnatak coast is characterized by having drowned bedrock valleys or 'ria'-type shores forming a unique physiographic setting for the growth and distribution of mangrove vegetation. Such a physiographic setting is found to be tectonically unstable in nature owing to the occurrences of earthquakes and tsunamigenic impacts. The unstable coast experience of shifting of intertidal, subtidal, and supratidal zones is due to the uplift and subsidence events. This coastal tract supports both the emerged and submerged forests. During high tides, the hydraulic head of saltwater forcibly moves in a landward direction and facilitates transmission of pore water salinity during tidal pumping to support mangroves in the emerged platforms of carbonate sediments. However, the freshwater breaches the modern platform from the landward sides during low tide to reduce the pore water salinity of the shore fringes along the islands, thereby enabling freshwater-loving mangroves to grow in the upper middle parts of the carbonate shore platform of the Karnatak coast. The rivers of western Karnataka meet with the Arabian Sea at the Malabar coast, forming tide-dominated estuaries that spread over the coastal basin receiving the supply of fluviomarine sediments loaded with nutrients, which support the mangroves in the coastal zones of Karnataka State. Out of 11 species of halophytic mangrove species recorded from this zone, salt-tolerant mangroves such as *Kandelia* spp., *Ceriops* spp., *Rhizophora* spp., and *Avicennia alba* occur in a higher density.

Goa coast: Goa coast is characterized by drowned bedrock valleys or 'ria'-type shores influenced by the intrusion of seawater during high tides and swelling waves transfer marine sediments into the water-accumulated valleys to fill the shallow basins generated by subsidence in the coastal belts. However, breaching of fresh water from the landward sides during low tide reduces the pore water salinity of the shore fringes and thereby enables mangroves to grow in the upper middle parts of the carbonate shore platform in the Andaman and Nicobar group of islands in less saline conditions on the Goa Coast. The total mangrove coverage of the Goa coast accounts for 339.04 km^2, of which only 200 km^2 come under the purview of dense and healthy mangrove forest, and a recent report has indicated that an area of 5.48 km^2 of mangrove zones have been degraded (MOEF, 2012).

The mangroves of this region of western India are also vulnerable to the sea-level change of the Bay of Bengal. An area of 366.82 km^2 of the Goa coast is occupied by mangroves, of which 339.04 km^2 are scrub land and 27.78 km^2 are dense mangrove forest, but 5.48 km^2 of the mangrove zones are degraded (MOEF, 2012). Out of seven major estuaries on the Goa coast, two estuaries, viz. Mondovi and Zuari, with their interconnecting Camberjua estuary, support more than 75% of mangroves on the coast. An estimate indicates that a total area of over 200 km^2 of healthy mangroves was distributed in different coastal areas on the Goa coast in the past. About 20 species of mangroves and nonmangrove halophytes are reported from the coastal settings of Goa.

2.3.2.3 Trend of Depletion of Mangroves in India from Remote Sensing Data

There has been a drastic decline in the extent of mangroves across different parts of the tropical world including Sundarbans because of so many reasons like nonjudicious exploitation of biological resources, conversion of mangroves for agriculture, aquaculture and other modes of land use changes, pollution of different kinds, ecotourisms, bioinvasion, salinity invasion and last but not the least global warming. In the 1960s, the total area of Indian mangroves was estimated to be 681,976 ha, of which nearly 45% was shared by the deltaic islands of Sundarbans near the Bay of Bengal (Blasco 1975, 1977). Afterward, the total mangrove area was recorded as 356,500 ha (Saengar et al., 1983) and in another survey conducted in 1992, the total area of Indian mangroves was estimated to be 437,400 ha, which included the mangroves of the Andaman and Nicobar islands. Over-exploitation of mangroves, mostly through land-use changes and the deforestation process have resulted in the formation of open marshy land of approximately 1, 00,000 ha. Mangroves along the west coast of India suffered a lot with the conversion of mangroves into highly degraded areas (Blasco 1975, 1977). The coastal areas with mangroves such as the Gulf of Kutch, the Bombay coast, and the Cochin backwaters are depicted as glaring examples of reclamation, deforestation, population pressure, and pollution (Untawale, 1984).

In information gathered from remote sensing data provided by the Forest Survey of India, Dehradun, it was observed that the extent of mangrove coverage in 1987 was 4046 km^2, which subsequently increased to 4921 km^2 in 2017. However, the mangrove cover had experienced a declining trend of 0.61% per year within the past three decades, in Andhra Pradesh, and a decrease of 0.34% per year in the Andaman and Nicobar Islands during that period. In contrast, other regions of India with mangroves showed an upsurge in the extent of mangrove coverage. Notably, Gujarat recorded a growth of 168%, Maharashtra of 117%, Tamil Nadu of 113%, Odisha of 22%, Goa 26-fold, Karnataka ten-fold, Kerala nine-fold, Daman and Diu three-fold, and Puducherry of two-fold. The Sundarbans, on the other hand, exhibited a marginal increase of merely 1.83% at a rate of 0.061% per year. Summing of all this information, it can be postulated that India achieved an overall increase in the growth and propagation of mangrove cover of 21.63% over a period of 30 years, corresponding to an annual increase of 0.72%. In other words, it can be claimed that the extent of mangrove cover in India witnessed an increase of 181 km^2, amounting to an annual growth rate of 1.9% between the years 2013 and 2017. Contrary to this, mangrove coverage across the globe recorded a declining trend at a rate of 0.66%

during the same period. In addition, it is to be noted that the open sparse mangrove cover in India presently represents 40% of the total mangroves of the country, which is required to be transformed into denser mangrove cover in the near future (SRF, 2015). Moreover, two globally threatened species of mangroves of India, namely *Heritiera fomes* and *Sonneratia griffithii*, are among the 11 species on the IUCN Red List. These two species face the imminent risk of local extinction owing to several eco-bio-ethological factors such as the dearth of a freshwater supply, low-seed viability, and slow growth rates.

The Indian Sundarbans, forming the world's largest contiguous mangrove ecosystem, has been playing crucial roles as a buffer against tropical cyclones in the Bay of Bengal, representing a dynamic landscape after being constantly influenced by both natural and anthropogenic forces. It encompasses a mosaic of ecological compartments, including high and low mangroves, high and low marshes, mangrove associates, and salt pans. These compartments are distinguished not just by vegetation and hydrological regimes but also by elevation, salinity tolerance, tidal inundation patterns, and anthropogenic intervention (Rahman et al., 2010; Kuenzer et al., 2011; Ray, 2013).

Using the area-based dataset and decadal rate of change data for 1995, 2005, 2015, and 2025, this analytical discussion explores the physical nature, spatial dynamics, and ecological implications of these land-cover types in response to climatic variability, sea-level rise, and extreme meteorological events, such as cyclones and storm surges (Tables 2.1, 2.2, 2.3, and 2.4; Figs. 2.15, 2.16, 2.17, 2.18, and 2.19).

Table 2.1 Areal dynamics of the ecological compartments of the Indian Sundarbans (from 1995 to 2025)

Ecological compartments	Rate of change in % (1995 to 2005)	Rate of change in % (2005 to 2015)	Rate of change in% (2015 to 2025)
Salt pan	2.59	9.25	−0.38
Low marsh	0.58	19.70	0.57
High marsh	2.47	12.55	−0.66
Mangrove associate	−1.51	−0.34	−0.46
Low mangrove	3.49	−4.55	−0.25
High mangrove	−2.24	−4.19	−0.44

Table 2.2 Rates of changes in the area of the ecological compartments of the Indian Sundarbans: low mangrove trajectory—short-term gain, long-term decline

Ecological compartments	1995 (Area in km^2)	2005 (Area in km^2)	2015 (Area in km^2)	2025 (Area in km^2)
Salt pan	145.77	183.54	353.34	339.89
Low marsh	103.62	109.60	325.50	344.12
High marsh	141.48	176.46	397.86	371.53
Mangrove associate	431.92	366.84	354.24	337.85
Low mangrove	528.92	713.41	388.67	378.87
High mangrove	765.00	593.37	344.54	329.54

Table 2.3 Ecological transformation through areal transitions: transition matrix

From 1995 to 2025	Salt Pan	Low Marsh	High Marsh	Mangrove Associate	Low Mangrove	High Mangrove	Loss
Salt Pan	73.06	12.23	5.55	3.75	4.16	12.20	37.92
Low Marsh	52.32	21.78	8.87	4.21	3.13	5.23	73.77
High Marsh	51.08	37.61	47.59	9.52	7.23	8.14	113.61
Mangrove Associate	79.03	129.45	82.83	47.59	39.27	34.08	364.68
Low Mangrove	20.54	101.93	131.09	97.24	87.74	72.17	422.99
High Mangrove	9.41	34.72	117.92	169.24	227.13	168.60	558.43
Gain	212.39	315.97	346.28	283.98	280.93	131.85	

Table 2.4 Index of change

From 1995 to 2025	Loss	Gain	Unchanged	Change index
Salt pan	37.9215	212.3964	73.0611	0.1328
Low marsh	73.7748	315.9711	21.7818	0.2184
High marsh	113.6151	346.2858	47.5965	0.2884
Mangrove associate	364.6872	283.9851	47.5965	1.0998
Low mangrove	422.9919	280.9359	87.7464	1.1473
High mangrove	558.4338	131.85	168.6087	1.8586

2.3.2.3.1 High Mangrove Dynamics: Alarming Decline of Mature Forests

High mangroves are indicative of mature, climax mangrove communities growing in elevated zones tolerating moderate salinity and less frequent tidal inundation. These areas are typically dominated by halophytic mangrove species such as *Excoecaria agallocha*, *Avicennia officinalis*, *Sonneratia apetala*, *Ceriops decandra*, and *Xylocarpus mekongensis*, and all of these mangrove plants form the core of biodiversity-rich patches of the Sundarbans.

In 1995, high mangroves covered an area of 765.00 km^2, which has declined sharply to 329.54 km^2 in 2025, reflecting a loss of over 57% of the area in three decades. The rate of decline has shown an intensifying trend in tune with −2.24% during 1995–2005, to −4.19% during 2005–2015, and then to −0.44% during 2015–2025.

The most substantial reduction, noticed between 2005 and 2015, corresponded with post-cyclone degradation effects. Cyclone Aila of 2009 resulted the inundation of a vast area of the Sundarbans, India, with saline water, which led to the massive destruction of mature mangrove stands. The high biomass and low regeneration capacity of high mangrove species make them particularly vulnerable to prolonged saline flooding, physical uprooting, soil erosion, and sand deposition. These kinds

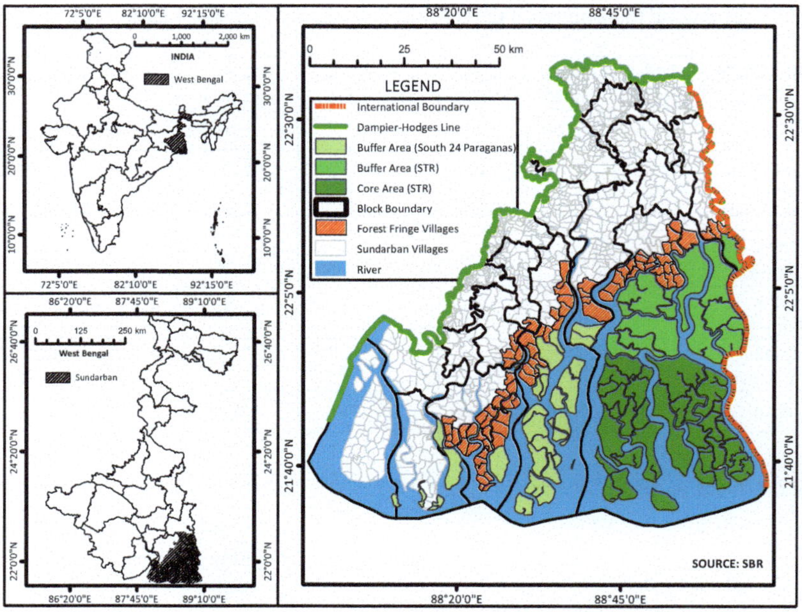

Fig. 2.15 Map of Sundarbans pointing out International Boundary, Dampier-Hodges Line, core and buffer areas of the Sundarbans Biosphere Reserve, estuarine networks, and mangrove forest with fringe areas

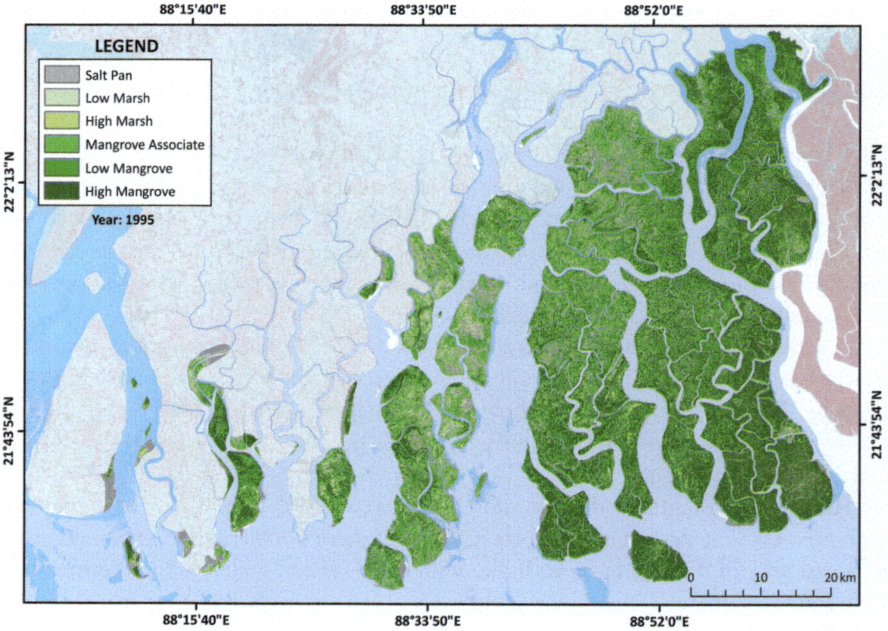

Fig. 2.16 Satellite imagery to show different ecological settings within Sundarbans, India, which include salt pan, low marsh, high marsh, mangrove associate, low mangrove, and high mangrove during 1995

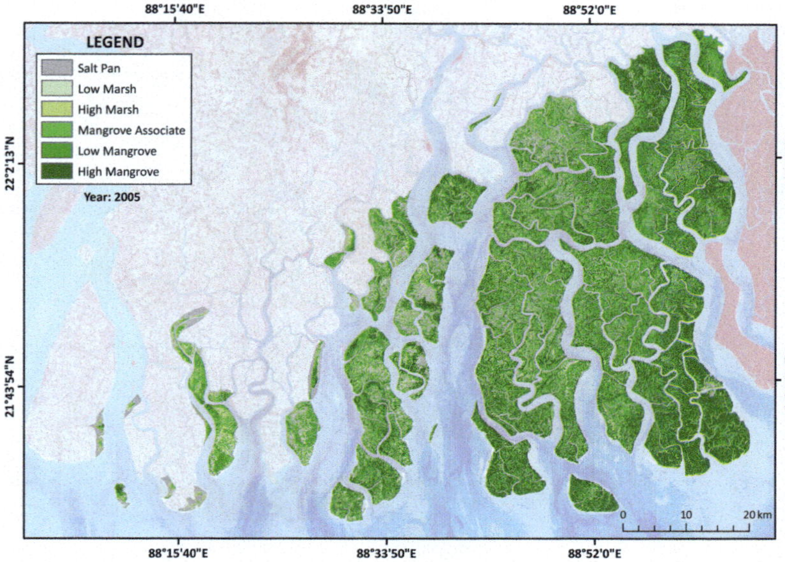

Fig. 2.17 Satellite imagery to show different ecological settings within Sundarbans, India, which include salt pan, low marsh, high marsh, mangrove associate, low mangrove, and high mangrove during 2005

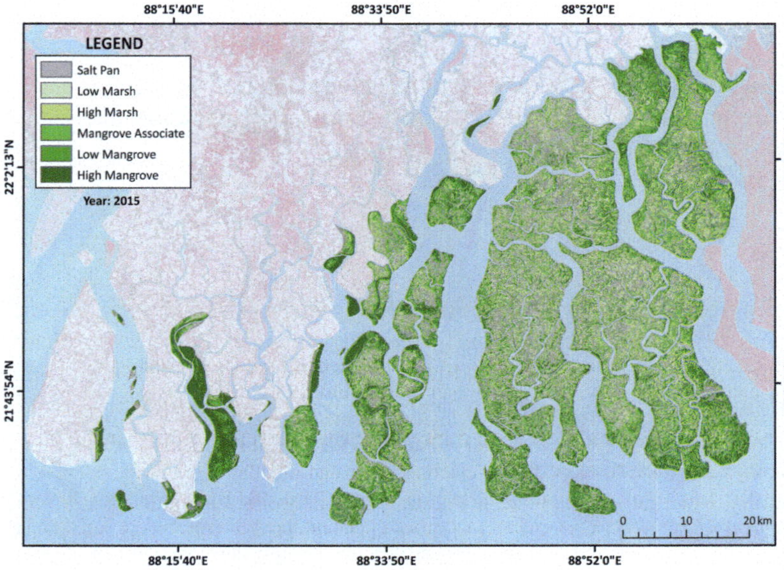

Fig. 2.18 Satellite imagery to show different ecological settings within Sundarbans, India which include salt pan, low marsh, high marsh, mangrove associate, low mangrove, and high mangrove during 2015

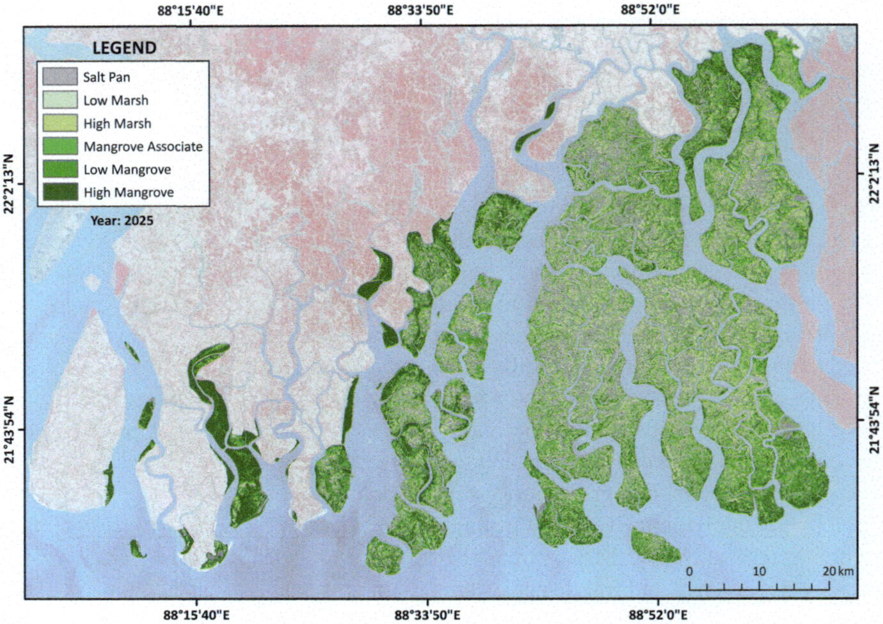

Fig. 2.19 Satellite imagery to show different ecological settings within Sundarbans, India, which include salt pan, low marsh, high marsh, mangrove associate, low mangrove, and high mangrove during the period 2019

of negative impacts are reflected both by resulting ecological degradation and reduction in the normal ecological resilience in the face of extreme climatic events mediated by sea-level rise.

2.3.2.3.2 Low Mangrove Trajectory: Short-Term Gain, Long-Term Decline

Low mangroves, characterized by pioneering species such as *Porteresia coarctata*, *Avicennia alba*, *Acanthus ilicifolius*, *Phoenix paludosa*, *Aegiceras corniculatum*, and *Rhizophora apiculata*, thrive at lower levels of intertidal zones, sustaining prolonged inundation pressure with greater salinity. These species often colonize in newly accreted lands and degraded mangrove patches, representing an early successional stage in mangrove ecology.

Initially, low mangroves were found to expand significantly during 1995 to 2005 from 528.92 km^2 to 713.41 km^2, with a growth rate of +3.49%. This early phase of expansion was supposed to be due to colonization in response to sediment accretion and demands to fill up temporary ecological vacuums left by degraded high mangrove zones. However, this trend of mangrove settlements was reversed sharply post-2005, with a shrinkage of mangrove zones to 388.67 km^2 by 2015 and

378.87 km^2 by 2025, accompanied by negative growth rates of −4.55% and − 0.25% respectively.

This pattern of fluctuation indicates the development of increased habitat instability. Although low mangroves are more tolerant of salinity and flooding, they are least immune to sustaining greater salinity intrusion, soil erosion, vertical accretion, and greater deposition of sands after being shifted and carried by excessive wind flows as an impact of repeated cyclones. The decline of mangrove coverage during this period also hints at a failure in ecological succession toward the development of more stable mangrove communities, mainly due to suboptimal regeneration conditions or conversion of these lands into nonvegetated saline mudflats, salt pans, or sandy habitats.

2.3.2.3.3 Mangrove Associates: Depletion of Transitional and Edge Vegetation

Associated plants of true mangroves constitute those plant communities that coexist with mangroves of the border and core areas in mangrove forests that do not exhibit the full suite of morphological adaptations like true mangrove plants such as pneumatophores, stilt roots, prop roots, sunken stomata, etc. These may include shrubs and herbaceous halophytes such as *Phoenix paludosa*, *Clerodendrum inermis*, or grasses such as *Porteresia coarctata*. They function as ecotonal buffers, mitigating salinity ingress and aiding species migration.

The area covered by mangrove associates in 1995 was 431.92 km^2. This area started declining progressively to 366.84 km^2 in 2005, to 354.24 km^2 in 2015, and to 337.85 km^2 in 2025. Although the rate of decline appeared modest (−1.51%, −0.34%, and − 0.46% over the three decades), the trend was found to be consistent and worrisome.

The gradual loss of area with mangrove-associated plants indicates that the first zones of edge habitats were found to be impacted by cyclonic saltwater intrusion, sand deposition, tidal embankment breaches, and soil erosion. This trend of reduction was considered to mirror ecological compression, where the narrowing of vegetational gradients led to the loss of both species diversity and ecological resilience. The continued decline of mangrove associates suggests that this was not only considered vegetation loss but also eco-degradation of vital microhabitats through soil erosion weakening the transitional ecological functionality.

2.3.2.3.4 High Marsh Expansion Followed by Saturation

High marshes supported by salt-tolerant grasses, sedges, and shrubs such as *Scirpus triqueter*, *Sesuvium portulacastrum*, *Phragmites karka*, *Salicornia brachiata*, and *Suaeda maritima* represent the transitional zone between mangroves and terrestrial ecosystems and experience less frequent inundation. They serve as sediment trappers and are vital in maintaining the ecological connectivity of coastal zones. The area under high marsh increased from 141.48 km^2 in 1995 to a peak of 397.86 km^2 in 2015, an almost three-fold increase, with impressive growth rates of +2.47% and + 12.55% in the first two decades. However, this expansion reversed slightly in the following decades, with a decrease to 371.53 km^2 by 2025 (−0.66%).

This expansion phase may reflect sediment deposition and mangrove retreat due to storm surge and saline-water inundation. High marshes may have proliferated in areas where mangroves failed to regenerate but where some halophytic vegetation could become established. However, the slight decline in the last decade may indicate increasing salinity, shifts in tidal hydrology, or even human-induced conversion of mangrove zones.

2.3.2.3.5 Low Marsh: A Rapidly Expanding Zone of Ecological Flux

Low marshes are situated in the intertidal zone, and are inundated daily by brackish water. These zones support salt-tolerant herbaceous vegetation and are often considered ecotonal or successional landscapes. Low marshes have expanded from 103.62 km^2 in 1995 to a significant proportion of 344.12 km^2 in 2025.

The rate of change in the decade 2005–2015 was exceptionally high (+19.70%), which slowed down to +0.57% in the next decade. This expansion likely results from both the dieback of mangroves—creating open, saline mudflats—and sedimentation, which allows colonization by marsh grasses.

The proliferation of low marshes could be interpreted as a sign of great ecological simplification, where complex, multi-tiered forest systems tend to collapse paving the way for the formation of flat and uniform marshlands. This transformation is supposed to be considered a precursor to conversion into salt pans or barren saline flats if salinization continues unabated.

2.3.2.3.6 Hypersaline Patch Expansion: Ecological Degradation and Salinity-Induced Habitat Shift

The growing presence of hypersaline patches in the Indian Sundarbans over the last three decades signals a concerning trend of ecological degradation. Unlike conventional or anthropogenically maintained salt pans, hypersaline patches refer to naturally evolved or ecologically transitioned zones characterized by extremely high salinity levels, resulting from prolonged saline-water stagnation, poor freshwater flushing, high evaporation rates, and soil salinization. These areas, although lacking structured human infrastructure such as traditional salt pans, emerge as ecologically inert zones with diminished biological productivity and altered micro-topography.

Between 1995 and 2025, the spatial extent of areas classified under this compartment increased from 145.77 km^2 to 339.89 km^2, registering an overall growth of more than 133%. This expansion reflects a steady upward trajectory, particularly sharp between 2005 and 2015 (+9.25%), with a marginal decline (−0.38%) observed between 2015 and 2025.

This transformation is indicative of a cascading ecological failure rather than mere land-use change. The hypersaline zones have primarily emerged in areas that once supported mature mangrove stands or vegetated marshlands. Following repeated cyclonic storms, events such as Aila and Amphan, these areas were inundated with high-salinity tidal water, and in the absence of adequate flushing by freshwater rivers, salt accumulated in the soil profile. Consequently, native halophytic vegetation failed to regenerate, leading to the development of bare or sparsely vegetated hypersaline flats.

Ecologically, the emergence of hypersaline patches mark a terminal stage of degradation in the coastal wetland succession pathway. These zones exhibit reduced organic content, high electrical conductivity, and often become unsuitable for even the most salt-tolerant marsh or mangrove flora. They interrupt habitat continuity for fauna and impair ecosystem services such as nutrient cycling, carbon sequestration, and sediment stabilization.

From a hydrological perspective, hypersaline patches often form in micro-basins or depressions with limited connectivity to dynamic tidal flows. Their persistence is reinforced by topographic trapping of saline water and a regional decline in freshwater discharge due to upstream damming and diversion of the Ganges and its distributaries. The consequent rise in soil salinity transforms once productive tidal flats or brackish wetlands into ecological dead zones.

Anthropogenic stressors indirectly accelerate this transformation. The conversion of adjacent wetlands into aquaculture ponds, embankment-induced waterlogging, and deforestation of buffer mangroves all contribute to the expansion of salinized terrain. Furthermore, climate-change-induced sea-level rise and enhanced evapotranspiration rates intensify the desiccation and salinization of already stressed zones, further contributing to the formation and persistence of hypersaline patches.

The ecological compartments of the Indian Sundarbans exhibit a clear, quantifiable trend of habitat conversion, successional arrest, and land degradation over the 30-year period from 1995 to 2025. High and low mangrove zones—critical for biodiversity, coastal protection, and climate resilience—have shrunk drastically. Simultaneously, low marsh and hypersaline (saline blank) areas have expanded, signaling ecosystem simplification and loss of ecological complexity.

The mangrove associates and high marsh zones show transitional behavior, but their roles are being overwhelmed by the combined effects of extreme climatic events, particularly cyclones such as Sidr, Aila, Amphan, and Yaas, along with sea-level rise, tidal surges, and freshwater deprivation.

In essence, the ecological restructuring observed is not simply a product of natural change, but a multidimensional response to intersecting pressures. The Sundarbans stands at a precarious ecological threshold. Unless proactive, science-driven restoration and freshwater management strategies are implemented, large parts of this global biodiversity hotspot risk transitioning into irreversible hypersaline wastelands.

2.3.2.3.7 Unraveling the Drivers of Ecological Transformation in the Indian Sundarbans

The dramatic transformations in the spatial and functional characteristics of ecological compartments in the Indian Sundarbans, ranging from the degradation of high mangroves to the expansion of saline blanks, cannot be adequately explained without considering the broader climatic and environmental context. This low-lying intertidal deltaic region is not only shaped by tidal fluxes and fluvial sedimentation but is also acutely sensitive to a convergence of macro- and micro-environmental stressors, particularly those associated with anthropogenic climate change, regional hydro-engineering, and geomorphic instability. These are as follows:

1. **Sea-level rise and delta submergence**

   The Indian Sundarbans lie within one of the most vulnerable coastal zones globally in the context of sea-level rise. The region is witnessing a relative sea-level rise at a rate of approximately 3.14 mm/year, which is significantly higher than the global average owing to land subsidence, tidal amplification, and tectonic sinking. When coupled with the natural sediment compaction in the Bengal Basin, the effective vertical land loss exacerbates tidal ingress and pushes saline water further inland, altering the salinity gradients that are vital for zonation and regeneration of mangrove and marsh communities.

   As the sea level rises, even slightly elevated high mangrove zones are increasingly submerged or waterlogged, leading to root suffocation and dieback of mature species such as *Heritiera fomes*. Likewise, marginal areas with thinner soils and poor drainage tend to become hypersaline over time, rendering them unsuitable for regrowth, thus accelerating the emergence of saline blanks or hypersaline patches.
2. **Cyclonic storm surges and recurrent coastal inundation**

   The last two decades have seen a sharp rise in the frequency, intensity, and destructiveness of tropical cyclones in the Bay of Bengal, many of which have directly impacted the Indian Sundarbans (Paul, 2002b). Notable among them are:

   (a) Cyclone Sidr: Although it primarily impacted Bangladesh, it induced strong tidal waves that eroded mangrove fringes in the Indian sector.
   (b) Cyclone Aila: It caused massive embankment breaches, leading to persistent saline flooding in both agricultural and forested areas. The ecological impact of Aila was perhaps the most profound in recent memory.
   (c) Cyclone Amphan: A super cyclone with wind speeds exceeding 190 km/h, Amphan devastated nearly 28% of the forest area in the Indian Sundarbans.
   (d) Cyclone Yaas and others continue to contribute to soil salinization and embankment destabilization.

   These events not only destroy above-ground vegetation but also change hydro-edaphic conditions, i.e., salinity, soil structure, pH, and microbial balance. The effects are long term: many areas affected by Aila in 2009 had not fully regenerated even a decade later. The recurrence of such events—sometimes at intervals of just 2–3 years—does not allow time for ecological recovery, pushing the system into a regressive successional loop.

3. **Declining freshwater inflows and salinity encroachment**

   One of the defining features of the Indian Sundarbans is its brackish water regime, formed by the mixing of tidal saline influx from the Bay of Bengal and freshwater discharge from the Ganges–Bhagirathi–Hooghly river system. However, the depletion of upstream freshwater flows due to extensive damming (such as the Farakka Barrage) and diversion for agriculture and urban use has significantly reduced the volumetric input of freshwater into the estuarine system.

   This hydrological deficit has allowed saline tidal intrusion to extend further inland, pushing the salt wedge deeper into areas that were historically dominated by freshwater or moderately brackish flora. As a result, many species that require lower salinity thresholds for regeneration—such as *Nypa fruticans* or *Excoecaria agallocha*—are failing to reproduce, and even resilient species such as *Avicennia marina* are showing signs of stress. The transformation of once-productive mangrove and marsh zones into saline blanks is directly linked to this hydro-salinization of the deltaic landscape.

4. **Sediment starvation and coastal erosion**

   The Sundarbans delta is fundamentally a sediment-dependent system. Sediment deposition not only builds land but also provides the substrate for mangrove establishment and marsh succession. However, upstream damming, soil-conservation measures, and changes in river course have substantially reduced the sediment load reaching the delta (Allison & Kepple, 2001).

   The result is a negative sediment budget, wherein erosion by wave action and tidal scouring exceeds deposition. This is especially apparent in the retreating southern and southwestern fringes of the delta. When sediment input falls short of compensating for land subsidence and sea-level rise, the land literally sinks, often converting to mudflats or eventually hypersaline zones owing to water stagnation and salt accumulation.

5. **Anthropogenic pressures and landscape fragmentation**

   Human-induced landscape modifications further complicate the region's ecological balance. The construction of embankments, although aimed at flood control, has disconnected the natural tidal flow in many areas, leading to water-logging and salt entrapment. Agricultural encroachment, illegal aquaculture, brick kilns, and settlement expansion have consumed large swathes of marsh and mangrove areas, reducing habitat connectivity and fragmenting biodiversity corridors.

   Additionally, unsustainable practices such as the clearing of mangrove associates for fuelwood and the over-extraction of fish and crabs have affected both the structure and function of the estuarine system. Fragmented ecosystems are less resilient to environmental shocks, and small, isolated mangrove patches often fail to survive long-term stress, especially when disconnected from sediment sources or freshwater inflows.

6. **Climate-change feedback loops and future risk**

   The interplay of climate-change-induced stressors in the Indian Sundarbans has triggered complex feedback mechanisms that are accelerating ecological degradation. As mangroves decline owing to rising salinity, tidal inundation, and cyclone damage, the land loses its vegetative cover, which in turn exposes soil to erosion and enhances tidal scouring. This further deepens tidal channels and facilitates greater saline-water intrusion, worsening the very conditions that caused the initial vegetation loss. The diminishing vegetation also reduces the region's carbon-sequestration capacity, weakening its role as a climate regulator. This not only contributes indirectly to global warming but also amplifies local vulnerability by fostering warmer surfacetemperatures, which intensify cyclone formation and rainfall variability. These self-reinforcing loops are shifting the Sundarbans from a relatively stable ecosystem to a threshold-driven, nonlinear regime—one in which future risks are unpredictable and potentially irreversible. Without urgent mitigation and restoration efforts, the region may move beyond its ecological tipping point.

   The broader climatic and environmental context paints a sobering picture of the Indian Sundarbans. The observed land-cover changes are noted as merely fluctuations in natural dynamics, which are symptomatic of a deeper systemic shift in the process of maintaining coastal ecological equilibrium. The interplay of sea-level rise, cyclone-induced saltwater intrusions, freshwater depletion, and human-induced landscape alterations are together orchestrating a slow-motion ecological collapse.

   If this trend continues unchecked, the region may face not only loss of biodiversity and ecosystem services but also a humanitarian crisis, given the dependence of millions of people on its fisheries, forest products, and protection against storms. Understanding these drivers in detail is the first step toward adaptive management, ecosystem restoration, and climate-resilient planning for the Indian Sundarbans.

2.3.2.3.8 Areal Transition of Ecological Compartments

The areal transition matrix and change index data from 1995 to 2025 for the Indian Sundarbans offer a detailed quantitative insight into the transformation of its ecological compartments: high mangrove, low mangrove, mangrove associates, high marsh, low marsh, and hypersaline patch (previously labeled salt pan). The dataset reflects not just the spatial reshuffling of land cover but also reveals the deeper ecological stress and systemic vulnerability of the region to climatic extremes and anthropogenic pressures.

2.3.2.3.9 Ecological Transformation Through Areal Transitions

Over the 30-year period from 1995 to 2025, the Sundarbans have undergone significant land-cover transition across all ecological compartments. The transition matrix quantifies both the area gained, area lost, and net changes in square kilometers, whereas the Change Index (calculated as the sum of gains and losses divided by the unchanged area) provides a normalized measure of ecological volatility.

The high mangrove zone, which serves as the most mature and ecologically stable component of the deltaic forest, suffered the most substantial decline. A total loss of 558.43 km^2 is recorded—substantially outweighing the gain of 131.85 km^2. Only 168.61 km^2 remained unchanged. The resulting Change Index of 1.8586, the highest among all compartments, signifies extreme ecological instability. This reflects widespread degradation likely triggered by intense cyclonic activity—notably Aila, Amphan, and Yaas—which devastated mature mangrove tracts through wind uprooting and prolonged saline inundation. These disturbances not only stripped vegetative cover but also increased soil salinity beyond the regeneration threshold of species such as *Heritiera fomes* and *Xylocarpus* spp. Furthermore, the decline was compounded by reduced freshwater flushing from the Ganges distributaries, which is crucial for seedling recruitment and salinity buffering.

Similarly, low mangroves, which usually act as buffer zones with pioneer species such as *Avicennia marina*, also recorded a net loss of 422.99 km^2, despite gaining 280.93 km^2 through expansion into other categories, particularly from degraded mangrove associates and marshlands. The compartment's Change Index stands at 1.1473, indicating substantial flux. These areas likely expanded briefly following earlier disturbance events but subsequently failed to stabilize, potentially because of sediment instability, hydrodynamic changes, and frequent saline incursions, eventually transitioning into less productive or hypersaline terrain.

Mangrove associates, which function as ecotonal buffers between mangrove cores and nonforested wetlands, demonstrated interesting dynamics. Although they gained 283.99 km^2, mainly from high and low marshes, the loss was even more significant—364.69 km^2. The relatively high Change Index of 1.0998 indicates a landscape in flux. These communities are particularly susceptible to both tidal scouring and human-induced edge degradation, such as conversion for aquaculture or fuelwood harvesting. The net result is a contraction of transitional zones essential for species dispersal and buffer functions.

The high marsh zones exhibited a relatively moderate Change Index of 0.2884, reflecting both expansion and attrition. These zones gained 346.29 km^2, mainly from mangrove associates and low mangroves, and lost 113.62 km^2, possibly because of elevation shifts, increased submergence from sea-level rise, and delayed drainage post-cyclones. The gains suggest areas where sediment deposition permitted marsh colonization, but losses indicate the regressive transformation of high marshes into hypersaline patches or barren zones owing to hydrological imbalance.

Low marsh areas experienced the most dynamic expansion, gaining 315.97 km^2 while losing 73.77 km^2, resulting in a moderate Change Index of 0.2184. These areas likely expanded into zones vacated by degraded mangroves and associates, signifying a shift in ecological dominance toward herbaceous salt-tolerant communities. However, these are typically early-successional stages and may reflect stalled regeneration, particularly where recurrent saline flooding inhibits further vegetative maturity.

The hypersaline patches gained 212.40 km^2, while losing 37.92 km^2, with 73.06 km^2 remaining unchanged, and had a relatively low Change Index of 0.1328. Despite its lower index, this transformation is ecologically significant. These areas likely emerged because of long-term salt accumulation, often following cyclone-induced breaches of embankments and persistent saline waterlogging in low-lying areas. Their increase reflects terminal ecological states, where neither mangroves nor marshes are able to regenerate, and where soil conditions deteriorate to levels unsuitable for any significant biological function.

The change index values and transition data collectively underscore that the Indian Sundarbans is not merely experiencing marginal shifts in vegetation—it is undergoing a fundamental ecological reorganization driven by a convergence of extreme events, chronic environmental stress, and disrupted successional processes.

The repeated impact of tropical cyclones, with ever-increasing intensity, leads to widespread uprooting of forest stands, salinization of marshlands, and permanent inundation of low-lying areas. The sea-level rise, interacting with land subsidence and sediment starvation, accelerates tidal ingress and undermines soil anchorage and accretion processes necessary for mangrove establishment. In parallel, hydrological modifications upstream in the Ganges–Brahmaputra system deprive the Sundarbans of vital freshwater inputs, thus shifting the saline–freshwater equilibrium and pushing the ecosystem further toward salinity dominance.

Moreover, human pressures such as embankment construction, conversion of mangrove fringes for aquaculture, and deforestation of associated species create fragmented landscapes with diminished ecological resilience. These alterations not only reduce habitat integrity but also disable natural regenerative mechanisms, particularly in mangrove nurseries and seed banks.

The end result is a net conversion of biologically rich, structurally complex forest systems into simplified, stress-tolerant, and often ecologically stagnant states, such as low marshes and hypersaline patches. The transition data suggest that once areas are degraded, they seldom revert to their original states under current climatic and socio-environmental regimes—indicating permanent ecological loss if proactive interventions are not undertaken.

The transition matrix and Change Index data provide quantitative evidence of a rapidly evolving deltaic ecosystem under severe stress. High and low mangroves are contracting dramatically, mangrove associates are diminishing as transitional buffers, whereas low marshes and hypersaline zones expand into degraded and salinized land. These shifts are emblematic of a larger pattern of ecological destabilization triggered by climatic extremes, sea-level rise, freshwater scarcity, and unsustainable human practices. The Indian Sundarbans is thus caught in a critical phase—if these trends continue, the region risks crossing ecological thresholds beyond which recovery may be impossible. Only integrated climate adaptation, habitat restoration, and upstream–downstream water governance can halt or reverse this trajectory.

2.3.2.3.10 Methodological Approaches to the Estimation of Topographic and Structural Changes Through the Decades

In this study, the classification of the Indian Sundarbans forest floor into six major ecological compartments—namely high mangrove, low mangrove, mangrove associates, high marsh, low marsh, and hypersaline patch—has been achieved through a scientifically rigorous, multi-spectral remote-sensing-based methodology that integrates multiple spectral indices using principal component analysis (PCA). This approach is designed to overcome the limitations of single-index classification by capturing the complex interplay of vegetation structure, pigment content, soil salinity, and canopy greenness that characterizes each ecological unit in a dynamic tidal estuarine environment.

The methodological foundation of this classification lies in the use of a suite of biophysical and spectral indices derived from multi-temporal and multi-sensor satellite data, which were strategically chosen to enhance the detection and differentiation of vegetation types and saline conditions across the landscape. These indices include the Mangrove Index and the Mangrove Vegetation Index, which specifically target mangrove physiognomic traits by utilizing the shortwave infrared (SWIR) and near-infrared (NIR) reflectance properties to distinguish between true mangroves and associated vegetation.

$$\textbf{Mangrove Index} = \left(\textbf{NIR}\,\textbf{SWIR1} / \textbf{NIR}\,\textbf{SWIR1}\right)\textbf{10000} \quad (1)$$

(Winarso et al., 2014)

$$\textbf{Mangrove Vegetation Index} = \textbf{NIR}\,\textbf{Green} / \textbf{SWIR1} + \textbf{Green} \quad (2)$$

(Baloloy et al., 2020)

The Pigment Index (PI), sensitive to chlorophyll concentration, which is essential in differentiating between healthy high mangrove stands and degraded or early successional forms such as mangrove associates and low mangroves. The spectral reflectance in the red-edge and green bands captures subtle pigment variations in canopy cover.

$$\textbf{Pigment Index} = \left(\textbf{NIR} - \textbf{BLUE}\right) / \left(\textbf{NIR} + \textbf{RED}\right) \quad (3)$$

(Dharmawan & Pramudji, 2020)

The Vegetation Soil Salinity Index (VSSI) and the Normalized Difference Salinity Index (NDSI) are both designed to detect the salinity stress on vegetation and bare soil. These indices incorporate band ratios involving red, SWIR, and thermal bands to indicate areas suffering from salt accumulation—an especially critical factor in identifying hypersaline patches and salinity-stressed marshes.

$$\textbf{Vegetation Soil Salinity Index} = \textbf{2}^{*}\textbf{GREEN} - \textbf{5}^{*}\left(\textbf{RED} + \textbf{NIR}\right) \quad (4)$$

(Dehni & Lounis, 2012)

The Tasseled Cap Greenness Index (TCGI), which emphasizes vegetation health and photosynthetic activity by projecting reflectance data into a transformed brightness–greenness–wetness space. The TCGI is particularly useful for distinguishing dense mangrove canopies (typical of high mangroves) from sparse vegetation or degraded zones.

$$\begin{aligned}&\left(\mathbf{Band1}^{*}-\mathbf{0.28}\right)+\left(\mathbf{Band2}^{*}-\mathbf{024}\right)+\left(\mathbf{Band3}^{*}-\mathbf{0.54}\right)\\&+\left(\mathbf{Band4}^{*}\,\mathbf{0.72}\right)+\left(\mathbf{Band5}^{*}\mathbf{0.08}\right)+\left(\mathbf{Band7}^{*}-\mathbf{0.18}\right)\end{aligned}\tag{5}$$

(Baig et al., 2014).

$$\begin{aligned}&\left(\mathbf{Band1}^{*}-\mathbf{0.28}\right)+\left(\mathbf{Band2}^{*}-\mathbf{024}\right)+\left(\mathbf{Band3}^{*}-\mathbf{0.54}\right)\\&+\left(\mathbf{Band4}^{*}\,\mathbf{0.72}\right)+\left(\mathbf{Band5}^{*}\mathbf{0.08}\right)+\left(\mathbf{Band7}^{*}-\mathbf{0.18}\right)\end{aligned}\tag{6}$$

(Healey et al., 2005).

The Soil Adjusted Vegetation Index (SAVI), which adjusts for soil background reflectance, enhances vegetation detection in low-cover areas. It is particularly effective in demarcating marsh zones, early successional plant communities, or mixed soil–vegetation patches common to low marshes and mangrove associates.

$$\mathbf{SAVI}=\left(\left(\mathbf{NIR}-\mathbf{R}\right)/\left(\mathbf{NIR}+\mathbf{R}+\mathbf{L}\right)\right)^{*}\left(\mathbf{1}+\mathbf{L}\right),\mathbf{Where\ L\ is\ constant}\tag{7}$$

(Huete, 1988)

Each of these indices, although individually informative, only captures certain spectral or ecological dimensions. To integrate the complex, multidimensional data into a robust classification framework, PCA was applied. PCA serves two primary functions in this context: dimensionality reduction and feature extraction. By transforming the correlated indices into a new set of orthogonal components (principal components), PCA isolates the most statistically significant variance directions in the data. The eigenvectors associated with these components—essentially the weights that define how much each original index contributes to the new composite dimensions—form the mathematical basis for constructing the final classification model.

The linear equation derived from the eigenvectors represents a composite index or integrative classifier that synthesizes spectral responses across vegetation type, chlorophyll health, salinity stress, and soil influence. By applying this equation pixel-wise across the raster domain, each pixel is assigned a value that reflects its weighted spectral composition. Thresholds, determined through field-verified training samples and ancillary ecological knowledge, are then used to segment the continuous index values into discrete land-cover categories corresponding to the six ecological compartments.

This integrated methodology ensures that the classification process is not only spectrally nuanced but also ecologically grounded. For example, high mangroves are identified where the TCGI and PI values are high, salinity indices are low, and the SAVI reflects dense canopy with minimal soil exposure. In contrast, hypersaline patches manifest as areas with high VSSI and NDSI values, low or negative SAVI and PI, and minimal greenness. Mangrove associates and marsh types occupy intermediate spectral niches, marked by moderate pigment levels, higher soil exposure, and fluctuating salinity stress depending on tidal influence.

The final classified map thus offers a high-resolution, ecologically meaningful delineation of the Sundarbans forest floor. By combining data-driven dimensional reduction with biophysical interpretation, the methodology not only captures current land-cover patterns but also allows for tracking temporal ecological transitions—making it an essential tool for change detection, degradation monitoring, and conservation planning in this globally critical mangrove ecosystem.

2.3.3 Mangroves of Sundarbans on the Indian Subcontinent

Mangroves, being woody halophytes characterized by unique biological adaptive attributes such as tidal root systems, e.g., pneumatophores, stilt roots, a viviparous germination system, sunken stomata, thick wax-coated leaves, exhibit stunted growth withstanding stresses from saltwater spray, shortages of nutrients within intertidal belts in the Indian Sundarbans experiencing tidal inundation twice a day, and are fringed with specific geomorphological and physiographic settings and coastal processes facilitating influx and outflux of sediment supply and deposition under definite geological settings to result in definite coastal configurations (Paul, 1999, 2002a). Three major ecological attributes pertaining to physical mangrove settings have been identified in lower latitudes in Indo-Pacific and Atlantic systems, especially between 32 °N and 38 °S, which are influenced by a carbonate setting under tides dominated by river and estuaries.

So many climatological (rainfall, temperature, atmospheric density, low wind, humidity, etc.), geophysical (tidal amplitudes, intensity of water currents and flows, relative sea-level variations, etc.), and geomorphological (sedimentation, erosion, accretion, etc.) factors and processes drive the development of mangrove-based physiographic components. Subsequently, based on the formation and expansion of land forms, biological components, through their interactions among themselves, through food chain–food web dynamics, as well as with the nonliving components through the bioturbation process, also determine the settlement, growth, propagation, and ecological succession of mangrove species of a particular eco-region.

The combined action of all those ecological parameters and processes ultimately determine physiographic settings of particular eco-zone where specific mangrove species grow in response to prevailing ecological parameters of different coastal zones in India and other tropical parts of the world. Coastal India includes four different kinds of physiographic settings, which are (i) deltaic depositional low-lying areas with intervening distributaries and tidal creeks; (ii) tidal river-estuaries with deltaic islands and subtidal shoals; (iii) lagoons with diminishing mixing and coastal embayment with backwater influence with increased salinity; (iv) carbonate banks in the shores and drowned bedrock valleys. Mangroves, after being settled in a specific physiographic setting, are adjusted with supply and depositional patterns of sediment, a quantum of annual freshwater flows and mixing with saline seawater, and different favorable environmental gradients promoting growth and diversity of not only halophytic vegetation but other associated plant and animal species. The environmental triggering factors include the nutrients and textural components of soil, the downstream supply of sediments, depositional patterns of alluvium, freshwater influx, salinity of flowing water and interstitial water within soil, hot and humid tropical climatic factors, tidal fluxes, and amplitudes. All these environmental gradients control the mangrove growth and distribution. The carbonate banks formed by coralline debris, micrites or chemogenic cements, siliciclastic materials, and other bioclasts support a wide range of mangroves with evergreen characteristics in the Andaman and Nicobar Islands. Drowned bedrock shores of Goa, Karnataka, and the Andaman and Nicobar Islands provide the required physiographic setting for the growth and propagation of mangroves. Physiographic settings developed tectonically owing to occurrences of earthquakes and tsunamis result in uplifting and subsidence events, which in turn result in the shifting of intertidal, subtidal, and supratidal zones of the unstable coasts.

The deltaic complex of Sundarbans, along with its surrounding ecological set-ups, bears a dynamic physiographical and geomorphological evolutionary history. The eco-region of Sundarbans constituting an integral part of the Bengal Basin, represents coalesced multi-generation deltas that have been prograded into several phases characterized by distinctive multilevel delta surfaces, terraces, paleo-channels on continuously experiencing the positive pushing forces from the changes of interglacial eustatic sea level that occurred during the Plio-Pleistocene period at the Bay of Bengal. Several archipelagic islands of different shapes and sizes have developed between the sea and mainland, some of which persisted throughout recorded history whereas others could not raise their heads over the surface of water to depict their existence.

The total Sundarbans area of both the Indian and the Bangladesh parts lying south of the Tropic of Cancer is bounded by the Baleshwari River flowing through the eastern sector of Bangladesh and the Hooghly estuary on the western sector of eastern India. Biogeographically, mangrove forests are designated as **'tidal forests'** and the entire ecosystem combining forest and estuarine networks is called '**coastal wetlands.'**

The Bengal Delta complex, supported by mangrove vegetation, popularly known as the Sundarbans Delta Complex, is developed by the filling up of the Bengal Basin through shelf sedimentation and deposition of tertiary marine geosynclinals. The Sundarbans Delta Complex followed by gradual progradation of the Quaternary Ganges–Brahmaputra delta fronts toward the southern sea. This deltaic complex is characterized by luxuriant growth and prolific-diversity mangrove vegetation overlying a huge thickness of tertiary marine sediments, forming an integral coastal part of the active subsidence of the Bengal Delta Complex (Inman and Nordstrom 1971). The opposite continental coast is considered a zone of collision with plate tectonics, which imposes an overriding control of the evolution of the coastline of West Bengal. Numerous faults have developed owing to tectonic control of the basin, mostly during the Quaternary period. The subsurface southerly tilt of the Bengal Basin is manifested by the trend of flows of the all-important river systems of Sundarbans from north to south and this southerly tilt is supposed to have been effectively viewed since the late seventeenth century (Sengupta 1966, 1972).

The existing drainage pattern of the coastal West Bengal with Sundarbans has resulted from two distinctly noticeable factors through the tertiary history of the Bengal Basin, which are (i) the increasing rate of southerly tilt of the West Bengal part of the Bengal Basin to a relatively greater rate of subsidence of the southwestern part of the hinge and (ii) the southerly slope of the Bengal Basin resulted from the movement on the hinge zone at the edge of the shelf (Sengupta, 1966). The physiographic changes over centuries have largely controlled the geographical distribution of mangrove plants in the tropical–subtropical coastal land masses after the break down of Gondwanaland owing to continental drift, since the advent of mangrove ecosystems on the earth.

The Sundarbans mangrove ecosystem in its present form does have their ancient counterparts buried under deltaic but inland sediment under the influence of freshwater cover further inland. Owing to the location of Sundarbans on the Ganges–Brahmaputra deltaic system flanked by older rocks from tertiary and older sediments to younger Pleistocene terraces, the regional delta-building geological processes take place involving both different fluvial and coastal accretion and erosion processes. Moreover, a higher amplitude during spring tide (4.6–5.5 m) with strong tidal currents in the macrotidal coast in West Bengal with the formation of an overlapping tidal plain across the intermittently subsiding Bengal delta having a network of deeply scoured (>30 m) tidal rivers and inlets contributed in the delta building process.

2.3.3.1 Bengal Delta, the Foundation of the Sundarbans Mangrove Ecosystem

The Bengal delta, being one of the largest tide-dominated river deltas in the world was developed by the deposition of the transported terrigenous sediments by the Ganges–Brahmaputra River systems to the southern margin of the deltaic system followed by continuous working, reworking, and redistribution by the strong tidal currents enunciated from the Bay of Bengal. The combined operational impact of two geo-physical processes resulted in the formation of a massive Bengal delta throughout the Plio-Pleistocene period: (i) discharge of great volumes of clastic sediments by the Ganges–Brahmaputra drainage system through the Bengal trough as an outcome of denudation of the rapidly rising Himalayan chain in the northern parts of India and (ii) a high rate of subsidence of the Bengal Basin floor along a north–south axis facilitating selective deposition of coarser clastics in the Bengal Basin adjacent to the Indo-Burmese thrust front, from where the onward transport of predominantly finer terrigenous clastics was triggered to be deposited in a large part of the Bengal delta.

2.3.3.2 Deltaic Flood Plain and Lower Delta Plain in Sundarbans

The upper deltaic flood plain of the Ganges–Brahmaputra river basin with freshwater-dominated marshes and poorly drained swamps was formed as a result of delta progradation due to both the higher-elevation and freshwater flow into the upper delta plain and brackish water drainage to the lower delta plain form sub-aerial facie. A sizeable part of the Sundarbans mangrove estuarine complex with Holocene alluvial stratigraphy with an average thickness of 40 m represents a deltaic system with a low flat alluvial flood plain.

The maximum sea-level transgression at the western part of the lower delta plain (tidal) of the present-day Sundarbans on the Hugli–Matla estuaries occurred at about 6500 BP by the discharge of the Ganges–Brahmaputra rivers when the shoreline was 100–300 km inland of the present shoreline. This fact has been substantiated by the documentation of floral and faunal assemblages on the sub-surface Holocene sediments around Calcutta about 40 years back, indicative of the tidal mangrove forest flourishing further north of the present Sundarbans about 6000–7000 years BP (Banerjee and Sen 1988). Consequently, shoreline progradation coupled with basin infilling account for about 30% growth and development of the modern delta of the Ganges–Brahmaputra rivers where the paleo-shoreline followed an approximately 3-m elevation contour of the present condition (Allison 1998).

2.3.3.3 Geo-Morphological Uniqueness of Sundarbans

The West Bengal coast of India is characterized by the presence of the largest tide-dominated Hooghly estuary with numerous channels, creeks, and impressive digitized architecture, supposed by the extensive mangroves in terms of available evidence indicating that maximum sea-level transgression of this coastal stretch lying beyond the paleo-shoreline during Holocene time occurred at about 6500 BP when the shoreline was low to 300 km inland of the present shoreline. Shoreline progradation and basin infilling by the Ganges–Brahmaputra rivers subsequent to this time accounts for around 30% of the modern delta plain. However, the roles of different environmental factors determining the extent of process and nature of supply and deposition of sediment from the river Ganges and its distributaries to the lowland flood plain, lower delta plain, and marine end, or from the marine side in different parts of coastal West Bengal remain unknown.

The fluctuation of tides in the form of the rise and fall of the water level causes migration of energy zones, which results in distinct geomorphological and sedimentary structures on the surface of bottom sediments in both the intertidal and subtidal zones, are dependent on several factors such as beach slope, the textural attributes of the sediments, local geomorphological units, the depth of the water, flow direction, and flow velocity.

Different sub-zones of the intertidal belt within the foreshore of the coastal belts of Sundarbans are characterized by different geomorphological settings, especially by the presence of different forms of ripples (sinuous trains of backwash ripple marks, swash marks, and rhomboid marks at the upper foreshore zone; ripples named lignoid, polygonal interference, ladder back type, rill marks, rhomboid marks, scale ripple marks at the middle foreshore zone with runnels and ridges). However, they are characterized by the double-crested ripple, capped-off, ripple–rill complex at the lower foreshore zone, which indicate the strength of the flow regime from the lower to the upper foreshore zone. Mangrove forests usually growing up at the land–sea interface and experiencing a relatively stable period of sea level is appeared to be a prerequisite for the development of old forests, which indicates past changes in coastal conditions in respect of sea level and the simultaneous the response of mangroves to environmental change (Berger et al., 2006). Considerable insights in respect of response of mangroves to climate change in the future can be generated after comparing species of the present forest with paleoecological information with regard to the response of mangroves to past sea-level changes,

2.3.3.4 Neotectonic Settings and the Sundarbans Mangrove Ecosystem

Reconstruction of the earth history depends on the interpretation of ancient sedimentary environments, which in turn rely on sedimentary structures in the specific environmental zones of the intertidal beach-face within the surface structures.The deltaic islands of Sundarbans extends geographically along the easternmost coast of India in between the sea and plains of the lower Bengal basin transgressing Bangladesh with a geo-genetic link along the southern fringe. The tectonic evolution of the greater Bengal Basin is fundamentally dependent on the collision pattern of the Indian plate with the Tibetan (Eurasian) and Burma plates. The collision of these plates are seen in two different forms: (i) the oblique subduction of the oceanic crust beneath the Burma plate, resulting in the development of accretionary wedges, which together with thrusting and folding have subsequently uplifted the Indo–Burman Ranges; (ii) continent–continent collision of the Indian plate into the Tibetan plate, as expressed by thrusting, lateral displacements, and uplift at the north to northeasterly direction resulting in the development of the eastern Himalayas (Sikder and Alam, 2003). The Bengal basin, being one of the world's widest, deepest and tectonically active basins, centering below the present Ganges–Brahmaputra river mouths, has two parts, which are separated by a hinge zone marked by high gravity and magnetic anomalies that is thought to represent the dividing line between the Indian platform with full thickness of the continental crust and the Bengal Foredeep (Bilham and England, 2001).

Extensive fluvio-marine deposits though millions of years of the river Ganges and Bay of Bengal led to the geological development of Sundarbans. Lithological logs recovered from the boreholes at Sundarbans include the thick and effective clay blanket (15–75 m) at the top with the underlying presence of very coarse sediment, gravels of rock fragments, and quartz, which help to trace different phases of tectonic history of the Quaternary age of the region (Allison, 1998). The character of the sediments of the deltaic islands at seaward zones, facing the Bay, contain more sand and a higher percentage of silty clay, and the clay components of the sediments exhibit an increasing trend from the south to the north.

The Bengal Basin evolved in the northeastern parts of the Indian subcontinent from a passive continental margin during the pre-Oligocene geological era to a remnant ocean basin formed during the beginning of the Miocene period and comprised three geotectonic provinces as an outcome of the regional tectonic plate scenario, especially owing to the collision pattern of the Indian plate with the Burma and Tibetan (Eurasian) plates. These geotectonic provinces are: (i) the Chittagong-Tripura Fold Belt in the east by the collision and sub-duction-related orogeny; (ii) passive to extensional cratonic margin in the west, the stable shelf; and (iii) the central deep basin or remnant ocean basin.

The developmental process of the Bengal basin was initiated around 126 million years back with the Alpine Himalayan Orogeny bounded by Mikir Hill and the Meghalaya hills in the north of Assam–Nagaland on the north coast of the Cachar Surma Valley. The Bengal geo-province represents the westernmost portion of this entire basin, which also comprises the western part of Bangladesh. The initial delta building continued up to 49.5 million years BP. During the Oligocene period, at the time of the initial **'proto delta'** formation process marked by, a strong regression of sea with the shifting of the coastline southward took place. Afterward, during the tertiary period, this **'transitional delta'** experienced rapid deposition of sediments and subsequently the **'modern delt,'** was formed.

The late Pleistocene mangrove strand was situated on the west of the river Ganges. Between the sixteenth and eighteenth century, the Bengal basin was tilted easterly along a hinge zone because of neo-tectonic movement and then curved toward Bangladesh, when the flows of the Ganges were directed to the river Padma within Bangladesh (Sanyal et al., 1984). McDowell estimated the sediment load of the Ganges–Brahmaputra river system, which was found to be second in the world, next to the Yellow river of China, which amounted to 1670 million tonnes per year. In Sundarbans in India, the presence of mangrove plants extended to the presently Kolkata metropolitan area, as several freshwater-loving mangrove plant species travelled 60 km inland in the inner estuary (Banerjee and Sen, 1984, 1987/1985).

2.3.3.5 Delta-Building Process in Sundarbans

An understanding of the processes of a long depositional history of the delta-building process spanning a period of more than 70 million years is needed to undertake proper eco-management of the Sundarbans mangrove ecosystem, especially in view of the sediment dispersal mechanism in the face of fluctuating tidal impacts during high tides and low tides. The deltaic islands are the areas where fluvial and marine land-building processes have worked simultaneously with the cyclical retreat and advancement of the sea during past geological ages. The retreats usually take place at the sea face, gradually but steadily in a southward direction where continuous sedimentation takes place to build up new land on the continental shelf. Constantly built-up topographic zones within Sundarbans exhibiting vast new alluvial plains intersected by a network of innumerable creeks and tidal channels along with several estuaries possessing soil, water, air, flora, and fauna with unique characteristics, making the entire ecosystem vibrant.

The deltaic zones are divided into three parts based on the process of their development, such as moribund, (formed in the decayed riverine zones); the mature (elevated parts of the delta above the highest water level); and the active deltas (deposition of sediment takes place only during high tide with sediment-loaded flood water). The maximum amount of sediments reaches to the sea through major estuaries such as Hooghly, Mooriganga, and Matla, but tidal actions push a certain portion of this sediment load back through creeks and rivers. The sediment-loaded water during high tide reaches and spills over flood plains, whereas receding water during low tide helps the settlement of those deposited sediments elevating land along the flood plains of these estuarine-rivers. Thus, the Sundarbans area is identified. The tidally active delta part of Sundarbans has still been undergoing the land-building process intersected by interlacing channels and intervening deltaic islands with the elevation of around 3 m above the mean sea level and the occurrence of dense mangrove forests gives this littoral tract a distinctive geographical identity. An estimate has revealed that the Ganges–Brahmaputra riverine system carries sediments amounting of 1.67 billion tonnes. However, no appreciable growth of new land has been noted along the coast during last three centuries, despite receiving this greater sediment load. A number of geo-physical factors have been playing their roles in the delta-building process, which include (i) erosion of the southern littoral tracts owing to the continuous attack by destructive waves; (ii) the siphoning of sediment away from the coast through the submarine canyon of the Bay of Bengal; (iii) slow subsidence due to auto-compaction of newly deposited sediment, surpassing the effect of accretion.

Some parts of the Bengal Basin with the deltaic plains have been tilting gradually toward the east since the period of their formation, mainly because of the discharge of freshwater, which shifted through Bangladesh, imposing severe stress on the freshwater budget for the Hooghly–Matla estuarine networks (Morgan and McIntire, 1959). The Hooghly–Matla estuarine system experiences a higher rate of freshwater discharge, having comparatively less well-mixed saline water than its eastern counterpart.

2.4 Geological Past and Estuarine Networks of India

The Sundarbans mangrove–estuarine system is geologically considered recent in origin where formation of different deltaic islands of the Ganges–Brahmaputra system was supposed to have been initiated at the end of Miocene and reached its present condition around 10,000 years back, spanning a period from the Pleistocene era to recent times. In this tidal forest region both marine land and fluvial processes have been working simultaneously during past geological ages resulting in cyclical advancement of the land mass and gradual retreat of the sea southward alongside a continuous sedimentation process to build up new land on the continental shelf.

The continuous tectonic movements, especially in northwestern Punjab and also in the southeastern flow of the river Ganges with deposition of sediments, brought about several geomorphological changes in the Bengal Basin since the Tertiary period and resulted in the development of the deltaic islands in Sundarbans (Wadia, 1961; Gopal & Chauhan, 2006). The huge amount of terrestrial sediments brought in by the Ganges–Brahmaputra system during the post-Pleistocene period, probably bypassing the deltaic regions to a considerable extent, ultimately contributed to the rapid growth of the Bengal Deep Sea Fan. The geotectonic movements during twelfth and sixteenth centuries, in and around the Bengal Basin, forced the Ganges river to change its course eastward and subsequently joined with the Brahmaputra riverine flows, which pushed the deltaic portion of the Bengal Basin to shift in an eastward direction (Morgan and McIntire, 1959). The hydrology and sedimentation patterns of the Bengal Basin were significantly altered because of these continuous tectonic activities, which resulted in shifting of the freshwater flow eastward toward Bangladesh, and thereby imposing salinity invasion in the Hooghly–Matla estuarine complex of Indian Sundarbans (Gopal & Chauhan, 2006).

The tropical estuarine complex of the entire Sundarbans mangrove ecosystem, constituted by a network of some major estuaries such as Hooghly, Mooriganga, Saptamukhi, Bidyadhari, Thakuran, Raimangal, Bartala, Gosaba, Haribhanga, and Matla, which occupy an important place on the global map and can be categorized as a river–delta estuary, was advocated by Odum in addition to Pritchard's (1967) four categories of estuarine systems. Out of these estuaries, six estuaries (Mooriganga, Saptamukhi, Thakuran, Gosaba along with Ajmalmari, and Matla) became disconnected from the main flow of the river Ganges, because of siltation and their estuarine character, and are now being maintained by the monsoonal runoff alone (Cole and Vaidyaraman, 1966). Therefore, the eco-geological processes involved in the formation of deltaic systems of Sundarbans has nearly ceased in the west parts but has assumed an accelerated pace in the eastern sector (Gopal & Chauhan, 2006). In addition to ten major estuaries, 31 rivers form a network of intricate freshwater–salt-water courses intersecting the length and breadth of the Indian Sundarbans and serving two-fold purposes of freshwater discharge along with facilitating the to-and-fro movement of saline water in between the freshwater regime with the open sea. Several dominant estuarine physical processes such as tidal flooding, waves, sea-level change, river fluxes, catastrophic storms etc. play their roles in determing the hydrodynamics of the region. These so-called 31 rivers have tidal influences within the low-lying coastal deltaic plain of Indian Sundarbans with sources of freshwater from upstream, most of which are the result of tidal incursion and retreat. Most of these rivers during the process of to-and-fro movement of tidal water on a gently sloping plain are also joined by many other smaller channels and water flows of lesser magnitude are locally called khal, gang, etc. The hydrodynamical orientation of all these water flows are determined by geomorphic setting, patterns of sedimentation, complex interaction of several geological, climatological, geomorphological factors such as the tectonic framework of the Bengal basin, and biological processes.

2.5 Administrative History of Sundarbans Biosphere Reserve, Especially during the Colonial Period

The deltaic Sundarbans on the Indian subcontinent started originating around 5000 years ago by the silt carried by three mighty rivers, the Ganges, the Brahmaputra, and the Meghna, and is presently located across two neighboring countries, India and Bangladesh, accommodating one of the largest mangrove forests in the world (Allison et al., 2003). The entire Sundarbans spreads over about 9630 km^2 between 21° 40′ N and 22° 10′ N latitude, and 88° 2′ E and 89°6′ E longitude, and is the part of the tide-dominated lower deltaic plain.

Sundarbans was first recorded and declared as lands of importance because of forest wealth at the district of 24 Parganas, of the undivided Bengal province, India, by the then Jaigir, Lord Clive of the East India Company in the 1757. After that, the history of the Sundarbans was only about the change of land-use pattern converting the forest areas into agricultural lands and that was introduced and supervised by the then Collector General of 24 Parganas, Mr. Claude Russel in 1770. The forest tracts of the then Sundarbans came under well-planned and proper eco-management following scientific guidelines since the 1970s. Earlier, Heimig in the British era introduced the first working plan for the Sundarbans forest during the period 1893–1894 to 1902–1903 where a permit was given strictly for the cutting of mangrove trees, mostly *Heritiera fomes*, locally named Sundari tree, only after plants attained considerable growth with a diameter of more than 3 inches.

Sundarbans is located in the Gangetic deltaic system of the coastal region of lower Bengal with close proximity to the southern corner of the sea, Bay of Bengal (21° 31′ to 22° 38′; N and 88°5′/–90° 28′ E), with its expansion from the mohana of the Bhagirathi river to the mohana of the Meghna river in the west, to tracts of mangrove forests from the districts of Noakhali, Chattogram, etc., in the east, and also extending far to the southern parts of the deltaic islands of Hatiya, Sandwip, etc. All of these areas were under the territories of three major districts of the undivided Bengal Province, viz. 24 Parganas, Bakharganj, and Khulna (Roy, 1949; Chakraborty, 2009).

The history of Sundarbans, a unique landscape with a vast expanse of evergreen forests intermingling with a vast sheet of tide-dominated brackish water bodies, mostly in the form of estuaries associated with thousands of tidal creeks and channels, revolves around the interaction of human beings with the everchanging natural settings constituted by half water and half forested land. In this terrain, deltaic islands have developed through the ages because of complex interactions among marine and riverine water, flora, and fauna. These deltaic islands have unique halophytic mangrove forests that had been alternately inhabited and deserted, not only because of man–wildlife interactions (poisonous snakes, tigers, crocodiles, etc.) but from time to time the occurrence of so many natural disasters such as catastrophic storms, cyclones, tidal waves, reduced flows of freshwater, etc. The Sundarbans is the only mangrove–tiger land in the world where the tiger stands at the pinnacle of both the aquatic and the terrestrial food chain. The entire area of Sundarbans covering an area of around 10,000 km^2 shared by two neighboring countries, India and Bangladesh, have been designated a World Heritage site in India and a Ramsar site in Bangladesh. Tracing the history of this unique natural landscape, it was noted that Hiuen-Tsang described this eco-region as a low-lying area bordering the sea with highly productive green coverage mangrove forests, considering it as part of the land of Samatata in the seventh century. This area of lower Bengal has been repeatedly called *Bati* or *Bhati* in medieval texts. In traditional oral folk songs, the entire land mass extending from the lower Gangetic plains in the west to the estuarine networks of Chittagong in the east was referred to as *Bangala* or *Bhati* and characterized by dense tidal forests. In the famous book of Bengali history (**'Bangalir Itihas'**), written by Niha Ranjan Roy, several sources of evidence were cited where many uninhabited parts of the Sundarbans were cultivated by the people who invaded those deltaic zones after clearing the mangrove forests in the Middle Ages. Another historical document was published by William W. Hunter in Hunter, 1875, who clearly mentioned the presence of several human settlements in several parts of the Sundarbans. Historians also opine that king Pratapaditya was the actual ruler of the Sundarbans at the end of the sixteenth century, when the possessor of this land was one of a group of chiefs locally named *Barah Bhuiyas* (the Twelve Chiefs), who were liegemen of the Mughal emperors. The king Pratapaditya infringed the authority of the Mughal emperor, Akbar, all-time great king of India, and set an independent stronghold on the marshy zones of lower Bengal.

It is thought that several zones in coastal Bengal were abandoned and converted into mangrove forests because of the impacts of several adverse environmental conditions without getting any support from society and the rulers due to their lack consciousness. The devastating flood of 1584 considerably destroyed the available resource bases and dislodged several thousands of inhabitants of the region. These problems emanating from nature were compounded by the Portuguese and Magh incursions. In the map prepared by Rennell in 1761, the entire coastal tract to the south of Bakharganj was highlighted as "**country depopulated by the Maghs.**" This invasion by external people such as the Portuguese coupled with the impacts of cyclones and storm waves played pivotal roles in the depopulation of the entire coastal area. The Portuguese were seen to occupy Tardaha on the Bidyadhari estuary up to the end of sixteenth century, with their practice of attacking and robbing with trade. Therefore, during that period of time, the whole of the Sundarbans area became a haven for Portuguese and Magh pirates. The drastic disappearance of human settlement in the Sundarbans towards the end of sixteenth century was thought to be due to certain alterations in the functioning of the mangrove ecosystem with the shifting of the course of riverine flows reducing the freshwater flows and bringing forth hypersaline conditions. During the eighteenth century, travelers from their ships occasionally noted **"reddish brown turbid waters," "mud-flats," "inundated low-lying forests,"** and **"sand banks"** in and around the then Sundarbans. By the beginning of the nineteenth century, the application of knowledge of the British rulers about the Indian landscape brought about strong desire regarding the ecological and economic potential of the Sundarbans landscape, which was at that point in time the favorite habitats for ferocious wildlife, such as poisonous snakes, tigers, and crocodiles, and also as a place with the "**barbaric**" custom of infanticide practiced by Hindu devotees.

The reclaiming phases of the Sundarbans mangrove forest coincided from the nineteenth century with the rise to power of the British East India Company. The main administrative center for their Governance was set up in the then Calcutta, only around 100 km away from the Sundarbans.

However, previous efforts for the reclamation of Sundarbans in the eighteenth century had been unsuccessful. The second phase of reclamation work began in full swing in the second decade of the nineteenth century, but it also witnessed little success owing to the depredations of man-eaters.

William W. Hunter, in his 60-page report published in 1875, portrayed the Sundarbans as an alluvial plain, covered by mangrove forests and the entire forest area was surrounded by a dozen large estuaries and their associated hundreds of tide-dominated creeks and channels crisscrossing the entire region.

The extent of the then Sundarbans up to middle of the eighteenth century was from the Hooghly river estuary in the west to Jamuna river at Bakharganj in the East, mainly restricting its territory within the district 24 Parganas. At that time, the entire area within Sundarbans was virgin, without experiencing harmful human intervention, especially no agricultural activity on the noncultivable land. However, this region was fragmented into different landmasses and large chunks of those fragmented lands had been distributed among the local zamindars and affluent farmers since 1773 on a lease basis. This venture by the then ruling British Colonial Administrators was undertaken in order to earn some revenue from that wildlife-infested tidal forest.

Lloyd, Farrington, and Bart developed the first working plan of this and maintained the same execution of this policy of felling trees each year from 1903 to 1912. Trafford placed the second working plan for the policy of felling trees for a period spanning from 1912 to 1932 and the third working plan for the same purpose was made by Curtis from 1931 to 1952 but with a gap from 1937 to 1938. After discussion, another working plan was chalked out by Mr. S. Chowdhury, which emphasized assembling all circles into a single working circle from 1937–1938 to 1948–1949. This scheme was continued in 20-year felling cycles under the then forest management system. The first working plan for managing the forests of Sundarbans in the independent India was introduced during the years 1949–1950 and was continued for the next 10 years. No working plan was prepared after the end of this plan in 1958–1959 and the same plan was followed up to 1976–1977 for felling operations in the annual coupe. The refuge problem at Marichjhapi created inconvenience for the implementation of this plan during the consecutive years from 1977 to 1980.

The entire Sundarbans region is at present distributed between the Sundarbans Tiger Reserve and the 24 Parganas South Forest Division, which are geographically separated by the Matla River. In 1989, the Ministry of Environment and Forests, Government of India, declared the entire 9630 km^2 of Sundarbans the Sundarbans Biosphere Reserve (**SBR**) for conservation, management, research, training, and maintaining harmony between man and the environment of mangrove forest lands, Sundarbans (Anonymous, 2003).

In 1770, agricultural activities were introduced within the territory of the then Sundarbans by the then British rulers under the leadership of Clod Russel but to a very meager extent. Later, in 1784, the forest tracts of Sundarbans were given importance as a potential landscape, with its ecological uniqueness, by Tillman Henkel, and all of such administrative initiatives ultimately helped to form a special administrative task force with the subsequent creation of a post, named **"Commissioner of Sundarbans"** to impose regulatory control over the entire eco-region. Alongside the initiatives taken by Leftenant, Hages, and Dampier to earmark the entire area of Sundarbans with definite territory during the period 1822–1830, which resulted in preparation of an acceptable map of Sundarbans by Ellison pointing out the presence of several sub-divisional areas such as Barasat, Basirhat, and Bangram, surrounded and intersected by a network of river estuaries such as Ichamati, Bidyadhari, Kalindi, and Raimangal, along with innumerable creeks and tributaries. The map of the entire Sundarbans prepared by Ellison, covering an area of 9629.9 km^2 included 15 different thanas: (1) Sagar; (2) Patharpratima; (3) Kakdwip; (4) Namkhana; (5) Mathurapur; (6) Kultali; (7) Joynagar; (8) Basanti; (9) Harowa; (10) Canning (erstwhile named Matla); (11) Minakhan; (12) Gosaba; (13) Sandeshkhali; (14) Hasnabad, and (15) Hingalgunj. The erstwhile 24 Parganas district was split into two districts, North 24 Parganas and South 24 Parganas in 1986. The district North 24 Parganas now has six thanas (Harowa, Gosaba, Sandeshkhali, Minakhan, Hasnabad, and Hingalgunj under the Basirhat subdivision) and the district South 24 Parganas included the remaining nine thanas within two subdivisions, Diamond Harbor and Alipur (District Headquarters). The total area of the then Sundarbans was scientifically delimited by Dampier and Hages by outlining the area that is still considered the **"Dampier–Hages Line."** Currently, the headquarters for Sundarbans is set up at Alipur (Roy, 1949).

Since 1793, with the initiation and implementation of the policies pertaining to permanent settlement, personal land with no stable demarcating line within the forest did not have property value along the margin of Sundarbans as the forest was the property of the Government and did not come under the purview of laws pertaining to the permanent settlement. Way back, up to the eighteenth century, there was no idea about the total area of the then Sundarbans, which was thought to be around 4500 km^2. In 1817, this was brought to the notice of the Land Revenue Department that local Zamindars used to earn a certain amount of revenue from the people residing near the forest as water tax and salt tax, In 1818, the British Government enacted and imposed three laws in order to establish total control over the forest resources. In 1853, the Government declared its policy for developing principles with the aim of clearing the forests to facilitate agricultural activities toward more food production and health management.

2.6 Uniqueness of the Mangrove Ecosystem with Special Reference to Sundarbans, India

The Ganges–Brahmaputra deltaic systems of India and Bangladesh are ecologically unique for possessing diverse coastal habitats such as salt marsh, grass belts, estuaries, creeks, backwater, inland flood plains, dunes, intertidal and subtidal zones, all of which together are considered to be the Sundarbans mangrove ecosystem, the largest among all the mangrove areas of the world, which functions and is sustained through an interactive process of all such subsystems exchanging different biological and nonbiological components. Most of this mangrove forest is located on the low-lying deltaic plains at southern ends, with an elevation of about 2 m above the present mean sea level (Hussain and Acharya, 1994; Sanyal, 1996; Naskar and Mandai, 1999). This mangrove ecosystem of Sundarbans renders lot of ecosystem services for the benefit of the environment in general and people of the region in particular (Chakraborty, 2019), but appropriate conservation strategies are required to save this sensitive and fragile ecosystem, maintaining the normal ecological qualities of the soil and water that determine the growth, development, richness, diversity, and zonation of the mangroves on a local scale (Lugo and Snedaker, 1974; Hutchings and Saenger, 1987; Naskar & Guha Bakshi, 1987; Woodroffe, 1990; Ellison & Stoddart, 1991; Duke et al., 1998; Saenger, 2002; Syvitski and Saito, 2007; Braud, 2008; Saito et al., 2007; Sarkar, 2009; Vorosmarty, 1998; Paul & Paul 2023). Owing to their existence within intertidal habitats, and because of their sensitiveness and adaptability in the unstable and oscillating ecological conditions, mangroves are considered an indicator of sea-level change. In order to cope with the changing ecological conditions, mangroves move landward or seaward in accordance with sea regression or transgression and this ecosystem is often referred to as a dancing ecosystem (Sanyal 1990).

2.7 Coastal Environment of West Bengal, India

The Indian coast with a coastline of 7500 km, including the two island territories, exhibits some common geomorphological features of coastal zones such as tide-dominated mudflats and sand flats, varied forms of ripple markings, long-shore currents, occasional occurrence of dunes, the presence of innumerable creeks and estuaries with brackish water, lagoons with back water, mangroves, salt marsh, and saline grass belts. However, the Indian coastline, with lots of diversity in respect of both geomorphological and biodiversity, has been under tremendous human-mediated environmental stresses, which have resulted in substantial eco-degradation of different structural components of this ecosystem (Qasim et al., 1988). The coastal area of West Bengal extends over 0.28 million hectares and 220 km of coastal line shared by two districts viz. Midnapore (East) and South 24

Parganas. The coastal zones of West Bengal highlight a long history of deposition and seaward advancement over the Bengal Basin because faulty movements of the tectonic settings by geotectonic process, which results in concomitant sea-level changes in multidimensional domains of physical states (Paul, 2002; Chakraborty, 2011, 2013, 2018). A massive erosion process operating on the coastline of West Bengal results in the shifting of sedimentation back to inland areas, filling up their own valleys and coastal plain streams with sandy sediments in response to a recent, slow, and small rise in sea level in the region. Throughout the length and breadth of this coast, the back shore and beaches are characterized by sand cliffs and clipped clay banks and experience an alarming rate of erosion (Paul, 2002).

2.7.1 The Coastal Environment of Midnapore (East) District, West Bengal, India, continual Part of Sundarbans

The coastal tracts of Midnapore (East) District, West Bengal, India, sharing about 27% (60 km) of the total length of the coastal tract of West Bengal, India, is a contiguous part of the deltaic Sundarbans mangrove ecosystem, mostly restricted within the South 24 Parganas district, West Bengal, India, and extends along the Hooghly estuary from New Digha (at the confluence of the Subarnarekha estuary with the Hooghly) to the extreme south-western point of the Midnapore district and then curves around Junput, Rasulpur, Khejuri, and Haldia east to further north east up to Tamluk or even Kolaghat, located on the bank of the Rupnarayan river (Chakraborty, 2010). The coastal tract of Midnapore is endowed with characteristic geomorphological settings such as sand dunes, longshore currents, high salinity, less turbidity, and less vegetational coverage (Chakraborty, 2010) than its counterpart, the South 24 Pargana district, accommodating the Sundarbans mangrove ecosystem. The inner zones within the littoral tracts of coastal plains of Midnapore (East) District, West Bengal, India, represent the corresponding shorelines of the post-Pleistocene period (Paul, 2002).

This macro-tidal coastal environment is characterized by having long waves with high energy produced across the Bay of Bengal, coupled with predominantly southwesterly monsoon winds blowing onshore and also easterly and southerly winds strengthened by the occasional occurrence of cyclones over the Bay-head coast during the summer months. The littoral tracts of Midnapur (East) coast with sandy sediments have been confronted by a number of coastal incidents such as sea-level rise, coastal erosion, shifting of dislodged and loosened sediments from one part of the coast to other coastal regions, etc. In order to control those ongoing problems, mitigation measures have been adopted against storm hazards, coastal pollution, industrial developments, aquaculture activities, navigability, coastal recreational activities, exploitation of natural resources, undertaking appropriate rehabilitation measures for environmental refugees and the introduction of invasive species, etc. All of those issues have created conflicts among different resource users: local people, landowners and other interested groups, developers and environmental planners, ecologists, engineers, geoscientists, and economists (Paul, 2002). The Subarnarekha river estuary, representing the prime estuary of the Midnapore (East) coast, locally named **"the streak of gold"** and traversing about 460 km through the picturesque terrain of varied landscapes of three states of Eastern India, viz. Jharkhand, Odisha, and West Bengal, after having its origin from the hilly tracts near Ranchi city of the Jharkhand state and before opening out to the Bay of Bengal.

2.7.2 Coastal Environment of Sundarbans, West Bengal, India: Evolution of Conservation Strategies

The deltaic systems of the Indian Sundarbans is surrounded by the Ichamati–Raimangal river estuaries in the east, by the Hooghly estuary in the west, by the sea of the Bay of Bengal in the south, and the Dampier–Hodges line drawn in 1829–1830 in the north. The entire Sundarbans eco-region, shared by the two neighboring Asian countries, India and Bangladesh, with a total area of 25,500 km^2, of which the total area of Indian Sundarbans is about 9630 km^2 and the remaining part remains within the territory of Bangladesh Sundarbans. The length of shores of Indian Sundarbans is 130 km out of the total 260 km coastal length of West Bengal, whereas the mangrove forest at the Bangladesh part of Sundarbans is included within a coast line of about 230 km out of 710 km of the total coastal length.

The Sundarbans of India is famous for its healthy mangrove ecosystem, on which a number of Reserve Forests of several kinds (tiger projects, crocodile projects, sanctuaries, and biosphere reserves) are present, occupying nearly 4260 km^2 (Project Tiger: 2585 km^2; Sundarbans National Park: 1330 km^2; Sajnekhali Wildlife Sanctuary: 362 km^2; Holiday Island Wildlife Sanctuary: 6 km^2; Lothian Wildlife Sanctuary: 3859 km^2; buffer zone: 893 km^2).

Currently, 54 deltaic islands out of 104 such islands in the Sundarbans region have been reclaimed for human settlements, with a human population of more than one million spreading over two administrative districts, namely South 24 Parganas with 13 blocks and North 24 Parganas with 6 blocks. The estuarine mangrove complex of Indian Sundarbans possesses several major river estuaries such as the Hooghly (Ganges and Bhagirathi), Mooriganga, Saptamukhi, Bidyadhari, Kalindi, Bartala, Roymangal, Thakuran, Matla, Ajmalmari, Haribhanga, and Gosaba, which flow from west to east. Out of these estuaries, five (Mooriganga, Saptamukhi, Gosaba, Thakuran, and Matla) have lost their connectivity with the main flow of the river Ganges owing to the continuous siltation process and reduced freshwater supply, which modify their estuarine character, depending mostly on the monsoonal runoff (Cole and Vaidyaraman, 1966; Rahman et al, 1979, 2010). The Sundarbans mangrove ecosystem is endowed with several characteristic geomorphological settings such as mudflats, sand beaches, coastal dunes, grass lands, mangrove swamps, estuarine networks, shallow creeks, and riverine channels (Chaudhuri & Choudhury, 1994; Paul, 2002; Chakraborty, 2011). The Ministry of Environment and Forests, Government of India, declared the entire 9630 km^2 of Sundarbans within Indian territory in the state of West Bengal as the SBR in 1989 for conservation, eco-management, research, training, and maintaining the harmonious relationships of human beings with mangrove forest lands intermingling with estuarine networks of Sundarbans, India (Anonymous, 2003).

Indian Sundarbans (latitude between 21′ 31° N and 22′ 30° N and longitude between 88′ 10° E and 89′ 51° E) with its counterparts of mangrove belts in Bangladesh constitute two-thirds of the total mangrove coverage of the world and the largest deltaic plain of fluvio-marine deposits contributed by continuous sedimentation from three major riverine systems (Ganges, Brahmaputra, Meghna) at the confluence of the Bay of Bengal. The geographical limit of Indian Sundarbans is decided by an imaginary line, known as the Dampier and Hodges line, drawn in 1831 and representing the international boundary between India and Bangladesh, and the Bay of Bengal in the south. The river Hooghly lies in the north west, the rivers Ichamati and Kalindi remain in the west, and the river Haribhanga flows in the east. The mangroves of Sundarbans in India encompasses parts of the districts of South 24 Parganas, southern parts of the North 24 Parganas and also the coastal belts of Midnapore (East) in the state of West Bengal, India.

In order to ensure conservation of the biodiversity of the mangrove ecosystem, several attempts have been made, which include the declaration of Sundarbans Tiger Reserve in 1973; the Sundarbans Wildlife Sanctuary in 1977, the World Heritage Site by UNESCO in 1987; the Global Biosphere Reserve by UNESCO in 2001; assignment of the Status of National Park in 1984.

2.8 Estuarine Networks of Sundarbans, India: Present Day Drainage Patterns

2.8.1 Estuarine Networks in Indian Sundarbans: Different River Estuaries of Sundarbans

An estuary is a semi-enclosed coastal body of water, where the seawater is considerably diluted by the influx of freshwater from land runoff through adjoining freshwater rivers and always experience tidal interplays (Pritchard, 1967; Fairbridge 1980). Estuaries of Sundarbans categorized as coastal plain estuaries are formed from rising sea levels invading into existing river valleys.

Tidal courses of rivers and river channels with saline estuarine environment are endowed with important ecological values, which are extremely sensitive to changes because of the influence of several natural factors as well as anthropogenic interferences (Ginsberg and Perillo, 2004).

The Sundarbans mangrove ecosystem, characterized by a low flat alluvial plain with dense assemblage of halophytic mangrove plant species is intersected by innumerable tidal rivers, estuaries, creeks, and saltwater channels (Bagchi, 1972; Paul, 1999, 2002; Chakraborty, 2009, 2010, 2011, 2017, 2018). The dynamic flow patterns through the distinct system of drainage networks of tidal waters and the ongoing process of erosion and accretion of lands make complex geomorphic setups throughout the length and breadth of the entire area. The estuaries within Indian Sundarbans flow in an essentially southward direction and end at the Bay of Bengal forming a funnel-shaped opening. Both banks of these estuaries and their interconnected creeks exhibit different forms of slumping and terrace structures formed because of the compaction of sediments on the one hand and neo-tectonic movements of the Bengal basin-mediated subsidence of marginal bank materials on the other (Umitsu, 1993).

Flows in different connecting channels or creeks with the rivers and estuaries are disrupted because of the deposition of marine sediments in the marine or brackish water zones in the estuarine networks rather than the fresh connection, development of aquaculture and agricultural lands, land-use changes, and waste disposal from upstream of the adjacent metropolitan city of Kolkata and upstream activities have contaminated the river basin with other forms of pollution in and around the Sundarbans, affecting the natural drainage system. Upstream and other activities carry large amounts of waste from urban, rural agricultural land, surface run-off, and industrial effluents from their vast drainage system (Paul, 1999).

Both banks of the river estuarine systems fringed with mangroves determine the nature of stratifications of the banks, disturbed heavily by pneumatophores and other forms of roots of mangroves, and also by the biological activities of benthic infauna. With the strengthening of banks by the construction of embankments against ongoing erosion pressure, mangroves become restricted in the intertidal zone, limiting marsh vegetation behind the embankments.

Supratidal levee deposits possess the following features:

(i) Dominance of mud in the upward direction of bank deposits with the rapid thinning of sand laminae.
(ii) Fine obscured laminae of silt and clay disturbed by roots of marsh vegetation characterize profusely marshy areas.
(iii) Mostly the upper portion of bank deposits are endowed with plant remains and plant roots.
(iv) Patches of salt encrustations are observed on the levee surface owing to evaporation of saline water during overbank flooding.
(v) Mud cracks of different shapes on the upper parts of the estuarine banks are formed owing to suspensional mud deposits, indicating sub-aerial exposure.

The catchment areas of different estuaries, creeks, channels, and connecting rivers of Sundarbans have been losing their life rapidly owing to different human activities involving inland fisheries, brick making, the tannery industry, etc., which have further worsened the ecological condition of the drainage basin because of the escalating rate of siltation on the river bed through transportation of materials, land reclamation, agricultural activities, cultivation, dumping of sewage sludge, and fisheries' waste.

Thus, the Bidyadhari river catchment is losing life rapidly. Along with a stupendous change in its ecological behavior, its capacity for holding sewage is also coming to a halt. Centuries back, estuarine networks and associated drainage systems of Sundarbans were fed by the upland freshwater discharge from the Ganges, Padma, Brahmaputra, and Meghna river systems, but with the passage of time, changing patterns of freshwater flows from the headwater discharge have made the most of the parts of Sundarbans freshwater deficient because all these river systems were cut-off during this period and become active during the rainy season, after being fed by the precipitation in the catchment areas.

A network of tide-dominated lotic aquatic systems throughout the length and breadth of Sundarbans, India, covering all the deltaic islands (54 reclaimed and 50 virgin), which include Hooghly, Mooriganga, Saptamukhi, Bartala, Bidyadhari, Raimangal, Gosaba, Herobhanga, Thakuran, Matla, and also so many other tide-dominated rivers such as Kalindi, Ichamati, Kalagachia, Gomal, Piyali, Karatal, Jamira, and Jagaddal.

Described in the following sections are some of the rivers of both ecological and historical importance.

2.8.1.1 The Bidyadhari River and Historical Connectivity with the Kolkata Metropolitan Area

The Bidyadhari river estuary, with a very circuitous tidal course of about 95 km, had its origin a little north-east of Dogachhia as a drainage channel, called Nona Gang, occupying low interfluves between the rivers Sunti and Padma. Bidyadhari river flows through both districts of West Bengal, North 24 Parganas and South 24 Parganas, toward the south past the Jessore road and Barasat–Basirhat rail track. It also received sewage and storm water from erstwhile Calcutta to be canalized through two different channels into the central lake canal emptying into another branch of the Bidyadhari south of Bamanghata (88′ 29° E and 22′ 31° N). Four different names have been assigned to the main Bidyadhari river system from the Haroa: Haroa River, Malancha River, Chaumuha River, and Kalagachia River from north before it reaches Dhamakhali for a total stretch of 120 km.

The freshwater discharge potential of the rivers has been considerably reduced, mainly because of the largely silted up head water portion of this riverine branch, which subsequently lost upstream connections with the Ganges river, and also the blockage of many other narrow or wide freshwater tributaries because of different land-use changes and a continuous heavy siltation process. The Bidyadhari river meets the Matla river in the eastern and northern sectors of the Sundarbans with different reserve forests. After entering within the tidal zones of Sundarbans, the Bidyadhari river, which flowed north-east past Haroa, where it was known as the Haroa gang, then bent westward to join the Nona khal. This river, while flowing in the south east direction, receives effluents from

Kolkata urban belts from Tolly's Nullah from the east on one hand and bidding of another tide-dominated Piyali river flowing south before reaching one of the gateways of Sundarbans, India, at Port Canning. This south flowing river gives off other smaller branches like Kuriabhanga, Kartoya, Khurti, and Karats. The united streams form the Matla River or rather estuary, which is navigable up to Canning. The Bidyadhari river flowed south-westwards to the junction of the Tolly's Nullah, and Beliaghata canal within Kolkata metropolitan city and then it joined the Karatoya and Atharobanki rivers southeast to Matla near Canning, the beginning of Sundarbans. The united flows of different creeks and tributaries formed from the Matla River reached the sea to the south and on receiving tidal flux from the marine zone became navigable up to Canning. Recently, the derelict portion of the Bidyadhari river near Kolkata has been under profuse siltation at an increasing rate and is being used as an outfall channel for the storm water and sewage of the city. The siltation process has been accelerated because of the development of local fisheries and some parts after being reclaimed for rice cultivation, the effect of which synergistically decreased the spill of river water over the adjoining land and, consequently, increased the deposit of silt in the river bed. Other contributory causes have been the construction of the Dhapa lock, the closing of tributaries, and restricting its flow, resulting in the canalization of the Bhangar khal.

2.8.1.2 Decline of the Bidyadhari River

It was observed during the years 1901–1912 that the bed of the Bidyadhari river had risen about 8 m in 8 years 1.6 km below Bamanghata, whereas the cross-sectional area had been reduced from 715 m^2 giving a mean rate of contraction of 45 m^2. It was concluded by the Hooghly–Bidyadhari Canal Enquiry Division that the Bidyadhari had a very short remaining lease of life, and that in 6 years' time it would be useless as an outfall channel for the sewage of Kolkata unless remedial measures were taken. It was also suggested that the only practicable way of dealing with the situation was to canalize the channel of Bidyadhari from Dhapa to the off shoot of the Piali river, 15 km below Bamanghata, and the channel of the Piali river from its head to its outfall into the Matla river.

The Government of Bengal, in 1928, declared the Bidyadhari river useless based on the opinions expert members and it was decided to stop maintaining the river any further. The decline of the Bidyadhari river is thought to be due to various manmade and natural causes such as the development of fishery and aquaculture, constructions of embankments, and other forms of land-use changes. The water quality of the main flows of the Bidyadhari river has been gradually but steadily deteriorating owing to the release of industrial effluents with domestic sewage, some of which are being productively utilized in the adjoining wetlands for aquaculture and also for irrigating the agricultural efforts of the adjoining lowland.

Different creeks connected to the Bidyadhari river are Katakhali, Pathankhali, Radhanagar Durga Mandir, Nabanki, Kayer, Dundul Khali, Netidhopani,

Mocumberiya, etc., and several tidal channels are Gumdi and Durga Doania near the Sajnekhali Sanctuary; Kartal gang, and Moukhali nala.

2.8.1.3 Ichamati River

Ichamati river, after having its origin in the district of Nadia, is bifurcated into two branches at Hingalganj, as the western smaller channel known as Sahebkhali and the main flowing river at the eastern side known as Kalindi. The Bidyadhari and Sahebkhali rivers are joined at Sandeshkhali Block 2 and form the course of the Raimangal river. The river Kalindi joins with the main flow of the Bidyadhari river downstream near Sandeshkhali II and the combined stream flows as Raimangal estuary south in the Sundarbans Reserve Forest as Haribhanga estuary and opens up in the Bay of Bengal. The combined flows of Ichamati–Bidyadhari–Rainmangal–Haribhanga act as the international boundary between India and Bangladesh, between the districts of Nadia and North 24 Parganas just before Bagdah PS and it enters at Hingalgunj block into the Dampier–Hodges imaginary boundary line demarcating the Sundarbans. The river Ichamati is connected with other rivers and estuaries of Sundarbans, namely Jamuna, Raimangal, Bidya, Jhilla, Kalindi, etc. Recently, waters of the Ichamati river overflowed after a short spell of rain, resulting in flooding during the monsoon every year, owing to heavy siltation, the setting up of fisheries by blocking the river with wooden rafts, the development of so many brick fields, the building up of houses on the riverbed, the luxuriant growth of water hyacinth blocking the river flow, increasing activities in local markets, etc.

2.8.1.4 Thakuran Estuary

Thakuran estuary, being a notable tide-dominated river estuary located in the buffer zone of Sundarbans Biosphere Reserve, flows between the Matla and Saptamukhi estuaries, maintaining a connection with the Bidyadhari and Matla estuaries in its stretch of about 80 km from north to south. This estuary is also a freshwater-starved estuary, mainly receiving freshwater in the rainy season through precipitation, but some amount of freshwater comes through the interconnecting flows from the Ichamati–Bidyadhari to the Raimangal estuary. The Thakuran river, representing an important drainage basin covering an area of 900 km^2 over the basement of alluvial and tidal sediments, belongs to a macro-mesotidal regime from its mouth (4.6 m to 5.5 m during spring tide) further upstream (2.8 m to 3.2 m during spring tide). The middle stretch of the river, after being bifurcated gives the origins of two separate channels, the main flow in the east as the Thakuran river, and the smaller channel in the western direction as the Jagaddal river. Beyond that point of bifurcation further north, the width of the Thakuran river is gradually reduced and connects to the Matla river through the Nonapukur-Piali, a distributary of the Matla. The Jagaddal river at its upper end meets with two creeks, the Chirapat creek and the Pakhiralay creek. A network of tide-dominated channels and creeks connected from north to south with the Thakuran river estuary are the Moni river, the Pukchara river, the Raidighi river, the Chatua river, the Kankalmari–Ajmalmari–Suia rivers, the Jagaddal river, Kadrakhali Khal, Bainchapi Khal, Guru Khal, Kamari Khal, Kalcher–Curzon creeks, Chirapat or Pakhiralay creeks, Chulkati Gang, Dhulibasani Gang, and Gokul Tali Gang.

2.8.1.5 Hooghly Estuary

The main flow of the river Ganges on entering the Sundarbans region is named the Hooghly estuary, which bridges the gap between the many freshwater rivers and the sea, the Bay of Bengal and experiences all through its stretch the meso-macro tidal influence. Of the two major water courses of the Hooghly estuary, the major one flows through the western side of the largest Gangetic delta of the world, the Sagar Island, and ends up in the Bay of Bengal. The other branch of the Hooghly estuary called the Mooriganga estuary flows southward by the sides of the deltaic islands such as Kakdwip, Namkhana, Bakkhali, Jamboodeep, and Mousuni islands. Many of the tributaries of the Hooghly estuary, once flowing through the deltaic islands of the region, have been disconnected and presently experience only reduced freshwater flow to the central part of Sundarbans in comparison with the Ichhamati–Rainmangal estuaries, carrying comparatively more freshwater discharge to the eastern sector of the Sundarbans region.

2.8.1.6 Matla Estuary

The Matla river with its many branches and sub-branches flows south to the sea with navigable tracts for steamers/burges up to Canning Port around a century ago. However, both the drainage system and the depth of this large estuary of Sundarbans have been drastically reduced. The Matla estuary appears to be the widest estuary in the Sundarbans region, next to the Hooghly estuary. After originating at Canning of the North 24 Parganas, by the combined flows of Karati in the north and Amjhara in the east of the Bidyadhari estuaries. The Matla estuary flows southward and joins the Bidyadhari estuary near Herobhanga reserve forest and opens up into the Bay of Bengal, covering a stretch of 120 km.

The network of different channels connected to the Matla estuary consists of the Piyali river, Mayadwip river, Pathankhali river, Nabipukur creek, Chota Matla creek, Chota Bidya creek, Sialfeli creek, Bainchapi creek, Jhara creek, Herobhanga creek, Kultala Gang, Kankalmari Ajmalmari Suia Gang, Gura Khal, Dulibhasani Gang, Bidya Khal Kamli, and Dulibhasani Gang named from Suia Gang, Gokultali Gang, and numerous tidal inlets such as Netidhopani, Storkhali, Suryamukhi, Havatidunia, Haldi, and Gobdi creeks, Choubanka, Lakshmi Banka, Chhaimari, Chota, and Bara Arhbhanga Khal are in the extreme south.

2.8.1.7 Mooriganga Estuary

The Mooriganga estuary is thought to have originated by the extension of the Bhagirathi river. This estuary ultimately meets the Saptamukhi estuary via a narrow channel, known as the Hatania Dowania river.

2.8.1.8 Saptamukhi Estuary

The Saptamukhi estuary is formed by the combination of several creeks, namely Thikara Khal in the north, and the Banstala Khal and Ghugudanga Khal in the South. The major branches of this estuary in the west are Hatania–Doania creek, Edward's creek, Blind creek, Patibunia creek, the Kalnagini Khal, Ghugudanga Gang, Chandanpiri Khal, and Bakkhali river around Fraserganj. The Saptamukhi estuary, with its vast expanse but shallow depth, is encircled by a number of deltaic islands such as Namkhana, Bakkhali, Fraserganj, Henry island, Prentice island, Lothian island, etc. Several mid-channel bars emerge from the riverbeds at different locations in this estuary.

2.9 Environmental Changes and Impacts on Mangroves: Assessment Through Land-Use/Land-Cover Changes

It is evaluated with the help of remote-sensing data and the LULC map that the extent of mangroves in the Indian Sundarbans has decreased to a large extent at the rate of 4.7197 ha/year, because of the clearing of open-forest land for conversion to other human activities such as agriculture, fisheries, and human settlement. Increase of the water areas under canals and rivers by 444.12 ha and 285.60 ha respectively indicates the submergence of low-lying land as the interconnecting drainage network undergoes large-scale siltation or sediment deposition. A large-scale decrease in the wetland area (1064.58 ha) and an increase in aquaculture area (3728 ha) have been observed over a period of ten years (2009 to 2019), where the probable cause is conversion of wetlands for aquaculture practices or intensive fish farming (Tables 2.1, 2.2, 2.3, and 2.4). Conversion of open-forest land has taken place by human activities to cultivable agricultural land, aquaculture, human settlements, development of small industries (brick kiln), and any other economic usage. The decline in agricultural land (2905.9 ha) over the course of one decade also indicates the general conversion of agricultural land for aquaculture purposes with the motive of gaining short-term economic benefit. Simultaneously, rural settlements have increased at expected rates with greater population pressure. Remote-sensing data studies have also revealed the large-scale conversion of agricultural fields to dense human settlements and fish farming ponds in Sundarbans. Land-use/land-cover changes depending on spatial, social, and economic conditions occur if augmentation seeks attention on a global scale with the reinforcement of land for cultivation and livestock management (Lambin and Geist, 2006). The impact of human activities on land has grown massively in the last two centuries, eventually affecting the earth's nature-shifting, all-encompassing landscapes. The results of these include intensified agriculture, biodiversity loss, reduced forest land, vast land degradation, and soil erosion. Coastal areas with greater biodiversity and high productivity are highly sensitive ecological areas and increasing population density in such coastal zones rapidly prompted an increase in industrialization and urbanization, thereby inviting eco-degradation (Clark, 1993).

Mangrove ecosystems across the world have been under stress and have become vulnerable owing to human population growth, loss of livelihoods, migration of environmentally displaced people for greater living resources, trades, tourism, etc., along with different forms of pollution, including climate change (Lakshmi and Edward, 2010; Chakraborty et al., 2014). By studying the Rennel Map I and XX, it has been observed that the inhabitants of the Sundarbans region, extensively in the blocks of Hasnabad, Canning, Haroa, Kultali, Bhangarand, and Patharpratima extended their land areas by embankment construction along the low tide line, which in turn have raised the tidal channels or creeks as the huge amount of silt carried by a network of channels and creeks could not be deposited on the tidal mud flat. Both the living and nonliving structural components in different eco-zones of Sundarbans, India, have been in the process of changes in tune with the climate-change-mediated drastic alterations in the coastal environment. Estuarine networks in close proximity to the Sundarbans Reserve forest have undergone remarkable changes in land-use patterns, pushing the interests of the local population to the corners as land use describes the right to the land for obtaining economic benefits by drastically deviating from the century old traditional resource management practices. Moreover, increasing soil and water salinity have become the leading environmental problems in the coastal deltaic islands of Sundarbans, where both of these environmental parameters have been markedly increased during the period 1975 to 2000. This became particularly alarming for soil health in the aftermath of cyclone Aila in 2009.

2.10 Sediment Budget in the Ganges–Brahmaputra–Meghna Deltaic System: Trends in Deltaic physiographic Change Over Three Decades in the Asia-Pacific Region

Massive eco-degradation and deforestation of mangrove systems across the Asia-Pacific region, especially in the Sundarbans formed in lower Bengal on the Ganges–Brahmaputra river system, along with the Irrawaddy and Mekong deltas, have been noted during the past few decades (Coleman et al., 2008; Saito et al., 2007). All of these depletions coupled with the eco-degradation of the mangrove ecosystem have been the outcomes of strong demographic pressures, large-scale deforestation, followed by deposition of sediments in parts of this coastal environment because of upstream activities.

A thorough assessment through remote sensing and satellite images from 1973 to 2000 had shown the highest rates of erosion at the southern edges of Mayadwip, Bulcherry Island, and Bhangaduni Island, on the southernmost parts of the Sundarbans delta. However, through fresh sedimentation followed by progradation, landmass has been continuously emerged and developed in the Sundarbans, counterbalancing a large part of the losses, which is revealed by a gain in mangrove forests (2925 ha) between the 1970s and 1990s, which almost equaled the mangrove forest lost owing to erosion (3157 ha). Moreover, an estimate during the 1990s to 2000s, the rate of erosion along the banks of major river channels has been claimed to be seven times greater than in the core zones of the mangrove forest, at the coast of the Bay of Bengal. Giri et al. has suggested that the construction of the Farakka barrage in India disrupted the balance between land creation and loss, as that had existed before, alongside alternation of the local hydrology. These changes have had an immense impact on the erosion-prone western part of the Sundarbans, India.

Reduction in sediment contributions (20%), decreased freshwater flow as a consequence of the construction of dams in the upstream, and massive shrinkage of mangrove cover are not reflected as much as erosion in the mangrove ecosystems in the deltaic islands of Bengal or Papua New Guinea, where an increase in the expansion of the net area (2.7%) was observed.

The Bengal basin formed by the depositional environment contributed principally by the rivers, including the Ganges–Brahmaputra–Meghna (GBM) system. An estimation based on geo-chronological and geo-physical investigations has revealed the sediment load from the Ganges–Brahmaputra riverine systems as to be 1 billion metric tons at a site around 300 km inland of the coast of the Bay of Bengal.

A sediment budget, of the sediment load carried by these two rivers, indicates the following partitioning:

(i) 21% of sediment is found to be deposited in the top set of beds of the subaqueous delta near the estuarine mouth (Allison 1998);
(ii) 30–39% of the sediment is sequestered in the huge flood plain landward of the ocean (Allison 1998);
(iii) Seaward progradation of the subaqueous delta occurred with the 20% deposition sediments on the river.
(iv) 20%–29% may reach the Swatch of No Ground.

Since 1840, net erosion has been increasing at the western part of the shoreline of the Indian Sundarbans, resulting in 3–4 km of retreat, making this zone sediment starved, which promotes eastward migration of the river mouth experiencing decreased sediment delivery by the river Ganges via local distributaries (Allison, 1998). Erosion coupled with saline intrusion accompanied by regional subsidence through tectonic and compaction processes brings about changes in coastal geomorphic features of recent times impacting the sedimentary environment of the estuary in the lower part of the Ganges delta surrounding the total Sundarbans area. This analysis has shown that total wetland conversion to open water and agriculture lands amounted to 4920 km^2 between 1981 and 2001.

Considerable changes in mangrove and deltaic systems have taken place during the last few decades around the world and counteract the process of several natural and anthropogenic efforts such as undertaking afforestation programs against deforestation, freshwater supply, abatements of the after effects of storms, and enhancing subsidence through sediment supply. Incidentally, mangroves can quickly colonize along many coastlines after being planted in the deforested deltaic systems and steadily provide suitable ecological conditions for the reestablishment of mangroves (Richards, 1993; Ganguly and Chakraborty, 2026). Five major factors play important roles in determining mangrove distribution in the seaward extent of deltaic systems: (i) input of sediments in fluvially controlled delta; (ii) supply of sediment causing formation of deltas associated with mangroves; (iii) newly formed deltaic islands extending seaward into deeper waters; (iv) keeping pace with subsidence in the absence of sufficient sediment; (v) transgressing the original shoreline (Syvitski et al., 2009). Several other factors influence the rates of sediment supply, including land-use practices, mountain elevations, and uplift rates.

The rate of subsidence is determined by sediment compaction and geologically crustal movement such as tectonic movement (Ericson et al., 2006; Syvitski et al., 2009). Extraction of groundwater and fossil fuels can also affect the local rate of subsidence. Also, decadal variations in wave conditions and sea-level positions have been suggested to be important to coastal geomorphological changes. The final factor governing the subaerial elevation of deltas and ultimately the distribution of mangroves, is change in eustatic (global) rise in sea level, which, combined with the other factors, determines the vertical rate of accretion needed to maintain the land-surface elevation through the natural rates of fluvial sediment deposition, coastal currents with higher intensities, and the accelerated subsidence process because of groundwater and hydrocarbon extraction (Ericson et al., 2006). These events set a threshold for the establishment of mangrove seedlings, which can potentially help consistent patterns of mangrove change, deciphering the interplay of these factors for determining the spatial extent of deltaic coastal systems, including mangroves, or experience recession or progradation (Ericson et al., 2006).

2.11 Mangroves of Bangladesh and Their Ecological Status

The Sundarbans mangrove ecosystem is famous internationally not only for its large green expanse with higher biological productivity but also for its ecological and economic potential for rendering valuable ecosystem services (Figs. 2.20, 2.21, 2.22, 2.23, and 2.24). In this context, research for a proper understanding of the ecosystem dynamics of this mangrove forest simultaneously together both in the India and Bangladesh can play great roles in changing the national economy, and can also open up avenues for livelihood generation of the local people through fishing, aquaculture, tourism, wood, and nonwood products. However, despite the formulation of several laws, regulations, and policies for undertaking appropriate eco-management plans, the forest has been in the process of eco-degradation along

Fig. 2.20 Mangrove landscape of the Bangladesh Sundarbans (1). (**A**, **B** and **C**)The Rupsa river estuary, one of the major freshwater-supplying and -strengthening water-drainage systems of the Bangladesh mangrove areas of Sundarbans is under multiple pollution threats. (**D**) Entry point of tourists and peoples to the Sundarbans of Bangladesh at Mongla of Khulna district is under the harmful impact of human activities; (**E**, **F**, **G** and **H**) Different industrial developments in the very close vicinity of the Sundarbans of Bangladesh

Fig. 2.21 Mangroves in Bangladesh under the process of eco-degradation (2). (**A**, **B**, **C**, **D**, **E**, **F** and **J**) All of these images clearly depict the poor growth of mangroves and destabilization of the floor of the mangrove forest by different forms of environmental perturbation, especially soil erosion, and various mangrove species demonstrate the roles of their root systems in their own existence

Fig. 2.22 Mangrove landscapes of the Bangladesh Sundarbans (3). (**A**) Mangroves and mudflats, but the growth and biomass of mangroves indicate a lack of proper ecological health as per the manifestation of the feeble canopy structure and the biomass of different mangrove plants. (**B** and **C**) Narrow mudflats, but healthy growth of mangroves belonging to the species *Aegiceras corniculatum* (**D**). (**C**, **E**, **F**, **G**, and **H**) The charismatic mangrove plant species, *Nypa fruticans*, locally named **'Golpata,'** demonstrate very healthy growth in terms of body size and biomass indicating ecologically healthy conditions, which command special distinction of the mangrove ecosystem of Bangladesh

Fig. 2.23 Aquaculture and fishery activities in the Bangladesh mangrove environment (4). (**A**) Capture fisheries with the involvement of various fishing gear and crafts, especially mechanized trawlers, still bring in a lot of revenue for the entire country as well as providing sustenance to a good number of local people through livelihood generation. (**B**, **C**, **D**, **E**, **F** and **G**) Aquaculture activities by the local people in and around the mangrove fringe areas of the mangrove belts of the Bangladesh Sundarbans following traditional extensive aquaculture systems. (**H** and **I**) Traditional capture fishing by the local fishermen in the mangrove belts of the Sundarbans of Bangladesh, where the fishermen were found to be catching a shallow-water-inhabiting catfish, *Plotosus canius*. (**J**) Local fish market in very close proximity to the Bangladesh Sundarbans

Fig. 2.24 Mangrove tourism in the Bangladesh Sundarbans (5) (**A**) Different water vessels for providing water transport for the tourists to visit different interior parts of the Sundarbans in Bangladesh. (**B**, **C**, **D**, **E**, and **F**): Different spots in the Bangladesh Sundarbans with modest infrastructural facilities for helping visitors. (**G** and **H**): These images are for acquainting visitors with the wealth of prime biodiversity of the Sundarbans in Bangladesh

with depletion of its areal extent (Tables 2.1, 2.2, 2.3, and 2.4; Figs. 2.15, 2.16, 2.17, 2.18, and 2.19). The plant disease **"top dying"** has been identified since the1980s as one of the main causes of deterioration of the abundance of *Hereteria fomes* (Sundari trees) in the Sundarbans of Bangladesh (Sarwar, 2015). Moreover, this has been aggravated by the increase in the salinity and both of these problems have reduced the abundance of Sundari trees to almost 50% (Rahman et al., 2010).

The wealth of biodiversity in the mangroves on the Asian subcontinent represents over 41% of the world's mangroves in South and Southeast Asia, where Indonesia alone accommodates 23% (Malik et al., 2017). A further 20% of the total mangrove area exists in Australia, Brazil, and Nigeria (Gopal & Chauhan, 2006). However, the mangroves located in the GBM delta shared between Bangladesh and India are the largest mangrove forests in the world, covering around 6017 km^2 and representing a unique ecosystem, which is the habitat of the endangered Royal Bengal tiger (*Panthera tigris tigris*) (Choudhuri and Choudhury, 2004; Iftekhar & Saenger, 2008). According to the recent estimation, 62% of the entire area of the Sundarbans is in Bangladesh (Rahman et al., 1979) and the rest of the area (38%) is in the state of West Bengal in India (Sanyal, 1983). The southwest part of the Sundarbans forest in Bangladesh was declared the Ramsar site (Islam, 2011). The economic potential of mangroves in Bangladesh enables it to earn 41% of the total revenue generated from the forests (Roy & Hossain, 2015; Hassan et al, 2022). However, currently, 50% of the covered area of the erstwhile Sundarbans mangrove forests in Bangladesh around a couple centuries ago have been depleted (Das & Mandal, 2016). In recent decades, the ecological set-ups and biodiversity of Sundarbans in Bangladesh has been threatened owing to several natural and human-induced pressures, including overexploitation of forest resources, changes in coastal land use, climate change, rise in sea level, natural disasters, pollution by heavy metals, micro-plastics, agricultural waste (pesticides, chemical fertilizers), effluent discharges from industry and aquaculture ponds, a decrease in freshwater flow, oil spillage from nearby sea-ports, imposing severe negative effects on biodiversity, upstream waterflow reduction, increased erosion followed by vertical accretion, salinity invasion, disease outbreaks, lack of knowledge about forest conservation, uncontrolled tourism, a lapse in executing conservation rules, lack of adequate planning and management, and an increase in wildlife–human conflicts (MacFarlane et al., 2007; Islam & Bhuiyan, 2016; Chowdhury & Maiti, 2016; Das & Mandal, 2016). Out of all of these, several diseases, natural disasters, rises in sea level, shrimp farming, and insufficient regeneration also affect the Sundarbans of Bangladesh profusely (Hossain et al., 2013, 2015). The development and ongoing activities at the Mongla export processing zone and sea-ports in Khulna, for exports and imports, have aggravated the process of eco-degradation of the mangrove ecosystem in Bangladesh (Hossain et al., 2015). The harmful impacts coupled with anthropogenic pressure and releasing of contaminants from multiple sources may reduce the ecological stability of the mangroves in the long run (Hossain & Bhuiyan, 2016).

References

Allison, M. A. (1998a). Geologic framework and environmental status of the Ganges-Brahmaputra Delta. *Journal of Coastal Research, 14*, 826–836.

Allison, M. A. (1998b). Historical changes in the Ganges-Brahmaputra Delta front. *Journal of Coastal Research, 14*(4), 1269–1275.

Allison, M. A., & Kepple, E. B. (2001). Modern sediment supply to the lower delta plain of the Ganges-Brahmaputra River in Bangladesh. *Geo-Marine Letters, 21*(2), 66–74.

Allison, M. A., Khan, S. R., Goodbred, S. L., Jr., & Kuehl, S. A. (2003). Stratigraphic evolution of the late Holocene Ganges–Brahmaputra lower delta plain. *Sedimentary Geology, 155*(3-4), 317–342.

Annon (2003) *Mangrove ecosystem: Biodiversity and its influence on the natural recruitment of selected commercially important finfish and shellfish species in fisheries*. National Agricultural Technology Project (NATP). Indian Council of Agricultural Research (ICAR). Principal Investigator: George, J. P.; Co–PI: Chakraborty, S.K. and Damory, S. N. (pp. 1–514).

Bagchi, K. (1944). *The Ganges Delta*. University of Calcutta.

Baig, M., Zhang, L., Shuai, T., & Tong, Q. (2014). Derivation of a tasselled cap transformation based on Landsat 8 at-satellite reflectance. *Remote Sensing Letters, 5*, 423–431.

Baloloy, A. B., Blanco, A. C., Ana, R. R. C. S., & Nadaoka, K. (2020). Development and application of a new mangrove vegetation index (MVI) for rapid and accurate mangrove mapping. *ISPRS Journal of Photogrammetry and Remote Sensing, 166*, 95–117.

Banerjee, L. K. (1984). Vegetation of the Bitarkanika sanctuary. Osissa State. *Journal Economic Taxonomy Botany, 5*, 1065–1079.

Banerjee, L. K. (1987/1985). Comparative study on mangroves of the Sundarbans and that of the Mahanadi Delta in Eastern India. In L. J. Bhosale (Ed.), *The Mangroves: Proceedings of National Symposium. Biology. Utilization and Conservation of Mangroves. Kolahpur, India* (pp. 527–530).

Banerjee, L. K., Sastry, A. R. K., & Nayar, M. P. (1989). *Mangroves in India: Identification manual* (p. 113). Botanical Survey of India.

Berger, U., Adams, M., Grimm, V., et al. (2006). Modelling secondary succession of subtropical mangroves: Causes and consequences of growth reduction in pioneer species. *Perspectives in Plant Ecology, Evolution and Systematics, 7*, 243–252.

Bilham, R., & England, P. (2001). Plateau "pop-up" in the great 1897 Assam earthquake. *Nature, 410*, 806–809.

Biswas, H., Mukhopadhyay, S. K., & De, T. K. (2004). Biogenic controls on the air-water carbon-dioxide exchangein the Sundarban mangrove environment, northeastcoast of Bay of Bengal, India. *Limnology and Oceanography, 49*(1), 95–101.

Blasco, F. (1984). Climate factors and the biology of mangrove plants. In S. C. Snedaker & J. G. Snedaker (Eds.), *The mangrove ecosystem: Research methods* (pp. 18–35). UNESCO/SCOR.

Blasco, F. (1975). *The mangroves of India*. Inst. Francais de Pondichery. Trans.Sec. Sci. Tech Vol 14, 1–175.

Blasco, F. (1977). Outlines of the ecology, botany and forestry of the Mangals of the Indian subcontinent. In V. J. Chapman (Ed.), *Wet ecosystem of the world* (pp. 241–262). Elsevier.

Chakraborty, P. K., & Naskar, K. R. (1988). Role of mangrove in estuarine fisheries development. *Bulletine Central Inland Capture Fishery Research Institute, Barrackpore, 57*, 229–233.

Chakraborty, R. (2009). Local people and the global tiger: An environmental history of the Sundarbans. *Global Environment, 3*, 72–95.

Chakraborty, S. K. (2009). Biological diversity of Mindapore Coast, West Bengal. *Environmental Forensics, 2*(2), 6.

Chakraborty, S. K. (2010). Coastal environment of Midnapore, West Bengal: Potential threats and management. *Journal of Coastal Environment, 1*(1), 28–40.

Chakraborty, S. K. (2011). Mangrove ecosystem of Sundarbans, India: Biodiversity, ecology, threats and conservation. In J. N. Metras (Ed.), *Mangroves: Ecology, biology and taxonomy* (pp. 83–112). NOVA Publisher.

Chakraborty, S. K. (2017). Ecological services on intertidal benthic fauna and the sustenance of coastal wetlands along the Midnapore(East) coast, West Bengal, India. In *Coastal wetlands: Alteration and remediation* (pp. 777–886). Springer.

Chakraborty, S. K. (2018). Bioinvasion and environmental perturbation: Synergistic impact oncoastal—Mangrove ecosystems of West Bengal, India. In C. Makowski & C. Finkl (Eds.), *Impacts of invasive species on coastal environments: Coast in crisis* (pp. 171–245). Springer.

Chakraborty, S. K., Maiti Dutta, S., Ghosh, P. B., Ray, R., & Paul, A. K. (2014). Impact of global warming on Sundarbans mangrove ecosystem, India: Role of different assessment tools from ecosystem monitoring to molecular markers. In *Proceedings of the international conference on green India: Strategic knowledge for combating climate change: Prospects and challenges* (pp. 181–200).

Chaudhuri, A. B., & Choudhury, A. (1994). *Mangroves of the Sundarbans. Vol. 1, India* (pp. 1–247). IUCN.

Chowdhury, A., & Maiti, S. K. (2016). Identifying the source and accessing the spatial variations, contamination status, conservation threats of heavy metal pollution in the river waters of Sundarban biosphere reserve. *Journal of Coastal Conservation, India, 20*, 257. https://doi.org/10.1007/s11852-016-0436-x

Clark, P. J. (1993). Dispersal of grey mangrove (Avicennia marina) propagulesin southeastern Australia. *Aquatic Botany, 45*, 195–204.

Cole, C. V., & Vaidyaraman. (1966). Salinity distribution and effect of fresh water flows in the HooglyRiver. In *Proceedings of Tenth Conference on Coastal Engineering, Takyo* (pp. 1312–1434). American Society of Civil Engineers.

Coleman, J. M., Huh, O. K., & Braud, D. (2008). Wetland loss in world deltas. *Journal of Coastal Research, 24*, 1–14.

Dam Roy, S., & Krishnan, P. (2005). Mangrove stands of Andaman vis-à-vis tsunami. *Current Science, 89*, 1800–1804.

Das, C. S., & Mandal, R. N. (2016). Coastal people and mangroves ecosystem resources vis-à-vis management strategies in Indian Sundarban. *Ocean and Coastal Management, 134*, 1–10.

Dehni, A., & Lounis, M. (2012). Remote sensing techniques for salt affected soil mapping: Application to the Oran region of Algeria. *Procedia Engineering, 33*, 188–198.

Dharmawan, I. W. E., & Pramudji. (2020). Mangrove community structure in Papuan Small Islands, Case study in Biak Regency. In *Proceeding The IOP Conference Series: Earth and Environmental Science, Purwokerto, Indonesia, 21–23 August 2019* (pp. 1–8).

Didar, U. L., Islam, S. M., & Hossain, B. M. A. H. (2018). Sundarbans mangrove forest of Bangladesh: Causes of degradation and sustainable management option. *Environmental Sustainability, 1*(3), 113–131.

Duke, N. C., Meynecke, J.-O., Dittmann, S., Ellison, A. M., Anger, K., Berger, U., Cannicci, S., Diele, K., Ewel, K. C., Field, C. D., Koedam, N., Lee, S. Y., Marchand, C., Nordhaus, I., & Dahdouh-Guebas, F. (2007). A world without mangroves? *Science, 317*, 41–42.

Ellison, J. C., & Stoddart, D. R. (1991). Mangrove ecosystem collapse during predicted sea-level rise: Holocene analogues and implications. *Journal of Coastal Research, 7*, 151–156.

Fairbridge, R. W 1980. The estuary: its definition and geodynamic cycle. InOlausson, E., and Cato, I (eds), *Chemistry and Biogeochemistry of Estuaries*. New York: Wiley.

Ericson, J. P., Vörösmarty, C. J., Dingman, S. L., Ward, L. G., & Meybeck, M. (2006). Effective sea-level rise and deltas: Causes of change and human dimension implications. *Global and Planetary Change, 50*, 63–82.

FSI. (1999). *Final SFR, 2003* (pp. 39–43). http://fsi.nic.in/documents/sfr_2003_hindi.pdf

Ganguly, A., & Chakraborty, S. K. (2026). Synergistic impact of urbanization of Kolkata Metropolitan and East Kolkata Wetlands on the mangrove-estuarine complex of Sundarbans, India. In *Wetlands and urban resilience: A global outlook* (Sustainable Development Goals Series) (pp. 375–398). Springer.

Gopal, B., & Chauhan, M. (2006). Biodiversity and its conservation in the Sundarban Mangrove Ecosystem. *Aquatic Sciences, 68*, 338–354.

Hasan, J., Shaha, D. C., Kundu, S. R., Md Yusoff, F., Cho, Y.-K., Haque, F., Abdus Salam, M., Ahmed, S., Wahab, A., Ahmed, M., Hossain, I., & Afrad, S. I. (2022). Phytoplankton Community in Relation to Environmental Variables in the Tidal Mangrove Creeks of the Pasur River Estuary. *Bangladesh Conservation, 2*, 587–612.

Healey, S., Cohen, W., Yang, Z., & Krankina, O. (2005). Comparison of Tasseled Cap-based Landsat data structures for use in forest disturbance detection. *Remote Sensing of Environment, 97*, 301–310.

Hossain, G. M., & Bhuiyan, M. A. H. (2016). Spatial and temporal variations of organic matter contents and potential sediment nutrient index in the Sundarbans Mangrove Forest. Bangladesh. *KSCE Journal of Civil Engineering, 20*(1), 163–174.

Hossain, M. S., Uddin, M. J., & Fakhruddin, A. N. M. (2013). Impacts of shrimp farming on the coastal environment of Bangladesh and approach for management. *Reviews in Environmental Science and Biotechnology, 12*, 315–332.

Hossain, M. S., Dearing, J. A., Rahman, M., & Mand Salehin, M. (2015). Recent changes in ecosystem services and human well –being in the Bangladesh coastal zone. *Regional Environmental Change, 16*, 429. https://doi.org/10.1007/s10113-014-0748-z

Huete, A. R. (1988). A soil-adjusted vegetation index (SAVI). *Remote Sensing of Environment, 25*, 295–309.

Hunter, W. W. (1875). *A statistical account of Bengal, Vol. I, District of 24 Parganas and Sundarbans* (pp. 285–346). Truebner & Company.

Hussain, Z., & Acharya, G. (Eds.). (1994). *Mangroves of the Sundarbans. Vol. 2: Bangladesh* (pp. 1–257). The IUCN Wetlands Programme Report. World Conservation Union.

Hutchings, P. A., & Saenger, P. (1987). *Ecology of mangroves*. University of Queensland Press.

Iftekhar, M. S., & Saenger, P. (2008). Vegetation dynamics in the Bangladesh Sundarbans Mangroves: A review of forest inventories. *Wetlands Ecology and Management, 16*, 291–312.

Inman, D. L., & Nordstrom, C. E. (1971). On the tectonic and morphologic classification of coasts. *Journal of Geology, 79*, 1–21.

Islam., M. S. (2011). *Biodiversity and livelihoods: A case study in Sundarban forest, world Heritage and Ramsar Site (Bangladesh).* Unpublished Master's Thesis, University of Klagenfurt, Austria.

Islam, S. M. D., & Bhuiyan, M. A. H. (2016). Impact scenarios of shrimp farming in coastal region of Bangladesh: An approach of an ecological model for sustainable management. *Aquaculture International, 24*(4), 1163–1190.

Kathiresan, K. (2000). A review of studies of Pichavaram mangrove, Southeast India. *Hydrobiologia, 430*, 185–205.

Kathiresan, K., & Bingham, B. L. (2001). Biology of mangroves and mangrove ecosystems. *Advances in Marine Biology, 40*, 81–251.

Kathirensan, K., & Quasim, S. Z. (2005). *Biodiversity of mangrove ecosystem* (p. 251). Hindustan Publishing Corporation (India).

Kuenzer, C., Bluemel, A., Gebhardt, S., Vo Quoc, T., & Dech, S. (2011). Remote sensing of mangrove eco systems: A review. *Remote Sensing, 3*, 878–928.

Lakshmi, S. A., & Edward, J. K. P. (2010). *Coastal issues and management strategy for Sagar Island in Bay of Bengal*. PhD thesis 2, 96–101.

Lambin, E. F., & Geist, H. J. (2006). *Book on land use and landcover change: Local processes and global impacts*. Springer.

Lugo, A. E., & Snedaker, S. C. (1974). The ecology of mangroves. *Annual Review of Ecology and Systematics, 5*, 39–64.

MacFarlane, G. R., Koller, C. E., & Blomberg, S. P. (2007). Accumulation and partitioning of heavy metals in mangroves : A synthesis of field based studies. *Chemosphere, 69*, 1454–1464.

Malik, A., Mertz, O., & Fensholt, R. (2017). Mangrove forest decline: Consequences for livelihoods and environment in South Sulawesi. *Regional Environmental Change, 17*, 157–169.

MOEF. (2012). *Ministry of forest and Climate Change, Government of India.*

Morgan, J. P., & McIntire, W. G. (1959). Quaternary geology of the Bengal Basin, East Pakistan and India. *Bulletin of the Geological Society of America, 70*, 319–342.

Naskar, K., & Guha Bakshi, D. N. (1987). *Mangrove swamps of the Sundarbans* (p. 282). Naya Prakash.

Naskar, K. R., & Mandal, R. N. (1999). *Ecology and biodiversity of Indian Mangroves* (pp. 1–754). Daya Publishing House.

National Cyclone Risk Mitigation Project (NCRMP). (2016). *Implemented through World Bank Assistance (Map showing coastal areas affected by cyclones).* Based on IMD data on cyclone hazards.

Paul, A. K. (1999). Geomorphic significance of mangrove Swamps—a study in the Sundarban. In K. R. Naskar & P. Sanyal (Eds.), *Mangal of the Sundarban* (pp. 39–52). Naya Prokash.

Paul, A. K. (2002a). *Coastal geomorphology and environment* (p. 582). ACB Publications.

Paul, A. K. (2002b). *Tsunami the disaster in Indian Ocean Nations* (p. 125). ACB Publications.

Paul, A., & Paul, A. K. (2023). Geomorphological analysis of the coral fringed coasts of Andaman and Nicobar Islands using geospatial techniques. In *Crisis on the Coast and Hinterland: Assessing India's East Coast with geomorphological, environmental and remote sensing and GIS approaches* (pp. 303–325). Springer.

Paul, A. K., Ray, R., Kamila, A., & Jana, S. (2017). Mangrove degradation in the Sundarbans. In *Coastal wetlands: Alteration and remediation* (Vol. 21, pp. 357–392).

Pritchard, D. W. (1967). What is an estuary: Physical viewpoint. In G. H. Lauff (Ed.), *Estuaries* (Vol. 1, pp. 149–117). American Association for the Advancement of Science.

Rahman, N., Billah, M. M., & Chaudhury, M. U. (1979). Preparation of an up-to date map of Sundarban forests and estimation of forest areas of the same by using Landsat imageries. In *Second Bangladesh National Seminar on remote sensing, Dhaka, 9–15 (December, 1979).*

Rahman, M. M., Rahman, M. M., & Islam, K. S. (2010). The causes of deterioration of Sundarban mangrove forest ecosystem of Bangladesh: Conservation and sustainable management issues. *AACL Bioflux, 3*(2), 77–90.

Ray, R., & Mondal, S. (2014). Multi criteria based forest cover type mapping using Landsat ETM+ digital data—A case study on Eastern part of Jungle Mahal Rgion, West Bengal. *Indian Journal of Regional Science, XLVI*, 77–84.

Ray, R., Paul, A. K., & Basu, B. (2013). Application of supervised enhancement technique in monitoring the mangrove forest cover dynamics—A study on Ajmalmari Reserve Forest, Sundarban, West Bengal. *International Journal of remote Sensing and Geoscience, 2*(1), 16–21.

Richards, J. F. (1993). Agricultural impacts in tropical wetlands: Rice paddies for mangroves in South and Southeast Asia. In M. Williams (Ed.), *Wetlands: A threatened landscape* (pp. 217–233). Blackwell.

Roy, N. (1949). *Bangalir Itihas (The history of the Bengalis)* (p. 105).

Roy, T. K., & Hossain, S. T. (2015). Attitudes towards current and alternative management of the Sundarbans mangrove forest, Bangladesh to achieve sustainability. *Journal of Environmental Planning and Management, 58*(2), 213–228.

Sanyal, P. (1983). Mangrove tiger land: The Sundarbans of India. *Tiger Paper, 10*(3), 1–4.

Saenger, P. (2002). *Mangrove ecology, silviculture and conservation* (pp. 1–361). Springer.

Saenger, P., & Siddique, N. A. (1993). Land from the sea: The afforestation program of Bangladesh. *Ocean and Coastal Management, 20*, 23–39.

Saenger, P., & Snedakar, S. C. (1993). Pantropical trends in mangrove above-ground biomass and annual litter fall. *Oecologia, 96*, 293–299.

Saenger, P., Heger, E. J., & Davie, J. D. S. (1983). Global status of mangrove ecosystem. Commission on ecology papers No#. International Union for conservation of nature and Natural Resources. The Environmentalist Supp. No.3: 1–88.

Saito, K., Atlin, G. N., Linquist, B., Phanthaboon, K., Shiraiwa, T., & Horie, T. (2007). Performance of traditional and improved upland rice cultivars under nonfertilized and fertilized conditions in Northern Laos. *Crop Science, 47*.

Sarkar, A., Sengupta, S., McArthur, J. M., Ravenscroft, P., Bera, M. K., Bhushan, R., Samanta, A., & Agrawal, S. (2009). Evolution of Ganges–Brahmaputra western delta plain: Clues from sedimentology and carbon isotopes. *Quaternary Science Reviews, 28*, 2564–2581.

Sarwar, M. G. M. (2015). *Sundari protecting the biodiversity of the Sundarbans by reducing human pressure* (p. 1212). Concern Worldwide Bangladesh.

Selvam, V. (2003). Environmental classification of mangrove wetlands of India. *Current Science, 84*(6), 757–765.

Syvitski, J. P. M., Kettner, A. J., Overeem, I., Hutton, E. W. H., Hannon, M. T., Brakenridge, G. R., Day, J., Vörösmarty, C., Saito, Y., Giosan, L., & Nicholls, R. J. (2009). Sinking deltas due to human activities. *Nature Geoscience, 2*, 681–686.

Venkatesan, K. R. (1966). The mangroves of Madras state. *Indian Forester, 92*(1), 27–34.

Vorosmarty, C. J., Li, C., Sun, J., & Dailn, Z. (1998). Drainage basins, river systems, and anthropogen river systems, and anthropogenic change: The Chinese example. In *Asian change in the context of global climate change*.

Winarso, G., Purwanto, A. D., & Yuwono, D. M. (2014). New mangrove index as degradation/health indicator using Remote Sensing data: Segara Anakan and Alas Purwo case study. In *Proceedings of the 12th Biennial Conference of Pan Ocean Remote Sensing Conference, Bali, Indonesia, 4–7 November 2014*.

Chapter 3
Ecology of Mangrove Ecosystem of Sundarbans, India

Abstract Mangroves of the estuarine deltaic complexes are adapted with an extreme supply of sediments coupled with a differential rate of dilution of saline tidal seawater with a good quantity of annual freshwater flows from the riverine networks, which in turn support favorable ecological setups for the growth and diversity of mangroves with evergreen characteristics alongwith their associated faunal components. The nutrient load of deposited alluviums under the impacts of marine forcing factors and land-contributed sediments of floodwater in the hot and humid climatic conditions with brackish water-level salinity, a specific range of tidal exposure and inundation, textural and mineral components of sediments play determining roles in promoting the distribution and diversity of mangroves in low-lying deltaic areas of Ganges delta, Bhitarkanika delta, Mahanadi delta, Godavari delta, and Krishna delta of the coast of India around the Bay of Bengal. Estuaries are the traditional geological sinks for the downstream supply of sediments enjoying bidirectional flows of waters and sediments, which result in wide shore-fringed tidal flats and islands, which in couse of time become suitable suitable habitats for mangroves and salt marshes to settle and propagate in India.

The mangrove ecosystem is a holistic approach with the combination of two major subsystems: the mangrove forest subsystem and the adjoining aquatic subsystem, which are connected to one another by the interplay of high tide and low tide and the continuous and steady interactions among many structural components constituting these sub-ecosystems, ultimately ensuring the functionality of the mangrove ecosystem and making it the most productive ecosystem in the world. Different natural processes such as tidal cycles, delta-forming processes, erosion followed by accretion, monsoonal storms, cyclones, etc., acting independently or in combination, have made the Sundarbans mangrove ecosystem most fragile and harsh. Large quantities of mangrove contributing organic detritus are formed by the decomposition of plenty of mangrove plant litter through the continuous interactions among mangrove litter, benthic fauna, and microbes after being facilitated by the moist sediments, a plentiful supply of oxygen and microbes–nutrients–plants-faunal interaction acting as a mechanism for recycling and conserving nutrients in the man-

S. K. Chakraborty, S. K. Acharya, *Eco-dynamics and Environmental Perturbations in Mangrove Ecosystems*,
https://doi.org/10.1007/978-3-032-08710-2_3

grove ecosystem. Unique hydrological, bio-geochemical, and ecological functioning of mangrove ecosystems have not only resulted in the ecological stability and power for ecological resilience so that this biologically productive, economically potential, and aesthetically celebrated mangrove ecosystem serve as the breeding ground of many commercially important shell fish, finfish, perform as buffer against natural disasters, protect shorelines by resisting erosion and contribute profusely for the wellbeing of human society by rendering many ecosystem services. This chapter discusses relevant components of the subject of ecology emphasizing on nutrient cycling processes, which are supposed to help the development of a proper eco-management strategy.

Keywords Mangrove ecosystem · Sundarbans · Biologically productive · Nutrient cycle · Microbial interaction

3.1 Mangroves and Mangrove Ecosystem: Ecology, Distribution, and Diversity

Ecologically, the mangrove ecosystem can be defined as an assemblage of woody halophytic tropical and semi-tropical plants (trees, climbers, and shrubs) that inhabit the intertidal belts of the estuarine-dominated coastal zone. The plant community constituting the mangrove ecosystem possesses the special power of adaptation, which enables them to survive in the very unstable and oscillating coastal environment structured by variable ecological parameters, mainly because of daily tidal inundation and exposure. In this context, the mangrove-dominated ecosystem is more unique owing to its ecology in place of its taxonomy. The global diversity of mangrove plants is represented by 20 plant families containing mangrove species, of which only two families viz. Pellicieraceae and Avicenniaceae, are composed only of mangroves. In another family, Rhizophoraceae, only four genera out of a total of 16 live in mangrove ecosystems (Duke, 1992). An approximate 150,000 km^2 of tropical and subtropical areas between 25°N and 25°S of sheltered coastal zones of the world are occupied by mangroves, representing 50% of their historic range. However, the latitudinal distribution of mangroves is mostly restricted by temperature because perennial mangrove species can hardly withstand freezing conditions (Blasco, 1975, 1977; Broecker et al., 1993).

The coastal zone, being the transitional zone between the land and sea, represents one of the most eco-dynamically vibrant ecosystems, which always remains under the influence of both terrestrial and saline water influences. Being the meeting site of the three spheres of global environment, hydrosphere, atmosphere, and lithosphere, the coastal zone acts not only acts as the repositories of most of the Earth's natural resources but also as a generating ground for major pollution as well as acting as the battleground in the fight against climate change. Several unique ecosystems of the world such as mangroves, corals, intertidal, estuaries, lagoons, flood-plain wetlands, and dunes occur within the coastal environment and thereby lots of ecosystem goods and services.

Mangrove ecosystems functioning at the edge of the land–sea are very idiosyncratic from the point of view of energy flows and nutrient cycling through an intricate process of trophic interactions between two subsystems, halophytic forest and estuaries, both of which remain connected by another unique physical process, the tide, with its diurnal and seasonal rhythms. The eco-geomorphodynamics of the estuarine-mangrove ecosystem of the Sundarbans, India, is very intricate, with the involvement of many ecological processes such as biological productivity, food chain and food web, reproduction, settlements of flora and fauna, and ecological succession, where innumerable structural components of this ecosystem (soils, brackish waters, microbes, planktons, benthos, nektons, mangrove vegetations) play their roles in order to make the ecosystem function effective, vibrant, and productive (Chakraborty et al. 1986; Chakraborty, 1996; Chakraborty & Choudhury 1989; Chakraborty & Chowdhury 1993; Chakraborty & Chowdhury, 1994; Chakraborty, 2009; Chakraborty, 2010; Chakraborty et al., 2012).

In order to understand the ecosystem dynamics of the mangrove ecosystem as well as the prospective impact of ongoing ecological perturbation on the structure and function of this ecosystem, proper knowledge of the basic ecological principles is thought be a prerequisite to explain both phenomena, eco-dynamics and environmental perturbation, in respect of the mangrove-estuarine ecosystem. Much emphasis has to be placed on developing an insight into the seasonal dynamics of trophic relationships, mode of litter decomposition, nutrient transport and cycling processes, monitoring of ecological changes, eco-restoration efforts, conservation strategies, etc. (Chakraborty, 2011, 2013, 2018).

The following are major characteristics of mangroves:

1. The mangrove ecosystem, by virtue of its functioning, ensures unidirectional flows of energy from three major sources of autotrophs, primary producers (green mangrove plants, benthic algae, and aquatic phytoplankton) resulting in higher biological productivity.
2. Mangrove plants with their pneumatophores and trap roots stabilize the coastal belt and artificial embankments and resist soil erosion.
3. Mangrove ecosystem recharge with nutrients and help salinity retention in water and soil.
4. With the influx of a huge amount of sea water and also river water, the mangrove ecosystem not only acts as a sink but also helps to dilute the dumping and release of waste and improve water quality.
5. The ecological function of mangroves results in the development of nursery and fish breeding grounds by enabling the growth and propagation of plants, animals, and microorganisms, maintaining biodiversity, protecting landscapes, and safeguarding structural diversity.
6. Maintenance of healthy ecological health of the mangrove ecosystem by virtue of the intricate interactions among all of its structural components have a direct impact on the socioeconomic benefits such as tourism, recreational activities, fishing, and aquaculture.
7. The assemblages of mangroves plants along with the soil help to trap and mineralize debris.

8. The mangroves ecosystem accelerates the process of the sorting of the nutrient from the rich organic materials by the influx of a large volume of tidal water and the wave action of the estuarine–marine system.
9. Representatives of almost of kinds of animal phyla—lower invertebrates, mollusks, arthropods, reptiles, birds, and mammals—are found in this ecosystem and those organisms such as plankton, benthos, nektons, and wild animals contribute to enriching the productivity of the ecosystem.
10. Mangroves act as natural bio-engineered wall buffering against the marine-forcing factors such as tidal waves, tidal currents, littoral currents, storms, cyclones, and tsunamis in the coastal zones of the Sundarbans mangrove-estuarine networks, deltaic islands, both virgin and reclaimed, interdistributary flood basins, and carbonate platforms.

3.2 Ecosystems and Basic Concept: A Prerequisite for Assessing Ecosystem Functioning and Driver of Eco-Degradation

An ecosystem, being a complex entity, is composed of living and nonliving component beings and exhibits nonlinear responses brought about by crossing thresholds that alter their composition and key processes that involve the interactions within and between these two components Such responsive changes in turn affect ecosystem stability, resilience, and functions. On crossing this critical threshold, even a relatively small stress may trigger rapid ecosystem degradation and loss of integrity. Climate change may very likely be the factor pushing some ecosystems over the limit. Along with so many, climate change acts as a global driver of change that through the process of interactions exacerbates the impacts of prevailing environmental and human determinants of the natural world.

In response to several ways of large-scale environmental change, the constituent biotic components of an ecosystem benefit or remain unchanged with the continuation of original distribution range, modes of adaptations, dispersal to new regions, decline in numbers, or become extinct. As species respond individually and collectively, chain reactions operate in the food webs of coastal and marine areas. Any change in the physiology, reproduction, and growth patterns of a species ultimately affect other species and even human beings, as they depend on all other biota of the ecosystem.

The extensive influence of inputs from green biomass mainly from mangrove forest based terrestrial sub-system and exchanges with the adjoining terrestrial landscape. Food-webs of an aquatic ecosystem are determined by allochthonous organic inputs from the adjoining terrestrial systems or by the primary producers along with benthic communities. However, any change in terrestrial ecosystems affecting the transport of organic matter or sediments into the estuarine system becomes an important issue toward structuring aquatic biological communities in the region.

In the Sundarbans mangrove ecosystem, estuaries play dominant roles in the region's long and broad coastal plain by virtue of their extensive interactions with the surrounding mangrove ecosystem mostly by the tidal interplays and also with the upstream freshwater ecosystems by way of mixing saline water with riverine flows. The estuarine networks of the region are unique in respect of their high biological, physiographical, and habitat diversity across a wide range of tropical climatic conditions.

3.3 Ecological Requirements of Mangroves: An Interpretation

Mangroves are collectively referred to as woody halophytic angiospermic trees inhabiting within the highest and lowest tidal levels, scientifically known as the intertidal zone of coastal–estuarine belts of the tropical and subtropical coasts of the world, between 25° N and 25° S, with a modest temperature (more than 20° C), higher humidity, a wide range of fluctuating salinity, and considerable annual rainfall. Therefore, both the diversity and the density of mangroves decline gradually and steadily with increasing latitudes (Chaudhuri & Choudhury, 1994). The vegetation of these mangroves with typical halophytic adaptive features form the prime structural component of the tide-influenced mangrove–estuarine ecosystem of Sundarbans, which is both dynamic and open. In addition, this mangrove ecosystem is fragile where the ecological balance is disrupted after being impacted by even feeble threats. The deltaic Sundarbans, lying within 30° N of the Equator, has plenty of solar radiation (more than 500 kWh/day) and modest windflows, which accelerate the rates of decomposition of mangrove litter and speedy transformation of organic matter in soil. The textural composition of the deltaic soil helps to fasten the process of desalination in the water-logged areas after experiencing the impact of precipitation. The tidal amplitudes of the estuaries and connected network of creeks experience a wide range of oscillation, which is reflected by the variability of inundation and exposure pressures twice daily. All of these natural phenomena help to promote the diversification of the micro and macro biota, both flora and fauna in the Sundarbans mangrove ecosystem that could only be sustained under the unusual changing eco-geomorphological dynamics. Continuous interactions among different structural components of two subsystems—the mangrove forest ecosystem and the estuarine ecosystem—ensure the exchange of materials between the two subsystems by transporting litter-contributed detritus from the forest floor to the estuarine flows during low tide and the complete inundation of the forest by the high tides, making this ecosystem an open system. In addition, ecosystem functioning also controls the flow characteristics and pathways of the transport of nutrients, in respect of both the amount and rate by the synergistic influence of rainfall, tide, litter production, decomposition, terrestrial runoff, etc., and also the storage of accumulated materials. These kinds of flows and modes of transformations are meant to import the organic matter from the landward forest ecosystem and export of the dissolved organic matter and particulate matters to the adjacent estuarine ecosystems (Lugo, 1980; Ketchum, 1983; Odum, 1984; Ong, 1984; Karimn, 1988; Jiménez, 1992; Chakraborty, 2013).

3.4 Why Mangroves Are the Most Productive Ecosystem of the World

The mangrove ecosystem, also designated '**tidal forest**' or '**coastal wetland**,' is very important from the ecological, economic, and social perspectives and plays a significant role as a stabilizer of coastal belts by arresting erosion, protecting embankments, promoting land accretion, acting as nursery grounds for aquaculture and fisheries, performing important links between the land and the sea, particularly along the deltaic estuaries, serving as a buffer against natural disasters by acting as the green wall against waves and tsunamis, producing plant-based products such as tannin, timber, paper and pulp, and charcoal, and stabilizing embankments.

The mangrove ecosystem represents one of the most biologically productive coastal wetlands occurring in the intertidal belts of the tropical and subtropical parts of the globe (Macnae, 1969; Chaudhuri & Choudhury, 1994). This ecologically unique ecosystem is dominated by salt-tolerant intertidal halophytic vegetation enjoying the impacts of exposure and inundation of two high tides and two low tides a day respectively, and thereby constitute a dynamic ecosystem for the generation of biodiversity on the one hand and the perpetuation of a naturally developed ecological balance on the other (Chakraborty, 2011). This coastal ecosystem, having a blend of both grazing and detritus food chains, becomes highly productive, experiencing an average productivity of 2500 mg cm^2/day and governing the biogeochemical cycles in general and the nutrient cycle in particular of the coastal environment (Jennerjahn & Ittekkot, 1999). The worldwide recognition of the mangrove ecosystem in terms of its ecological potential, economic benefits, and aesthetic significance is because of two important attributes. **First**, a large quantity of energy in the form of mangrove plant-contributed detritus is exported to open water bodies from the mangrove forest (Odum & Heald, 1975) and a positive correlation exists between the total yield of fisheries and the extent of mangroves from adjacent water (Macnae, 1974; Lee, 1995), **Second**, profitable regional and international markets have developed, enjoying the steady supply of high-quality aquaculture products that sustain the livelihoods of a considerable number of local people.

As the mangrove ecosystem is composed of an array of both living and nonliving components that constitute the combination of two subsystems, the mangrove forest subsystem and its adjoining estuarine aquatic subsystem, which are connected through continuous interactions among themselves by the presence of a unique physical process, the tidal interplays, which by virtue of their inundation and exposure pressures ensure the exchange of materials—supplying a huge volume of water along with anionic components during high tide and supplying million tons of detritus-based nutrients to the estuarine flows during low tide, thereby helping continuous nutrient cycling. In addition, the variability of such exposure and inundation processes during diurnal, seasonal, and annual tidal cycles also plays a considerable role in determining the transporting of nutrients, controlling of soil salinities, and dispersing of propagules (Chakraborty, 2011). The schematic diagram depicting the conceptual model of the functioning of the mangrove ecosystem highlights the interactions and the interconnection among riverine water inflow, tidal interplays, exchange of nutrients, autotroph–heterotroph interactions, oscillation of physico-chemical parameters, and the activities of sediment-dwelling faunal components.

Major parts of the forest subsystem are inundated during high tide and benefit from receiving major inputs in the form of minerals and also tiny organisms such as phytoplankton, zooplankton, and other free-living microbiota from estuarine water, which are taken by so many suspension feeders and filter feeders as food. Besides, during the period of inundation, the entire green biomass and the bottom soil are recharged with moisture, and promote deposition of sediments and supply of nutrients (macro, micro, and trace elements) for the forest subsystem. The receding water during low tide takes away enormous amounts of the mangroves plants' contributed detritus to the adjoining aquatic subsystem. The nutrients released from the detritus are utilized by phytoplankton along with plenty of water and sunlight available in the open water system. The evergreen and perennial mangrove forest generates a very large number of mangrove leaves throughout the year, which on silt–clay-loaded bottom moist soil are broken down by the foraging behavior of an array of macrobenthic fauna (crabs, gastropods, insects, etc.) into smaller pieces of leaves. These are consequently interacted with by microarthropods (Collembola, Thysanura, Acarina, Isopoda, Hymenoptera, etc.) followed by the exo-enzymatic actions of the microbial communities (bacteria, fungi, protozoa) on the broken plant litter for the production of detritus through the decomposition process. The bottom dwelling deposit feeders such as crabs, mollusks, polychaetes, nematodes, etc., through their feeding activities, reshuffle the sediments of the forest floor, bringing up the soils from the lower strata of the sediment to the upper surface and soils from the surface layer move down to the lower strata of sediment and thereby not only continuously alter the texture of the bottom soils of the forest but also reposition the nutrients. Three prime processes involved in the decomposition processes are the (1) respiration with the release of carbon in gaseous form by micro flora and fauna; (2) leaching of soluble substances; and (3) the physical breakdown (corrosion) of decaying materials by fauna assisted by physical factors (Heal & French, 1974).

In this context, the underlying principle of the functioning of the mangrove ecosystem revolves on the continuous interaction of an array of structural components, both living and nonliving, constituting both subsystems (forest and estuarine), which are interlinked, and exchange materials between them in order to make this unique ecosystem highly biologically productive. The shallow water forest subsystem exports nutrients and energy-producing substances to the adjoining estuarine–coastal system. The biological components such as plankton, nekton, aquatic birds, and reptiles move freely between the two fixed subsystems and thereby produce, transport, convert, and utilize energy and nutrients after being facilitated by diurnal and seasonal tidal periodicities (Odum, 1971).

Heretofore, the biological productivity of the mangrove ecosystem resulted because of the combined or independent influence five attributes: (1) three major sources of primary production (mangrove plants, benthic algae, and phytoplankton); (2) continuous movement of water with moderate to higher water currents and tidal amplitudes during spring and neap tides orchestrated by the diurnal, seasonal, and annual tidal interplays; (3) a plentiful supply of nutrients is ensured by the continuous production and transport of mangrove litter-contributed detritus; (4) annual supply of sediment carried by riverine flows of the Ganges and Brahmaputra riverine networks and also by river floods into the coastal belts to help the enhancement of productivity of deltaic mangrove forests, which simultaneously also reduce salinity through the overdilution of sea water within the estuarine networks, resulting in conducive ecological setups for promoting the growth and diversity of mangroves in the Sundarbans areas; and (5) rapid utilization followed by regeneration and conservation of nutrients by activities of diversified forms of microorganisms and filter feeders. In addition, two different food chains operate for the all-round functioning of this ecosystem, which are (1) the grazing food chain, which begins with the direct consumption of live plants, both phytoplankton and mangrove vegetation by herbivorous consumers, followed by the consumption of omnivorous and carnivorous fauna; (2) the detritus food chain, which after initiating from the dead organic matter ultimately reaches the detritivores and their predators, facilitated by the all-round activities of microorganisms. All of these food chains are interconnected and interwoven with one another to develop a very intricate mangrove estuarine food web. A cluster of biological, chemical, and physical processes operating in mangrove ecosystems ultimately result in higher levels of productivity on undertaking a wide range of interactions among different structural components **viz**. soil, water, flora, and fauna of the ecosystem.

The intertidal and subtidal zones at the mangrove forest and estuary, being parts of the littoral zone with the penetration of sunlight almost up to the bottom layer of the water column, are unique from the geomorphological points of views and also in the energy-cycling processes of the mangrove ecosystem. The intertidal zone of the Sundarbans mangrove ecosystem, being a part of the littoral zone (shallow zone receiving sunlight up to the bottom layers of a water column) experiences continuous inundation and exposure twice a day by the rising and falling tides during high and low tides respectively. This zone marks the initial point of the marine ecosystem at the juncture of the land and sea, where harsh ecological exchange processes occur owing to tidal fluctuations receiving and accumulating the discharge from the terrestrial ecosystem with the results of changes in the volume of overlying water, currents, and waves of water, etc.

3.4.1 The Intertidal and Subtidal Zones

The intertidal zone of the Sundarbans mangrove ecosystem, being a part of the littoral zone (shallow zone receiving sunlight up to the bottom layers of the water column) experience continuous inundation and exposures twice a day by the rising and falling tides during high and low tides respectively. This zone marks the initial point of the marine ecosystem at the juncture of land and sea where harsh ecological exchange processes occur owing to tidal fluctuations receiving and accumulating the discharge from the terrestrial ecosystem with the results of changes in the volume of overlying water, currents, and waves of water, etc. The variability of ecological factors and processes offers very challenging ecological conditions in this zone for the residing organisms who spend their entire life or phases of life as the intertidal biota and develop adaptabilities to withstand and overcome the changing and even adverse ecological states, adopting different behavioral activities such as varied modes of feeding mechanisms (filter feeding, deposit feeding, etc.) to tackle the problems of an uncertain supply of food, morphological and ecophysiological adaptabilities to sustain against fierce currents of water, etc. The upper intertidal zone, known as the supralittoral zone, is subjected to intermittent submergence and quite frequently drenched in spray from crashing waves. The mid-littoral zone within the intertidal belt gets submerged during high tides on all days of the lunar tidal cycle and provide more stable and consistent ecological conditions for the biota to flourish. The lowest mark of the intertidal zone within low littoral zone merges with the subtidal zone, which is also a part of the littoral zone but always remain under shallow sea water (Flow Chart 3.1).

The variability of ecological factors and processes offer very challenging ecological conditions in this intertidal zone for the residing organisms who spend their entire life or phases of life for as the intertidal biota and develop adaptabilities to withstand and overcome the changing and even adverse ecological states, adopting different behavioral activities such as varied modes of feeding mechanisms (filter feeding, deposit feeding, etc.) to tackle the problems of an uncertain supply of food, and morphological and ecophysiological adaptabilities to sustain against fierce currents of water. The upper intertidal zone, known as the supralittoral zone, is subjected to intermittent submergence and is quite frequently drenched in spray from crashing waves. The midlittoral zone within the intertidal belt gets submerged during high tides on all days of the lunar tidal cycle and provides more stable and consistent ecological conditions for the biota to flourish. The lowest mark of the intertidal zone within low littoral zone merges with the subtidal zone, which is also a part of the littoral zone but always remain under shallow sea water (Flow Charts 3.2, 3.3, 3.4, 3.5, 3.6, 3.7, 3.8, 3.9 and 3.10).

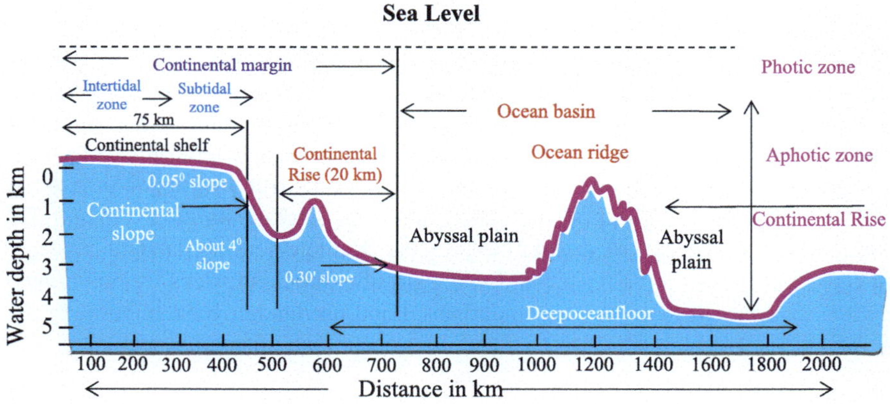

Flow Chart 3.1 Diagrammatic representation of the major zones of the continental margin and the ocean basin

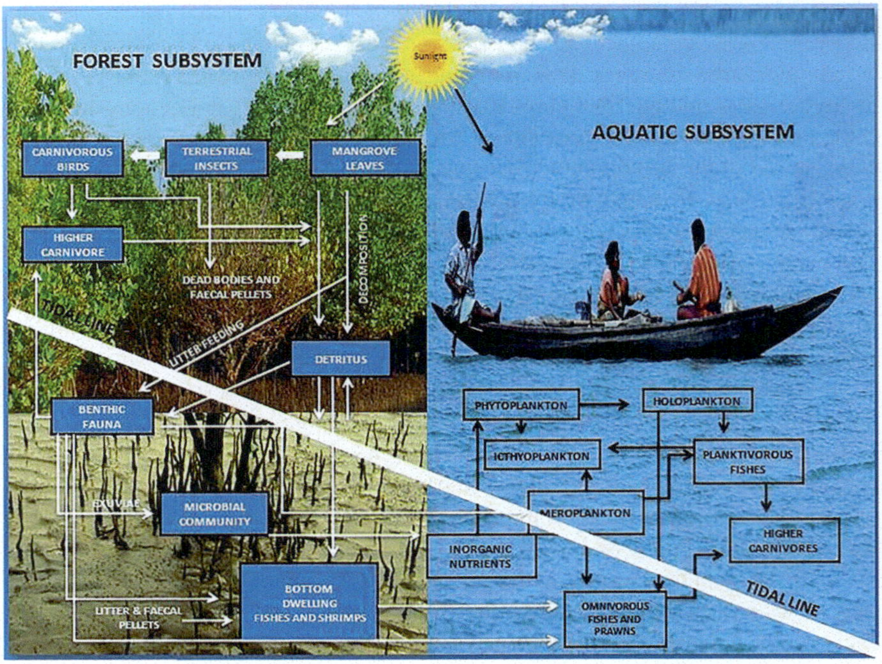

Flow Chart 3.2 Diagramatic representation of ecosystem functioning of the mangrove ecosystem of Sundarbans, India, where both subsystems—forest and estuarine—are interconnected and interlinked by tidal interplay (represented by the white line), facilitating the exchange of materials that support the growth and propagation of galaxy flora and fauna in the form of phytoplankton, benthic algae, zooplankton, benthic fauna, nekton, wild animals, and different arborial faunal components such as birds and insects

The degrading of plant litter, after being interacted with and consumed by a wide variety of faunal components (deposit feeders, grazers, detritivores, etc.), releases a considerable amount of nutrients to the mangrove ecosystem for recycling and regeneration. One estimate from the mangroves at Rookery Bay indicates that the litter fall in the Sundarbans mangrove forest exceeds 550 g dry wt./m^2/year (Lugo & Snedaker, 1974), which is less than the estimate (876 g dry wt./m^2/year) made by Heald (1971) of mangroves at the Everglade Park area in Florida (Jadav & Choudhury, 1985). Similarly, research study both on the field and laboratory have revealed different phases of the decomposition process of the mangrove plant litter through the interaction of a hoard of microarthropod fauna in their different phases of succession of arrival in the decomposing mangrove plant litter (Dey et al., 2008).

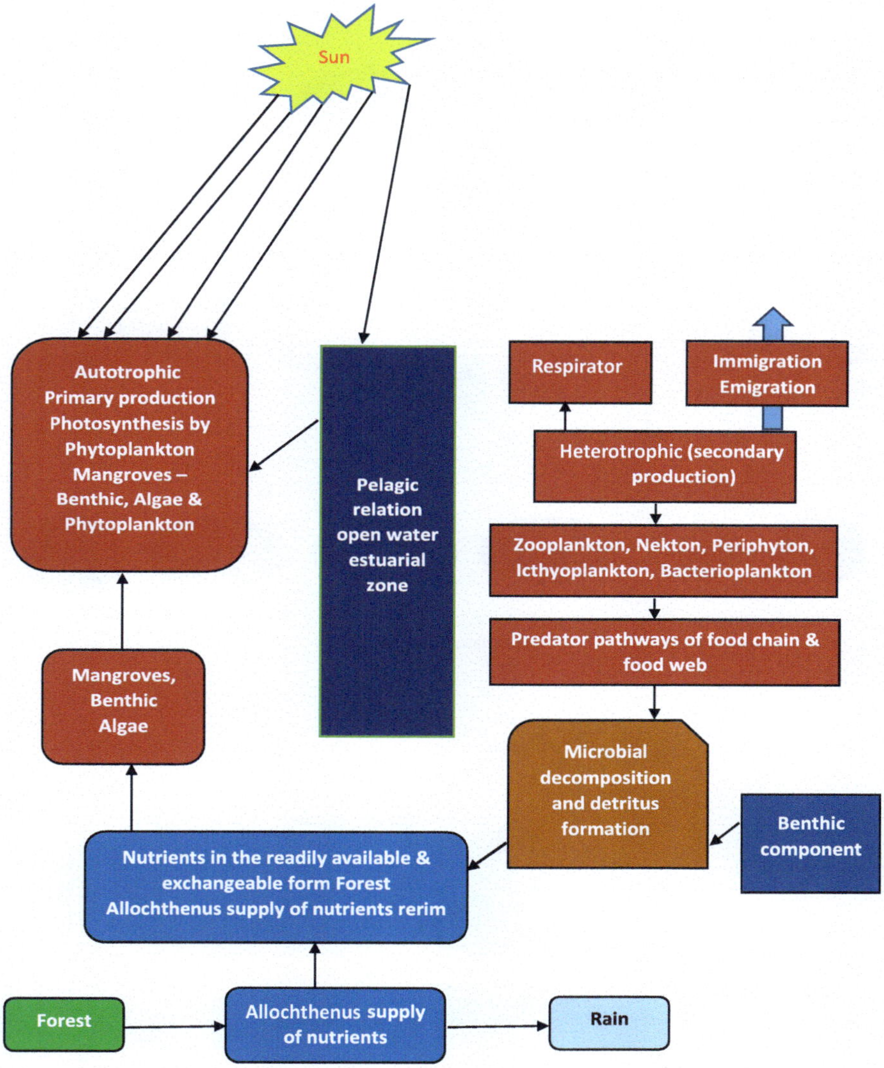

Flow Chart 3.3 Food-chain-food-web dynamics of the mangrove-estuarine complex of Sundarbans involving both autotrophic and heterotrophic biological productive systems

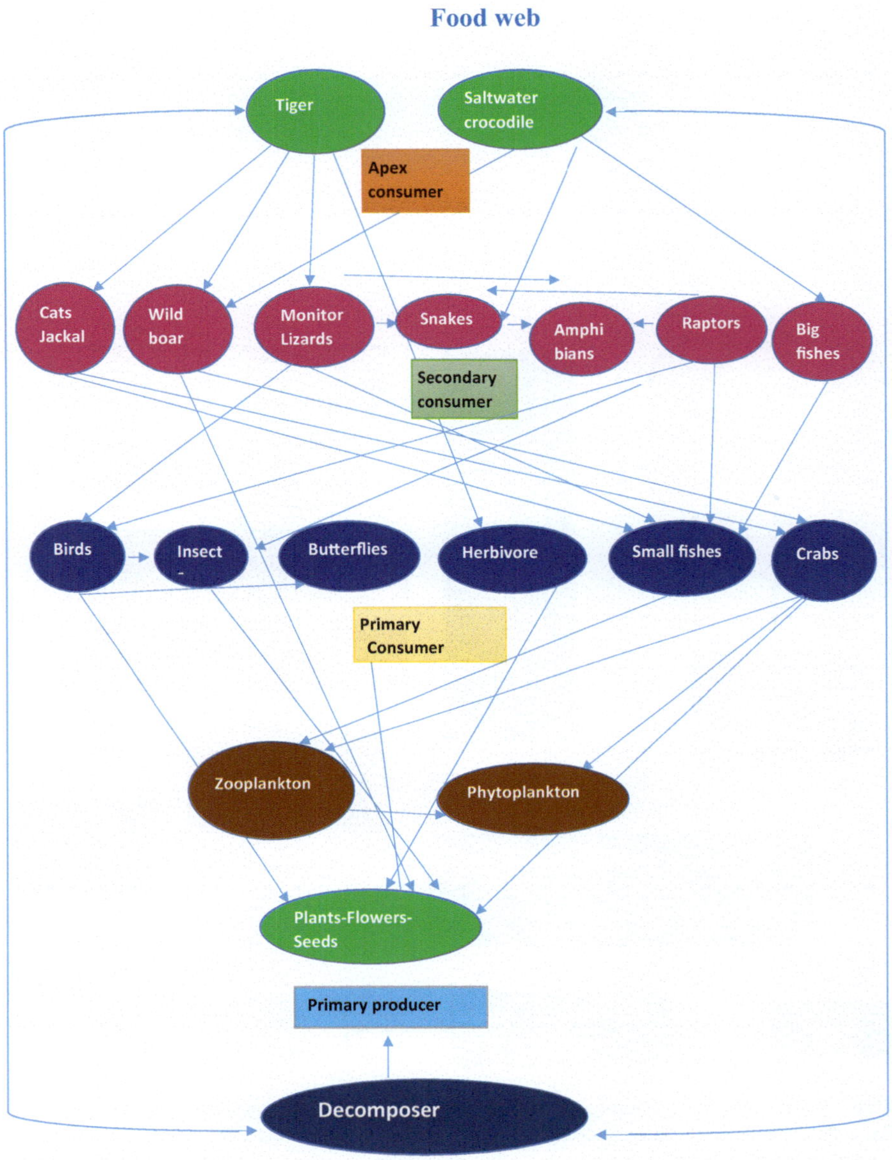

Flow Chart 3.4 Intricate food web in the Sundarbans mangrove ecosystem through the trophic interactions of different groups of flora and fauna, with tiger and crocodile as the top predators

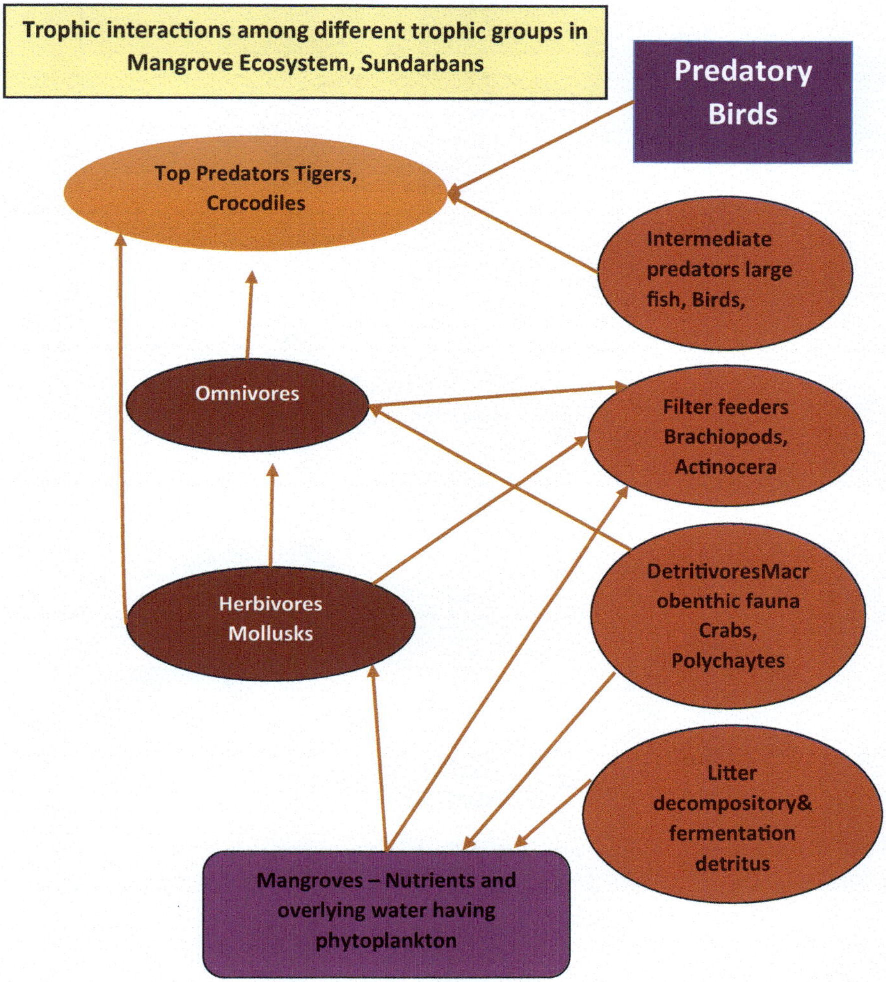

Flow Chart 3.5 Trophic interactions among different groups of biota that have different feeding habits in the Sundarbans mangrove ecosystem

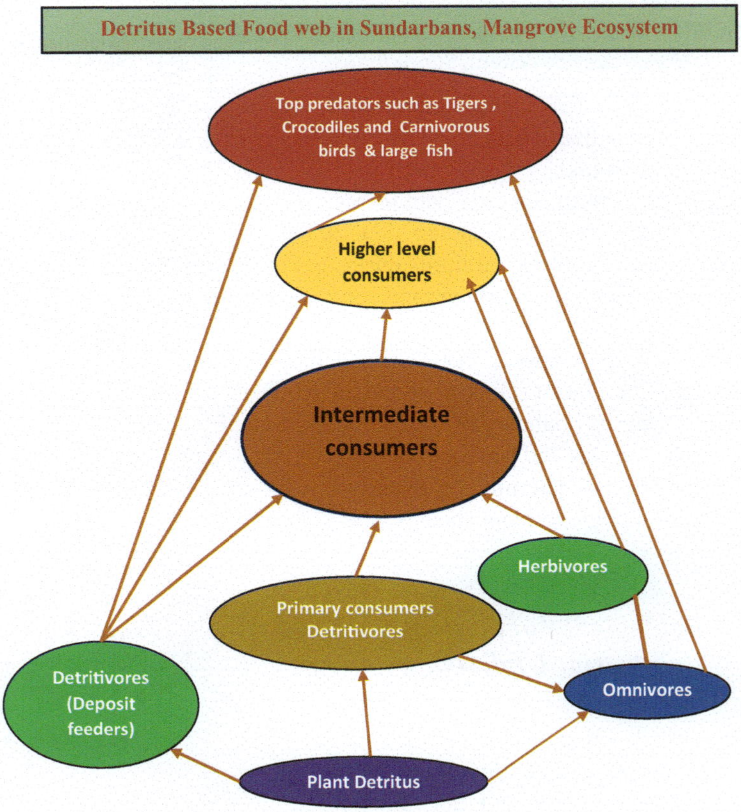

Flow Chart 3.6 Detritus-based food web in the Sundarbans mangrove ecosystem which integrates both grazing and autotrophic food chains but culminate in the top predators

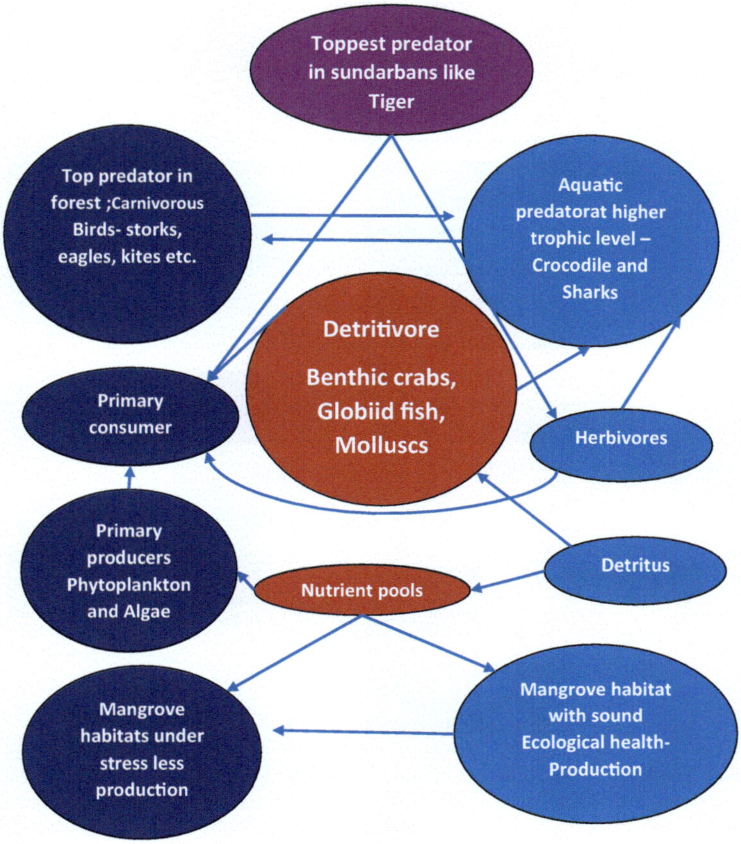

Flow Chart 3.7 Exchange of nutrients among different groups of organisms within the Sundarbans mangrove ecosystem, India, starting from detritus-inhabiting organisms to top predators in both stressed and ecologically healthy mangrove belts

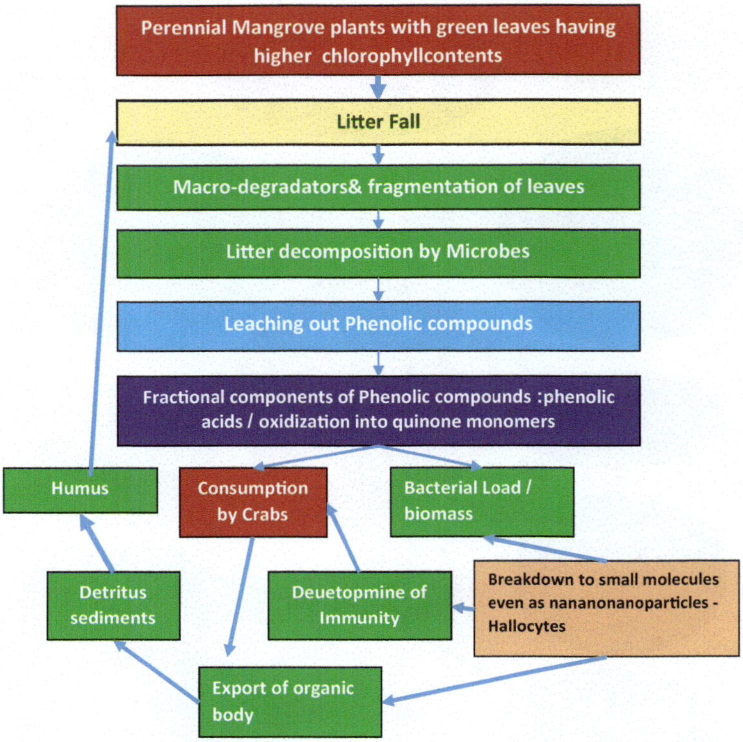

Flow Chart 3.8 Bioaccumulation and biotransformation of phenolic compounds of mangrove-plant leaves through litter decomposition by the active involvement of macrobenthic fauna (intertidal crabs) and microbes producing humus-based nutrients and ultimately nanoparticles, such as halloysite

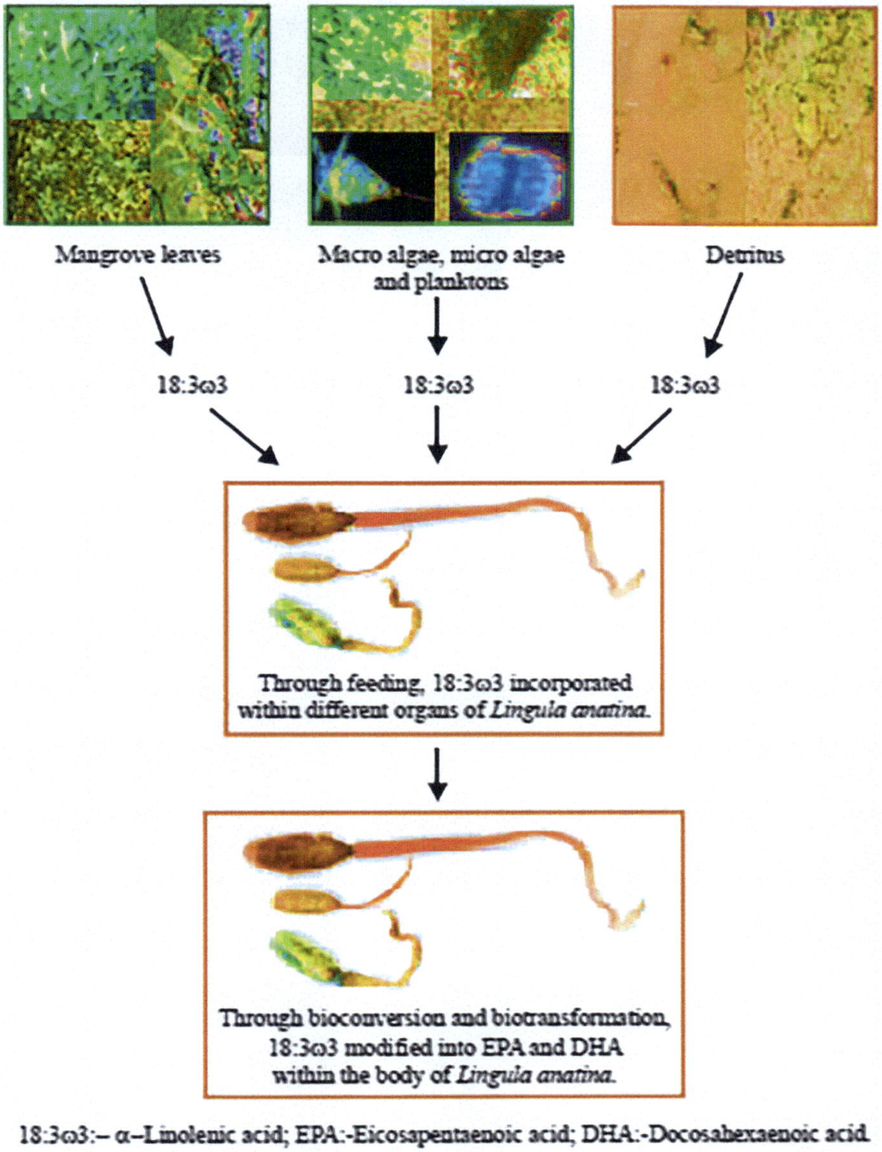

Flow Chart 3.9 Bioconversion and bio-transformation of fractional components of lipids of mangroves through the activities of one major intertidal macrobenthic fauna of Sundarbans, *Lingula anatina*, belonging to the minor phylum, Brachyopoda

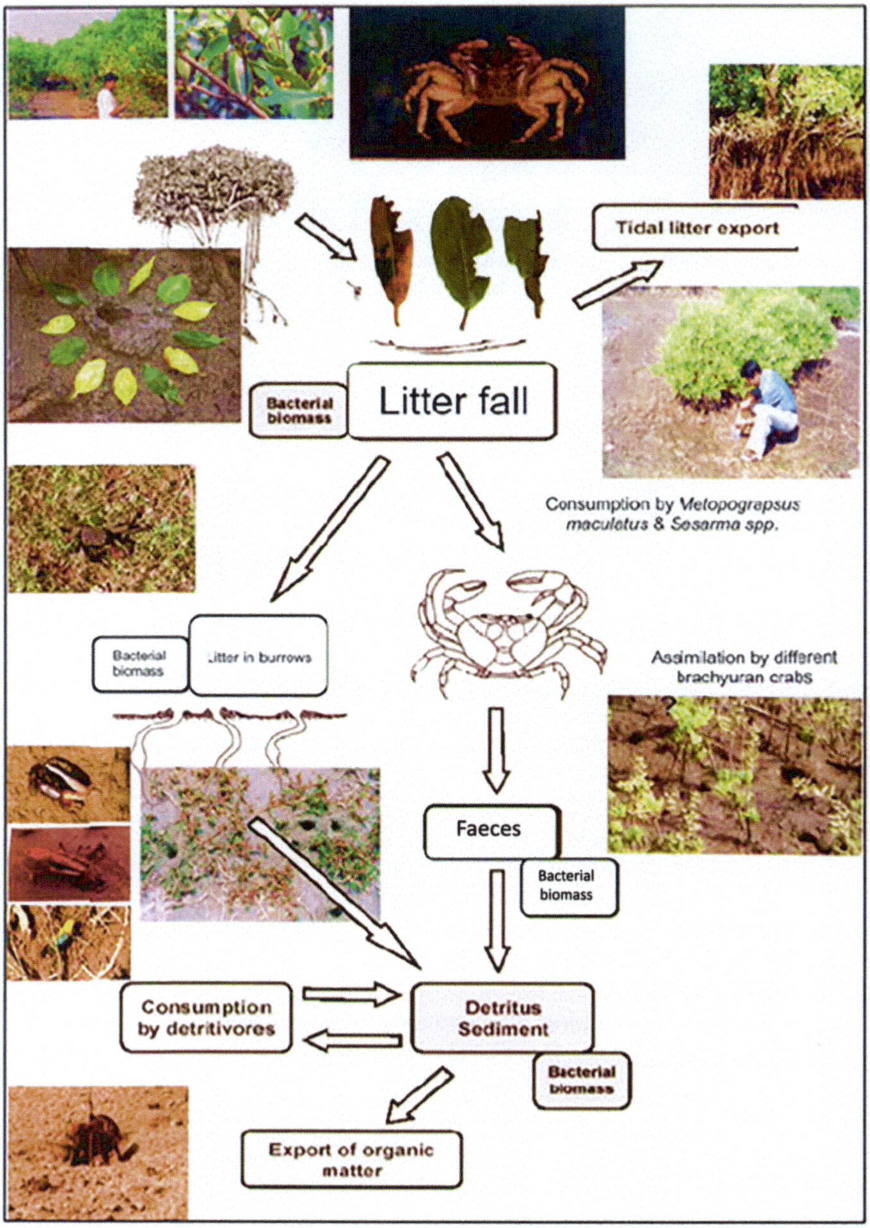

Flow Chart 3.10 Production and transport of nutrients from the mangrove forest subsystem to the estuarine subsystem of the Sundarbans through a series of trophic interactions among the constituent biota by virtue of their survival strategies with different forms of growth, proliferation, biotic interactions, and behavioral manifestations such as bioturbation

3.5 Eco-Biological Characteristics of Mangroves

3.5.1 Characteristics of Reproduction and Growth in the Lifecycle of Mangroves

Most of the species of mangrove ecosystem are hermaphroditic (presence of both sexes in an individual) where pollinations are made by different groups of faunal components such as insects and birds), except for the self-pollinated species under the genus *Rhizophora*. Viviparous germination is a unique biological characteristic of true mangrove species germination taking place within an embryo, which still remains attached to the parent's body. The embryo does not possess a dormant stage but grows out of the seed coat and also from the burst out fruit prior to detaching from the plant followed by dropping down to the bottom soil in the forest floor. Therefore, propagules of mangroves are designated as seedlings, not as seeds. Achieving vivipary in the life history of mangrove plant species helps mangroves not only to combat the problems of varying ecological parameters such as differential grades of impacts from the oscillating seasonal salinities, round-the-clock inundation and exposure pressures characteristic of intertidal environments but also help to prolong the survivability of seedlings. Moreover, in order to ensure the long range of dispersal of propagules in terms of sizes, shapes, weight, etc., a considerable amount of nutrients are stored within them to float for an extended period (Tomlinson, 1986). Seedlings are deposited and settle down depending on the power of buoyancy within the soft forest floor in the intertidal belt in the face of tidal amplitudes and currents of water. After settlement, many seedlings face stresses from the oscillation of salinity, variable intensities of flooding, the availability of light, and even grazing pressure from the ground-dwelling benthic fauna (Smith, 1992). All of these factors determine the rates of mortality as well as the survival rates of new seedlings and saplings (Smith, 1992).

3.5.2 Adaptations to Salinity

Mangroves inhabiting in the intertidal zone have to experience a relatively wide range of flooding and salinity conditions during their growth and proliferation facing less competition with freshwater plants. Mangroves are endowed with a number of morphological and physiological adaptive characteristics that enable those halophytic vegetational assemblages to successfully colonize under those environmentally challenged unstable ecological conditions with highly fluctuating ecological parameters. Mangroves do not require saline water for their growth and survivability but prefer to grow in brackish water, reducing competition with freshwater vegetation. The level of salinity in mangrove-dominated coastal environments is controlled by many geo-hydrological factors and ecological processes where saline water from the sea intrudes via the periodic tidal flows and mixes with freshwater coming from rivers, rainfall, land runoff, and even from groundwater. Higher intensities of evapotranspiration (water loss through the soil and plant leaves) in the tropical and subtropical environment along with restricted flows of freshwater are considered a probable cause of the enhancement of salinity over both time and space.

Mangroves have evolved different mechanisms in order to tolerate high salinities such as salt secretion and salt exclusion by plant tissues of most mangroves but to varying extents (Rützler & Feller, 1996). For the purpose of salt secretion, salt-secreting glands are developed to remove salts from plant tissues that cause salt crystals to form on the leaf surfaces. Afterward, all of these crystals are washed away by the rain or even blown away by windflows. Falling of matured leaves after accumulating a considerable amount of salts represents another mechanism for eliminating excess salt in mangroves (Kathiresan & Bingham, 2001).

An experimental study on four common mangrove plant species in Sri Lanka **viz.** *Ceriops tagal, Avicennia marina, Rhizophora apiculata*, and *Rhizophora mucronata,* to note the salt-retention capacity of those species through in situ and ex situ conditions, which have revealed *A. marina* to be the most efficient at retaining salt within plant tissues, whereas *C. tagal* is superior to *R. mucronata* but inferior to *A. marina* in this respect. It has also been observed that increased salinity of substrate imposes significant negative impacts on photosynthesis and growth performance of mangrove seedlings. Experimental mangrove species also regulate salt concentration through subsequent shedding of salt-laden leaves as part of the salt-regulation strategy of this mangrove species (de Silva & Amarasinghe, 2021). It was also observed that the amount of proline within the mangrove cells becomes significantly high as an adaptive means of balancing high concentrations of sodium and chloride ions. The cellular osmotic balance is maintained by *C. tagal* through the production and accumulation of cyclitols along with Na^+ and Cl^-.

3.5.3 Adaptations to Flooding

Mangrove forests always experience periodic flooding, the magnitude and frequency of which are determined by local topography, geology, tidal action, groundwater, freshwater inputs, associated riverine networks, meteorological parameters such as rainfall, windflows, temperature, humidity, etc., surface runoff, and water-quality parameters, mainly salinity and evapotranspiration. All of these attributes are directly connected to the ability of mangroves to tolerate flooding pressures, but which varies in time and space. In the intertidal belts, the intensity, duration, and frequency are reduced in a landward direction. Zonation, distribution, and succession of the mangroves are also driven by the gradient of sloping, which ultimately determines the species composition of mangrove assemblages in lower (maximum duration of tidal inundation), middle (moderate duration of tidal inundation) and upper intertidal zones (least duration of tidal inundation) (Robertson & Alongi, 1992). The impacts of flooding result in the reduction of soil oxygen, deformities in root tissues requiring the necessary amount of oxygen to continue metabolic activities, and enhancement of the accumulation of toxic substances such as sulfides.

3.5.4 *Mass Mortality of Mangroves*

Large-scale mortality of mangroves is brought about by several biological causes after being inflicted by severe environmental disturbances within the forests. All of these imposed disturbing factors include natural disasters such as cyclone, hurricanes, and other forms of storms that cause direct physical damage to mangroves and also the heavy deposition of sediments leading to drastic alteration of the textural composition of the mangrove forest floor, thereby disturbing nutrient availability.

3.5.5 *Shoreline Stabilization and Protection*

As mangroves represent biotic assemblage in a coastal environment, they play a significant role in the process of soil formation and shoreline stabilization. The hydrological regime of the shallow marine-estuarine ecosystems is profusely regulated by extensive and robust above-ground root structures of mangrove forests in the form of prop roots, drop roots, and pneumatophores by acting as a sieve for reducing current velocities and enhancing sedimentation and sediment retention (Carlton, 1974; Augustinus, 1995). Mangroves not only act for trapping sediments through the intricate matrix of fine roots but also produce sediment by promoting accumulation of organic matter derived from mangrove plants. In addition, roots and leaves of mangroves help to accelerate the soil elevation, especially in areas with low sediment delivery. Enhanced rates of soil formation, sedimentation, and sediment retention help mangroves to stabilize soils, which are a prerequisite for tackling the risk of soil erosion, mostly under high-energy conditions such as tropical storms. The topographic condition, having distinct zonation patterns with specific species composition of mangroves in the intertidal zone, play important roles in ensuring protection of the coastal zone. Mangroves are endowed with the ability to lessen the impacts of the winds with higher intensity and tropical storms, tidal waves with the strong force of currents, floods with an excessive volume of water that accompany them, thereby protecting the uplands from more severe damage (Tomlinson, 1986). Even after damage by the devastating impacts of all those forces, the mangrove forests in general exhibit great aptness in recovering the loss, as they can produce plenty of propagules from which seedlings grow rapidly and attain sexual maturity quickly, thereby accelerating the pace of natural regeneration.

3.5.6 *Purification of Water and Nursery Ground for Finfish and Shellfish*

The intricate root system of mangroves traps not only sediment but in that process they can also filter the turbid water by facilitating the steady deposition and trapping of sediments and by taking up excess nutrients, thereby maintaining the water quality. They also have the ability of phytoremediation by extracting and accumulating toxic materials such as heavy metals from the water. The improved water quality loaded a good amount of nutrients to create habitats for the growth and proliferation of diversified forms of commercially important shellfish and finfishes. Both the anadromous and catadromous fishes select this mangrove habitat to lay their fertilized eggs so that the hatched-out seedlings and fingerlings can find plenty of nutrients mediated by planktonic organisms as their novel food sources.

3.6 Uniqueness of the Mangrove Ecosystem: Trophic Interactions in the Estuarine-Mangrove Ecosystem of West Bengal, India

The '**Marine Forest**' or '**Mangals**' are the terms used to portray the mangrove community, an assemblage of different species of mangrove plants that occur in the intertidal belt of tropical and subtropical coasts preferably at a temperature of more than **20° C** and are equipped with special halophytic adaptive characteristics such as pneumatophores, stilt roots, sunken stomata, salt gland, root buttress, and viviparous germination, etc. Detritus of the mangrove ecosystem utilizes the food materials within its periphery. The intricate arrangement of mangrove roots bewilder sediments, reduce wave action, promoting more deposition of silty clay sediments, minimize the coastal erosion, diminish atmospheric depression, resist storms, cyclones, and tsunamis, tidal bores and surges, and provide a suitable habitat for the growth and reproduction of benthic fauna and finfishes.

The functioning of the mangrove ecosystem involves three different types of green plants (mangrove plants, benthic algae, and phytoplankton) representing the first trophic level as primary producers that act as a novel source of food to the primary consumers, which are considered representatives of the secondary trophic level (insects, different kinds of benthic fauna, herbivorous fishes, and forest-inhabiting wildlife such as deer, wild boars, birds, monkeys, etc.). Mainly carnivores, along with omnivorous fauna (larger finfishes, dolphins, leopard cats, turtles, water monitors, and various birds) belong to the tertiary trophic levels, which facilitate through trophic interactions a unidirectional flow of energy that ultimately culminates in the highest trophic level, represented by top carnivores such as the Royal Bengal tiger (*Panthera tigris tigris*) and the estuarine crocodile (*Crocodylus porous*) (Alexander, 1977; Blasco, 1977; Alongi, 1988a, b; 2009; Bouillon, et al., 2008; Bouillon, 2011; Andersen and Kristensen 1988; Annon, 2003, 2005; Alongi and Mckinnon, 2005; Aziz and Paul, 2015).

3.7 Significance of the Mangrove Ecosystem

The biological productivity within the mangrove ecosystem occurs at the dynamic interface between estuary and land. This ecosystem withstands continuous disturbances because of inundation and exposure pressures, fluctuating temperature and salinity, variable degrees of anoxia, turbid water loaded with dislodged silts and clays as erosion pressure, etc., is directly connected to the cycling of essential nutrients. The gross and net primary production of the mangrove forest amounts to 4.6 kg carbon/ha/year and 1.9 kg carbon/ha/year respectively (Alongi, 2009). These rapid rates of productivity commands a required supply of nutrients and other trace elements to promote the growth, physiological maintenance, and production, which are translated into rapid uptake of soil nutrients with the active help of bacteria, fungi, archaea, protozoa, and microalgae. This cycling of nutrients depicts strong linkage between nutrient cycling and carbon fixation (Alongi, 2014). A mass balance shows that the ecological health of the mangrove ecosystem depends on the ammonification and nitrification of soils. Nutrient availability and tree photosynthesis are intimately linked in the mangrove ecosystem, which therefore is considered nutrients limited. This relationship-mediated forest productivity is disrupted by various ecological attributes and processes such as deforestation, hydrological alteration, land-use change, increasing intensity and frequency of storms, enhanced temperatures, higher concentration of atmospheric CO_2, and a rising sea-level.

The mangrove ecosystem is composed of two major subsystems—forest and associated estuaries represents one of the most productive ecosystems in the subtropics and tropics constituting an ecological lifeline of most of the world's coastlines at low latitudes. Mangroves, the tidal woody and salt-tolerant forests that occur over 152,000 km^2 of coastal area occupy <1% of the world's coastal zones (Alongi, 2002) and renders an array of ecological services and economic goods (Chakraborty, 2018). Innumerable species of flora and fauna act as ecosystem engineers contributing to structure and shape and maintaining the ecological uniqueness of intertidal zones, developing it as prime accumulation sites for soils, sediment particles, and associated elements required for nutrient recycling.

The morpho-anatomical excellence of mangroves as manifested by their robust body structure associated with sunken stomata, thick wax covering the leaves, a unique system of roots as pneumatophores and proof roots, viviparous germination, etc., have made them resilient to outer disturbances (Alongi, 2008).

Striking features of mangroves and the mangrove ecosystem in respect of a nutrient reservoir and nutrient cycling include the following:

1. Below-ground reservoir of nutrients that help to drive the process of ecosystem functioning.
2. Faster nutrient flux and microbial decomposition facilitating rapid biotic turnover.
3. Efficient rates of water and nutrient use help to augment ecological recovery.
4. Body architectural design leads to rapid reconstruction, rehabilitation, and eco-restoration of the ecodegraded environment.
5. Functional roles of keystone species help restoration and recovery of the ecodegrading mangrove ecosystem.
6. Mangrove-ecosystem stability is maintained by the negative and positive feedback mechanism that helps to dampen changes in the ecological steady-state (Alongi, 2008).

7. Restriction of mangrove occurrence in the intertidal zones is associated with five different ecotypes across the world, enjoying a distributional range of an approximate area of 150,000 km^2 of sheltered coastlines in the tropics and subtropics that include fringe, riverine, basin, overwash, and dwarf forests.

All of the above-mentioned characteristics are instrumental for the inbuilt resilience of the mangrove ecosystem to resist and combat the imposing ecological and anthropogenic stresses, including deforestation and creation of impoundments for aquaculture, large-scale erosion followed by accretion, increased sedimentation and nutrient recharging, occurrence of natural disasters such as storms with increased frequency and intensity, steadily rising temperatures, increasing levels of atmospheric CO_2, and climate change-induced sea-level rise (Alongi, 2008, 2015; Gilman et al., 2008).

3.8 Dunes and Their Significance Along the Sundarbans Mangrove-Estuarine Complex, India

A sand dune is formed by a mound of sand deposited by winds that rises to a summit and possesses a slip face. In the Sundarbans, dunes run parallel to the southern coastal tract line without any definite shapes during the initial stages of formation. These eolian sand deposits convert a sandy area to a "**dune field**." Wind-mediated formation of sand ripples also make many areas a dune field. In Sundarbans, sand dunes are generally oriented parallel to the prevalent wind as barchans or transverse dunes. Barchan dunes are crescent-shaped structures formed by the deposition of sands with unidirectional wind flow which move advance by avalanching sand on the slip face. In contrast, elongated transverse dunes appear nearly straight with regularly spaced sand ridges. The windward sides of the dunes contain superimposed eolian ripples but hardly support that much vegetation. The formation of dunes follows a basic sequence of dune-formation stages.

During the initial stage of dune formation, the perennial grass *Paspalum vaginatum* of the family Poaceae forms an arc-shaped train in the sand with the convexity of the arc pointed along the direction of the wind. Different species of grass help to trap sand particles, triggering the initiation of the growth of mounds of sand. The growth of the vegetation continues, which further accelerates the accumulation of sand, which, in turn, results in the growth of the dunes (Fig. 3.1).

Dunes are an important landform of the sea-facing delta formed at the water–land interface at high-tide levels of coastal environment by either wind- or even water-driven sand materials from foreshore sands to the supratidal backshore zone and physically appear as mounds or ridges. Each dune consists of a windward side and also a slip face. The dunes are structured by a number of factors such as wind flows, direction of winds, and longshore currents, and support the lives of some specialized flora and fauna, which in turn help to stabilize the dunes. The coastal dunes of the Sundarbans occur mostly in close proximity to the sea and physically

Fig. 3.1 Dunes in the coastal belt of West Bengal, India. (Images of dunes in the coastal belt of West Bengal, India, depicting their geomorphological features with dune-stabilizing creeper plants, *Ipomoea pes-caprae*, saline grass, and eco-assessment processes to record ecological changes. A large dune in the coastal belt of Midnapore (East) is in the process of destruction. (**A**, **B** and **C**) Some of these natural dunes are still in existence in the coastal belt of Midnapore (East), West Bengal, India; (**D**) A natural dune is stabilized by creepers and saline grass; (**E**) Along with the destruction of winds, natural plants of cashew nuts are uprooted; (**F**) and (**G**) Different geomorphological attributes are being collected by researchers; (**H**) and (**I**) The most efficient sand binders, the creeper, *Ipomoea pes-caprae,* of the family Convolvulaceae play vital roles in arresting coastal soil erosion and ensuring the stability of the dunes)

appear with shapes such as dimensionless mounds and also sometimes possess sinuous crests running parallel to the coastline and represent transverse and barchan types. Three prime factors and environmental processes are found to become responsible for the formation of dunes, which are (1) intensity of the wind flow and pattern; (2) deposition of sand particles in a particular site, forming a mound-like structure; (3) settlement of floral components such as climbers, creepers, herbs, and shrubs; (4) a number of ecological parameters such as rainfall, salinity, moisture content, and substrate quality play important roles in determining the distribution of settled plant species, which in turn play important roles such as the stability of the dunes and arresting of the migration of the dunes (Bhattacharya, 2000). The dunes-inhabiting plants include *Suaeda maritima*, *Salicornia* spp., *Paspalum vaginatum*, *Cynodon dactylon*, *Aeluropus lagopoides*, *Sesuvium portulacastrum*, and *Ipomoea pes-caprae*. Among all these creepers *Ipomoea pes-caprae* play important roles in the formation, growth, stabilization, binding of sands, and arresting migration of the coastal dunes along the upper parts of the intertidal zones of Sundarbans (Bhattacharya, 2000). Vegetation assemblages of dunes and their combined roles are connected with the geomorphology and evolution of landforms in a specific coastal tract. The independent roles of plants also vary as their roots and rootlets have different functional roles as the of herbaceous and grassy vegetations act as sediment binders, whereas the leaves and stems of shrubs such as *Suaeda maritima* and *Salicornia* spp. help fall out of suspensional sands. Simmilarly creepers such as *Ipomoea pes-caprae* arrest dune migration to a considerable extent, and all dune-inhabiting plants by their biomass help the stabilization of the dunes (Bhattacharya, 2000; Vila' et al., 2000, 2006).

Coastal dunes are spectacular geomorphological structures occurring at the sea-facing deltaic islands of Sundarbans. The mound-shaped dunes are provided with transverse sinuous crests. They are formed by the continuous deposition of sands carried by the wind flow at the supratidal backshore zone and stabilized by some salt-tolerant plants such as *Ipomya bilobata*, *Pandanus tectorius*, and *Spartina alterniflora*. Three major parameters responsible for the formation of dunes are as follows (Bhattacharya, 2000).

1. Intensity, vigor, and direction of wind flow.
2. Total rainfall on that area of formation of dunes.
3. Types of dunes stabilizing the floral community with the species *Ipomoea bilobata*, *I. pes-caprae*, *Salicornia* spp., *Paspalum vaginatum*, *Suaeda maritima*, *Cynodon dactylon*, *Aeluropus lagopoides*, *Sesuvium portulacastrum*, etc.
4. Physiography of the sea-facing islands.
5. Roles of physico-chemical parameters such as salinity, pH, moisture content, textural, and nutrient concentration of sediment promoting growth of dune stabilizing plants.
6. Roles of creepers and shrubs helping in the creation of these coastal dunes by arresting sand grains behaving as barriers.
7. The rootlets of these creepers help sand binding, thereby arresting the migration of newly formed dunes (Bhattacharya, 2000; Van der Maarel, 2003).

3.9 Mangroves and Nutrient Cycles

Kinetic energy-loaded solar energy, on entering the planet, drives the ecosystem processes after being primarily arrested and thereafter converted by autotrophic organisms such as phytoplankton, benthic algae, sea grasses, mangroves, etc., and subsequently transformed into heat energy following a unidirectional pathway of energy flow. Organic matter through the decomposer cycle is converted into chemical elements, which are utilized by almost all plants to form their biomass, which enjoys distribution to every corner of the oceanic ecosystem. The nutrient cycle depicts the cyclical process of prime chemical transformations of varied forms of chemical elements such as carbon, nitrogen, silicon, phosphorus, sulfur, and other types of macroelements, which are recycled to ensure the functioning of the ocean ecosystem (Walsh, 1991; Broecker et al., 1993; Zweifel et al. 1996; Khan et al. 2005; Koch et al. 2006; Koch et al. 2009).

Nutrients, through their cycling and retention mechanisms within the mangrove ecosystem, determine the growth, development, and propagation of mangroves and thereby ensure higher rates of biological productivity (Ball, 1988; Reef, 2010). The survivability of mangroves and the sustainability of the mangrove ecosystem are nutrient limited, especially by nitrogen and phosphorus, whereas other forms of micronutrients such as sulfur, potassium, magnesium, boron, and sodium hardly influence the functioning of the mangrove ecosystem in a saline water-dominated environment (Warren & Underwood 1986; Zweifel et al. 1996; Alongi, 2017). The interactive effects between a variety of nutrients with different species of mangrove plants are determined by several ecological factors such as temperature, pH, turbidity, salinity, soil texture, intensity of tidal waves, and duration of both tidal inundation and exposure. The photosynthetic potential of mangroves is very much regulated by the supply of nutrients, which is reflected by higher nutrient use efficiencies along with enhanced rates of resorption of leaves (Alongi, 2009). Mangroves, by virtue of their adaptability to surviving in anoxic soils, can conserve nutrients and water very effectively, mostly experiencing oligotrophic conditions during the period of inundation under tides (Ball, 1988). Sclerophylly in mangrove leaves being an adaptive manifestation for ensuring survivability in water-scarce habitats exhibits extremely low leaf-water potentials by way of eliminating much of their water content. The evergreen and perennial mangrove forest, having sclerophyllous leaves, possess high root/shoot ratios in comparison with tropical terrestrial forests (Ball, 1988). However, such high root/shoot ratios, being an adaptation to saline ecological setups, demonstrate variation among different mangrove species over time and with forest structure. The higher biomass of mangrove roots can be treated as another adaptive manifestation in capturing nutrients from soils loaded with poor nutrients (Ball, 1988; Reef, 2010). Several research studies on litter decomposition and nutrient cycling have been undertaken by the researcher in the back drop of global perspectives.

The grazing food chain of the mangrove-estuarine complex of Sundarbans are dependent on the breakdown of mangrove litter and their subsequent inorganic transformations through decomposition process. The composition of the litter, abiotic factors and microbial communities as decomposer contribute in the nutrient cycling within mangrove ecosystem through "feed-back mechanisms" with one another, and thereby increasing or decreasing the decomposition rates (Swift et al, 1979). Leaf breakdown occurs due to physical fragmentation, animal feeding, microbial activity and leaching (Stewart and Davies, 1989). The chemical composition of mangrove detritus undergo massive changes during the processes of decomposition and aging (Rice and Tenore, 1981). Rapid leaching of dissolved organic matters take place when the mangrove leaves fall down on water-sediment interphase of mangrove forests. Resulting the production and releasing of dissolved organic nitrogen (DOM) and also inorganic organic carbon, and nitrogen (Robertson, 1988). A study on mangrove litter contributed nutrient cycling from the Hooghly-Matla estuarine complex of Sundarbans, India has shown that production and biotransformation of nitrogen (N) in the form of dissolved organic nitrogen (DON) and dissolved inorganic nitrogen (DIN) which intimately connected with the litter production and decomposition to detritus involving different pathways of conversions of soil total nitrogen (STN), soil organic nitrogen (SON), soil inorganic nitrogen (SIN), total organic nitrogen of water (TON), dissolved organic nitrogen of water (DON), particulate organic nitrogen of water (PON) and dissolved inorganic nitrogen of water (DIN) (Mandal et al., 2011).

3.9.1 Roles of Nitrogen Mass Balance in Restricting Distribution of Mangroves

Nitrogen cycling in the mangrove ecosystem becomes complex because of mainly low dissolved nitrogen concentrations but slow turnover of particulate nitrogen. Most of the nitrogen, after having their allmost similar amount of production from mangrove plants and benthic algae, remains stored in the soil. Assessment and documentation of nitrogen mass balance help to identify global change in mangroves for the reasons mentioned below.

First, the primary production of the mangrove ecosystem is controlled by the combined effect of nitrification, ammonification, and/or dissimilatory nitrate reduction to ammonium in the soil. Disruption of ammonification or nitrate reduction processes by any kind of disruptive forces because of global changes such as deforestation followed by soil erosion will certainly have negative impacts on mangrove productivity.

Second, different microbial processes on soil such as nitrification, denitrification and ammonification do not play much of a role in the nitrogen dynamics of the mangrove ecosystem and therefore the factors that change the rates of all such processes seem to have little impact on the nitrogen dynamics of the mangrove ecosystem.

Third, as burial of a considerable amount of mangrove leaf litter along with its conversion products as a substantial fraction of total nitrogen takes place in mangrove forest soils, it acts as a significant nitrogen conservation mechanism.

Fourth, the production of mangrove plant litter and below-ground root biomass being affected by any kind of global issues will have a substantial impact on net primary productivity.

Fifth, any kind of global changes affecting tidal exchange such as construction of impoundments leading to alteration of hydrology will likely have considerable impacts on the balance of the nitrogen dynamics of the mangrove forests.

Sixth, nitrogen fluxes tend to make an overall balance in the mangrove ecosystem and disruption by any kind of global change such as deforestation appears to impose a negative effect on this balance and thereby on the survival of the mangrove ecosystem.

In contrast to nitrogen cycling, the cycling of phosphorus and other limiting nutrients such as iron and copper play small roles in the soil and water of the mangrove ecosystem. The phosphorus cycle, being tightly linked to soil geochemical processes, the lower availability of phosphorus concentrations (phosphate, organic phosphorus) focus on low nutrient availability (Vitousek & Matson 1987; Mann & Steinke, 1989; Alongi et al, 1992; van de Graaf et al., 1995; Pelegrí & Twilley 1998; Howarth et al., 2002; Luizão et al., 2004; Alongi, 2018).

3.9.2 Increased Nutrient Enrichment and Adverse Effects on the Mangrove Ecosystem

Enhanced nutrient availability is the leading cause of higher nutrient enrichment of the coastal environment and puts intense pressure on the biotic community, especially on the accelerated growth of intertidal mangrove forests. Mostly, such nutrient enrichment favors the growth of shoots relative to roots and thereby enhances the vulnerability to environmental stresses that adversely affect plant–water relations (Morell & Corredor, 1993; Middelburg et al., 1995; Rabouille, 1997). Two major ecological parameters, such as high salinity and low humidity, trigger greater investment of nutrients in roots, impairing the demand for water by the shoots. Higher salinity of sediment results in low humidity and low rainfall. However, the increased growth of mangroves in response to coastal eutrophication is counterbalanced by the costs of decreased resilience due to mortality, and increased salinity in soil and water along climatic gradients (Lovelock et al., 2009). Adverse ecological consequences of nutrient enrichment are noted in coastal ecosystems (Cloern, 2001), which include the occurrence of algal blooms, destruction of coral reefs, along with their associated biodiversity components (Hughes et al., 2003), reduce ecosystem resilience, leading to drastic loss of biodiversity (Levine et al., 1998; Scheffer, et al., 2001) and sometimes resulted in the development of "dead zones" (Rabalais et al., 2002). However, in contrast to all those negative consequences, nutrient enrichment, even by the sewage- and aquaculture-based waste, was found to enhance tropical mangroves to some extent (Clough et al., 1983; Boto & Wellington, 1983; Robertson & Phillips, 1995; Feller et al., 2002, 2003; Lovelock et al., 2007).

Harmful effects of increased nutrients on mangrove forest cause less investment in terms of nutrients support in roots and higher in above ground parts of plants resulting in higher growth rates (Grime, 1979; Tilman, 1991). Prolonged exposure to increased amounts of nutrients prompts mangroves to invest more nutrients in the roots for tolerance against draught conditions because of greater susceptibility of mangrove plants toward environmental stressors (Chapin, 1991). Enhanced salinity triggers mangrove vegetation to allocate more carbon to the growth of roots in comparison with shoots and this pattern of growth is also amplified by the decrease of humidity (Ball, 1988a, 1988b; Ball et al., 1997). Instability developed within mangrove forests owing to increased availability of nutrients lowers their resilience to the oscillation of ecological variables and the roles of particular nitrogen-based nutrients for the growth of shoots relative to roots (Naidoo, 2009).

Nutrients that have phosphorus control the improvement of water relations under the influence of greater salinity (Clarkson et al., 2000) by regulating the water contents of leaves, photosynthetic efficiency at using water (Lovelock et al., 2007; Lovelock & Feller, 2003), and hydraulic conductivity within stems (Lovelock et al., 2006, 2007). Thus, the ratio of available nitrogen:phosphorus at sites enriched in nutrients may be important in determining the resilience of mangrove forests to periodic development of hypersaline soils under conditions of low rainfall and humidity. The mortality of mangrove trees alongside reduction of canopy size results in retardation of primary productivity and a reduction of the power of carbon sequestration relatively impacts more on ecosystem functioning, especially in drought-affected and scrub-forest-dominated mangrove forest cover (Cintron et al., 1978).

The aforementioned research outcomes bring forth two hypotheses: (1) mangroves invest less in roots, resulting in more mortality after being exposed to higher availability of nutrients in contrast to maximum investment in aboveground biomass canopies; and (2) higher mortality of mangroves due to differential allocation of nutrients is very much site specific because of in-site fluctuating rainfall of shorter duration, low humidity, and higher salinity in the soil and water. These hypotheses can thereby explain differential patterns of mortality of mangroves in seaward and landward forests. Nutrient-mediated instability mainly because of coastal eutrophication tends to have far-reaching ecological consequences by disrupting the natural process of functioning of the mangrove ecosystem in view of ongoing and future climatic conditions.

Several ecological attributes and plant-mediated processes regulate nutrient cycling within mangrove forests, which include primary productivity, nutritional quality of plant tissue, efficiency of plants for nutrient utilization, nutrient resorption, sclerophylly, decomposition, and investment of nutrients for plant defense. Inherent functional abilities of plants influence nutrient cycling through uptake, allocation, utilization, and loss of nutrients (Hobbie, 1992). Spatial variation of the species composition shows differential efficiencies of nutrient-conservation mechanisms, as revealed from research information from temperate and tropical latitudes (Aerts, 1990).

Some mangrove forests with specific plant species in low-nutrient environments become more competitive because they can show more nutrient conservation mechanisms by acquiring the ability to use, retain, and recycle nutrients effectively for the better functioning of the ecosystem (Aerts & Berendse, 1988; Aerts, 1990). In mangrove forest soil with low fertility, several plant species are found to develop the power of adaptation to overcome problems of nutrient deficiency, including low photosynthetic capacity, being evergreen, sclerophylly, high shoot:root ratio, a lower amount of leaching from plant tissue, long life-span, greater investment in structural and chemical defense, minimum losses to herbivory, sluggish growth rates, higher rates of nutrient-uptake, nutrient-resorption and nutrient-use efficiencies, symbiotic mycorrhizal relationships, and the low decomposability of litter (Beadle, 1967; Grubb, 1977; Shaver & Melillo, 1984; Medina et al., 1990; Hobbie, 1992; Aerts, 1995).

3.9.2.1 The Microbial Loop in the Nutrient Cycle

Most of the microbes (bacteria and fungi) in the marine ecosystem play important roles as decomposers, which decompose the waste organic matter and dead organisms by breaking them exo-enzymatically and thereby recycling most of the chemical elements that remain within them, forming a microbial loop, which derives its nomenclature because of its connections with the classical food chain and food web processes through complex relationships. A large amount of suspended particulate organic matter (**POM**) is also produced by the breakdown of dead organisms, after experiencing the shedding followed by breaking of exoskeletons, release of mucus, and dumping of solid waste matter. The excretion of zooplankton and other faunal components within the mangrove ecosystem produce dissolved organic matter (**DOM**) along with the leaking out of phytoplankton. The metabolic processes of microbial components (bacteria, fungi, and protozoa) utilize **POM** and **DOM**, which in turn are consumed by zooplankton and complete the connectivity among different components of the microbial loop in the normal food chain.

3.10 Mangroves and Man: Construction of Boats, Harvesting of Honey and Catching of Fishes

The diversity, density, and extent of the distribution of mangroves of Sundarbans have been dwindling because of nonscientific aquacultural practices, the colossal loss of green coverage through indiscriminate felling of the mangrove plant species for food (honey and fruits), timbers, fuel-shelter materials, branches for erecting frames for beetle leaf cultivation, etc., threatening the very existence of important species of mangroves. Different species of mangroves growing in a habitat with specific physico-chemical attributes that are now on the verge of extinction because of changing land-use patterns, i.e., conversion of agriculture, for aquaculture, and reclamation of mangrove delta for residential requirements for the island-dwellers in the Sundarbans.

The significant contribution of mangroves by acting as a "**Nursery Ground**" for so many commercially important shell fish and finfish mainly takes place by the autochthonous supply of nutrients in the form of decomposed products of mangrove leaves. This kind of detritus-based substance loaded with nitrogen-based nutrients also plays an important role in nitrogen enrichment in the global context. The higher biological productivity of the mangrove ecosystem generates many ecosystem goods and renders many ecosystem services to society, as mentioned below, that are directly connected to the availability, enrichment, and cycling of nitrogen.

1. Help to develop coastal fisheries by way of providing materials for food, shelter, and medicine and also acting as nursery grounds for a number of both permanent inhabitants and also temporarily commercially important species that spend some phases of their life cycle among mangroves (Crona & Rönnbäck, 2005; Aburto-Oropeza et al., 2008; Chakraborty, 2018);
2. Provide access to varieties of food sources (crabs, fish, mollusks, fruits, leaves, and honey), wood products (firewood, charcoal, and timber for construction), and nonwood products (tannins, wax, herbal remedies dyes, thatch, fodder), which are closely associated with the livelihood of human beings living in or near mangroves (Bandaranayake, 1998; Glaser, 2003; Walters et al., 2008; Hussain & Badola, 2010; Warren-Rhodes et al., 2011
3. Act as phytoremediators by extracting toxic substances such as heavy metals and trapping followed by subsequent storage of sediment, organic matter, nutrients, from surrounding soils and waters (Alongi & Mckinnon, 2005)
4. Potential agent for efficient carbon sequestration because of allocation of high carbon in below-ground biomass, ensuring higher primary productivity, reduced sediment respiration rates, promoting long-term organic carbon burial within sediments, and providing substantial exportation of refractory dissolved organic carbon to the adjoining marine waterbodies (Komiyama et al., 2008; Kristensen et al., 2008; Donato et al., 2011; Duarte et al., 2013).

Ecological succession points out that different mangrove-plant species display different preference of their habitat choice as revealed by the presence of mangrove grass, *Porteresia coarctata*, as a pioneer species to stabilize the newly deposited sediment caused by the resultant forces developed because of the counteracting effects of maritime flows with riverine flows. Mangrove trees such as *Avicennia*, *Ceriops*, and *Excoecaria* stand at the confluence of the river mouth, whereas mangrove palms *Phoenix* and *Nypa* prefer to inhabit in the bank of tidal inlets, creeks, and rivers that have a fresh water influence. Similarly, mangrove plants such as *Sonneratia, Bruguiera*, and *Ajiceros* in the deeper parts of the Sundarbans mangrove ecosystem with their thousands of flowers stamens are exposed at night to tempt the bats and insects for the process of pollination. Bees prefer to build their nests on the branches of *Ajiceros* spp. for the production of honey and wax.

The life of the peoples of Sundarbans is intermingled with the catching of fishes utilizing their own traditional knowledge and locally available fishing appliances.

They are well acquainted with how to construct their fishing boats using the trunks and branches of mangrove tree plant species. The keels of the boats are constructed by the trunks of large and mature *Avicennia* spp., which can tolerate higher salinity. Similarly, trunks of mangrove trees such as *Ceriops*, *Rhizophora*, and *Xylocarpus* are used for creating knees, masts, and planks respectively. In addition, good-quality smokeless charcoals with high calorific value are produced by the local peoples from the mangrove-plant species *Rhizophora*. In order to stop the problem of over-exploitation of mangroves, dwellers of this region should adopt an integrated approach through aquaculture–/agriculture–/silviculture systems for the sustainable usage of mangrove resource bases alongside fulfilling the basic requirements of local peoples in respect of food, fodder, shelter materials, and fuels.

The products and services provided by mangroves are not all easily quantified in economic terms. However, products are harvested directly such as wood from the forest and other derivative food items such as crabs, gastropods, bivalves, shrimp, fish, honey, etc. (Pauly & Ingles, 1999). Another important commodity is fuels derived from mangrove wood, either as charcoal or firewood. The forest products are also used as the primary material for the construction of boats, the small huts of rural people, and even some furniture. Different species under the genus *Rhizophora* are used commercially around the world, especially in the different countries of Asia (Bandaranayake, 1998). Besides wood, other parts of mangrove trees such as bark have traditionally been used as a major source of tannins, which are used as preservatives as well as dyes. The pneumatophores of several mangrove species are used for making fishing floats and corks, and as a soap substitute. Mangrove-derived organic substances are used to produce cosmetics, soaps, and synthetic fibers. The fruits, seeds, seedlings, and tender leaves of *Avicennia marina* and the vegetative parts of other mangrove species are consumed by humans as food (Bandaranayake, 1998). Moreover, the potential of several species of mangroves as medicinal products, especially for treating rheumatism, dysentery, and malaria, have been documented. The toxic substances derived from mangroves have been used as pesticides, and antimicrobial substances (Bandaranayake, 1998). The mangrove forest and its adjoining landscape, with hundreds of charismatic wild faunal components (reptiles, birds, and large carnivores) make it a place that is attractive to humans and therefore paves the ways for promoting ecotourism.

3.11 Ecology of Mangroves: Zonation and Succession of Mangroves of West Bengal Coast, India

The coastal mangrove-estuarine ecosystem of Sundarbans supports the lives of a galaxy of biodiverse components in the form of halophytic mangrove plants and their associated vegetation, plankton, benthos, wildlife, and microbes, which through trophic interactions among themselves and also with their ecological variables demonstrate distinct ecological succession in order to help the proper functioning of the ecosystem and ensure the eco-dynamics of the said ecosystem in tune with the changing ecological gradients (Bunt, 1991, 1996, 1999; Bunt & Bunt, 1999; Chakraborty, 2011, 2017). The zonation and succession patterns are determined by nature's own mode of activities harmonizing and synchronizing the morpho-anatomical excellence and biological attributes of mangrove plants, ecological characteristics of sediment and water, tidal amplitude and intensity of water currents, the depth and volume of water, sloping gradients of the intertidal and subtidal bottoms, duration of tidal inundation and exposure periods, etc., the combined impact of all those ecological attributes tend to determine the distribution, diversity, density, and zonation of both flora and fauna through ecological succession.

3.11.1 Mangrove Zonation and Development of Ecotypes

Mangroves colonize in specific ecological zones along the coast such as on estuaries, deltas, lagoons, and islands after being determined by geomorphological, topographical, and hydrological features within each of these eco-settings, pointing out some mangrove ecotypes. Five of the most familiar ecotypes are the riverine zone, the basin, the fringe, the overwash, and dwarf-forest ecotypes (Lugo & Snedaker, 1974; Twilley, 1988). A riverine forest occurs along the estuarine tracts and is structured in terms of dimension and density by the influence of periodical tidal inundation and exposure supplying mostly nutrient-rich brackish water. Drainage of shallow depressions formed in the interior of mangrove-forest zones because of continuous interaction of the forest floor with tidal actions, called drainage basin forests, are characterized by slow-flowing stagnant water. A fringe forest bordering the protected shorelines, tidal canals, creeks, and lagoons experience total inundation for at least half a day by daily tides because of its occurrence at lower parts of intertidal zone. Overwash forests occurring on those islands that get frequently washed over or inundated by tides. Dwarf forests, also known as scrub forests, mainly grow in the upper parts of intertidal zones getting prolonged exposure and higher evaporation, where hydrology is characterized by high salinity and oligotrophic conditions. Each of these mangrove ecotypes is endowed with different patterns of plant assemblages, biogeochemistry, and biological productivity, which are determined by a combination of different ecological factors and processes such as temperature, salinity, soil properties, tides, rainfall, freshwater discharge, biological interactions, and other forms of natural disturbances such as storms, floods, freshwater flooding, etc. (Lugo, 1980; Jiménez, 1992; Ellison et al. 2002; Macintosh et al., 2002; Dahadough-Guebas et al., 2005).

Ecological zonation patterns of several plant species exhibiting linear patches occur in many parts of the Indian Sundarbans such as Lothian, Bhagbatpur, Jambu island, Patibania, Saznakhali, Sajelia, and Kultali Gang, whereas other species are found to display nonlinear zonation patterns along the shore fringes of the south-eastern part of the Indian Sundarbans (Matla Darbhanga complex). Dispersal of seedlings over long distances by floodwater during spring tides, coastal erosion, cyclonic storms, changing the shore profile with the development of some adversities within soil conditions and the non-availability of proper protection measures against tidal currents, etc., thereby result in a wide range of distribution of these species in the mangrove swamps.

3.11.2 Dispersal of Propagules and Establishment of Seedlings: Prerequisite for Zonation and Succession of Mangroves

Spatial and temporal variation in the availability of propagules tend to have a striking influence on shaping the community structure through post-settlement biotic interactions and this phenomenon is designated supply-side ecology. The tidal sorting hypothesis (TSH) as proposed by Rabinowitz's (1978) call for supply-side dynamics, which explains the significance of tidal-zonation patterns of different species of mangrove-tree species experiencing and withstanding the interacting effects of the size of the propagules and the depth of the water in respect of the dispersal and establishment of mangrove plants. According to this hypothesis, smaller propagules are carried by the tidal water more toward the inland regions than the large propagules determining the settlements of small propagules in higher numbers at the landward direction or supratidal zones. The larger propagules settle down in areas more toward the seaward direction as they are able to withstand troubles from the higher intensity of the water current and finding greater access to the soil surface. In accordance with the TSH, the large propagules of *Rhizophora* rooted and persisted better in this turbulent habitat than did the smaller propagules of the other species.

Dispersal patterns of propagules not only determine the zonation, multispecies assemblage in contrast to the population of individual species of mangrove plants across the sloping gradients as well as the duration of tidal inundation but also effectively control population dynamics along with structural entities of the biotic community (Hutchinson, 1951; Skellam, 1951; Andrewartha & Birch, 1954; Huffaker, 1958; MacArthur & Wilson, 1967; Krebs et al., 1969; Morgan, 2001; Underwood & Keough, 2001). Research evidence points out the importance of the juvenile recruitment of intertidal benthic fauna and demersal fish in determining larval production and release, dispersal, settlement and metamorphosis, population dynamics, and biotic community interactions (Young, 1987; Underwood & Fairweather, 1989).

The term "**supply-side ecology**" was coined for the development of a hypothesis that highlights the ecological systems that have a specific community structure comprising local assemblages of different species linked by dispersal and influenced by the supply of new individuals arriving and recruiting to a site (dispersing larvae or diaspores) rather than by post-recruitment biotic interactions (Lewin, 1986). This idea was formalized in a number of mathematical models taking into consideration (1) measurement of seed-dispersal patterns, (2) correlation of these patterns with spatial or temporal variation in the species composition and structure of vegetation, and iii) experimentally manipulate seed shadows to confirm a causal relationship between seed-dispersal patterns and the distribution and structure of plant assemblages. Earlier supply-side theories of community structure as postulated by Rabinowitz's (1978) TSH, which attributes the differential distribution of mangrove-tree species with tidal elevation (i.e., zonation) to the interacting effects of water depth and propagule size on propagule dispersal and seedling establishment. This hypothesis invoked species-specific physiological specialization to differing edaphic conditions, which explains mangrove zonation (Rabinowitz, 1978; Watson, 1928; Macnae, 1969; Chapman, 1976; Smith, 1992). This hypothesis also postulates that the species were physiologically adapted to grow and survive under distinct regimes of tidal inundation as well as soil- and water-quality parameters. The duration of maintaining the buoyancy of mangrove propagules varies among different species of mangroves and different intensities of soil and water with water parameters (Rabinowitz, 1978). During floating, experiencing directional dispersal in the tidal currents, the propagules accumulate moisture and are deposited as per their sizes and weights and result in mangrove zonation. It was hypothesized that tidal currents during higher spring tides cause dispersal of smaller propagules farther inland than larger propagules and larger propagules are physically restricted from dispersing into these shallower areas. Their mangrove assemblages in the shallower parts of inland areas are characterized by having smaller propagules, whereas the larger sized propagules are produced by those mangrove plants that are settled and grow in deeper seaward zones.

Research investigations have revealed in respect of TSH about the sizes of propagules of mangrove-plant species is that the order of ranking is *Laguncularia* > *Avic ennia* > *Rhizophora*. The enhanced buoyancy hardly increases the distance for effective dispersal of the propagules because of their viability, which has declined over time because of several factors, including the biotic interactions such as insect feeding and tissue decay (Wayne, 2007). The predominance of local establishment with specific life-history attributes coupled with long-distance dispersal renders an important connotation toward shaping the ecology and evolution trends of mangroves. The rooted seedlings of most dispersing propagules settle and grow nearer to the parent mangrove plants and thereby determine species composition of the mangrove, which in turn strongly affects the patterns and rates of light-gap regeneration as a part of the primary mode of canopy-tree replacement. The changing patterns of longer distance dispersal of different mangrove species overcoming ecological constraints have appeared to structure the biogeographic distribution, diversity of species, and evolution of mangroves (Duke, 1995; Duke et al., 1998, 2002; Ellison et al., 1999; Cain et al., 2000; Ellison, 2002). In this context, it can be concluded that the salinity and size of the propagules coupled with buoyancy determine the direction of their movement and range of dispersal, especially in the estuarine environment, which experiences salinity gradients and tidal interplays.

However, TSH, after its first publication (Rabinowitz, 1978) to explain the mechanism of zonation of mangroves also encountered a number of criticisms, which mainly belong to three categories. **First**, physiologists challenging the ability of mangrove species to tolerate the edaphic conditions in explaining zonation patterns do not find strong support.

Second, the TSH advocates the domination of mangrove species in inland zones with smaller propagules, whereas species with relatively large propagules settle and grow in seaward stands.

Third, the TSH has been criticized for its assumption that the shallow water of the estuarine networks experiencing tidal influence with differential elevations restricts the upper distributional limits during landward movement of larger propagules carried by rising tides, which is in contrast to the dispersal of smaller propagules toward inland areas.

In this context it is opined that dispersal limitation ultimately determines the structuring of inland forest stands at higher elevation. The settlement of larger propagules at the seaward boundary of the forest where there is deeper water encounter the intensity of currents and waves. In addition, interacting biotic and abiotic factors and processes tend to determine species composition, distribution, and modes of regeneration of mangrove-plant species spatially and temporally.

3.11.3 Ecological Zonation in the Sundarbans Mangrove Ecosystem: Underlying Scientific Background

Determining underlying mechanisms for species composition, abundance, and distribution have been, and continue to be, one of the major concerns of community ecologists for identifying and documenting the scientific facts in this regard (Connell, 1961; Pielou, 1977; Bertness & Ellison, 1987; Dale, 1999). Distinct and pronounced zonation patterns displayed by the halophytic shrubs and trees of the tropical mangrove ecosystem exhibit a definite ordering of location of one species, or groups of species (Watson, 1928; Chapman, 1944, 1976, 1979; Macnae, 1969; Snedaker, 1982; Mukherjee, 1992) in accordance with the gradient of sloping and elevation of intertidal zones restricting a group of species at the lower limit whereas the upper limit is marked by the presence of other groups of vegetation where some overlapping of the territory of some species are encountered (Whittaker, 1967). Such a well-designed zonation pattern restricting one group of mangrove species to a definite zone or territory and replaced by another group enjoying another territory or zone is attributed to the ability of one particular species to respond, adjust, and adapt to different eco-biological attributes such as salinity, duration, and intensity of tidal inundation and exposure, edaphic gradients, elevation, sloping gradients across different zones of intertidal belts (Snedaker, 1982; Smith, 1992), but biotic factors only contribute in structuring the zonation to the least possible extent (Ball, 1980).

The Sundarbans mangrove forest, covering an area of 10,000 km^2 in the Ganges Delta of India and Bangladesh supports the world's most diversified biodiversity components, unique in their adaptability and charismatic appeal such as the Royal Bengal tiger (*Panthera tigris tigris*) and the '**Sundari tree**' (*Herriteria fomes*) (Khan et al., 1990; Blasco et al., 1992; Chaudhuri & Choudhury, 1994; Choudhury et al., 1994; Hussain & Acharya, 1994). Different geomorphological processes have been regulating the extent and structure of mangrove forests (Thom, 1967, 1982) and other factors such as salinity, pH, temperature, etc., with their spatial or temporal variation tend to limit distribution or abundance of different species (Ball, 1980, 1988a, 1988b; Ball & Pidsley, 1995). The Sundarbans mangrove forest is characterized by a complex network of branching and meandering distributaries and rivers varying from a few meters to several kilometers in width, and a collection of low-lying, shifting islands. The mangrove forest has been grown on the deltaic islands of the geologically recently developed Bengal basin, which was established by the deposition of quaternary sediments by the three major river systems of the Indian Subcontinent: the Ganges, Brahmaputra, and Meghana rivers (Bagchi, 1944; Morgan & MacIntire, 1959). However, the modes of subsidence in different parts of the Bengal basin did not follow the uniformity across the delta because of the uneven rates of sediment deposition as well as differential tectonic activity. The eastern parts of the deltaic Sundarbans experienced more rapid subsidence than the western side, which resulted in an apparent western-to-eastern downward tilt, facilitating the flows of water through the estuarine networks of the Sundarbans (Deb, 1956). This kind of drainage pattern shifts more toward an eastward direction along with the western-to-eastern tilt, resulting in a considerable decline in freshwater flow into the western side of the Sundarbans, resulting in higher salinity gradients at the deltas located in the western sector.

Different species of flora and fauna display different zonation patterns within the mangrove ecosystem of Sundarbans along the intertidal slope, reflecting their adaptation to different degrees of territoriality in tune with changing ecological parameters, viz. sediment texture, nutrient load, salinity gradients, turbidity, duration of tidal inundation and exposure, sizes and shapes of propagules, ability to withstand anoxic conditions, food and feeding habits of fauna, biotic interactions, etc. (Chakraborty & Choudhury, 1992a, 1992b). Mudflats are developed because of the continuous deposition of sediments due to the interaction of incoming seawater with turbid freshwater of riverine flows in some sheltered areas with shallow water coupled with low velocity of water movement, leading to higher deposition of soil particles, thereby causing formation of the deltaic islands of the Sundarbans. The finer sediment, rich in organic content, mainly composed of small silts and clays, constitute the thick mudflats, which enable the roots of the newly settled plant species to penetrate the soil. At the beginning, thick algal mats settle on the surface of these newly developed mudflats followed by the settlement, growth, and proliferation of mangrove-associated saline grasses, which can penetrate very easily their adventitious roots on the soft bottom and arrest the soil at the site of their settlements, avoiding soil erosion and stabilizing the mudflats (Ginsburg & Lowenstam, 1958). Consequently, luxuriant growth of other mangrove plant species occurs, displaying distinct zonation and succession patterns, replacing one group of plants with others with the passage of time. The replacement of one species of mangroves with another in a particular mudflat brings about distinct patterns of mangrove species assemblage, characteristic of a particular eco-zone and this kind of development of vegetational stands consolidates, elevates the sloping grades of the substratum with the development of shades, facilitating definite biogeochemical cycling, creating new ecological habitats and niches, making the way for other fauna and flora to survive in the Sundarbans mangrove ecosystem (Chakraborty, 2013). In Sundarbans, the zonation of mangrove plant species along with their associated vegetation demonstrated the occurrence of the saline grass, *Porteresia coarctata* at the seaward edge of the forest, followed by the landward settlements of *Avicennia officinalis*, *Suaeda maritima*, *Aegiceras corniculatum*, *Acanthus ilicifolius*, *Sonneratia apetala*, *Ceriops roxburghiana*, *Bruguiera gymnorhiza*, etc. Two important mangrove plant species, viz. *Heritiera fomes* and *Nypa fruticans*, require more freshwater and generally occur above the mid-tidal level together with *Phoenix paludosa*, a major mangrove palm species. Mangrove species such as *Excoecaria agallocha*, *Ceriops decandra*, *Rhizophora apiculata*, and *Bruguiera gymnorhiza* are found to inhabit in a relatively landward direction, experiencing a higher intensity of currents and longer exposure. The zonation patterns in mangroves are considered more steady-state adjustment rather than displaying successional stages (Lugo & Snedaker, 1974).

The succession of the group of vegetation as observed in the Sundarbans, covering around 70% of the forest area is *Porteresia coarctata*, *Phragmites karka*, *Sesuvium portulacastrum*, *Avicennia officinalis*, *Acanthus ilicifolius*, *Aegiceras corniculatum*, *Avicennia alba*, *Avicennia marina*, *Sonneratia apetala*, *Bruguiera gymnorhiza*, *Salicornia brachiata*, *Ceriops decandra*, *Heritiera fomes*, *Rhizophora apiculata*, *Xylocarpus mekongensis*, *Ceriops decandra*, *Excoecaria agallocha*, *Phoenix paludosa*, and *Suaeda maritima* (Chakraborty, 2011). Smith (1992) reviewed and inferred that determining ecological factors and processes resulting the ecological succession of different species as well as the forest structure of mangroves, highlighting the following: (1) response of species to geomorphological influences such as sea-level rise; (2) land-building processes driving the autogenic succession of mangroves; (3) physiological adaptations to stress gradients such as soil, salinity in the intertidal zone; (4) dispersal patterns of mangrove propagules in accordance with size and rooting time; and (5) post-disposal processes determining the establishment of propagules, such as predation and competition.

Different mangrove species have different optimal requirements of salinity, nutrient support, and soil textural components (Hatchings & Saenger, 1987), depicting correlations between properties of soils and recruitment, settlement, establishment, and dominance of different species (McKee, 1995). Tidal dispersal and retention of propagules in relation to their proximity to the parent trees also influence the initial community structure (McKee, 1995). Different kinds of paleogeological events determining the deltaic formation through differential subsidence as well as directing the flow patterns of different estuaries of the Sundarbans mangrove ecosystem resulted in a shift of *Heritiera fomes*, *Nypa fruticans*, and *Excoecaria agallocha*, and to a lesser extent *Ceriops decandra* and *Sonneratia apetala*, principally reflecting the west-to-east and south-to-north gradients of freshwater flow patterns, which have a considerable impact on deciding the salinity profiles. These observations indicate that edaphic variables are not that important in deciding the distribution and zonation of mangrove plants in the Sundarbans, but propagule dispersal coupled with interspecific competition appear to be more important in deciding the relative abundance of co-occurring species, the size and structure of the forest, and the distribution of different species inhabiting the mangrove ecosystem. In the case of overlapping zonation, where several species occur in a certain part of the intertidal zone, forming an assemblage of mangrove vegetation, displaying an ideal example of species co-existence, but varying degrees of tidal elevation and changes in edaphic factors bring about changes in modes of settlement, establishment, interspecific interactions, and zonation patterns in different eco-zones across the coastal environment (Gleason, 1926; Bunt & Stieglitz, 1999).

3.12 Mangrove Fauna, Food Chain, and Food Web: Roles of Organic Matters

The food chains and food web dynamics of the mangrove ecosystem are driven profusely by the production, transport, and modes of utilization of mangrove-derived organic matter (OM) which wide range of variations with forest type, the availability of amount of clear open waters, as mangrove ecosystems possess a high proportion of forest floor to open but turbid water (Blader et al., 1985). In the mangrove ecosystem, carbon, after being fixed by mangrove plants, contributes profusely to the production of OM and ultimately make it available to the mangrove food web. Other primary producers such as planktonic and benthic algae, also add to the production of nutrients in addition to decomposed products of mangrove leave litter and OM supplied by riverine flows, and thereby impact the total productivity of the ecosystem (Warren & Underwood, 1986; Samanta et al. 2018).

Food chains and their interconnections in the mangrove help the transfer of energy and materials from primary producers (autotrophs) through a sequence of trophic levels (a number of herbivores, omnivores, and carnivores) and thereby result higher biological productivity. Two types of food chains are present, the first one is one grazing food chain, in which energy flows start from the primary producers through repeated eating and being eaten processes and the second one, is the detritus food chain, which starts from dead and microbiologically decomposed organic substances through feeding by detritivores and subsequently by carnivores. The food web, on the other hand, represents the interlinkage of several food chains within an ecosystem and ensures the functioning of that ecosystem through energy transfer and the cycling of matter. The complex food web of the Sundarbans mangrove ecosystem interconnects the forest subsystem and estuarine subsystem by continuous inundation and exposure during high tide and low tide, which ensures the continuous exchange of materials between the two subsystems that constitute the mangrove ecosystem, making this ecosystem one of the most productive in the world. This productive ecosystem harbors a galaxy of biodiversity components, including endangered wild charismatic faunal species such as the Royal Bengal tiger, the estuarine crocodile, the olive ridley turtle, gangetic and irrawabati dolphins, otters, fishing cats, different species of snakes (pythons, cobra, etc.), and birds (Chandra et al., 2020). All of these wild species inhabit different ecological niches of the mangrove habitats, including different parts of mangrove trees –trunk, branches, leaves, flowers, and fruits.

The Sundarbans mangrove ecosystem, along with its food web, is endowed with a vast expanse of green perennial mangrove forest with an extensive plant community of sea grasses, salt marshes, mangroves, and macro algae, and a vast sheet of estuarine water that has its own structural biological components such as phytoplankton, zooplankton, shell fishes and finfishes, reptiles, and aquatic birds.

The input of energy from the forest to water is in the form of mangrove litter-contributed detritus and promotes the growth and flourishing of detritivores (brachyuran crabs, gobiid fish, etc.), herbivores (meiobenthic fauna, copepods adult and larval insects, gastropods, etc.), omnivores (benthic arthropods including crustaceans and amphipods, polychaetes, and demersal fish), primary carnivores (polychaetes, larvae of fish, turtles, etc.), secondary carnivores (aquatic birds, catfish, etc.), tertiary carnivores (sharks, catfish, predatory fish, reptiles, etc.), and decomposers (protozoa, bacteria, fungi, etc.), respectively representing each of the feeding groups.

3.13 Structural Components of the Mangrove Ecosystems

Green plants, being primary producers, are able to synthesize complex organic compounds from inorganic substances by photosynthesis in the presence of sunlight, and this group of biotic components include phytoplanktons, benthic algae, saline grasses, and mangrove vegetation. Autotrophic algae play key roles in contributing a major portion of the biological production of the mangrove ecosystem followed by halophytic mangrove plant species. Ecological succession, as well as vertical and horizontal distribution patterns of mangrove plant species, including algae, are governed by several environmental processes and factors such as tidal influences, intensity of wave action, salinity profiles, temperature condition, range of light penetration, moisture recharging, levels of desiccation, and sediment load in the form of turbidity. (Farnsworth & Ellison, 1996).

The trophic level represented by with the group of animals as consumers take their nutritional support from eating other organisms as food and they are classified into three different categories based on their food preferences: (1) herbivores; (2) omnivores; and (3) carnivores.

Herbivores are those group of animals that depend entirely on plant species for their nourishment with food and energy. Different groups of meiobenthic fauna (nematodes, copepods, ostracods, etc.), insect larvae, bivalves, gastropods, polychaetes, sipunculids, echiurids, etc., belong to this group. Omnivores represent the type of trophic level where the animals feed on heterogeneous types of food items including both plants and animals. Different species of intertidal macrobenthic fauna such as brachyuran crabs, polychaetes, amphipods, shrimps, mollusks, fish, and a number of meiobenthic fauna are examples of omnivores. Carnivores eat the flesh of other animals by predating them. They are also categorized into three different types on the basis of their food habits: (1) primary carnivores such as small fin fish, fin-fish larvae, eels, catfish, and soles; (2) secondary carnivores such as birds including kingfisher, heron, wader, sea gull, as well as large catfish; (3) tertiary carnivores including large and highly energetic predators, for example, fin fishes such as perch, shark, tarpon, as well as birds such as sea-gull, eagle, etc. Out of several species of fauna dwelling in the mangrove ecosystem fish and birds are considered important consumers in the mangrove ecosystem, consuming algae, plankta, both meio- and macrobenthic fauna, and other species of fin fishes (Brenchley, 1982; Berendse & Aerts,1984; Williamson et al., 1994). Decomposers represent those organisms that play their functional roles in converting the biomass of the mangrove ecosystem into organic and inorganic matter. They include fungi, bacteria, and protozoa.

3.14 Ecological Resilience and the Mangrove Ecosystem

Ecological processes involving different feedback from interaction among biota and their surrounding ecological variables have important consequences for the ecodynamics of the ecosystem (Rietkerk & van de Koppel, 2008). Several self-organized patterns are observed to develop from such interactions operating from the molecular to the ecosystem level, involving biological, geo-morphological, chemical, and physical processes, which act as prerequisites in maintaining ecological resilience.

Ecological resilience refers to the amount of disturbance that an ecosystem can withstand without hampering the normal ecosystem functioning by bringing back the stability of the ecosystem after being disturbed for some time. The roles of interrelationships and interconnectivities within the ecosystem in keeping the feedback loops going ensure the maintenance of the ecosystem's resilience. However, any kind of environmental change impacting the biotic components involved in the ecosystem trigger them to exhibit threshold responses that tend to shift the natural processes of interactions within an ecosystem. The interactions among biological (abundance of large bioturbating macrofauna, micro phyto-benthos, and detrital matter), geological (sediment grain size), and physical processes (tides, waves, currents, density across environmental gradients, and ecological feedback in order to understand the potential to improve our empirical understanding of threshold responses to change in ecological systems. In order to understand the risk of change to ecosystem function or resilience in respect of ecosystem dynamics, assessment of threshold responses implies adjudging the potential for cumulative effects of multidimensional stresses or disturbances. Such underlying ecological principles help to chalk out strategies for environmental management, restoration, and conservation policies by emphasizing the maintenance or enhancement of ecological resilience.

The ecological consequences of chronic multiple stressor impacts can be understood by assessing feedback responses developed in the process of interactions, especially positive (facilitatory) processes across scales of space and time that generate ecological resilience through buffering the capacity of a system to respond to change. Mangrove ecosystems are subjected to multiple stress factors, which are derived from the alteration of natural set-ups of both land-based forests (sediment and nutrient loading) and the sea (ecosystem effects of nutrient loading) and estuaries (ecosystem effects of fishing and climate change) along strong environmental gradients, which promote environmental variation, leading to modified stress responses.

3.15 Driving Forces Controlling the Ecological Quality of Natural Water: Chemicals and Growth Factors

The chemical properties of natural waters are controlled by a number of natural age-old processes such as rainfall, evaporation, erosion, sedimentation, continuous inputs of terrestrial materials, and utilization of materials within the water bodies. The diurnal, seasonal and long term annual cycling of the major elements within the water bodies are determined by the lively interactions of biological components with chemical substances.

Water, being the universal solvent, hardly occurs in pure form but accommodates a wide range of inorganic and organic substances in the form of dissolved gases and solids and ultimately provides the growth factors for the survivability of aquatic organisms, which in turn excrete in the form of ammonia, phosphate, carbon dioxide, and a large percentage of everything that has been consumed by the aquatic biota.

3.15.1 Organic and Inorganic Compounds

The organic matter, after being produced, is stored and transported by the influence of a number of eco-geological attributes of the water bodies, factors such as, **first**, by the interactions of the structure of water channels and floodplains involving the prevailing inorganic constituents of sediments. **second**, POM, being a key basal resource in lotic water bodies, affects the biological productivity of the local ecosystem. However, increasing complexity in channel morphology limits the power for carrying organic matter at a higher speed, allowing it to be retained, processed, and consumed within reach. **Last**, filter-feeding organisms trap fine organic particulates, which are processed through the food web. Nutrients with inorganic chemical constituents provide the chemical support to the entire food chain where the nutrients are utilized by the biotic organisms but certain ones are converted by microorganisms to organic growth factors such as vitamins.

A vast array of dissolved gases and inorganic materials that are present in natural estuaries-mangrove forests can be divided into two major categories: (1) detritus, semi-decomposed derivatives from the organic substances deposited on the bottom of the mangrove forest as well in as the estuaries; (2) others are composed of fractional elements derived from dead organic biota coated with bacteria, fungi, and small protozoa. These bio-assemblages, having bio-active substances within the lotic ecosystem, impart considerable regulatory impacts on the water and soil characteristics by virtue of their distribution in relation to depth, volume, density, current, substrate, and food availabilities in different aquatic habitats and niches (Pradhan et al., 2008; Chakraborty, 2013; Chakraborty et al., 2014, 2021; Annon, 2010).

3.15.2 Nutrients Dynamics and the Mangrove Ecosystem

The nutrients, which are present in small amounts, appear to be important for ensuring the growth of plants and animals. They render threefold contributions within the aquatic ecosystem: (1) as both major and minor nutrients for biota; (2) impart an ionic effect as growth factors; (3) as toxicants causing harms to both the biota and the ecosystem. The common nutrients required in higher amounts for the development of cells and the growth of organisms include $\mathbf{O_2}$, $\mathbf{CO_2}$, **Fe**, $\mathbf{NH_4}$, $\mathbf{NO_3}$, $\mathbf{PO_4}$, $\mathbf{SiO_2}$, and $\mathbf{SO_4}$. Important minor nutrients, which are needed in the living organisms in smaller amounts, include **Mn**, **Co**, **Mo**, **Cu**, and **Zn**, whereas sodium (**Na**), potassium, and chloride (**Cl**) ions are often required in larger amounts in aquatic ecosystems, but their ionic effect is restricted only at the surface of delicate membranes through which ions are exchanged. Other inorganic elements, such as copper or zinc, can act as either toxicants or as growth stimulators. A group of organic compounds produced by algae and some animals are also found, which alter the chemical state of the natural waters and in some cases act as chelating agents, changing the ionic state of metals, which may otherwise become toxic.

Nutrient dynamics are governed by many intricate trophic interactions coupled with hydro-geographic and ecological processes in the aquatic ecosystems, which also include bio-geological processes such as erosion and accretion, formation of landforms, precipitation, and runoff, development of vegetative cover, as well as bioturbation processes by faunal components.

The organic compounds produced from the plants as unwanted excretory materials stimulate or inhibit their own growth or that of their competitors. Aquatic organisms unable to synthesize growth factors such as vitamins depend on these materials for their metabolic needs from the products from organisms. Some organic compounds produced by freshwater blue-green algae or red tides of marine and brackish-water dinoflagellates also act as toxic substances. Aquatic fauna are often observed to communicate chemically in water as they do on land (Kremling, 1983; Hayes & Barry, 2008; Hasler, 1996).

3.16 Nutrient Cycle and the Sundarbans Mangrove Ecosystem

The Indian coastline, extending over 7000 km, in comparison with the world coastline, which spreads over about 350,000 km, displays a wide range of geomorphological types, ecosystems, and biodiversity. The coastal–estuarine–mangrove complex represents that part of the earth where land, estuaries, and atmosphere interact with each other, creating a shallow sheet of water bodies covering around 7% (26 × 106 km^2) of the surface of the global ocean, with a depth of <200 m. Representing only a very small part of the global oceanic environment, the coastal zone, with its estuaries, backwater bodies, and mangroves contribute considerably to the biogeochemical cycles of the globe, for three reasons: (1) a huge amount of nutrients, both organic and inorganic, are supplied to the aquatic systems from the land-based forest subsystem through runoff and groundwater discharge; (2) exchanges of a considerable amount of energy and matter with the open ocean; and (3) develops through an continuous interactive process as one of the most geomorphologically diverse, geochemically complex and biologically productive eco-zones of the biosphere. As per the latest estimates, it amounts to 14–30% of the primary production of the ocean, 80% of organic-matter deposition, 90% of mineralization in the sediments, 75–90% sink within the ocean for a suspended river load, and for ensuring 50% of the total calcium carbonate deposition (Seitzinger, 1990; Thamdrup & Dalsgaard, 2002). Moreover, being recognized as the breeding ground of so many shell fish and fin fish, this ecotonic environment produces 90% of the total fish catch of the world (Mulder et al., 1995). Recent estimates also reveal that this sensitive but productive area renders ecosystem goods and services to the tune of 43% of global values (Meybeck, 1982).

Mangrove forests being one of the most productive eco-zones of the world, occurring in the intertidal zones of the tropics and sub-tropics (31° N to 39° S), play important roles in coastal estuarine biogeochemistry by acting as nutrient filters bridging the gap between forest land and the adjoining estuary (Bouillon et al., 2008; Keuskamp et al., 2015). The nutrient cycles operating in the mangrove forest-based ecosystem also play significant roles in the total organic carbon (TOC) oceanic cycle along with nitrogen and phosphorus cycles, after being linked with the adjacent coastal water because of their ability to efficiently trap, uptake, and recycle suspended litter in the water column (Alongi, 1996; Twilley et al., 1986). The primary producers of the mangrove ecosystem includes mangrove trees, macroalgae, periphyton, and phytoplankton, contributing to total mangrove primary production, which varies with their relative surface cover, delivery, and turbidity. Community metabolism of mangrove trees displays considerable variation on both the local and the regional scales, primarily as a response to environmental processes and parameters such as tide, climate, water-quality parameters such as salinity, pH, turbidity, etc., soil texture and nutrients. Primary production in the water column is often limited by high turbidity and fluctuation of salinity profiles.

Research information pertaining to decomposition processes in mangroves appears to be critical for assessing the potential of mangroves to act as blue carbon sinks, both spatially and temporally. The roles of the types of vegetation, tidal interplays, faunal community structure, microbial processes, and several water and soil quality factors play an important role in decomposition. The significance of phenolic compounds in influencing the rate of decomposition in mangroves needs more attention. The formation of peat in coastal zones is expected to throw light on the better understanding of the carbon budget within mangroves, which tends to open up new avenues for the proper assessment of carbon sequestration through eco-engineering and therefore the suppression of the decomposition of organic matter through the strengthening of the enzyme latch, allowing mangroves to be used as a form of bio-geoengineering (Freeman et al., 2012). Moreover, the ever-increasing threats (oil spills, agricultural waste, industrial discharge, and global warming) to the mangrove ecosystem, disrupting the food chain–food web dynamics along with microbial processes, have become a serious concern in respect of the fate of mangroves (McKee & Faulkner, 2007).

Mangrove forests receiving inputs in the form of dissolved nutrients from land and sea barely maintain high productivity in the mangrove ecosystem. Only the autochthonous production of nutrients through the recycling of mangrove plants-derived organic matter help to meet this high nutrient demand (Holguin et al., 2001). Mangrove leaves containing up to 40% water-soluble components play a key role through their conversion into bacterial biomass in less than 8 h after falling onto the moisture-enriched muddy mangrove floor (Benner et al., 1986). The macrobenthic fauna such as crabs, mollusks, insects, etc., act as macrodegraders causing fragmentation of leaves and thereby increasing the scopes for microbes to act upon the degrading leaves for ready conversion to detritus, which are exported to the adjacent coastal waters. Fungi initially invade within the mesophyll tissue of fragmented leaf litter and bacteria thereafter complete decomposition through exoenzymatic processes, facilitating the cycling and transformation of nutrients. These kinds of material exchanges occur both in particulate and dissolved forms temporally and spatially and support coastal fisheries (Twilley, 1985; Wattayakorn et al., 1990; Wolanski et al., 1992; Furukawa et al., 1997).

Estuaries along with networks of brackish water creeks in the Sundarbans mangrove ecosystem act as a link between the marine environment of the Bay of Bengal and more than a hundred deltaic islands, both reclaimed and virgin, and receive a good amount of organic and inorganic nutrients. This ecotone, loaded with a lot of autochthonous and allochthonous nutrients, help contribute directly to enhance primary production, modify changes in habitat structure, and determine trophic dynamics of the entire ecosystem.

The modes of production, distributional patterns, and fate of nutrients produced in the water column are found to exhibit differences among the open ocean, estuaries, and mangrove-dominated deltaic islands. In addition, quantum of production and exchange of nutrients determine the trophic dynamics in a considerable way, decreasing the number of trophic levels and thereby markedly increasing primary production. This is substantiated from a research outcome where six trophic levels were noticed in oligotrophic waters, whereas in upwelling areas there were three trophic levels with more potential for primary production (Boto, 1982). Moreover, the prevalence of large cells among coastal phytoplankton in comparison with micro-phytoplankton and pico-phytoplankton dominating in the open sea and faunal components grazing over them produces large fecal pellets in comparison with small phytoplankton-grazing fauna releasing small fecal pellets, which are not exported from the photic zone efficiently as a result of low settling velocities (Boto, 1982). Production of large fecal pellets by the herbivorous and omnivorous fauna in the coastal –mangrove complex with a shallow water depth promotes primary production in the mangrove forest subsystem alongside simultaneous detritus production, which in turn triggers biological activity in the sediments. Out of the total biological production, around 30% is utilized for the respiration of organisms in the water column and an equal amount undergoes mineralization in the sediments.

The productivity of the coastal estuarine mangrove complex is mainly controlled by mangrove plants, macro algae, and phytoplankton on getting the supply of nitrogen or phosphate-based nutrients. The nutrient-rich water exported from the mangroves by a receding low tide when a comparatively well-mixed water body leaves the mangroves but is transported back at the initiation of the flood tide, which ensures the mixing up of the nutrient-poor estuarine water to spread out into the mangroves at high tide. However, several environmental problems within the coastal water are attributed to the introduction of excess nitrogen, leading to eutrophication, which results in toxic production of algal biomass as red and brown tides, destruction of the habitats of seagrass beds, changes in the species composition in the prevailing biotic community, loss of biodiversity and altered distribution of species, enhanced sedimentation, and drastic depletion of dissolved oxygen, leading to hypoxia and anoxia.

3.16.1 Detritus and Mangrove Litter: Detritus Food Chain and Food Web

Detritus, after having formed from the decomposition of mangrove litter, constitutes the primary food source in the detritus food chain in the mangrove ecosystem. The initial fragmentation and disintegration of freshly fallen mangrove leaves on moisture-enriched and microbe-loaded soft sediments on the floor of the mangrove forest by the macrobenthic and periphytonic faunal components make way for production of semi-decomposed organic matter in the microbial (fungi and bacteria) decomposition process, with the result of the production of detritus.

The nutrient-rich detritus-enriched mangrove forest floors promote the growth and reproduction of a hoard of benthic fauna (macrobenthos such as crabs, gastropods, bivalves, polychaetes, sipunculids, brachiopods, actinarians, etc., and a number of bottom-dwelling finfish such as gobiid fish, mullets, etc., which consume a good amount of detritus for generating energy for their behavioral activities. Higher production of mangrove plant-derived detritus and its simultaneous accumulation within the mangrove forest floor act as the prime source of energy in order to drive the eco-dynamics of the mangrove ecosystem.

Detritus, being the products resulting from the disintegration of dead organic matter, forms the main constituent of the detritus food webs in the Sundarbans mangrove ecosystem through some intricate interactive pathways involving phytoplankton, sea grasses and mangroves, and benthic and epiphytic algae, which constitute the major portion of primary producers in this estuarine system. Although herbivores depend entirely on primary producers, only a minor portion of the primary production in mangrove ecosystems is consumed directly by herbivores. Therefore, an enormous amount of plant detritus acts as the major source of energy for an array of faunal components, mostly detritivorous benthic fauna and demersal fishes, which drive the biological machinery of the mangrove ecosystem. After being formed from the mangrove litter, primary detritus is exported to adjacent aquatic systems by tidal waters. A major part of the living biomass does not undergo decomposition because it contains the complex organic matters lignin and cellulose, which even the macro-consumers hardly digest and are therefore are avoided as food items.

Detritus is considered the largest pool of organic matter in the biosphere (Hedges, 1992). The simplistic definition of detritus is that it accommodates all the particulate organic matter that was there to constitute living organisms (Odum, 1971). In another way, detritus can be defined as a conglomeration of all dead organic carbon, distinguishable from living organic and also from inorganic carbon and is depleted by nonpredatory means from many trophic levels (egestion, excretion, secretion, etc.) or from other external sources that enter and cycle in the ecosystem (allochthonous organic carbon).

Out of all biomolecules acting as potential sources of detritus, cellulose, and lignin represent the most abundant biomolecules in living organisms of the coastal mangrove ecosystem (Hedges, 1992). Particulate matter, especially detrital carbon, remain in the water column in a dissolved form. The proportionate ratio of dissolved to particulate organic carbon is 10:1 in lakes and oceans (Wetzel, 1984). The other constituents of dissolved detritus (generally <50%) consists of clearly defined biomolecules such as carbohydrates, peptides, lipids, nucleic acids, lignin, and their monomers (Münster, 1993).

Humic substances as naturally occurring, biogenous organic substances can generally be characterized as black to yellow in colored substances with a high molecular weight and refractory properties (Aiken et al., 1985). Besides, the chemical properties of humic substances containing chromophoric moieties endowed with non-ionic macroporous resins, can bind large (>500 g/mol) molecules with acidic functional groups at a low **pH**. The formation of the humic substances takes place through degradative pathways from those organic materials that could not readily be biodegraded by detrivores. Moreover, condensation of simple monomeric compounds sometimes leads to the production of humic substances.

3.16.2 Decomposition Process and Mangrove Litter

Organic matter in the form of detritus and other elemental form components of nutrients in the form of macro- (nitrogen, carbon, phosphorus, etc.) and micronutrients (Si, Fe, Mg, Br, etc.) are produced from the leaf litter of mangrove forests by the activities of both macro- and micro-decomposers and transferred from the mangrove forest subsystem to the adjoining estuarine subsystem. The entire process take place involving at least eight basic steps: (1) breaking up of the fallen leaves into several smaller parts, mostly by the macrobenthic and microarthropods; (2)

mixing up of the fragment parts of the leaves with water, especially during the period of tidal inundation; (3) exposed mesophyll tissue of soft and partially decomposed leaves and other parts of mangrove plants are attacked by the mycelia of fungi and other microbes such as bacteria; (4) leaching out of soluble nutrients from the soft leaves under the process of decomposition by water; (5) colonization and proliferation of microorganisms with their enzymatic activities to facilitate the process of decomposition; (6) consumption of tough and hard vegetative parts of the plant material by herbivores; (7) defecation and other bioturbatory activities of benthic fauna add more nutrients and also reshuffle the nutrients into different strata of the sediment on the mangrove forest floor; and (8) combined action of several physical forces such as tidal flows, especially during low tides, water-current movement, erosion pressure, and deposition patterns ensures the transportation of dislodged sediments and loads the newly generated nutrients from the surface of the mangrove forest floor (Ashton et al., 1999; Kathiresan & Bingham, 2001; Swift et al., 1979).

In this context, it is an established fact that the mangrove ecosystem is an amalgamation of two subsystems, the mangrove forest subsystem and estuarine subsystem linked up with each other by virtue of the ebbs and flows of tides exchanging materials essential for the sustenance of both of the subsystems including adjoining ecosystems such as seagrass, coral reef, and other marine ecosystems by way of maintaining ecological stability. All of these coastal ecosystems provide habitats for the growth and propagation of juveniles of so many shell fish and fin-fish by developing a coupling interactive system for phytoplankton, zooplankton, and benthos (Alongi, 2009, Alongi et al., 1992), marine inputs, and water-column production (Bouillon et al., 2008). All of these interlinked and interdependent components and ecological processes lead to development of intensive biogeochemical processes with the potential for global element cycles (Dittmar et al., 2006; Bouillon et al., 2008). The species composition, size, structure, and biomass are important from the point of view of primary productivity in the mangrove ecosystem as the intensity and quantity of litter act as a vital index of potential carbon export from the mangrove forest to the adjoining coastal estuarine ecosystems (Twilley & Day, 1999; Twilley et al., 1996) and a valuable indicator of mangrove productivity (Clough, 1992). The total amount of litter produced in a particular mangrove forest is also considered an important indicator influencing directly the ecological functioning of other adjoining ecosystems through the process of outwelling of POM between mangrove forests and other coastal habitats. Ambient temperature, along with other ecological parameters such as salinity, pH, etc., act as regulating factors directing the dynamics and cycling of leaf litter, mediating nutrient cycling in the mangrove ecosystem (Mfilinge et al., 2002).

A range of natural products that contribute to the production of organic matter in the mangrove ecosystem are fallen litter from mangrove trees that is composed primarily of leaves, propagules, tree bark and branches, subsurface roots, as well as dead biomass of all inhabiting biota (Tam & Wong, 1998; Bouillon et al., 2004a, 2004b; Kristensen et al., 2008). The decomposition process in the mangrove ecosystem regulates the flux of nutrients in the form of nitrogen, phosphorus, and carbon, and nutrients in mangroves (McKee & Faulkner, 2000). Maintaining a balance between rates of decomposition and the production of organic matter (OM), mangroves and other vegetated coastal mangrove estuarine complex maintain the global carbon sink (Donato et al., 2011; Yuan et al., 2015). In this context, nutrient cycling within mangroves is attracting global attention in view of the ongoing issues on climate change, as mangroves have been identified as the site for "**blue carbon**" sequestration. Therefore, continuous and steady reduction (30–50%) in global mangrove coverage during the last few decades is a matter of great concern, as reported in Donato et al. (2011).

Different ecological parameters and attributes (temperature, rainfall, soil texture, tidal amplitude, water current, species composition, bioturbation intensity, abundance of wild fauna, redox state) tend to control the quantity, magnitude, and fate of OM production through decomposition in mangroves, both temporally and spatially (Middelburg et al., 1996; Ashton et al., 1999). The decomposition process taking place in the mangrove forest includes many other eco-biological processes such as grazing, cutting and fragmentation of fallen mangrove litter by macro-invertebrates, and leaching of semi-decomposed substances and microbial decomposition, expediting nitrogen, carbon, and phosphorous cycles. Placing a greater emphasis on proper understanding for developing a more comprehensive understanding of decomposition processes as the gained knowledge base certainly can contribute to sequestering blue carbon as a effective measure for the mitigation of climate change-mediated environmental problems.

Although decomposition of plant materials is accelerated by an increase in temperature, decomposition of leaf litter of mangrove plants is slower in drier parts of tropical countries, mainly because of the influence of higher salinity and climate-change-associated factors (Alongi, 2009; Keuskamp et al., 2015). Other ecological factors such as the textural composition of sediment and the intensity of light and humidity also play important roles in the association of changing salinity to expedite the process of mangrove leaf litter decomposition by boosting the diversity and density of the decomposing biota.

The rate of decomposition is observed to have been influenced by tidal action, as a number of studies have found the highest rates of decomposition in lower intertidal and subtidal zones, which experience frequent flooding (Twilley et al., 1986). However, Feller et al. (2002) found faster rates of decomposition in the part of intertidal zones that are exposed and inundated frequently in contrast to permanently inundated subtidal zones and upper-part intertidal zones. Tidal flushing develops conducive ecological conditions with specific physico-chemical circumstances for accelerating the decomposition process (Middleton & McKee, 2001).

3.16.3 Mangrove Litter and Decomposition

A major driver of nutrient cycles through detrital food chains is the decomposition of mangrove leaves, which are the reservoir of a large amount of complex organic matter along with nitrogen, phosphorus, and carbon within mangrove forests and act as a source of energy for the onward transfer to adjacent coastal waters. However, this apparently simple issue poses difficulty for proper explanation and justification because the of complex nature of composition of organic matter, which varies from species to species and also exhibits differential patterns of decomposition processes alongside the biotransformation of decomposition products from mangrove forests to adjoining coastal water bodies.

One of the reasons for the slower decomposition rates of mangrove plant litter of particular species is the higher concentration of hydrolyzable tannins, cellulose, and lignins, which hinders the process of decomposition (Robertson, 1988). All those complex organic compounds work against crabs, most of which are detritovores or herbivores (Chakraborty et al., 2013). The hydrolyzable forms of tannins are lost owing to the process within the initial days of decomposition (Robertson, 1988).

Decomposition of carbon-rich compounds in mangrove sediment was found to be slow because of the presence of high levels of recalcitrant compounds, e.g., tannins, lignin, and cellulose) (Kristensen, 2008). Although they can be readily degraded, this happens slowly under a predominantly anaerobic situation, which which ultimately lead to slow down the degradation of cellulose and lignin even under aerobic conditions (Hawkins & Freeman, 1994). Lignin in leaves of a mangrove plant species, *Rhizophora mangle*, has a half-life of 150 years in the upper surface of mangrove sediment. This slow trend of decomposition of mangroves acts as contributing factor toward the accumulation of OM in some mangrove forests (Jennerjahn & Ittekkot, 1999). Moreover, fermenting prokaryotes achieving anaerobic conditions split initially large molecules into smaller ones, after which they are oxidized to CO_2 by a range of microorganisms adapted to such anaerobic conditions by utilizing electron acceptors in the sequence of Mn^{++++}, NO^{---}, Fe^{+++}, and SO_4^{----} to yield energy (Kristensen et al., 2008).

Research information showed faster rates of decomposition of the leaves of *Avicennia marina* in comparison with those of *Ceriops tagal* and *Rhizophora stylosa* in Australia (Robertson, 1988). Similar studies conducted in India revealed the occurrence of much lower tannin concentrations in the leaves of *Avicennia marina* than in those of *Rhizophora. apiculata* during the first 3 weeks of decomposition, whereas tannin concentrations became almost similar in both species at the very end of the decomposition process. All of this research information suggests that leaf decomposition is dependent on the amount, chemical composition, location, seasonal ecological parameters, and duration of inundation and exposure. The quantity and composition of amino acids and hexosamines, the products of decomposition, vary from species to species and also from season to season, which indicates the process of the transformation of nitrogenous compounds (production, preservation, immobilization) during decomposition. This kind of research information assumes greater significance in the eco-management of the mangrove ecosystem in view of the present pace of global warming, which brings about changes in the species composition of the mangrove community, which in turn modify the cycling of nutrients, especially the essential ones such as nitrogen in mangrove ecosystems.

3.16.4 Mangrove Litter and Conversion into Detritus

The very productive mangrove ecosystem is endowed with a high biomass of flora and fauna (Chakraborty & Choudhury, 1992a, 1992b; Koch & Wolff, 2002; Alongi, 2014; Hutchison et al., 2014; Ray et al., 2014). Threefold characteristics of the mangrove ecosystem are high primary production, rapid pace of decomposition of litter to detritus through the processes of fragmentation, autolysis, and photo-degradation, leaching with an active contribution of macro- and micro-decomposers and prompt utilization of nutrients promoting the maintenance of the nutrient cycle (Benner & Hodson, 1985; Ashton et al., 1999; Gessner et al., 1999; Komiyama et al., 2008; Treplin & Zimmer, 2012; Hutchison et al., 2014). The estimated aboveground net primary production of mangrove forests (11.1 tonnes carbon/ha/year) exceeds that of terrestrial moist deciduous forests (10.8 tonnes carbon/ha/year) (Alongi, 2014). The functioning of the mangrove ecosystem, especially through the production of nutrient-enriched detritus, is governed by the species composition of mangrove plants, chemical constituents of litter, seasonal impacts of rainfall, temperature, humidity, supply of oxygen, range of the duration of tidal inundation and exposure, and the species composition of litter-dwelling fauna and microbes (Robertson, 1988; Tam et al., 1990; Mackey & Smail, 1996; Woitchik et al., 1997; Twilley et al., 1997; Mfilinge et al., 2002; Bosire et al., 2005; Sanchez-Andres et al., 2010). Most of the macrobenthic fauna initially act upon the fallen leaves and break down and bury the fragmented litter, which enhances the litter quality because microbes find a larger surface area to exo-enzymatically decompose litter with higher decomposition rates, leaching out the nutrients. Therefore, macrobenthic litter-consuming brachyuran crabs boost litter decomposition through litter consumption and burial where a large part of the organic production is recycled within the mangrove forest (Lee, 1998). The presence of higher levels of different recalcitrant compounds, including lignocellulose and tannins in the leaves of mangroves makes them highly resistant to degradation by most of the microorganisms (Robertson, 1988). However, identified several phylogenetically diverse species of marine fungi that have been recorded that possess the ability to degrade lignocellulose owing to their power to tolerate low oxygen concentrations.

An enormous amount of nutrients and dissolved organic compounds (DOCs) is exported to the nearby estuarine systems by tidal movements during low tides triggering the growth and propagation of phytoplankton, whereas a sufficient amount retained within the forest floor also elicits the productivity of trees (Steinke et al., 1993). However, the presence of higher concentrations of phenolic substances within the litter slows down the microbial decomposition rate. The building up of phenolic compounds within the mangrove ecosystem as well as inhibiting the microbial activity results in the accumulation of OM on the mangrove forest floor.

Decomposition rates within the mangrove ecosystem are determined by ecological processes (tidal amplitude along with duration of tidal inundation and exposure, intensity of water currents, supply of oxygen, etc.), but mostly by the species composition of mangrove plants, constituting a particular eco-zone. The mangrove forest is dominated by plant species such as *Avicennia* exhibited three times more anaerobic respiration than *Rhizophora* forests. It was experimentally determined that the leaves of *Sonneratia alba* underwent faster decomposition than the leaves of *Rhizophora apiculata* (Ashton et al., 1999).

3.16.4.1 Outwelling and Transport of Mangrove Organic Matter

Biological production, both pelagic primary production and offshore secondary production, in the adjoining water bodies of the mangrove forest is directly connected to the production potential and quantity of the export of detritus and faunal biomass, and has been widely used as an argument for mangrove conservation. The exchange of materials between waterways and mangrove forests is also dependent on tidal flooding (Dittmar & Lara, 2001a). Alongside decomposition, other ecological processes such as mineralization or burning of leaf litter toward the production of OM and their transport is influenced not only by the rate of biological production but also by the hydrological and geomorphological characteristics of each ecosystem (Jennerjahn & Ittekkot, 1999). Greater fluxes of OM from mangroves to adjacent coastal waters directly determine the estuarine (brackish) water food webs, mainly because of the outwelling-mediated transport of materials (Alongi, 1990; Dittmar & Lara, 2001c). A large quantity of nutrients is found to have been released from detritus-rich mangrove floors to water columns of the tidal canals, creeks, and large estuaries because of tidal pumping during the ups and downs of high and low tides and thereby result in the fluctuation of nutrient concentrations, which in turn regulate both gross and net productivities in the water column (Dittmar & Lara, 2001a). Therefore, the decline in mangrove-forest cover during last few decades certainly will negatively impact the flux of terrigenous DOM and export of organic carbon from mangroves to coastal waters (Alongi, 2008; Kristensen et al., 2008).

3.16.4.2 Phenolic Compounds and Mangrove Leaf Litter Decomposition

The nutrient cycling in the mangrove ecosystem from organic to inorganic form through different trophic levels is facilitated by the food chain–food web dynamics involving an array of faunal components as well as microbes (Costanza et al., 1997; Chapin et al., 2000; Chakraborty, 2017). Alongside different species of fungi and bacteria, intertidal benthic fauna play important roles in the physiological transformation of nutrients from one form to another by way of their feeding behavior and also by releasing their nutrient-enriched fecal matter, which helps the growth of several microbes for expediting the decomposition of mangrove litter (Chakraborty, 2011, 2017).

Polyphenols, organic chemical compounds, occur in nature as secondary metabolites of plants. The functional roles played by those organic compounds fall into three categories: as antibacterial agents (Coppo & Marchese, 2014), as antifungal compounds (Giovanna et al., 2020), in nutrient cycling (Hättenschwiler & Vitousek,

2020). Moreover, phenolic compounds present in decomposed leaf litter play important roles in determining species composition and eco-dynamics of forest vegetation by rendering an allelopathic influence. As antioxidants, phenolic compounds play a considerable role in scavenging endogenous free radicals. The free radicals and oxidative stress are connected with conditions such as neurodegenerative disorders, aging, cancer, diabetes, atherosclerosis, etc. The anti-inflammatory properties of these compounds also boost the immune system, which also helps to form body structures of plants and animals (Wigglesworth, 1988) and coloration. Phenolic acids, acting as important precursors of humic substances in soils (Ziolkowska et al., 2020), play valuable roles in soil formation and pedogenesis by influencing cycling of minerals and organic matter. Through the production of organic metal complexes, the scopes of availability of micro and macro nutrients are increased.

However, the concentration of phenolic compounds within mangrove ecosystems does not only determine the accumulation of organic matter. Through an experimental study, Saraswati et al assessed the role of an '**enzymic latch**' in causing the accumulation of carbon stores in peatlands (Freeman et al., 2001, 2004), which is also found to be present in the mangrove ecosystem. An 'enzymic latch' is a phenomenon by which the activity of phenol oxidase, the enzyme responsible for the decomposition of phenolic compounds, is inhibited and thereby facilitates the building up of phenolics under anaerobic conditions in the soil (Freeman et al., 2001). However, the '**enzymic latch**' hardly occurs in mangroves with litter containing lower phenolic concentrations.

The very presence of mangrove plant-derived polyphenols are found to occur within the detritus of saline conditions of estuarine soil against the fierce action of the ocean tides most of and those polyphenols are bio-transformed and recycled between producers and consumers, because the benthic fauna of the soil serves as an important link for the eco-biological cycling of polyphenols. The benthic fauna on the mangrove–estuarian complex not only get their primary source of nutrition from detritus-loaded bottom soils but also lend support to other primary producer communities such as mangrove plants, macro- and microalgae, etc., for their own nutrition supply. Most of the phenolic compounds during the process of decomposition are transformed into different fractional components, which are taken up by different biotic components through estuarine trophic interactions, especially by the functional roles of benthic fauna, ensuring the biotransformation of plant biochemicals such as polyphenols, which can be cited as an important example of co-evolution with the complementary dependence of so many actors (plants, animals, and detritus) operating in their own laboratory of nature.

A recent study on the bio-cycling of polyphenols in the estuarine mangrove ecosystem of the North Eastern coastal belt of India has revealed that the antioxidative potential and free-radical scavenging activity of this organic compound moves from a mangrove shrub (leaves of *Acanthus ilicifolius*) to detritus and from there to benthic organism, *Uca lactea annulipes* (fiddler crab), of the estuarine mangrove ecosystem through the processes and pathways of eco-biological transformation. The presence of deprotonated polyphenols after being released from the decomposing plant litter to detritus are bio-transformed to a few fragmented parent ions of phenolic acids in the crab hemolymph of *Uca lactea annulipes*. The studied fiddler crabs act as agents in the bio-cycling of polyphenols, which depicts the interdependence of plants, animals, and soil exchanging essential life-supporting ingredients as an outcome of the co-evolutionary process (Ghosh et al., 2023). Earlier research studies on bioturbatory activities of intertidal brachyuran crabs effectively help to remineralize the detritus to CO_2 and dissolved organic carbon, which tends to increase labile organic carbon in the sediment (Fanjul et al., 2015). It was also observed that the calcareous carapace made up of $CaCO_3$ of the fiddler crabs is associated with microbes forming microbial biofilms to expedite microbial nitrogen transformation (Zilius et al., 2020).

Research information based on the estimation of polyphenols in different compartments of the mangrove–estuarine ecosystem from the coastal belt of West Bengal–Odisha of India has revealed that the maximum amount of that compound remains present in the leaves of *Acanthus ilicifolius*, a mangrove shrub, than detritus, whereas the studied intertidal detritivorous fiddler crab, *Uca annulipes annulipes* showed the presence of smaller quantities (Ghosh et al , 2023). This occurrence of differential quantities of polyphenols in the order of highest to lowest is the leaves of the mangrove plant to detritus to detritivorous benthic fauna, and this trend is explained by different facets of decomposition process of mangrove leaves, where macro-decomposers (crabs, mollusks, insects, etc.), and bacteria convert leaves to detritus. Ecological integrity is maintained by the continuous interactions among different structural components of the ecosystem resulting from the proliferation of biodiversity components that ensure the survivability of mangrove flora and fauna alongside rendering many ecosystem services (Chakraborty, 2011, 2017). After becoming morphologically, behaviorally, and physiologically adapted to their respective habitats, different floral, microbial and faunal components of the biotic community undergo continuous and steady interactions among themselves, which accelerates the decomposition process of mangrove litter and thereby influences the ecosystem functioning to a large extent (Kristensen, 2008). The microorganisms inhabiting the mangrove floor metabolize the nutrients present within the detritus for their own metabolic activities and convert a major proportion of detritus to CO_2 and in the process reduce the efficiency of the transfer.

This process ensures biotransformation and recycling of the nutrients and energy in the mangrove ecosystem. Most of the polyphenols identified in the same study from the West Bengal–Odisha coastal belts are gallic acid (*m/z* 170), bergapten (*m/z* 216), catechol (*m/z* 110), p-coumaroylquinic acid (*m/z* 338), quercetin 3-o glucoside (*m/z* 478), peonidin 3-o arabinoside (*m/z* 449), rosmarinic acid (*m/z* 360), dihydromyricetin 3-o rhamnoside (*m/z* 466), avenanthramide K (*m/z* 315), petunidin 3-o glucoside (*m/z* 479), and (+) catechin 3-o gallate (*m/z* 442). The acanthus leaf, detritus, and crab's hemolymph showed the presence of a complex mixture of several acids containing petunidin 3-o along with glucoside (*m/z* 479) carboxyl (wavenumber 1500–2000 cm^1) and that mixture was found to demonstrate functional behavior as a dibasic, or occasionally as a tribasic acid.

The assessment of antioxidant activities using the Folin–Ciocalteu, DPPH, ABTS, and FRAP techniques of three major biological components of the mangrove–estuarine ecosystem such as the leaves of a mangrove shrub *Acanthus ilicifolius,* detritus from intertidal belt and hemolymph of one major intertidal macrobenthic fauna, the fiddler crab (*Uca annulipes annulipes)* revealed maximum antioxidant activity in the leaves of mangrove plants followed by detritus and then hemolymph of intertidal benthic crabs. Analysis by infra-red spectroscopy indicated the presence of aromatic –C=C– groups, carboxylic –COOH, and phenolic –OH groups. Polyphenols extracted from the mangrove leaves exhibited the occurrence of dihydromyricetin and 3-o rhamnoside rosmarinic acid, whereas catechin 3-o gallate and quercetin 3-o glucoside were found in the detritus. Diversified forms of polyphenols such as catechol, bergapten, avenanthramide, gallic acid, p-coumaroylquinic acid, petunidin 3-o glucoside, and peonidin 3-o arabinoside were recorded from the bodies of crabs. Interestingly, spectroscopic data showed the presence of some parent ions of the phenolic acids in the hemolymph of crabs that were also present in the leaves. This bio-transformation of organic components from one compartment to others of the mangrove ecosystem appears to represent a critical part of nutrient cycling through which nutrients move and disperse across the ecosystem. This further establishes the role of the fiddler crabs as agents participating in the bio-cycling of polyphenols. Fiddler crabs, an important benthic fauna of the mangrove ecosystem, play important roles as a bio-transferring agent facilitating the bio-cycling of polyphenols because they consume detritus from the floor of the mangrove forest and in that way gain biomass and also lend immune boosting support for their fruitful survivability. The crabs help to transform polyphenols and their fractional components, having antioxidant potential from the detritus to their shells, muscles, and body fluids, which participate in the eco-biological cycling through intricate trophic interactions (Soundarapandian et al., 2014) This process of eco-biological cycling of polyphenols from mangroves through detritus and fiddler crabs brings about a striking shift of the pathway of nutrient cycling from mineral to organic forms, which tends to minimize potential nutrient losses from the ecosystem and maximize litter nutrient recovery by the symbiotic microflora (Ghosh et al., 2023).

Although the mangrove forest is a perennial evergreen forest, the highest fall of mangrove litter (leaves, barks, twigs, seeds, and flowers) is found during the monsoon period. Out of the several parts of mangrove plants constituting mangrove litter, leaves comprise around 80% to 90%, whereas other reproductive parts such as flowers, seeds, and fruits comprise about 10% in the decomposition process, which ultimately recycles and helps the exchange of nutrients within the mangrove ecosystems. The total amount of the litter fall varies from one species to other depending on the ecological variables of different habitats (Imbart & Menard, 1997). On average, 900 g carbon/m/year of litter falls on the mangrove forest, of which 10% to 15% is transported to the adjacent estuarine systems by the receding flows of water during low tide, whereas 250 g carbon/m/year leaches out as soluble organic matter leaving 200 g carbon/m/year to remain as particulate organic detritus. The concentration of nutrients in the estuarine subsystem is determined by microbial immobilization of mangrove litter resulting from the loss in carbon content (40% to 70%) and the increase in nitrogen content (0.1% to 0.4%) within the mangrove ecosystem (Das, 2010).

The nutrient cycles in the mangrove ecosystems are governed by several factors (Middelburg et al., 1996) and the process of nutrient recycling also determines the roles of the mangrove ecosystem as either a source of nutrients or a sink for nutrients (McKee & Faulkner, 2000). Both the vegetative and reproductive parts of mangrove plants as litters undergo the decomposition process on the way to forming detritus, where a number of biological, chemical, and physical processes govern the entire decomposition process starting from the falling of litter to the muddy bottom of the mangrove forest floor, the breaking up of larger parts into smaller ones by macrodegraders (crabs, insects, mollusks, etc.) and microbial decomposition. Biological processes include the foraging behavior of larger animals whereas chemical processes involve the microbial decomposition through the exoenzymatic process of digestion coupled with mineral uptake. Physical processes include the actions of tides, the oscillation of waves and the intensity of water currents, the impacts of rhythmic inundation and exposure, rainfall, and the oscillation of the temperature. Diurnal, seasonal, and annual fluctuations of these processes tend to control the rate of nutrient production, export, import, and storage in this ecologically potential ecosystem. On falling to the ground, a portion (2% to 14%) of mangrove litter is taken away by the receding low tides from the mangrove forest zones and drifts to the adjacent estuarine ecosystems where it becomes moistened and settles down to the bottom of the sub-tidal shallow water and is used by demersal shellfishes and finfishes directly as their food. However, the presence of lignin and cellulose within mangrove litter poses a difficulty for the digestion of macro-consumers and therefore for the production of ideal food for the mangrove faunal community. Most of the mangrove litter is required to undergo complete decomposition. In the Sundarbans mangrove ecosystem, three kinds of breakdown processes operate on this mangrove litter, which are: (1) mechanical fragmentation at the time of falling by tidal and wave actions; (2) fragmentation of larger parts into smaller pieces by macrobenthic fauna; (3) leaching of soluble organic residues; (4) decomposition by fungi and bacteria.

Mangrove vegetation is the principal source of produce of mangrove forest litter that contributes most of the primary production in the mangrove ecosystem after degradation by microbial activity. Aquatic organisms in the mangrove ecosystems of the Sundarbans depend on the detritus derived from the decomposition of the mangrove forest litter. The lengthy conversion of mangrove litter into detritus by way of transforming complex organic matter from this litter to simpler soluble organic matter is caused by the very slow decomposition process. The sequential process involves events such as: (1) the mature leaves fall down at a regular intervals; (2) fallen leaves become moist on falling onto the muddy mangrove floor, and the color of the leaves becomes pale within 3 days; (3) several colored spots appear on the soft leaves; (4) macrodegraders start to chew the leaves and pull the larger leaves into pieces; (5) mycelia of fungi start to intrude on the mesophyll tissue of the leaves, making all parts of the decomposing leaves soft; (6) the entire decomposition process is completed with 4 weeks depending on the ecological variables, exposing only the inner veins of the leaves.

3.16.5 *The Roles of Microbial Activity in Sedimentary Decomposition*

The mangrove plant litter, on falling to the ground, softens on obtaining moisture from the muddy sediment and then it is broken into smaller pieces by the masticating activities of macrodegraders such as benthic fauna, which, through fragmentation of leaf litter, increase the surface area so that fungi can invade the mycelia of semi-decomposed plant litters. At the same time bacteria play an important role in the decomposition process by way of producing extracellular enzymes that convert fatty acids, phosphate, etc. (Alongi et al., 1993; Holguin et al., 2001; Keuskamp et al., 2015). Although proper decomposition starts with the colonization of bacteria and fungi (Holguin et al., 2001), prior to that macrobenthos and several groups of microarthropods also initiate the process of decomposition by breaking the newly fallen mangrove leaves into pieces, providing scope for the fungi to invade the mesophyll tissue of the fallen leaves. In mangroves of tropical coastal belts, 91% of the microbial biomass are represented by bacteria and fungi, whereas protozoa and algae only make up 2% and 7% respectively (Alongi, 1988a; Holguin et al., 2001). Five major groups of bacteria are found to be responsible for decomposition in mangrove ecosystems, which are nitrogen-fixing, sulfate-reducing, and methane-producing microbes. In addition, some other types such as iron-reducing bacteria are observed to occur in mangroves located in areas receiving inputs from mines and metals-based industries. A manglicolous group of fungi is found in the mangrove ecosystem, and play important roles in nutrient cycling by synthesizing enzymes that are required to degrade lignin, cellulose, tannins, and other such plant compounds. Different electron acceptors are involved in the decomposition process under both aerobic and anaerobic conditions.

The aerobic decomposition process to convert the mangrove plant litter into OM occurs on the surface of the forest floor involving fresh litter where detritus is deposited on the surface of the sediment (Holguin et al., 2001; Kristensen et al., 2008). The mangrove leaf litter buried at a certain depth below the sediment undergoes anaerobic decomposition, because at that depth O_2 remains in small amounts as respiration of ground-dwelling fauna and pneumatophores of mangrove plants consume so much of the oxygen, leaving only a small fraction available to a depth greater than 2 mm below the surface (Holguin et al., 2001; Kristensen, 2008). However, exception persists in the aerobic zone at a certain depth, which has been created by roots and the burrows of macrobenthic fauna (Kristensen et al., 2008).

3.16.6 *The Role of Detritus in Food Webs*

Food webs referred to as interlinkages of several food chains are of two types: grazer and detritus food webs on the basis of nutrients and energy sources (Wetzel et al. 1972). Detritus is used as an important source of food for all those faunal components in the nutrient-enriched mangrove ecosystem. Grazing food chains and food webs start from and depend on living primary producers, whereas in the detritus food chains and food webs heterotrophic bacteria and fungi constituting the major parts of the detritus are consumed by detritus-consuming faunal components.

The autochthonously generated detritus also plays important roles in carbon cycling by separating grazer and detritus food webs and thereby determining the trophic levels of organisms by governing internal carbon transfer processes such as exudation, (Chen & Wangersky, 1996); mastication and egestion (Viitasalo et al., 1999); and feeding (Strom et al., 1997); viral lysis (Middelboe et al., 1996). Microbial-nutrient relationships (Alongi et al., 1993; Polis & Strong, 1996; Hessen, 1998). In contrast, the microbial loop, being the detritus food web in the water column, operates partially by autochthonous detritus but depends on imported allochthonous detritus, making the aquatic ecosystem heterotrophic (Mann, 1988; Salonen et al., 1983; del Giorgio et al., 1997).

Carbon cycling is directly connected to the production of detritus through the decomposition process as it facilitates mineralization of organic carbon to make it readily available for the primary producers and thereby supports the detrital food web. Detrivores as important degraders and mineralizers in the process of production of detritus use part of the detrital carbon for biosynthesis and utilization of the chemical energy in order to derive their need for energy, which is released from the cleavage of covalent bonds only inside the cells. Thus, molecules of detrital materials penetrate the cell membrane, including two types: (1) actively with a polarized state of small molecules (Page et al., 1989; Liu et al., 1993) and (2) passively with hydrophobic molecules.

For the biological mineralization of detritus, suitable environmental conditions (30 °C to 113 °C, at a pH between 2 and 11) are required in the aquatic ecosystem for the growth and propagation of microorganisms, including the bioavailability of detritus prior to uptake, and cleavage of large (>500 g/mol) polar molecules is required. To ensure the successful transportation of cleaved products, cellular enzymes must process the detrital molecules (Schwarzenbach et al., 1993; Alexander, 1999). Bioavailability of detritus occurs at the time of the development of direct physical contact between detritus and detritus-processing enzymes (Alexander, 1999). Several components in the degradable biomass such as cell walls, thick leaves, bark, and other woody structures act as a structural barrier to protect the given parts of detritus and thereby limit their bioavailability (Enríques et al., 1993). However, all those structural barriers mainly composed of complex organic substances such as lignin and cellulose are broken down at the initial phase of decomposition toward the production of detritus by macroscopic detrivores (shredders, grinders) (Fenchel, 1970; Robertson & Mann, 1980; Harrison, 1982). Extracellular enzymes of detrivores help to cleave detritus with a high molecular weight (500 g/mol) and is one of the most important factors controlling the use of detritus. Biopolymers such as nucleic acids, polysaccharides, proteins, and lipids are cleaved by hydrolytic extracellular enzymes. Hydrolytic enzymes hardly require any external energy for their catalytic action as the catalytic site of an enzyme, on developing contact with the substrate, promotes thermodynamically favorable hydrolysis. Heterogeneous polymers of degrading litter such as lignin, cellulose, and other humic substances are cleaved by some of the nonspecific oxidizing extracellular enzymes, which activate oxygen or manganese (II) cation in order to give them the potential to attack molecules of detritus with double bonds and aromatic structures having electron-rich regions. Several microbes (fungi and bacteria) that have oxidizing ligninolytic enzymes are important in the decomposition of particulate detritus in the littoral (Kirk & Farrell, 1987). The detritivores, by virtue of their foraging activities such as searching, holding, and engulfing detritus, accelerate the process of biodegradation. Intraspecific and interspecific competition between detrivores plays an important role in determining the biodegradation process as evidenced by the fact that functional roles of predator, environmental heterogeneity, and disturbances help to limit potential competition (Leibold, 1996; Flöder & Sommer, 1999).

3.16.6.1 Physical and Chemical Factors Affecting the Decomposition of Detritus

Detritus is found in nature in three phases: solid (particulate), dissolved, and gaseous phases. The movement of transport among these three phases ensures the redistribution of detritus and is regulated by physico-chemical factors such as sorption—desorption and dissolution—volatilization). Changing ecological conditions that have a specific range of different ecological parameters (lower alkaline pH, oxygen saturation, wind, water movements, photochemical reactions during the day, freezing, and drought) govern the rate of mangrove litter decomposition by fragmenting particulate detritus with the production of detritus followed by alkaline hydrolysis of a detritus molecule without breaking covalent bonds (Robertson & Mann, 1980; Peduzzi & Herndl, 1991; Mateo & Romero, 1996). The physical fragmentation process helps the biological degradation of detritus by promoting close contact between degradative enzymes, which is followed by chemical processes for modifying and transporting detritus.

3.17 Biogeochemical Cycles in the Perspectives of the Mangrove Ecosystem

The 40 prime chemical elements required to drive life processes and the growth of living organisms are of two types: micronutrients, constituting less than 0.2% of dry organic matter, are required in trace amounts, and macronutrients (oxygen, hydrogen, carbon, nitrogen, and phosphorous), constituting more than 1% each of dry organic weight, which are present in greater amounts. However, many micronutrients such as boron, bromine, aluminum, cobalt chromium, iodine, and fluorine, do not occur in many species. The cyclical movement of all these nutrients in the ecosystem and their transformation from one trophic level to another ultimately sustain the functioning of the ecosystem. Hence, the cycle is referred to as the biogeochemical cycle. The two types of biogeochemical cycles are the gaseous cycle (cycles of carbon, nitrogen, and oxygen gases) and the sedimentary cycle, occurring in the lithosphere, constitute the major reservoir of different elements such as sulfur and phosphorus, which are released by the process of weathering.

Different coastal environmental settings across the globe have different types of geomorphological processes, soil texture, biological diversity, nutrient loads, rates of organic matter decomposition, potential for carbon storage, in the biomass both above- and belowground, the stoichiometric ratio of nitrogen to phosphorus [nitrogen:phosphorus]) (Aller, 1982, 1988, 2001; Twilley et al., 2017). All of these eco-bio-geological attributes promote nutrient exchange, microbiologically controlled decomposition processes, aeration of soil layers, and trigger the growth of biological entities in the mangrove ecosystem (Vitousek, 1984).

3.17.1 Nitrogen Cycling in the Mangrove Ecosystem

Nitrogen, being one of major essential major macroelements, in both inorganic and organic forms as well as in oxidation states, undergoes cycling, which involves five processes (nitrogen fixation, nitrogen uptake, nitrogen mineralization (decay), nitrification, and denitrification) through the biosphere, atmosphere, and geosphere. Nitrogen is an important nutrient for all organisms and a major ingredient in all amino acids and hexosamines, in addition to constituting the building blocks of proteins, which form a major part of OM in both the aquatic and forest subsystems (Lee, 1989). Different forms of proteins making up a significant proportion of mangrove leaves undergo the decomposition process releasing amino acids and also elemental forms of nitrogen (Jennerjahn & Ittekkot, 1997). Amino acids represent a prime source of nitrogen and are immobilized by tannins within the mangrove ecosystem (Hernes et al., 2001).

Both soil and the planktonic microbial community represented mostly by bacteria and archaea and to a lesser extent by fungi and protists, in mangrove ecosystems, with their diversity and abundance, rapidly process, transform, turn over, and conserve various forms of nitrogen through a series of transformation processes. However, rates of individual transformation processes vary greatly depending on the microbial composition in the forest type and the status of soil nutrients and other eco-biological attributes such as temperature, soil fertility, microbial composition, root activities, bioturbation, redox, intertidal position, and salinity (Alongi, 1988a, 1989, 1990, 1996, 2002). nitrogen immobilization can be assessed by the gross rates of ammonification and nitrification, which always remain higher than net rates. N2-fixation rates are found to be highest on algal, fungal, or bacterial crusts grown on the bark of tree stems, pneumatophores, and prop roots in comparison with on the soil surface, litter, leaves, below-ground roots, and rhizomes. The direction and magnitude of exchange depends on many factors, such as geomorphology, tidal amplitudes, climatic condition, the ratio of waterway area to forest, temperature, pH, dissolved O_2 concentration, salinity, and plankton metabolism.

3.17.2 *Mineralization and Immobilization of Nitrogen in Mangrove Forests*

Relative nitrogen mineralization refers to the mineralization of nitrogen per unit of total nitrogen and acts as a useful index for comparing the soil quality among different zones of mangrove forests (Alongi et al, 1992; Reis et al, 2017). Immobilization of nitrogen takes place via the microbial conversion of inorganic nitrogenous compounds such as NH_4 and NO_4 into organic forms in the process of the decomposition of organic matter (Lindau et al., 1994). The relative nitrogen mineralization was found to decrease exponentially toward the landward direction under the influence of inland water bodies in comparison with the mangrove belts at the mouth of the estuary (Chen & Twilley, 1999). The deposition and accumulation of mangrove leaf litter tend to determine the carbon:nitrogen ratio of soil organic matter, which is reflected by higher carbon:nitrogen ratios at the litter horizon of soils low in nitrogen in comparison with litter in nitrogen-rich sites (Vitousek, 1982; Morris, 1991; Bridgham et al., 1998).

3.17.3 *Nitrogen Cycle Through Nitrification and Denitrification in the Mangrove Ecosystem*

The nitrogen cycle in the coastal–estuarine–mangrove complex plays vital roles within the overall biogeochemical cycles. An unprecedented increase in human-mediated global pollution has been considerably impacted by the climate-change-induced global nitrogen cycle, which also indirectly imposes cascading effects to alter normal processes sparking off positive feedback in the climate system. The enhanced nitrogen loading has been observed to cause long-term decline in coastal fisheries by influencing major processes such as fluxes and transformation of nitrogen in the coastal mangrove ecosystem.

Microorganisms, especially bacteria, play major roles in all of the principal nitrogen transformation processes involved in dynamic nitrogen cycling subject to the regulatory impacts of environmental factors (temperature, pH, oxygen availability, rainfall, moisture, etc.) on microbial activity. The exchange of organic matter (OM) and nutrients among different subsystems of mangrove–estuarine ecosystems emphasizes the recycled production of nutrients regenerated by the bacterial decomposition of dead biomass alongside new production, even obtaining a sustained supply of nutrients from external inputs.

Despite having a rich concentration of OM, which is considered an important source of nutrients (McKee & Faulkner, 2000), the mangrove ecosystem is still considered to be deficient in nutrients, especially nitrogen along with phosphorous (Sengupta & Chaudhuri, 2001; Alongi, 1996, Alongi, 2014; Middelburg et al., 1996; Vazquez et al., 2000; Holguin et al., 2001). This is supported by the fact that mangrove ecosystems possess different eco-zones that have a varying amount of nutrient content (Tam & Wong, 1998) as the sediments are rich in nitrogen content compared with the nitrogen concentration in their litter (Middelburg et al., 1996). In spite of exhibiting this kind of difference in respect of nutrient content within the mangrove ecosystem, it is still considered to be a highly productive ecosystem because it possesses an effective nutrient recycling system, which alongside retaining rare essential nutrients, generates new nutrients from the process of litter decomposition (Holguin et al., 2001, Jennerahn and Ittekkot, 2002).

In the open ocean the main source of nitrogen is the reservoir of deposited materials in the deeper zones, which comes to the surface and sub-surface strata by the upwelling process and is also redistributed by vertical mixing resulting from the churning actions of oceanic currents, waves, and atmospheric interactions. This supplying of nutrients ultimately sustains the high productivity of the coastal zone. However, the origin of nutrients, especially, is much more complex in coastal zones, where inputs from the river account for almost half of the total, which is in contrast to negligible contributions from atmospheric deposition (8%) and nitrogen fixation (1%) (Galloway, 1996; Howarth, 1996; Twilley, 1988; Wollast, 1993). The sources of nitrogen to the coastal ecosystems therefore include deposition from the atmosphere, influx by freshwater rivers and estuaries, drainage of chemical fertilizers from agriculture, and the release of wastewater from industrial and municipal sewage. In the surface water of estuaries, nitrogen remains mostly in the form of dissolved organic nitrogen (DON), but despite decades of research, only a small fraction of the DON in surface ocean waters has been chemically characterized. Coastal mangrove deltaic ecosystems make a major contribution to the nitrogen cycling processes because of the great influence of benthic primary producers on nutrient transformations and ensure exchanges of materials across the sediment–water interface. This nitrogen cycling plays a key role in overall nutrient cycling in a well-mixed water column because nitrogen plays a prime role as a primary nutrient for plants, limiting growth, propagation, and production by plants, algae, and microbes. Higher atmospheric deposition rates of nitrogen into coastal waters are caused by the cycling of complex ammonia in the atmosphere resulting in higher NH_3 concentrations, which facilitate greater exchange of fluxes for ammonia into coastal water.

The nitrogen cycle in the mangrove ecosystem is operated more by microbial activity than by chemical processes (Alongi et al., 1992). The rates of various transformations of the nitrogen pool within the mangrove forest are dependent on a wide range of physical–chemical parameters of the soil. Immobilization of a good amount of nitrogen as plant tissue is caused by the uptake of nitrogen (Clough et al., 1983). The differential concentrations of nitrogen within mangrove leaves and other parts of the mangrove plants of different species indicate several other ecological attributes such as species composition and zonation patterns of mangrove plants (Lacerda et al., 1986; Sah et al., 1989), the nutrient status of the sediment in respect of their position within the forest (Karl et al., 2002; Boto & Wellington, 1983), the structural uniqueness of the forest (Vorosmarty et al., 1998; Karl et al., 2002), the maturity of the leaves (Clough et al., 1983; Clough & Attiwill, 1975), and the nitrogen content in fruit, propagules, small fibrous roots, and small branches (Wong et al., 1995; Gong & Ong, 1990; Clough & Attiwill, 1975). The nitrogen pool within the sediment of the mangrove ecosystem is composed mostly of remains in organic form, which is not readily available in mangrove plants (Clarke, 1985). Concentrations of dissolved inorganic form, especially nitrate (NO_3), are found in lower amounts in estuarine mangrove zones in comparison with sediments of other parts of the marine environment (Alongi et al., 1992; Clarke, 1985). Other abundant forms of dissolved inorganic nitrogen occurs as ammonium (NH_4) but nitrogen in the form of NO_2 is found consistently in mangrove sediments in smaller concentrations (Alongi et al., 1992; Clarke, 1985; Boto & Wellington, 1984; Morell & Corredor, 1990; Purvaja, 1995; Alongi, 1996).

One of the prime ecosystem services rendered by mangrove ecosystems in the global context is the enhancement of nitrogen (N) availability through nitrogen enrichment (Reis et al., 2017). Nitrogen dynamics in both conserved and non-conserved mangrove ecosystems are dependent on both anthropogenic nitrogen enrichment and the nitrogen fluxes, as well as transformation processes in the interface between sediments of forest and estuaries (biological nitrogen fixation, mineralization, ammonification, nitrification, denitrification, and volatilization). The biological nitrogen fixation (BNF) process representing as one of the major pathways involves the entry of nitrogen into the mangrove ecosystem with the simultaneous reduction of atmospheric N (N_2) into ammonia (NH_3) by the activities of microorganisms—possessing nitrogenase enzyme complex—which are termed as diazotrophs (Alongi, 2002, 2009). Biological nitrogen fixation (BNF) occurs both in mangrove stems and roots (pneumatophores) by the cyanobacterial crusts growing on living tissues, sediments, leaf litter, and microbial mats. Several ecological factors play a role in determining BNF rates in aboveground mangroves, such as light intensity, photorespiration, water content, and phosphorus (P), along with endogenous oxygen (O_2) production and reduced dissolved organic C (DOC) availability. In addition, high salinity and a low percent of moisture also have been found to regulate N fixation by the microbes settled on trunks of *Avicennia germinans* (L.) L. in the mangrove forest of Guadeloupe, Lesser Antilles (Sheridan, 1991, 1992). The intensity of light, endogenous O_2 production, water content, sulfide (H_2S) concentration, and the availability of dissolved organic carbon (DOC) also regulate N fixation rates in mangrove microbial mats in Twin Cays, Belize (Joye & Lee, 2004; Lee & Joye, 2006). Nitrogen (N) fluxes across the sediment-water interface represent another major pathway by which nitrogen (N) enters into the mangrove ecosystem via tidal exchange (Alongi, 2002, 2009). However, the role of the mangrove ecosystem as a nitrogen (N) sink for coastal water bodies appears to be unclear, even for the dissolved forms of nitrogen (N) (Adame & Lovelock, 2011). The prevailing concentration gradients between pore-water and overlying water establish the proposition of the release of dissolved nitrogen (DN) into the overlying water, but net fluxes of DN were found to be negligible into the sediment, which is supposed to be due to the rapid uptake of DN by plants and benthic microalgae coupled with the efficient conservation activities of microbial assemblages (Kristensen et al., 1988; Alongi et al., 1993). Nitrogen (N) mineralization representing a microbially mediated process for the conversion of organic N to inorganic forms involves several other biochemical processes, such as ammonification (the production of ammonium, NH_4) and nitrification (the oxidation of NH_4 to nitrite, NO_2, and nitrate,

NO_3). Some N transformations processes by the microbial activities also result in gaseous N losses from mangroves to atmosphere; these processes include denitrification, which involves the reduction of NO_3 and NO_2 and their transformation to nitrous oxide (N_2O) and nitrogen gas (N_2). Nitrogen enrichment of the mangrove ecosystem by the anthropogenic activities also influences the patterns of nitrogen dynamics, which has been revealed by several research studies (Alongi et al., 2002, 2005; Whigham et al., 2009; Romero et al., 2012; Fernandes et al., 2012).

In view of the above, research studies undertaken in the Sundarbans mangrove ecosystem, India, have pointed out that despite differences among different ecozones within the mangrove forest in respect to the rates of nitrogen transformation and flux, mangroves play a considerable role in regulating and boosting these rates in tropical and subtropical regions. Anthropogenic activity-mediated nitrogen enrichment imposes an extensive impact on the structural components as well as the functioning of the mangrove ecosystem, especially on N cycling, thereby triggering an increase in N_2O fluxes to the atmosphere and acting as an important contributor to global warming (Sengputa & Chaudhuri, 1991). Fringe forests in response to nitrogen enrichment have been shown to possess higher rates of nitrogen transformation in sediments on shorter timescales and higher N losses to the atmosphere compared to other mangrove types (Reis et al., 2017).

3.17.4 Nitrification (N_2-Fixation) and the Mangrove Ecosystem

Nitrogen fixation is the process that converts the gaseous forms of nitrogen into combined forms such as ammonia and organic nitrogen by microbial (bacteria, cyanobacteria, and fungi) activities. Biological nitrogen fixation represents a process of the reduction of nitrogen gas to elemental forms of nitrogen in a variety of ways involving an array of microbes and archaea with the help of an enzyme, nitrogenase (Postgate, 1978). This functional process in mangrove ecosystems supplies a usable form of nitrogen to plants, which is considered a limiting attribute for the productivity of plants (Alongi, 1996), mangrove roots (Boto et al., 1985), and the oxidized lining of animal burrows (Capone, 1982).

The initial phase of decomposition of mangrove leaves on the soil surface exhibited a decrease in nitrogen and phosphorus concentrations, followed by a marked increase in these concentrations in the decomposed products. It has been recorded that nitrogen immobilization of about 13% to 21% was done by the microbial community, which took active roles in decomposing the leaves of *Rhizophora mucronata* (Woitchik et al., 1997). The high rates of nitrogen fixation linked to mangrove ecosystems are associated with the higher rates of accumulation of biodegradable OM, which were observed to have a close link with the enhanced rates of nitrogen fixation in the mangrove ecosystem (Holguin et al., 2001). Research information has established the fact that nitrogen fixation is the second major form of bacterial activity in mangroves after carbon decomposition by sulfate-reducing bacteria

(Holguin et al., 2001; Zuberer & Silver, 1978; van der Valk & Attiwill, 1984). In the densely vegetated stands of the mangrove forest, nitrification rates generally increase with depth, reflecting greater activity of nitrifying bacteria with the production of enzymes where nitrifiers are exposed to more oxic conditions. The penetrability of oxygen within the sediment serves an indicator of the extent of the nitrification zone, providing an estimate of the nitrification rate within the sediment. Also, the presence of NO_3 in the porewater down to a depth of at least 30 cm in the vegetated sediment indicates active nitrification below the oxic surface layer. The presence of NO_3 through active subsurface nitrification within the root mats of mangrove vegetation is mainly caused by downward translocation coupled with the steady release of O_2 by the roots of the mangroves (Reddy et al.,1989; Caffrey & Kemp, 1902). It is postulated that extensive uptake of production followed by uptake of ammonia in mangrove forests slows down the process of nitrification. A research study has pointed out that the halloysite, a nanoparticle derived from the decomposition of the mangrove shrub *Acanthus ilicifolius* inhibits the urase activities of the detritus and thereby probits the production of ammonia from the breakdown of urea (Ghosh et al., 2024).

3.17.5 Biological Nitrogen Fixation and the Mangrove Ecosystem

Biological nitrogen fixation represents one of the main pathways ensuring the entry of nitrogen within mangrove ecosystems by reducing the atmospheric nitrogen (N_2) to ammonia (NH3+) by the activities of a group of microorganisms that possess the nitrogenase enzyme complex, scientifically designated as diazotrophs (Alongi, 2002, 2009). Biological nitrogen fixation occurs in mangrove forests, preferably on the plant roots of the pneumatophores, on open floors on sediments, microbial mats, leaf litter, and cyanobacterial crusts growing on trunks.

Several factors influencing the process of biological nitrogen fixation include reduced light intensity, water content, photorespiration, endogenous oxygen (O_2) production, phosphorus availability, reduced availability of dissolved organic carbon, and retarded rates of nitrogen fixation.

3.17.6 Nitrogen Fluxes Across the Sediment–Water Interface

In nutrient cycling, nitrogen enters and leaves the mangrove ecosystem through the tidal exchange processes (Alongi, 2002, 2009). Mangroves acting as sinks or a source of nitrogen in the coastal environment exhibit variabilities in respect of time and space and also reveal variation in dissolved forms of nitrogen (Adame & Lovelock, 2011).

3.17.7 Nitrogen Losses to the Atmosphere

Alongside losses of nitrogen through tidal exchange, another means of loss of nitrogen from mangroves to the atmosphere is microbial-mediated nitrogen transformations in sediment. Denitrification representing a reduction process for converting NO_3− and NO_2− to N_2O and N_2 are of two main types of direct denitrification supported by NO_3− diffusing from overlying water into sediment, and coupled denitrification, where NO_3− is supplied from nitrification. Taking into consideration the magnitude of differences in respect of rates between denitrification, ammonification, and nitrification within mangrove forest stands, it can be inferred that direct denitrification is considered an important pathway for the loss of gaseous nitrogen to the atmosphere from the mangrove forest. The maximum rates of denitrification within mangroves may in part reflect the roles of NO_3− supplied from flood tides toward enhancing direct denitrification in mangroves (Rivera-Monroy & Twilley, 1996). However, the wide ranges of rates of N_2O fluxes and denitrification within mangrove sediments indicate that mangroves exhibit high rates of nitrogen loss to the atmosphere (Alongi, 2013). In addition, increased availability of nitrogen and rates of cycling of nitrogen in mangrove sediment are directly connected to net primary productivity and assimilation of extra nitrogen.

3.17.8 Denitrification in Mangroves

Denitrification is a process by which different nitrogenous compounds are converted to free nitrogen by the activities of specific bacteria, designated as denitrifying bacteria, which help chemical reduction of soil nitrites or nitrates to gaseous end products. Several biogeochemical processes such as mineralization, nitrification, and denitrification alternating with anoxic and oxic conditions coupled with tidal inundation and exposure within the mangroves tend to slow the pace of denitrification in comparison with the estuarine counterpart (Rivera-Monroy & Twilley, 1996). Two types of denitrification are observed in the mangrove ecosystem, direct and coupled, depending on the source of NO_3 The direct denitrification in mangrove systems is driven by the diffusion of NO_3 into the sediment, whereas coupled denitrification occurs with the help of nitrate products generated by the process of nitrification (Rivera-Monroy et al., 1995). This interpretation is validated by the incidence of a slower rate of natural denitrification in mangrove sediment depending on the availability of NO_3. Mangrove forest tracts exhibit higher denitrification rates in comparison with the naturally occurring ones on receiving a large amount of NO_3 from the sewage treatment plants (Nedwell, 1975; Corredor et al., 1999). A hypothesis has been put forward by Nixon et al. (1996), which states that denitrification in the sediments of the continental shelf zones of the North Atlantic Ocean must be balanced by an equivalent flux of nitrification in the open ocean in order to maintain a steady-state condition in the marine ecosystem.

3.17.9 Solar Radiation and Radiative Transfer for Nutrient Cycling in Water

Detritus, being the largest pool of organic matter in the biosphere, involves decomposition of detrital carbon to the level of 50% to 100% of the carbon flux in the aquatic ecosystem, mainly through biological as well as photochemical reactions. The decomposition of detritus has two major impacts on carbon cycling in the aquatic ecosystem: (1) it directs the microbial processes accelerating the decomposition of detrital carbon and strengthening the stability of the detrital food web and (2) it helps the mineralization of organic carbon.

It has been noted that more dissolved nitrogen and phosphorus are converted into particulate form after being exposed to solar radiation in open water than in non-exposed water. The nutrient dynamics indicated increased production of microbial biomass in the solar radiation-exposed water. The leaves of plants suffer from the loss of organic matter after prolonged exposure to solar radiation and this also triggers the fragmentation of leaves. It is also found that providing detrital leaves with exposure to radiation enhances the incorporation of 14C-leucine, making the introduction of microbial biofilm two to four times higher than without solar radiation, justifying the significance of solar radiation in the decomposition of detritus.

After emitting a constant flux of radiation (1450 W m^2) from the sun, solar radiation triggers photochemical reactions on reaching the outer edge of the Earth's atmosphere. However, such a triggering impact on the visible range of the spectrum (i.e., in the wavelength range 290 to 3000 nm) and the proportion of UV-B radiation vary in respect of latitude, longitude, time, atmospheric density, water vapor, cloud coverage, wind flows, and load of atmospheric pollutants such as aerosols, particulate matter, ozone, and gases (Molina & Molina, 1986; Frederick & Lubin, 1988; Zepp & Cline, 1977; Miller et al., 1999; Blumthaler et al., 1992; Frederick et al., 1989; Madronich et al., 1995).

As the absorption of a single photon represents the primary event in photochemistry, the solar radiation tends to influence the photon flux density (number of photons reaching the surface of the Earth per unit area (during a normal summer period enjoying a higher global radiation of 600 W m^{-2}, (1024 photons below 500 nm are received by a square meter of the Earth's surface). This is in contrast to the period at night when the photon flux density reaches close to zero, pointing out variation of the temporal photon flux density of about 1024-fold.

Proper assessment of photochemical reactions requires an accurate understanding of the intensity, the spectral distribution, and the incidence of photon flux. Thus, in different aquatic environments, the attenuation of solar radiation may vary by over five orders of magnitude.

1. *Absorption of radiation*

   Absorption of radiation with photon particles triggers a photochemical reaction, which in turn results in electronic excitation of a detritus molecule (Stumm, 1992). The humic substances such as those chromophoric in nature are found to be the most important group of detritus and contribute most to absorption. In an open-water ecosystem, mild absorbers are the water (Arrigo, 1994; Smith & Baker, 1978), nitrate and nitrite (Zuo & Deng, 1998), suspended particulate matter such as chlorophylls, carotenoids, and polyunsaturated fatty acids (Zafiriou et al., 1984), and degradation products of pigments.

2. *Photochemical reactions*

   Energy from the excited molecules is lost via physical or chemical quenching (Mialocq, 1997). Most of the excitation energy is lost either as heat (Haag & Mill, 1990) or by emitting a photon, or as sensitization by transferring energy to another chemical species (Mialocq, 1997). Photochemical reactions occur either through direct or indirect reactions (Hoigné, 1990; Schwarzenbach et al., 1993). Direct reaction takes place through a chemical reaction involving excited molecules, such as intramolecular rearrangement, isomerization, fragmentation, and abstraction of a hydrogen atom (Schwarzenbach et al., 1993). An indirect photochemical reaction, also designated photosensitization, undergoes the chemical reaction by initially transferring energy from the excited molecule to another molecule (Zepp et al., 1985; Hoigné, 1990; Schwarzenbach et al., 1993; Mialocq, 1997). Humic substances as the dominant absorbers act as important sensitizers for photo-chemical reactions (Zepp et al., 1985).

3. *Prediction of the rate of photochemical reactions*

   A photo-chemical reaction is similar to that of the mineralization of organic carbon and proceeds through several reaction mechanisms, each of which has its own properties.

   The prediction of the rates of photochemical reactions can be estimated by taking into consideration three parameters: (a) the absorption of photons by the medium; (b) scalar photon flux density; and (c) the quantum yield of the reaction (Zepp & Cline, 1977). The product of (a) and (b) provides an estimate of the rate of photon absorption. The total number of absorbed photons generated from the photochemical reaction can be assessed by the quantum yield.

4. *Photochemical reactions for the processing of detritus*

   The aromaticity of the humic compounds is decreased by the photochemical reactions and is increased by the number of aliphatic structures, carboxylic and carbonyl groups (Kulovaara et al., 1996). The molecular weight of dissolved detritus is found to be increased by photochemical reactions, whereas usually the molecular weight of dissolved detritus is decreased (Strome & Miller, 1978; Geller, 1986; De Haan, 1993; Hongve, 1994). The common photoproduct is CO_2 (Gjessing & Gjerdahl, 1970; Chen et al., 1978; Miller & Zepp, 1995; Granéli et al., 1996, 1998; Gao & Zepp, 1998; Miller & Moran, 1997; Moran & Zepp, 1997), whereas other organic photoproducts acting as suitable substrates for bacterial proliferation are subject to bacterial biodegradation. Photochemical reactions help to mobilize and release detritus-bound nutrients in the form of inorganic phosphorus and nitrogen from complex organic compounds (humic substance–iron–phosphorus complexes) (Manny et al., 1971; Cotner & Heath, 1990; Bushaw et al., 1996). After being released from photochemical reactions, the nutrients are provided to detrivorous fauna, which in turn expedite the process of biodegradation (Bushaw et al., 1996).

Photodegradation of senescent phytoplankton results in the chain-wise degradation of chlorophylls, chlorophyll phytyl chains, sterols, unsaturated fatty acids, carotenoids, and cyanobacterial toxins (Rotani, 1998; Choswell et al., 1999). Photochemical reactions on lignin-based compounds bring forth cleavage of methoxy groups resulting in the formation of hydroxyl, carboxyl, quinones, and keto groups (Weir et al., 1994, 1996; Argyropoulos & Sun, 1996). In the process of photochemical reactions different bonds within a polymer break down releasing phenoxy monomers of low-molecular-weight compounds (Gold et al., 1983; Castellan et al., 1987; Argyropoulos & Sun, 1996). Twenty-six aliphatic acids were documented after the photo-degradation of lignin (Sun et al., 1998).

5. *Effects of solar radiation on detrivores*

The ecological process leading to the biodegradation of protein and several other forms of labile detritus is slowed down by solar radiation (Naganuma et al., 1996; Benner & Biddanda, 1998; Tranvik & Kokalj, 1998; Anesio et al., 1999; Obernosterer et al., 1999). Solar radiation helps the decomposition of lignin, as evident in the process of the conversion of the leaves of seagrass into detritus and also brings about changes in the chemical properties of lignin residues in the water column (Opsahl & Benner, 1993). Photochemical reactions within cells are noxious for the well-organized biochemical machinery, involving many reactions that take care of the functions of life. Solar UV radiation is observed to cause photohydration of cytosine and photo dimerization of adjacent bases (Jeffrey et al., 1996). In addition, solar radiation is found to increase enzyme activity in waters loaded with humic compounds by breaking the complexes between enzymes and humic substances (Wetzel et al., 1995). Moreover, the impact of solar radiation results in the loss of selected sensitive species, bringing forth the changes in the structure of the biological community of UV-sensitive populations (Bothwell et al., 1994). However, organisms are able to protect themselves and recover from the negative effects of solar radiation (Herndl et al., 1993; Müller-Niklas et al., 1995; Pakulski et al., 1998). Antioxidants such as carotenoids, mycosporine, melanin, and flavonoids, play very effective roles in protecting the species by absorbing harmful UV radiation with the help of antioxidant enzymes such as catalase, superoxide dismutase, glutathione reductase, and glutathione peroxidase (Krasnovsky, 1994; Zellmer, 1995).

3.18 Photodegradation of Structure Aligned: Impact of Solar Radiation with Implications for Bacteria

Solar radiation is found to increase microbial activity by liberating nutrients and labile substrates for microbes (Cotner & Heath, 1990; Bushaw et al., 1996). However, photochemical degradation of lignin requires enhanced activity of microbes on the leaves (Opsahl & Benner, 1993). Loss of lignin through the photochemical degradation process results in structural deformities of the leaves and thereby leads to the fragmentation of the leaves, which in turn enables microbes, especially fungi, to invade the mesophyll tissues of the leaves and promote colonization of the leaves by bacteria to facilitate the completion of the decomposition process.

3.19 Mangroves and Methane Emission

Methane (CH_4), being an important trace gas in the global atmosphere, for example, acts as a sink for OH, a source of CO_2, O_3, and CO in the troposphere and represents one of the greenhouse gases causing global warming by absorbing outgoing radiation in the infrared spectrum of wavelength. Recent research information has established the fact that the global emission of concentration of CH_4 in the atmosphere has been increasing by 1–2% a year (Graedel & McRae, 1980). The wetlands receiving sewage effluent (Harriss & Sebacher, 1981) and nitrogen-based fertilizers tend to produce more emissions than other wetlands without such substances. The macrophytic plants of the aquatic bodies influence the rates of CH_4 gas from the soil to the atmosphere and especially in marine habitats with higher sulfate concentration result in more emission of CH_4 (Reeburgh & Heggie, 1977) by regulating the relative activities of sulfate-reducing bacteria in the overall magnitude and distribution of CH_4 concentrations. Temperature, salinities, and organic matter, being important environmental factors, exert ultimate controls over the complex biological influences on these soils, which are crucial to prevailing atmosphere–biosphere links. Salinity gradients within the mangrove ecosystem influence the changing flux of methane, both seasonally and spatially, demonstrating largest peak fluxes occurring at the site of lowest salinity. The conjunct roles of methanogenic and sulfate-reducing bacteria both in the soils and pore water, exhibiting an inverse relationship, act as a primary control on CH_4 flux.

Owing to the reduction of CO_2, and fermentative disproportionation of low-molecular-weight compounds such as acetate, methane production has been found to be relatively low in mangrove ecosystems, with a range of emissions of 0 to 5 mmol m^{-2} day^{-1} (Kristensen et al., 2008). However, it is perceptible that methane emission is drastically increased from the mangrove ecosystem because of the loading of organic compounds including nutrients from different anthropogenic activities, which in turn induce grave oxygen stress as well as supplying labile organic carbon (Kristensen et al., 2008).

3.19.1 Production Processes and Sources of Methane: Variability Fluxes of Concentration

Methane (CH_4) is one of the prime greenhouse gases causing greater global warming and thereby has the potential to influence the global climate directly or indirectly. In the marine–estuarine environment, CH_4 is produced under both the direct and indirect influence of anaerobic conditions in sediments (Holmes et al., 2000) and also under aerobic influence in surface waters with nutrient-limiting conditions. The higher organic loads, coupled with the prevalence of anoxic conditions in estuarine sediments, accelerated the production process of CH_4. Sediments of shallow coastal water zones endowed with grasses also act as a significant source of CH_4 to the water column (Anderson & Bryant, 1990; Mau et al., 2015; Treude & Ziebis, 2010).

Methane (CH_4) representing the second most abundant greenhouse gas contributes profusely to global warming-mediated climate change. This gas, which has a pungent smell, is produced under anaerobic condition in the mangrove ecosystem (Foster et al., 2007; Dutta et al., 2017). Moreover, microbial methanogenesis is instrumental for the conversion of about 1% of the annually fixed CO_2 through the process of photosynthesis, thereby helping in the cycling of an estimated 1 billion tons of methane (CH_4) each year.

The mangrove ecosystem producing methane (CH_4) via a diffusion process across the air–soil interface of the intertidal zone acts as one of the major sources and exchanges the same across the forest–water–air continuum of the coastal environment. About 75% of all oceanic methane emissions take place from the coastal shelf area, which covers about 16% of the world's oceans and stands as one of the most biologically productive eco-regions of the world (Bange et al., 1994). Tropical coastal mangrove ecosystems being among the most productive ecosystems in the world are characterized by high turnover rates of organic matter and nutrient recycling between the marine and terrestrial ecosystems (Odum & Heald, 1972; Nedwell et al., 1994; Chakraborty, 2011; Kathiresan, 2011). Therefore, mangroves have assumed great importance in driving the biogeochemical cycling of carbon and associated elements along the tropical continental margins, especially because of the bioconversion of mangrove litter into methane (CH_4) under the anoxic conditions of the sediments, facilitated by the higher temperatures in humid conditions and thereby forming a prime source of atmospheric methane (Purvaja et al., 2004). Lateral movement of the supersaturated pore-water with Methane (CH_4) carried to the estuary through tidal creeks could also act as a contributor of methane (CH_4) to the estuarine water (Bartlett et al., 1985). The hydrochemical parameters, including the tidal oscillation of the water table level, temperature, turbidity, and salinity, regulate the emission of methane (CH_4) from the estuarine water (Kang & Freeman, 2002).

A higher emission rate of methane was noted in the Hooghly estuary of the Sundarbans, India, during the post-monsoon season, which coincided with the maximum blooming of phytoplankton due to the maximum transport of nutrients from the mangrove forest, coupled with the transparency of the water resulting from lower inputs from river flows. Dissolved methane concentrations in the Hooghly estuary, one of the major estuarine flushing channels exhibited lower values ranging between 10.3 and 59.25 nmol/L with emission rates fluctuating between 0.88 and 148.6 μmol/m^2/day. The concentration of dissolved methane in the pore water of the mangrove forest tracts of the Sundarbans, India, reached a maximum concentration of 5769 nmol/L, which was found to be lower in comparison to those of coastal salt marshes and freshwater wetlands under tidal influence (Dutta et al., 2017).

A research study dealing with the biogeochemical cycling of methane (CH_4) in the highly anoxic but organic carbon-rich sediments of the intertidal zone at Lothian Island, a virgin deltaic island in the Saptamukhi Estuary of the Indian Sundarbans, during a period of 2 years (June 2010–December 2012), has revealed mean rates of methanogenesis in the range of 3547 μmol/m^3 and 48.88 μmol/m^3 in wet sediment per day within the subsurface sediment (up to a depth of 25 cm) and sediments from the shallower parts of the subtidal zone (up to a depth of 5 cm), respectively. The concentration of methane (CH_4) in the pore-water was observed to be 53.4 times more supersaturated than that in the nearby estuarine water, indicating a significant efflux of methane (CH_4) from the sediments of the mangrove forest floor to the adjoining estuarine water through the processes of advective and diffusive transport. About 8.2% of the total CH_4 produced in the intertidal mangrove sediments was transported to the adjacent estuary. The advective flux of total methane (CH_4) produced in the sediments of the intertidal mangrove forests to the estuarine water was found to be 20 times higher than the diffusive CH_4 flux. The total amount of methane (CH_4) released from the mangrove-estuarine complex of the Sundarbans to the regional atmosphere was estimated to be 0.09 Gg per year (Dutta et al., 2017).

3.20 Carbon and Carbonate Metabolism in Coastal Aquatic Ecosystems

Out of several ecosystems in the realm of the oceanic environment, the coastal continental shelf region enjoys the major share of the net ecosystem production (NEP) (75%) followed by macrophyte-dominated ecosystems (16%), mangroves (7%), marshes (3%), and coral reefs (2.6%) (Gattuso et al., 1998; del Giorgio et al., 1997). The coastal regions of the ocean ecosystem contribute more than 40% of marine calcium carbonate production (Cebridn et al., 1997; Milliman, 1993), which is reflected by the higher net sinks and sources of CO_2 for the atmosphere (Kempe, 1995). This is because of the transfer of large fluxes of nutrients from the deep ocean to the coastal region by upwelling, vertical mixing, and outwelling.

Both aerobic and anaerobic modes of sulfate reduction contribute significantly to the respiration process in mangrove sediments (Kristensen et al., 2008), which accounts for 75% to 125% of total mineralization in the functioning of the mangrove ecosystem. However, sulfate reduction undergoes variation because of the variation of species composition, age, stand, and biomass of the plants. In a 60-year-old stand, sulfate reduction accounted for around 75% of carbon oxidation, which could lend support to all carbon oxidation of a 15-year-old stand. It was found through a research study that assemblage of mangrove plants within mangrove forest dominated with the plant species: *Rhizophora* exhibited higher rates of sulfate reduction than *Avicennia* spp.

The accelerating magnitude of ecological changes brings forth altered sediment transport, freshwater runoff, and fluxes of nutrients to coastal systems, mainly because of anthropogenic intervention through altered land-use patterns and a reduction in the extent of vegetation cover (Walsh, 1991). Global carbon cycles are influenced by the transfer of organic matter from the land to the ocean via estuarine and riverine networks (Richey et al., 1990). Estuaries have become heterotrophic ecosystems because of ever-escalating, intensive anthropogenic disturbance, serving as a major source of carbon dioxide (CO_2) generation (Frankignoulee et al., 1998; Mukhopadhyay et al., 2002a, b; Biswas et al., 2004). Efficient carbon (C) sequestration is connected with higher primary productivity, the allocation of carbon (C) in belowground biomass, the burial of long-term organic carbon (C), lower rates of sediment respiration, and the substantial transportation of refractory dissolved organic carbon (C) to the ocean, first to open waters and then burial in bottom sediments (Kristensen, 2008; Donato et al., 2011; Duarte et al., 2013).

Tidal mangrove forests by virtue of their unique ecological characteristics can efficiently utilize limiting nutrients and sequester carbon effectively by balancing carbon allocation within biomass and releasing respiratory CO_2 (Alongi, 2009, 2012, 2014). The mangrove ecosystem being one of the most productive ecosystems of the world with the richest carbon stocks can ensure higher exchange of nutrients between forests and estuaries through high turnover rates of organic matter (Donato et al., 2011, 2012; Alongi, 2014).

Globally, despite accounting for only 0.7% of the tropical forest area, mangroves play a vital role in carbon cycling and their destruction leads to 10% of global CO_2 emissions (Alongi & Mukhopadhyay, 2015). The productive and biodiverse mangrove ecosystems in tropical and subtropical coastal estuarine zones are highly threatened ecosystems, as almost one-third of the world's mangroves have been lost in the past 50 years (Hutchison et al., 2014; Alongi, 2002). It is believed that all mangroves across the world might become extinct by the end of the present century (Duke et al., 2007). However, the rates of such disappearance have declined in the recent past because of an increasing awareness of the importance of this ecosystem (FAO, 2007; Kauffman et al., 2003) and the adoption of proper conservation strategies (Grabowski & Chazdon, 2012).

Land-use change coupled with deforestation driven by anthropogenic processes impairs carbon stocks and the strength of carbon sinks within the plant biomass, water, and soils of the Sundarbans mangrove ecosystem, thereby paving the way for the conversion of a productive estuary into a heterotrophic one (De et al., 1991; Frankignoulee et al., 1998; Mukhopadhyay et al., 2002a, b, 2006). The Hooghly estuary emerged as a deltaic offshoot of the mighty Ganges River, which drains much of the sediment from the southern slopes of the Himalayas while flowing through the western part of the mangrove-estuarine complex of Sundarbans, providing inorganic nutrients to the ecosystem. Most of the nutrients, organic detritus, and inorganic elements (N, P, C, K, S, etc.) are produced because of the decomposition of mangrove litters falling on the floors of mangrove forests. Through interacting with the water, sediments, and microbes, these nutrients are released during estuarine transport in the forms of dissolved inorganic nitrogen (DIN), dissolved reactive phosphate (DRP), dissolved inorganic carbon (DIC), silicates, and ammoniacal nitrogen. A study has reported that monsoonal runoff could account for 64% of

DIC, 71% of DIN, and 76% of both DRP and silicates of the total transport of such nutrients in the Sundarbans, India (Mukhopadhyay et al., 2006).

Minerals in association with organic matters tend to rest temporarily on banks, floodplains, and riverbeds and their resuspension during turbulent conditions could release nutrients into the water column. The Ganges River estuary, representing one of the nine major rivers in Asia, plays a vital role in transporting a huge amount of sediment materials, especially during the monsoon season, accumulating a vast amount of nutrients and organic material.

3.20.1 Roles of Decomposition of Detritus in the Cycling of Carbon and Nitrogen in the Tropical Mangrove Forest

The elemental composition of nutrients derived from the decomposition of detritus of *Rhizophora apiculata* and *Enhalus acoroides* have shown particulate organic carbon and nitrogen in detritus and sediments that help the exchange of oxygen, CO_2, DOC, and inorganic nitrogen across the sediment-water interface. The rates and duration for completion of the decomposition of seagrass leaves were up to 60% after 43 days, without any decomposition of the seagrass rhizome detritus. The impacts of CO_2 production after the addition of detritus promote an immediate increase in sediment oxygen uptake (SOU) in the seagrass sediments, which is in contrast to the mangrove plants-dominated sediments, where SOU remained low under similar conditions. The carbon-rich detritus produced by the decomposition of both mangrove leaves and seagrass result in a decrease in carbon:nitrogen ratios, but the decomposition of seagrass leaves in the less carbon-rich detritus promotes an increase in carbon:nitrogen ratios. The colonization of bacteria promptly takes place because of the decrease in carbon:nitrogen ratio, which tends to incorporate nitrogen from both the overlying water column and from the particulate organic nitrogen reservoir of sediments.

The potential for nitrogen enrichment in mangrove sediments diminishes owing to lower nitrogen fixation during decomposition, whereas a greater nutrient reservoir enhances the potential of the direct incorporation of bacterial nitrogen from porewater pools (Holmer et al., 2001). The amendment of organic matter reduces nitrification and enhances denitrification. In addition, low rates of mineralization in the sediments and low concentrations of ammonia (NH_4) in the overlying water column tend to regulate the process in tropical mangrove forest concentrations. The chain-wise production processes in the mangrove ecosystem reveal that an increase in production of CO_2 and SOU indicates that the addition of detritus enhances the process of mineralization and thereby opens up avenues for the increased sources of NH_4 and thereby promotes greater nitrification, but reduced denitrification (Rivera-Monroy & Twilley, 1996).

Significant immobilization of nitrogen takes place during the degradation of the nitrogen-poor mangrove detritus. The carbon:nitrogen ratios of the mangrove detritus decrease because of colonization by bacteria and greater loss of carbon than nitrogen in the mangrove leaves (Steinke et al., 1993; Wafar et al., 1997). Occurrence of higher concentrations of tannins, as enzymatic inhibitors and anti-microbial agents, is the cause of lower rates of mineralization in mangrove sediments. The mineralized nitrogen remains stored in the sediments of the mangrove floor, where the early phase of degradation of mangrove leaves enhances the availability of nutrients to the mangrove plants and also to detritivores.

3.20.2 Significance of the Coastal Ecosystem in the Context of the Global Oceanic Carbon Cycle

In the global context, all coastal ecosystems represent net autotrophs in contrast to estuaries, which are net heterotrophs demonstrating a negative NEP. Out of several eco-regions of the oceanic environment, the continental shelf proper is the prime contributor to NEP of the coastal zone (75%) followed by macrophyte-dominated ecosystems (16%), mangroves (7%), marshes (3%), and coral reefs (2.6%) (del Giorgio et al., 1997; Hazra and Sanyal, 1996; Howarth et al., 2002; Mumby & Hastings 2008).

Out of the total production of calcium carbonate ($CaCO_3$) in the marine environment, the coastal belt contributes more than 40%, which is produced in the continental shelf region and transferred and deposited on continental slopes, raising parts of the ocean. In addition, the delivery and accumulation of carbon are increased by human activities (Milliman, 1993; Alongi et al., 1989; Meybeck, 1993). This accumulated carbon is deposited in the sediment or oxidized to CO_2, or transported to the deep open ocean, playing a major role in the global carbon cycle. The coastal belts in the oceanic realm, being heterotrophic owing to a balance between increased primary production and increased respiration, act as a source of CO_2 to the atmosphere (Smith & Hollibaugh, 1993; Wollast & Mackenzie, 1989). Remineralization also releases nutrients from the organic carbon and promotes primary production. Alongside a higher pace of sedimentation in the coastal zone, mainly because of increased continental erosion due to deforestation, altered land uses and modern technology-based agricultural practices accelerate the rate of carbon burial in coastal sediments (Wollast, 1991).

3.20.3 Mangroves and Carbon Cycling with Special Reference to Eutrophication

A major portion of dissolved and particulate materials from the continent to the marine system is transferred by estuarian pathways. A wide range of biodiversity, ecological variables, geomorphological settings, physical processes such as tidal amplitudes, currents of water, stratifications, waves, etc., intense sedimentation, resuspension, and geochemistry have occurred in estuaries, making these brackish water bodies extremely dynamic systems (Odum & Heald, 1975; Twilley, 1998). The relationships between the biological system (population, community, interspecific competition) and the global carbon cycle depends on eco-biological attributes such as (1) the balance between organic carbon production, primary production, and consumption; (2) the balance between precipitation of calcium carbonate followed by dissolution; (3) climatic conditions (temperature and light), nutrient availability, etc. Higher production than consumption makes any system autotrophic in respect of organic carbon, which is contrast to a heterotrophic system, when consumption exceeds production.

In recent decades, the number of research studies highlighting local and regional carbon stocks have increased greatly but only a few have attempted to provide estimates of global mangrove carbon budgets (Twilley et al., 1982; Bouillon et al., 2008), even though the fact remains that mangroves are the most carbon-dense tropical forests (Donato et al., 2011).

Mangroves act as carbon sinks because of the production and decomposition of litter, but around 50% of mangroves are being increasingly cleared by human activities (Hatcher et al., 1989), which led not only to a reduction in the fixation of atmospheric CO_2 along with carbon burial but also the oxidation of the stored organic carbon in sediments for releasing CO_2 into the atmosphere (Springfield, 1996). All of these incidences vary both spatially and temporally because of the many ecological factors and processes such as the species composition of the forest, the diversity and density of faunal components belonging to different trophic levels, temperature, rainfall, salinity, and microbial activity. Mangroves play crucial roles in blue carbon sequestration by acting as carbon sinks, which are crucial in mitigating greenhouse gas-induced climate change.

The litter fall of mangroves acts as a major source of carbon for benthic heterotrophs, followed by phytoplankton, microbenthos, and macrophytes, and also increases benthic bacterial productivity (Alongi, 1988a; Boto et al., 1989; Moriarty, 1986; Stanley et al., 1987). The power of consumption and hiding by crabs enhance the retention and processing of litter within the mangrove system (Robertson et al., 1992). Moreover, microbial pathways enhancing a large proportion of the total carbon flow in mangrove ecosystems support the '**carbon sink hypothesis**' (Alongi, 1989, 1990), which addresses the fact that only a small proportion of major bacterial biomass is consumed in the sediment, leaving the remaining bacterial carbon to be recycled very efficiently by way of carbon turnover and natural mortality within the microbial food web operating within the mangrove floor. The amount of organic carbon accumulated, sequestered, and exchanged between the forest and aquatic subsystems exhibits a wide range of variation based on the species composition, geomorphological characteristics, intensity and frequency of tidal inundation, wind speed and direction, freshwater discharge through river flows), and biological interactive attributes (leaf-litter fall followed by fragmentation, burial, and consumption by detritivores) (Middelburg et al., 1996; Twilley et al., 1992; Robertson et al., 1991).

The balance between autotrophy and heterotrophy within the mangrove estuarine ecosystem is influenced by eutrophication caused by an increased nutrient load, which in turn enhances NEP, and shifts the system toward increased autotrophy (Boynton et al., 1996; Heip et al., 1995; Kempe, 1995; Middelburg, 1996; Schlesinger, 1997; Smith & Hollibaugh, 1993; Kempe, 1982; Lohrenz et al., 1997). In contrast, heterotrophy becomes more prevalent in the macrotidal zone, with a wide range of salinity fluctuation from almost freshwater level to 30% because of respiration of the organic carbon, leading to increased heterotrophy with CO_2 supersaturation (Kempe, 1982).

3.20.4 Carbon Dynamics During the Process of Decomposition

The loss of carbon from the mangrove forest was generally higher because of the difference in the amount of flux incubation than carbon recovered in the flux, and the deficit in the recovery of carbon is supposed to be due to a number of factors:

1. The loss of carbon is required to be estimated properly because difference in the the mangrove leaves can recover carbon more efficiently than other plants in terrestrial ecosystem.
2. Changes in pH during decomposition influence the precipitation or dissolution of carbonate, which in turn affects CO_2 concentrations.
3. It is difficult to estimate CO_2 production properly owing to the low frequency of CO_2 flux, mostly during the initial decomposition.

A major source of inorganic carbon is respiratory CO_2 derived from mangrove detritus (Hemminga et al., 1994; Lin et al., 1991). The roles huge amount of mangrove litter in the decomposition processes within the sediment of mangrove forests through the tight coupling with adjacent ecosystems in the vicinity of the forests are enormous to sustain the ecosystem functioning.

3.21 Phosphorus Cycling in the Mangrove Ecosystem

Deficiency of phosphorus is a major factor limiting primary productivity in the mangrove ecosystem. Particulate phosphorus occurs in the mangrove soil as detrital phosphate bound with several metallic elements (Fe-P, Ca-P, Al-P, etc.), and also as organic phosphorus and water-soluble phosphorus (Cotner Jr. & Heath, 1990, Alongi et al, 1992). Organic phosphorus in the phytate form remains bound to humic compounds and therefore is not available for plants and microbial populations (Alongi et al, 1992; Nielsen and Anderson, 2003; Lovelook et al, 2006). The major sources of phosphorus in the mangrove ecosystem is derived from different sources such as (1) erosion and weathering of phosphate minerals, (2) dry deposition from the atmosphere, (3) in water metal phosphates are solubilized, (4) absorption of phosphate onto clays, (5) from microbial excretion and also excretion from larger faunal components, and (6) release of phosphorus-based compounds from human activities (industrial effluents, sewage, agricultural runoff). Phosphorus-based substances are always found in oxidized form such as inorganic orthophosphates and pyrophosphate. Phosphorus undergoes a complex biogeochemical transformation process after being taken up by mangrove roots and involved in a number of metabolic processes leading to the generation of adenosine triphosphate. Many phosphate-solubilizing bacteria (PSB), which are associated with rhizospheres and roots of mangroves, direct major pathways for mangroves to acquire sufficient nutrition (Cotner Jr. & Heath 1990; Alongi, 2008). Production of organic acids, H+ protonation, and alkaline phosphatases within mangroves takes place because of phosphorus solubilization (Silva et al., 2007). Alleviation of phosphorus limitation alongside stimulating the growth of mangroves are due to the greater colonization and activities of these microbes. After being stimulated by PSB and probably arbuscular mycorrhizal fungi, hydrolyzing enzymes for phosphorus esters such as soil phosphatases play a crucial role in mineralizing organic phosphorus in mangrove soils. The possibility of increasing the amount of phosphorus enhances considerably in the mangrove ecosystem by the activities of phosphate-solubilizing bacteria and solubilizing arbuscular mycorrhizal fungi (Howarth et al., 2002; Alongi, 2008).

The concentrations of phosphorus is found to have been reduced considerably in detritus because of the precipitation of phosphates having a greater abundance of cations in the interstitial water of intertidal belts of the mangrove ecosystem from which plants are unable to take phosphates (Holguin et al., 2001). However, the prevailing anaerobic conditions in mangrove sediments tend to promote the dissolution of nonsoluble phosphate through the production of sulfide (Holguin et al., 2001).

As a part of greenhouse-gas mitigation strategies, mangrove afforestation alongside protection of existing mangrove forests on a global scale has been given more emphasis during the last few decades because of the greater ability of mangroves for carbon sequestration for mitigating atmospheric greenhouse-gas emissions and also their potential for storing greater amounts of carbon per unit area in comparison with any other forest ecosystem (Ewel et al., 1998; Twilley et al., 2017; Twilley & Pablo Riul, 2018). A framework emphasizing the diverse coastal geomorphology in different coastal environmental settings after being determined by freshwater influx through riverine flows, tidal cycles with different amplitudes and duration, the intensity of water currents, dimension, and direction of waves and meteorological parameters such as rainfall, temperature, etc., have shaped the structure and expanse of mangroves, which in turn explains global variations in the potential and contribution of mangrove carbon storage and global carbon dynamics.

3.22 Influence of Animals on the Ecosystem Dynamics of the Mangrove Ecosystem

A diverse groups of animals inhabiting the mangrove ecosystem has support in terms of habitats and food materials. Some animals spend their entire lives within the mangrove ecosystem, whereas other groups of animals enjoy the support only for certain stages of their life cycle, mostly for their reproduction and nurturing of juvenile stages (Yañez-Arancibia et al., 1988). The intricate arrangements of pneumatophores and the trap roots of mangroves growing up in the intertidal zone provide space for the settlement, colonization, and growth of algae, periphytons such as sponges, mollusks, and crustacean fauna in the form of wood borers, and fouling organisms, which also play important roles in the functioning of mangrove ecosystems as macro decomposers of mangrove leaves through the process of grazing. Several species of crabs and insects cause profuse damage of propagules or seedlings as predators. The mangrove ecosystem, by virtue of enjoying the to and fro movement of materials between two subsystems, forest and aquatic, serves as the nursery ground for several commercially important shell fish and finfish. Many semi-aquatic and even terrestrial animals, mostly insect and avian species, are found in mangrove forests and play crucial roles as herbivores, pollinators, predators, and also acting as an important food source for other animals (Hogarth, 2007; Ewel et al., 1998). Several species of mammals living in mangrove forests include monkeys, bats, rodents, raccoons, mink, otter, wild pigs, deer, black bear, and the only tidal large carnivore, the tiger. Large aquatic species living in mangrove-dominated estuaries include sea turtles, manatees, and dolphins.

Different large animals, owing to their size, longevity, and habitat requirements in the mangrove ecosystem, are responsible for biogeochemical, successional, and also for landscape alterations because of the interactions among themselves and with the plant community through the process of herbivory or predation, which influences their food supply, consumer population dynamics reverberating throughout all trophic levels, causing alterations to the ecosystem and the biotic community structure., life history strategy, etc., which in turn induces physical and chemical alterations to the environment affecting the biogeochemical cycling of nutrients and ions in soils, sediments, and water. Bioinvasion by exotic plants and animals causes disruption of ecosystem functioning (Elton, 1958; Vitousek, 1986).

3.22.1 Benthic Fauna as Ecosystem Engineers: Bioturbation and Nutrient Cycling

Ecosystem engineering is the direct or indirect controlling process of generating and utilizing resources by an organism's ability, which in turn tends to result in changes in the physical state of the abiotic or biotic condition of a habitat with creation, destruction, or modification (Jones et al., 1997). Different ecological parameters determine the abilities of ecological engineers alongside facilitating the processes of creating new avenues for resource generation and utilization.

A habitat includes temperature, space to act, humidity, moisture, light texture of sediment, etc., which together help to modulate the abiotic parameters such as currents, sediment erosion and deposition, nutrient reshuffling, etc.

Bioturbation, as an important behavioral manifestation of ecological engineers, refers to a process that affects sediments through the formation of several biogenic structures and also by ensuring the redistribution of elemental components of soil particles in marine, estuarine, mangrove, and terrestrial ecosystems across the world. These bioturbators, also designated as ecological engineers, bring forth through their biogenic activities biogeochemical heterogeneity by way of modulating carbon and nutrient availability, altering the process of oxygen supply, changing redox potential, pH, and temperature in water and sediments, and causing changes in microbial communities (Booth et al., 2019; Fusi et al., 2022).

Bioturbation involves different biological activities of benthic faunal components such as feeding, mobility, burrowing, etc., which in turn disrupt the geochemistry of sediments and their interstitial water by disturbing through the processes of stirring or mixing of sediment layers (Woodin et al., 1998). In the Sundarbans mangrove ecosystem, major bioturbators include brachyuran crabs, polychaetes, and gobiid fishes, which can burrow continuously and produce several bioturbatory structures such as sand balls, mud balls, pseudo-pellets, sand pyramids, semidome, chimney, hood, etc., which reflect seasonal variations in the rate of sediment destabilization, pumping of water and oxygen within soil, intensity of soil excavation, etc. (Annon, 2003; Chatterjee et al., 2008). Several species of soil-inhabiting microarthropods also demonstrate bioturbatory activities that expedite the process of the microbial degradation rate of sediment organic materials by the microbial and benthic community (Dey et al., 2010; Chakraborty, 2017).

Intertidal macrofaunal groups demonstrating diverse forms of diet and feeding habits such as herbivores, predators, detritivores, and omnivores impact consequential effects on the sediment environment. Through the process of burrowing by benthic fauna (crabs, polychaetes, actinarians, etc.) especially in the intertidal belt during low tide, reshuffling of nutrients, alteration of textural components, downward vertical transfer of gases, pumping of oxygen, enhancement of the surface area to promote gaseous exchange, and facilitating redox reactions (Kristensen, 2008; Chatterjee et al., 2008), whereas at high tide, they provide sites with steep geochemical gradients. However, during the high tide, the benthic fauna through their bioturbatory activities, develop boundaries between oxygen-loaded water with anoxic but metabolites-enriched sediment (Furukawa, 2005). Enhancement of oxygenation in the sediment by constructing burrows through the reworking of the substrate in intertidal belts triggers cascades of biogeochemical cycling at anoxic and nutrient-poor sediment, which in turn expedite the process of flourishing of organisms. Bioturbatory activities such as burrowing and sediment mixing by macrobenthic fauna increase habitat heterogeneity, which traps organic matter and promotes functional redundancy in microbial communities, which in turn increase the ecological resilience of the entire ecosystem by their synergistic roles with other organisms (Kristensen, 2008; Chatterjee et al., 2008; Furukawa, 2005; Bertics & Ziebis, 2009; Kristensen & Kostka, 2005; Fenchel, 1996; Botto & Iribarne, 2000; Lee, 1998; Icely & Jones, 1978).

Intertidal macrobenthic fauna feeding on mangrove litter promotes colonization of the diverse microbial community (bacteria, archaea, and fungi) through the process of detritus (Lee, 1998). Different feeding habits reduce organic carbon of the sediments into which this carbon would leach. The detritivorous crabs feeding on surface sediment harvest bacteria, meiofauna, and microalgae from the surface layer and deposit pellets triggering biochemical reactions and oxidation (Icely and Jones, 1978; Botto & Iribarne, 2000). Grazers such as gastropods, through their bioturbation activities by grazing over the algal mats across the surface of the sediments and also from the mangrove trunks expedite litter degradation and displacement of sediment on the mangrove floor. The brachyuran crabs under the family Grapsidae (Sesarmidae crabs), as leaf eaters and litter consumers, also supplement their diets with nitrogen-enriched components, taken from the mud loaded with bacteria (Kristensen, 2008; Skov and Hartnoll, 2002). These crabs displaying arboreal behavior collect mangrove leaves and propagules and store this collected food in their burrows (Andreetta et al., 2014; Lee, 1998). All of these crabs, by exhibiting their biological activities, exert a considerable impact on the structure and function of the mangrove ecosystem through consumption of mangrove leaf litter, thereby significantly reducing the export of detritus and enhancing the retention of organic matter, and their fecal matter also supports coprophagous organisms (Lee, 1999a, 1999b; Kristensen, 2008). In addition, bioturbatory macrobenthic fauna alter the soil texture, surface topography, chemistry of the soils, degree of anoxia, and abundance of other fauna and microbial populations (Alongi, 2009). The presence of the burrows networks alter the movement of tidal water by reducing and also speeding up the flow of water, enabling the redistribution of dissolved and particulate materials in between forest and adjacent aquatic subsystems (Chatterjee et al., 2008).

Out of so many benthic faunal components displaying different forms of bioturbatory activity, deposit feeders represent a prime faunal group of bioturbators because of their abilities to constantly process sediment for food, which results in the horizontal and vertical movement of particles in the sediment. However, in some cases biogenic alterations through the activities of bioturbators speed up the process of erosion of the sediments (Luckenbach, 1986), the rearrangement of sediment particles (Lim, 2006), and modification of concentrations of constituents of pore water (Aller, 1982), which in turn regulate the bioturbation process individuals to burrow through the sediment (Brenchley, 1982). Habitat heterogeneity brought about because of the bioturbation process influences the distribution and abundance of organisms with higher densities, which in turn lead to cause modification of physical and biogenic structures in much more effective way through their combined activities (Lee, 2008).

Morphology and sediment qualities in and around burrows of macrofauna tend to determine biogeochemical gradients and the ecology of the water–sediment interface as diffusion of oxygen from air or water into sediment varies as per the architectural plans of burrows (Bertics & Ziebis, 2009; Kristensen & Kostka, 2005; Fenchel, 1996).

Soil microarthropods also play their functional roles in the process of mangrove plant litter decomposition and are considered to be among the strongest determinants of plant-litter decomposition. In this process, these faunal components play an important role in nutrient cycling and enriching the system's wealth of biodiversity. The peak population of microarthropods coincides with the highest availability of the nutrient status of the ecosystem (Chakraborty & Chowdhury, 1985, 1994; Bhattacharya, 2000; Kremen & Ostfeld, 2005; Bouillon, 2011; Kristensen et al, 2008, 2012; Dey et al., 2010).

3.22.2 Functional Roles of Mangrove Intertidal Benthic Communities in the Sustenance of the Ecosystem: Their Roles in the Bio-Cycling of Fractional Fatty Acid Components of Lipids

Intertidal macrobenthic fauna play important roles through nonlinear interactions among themselves and also with other animals in search of their food and shelter, utilizing the environmental setups constituted by variable ecological parameters and functions for the sustenance and stability of the marine-estuarine ecosystems. By virtue of their differential feeding strategies (deposit feeders, suspension feeders, grazers, filter feeders, etc.), the foraging behaviors and excretory processes of mangrove-associated faunal components, especially benthic fauna, play important roles in nutrient cycling in mangrove–estuarine ecosystems, which convert complex organics into simpler ones (carbon, nitrogen, phosphorus, etc.), which contribute substantially to meet the nutrient demands of primary producers (Hedges, 1978; Kristensen et al., 1988; Lee & Kneib, 1994; Kristensen et al., 1995; Hazra & Choudhuri, 1990; Geller, 1986; Hazra & Sanyal, 1996; Hedges, 1978; Kristensen & Kostka, 2005; Kristensen, 2008; Lerberg et al., 2000; Leme, 2010; Fanjul, 2015). Among several other organic components, different types of fatty acids are considered an important determinant for the stability and ecological health of the mangrove ecosystem, which has been established from a study on the functional roles rendered by selected intertidal macro benthos, at an ecotone (Talshari), the confluence of an estuary, Subarnarekha with an open sea, Bay of Bengal, near Midnapore (East), coast, West Bengal, India.

The functioning of benthic assemblages in marine–estuarine ecosystems are dependent on the biological traits (feeding methods, body size, flexibility, mobility, and burrowing activity) exhibited by constituent species, which helps to determine the ecological processes by virtue of their behavioral manifestations such as feeding (deposit, suspension, grazing, and filter feeding) and movement (crawling, burrowing, etc.) activities and thereby disrupt and dislodge sediments, pump oxygen and water into the deeper parts of the sediment, and speed up organic-matter decomposition (Webb & Eyre, 2004; Hall-Spencer et al., 1999; Wassenberg et al., 2002).

Macrobenthic faunal groups help the ecosystem with nutrient regeneration with an water–sediment interphase. Therefore, the new addition and removal of such bottom-dwelling faunal components govern the composition of the biotic community, primary and secondary productivity, but biogeochemical cycles by controlling the flux of nitrogen, phosphorus, organic carbon, and oxygen, alongside maintaining the textural composition of the sediment–surficial characteristics, and community composition (Austen et al., 2002; Herman et al., 1999; Levin et al., 2001). However, the soft-bottom habitats of all those benthic fauna are subjected to a wide range of both anthropogenic and natural stresses stemming mostly from the changing patterns of resource use, pollution and nonjudicious modes of exploitation (Gray, 1997). The pressing need of the moment is to gain a proper understanding of the intricate relationships among the diversified biodiversity components in relation to the changing ecosystem processes of shallow coastal zones to undertake proper strategies for the conservation of species with a low potential for survival after disturbance, and to combat long-term degradative change (Bolam et al., 2002; Lohrer et al., 2004; Solan et al., 2004; Beukema et al., 1999; Thrush et al., 2003; Thrush & Whitlatch, 2001).

Understanding of ecophysiology, functional relationships and the natural history of the biotic community interactions, trophic linkages, hydrodynamics, and biogeochemical gradients, are a prerequisite for identifying the causes of the extinction of large and high-trophic-level predators in marine–estuarine ecosystems (Dayton, 2003; Grime, 1997; Wardle et al., 1997; Dayton et al., 1998; Duffy, 2003; Jackson, 2001). The scientific principles as outlined by **the mass ratio theory** dealing with the immediate effects of biodiversity on ecosystem functioning in a terrestrial ecosystem can be evaluated by the contribution of individual species to total plant biomass (Grime, 1998) and that is also true for coastal macrobenthic fauna by influencing the processes of fluxes of energy and matter (Green et al., 1998; Sandnes et al., 2000).

The functional relationships between benthic fauna and fluxes of energy determine spatial and temporal changes in the distribution and behavior of species across the sediment–water interface, leading to bioturbation and mineralization processes after being influenced by strong feedback loops in respect of their life-support resources (King, 1986; Hobbs & Humphries, 1995; Kim et al., 2006; Hayes & Barry, 2008; Herman et al., 1999). In the shallow-water benthic habitats macrophytes function to determine the flux of nutrients, energy, and oxygen, where grazers are found to reduce the abundance of microphytes destabilizing the surface sediments (de Deckere et al., 2001), lowering nutrient uptake, enabling oxygen to penetrate into the deeper strata of the sediment, and thereby increasing tidal flow-mediated hydraulic flushing by the tidal flow (Munksby et al., 2002). Similarly, deposit feeders help the penetration of organic matter into the deeper parts of the sediment and speed up microbial activities for the decomposition and mineralization of organic matter within deeper layers of the sediment, where anaerobic pore water get subjected to requires greater consumption of oxygen and simultaneously produces ammoniacal nitrogen ($NH_4 - N$). By way of nonlocal transport to drive the decomposition process, pore water loaded with $NH_4 - N$ is advected in aerobic zones triggering nitrifying bacteria to convert $NH_4 - N$ to nitrate-nitrogen ($NO_3 - N$) to photosynthesizing microphytes or undergoes denitrification with production of N_2 gas. However, several eco-biological factors such as movement, burrowing, biodeposits, grazing, production of feces, and bioturbation control the complex interactions among benthic macrofauna, microphytes, and chemical fluxes by the feedback mechanisms driving ecosystem function in soft sediments (Lohrer et al., 2004). The suspension feeders are observed to play more crucial roles by depositing organic materials at the sediment surface as bio-deposits than deposit feeders, which inject organic matter into deeper parts of the sediment. Ecosystem functioning of intertidal macrobenthic communities regulates biogeochemical reactions and other biotic interaction processes integrating over small-scale heterogeneity in the sediment, highlighting the eco-potential of different functional groups of large macrofauna depending on the flux of textural components of sediments, organic matter, and other small surface-dwelling organisms, attenuating the effects of within-patch dynamics (Englund & Cooper, 2003; Hewitt et al., 1997). However, large infaunal benthic fauna acting as ecosystem engineers in shaping the benthic community structure, regulating nutrient flux, and thereby causing system shifts owing to changes in disturbance regimes, affecting the local productivity of estuaries but also impacting nutrient transformation in the larger ecosystem (Thrush et al., 2003).

3.23 Bioaccumulation and Biotransformation of Lipid and Its Fractional Components Through Different Structural Components of the Mangrove–Estuarine Ecosystem

In the marine ecosystem, lipid is an important biochemical component of living organisms and functions as a denser form of energy (Sargent et al., 1976; Ackman, 1989; Alikunhi et al., 2010). Lipid containing certain forms of essential fatty acids and sterols also acts as major determinants of ecosystem stability (Alikunhi et al., 2010; Parrish, 2013). Their heterogenous chemical properties make them important biomarkers in assessing the carbon cycle and the trend of exchange of materials under the influence of other ecological parameters such as temperature, salinity, and pressure (Parrish, 2013). Fatty acid, a carboxylic acid with an aliphatic, represents the structural entities of fat and lipid, where lipids represent a broad group of macronutrients whereas fat is a constituent of lipids. Lipids are one of the main storehouses of the body, containing triglycerides, steroids, and phospholipids. Fatty acids are important, not only for their impact on animal growth but also for many other facets of functions such as reproduction, immunity, ionic balance regulation, and even buoyancy regulation and buoyancy control.

Recent investigation has revealed that muscles of the studied animal, *L. anatina*, stored a large amount of all fatty acids.

Proper estimation and evaluation of structural components and their changes of fatty acids act as useful tools for studying trophic ecology, which throws light on the food web connections, in contrast to those studies emphasizing traditional gut-content analysis in order to obtain information on dietary intake, food constituents, and sequestration of lipid reserves for a prolonged period (Kapoor & Patil, 2011).

In a marine–estuarine ecosystem, fatty acids within the organisms occurring at different trophic levels show generally qualitative similarities in respect of their composition. Fatty acids produced within primary producers of the mangrove ecosystem, phytoplankton in estuarine ecosystems, and halophytic mangrove plant species in the forest ecosystem undergo marked changes in their biochemical constituents in the process of decomposition of the biomass of all those primary producers that are deposited and stored in the tidal basin (Ackman, 1989). All of the fractional components of bio-transformed fatty acids enter the body of macrobenthic fauna feeding upon detritus over the surface of mudflats and enjoying a position as faunal components at a higher trophic level of the mangrove ecosystem. A recent research study at an ecotone (at the confluence of an estuary, Subarnarekha with open sea, Bay of Bengal [longitude 87°5′E to 88°5′E and latitude 20°30′N to 22°2′N] near New Digha, West Bengal, India) has highlighted the functional roles of three major benthic fauna (*Uca acuta acuta*, a fiddler crab; *Lingula anatina*, a brachiopod; and *Cerebratulus bengalensis*, a nemertine) in the bioconversion and biotransformation of lipid components in this ecosystem on a comparative basis (Samanta et al., 2014) (Tables 3.1, 3.2, 3.3, 3.4, 3.5, and 3.6).

Detailed analyses of fatty acids were made during the pre-monsoon period 2014–2015 (from the individuals of three major macrobenthic fauna, *Lingula anatina* (brachiopod), *Uca acuta acuta* (fiddler crab), and *Cerebratulus bengalensis* (nemertine) just after their collection from the intertidal mudflats of an ecotone (Talshari), along with the simultaneous collection of detritus and mangrove leaves from three major mangrove plants, viz. *Avicennia marina*, *Acanthus ilicifolius*, and *Suaeda maritima*. The collection, extraction, preparation estimation, and identification of total lipids and fractional components of fatty acids were done following the methods of several researchers (Ackman, 1989; Sargent et al., 1976; Alikunhi et al., 2010; Gunstone, 2002; SanGiovanni & Chew, 2005; Kapoor & Patil, 2011).

Lipid contents of selected food sources of three studied benthic fauna have shown that zooplankton (S1) contained the highest amount (0.84%) of total lipids, whereas detritus exhibited the lowest amount (0.15%). In the results of analyzing different organs of *Lingula anatina*, muscles (S12) were found to possess the highest amount (2.95%) of lipid components, whereas pedicle (S1) contained the lowest (0.35%) amount of total lipids (Tables 3.1, 3.2, and 3.3). The percentages of phospholipid (PL), neutral lipid (NL), and glycolipid (GL), which constituted the total lipid (TL) in the body flesh of *Uca acuta acuta* and the muscles of *Lingula anatina* are presented in Table 3.2). Fractional analyses of lipids from different parts of the studied faunal materials have revealed the occurrence of the highest amount of saturated fatty acids (SAFAs) from the lophophore and detritus of *Lingula anatina* as

Table 3.1 Percentage of total lipid (TL, % w/w) obtained from various body parts of studied animals and their food sources

Samples	Amount taken (g)	Total lipid obtained (mg)	Percentage of total lipid (w/w)
Planktons (S1)	2.80	23.6	0.84
Mangrove leaves			
Avicennia marina (S2)	15	94.8	0.63
Acanthus ilicifolius (S3)	15	29.2	0.19
Suaeda maritima (S4)	15	38.0	0.25
Detritus (S5)	9.63	14.2	0.15
***Uca acuta acuta* (Fiddler crab)**			
Body flesh (S6)	30	240.20	0.81
Big chela flesh (S7)	5	33	0.66
Gut content (S8)	5	59.5	1.19
Hepatopancreas (S9)	11.41	161.5	1.42
***Lingula anatine* (Brachyopoa)**			
Lophophore (S10)	1.08	8.5	0.79
Pedicle (S11)	3.20	11.2	0.35
Muscle (S12)	4.27	126.1	2.95
Gut content (S13)	0.56	11.7	2.09
***Cerebratulus bengalensis* (Nemertean)**			
Muscles (S14)	5.23	125.54	2.39

Table 3.2 Percentage of neutral lipid, glycolipid, and phospholipid obtained from total lipids of the body flesh of *Uca acuta acuta* and the muscle of *Lingula anatina*

Uca acuta acuta (Fiddler crab), % w/w		*Lingula anatina* (Brachiopoda), % w/w	
Neutral lipid	38.90	Neutral lipid	34.55
Glycolipid	8.40	Glycolipid	11.41
Phospholipid	52.66	Phospholipid	54.03

53.3% and 53.2% respectively. Similarly, muscles of *Cerebratulus bengalensis* (S14), a nemertean species, and *Acanthus ilicifolius* (S3), a mangrove shrub, displayed the lowest amount of SAFAs at 20.1% and 30.2% respectively (Tables 3.3 and 3.4). Out of a total of 14 studied samples, the largest amount (27.9%) of monounsaturated fatty acid (MUFA) was recorded from detritus, whereas the largest amount (57.54%) of polyunsaturated fatty acid (PUFA) was recorded from another mangrove shrub, *Suaeda maritima* (S4; Tables 3.3, 3.4, 3.5, and 3.6). Only arachidic acid (20:0) was found to occur in *Cerebratulus bengalensis* but not in any of the other 14 analyzed for SAFA. This nemertean fauna also showed the presence of a good amount of gondoic acid (20:1).

The maximum amount (39.6%) of α-linolenic acid (ALA, 18:3ω3) was estimated in the leaves of mangrove plant species, *Avicennia marina*, whereas the total

Table 3.3 Fractional compositions of fatty acids of total lipids in different food samples of studied animals as determined by gas-liquid chromatography of methyl esters (% w/w of each component in total fatty acids)

Components	Names of fatty acids	Plankton	Mangrove leaves			Detritus
		S1	S2	S3	S4	S5
14:0	Myristic	9.0	0.5	1.1	0.5	6.5
15:0	Pentadecanoic	1.3	1.4	1.0	1.8	2.4
16:0	Palmitic	20.4	35.2	22.1	29.7	38.2
17:0	Margaric	1.1	0.9	0.8	0.4	0.6
18:0	Stearic	9.5	3.5	4.9	5.1	4.4
20:0	Arachidic					
22:0	Behenic	0.3	0.1	0.1		0.4
24:0	Lignoceric	1.3		0.2	0.2	0.7
Total SAFA		**42.90**	**41.60**	**30.2**	**37.7**	**53.2**
14:1	Tetradecenoic	0.3				2.7
15:1	Pentadecenoic	0.1		0.3	0.5	0.6
16:1	Palmitoleic	10.8	1.5	1.8	1.3	11.0
17:1	Heptadecenoic	1.7	0.4	0.3		0.3
18:1ω9	Oleic	5.3	14.4	13.1	2.3	13.20
20:1	Gondoic					
22:1	Erucic	0.2				0.1
24:1	Tetracosanoic	0.2				
Total MUFA		**18.6**	**16.3**	**15.5**	**4.1**	**27.9**
16:2	Hexadecadienoic	0.2		0.4		1.1
17:2	Heptadecadienoic	2.2				
18:2ω6	Linoleic	1.9	10.3	21.9	17.3	9.0
18:3ω6	γ linolenic	0.3	1.1	0.1	0.2	1.5
18:3ω3	**α linolenic**	**3.1**	30.1	31.0	39.6	2.7
20:3ω3	Eicosatrienoic	0.1				0.3
20:4ω6	Arachidonic	2.6				**0.2**
20:4ω3	Eicosatetraenoic	1.9	0.1	0.3	0.4	**1.3**
22:4ω6	Adrenic	0.6	0.04	0.1	0.04	
20:5ω3	**Eicosapentaenoic**	**12.5**				1.2
21:5ω3	Heneicosapentaenoic	0.1				
22:5ω6	Osbond, ω6DPA	0.2				0.04
22:5ω3	Clupanodonic, DPA	1.0				**0.02**
22:6ω3	**Docosahexaenoic**	**11.1**				0.20
Total PUFA		**37.8**	**41.64**	**53.8**	**57.54**	**17.56**
Total –ω3		**29.8**	**30.20**	**31.30**	**40**	**5.72**
Total –ω6		**5.6**	**11.44**	**22.1**	**17.54**	**10.74**
PUFA/SAFA		**0.89**	**1.00**	**1.78**	**1.52**	**0.33**

Table 3.4 Fractional compositions of fatty acids of total lipids in different body parts of the studied animals as determined by gas-liquid chromatography of methyl esters (% w/w of each component in total fatty acids)

Components	*Uca acuta acuta*(Fiddler crab)				*Lingula anatine* (Brachiopoda)				*Cerebratulus bengalensis* (Nemertean)
	S6	S7	S8	S9	S10	S11	S12	S13	S14
14:0	2.2	4.3	4.0	6.9	1.4	0.9	4.5	3.1	1.0
15:0	2.3	4.5	1.8	0.4	0.9	1.2	1.3	1.3	1.26
16:0	21.6	18.7	25.5	13.6	23.0	23.4	14.7	23.2	7.96
17:0	2.5	1.3	2.6	3.0	3.3	2.9	1.8	1.9	1.71
18:0	9.4	12.4	10.2	10.7	21.3	19.8	11.7	10.1	5.32
20:0									2.01
22:0	0.4	0.1	0.6	0.3	0.7	0.8	0.5	1.0	0.62
24:0	0.7	0.5	0.3	1.0	2.7	2.8	1.2	0.6	0.13
Total SAFA	**36.9**	**41.8**	**45**	**35.9**	**53.3**	**51.8**	**35.7**	**41.2**	**20.01**
14:1				0.1					0.86
15:1	0.7		0.3	0.3		1.0			0.20
16:1	9.9	8.0	17.4	10.3	4.5	1.5	8.3	8.7	1.66
17:1	0.6	2.6	1.1	3.9	0.7	5.2	1.3	1.3	0.80
18:1ω9	6.3	7.2	7.3	9.6	4.2	3.4	8.5	7.0	3.02
20:1									0.29
22:1	0.1	1.0	0.2	0.3	0.5	0.7	1.1	0.2	0.57
24:1	0.4	1.0	0.6	0.7	0.6	0.9	0.4		0.66
Total MUFA	**18**	**19.8**	**26.9**	**25.2**	**10.5**	**12.70**	**19.6**	**34.4**	**8.06**
16:2	0.2		2.0	2.5	0.1	0.6	0.1	0.2	
17:2					2.3	0.5	0.1	0.1	
18:2ω6	2.9	4.7	3.9	4.4	1.2	1.2	1.8	1.9	1.84
18:3ω6	0.5	0.4	0.5	1.2	0.1			0.2	0.46
18:3ω3	0.3	0.3	0.4	1.0	**1.0**	**0.7**	**3.5**	**2.5**	1.36
20:3ω3	0.2	1.5	0.3	0.1	0.1	0.5	0.1	0.1	4.48
20:4ω6	0.1		0.1	0.7	10.4	9.6	7.6	5.3	5.00
20:4ω3	6.2	8.4	7.6	9.8	2.4	2.3	2.0	2.1	
22:4ω6	0.1	0.1	0.1	0.2	0.5		0.8	0.8	
20:5ω3	2.3	14.3	10.3	13.3	**10.1**	**12.7**	**14.4**	**15.8**	
21:5ω3	0.1			0.2	0.2		0.04		
22:5ω6				0.007	0.7	0.9	0.6	2.1	
22:5ω3	0.1	0.2	0.1	0.2	1.3	1.6	2.4	1.6	
22:6ω3	9.3	7.0	2.5	4.8	**5.2**	**4.5**	**10.3**	**7.9**	2.89
Total PUFA	**22.3**	**36.9**	**27.8**	**38.40**	**35.6**	**35.1**	**43.74**	**40.60**	**16.03**
Total –ω3	18.5	31.7	21.2	29.4	20.3	22.3	32.74	30	8.73
Total –ω6	3.6	5.2	4.6	6.50	12.9	11.7	10.8	10.3	7.3
PUFA/ SAFA	**0.60**	**0.88**	**0.60**	**1.06**	**0.66**	**0.67**	**1.22**	**0.98**	**0.80**

Table 3.5 Fatty acid compositions of neutral lipid (NL), glycolipid (GL), and phospholipid (PL) from body flesh of *Uca acuta acuta* (fiddler crab) as determined by gas-liquid chromatography of methyl esters (% w/w of each component in total fatty acids)

Components	Body flesh/NL	Body flesh/GL	Body flesh/PL
14:0	10.4	3.6	0.4
15:0	0.5	0.9	0.2
15:1		1.5	0.2
16:0	48.9	32.6	20.0
16:1	19.0	7.9	7.4
16:2	0.9		0.1
17:0	0.6	3.0	0.5
17:1	2.2	2.6	0.1
17:2		0.1	
18:0	4.5	25.0	13.2
18:1ω9	3.2	4.4	7.2
18:2ω6	1.7	2.2	3.3
18:3ω6	0.6	0.3	0.7
18:3ω3	0.3	3.6	0.3
20:3ω3	0.04		0.3
20:4ω6	0.6		0.1
22:0	0.1		0.6
22:1		0.6	0.4
20:4ω3	1.2	4.1	8.2
22:4ω6	0.1	0.6	0.1
20:5ω3	3.6	1.1	22.0
21:5ω3	0.009	1.4	0.2
22:5ω6	0.1		
24:0	0.05	0.2	1.0
24:1	0.2	2.5	0.7
22:5ω3	0.01	0.4	0.3
22:6ω3	0.5	0.6	11.9

The –ω values represent the methyl end chain from the center of the double bond furthest removed from the carboxyl end

quantity of this fatty acid declined abruptly in the already studied macrobenthic species (Tables 3.2 and 3.3).

Out of 30 fatty acids, 13 categories were found to occur in mangrove leaves, whereas only seven viz. 17:2, 20:3ω3, 20:4ω6, 20:5ω3, 21:5ω3, 22:5ω6, 22:5ω3, and 22:6ω3 were recorded from the macrobenthic samples (Tables 3.1, 3.2, 3.3, and 3.4). Both arachidonic acid (AA) and eicosapentaenoic acid (EPA) were recorded in higher amounts (10.4% and 15.8%) in the macrobenthic brachiopod species, *Lingula anatina* (Table 3.3 and 3.4). Fractional components of fatty acids in the food components of *Cerebratulus bengalensis* showed the presence of a considerable amount of ALA (Tables 3.3, 3.4, 3.5, and 3.6). Nine different types of MUFAs and 12 different types of PUFAs were estimated within the muscles of *Cerebratulus*

Table 3.6 Fatty acid compositions of total lipid obtained from L1 (*Avicennia marina*), L2 (*Acanthus ilicifolius*), L3 (*Suaeda maritima*), and detritus samples as determined by gas-liquid chromatography of methyl esters (% w/w of each component in total fatty acids)

Components	L1	L2	L3	Detritus
14:0	0.5	1.1	0.5	6.7
14:1				2.6
15:0	1.4	1.0	1.8	2.6
15:1		0.3	0.5	0.5
16:0	35.2	22.1	29.7	38.3
16:1	1.5	1.8	1.3	10.9
16:2		0.4		1.0
17:0	0.9	0.8	0.4	0.4
17:1	0.4	0.3		0.2
18:0	3.5	4.9	5.1	4.5
18:1ω9	14.4	13.1	2.3	13.1
18:2ω6	10.3	21.9	17.3	9.0
18:3ω6	1.1	0.1	0.2	1.6
18:3ω3	30.1	31.0	39.6	2.7
20:3ω3				0.3
20:4ω6				0.2
22:0	0.1	0.1		0.3
22:1				0.1
20:4ω3	0.1	0.3	0.4	1.2
22:4ω6	0.04	0.1	0.04	
20:5ω3				1.1
22:5ω6				0.03
24:0		0.2	0.2	0.7
22:5ω3				0.02
22:6ω3				0.2

bengalensis. The highest amount of AA (20:4ω6; 5.00%) was registered followed by eicosatrienoic acid (ETE, 20:3ω3, 4.48%), docosahexaenoic acid (DHA, 22:6ω3, 2.89%), and so on, out of several estimated PUFAs. However, all those fractional components of lipid in the form of fatty acids (AA, eicosatrienoic acid [ETE]) and DHA in mangrove leaves get biotransformed in the decomposition process and through consumption by benthic faunal components. The macrobenthic fiddler crab species, *Uca acuta acuta*, exhibited the occurrence of seven different types of SAFAs and MUFAs and 13 different types of PUFAs. Out of 30 fatty acids, seven types of SAFAs, six categories of MUFAs, and 14 categories of PUFAs were recorded from different body parts of the brachiopod species, *Lingula anatina* (Table 3.3 and 3.4). The ratio PUFA/SAFA always displayed very low values. Two PUFAs (21:5ω3 and 22:5ω6) were not observed in two of the macrobenthic faunal components studied, *Cerebratulus bengalensis and Uca acuta acuta*. The generated qualitative and quantitative research information with regard to estimated and reported fatty acids in the studied samples have been considered to be biomarkers.

In the process of undertaking strategies toward ecosystem management, background ground truth information pertaining to the functioning of the natural ecosystem with special emphasis on the diversity and functional roles of biodiversity components for rendering ecosystem goods and services are required (Chakraborty, 2017). As different eco-zones within coastal mangrove ecosystems (intertidal belt, estuary, creeks, mangrove forest, etc.) are linked through biogeophysical interactions, exhibiting connectedness via nutrient fluxes, floral and faunal successions, biotic interactions, dispersals, migrations, etc., one function in any system is expected to influence the functions of other systems, reflected by energy transfer, the efficiency of nutrient cycling, sediment deposition, and sediment formation, thereby regulating the biological productivity of a coastal ecosystem (Chakraborty, 2017). A recent research study has highlighted the mode of occurrence of several types of fatty acids in body muscles of three mangrove–estuarine benthic macrofauna, namely *Cerebratulus bengalensis,* a nemertean species, *Uca acuta acuta*, a fiddler crab species, and *Lingula anatina*, a brachiopod *species* inhabiting the same ecotone, at the confluence of an estuary (Subarnarekha) with the Bay of Bengal (longitude 87°5′E to 88°5′E and latitude 20°30′N to 22°2′N) on the North-East coast of India. Biochemical fractional components were also evaluated from the other structural components of the same habitat of the studied species, mostly their food sources (planktons, detritus, and mangrove leaves). The presence of different amounts of diversified forms of fatty acid components in different body parts of three studied macrobenthic species, along with their food sources, have enabled a conclusion to be reached on the process of bioconversion and biotransformation of several bioactive substances between interlinked ecosystems.

Fatty acids, being qualitative markers, help to trace or confirm intricate prey–predator relationships in the marine estuarine environment (Emig, 1997; Auel et al., 2002). Qualitative similarities are observed in the fatty acid composition of different organisms occupying different trophic levels in a marine ecosystem. Phytoplankton, being the first link of the food chain, synthesize diversified forms of fatty acids including ω–3 fatty acids, a unique fatty acid of all marine life, and the composition of all such fatty acids undergo significant alteration in the decomposing leaves of mangroves (Tables 3.5 and 3.6). Different organs of the brachiopod species *Lingula anatina*, an intertidal detritivorous macrobenthic animal, showed the presence of a very good amount of 20:5ω3 fatty acids associated with other PUFAs of the ω–3 series and with other polyunsaturated fatty acids of the ω–3 series (Table 3.4). This observation is interpreted by the intake of a large quantity of precursor fatty acid (18:3ω3) through the consumption of phytoplankton and their consequent chain elongations followed by desaturation processes leading to the generation of ω–3 unsaturated fatty acids. This kind of diet enriched with ω–3 acids is supposed to explain the trend of accumulation of all such polyunsaturated fatty acids in required amounts in the different body parts of the studied benthic fauna. ALA (18:3ω3), the primary precursor of the biochemical entity for the synthesis of ω–3 family of fatty acids (Heald & Odum, 1970) in the bodies of animals, is derived from food materials. The major pathways for the synthesis of EPA (20:5ω3) and DHA (22:6ω3) involve acyl-coenzyme-A esters as substrates on which the sequence of chain elongation and desaturation steps (Δ5 and Δ6 desaturases) takes place (Gunstone, 2002; SanGiovanni & Chew, 2005; Kapoor & Patil, 2011). In this research study, the bioconversion of ALA within the studied faunal components is supposed to take place with the help of linolenate (18:3ω3), which is contributed to the studied benthic fauna by their primary food sources.

Lipids as a biochemical component not only have important physiological functions but also regulate the flow of organic materials through food chains and food web dynamics in the ecosystem. Several biochemical constituents undergo conversion in the process of ecosystem functioning in the mangrove–estuarine ecosystem. A recent study conducted in the coastal Midnapore (East) district has highlighted the mode of bioconversion and biotransformation of lipids in the following pathway: mangroves → soil → water → macrodecomposers → microdecomposers → detritus → detritivores (benthos) → detritus, etc. This study involving three different macrobenthos (fiddler crab, brachiopod, and nemertine) inhabiting an ecotone for the detailed fractional analysis of lipids and their bioconversion processes revealed the modes of occurrence and characteristics of different fractional components of fatty acids in different parts of the studied fauna and the mode of bioconversion of such biochemical entities. The steady decomposition of mangrove litter ultimately ensures the bioaccumulation of decomposition products (nutrients) that also render benefit to all the biodiversity components through trophic interactions (Bligh & Dyer, 1959). A higher occurrence of EPA and DHA within several organs of the bodies of the studied faunal components is supposed to have been derived because of the bioconversion of ALA from the primary food sources and EPA and DHA as the precursors of several metabolites act as potent lipid mediators. The biomedical importance of these two fatty acids has been recognized by several researchers, and can be used for the prevention or treatment of several diseases in humans (Ackman, 1989; Barlocher & Corkum, 2003).

The muscles of the studied animal *Cerebratulus bengalensis* stored a major amount of all fatty acids and the presence of EPA and DHA in plankton samples indicated that this animal obtains these fatty acids from plankton as a food source. The leaves of the mangrove plants of the studied area have been found to store moderate to high amounts of ALA (18:3ω3), which appears to be the precursor of long-chain PUFAs, viz. EPA (20:5ω3) and DHA (22:6ω3). The occurrence of higher amounts of carnivorous markers of the studied species such as oleic acid (18:1ω9) was supposed to have been derived from animal sources such as zooplankton, animal detritus, etc. The basal part of the trophic pyramid in this complex estuarine mangrove ecosystem by macroalgae, diatoms, and dinoflagellates, is where animals derived their higher amounts of 22:6ω3 and 20:4ω6 (Christie & Han, 2010). This research investigation on the mode of occurrence of different classes of fatty acids in different structural components of the mangrove–estuarine ecosystem had enabled a conclusion on the mode of biotransformations and bioconversions of these bioactive substances. The presence of high levels of carnivorous markers of the studied species, i.e., oleic acid (18:1ω9, derived from animal sources because of the consumption of zooplankton, animal detritus, etc., that occurred in the intertidal belts) in muscles of *Uca acuta acuta, Cerebratulus bengalensis, Lingula anatina* have indicated that they are inhabitants of the studied ecotone. This study depicting the flows of lipid among different trophic levels forms an important model in respect of population dynamics, emphasizing the bioaccumulation of hydrophobic chemicals (Kelly & Scheibling, 2012; Loreau et al., 2001).

The present investigation highlights the following points with regard to the bioaccumulation and biotransformation of plant biochemicals, especially of lipid and fatty acid components from plants, to detritus, and from there to selected intertidal benthic fauna of the Midnapore (East) coastal tract, West Bengal, India:

1. Mangrove plant leaves possess high levels of ALA (18:3ω3), which acts as a precursor of long chain PUFAs, viz. EPA (20:5ω3) and DHA (22:6ω3).
2. The fiddler crab species, *Uca acuta acuta*, being an important bio-energetically significant benthic faunal component, feeds on leaf detritus, and thereby consumes a considerable amount of ALA (18:3ω3).
3. The ALA thus consumed by this abundant detritivore in the estuarine mud flat is converted in vivo to long-chain PUFAs of the –ω3 series, viz. EPA and DHA.
4. Considerably higher levels of EPA and DHA have been found in the various body parts of the studied benthic fauna, which are capable of biosynthesizing long-chain PUFAs efficiently (Prato et al., 2012; Moreno et al., 1979).

3.24 Functional Role of Microarthropods in Nutrient Cycling of Mangrove–Estuarine Ecosystem of the Midnapore Coast of West Bengal, India

The litter-rich sediments in the floor of the mangrove forest is found to have been packed with a diversified form of microarthropods, which, by virtue of their abundance, succession, and community interactions, render valuable services to the sustenance of food chain–food web dynamics and also as functional contributors facilitating mangrove leaf-litter decomposition. The functional roles played by these soil inhabitants toward maintaining nutrient cycling by way of the decomposition of mangrove litters are controlled by a number ecological parameters such as temperature, salinity, pH, texture of the soils, and the availability of their foods.

3.24.1 Diversity and Population Density of Microarthropods at Nayachar Island

From a comprehensive research study at the mangrove belts of Nayachar Island, a newly emerged virgin deltaic island located on the northeastern part of Midnapore (East) coast of West Bengal, India, an extending eco-zone of the Sundarbans Mangrove ecosystem, India, seven major groups of mangrove litter-inhabiting microarthropods were recorded in the mangrove forest tracts of Nayachar deltaic islands, which occurred in the following order of abundance as Acarina (36.3%), Collembola (27.2%), Coleoptera (10%), Diptera (5.7%), Isopoda (5.1%), Hymenoptera (5%), and other microarthropod fauna (10.6%). Forty-four recorded microarthropod species belonging to those seven groups were found to display distinct seasonal fluctuation in respect of their population density (minimum during monsoon seasons and maximum during post-monsoon seasons) and community structure, viz. species diversity index, species richness index, species evenness index, and Simpson's Reciprocal Index (all lowest during pre-monsoon seasons and highest during post-monsoon seasons). There was a distinct pattern of succession in terms of their occurrence during different time intervals in the decomposition process of different groups of microarthropods in mangrove plant litter, showing the highest decomposition products at the end of 12 months (70%) and the lowest (34%) at the end of 3 months. Similarly, the highest density and diversity of those fauna were observed during the period of 6–9 months of decomposition. The sub-tropical climate in this region has three different seasons, each of 4 months' duration, viz. pre-monsoon (March–June), monsoon (July–October), and post-monsoon (November–February) (Chaudhuri & Choudhury, 1994). The abundance of all the soil- and litter-inhabiting microarthropods exhibited distinct seasonal fluctuation, which varies from one group to another. The populations of microarthropods were found to have maximum abundance during the post-monsoon season, which was thought to be due to the higher abundance of species belonging to Collembola (*Isotomiella minor*, *Isotomurus balteatus*, *Isotoma* spp., *Proisotoma* spp., *Entomobrya* spp., *Sinella* spp., *Calx* spp., *Lepidocyrtus medius*, *Lepidocyrtus* spp., *Cephoderus* spp., *Sminthurides appendiculatus*, *Sminthurides aquaticus*, *Sminthurides* spp., *Mesaphorura choudhuri*, and *Cryptopygus* spp.); Acarina (*Scheloribates parvus*, *Scheloribates praeincisus*, *Scheloribates thermophilus*, *Xylobates seminudus*, *Galumna flabellifera*, *Oppia* spp., *Tectocepheus velatus*, *Tectocepheus* spp., *Allonothrus* spp., *Masthermannia* spp., *Metabelba obtusus*, and *Hypozetes* spp.); Isopoda (*Porcellionides* spp. and *Philoscin* spp.) (Dey et al., 2008). The abundance as well as distribution of different groups of microarthropods were supposed to be due to the association of different groups with plenty of mangrove litter, which acted as suitable habitats endowed with a considerable amount of nutrients and by virtue of differential grades of tolerability of different species belonging to different groups against varying ecological conditions.

The succession of soil microarthropods was observed to have been initiated with the occurrence of Collembola followed gradually by Acarina, Coleoptera, Diptera, Isopoda, Hymenoptera, and other arthropod populations, all of which together played their determining roles to drive the decomposition process, producing a huge amount of nutrients after 6 months of decomposition of mangrove litter. Analysis of relative abundance based on species composition and modes of occurrence of different species in the studied microarthropod community has revealed that maximum diversity and density of Collembola (45.45%), Acarina (37.5%), Coleoptera (14.28%), and Isopoda (19.44%) were observed on the third, sixth, ninth, and 12th months of the entire period of decomposition in the field-based experimental set-ups. Different physico-chemical parameters of soils such as pH, total salinity (ppt), organic carbon (%), nitrogen (ppm), total phosphorus (ppm), and available potassium (ppm) during the process of litter decomposition also demonstrated distinct seasonal variation in tune with the changing pattern of species composition and fluctuating population density of microarthropods (Dey et al., 2008, 2010).

3.24.2 Seasonal Community Analysis of Microarthropods in the Mangrove Habitat

Detailed analysis of the diversity and population density of microarthropods revealed that out of all the microarthropod faunal components, the most abundant groups were Collembola (15 species) followed by Acarina (two species), Hymenoptera (four species), Coleoptera (three species), Diptera (two species), and Isopoda (two species). The species diversity index (H′), representing a value of relationship between the number of individuals and the number of species, fluctuated from a minimum of 2.53 (monsoon) to a maximum of 3.53 (post-monsoon) and displayed significant positive correlation with total nitrogen but a significant negative correlation with rainfall, relative humidity, and soil temperature (Table 3.2). The values of this community index tends to be low in physically controlled ecosystems and high in biologically controlled ecosystems (Odum, 1971; Odum & Heald, 1975). The monsoon registered a minimum evenness index of the soil microarthropod community of 0.93 to a maximum of 0.96) during the post-monsoon season, and exhibited a significant negative correlation with the potassium and nitrogen contents of mangrove sediment. This index of the microarthropod community indicates the unequal representation against a hypothetical community in which species are equally common or not. In the same study, the species richness, highlighted simply as the ratio between total species and total numbers of individuals, varied from a minimum of 21.25 during the monsoon season to a maximum of 37.25 during the post-monsoon season, and exhibited a significant positive correlation with nitrogen and a negative correlation with temperature, relative humidity, and rainfall. Simpson's Reciprocal Index, based on the abundance of soil microarthropod community, displayed seasonal fluctuation with a minimum of 10.98 during the monsoon season to a maximum of 29.35 during the post-monsoon season and also revealed a significant negative correlation with temperature, pH, and rainfall.

Canonical analysis (CCA), a multivariate statistical analytical tool, has been used to evaluate species assemblage structure (Ysebaert et al., 2002). The CCA analysis in the present research study showed that several physico-chemical parameters of soil and major meteorological parameters such as temperature and rainfall, imposed impacts with different intensities on soil microarthropodal faunal abundance and distribution. CCA shows inter-relationships of different taxonomic groups of microarthropods with the different physico-chemical parameters of soil. The concentrations of total nitrogen and available potassium displayed a positive correlation with the microarthropods, whereas other parameters were found to exhibit an inverse relationship with different groups of microarthropods. Extreme levels of inverse relationships were found to exist between the entire population of microarthropods and salinity and rainfall. Different ecological parameters of soil revealed different decomposition rates during litter decomposition by controlling the abundance of soil microarthropods as revealed by CCA in this decomposition study (Dey et al., 2008, 2010).

3.25 Biological Traits Analysis, A Tool for Assessing Mangrove Ecosystem Functioning

Biological traits analysis (BTA) highlights biological characteristics in the form of traits, which are expressed by biotic components, not in an isolated or independent manner but in the form of biotic assemblages, termed biological trait composition. It is very unique from the point of view of incorporating information on the prime eco-biological attributes of all the constituents of the biotic assemblage, instead of depending on the taxonomic identity and also on the trophic/functional group by employing a wide range of information on attributes pertaining to biotic interactions with physical and chemical environments alongside intraspecific variation in trait expression (Kinne,1963; Kirubakaran et al., 2011).

As the undertaking of an ecosystem approach as a conservation management strategy for marine-estuarine–mangrove ecosystems, the development of proper methodologies toward investigating the ecological functioning of ecosystems commands the proper attention of environmental managers and conservationists. BTA has been considered as a novel approach for this purpose using a number of eco-biological attributes such as species composition and abundance of biotic community, different phases of life history, the morphological, anatomical, and behavioral excellence of the species, along with their functional potential for the sustenance of ecosystem stability. Both the anatomical and eco-physiological adaptabilities against the sharp variation of different ecological parameters throughout different months, seasons, and years bring about a complex relationship and mode of associations between traits and distinct environmental factors. In addition, multidimensional impacts of different anthropogenic activities determine benthic assemblage and functional efficiency.

Ecological functions performed by arine benthic faunal assemblages of marine–estuarine ecosystems can be explained more efficiently by biological traits analysis, providing more information than both the relative taxon composition and trophic group approaches. BTA offers a number of advantages over relative taxon composition analysis in forming the underlying structure of the epibenthic assemblages depicting consistently spatial heterogeneity in different eco-zones to describe epibenthic functioning by relating changes in composition to the characteristics of selected taxa. Similarly, comparison of BTA highlighting more small-scale heterogeneity with the trophic group approach analyses, is appeared to be prerequisite to describe epibenthic functioning by relating changes in composition to the characteristics of selected taxa.

3.26 Integrative Approaches for the Eco-Assessment of the Estuarine–Mangrove Ecosystem

The need of the hour is to devise integrated and holistic approaches combining pertinent basic information on the complex interactive condition of the ecosystems in the light of impacts from anthropogenic activities. It requires proper recording of both the biological and the chemical components involved in the ecological processes of an ecosystem alongside addressing the roles of organisms and the physical environment in these processes.

Different forms of anthropogenic activities for harvesting food resources (fishes, honey, fodders, fruits, flowers, etc.), shelter materials (branches and stems of mangrove plants), and the release of different chemical pollutants and allochthonous forms of nutrients exert intense pressure on marine ecosystems. Little is known about the extent of all those effects on human-mediated stress on marine ecosystems along with the power of tolerability and prevailing legislation and international agreements on these problems (Gislason et al., 2000). Different cost-effective biomonitoring tools based on sound underlying principles, suitable for different geographical settings, have emerged to assess the intensity of the human impact on the mangrove ecosystem and to evaluate the success of management techniques initiated to remediate it.

As an ecosystem does not exist in isolation, the devising of biomonitoring and eco-assessment tools requires ground truth information of the structure of biotic assemblages, modes of interactions among biotic and abiotic components in marine–estuarine ecosystems along with their responses against anthropogenic activities (Lindeman, 1942). Marine–estuarine–mangrove ecosystems are composed of three major categories of structural components: (1) the biotic environment (biotic community constituted by an assemblage of living organisms such as microbes, phytoplankton, microalgae, benthos, nekton, large wild reptiles, birds, and mammals); (2) the physical environment (intertidal and subtidal zones, sediment with distinct textural components, waves, currents, tidal interplay, and water temperature); (3) the chemical environment (nutrient loads with nitrogen, phosphorus carbon, oxygen, water-quality parameters such as salinity, turbidity, pH, etc.). An alternative approach integrates combined information pertaining to the impacts on biological communities o nonbiological ones at ecosystem level and controlling ecological processes such as energy flows and cycles of substances in the functioning of the marine ecosystem to achieve ecosystem stability.

3.26.1 Discrete Approaches to Assessment of Ecological Changes

Identification of species using taxonomic keys appears to be a very useful tool for the proper documentation of species composition where not all species could be identified, limiting the usefulness of species-level studies (Vyerman et al., 1996). In an approach to recording the composition, abundance, and distribution of species inhabiting a studied ecosystem, inconsistency in the perfect recording because of misidentification can be very common (Vecchione et al., 2000). Taxonomic assessment emphasizing higher levels of identification, such as family-level assessment, can help to overcome such problems because the potential of this approach appears to be more sensitive than species-level assessments (Jennings et al., 1996; Wantiez, et al., 1997), Focusing at higher taxonomic levels in tune with the natural variability of intertidal and subtidal benthic faunal assemblage in marine and estuarine environments can lessen the problem because it offers more stable analyses of the biotic community (Underwood & Chapman; Olsgard et al., 1998; Weslawki et al., 1999).

The taxon richness represents the simplest and quickest measure for the assessment of the changing biological assemblages as it combines other univariate measures such as diversity indices emphasizing richness, abundance, and distribution of taxa (Magurran, 1988). However, as stress causes a reduction of taxon diversity, low diversity indicates an impacted system. Sometimes the imposition of human-mediated stress can have a positive impact, resulting in an increase in diversity and may not be considered as the contributor for the reduction of diversity. A number of research studies have assessed the impacts of fishing on the abundance and species assemblage of pelagic and demersal fishes (Jennings & Kaiser, 1998), on fishery management practices (Edgar & Barrett, 1999), on nutrient enrichment, and on dredging-related problems.

Ecosystem responses refer to the manifestation of complex interactions among biotic assemblages in response to enrichment-mediated impacts. Ecological functioning deals with the maintenance and regulation ecological stability of ecosystem processes. Ecological functioning involves interactions among organisms and their surrounding ecological parameters in order to project the roles of the taxonomic composition of assemblages at the ecosystem level.

Ecological functioning depends on the species composition within the biotic community of the coastal ecosystem and therefore incorporates, both implicitly and explicitly, different information on the eco-biological attributes and the ecological roles of different taxa that constitute the biotic communities. The approach in ecological functioning depends on taxonomic assessment with the biological characteristics of selected taxa that interact with their physical and chemical environment in variable ways depending on the characteristics they express, which is reflected by changes in their occurrence and abundance.

Different species that have a similar habitat or share the same habitat do not respond in the same way and thereby make it difficult to determine the responses owing to anthropogenic effects at the ecosystem level. This limitation has necessitated the analyses of functional groups belonging to different trophic levels incorporating more information on additional characteristics on morphology, anatomy, biology, and behavior and can be used to monitor the impacts of several human-mediated stresses such as aquaculture practices, capture fisheries (Garrison & Link, 2000), habitat fragmentation, bioinvasion, and pollution (Hedges, 1978; Gaston et al., 1998; Hazra et al., 2002, Maiti Dutta et al., 2014).

References

Aburto-Oropeza, O., Ezcurra, E., Danemann, G., Valdez, V., Murray, J., & Sala, E. (2008). Mangroves in the gulf of California increase fishery yields. *Proceedings of the National Academy of Sciences of the United States of America, 105*, 10456–10459. https://doi.org/10.1073/pnas.0804601105

Ackman, R. G. (1989). Fatty acids. In R. G. Ackman (Ed.), *Marine biogenic lipids. Fats and oils* (Vol. 2, pp. 145–178). CRC Press.

Adame, M. F., & Lovelock, C. E. (2011). Carbon and nutrient exchange of mangrove forests with the coastal ocean. *Hydrobiologia, 663*, 23–50. https://doi.org/10.1007/s10750-010-0554-7

Aerts, R., & Berendse, F. (1988). The effect of increased nutrient availability on vegetation dynamics in wet heathlands. *Vegetatio, 76*, 63–69.

Aerts, R. (1990). Nutrient use efficiency in evergreen and deciduous species from heathlands. *Oecologia, 84*, 391–397.

Aerts, R. (1995). The advantages of being evergreen. *Trends in Ecology & Evolution, 10*, 402–407.

Alexander, M. (1977). *Introduction to soil microbiology*. Wiley.

Aller, R. C. (1982). The effects of macrobenthos on chemical properties of marine sediment and overlying water. In P. L. McCall & M. J. S. Tevez (Eds.), *Animal–sediment relations*. Plenum Press.

Aller, R. C. (1988). Benthic fauna and biogeochemical processes in marine sediments: The role of burrow structures. In T. H. Blackburn & J. Sorensen (Eds.), *Nitrogen cycling in coastal marine environments*. Wiley.

Aller, R. C. (2001). Transport and reactions in the bioirrigated zone. In B. P. Boudreau & B. B. Jørgensen (Eds.), *The benthic boundary layer: Transport processes and biogeochemistry*. Oxford University Press.

Alongi, D. M. (1988a). Bacterial productivity and microbial biomass intropical mangrove sediments. *Microbial Ecology, 15*, 59–79.

Alongi, D. M. (1988b). Microbial–meiofaunal interrelationships in some tropical intertidal sediments. *Journal of Marine Research, 46*, 349–365.

Alongi, D. M. (1989). The role of soft-bottoms benthic communities in tropical mangrove and coral reef ecosystems. *Reviews in Aquatic Sciences, 1*, 243–280.

Alongi, D. M. (1990). The ecology of tropical soft-bottom benthic ecosystems. *Oceanography and Marine Biology: An Annual Review, 28*, 381–496.

Alongi, D. M. (1996). The dynamics of benthic nutrient pools and fluxes in tropical mangrove forests. *Journal of Marine Research, 54*, 123–148.

Alongi, D. M. (2002). Present state and future of the world's man grove forests. *Environment Conservation, 29*, 331–349.

Alongi, D. M. (2008). Mangrove forests: Resilience, protection from tsunamis, and responses to global climate change. *Estuarine, Coastal and Shelf Science, 76*, 1–13.

Alongi, D. M. (2009). Paradigm shifts in mangrove biology. In *Coastal wetlands: An integrated ecosystem approach* (pp. 628–629). Elsevier.

Alongi, D. M. (2009). *The energetics of mangrove forests*. Springer.

Alongi, D. M. (2012). Carbon sequestration in mangrove forests. *Carbon Management, 3*, 313–322.

Alongi, D. M. (2013). Mangrove-microbe-soil relations. In E. Kristensen, R. R. Haese, & J. E. Kostka (Eds.), *Interactions between macro- and microorganisms in marine sediments* (pp. 85–103). American Geophysical Union. https://doi.org/10.1029/CE060p0085

Alongi, D. M. (2014). Carbon cycling and storage in mangrove forests. *Annual Review of Marine Science, 6*, 195–219.

Alongi, D. M. (2015). The impact of climate change on mangrove forests. *Current Climate Change Reports, 2015*(1), 30–39.

Alongi, D. M., & Mukhopadhyay, S. K. (2015). Contribution of mangroves to coastal carbon cycling in low latitude seas. *Agricultural and Forest Meteorology, 213*, 266–272.

Alongi, D. M. (2018). Impact of global change on nutrient dynamics in mangrove forests. *Open Journal of Ecology, 11*(8).

Alongi, D. M., & McKinnon, A. D. (2005). The cycling and fate of terrestrially derived sediments and nutrients in the coastal zoneof the Great Barrier Reef Shelf. *Marine Pollution Bulletine, 51*, 239–252.

Alongi, D. M., Boto, K. G., & Tirendi, F. (1989). Effect of exported mangrove litter on bacterial productivity and dissolved organic carbon fluxes in adjacent tropical nearshore sediments. *Marine Ecology Progress Series, 56*, 133–144.

Alongi, D. M., Boto, K. G., & Robertson, A. I. (1992). Nitrogen and phosphorus cycles. In A. I. Robertson & D. M. Alongi (Eds.), *Tropical mangrove ecosystems* (pp. 251–292). American Geophysical Union.

Alongi, D. M., Christoffersen, P., & Tirendi, F. (1993). The influence of forest type on microbial-nutrient relationships in tropical mangrove sediments. *Journal of Experimental Marine Biology and Ecology, 171*, 201–223. https://doi.org/10.1016/0022-0981(93)90004-8

Alongi, D. M., & Mckinnon, A. D. (2005). The cycling and fate of terrestrially-derived sediments and nutrients in the coastal zone of the great barrier reef shelf. *Marine Pollution Bulletin, 51*(239), 252.

Alongi, D. M., Tirendi, F., & Clough, B. F. (2000). Below-ground decom position of organic matter in forests of the mangroves *Rhizophora stylosa* and *Avicennia marina* along the arid coast of Western Australia. *Aquatic Botany, 68*, 97–122.

Alongi, D. M., Trott, L. A., Wattayakorn, G., & Clough, B. F. (2002). Below ground nitrogen cycling in relation to net canopy production in mangrove forests of southern Thailand. *Marine Biology, 140*(855), 864.

Alongi, D. M., Pfitzner, J., Trott, L. A., Tirendi, F., Dixon, P., & Klumpp, D. W. (2005). Rapid sediment accumulation and microbial mineralization in forests of the mangrove *Kandelia candel* in the Jiulongjiang estuary, China. *Estuary and Coastal Shelf Science, 63*(605), 618. https://doi.org/10.1016/j.ecss.2005.01.004

Andersen, F. O., & Kristensen, E. (1988). The influence of macrofauna on estuarine benthic community metabolism: A microcosm study. *Marine Biology, 99*, 591–603.

Andreetta, A., Fusi, M., Cameldi, I., Cimo, F., Carnicelli, S., & Cannici, S. (2014). Mangrove carbonsink . Do burrowing crabs contribute to sediment carbon storage ? Evidence from Kenyan mangrove system. *Journal of Sea Research, 85*, 524–533.

Andrewartha, H. G., & Birch, L. C. (1954). *The distribution and abundance of animals*. University of Chicago Press.

Anesio, A. M., Denward, C. M., Tranvik, L. J., & Granéli, W. (1999). Decreased bacterial growth on vascular plant detritus due to photochemical modification. *Aquatic Microbial Ecology, 17*, 159–165.

Annon. (2003). *Mangrove ecosystem: Biodiversity and its influence on the natural recruitment of selected commercially important finfish and shellfish species in fisheries* (pp. 1–514). National Agricultural Technology Project (NATP). Indian Council of Agricultural Research (ICAR). Principal Investigator: George, J. P.; Co–PI: Chakraborty, S.K. and Damory, S. N.

Annon. (2005). *Studies on bio-resource assessment and management of degraded mangrove ecosystemof Midnapore Coast, West Bengal* (pp. 1–99). Research project report, Ministry of Environment and Forests, Govt. of India. Principal Investigator: Susanta Kumar Chakraborty [Sanction No.3/6/2001-CSC (M) Dated 5th Nov.2001].

Argyropoulos, D. S., & Sun, Y. (1996). Photochemically induced solid-state degradation, condensation, and rearrangement reactions in lignin model compounds and milled wood lignin. *Photochemistry and Photobiology, 64*, 510–517.

Arrigo, K. R. (1994). The impact of ozone depletion on phytoplankton growth in the Southern Ocean: Large-scale spatial and temporal variability. *Marine Ecology Progress Series, 114*, 1–12.

Ashton, E. C., Hogarth, P. J., & Ormond, R. (1999). Breakdown of mangrove leaf litter in a managed mangrove forest in peninsular Malaysia. *Hydrobiologia, 413*, 77e88.

Augustinus, P. G. E. F. (1995). Geomorphology and sedimentology of mangroves. In G. M. E. Perillo (Ed.), *Geomorphology and sedimentology of estuaries. Developments in sedimentology 53* (pp. 333–357). Elsevier Science.

Austen, M. C., Lambshead, P. J. D., Hutchings, P. A., Boucher, G., Snelgrove, P. V. R., Heip, C., King, G., et al. (2002). Biodiversity links above and below the marine sediment–water interface that may influence community stability. *Biodiversity and Conservation, 11*, 113–136.

Aziz, A., & Paul, A. (2015). Bangladesh sundarbans: Present status of the environment and biota. *Diversity, 7*, 242–269. https://doi.org/10.3390/d7030242

Bagchi, K. (1944). *The Ganges Delta*. University of Calcutta.

Bange, H. W., Bartell, U. H., Rapsomanikis, S., & Andreae. (1994). Methane in the Baltic and North Seas and a Reassessment of the marine emissions of methane. *Global Biogeochemical Cycles, 8*(4), 465–480.

Ball, M. C. (1980). Patterns of secondary succession in a mangrove forest of southern Florida. *Oecologia, 44*, 226–235.

Ball, M. C. (1988). Salinity tolerance in the mangroves, Aegiceras corniculatum and Avicennia marina. I. Water use in relation to growth, carbon partitioning and salt balance. *Australian Journal of Plant Physiology, 15*, 447–464.

Ball, M. C. (1988a). Ecophysiology of mangroves. *Trees, 2*, 129–142.

Ball, M. C. (1988b). Salinity tolerance in the mangroves Aegiceras corniculatum and Avicennia marina. I. Water use in relation to growth, carbon partitioning, and salt balance. *Australian Journal of Plant Physiology, 15*, 447–464.

Ball, M. C., & Pidsley, S. M. (1995). Growth responses to salinity in relation to distribution of two mangrove species, Sonneratia alba and S. Lanceolata, in northern Australia. *Functional Ecology, 9*, 77–85.

Ball, M. C., Cochrane, M. J., & Rawson, H. M. (1997). Growth and water use of the mangroves, Rhizophora apiculata and R. Stylosa, in response to salinity and humidity under ambient and elevated concentrations of atmospheric CO_2. *Plant, Cell & Environment, 20*, 1158–1166.

Bandaranayake, W. M. (1998). Traditional and medicinal uses of mangroves. *Mangroves and Salt Marshes, 2*, 133–148.

Bartlett, K. B., Harris, R. C., & Sebacher, D. I. (1985). Methane flux from coastal salt marshes. *Journal of Geophysical Research, 90*, 5710–5720.

Barlocher, F., & Corkum, M. (2003). Nutrient enrichment overwhelms diversity effects in leaf decomposition by stream fungi. *Oikos, 101*(2), 247–252.

Beadle, N. C. W. (1967). Soil phosphate and its role in molding segments of Australian flora and vegetation with reference to xeromorphy and sclerophylly. *Ecology, 47*, 992–1007.

Benner, R., & Biddanda, B. (1998). Photochemical transformation of surface and deep marine dissolved organic matter: Effects on bacterial growth. *Limnology and Oceanography, 43*, 1373–1378.

Benner, R., & Hodson, R. E. (1985). Microbial degradation of the leachable and lignocellulosic components of leaves and wood from Rhizophora mangle in a tropical mangrove swamp. *Marine Ecology Progress Series, 23*, 221–230.

Benner, R., Peele, E. R., & Hodson, R. E. (1986). Microbial utilization of dissolved organic matter of the red mangrove, *Rhizophora mangle*, in the Fresh Creek estuary, Bahamas. *Estuarine, Coastal and Shelf Science, 23*, 607–619.

Berendse, F., & Aerts, R. (1984). Competition between *Erica tetralix* L, and *Molinia caerulea* (L.) Moench as affected by the availibility of nutrients. *Acta Oecological Oecologica Plantarum, 5*, 3–14.

Bertness, M. D., & Ellison, A. M. (1987). Determinants of pattern in a New England salt marsh plant community. *Ecological Monographs, 57*, 129–147.

Beukema, J. J., Flach, E. C., Dekker, R., & Starink, M. (1999). A long-term study of the recovery of the macrozoobenthos on large defaunated plots on a tidal flat in the Wadden Sea. *Journal of Sea Research, 42*, 235–254.

Bhattacharya, A. (2000). Some geomorphic observations indicating shoreline variation in the coastal tract of West Bengal (India)—A case study around Bakkhali in Sunderbans. In *Proceedings of the International Quaternary Seminar on INQUA Shoreline* (pp. 32–37). Indian Ocean Sub-Commission, New Academic Publishers.

Biswas, H., Mukhopadhyay, S. K., & De, T. K. (2004). Biogenic controls on the air–water carbon dioxide exchange in the Sundarbans mangrove environment, northeast coast of Bay of Bengal, India. *Limnology and Oceanography, 49*(1), 95–101.

Blader, S. J. M., Young, S. W., & Dunning, M. C. (1985). Community structure and zoogeographic affinities of coastal fishes of the Dampier region of North-Western Australia. *Australian Journal of Marine and Freshwater Research, 36*, 247–266.

Blasco, F. (1975). *The mangroves of India* (Vol. 14, pp. 1–175). Institut Français de Pondicherry, Pondicherry Travaux de la Section, Scientific et Technique.

Blasco, F. (1977). Outlines of the ecology, botany and forestry of the Mangals of the Indian subcontinent. In V. J. Chapman (Ed.), *Wet ecosystem of the world* (pp. 241–262). Elsevier.

Blasco, F., Bellan, M., & Chaudhury, M. (1992). Estimating the extent of floods in Bangladesh using SPOT data. *Remote Sensing of Environment, 39*, 167–178.

Blumthaler, M., Ambach, W., & Rehwald, W. (1992). Solar UV-A and UV-B radiation fluxes at two Alpine stations at different altitudes. *Theoretical and Applied Climatology, 46*, 39–44.

Bolam, S. G., Fernandes, T. F., & Huxham, M. (2002). Diversity, biomass, and ecosystem processes in the marine benthos. *Ecological Monographs, 72*, 599–615.

Bosire, J. O., Dahdouh-Guebas, F., Kairo, J. G., Kazungu, J., Dehairs, F., & Koedam, N. (2005). Litter degradation and CN dynamics in reforested mangrove plantations at Gazi Bay, Kenya. *Biological Conservation, 126*, 287–295.

Bothwell, M. L., Sherbot, D. M. J., & Pollock, C. M. (1994). Exosystem response to solar ultraviolet-B radiation: Influence of trophic-level interactions. *Science, 265*, 97–100.

Botto, F., & Iribarne, O. (2000). Contrasting effects of two burrowing crabs (*Chasmagnathus granulata* and *Uca uruguayensis*) on sediment composition and transport in estuarine environments. *Estuarine, Coastal and Shelf Science, 51*, 141–151.

Boto, K. G. (1982). Nutrient and organic fluxes in mangroves. In B. F. Clough (Ed.), *Mangrove ecosystems of Australia: Structure, function and management* (pp. 239–257). Australian Institute of Marine Science and Australian National University Press.

Boto, K. G., & Wellington, J. T. (1983). Phosphorous and nitrogen nutritional status of a northern Australian mangrove forest. *Marine Ecology Progress Series, 11*, 63–69.

Boto, K. G., & Wellington, J. T. (1984). Soil characteristics and nutrient status in a northern Australian mangrove forest. *Estuaries, 7*, 61–69.

Boto, K., Saffigna, P., & Clough, B. (1985). Role of nitrate in nitrogen nutrition of the mangrove Avicennia marina. *Marine Ecology Progress Series, 21*, 259–265.

Boto, K. G., Alongi, D. M., & Nott, A. L. J. (1989). Dissolved organic carbon-bacteria interactions at sediment-water interface in a tropical mangrove system. *Marine Ecology Progress Series, 51*, 243–251.

Bouillon, S. (2011). Storage beneath mangroves. Nature. *Geosciences, 4*, 1–2.

Bouillon, S., Borges, A. V., Castañeda-Moya, E., Diele, K., Dittmar, T., Duke, N. C., Kristensen, E., Lee, S. Y., Marchand, C., Middelburg, J. J., Rivera-Monroy, V. H., Smith, T. J., III, & Twilley, R. R. (2008). Mangrove production and carbon sinks: A revision of global budget estimates. *Global Biogeochemical Cycles, 22*. https://doi.org/10.1029/2007GB003052

Bouillon, S., Koedam, N., Baeyens, W., Satyanarayana, B., & Dehairs, F. (2004a). Selectivity of subtidal benthic invertebrate communities for local microalgal production in an estuarine mangrove ecosystem during the post–monsoon period. *Journal of Sea Research, 51*, 133–144.

Bouillon, S., Moens, T., Overmeer, I., Koedam, N., & Dehairs, F. (2004b). Resource utilization patterns of epifauna from mangrove forests with contrasting inputs of local versus imported organic matter. *Marine Ecology Progress Series, 278*, 77–88.

Boynton, W. R., Murray, L., Hagy, J. D., Stokes, C., & Kemp, W. M. (1996). A comparative analysis of eutrophication patterns in a temperate coastal lagoon. *Estuaries, 19*, 408–421.

Brenchley, G. A. (1982). Mechanisms of spatial competition in marine soft-bottom communities. *Journal of Experimental Marine Biology and Ecology, 60*(1), 17–33.

Bridgham, S. D., Updegraff, K., & Pastor, J. (1998). Carbon, nitrogen and phosphorus minerali zation in northern wetlands. *Ecology, 79*(5), 1545–1561.

Broecker, W. S., Lao, Y., Klas, M., Clark, E., Bonani, G., et al. (1993). A search for an early Holocene CaCO3 preservation event. *Paleoceanography, 8*, 333–339.

Bunt, J. S. (1996). Mangrove zonation: An examination of data from seventeen riverine estuaries in tropical Australia. *Annals of Botany, 78*, 333–341.

Bunt, J. S. (1999). Overlap in mangrove species zonal patterns: Some methods of analysis. *Mangroves and Salt Marshes, 3*, 155–164.

Bunt, J. S., & Bunt, E. D. (1999). Complexity and variety of zonal pattern in the mangroves of the Hinchinbrook area, northeastern Australia. *Mangroves and Salt Marshes, 3*, 165–176.

Bunt, J. S., & Stieglitz, T. (1999). Indicators of mangrove zonality: The Normanby river, N.E. Australia. *Mangroves and Salt Marshes, 3*, 177–184.

Bunt, J. S., Williams, W. T., Hunter, J. F., & Clay, H. J. (1991). Mangrove sequencing: Analysis of zonation in a complete river system. *Marine Ecology Progress Series, 72*, 289–294.

Bushaw, K. L., Zepp, R. G., Tarr, M. A., Schultz-Jander, D., Bourbonniere, R. A., Hodson, R. E., Miller, W. L., Bronk, D. A., & Moran, M. A. (1996). Photochemical release of biologically available nitrogen from aquatic dissolved organic matter. *Nature, 381*, 404–407.

Cain, M. L., Milligan, B. G., & Strand, A. E. (2000). Long distance seed dispersal in plant populations. *American Journal of Botany, 87*, 1217–1227.

Carlton, J. M. (1974). Land-building and stabilization by mangroves. *Environmental Conservation, 1*(4), 285–294.

Castellan, A., Vanucci, C., & Bouas-Laurent, H. (1987). Photochemical degradation of lignin through C-O bond cleavage of non phenolic benzyl arylether units. A study of the photochemistry of (2′,4′,6′-trimethyl-phenoxy)-3,4 dimethoxy toluene. *Holzforschung, 41*, 231–238.

Cebridn, J., Duarte, C. M., Marb, N., & Enriquez, S. (1997). Magnitude and fate of the production of four co-occurring western Mediterranean seagrass species. *Marine Ecology Progress Series, 155*, 29–44.

Chakraborty, S. K. (1996). Aquaculture potential of mangrove ecosystem of Sundarbans, India. In *Proceedings of National Seminar on Fisheries—A Multimillion Dollar Industry, Aquaculture* (pp. 72–83). Foundation of India (AFI).

Chakraborty, S. K. (2009). Biological diversity of Mindapore Coast, West Bengal. *Environmental Forensics, 2*(2), 6.

Chakraborty, S. K. (2010). Coastal environment of Midnapore, West Bengal: Potential threats and management. *Journal of Coastal Environment, 1*(1), 28–40.

Chakraborty, S. K. (2011). Mangrove ecosystem of Sundarbans, India: Biodiversity, ecology, threats and conservation. In J. N. Metras (Ed.), *Mangroves: Ecology, biology and taxonomy* (pp. 83–112). NOVA Publishers.

Chakraborty, S. K. (2013). Interactions of environmental variables determining the biodiversity of coastal mangrove ecosystem of West Bengal, India. *The Ecoscan Special Issue, 3*, 251–265.

Chakraborty, S. K. (2017). Ecological services on intertidal benthic fauna and the sustenance of coastal wetlands along the Midnapore (east) coast, West Bengal, India, p. 777-886. In: Finki, C. W., and Chapin FS (1991) integrated responses of plants to stress: A centralized system of physiological responses. *BioScience, 41*, 29–36.

Chakraborty, S. K. (2018). Bioinvasion and environmental perturbation: Synergistic impact on coastal—Mangrove ecosystems of West Bengal, India. In C. Makowski & C. Finkl (Eds.), *Impacts of invasive species on coastal environments: Coast in crisis* (pp. 171–245). Springer.

Chakraborty, S. K., & Choudhury, A. (1989). Ecological studies on the zonation of brachyuran crabs in a virgin mangrove Island of Sundarbans, India. *Journal of the Marine Biological Association of India, 34*(1 & 2), 189–194.

Chakraborty, S. K., & Choudhury, A. (1992a). Population ecology of fiddler crabs (*Uca.* spp.) of the mangrove estuarine complex of Sundarbans, India. *Tropical Ecology, 33*(1), 78–88.

Chakraborty, S. K., & Choudhury, A. (1992b). Ecological studies on the zonation of brachyuran crabs in a virgin mangrove Island of Sundarbans, India. *Journal of the Marine Biological Association of India, 34*(1&2), 189e194.

Chakraborty, S. K., & Choudury, A. (1985). Distribution of fiddler crabs in Sundarbans mangrove estuarine complex, India. In *The Mangroves. Proceedings of National Symposium Biology Utilization, conservation of mangroves* (pp. 467–472).

Chakraborty, S. K., & Chowdhury, A. (1993). Community organization of macrobenthic mollusks of Sagar Island, Sundarbans, India. *Indian Biologist, 25*(1), 33–36.

Chakraborty, S. K., & Chowdhury, A. (1994). Community structure of macrobenthic polychaetes of intertidal region of Sagar Island, Hooghly estuary, Sundarbans, India. *Tropical Ecology, 35*(1), 97–104.

Chakraborty, S. K., Choudhury, A., & Deb, M. (1986). Decapod Brachyura from Sundarbans mangrove estuarine complex, India. *Journal of the Bengal Natural History Society, 5*(1), 55–68.

Chakraborty, S. K., Chatterjee, S., Dey, M. K., et al. (2012). Ecological changes of an eco-rehabilitated delta of Hooghly estuary. *Journal of Coastal Environment, 3*(1), 9–27.

Chakraborty, S. K., Maiti Dutta, S., Ghosh, P. B., et al. (2014). Impact of global warming on Sundarbans mangrove ecosystem, India: Role of different assessment tools from ecosystem monitoring to molecular markers. In *Proceedings of the International Conference on Green India: Strategic knowledge for combating climate change: Prospects and challenges* (pp. 181–200).

Chapin, I. I. I., Zavaleta, F., & Eviner, E. V. (2000). Consequences of changing biodiversity. *Nature, 405*, 234–242. https://doi.org/10.1038/35012241

Chapman, V. J. (1944). The Cambridge university expedition to Jamaica. Part 1. A study of the botanical processes concerned in the development of the Jamaican shore line. *Botanical Journal of the Linnean Society, 52*, 407–447.

Chatterjee, S., Bhunia, G., & Chakraborty, S. K. (2008). Bioturbation of Brachyuran crabs and its impact on coastal ecosystem of Midnapore district, West Bengal, India. In *Proceedings of Fourth InternationalConference on Environmental Science and Technology at Houston, Texas U.S.A* (pp. 133–148).

Connolly, R. M., & Jones, G. K. (1996). Determining effects of an oil spill on fish communities in a mangrove-seagrass ecosystem in southern Australia. *Australasian Journal of Ecotoxicology, 2*, 3–15.

Chapman, V. J. (1976). *Mangrove vegetation*. J. Cramer, Vaduz, Liechtenstein.

Chapman, A. R. O. (1979). *Biology of seaweeds*. Edwin Arnold.

Chaudhuri, A. B., & Choudhury, A. (1994). *Mangroves of the Sundarbans. Vol. 1, India*. IUCN.

Chen, R., & Twilley, R. R. (1999). A simulation model of organic matter and nutrient accumulation in mangrove wetland soils. *Biogeochemistry, 44*, 93–118.

Chen, W., & Wangersky, P. J. (1996). Production of dissolved organic carbon in phytoplankton cultures as measured by high-temperature catalytic oxidation and ultraviolet photo-oxidation methods. *Journal of Plankton Research, 18*, 1201–1211.

Chen, Y., Khan, S. U., & Schnitzer, M. (1978). Ultraviolet irradiation of dilute fulvic acid solutions. *Soil Science Society of America Journal, 42*, 292–296.

Choswell, R. K., Shawn, G. R., Eaglesham, G., Smith, M. J., Norris, R. L., Seawright, A. A., & Moore, M. R. (1999). Stability of cylindrospermopsin, the toxin from the cyanobacterium, *Cylindrospermopsis raciborskii*: Effect of pH, temperature, and sunlight on decomposition. *Environmental Toxicology, 14*, 155–161.

Choudhury, A. M., Quadir, D. A., & Islam, M. D. J. (1994). Study of Chokoria Sundarbans using remote sensing techniques. *ISME Mangrove Ecosystem Technical Reports, 4*, 1–34.

Christie, W. W., & Han, X. (2010). *Lipid analysis-isolation, separation, identification and lipidomic analysis* (4th ed., pp. 1–466). Oily Press, Elsevier.

Cintron, G., Lugo, A. E., & Pool, D. J. (1978). Mangrove of arid environments in Puerto Rico and adjacent islands. *Biotropica, 10*, 110–121.

Clarke, P. J. (1985). Nitrogen pools and soil characteristics of a temper ate estuarine wetland in eastern Australia. *Aquatic Botany, 23*, 275–290.

Clough, B. F. (1992). Primary productivity and growth of mangrove forests. In A. I. Robertson & D. M. Alongi (Eds.), *Tropical mangrove ecosystem, coastal and estuarine studies* (pp. 225–250). American Geophysical Union.

Clough, B. F., & Attiwill, P. M. (1975). Nutrient cycling in a community of Avicennia marina in a temperate region of Australia. In G. E. Walsh, S. C. Snedaker, & H. J. Teas (Eds.), *Proceedings of the International Symposium on Biology and Management of Mangroves* (Vol. 1). Institute of Food and Agricultural Science, University of Florida.

Clough, B. F., Boto, K. G., & Attiwill, P. M. (1983). Mangroves and sewage: A re-evaluation. In H. J. Teas (Ed.), *Biology and ecology of mangroves* (pp. 151–161). Dr. W. Junk Publishers.

Connell, J. H. (1961). Effects of competition, predation by Thiais lacpillius and other factors on natural populations of the barnacle Ba/anuiis balacnoides. *Ecological Monographs, 31*, 61–104.

Coppo, E., & Marchese, A. (2014). Antibacterial activity of polyphenols. *Current Pharmaceutical Biotechnology, 15*(4), 380–390. https://doi.org/10.2174/138920101504140825121142

Costanza, R., D'Arge, R., & de Groot, R. (1997). The value of the world's ecosystem services and natural capital. *Nature, 387*, 253–260. https://doi.org/10.1038/387253a0

Cotner, J. B., Jr., & Heath, R. T. (1990). Iron redox effects on photosensitive phosphorus release from dissolved humic materials. *Limnology and Oceanography, 35*, 1175–1181.

Dahadough-Guebas, F., Hettiarachchi, S., Leo Seen, D., Batelaan, O., Sooriyarachchi, S., Jayatissa, L. P., & Koedam, N. (2005). Transition in ancient inland freshwater resource management in Sri Lanka affect biota and human populations in and around coastal lagoons. *Current Biology, 15*, 579–586.

Dale, M. R. T. (1999). *Spatial pattern analysis in plant ecology*. Cambridge University Press.

Das, G. K. (2010). Biodiversity in Sunderbans. In A. K. Dwivedi & S. C. Tripathi (Eds.), *Environment and biodiversity, environmental pollution and risk to biodiversity* (pp. 1–34). Lambert Academic Publishing AG & Co.

Das, S., Jana, T. L., & De, T. K. (2014). Vertical profile of phosphatase activity in the Sundarbans mangrove forest, North East coast of Bay of Bengal, India. *Geomicrobiology Journal, 31*, 716–725.

Dayton, P. K. (2003). The importance of the natural sciences to conservation. *The American Naturalist, 162*, 1–13.

Dayton, P. K., Tegner, M. J., Edwards, P. B., & Riser, K. L. (1998). Sliding baselines, ghosts, and reduced expectations in kelp forest communities. *Ecological Applications, 8*, 309–322.

de Deckere, E. M. G. T., Tolhurst, T. J., & de Brouwer, J. F. C. (2001). Destabilization of cohesive intertidal sediments by infauna. *Estuarine, Coastal and Shelf Science, 53*, 665–669.

De, T. K., Ghosh, S. K., Jana, T. K., & Choudhury, A. (1991). Phytoplankton bloom in the Hooghly estuary. *Indian Journal of Marine Sciences, 20*, 134–137.

Donato, D. C., Kauffman, J. B., Murdiyarso, D., Kurnianto, K., Stidham, M., & Kanninen, M. (2011). Mangroves among the most carbon-rich forests in the tropics. *Nature Geoscience, 4*, 293–297.

Donato, D. C., Kauffman, J. B., Mackenzie, R. A., Ainsworth, A., & Pfleeger, A. Z. (2012). Whole-island carbon stocks in the tropical Pacific: Implications for mangrove conservation and upland restoration. *Journal of Environmental Management, 97*, 89–96.

Duarte, C. M., Losada, I. J., Hendriks, I. E., Mazarrasa, I., & Marba, N. (2013). The role of coastal plant communities for climate change mitigation and adaptation. *Nature Climate Change, 3*, 961–968.

Duke, N. C., Meynecke, J. O., Dittmann, S., Ellison, A. M., Anger, K., Berger, U., Cannicci, S., Diele, K., Ewel, K. C., Field, C. D., Koedam, N., Lee, S. Y., Marchand, C., Nordhaus, I., & Dahdouh-Guebas, F. (2007). A world without mangroves? *Science, 317*, 41–42.

Dutta, M. K., Mukherjee, R., Jana, T. K., & Mukhopadhyay, S. K. (2015). Biogeochemical dynamics of exogenous methane in an estuary associated to a mangrove biosphere; The Sundarbans, NE coast of India. *Marine Chemistry, 170*(20), 1–10.

Dutta, M. B., Bianchi, T. S., & Mukhopadhyay, S. K. (2017). Mangrove methane biogeochemistry in the Indian Sundarbans: A proposed budget. *Frontiers in Marine Science, 4*, 187.

De Haan, H. (1993). Solar UV-light penetration and photodegradation of humic substances in peaty lake water. *Limnology and Oceanography, 38*, 1072–1076.

de Silva, W., & Amarasinghe, M. (2021). Response of mangrove plant species to a saline gradient: Implications for ecological restoration. *Acta Botânica Brasílica, 35*(1), 151–160.

Deb, S. C. (1956). Palaeoclimatology and geophysics of the Gangetic Delta. *Geological Reviews of India, 18*, 11–18.

del Giorgio, P. A., Cole, J. J., & Cimbleris, A. (1997). Respiration rates in bacteria exceed phytoplankton production in unproductive aquatic systems. *Nature, 385*, 148–151.
Dey, M. K., Hazra, A. K., & Chakraborty, S. K. (2008). Diversity of microarthropods and their role in the plant litter decomposition in the coastal tract of east Midnapore District West Bengal, India. *Zoological Research in Human Welfare, 20*, 207–226.
Dey, M. K., Hazra, A. K., & Chakraborty, S. K. (2010). Functional role of microarthropods in nutrientcycling of mangrove-estuarine ecosystem of Midnapore coast of West Bengal, India. *International Journal of Environmental Technology and Management, 12*(1), 67–84.
Dittmar, T., & Lara, R. J. (2001a). Driving forces behind nutrient and organic matter dynamics in a mangrove tidal creek in north Brazil. *Estuarine, Coastal and Shelf Science, 52*, 249–259.
Dittmar, T., & Lara, R. J. (2001b). Molecular evidence for lignin degradation in sulfate reducing mangrove sediments (Amazônia, Brazil). *Geochimica et Cosmochimica Acta, 65*, 1403–1414.
Dittmar, T., & Lara, R. J. (2001c). Do mangroves rather than rivers provide nutrients to coastal environments south of the Amazon River? Evidence from long term flux measurements. *Marine Ecology Progress Series, 213*, 67–77.
Dittmar, T., Hertkorn, N., Kattner, G., & Lara, R. J. (2006). Mangroves, a major source of dissolved organic carbon to the oceans. *Global Biogeochemical Cycles, 20*(1).
Donato, D. C., Kauffman, J. B., Murdiyarso, D., Kurnianto, S., Stidham, M., & Kanninen, M. (2011). Mangroves among the most carbon-rich forests in the tropics. *Nature Geoscience, 4*(5), 293–297.
Duarte, C. M., Dennison, W. C., Orth, R. J. W., & Carruthers, T. J. B. (2008). The charisma of coastal ecosystems: Addressing the imbalance. *Estuaries and Coasts, 31*, 233–238.
Duffy, J. E. (2003). Biodiversity loss, trophic skew and ecosystem functioning. *Ecology Letters, 6*, 680–687.
Duke, N. C. (1992). Mangrove floristics and biogeography. In A. I. Robertson & D. M. Alongi (Eds.), *Tropical mangrove ecosystems* (pp. 63–100). American Geophysical Union.
Duke, N. C. (1995). Genetic diversity, distributional barriers and rafting continents? More thoughts on the evolution of mangroves. *Hydrobiologia, 295*, 167–181.
Duke, N. C., Ball, M. C., & Ellison, J. C. (1998). Factors influencing biodiversity and distributional gradients in mangroves. *Global Ecology and Biogeography Letters, 7*, 27–47.
Duke, N. C., Lo, E. Y. Y., & Sun, M. (2002). Global distribution and genetic discontinuities of mangroves? Emerging patterns in the evolution of Rhizophora. *Trees, 16*, 65–79.
Edgar, G. J., & Barrett, N. S. (1999). Biotic affinities of rocky reef fishes, invertebrates and macroalgae in different zones of the Port Davey marine protected area, south-western Tasmania. *Aquatic Conservation : Marine and Freshwater Systems, 20*, 282–296.
Ellison, A. M. (2002). Macroecology of mangroves: Large-scale patterns and processes in tropical coastal forests. *Trees, 16*, 181–194.
Ellison, A. M., Farnsworth, E. J., & Merkt, R. E. (1999). Origins of mangrove ecosystems and the mangrove biodiversity anomaly. *Global Ecology and Biogeography, 8*, 95–115.
Ellison, A. M., Mukherjee, B. B., & Karim, A. (2000). Testing patterns of zonation in mangroves: Scale dependence and environmental correlates in the Sundarbans of Bangladesh. *Journal of Ecology, 88*, 813–824.
Elton, C. S. (1958). *The ecology of invasions by animals and plants*. Methuen.
Englund, G., & Cooper, S. D. (2003). Scale effects and extrapolation in ecological experiments. *Advances in Ecological Research, 33*, 161–213.
Ewel, K. C., Twilley, R. R., & Ong, J. E. (1998). Different kinds of mangrove forests provide different goods and services. *Global Ecology and Biogeography, 7*, 83–94.
Fanjul, E. (2015). Effect of crab bioturbation on organic matter processing in south West Atlantic intertidal sediments. *Journal of Sea Research, 95*, 206–216. https://doi.org/10.1016/j.seares.2014.05.005
FAO [Food and Agriculture Organization of the United Nations]. (2007). *The world's mangroves 1980–2005. FAO Forestry Paper 53*. FAO.
Farnsworth, E. J., Ellison, A. M., & Gong, W. K. (1996). Elevated CO_2 alters anatomy, physiology, growth,and reproduction of red mangrove (Rhizophora mangle L.). *Oecologia, 108*, 599–609.

Farnsworth, E. J., & Ellison, A. M. (1991). Patterns of herbivory in Beliean mangrove swamps. *Biotropica, 23*, 555–567.
Farnsworth, E. J., & Ellison, A. M. (1997). Global patterns of pre-dispersal propagule predation in mangrove forests. *Biotropica, 29*(3), 318–330.
Feller, I. C., Whigham, D. F., McKee, K. L., & O'Neill, J. P. (2002). Nitrogen vs. phosphorus limitation across an ecotonal gradient in a mangrove forest. *Biogeochemistry, 62*, 145–175.
Feller, I. C., Whigham, D. F., McKee, K. L., & Lovelock, C. E. (2003). Nitrogen limitation of growth and nutrient dynamics in a mangrove forest, Indian River lagoon, Florida. *Oecologia, 134*, 405–414.
Fernandes, S. O., Michotey, V. D., Guasco, S., Bonin, P. C., & LokaBharathi, P. A. (2012). Denitrification prevails over an ammox in tropical mangrove sediments (Goa, India). *Marine Environmental Research, 74*, 9–19.
Foster, R. A., Subramaniam, A., Mahaffey, C., Carpenter, E. J., Capone, D. G., & Zehr, J. (2007). Influence of the Amazon River plume on distributions of free-living and symbiotic cyanobacteria in the western tropical North Atlantic Ocean. *Limnology and Oceanography, 52*, 517–532.
Frankignoulee, Abril, G., Borges, A., Bourge, I., Canon, C., Delille, B., & Libert, E. (1998). Carbon dioxide emission from European estuaries. *Science, 282*, 434–436.
Frederick, J. E., & Lubin, D. (1988). The budget of biologically active ultraviolet radiation in the earth atmosphere system. *Journal of Geophysical Research, 93*, 3825–3832.
Frederick, J. E., Snell, H. E., & Haywood, E. K. (1989). Solar ultraviolet radiation at the earth's surface. *Photochemistry and Photobiology, 50*, 443–450.
Furukawa, K., Wolanski, E., & Mueller, H. (1997). Currents and sediment transport in mangrove forests. *Estuarine, Coastal and Shelf Science, 44*, 301–310.
Gao, H., & Zepp, R. G. (1998). Factors influencing photoreactions of dissolved organic matter in a coastal river of the southeastern United States. *Environmental Science & Technology, 32*, 2940–2946.
Gattuso, J. P., Frankignoulle, M., & Wollast, R. (1998). Carbon and carbonate metabolism in coastal aquatic ecosystems. *Annual Review of Ecology and Systematics, 29*, 405–434.
Geller, A. (1986). Comparison of mechanisms enhancing biodegradability of refractory lake water constituents. *Limnology and Oceanography, 31*, 755–764.
Gessner, M. O., Chauvet, E., & Dobson, M. (1999). A perspective on leaf litter breakdown in streams. *Oikos, 85*(2), 377–384.
Ghosh, S. K., Jana, T. K., Singh, B. N., & Choudhury, A. (1987). Comparative study on carbon dioxide system in virgin and reclaimed mangrove waters of Sundarbans during freshet, Mahas agar. *Bulletin of the National Institute of Oceanography, 20*, 155–161.
Ghosh, P. B., Singh, B. N., Chakraborty, C., Saha, A., Das, R. I., & Choudhury, A. (1990). Mangrove litter production in the tidal creek of Lothian island, Sundarbans, India. *Indian Journal of Marine Science, 19*, 292–293.
Grabowski, Z. J., & Chazdon, R. L. (2012). *Beyond carbon: Redefining forests and people in the global ecosystem services market*. S.A.P.I.EN.S. [Online], 5.1 (available online since 07 April, 2012). Retrieved September 18, 2016, from http://sapiens.revues.org/1246
Gilman, E. L., Ellison, J., Duke, N. C., & Field, C. (2008). Threats to mangroves from climate change and adaptation options: A review. *Aquatic Botany, 89*, 237–250.
Ginsberg, R. N., & Lowenstam, H. A. (1958). The influence of marine bottom communities on the depositional environment of sediments. *Journal of Geology., 66*, 310–318.
Hawkins, J. E., & Freeman, C. (1994). Rising sea levels—Potential effects upon terrestrial greenhouse gas production. *Soil Biology and Biochemistry*.
Giovanna, S., Brasili, E., & Pasqua, G. (2020). Antifungal activity of phenolic and polyphenolic compounds from different matrices of *Vitis vinifera* L. against human pathogens. *Molecules, 25*(16), 3748. https://doi.org/10.3390/molecules25163748
Gjessing, E. T., & Gjerdahl, T. (1970). Influence of ultra-violet radiation on aquatic humus. *Vatten, 2*, 144–145.
Gleason, H. A. (1926). The individualistic concept of the plant association. *Bulletin of the Torriey Botanical Club, 53*, 3–26.

Gold, M. H., Kutsuki, H., & Morgan, M. A. (1983). Oxidative degradation of lignin by photochemical and chemical radical generating systems. *Photochemistry and Photobiology, 38*, 647–651.

Gong, W., & Ong, J. (1990). Plant biomass and nutrient flux in a managed forest in Malaysia. *Estuarine, Coastal and Shelf Science, 31*, 519–530.

Granéli, W., Lindell, M., & Tranvik, L. (1996). Photo-oxidative production of dissolved inorganic carbon in lakes of different humic content. *Limnology and Oceanography, 41*, 698–706.

Granéli, W., Lindell, M., Marcal de Faria, B., & de Assis Esteves, F. (1998). Photoproduction of dissolved inorganic carbon in temperate and tropical lakes—dependence on wavelength band and dissolved organic carbon concentration. *Biogeochemistry, 43*, 175–195.

Gray, J. S. (1997). Marine biodiversity: Patterns, threats and conservation needs. *Biodiversity and Conservation, 6*, 153–175.

Green, M. O., Hewitt, J. E., & Thrush, S. F. (1998). Seabed drag coefficients over natural beds of horse mussels (Atrina zelandica). *Journal of Marine Research, 56*, 613–637.

Grime, J. P. (1997). Biodiversity and ecosystem function: The debate deepens. *Science, 227*, 1260–1261.

Grime, J. P. (1998). Benefits for plant diversity to ecosystems: Immediate, filter and founder effects. *Journal of Ecology, 86*, 902–910.

Hatcher, B. G., Johannes, R. E., & Robertson, A. I. (1989). Review of research relevant to the conservation of shallow tropical marine ecosystems. *Oceanography and Marine Biology: An Annual Review, 27*, 337–414.

Hättenschwiler, S., & Vitousek, P. M. (2020). The role of polyphenols in terrestrial ecosystem nutrient cycling. *Trends in Ecology & Evolution, 15*(6), 238–243. https://doi.org/10.1016/S0169-5347(00)01861-9

Hayes, K. R., & Barry, S. C. (2008). Are their any consistent prediction of invasion success? *Biological Invasions, 10*(4), 483–506.

Hazra, A. K., & Choudhuri, D. K. (1990). Ecology of subterranean macro and micro arthropods fauna in different degraded and polluted soil environment of West Bengal. India. In *Records of the Zoological Survey of India. Occasional Paper, No. 120* (pp. 1–280).

Hazra, A. K., & Sanyal, A. K. (1996). Ecology of collembola in a periodically inundated newly emerged alluvial Island in the river Hooghly, West Bengal. *Proceedings of the Zoological Society (Calcutta), 49*(2), 157–169.

Hazra, S., Ghosh, T., DasGupta, R., & Sen, G. (2002). Sea level and associated changes in the Sundarbans. *Science and Culture, 68*, 309–312.

Heal, O. W., & French, D. D. (1974). Decomposition of organic matter in tundra. In A. J. Holding et al. (Eds.), *Soil organisms and decomposition in tundra* (pp. 279–309). Tundra Biome Steering Committee.

Heald, E. J., & Odum, W. E. (1970). The contribution of mangrove swamps to Florida fisheries. *Proceedings of the Gulf and Caribbean Fisheries Institute, 22*, 130–135.

Hedges, J. I. (1978). The formation and clay mineral reactions of melanoidins. *Geochimica et Cosmochimica Acta, 42*, 69e76.

Heip, C. H. R., Goosen, N. K., Herman, P. M. J., Kromkamp, J., Middleburg, J. J., et al. (1995). Production and consumption of biological particles in temperate tidal estuaries. *Oceanography and Marine Biology: An Annual Review, 33*, 1–149.

Herman, P. M. J., Middelburg, J. J., Van de Koppel, J., & Heip, C. H. R. (1999). Ecology of estuarine macrobenthos. *Advances in Ecological Research, 29*, 195–231.

Herndl, G. J., Müller-Niklas, G., & Frick, J. (1993). Major role of ultraviolet-B in controlling bacterioplankton growth in the surface layer of the ocean. *Nature, 361*, 717–719.

Hernes, P. J., Benner, R., Cowie, G. L., Goni, M. A., Bergamaschi, B. A., & Hedges, J. I. (2001). Tannin diagenesis in mangrove leaves from a tropic estuary: A novel molecular approach. *Geochimica et Cosmochimica Acta, 65*, 3109–3122.

Hewitt, J. E., Pridmore, R. D., Thrush, S. F., & Cummings, V. J. (1997). Assessing the short-term stability of spatial patterns of macrobenthos in a dynamic estuarine system. *Limnology and Oceanography, 42*, 282–288.

Hobbs, R. J., & Humphries, S. E. (1995). An integrated approach to the ecology and management of plant invasions. *Conservation Biology, 9*, 761–770.
Hutchings, P. A., & Saenger, P. (1987). *Ecology of mangroves*. University of Queensland Press.
Hoigné, J. (1990). Formulation and calibration of environmental reaction kinetics; oxidations by aqueous photooxidants as an example. In W. Stumm (Ed.), *Aquatic chemical kinetics* (pp. 43–70). Wiley.
Holguin, G., Vazquez, P., & Bashan, Y. (2001). The role of sediment microorganisms in the productivity, conservation, and rehabilitation of mangrove ecosystems: An overview. *Biology and Fertility of Soils, 33*, 265–278.
Hongve, D. (1994). Sunlight degradation of aquatic humic substances. *Acta Hydrochimica et Hydrobiologica, 22*, 117–120.
Howarth, R. W., Sharpley, A. W., & Walker, D. (2002). Sources of nutrient pollution to coastal waters in the United States: Implications for achieving coastal water quality goals. *Estuaries, 25*, 656–676.
Huffaker, C. B. (1958). Experimental studies on predation: Dispersion factors and predator-prey oscillations. *Hilgardia, 27*, 343–383.
Hussain, Z., & Acharya, G. (Eds.). (1994). *Mangroves of the Sundarbans* (Vol. 2). IUCN.
Hutchinson, G. E. (1951). Copepodology for the ornithologist. *Ecology, 32*, 571–574.
Hutchison, J., Manica, A., Swetnam, R., Balmford, A., & Spalding, M. (2014). Predicting global patterns in mangrove forest biomass. *Conservation Letters, 7*, 233–240.
Jackson, J. B. C. (2001). What was natural in the coastal oceans? *Proceedings of the National Academy of Sciences of the United States of America, 98*, 5411–5418.
Jeffrey, S. W. (1997). Preparation of chlorophyll standards. In S. W. Jeffrey, R. F. C. Mantoura, & S. W. Wright (Eds.), *Phytoplankton pigments in oceanography: Guidelines to modern methods* (pp. 207–238). UNESCO.
Jennerjahn, T. C., & Ittekkot, V. (1997). Organic matter in sediments in the mangrove areas and adjacent continental margins of Brazil: I. Amino acids and hexosamines. *Oceanologica Acta, 20*, 359–369.
Jennerjahn, T. C., & Ittekkot, V. (1999). Changes in organic matter from surface waters to continental slope sediments off the Sao Francisco River, Eastern Brazil. *Marine Geology, 161*, 129.
Jiménez, J. A. (1992). Mangrove forest of the Pacific Coast of Central America. In U. Seelinger (Ed.), *Coastal plant communities of Latin America* (pp. 259–267). Academic Press.
Joye, S. B., & Lee, R. Y. (2004). Benthic microbial mats: important sources of fixed nitrogen and carbon to the twin cays, Belize ecosystem. *Atoll Research Bulletin, 528*, 1–26.
Kang, H., & Freeman, H. (2002). The influence of hydrochemistry on methane emissions from two contrasting northern wetlands. *Water, Air, Soil Pollution, 141*, 263–272.
Kapoor, R., & Patil, U. K. (2011). Mini review importance and production of omega-3 fatty acids from natural sources. *International Food Research Journal, 18*, 493–499.
Karimn, A. (1988). *Environmental factors and the distribution of mangroves in Sundarbans with special reference to Heritiera fomes Buch.-ham*. PhD Thesis, University of Calcutta, Calcutta, India.
Karl, D., et al. (2002). Dinitrogen fixation in the world's oceans. *Biogeochemistry, 57/58*, 47–98.
Kathiresan, K. (2011). Ecobiology of mangroves. In J. N. Metras (Ed.), *Mangroves: Ecology, biology and taxonomy* (pp. 1–50). NOVA Publishers.
Kauffman, J. B., Steele, M. D., Cummings, D. L., & Jaramillo, V. J. (2003). Biomass dynamics associated with deforestation, fire, and conversion to cattle pasture in a Mexican tropical dry forest. *Forest Ecology and Management, 176*, 1–12.
Kristensen, E. (2008). Mangrove crabs as ecosystem engineers; with emphasis on sediment processes. *Journal of Sea Research, 59*(1–2), 30–43.
Kristensen, E., Andersen, F. Ø., & Kofoed, L. H. (1988). Preliminary assessment of benthic community metabolism in a south-east Asian mangrove swamp. *Marine Ecology Progress Series, 48*, 137–145.
Kathiresan, K., & Bingham, B. L. (2001). Biology of mangroves and mangrove ecosystems. *Advances in Marine Biology, 40*, 81–251.

Kelly, J. R., & Scheibling, R. E. (2012). Fatty acids as dietary tracers in benthic food webs. *Marine Ecology Progress Series, 446*, 1–22.

Kempe, S. (1995). Coastal seas: a net source or sink of atmospheric carbon dioxide? *LOICZ Reports and Studies, 1*, 1–27.

Ketchum, B. H. (Ed.). (1983). *Estuaries and enclosed seas* (p. 500). Elsevier.

Khan, F. A., Choudhury, A. M., & Islam, M. D. J. (1990). *Timber volume inventory in the Sundarbans using aerial photography and other remote sensing techniques.*

Khan, M. N. I., Suwa, R., & Hagihara, A. (2005). Carbon and nitrogen pools in a mangrove stand of Kandelia obovata (S.L.). *Wetlands Ecology and Management, 15*, 141–153.

Kim, T. W., Tae, K. K., Hong, S. K., & Choe, J. C. (2006). Surface mating as an alternative mating strategy in the fiddler crab *Uca lactea. Journal of Ecology and Field Biology, 29*(1), 49–53.

King, G. M. (1986). Inhibition of microbial activity in marine sediments by a bromophenol from a hemichordate. *Nature, 323*, 257–259.

Kinne, O. (1963). The effect of temperature and salinity on marine and brackish water animals. *Oceanography and Marine Biology: An Annual Review, 1*, 301–340.

Kirubakaran, P., Kothapalli, R., Dhanachandra, S. K. H., et al. (2011). Silico studies on marine actinomycetes as potential inhibitors for *glioblastoma multiforme. Bioinformation, 6*, 100–106.

Koch, V., & Wolff, M. (2002). Energy budget and ecological role of mangrove epi-benthos in the Caete estuary, North Brazil. *Marine Ecology Progress Series, 228*, 119–130.

Koch, E. W., Ackerman, J., van Keulen, M., & Verduin, J. (2006). Fluid dynamics in seagrass ecology: From molecules to ecosystems. In A. W. D. Larkum, R. J. Orth, & C. M. Duarte (Eds.), *Seagrasses: Biology, ecology and conservation* (pp. 193–225). Springer-Verlag.

Koch, E. W., et al. (2009). Non–linearity in ecosystem services: Temporal and spatial variability in coastal protection. *Frontiers in Ecology and the Environment, 7*, 29–37.

Komiyama, A., Ong, J. E., & Poungparn, S. (2008). Allometry, biomass, and productivity of mangrove forests: a review. *Aquatic Botany, 89*, 128–137.

Krasnovsky, A. A., Jr. (1994). Singlet molecular oxygen and primary mechanisms of photooxidative damage of chloroplasts. Studies based on detection of oxygen and pigment phosphorescence. *Proceedings of the Royal Society of Edinburgh, 102B*, 219–235.

Krebs, C. J., Keller, B. L., & Tamarin, R. H. (1969). Microtus population biology: demographic changes in fluctuating populations of M. ochrogaster and M. pennsylvanicus in Southern Indiana. *Ecology, 50*, 587–607.

Kremen, C., & Ostfeld, R. S. (2005). Measuring, analyzing and managing ecosystem services. *Frontiers in Ecology and the Environment, 3*(10), 540–548.

Kremling, K. (1983). Trace metal fronts in European shelf waters. *Nature, 303*, 225–227.

Kristensen, E. (2008). Mangrove crabs as ecosystem engineers; with emphasis on sediment processes. *Journal of Sea Research, 58*, 30–43.

Kristensen, E., & Kostka, J. E. (2005). Macrofaunal burrows and irrigation in marine sediment: microbiological and biogeochemical interactions. In E. Kristensen, R. R. Haese, & J. E. Kostka (Eds.), *Coastal and estuarine studies* (pp. 125–158). America Geophysical Union.

Kristensen, E., Andersen, F. O., & Kofoed, L. H. (1988). Preliminary assessment of benthic community metabolism in a south-east Asian mangrove swamp. *Marine Ecology Progress Series, 48*, 137–145.

Kristensen, E., Holmer, M., Banta, G. T., Jensen, M. H., & Hansen, K. (1995). Carbon, nitrogen and sulfur cycling in sediments of the Ao Nam Bor mangrove forest, Phuket, Thailand: a review. *Phuket Marine Biological Center Research Bulletin, 60*, 37–64.

Kristensen, E., Bouillon, S., Dittmar, T., & Marchand, C. (2008). Organic carbon dynamics in mangrove ecosystems: A review. *Aquatic Botany, 89*, 201–219.

Kristensen, E., Penha-Lopes, G., Delefosse, M., Valdemarsen, T., Quintana, C. O., & Banta, G. T. (2012). What is bioturbation? The need for a precise definition for fauna in aquatic sciences. *Marine Ecology Progress Series, 446*, 285–302.

Lacerda, L. D., Jose, D. V., de Rezende, C. E., Francisco, M. C. F., Wasserman, J. C., & Martins, J. C. (1986). Leaf chemical characteristics affecting herbivory in a New World mangrove forest. *Biotropica, 18*, 350–355.

Lee, S. Y. (1998). Ecological role of grapsid crabs in mangrove ecosystems: a review. *Marine and Freshwater Research, 49*, 335–343.

Lee, S. Y. (1999a). The effect of mangrove leaf litter enrichment on macrobenthic colonization of defaunated sandy substrates. *Estuarine, Coastal and Shelf Science, 49*(5), 703–712.

Lee, S. Y. (1999b). Tropical mangrove ecology: Physical and biotic factors influencing ecosystem structure and function. *Australian Journal of Ecology, 24*, 355–366.

Lee, R. Y., & Joye, S. B. (2006). Seasonal patterns of nitrogen fixation and denitrification in oceanic mangrove habitats. *Marine Ecology Progress Series, 307*, 127–141.

Lee, S. Y. (2008). Mangrove macrobenthos: Assemblages, services, and linkages. *Journal of Sea Research, 59*, 16–29.

Lee, S. Y., & Kneib, R. T. (1994). Effects of biogenic structure on prey consumption by the Xanthid crabs *Eurytium limosum* and *Panopeus herbstil* in a salt marsh. *Marine Ecology Progress Series, 104*(1&2), 39–47.

Leme, M. H. A. (2010). Seasonal changes in reproductive traits of the crab Sesarma rectum (Grapsoidae: Searmidae) on the northern coast of Sao Paulo state, Brazil. *Journal of Crustacean Biology, 26*(2), 141–147.

Lerberg, S. B., Frederick Holland, A., & Sanger, D. M. (2000). Responses of tidal creek macrobenthic communities to the effects of watershed development. *Estuaries, 23*(23), 838–853.

Lim, S. L. S. (2006). Fiddler crab burrow morphology: how do burrow dimensions and bioturbative activities compare in sympatric populations of *Uca vocans* (Linnaeus, 1758) and *U. annulipes* (H. Milne Edwards, 1837)? *Crustaceana, 79*(5), 525–540.

Lindau, C. W., DeLaune, R. D., & Pardue, J. H. (1994). Inorganic nitrogen processing and assimilation in a forested wetland. *Hydrobiologia, 277*, 171–178.

Lindeman, R. L. (1942). The trophic–dynamic aspect of ecology. *Ecology, 23*, 394–418.

Lohrenz, S. E., Fahnenstiel, G. L., Redalje, D. G., Lang, G. A., Chen, X., et al. (1997). Variations in primary production of northern Gulf of Mexico continental shelf waters linked to nutrients inputs from the Mississippi River. *Marine Ecology Progress Series, 155*, 45–54.

Loreau, M., Naeem, S., Inchausti, P., Bengtsson, J., Grime, J. P., et al. (2001). Ecology-biodiversity and ecosystem functioning: current knowledge and future challenges. *Science, 294*(5543), 804.

Lovelock, C. E., & Feller, I. C. (2003). Photosynthetic performance and resource utilization of two mangrove species coexisting in hypersaline scrub forest. *Oecologia, 134*, 455–462.

Lovelock, C. E., Ball, M. C., Feller, I. C., Engelbrecht, B. M. J., & Ewe, M. L. (2006). Plant function in nitrogen and phosphorus limited mangrove ecosystems. *New Phytologist, 172*, 514–522.

Lovelock, C. E., Ball, M. C., Martin, K. C., & Feller, I. C. (2009). Nutrient enrichment increases mortality of mangroves. *PLoS One, 4*(5), e5600.

Luckenbach, M. W. (1986). Sediment stability around animal tubes: The roles of hydrodynamic processes and biotic activity. *Limnology and Oceanography, 31*, 779–787.

Lugo, A. E. (1980). Mangrove ecosystems Successional or steady state? *Biotropica, 12*(2), 65–72.

Luizão, R. C. C., Luizão, F. J., Paiva, R. Q., Monteiro, T. F., Sousa, L. S., & Kruijt, B. (2004). Variation of carbon and nitrogen cycling processes along a topographic gradient in a central Amazonian forest. *Global Change Biology, 10*, 592–600. https://doi.org/10.1111/j.1529-8817.2003.00757.x

MacArthur, R. H., & Wilson, E. O. (1967). *The theory of island biogeography*. Princeton University Press.

Macintosh, D. J., Ashton, E. C., & Havanon, S. (2002). Mangrove rehabilitation and intertidal biodiversity: A study in the Ranong mangrove ecosystem, Thailand. *Estuarine, Coastal and Shelf Science, 55*, 331–345.

Macnae, W. (1969). A general account of the fauna and flora in mangrove swamps and forests in the Indo-West-Pacific region. *Advances in Marine Biology, 6*, 73–270.

Macnae, W. (1974). *Mangrove forest and fisheries. Indian Ocean Programme Publication Number-34* (p. 351). Indian Ocean Fishery Commission. OFC/DEV/74/34.

Madronich, S., McKenzie, R. L., Caldwell, M. M., & Björn, L. O. (1995). Changes in ultraviolet radiation reaching the earth's surface. *Ambio, 24*, 143–152.

Magurran, E. A. (1988). *Ecological diversity and its measurement*. Princeton University Press.

Maiti Dutta, S., Banerjee Mustafi, S., Raha, S., & Chakraborty, S. K. (2014). Assessment of thermal stress adaptation by monitoring Hsp70 and MnSOD in the freshwater gastropod, Bellamya bengalensis (Lamark, 1882). *Environmental Monitoring and Assessment, 186*(12), 8961–8967.

Mandal, S., Ray, S., Sarkar, A., & Ghosh, P. B. (2011). Degradation of mangrove litter and its contribution to the adjacent estuary of Sagar island, Sundarban mangrove ecosystem, India. In J. N. Metras (Ed.), *Mangrove ecology, biology and taxonomy* (pp. 241–250). Nova Science Publications.

Mann, F. D., & Steinke, T. D. (1989). Biological nitrogen fixation (acetylene reduction) associated with green algal (cyanobacterial) communities in the Beachwood Mangrove Nature Reserve. 1. The effect of environmental factors on acetylene reduction activity. *South African Journal of Botany, 55*, 438–444.

Manny, B. A., Miller, M. C., & Wetzel, R. G. (1971). Ultraviolet combustion of dissolved organic nitrogen compounds in lake waters. *Limnology and Oceanography, 16*, 71–85.

McKee, K. L., & Faulkner, P. L. (2000). Restoration of biogeochemical function in mangrove forests. *Restoration Ecology, 8*(3), 247–259.

Meybeck, M. (1982). Carbon, nitrogen, and phosphorus transport by World Rivers. *American Journal of Science, 282*, 401–450.

Meybeck, M. (1993). C, N, P and S in rivers: From sources to global inputs. In R. Wollast, F. T. Mackenzie, & L. Chou (Eds.), *Interactions of C, N, P and S biogeochemical cycles and global change* (pp. 163–193). Springer-Verlag.

Mfilinge, P. L., Atta, N., & Tsuchiya, M. (2002). Nutrient dynamics and leaf litter decomposition in a subtropical mangrove forest at Oura Bay, Okinawa, Japan. *Trees, 16*, 172–180.

Mialocq, J. C. (1997). Fundamentals of interaction between light and matter. In M. Chanon (Ed.), *Homogenous photocatalysis* (pp. 16–51). Wiley.

Middelburg, J. J., Klaver, G., Nieuwenhuize, J., & Vlug, T. (1995). Carbon and nitrogen cycling in intertidal sediments near Doel, Scheldt estuary. *Hydrobiologia, 311*, 57–69.

Middelburg, J. J., Nieuwenhuize, J., Slim, F. J., & Ohowa, B. (1996). Sediment biogeochemistry in an East African mangrove forest (Gazi Bay, Kenya). *Biogeochemistry, 34*, 133–155.

Miller, W. L., & Moran, M. A. (1997). Interaction of photochemical and microbial processes in the degradation of refractory dissolved organic matter from a coastal marine environment. *Limnology and Oceanography, 42*, 1317–1324.

Miller, W. L., & Zepp, R. G. (1995). Photochemical production of dissolved inorganic carbon from terrestrial organic matter: Significance to the oceanic organic carbon cycle. *Geophysical Research Letters, 22*, 417–420.

Miller, R. V., Jeffrey, W., Michell, D., & Elasri, M. (1999). Bacterial responses to ultraviolet light. *ASM News, 65*, 535–541.

Milliman, J. D. (1993). Production and accumulation of calcium carbonate in the ocean: budget of a nonsteady state. *Global Biogeochemical Cycles, 7*, 927–957.

Molina, L. T., & Molina, M. J. (1986). Absolute absorption cross sections of ozone in the 185- to 350-nm wavelength range. *Journal of Geophysical Research, 91*, 14501–14514.

Moran, M. A., & Zepp, R. G. (1997). Role of photoreactions in the formation of biologically labile compounds from dissolved organic matter. *Limnology and Oceanography, 42*, 1307–1316.

Morell, J. M., & Corredor, J. E. (1993). Sediment nitrogen trapping in a mangrove lagoon. *Estuarine, Coastal and Shelf Science, 37*, 203–212.

Moreno, V., de Moreno, J. E., & Brenner, R. R. (1979). Biosynthesis of unsaturated fatty acids in diatom Phaeodactylum tricornutum. *Lipids, 14*(1), 15–19.

Morgan, S. G. (2001). The larval ecology of marine communities. In M. D. Bertness, S. D. Gaines, & M. E. Hay (Eds.), *Marine community ecology* (pp. 159–181). Sinauer Associates.

Morgan, J. P., & MacIntire, W. G. (1959). Quaternary geology of the Bengal Basin, East Pakistan and India. *Bulletin of the Geological Society of America, 70*, 319–342.

Moriarty, D. J. W. (1986). Measurement of bacterial growth rates in aquatic systems from rates of nucleic acid synthesis. *Advances in Microbial Ecology, 8*, 245–292.

Morris, J. T. (1991). 1991 Effects of nitrogen loading on wetland ecosystems with particular reference to atmospheric deposition. *Annual Review of Ecology and Systematics, 22*, 257–279.

Mukherjee, B. B. (1992). On the ecology of the mangroves with special reference to those in Sundarbans in the territory of India and Bangladesh. *Journal of the Natural History Museum and Institute, Chiba, 2*, 77–81.

Mukhopadhyay, S. K., Biswas, H., De, T. K., Sen, S., & Jana, T. K. (2002a). Seasonal effects on air water of carbon dioxide exchange in the Hooghly estuary, NE coast of Bay of Bengal, India. *Journal of Environmental Management, 4*, 549–552.

Mukhopadhyay, S. K., Biswas, H., De, T. K., Sen, B. K., Sen, S., & Jana, T. K. (2002b). Impact of Sundarban mangrove biosphere on the carbon dioxide and methane mixing ratio at the NE coast of Bay of Bengal, India. *Atmospheric Environment, 36*, 629–638.

Mukhopadhyay, S. K., Biswas, H., De, T. K., & Jana, T. K. (2006). Fluxes of nutrients from the tropical River Hooghly at the land–ocean boundary of Sundarbans, NE Coast of Bay of Bengal, India. *Journal of Marine Systems, 62*(1–2), 9–2.

Mulder, A., van de Graaf, A. A., Robertson, L. A., & Kuenen, J. G. (1995). Anaerobic ammonium oxidation discovered in a denitrifying fluidized-bed reactor. *FEMS Microbiology Ecology, 16*, 177–183.

Müller-Niklas, G., Heissenberger, A., Puškaric, S., & Herndl, G. J. (1995). Ultraviolet-B radiation and bacterial metabolism in coastal waters. *Aquatic Microbial Ecology, 9*, 111–116.

Mumby, P. J., & Hastings, A. (2008). The impact of ecosystem connectivity on coral reef resilience. *Journal of Applied Ecology, 45*, 854–862.

Naganuma, T., Konishi, S., Inoue, T., Nakane, T., & Sukizaki, S. (1996). Photodegradation or photoalteration? Microbial assay of the effect of UV-B on dissolved organic matter. *Marine Ecology Progress Series, 135*, 309–310.

Naidoo, G. (2009). Differential effects of nitrogen and phosphorus enrichment on growth of dwarf Avicennia marina mangroves. *Aquatic Botany, 90*, 184–190.

Nedwell, D. B. (1975). Inorganic nitrogen-metabolism in a eutrophicated tropical mangrove estuary. *Water Research, 9*, 221–231.

Nedwell, D. B., Blackburn, T. H., & Wiebe, W. J. (1994). Dynamic nature of the turnover of organic carbon, nitrogen and sulphur in the sediments of a Jamaican mangrove forest. *Marine Ecology Progress Series, 110*(2/3), 223–231.

Nixon, S. W., Ammerman, J. W., Atkinson, L. P., Berounsky, V. M., Billen, G., et al. (1996). The fate of nitrogen and phosphorus at the land-sea margin of the North Atlantic Ocean. *Biogeochemistry, 35*, 141–180.

Obernosterer, I., Reitner, B., & Herndl, G. J. (1999). Contrasting effects of solar radiation on dissolved organic matter and its bioavailability to marine bacterioplankton. *Limnology and Oceanography, 44*, 1645–1654.

Odum, E. P. (1971). *Fundamentals of ecology* (pp. 1–574). W.B. Saunders.

Odum, W. E., & Heald, E. J. (1972). Trophic analyses of an estuarine mangrove community. *Bulletin of Marine Science, 22*, 671–738.

Odum, E. P. (1984). The status of three ecosystem-level hypotheses regarding salt marsh estuaries: Tidal subsidy, outwelling and detritus-based food chains. In V. S. Kennedy (Ed.), *Estuarine perspectives* (pp. 485–495). Academic Press.

Odum, E. P., & Heald, E. J. (1975). Mangrove forest and aquatic productivity. In A. D. Haster (Ed.), *Coupling of land and water System, Ecological studies* (Vol. 10, pp. 129–136). Springer-Verlag.

Ong, J. E. (1984). Mangrove outwelling? In J. E. Ong & W. K. Gong (Eds.), *Productivity of the mangrove ecosystem: Management implications* (pp. 30–36). Universiti Sains Malaysia.

Opsahl, S., & Benner, R. (1993). Decomposition of senescent blades of the seagrass *Halodule wrightii* in a subtropical lagoon. *Marine Ecology Progress Series, 94*, 191–205.

Pakulski, J. D., Aas, P., Jeffrey, W., Lyons, M., Von Waasenberger, L., Mitchell, D., & Coffin, R. (1998). Influence of light on bacterioplankton production and respiration in a subtropical coral reef. *Aquatic Microbial Ecology, 14*, 137–148.

Pauly, D., & Ingles, J. (1999). The relationship between shrimp yields and intertidal vegetation (mangrove) areas: a reassessment. In A. Yanez-Arancibia & A. L. Lara-Dominguez (Eds.), *Mangrove ecosystems in Tropical America* (pp. 311–316). Institute de Ecologia; UICN/ORMA, NOAA/NMFS.

Pelegrí, S. P., & Twilley, R. R. (1998). Heterotrophic nitrogen fixation (acetylene reduction) during leaf-litter decomposition of two mangrove species from South Florida, USA. *Marine Biology, 131*, 53–61. https://doi.org/10.1007/s002270050296

Prato, E., Danieli, A., Maffia, M., & Biandolino, F. (2012). Lipid contents and fatty acid compositions of Idotea baltica and Sphaeroma serratum (Crustacea: Isopoda) as indicators of food sources. *Zoological Studies, 51*(1), 38–50.

Purvaja, R. (1995). *Interactions between greenhouse gas, nutrient, major and trace elements in the Pichavaram mangroves of South India* (pp. 1–256). PhD thesis, Anna University, Chennai.

Purvaja, R., Ramesh, R., & Frenzel, P. (2004). Plant-mediated methane emission from an Indian mangrove. *Global Change Biology, 10*, 1–10. https://doi.org/10.1111/j.1365-2486.2004.00834.x

Rabinowitz, D. (1978). Early growth of mangrove seedlings in Panama and an hypothesis concerning the relationship of dispersal and zonation. *Journal of Biogeography, 5*, 113–133.

Rabouille, C. (1997). Human perturbation on carbon and nitrogen cycles in the global coastal ocean. In *Presented at LOICZ Open Science Meeting Third, The Netherlands*.

Ray, R., Majumder, N., Das, S., Chowdhury, C., & Jana, T. K. (2014). Biogeochemical cycle of nitrogen in a tropical mangrove ecosystem, east coast of India. *Marine Chemistry, 167*, 33–43.

Reis, C. R. G., Nardoto, G. B., & Oliveira, R. S. (2017). Global overview on nitrogen dynamics in mangroves and consequences of increasing nitrogen availability for these systems. *Plant and Soil, 410*, 1–19.

Rice, D. L., & Tenore, K. R. (1981). Dynamics of carbon and nitrogen during the decomposition of detritus derived from estuarine macrophytes. *Estuarine, Coastal and Shelf Science, 13*, 681–690.

Richey, J. E., Hedges, J. I., Devol, A. H., Quay, P. D., Victoria, R., Martinelli, L., & Forsberg, B. R. (1990). Biogeochemistry of carbon in the Amazon River. *Limnology and Oceanography, 35*, 352–371.

Rivera-Monroy, V. H., & Twilley, R. R. (1996). The relative role of denitrification and immobilization in the fate of inorganic nitrogen in mangrove sediments (Terminos Lagoon, Mexico). *Limnology and Oceanography, 41*, 284–296.

Rivera-Monroy, V. H., et al. (1995). Direct denitrification in mangrove sediments in Terminos Lagoon, Mexico. *Marine Ecology Progress Series, 126*, 97–109.

Robertson, A. I. (1988). Decomposition of mangrove leaf litter in tropical Australia. *Journal of Experimental Marine Biology Ecology, 116*, 235–247.

Robertson, A. I., & Alongi, D. M. (Eds.). (1992). *Tropical mangrove ecosystems* (p. 329). Coastal and Estuarine Studies Series, American Geophysical Union.

Robertson, A. I., Daniel, P. A., & Dixon, P. (1991). Mangrove forest structure and productivity in the Fly River estuary, Papua New Guinea. *Marine Biology, 11*(1), 147–155.

Robertson, A. I., Alongi, D. M., & Boto, K. G. (1992). Food chain and carbon fluxes. In A. I. Robertson & D. M. Alongi (Eds.), *Tropical mangrove ecosystems* (pp. 293–326). Am. Geophys. Union.

Romero, I. C., Jacobson, M., Fuhrman, J. A., Fogel, M., & Capone, D. G. (2012). Long-term nitrogen and phosphorus fertilization effects on N_2 fixation rates and nifH gene community patterns in mangrove sediments. *Marine Ecology, 33*, 117–127.

Rotani, J. F. (1998). Photodegradation of unsaturated fatty acids in senescent cells of phytoplankton: Photoproduct structural identification and mechanistic aspects. *Journal of Photochemistry and Photobiology A: Chemistry, 114*, 37–44.

Roy, S., Ulanowicz, R. E., Majee, N. C., & Roy, A. B. O. Y. (2000). Network analysis of a benthic food web model of a partly reclaimed island in the Sundarbans mangrove ecosystem, India. *Journal of Biological Systems, 8*, 263–278.

Rützler, K., & Feller, I. C. (1996). Caribbean mangrove swamps. *Scientific American, 1996*, 94–99.

Sah, K. D., Sahoo, A. K., Gupta, S. K., & Banerjee, S. K. (1989). Mangrove vegetations of the Sundarbans and their effect on the physicochemical and nutrient status of the soils. *Proceedings of the National Academy of Sciences, India Section B, 2*, 125–132.

Samanta, S., Das, T., Ghorai, S., & Chakraborty, S. K. (2018). *Biocycling of fractional fatty acid components of lipids in mangrove-benthic systems, in Midnapore (east) coast, India*. Examines of Marine Biology & Oceanography. https://doi.org/10.31031/EIMBO.2018.01.000510

Sanchez-Andres, R., Sanchez-Carrillo, S., Alatorre, L. C., Cirujano, S., & Alvarez-Cobelas, M. (2010). Litterfall dynamics and nutrient decomposition of arid mangroves in the Gulf of California: their role sustaining ecosystem heterotrophy. *Estuarine, Coastal and Shelf Science, 89*(3), 191–199.

Schlesinger, W. H. (1997). *Biogeochemistry. An analysis of global change* (p. 588). Academic.

Schwarzenbach, R. P., Gschwend, P. M., & Imboden, D. M. (1993). *Environmental organic chemistry*. Wiley.

Seitzinger, S. P. (1990). Denitrification in aquatic sediments. In N. P. Revsbach & J. Sorenson (Eds.), *Denitrification in soil and sediment* (pp. 301–322). Plenum.

Sengputa, A., & Chaudhuri, S. (1991). Ecology of heterotrophic dinitrogen fixation in the rhizosphere of mangrove plant community at the Ganges River estuary in India. *Oecologia, 87*, 560–564.

Sheridan, R. P. (1991). Nitrogen-fixing micro epiphytes in a tropical mangal community, Guadeloupe, French West Indies. *Biotropica, 23*, 530–541.

Sheridan, R. P. (1992). Nitrogen fixation by epicaulous cyanobacteria in the pointe de la saline mangrove community, Guadeloupe, French West Indies. *Biotropica, 24*, 571–574.

Sengupta, A., & Chaudhuri, S. (2001). Ecology of heterotrophic dinitrogen fixation in the rhizosphere of mangrove plant community at the Ganges river estuary in India. *Oecologia, 87*, 560–564.

Silva, C. A. R., Oliveira, S. R., Rêgo, R. D. P., & Mozeto, A. A. (2007). Dynamics of phosphorus and nitrogen through litter fall and decomposition in a tropical mangrove forest. *Marine Environmental Research, 64*(4), 524–534.

Skellam, J. G. (1951). Random dispersal in theoretical populations. *Biometrika, 38*, 196–218.

Smith, R. C., & Baker, K. S. (1978). Optical classification of natural waters. *Limnology and Oceanography, 23*, 260–267.

Smith, S. V., & Hollibaugh, J. T. (1993). Coastal metabolism and the oceanic organic carbon balance. *Reviews of Geophysics, 31*, 75–89.

Smith, T. J., III. (1992). Forest structure. In A. I. Robertson & D. M. Alongi (Eds.), *Tropical mangrove ecosystems* (pp. 101–136). American Geophysical Union.

Snedaker, S. C. (1982). Mangrove species zonation: Why? In D. N. Sen & K. S. Rajpurohit (Eds.), *Contributions to the ecology of halophytes* (pp. 111–125). Dr. W. Junk.

Soundarapandian, P., Roy, S., & Varadharajan, D. (2014). Antioxidant Activity in Hard and Soft Shell Crabs of *Charybdis lucifera* (Fabricius, 1798). *Journal of Aquaculture Research & Development, 5*, 288. https://doi.org/10.4172/2155-9546.1000288

Springfield, V. A. (1996). U.S. Atomic Energy Commission 160. In *World Resources 1996–1997* (p. 384). Oxford University Press.

Stanley, S. O., Boto, K. G., Alongi, D. M., & Gillian, F. T. (1987). Composition and bacterial utilization of free amino acids in tropical mangrove sediments. *Marine Chemistry, 22*, 13–30.

Steinke, T. D., Holland, A. J., & Singh, Y. (1993). Leaching losses during decomposition of mangrove leaf litter. *South African Journal of Botany, 59*(1), 21–25.

Stewart, B. A., & Davies, B. R. (1989). The influence of different litterbag designs on the breakdown of leaf material in small mountain stream. *Hydrobiologia, 183*, 173–177.

Strome, D. J., & Miller, M. C. (1978). Photolytic changes in dissolved humic substances. *Verhandlungen des Internationalen Verein Limnologie, 20*, 1248–1254.

Stumm, W. (1992). Chemistry of the solid-water interface. Chap. 10. In *Heterogeneous photochemistry* (pp. 337–367). Wiley.

Sun, Y.-P., Nguyen, K. L., & Wallis, A. F. A. (1998). Ring-opened products from reaction of lignin model compounds with UV-assisted peroxide. *Holzforschung, 52*, 61–66.

Swift, M. J., Heal, O. W., & Anderson, J. M. (1979). *Decomposition in terrestrial ecosystems*. University of California Press.

Tam, N. F. Y., & Wong, Y. (1998). Variations of soil nutrient and organic matter content in a subtropical mangrove ecosystem. *Water, Air, and Soil Pollution, 103*(1), 245–261.

Tam, N. F. Y., Vrijmoed, L. L. P., & Wong, Y. S. (1990). Nutrient dynamics associated with leaf decomposition in a small subtropical mangrove community in Hong Kong. *Bulletin of Marine Science, 47*(1), 68e78.

Thamdrup, B., & Dalsgaard, T. (2002). Production of N_2 through anaerobic ammonium oxidation coupled to nitrate reduction in marine sediments. *Applied and Environmental Microbiology, 68*, 1312–1318.
Thom, B. G. (1967). Mangrove ecology and deltaic geomorphology: Tabasco, Mexico. *Journal of Ecology, 55*, 301–343.
Thom, B. G. (1982). Mangrove ecology: a geomorphological perspective. In B. F. Clough (Ed.), *Mangrove ecosystems in Australia. Structure, function and management* (pp. 3–17). Australian National University Press.
Thrush, S. F., & Whitlatch, R. B. (2001). Recovery dynamics in benthic communities: balancing detail with simplification. In K. Reise (Ed.), *Ecological comparisons of sedimentary shores* (pp. 297–316). Springer-Verlag.
Thrush, S. F., Hewitt, J. E., Norkko, A., Cummings, V. J., & Funnell, G. A. (2003). Catastrophic sedimentation on estuarine sandflats: Recovery of macrobenthic communities is influenced by a variety of environmental factors. *Ecological Applications, 13*, 1433–1455.
Tomlinson, P. B. (1986). *The botany of mangroves* (p. 433). Cambridge University Press.
Tranvik, L., & Kokalj, S. (1998). Decreased biodegradability of algal DOC due to interactive effects of UV radiation and humic matter. *Aquatic Microbial Ecology, 14*, 301–307.
Treplin, M., & Zimmer, M. (2012). Downed or dry: a cross-habitat comparison of detrital breakdown processes. *Ecosystems, 15*, 477–491.
Twilley, R. R. (1985). The exchange of organic carbon in basin man grove forests in a southwest Florida estuary. *Estuarine, Coastal and Shelf Science, 20*, 543–557.
Twilley, R. R. (1988). Coupling of mangroves to the productivity of estuarine and coastal waters. In B. Jansson (Ed.), *Coastal offshore ecosystem interactions* (pp. 155–180). Springer-Verlag.
Twilley, R. R., & Day, J. W. (1999). The productivity and nutrient cycling of mangrove ecosystem. In A. Yanez-Arancibia & A. L. Lara Domınguez (Eds.), *Ecosistemas de Manglar en America Tropical* (pp. 127–152). NOAA/NMFS.
Twilley, R. R., Lugo, A. E., & Patterson-Zucca, C. (1986). Litter production and turnover in basin mangrove forests in southwest Florida. *Ecology, 67*(3), 670–683.
Twilley, R. R., Chen, R. H., & Hargis, T. (1992). Carbon sinks in mangroves and their implications to carbon budget of tropical coastal ecosystems. *Water, Air, & Soil Pollution, 64*, 265–288.
Twilley, R. R., Pozo, M., Garcia, V. H., Rivera-Monroy, V. H., Zambrano, R., & Bodero, A. (1997). Litter dynamics in riverine mangrove forests in the Guayas River estuary, Ecuador. *Oecologia, 111*, 109–122.
Underwood, A. J., & Fairweather, P. G. (1989). Supply-side ecology and benthic marine assemblages. *Trends in Ecology & Evolution, 4*, 16–20.
Underwood, A. J., & Keough, M. J. (2001). Supply-side ecology: the nature and consequences of variations in recruitment of intertidal organisms. In M. D. Bertness, S. D. Gaines, & M. E. Hay (Eds.), *Marine community ecology* (pp. 183–200). Sinauer Associates.
van de Graaf, A. A., Robertson, L. A., & Kuenen, J. G. (1995). Anaerobic ammonium oxidation discovered in a denitrifying fluidized-bed reactor. *FEMS Microbiology Ecology, 16*, 177–183.
Vitousek, P. M. (1982). Nutrient cycling and nutrient use efficiency. *American Naturalist, 119*, 553–572.
Vitousek, P. M., & Matson, P. A. (1987). Nitrogen transformations in a range of tropical forest soils. *Soil Biology and Biochemistry, 20*, 361–367. https://doi.org/10.1016/0038-0717(88)90017-X
Vorosmarty, C. J., Li, C., Sun, J., & Dailn, Z. (1998). Drainage basins, river systems, and anthropogenic change: The Chinese example. In J. N. Galloway & J. M. Melillo (Eds.), *Asian change in the context of global change*. Cambridge University Press.
Walsh, J. J. (1991). Importance of continental margins in the marine biogeochemical cycling of carbon and nitrogen. *Nature, 350*, 53–55.
Warren, J. H., & Underwood, A. J. (1986). Effects of burrowing crabs on the topography of mangrove swamps in New South Wales. *Journal of Experimental Marine Biology and Ecology, 102*, 223–235.
Watson, J. G. (1928). Mangrove forests of the Malay Peninsula. *Malay Forest Records, 6*.

Wattayakorn, G., Wolanski, E., & Kjerfve, B. (1990). Mixing, trapping and outwelling in the Klong Ngao mangrove swamp, Thailand. *Estuarine, Coastal and Shelf Science, 31*, 667–688.
Weir, N. A., Arct, J., Ceccarelli, A., & Siren, A. (1994). Some photo-reactions of substituted poly(acrylophenones). *European Polymer Journal, 30*, 701–706.
Weir, N. A., Arct, J., Ceccarelli, A., & Puumala, D. (1996). Long-wave photochemistry of dimethoxylated compounds related to lignin. *Polymer Degradation and Stability, 52*, 119–129.
Wetzel, R. G. (1984). Detrital dissolved and particulate organic carbon functions in aquatic ecosystems. *Bulletin of Marine Science, 35*, 503–509.
Wetzel, R. G., Hatcher, P. G., & Bianchi, T. S. (1995). Natural photolysis by ultraviolet irradiance of recalcitrant dissolved organic matter to simple substrates for rapid bacterial metabolism. *Limnology and Oceanography, 40*, 1369–1380.
Whittaker, R. H. (1967). Gradient analysis of vegetation. *Biological Reviews, 42*, 207–264.
Whigham, D. F., Verhoeven Jos, T. A., Samarkin, V., & Megonigal, P. J. (2009). Responses of *Avicennia germinans* (black mangrove) and the soil microbial community to nitrogen addition in a hypersaline wetland. Estuaries and Coasts, 32, 926–936.
Woitchik, A. F., Ohowa, B., Kazungu, J. M., Rao, R. G., Goeyens, L., & Dehairs, F. (1997). Nitrogen enrichment during decomposition of mangrove leaf litter in an east African coastal lagoon (Kenya): relative importance of biological nitrogen fixation. *Biogeochemistry, 39*, 15–35.
Wolanski, E., Mazda, Y., & Ridd, P. (1992). Mangrove hydrodynamics. In A. I. Robertson & D. M. Alongi (Eds.), *Tropical mangrove ecosystems, coastal and estuarine studies* (pp. 436–462). AGU.
Wollast, R. (1991). The coastal organic carbon cycle: fluxes, sources, and sinks. In R. F. C. Mantoura, J.-M. Martin, & R. Wollast (Eds.), *Ocean Matgin processes in global change* (pp. 365–381). Wiley.
Wollast, R. (1993). Interactions of carbon and nitrogen cycles in the coastal zone. In R. Wollast, F. T. Mackenzie, & L. Chou (Eds.), *Interactions of C, N, P and S biogeochemical cycles and global changes* (pp. 195–210). Springer-Verlag.
Wollast, R., & Mackenzie, F. T. (1989). Global biogeochemical cycles and climate. In A. Berger, S. Schneider, & J.-C. Duplessy (Eds.), *Climate and Geo-Sciences* (pp. 453–473). Kluwer.
Wong, Y. S., Lan, C. Y., Chen, G. Z., Li, S. H., Chen, X. R., Liu, Z. P., & Tam, N. F. Y. (1995). Effect of wastewater discharge on nutrient contamination of mangrove soils and plants. *Hydrobiologia, 295*, 243–254.
Ysebaert, T., Maeire, P., Herman, P. M. J., & Verbeek, H. (2002). Macrobenthic species response surface along estuarine gradients; Prediction by logistic regression. *Marine-Ecological Progress Series, 225*, 79–95.
Yuan, J., Ding, W., Liu, D., Kang, H., Freeman, C., Xiang, J., & Lin, Y. (2015). Exotic Spartina alterniflora invasion alters ecosystem–atmosphere exchange of CH4 and N2O and carbon sequestration in a coastal salt marsh in China. *Global Change Biology, 21*, 1567–1580.
Zellmer, I. D. (1995). UV-B-tolerance of alpine and arctic *Daphnia*. *Hydrobiologia, 307*, 153–159.
Zepp, R. G., & Cline, D. M. (1977). Rates of direct photolysis in aquatic environment. *Environmental Science & Technology, 11*, 359–366.
Zepp, R. G., Schlotzhauer, P. F., & Sink, R. M. (1985). Photosensitized transformations involving electronic energy transfer in natural waters: Role of humic substances. *Environmental Science & Technology, 19*, 74–81.
Zilius, M., et al. (2020). N2 fixation dominates nitrogen cycling in a mangrove fiddler crab holobiont. *Scientific Reports, 10*, 13966. https://doi.org/10.1038/s41598-020-70834-0
Ziolkowska, A., Debska, B., & Banach-Szott, M. (2020). Transformation of phenolic compounds in meadow soils. *Scientific Reports, 10*, 19330. https://doi.org/10.1038/s41598-020-76316-7
Zuberer, D. A., & Silver, W. S. (1978). Biological dinitrogen fixation (acetylene reduction) associated with Florida mangroves. *Applied and Environmental Microbiology, 35*, 567–575.
Zweifel, U. L., Blackburn, N., & Hagström, Å. (1996). Cycling of marine dissolved organic matter. I. An experimental system. *Aquatic Microbial Ecology, 11*, 65–77.

Chapter 4
Biodiversity of the Sundarbans Mangrove Ecosystem, India

Abstract Biodiversity (or biological diversity) is formally defined at three specific levels—genetic variations within each species or population, the variety and abundance of different species within a community or region, and the range of ecosystems within the region under consideration. Biodiversity delivers valuable ecosystem services and provisions of goods such as food, fodder, fuel, medicines, and building materials vital to human survival. Biodiversity worldwide has been threatened by both natural and anthropogenic drivers. Major pressures include habitat destruction, invasive alien species, shifts in species composition, disruption of food-web dynamics, chemicalization of the environment, altered geomorphic processes, changing rainfall patterns, extreme events (floods and droughts), rising temperatures, and saltwater intrusion. The horizontal and vertical growth of mangrove swamps, as well as their diversity, vary regionally according to environmental setting, resulting in three distinct types in the Indian Sundarbans: estuarine, deltaic, and lagoonal systems.

The Indian Sundarbans has a hot, humid tropical monsoon climate with three distinct seasons: pre-monsoon (March–May), monsoon (June–October), and post-monsoon (November–February). Over the eight months of pre-monsoon and monsoon, Sundarbans remains very hot, sultry, and humid, with more turbulent freshwater flows. During this period, the region experiences frequent tropical depressions originating in the Bay of Bengal, associated with fierce cyclonic storms that often result in severe biodiversity loss and ecosystem damage.

Changing environmental conditions force flora and fauna either to adapt or to migrate to more suitable habitats, often abandoning former ranges that become environmental refuges. These shifts alter species composition of communities and disrupt ecosystem functions and services, reducing productivity and ecosystem resilience. Over time, these changes adversely affect human well-being through diminished food security, loss of productive agricultural lands, increased geographic spread of vector-borne diseases, eutrophication of water bodies, and disruption of seasonal cycles of plant and animal life leading to disruption of environmental services.

S. K. Chakraborty, S. K. Acharya, *Eco-dynamics and Environmental Perturbations in Mangrove Ecosystems*,
https://doi.org/10.1007/978-3-032-08710-2_4

The Sundarbans mangrove ecosystem renders numerous goods and services for human beings in the forms of forestry, aquaculture, honey, agriculture, fish, and wildlife—reptiles, birds, and mammals. Natural production of honey synchronizes with the blossoming of flowers in different mangrove plants, mostly during the months of late February, March, April, and early May, when atmospheric temperatures are increasing steadily from 26°–38° C. During this peak honey-collection season, traditional harvesters—known locally as mawals—enter the forest in organized groups led by an experienced guide. Moving in a systematic pattern, they track the flight paths of honeybees to locate the beehives high in the mangrove canopy. Honey from Sundarbans mangrove trees are mostly collected from *Aegiceras carniculatum* (Khalsi), *Sonneratia apetala* (Keora), and *Ceriops decandra* (Goran). The diverse assemblage of halophytic plants and faunal communities—including zooplankton, benthic invertebrates, and nekton—inhabiting in the dense networks of estuaries, rivers, creeks, and channels in Sundarbans rendering valuable ecosystem services which are vital for sustaining local livelihoods. From mammals to different forms of nonchordates, emphasizing on their seasonality of occurrence, distribution, community interractions etc. and on plants-insects interraction, Sundarbans has massive potential in its faunal diversity alongside unique assemblages of true mangroves and mangrove-associated floral components. Sundarbans is widely recognized as one of the world's premier biodiversity hotspots, distinguished by its exceptionally rich and characteristic faunal diversity. This rich and unique setting has continuously attracted researchers in ecology, biodiversity conservation, and systematics ever since the pioneering work of Stoliczka (1869) on the malacofauna of lower Bengal's brackish-water systems. Most importantly, alongside documenting and interpreting the ecological perspectives of the biodiversity wealth including newly discovered species, mathematical tools are applied to explain man-wildlife conflicts.

Keywords Biodiversity · Sundarbans · Tropical climate · Mangrove plants · Faunal diversity · Man-wildlife conflicts

4.1 Biodiversity in General and Sundarbans in Particular

Biological diversity is defined by the United Nations Convention as the "variability among living organisms from all sources, including, inter alia, terrestrial, marine and other aquatic ecosystems and the ecological complexes of which they are a part; this includes diversity within species, between species and of ecosystems" (Chakraborty, 2023).

Over 300,000 marine species are currently recorded across virtually all oceanic and coastal habitats—continental shelves and slopes, the intertidal and subtidal littoral zones, coral reefs, mangroves, seagrass meadows, rocky and sandy shores, underwater seamounts, and the deep oceans—encompassing corals, sponges, anemones, benthic invertebrates, pelagic fauna like plankton, and fish representing nektonic forms. Voluminous water with considerable depth display stratification in respect of density, temperature, salinity and nutrients because of upwelling and outwelling processes which in turn determine the growth and propagation of various forms of biota starting from tiniest plankton to many larger species like dolphins, sea turtles, whales, seals, walrus, seabirds, and a hoard of large fish like, sardines, tuna, and mackerel. A major physical process, known as upwelling—upward movement of deep and cooler water loaded with nutrients to the ocean surface—appears to be the major source of life support ingredients for the development of pelagic fisheries. Marine, coastal, and oceanic ecosystems collectively host the representative faunal components of 32 out of the 36 recognized animal phyla, 15 of which are exclusively marine, and five almost exclusively so (Thom, 1967; Snedaker, 1984; Smith et al., 1992; Spalding et al., 1997).

Highly productive mangrove ecosystem areas act as nursery and feeding grounds for so many finfish and shellfish species in the forms of species of fish, invertebrates, higher chordates like reptiles, birds, and mammals. These faunal communities, both independently an collectively, deliver critical ecosystem services by protecting coastlines from erosion, sedimentation, and the destructive impacts of tropical cyclones and storm surges, while also supporting coastal fisheries (Boesch & Turner, 1984; Kathiresan & Qasim, 2005). Improved understanding of the geographical distributional patterns of mangrove flora and fauna has enabled greater utilization of mangrove-associated fisheries during over the last couple of decades (Odum & Heald, 1975c; Lugo, 1980; Ellison et al., 1999; Annon, 2003; Anonymous, 1987; Macintosh et al., 2002; Lovelock et al., 2006; Blasco, 1977; Blasco & Aizpuru, 2002; Gray and Elliot, 2009; McKee et al., 2007).

Mangroves colonize areas of low-wave–energy environment such as protected shorelines, river floodplains, and deltaic islands, where fine mud and clay sediments accumulate, thriving across a wide range of high-to-low nutrient concentrations (Odum et al., 1982). Continuous nutrient cycling in mangrove ecosystems happens by virtue of the presence of a unique physical process of tidal flushing, linking the mangrove subsystem to the adjacent estuarine subsystem. These twice daily tidal exchanges facilitate constant interaction among living and nonliving components, resulting in highly efficient nutrient retention, transformation, and recycling (Naskar & Guha Bakshi, 1987; Mukherjee et al., 2013; Chakraborty, 2011, 2018). Vast sheets of green coverage in mangrove forests along estuaries with fluctuating tidal water act as important nutrient sinks, controling soil salinities, and helping to disperse propagules.

Of India's nine maritime states, which together have a coastline of approximately 7500 km, West Bengal occupies a unique geographical location possessing the mighty the Sundarbans mangrove ecosystem. This vast mangrove forest forms part of Hooghly–Matla estuarine complex which is shared with the neighboring country, Bangladesh. The biodiversity of the coastal area of West Bengal extends over 0.82 million ha and along 220 km of coastline shared by two coastal districts—South 24 Parganas and Midnapore (East). The area contains a good number of mangroves and their associated species of flora, including various species of phytoplanktonic and benthic algae, fungi, lichen, zooplankton, ichthyoplankton, benthos, soil inhabiting arthropods, fish, amphibians, reptiles, birds, and mammals.

Mangrove colonization and growth initiate sediment accretion, which in turn promotes the formation of deltaic islands along the West Bengal coast. This process is accompanied by the simultaneous settlement and flourishing of associated flora and microbes fauna, regulated by the complex interplay of ecobiophysical and physiochemical factors (Boesch, 1977; Bhowmik et al., 1985; Biswas et al., 1985; Annon, 2003; Alcock, 1985; Paterson et al., 1998; Alfred & Ramkrishna, 2004; Chakraborty, 2011, 2018).

The mangrove–estuarine complex of Sundarbans is one of Earth's most dynamic, diverse, and productive ecoregions, driven by continuous interactions among numerous ecological factors and processes at the interface between of halophytic forests and estuarine networks. Through the functional ecodynamics, the mangrove ecosystem supports the growth and propagation of a myriad of living organisms in the form of plankton, benthos, periphyton, and nekton. The shallow estuarine and near-shore habitats with submerged aquatic vegetation and large woody debris are structurally complex, exhibiting ecological heterogeneity and a high degree of biodiversity. This biologically productive transitory ecological habitat acts as the nursery grounds for the juveniles of many commercially important finfish and shellfish. These conditions supply food, refuge from predators, and the essential natural ingredients required for spawning and physiological adaptation to saltwater life. The morphodynamics of near-shore regions sustain a highly productive food web based upon the exchange of materials in the form of mangrove litter, decomposed detritus, and elemental nutrients produced by autotrophic organisms—primarily vascular halophytic plants, benthic algae, and phytoplankton—thriving in these shallow-water estuarine habitats. Mangrove ecosystems deliver a wide variety of biological provisions that fulfill commercial, cultural, aesthetic, and recreational requirements of local human population. Globally, the total economic value of services rendered by coastal wetlands is estimated at more than $2 trillion per year. An understanding of the prevailing linkage between biotic and non-biotic components in respect to the mutual interdependences of different coastal habitats will help in the development strategies to sustain these ecosystems in an integrated manner.

Estuaries are unique transitional ecosystems situated between relatively stable marine and limnetic environments. They are characterized by continually changing ecological conditions across a broad range in hydrographical parameters (Kinne, 1967). Natural ecosystems support diverse populations, which remain associated and interact in a very interesting and intriguing manner, forming a *biotic community,* which is defined as follows: "A biotic community is an assemblage of populations living in a prescribed area or physical habitat representing an organized unit having specific characteristics of an individual and population and functions as a unit through coupled metabolic transformations" (Odum, 1977). Species density and diversity—two important criteria of ecological communities—not only reflect the nature of habitats but also revela the mode of interactions among an array of abiotic and biotic components (Chakrabarti, 1987, 1995, 1998, 2003, 2009, 2010; Chakraborty and Chowdhury,

1992; Goudha & Panigrahy, 1996; Chakraborty et al., 2005, 2012, 2014, 2005, 2012, 2014, 2019, 2021; Chakravarty et al., 2004, 2023).

Estuaries, as unique ecosystems occupying the transition zones between the more stable marine ecosystem and limnetic regions of the freshwater riverine areas, possess continually changing ecological conditions within a wide range of hydrobiological and ecographical conditions (Kinne, 1967). Sundarbans mangroves represent a highly dynamic, diverse, and productive ecoregion on Earth because of the continuous interaction among numerous ecological factors and processes at the interphase of halophytic forest covered land and estuarine networks. This complex ecosystem supports the growth and propagation of a multitude of living organisms in the form of plankton, benthos, periphyton and nekton. The shallow estuarine habitats of phytoplankton, zooplankton, and fish with adjoining mangrove vegetation are structurally complex, exhibiting ecological heterogeneity, which promotes higher biodiversity. This biologically productive ecozone acts as the nursery grounds for the juveniles of many commercially important finfish and shellfish by providing food, protection from predators and other natural ingredients required for spawning. The flora and fauna inhabiting this ecosystem tolerate fluctuating environmental conditions and survive through a combination of morphological adaptations and physiological flexibility to overcome the challenges of such adversity. Nearshore ecodynamics develop unique food webs involving biotic components produced by the exchange of materials in the form of mangrove litter which contributes nutrients formed by microbial decomposition and transported to the nearby estuarine subsystem. Continuous exchange of nutrients between forest and estuarine subsystems within Sundarbans mangrove ecosystem are used by autotrophic organisms like vascular halophytic plants, benthic algae and phytoplankton which grow in highly productive shallow-water estuarine habitats. Biological productivity is boosted because the mangrove ecosystem produces a wide variety of biological organisms and biological products to fulfill commercial, cultural, aesthetic, and recreational requirements of human population of the region (Chakraborty et al., 1992, 2005, 2013, 2019; Chandra et al., 2003; Khalua et al., 2003, 2008; Chakravorty et al., 2005; Gopal & Chauhan, 2006; Chatterjee et al., 2008, 2014; Datta et al., 2017, 2018; Chakrabarty, et al., 2023). An estimation has revealed that the total global economic value of all the services rendered by coastal wetlands exceeds $2 trillion per year. Therefore, an understanding of the prevailing linkage between biotic and abiotic components in respect of mutual interdependences of different coastal habitats help not only sustaining the ecosystems but also generate ecological goods in an integrated manner. Natural mangrove ecosystem provide shelter for a variety of living organisms which interact with each other in intriguing manner forming a *biotic community*, which is defined as "an assemblage of populations living in a prescribed area or physical habitat representing an organized unit having specific characteristics of an individual and population and functions as a unit through coupled metabolic transformations" (Odum, 1977; Chakrabarti, 1987; Chakraborty, 2017). Density and diversity of species being the two important attributes within a biotic community could assess not only the nature of habitat but also tend to reflect the mode of interactions among an array of abiotic and biotic components (Chapman, 1976; Camilleri, 1992; Chakravarty et al., 2005).

4.2 The Evolution of Life in Marine Ecosystems

The evolution of living organisms in marine–oceanic ecosystems occurs through the gradual change within genetic constitution of a given species, mostly over time resulting in the flourishing of biodiversity in time and space. In biotic communities, the adaptive advantages of one species over another can be explained with the help of the concept of natural selection, as put forward by Charles Darwin. Superior abilities yield greater success in acquiring environmental resources in the form of food, shelter, and sexual partners, as well as adaptive advantages in avoiding predators In mangrove–estuarine ecosystem, individuals exhibiting superior adaptive traits are better able to survive within changing ecological conditions. This ensures perpetuation of their race by transferring these advantageous genetic traits to subsequent generations, leaving more individuals as descendants within the same population of a particular species.

4.3 Biodiversity–Ecosystem Functional Relationships with Special Reference to the Sundarbans Mangrove Ecosystem

Biodiversity encompasses variation at three hierarchical levels—genes, species, and ecosystems. At the species level, the focus is on species richness—the total number of different species within a community. Species richness serves as the key indicator of ecosystem functioning and reflects prevailing biotic and abiotic conditions. The relationship between species richness and ecosystem functioning is explained by three major hypotheses.

Hypothesis 1: The linear (rivet) hypothesis suggesting that all species equally and critically contribute to the functioning of ecosystem (Lawton, 1994).

Hypothesis 2: The redundancy hypothesis highlighting the cause of loss of species in ecosystems as a consequence of their nonperformance as long as the major functional groups are still present which in other words point out that the functional traits of the species is more important than the number of species but the redundant species play important roles in ensuring ecosystem resilience to perturbation (Lawton, 1994).

Hypothesis 3: The idiosyncratic hypothesis stating the roles of species diversity in regulating the ecosystem functioning but not in a predictable direction as the individual species contributes in context-dependent and complex manners (Lawton, 1994).

These three hypotheses, taken either independently or in a combined manner, depict the functional relationships of biodiversity within an ecosystem; however, additional important components of biodiversity, such as density of a species, also affect these relationships. Certain keystone species, rather than species diversity, may have a significant influence on ecosystem processes such as productivity, nutrient cycling, and food-web dynamics (Thrush et al., 2006). Owing to the unique fit of a keystone species within a particular ecosystem, the loss of it will likely result in the disruption of ecosystem functioning in tune with the rivet hypothesis as the functioning of this species can hardly be taken up by another. Changes in the absolute density and relative abundance of species tends to bring changes related to altering resource use or sharing patterns facilitation (Karlson et al., 2007; McKie et al., 2008).

Sundarbans estuarine delta complex—the world's largest mangrove forest with luxuriant mangrove plants—sustains its estuarine components through the continuous interaction between terrestrial and marine ecosystems (Vannucci, 1989). An eco-hydrological approach to understanding the functioning of the mangrove ecosystem emphasizes the interactive pathways of different environmental variables with biodiversity components and appears to be a prerequisite for sustainable biodiversity management and ecosystem restoration. The biocomplexity of mangrove ecosystems distinguishes them from other ecosystems due to the prevailing interplay of highly specialized adaptations and extreme trail plasticity that enables mangroves to thrive in the oscillating ecological conditions of the intertidal environment.

Floral diversity of the deltaic Sundarbans having heterogeneous species composition flourishes along the gentle slopes of the tidal channels of mangroves and salt marsh of coastlines, river flood plains, mouth of the river and delta fronts of the estuarine–mangrove complex of Sundarbans. The unique mosaic of habitats that occurs within the mangrove ecosystem supports an extremely rich diversity, not only of flora, but also fauna in the form of zooplankton, benthos, periphyton, and nekton. These biotic components, either independently or in a combined manner, serve as the driving force of ecodynamics through trophic interaction (Meadows & Tait, 1989; Saberwal et al., 1994; Chapin et al. 1997; Das & Bhattacharya, 1999a, 1999b; Perrings et al., 2011). They serve to filter pollutants, act as phytoremediators for toxic persistent substances, and protect coastal regions by serving as a buffer against natural disasters. They also supply nutrients to adjacent ecosystems (Figs. 4.1, 4.2, 4.3, 4.4, 4.5, 4.6, 4.7, 4.8, 4.9, 4.10, 4.11, 4.12, 4.13, 4.14, 4.15, 4.16, 4.17, 4.18, 4.19, 4.20, 4.21, 4.22, 4.23, 4.24, 4.25, 4.26, 4.27, 4.28, 4.29, 4.30, 4.31, 4.32, 4.33, 4.34, 4.35, 4.36, 4.37, 4.38, 4.39, 4.40 and Tables 4.1, 4.2, 4.3, 4.4, 4.5, 4.6, 4.7, 4.8, 4.9, 4.10, 4.11, 4.12, 4.13, 4.14, 4.15, 4.16, 4.17, 4.18).

Fig. 4.1 Mangrove plant species in the Sundarbans mangrove ecosystem, India (1): (**A** and **B**) Beds of a dominant saline grass, *Portarasia Coarctata,* a pioneering grass species for the delta formation though sediment stabilization.; (**C** and **D**), *Excoecaria agallocha*, a dominant and abundant mangrove tree species, locally referred to as Gneio, attaining a height of about 6 meters inhabiting mainly at supralittoral zones sustaining no tidal inundation for a prolonged period; (**E**, **F** and **G**) the most dominant shrub mangrove species, *Acanthus illicifolius*, characterized by thorny thick leaves and colorful magenta flowers, inviting several honey-producing and pollinator insect species; (**H**) Commonly-occurring fern species, *Acrostichum aureum*, by the side of freshwater-dominated estuarine tracts

Fig. 4.2 Mangrove plant species in the Sundarbans mangrove ecosystem, India (2): (**A**) *Salicornia brachiata*, under the family Chenopodiaceae being a succulent herb species covering the surface of the tidal flats alongside resisting soil erosion as well as providing an ecological niche for benthic fauna like brachyuran crabs, molluscs, etc.; (**B** and **C**) *Suaeda nudiflora*, emerging from the intertidal saline grass mat, *Cyperus javanicus*; (**C** and **D**) *Suaeda maritima*, profuse growth which, in the upper tidal levels, renders stability of the sediment and maintains microclimatic conditions to provide suitable habitats for the benthic fauna; (**E**) *Pandanus tectorius*, locally referred to as "Keya" growing on the upper parts of the sandflats act as a stabilizer of sand dunes; (**F**) *Ceriops decandra* (Goran), a strong mangrove tree, providing shelter for so many arboreal faunal components alongside ensuring stability of mangrove floor by resisting soil erosion; (**G**) *Bruguiera gymnorrhiza* (Kankra), a large mangrove plant species renders so many ecosystem benefits, starting from enhancing carbon sequestration, to resisting soil erosion, and providing shelter to arboreal faunal species, etc.; (**H**) *Phoenix pelludosa,* mangrove palm (Hetal), which stabilizes the soil of the mangrove forest floor, helps with accretion, and provides habitats for wild animals, like the tiger

Fig. 4.3 Mangrove plant species in the Sundarbans mangrove ecosystem, India (3): (**A**) *Aegiceras corniculatum* (Khalsi); (**B**) *Aegiceras corniculatum*; (**C**) *Avicennia officinalis* (Sadabain); (**D**) *Kandelia candel* (Garia); (**E**) *Rhizophora decandra* (Garjan); (**F**) *Avicennia officinalis* (Sadabain); (**G**) *Avicennia marina* (Kalobain); (**H**) *Avicennia alba* (Pairabain)

Fig. 4.4 Mangrove plant species in the Sundarbans mangrove ecosystem, India (4): (**A**) *Brownlowia tersa* (Lata); (**B**) *Avicennia alba* (Pairabain); (**C** and **D**) *Xylocarpus granatum* (Dhudul); (**E** and **F**) *Nypa fruticans* (Gol Pata); (**G**, **H** and **I**) *Heritiera fomes* (Sundari)

Fig. 4.5 Mangrove plant species in the Sundarbans mangrove ecosystem, India (5): (**A** and **B**) *Rhizophora mucronata* (Garjan); (**C**) *Aegialites rotundifolia* (Tora); (**D**) *Rhizophora apiculata* (Garjan); (**E** and **F**) *Sonneratia caseolaris* (Ora); (**G**) *Ceriops tagal* (Math goran); (**H**) *Xylocarpus mekongensis* (Passur)

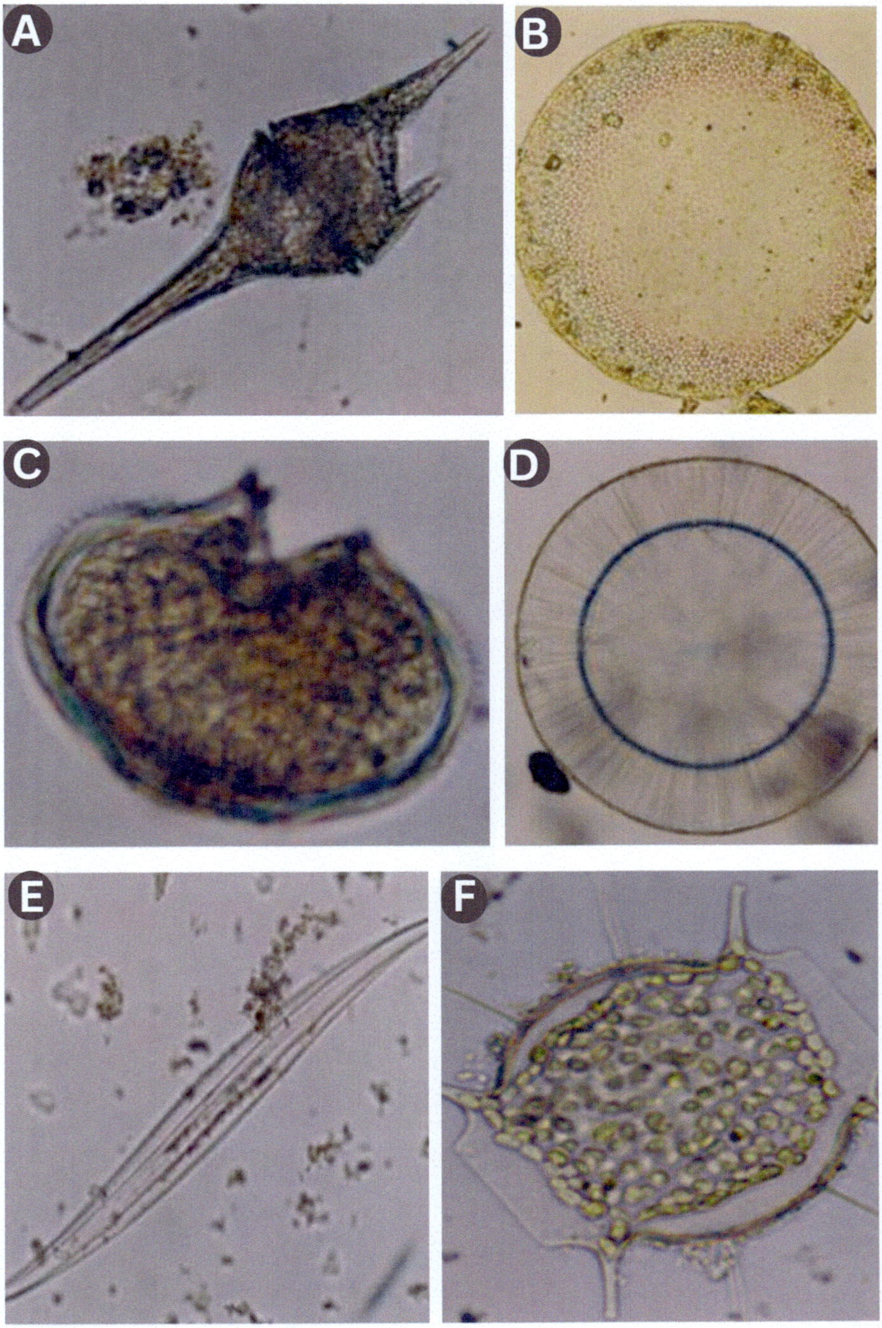

Fig. 4.6 Phytoplanktons of the mangrove ecosystem of Sundarbans, India: (**A**) *Ceratium* sp.; (**B**): *Coscinodiscus* sp.; (**C**) *Peridinium* sp.; (**D**) *Gossleriella* sp.; (**E**) *Biddulphia* sp.; (**F**) *Pleurosigma* sp.

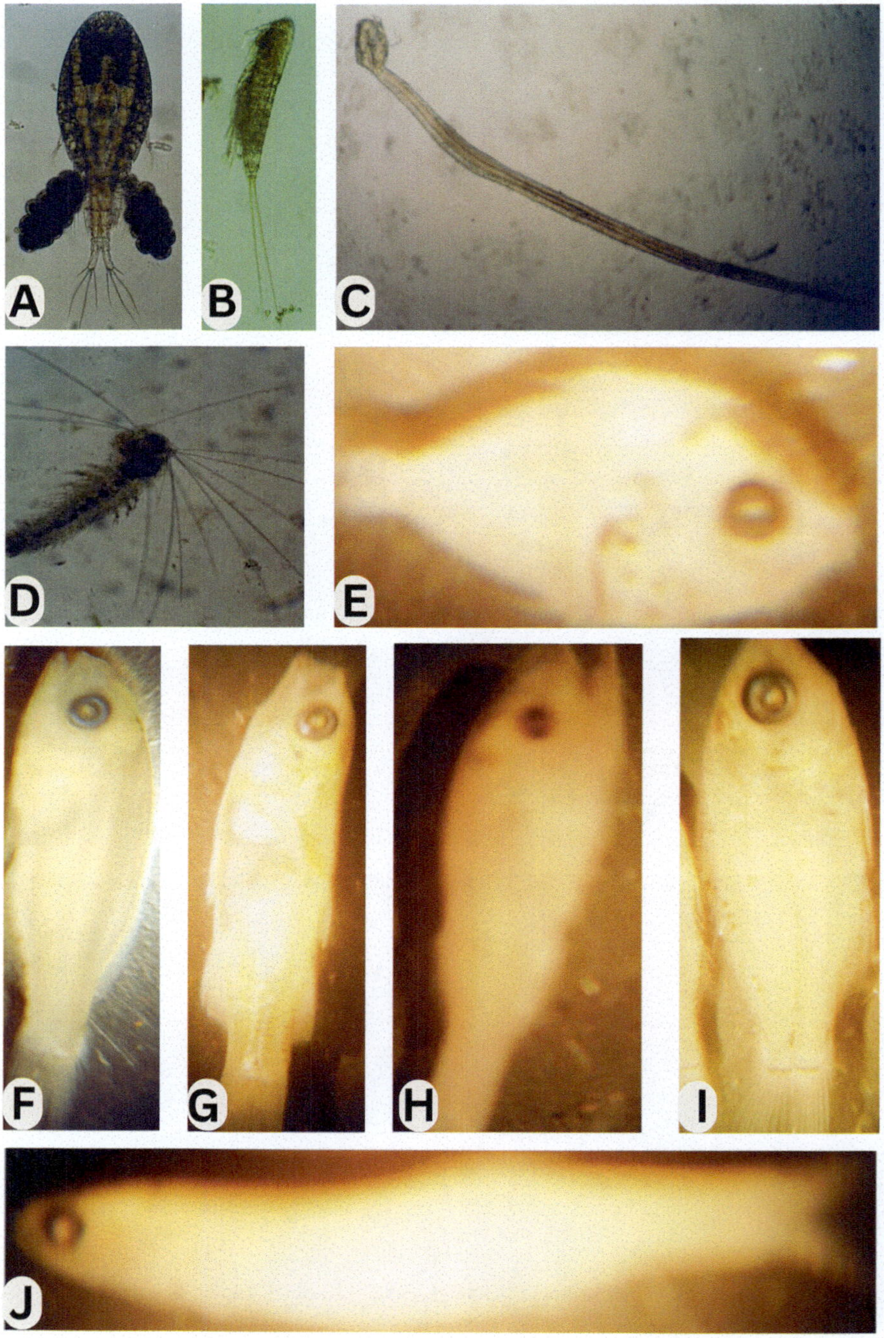

Fig. 4.7 Zooplankton and Ichthyoplankton of the Sundarbans mangrove ecosystem, India: (**A**) Cyclopoid Copepoda; (**B**) Harpacticoid Copepoda; (**C**) Chaetognatha; (**D**) Polychate larva as meroplankton; (**E**) Larva of the fish family Mugilidae; (**F**) Larva of the fish family, Leiognathidae; (**G**) Larva of fish family Scaenidae; (**H**) Larva of fish family Stolephorus; (**I**) Larva of fish family Ambassidae; (**J**) Larva of fish family Gerridae

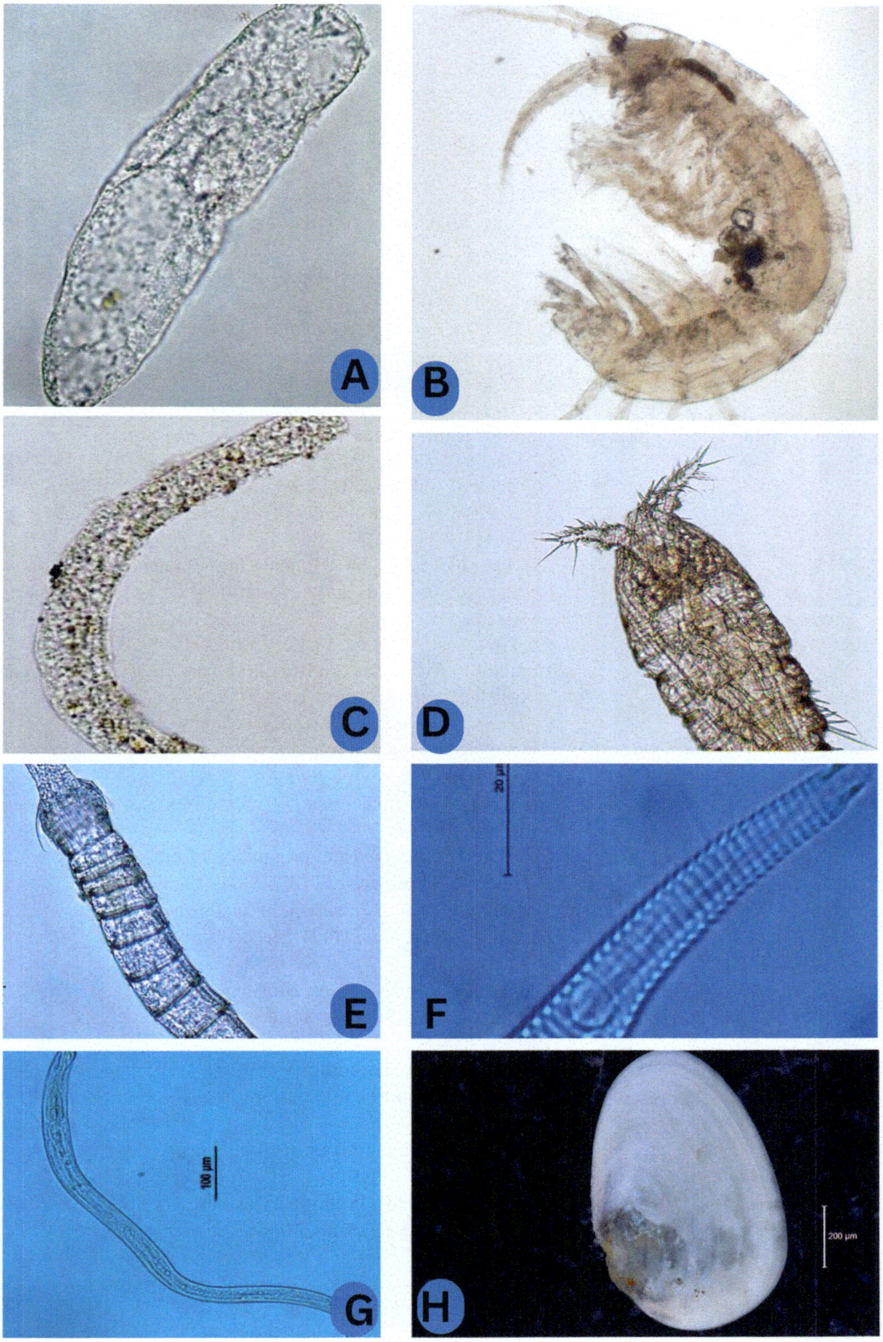

Fig. 4.8 Diversity of meiobenthic fauna in the Sundarbans mangrove ecosystem, India: (**A**) Ciliophora; (**B**) Copepoda; (**C**) Gastrotricha; (**D**) Herpacticoid Copepoda; (**E**) Kinorinka; (**F**) Nematoda, anterior part; (**G**) Nematoda, full body; (**H**) Bivalvia

Fig. 4.9 Diversity of brachyuran crabs in the Sundarbans mangrove ecosystem, India: (**A**) *Ocypoda macrocera*; (**B**) *Uca acuta acuta* (male); (**C**) *Uca acuta acuta* (female); (**D**) *Uca triangularis triangularis* (male); (**E** and **F**) *Uca triangularis triangularis* (female); (**G**) *Uca annulipes annulipes*; (**H**) *Dotilla blanfordi*; (**I**) *Metaplax intermedia*; (**J**) *Uca dussumieri dussumieri* (female, juvenile); (**K**) *Sesarma chromantes bidens*; (**L**) *Scylla serrata*, the only edible crab in the Sundarbans mangrove ecosystem

Fig. 4.10 Diversity of habitats of intertidal brachyuran crabs in the Sundarbans mangrove ecosystem, India: (**A**) Habitat of *Dotilla blanfordi* with sand pellets produced by this tiny crab; (**B**) Assemblage of ghost crabs, *Ocypoda macrocera*, on their way to low tidal mark for releasing fertilized ovum at the time of spring tide; (**C**) Mud balls produced by male fiddler crab, *Uca acuta acuta;* (**D**) Claws waving by male *U. acuta acuta* to mark their territory and also send signals to their female counterpart as a part of sexual behavior; (**E**) Ideal habitat for the detritivorus macrobenthic fauna including *U. acuta acuta*; (**F**) *Uca. annulipes annulipes* (male) constructing burrow; (**G** and **H**) Congregation of males, *Uca. annulipes annulipes*, mainly males waiving their large chela; (**I**) *Uca triangularis bengali* (female) near burrow; (**J**) *Uca triangularis bengali* (male) near burrow; (**K**) Habitat of *Sesarma chiromantes bidens* outside its burrow; (**L**) *Scylla serrata* outside its burrow for the purpose of foraging

Fig. 4.11 Diversity of Molluscs (Bivalves) in the Sundarbans mangrove ecosystem, India: (**A**) *Meretrix casta*; (**B**) *Macoma barmanica*; (**C**) *Nerita articulata*, a common gastropod; (**D**) *Meretrix meretrix*; (**E**, **F** and **G**): *Anadora granosa*: (**F**) *Sanguilaria (Solalillina) accuminata*; (**H**) *Neo-Solen kampi,* a rare bivalve

Fig. 4.12 Diversity of Molluscs (Gastropods) in the Sundarbans mangrove ecosystem: (**A**) An identified large sized gastropod; (**B**) *Cerithidea obtusa*; (**C**) *Cerithidea (Cerithideopsis) cingulata*; (**D**) *Littorina (Littoraria) melanostoma*; (**E**) *Littorina (Littorinopsis) scabra scabra*; (**F**) *Haminea crocata*; (**G**) *Nerita articulata*; (**H**) *Assiminea beddomeana*; (**I**) *Telescopium telescopium*; (**J**) *Assiminea brevicula*; (**K**) *Ellobium (Auricula) gangeticum*

Fig. 4.13 Diversity of habitats of different Molluscs (Gastropods): (**A** and **B**) Habitats of Gastropod, *Telescopium Telescopium*, on the detritus enriched mudflat; (**C**) External openings of the burrows of the bivalve, *Solen kampi*, the razor clam; (**D**) The tiny gastropod, *Littorina melanostoma* on the leaves of *Avicennia alba*, an abundant mangrove plant in the Sundarbans mangrove ecosystem; (**E**) A number of gastropods, belonging to the species, *Cerethedia obtusa*, crawling over a mangrove trunk; (**F**) Mangrove plants associated mussel, *Anodontaanatina*; (**G**) Mangrove-associated Gastropoda, *Nerita semirugosa;* (**H**) *Telescopium telescopium*, a larger Gastropoda attached with thorny mangrove shrub, *Acanthus ilicifolius;* (**I**) *Cerithedia cingulata*, the most abundant grazing gastropod on the mudflat

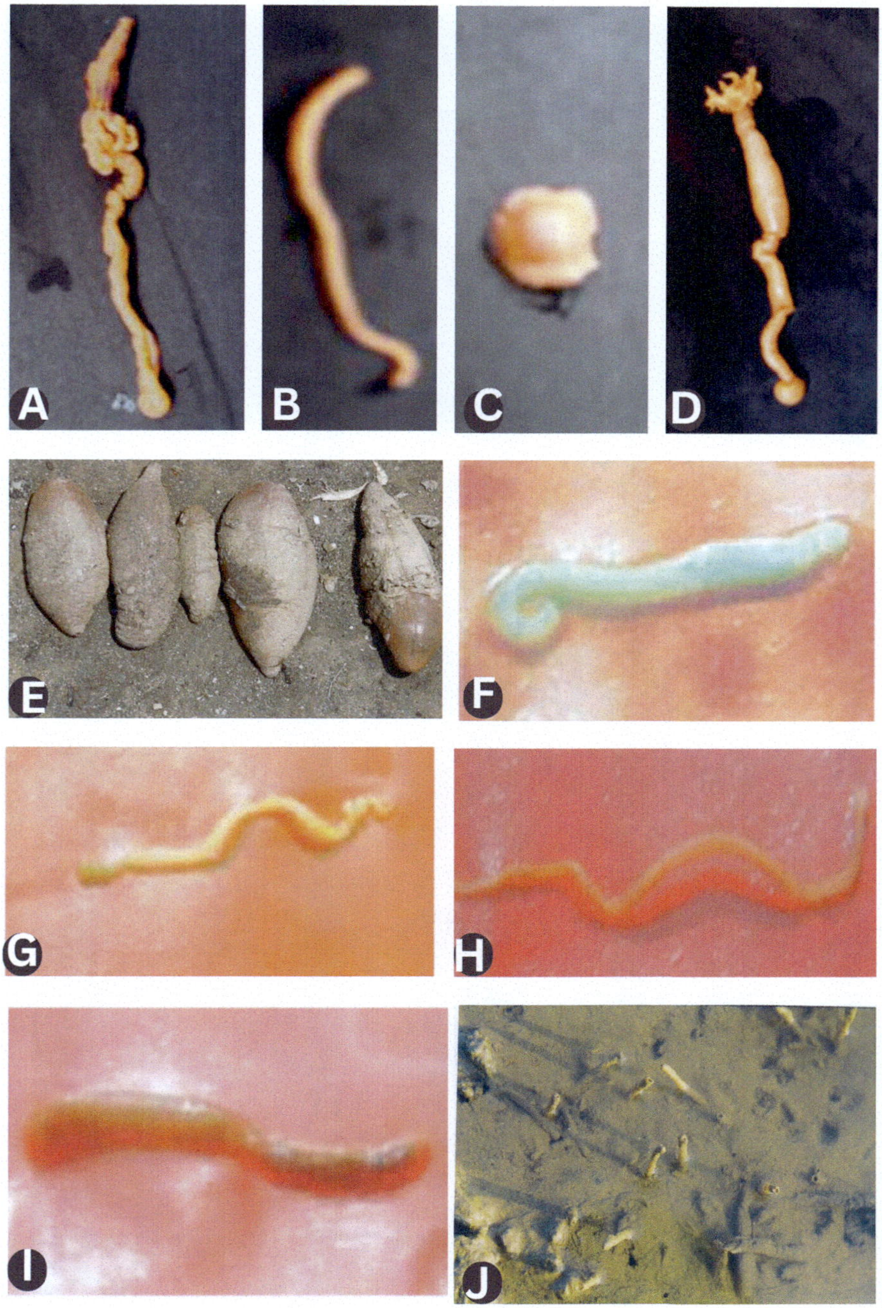

Fig. 4.14 Diversity of Actinarians, under phylum Cnidaria and Diversity of Polychaetes under phylum Annelida (**A**) *Phytocoetopsis ramunii;* (**B**) *Edwardsia jonesii;* (**C**) *Diaduminis chelirean;* (**D**) *Paracondylactis indica;* (**E**) Holothuroids, Sea Cucumber, *Holothuria* sp. (**F**) *Neanthes chingrighattensis*; (**G**) *Glycera tesselata*; (**H**) *Talesapia annandalei*; (**I**) *Mastobranchus indicus*; (**J**) Burrow structures of *Diopatra cuprea cuprea, a polychaeta as ecological indicator*

Fig. 4.15 Macrobenthic fauna from the Sundarbans mangrove ecosystem, India: (**A**) Assemblage of globiid fish, *Boleopthalmus boddarti* on the mud flat; (**B**) Globiid fish, *Boleopthalmus boddarti*; (**C**) Jellyfish, phylum Cnidaria, *Acromitus flagellates*; (**D**) Sipunculida, Phascolosoma sp.; (**E**) Hermit crab; (**F** and **G**) Brachiopods, *Lingula anatima*; (**H**) Nemertea: Heteronemertea; *Polydendrorhynchus amaleshi*; (**I**) Nemertean *Polydendrorhynchus amaleshi* with extended proboscis

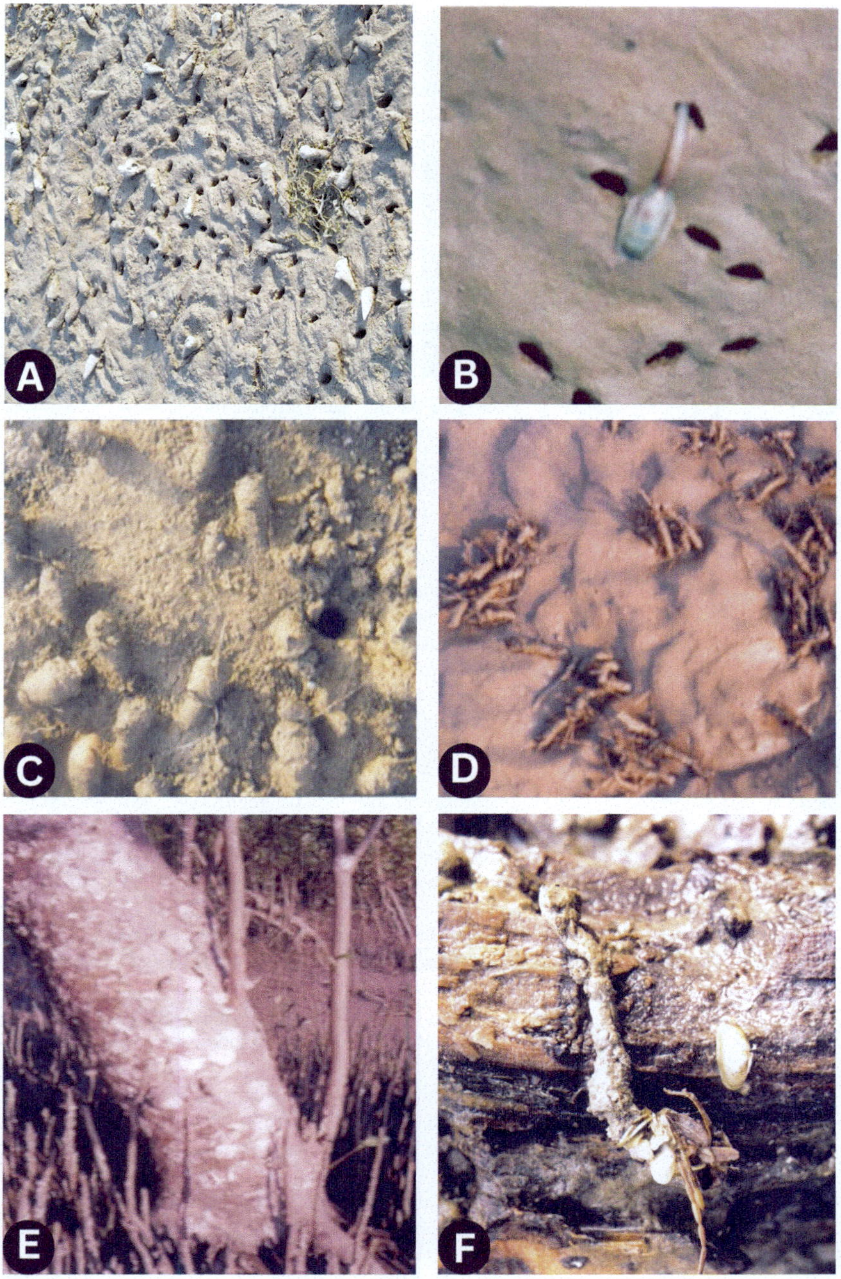

Fig. 4.16 Diversity of habitats of macrobenthos in various parts of Sundarbans, India: (**A** and **B**) Habitats of brachiopods, *Lingula anatina*, with specific markings and impressions of their presence in the mudflats; (**C**, **D** and **F**): Projected tubes over the surface of mudflats constructed by polychaete species like *Diopatra cuprea cuprea* and *Mastobranchus indicus*, not only as the mask of their presence, but also indicates the trend of ecodegradation of their habitats; (**E**) Mangrove trunks and roots serve as important sites for the settlement of periphyton

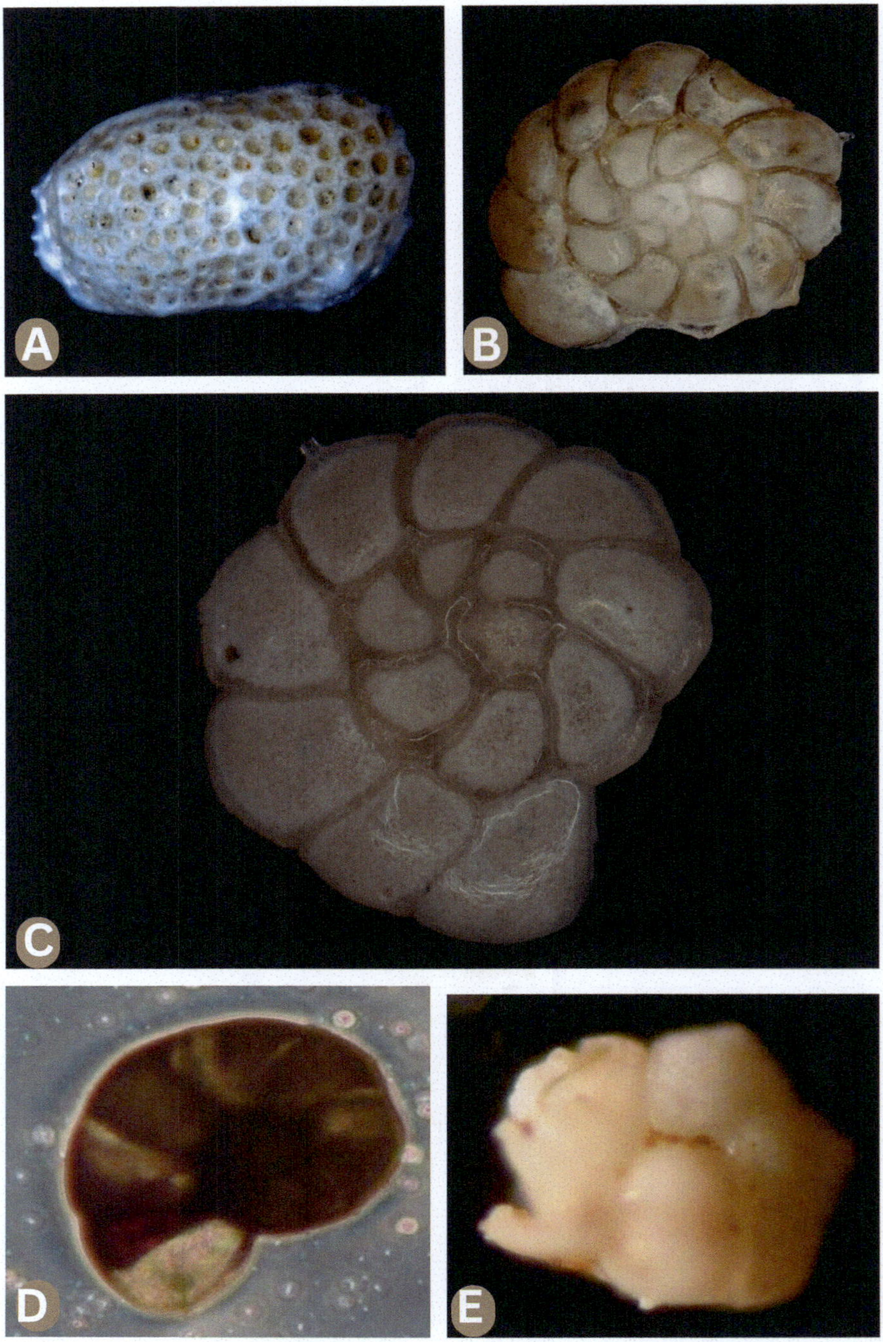

Fig. 4.17 Some representatives of Foraminifera in West Bengal Coast, India: (**A**, **B**, **C**, **D** and **E**) Different for aminifera collected from the estuarine and marine zones in and around Sundarbans, India

Fig. 4.18 Horseshoe crabs with their habitats, sites of collection in the West Bengal coast, India: (**A**, **B**, **C**, **D**, **E**, **F** and **G**) All those images have attempted to depict the images of two major species of horseshoe crabs (*Carcinorpius rotandicaudata* and *Tacypleus gigas*) inhabiting in mid-littoral zones of the coastal belts at the junction of the states of West Bengal and Odisha, in India. Assessment of the ecological conditions of their habitats helps taking up proper in situ conservation measures

Fig. 4.19 Diversity of entomofauna of the Sundarbans mangrove estuarine complex, India: (**B** to **L**) represent different mangrove plants dwelling insects which serve multifold functions like pollinators, ecological indicators, predators and food for birds and reptiles, etc.: (**A**) *Apis mellifera* (Order: Hymenoptera, Family: Apidae); (**B**) *Apis dorsata* (Order: Hymenoptera, Family: Apidae); (**C**) *Dysdercus* sp. (Order: Hemiptera: Family: Pyrrhocoridae); (**D**) *Euploea core* (Order Lepidoptera, Family Nymphalidae); (**E**) Mass aggregation of newly laid insect eggs on mangrove plant's leaf; (**F**) *Micraspis discolor* (Order Coleoptera, Family Coccinellidae); (**G**) Order Hemiptera, Family Cicadidae, common name Cicada; (**H**) *Sarcophaga* sp. (Order Diptera, Family Sarcophagidae); (**I**) Solitary bee (Order Hymenoptera); (**J**) Order Coleoptera, Family Cerambycidae; (**K**) Midge (Order Diptera); (**L**) *Utetheisa* sp. (Order Lepidoptera, Family Erebidae)

Fig. 4.20 Diversity of Shellfish (Prawns and Shrimps) from different brackish water bodies of Sundarbans, India (1): (**A**) Lobster (Crustacea, Malacostraca, Nephropidae) *Homarus gammarus*; (**B** and **I**) *Macrobrachium rosenbergii*; (**C** and **D**) Tiger shrimp, *Paeneus monodon*; (**E**) *Solenocera crassicornis;* (**F**) *Penaeus indica*; (**G**) Red marine prawns (Kori); (**H**) *Fenneropenaeus indicus*

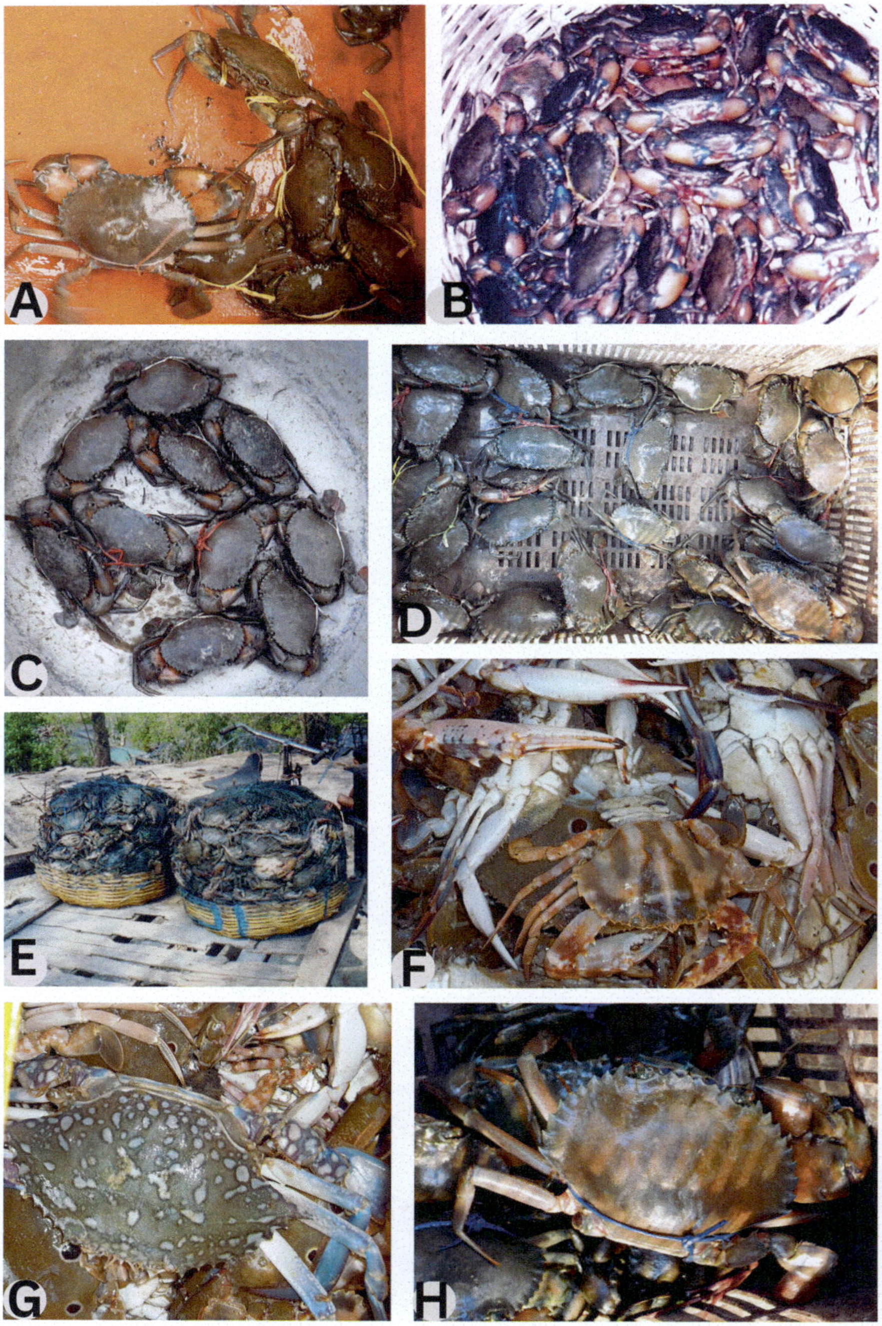

Fig. 4.21 Diversity of Shellfish, edible Crabs in the Sundarbans mangrove ecosystem, India (2): (**A**, **B**, **C**, **D** and **E**) The only edible intertidal macrobenthic crab in the marine coastal belts of West Bengal, India, including the Sundarbans mangrove ecosystem, the mud crab, *Scylla serrata*; (**F**, **G**, and **H**) Subtidal large flattened and colorful crab, *Portunus pelagicus*, occasionally used as human food

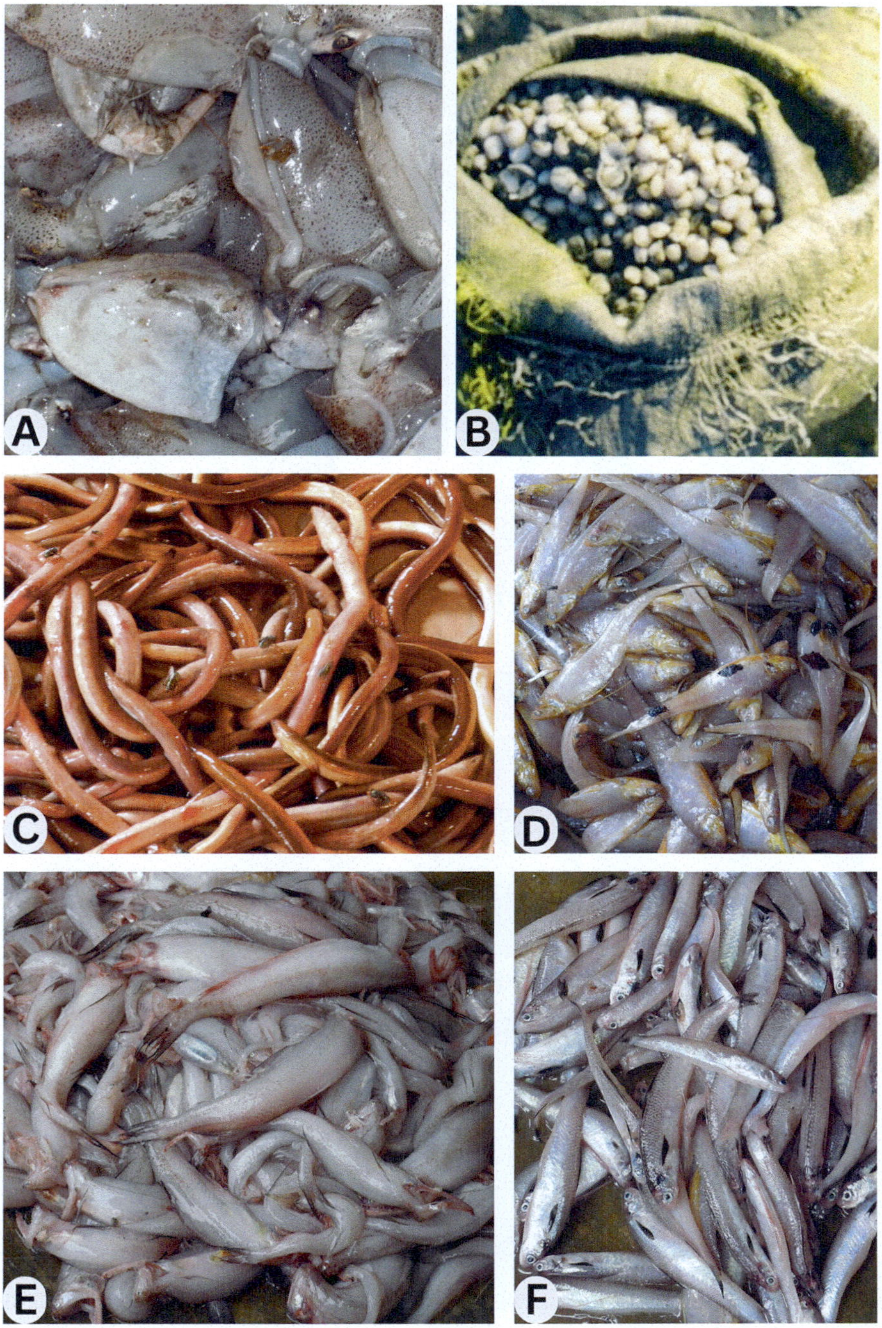

Fig. 4.22 Diversity of Shellfish and Finfish in the Sundarbans mangrove ecosystem (3): (**A**) Cuttle fish, *Sepia officinalis*; (**B**) Bivalves, *Meretrix meretrix*; (**C**) Burmese goby eel, *Taenioides buchanani*; (**D**) Amudi Bhetki, *Coilia dussumieri*; (**E**) Bombay Duck, *Herpodon neherrius*; (**F**) Nona bele, *Glossogobius giuris*

Fig. 4.23 Diversity of Finfish in mangrove estuarine complex of Sundarbans, India (1): (**A** and **B**) The most abundant commercially important finfish in the estuarine environment of Sundarbans, India, known as Asian seabass, is *Lates calcarifer* (Bhetki); (**C**) Locally named as Telia Bhetki, Tiger tooth croaker, *Otolithes ruber*; (**G**) Gang Mourala, *Stolephorus indicus*, *Protonibeadia canthus*; (**D**) Hilsa fish, *Tenualosa ilisha*; (**E**) Phasa, *Setipinna phasa*; (**F** and **G**) Pankhaki, *Toxotes chatareus*; (**H**, **I** and **J**) Mackerel, *Scombers combrus*

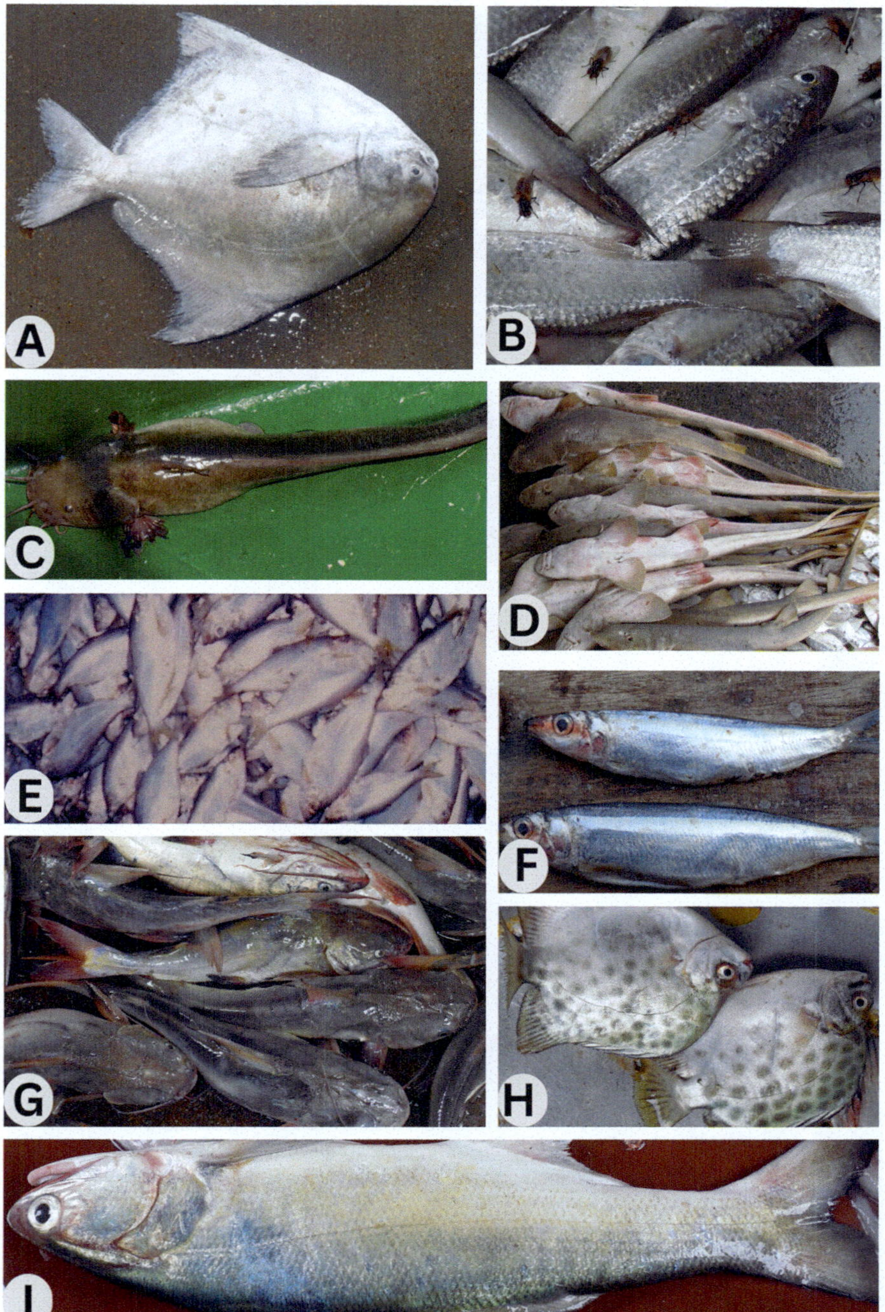

Fig. 4.24 Diversity of Finfish in mangrove estuarine complex of Sundarbans, India (2): (**A**) Pomfret, *Pampus argenteus*; (**B**) Parse, *Chelonsub viridis*; (**C**) Kat Magur, *Plotosus canius;* (**D**) Juvenile sharks, dogfish, *Scoliodon laticaudus*; (**E**) Phasa, *Setipinna phasa;* (**F**) Rainbow sardine, Turturi, *Dussumieria acuta*; (**G**) Nona Tangra, *Arius arius*; (**H**) Paira Chanda, *Scatophagus argus*; (**I**) Bhangor, *Mugil cephalus*

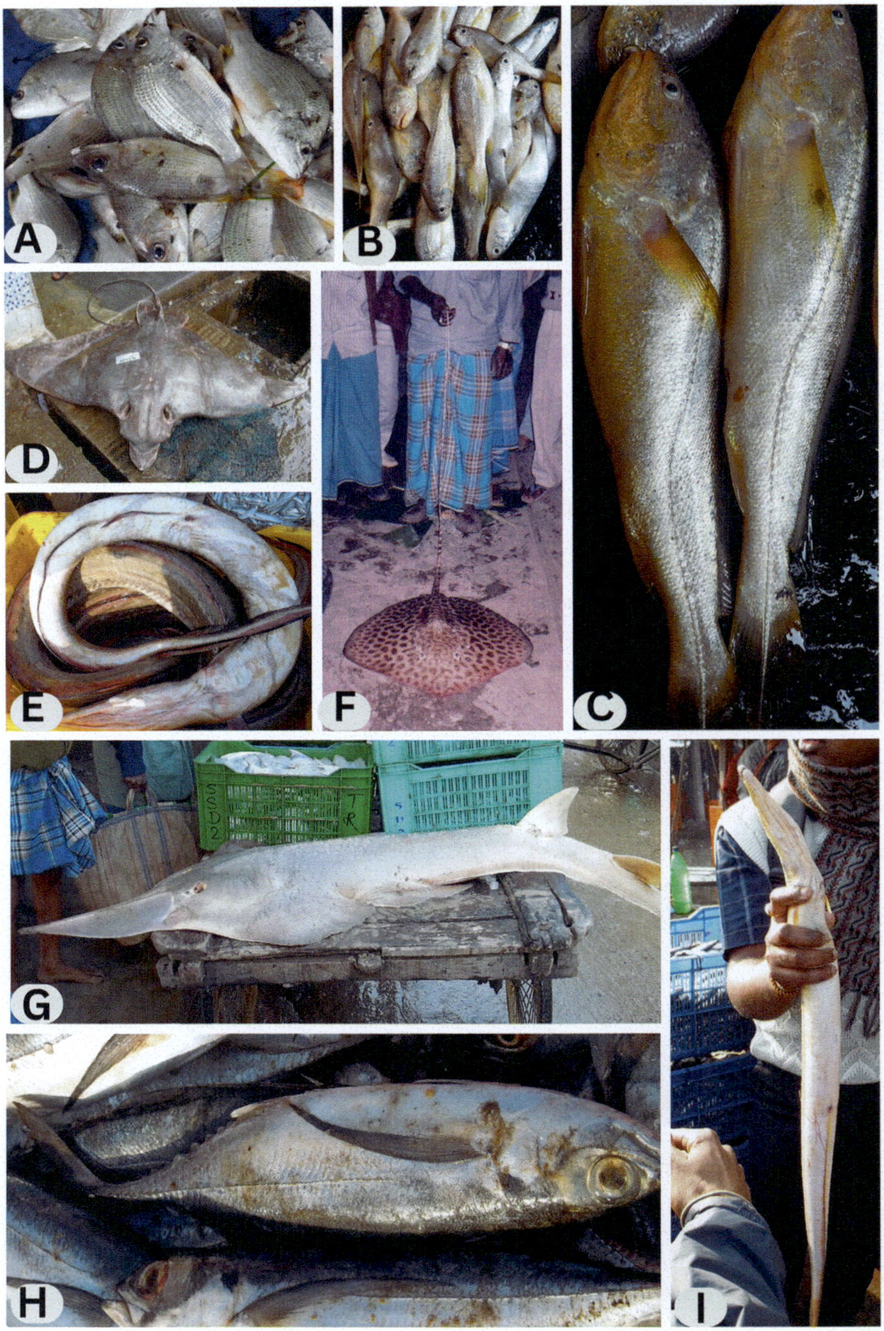

Fig. 4.25 Diversity of Finfish in the mangrove estuarine complex of Sundarbans, India (3): (**A**) Datne match (Fish), *Acanthopagrus datnia*; (**B**) Amudi bhetki, *Coilia dussumieri*; (**C**) Bhola bhetki; (**D** and **F**) Sting rays, *Dasyatis pastinaca*; (**E** and **I**) Eel fish, *Anguilla bengalensis*; (**G**) Sword-headed shark, *Pristiophorus cirratus*; (**H**) Sardin, *Sardina pilchardus*

Fig. 4.26 Fishing activities in Sundarbans, India: Crafts and Gears with traditional approaches: (**A**) Fishing jetty at Fresergaunj near Bakhkhali Tourist Complex of Namkhana block of Sundarbans with lot fishing activities; (**B**) Gears are operated from indigenously constructed large sized craft; (**C** and **D**) Local fishermen making surveillance with their traditional modes of fishing activities in search of fish and crabs; (**E**, **F**, **G** and **H**) Arrangement of fishing gear with the harvest of shellfish by local fishermen

Fig. 4.27 Round the clock fishing activities of local people of Sundarbans with the help of indigenously prepared fishing crafts and gears using their traditional knowledge brought down to them through generations to collect and harvest both shellfish and finfish from the tidal mudflats, sandflats, creeks, channels and main flows of estuaries: (**A**) A young person is competently unearthing a mud crab, *Scylla serrata*, from its burrow with indigenous skills; (**B**, **C** and **D**) Fishing by manually operated indigenously prepared nets; (**E**, **F** and **G**) fishing simply by applying plant products and fish attractants

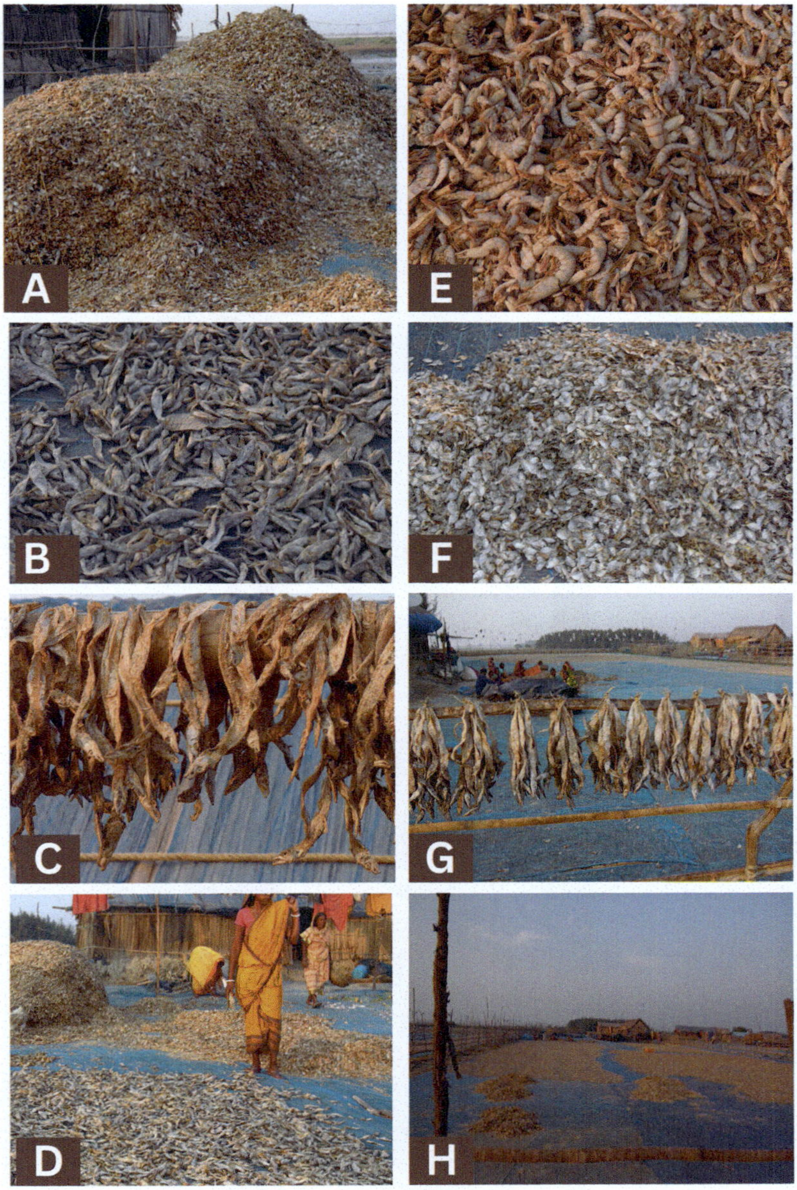

Fig. 4.28 Drying of fish in different parts of coastal belts of West Bengal including Sundarbans, India: (**A**, **B**, **C**, **D**, **E**, **F**, **G** and **H**) the Sundarbans mangrove estuarine complex enriched with biological wealth mostly by virtue of the occurrence of diversified forms of finfish and shellfish, many of which are not directly consumable by human beings; a huge amount of harvested, but not directly consumable, fishery resources are processed, mostly by natural processes, exposing them to open sunlight with the direct involvement of local people; all the dried-up fishery resources are used as fish feeds and even as human foods

Fig. 4.29 Captive Breeding Program of the Crocodile Conservation Project (1): (**A**, **B** and **C**) Depiction of different culture protocols and ingredients required for such culture of the estuarine crocodile, *Crocodilus porosus*, by scientific hoardings, posters, models, and published documents: (**D**, **E**, **F** and **G**) Collection, preservation, and maintenance of eggs, juveniles, and other developmental phases of this charismatic wild species; (**H**) Growing up of crocodiles at early stages of their lifecycle in captive conditions, maintaining suitable ecological condition

Fig. 4.30 Captive Breeding Program as Crocodile Conservation Project (2): (**A** and **B**) Awareness-building process by exhibiting several posters and publications highlighting necessity of the conservation of estuarine crocodile, *Crocodilus porosus* at the Bhagabatpur Crocodile Conservation Project located on the Saptamukhi estuary of Sundarbans, India; (**C** and **D**) Keeping sub-adult estuarine crocodiles in captive condition providing necessary foods, water, and maintaining temperature, humidity, and wind circulation in the artificial ponds supplied with natural estuarine water; (**E** and **F**) Regular cleaning of the captive ponds by trained personnel as a maintenance drive to keep the maturing estuarine crocodiles healthy and away from any scope of infection under controlled ecological conditions

Fig. 4.31 Captive Breeding Program for conservation of the turtle. Conservation for ensuring protection of all sea turtles was taken up very seriously by the Government of India by the Wildlife Protection Act under Schedule 1. Initially, during the early 1980s, attempts were made to rear and release to their natural home in the surrounding estuaries as a rehabilitation program for the Olive Ridley Turtle, *Lepidochelys olivacea*, but afterwords, more attention was paid to undertaking conservation ventures by rearing and rehabilitating river terrapin, *Batagarbaska*, within the premises of the Sajjnekhali Mangrove Interpretation center and the Sajjnekhali Tourist Complex. Images: (**A**, **B**, **C**, **D** and **E**) depict different infrastructural components for this purpose; (**F**) Photo of a river terrapin, *Batagar baska*

Fig. 4.32 Some representative species of reptiles in the Sundarbans mangrove ecosystem, India: (**A**, **B**, **C**, and **D**) Estuarine crocodiles in different poses and postures on mid-littoral zones of different mudflats in the core areas of the Sundarbans Biosphere Reserve, India depicting the healthy physical body anatomy of all these crocodiles, indicating the healthy ecological conditions of their habitats in Indian Sundarbans; (**E**) Water monitor, *Varanus salvator;* (**F**) River terrapin, *Batagar baska;* (**G**) Olive Ridley turtle, *Lepidochelys olivacea;* (**H**) Viper snake, *Deboia russelii*; (**I** and **J**) Cobra, *Ophiophagus Hannah*

Fig. 4.33 Diversity of birds in the Sundarbans mangrove ecosystem, India (1): (**A** to **I**) Different aquatic birds at the Sundarbans mangrove ecosystem enjoying different aquatic habitats; (**A**) Flocks of the common pochard, *Aythya ferina*; (**B**) Flocks of the little cormorant, *Microcarbo niger;* (**C**) Flocks of the lesser whistling duck, *Dendrocygnaja vanica;* (**D**) Flocks of the Indian cormorant, *Phalacrocorax fuscicollis;* (**E**) Great egret, *Ardea alba;* (**F**) Flocks of the intermediate egret, *Ardea intermedia,* and the little heron, *Butorides striata;* (**G**) Wooly-Necked Stork, *Ciconia episcopus;* (**H**) flocks of the spotted dove, *Streptopolia chinensis;* (**I**) Red-wattled lapwing, *Vanellus indicus*

Fig. 4.34 Diversity of birds in the Sundarbans mangrove ecosystem, India (2): (**A**) Bully Fish Owl, *Ketupa ketupa;* (**B**) Red Jungle Cock, *Gallus gallus;* (**C**) Paddy field warbler, *Acrocephalus Agricola;* (**D**) Booted wabler, *Induna caligata*; (**E**) Brown-winged Kingfisher, *Pelargopsis amauroptera*; (**F** and **G**) Black Capped Kingfisher, *Halcyon pilieta*; (**H**) Collard Kingfisher, *Todiramphus chloris*; (**I**) Pied Kingfisher, *Cryle rudis;* (**J**) Common kingfisher (*Alcedo atthis*); (**K**) Small blue kingfisher (*Alcedo coerulescens*); (**L**) Stork-billed kingfisher (*Pelargopsis capensis*); (**M**) Brahmini Kite, *Hallastur indus*; (**N**) Hen Harrier, *Circus cyaneus*

Fig. 4.35 Diversity of birds in the Sundarbans mangrove ecosystem, India (3): (**A**) Great Egret, *Ardea alba;* (**B**) Intermediate Egret, *Ardea intermedia;* (**C**) Lesser Adjutant, *Leptoptilos javanicus*; (**D**) Common Snipe, *Callinago gallinago*; (**E**) Eurasian Curlew, *Numenius arquata*; (**F**) Spotted Red shank, *Tringa erythropus;* (**G**) Common Sandpiper, *Actitis hypoleucos;* (**H**) Yellow Bittern, *Botaurus sinensis;* (**I**) Black bittern, *Ixobrychus flavicolis;* (**J**) Broad-Billed Sandpiper, *Calidris falcinellus*; (**K**) Little Stint, *Calidris minuta*

Fig. 4.36 Representative mammalian fauna of the Indian Sundarbans: (**A**) Deer, *Axix axis*; (**B**) Wild Boar, *Sus scrofa*; (**C**) Monkey, *Maccaca mulata*; (**D**) Bat Indian Flying Fox, *Petropus giganticus*; (**E**) Smooth-coated Otter, *Lutrogale perspicilata*; (**F**) Jungle cat, *Felis chaus*; (**G**) Leopard cat, *Felis bengalensis*; (**H**) Irrabati Dolphin, *Orcella brevirostris*; (**I**) Gangetic river Dolphin, *Platanista gangetica*

Fig. 4.37 Tigers of Sundarbans, named 'Royal Bengal Tiger' are considered to be the most mysterious wildlife of the world because of their intelligence, agility, power capturing prey with remarkable strength of body, tolerability against the ecological challenges of their habitats, including their ability to endure salinity, determination in protecting jurisdiction of territory, notorious habits and ferociousness. All morphological, anatomical, and physiological adaptations to maintain their unique behavioral patterns are supposed to have been refined in the process of evolution over centuries through modification of their biological systems in comparison to other subspecies of tiger distributed across different parts of the world. (**A**, **B**, **C**, **D** and **E**) Tiger in Sundarbans, *Pantheratigristigris*, in different forms of its activities; (**A**) Tiger crossing the creeks on its way from one deltaic island to other in search of pray; (**B** and **C**) Tigers in the core of the mangrove forest; (**D**) Waiting within the tunnels of mangrove bushy palms, *Phoenix pelludosa*, tiger of Sundarbans concentrated on observing its surroundings; (**E**) Tiger of Sundarbans in camouflaging mode within a groove of mangrove plants, *Ceriops decandra*

Fig. 4.38 Creation of new infrastructural facilities like open fields for foraging of wild animals: (**A**, **B**, **C** and **D**) interpretation centre to promote tourism and build awareness for the sustainable conservation of nature, alongside protecting the natural geo-morphological settings like dunes; (**E**) freshwater ditches, wetlands; (**F**) for the protection of coastal belts from salinity ingress, flood control and ground water recharging and biodiversity development in both the core and buffer areas of the Indian Sundarbans

Fig. 4.39 Development museum (**A**, **B** and **C**); interpretation centres (**E** and **F**); erecting of fencing to restrict the movement of wild animals only within the mangrove forests and thereby avoiding man-wild animal conflicts; (**D** and **F**); construction of tourist lodges and concrete pathways for the tourists (**G** and **H**); raising of mangrove saplings in the nursey beds as prerequisites for mangrove afforestation programs and all those well-managed activities of the Forest Department of Government of West Bengal, India, which will certainly strengthen biodiversity conservation

Fig. 4.40 All images depict different religious activities of different parts of Sundarbans, India, promoting the influx of tourists to religious grounds, to understand the ethnic culture of the region and to gain an insight about the religious beliefs of the people of Sundarbans for their own well-being and protection from the wild animals and natural disasters: (**A**) Kapil Munir Temple, a famous Hindu temple located at the extreme southern corner on the largest Gangetic Delta of Sundarbans, Sagar Island, where millions of peoples from different parts of India throng to satisfy their religious beliefs with the holy baths in the confluence of the Hooghly estuary with the Bay of Bengal; (**B**) Rituals after the death of a local person at the bank of an estuary; (**C**, **D**, **E**, **F**, **G** and **H**) Different images of the most famous idols of the Goddess, 'Ban-Bibi'

Table 4.1 Mangrove plant species in the Indian Sundarbans (major and minor elements) along with different mangrove plant associates

Families & genera of mangroves & the mangrove associates	Species Names of the Mangroves/Mangrove Associates, etc.	Local Name\ Vernacular Names	Occurrence Abundance Frequency
Major elements of mangrove plants in Sundarbans, India			
A. Rhizophoraceae			
1. *Rhizophora* L. (all are trees)	*R. apiculata* Blume	Bhora, Garjan	++
	R. mucronata Lamk	Khamu, Garjan	+++
2. *Bruhuiera* Lamk (Most are trees/few are shrubs)	*B. gymnorrhiza* (L.) Lamk	Khakra	+++
	B. sexangula (Lour.) Poir	Khakra	++
	B. cylindrical (L.) Blume	Bakul Kakra	++
	B. parviflora W. & A. ex. Griff.	Bakul Kakra	+
3. *Ceriops* Arnold (small trees/shrubs)	C. decandra (Griff.) ding Hou	Jhamti Goran	+++
	C. *tagal* (Perr.) Robin	Mat Goran	+++
4. *Kandelia* W.& A. (small tree)	*K. candel* (L.) Druce	Goria	++
B. Avicenniaceae			
5. *Avicennia* L. (small-medium tree)	*A. alba* Blume	Kal-ban/bain/bani	+++
	A. officinalis L.	Jat-ban/bain/bani	+++
	A. marina	Peyara-ban/bain/bani	+++
C. Sonneratiaceae			
6. *Sonneratia* Linn. f. (medium-tall tree)	*S. alba* J. Smith	Sada Chak-Keora	?
	S. casolaris (L) Engle	Chak-Keora	+
	S. griffuthii Kurz.	Ora	++
	S. apetala Buch. Ham.	Keora	+++
D. Combretaceae			
7. *Lumnitzera* Willd (shrub/small tree)	*L. racemosa* Willd.	Kripa/Kripal	+++
E. Arecaceae (nom. Alt. Palmac)			
8. *Nypa* Steck	*N. fruticans* (Thunb.)	Golpata	++
Minor elements of mangrove plants in Sundarbans, India			
A. Meliaceae			
1. *Xylocarpus* Roem (medium-tall tree)	*X. granatum* Koenig	Dhundul	+++
	X. mekongensis Pierre	Pasur	+++
2. *Aglaia* Lour.	*A. cucullata* (Roxb.) Pelle	Amur	+
B. Myrsinaceae			
3. *Aegiceras* Gaertn (shrub-small tree)	*A. corniculatum* (L.) Blanco	Khalsi	+++
C. Euphorbiaceae			
4. *Excoecaria* L. (medium tree)	*E. bicolor* Hassk	Lata-Gneo	**?**
	E. agallocha L.	Gneo/ Gneoa	+++

Table 4.1 (continued)

Families & genera of mangroves & the mangrove associates	Species Names of the Mangroves/Mangrove Associates, etc.	Local Name\ Vernacular Names	Occurrence Abundance Frequency
D. Aegialitidaceae (nom. Alt. Plumbaginaceae)			
5. *Aegialialitis* Brown (shrub)	*A. rotundifolia* Roxb.	Tora	+++
E. Sterculiaceac			
6. *Heritiera* Aiton (tall tree)	*H. fomes* Buch. Hamilton	Sundari	+
	H. lottoralis Dryander	Sundari	?
F. Rubiaceae			
7. *Scyphiphora* Gaert.	*S. hydrophyllacea* Gaer.f.**	Tagri Bani	++
G. Pteridaceae			
8. *Acrostichum* L.	*A. aurewm* L	Hudo fern	++
H. Acanthaceaf			
9. *Acanthus* L. (shrub/twiner)	*A. Ilicifolius* L.	Harkoch	+++
	A. volubilis wall*	Lata-Hargoja	+
I. Rutaceae			
10. *Atalantia* M. Roem (shrub)	*A. correa* M. Roem**	Ban Lebu	+
J. Tillaceae			
11. *Brownlowia* Roxb. (shrub)	*B. tersa* (L.) Kosterm (= B. laceolata Benth.)	Lata-Sundari	++
Mangrove associates, as the back mangroves of Sundarbans, India			
A. Malvaceae			
1. *Hibiscus* L. (medium tree)	*H. tiliaceus* L.	Ban-Kapas	++
	H. tortuosus Roxb.	Ban-Kapas	+++
2. *Thespesia* Soland ex Corr (Midium tree)	*T. populneoides* (L.) sold	Paras	++
	T. populnea (Roxb)		
	Kostel	Paras	++
	T. lampus (Cav.) Dalz. & Gibs		
B. Fabaceae			
(a) Caesalpiniaceae			
3.*Cynometra* L. (twiner)	*C. ramiflora* L.	Gila	++
4. *Caesalpinia* L. (twining shrub)	*C. bonduc* (L.) Rocb.	Nata-Karanjan	+++
	C. crista L.	Shingri-Lata	+++
(b) Papilionaceae			
5. *Dalbergia* L. f. (small tree)	*D.spinose* Roxb.	Chulia Kanta	++
6. *Derris* Lour.	*D. scandens* Benth.	Noa-Lata	+++
	D. trifoloata Lour.	Pan-lata	+++
	D. indica bennet	Karanja	+++
C. Arecaceae (noom. Alt. Palmac)			
7. *Phoenic* L.	*P. palucdosa* Roxb.	Hental	+++
D. Lecythidaceae			
8. *Barringtonia* Forst. (tree)	*B.asiatida* (L.) Kurz.	Hijal	++
	B. racemose Roxb.	Samudra	++

(continued)

Table 4.1 (continued)

Families & genera of mangroves & the mangrove associates	Species Names of the Mangroves/Mangrove Associates, etc.	Local Name\ Vernacular Names	Occurrence Abundance Frequency
E. Bignoniaceae			
9. *Dolichandrone* seem. (tree)	*D. spathaceae* (L.f.) Sch.	Gorshingah	++
F. Combretaceae			
10. *Dolichandrone* seem. (tree)	*B. catappa* L.	Kat Badam	+++
G. Ebenaceae			
11. *Diospyros* L. (tree of shrub)	*D. ferrea* (will.) Roxb.	Ban gab	+++
H. Cluciaceae (noom. Alt. Guttiferac)			
12. *Calophyllum* L. (tree)	*C. inophyllum* L.	Kath Champa	++
I. Sapindaceae			
13. *Allophyllus* L.	*A. cobbe* L.	—	++
J. Pandanaceae			
14. *Pandanus* Parkin. (shrubs-tree)	*P. tectorius* sol. Ex Park.	Keya	++
	P. foetidus Roxb.	Keya	++
K. Sapotaceae			
15. *Manilkara* Adans. (tree)	*M. hexandra* (Roxb.) dub.	Bilati bakul (Pernaps introduced)	+
L. Asclepiadaceae			
16. *Sarcolubus* R. Br. (twiner/shrub)	*S. carinatus* wall.	Baoli Lata	+++
	S. globosus wall	Baoli Lata	+++
17. *Pentalropis* Wight & Arn.	*P. capensis* (Linn.f) bullock	—	+++
M. Chenopodiaceae			
18. *Suaeda* Forssk. (herbaceous)	*S. maritima* Dumort.	Gire-Sak	+++
	S. nudiflora Moq.	Gire-Sak	+++
19. *Salicornia* L. (herbaceous)	*S. brachiate* Roxb.	—	++
N. Amaryllidaceae			
20. *Crinum* L. (herbaceous)	*C. defixum* Kar-Gawl.	Sukh-Darshan	++
O. Araceae			
21. *Cryptocoryne* Fisch.	*C. cilliata* (Roxb.) Fisch.	Kerali	++
P. Verbenaceae			
22. *Clerodendrum* L. (shrub)	*C. inerme* (L.) Gaertn.	Ban Jui	+++
Q. Ruppiaceae			
23. *Ruppia* L.	*R. maritima* L.	Nona Jhanji	+++
R. Tanaricaceae			
24. *Tanarix* L.	*T. dioca* Roxb.	Nona Jhau	+++
	T. gallica L.	Noona Jhau	+++
S. Boraginaceae			
25. *Heliotrophium* L.	*H. curassavicem* L.	NoonaHatisur	+++

Table 4.1 (continued)

Families & genera of mangroves & the mangrove associates	Species Names of the Mangroves/Mangrove Associates, etc.	Local Name\ Vernacular Names	Occurrence Abundance Frequency
T. Poaceae			
26. *Poaceae*	*P. coarctata* (Roxb.) Takeoka	Dhani Ghas	+++
27. *Myriostachya* Hook.f.	*M. wightiana* Hook.f.	Nalai	+
28. *Aeluropus* L.	*A.lagopoides* (L.) tin.	Nona Durba	
U. Opuntiaceae			
29. *Oputia* L.	*O. dillenii* hew.	Nag-Phana	+++
V. Solanaceae			
30. *Solanum* L.	*S. trilobatum* L.	Nona begum/Kata begum	++
W. Lauraceae			
31. *Cassytha* L.	*C. filiformis* L.	Akash bel	+
X. Loranthaceae			
32. *Macrosolen* (Bl.) Reichenb.	*M. cochinchinensis* (Lour.) Van Tiegn.	Chotomanda	+++
33. *Viscum* L.	*V. monoicum* Roxb.	Manda	++
Y. Rubiaceae			
34. *Hydrophyllax*	*H. maritima* L.	–	+
Z. Convolvulaceae			
35. *Ipomoea* L.	*I. Pes–* caprae (L.) sweet	Chagal Kuri	+++

Table 4.2 List of mangrove plants as recordedfrom four ecoZones with contrasting ecological characteristics in the Indian Sundarbans

SL. No	Name of the plant species	S-I	S-II	S-III	S-IV
	True mangrove				
1	*Nypa fruticans* Wurmb.	+	–	–	+
2	*Phoenix paludosa* Wurmb	+	–	+	+
3	*Avicennia alba* Blume.	+	+	+	+
4	*Avicennia marina* (Forsk)Vierh.	+	+	+	+
5	*Avicennia officinalis* L.	+	+	+	+
6	*Lumnitzera racemosa* (L.) Sweet.	+	–	+	–
7	*Excoecaria agallocha* L.	+	+	+	+
8	*Xylocarpus granatum* Koening.	+	–	–	–
9	*Xylocarpus mekongensis* Pierre.	+	–	+	–
10	*Aegiceras corniculatum* (L.) Blanco.	+	–	+	–
11	*Bruguiera cylindrica* (L.) Bl.	+	–	+	+
12	*Bruguiera gymnorrhiza* (L.) Lamk.	+	–	+	+
13	*Ceriops decandra* (Griff.) Ding Hou	+	–	+	+
14	*C. tagal*	+	–	–	+
15	*Sonneratia apetala* Buch. -Ham.	+	–	+	+
16	*Sonneratia griffithii* Kurz.	+	–	+	+
17	*Heritiera fomes* Buch. -Ham.	+	–	–	+

Table 4.2 (continued)

SL. No	Name of the plant species	S-I	S-II	S-III	S-IV
18	*Rhizophora apiculata* Lamk.	+	–	+	+
19	*Rhizophora mucronata* Bl.	+	–	+	+
20	*Aegialitis rotundifolia* Roxb.	+	–	–	–
21	*Acanthus ilicifolius* L.	+	+	+	+
22	*Acanthus volubilis*	+	+	–	+
23	*Aglaia cucullata*	+	–	–	–
24	*Kandelia candle* (L.) Druce.	+	–	–	–
	Mangrove-associated plants and mesophytes				
1	*Sarcolobus carinatus* wall	+	–	–	–
2	*Sarcolobus capensis* (L.) bull	+	–	–	+
3	*Sarcolobus globossus* wall	+	–	+	–
4	*Sesuvium portulacastrum* L.	+	–	–	–
5	*Heliotrophium curassavicum* L.	+	–	–	+
6	*Caesalpinia crista* L.	+	–	–	–
7	*Caeslpinia bonduc* (L.) Roxb	+	–	–	–
8	*Salicornia brachiata* Roxb	+	–	–	–
9	*Suaeda maritima* Dumort	+	+	+	+
10	*Suaeda nudiflora* Roxb	+	+	–	+
11	*Ipomoea pescaprae* (L.) sweet	+	–	+	+
12	*Macrosolen cochinchinensis* (Lour.) Tiegh	+	–	–	–
13	*Thespesia populnea* (L.) Soland ex. Corr	+	–	+	–
14	*Pandanus tectorius* Soland	+	–	–	–
15	*Derris trifoliata* Lour	+	–	+	+
16	*Derris indica* bennet	+	–	–	+
17	*Premna integrifolia*	+	–	+	–
18	*Premna latifolia*	+	–	+	–
19	*Porteresia coarctata* (Roxb) Takeoka	+	+	+	–
20	*Myriostachya wightiana* Hook	+	–	–	–
21	*Tamarix gallica L.*	+	–	–	–
22	*Cynodon* sp.	–	–	–	+
23	*Paspalum scorbiculatum*	–	–	–	+
24	*Cyperus* sp.	–	+	–	+
25	*Cryptocorynae cliata*	–	–	–	+
26	*Hyptis suaviolenae*	–	+	–	+
27	*Dodonia vispasum*	–	+	–	+
28	*Prosopis julifera*	–	+	–	+
29	*Pongomia pinata*	–	–	–	+
30	*Eleocharis* sp.	–	–	–	+
31	*Glycosmis pentaphyla*	–	–	–	+
32	*Opantia* sp.	–	+	–	+
33	*Datura metal*	–	–	–	+
	Mesophytic invasive plants				
1	*Alternanthera* sp.	+	+	–	+
2	*Antigonon leptopus*	–	–	–	+

Table 4.2 (continued)

SL. No	Name of the plant species	S-I	S-II	S-III	S-IV
3	*Casuarina equisetifolia*	+	−	−	+
4	*Eichhornia crassipes*	+	−	−	+
5	*Eucalyptus* sp.	+	+	−	+
6	*Eupatorium odoratum.*	+	−	−	+
7	*Lantana camar*	+	+	−	+
8	*Leucaena leucocephala*	+	−	−	+
9	*Mikania scandens*	+	−	−	+
10	*Parthenium hysterophours*	+	+	−	+

S-I = Jharkhali, S-II = Canning, S-III = Saptamukhi, S-IV = Bokkhali
Note: + = (Present); − = (Absent)

Table 4.3 Different species of algae as recorded from the Sundarbans, India

Name of Devision, Class Order and family Chlorophyta	Name of the Species	
Chrysophyta	*1. Enteromorpha clathrate*	Periphytic
	2. Enteromorpha compressa	"
	4. Enteromorpha prolifera	"
	5. Ulva lactuca	"
	6. Ulva patengensis.	"
	7. Ulva ap.	Benthos
	8. Chaetomorpha aerea	"
	9. Chaetomorpha aerea	"
	10. Lola capillaris	"
	11. Lola imlexa	"
	12. Lola tortusa	"
	13. Rhizoclonium grande	"
	14. Rhizoclonium hookeri	"
	15. Rhizoclonium hookeri	"
	16. Rhizoclonium riparium	"
	17. Cladophora echinus	"
	18. Spirogyra sp.	"
	19. Coleochaete sp.	Epiphytic
	20. Ulothrix sp.	Periphytic
	21. Vaucheria prescottii	Periphytic
Phaeophyta	*22. Colpomenia sinusa*	Benthos
Rhodophyta	*23. Catenella impudica*	Benthos
	24. Catenella repenlla repens	Benthos
	25. Polysiphonia denudate	Periphytic
	26. Polysiphonia mollis	"
	27. Compsopogon coerleus	"
	28. Catenella impudica	"

(continued)

Table 4.3 (continued)

Name of Devision, Class Order and family Chlorophyta	Name of the Species	
Cyanophyta	*29. Xenococcus chaetomouphae*	"
	30. Xenococus cladophorae	"
	31. Aphanotheca castagnii	"
	32. Merismopedia glauca	Periphytic
	33. Gloeocapsa calcarea	"
	34. Gloeocapsa mantana	"
	35. Gloeocapsa kuetzingiana	"
	36. Gloeocapsa rupestris	"
	37. Gloetheca samoensis	"
	38. Gloethaeca membranacea	"
	39. Merismopedia tennissium	"
	40. Aphanocapsa thermalis	"
	41. Aphanocapsa roeseana	"
	42. Aphanocapsa pulehra	"
	43. Micrcystiss litoralis	"
	44 Micrcystis bengalensis	"
	45. Micrcystis pulverea	"
	46. Myxosareina spectabilis	"
	47. Johannesbaptisti a pellucida	"
	48. Chamaesiphon curvatus	"
	49. Dermocarpa hemisphaerica	"
	50. Stichosiphon sansibaricus	"
	51. Raphidiopsis indica	"
	52. Oscillatoria amoena	"
	53. Oscillatoria bonnemaisonii	"
	54. Oscillatoria margaretifera	"
	55. Oscillatoria martini	"
	56. Oscillatoria subbrevis	"
	57. Oscillatoria tennis	"
	58. Lyngbya allorgei	"
	59. Lyngbya conferviodes	"
	60. Lyngbya hieronymusii	"
	61. Lyngbya lutea	"
	62. Lyngbya magnifica	"
	63. Lyngbya semiplana	"
	64. Hydrocoleu, m cantharidisum	"
	65. Schizothrix lamyi	"
	66. Microcoleus chthonoplastes	"
	67. Plectonema wollei	"
	68. Anabaena variabilis	"
	69. Crinalim magnum	"
	70. Scytonema siculum	Periphytic
	71. Calothrix confrrviflora	"
	72. Calothrix parasitica	"
	73. Calothrix cotarenii	"

Table 4.4 Distribution of different phytoplankton atfour ecozones in the Indian Sundarbans with contrasting ecological characters

Sl. No.	Species name	S-I	S-II	S-III	S-IV
	Bacillariophyta (Diatomeae)				
	Order: Centrales				
	Sub-order: Discoideae				
	Family: Coscinodisceae				
	Sub-family: Melisirineae				
1	*Melosira* sp.	+	–	+	+
2	*Pyxidicula* sp.	+	–	+	+
3	*Sceletonema* sp.	+	+	+	+
4	*Thalassiosira* sp.	+	+	+	+
	Sub-family: Coscinodiscineae				
5	*Coscinodiscus* sp.	+	+	+	+
6	*Planktoniella* sp.	+	–	+	+
	Sub-order: Solenoideae				
	Family: Solenieae				
	Sub-family: Lauderiineae				
7	*Lauderia* sp.	–	–	+	+
	Sub-family: Rhizosoleniinae				
8	*Rhizosolenia* sp.	+	+	+	+
	Sub-order: Biddulphioideae				
	Family: Chaetocereae				
9	*Chaetoceros* sp.	+	–	+	+
	Family: Biddulphieae				
	Sub-family: Triceratineae				
10.	*Ditylum* sp.	+	+	+	+
11	*Triceratium* sp.	+	+	+	+
12	*Biddulphia* sp.	+	+	+	+
	Order: Pennales				
	Sub-order: Araphidineae				
	Family: Fragilarioideae				
	Sub-family: Tabellarieae				
13.	Grammatophora	+	–	+	–
	Sub-family: Fragilarieae			+	
14	*Fragilaria* sp.	+	–	+	+
15	*Thallassionema* sp.	+	+	+	+
16	*Thallassiothrix* sp.	+	+	+	+
17	*Asterionella* sp.	–	–	+	+
	Sub-order: Biraphideae				
	Family: Naviculoideae				
	Sub-family: Naviculeae				
18	*Gyrosigma* sp.	+	–	+	+
19	*Pleurosigma* sp.	+	+	+	+
20	*Diploneis* sp.	+	–	+	+

(continued)

Table 4.4 (continued)

Sl. No.	Species name	S-I	S-II	S-III	S-IV
21	*Navicula* sp.	+	+	+	+
	Sub-family: Amphiproroideae				
22	*Amphora* sp.	–	–	+	+
23	*Cymbella* sp.	–	+	+	–
	Family: Nitzschiaceae				
	Sub-family: Nitzschioideae				
24	*Bacillaria* sp.	+	–	+	–
25	*Nitzschia* sp.	+	+	+	+
	Family: Surirellaceae				
	Sub-family: Surirelloideae				
26	*Surirella* sp.	–	–	+	–

S-I = Jharkhali, S-II = Canning, S-III = Saptamukhi, S-IV = Bokkhali
Note: + = (Present); – = (Absent)

Table 4.5 Faunal diversity of Sundarbans, India

Sl. Numbers.	Kingdoms	Phylum	Species
A	Protista		
		Euglenozoa	1
		Sarcomastigophora	22
		Metamonada	7
		Dinoflagellata	8
		Chromista	1
		Cercozoa	2
		Amoebozoa	8
		Myxozoa	24
		Apicomplexa	32
		Ciliophora	34
		Foraminifera	1
		Total	140
B	Animalia		
		Arthropoda	1326
		Cnidaria	14
		Mollusca	173
		Ctenophora	1
		Porifera	1
		Nematoda	48
		Sipuncula	1
		Brachiopoda	1
		Echiura	3
		Chordata	823
		Annelida	74
		Bryozoa	3
		Chaetognatha	4
		Echinodermata	14
		Total	2487

Table 4.6 Distribution of different zooplankton at four ecozones in the Indian Sundarbans with contrasting ecological characters

Sl. No.	Group Name	S-I	S-II	S-III	S-IV
1.	Copepods	+	+	+	+
2.	Cladocera	+	+	+	+
3.	Chaetognatha	+	+	+	+
4.	Sergestidae	+	+	+	+
5.	Appendicularians	+	+	+	+
6.	Hydromedusae	–	–	–	+
7.	Ctenophora	+	–	–	+
8.	Decapode larvae	+	+	+	+
9.	Molluscan larvae	+	+	+	+
10.	Polychaetes larvae	+	+	+	+

S-I = Jharkhali, S-II = Canning, S-III = Saptamukhi, S-IV = Bokkhali
Note: + = (Present); – = (Absent)

Table 4.7 Diversity of species under the phylum Cnidaria

Sl. No.	Phylum	Class	Order	Family	Species
1	Cnidaria	Hydrozoa	Anthomedusae	Bougainvilliidae	*Bougainvillia fulva* Agassiz & Mayer, 1899
2			Leptomedusae	Campanulariidae	*Obelia* sp.
3				Eucopidae	*Blackfordia virginica* Mayer, 1910
4				Phialuciidae	*Octophialucium indicum* Kramp, 1958
5				Eirenidae	*Eirerie ceylonensis* Browne, 1905
6					*Eirene menoni* Kramp, 1953
7				Aequoreidae	*Aequorea pensilis* (Eschscholtz, 1829)
8			Limnomedusae	Moerisiidae	*Moerisia gangetiea* Kramp, 1958
9				Limnocnididae	*Limnocnida indica* Annandale, 1912
10			Trachymedusae	Geryoniidae	*Liriope tetraphylla* (Chamissio & Eysenhardt, 1821)
11			Cubomedusae	Carybdeidae	*Tamoya gargantuan* Haeckel, 1880
12			Coronatae	Atollidae	*Atolla wyvillei* Haeckel, 1880

(continued)

Table 4.7 (continued)

Sl. No.	Phylum	Class	Order	Family	Species
13			Rhizostomeae	Cassiopeidae	*Cassiopea Andromeda* (Forskal, 1775)
14				Cepheidae	*Netrostoma typhlodendrium* Schultz, 1898
15				Catostylidae	*Acromitus flagellatus* (Maas, 1903)
16				Versurigidae	*Versuriga anadyomene* (Maas, 1903)

Table 4.8 Distribution of different ichthyoplankton families at four ecozones in the Indian Sundarbans with contrasting ecological characters

Sl. No.	Name of the ichthyoplankton families	S-I	S-II	S-III	S-IV
1.	Anguillidae	+	+	+	+
2.	Elopidae	–	–	+	+
3.	Ophichthidae	+	–	+	+
4.	Muraenidae	+	–	+	+
5.	Gobiidae	+	+	+	+
6.	Sciaenidae	+	–	+	+
7.	Mugilidae	+	+	+	+
8.	Ambassidae	+	+	+	+
9.	Clupeidae	+	+	+	+
10.	Lutjanidae	–	–	+	+
11.	Scatophagidae	+	+	+	+
12.	Centropoidae	–	–	–	+
13.	Engraulidae	+	+	+	+
14.	Carangidae	+	–	+	+
15.	Teraponidae	+	+	+	+
16.	Leiognithidae	+	–	–	+
17.	Sillaginidae	+	+	+	+
18.	Scombridae	–	–	–	+
19.	Polynemidae	+	+	+	+

S-I = Jharkhali, S-II = Canning, S-III = Saptamukhi, S-IV = Bokkhali
Note: + = (Present); – = (Absent)

Table 4.9 Occurrence of different macrozoobenthic species recorded from different tidal levels of four study sites

	Species	S I			S II			S III			S IV		
		LTL	MTL	HTL	LTL	MTL	HTL	LTL	MTL	HTL	LTL	MTL	HTL
	Phylum—Cnideria Class—Anthozoa Order—Actiniaria Family—Edwardsiidae												
1	*Edwardsia jonesii* Seshaiya & Cuttress	+	+	–	+	+	–	+	+	–	+	–	–
2	*Pelocoetes exul* Annandale	+	+	–	–	–	–	+	+	–	–	+	–
3	*Phytocoeteopsis ramunii* Panikkar	+	+	–	–	–	–	–	–	–	–	–	–
	Family-Diadumenidae												
4	*Diadumene schilleriana* (Stoliczka)	–	+	+	–	–	–	–	–	–	–	+	+
	Family—Actiniidae												

(continued)

Table 4.9 (continued)

	Species	S I			S II			S III			S IV		
		LTL	MTL	HTL	LTL	MTL	HTL	LTL	MTL	HTL	LTL	MTL	HTL
5	*Paracondylactis indica* Dave	+	+	–	+	+	–	+	+	–	+	+	–
	Phylum—Annelida												
	Class—Polychaeta												
	Sub-class-Errantia												
	Family—Hesionidae												
6	*Talehsapia annandalei* Fauvel	+	+	+	–	–	–	+	+	–	–	–	–
	Family—Phyllodocidae												
7	*Eteone oranata* Grube	–	–	–	–	–	–	–	–	–	+	+	–
8	*E. barantollae* Fauvel.	–	–	–	–	–	–	+	+	–	+	+	–
	Family-Nereidae												
9	*Neanthes chingrighattensis* Fauvel	–	–	–	+	+	+	+	+	+	+	+	–

10	*Perinereis cultrifera* Grube	+	+	+	+	+	−	−	−	−	−	−	−
11	*P. nigropunctata* Horst	+	+	−	+	+	+	+	+	+	−	−	−
12	*Nemalycastis indica* southern	+	+	+	+	+	−	+	+	+	+	+	−
13	*N. fauveli* Rao	+	+	−	+	+	−	−	−	−	+	+	−
14	*Dendronereides gangetica* Misra	−	−	−	+	+	+	−	−	−	+	+	−
15	*D. heteropoda* southern	−	−	−	−	+	+	−	−	−	+	+	−
16	*Dendronereis aestuarina* southern.	+	+	+	+	+	−	−	−	−	−	−	−
	Family—Glyceridae												
17	*Glycera tesselata* Grube.	−	−	−	+	+	+	+	+	−	−	−	−
18	*G. alba* Rathake.	−	−	−	−	+	+	−	−	−	+	+	+

S-I = Jharkhali, S-II = Canning, S-III = Saptamukhi, S-IV = Bokkhali
Note: + = (Present); − = (Absent)

Table 4.10 Diversity of brachyuran crabs in Sundarbans, India

Sl. No.	Family and species	Diversity of Brachyuran Carbs in Sundarbans India through	
		Molay Dev Roy	Chakraborty et al. (1986)
	Infraorder BRACHYURA Linnaeus, 1758		
	Section **Podotremata** Guinot, 1977		
	Superfamily DROMIOIDEA DeHaan, 1833		
	Family DROMIIDAE DeHaan, 1833		
	Genus ***Conchoecetus*** Stimpson, 1858		
1.	*C. artificiosus* (Fabricius, 1798) (M)	+	
	Superfamily RANINOIDEA DeHaan, 1839		
	Family RANINIDAE DeHaan, 1839		
	Genus ***Raninoides*** H. Milne Edwards, 1837		
2.	*R. personatus* Henderson, 1888 (M)	+	
	Section **Eubrachyura** SaintLaurent, 1980		
	Subsection **Heterotremata** Guinot, 1977		
	Superfamily AETHROIDEA Dana, 1851		
	Family AETHRIDAE Dana, 1851		
	Genus ***Drachiella*** Guinotin Serèneand Soh, 1976		
3.	*D.morum* (Alcock, 1896) (M)	+	
	Superfamily CALAPPOIDEA DeHaan, 1833		
	Family CALAPPIDAE DeHaan, 1833		
	Genus ***Calappa*** Weber, 1795		
4.	*C.lophos* (Herbst, 1790) (M)	+	
5.	*C.pustulosa* Alcock, 1896	+	
	Family MATUTIDAE DeHaan, 1835		
	Genus ***Ashtoret*** Galiland Clark, 1994		
6.	*A.lunaris* (Forskål, 1775) (E)	+	
	Genus ***Matuta*** Weber, 1795		
7.	*M. planipes* Fabricius, 1798 (M, E)	+	
8.	*M. victor* (Fabricius, 1781)	+	
	Superfamily CORYSTOIDEA Samouelle, 1819		
	Family CORYSTIDAE Samouelle, 1819		
	Genus ***Jonas*** Jacquinot, 1853		
9.	*Jonasindica* (Chopra, 1935) (M)	+	
	Superfamily DORIPPOIDEA MacLeay, 1838		
	Family DORIPPIDAE MacLeay, 1838		
	Genus ***Dorippe*** Weber, 1795		
10.	*Dorippequadridens* (Fabricius, 1793) (M)	+	
	Genus ***Dorippoides*** Serèneand Romimohtarto, 1969		
11.	*D. facchino* (Herbst, 1785) (M)	+	
	Genus ***Neodorippe*** Serèneand Romimohtarto, 1969		
12.	*N.callida* (Fabricius, 1798) (E)	+	
	Family ETHUSIDAE Guinot, 1977		

Table 4.10 (continued)

Sl. No.	Family and species	Diversity of Brachyuran Carbs in Sundarbans India through Molay Dev Roy	Chakraborty et al. (1986)
	Genus ***Ethusa*** Roux, 1830		
13.	*E. indica* Alcock, 1894 (M)	+	
	Superfamily ERIPHIOIDEA MacLeay, 1838		
	Family MENIPPIDAE Ortmann, 1893		
	Genus *Menippe* DeHaan, 1833		
14.	*M. rumphii* (Fabricius, 1798) (M)	+	
	Genus ***Myomenippe*** Hilgendorf, 1879		
15.	*M. hardwickii* (Gray, 1831) (M, E)	+	
	Superfamily GONEPLACOIDEA MacLeay, 1838		
	Family SCALOPIDIIDAE Stevcic, 2005		
	Genus ***Scalopidia*** Stimpson, 1858		
16.	*S. spinosipes* Stimpson, 1858 (M)	+	
	Superfamily LEUCOSIOIDEA Samouelle, 1819		
	Family IPHICULIDAE Alcock, 1896		
	Genus ***Iphiculus*** Adamsand White, 1848		
17.	*I. spongiosus* Adamsand White, 1848	+	
	Genus ***Pariphiculus*** Alcock, 1896		
18.	*P. mariannae* (Herklot, 1852) (M)	+	
	Family LEUCOSIIDAE Samouelle, 1819		
	Subfamily EBALIINAE Stimpson, 1871		
	Genus ***Arcania*** Leach, 1817		
19.	*A. erinaceus* (Fabricius, 1793) (M)	+	
20.	*A. septemspinosa* (Fabricius, 1787) (M)	+	
	Genus ***Ixa*** Leach, 1815		
21.	*I. cylindrus* (Fabricius, 1787) (M)	+	
22.	*I. inermis* Leach, 1817 (M)	+	
	Genus ***Myra*** Leach, 1817		
23.	*M. elegans* Bell, 1855	+	
24.	*M. fugax* (Fabricius, 1798)	+	
	Genus ***Philyra*** Leach, 1817		
25.	*P. globus* (Fabricius, 1775) (M)	+	
	Subfamily LEUCOSIINAE Samouelle, 1819		
	Genus ***Euclosia*** Galil, 2003		
26.	*E. rotundifrons* Chopra, 1933 (M)	+	
	Genus ***Leucosia*** Weber, 1795		
27.	*L. craniolaris* (Linnaeus, 1758) (M)	+	
	Genus ***Seulocia*** Galil, 2005		
28.	*S. rhomboidalis* DeHaan, 184 (M)	+	
	Superfamily MAJOIDEA Samouelle, 1819		
	Family EPIALTIDAE MacLeay, 1838		
	Subfamily PISINAE Dana, 1851		

(continued)

Table 4.10 (continued)

Sl. No.	Family and species	Diversity of Brachyuran Carbs in Sundarbans India through	
		Molay Dev Roy	Chakraborty et al. (1986)
	Genus ***Doclea*** Leach, 1815		
29.	*D. armata* DeHaan, 1839 (M)	+	
30.	*D. canalifera* Stimpson, 1857 (M)	+	
31.	*D. hybrida* (Fabricius, 1798) (M)	–	
32.	*D. muricata* (Fabricius, 1787) (M)	+	
33.	*D. ovis* (Fabricius, 1787) (M)	+	
34.	*D. rissonii* Leach, 1815 (M)	+	
	Genus ***Hyastenus*** White, 1847		
35.	*H. diacanthus* (DeHaan, 1839) (M)	+	
	Genus ***Phalangipus*** Latreille, 1825		
36.	*P. longipes* (Linnaeus, 1758) (M)	+	
	Family HYMENOSOMATIDAE MacLeay, 1838		
	Genus ***Neorhynchoplax*** Sakai, 1938		
37.	*N. inachoides* (Alcock, 1900) (E)	+	
38.	*N. nasalis* (Kemp, 1917) (E)	+	
39.	*N. woodmasoni* (Alcock, 1900) (E)	+	
	Genus ***Hymenicoides*** Kemp, 1917		
40.	*H. carteri* Kemp, 1917 (E)	+	
	Genus ***Trigonoplax*** H. Milne Edwards, 1853		
41.	*T. unguiformis* (DeHaan, 1839) (E)	+	
	Superfamily PARTHENOPOIDEA MacLeay, 1838		
	Family PARTHENOPIDAE MacLeay, 1838		
	Subfamily PARTHENOPINAE MacLeay, 1838		
	Genus ***Cryptopodia*** H. Milne Edwards, 1834		
42.	*C. angulata* H. Milne Edwards & Lucas 1841 (M)	+	
	Genus ***Enoplolambrus*** A. Milne Edwards, 1878		
43.	*E. pransor* (Herbst, 1796) (M) (M)	+	
	Superfamily PILUMNOIDEA Samouelle, 1819		
	Family GALENIDAE Alcock, 1898		
	Subfamily GALENINAE Alcock, 1898		
	Genus ***Galene*** DeHaan, 1833		
44.	*G. bispinosa* (Herbst, 1783) (M)	+	
	Subfamily HALIMEDINAE Alcock, 1898		
	Genus ***Halimede*** DeHaan, 1835		
45.	*H. fragifer* DeHaan, 1835 (M)	+	
46.	*H. tyche* (Herbst, 1801)	+	
	Family PILUMNIDAE Samouelle, 1819		
	Subfamily PARAPANOPINAE Stevcic, 2005		
	Genus ***Parapanope*** DeMan, 1895		
47.	*P. euagora* DeMan, 1895 (M)	+	
	Subfamily PILUMNINAE Samouelle, 1819		

Table 4.10 (continued)

Sl. No.	Family and species	Diversity of Brachyuran Carbs in Sundarbans India through Molay Dev Roy	Chakraborty et al. (1986)
	Genus ***Benthopanope*** Davie, 1989		
48.	*B. indica* Stimpson, 1858 (E)	+	
	Genus ***Eurycarcinus*** A. Milne Edwards, 1867		
49.	*E. bengalensis* Deb, 1998 (E)	+	
50.	*E. natalensis* (Krauss, 1843) (M, E)	+	
51.	*E. orientalis* A. Milne Edwards, 1867 (M, E)	+	
	Genus ***Heteropanope*** Stimpson, 1858		
52.	*H. glabra* Stimpson, 1858 (E)	+	
53.	*H. neolaevis* Deb, 1998 (E)	+	
	Genus ***Heteropilumnus*** DeMan, 1895		
54.	*H. ciliatus* (Stimpson, 1858) (E)	+	
	Subfamily RHIZOPINAE Stimpson, 1858		
	Genus ***Typhlocarcinus*** Stimpson, 1858		
55.	*T. nudus* Stimpson, 1858 (M)	+	
	Superfamily PORTUNOIDEA Rafinesque, 1815		
	Family PORTUNIDAE Rafinesque, 1815		
	Subfamily CAPHYRINAE Paul'son, 1875		
	Genus ***Lissocarcinus*** Adams & White, 1848		
56.	*L.arkati* Kemp, 1923 (M)	+	
	Subfamily PORTUNINAE Rafinesque, 1815		
	Genus ***Portunus*** Weber, 1795		
	Subgenus ***Achelous*** DeHaan, 1833		
57.	*P. (A.) granulatus* (H. Milne Edwards, 1834)	+	
	Subgenus ***Lupocycloporus*** Alcock, 1899		
58.	*P. (L.) gracilimanus* (Stimpson, 1858) (M)	+	
	Subgenus ***Monomia*** Gistel, 1848		
59.	*P. (M.) gladiator* Fabricius, 1798 (M)	+	
	Subgenus ***Portunus*** Weber, 1795		
60.	*P. (P.) pelagicus* (Linnaeus, 1758) (M)	+	
61.	*P. (P.) sanguinolentus* (Herbst, 1790) (M)	+	
62.	*P. (P.) pubescens* (Dana, 1852) (M)	+	
	Subgenus ***Xiphonectes*** A. Milne Edwards, 1873		
63.	*P. (X.) hastatoides* Fabricius, 1798 (M)	+	
64.	*P. (X.) pulchricristatus* (Gordon, 1931) (M)	+	
	Genus ***Scylla*** DeHaan, 1833		
65.	*S. serrata* (Forskål, 1775) (M, E)	+	
66.	*S. tranquebarica* (Fabricius, 1798) (E)	+	
	Subfamily ***Thalamitinae*** Paul'son, 1875		
	Genus ***Charybdis*** DeHaan, 1833		
	Subgenus ***Charybdis*** DeHaan, 1833		

(continued)

Table 4.10 (continued)

Sl. No.	Family and species	Diversity of Brachyuran Carbs in Sundarbans India through	
		Molay Dev Roy	Chakraborty et al. (1986)
67.	*C. (C.) affinis* Dana, 1852 (M)	+	
68.	*C. (C.) callianassa* (Herbst, 17) (M)	+	
69.	*C. (C.) feriatus* (Linnaeus, 1758) (M)	+	
70.	*C. (C.) helleri* (A. Milne Edwards, 1867) (M)	+	
71.	*C. (C.) lucifera* (Fabricius, 1798) (M)	+	
72.	*C. (C.) miles* (DeHaan, 1852) (M)	+	
73.	*C. (C.) orientalis* Dana, 1852 (M)	+	
74.	*C. (C.) rostrata* (A. Milne Edwards, 1861) (M, E)	+	
75.	*C. (C.) variegata* (Fabricius, 1798) (M)	+	
	Subgenus ***Goniohellenus*** Alcock, 1899		
76.	*C. (G.) truncata* (Fabricius, 1798) (M)	+	
77.	*C. (G.) vadorum* Alcock, 1899 (M)	+	
	Genus ***Thalamita*** Latreille, 1829		
78.	*T. crenata* (Latreille, 1829) (M)	+	
	Superfamily XANTHOIDEA MacLeay, 1838		
	Family XANTHIDAE MacLeay, 1838		
	Subfamily XANTHINAE MacLeay, 1838		
	Genus ***Liagore*** DeHaan, 1833		
79.	*L. erythematica* Guinot, 1971 (M)	+	
	Genus ***Orphanoxanthus*** Alcock, 1898		
80.	*O. microps* (Alcockand Anderson, 1894) (M)	+	
	Superfamily GRAPSOIDEA MacLeay, 1838		
	Family GRAPSIDAE MacLeay, 1838		
	Genus ***Metopograpsus*** H. Milne Edwards, 1853		
81.	*M. latifrons* (White, 1874) (E)	+	
82.	*M. messor* (Forskål, 1775) (E)	+	
	Genus ***Pachygrapsus*** Randall, 1840		
83.	*P. porpinquus* DeMan, 1908 (E)	+	
	Subsection **Thoracotremata** Guinot, 1977		
	Family SESARMIDAE Dana, 1851		
	Genus ***Clistocoeloma*** A. Milne Edwards, 1873		
84.	*C. merguiense* DeMan, 1888 (E)	+	
	Genus ***Episesarma*** DeMan, 1895		
85.	*E. mederi* (H. Milne Edwards, 1853) (E)	+	
	Genus ***Metasesarma*** H. Milne Edwards, 1853		
86.	*M. rousseauxii* H. Milne Edwards, 1853 (E)	+	
	Genus ***Muradium*** Serèneand Soh, 1970		
87.	*M. tetragonum* (Fabricius, 1798) (E)	+	
	Genus ***Neosarmatium*** Serène & Soh, 1970		
88.	*N. smithi* (H. Milne Edwards, 1853) (E)	+	

Table 4.10 (continued)

Sl. No.	Family and species	Diversity of Brachyuran Carbs in Sundarbans India through	
		Molay Dev Roy	Chakraborty et al. (1986)
	Genus ***Parasesarma*** (DeHaan, 1835)		
89.	*P. pictum* (DeHaan, 1835) (E)	+	
90.	*P. plicatum* (Fabricius, 1798) (E)	+	
	Genus ***Perisesarma*** DeMan, 1895		
91.	*P. bidens* (DeHaan, 1835) (E)	+	
	Genus ***Pseudosesarma*** Serène & Soh, 1970		
92.	*P. edwardsi* (DeMan, 1887) (E)	+	
	Genus ***Sesarmoides*** Serène & Soh, 1970		
93.	*S. kraussi* (DeMan, 1888) (E)	+	
94.	*S. longipes* (Krauss, 1843) (E)	+	
	Genus ***Sesarmops*** Serène & Soh, 1970		
95.	*S. impressum* (H. Milne Edwards, 1837) (E)	+	
96.	*S. intermedius* (DeHaan, 1835) (E)	+	
	Family VARUNIDAE H. Milne Edwards, 1853		
	Subfamily CYCLOGRAPSINAE H. Milne Edwards, 1853		
	Genus ***Metaplax*** H. Milne Edwards, 1852		
97.	*M. crenulata* (Gerstaecker, 1856) (E)	+	
98.	*M. dentipes* (Heller, 1865) (E)	+	
99.	*M. distincta* H. Milne Edwards, 1852 (E)	+	
100.	*M. indica* H. Milne Edwards, 1852 (E)	+	
101.	*M. intermedia* DeMan, 1888 (E)	+	
	Subfamily VARUNINAE H. Milne Edwards, 1853		
	Genus ***Parapyxidognathus*** Ward, 1941		
102.	*P. deianira* (DeMan, 1888) (E)	+	
	Genus ***Ptychognathus*** Stimpson, 1858		
103.	*P. dentatus* DeMan, 1892 (E)	+	
104.	*P. onyx* Alcock, 1900 (E)	+	
	Genus ***Pyxidognathus*** A. Milne Edwards, 1878		
105.	*P. fluviatilis* Alcock, 1900 (E)	+	
	Genus ***Varuna*** H. Milne Edwards, 1830		
106.	*V. litterata* (Fabricius, 1798) (E, M, F)	+	
	Superfamily OCYPODOIDEA Rafinesque, 1815		
	Family DOTILLIDAE Stimpson, 1858		
	Genus ***Dotilla*** Stimpson, 1858		
107.	*D. blanfordi* Alcock, 1900 (E)	+	
108.	*D. intermedia* DeMan, 1888 (E)	+	
	Genus ***Dotillopsis*** Kemp, 1919		

(continued)

Table 4.10 (continued)

Sl. No.	Family and species	Diversity of Brachyuran Carbs in Sundarbans India through	
		Molay Dev Roy	Chakraborty et al. (1986)
109.	*D.brevitarsis* (DeMan, 1888) (E)	+	
	Genus ***Ilyoplax*** Stimpson, 1858		
110.	*I.gangeticus* (Kemp, 1919) (E)	+	
111.	*I.stapletoni* (DeMan, 1908) (E)	+	
	Genus ***Scopimera*** DeHaan, 1835		
112.	*S.globosa* DeHaan, 1835 (E)	+	
113.	*S.investigatoris* Alcock, 1900 (E)	+	
114.	*S.proxima* Kemp, 1919 (E)	+	
	Family MACROPHTHALMIDAE Dana, 1851		
	Genus ***Macrophthalmus*** Latreille, 1829		
	Subgenus ***Macrophthalmus*** Latreille, 1829		
115.	*M.(M.) brevis* (Herbst, 1804) (E)	+	
116.	*M.(M.) crassipes* H.MilneEdwards, 1834 (E)	+	
117.	*M.(M.) sulcatus* H.MilneEdwards (M)	–	
118.	*M.(M.) transversus* (Latreille, 1817) (M, E)	+	
	Subgenus ***Mareotis***		
119.	*M.(M.) depressus* Rüppell, 1830 (E)	+	
120.	*M.(M.) pacificus* Dana, 1851 (E)	+	
121.	*M.(M.) teschi* Kemp, 1919 (E)	+	
122.	*M.(M.) tomentosus* Souleyet, 1841 (E)	+	
	Subgenus ***Paramareotis*** Komai, Goshima and Murai, 1995		
123.	*M.(P.) erato* DeMan, 1888 (E)	+	
	Subgenus ***Venitus***		
124.	*M.(V.) dentipes* LucasinGuérin-Méneville, 1838 (E)	+	
	Family OCYPODIDAE Rafinesque, 1815		
	Subfamily OCYPODINAE Rafinesque, 1815		
	Genus ***Ocypode*** Weber, 1795		
125.	*O.ceratophthalma* (Pallas, 1772) (M)	+	
126.	*O.macrocera* H.Milne Edwards, 1837 (M)	+	
127.	*O.platytarsis* H.Milne Edwards, 1852 (M)	–	
	Subfamily UCINAE Dana, 1851		
	Genus ***Uca*** Leach, 1815		
128.	*U.dussumieri* (H.Milne Edwards, 1852) (E)	+	
129.	*U.rosea* (Tweedie, 1937) (E)	+	
130.	*U.vocans* (Linnaeus, 1758) (E)	+	
131.	*U.lactea* (DeHaan, 1835) (E)	+	

Table 4.10 (continued)

Sl. No.	Family and species	Diversity of Brachyuran Carbs in Sundarbans India through	
		Molay Dev Roy	Chakraborty et al. (1986)
132.	*U. triangularis* (A. Milne Edwards, 1873) (E)	+	
	Superfamily PINNOTHEROIDEA DeHaan, 1833		
	Family PINNOTHERIDAE DeHaan, 1833		
	Subfamily PINNOTHERINAE DeHaan, 1833		
	Genus ***Nepinnotheres*** Manning, 1993		
133.	*N. cardii* (Bürger, 1895) (M)	–	
	Genus ***Pinnotheres*** Bosc, 1802		
134.	*P. boniensis* Stimpson (1858) (M)	+	
135.	*P. mactricola* Alcock, 1900 (E)	+	
	Total = 135 species	**131**	

Table 4.11 Diversity of mollucs in Sundarbans, India

SL. NO.	Class	Order	Family	Species
1	Gastropoda	Archaeogastropoda	Acmaeidae	*Potamacmaea fluviatilis* (Blanford)
2			Cyclostrematidae	*Tubiola microscopica* (Nevill)
3			Neritidae	*Nerita articulata*Gould
4				*Neritina smithi*Wood
5				*Neritina* (*Dostia*) *violacea* (Gmelin)
6				*Pseudonerita obtusa* (Benson)
7				*P. sulculosa* (von martens)
8				*Clithon reticularis* (Sowerby)
9				*Septaria caerulescens* (Sowerby)
10				*S. lineata* (Lamarck)
11			Viviparidae	*Bellamya bengalensis* f. *typica* (Lamarck)
12				*Bellamya dissimilis* (Mueller)
13				*B. crassa* (Benson)
14			Pilidae	*Pila globosa* (Swainson)
15		Mesogastropoda	Litiorinidae	*Littorina* (*Littorinopsis*) *melanostoma* (Gray)
16				*L.* (*Littorinopsis*) *scabra scabra* (Linnaeus)
17				*L.* (*Linoraria*) *undulata* (Gray)

(continued)

Table 4.11 (continued)

SL. NO.	Class	Order	Family	Species
18			Iravadiidae	*Iravadia omata* (Blanford)
19			Stenothyridae	*Stenothyra blanfordiana* (Nevill)
20				*S. deltae* (Benson)
21				S. *nana* (Annandale and Prashad)
22				*S. ornata* (Annandale and Prashad)
23				*S. soluta* (Annandale and Prashad)
24				S. woodmasoniana (Nevill)
25				*Gangetia miliacea* (Nevill)
26			Bithyniidae	*Digoniostoma ceranleopoma* (Benson)
27			Assimineidae	*Assinlinca* beddomeana (Nevill)
28				*A. brevicula* (Pfeiffer)
29				*A. francesiae* (Wood)
30				*A. hungerjordiana* (Nevill)
31				*A. microsculpta* (Nevill)
32				*A. theobaldiana* (Nevill)
33				*A. woodmasoniana* (Nevill)
34			Architectonicidae	*Architectonica perspectiva* (Linnaeus)
35			Thiaridae	*Thiara (Thiara) scabra* (Mueller)
36				*T. (Melanoides) tuberculata* (Mueller)
37				*T. (Tarebia) granifera* (Lamarck)
38				*T. (Tarebia) lineatus* (Gray)
39				*T. (Mainwaringia) paludomoidea* (Nevill)
40				*Broda costula* (Rafinesque)
41				*Larina burmana* (Blanford)
42			Potamididae	*Telescopium (T) telescopium* (Linnaeus)
43				*Cerithidea (Cerithideopsilla) cingulata* (Gmelin)
44				*C. (Cerithidea) obtusa* (Lamarck)

Table 4.11 (continued)

SL. NO.	Class	Order	Family	Species
45			Epitoniidae	*Acrilla acuminata* (Sowerby)
46			Naticidae	*Natica gualteriana* (Recluz)
47				*Natica* tigrina (Roeding)
48		Neogastropoda	Muricidae	*Thais lacera* (Born)
49			Pyrenidae	*Columbella duclosiana* (Sowerby)
50			Melongenidae	*Pugilina cochlidium* (Linnaeus)
51			Nassariidae	*Nassarius foveolatus* (Reeve)
52				*N. (Pygmaeonassa) orissaensis* (Preston)
53				*N. stolatus* (Gmelin)
54			Olividae	*Arnalda ampla* (Gmelin)
55			Turridae	*Asthenotoma vertebrata* (Smith)
56			Ringiculidae	*Ringicula propinquans* (Hinds)
57			Cylichnidae	*Tornatina estriata* (Preston)
58			Aglajidae	*Melanoclamys* sp.
59			Atyidae	*Haminea crocata* (Pease)
60		Basommatophora	Ellobiidae	*Cassidula nucleus* (Gmelin)
61				*Ellobium aurisjudae* (Linnaeus)
62				*E. (Auricula) gangeticum* (Pfeiffer)
63				*Melampus pulchella* (Petit)
64				*Pythia plicata* (Gray)
65			Lymnaeidae	*Lymnaea (Pseudosuccinea) acuminata f. Typica* (Lamarck)
66				*L. (P.) acuminata* f. *Rufescens* (Gray)
67				*L. (P.) luteola* f. *Typica* (Lamarck)
68			Planorbidae	*Indoplanorbis exustus* (Deshayes)
69		Soleolifera	Onchidiidae	*Onchidium pallidum* (Stoliczka)
70				*O. tenerum* (Stoliczka)

(continued)

Table 4.11 (continued)

SL. NO.	Class	Order	Family	Species
71				*O. tigrinum* (Stoliczka)
72				*O. typhae* (Buchannan)
73	Scaphopoda		Dentaliidae	*Dentalium octangulatum* (Donovan)
74	Bivalvia	Arcoida	Arcidae	*Scaphula celox* (Benson)
75				*Scaphula deltae* (Blanford)
76				*Anadara granosa* (Linnaeus)
77			Noetiidae	*Striarca lactea* (Linnaeus)
78		Mytiloida	Mytilidae	*Modiolus striatulus* (Hanley)
79				*M. undulatus* (Dunker)
80		Pterioida	Anomiidae	*Placuna (P.) placenta* (Linnaeus)
81				*Enigmonia aenigmatica* (Holten)
82			Ostreidae	*Crassostrea cuttackensis* (Newton and Smith)
83				*C. gryphoides* (Schlotheim)
84				*Saccostrea cucullata* (Born)
85		Unionoida	Unionidae	*Parreysia favidens* (Benson)
86		Veneroida	Lucinidae	*Eamesiella philippinarum* (Hanley)
87			Mactridae	*Mactra luzonica* (Deshayes)
88			Solenidae	*Solen brevis* (Gray)
89			Cultellidae	*Cultellus subelliptica* (Dunker)
90				*Neosolen aquaedulcioris* (E.N. Ghosh)
91				*Pharella javanicus* (Lamarck)
92				*Siliqua albida* (Dunker)
93				*Tanysiphon rivalis* (Benson)
94			Tellinidae	*Tellina corbuloides* (Hanley)
95				*T. irridescens* (Benson)
96				*T. sinuata* (Spengler)
97				*Sirigilla splendida* (Anton)
98				*Macoma birmanica* (Philippi)

Table 4.11 (continued)

SL. NO.	Class	Order	Family	Species
99			Donacidae	*Donax incarnatus* (Gmelin)
100			Psammobiidae	*Sanguinolaria* (Soletellina) *acuminata* (Deshayes)
101			Semelidae	*Theora opalina* (Hinds)
102			Solecurtidae	*Novaculina gangetica* (Benson)
103			Corbiculidae	*Corbicula striatella* (Deshayes)
104				*C.bensoni* (Deshayes)
105				*Polymesoda* (Geloina) *bengalensis* (Lamarck)
106			Trapeziidae	*Trapezium sublavigatum* (Lamarck)
107			Veneridae	*Gafrarium pectinatum* (Linnaeus)
108				*Meretrix meretrix* (Linnaeus)
109				*Pelecyora trigona* (Reeve)
110			Glauconomidae	*Glauconome cerea* (Reeve)
111				*Sculpta* (Sowerby)
112		Myoida	Myidae	*Sphenia perversa* (Blanford)
113			Corbulidae	*Corbula abbreviata* (Preston)
114				C. *calcarea* (Preston)
115				C. *gracilis* (Preston)
116			Pholadidae	*Martesia fragilis* (Verrill and Bush)
117				*Lignopholas fluminalis* (Blanford)
118				*Bamea candida* (Linnaeus)
119			Teredinidae	*Bactronophorus thoracites* (Gould)
120				*Dicyathifer manni* (Wright)
121				*Bankia carinata* (Gray)
122				*B. campanellata* (Moll and Roch)
123				*B. nordi* (Moll)

(continued)

Table 4.11 (continued)

SL. NO.	Class	Order	Family	Species
124				*B. rochi* (Moll)
125				*Nausitora dunlopei* (Wright)
126			Laternulidae	*Laternula truncata* (Lamarck)
127			Cuspidariidae	*Cuspidaria chilkaensis* (Preston)
128	Cephalopoda	Sepioidea	Sepiidae	*Sepia aculeata* (d'Orbigny)
129				*Sepiella inermis* (d' Orbigny)
130		Teuthoidea	Loliginidae	*Loliolus investigatoris* (Goodrich)

Table 4.12 Diversity of polychaetes in Sundarbans, India

Sl. No.	Family	Subfamily	*Species*
1	Polynoidae	Lepidonotinae	*Lepidonotus tenuisetosus* (Gravier, 1901)
2			*Lepidonotus* sp.
3		Harmothoinae	*Gattyana fauveli* (Misra)
4			*Gaudichaudius cimex* (Qua'refages, 1866)
5	Amphinomidae		*Chloeia parva* (Baird, 1870)
6	Phyllodocidae		*Anaitides madeirensis* (Langerhans, 1880)
7			*Eteone barantollae* (Fauvel, 1932)
8			*Eteone ornata* (Grube, 1878)
9	Hesionidae		*Hesione splendida* (Savigny, 1818)
10	Pilargidae		*Sigambra constricta* (Southern, 1921)
11			*Sigatargis commensalis* (Misra)
12	Talehsapiidae		*Talehsapia annandalei* (Fauvel, 1932)
13	Nereididae	Namanereidinae	*Namalycastis fauveli* (Rao, 1981)
14			*Namalycastis indica* (Southern, 1921)
15			*Namalycastis meraukensis* (Horst, 1918)
16		Nereidinae	*Ceratonereis burmensis* (Monro, 1937)
17			*Dendronereides gangetica* (Misra)
18			*Dendronereides heteropoda* (Southern, 1921)
19			*Dendronereides heteropoda* (Southern, 1921)
20			*Dendronereis dayi* (Misra)
21			*Ganganereis sootai* (Misra)
22			*Lycastonereis indica* (Rao, 1981)
23			*Neanthes chingrighattensis* (Fauvel, 1932)
24			*Neanthes glandlcincta* (Southern, 1921)
25			*Neanthes meggitti* (Monro, 1931)
26			*Perinereis cavifrons* (Ehlers, 1920)
27			*Perinereis cultrifera* (Grube, 1840)
28			*Perinereisnigropunctata* (Horst, 1889)

Table 4.12 (continued)

Sl. No.	Family	Subfamily	*Species*
29	Nephtyidae		*Nephtys oligobranchia* (Southern, 1921)
30			*Nephtys polybranchia* (Southern, 1921)
31	Glyceridae		*Glycera eonvolula* (Keferstein, 1862)
32			*Glycera lancadivae* (Schmarda, 1861)
33			*Glycera rouxii* (Audouin & Milne Edwards, 1878)
34			*Glycera tesselata* (Grube, 1863)
35	Goniadidae		*Glycinde oligodon* (Southern, 1921)
36			*Gonilula emerita* (Audouin & Milne Edwards, 1833)
37	Onuphidae		*Diopatra cuprea* (Bosc, 1802)
38			*Hyalinoecia tubicola* (Miller, 17**76)**
39	Eunicidae		*Eunice aphroditois* (Pallas, 1788)
40			*Lysidice natalensis* (Kinberg, 1865)
41			*Marphysa mossambica* (Peters, 1854)
42			*Marphysa sanguinea* (Montagu, 1807)
43			*Marphysa* sp.
44	Lumbrineridae		*Lumbrineris bilabiata* (Misra)
45			*Lumbrineris heteropoda* (Marenzeller, 1879)
46			*Lumbrineris notocirrata* (Fauvel, 1932)
47			*Lumbrineris polydesma*' (Southern, 1921)
48	Orbiniidae	Orbiniinae	*Leitoscoloplos* sp.
49			*Scoloplos (Scolopolos) sagarensis* (Misra)
50	Spionidae		*Minuspio cirrifera* (Wiren, 1883)
51			*Polydora normalis* (Day, 1957)
52			*Polydora* sp.
53			*Pseudopolydora kempi* (Southern, 1921)
54			*Spio bengalensis* (Willey, 1908)
55	Flabelugeridae		*Pherusa bengalensis* (Fauvel, 1932)
56	Capitellidae		*Barantolla seulpta* (Southern, 1921)
57			*Capitella capitata* (Fabricius, 1780)
58			*Mastobranchus* sp. *[cf. Indicus]* (Southern, 19**21)**
59			*Parlleteromastus tenuis* (Monro, 1937)
60	Maldanidae	Maldaninae	*Asychis gangeticus* (Fauvei, 1932)
61	Oweniidae		*Owenia fusiformis* (delle Chiaje, 1841)
62	Sternaspidae		*Sternaspis scutata* (Renier, 1807)
63	Sabellariidae		*Sabellaia pectinata* (Fauvel, 1932)
64			*Sabellaria alcocki* (Gravier, 1906)
65	Terebellidae	Amphitritinae	*Loimia medusa* (Savigny, 1818)
66			*Pista typha* (Grube, 1878)
67	Ampharetidae	Melinninae	*Isolda pulchella* (Muller, 1858)
68	Sabellidae	Sabellinae	*Potamilla leptochaeta* (Southern, 1921)
69	Serpulidae	Serpulinae	*Ficopomatus macrodon* (Southern, 1921)

(continued)

Table 4.13 Major brackish water and freshwater finfish and shellfish of Sundarbans, India

Common Name	Local Name	Scientific Name
Freshwater finfish		
Catla	Catla	*Catla catla*
Rohu	Rui	*Labeo rohita*
Mrigal	Mrigal	*Cirrhinus mrigala*
Bata	Bata	*Labeo bata*
Calbasu	Kalbos	*Labeo calbasu*
Olive barb	Saral Punti	*Puntius sarana*
Ticto barb	Punti	*Ticto ticto*
Tint Punti	Tint Punti	*Puntius puntius*
Japani Punti	Java Punti	*Puntius javanicus*
Grass carp	Gheso rui	*Ctenopharyngodon idella*
Common carp	Ameriacan rui	*Cyprinus carpio*
Silver carp	Silver carp	*Hypophthalmichthys molitrix*
Bighead carp	Bighead	*Aristichthys nobilis*
Pangas	Pangas	*Pangasius sutchi*
Rupchanda	Rupchand	*Pygocentrus nattereri*
Chitol (feather back)	Chitol	*Chitala chitala*
Tilapia	Tilapia	*Oreochromis niloticus O. mossambicus*
Air breathing walking catfish	Magur	*Clarias batrachus*
Air breathing stinging catfish	Singi	*Heteropneustes fossilis*
Climbing perch	Koi	*Anabas testudineus*
Gaint Murrel	Sall	*Channa marulius*
Striped Murrel	Sol	*Channa striata*
Spotted Murrel	Lata	*Channa punctata*
Barca snakehead	Chang	*Channa gachua, C. orientalis*
Pholui	Pholui	*Notopterus notopterus*
Goby	Bele	*Glossogobius* sp.
Gangetic leaf fish	Nadosh	*Nandus nandus*
Gangetic striped dwarf catfish	Tangra	*Mystus cavasius, Mystus vittatus*
Gaint river catfish	Tangra	*Sperata seenghala*
Long whiskered catfish	Aar	*Sperata aor*
Wallago	Boal	*Wallago Attu*
Pabdah catfish	Pabda	*Ompok pabda, Ompok pabo*
Freshwater shellfish		
Gaint freshwater prawn	Golda chingri	*Macrobrachium rosenbergii*
Indian freshwater prawn	Sada chingri	*Macrobrachium malcomsonii*
Ganga River prawn	Chapra chingri	*Macrobrachium choprai*
Brackishwater finfish		
Indian seabass	Bhetki	*Lates calcerifer*
Gold spot mullet	Parse	*Liza parsia*
Large scale mullet	Rupoli parse	*Liza macrolepis*
Tade mullet	Bhangan	*Liza tade*
Flathead mullet	Ansh Bhangan	*Mugil cephalus*

Table 4.13 (continued)

Common Name	Local Name	Scientific Name
Corsula	Khorsula	*Rhinomugil corsula*
Pearl spot	Mukta Gachha	*Etroplus suratensis*
Tank goby	Bele	*Glossogobius giuris*
Long whiskers catfish	Nona tengra	*Mystus gulio*
Paradise threadfin	Topse	*Polynemus paradiseus, P. indicus*
Spotted scat	Pairachanda	*Scatophagus argus*
Milk fish	Milk fish	*Chanos chanos*
Yellow fish	River pangus	*Pangasius pangasius*
Brackishwater shellfish		
Gient Tiger prawn	Bagda chingri	*Peneaus monodon*
Ginger prawn	Honnye chingri	*Metapenaeus monoceros*
Banana prawn	Kola chingri	*Fenneropenaeus merguiensis*
Yellow prawn	Chamne chingri	*Metapenaeus brevicornis*
Kiddi shrimp	Rosna chongri	*Parapenaeopsis stylifera*
Indian white shrimp	Chapra chingri	*Fenneropenaeus indicus*
Mud crab	Kalo kankra	*Scylla serrata*
Mud crab	Neel kankra	*Scylla tranqueberica*

Table 4.14 Diversity of lizards in Sundarbans, India

No.	Species	Common English Name
1	*Varanus salvator*	Water monitor lizard
2	*Varanus bengalensis*	Bengal land monitor lizard
3	*Gekko gecko*	Tokay
4	*Calotes versicolor*	Indian garden lizard
5	*Varanus flavescens*	Yellow monitor lizard
6	*Hemidactylus flaviviridis*	Yellow-green house gecko
7	*Hemidactylus brooki*	Brook's house gecko
8	*Hemidactylusl eschenaulti*	Bark gecko
9	*Hemidactylus frenatus*	Asian house gecko
10	*Lygosoma punctata*	Spotted supple skink
11	*Lygosomaalbo punctata*	White-spotted supple skink
12	*Eutropis carinata*	Keeled grass skink
13	*Eutropisma cularia*	Bronze grass skink

Table 4.15 Diversity of snakes in Sundarbans, India

No.	Species	Common English name
1	*Oligodon arnensis*	Banded kukri snake
2	*Enhydrissi eboldii*	Siebold's smooth water snake
3	*Lycodon jara*	Yellow-speckled wolf snake
4	*Ahaetulla nasuta*	Common vine snake
5	*Amphiesmas tolatum*	Buff-striped keelback
6	*Enhydrisen hydris*	Common smooth water snake
7	*Atretiumschis tosum*	Olive keelback water snake
8	*Boigatrig onata*	Common Indian cat snake
9	*Cerberus rynchops*	Dog-faced water snake
10	*Ramphotyphlops braminus*	Common worm snake / Brahminy blind snake
11	*Chrysopelea ornata*	Ornamental flying snake
12	*Coelognathus radiatus*	Copperheaded trinket snake
13	*Acrochordatus granulatus*	File/wart snake
14	*Fordonialeucobalia*	White-bellied mangrove snake
15	*Gerardiaprevostianus*	Glossy marsh snake
16	*Lycodon aulicus*	Common wolf snake
17	*Dendrelaphis pictus*	Painted bronze-back tree snake
18	*Dendrelaphis tristis*	Common bronze-back tree snake
19	*Ptyas mucosa*	Rat snake
20	*Sibynophis subpunctatus*	Black-headed snake
21	*Xenochrophis cerasogaster*	Painted keelback
22	*Xenochrophis piscator*	Checkered keelback
23	*Xenochrophis chrurrenbergi*	Bar-necked keelback
24	*Bungarus caeruleus*	Common krait
25	*Bungarus fasciatus*	Banded krait
26	*Naja kaouthia*	Monocellate cobra
27	*Hydrophiscyano cinctus*	Annulated sea snake
28	*Ophiophagus hannah*	King cobra
29	*Enhydrinas chistosa*	Hook-nosed sea snake
30	*Hydrophisc aerulescens*	Malacca sea snake
31	*Daboiarusselii*	Russell's viper
32	*Trimeresuruserythrurus*	Spot tailed pit viper
33	*Hydrophisfasciatus*	Fasciated sea snake
34	*Python morulus*	Indian python
35	*Laticauda colubrina*	Colubrine amphibious sea snake
36	*Laticauda laticauda*	Common amphibious sea snake
37	*Microcephalophis cantoris*	Cantor's narrow-headed sea snake

Table 4.16 Diversity of turtles in Sundarbans, India

No.	Species	Common English Name
1	*Geoclemys hamiltonii*	Spotted pond turtle
2	*Pangshur atectum*	Indian roofed turtle
3	*Eretmochelys imbricata*	Hawksbill sea turtle
4	*Batagur baska*	River terrapin
5	*Lepidochelys olivacea*	Olive Ridley sea turtle
6	*Batagur dhongoka*	Three-striped roofed turtle
7	*Nilssoni ahurum*	Peacock softshell turtle
8	*Nilssoni agangetica*	Indian softshell turtle
9	*Lissemys punctata*	Indian flapshell turtle
10	*Pelochelys cantorii*	Asian giant softshell turtle
11	*Chitra indica*	Narrow-headed softshell turtle

Table 4.17 Diversity of birds in Sundarbans, India

Sl No.	Name	Sc. name
1.	Red jungle fowl	*Gallus gallus* (Linnaeus, 1758)
2.	Common quail	*Coturnix coturnix* (Linnaeus, 1758)
3.	Rain quail	*Coturnix coromandelica* (J. F. Gmelin, 1789)
4.	Lesser whistling-duck	*Dendrocygna javanica* (Horsfield, 1821)
5.	Fulvous whistling-duck	*Dendrocygna bicolor* (Vieillot, 1816)
6.	Ruddy shelduck	*Tadorna ferruginea* (Pallas, 1764)
7.	Common shelduck	*Tadorna tadorna* (Linnaeus, 1758)
8.	Greylag goose	*Anser anser* (Linnaeus, 1758)
9.	Bar-headed goose	*Anser indicus* (Latham, 1790)
10.	Knob-billed duck	*Sarkidiornis melanotos* (Pennant, 1769)
11.	Cotton pygmy-goose	*Nettapus coromandelianus* (J. F.Gmelin, 1789)
12.	Gadwall	*Mareca strepera* (Linnaeus, 1758)
13.	Falcated duck	*Mareca falcata* (Georgi, 1775)
14.	Eurasian wigeon	*Mareca Penelope* (Linnaeus, 1758)
15.	Northern pintail	*Anas acuta* (Linnaeus, 1758)
16.	Mallard	*Anus platyrhynchos* (Linnaeus, 1758)
17.	Spot-billed duck	*Anus poecilorhyncha* (Forster, 1781)
18.	Common teal	*Anas crecca* (Linnaeus, 1758)
19.	Northern shoveler	*Spatula clypeata* (Linnaeus, 1758)
20.	Garganey	*Spatula querquedula* (Linnaeus, 1758)
21.	Red-crested pochard	*Netta rufina* (Pallas, 1773)
22.	Common pochard	*Aythya farina* (Linnaeus, 1758)
23.	Ferruginous pochard	*Aythya nyroca* (Güldenstädt, 1770)
24.	Baer's pochard	*Aythya baeri* (Radde, 1863)
25.	Tufted duck	*Aythya fuligula* (Linnaeus, 1758)
26.	Greater scaup	*Aythya marila* (Linnaeus, 1761)
27.	Common golden eye	*Bucephala clangula* (Linnaeus, 1758)

(continued)

Table 4.17 (continued)

Sl No.	Name	Sc. name
28.	Red-breasted merganser	*Mergus serrator* (Linnaeus, 1758)
29.	Red-billed tropicbird	*Phaethon aethereus* (Linnaeus, 1758)
30.	Little grebe	*Tachybaptus ruficollis* (Pallas, 1764)
31.	Great White pelican	*Pelecanus onocrotalus* (Linnaeus, 1758)
32.	Spot-billed pelican	*Pelecanus philippensis* (Gmelin, 1789)
33.	Darter	*Anhinga melanogaster* (Pennant, 1769)
34.	Little cormorant	*Microcarbo niger* (Vieillot, 1817)
35.	Indian cormorant	*Phalacrocorax fuscicollis* (Stephens, 1826)
36.	Great cormorant	*Phalacrocorax carbo* (Linnaeus, 1758)
37.	Little egret	*Egretta garzetta* (Linnaeus, 1766)
38.	Western reef-egret	*Egretta gularis* (Bosc, 1792)
39.	Cattle egret	*Bubulcus ibis* (Linnaeus, 1758)
40.	Indian pond heron	*Ardeola grayii* (Sykes, 1832)
41.	Chinese pond heron	*Ardeola bacchus* (Bomaparte, 1855)
42.	Great egret	*Ardea alba* (Linnaeus, 1758)
43.	Intermediate egret	*Ardea intermedia* (Walger, 1829)
44.	Gray heron	*Ardea cinerea* (Linnaeus, 1758)
45.	Goliath heron	*Ardea goliath* (Cretzschmar, 1827)
46.	Purple heron	*Ardea purpurea* (Linnaeus, 1766)
47.	Little heron/striated heron	*Butorides striata* (Linnaeus, 1758)
48.	Black-crowned night heron	*Nycticorax nycticorax* (Linnaeus, 1758)
49.	Yellow bittern	*Ixobrychus sinensis* (Gmelin, 1789)
50.	Cinnamon bittern	*Ixobrychus cinnamomeus* (Gmelin, 1789)
51.	Black bittern	*Ixobrychus flavicollis* (Latham, 1790)
52.	Glossy ibis	*Plegadis falcinellus* (Linnaeus, 1766)
53.	Black-headed ibis	*Threskiornis melanocephalus* (Latham, 1790)
54.	Red-naped ibis	*Pseudibis papillosa* (Temminck, 1824)
55.	Eurasian spoonbill	*Platalea leucorodia* (Linnaeus, 1758)
56.	Painted stork	*Mycteria leucocephala* (Pennant, 1769)
57.	Asian openbill	*Anastomus oscitans* (Boddaert, 1783)
58.	Woolly-necked stork	*Ciconia episcopus* (Boddaert, 1783)
59.	Black-necked stork	*Ephippiorhynchus asiaticus* (Latham, 1790)
60.	Greater adjutant	*Leptoptilos dubius* (Gmelin. 1789)
61.	Lesser adjutant	*Leptoptilos javanicus* (Horsfield, 1821)
62.	Great frigatebird	*Fregata minor* (Gmelin, 1789)
63.	Lesser frigatebird	*Fregata ariel* (G. R. Gray, 1845)
64.	Wilson's storm-petrel	*Oceanites oceanicus* (Kuhl, 1820)
65.	Short-tailed shearwater	*Ardenna tenuirostris* (Temminck, 1836)
66.	Jouanin's petrel	*Bulweria fallax* (Jouanin, 1955)
67.	Masked finroot	*Heliopais personatus* (G. R. Gray, 1849)
68.	Slaty-legged crake	*Rallina eurizonoides* (Lafresnaye, 1845)
69.	Slaty-breasted rail	*Gallirallus striata* (Linnaeus, 1766)
70.	Water rail	*Rallus aquaticus* (Linnaeus, 1758)

Table 4.17 (continued)

Sl No.	Name	Sc. name
71.	White-breasted waterheron	*Amaurornis phoenicurus* (Pennant, 1769)
72.	Baillon's crake	*Zapornia pusilla* (Pallas, 1776)
73.	Ruddy-breasted crake	*Zapornia fusca* (Linnaeus, 1766)
74.	Watercock	*Gallicrex cinerea* (Gmelin, 1789)
75.	Gray-headed swamphen	*Porphyrio poliocephalus* (Latham, 1801)
76.	Common moorhen	*Gallinula chloropus* (Linnaeus, 1758)
77.	Common coot	*Fulica atra* (Linnaeus, 1758)
78.	Greater painted snipe	*Rostratula benghalensis* (Linnaeus, 1758)
79.	Eurasian woodcock	*Scolopax rusticola* (Linnaeus, 1758)
80.	Pintail snipe	*Gallinago stenura* (Bonaparte, 1831)
81.	Wood snipe	*Gallinago nemoricola* (Hodgson, 1836)
82.	Swinhoe's snipe	*Gallinago megala* (Swinhoe, 1861)
83.	Common snipe	*Gallinago gallinago* (Linnaeus, 1758)
84.	Jack snipe	*Lymnocryptes minimus* (Brunnich, 1764)
85.	Common sandpiper	*Actitis hypoleucos* (Linnaeus, 1758)
86.	Marsh sandpiper	*Tringa stagnatilis* (Bechstein, 1803)
87.	Common greenshank	*Tringa nebularia* (Gunnerus, 1767)
88.	Green sandpiper	*Tringa ochropus* (Linnaeus, 1758)
89.	Wood sandpiper	*Tringa glareola* (Linnaeus, 1758)
90.	Gray-tailed tattler	*Tringa brevipes* (Vieillot, 1816)
91.	Spotted redshank	*Tringa erythropus* (Pallas, 1764)
92.	Common redshank	*Tringa totanus* (Linnaeus, 1758)
93.	Terek sandpiper	*Xenus cinereus* (Güldenstädt, 1775)
94.	Black-tailed godwit	*Limosa limosa* (Linnaeus, 1758)
95.	Bar-tailed godwit	*Limosa lapponica* (Linnaeus, 1758)
96.	Whimbrel	*Numenius phaeopus* (Linnaeus, 1758)
97.	Eurasian curlew	*Numenius arquata* (Linnaeus, 1758)
98.	Ruddy turnstone	*Arenaria interpres* (Linnaeus, 1758)
99.	Asian dowitcher	*Limnodromus semipalmatus* (Blyth, 1848)
100.	Curlew sandpiper	*Calidris ferruginea* (Pontoppidan, 1763)
101.	Spoon-billed sandpiper	*Calidris pygmaea* (Linnaeus, 1758)
102.	Broad-billed sadpiper	*Calidris falcinellus* (Pontoppidan, 1763)
103.	Great knot	*Calidris tenuirostris* (Horsfield, 1821)
104.	Red knot	*Calidris canutus* (Linnaeus, 1758)
105.	Sanderling	*Calidris alba* (Pallas, 1764)
106.	Little stint	*Calidris minuta* (Leisler, 1812)
107.	Red-necked stint	*Calidris ruficollis* (Pallas, 1776)
108.	Temminck's stint	*Calidris temminckii* (Leisler, 1812)
109.	Long-toed stint	*Calidris subminuta* (Middendorff, 1853)
110.	Dunlin	*Calidris alpina* (Linnaeus, 1758)
111.	Ruff	*Calidris pugnax* (Linnaeus, 1758)
112.	Red phalarope	*Phalaropus fulicarius* (Linnaeus, 1758)
113.	Pheasant-tailed jacana	*Hydrophasianus chirurgus* (Scopoli, 1786)

(continued)

Table 4.17 (continued)

Sl No.	Name	Sc. name
114.	Bronze-winged jacana	*Metopidius indicus* (Latham, 1790)
115.	Indian thicknee	*Burhinus indicus* (Salvadori, 1865)
116.	Great thick-knee	*Esacus recurvirostris* (Cuvier, 1829)
117.	Eurasian oystercatcher	*Haematopus ostralegus* (Linnaeus, 1758)
118.	Black-winged stilt	*Himantopus Himantopus* (Linnaeus, 1758)
119.	Pled avocet	*Recurvirostra avosetta* (Linnaeus, 1758)
120.	Red-wattled lapwing	*Vanellus indicus* (Boddaert, 1783)
121.	Yellow-wattled lapwing	*Vanellus malabaricus* (Boddaert, 1783)
122.	White-tailed lapwing	*Vanellus leucurus* (Lichtenstein, 1823)
123.	River lapwing	*Vanellus duvaucelii* (Lesson, 1826)
124.	Gray-headed lapwing	*Vanellus cinereus* (Blyth, 1842)
125.	Pacific golden plover	*Pluvialis fulva* (Gmelin, 1789)
126.	Gray plover	*Pluvialis squatarola* (Linnaeus, 1758)
127.	Common ringed plover	*Charadrius hiaticula* (Linnaeus, 1758)
128.	Little ringed plover	*Charadrius dubius* (Scopoli, 1786)
129.	Kentish plover	*Charadrius alexandrines* (Linnaeus, 1758)
130.	Lesser sand plover	*Charadrius mongolus* (Pallas, 1776)
131.	Greater sand plover	*Charadrius leschenaultia* (Lesson, 1826)
132.	Oriental pratincole	*Glareola maldivarum* (Forster, 1795)
133.	Small pratincole	*Glareola lactea* (Temminck, 1820)
134.	Lesser black-backed Gull	*Larus fuscus* (Linnaeus, 1758)
135.	Pallas's gull	*Ichthyaetus ichthyaetus* (Pallas, 1773)
136.	Brown-headed gull	*Chroicocephalus brunnicephalus* (Jerdon, 1840)
137.	Black-headed gull	*Chroicocephalus ridibundus* (Linnaeus, 1776)
138.	Slender-billed gull	*Choicocephalus genei* (Breme, 1839)
139.	Black-legged kittiwake	*Rissa tridactyla* (Linnaeus, 1758)
140.	Gull-billed tern	*Gelochelidon nilotica* (Gmelin, 1789)
141.	Caspian tern	*Hydroprogne caspia* (Pallas, 1770)
142.	River tern	*Sterna aurantia* (J. E. Gray, 1831)
143.	Common tern	*Sterna hirundo* (Linnaeus, 1758)
144.	Lesser crested tern	*Thalasseus bengalensis* (Lesson, 1831)
145.	Great crested tern	*Thalasseus bergii* (Lichtenstein, 1823)
146.	Little tern	*Sternula albifrons* (Pallas, 1764)
147.	Whiskered tern	*Chlidonias hybrida* (Pallas, 1811)
148.	White-winged tern	*Chlidonias leucopterus* (Temminck, 1815)
149.	Black noddy	*Anous minutus* (bole, F., 1844)
150.	Indian skimmer	*Rynchops albicollis* (Swainson, 1838)
151.	Arctic skua	*Stercorarius parasisticus* (Linnaeus, 1758)
152.	Common kingfisher	*Alcedo atthis* (Linnaeus, 1758)
153.	Blue-eared kingfisher	*Alcedo meninting* (Horsfield, 1821)
154.	Oriental dwarf kingfisher	*Ceyx erithaca* (Linnaeus, 1758)
155.	Brown-winged kingfisher	*Pelargopsis amauroptera* (Pearson, 1841)
156.	Stork-billed kingfisher	*Pelargopsis capensis* (Linnaeus, 1766)

Table 4.17 (continued)

Sl No.	Name	Sc. name
157.	Ruddy kingfisher	*Halcyon coromanda* (Latham, 1790)
158.	White-throated kingfisher	*Halcyon smyrnensis* (Linnaeus, 1758)
159.	Black-capped kingfisher	*Halcyon pileate* (Boddaert, 1783)
160.	Collared kingfisher	*Todiramphus chloris* (Boddaert, 1783)
161.	Pied kingfisher	*Ceryle rudis* (Linnaeus, 1758)
162.	Indian roller	*Coracias benghalensis* (Linnaeus, 1758)
163.	Dollarbird	*Eurystomus orientalis* (Linnaeus, 1766)
164.	Green bee-eater	*Merops orientalis* (Latham, 1801)
165.	Blue-tailed bee-eater	*Merops philippinus* (Linnaeus, 1767)
166.	Chestnut-headed bee-eater	*Merops leschenaulti* (Vieillot, 1817)
167.	Rufous woodpecker	*Micropternus brachyurus* (Vieillot, 1818)
168.	Brown-capped pygmy woodpecker	*Dendrocopos nanus* (Vigors, 1832)
169.	Gray-capped pygmy woodpecker	*Dendrocopos canicapillus* (Blyth, 1845)
170.	Fulvous-breasted woodpecker	*Dendrocopos macei* (Vieillot, 1818)
171.	Yellow-crowned woodpecker	*Dendrocopos mahrattensis* (Latham, 1801)
172.	Lesser yellownape	*Picus chlorolophus* (Vieillot, 1818)
173.	Greater yellownape	*Chrysophlegma flavimucha* (Gould, 1834)
174.	Streak-throated woodpecker	*Picus xanthopygaeus* (Gray & Gray, 1846)
175.	Gray-headed woodpecker	*Picus canus* (Gmelin, 1788)
176.	Common flameback woodpecker	*Dinopium javanense* (Ljungh, 1797)
177.	Lesser flameback woodpecker	*Dinopium benghalense* (Linnaeus, 1758)
178.	Greater flameback woodpecker	*Chrysocolaptes guttacristatus* (Tickell, 1833)
179.	White-naped woodpecker	*Chrysocolaptes festivus* (Boddaert, 1783)
180.	Eurasian wryneck	*Jynx torquilla* (Linnaeus, 1758)
181.	Speckled piculet	*Picumnus innominatus* (Burton, 1836)
182.	Brown-headed barbet	*Psilopogon zeylanicus* (Gmelin, 1788)
183.	Lineated barbet	*Psilopogon lineatus* (Vieillot, 1816)
184.	Blue-throated barbet	*Psilopogon asiaticus* (Latham, 1790)
185.	Coppersmith barbet	*Psilopogon haemacephalus* (Muller, 1776)
186.	Common hoopoe	*Upupa epops* (Linnaeus, 1758)
187.	Pied cuckoo	*Clamator jacobinus* (Boddaert, 1783)
188.	Chestnut-winged cuckoo	*Clamator coromandus* (Linnaeus, 1766)
189.	Common hawk cuckoo	*Hierococcyx varius* (Vahl, 1797)
190.	Indian cuckoo	*Cuculus micropterus* (Gould, 1838)
191.	Eurasian/common cuckoo	*Cuculus canorus* (Linnaeus, 1758)
192.	Oriental/Himalayan cuckoo	*Cuculus saturatus* (Blyth, 1843)
193.	Lesser cuckoo	*Cuculus poliocephalus* (Latham, 1790)
194.	Gray-bellied cuckoo	*Cacomantis passerines* (Vahl, 1797)
195.	Plaintive cuckoo	*Cacomantis merulinus* (Scopoli, 1786)
196.	Asian koel	*Eudynamys scolopaceus* (Linnaeus, 1758)
197.	Green-billed malkoha	*Phaenicophaeus tristis* (Lesson, 1830)
198.	Greater coucal	*Centropus sinensis* (Stephens, 1815)

(continued)

Table 4.17 (continued)

Sl No.	Name	Sc. name
199.	Lesser coucal	*Centropus bengalensis* (Gmelin, 1788)
200.	Rose-ringed parakeet	*Psittacula krameri* (Scopoli, 1769)
201.	Alexandrine parakeet	*Psittacula eupatria* (Linnaeus, 1766)
202.	Plum-headed parakeet	*Psittacula cyanocephala* (Linnaeus, 1766)
203.	Red-breasted parakeet	*Psittacula alexandri* (Linnaeus, 1758)
204.	Asian palm swift	*Cypsiurus balasiensis* (J. E. Gray, 1829)
205.	House swift/little swift	*Apus affinis* (J. E. Gray, 1830)
206.	Fork-tailed swift	*Apus pacificus* (Latham, 1801)
207.	Large-tailed nightjar	*Caprimulgus macrurus* (Horsfield, 1821)
208.	Indian nightjar	*Caprimulgus asiaticus* (Latham, 1790)
209.	Savana nightjar	*Caprimulgus affinis* (Horsfield, 1821)
210.	Jungle nightjar	*Caprimulgus indicus* (Latham, 1790)
211.	Barn owl	*Tyto alba* (Scopoli, 1769)
212.	Brown fish owl	*Ketupa zeylonensis* (Gmelin, 1788)
213.	Buffy fish owl	*Ketupa ketupa* (Horsfield, 1821)
214.	Brown hawk owl	*Ninox scutulata* (Raffles, 1822)
215.	Oriental scops owl	*Octus sunia* (Hodgson, 1836)
216.	Indian scops owl	*Octus bakkamoena* (Pennant, 1769)
217.	Spotted owlet	*Athene brama* (Temminck, 1821)
218.	Short-eared owl	*Asio flammeus* (Pontoppidan, 1763)
219.	Common pigeon	*Columbidae livia* (Gmelin, 1789)
220.	Spotted dove	*Streptopelia chinensis* (Scopoli, 1786)
221.	Red-collared dove	*Streptopelia tranquebarica* (Hermann, 1804)
222.	Eurasian collared dove	*Streptopelia decaocto* (Frivaldszky, 1838)
223.	Oriental turtle dove	*Streptopelia orientalis* (Latham, 1790)
224.	Laughing dove	*Streptopelia senegalensis* (Linnaeus, 1766)
225.	Emarald dove	*Chalcophaps indica* (Linnaeus, 1758)
226.	Pin-tailed green pigeon	*Treron apicauda* (Blyth, 1846)
227.	Orange-breasted green pigeon	*Treron bicinctus* (Jerdon, 1840)
228.	Thick-billed green pigeon	*Treron curvirostra* (Gmelin, 1789)
229.	Yellow-footed green pigeon	*Treron phoenicopterus* (Latham, 1790)
230.	Osprey	*Pandion haliaetus* (Linnaeus, 1758)
231.	Black-shouldered kite	*Elanus caeruleus* (Desfontaines, 1789)
232.	Black-kite	*Milvus migrans* (Boddaert, 1783)
233.	Brahminy kite	*Haliastur indus* (Boddaert, 1783)
234.	White-bellied sea eagle	*Haliaeetus leucogaster* (Gmelin, 1788)
235.	Pallas's fish eagle	*Haliaeetus leucoryphus* (Pallas, 1771)
236.	Gray-headed fish eagle	*Icthyophaga ichthyaetus* (Horsfield, 1821)
237.	Short-toed snake eagle	*Circaetus gallicus* (Gmelin, 1788)
238.	Crested serpent eagle	*Spilornis cheela* (Latham, 1790)
239.	Eurasian marsh harrier	*Circus aeruginosus* (Linnaeus, 1758)
240.	Hen harrier	*Circus cyaneus* (Linnaeus, 1766)
241.	Pallid harrier	*Circus macrourus* (Gmelin, 1770)

Table 4.17 (continued)

Sl No.	Name	Sc. name
242.	Pied harrier	*Circus melanoleucos* (Pennant, 1769)
243.	Crested goshawk	*Accipiter trivirgatus* (Temminck, 1824)
244.	Northern goshawk	*Accipiter gentillis* (Linnaeus, 1758)
245.	Shikra	*Accipiter badius* (Gmelin, 1788)
246.	Oriental honey-buzzard	*Pernis ptilorhynchus* (Temminck, 1821)
247.	White-eyed buzzard	*Butastur teesa* (Franklin, 1831)
248.	Indian spotted eagle	*Clanga hastata* (Lesson, 1831)
249.	Greater spotted eagle	*Clanga clanga* (Pallas, 1811)
250.	Bonelli's eagle	*Aquila fasciata* (Vieillot, 1822)
251.	Steppe eagle	*Aquila nipalensis* (Hodgson, 1833)
252.	Booted eagle	*Hieraaetus pennatus* (Gmelin, 1788)
253.	Changeable hawk eagle	*Nisaetus cirrhatus* (Gmelin, 1788)
254.	White-rumped vulture	*Gyps bengalensis* (Gmelin, 1788)
255.	Slender-billed vulture	*Gyps tenuirostris* (Gray, 1844)
256.	Common kestrel	*Falco tinnunculus* (Linnaeus, 1758)
257.	Eurasian hobby	*Falco Subbuteo* (Linnaeus, 1758)
258.	Peregrine falcon	*Falco peregrinus* (Tunstall, 1771)
259.	Amur falcon	*Falco amurensis* (Radde, 1863)
260.	Red-necked falcon	*Falco chicquera* (Daudin, 1800)
261.	Indian pitta	*Pitta brachyura* (Linnaeus, 1766)
262.	Mangrove pitta	*Pitta megarhyncha* (Schlegel, 1863)
263.	Ashy woodswallow	*Artamus fuscus* (Vieillot, 1817)
264.	Common woodshrike	*Tephrodornis pondicerianus* (Gmelin, 1789)
265.	Golden-fronted leafbird	*Chloropsis aurifrons* (Temminck, 1829)
266.	Blue-winged leafbird	*Chloropsis cochinehinensis* (Gmelin, 1789)
267.	Brown shrike	*Lanius cristatus* (Linnaeus, 1758)
268.	Bay-backed shrike	*Lanius vittatus* (Valenciennes, 1826)
269.	Long-tailed shrike	*Lanius Schach* (Linnaeus, 1758)
270.	Gray-backed shrike	*Lanius tephronotus* (Vigors, 1831)
271.	Southern gray shrike	*Lanius meridionalis* (Temminck, 1820)
272.	Mangrove whistler	*Pachycephala cinerea* (Blyth, 1847)
273.	Indian Golden Oriole	*Oriolus kundoo* (Sykes, 1832)
274.	Black-naped Oriole	*Oriolus chinensis* (Linnaeus, 1766)
275.	Black-hooded Oriole	*Oriolus xanthornus* (Linnaeus, 1758)
276.	Large Cuckooshrike	*Coracina macei* (Lesson, 1831)
277.	Black-winged Cuckooshrike	*Lalage melaschistos* (Hodgson, 1836)
278.	Black-headed Cuckooshrike	*Lalage melanoptera* (Rupell, 1839)
279.	Rosy minivet	*Pericrocotus roseus* (Vieillot, 1818)
280.	Small minivet	*Pericrocotus cinnamomeus* (Linnaeus, 1766)
281.	Scarlet minivet	*Pericrocotus flammeus* (Forster, 1781)
282.	Bar-winged flycatcher-shrike	*Hemipus picatus* (Sykes, 1832)
283.	White-throated fantail	*Rhipidura albicollis* (Vieillot, 1818)

(continued)

Table 4.17 (continued)

Sl No.	Name	Sc. name
284.	Black drongo	*Dicrurus macrocercus* (Vieillot, 1817)
285.	Ashy drongo	*Dicrurus leucophaeus* (Vieillot, 1817)
286.	White-bellied drongo	*Dicrurus caerulescens* (Linnaeus, 1758)
287.	Bronzed drongo	*Dicrurus aeneus* (Vieillot, 1817)
288.	Hair-crested/spangled Drongo	*Dicrurus hottentottus* (Linnaeus, 1766)
289.	Greater racket-tailed Drongo	*Dicrurus paradiseus* (Linnaeus, 1766)
290.	Lesser racket-tailed drongo	*Dicrurus remifer* (Temminck, 1823)
291.	Rufous treepie	*Dendrocitta vagabunda* (Latham, 1790)
292.	House crow	*Corvus splendens* (Vieillot, 1817)
293.	Indian jungle crow	*Corvus macrorhynchos* (Wagler, 1827)
294.	Black-naped monarck	*Hypothymis azurea* (Boddaert, 1783)
295.	Asian paradise-flycatcher	*Terpsiphone paradisi* (Linnaeus, 1758)
296.	Common lora	*Aegithina tiphia* (Linnaeus, 1758)
297.	Orange-headed thrush	*Geokichla citrina* (Latham, 1790)
298.	Scaly thrush	*Zoothera dauma* (Latham, 1790)
299.	Eyebrowed thrush	*Trdus obscures* (Gmelin, 1789)
300.	Black-breasted thrush	*Turdus dissimilies* (Blyth, 1847)
301.	Black-throated thrush	*Turdus atrogularis* (Jarocki, 1819)
302.	Tickell's thrush	*Turdus unicolour* (Tickell, 1833)
303.	Siberian thrush	*Geokichla sibirica* (Pallas, 1776)
304.	Blue rock thrush	*Monticola solitarius* (Linnaeus, 1758)
305.	Blue-capped rock thrush	*Monticola cinclorhyncha* (Vigors, 1831)
306.	Red-throated flycatcher/ taiga flycatcher	*Ficedula albicilla* (Pallas, 1811)
307.	Red-breasted flycatcher	*Ficedula parva* (Bechstein, 1792)
308.	Little pied flycatcher	*Ficedula westermanni* (Sharpe, 1888)
309.	Verditer flycatcher	*Eumyias thalassinus* (Swainson, 1838)
310.	Paleblue flycatcher	*Cyornis unicolour* (Blyth, 1843)
311.	Blue-throated flycatcher	*Cyornis rubeculoides* (Vigors, 1831)
312.	Tickell's blue flycatcher	*Cyornis tickelliae* (Blyth, 1843)
313.	Gray-headed canary flycatcher	*Culicicapa ceylonensis* (Swainson, 1820)
314.	Asian brown flycatcher	*Muscicapa dauurica* (Pallas, 1811)
315.	Dark-sided flycatcher	*Muscicapa sibirica* (Gmelin, 1789)
316.	Spotted flycatcher	*Muscicapa striata* (Pallas, 1764)
317.	Siberian rubythroat	*Calliope calliope* (Pallas, 1776)
318.	Bluethroat	*Luscinia svecica* (Linnaeus, 1758)
319.	Indian blue robin	*Luscinia brunnea* (Hodgson, 1837)
320.	Golden bush robin	*Rarsiger chrysaeus* (Hodgson, 1845)
321.	Firethroat	*Calliope pectardens* (David, 1877)
322.	Oriental magpie robin	*Copsychus saularis* (Linnaeus, 1758)
323.	White-rumped Shama	*Copsychus malabaricus* (Linnaeus, 1758)
324.	Indian Robin	*Saxicoloides fulicatus* (Linnaeus, 1766)
325.	White-tailed blue robin	*Myiomela leucura* (Hodgson, 1845)

Table 4.17 (continued)

Sl No.	Name	Sc. name
326.	Black redstart	*Phoenicurus ochruros* (S. C. Gmelin, 1774)
327.	Siberian stonechat	*Saxicola maurus* (Pallas, 1773)
328.	White-tailed stonechat	*Saxicola leucurus* (Blyth, 1847)
329.	Pied bushchat	*Saxicola caprata* (Linnaeus, 1766)
330.	Chestnut-tailed starling	*Sturnus malabarica* (Gmelin, 1789)
331.	Brahminy starling	*Sturnus pagodarum* (Gmelin, 1789)
332.	Common starling	*Sturnus vulgaris* (Linnaeus, 1758)
333.	Asian pied starling	*Gracupica contra* (Linnaeus, 1758)
334.	Rosy starling	*Pastor roseus* (Linnaeus, 1758)
335.	Common hill myna	*Gracula religiosa* (Linnaeus, 1758)
336.	Common myna	*Acridotheres tristis* (Linnaeus, 1766)
337.	Bank myna	*Acridotheres ginginianus* (Latham, 1790)
338.	Jungle myna	*Acridotheres fuscus* (Wagler, 1827)
339.	Indian nuthatch	*Sitta castanea* (Lesson, 1830)
340.	Velvet-fronted nuthatch	*Sitta frontalis* (Swainson, 1820)
341.	Cinereous tit	*Parus cinereus* (Vieillot, 1818)
342.	Sand Martin	*Riparia riparia* (Linnaeus, 1758)
343.	Barn swallow	*Hirundo rustica* (Linnaeus, 1758)
344.	Wire-tailed swallow	*Hirundo smithii* (Leach, 1818)
345.	Streak-throated swallow	*Petrochelidon fluvicola* (Blyth, 1855)
346.	Red-rumped swallow	*Cecropsis daurica* (Laxmann, 1769)
347.	Red-whiskered bulbul	*Pycnonotus jocosus* (Linnaeus, 1758)
348.	Red-vented bulbul	*Pycnonotus cafer* (Linnaeus, 1766)
349.	Zitting cisticola	*Cisticola juncidis* (Rafinesque, 1810)
350.	Gray-breasted prinia	*Prinia hodgsonii* (Blyth, 1844)
351.	Yellow-bellied prinia	*Prinia flaviventris* (Delessert, 1840)
352.	Ashy prinia	*Prinia socialis* (Sykes, 1832)
353.	Plain prinia	*Prinia inornata* (Sykes, 1832)
354.	Common tailorbird	*Orthotomus sutorius* (Pennant, 1769)
355.	Oriental White-eye	*Zosteropes palpebrosus* (Temminck, 1824)
356.	Blyth's reed warbler	*Acrocephalus dumetorum* (Blyth, 1849)
357.	Blunt-winged warbler	*Acrocephalus concinens* (Swinhoe, 1870)
358.	Black-browned reed warbler	*Acrocephalus bistrigiceps* (Swainhoe, 1860)
359.	Large-billed reed warbler	*Acrocephalus orinus* (Oberholser, 1905)
360.	Clamorous reed warbler	*Acrocephalus stentoreus* (Hemprich &Ehrenberg, 1833)
361.	Thick-billed warbler	*Arundimax aedon* (Pallas, 1776)
362.	Paddyfield warbler	*Acrocephalus agricola* (Jerdon, 1845)
363.	Booted warbler	*Iduna caligata* (Lichtenstein, 1823)
364.	Sykes's warbler	*Iduna rama* (Sykes, 1832)
365.	Common chiffchaff	*Phylloscopus collybita* (Vieillot, 1817)
366.	Dusky warbler	*Phylloscopus fuscatus* (Blyth, 1842)

(continued)

Table 4.17 (continued)

Sl No.	Name	Sc. name
367.	Tickell's leaf warbler	*Phylloscopus affinis* (Tickell, 1833)
368.	Lemmon-rumped warbler	*Phylloscopus chloronotus* (Gray &Gray, 1847)
369.	Yellow-browed warbler	*Phylloscopus inornatus* (Blyth, 1842)
370.	Hume's leaf warbler	*Phylloscopus humei* (Brooks, 1878)
371.	Greenish warbler	*Phylloscopus trochiloides* (Sundevall, 1837)
372.	Large-billed leaf warbler	*Phylloscopus magnirostris* (Blyth, 1843)
373.	Blyth's leaf warbler	*Phylloscopus reguloides* (Blyth, 1842)
374.	Smoky warbler	*Phylloscopus fuligiventer* (Hodgson, 1845)
375.	Golden-spectacled/green-crown warbler	*Seicercus burkii* (Burton, 1836)
376.	Rusty-rumped warbler	*Locustella certhiola* (Pallas, 1811)
377.	Striated grassbird	*Megalurus palustris* (Horsfield, 1821)
378.	White-browed scimitar babbler	*Pomatorhinus schisticeps* (Hodgson, 1836)
379.	Striped tit-babbler	*Mixornis gularis* (Horsfield, 1822)
380.	Chestnut-capped babbler	*Timalia pileata* (Horsfield, 1821)
381.	Puff-throated babbler	*Pellorneum ruficeps* (Swainson, 1832)
382.	Yellow-eyed babbler	*Chrysomma sinense* (Gmelin, 1789)
383.	Striated babbler	*Argya earlei* (Blyth, 1844)
384.	Jungle babbler	*Argya striata* (Dumont, 1823)
385.	Bengal bushlark	*Mirafra assamica* (Horsfield, 1840)
386.	Ashy-crowned sparrow lark	*Eremopterix griseus* (Scopoli, 1786)
387.	Oriental skylark	*Alauda gulgula* (Franklin, 1831)
388.	Greater short-toed lark	*Calandrella brachydactyla* (Leisler, 1814)
389.	Sykes's short-toed lark	*Calandrella dukhunensis* (Sykes, 1832)
390.	Sand lark	*Alaudala raytal* (Blyth, 1845)
391.	Thick-billed flowerpecker	*Dicaeum agile* (Tickell, 1833)
392.	Orange-bellied flowerpecker	*Dicaeum trigonostigma* (Scopoli, 1786)
393.	Pale-billed flowerpecker	*Dicaeum erythrorhynchos* (Latham, 1790)
394.	Scarlet-backed flowerpecker	*Dicaeum cruentatum* (Linnaeus, 1758)
395.	Purple-rumped sunbird	*Nectarinia zeylonica* (Linnaeus, 1766)
396.	Purple sunbird	*Cinnyris asiatica* (Latham, 1790)
397.	Loten's sunbird	*Cinnyris lotenius* (Linnaeus, 1766)
398.	Crimson sunbird	*Aethopyga siparaja* (Raffles, 1822)
399.	Little spiderhunter	*Arachnothera longirostra* (Latham, 1790)
400.	House sparrow	*Passer domesticus* (Linnaeus, 1758)
401.	Forest wagtall	*Dendronanthus indicus* (Gmelin, 1789)
402.	White wagtail	*Motacilla alba* (Linnaeus, 1758)
403.	Citrine wagtail	*Motacilla citreola* (Pallas, 1776)
404.	Western yellow wagtail	*Motacilla flava* (Linnaeus, 1758)
405.	Gray wagtail	*Motacilla cinerea* (Tunstall, 1771)
406.	Richard's pipit	*Anthus richardi* (Vieillot, 1818)
407.	Paddyfield pipit	*Anthus rufulus* (Vieillot, 1818)

Table 4.17 (continued)

Sl No.	Name	Sc. name
408.	Tawny pipit	*Anthus campestris* (Linnaeus, 1758)
409.	Tree pipit	*Anthus trivialis* (Linnaeus, 1758)
410.	Olive-backed pipit	*Anthus hodgsoni* (Richmond, 1907)
411.	Rosy pipit	*Anthus roseatus* (Blyth, 1847)
412.	Baya weaver	*Ploceus philippinus* (Linnaeus, 1766)
413.	Black-breasted weaver	*Ploceus benghalensis* (Linnaeus, 1758)
414.	Streaked weaver	*Ploceus manyar* (Horsfield, 1821)
415.	Finn's weaver	*Ploceus megarhynchus* (Hume, 1869)
416.	Red avadavat	*Amandava amandava* (Linnaeus, 1758)
417.	Indian silverbill	*Euodice malabarica* (Linnaeus, 1758)
418.	Scaly-breasted munia	*Lonchura punctulata* (Linaeus, 1758)
419.	Black-headed munia	*Lonchura malacca* (Linnaeus, 1766)
420.	White-rumped munia	*Lonchura striata* (Linnaeus, 1766)
421.	Common rosefinch	*Carpodacus erythrinus* (Pallas, 1770)
422.	Chestnut-eared bunting	*Emberiza fucata* (Pallas, 1776)
423.	Yellow-browed bunting	*Emberiza Chrysophrys* (Pallas, 1776)
424.	Gray-necked bunting	*Emberiza buchanani* (Blyth, 1845)
425.	Ortolan bunting	*Emberiza hortulana* (Linnaeus, 1758)
426.	Little bunting	*Emberiza pusilla* (Pallas, 1776)
427.	Black-headed bunting	*Granativora melanocephala* (Scopoli, 1769)
428.	Crested bunting	*Melophus lathami* (Gray, 1831)

Table 4.18 Mammalian fauna of the Indian Sundarbans, India

No	Species	Common Name	Source
	Order CHIROPTERA		
	Family Pteropodidae		
1	*Rousettus leschenaultia* (Desmarest, 1820)	Fulvous fruit bats	Mallick (2011)
2	*Pteropus giganteus* (Brunnich, 1782)	Indian flying fox	Mandal and Nandi (1989), Mallick (2011)
3	*Cynopterus sphinx* (Vahl, 1797)	Greater short nosed fruit bat	Mandal and Nandi (1989), Mallick (2011)
	Family Rhinopomatidae		
4	*Rhinopomah ardwickii* (Gray, 1831)	Lesser mouse-tailed bat	Mandal and Nandi (1989), Mallick (2011)
	Family Emballonuridae		
5	*Taphozous longimanus* (Hardwickii, 1825)	Longed0winged tomb bat	Mandal and Nandi (1989), Mallick (2011)
	Family Vespertilionidae		
6	*Scotophilus kuhlii* (Leach, 1821)	Lesser Asiatic yellow house bat	Mandal and Nandi (1989), Mallick (2011)

Table 4.18 (continued)

No	Species	Common Name	Source
7	*Scotophilu sheathi* (Horsfield, 1831)	Greater Asiatic yellow house bat	Mandal and Nandi (1989), Mallick (2011)
8	*Pipistrellu smimus* (Wroughton, 1899)	Indian pygmy pipistrelle	Mandal and Nandi (1989)
	Family Megadermatidae		
9	*Megadermas pasma* (Linnaeus, 1758)	Lesser false vampire	Mandal and Nandi (1989), Mallick (2011)
10	*Megaderma lyra* (Geoffroy Saint-Hilarie, 1810)	Greater false vampire	Mandal and Nandi (1989), Mallick (2011)
	Family Rhinolophidae		
11	*Hipposider ospomona* (Anderson, 1918)	Anderson's leaf-nosed bat	Mallick (2011)
12	*Rhinolophuslepidus* (Blyth, 1844)	Blyth's horse shoe bat	Mandal and Nandi (1989), Mallick (2011)
13	*Coelops frithi* (Blyth, 1848)	Tailless leaf-nosed bat	Mallick (2011)
14	*Hipposiderosbicolor* (Temminck, 1834)	Bicolored round-leaf bat	Mandal and Nandi (1989), Mallick (2011)
15	*Hipposideros lankadiva* (Kelaart, 1850)	Indian leaf-nosed bat	Mandal and Nandi (1989), Mallick (2011)
	Order **SORICOMORPHA**		
	Family Soricidae	Shrews	
16	*Suncus murinus* (Linnaeus, 1766)	House shrew	Mandal and Nandi (1989), Mallick (2011)
	Order **PRIMATE**		
	Family Ceropithecidae		
17	*Macaca mulatta* (Zimmermann, 1780)	Rhesus macaque	Mandal and Nandi (1989), Mallick (2011)
	Order **LAGOMORPHA**		
	Family Leporidae		
18	*Lepus nigricollis* (Cuvier, 1823)	Indian rufous-tailed or black-naped hare	Mandal and Nandi (1989), Mallick (2011)
	Order **CARNIVORA**		
	Family Canidae		
19	*Vulpes bengalensis* (Shaw, 1800)	Indian fox	Mallick (2011)
20	*Canisaureus* (Linnaeus, 1758)	Jackal	Mallick (2011)
	Family Felidae		
21	*Felis chaus*	Jungle cat	Mandal and Nandi (1989), Mallick (2011)
22	*Prionailurus bengalensis* (Kerr, 1792)	Leopard cat	Mandal and Nandi (1989), Mallick (2011)
23	*Prionailurus viverrinus* (Bennett, 1833)	Fishing cat	Mandal and Nandi (1989), Mallick (2011)

Table 4.18 (continued)

No	Species	Common Name	Source
24	*Panther tigris* (Linnaeus, 1758)	Royal Bengal Tiger	Mandal and Nandi (1989)
	Family Herpestidae		
25	*Herpestes edwardsii* (Saint-Hilaire, 1818)	Gray mongoose	Mandal and Nandi (1989), Mallick (2011)
26	*Herpestes palustris* (Ghose, 1965)	Marsh mongoose	Mandal and Nandi (1989)
27	*Herpestesjavanicus* (Saint-Hilaire, 1818)	Small Asian mongoose	Mallick (2011)
	Family Mustelidae		
28	*Lutra lutra* (Linnaeus, 1758)	Common otter	Mandal and Nandi (1989)
29	*Lutrogalepers picillata* (Saint-Hilaire, 1826)	Smooth-coated otter	Mandal and Nandi (1989), Manjrekar and Prabu (2014)
30	*Aonyx cinereus* (Illeger, 1815)	Asian small-clawed otter	Mandal and Nandi (1989), Mallick (2011)
	Family Viverridae		
31	*Viver razibetha* (Linnaeus, 1758)	Large Indian civet	Mallick (2011)
32	*Viver riculaindica* (Saint-Hilaire, 1803)	Small Indian civet	Mandal and Nandi (1989), Mallick (2011)
33	*Paradoxurus hermaphrodites* (Pallas, 1777)	Common palm civet	Mandal and Nandi (1989), Mallick (2011)
	Order **CETACEA**		
	Family Platinistidae		
34	*Platanista gangetica* (Roxburgh, 1801)	Gangetic dolphin	Mandal and Nandi (1989), Mallick (2011)
	Family Phocoenidae		
35	*Neophoca enaphocaenoides* (Cuvier, 1829)	Little porpoise/ black finless porpoise	Mallick (2011)
	Family Delphinidae		
36	*Stenella attenuate* (Gray, 1846)	Pantropical spotted dolphin	Mallick (2011)
37	*Orcaella brevirostris* (Owen, 1866)	Irrawady dolphin	Mandal and Nandi (1989), Mallick (2011)
38	*Stanellamalayana* (Lesson, 1826)	Malay dolphin	Mandal and Nandi (1989)
39	*Sotaliap lumbea* (Cuvier)	Plumbeus dolphin	Mandal and Nandi (1989)
40	*Sousa chinensis* (Osbeck, 1765)	Indo–Pacific hump-backed dolphin	Mallick (2011)
	Order **ARDIODACTYLA**		

(continued)

Table 4.18 (continued)

No	Species	Common Name	Source
	Family Suidae		
41	*Sus scrofa* (Linnaeus, 1758)	Wild boar	Mandal and Nandi (1989), Mallick (2011)
	Family Cervidae		
42	*Axis axis* (Erxleben, 1777)	Cheetal or spotted deer	Mallick (2011)
	Order **PHOLIDOTA**		
	Family Manidae		
43	*Manis pentadactyla* (Linnaeus, 1758)	Chinese pangolin	Mandal and Nandi (1989), Mallick (2011)
	Order **RODENTIA**		
	Family Muridae		
44	*Bandicota indica* (Bechstein, 1800)	Greater bandicoot rat	Mandal and Nandi (1989), Mallick (2011)
45	*Mus booduga* (Gray, 1837)	Common Indian field mouse	Mandal and Nandi (1989), Mallick (2011)
46	*Bandicota bengalensis* (Gray& Hardwicke, 1833)	Lesser bandicoot rat	Mandal and Nandi (1989), Mallick (2011)
47	*Rattus norvegicus* (Berkenhout, 1769)	Norway rat	Mandal and Nandi (1989), Mallick (2011)
48	*Mus musculus* (Linnaeus, 1758)	House mouse	Mallick (2011)
49	*Rattusrattus* (Linnaeus, 1758)	House rat or black rat	Mandal and Nandi (1989), Mallick (2011)
	Family Hystricidae		
50	*Hystrix indica* (Kerr, 1792)	Indian crested porcupine	Mallick (2011)

4.4 Distribution and Extent of Mangroves Based Biodiversity Components

Mangrove forests, often known as *mangals* or *marine forests* are woody halophytes which grow mostly within intertidal belts of tropical, subtropical coastlines render significant roles for coastal protection, arrest soil erosion, resist natural disasters, promote biological enrichment and ensure socioeconomic development. Approximately 75% of the world's mangrove forests are found in just 15 tropical and subtropical countries. Asia accounts for the largest (≈42%), followed by African (≈20%), North and Central America (≈15%), Australia and the Pacific Islands (≈12%), and South America (≈11%).

In the deltaic Sundarbans mangrove ecosystem, out of a total area of 10 000 km^2, around 60% are found to in Bangladesh, whereas the remaining around 40% are in India. Sundarbans forms the largest continuous block of mangrove forest on Earth, developed along the major riverine networks of the Ganges, Brahmaputra, Meghna rivers, and their distributaries (including the Padma), which discharge into the Bay of Bengal.

4.5 Salt Marshes or Mangrove Swamp or Mangrove Ecosystem of Sundarbans, India

Both the mangrove forests and saltmarsh are brackish water wetlands, developed in the intertidal belts of sheltered coasts, and surrounded by either lagoon or estuaries. They differ in respect to the species composition of their respective biotic communities—the mangroves harbor woody halophytic plant species and their associated flora and fauna, whereas the salt marshes mainly contain herbs and grasses. Both landscapes experience the movement of tidal water through a network of tidal creeks carrying sediments loaded with organic matter. Vegetation in mangrove swamps and salt marshes obstructs and attenuates tidal flows, thereby reducing tidal current velocity and promoting the deposition of fine sediments. Sediment accretion creates stable habitats for the biotic community.

Salt marshes are periodically inundated saline wetlands dominated by herbaceous vegetation. They typically develop on relatively soft, muddy flat areas in the upper intertidal zone, generally occupying sheltered areas behind natural or artificial embankments designed to guarding inland fresh water bodies from saltwater intrusion. The sediments of the mangrove swamps and marsh do not exhibit much textural variation—the marsh sediments are composed of more than 90% silt with a small proportion of fine sand and clay.

The diverse forms of marsh vegetation play important ecological and economic roles both in nutrient cycling and as sediment baffles (Davis, 1983; Das & Bhattacharya, 1999b). In the Sundarbans mangrove ecosystem, India, the ground-layer mangrove vegetation includes creepers like *Suaedanudiflora*, *S. maritima*, *Sesuvium portulacastrum, Salicornia brachiata*, and *Salicornia scirpus. Hygrophyllux* spans grass species like in the *Porteresia coarctata* with a height of 30–60 cm which act as a pioneering species occurring in patches in the newly deposited mudflats. Out of the two species of *Suaeda* sp. and *Suaeda maritime* is found to survive even in the desiccated ground forming a stable matting whereas, *Suaeda mudiflora* is seen to grow profusely with greater density on unconsolidated loose sediments near embankments, attaining greater heights (60–75 cm). Several other types of vegetation in the marshy zones includes grasses like *Panicum* sp., *Paspalum* sp., *Aeluropus lagopoides*, *Sesuvium portulacastrum*, and the creeper, *Ipomoea pes-caprae*, sp. (Das & Bhattacharaya, 1999a; Table 4.1). A recent study (Chakravarti, 2008) recorded the distribution patterns of mangrove plant species within four ecozones with contrasting ecological patterns—study site I, called Jharkhali on Matla estuary; study site II, referred to as Canning on Matla estuary; study site III on Prentice and Luthian islands on the Saptamukhi estuary; and study site IV, known as Bokkhali on Hooghly estuary, which has shown differential distribution patterns based on prevailing ecological parameters and sustaining different environmental stress factors (Table 4.2).

4.5.1 Phenological Study of Mangroves in Sundarbans

Phenology refers to the study of the timing and duration of different events of life cycle of plants like leafing, senescence, flowering, and fruiting, in response to prevailing environmental parameters conditions. It is often evolutionarily adapted to maximize utilization of prevailing ecological conditions—abiotic conditions by way of plant fitness, community interaction, and ecosystem resilience. Phenology of mangrove plants depends on the oscillating tide-dominated ecological conditions of the interface between forest land and estuarine water, sustaining continuous threats from anthropogenic activities, and environmental consequences mediated by climate change.

In addition to recording the timing of four key phenophases—leaf flushing, senescence, flowering, and fruiting—additional quantitative parameters must be observed: duration, intensity, synchrony, frequency, skewness, and kurtosis. These metrics provide critical insight into the different traits that collectively shape species-specific strategies and ecosystem-level responses. Phenological variability across key phenophases plants to adapt to fluctuating environmental cues, thereby shaping life-history traits, pollinator and seed-disperser interactions, and resource-acquisition and utilization strategies of plant species in the fragile mangrove communities (Demaree & Rutishauser, 2011). In tropical mangrove ecosystems, temperature is not the main limiting parameter for plant phenology; instead, phenological events are mostly controlled by seasonal patterns of light and water availability (van Schaik et al., 1993). The timing of the flushing of some mangrove plant species in the tropical region exhibits a wide variety of vegetative and reproductive phenology, coinciding with the wet season (Coupland et al., 2005). Other mangrove plant species exhibit a dry-season flowering peak with fruiting in the wet season (Coupland et al., 2005; Kamruzzaman et al., 2017). However, several studies have pointed out higher interspecific variation instead of a community-level peak in vegetative and reproductive phenology, which demonstrated significant variability across different ecozones within a mangrove ecosystem (Duke et al., 1984; Nadia et al., 2012; De Alvarenga et al., 2017).

Recent studies undertaken in the Indian Sundarbans evaluated mangrove the phenological parameters focusing on relationships between duration, frequency, synchrony, and intraspecific variation (Chakraborty et al., 2023; Dey, 2009; Maiti Dutta et al., 2014; Datta & Chakraborty, 2017).

Most mangrove plant species in Sundarbans in India start flushing during the late monsoon period (September and October) and stop flushing between November and March. Mangrove plant species like *Sonneratia apetala* exhibit semiannual flushing activity. Most mangrove plant species, except *Bruguiera gymnorhiza,* start flowering in the pre-monsoon season (April to June), and continue until September. Three other species—*Ceriops decandra, Bruguiera cylindrica*, and *Bruguiera gymnorhiza*—showed flowering throughout the year. The flowering of mangrove plant species was followed by a fruiting phase, which occurs from May to September and disperses from October to January. Interestingly, mangrove plant species like *Rhizophora mucronata* and *Aegialitis rotundifolia* flower more than once a year, but their all-fruiting activity cycle occurred just once a year. Although *Aegialitis rotundifolia* had the highest synchrony, *Bruguiera gymnorhiza* exhibited the longest flowering duration. The duration of flushing was observed to be nearly the same among most of the species. Leaf senescence in mangrove species generally occurred at lower intensity but over longer durations than leaf flushing. Leaf shedding patterns varied markedly among species; for example, *Avicennia marina* shed leaves for about ten months of the year. In contrast, fruiting in *Bruguiera cylindrica* requires a lengthy duration coupled with highest degree of synchrony among individuals.

4.6 Functional Ecological Roles Marsh Sediment

The vegetation carpet within the marsh is formed by different creepers, herbs, and shrubs, along with the inhabiting benthic fauna including cnidarians, crabs, polychaetas, sipunculids, gastropods, and bivalves. These play important ecological roles like nutrient recycling of marsh sediments, as well as providing a baffling effect on the flood plains, protecting the freshwater-supplying river basin from waves and currents. Salt marshes serve as nursery grounds attracting both shellfish and finfish in the highly nutrient-rich water. Suspended particles in the form of fine sands, silt and clay along with different chemical elements including trace metals settle gradually and are trapped by marsh vegetation (Das & Bhattacharya, 1999a; Costa & Davy, 1992).

Ecological function of sediments of mangrove floors of Sundarbans are loaded with diversified forms bacteria and fungi which through the production of enzymes mostly for the decomposition of mangrove plant litter into detritus which enhance the fertility of both mangrove forest soil and estuarine water.

Higher microbial counts, particularly of bacteria were recorded in the upper littoral zones in undisturbed mangrove soils. They play great roles in the microbial degradation of mangrove litter. Cellulose, hemicellulose, cellubiose, pectin, lignin, chitin, lipid and protein are some of the essential litter substances which are exposed to microbial degradation (Biswas et al., 1986). A number of bacteria and fungi have been isolated from a wide variety of substrates within the Sundarbans mangrove ecosystem (Bhaumik et al., 1985). These include 48 bacterial strains which have been isolated from decomposed litter. Species of bacteria identified include *Brevibacterium lypolyticum, B. marinopiscosum, B. minutiferula, B.* sociovivum, *B.* stationis, Clostridium carnis, *C.* pectinivorum, Kurthiabes sonii, *Lactobacillus brevis, L. delbrueckii, L. fermenti*, Listiriamon ocytogenes, *Micrococcus agilis* and M. varians. Of these, the genera Brevibacterium and Micrococcus were found to be abundant.

Bacreia identified in decomposed litter have also been found within the gut tissuesofdetritivorus benthic fauna that rely on the substrate for their food sources. Bacteria have been identifies from the haemocoelomic fluid of Boleophthalmus spp and the gut of Mystusgulio, Uca spp.

Most of identified bacteria belong to ammonifying groups and some belong to nitrifying groups. A few strains, as chitinolytic bacteria suggesting these bacteria assist in the breakdown of chitin found in a variety of marine and estuarine arthropod sources.

4.7 Biodiversity Excellence in the Networks of Canals in the Indian Sundarbans

Within India's tropical setting are richly diversified forms of landscapes and ecosystems, and many different forms of wetlands, rivers, canals, and dams, serving as prime reservoirs of inland water resources. The network of canals in the Indian Sundarbans is being used not only for irrigation and flood control purposes, but also for culturing of fish, depending on seasonal prevailing fluctuation in salinity, turbidity, and pH. Three study sites were selected in the Indian Sundarbans, an extremely fragile mangrove ecosystem in the world. The canal network in the Indian Sundarbans covers approximately 907.33 ha, with a huge potential for fisheries development. Some fish species establish themselves as natural populations, entering the canals naturally from their source waters.

Two major types of canals having their own ecological uniqueness are present in the Indian Sundarbans which noted inundation canals having higher water levels receiving tidal water and perennial canals connected with dams and water reservoirs for supplying irrigation water round the year. Canal systems, networks of different connecting creeks and canals, both tides fed natural and excavated ones for freshwater in the Indian Sundarbans represent an important source for livelihood and household nutrition for the local populace.

Canals not only offer the huge potential for fish production, but also provide irrigation to 17 million ha—26% of India's total irrigated area. The Indian Sundarbans canal systems, covering about 907.33 ha, offer significant promise for fish production as seen in Egypt, Sudan, Thailand, and many other countries (Das Sarkar et al., 2019; Anon, 2003). Variations in salinity, temperature, pH, water transparency, and nutrients levels (NO_2, NO_3, SIO_3, HCO_3, PO_4, Ca^{++}, and Mg^{++}) strongly influence phytoplankton diversity (Agnieszka) and benthic faunal biomass.

The occurrence of diversified forms of natural flora and fauna in the canal is determined by allochthonous inputs of different forms of chemical nutrients like biota and toxic substances washed out from the irrigated agricultural land as well as from forested landscapes within the catchment area. The prime source of autochthonous nutrients is supplied through the decomposition of aquatic macrophytes, the biomass of which has been observed to be high in canals. However, all these inputs are not readily available to the fish and mostly they are removed from the canal system during the artificial management of canals. Therefore, the trophic status along with hydraulic characteristics (flow rates, depth, and water quality parameters) of the canal habitats significantly differ from that of the natural waters. The canals have flowing water (lotic water), even during the dry season, and exhibit a much lower plankton growth as contrasted with lentic water bodies. Continuous water flow flushes nutrients downstream and leaves minimal resources for the phytoplankton to thrive.

To address the underutilization of canal ecosystem resources for sustainable fish production, Central Inland Fisheries Research Institute (CIFRI), a research wing of Indian Council of Agricultural Research, has taken up an initiative to develop canal fisheries via ecomanagement and ecotechnological intervention strategies in selected Sundarbans canals—Bhetkimari, Bishalakhi, and Bharua—assessing their biological productivity and fish production potential. In view of this, the Indian Council of Agricultural Research (ICAR), through the Central Institute of Agricultural Research (CIFRI), initiated the *net partition systems* for carp polyculture in the selected Bishalakhi and Bharua canals. Canals provide the main source of water for irrigation as well as fish production. Fish of different forms and sizes enter these canals, forming the natural fish community. In several countries such as Thailand, Sudan and Egypt, bioresources of canals are effectively utilized for fish production (Annon, 2003). The participatory mode of scientific utilization of these resources with local fish farmers was used to carry out cultural activities involving fish. Alongside culturing Indian Major Carp fish (rohu, mrigal, and catla), a good number of small indigenous minor carp fish such as *Amblypharngyodon mola, Salmostomo bacaila*, *Puntius sophore, Puntius ticto*, and *Macrobrachium rosenbergii* (prawn) were taken into consideration. These initiatives not only generate considerable fishery yields, but also empower local communities to develop their own knowledge and practices for the sustainable use of these previously underutilized canal resources. Community-based development can be cited as the adaptive ecomanagement of peoples to adapt to the ongoing changes in the global environment.

Bishalakhi canal: The Bishalakhi canal (21°46′47.3″ N; 88°05′30.6″ E) is in Krishnanagar, within tidal dominated Sagar Island, the largest Gangetic delta of the world, which gets its supply of tidal water from its connection with the Muriganga estuary. This selected canal is with a total length of 2.5 km, the width of 22 m and a depth ranging from 2.4–3.0 m.

Bhetkimari canal: The Bhetkimari canal (21°36′08.9″ N 88°15′14.8″ E to 21°36′40.1″ N 88°15′22.5″ E) is in Madanganj, of the Namkhana block receiving tidal water from the estuaries, Hathenia and Dewania. The canal has a length of 1.5 km., width of 30 m, and depth ranging from around 1.2–2.7 m.

Bharua canal: The tide-fed canal at Shibpur, in Fraserganj block, receives tidal water from the Hooghly estuary. It has a length of 2 km, a width of 45 m, and a depth ranging from 1 m to 2.4 m.

The ecological parameters of these canals demonstrate seasonal fluctuations in salinity at Bhetkimari canal, which fluctuated from 1.26 ppt in monsoon season to

18.2 ppt in pre-monsoon. The pH was slightly alkaline (8.1–8.4), total alkalinity was moderate (110 mg/L), dissolved oxygen was 6.0–7.2 mg/L, and nitrate concentration was significantly diluted in the monsoon season (from 0.1 ppm to 0.01 ppm), whereas other nutrients such as phosphate-P in Bhetkimari canal increased significantly during monsoon season, probably from agricultural field run-off. The highest amount of silicate was observed during post-monsoon and in the pre-monsoon.

4.8 Biotic Components Constituting the Biotic Community in the Sundarbans Mangrove Ecosystem

The constituent biota in the biotic community highlights the status of the productivity of a water body. Diversity and density of varied species of phytoplankton indicate the biological wealth of any aquatic ecosystem which in turn governs the food web patterns. Any alteration in the interactions pathways of biotic and abiotic parameters could result in a shift in the biodiversity components (Ingole et al., 2010), especially in assemblages and the community of fish. In the coastal–estuarine ecosystem, population dynamics of fish are driven mainly by three major ecological variables—salinity, temperature, and nutrients—and how fish tolerate and adapt to these changing variables during the cyclical tidal inundations (Nsor & Obodai, 2016).

4.8.1 Marine Plants and Seaweeds in Marine Ecosystem

Seaweed growth is mostly in the shallow and rocky seashore of a temperate marine environment, depicting diverse colors of green, red, and brown because of the possession of diverse types of pigments. Seaweed has simple body architecture, more like large algae than higher plants, and they do not show any division of the body into roots, leaves, and stems. With the help of holdfasts, seaweed can anchor to the hard substratum and resist the force of currents of water by bending their bodies. The largest of the sea weeds, kelp, forms an underwater forest. Varied species of red filamentous algae (Rhodophyta and some mangroves) create a thin covering on rocky surfaces and actively take roles in consolidating the structure of shallow-water coral reefs in the tropical belts (Misra et al., 1984; Chakraborty, 2023).

4.8.1.1 Algal Groups in the Sundarbans Mangrove Ecosystem

A total of 73 species of benthic algae like *Oscillatoria princeps, Ditylum brightwelli, Chaetoceros loranzianus, Bacillaria paradoxa, Coscinodiscus centralis, Planktionella sol, Oithona* sp., and *Helediaptomus* sp. are encountered in different habitats and ecological niches of the four contrasting ecozones of the Sundarbans mangrove ecosystem in India (Chakraborty, 2008; Table 4.3).

4.8.2 Plankton in the Sundarbans Mangrove Ecosystem

Plankton is floating and drifting aquatic fauna, whose movement is at the mercy of the currents, as they cannot intentionally move horizontally, nor can they adjust their buoyancy to control vertical movement within the water column. The distribution of these planktonic organisms is not uniform across different estuaries of Sundarbans mangrove–estuarine complexes where bacterio-plankton (bacteria and cyanobacteria) are found to be the smallest planktonic forms with sizes of around two microns. Planktons are of two types, autotrophic (phytoplankton) and heterotrophic (zooplankton). Plankton dynamics are strongly influenced by physical processes such as upwelling, which supplies abundant nutrients that fuel phytoplankton growth and proliferation through photosynthesis. This bottom-up forcing enhances the diversity and density of herbivorous zooplankton, which in turn serve as prey for larger zooplankton, fish, and other higher faunal components like reptiles, birds, and mammals.

4.8.2.1 Plankton and Red Tides in Marine–Estuarine Environment

The bloom of planktonic fauna caused by a higher proliferation of both phytoplankton and zooplankton are referred to as red tides. This occurs at the surface of marine–estuarine water exhibiting various colors such as reddish brown, bluish, and yellowish as generated by the different species of phytoplankton (dinoflagellates) and zooplankton (Cloern et al., 1983; Pradhan & Chakraborty, 2008). These kinds of red tides occur mainly on the surface of water in maritime zones due to (1) mass mortality of planktonic forms from a sudden decrease of nutrient concentration, combined with the effect of dead organism decay, which also results in an explosion of bacterial diversity and density. Such proliferation of bacterial populations utilize profusely dissolved oxygen from the water to expedite the mass mortality other aquatic life forms including fish and other invertebrates; (2) Prolific growth of planktonic organisms prohibit the sunlight from penetrating into the subsurface of marine water, causing the impairment of primary production which in turn tends to deprive bottom-living organisms, seagrasses, seaweeds, corals and other associated biota; (3) Some species of dinoflagellates flourish within the toxic chemicals of the red tides, especially saxitoxin, which is tremendously harmful to the nervous system of higher chordates, but does not impart much injurious effects on most of the invertebrates. Red tides are caused by higher bloom of planktonic organisms, especially dinoflagellates, downwelling (convergence of currents), organic pollution, and global warming.

4.8.2.2 Phytoplankton in the Sundarbans Mangrove Ecosystem: Diversity and Distribution

The primary productivity is done by phytoplankton in combination of cyanobacteria along with other bacteria even in nutrient deficient waters by virtue of their efficiency to fix up atmospheric nitrogen gas instead of depending on nitrate as their main source of nitrogen. Dinoflagellates and diatoms representing two major groups of the phytoplankton of the marine ecosystem undertake major roles to account for about half of the planet's entire primary productivity and also for half of the planet's oxygen production. Diatoms containing a good amount of silica on their external skeleton occur in abundance in the upwelling of temperately generated nutrient-rich waters. The dinoflagellates constituting the major share of the microplankton occur replacing diatoms in warmer waters of the tropics and subtropics they are characterized with two whiplike flagella. The phytoplankton, which are the main source of autotrophic productivity, determine the ecosystem dynamics of marine ecosystem but restrict their function only into the upper strata of marine–estuarine ecosystem depending on the availability of substantial amount of light.

Totals of 62 species under 54 genera, 77 species under 66 genera, and 71 species under 63 genera of the phytoplankton belong to two major families, Cyanophyceae and Bacillariophyceae were recorded from three studied canals—Bishalakhi, Bhetkimari, and Bharua, respectively. The phytoplankton species under the family Cyanophyceae represent as a major microfloral component (> 35%) followed by Bacillariophyceae (> 28%) and Chlorophyceae (> 17%). The Shannon diversity indices were found to be > 2.95 in all canals indicating a moderately rich phytoplankton diversity of the systems. Other algal families like Euglenophyceaea, Conjugatophyceae, Trebouxiophyceae, Mediophyceae, Xanthophyceae, Coscinodiscophyceae, and Synurophyceae also accommodate several other phytoplankton species which exhibit distinct seasonal fluctuation. Although, diatoms are the most diverse group of phytoplankton across all the seasons in the studied canals, the least number of species (< 1%) belong to the family, Dinophyceae (Giri, 2009; Fig. 4.6 and Table 4.4).

4.8.2.2.1 Diversity and Population Dynamics of Different Phytoplankton

Phytoplankton is at the lowest stratum of the marine–coastal food-web, contributing almost 15% of global oceanic production. However, numerous environmental stresses are challenging the ecology and biology of phytoplankton, including drastic changes in land use and shrinkage of vegetation cover. These stresses also impact the natural process of sediment transport, freshwater flow and nutrient dynamics, resulting in changes in ecosystem structure and function (Walsh et al., 1981, 1991; Nixon & Pilson, 1983). The disruption of the ecosystem where phytoplankton blooms is interlinked with changes in estuarine water flow, stratification of the water column in terms of intensity of light and water density, grazing pressure from zooplankton, and nutrient dynamics (Cloern, 1996).

A two-year research study (from November 2001 to October 2003) investigated four different sites representing contrasting ecozones of Sundarbans through six seasons and 24 continuous months. The study recorded 26 genera of phytoplankton—*Melosira* sp., *Planktoniella, Thallassionema, Chaetoceros, Pyxidicula, Skeletonema, Coscinodiscus, Fragilaria, Rhizosolenia, Thalassiosira. Pleurosigma, Lauderia, Ditylum, Triceratium, Biddulphia, Grammatophora, Thallassiothri, Cymbella, Asterionella, Gyrosigma, Diploneis, Navicula Amphora, Bacillaria,*

Nitzschia, and *Surirella* sp. (Table 4.4). The population densities of phytoplankton ranged from a minimum of 893.3 (March 2002) to a maximum of 4163.6 (June 2003). All study sites record an increase in the abundance of phytoplankton during the spring.

A long-term study over multiple decades reveals the trending depletion of several major Sundarbans phytoplankton species like *Amphora hyalina, Asterionella japonica, Asterionellopsis glacialis, Amphiprora constrica, Bacillaria paradoxa, Bacillaria paradoxa, Bacteriostum hyalianum, Bacteriastrum delecatum, Biddulphia regia, Bacillaria paxilifer, Bellerochea malleus, Biddulphia sinensis, Ceratium furca, Chaetoceros didymus, Chaetoceros peruvianus, Chaetoceros holsaticum, Chaetoceros gracile, Chaetoceros decipiens, Ceratium tripos, Chaetoceros diversum, Chaetoceros lassionasum, Chaetoceros lorenzianus, Chaetoceros cinctum, Chaetoceros atlanticum, Chaetoceros lorenzianus, Chaetoceros curvisitum, Chaetoceros affinis, Ceratium furca, Ceratium fusus, Ceratium tripose, Cyclotell stylorum, Coscinodiscus concinnus, Coscinodiscus radiatus, Coscinodiscus concinnus, Coscinodiscus radiatus, Coscinodiscus eccentricus, Coscinodiscus gigas, Coscinodiscus lineatus Ceratualina pelagica, Coscinodiscus eccentricus, Corethron crinophyllum, Coscinodiscus concinnus, Coscinodiscus radiatus, Ditylum brightwelli, Dinophysis caudata, Ditylum brightwelli, Dinophysis caudate, Dinophysis caudata, Eucampia zodiacus, Eucampia zodiacus, Eucampia zodiacus, Guinardia flaccida, Halosphera viridis, Hemiaulus sinensis, Lauderia annulata, Leptocylindrius danicum, Leptocylindrus minimus, Melosira granulata, Melosira variance, Melosira numuloides, Navicula rhombica, Navicula rhombica, Nitzschia longissima, Nitzschia longissima, Nitzschia seriata, Nitzschia seriata, Nitzschia sigma, Nitzschia closterium, Nitzschia spatulata, Odentella regia, Odentella sinensis, Odentella mobilensis, Paralia sulcata, Planktoniella blanda, Paralia sulcata, Planktoniella blanda, Porocentrum micans, Pleurosigma elongatum, Pleurosigma elongatum, Rhizosolenia alata Rhizosolenia hebetate, Rhizosolenia shrubsolei, Rhizosolenia stolteriotn, Rhizosolenia alata, Rhizosolenia cylindrus. Thalassionema nitzchioides,* and *Thalassiosira decipiens* (Biswas et al., 2010).

4.8.2.3 Zooplankton in the Sundarbans Mangrove Ecosystem: Diversity and Population Dynamics

Zooplankton, being the secondary producer in the food chains of mangrove–estuarine ecosystem provide necessary inputs in the form of energy and nutrients to fish by offering themselves as food for many juvenile and adult zooplanktivorous fish species, thereby controlling the abundance and distribution of fish fauna in the ecosystem. Zooplankton are categorized into three major categories, based on their sizes—microzooplankton (rotiferra), mesozooplankton (copepoda, cladocera) and macrozooplankton (chaetognaths, euphausiids). There are two additional types, based on their life cycles as holoplankton, which spend their entire life as plankton, and meroplankton, which spend a part of their life cycle as plankton (Giri, 2009; Fig. 4.7 and Table 4.6).

The seasonal fluctuation of the zooplankton population are determined by a number of water quality parameters—salinity, pH, temperature, turbidity, physiographic conditions, intensity of water currents, and volume of water. Three algal groups—Cyanophyceae, Bacillariophyceae, and Chlorophyceae—were found to contribute to a major percentage of the total phytoplankton community. The post-monsoon season registered the highest abundance of phytoplankton (10 110 ± 2696), whereas the lowest abundance was recorded during the pre-monsoon season (6385 ± 1233). Out of the eight groups of zooplankton recorded, Rotifera constituted the most dominant group (45%), followed by Crustacea nauplii (20%), and Cladocera (12%). The faunal species composition of the zooplanktonic community of the canals is dominated by Crustacean nauplii (26%), followed by Calanoida (24%), Cyclopoida (14%), Cladocera (6%), and Rotifera (3%). Therefore, the order of zooplanktonic abundance is Crustacean nauplii > Rotifera > Copepoda > copepod eggs > Cladocera > Ostracoda.

4.8.2.3.1 Diversity and Population Dynamics of Different Zooplankton

A research study spanning a period of 2 years (November 2001 to October 2003) recorded ten groups of zooplankton from different sectors of the Indian Sundarbans—Copepoda, Cladocera, Appendicularia, Chaetognatha, Hydromedusae, Ctenophore, Sergestidae, decapod larvae, mollusc larvae, and polychaeta larvae, indicating very rich and diverse zooplanktonic fauna. Among different groups of zooplankton, species in the order Copepoda of the subphylum Crustacea have been found to be the most diverse and abundant group by virtue of their presence at all study sites, accounting for 57% of all zooplankton. The highest abundance of copepods (57%) included groups of zooplankton such as Cladocera (9%), decapod larvae (9%), molluscan larvae (8%), polychaete larvae (4%), Sergestidae (4%), Chaetognatha (4%), Appendicularia (2%), Ctenophore (2%), and Hydromedusae (1%).

The population densities of the various registered zooplankton fluctuates in tune with the tidal cycles and seasonal climatic conditions, with varying population density of copepods from 124.7/L (January 2003) at neap tide to 1017.7/L (June 2003) at spring low tide; population densities of Cladocera from 124.7/L (January 2003) at neap tide to 1017.7/L (June 2003) at spring low tide. The population density of chaetognath ranged from 11.3 (February 2002) at spring low tide to 90.7 (June 2003) at spring low tide; population density of *Sergestidae* ranged from 12.3 (March 2003) at spring low tide to 87.3 (July 2003) and at neap low tide; the population density of *Appendicularia* ranged from 4.7 (October 2003) at neap low tide to 8.3 (February 2002) to 52.3 (July 2002) at spring low tide; spring high tide, the population density of zooplanktonic *Hydromedusae* ranged from 15.0 (December 2002) at spring high tide to 22.3 (December 2002) to 97.3 (July 2002) at neap low tide; the population density of Ctenophora ranged from 12.3 (March 2003) at spring low tide to 64.3 (June 2003). at neap low tide, the population density of decapod larvae ranged from 17.0 (February 2003) at neap low tide to 151.0 (June 2002) at neap high tide, the population density of molluscan larvae fluctuated from 17.3 (February

2002) at spring low tide to 141.3 (July 2003) at neap high tide, the population density of polychaete larvae ranged from 8.3 (February 2002) at spring low tide to 97.3 (July 2003) during spring high tide.

4.8.2.4 Diversity and Population Dynamics of Different Ichthyoplankton

Larval and juvenile forms of finfish are named as ichthyoplankton, which constitute a critical component of pelagic biomass in estuaries, seas, and oceans. The pelagic and demersal fishery resources of any aquatic subsystem are directly connected with the structure and functions of the ichthyoplankton community. Although many ichthyoplankton may not be used as edible food, they supposedly play significant roles in maintaining and stabilizing the estuarine web–food chains. Some indirect applicability of the ichthyoplankton includes the manufacture of organic manure, poultry feed, and shrimp feed. However, this valued biotic component in the pelagic realm is gradually getting eroded because of the mass collection of wild tiger shrimp seeds, which has become a very common practice in the coastal zone of West Bengal, coinciding with the tidal cycles of an area. The reasons behind this ecologically adverse practice are thought to be due to ever escalating demand for the healthy wild seeds in the aquaculture farms grown up along the coastal belts in the state of West Bengal, India during last five decades which do not have their own shrimp hatcheries. The ichthyoplankton recorded in estuaries is usually dominated by larval forms of finfish and shellfish spawned within these systems (Neira et al., 1992). In coastal aquaculture operations, one of the common human practices is the collection of young stages of fish, popularly called *ichthyoplankton,* as a part of formation of natural seed-resources centers for stocking grow-out pond cages. To strengthen the process of ongoing aquaculture activities along the coastal environment, it is essential to identify the spawning grounds within spawning season—their recruitment, growth, and mortality. It is also imperative to develop skill for the proper identification of eggs, larvae, post larvae, and other developmental stages of the target species by establishing a balance among different components of capture fisheries like rational exploitation, conservation, and management. Undertaking pertinent research in the fields of distribution and abundance of ichthyoplankton can improve understanding of the interrelationships among various fish species in different phases of their life cycle and the effects on their adult spawning patterns.

A research study has recorded a total of 19 families of finfish producing ichthyoplankton within four different study sites (ecozones) of the Indian Sundarbans. Study site I is on the Hooghly estuary, at the deltaic islands, Fresergaunj and Bokkhali, with a fishing harbor and a sandy sea shore, respectively. Study site II is on the Saptamukhi estuary at two virgin deltaic islands, Lothian and Prentice, both characterized as having luxuriant mangroves coupled with extended mudflats. Study site III is the Matla estuary at the deltaic island, Jharkhali, a very busy tourist spot with dense mangrove but narrow strip of mudflat. Finally, study site IV is on the Matla estuary at Canning, a congested small town having excessive trawling activities, and recently developed semi-intensive aquaculture farms. The families of finfish with ichthyoplanktonic faunal components are Anguillidae, Mugilidae, Ophichthidae, Muraenidae, Gobiidae, Elopidae, Sciaenidae, Ambassidae, Clupeidae, Lutjanidae, Scatophagidae, Centropoidae, Engraulidae, Carangidae, Teraponidae, Leiognithidae, Sillaginidae, Scombridae, and Polynemidae. Ichthyoplanktons recorded from the study site I, Fresergunj and Bokkhali islands, harbored the highest number of families (19), followed by study site II on Saptamukhi estuary with 16 families, study site III at Jharkhali on Matla estuary with 15 families, and study site IV at Canning on the Matla estuary with ten families.

The population density of ichthyoplankton (no./haul) recorded at different study sites through 24 months and 6 seasons of 2 consecutive years (November 2001 to October 2003) reveals that the abundance of ichthyoplankton fluctuated from season to season as well as in between different tidal levels (at a maximum during spring high tide and minimum at neap low tide) and at the different study sites with the highest density during monsoon, followed by pre-monsoon, then post-monsoon. Computation of relative abundance of different groups of ichthyoplankton reveals that out of 19 ichthyoplankton families, the Clupeidae ranked first representing 14% of all recorded species, followed by Engraulidae (10%), Gobiidae (9%), Ambassidae (9%), Teraponidae (7%), Sciaenidae (6%), Carangidae (6%), Sillaginidae (6%), Mugilidae (5%), Lutjanidae (5%), Anguillidae (4%), Scatophagidae (4%), Polynemidae (4%), Leiognathidae (3%), Ophichthidae (2%), Muraenidae (2%), Elopidae (1%), Centropoidae (1%), and Scombridae (1%). The average population density (no./haul) of total ichthyoplankton varied from 2.19 (March 2003) to 15.40 (July 2003); 0.64 (March 2002) to 8.43 (August 2003); 0.03 (February 2002) to 15.33 (July 2003); and 0.39 (February 2003) to 22.64 (August 2002) at study sites I, II, III, and IV respectively. The average population density (no./haul) of total ichthyoplankton throughout the study period were recorded as 6.25, 3.26, 6.04 and 8.50 at study sites I to IV respectively (Table 4.8). Another estimate from a more recent research study noted that the annual average seed catch of finfish and shellfish is 10 077 nos./net/day of which 65–75% consists of shellfish, 3–5% tiger shrimp and 20–30% finfish. The peak period of seed availability particularly for the period from June to September recording averages a total catch between 9750 nos./net/day and 19 110 nos./net/day.

At study site I, Clupeidae represented 12% of the total catch composition, followed by Engraulidae (10%), Gobiidae (9%), Ambassidae (9%), Sciaenidae (8%), Teraponidae (7%), Leiognathidae (7%), Carangidae (7%), Sillaginidae (7%), Mugilidae (6%), Scatophagidae (5%), Anguillidae (4%), Muraenidae (4%) and Polynemidae (3%). At study site II, the percentage composition of different ichthyoplankton families were Engraulidae (15%), Clupeidae (13%), Gobiidae (13%), Teraponidae (11%), Ambassidae (10%), Sillaginidae (9%), Anguillidae (8%), Mugilidae (8%), Scatophagidae (5%), and Polynemidae (8%). At study site III, the percentage composition of different ichthyoplankton families were Anguillidae (3%), Elopidae (3%), Ophichthidae (3%), Muraenidae (3%), Gobiidae (8%), Sciaenidae (8%), Mugilidae (6%), Ambassidae (9%), Clupeidae (12%), Lutjanidae(9%), Scatophagidae (4%), Engraulidae (8%), Carangidae (8%), Teraponidae (6%), Sillaginidae (6%) and Polynemidae (3%). At study site IV, the percentage composition of different ichthyoplankton families were Clupeidae (16%), Engraulidae (10%), Lutjanidae (8%), Ambassidae (8%), Gobiidae (7%), Carangidae (7%), Teraponidae (5%), Leiognithidae (5%), Sillaginidae (5%), Sciaenidae (5%), Mugilidae (4%), Anguillidae (4%), Scatophagidae (3%), Polynemidae (3%), Elopidae (2%), Ophichthidae (2%), Muraenidae (2%), Scombridae (2%) and Centropoidae (1%; Fig. 4.7 and Table 4.8).

4.8.2.5 Periphyton in the Sundarbans Mangrove Ecosystem

Periphyton, as a heterogeneous group of biotic assemblages, remain attached to the stems of plants, woods, stones, and various other hard structures. It acts as primary producers like other phytoplankton and macrophytes in an aquatic ecosystem and is used directly as food for many aquatic organisms. The periphyton community, mostly composed of microorganisms, is also associated with different strains of bacteria, fungi, protozoa, algae, cyanobacteria, microcrustaceans, and molluscs after remaining attached to any natural, dead bodies and artificial structures acting as a food source to faunal components belonging to higher trophic levels in food-webs. It is important to understanding the growth of heterogeneity in the periphytic assemblage and their interaction with ongoing ecological changes because of the impact of several ecological attributes—including hydraulic conditions, ecological stressors, associated biodiversity components, and biotic interactions.

Studies of periphyton assemblages in a selected Sundarbans canal, Bishalakhi, investigated ecohydrological interactions between 16 different environmental variables and periphyton during pre-monsoon, monsoon, and post-monsoon seasons. Different ecological variables—salinity, total alkalinity, transparency, water temperature, specific conductivity, water depth, water velocity, Mg^{++}, PO4 – P, TP, and SiO4 – Si—showed significant differences ($p < 0.05$) across seasons. In total, 74 taxa of periphyton under 42 genera and six taxonomic groups were documented. Out of all periphyton species identified, diatoms dominated the periphyton community with respect to diversity and abundance; and the majority of which (36 species) revealed a significant positive correlation with dissolved oxygen and total nitrogen. Some macrobenthic fauna, including *Bellamya bengalensis, Pila globossa, Pila virens, Meiniplotica scabra, Tarebia granifera,* and *Thiara lineata*, were also found as periphyton in the canals.

4.8.2.5.1 Periphyton Abundance and Compositions in the Sundarbans, Mangrove Ecosystem

Higher species diversity of diatoms was found for those abundant species belonging to the orders Bacillariales and Naviculales. The largest number of species was found in the genus *Nitzschia* (16 species; Hassall, 1845), and *Navicula* (ten species; Bory, 1822). Other genera with a sizable number of species include *Fragilaria* (Lyngbye, 1819); *Gomphonema* (Ehrenberg, 1832); *Pinnularia* (Ehrenberg, 1843); *Gyrosigma*

(Hassall, 1845); *Amphora* (Ehrenberg ex Kützing, 1844); and *Amphipleura* (Kützing, 1844). Dominant diatom species include *Bacillaria paxillifera* (J. F. Gmelin) Linnaeus, 1791; *Cylindrotheca closterium* (Ehrenberg) W. Smith, 1853; *Cymbella lanceolata* (C. Agardh) Kirchner, 1878; *Synedra ulna* (Nitzsch) Ehrenberg, 1832; and *S. acus* (Kützing, 1844). The family Chlorophyceae includes species like *Schroederia* (Lemmermann, 1898); *Ankistrodesmus* (Corda, 1838; Scenedesmus Meyen, 1829); and *Monoraphidium* (Selenastrum Reinsch, 1867); while the family Cyanophyceae was represented by four genera—*Anabaenopsis Lyngbya* (C. Agardh ex Gomont, 1892); *Dolichospermum* (Bory ex Bornet and Flahault, 1886); and *Cylidrospermum* (Kützing ex É. Bornet and C. Flahault, 1886). The family Zygnematophyceae (Conjugatophyceae) was observed to have only one genus, *Spirogyra* (Link, 1820). From the community analysis, periphyton diversity was found to be moderate in the canal system with 22 species during the premonsoon, a higher abundance than during monsoon and post-monsoon seasons with only 12 species. The major species of periphyton include *Oscillatoria prolifica*, *Spirogyra* sp. *Cymbella cymbiformis*, *Schroederia indica*, *Amphora* sp., *Navicula protractoides*, *Nitzschia alpina*, *Surirella* sp., *Tetracyclus* sp., *Hantzschia* sp., and *Tribonema* sp.

4.8.2.5.2 Functional Significance of Periphyton

Despite some important functional roles in the mangrove ecosystem, periphytons do not play driving roles with regard to structuring. However, observing the effect of ecodynamics on their settlement, growth, and propagation offers useful information for restoring hydrology and for the purpose of monitoring the ecosystem (Iwaniec et al., 2006). Natural disturbances like floods and cyclone-induced incidents tend to influence the hydrological regime of respective aquatic ecosystems, and thereby, also the morphofunctional characteristics of periphyton communities.

Phytoplankton, a photo-autotrophic periphyton species, takes an active role in primary production, even in light-limited and nutrient-limited habitats because of their abilities to distribute ubiquitously. In the process of settlement and succession of periphyton, initial colonization of diatoms is observed first, followed by cyanobacteria, and filamentous green algae. Several ecobiological attributes govern the growth and colonization process, including texture and nutrients of substrates (Algarte et al., 2017), temperature (Francoeur et al., 1999), intensity of light, grazing pressure, nutrient availability, intensity of water currents, and duration of underwater inundation (Biggs & Thomsen, 1995). The increased production of periphyton in eutrophic waters is from excessive growth of filamentous green algae or cyanophytes, which reduces the availability of nutrients (Davis, 1983), which in turn depletes the concentrations of dissolved oxygen. As a majority of canals in the Indian Sundarbans remain underutilized in achieving sustainable fisheries, an ecofriendly approach with periphyton-based aquaculture can have a positive impact on the production of cultivable fish species, alongside improvement of water quality.

4.8.2.5.3 Periphyton Community Structure and Distribution in the Sundarbans Mangrove Ecosystem

Regarding the distribution and species diversity of periphyton in the Indian Sundarbans, previous studies like Naskar et al. recorded 22 taxa of epiphytic algae primarily dominated by Cyanophyceae (50%). The study conducted by Sinha et al. (2013) projected the dominance of Bacillariophytes (> 70%) across the seasons, which is aligned with earlier studies by Pandit et al., Dunck et al., and Kanavillil and Kurissery (2013). In the diatom assemblage, it was noted that the genus *Nitzschia* had the highest number of taxa, closely aligning with the research findings of Moresco et al., which showed that the genus *Navicula*, tallied the highest number of taxa, followed by *Nitzschia* and *Pinnularia* of the Guaiapó Stream in the Upper Parana River Floodplain, Brazil. The higher diversity and density of Bacillariophytes were found to be due to the bioavailability of silica in the environment. A higher abundance of the two dominant genera of periphyton, *Nitzschia* and *Navicula*, was correlated with a higher load of organic substances, along with moderate turbidity. The dominance of different periphytic epiphytic algal species under three major families Xanthophyceae, Chlorophyceae, and Cyanophyceae, especially during post-monsoon season, had direct relationships with salinity, transparency, and nutrients (Sinha et al., 2013). The considerable growth of filamentous algae in the periphyton community in different seasons can be explained by the availability of the huge amount of nutrients, a higher amount of light, and the reduced flow intensity of the water (Biggs et al., 1998).

4.8.2.6 Diversity of Entomofauna in Sundarbans, India

A rich diversity of insect fauna is found to occur in the ecologically important and climate-sensitive mangrove ecosystems, of Sundarbans Bioshere Reserve, India. The diverse forms of habitat heterogeneity, tend to support an array of coexisting ecosystems including pristine mangrove stands, tidal mudflats estuarine and freshwater creeks, brackish wetlands, reclaimed agricultural land, aquaculture ponds, and adjoining landscapes with lot anthropogenic activities. Such ecological complex ecological spectrum support insect faunal components in different niches suitable for the flourishing of different groups of insects.

Total 29 insect species under five orders have been documented from the Sundarbans Biosphere Reserve of which number of insect species under the order Hymenoptera was found to be maximum followed by Diptera, Lepidoptera and Coleoptera. Maximum abundance of all these species coincide with the flowering seasons of respective plant species. The insect faunal groups were registered in higher abundance during flowering seasons of the mangrove plants like *Avicennia marina*, followed by *Avicennia officinalis*, *Acanthus ilicifolius* and *Aegiceras corniculatum* where flowers provide important food sources for pollinators and thereby help maintaining pollinator diversity in mangrove community.

Insects plants interaction during flowering season depict higher possibility of successful pollination and seed-setting andthere such relationships play crucial role in reproductive success and propagation of mangroves. In Indian scenario, the major mangrove insect visitors belong to order Hymenoptera, Diptera, Lepidoptera and Coleoptera (Remadevi et al., 2019). Among them, hymenopterans, especially bees have higher abundance at dry seasons (d'Avila and Marchini, 2008), lepidopterans at early dry seasons (Mertens et al., 2021), whereas coleopterans and dipterans show their peak activity during wet season (Oliveira et al., 2021; Denlinger, 1980).

In a recent study, a total of **58 dipteran species** belonging to **19 families** were recorded in association with various mangrove plant species across the Sundarban Biosphere Reserve (SBR), India. Among these, **ten species represent new distributional records** for the region. These recorded dipteran species belong to 21 families and are categorised in several ecological guilds based on species diversity, richness, and relative abundance which are like flower visitors, frugivorous, leaf and trunk resting flies, carnivorous, saprophytes and hematophagous., of which flower visitors were found to be most dominant (Roy et al., 2016) (Fig. 4.19).

4.9 Benthic Fauna and Their Functional Contribution to the Sundarbans Mangrove Ecosystem

The intertidal belts in the form of sea shore, mudflats, and sandflats are endowed with a galaxy of life, in the form of benthic fauna, which dwell both in surface and subsurface zones. Their dominating presence is manifested by their highly dramatic activities, laid down on the sediment as a mass of changing contours. Traditionally, benthic fauna is classified into meiofauna and macrofauna by the use of defined sieve sizes, whereby organisms retained on a 0.5 or 1 mm mesh sieve are referred to as macrofauna (Rees, 1940; Mare, 1942; McIntyre, 1969; Bachelet, 1990; Schlacher & Woolridge, 1996). The distribution of biomass among different size classes of benthic fauna is considered to be a conservative structural feature which relates to the pre-dominant life-styles of different size groups, including microbes, meiofauna, and macrofauna (Schwinghamer, 1981, 1983). In tropical intertidal zones, the sizes of benthic fauna don't fit the conventional classification of either epifauna, surface-dwelling macrobenthos, or infauna, residing in the subsurface. Benthic fauna is often designated as mesofauna, having intermediate sizes between meiofauna and macrofauna (Dittman, 1995). The intertidal belts of the mangrove ecosystem is dominated by epifaunal macrobenthos like brachyuran crabs, molluscs, polychaetes, brachiopods, and cnidarian (Macnae & Kalk, 1962; Macnae, 1967, 1968; Sasekumar, 1974; Alongi & Sasekumar, 1992; Chaudhuri & Choudhury, 1994; Alfred & Ramakrishna, 2004; Chakraborty, 2018). In the estuarine ecosystem, macrobenthos is one of the most important structural elements of the food-web and it plays a significant role in ecosystem dynamics (Herman et al., 1999; Ysebaert & Herman, 2002).

Research on intertidal macrobenthic fauna has been centered around their functional roles in sustaining the mangrove–estuarine ecosystem, with special emphasis on Sundarbans, India (Chakraborty, 2018). Intertidal benthic fauna provide critical ecosystem services including bioturbation, which aerates and fertilizes sediment, enhances biodiversity, accelerates litter decomposition, and facilitates nutrient cycling. They also play a central role in food chains and food webs by transferring energy through trophic interactions, and they support capture fisheries by releasing millions of meroplanktonic larvae that serve as a vital food sources for both pelagic and demersal finfish (Anvar Batcha, 1984; Robertson, 1986; Micheli, 1993; Chakraborty, 2018). Intertidal macrobenthic fauna resides in an oscillating ecological condition experiencing repetitive tidal inundation and exposure, and are highly sensitive to environmental stress. Because of this sensitivity, they serve as excellent bioindicators for detecting the ecological changes and assessing the health of mangrove ecosystems (Kitamori, 1963; Washington, 1984; Anvar Batcha, 1997; Stark, 1998; Maity et al., 2014). The intertidal macrobenthic fauna of the Hooghly–Matla estuarine complex represents the most abundant faunal components in the mangrove ecosystems of Sundarbans, India, and exhibit tolerance through their adaptability to an array of dynamic physicochemical parameters in that environment (Sanders, 1968; Lie & Kisker, 1970; Chakraborty & Choudhury, 1992a, 1992b; Chakraborty, 2013). The macrobenthic fauna not only play a vital role in energy flow, but they are also responsible for the development of several biogenic structures (Chakraborty, 1995; Chatterjee & Chakraborty, 2014). Different groups of macrozoobenthos exhibit distinct zonation patterns in their dynamic estuarine–mangrove ecosystem settings. These patterns are shaped by various eco-biological factors and influenced by diurnal and seasonal tidal fluctuations. This dynamic setting reflects the varying abilities of each species to adapt to different levels of hydrodynamic stress in the coastal environment (Chakraborty & Choudhury, 1992a, 1992b; Ysebaert et al., 1998; Ravichandran et al., 2001; Giri et al., 2004).

The macrobenthic fauna of the mangrove ecosystem of Sundarbans, India demonstrate marked seasonal variations in respect to species composition, population density, and biomass, dependent on several oscillating ecological processes, such as beach configuration, the rate of accretion and erosion, soil characteristics, and hydrological parameters (Chakraborty & Choudhury, 1992a, 1992b, 1994; Chakraborty, 2013). The diversity and abundance of macrozoobenthos largely depend on the successful settlement of pelagic larvae, or meroplankton, produced by various species. These larvae must find suitable habitat at the land–water interface, primarily within intertidal zones, where they can tolerate the physical stresses imposed by strong waves and tidal currents. The ongoing interactions among biotic and abiotic parameters permit some limited selected species to settle down, grow, and propagate as diverse benthic faunal components in the tropical estuary. The high density and biomass of the benthic faunal community are interlinked with their breeding season, faster growth rate, and short life span (Parulekar et al., 1980). In the Sundarbans mangrove ecosystem, the peak population density of intertidal macrobenthic fauna is found during the monsoon months, albeit with lower biomass; whereas the highest biomass, coupled with lower population density, was observed during the pre-monsoon months of the year. Such seasonal oscillation of the population of macrobenthic fauna is likely due to independent and synergistic impacts of various ecological factors of water, textural components of soils, tidal processes, trophic relationship, reproductive potentiality. and recruitment process (Sanders, 1958; Masse, 1972; Chakraborty & Choudhury, 1992a, 1992b).

The macrobenthic fauna of this ecozone are represented by the major invertebrate faunal groups including Cnidaria, holothuroids, brachiopods, sipunculids, gastropods, bivalves, nemertine, polychaetes, amphipods, insects, euchirids, shrimps, isopods, stomatopods, decapods, hemichordate, and gobiid fish (Chaudhuri & Choudhury, 1994; Figs. 4.9, 4.10, 4.11, 4.12, 4.13, 4.14, 4.15, 4.16, 4.18 and Tables 4.5, 4.6, 4.9, 4.10, 4.11, 4.12).

4.9.1 Spatial and Temporal Variability of Benthic Fauna Emphasizing Habitat Diversity at Different Zones in Oceanic–Marine Ecosystems

The intertidal zones in the mangrove–estuarine complex of Sundarbans, representing one of the most productive natural wetlands in tropical and subtropical regions of the world, comprise diversified habitats with specific types of geomorphological settings, rich diversity, unstable and vulnerable ecological conditions, exposure to different types of environmental hazards (cyclone, flooding with tidal waves), and environmental perturbations (global warming, sea level rise, eutrophication, and ecological contamination); acting as a sink for sewage, oil, pesticides, and heavy metals. (Macnae, 1969; Chaudhuri & Choudhury, 1994). This unique ecosystem, dominated by intertidal salt-tolerant woody halophytic vegetation, supports high biological productivity, economic value, and ecological resilience. Influenced by two tidal cycles per day, it promotes bioresource enrichment while maintaining ecological balance through coastal protection (Chakraborty, 2011). This highly productive detritus-based coastal ecosystem achieves an average primary productivity of 2500 mg/cm^2 per day and contributes profusely to the biogeochemical cycles of the coastal environment (Jennerjahn et al., 2009). The significance of mangrove ecosystems for fisheries and aquaculture is widely recognized for two main reasons:

1. Mangroves export large quantities of energy-rich detritus to adjacent water bodies, fueling coastal food webs (Odum & Heald, 1975a, 1975b). Numerous studies have shown a positive correlation between the extent of mangroves and fishery yields in adjacent waters (Macnae, 1974; Lee, 1995).
2. Mangrove-based bioresources support profitable regional and international markets, providing livelihoods and income for millions of people in costal communities.

4.9.1.1 The Intertidal and Subtidal Zones

The intertidal zone of the Sundarbans mangrove ecosystem, being a part of the littoral zone (the shallow zone receiving sunlight to the bottom layers of the water column) experiences continuous inundation and exposure twice a day by the rising and falling of tides. This zone marks the initial point of the marine ecosystem at the juncture of land and sea where harsh ecological exchange processes occur due to tidal fluctuations receiving and accumulating of the discharge from the terrestrial ecosystem with the results of changes in the volume of overlying water, currents, and waves of water. The intertidal zone presents highly variable and often harsh ecological conditions. Organisms living there—either for their entire life or critical life phases—have adapted remarkably to survive the changing and often adverse ecological conditions. These adaptations include feeding mechanisms (e.g., filter feeding, deposit feeding) to cope with uncertain food supply, as well as morphological and ecophysiological adaptations to withstand strong tidal currents. The upper intertidal zone, often called the supra-littoral zone, is only occasionally submerged and is more frequently drenched by wave splash and spray. The midlittoral zone within the intertidal belt is submerged during high tides on all days of the lunar tidal cycle, providing more stable and consistent ecological conditions for the biota to flourish. The lower boundary of the intertidal zone—the low littoral zone—meets the subtidal zone, which remains permanently submerged under shallow seawater and is considered the deepest part of the littoral zone (Fig. 3.1).

4.9.1.2 The Abyssal Zone and Biota

The deeper part of the seafloor, which at a depth of about 4000 m represents *the Abyssal Zone*, often undergoes striking seasonal and spatial changes in ecological character. Decomposed detrital materials called phytodetritus, composed of dead phytoplankton, exoskeletons of Crustacea and fish, and fecal materials of pelagic animals, often sink from the surface layers to the abyssal seafloor. These organic materials, derived mostly from the living biota, are often carried along by bottom currents to accumulate in low-lying depressions and on the down-current side of slopes (Fig. 3.1).

The spatial and temporal variability of macrobenthic fauna are required to be treated together instead of separately because they occur concurrently in natural systems (Gupta, 1987; Dye, 2004; Gray and Elliot, 2009). Spatial and temporal variability are inexorably intertwined over the full range of time and space scales. Because ecological communities tend to exhibit variability across different spatial and temporal scales, understanding the factors and processes that determine such changes in relevant scales of space and time appears to become a prerequisite for assessing the interactions among biotic and abiotic components of ecosystems (Wiens, 1989; Levin, 1992; Lee & Williams, 2002; Bosire et al., 2004; Bryars & Havenhand, 2004; Hall Jr. & Killen, 2005). In order to undertake an integrated eco-management of estuaries, characterized by a high degree of spatial and temporal variability, the ecological information pertaining to these interactive processes helps draw better predictions about future environmental changes due to human impacts (Thrush et al., 1999).

4.9.2 Benthic Fauna of Sundarbans, India

4.9.2.1 Diversity, Zonation, Density, and Distribution of Macrobenthic Fauna in the Indian Sundarbans

The Hooghly–Matla estuarine complex is one of the largest and most extensive estuarine systems in the tropics (Chaudhuri & Choudhury, 1994). The intertidal belts of estuarine Sundarbans receiving plenty of mangrove-contributed detritus, and support a panorama of benthic life—throbbing with the cyclical tidal inundation and exposure (Frith et al., 1976; Chakraborty et al., 1986, 1992; Mitra & Dey, 1992; Chakraborty & Choudhury, 1993, 1994; Subba Rao et al., 1987, 1992, 1995a, 1995b; Dittman, 2000; Ellison, et al., 2000; Ellison, 2002; Dauvin et al., 2003). This ecoregion is renowned for its rich biodiversity, characterized by luxuriant mangrove forests and a highly diverse fauna with specialized adaptations, complex population dynamics, and strong community interactions. Among these, macrobenthos constitute one of the most important faunal groups. This ecologically significant ecoregion is endowed with several extended mudflat habitats which oscillate with tidal cycles fortnightly and seasonally. Although the overall drainage pattern of this ecosystem is found to be relatively stable, the detailed topography of the mudflats bordering the creeks demonstrate instability in their sedimentological characteristics due to periodical erosion and deposition processes. The complex root systems, twisted trunks, and branches of mangrove plants baffle tidal currents, promote sediment deposition, and gradually elevate intertidal mudflats. These structures also trigger the formation of diverse geomorphological structures—such as ripples, mounds, mud balls, sand bars, and dunes—through tidal redistribution of freshly deposited sediment within the swamps (Paul, 2002).

4.9.2.2 Diversity of Macrobenthic Fauna in the Indian Sundarbans

The faunal components of mangrove ecosystems involve terrestrial, estuarine, and marine species, including sessile forms (poriferans, cnidarians), wandering species (king crabs), surface-dwellers (gastropods, hermit crabs), and infauna in the subsurface sediment (actiniarians, polychaetes, bivalves, brachyuran crabs, wood-boring organisms, brachiopods, and gobiid fish). Other types of macrobenthos include biofouling faunal components like oysters and barnacles (Alfred & Ramakrishna, 2004; Chakraborty, 2018). Due to nutrient enrichment from mangrove detritus and enhanced upwelling, the coastal waters adjacent to luxuriant mangrove forests in the Bay of Bengal exhibit significantly higher species richness of both benthic and pelagic faunal, according to surveys conducted there, as compared to other coastal regions (Annon, 2003; Kathiresan, 2004). In terms of macrofaunal species richness, the Sundarban mangroves of West Bengal rank first in India with 1,434 faunal species belonging to 20 phyla, followed by 914 species of Andaman and Nicobar Islands, and 801 species of Tamil Nadu, among all coastal states of India endowed with mangroves (Alfred & Ramakrishna, 2004). Earlier research studies have revealed the occurrence of nine species of *Actiniaria* (Bairagi, 1998), 69 species of polychaetes, 26 species of brachyuran crabs (Chakraborty et al., 1986), and 120 species of molluscs (Subba Rao et al., 1995a, 1995b) in the Hooghly–Matla estuarine complex. Another comprehensive ecological survey documented a total of 85 different species of macrozoobenthic fauna from different tidal zones (low-tide level, mid-tide level, and high-tide level) within four different ecozones:

Bakhkhali–Freshergaunj islands on Hooghly estuary;
Prentice–Lothian islands on Saptamukhi estuary;
Jharkhali island on Matla estuary, within the tiger research project area; and
Canning port on Matla estuary, the gateway of the entry of Sundarbans to the north

The same research confirms the existence of a total of five species of sea anemones under four families and one order, 23 species of polychaetes under nine families and two subclasses, 18 species of brachyuran crabs under two families and one order, 33 species of molluscs under 21 families, and eight orders and six other macrobenthic fauna—one horse shoe crab, one gobiid fish, and one amphipod (Chakravarti, 2008). Other research studies have recorded additional species like nemerteans, brachiopods, and brachyuran crabs (Tables 4.5, 4.6, 4.9, 4.10, 4.11 and 4.12).

The most abundant macrobenthic fauna inhabiting in the intertidal belts of Sundarbans mangrove–estuarine complex are found under the phylum Mollusca as *Assiminea francesiae, Assiminea brevicula, Nerita articulata, Nassarius foveolatus, Cerithidea cingulata, Telescopium telescopium,* and *Littorina melanostoma;* under the subphylum Crustacea as *Uca acuta, U. triangularis bengali, Metaplax intermedia* (brachyuran crabs); under the class Polychaeta as *Mastobranchus indicus, Neanthes chingrighattensis, Talehsapia annandalei, Perinereis cultrifera, P. nigropunctata, Owenia fusifromes, Diopatra cuprea,* and *Paraheteromastus tenuis*; and under the phylum Cnidaria, as *Edwardsia jonesii* and *Paracondylactis indica.*

A striking feature of the Hooghly–Matla estuarine macrobenthic community is the unusually high diversity and dominance of gastropods species over all other groups. In most other estuaries worldwide, polychaetes are typically the most dominant and diverse group, followed by other macrobenthic fauna (Untawale & Parulekar, 1976; Fernando et al., 1983; Khan, 1995). It was assumed that relative species richness of brackish water fauna in the tropical estuaries was a matter of chance because of a higher preference of this faunal group for the tropical environment (Caspers, 1967). This is corroborated by the findings of Hartman (1974) who attributed the relatively high occurrence of polychaete fauna in the Indian Ocean to its tropical nature, where elevated temperatures facilitate easier osmoregulation (Panikkar, 1951).

4.9.2.3 Zonation of Macrobenthic Fauna in the Indian Sundarbans

Different macrobenthic fauna inhabit different tidal levels—high-tide level, HTL; mid-tide level, MTL; low-tide level, LTL—exhibiting different zonation patterns according to differing ecological variables which reflect their adaptation to different degrees of terrestriality. Several brachyuran crabs—*Illyoplax gangetia, Scylla serrata, Uca triangularis bengali, Uca dussumieri, Dotilla brevitarsis, Macrophthalmus pectinipes,* and *Metaplax crenulata*—have been found to inhabit the low-tide level, whereas *Metaplax intermedia, Metaplax indica, Dotilla blanfordi, Sesarma chiromentes bidens,* and *Sesarma taeniolatum* are the major inhabitants of the mid-tide level with the highest range of their activities within the mangrove forest floor. Several brachyuran crabs—*Uca acuta* and *Ocypoda macrocera*—restrict themselves within the high-tide level. Although several species of molluscs and polychaetes have been noticed in all the tidal levels of intertidal zone, most of the species under these two major groups occur in higher abundance within the midlittoral zone. The species classified under actiniarians, nemerteans and brachiopods occur within the low littoral zones.

Both biotic and abiotic factors, varying at spatial and temporal scales, influence the benthic distribution patterns of benthic fauna (Alongi, 1987; Fischer & Sheaves, 2003). The occurrence and activities of benthic faunal components tend to modify, stabilize, or destabilize the sediment. Animal-sediment interactions are closely linked to feeding modes. Based on these feeding activities, macrofauna are broadly categorized into two types of feeders: deposit feeders, which exhibit intense bioturbation; and suspension feeders, which rely on overflowing water on more stable substrates. Direct and indirect interactions among different species within a benthic community—such as predation, competition, mutualism, amensalism, and amelioration—shape the overall structure and function of coastal biotic communities. Among the many ecological factors, salinity and temperature are the primary drivers of animal distribution in brackish-water zones (Teal, 1958; Kinne, 1963), whereas sediment characteristics play the key role in structuring species composition across various coastal habitats (Abele, 1974). To reduce competitive interactions in a benthic community, a particular species may use the substrate as a source of nutrition, another uses it as a feeding site, while some others depend on the substrates for their shelter, resulting in a positive co-existence in the community. Higher densities and diversity of macrobenthic fauna are often concentrated in narrow intertidal belts due to the availability of numerous ecological niches where sediment characteristics play a decisive roles in determining species diversity (Sanders, 1968). Nutrient enrichment in the littoral zones of the mangrove ecosystem play crucial roles in facilitating the settlements, growth, population fluctuation, and biotic community interactions of fauna (Heald & Odum, 1970). Because of the highly buffered nature of sea water, pH does not play a limiting effect on marine organisms. In a research study on the seasonal fluctuation of macrobenthic community, polychaetes represent the highest percentage (46.44%), followed by molluscs (37.39%), brachyuran crabs (12.68%), actiniarians (3.48%), and other miscellaneous groups of benthic animals (0.07%). The highest abundance of total macrobenthic fauna was found during pre-monsoon, followed by monsoon and post-monsoon. However, polychaetes displayed their highest abundance during monsoon, followed by pre-monsoon and post-monsoon.

4.9.2.4 Meiobenthic and Microarthropods Fauna: Diversity, Distribution, and Ecological Services

A research study from the Midnapore (East) coastal tracts has recorded 60 species of meiobenthic fauna, dominated by nematoda, harpacticoid copepods, foraminifera, and polychaeta. These showed distinct seasonal variations with respect to their diversity and density that closely corresponds with fishery production and fish landings in the area (Datta & Chakraborty, 2015). These faunal components can be used to hypothesize the mutualistic dependence of meiobenthic fauna with fishery components and how such relationships may be used to determine the underlying principles for developing tools for fisheries management.

Meiofauna—which are haptosessile or mobile benthic invertebrates—distinguish themselves from macrobenthic fauna by virtue of their small sizes, though they share a sizable proportion of total benthic biomass in marine habitats. All of these tiny benthic faunal components within any estuarine and marine ecosystem play important functional roles: facilitating biomineralization; acting as important component in the food chain and food web processes; and serving as sensitive faunal components serving as excellent agents biomonitoring ecological change (Higgins & Thiel, 1988; Westheide, 1991; Giere, 1993). At least 22 of 33 metazoan invertebrate phyla have been recorded as meiofauna across the world. The taxa belonging to meiobenthic fauna which are Cnidaria, Turbellaria, Sarcomastigophora, Copepoda, Cladocera, Ostracoda, Ciliophora, Nemartina, Nematoda, Gastrotricha, Rotifera, Loricifera, Priapulida, Kinoryncha, Polychaeta, Oligichaeta, Sipuncula, Tardigrada, Mystacocarida, Syncarida, Thermosbaenacea, Isopoda, Tanaidacea, Amphipoda, Cumacea, Halacarida, Pycnogonida, Palpigradida, Insecta, Bryozoa, Entoprocta, Brachiopoda, Aplacophora, Gastropoda, Bivalvia, Holothuroidea, and Tunicata (Higgins & Thiel, 1988).

Meiofaunal organisms exert top-down control over trophic interactions within estuarine and marine ecosystems by grazing on bacteria, diatoms, detritus, protozoans, dissolved organic matter, and microalgae, or even preying on other meiofauna and juvenile stages of macrofauna (Tenore et al., 1977; Higgins & Thiel, 1988; Alongi, 1988). These meiobenthic faunal components are also used as novel sources of food for other macrofauna and demersal fish. Meiobenthic fauna act as sensitive organisms in indicating the ecological changes of their habitats in view different estuarine pollutions mostly because of their high species diversity, short span of life cycles with frequent generation times; and they also facilitate biomineralization (Kennedy & Jacoby, 1999; Datta & Chakraborty, 2015). Thirty-seven species of meiobenthic foraminifera (Ghosh et al., 2014), and more than 20 species of nematodes (Datta & Chakraborty, 2015), along with a number of oligochaetes, polychaetes, harpacticoid copepods, and kinorhynch have been found in this coastal belt (Datta & Chakraborty, 2015; Datta et al., 2016). Another research study has documented 51 species of soil microarthropods belonging to different insect orders, namely Collembola, Hymenoptera, Diptera, and Isoptera from different parts of the coastal belt where they actively contributed as macro decomposers in estuarine–mangrove nutrient cycling (Dey et al., 2007, 2008, 2010).

4.10 Fish Diversity: Fish Species Composition and Distribution

4.10.1 Mangroves Fisheries Connection

Estuaries and mangroves at the coastal zone act as important nursery grounds for littoral finfish and shellfish populations. These shallow aquatic habitats receive ample sunlight, which generates abundant phytoplankton and zooplankton as novel sources of food particles. These environments also provide shelter from predators and offer conducive ecophysiological conditions for the successful development of eggs, embryos, larvae, and juvenile fish (Jayaram, 1981; Blaber & Blaber, 1980; Blaber et al., 2000). Many coastal fish populations depend on these ecologically critical areas—rich with life-support ingredients—at least during specific phases of their life cycle (Weinstein, 1979; McHugh, 1985). Some studies have categorically pointed out positive relationships between the vast mangrove area and local fishery production (Martosubroto & Naamin, 1977; de Graaf & Xuan, 1998). These positive relationships follow the underlying principles that the higher quantity of organic nutrients, as well as dissolved and particulate chemical elements, are exported from the mangrove forest subsystem to the adjoining estuarine subsystem (Alongi, 1989; Hatcher et al., 1989; Chong et al., 1990; Lee, 1995). All these mangrove plant's nutrients are derived and found to stimulate biological productivity in the overall coastal environment, which is reflected by appreciably greater fishery yields off coasts with estuarine–mangrove forests than off non-mangrove coasts (Marshall, 1994). In the mangrove–estuarine belts, many faunal components get suitable shelter and plenty of food in the vegetated intertidal and subtidal habitats (Odum et al., 1982; Blaber, 1986; Robertson & Blaber, 1992).

Innumerable fish and shellfish thrive in the ecological habitats and niches of the water networks of Sundarbans—formed by rivers, estuaries, creeks, and channels. The rich water resources are crucial for ensuring ecosystem stability and enhancing livelihood generation for the people of the region. Almost 50% of the global commercial fish catch come from the sea and estuaries. These environments also support the lives of a large number of shellfish and demersal fish, which are the main food items sustained by benthic fauna (Figs. 4.20, 4.21, 4.22, 4.23, 4.24, 4.25, 4.26, 4.27 and Table 4.13).

4.10.2 Uniqueness of Sundarban Ecosystem with Respect to Fishery and Aquaculture

India serves as the nursery ground for almost 90% of the aquatic species on the eastern coast of India. The ecological health of Sundarban ecosystem is connected to the availability of important commercial species found on the continental shelf of India and neighboring countries. The food web dynamics depend heavily on the contribution of the Sundarban mangroves ecosystem to the steady supply of food materials in the form of both living and nonliving food resources. The large number of planktonic components of the deltaic Sundarbans mangrove–estuarine complex—including phytoplankton, zooplankton, and ichthyoplankton—flourish in a physiologically suitable environmental set ups with beneficial physicochemical attributes like temperature, pH, and salinity. This ecozone also receives an abundant supply of both autochthonous (large quantities of organic detritus from mangrove forest) and allochthonous nutrients (derived from land drainage). The arrival, growth, maturity, and propagation of many commercially important estuarine fish within the estuarine networks of Sundarbans make up a large part of the near-shore fishery of the northern Bay of Bengal. Several other finfish and prawns that spend most of their lives in freshwater reach the estuarine belts annually for spawning and rearing their juveniles in these brackish water zones. In such a context, many finfish and shellfish species that inhabit both marine and freshwater environments require these environmental supports to complete their life cycles, thereby constituting proper linkages between the two culture systems in this ecoregion. Sundarbans, India being the nursery ground of so many commercially important finfish and shell fish, boasts approximately 172 species of fish, belonging to 44 genera, 30 families, and 16 orders; 20 species of shrimp and prawns; and 34 species of crabs, including one commercially important edible species.

A total of 44 fish species belonging to 39 genera under 30 families and 16 orders have been recorded from selected canals of the Indian Sundarbans. Analysis of the fish assemblage reveals the higher relative abundance of species within the family Cyprinidae, followed by Danionidae, Bagridae, Ambassidae, Channidae, Notopteridae, Mastacembelidae, Anabantidae (3.02% Gobiidae, Oxudercidae, Mugilidae, Megalopidae, Schilbeidae, and Megalopidae respectively. The major species of fish fauna are *Boleophthamus boddarti*, *Oreochromismoss ambicus*, *Rasbora daniconius* and *Salmostomo bacaila*. *Puntius sophore*, *Puntius ticto*, *Amblypharngy odonmola*, *Megalops cyprinoides*, and *Chanos chanos*. Small indigenous fish species (SIFS) constitute a major share within the fish assemblage of Sundarbans canals, contributing > 75% of the total fish diversity. The species composition of the canal fish community is represented by *Amblypharyngodon mola*, *Parambassis ranga*, *Chanda nama*, *Puntius sophore*, *Pethia ticto*, *Trichogaster* spp., *Glossogobius giuris*, *Channa punctata*, *C. marulius*, *Salmophasia bacaila*, *Notopterus notopterus*, *Mystus gulio*, *Anabas testudineus*, *Macrognathus pancalus*. *Chelon parsia*, and *C. tade* were also major catches.

The fish assemblage showed only minor seasonal variation, largely due to the presence of sluice gates that strongly restrict both tidal exchange and the movement of fish species between the canal and the main river. Water quality parameters—including temperature, salinity, dissolved oxygen, pH, conductivity, transparency, turbidity, and nutrient concentrations—play crucial roles in structuring fish assemblages in the different brackish-water impoundments such as ponds, canals, and bheries.

4.10.3 Capture Fisheries in the Sundarbans Mangrove Ecosystem in Contrast to Other Estuaries

Fish landings data along the lower Sundarbans region of India revealed a major shift in catch composition over recent years. Landings are now increasingly dominated by trash fish along with some common fish species like *Harpadon nehereus*, sardines, and ribbon fish. A total of 84 finfish and shellfish species belonging to 38 families and 11 orders have been documented from the Narmada estuary, whereas a total of 59 species belonging to 27 families and 11 orders were recorded from the Korapuzha estuary. The flagship Hilsa program, based on analysis of catch composition, reproductive biology, sex ratio, size at first maturity, maximum sustainable yield (MSY), and spawning stock biomass of *Tenualosa ilisha* in the Hooghly estuary, revealed that the total Hilsa catch from the Hooghly–Bhagirathi river system in 2016–17 amounted to 48 922 t. Of this catch, 99.4% was from the marine sector. The MSY for the species was estimated at 32 953 t, which is close to the present average catch from the Hooghly–Bhagirathi system. The size group of Hilsa with sizes ranging from 341 mm to 360 mm had the highest reproductive potential in terms of fecundity.

4.10.4 Aquaculture Activities in Sundarban and Culturable Species

The South 24 Parganas district of West Bengal, home to more than 85% of the mangroves of the Indian Sundarbans, accounts for over 30% of the state's marine fish and prawn production. Culture and capture fisheries together constitute the economic backbone of Sundarbans and provide livelihood support to a large coastal population. In addition to coastal and brackish water aquaculture, freshwater aquaculture has been increasing steadily during last few decades. contributing to a developing parallel economy and enhancing livelihood security in Sundarbans. (Tzeng & Wang, 1993; Neira & Potter, 1994; Patra et al., 2005; Mishra et al., 2009; Nsor & Obodai, 2016).

A wide range of culturable finfish and shellfish species—both freshwater and brackish-water—are cultured in several backyard ponds across the Indian Sundarbans for aquaculture. The family members of the local people excavate a portion of their low-lying paddy field to serve three-fold purposes: collecting water for domestic and agricultural activities; construction of dwelling houses with the excavated earth materials; and for the culture of fish. These excavated areas, owned by local people, store rainwater in the monsoon season. Fish produced in these small ponds are used for the consumption of family members of the local people, while only a small amount is sold. Freshwater aquaculture is also conducted in some larger ponds, mostly owned by groups of families, and in land-shaping ponds, excavated for irrigation purposes in agriculture, as well as in low-lying inundated paddy fields. Brackish-water aquaculture is carried out by channeling tidal saline water through sluice gates from nearby estuaries or creeks into the agriculture fields or large artificial enclosures developed by erecting earthen dikes in coastal swamps, and are used for commercial pisciculture. In these bheries, Sundarban farmers culture a wide variety of freshwater and brackish-water fish species, including *Lates calcarifer, Mugil cephalus, Liza tade, Elops machnata, Oreochromis urolepis, Chanos chanos, Penaeus monodon, Penaeus indicus,* and *Scylla serrata* (Figs. 4.26, 4.27 and 4.28).

4.10.5 Shrimp Farming Status in Sundarban Region

In Sundarbans, brackish-water aquaculture is largely driven by shrimp farming, with *Peneaus monodon* (tiger shrimp) is the predominant species, because of its expanding export demand and high unit-value realization. A semi-intensive shrimp culture began at a good pace beginning in the middle of the '80s and continued through the '90s utilizing more than 34 600 ha in West Bengal. Production of shrimp through aquaculture increased from 12 500 t during 1990–1991 to 33 685 t in 2009–2010. Large-scale shrimp farming is usually practiced in low-lying areas and shallow-water bodies enclosed by constructed earthen bunds, which are filled with tidal water during lunar phases. Entry of the water is normally regulated by means of a sluice gate, made with wooden planks of bamboo poles and fitted with a screen for the filtration of the tidal water while allowing juveniles of fish and shrimp species to gain entry into these water bodies. The partial exchange of water is done during spring tides taking advantage of the tidal amplitudes in the lunar cycle. Such aquaculture practices enable the growth of filamentous algae along with other macrophytes in the shallow-water culture bodies, making an excellent natural rearing environment for the growth of shrimp. The rapid expansion of shrimp farming was halted in the mid-1990s, mostly in the year 1996–1997, when India's Supreme Court banned the indiscriminate increase in shrimp culture mainly over concern about several environmental problems.

West Bengal contributes the major share in aquaculture production of India, mainly because of the aquaculture activities in Sundarbans. A large proportion of Sundarbans population depends on culture and capture fisheries, which collectively form the backbone of the region's economy. Beyond brackish-water systems, freshwater aquaculture is of utmost daily importance as it contributes to supplemental income streams and livelihood security in Sundarbans. In addition to supporting the lives of more than 300 fish species, Sundarbans acts as the nursery ground for nearly 90% of the aquatic species of the eastern coast of India. The occurrence and abundance of commercially important fish species exploited by India and neighboring countries along the continental shelf of the Bay of Bengal depend heavily on the ecological health of the Sundarban mangrove ecosystem. This system provides critical nursery and feeding grounds, supplies a huge amount of essential nutrients from the mangrove forest and maintains key favorable environmental variables—temperature, salinity, transparency, and dissolved oxygen. These inputs constitute an important source of energy for a wide variety of estuarine consumers. Further, several commercially important estuarine fish grow to maturity in the ecotonic productive ecozones within Sundarbans, constituting a major share of the near-shore fisheries of the northern Bay of Bengal. A few freshwater fish and prawns descend to the estuary for spawning and to complete their life cycle. In this context, there exists linkages between inland and marine fisheries in Sundarbans, and commands an integrated approach to aquaculture development and management in this area.

Brackish water aquaculture in Sundarbans is centered mainly on shrimp farming. The production from these shrimp culture activities increased from 12 500 t during the year 1990–1991 to 33 685 t in 2009–2010 in the state of West Bengal, India. However, the growth of the semi-intensive shrimp farming suddenly halted during 1996–1997 because a sudden epidemic outbreak of white-spot disease. The annual production of shrimp from the traditional aquaculture fields like bheries or in low-lying areas and shallow-water bodies, influenced by tidal water, is incredibly low (200–350 kg/ha). Assisted by tidal lunar cycles, brackish water from the estuaries, rivers, creeks, backwater, and lagoons flows by gravity into large and shallow areas of 10–100 ha, enclosed by constructed earthen bunds on its periphery. Water depth of the rearing areas are kept in between 1 and 3 feet where partial exchange of water is made in the lunar phases taking advantage of the tidal amplitude of spring and neap tides. The entry of water is normally regulated by means of a wooden sluice made of planks and bamboo screens for filtration of the water. Larger predators are blocked by these screens while seeds of different fish and shrimp species find entry into these areas along with the tidal water. Farmers usually stock natural or hatchery-produced seeds of *Penaeus monodon* for better yield. In several cases, seeds of *Penaeus monodon* are initially released in specially-enclosed nursery areas. Thereafter, seeds are released into the main water bodies by cutting the earthen bund of the nursery pond at 2–3 separate locations. This aquaculture practice involves periodic stocking and harvesting of cultured species based on the intensity and amplitude of the tides.

4.11 Other Forms of Biodiversity Components Under Higher Chordates (Reptiles, Birds and Mammals)

After their origin and evolution during the Jurassic and Cretaceous periods, around 8000 species of living reptiles flourished on land, while only about 80 species inhabit the marine environment. There are only seven species of living marine turtles, two species of saltwater crocodiles, (the American crocodile, *Crocodylus acutus* and the Indo–Pacific crocodile, *Crocodylus porosus*), and one species of marine lizard (*Amblyrhynchus cristatus*) which occurs in the Galapagos Islands as a seaweed-eating iguana. Marine reptiles, by virtue of possessing salt-excreting glands instead of having kidneys, are able to keep up with the excess salt intake of marine life. The marine turtles demonstrate an amphibious mode of life cycle, spending some parts of their life cycle in the marine water, but employ land habitats for laying their eggs.

A few species of snakes are found as marine snakes dwelling in tropical warm waters of Indian and the Pacific Oceans, some of which represent the most venomous of all snakes. Sea birds—gulls, terns, puffins, ducks, waders, geese, plovers and sandpipers—prefer different zones of the marine–coastal ecosystem—seagrass, mangroves, lagoons, creeks, etc. They have developed adaptive features in the forms of webbed feet for swimming or paddling, and also evolved by the development of salt-excreting glands. Although, earth mammals appeared around 65 million years ago, there is a lesser diversity of mammals in mangrove–estuarine ecosystems, represented by Cetaceans (whales, dolphins, and porpoises).

4.12 Problems and Prospects of Biodiversity of the Indian Sundarbans, India

The halophytic plants within mangrove ecosystem can accumulate higher biomass at a rapid pace by means of sequestrating more carbon in comparison to other forest types. Conversion of mangrove forests to agriculture or aquaculture, results in the release of substantial carbon stocks from both aboveground biomass and extensive belowground root systems, making mangrove loss a major source of greenhouse gas (GHG) emissions. Mangroves and their associated organisms render a number of ecological services like arresting of erosion, releasing of huge amount of oxygen to the atmosphere, serving as breeding ground for so many commercially important shellfish and finfish, providing shelter for a good number of faunal components like birds and insects, supplying food and fodders as well as shelter materials for human beings (Spalding, 1997; Chakraborty, 2017).

A rich diversity of flora and fauna, inhabiting varied ecological niches and habitats, encounter all the moments of their life cycle within varying levels of ecological parameters (e.g., temperature, pH, salinity, rainfall, humidity, and nutrients), and ecological processes (e.g., erosion and accretion, volume and depth of water, duration of tidal exposure and inundation, freshwater inflow, and mixing with saline water) in salt marshes, creeks and tributaries, sandflats and mudflats of Sundarbans mangrove–estuarine complex (Anon, 2003; Chakraborty et al., 2009).

4.12.1 *Species Composition and Diversity in the Coastal–Estuarine Mangrove Complex of West Bengal, India*

The Sundarbans mangrove ecosystem and estuarine networks shared by two districts of West Bengal, India, South-24 Parganas and Midnapore (East) by virtue of offering ecological heterogeneity in form of back water, brackish water, intertidal belts, mangrove forest, etc support the lives of galaxy of flora and fauna which display different forms of ecological succession and community interractions (Chakraborty, 2011, 2017)

4.12.2 *Ecological Characteristics of Mangrove Habitats*

The tide-dominated coastal–estuarine environment at the low-lying Ganga delta provide a wide-range sheltered substratum, mostly muddy in nature, on which mangroves settle and grow (Hutchings & Richer, 1981; Krishnamurthy et al., 1987; Koch, 2000; Gray and Elliot, 2009). Exposed mudflats at the junction of sea and estuaries receive silt accumulation from the deposition of sediments which take place usually during high tide and through discharges by the floodwater of rivers during monsoons. Mangrove swamps are located above open mudflats and function as sediment sinks, analogous to coastal dunes. They are geo-morphologically significant to the mudflat–mangrove swamp interaction where the incoming rivers supply fresh silts settle down and deposit sediments to raise the level of the shore–fringe mudflats. Residual sediments are redistributed into the swamps by tidal drift at high tides. The anastomosing pneumatophores and crown of tips of the stilt roots interfere with the mobility of the sediments across the swamp surface. The colonization of mangroves and their luxuriant growth with intricate root-mat systems effect a process of rapid accretion, found to be as high as 12 cm per year in the recently-developed swamps of Prentice Island on the bank of the Saptamukhi estuary of Sundarbans, India. However, the muddy sediments at the floor of dense mangrove forest saturated with water at the tide-dominated intertidal environment, mostly maintain an anoxic condition with an acid-sulfate influence and bluish-gray color. Very selective plants can withstand this adverse ecological condition where the very top layers of the sediments are sometimes slightly oxidized and becomes brownish in color.

The ebb-dominated Sundarbans shore–fringe mudflats below the swamp level are scoured by a higher intensity of water flow. Along with the abrasive action of erosion, storm deposition and channel shifting often remove the surface and subsurface sediments of mudflats and thereby produce a significant swamp cliff that leaves the root zone exposed to wave and current attack. Research studies have shown the retreat of upper mudflat surface has been taking place at the rate of 30–43 cm per year along the intertidal belts of Dublahat, Beguakhali, Shibpur, Bakkhali, Henry's island, south of Lothian island, and Tar island, while the rate of such retreats in the cliffed swamp edge near Bhagbatpur crocodile project area, and along the banks of Lothian and Prentice islands at the Saptamukhi estuary, etc.) estuaries have been at 10–14 cm per year.

Many of the matured mangrove plants like *Avicennia alba, Avicennia marina, Sonneratia apetala, Exocoeria agalocha, Acanthus illicifelius and Ceriops* decandra have been uprooted, remaining in a hanging condition, and entirely removed from the cliffed banks of mangrove swamps along several creeks and the estuaries of Muriganga, Hooghly, Saptamukhi, Matla, Bidyadhari, Thakuran, Gosaba, and Raimangal. The mature plants under mangrove plant species like *Rhizophora mucronata* and *Phoenix pelludosa* can withstand being dislodged from the swamp edge by virtue of their root characters. This kind of instability of the mudflats due to unconsolidated sediments can be overcome by the viviparous habit of the seedlings of several mangrove plants and other tree species, like *Rhizophora apiculata* (Khamoo or Bhora) and *Aegialitis rotundifolia* (Satari or Tora). *Ceriops tagal* (Goran), *Rhizophora mucronata* Bhora, successfully adapt by virtue of their root system. The quick establishment of torpedo-shaped propagules on soft and saturated muds of the upper mudflats help settlements and colonization of mangrove plants along the tidal creeks. Establishment of mangrove seedlings and colonization of mangroves makes the forest floor, as well as the adjoining estuarine zones, habitable for innumerable faunal components by promoting decomposition cycles, nutrients enrichment, and oxygen recharging.

4.12.3 Diversity and Distribution of Floral Components of the Mangrove Ecosystem

Since the late eighteenth century, the landscape of the Indian Sundarbans has changed remarkably due to neotectonic movement that produced a tilting effect that drove a huge amount of freshwater to the major rivers and estuaries of the Bangladesh part of Sundarbans. This situation was aggravated due to the construction of the Farakka Barrage in 1975, for which a good amount of freshwater has been diverted to the Bangladesh side of the Indian Sundarbans. Owing to the shortage of freshwater flows along with large-scale erosion followed by vertical accretion, species within the mangrove ecosystem are threatened and many are already on the verge of extinction. Diversity of halophytic mangrove plants within mangrove ecosystems, in Sundarbans of the Indian subcontinent cover an area of about 66% of the total mangrove coverage in India. The total number of the major floral components within the mangrove ecosystem, categorized as mangrove associates, minor mangrove elements, and back mangroves, are reported to represent about 90% of the total mangrove species in the Indian subcontinent. An estimate has shown that Indian mangroves with all these major and minor elements are spread over the land areas of deltaic islands and estuarine mouths of about 6820 km^2. Another more recent estimate has shown that the total areas of the Indian Sundarbans inhabited by mangroves was 3554 km^2 (Naskar & Guha Bakshi, 1987; Iftekhar & Saenger, 2008).

A more recent estimate tallies the Indian mangrove areas at 39 699 km^2. Out of this, the total area of Sundarbans mangroves shares about 4.266.6 km^2, including estuarine rivers, creeks, other watersheds, and river banks. The largest area of mangroves in the deltaic systems of Sundarbans has attracted the attention of many naturists and ecologists worldwide since the late 18th century. Early botanical documentation of the lower Gangetic delta includes contributions from William Roxburgh (1814) in his *Hortus Bengalensis* catalogue of plants, followed by important scientific contributions on various plant species of the Sundarbans mangrove ecosystem by Master (1840), Voight (1845), Anderson (1862), Schlichs (1875), Clarke (1895), Prain (1903a, 1903b, 1903c, 1905) and Griffith (1933).

Both ground-truth surveys and analyses of satellite imagery have revealed continuous changes and alterations of the physiography of the Bengal basin in general and Sundarbans in particular, due to several natural changes—movement of tectonic plates, catastrophic cyclones, land subsidence, vertical erosion, geographical changes, changes in the course of estuarine flows, and massive soil erosion. Changes are also due to rapid population explosion and the overdependence of local people on mangrove-based natural resources (Naskar & Guha Bakshi, 1987). Review of literature has shown the contributions of different researchers to the taxonomic identification and diversity of Indian mangroves—Blasco (1977), Chanda, Mukherjee and Mukherjee (1978), Untawale (1986), Banerjee, Banerjee et al. (1989), Chaudhuri and Choudhury (1994), Naskar (1983) and Jagtap (1985, 1993).

Among the major species halophytic mangroves in India, the largest number of species (11) are included under genera Bruguiera, Rhizophora, Ceriops and Kandelia, of the family Rhizophoraceae in Sundarbans, India. Two more species have been recorded under the genus Rhizophoraviz., *R. lamarkii* and *R. stylosa* from the Andaman and Nicobar Islands (Jagtap, 1985; Banerjee et al., 1989), and Banerjee has also recorded the species *R. stylosa* from the Mahanadi Delta of Odissa coast in addition to the nine common species under the genus *Rhizophera* from Sundarbans, India. Another major mangrove genus *Avicennia*, includes three species: *Avicennia alba, A. marina*, and *A. officinalis*. The genus *Sonneratia* also has three mangrove species: *Sonneratia apetala, S. caseolaris*, and *S. griffithii*. Other mangrove species from the Indian Sundarbans are *Xylocarpus monogenesis, X. gronatum, Aglaia cuculata, Excoecaria agallocha, Lumnizera racemosa, Heritiera fomes, Aegiceras corniculatum, Aegialins rotundifolia,* and *Scyphiphora hydrophyllacea*.

The climbers growing the mangrove forest include *Brownlowia tersa, Salacia chinensis, Ipomea per-caprae, Derris scandans, D. trifoliata, Sarcolobus carinatus, Canavalia cathartica, C. lineata, Pentatropis capensis, Solanum surathense, S. trilobatum, Flagellaria indica*. Some grass species found to occur in higher density are *Phragmitis karka, Myriosachya wightiana,* and *Porteresia coarctata*. The herbs include *Aerva lanata, Ammonnia baccifera, Caesalpina crista, Selicornia brachiata*, and *Suaeda maritima*. One very common fern observed in mangrove–estuarine belts is *Acrostichum aureum*.

The adaptive abilities of different species of mangroves against the fluctuating ecological conditions determine the occurrence, distribution and succession of mangrove species (Chaudhuri & Choudhury, 1994). Globally 69 species of true mangroves belonging to 20 families and 26 genera have been documented by Duke, but Kathiresan and Bingham (2001) recorded and described 65 species under 22 genera and 16 families as to be true mangroves. Although, all those taxonomically separated halophytic plant species, trees and shrubs demonstrate special adaptations for their fruitful survivability in intertidal zone (Tomlinson, 1986). Such adaptability of mangrove plants to thrive well in the stressful ecological condition is ensured by the possession of viviparous germination, pneumatophores and trap roots, thick lignin coating on leaves, sunken stomata. However, almost 80 other species of herbaceous and woody plants, have been recorded as mangrove associates but showing no special adaptation for living in the intertidal environment and they bridge the gap between mangroves and terrestrial vegetation. The floral diversity of the Sundarbans mangrove ecosystem harbors 34 true mangrove species, and 40 species of mangrove associated plants (Anon, 2003), a total of 163 species of fungi (Chaudhuri & Choudhury, 1994), 150 species of algae (Naskar et al. 2004) and 32 species of lichens (Santra, 1998).

The species composition of mangrove vegetation depends on the geomorphological characteristics of the intertidal belts, mid-channel bars, levees, and river flood plain point bars along the estuarine networks of Sundarbans, which provide suitable habitats where the mangrove floral components flourish. Preferred salinity and availability of nutrients in the forms of organic carbon, soil texture, water retention capacity of soil to determine the ecological succession of mangrove assemblages. In the levees of the supratidal zone of Sundarbans are found the *Excoecaria—Avicennia—Phoenix* combination. The combinations of *Avicennia—Aegialitis—Rhizophora* and *Avicennia—Ceriops—Rhizophora* occur in the middle and lowest stretch of the rivers. Different mangrove species display varied modes of halophytic adaptations with the development of pneumatophores, prop roots, viviparous seedlings, salt exclusion, salt excretion, salt accumulation, and salt-balance mechanisms.

4.12.3.1 Mangrove Plant Species

In most of the areas of Sundarbans, mangrove forests are reclaimed for settlement and cultivation by felling mangrove trees and by protecting the hinterlands from saltwater menace through artificial damming. The mangrove plant species thrive in the habitats of intertidal zones of newly developed point bars and mid-channel bars, as well as on riverbanks and flood plan areas. In the Indian Sundarbans, the abundance of two important species of mangrove trees—*Heritiera fomes*, locally known as the sundari tree; and *Nypa fruticans*, locally known as golpata—have been declining steadily during over the last century, mainly because of salinity invasion. The mangrove trees of Sundarbans that produce timbers are *Excoecaria agallocha, Ceriops tagal, C. decandra, Avicennia officinalis, A. marina, A. alba, Bruguiera gymnorrhiza, Sonneratia apetala, S. griffithii, Rhizophora mucronata, R. decandra, Aegialitis rotundifolia, Aegiceras corniculatum,* and a palm species *Phoenix paludosa*, whereas, among the minor vegetation, *Salichornia* sp., *Sueada* sp., *Sarcolobus globosus, Derris trifoliata*, and *Acanthus ilicifolius* are the most important (Das & Bhattacharya, 1999b).

4.12.3.2 Diversity of Flora in Sundarbans, India

Of the 90 plant species found to occur in the mangrove forest of the Indian Sundarbans, 24 species are recognized as true mangroves, but 35 species are considered to be mangrove-associated species, while the remaining 31 species are termed 'Back' mangroves. According to the International Union for Conservation of Nature (IUCN) Red List, four mangrove plant species are likely to be under the threat of extinction in the wild. Out of the four species, *Heritiera fomes*, *Sonneratia griffithii*, locally named Ora, are categorized as critically endangered.

Several mangrove plant species such as *Rhizophora apiculata, Sonneratia apetala, Avicennia marina, Excoecaria agallocha, Bruguiera cylindrica,* and *Acanthus ilicifolius* are considered true mangrove plant species, whereas several other plant species are considered mangrove-associated plants as represented by *Sarcolobus carinatus, Suaeda maritima, Pandanus tectorius*. A number of mesophytic invasive plants like *Casuarina equisetifolia* and *Alternanthera* sp. are also found to invade several disturbed mangrove zones, putting ecological pressures on the existing mangrove plant assemblages (Bhakat et al., 2004). The primary producers in aquatic mangrove systems are represented by several phytoplankton species, including *Gosslorilla* sp., *Ceratium* sp., *Nitzschia* sp., *Fragilaria* sp., *Peridinium* sp., *Thalassiosira* sp., *Planktoniella* sp., *Diploneis* sp., *Rhizosolenia* sp., *Ditylum* sp., *Triceratium* sp., *Gyrosigma* sp., *Navicula* sp., *Dynophysis* sp., and *Oedegonium* sp. (Anon, 2003). In addition to phytoplankton, several autotrophic algae inhabit the mangrove forest subsystem as an epiphytic assemblage residing on the stems. Pneumatophores of mangrove trees are a benthic algal mat colonizing on the surface of the sediment and providing a novel source of nutrients to the whole ecosystem. An array of 48 bacterial strains from the decomposed litter of the mangrove ecosystem of Sundarbans, India play key roles in the total nutrient cycle of the ecosystem by their exoenzymatic activities leading to degradation of cellulose, hemicellulose, pectin, chitin, and lignin (Biswas et al., 1986). A total of 150 species of algae have been identified in the Sundarbans mangrove ecosystem. Two species belong to Pheophyta, 15 species belong to Rhotophyta, 50 species of algae belong to Cyanophyta, 39 species belong to Chlorophyta, 44 species belong to Chrysophyta of which 42 species are diatoms (Naskar et al., 2004; Table 4.3).

A total of 90 species of flora are recognized in the mangrove forest of the Indian Sundarbans, of which 24 species are designated as true mangroves, 31 species are back mangroves and 35 are associated-mangrove species. Interestingly, back mangrove species are so high in numbers that Sundarbans serves as home to 70% of all the mangrove species found in the world. Of the 24 species of true mangrove plants, four species are listed as IUCN RED LISTED species, which means that these four species face a threat of extinction in the wild. Of these, *Sonneratia griffithii*, commonly known as Ora, is listed as critically endangered and *Heritiera fomes* or sundari is listed as endangered flora throughout the world. Six species of back mangroves and eight species from the associated mangrove group are also listed in IUCN RED LISTED species (Figs. 4.1, 4.2, 4.3, 4.4, 4.5 and Tables 4.1, 4.2, 4.3).

4.12.3.3 Mangrove Species Under Threats

The mangrove species of Sundarbans are under serious threats, the causal factors of which are habitat degradation, conversion of mangrove areas for intensive aquaculture, unpredictable weather patterns, and increasing salinity.

Two important freshwater mangrove plants, *Heritiera fomes* and *Nypa fruticans*, are endangered in India. Similarly, some other mangrove species such as *Sonneratia acida, S. apetala, S. caseolaris, Aegiceras corniculatum, Kandelia kandel,* and *Rhizophoa* spp. require immediate conservation.

4.12.3.4 Habitats and Zonation Patterns of Mangrove Plants in the Sundarbans Mangrove Ecosystem, India

Intertidal zones of the accreting point bars and midchannel bars in the estuarine tracts of Sundarbans mainly support the settlement and growth of mangrove plants leading to the development of mangrove forests, whereas marsh vegetation is observed to grow mostly near the supratidal zones of natural levees on the estuaries and river flood plains. Tidal marshes in Sundarbans are developed along the subaqueous to subaerial levees and in distal flood plains.

Distinct zonation patterns of mangrove species along the intertidal slope of the Sundarbans mangrove ecosystem is driven by species-specific responses to prevailing ecological factors. These include sloping gradients of intertidal belts, temperature, turbidity, duration of tidal exposure and inundation, salinity, and textural composition of the sediment (Chakraborty & Choudhury, 1992a, 1992b). In some sheltered areas on the deltaic islands of Sundarbans, mudflats are more likely to develop due to higher sedimentation loads coupled with lower velocity of water. Rich organic substances are deposited on those mudflats where finer sediments, loaded with detritus, help the plant species to settle and penetrate their roots into the soft sediments. Initially, the surfaces of these bare mudflats are colonized by thick algal mats, which pave the way for the luxuriant growth of mangrove vegetation and initiate the well-defined sequential zonation patterns (Anon, 2003). Ginsburg and Lowenstam (1958) highlighted the role of mangrove vegetation—including benthic filamentous algae and marine phanerogams—in stabilizing sediments and trapping salt in shallow sheltered water bodies. The sequential settlement, growth, proliferation, and propagation of various mangroves species on newly accreting mudflats help consolidate and elevate the substratum, subsequently inviting successive species to grow and propagate (Paul, 2002). In Sundarbans, zonation patterns of mangroves reveal the presence of different mangrove species occur in tune with the prevailing ecological determinant, whereas landward zones with more freshwater influence are endowed with *Avicenna officinalis, Suaeda maritima, Bruguiera gymnorhiza, Sonneratia apetala*, and *Ceriops roxburghiana*. A number of mangrove plant species like *Heritiera fomes* and *Nypa fruticans* require more freshwater and in their habitat at the mid-tidal level. The mangrove species like *Excoecaria agallocha, Ceriops decandra, Rhizophora apiculata, Phoenix pelludosa* and *Bruguiera gymnorhiza* prefer to grow in relatively higher saline zones withstanding the thrust of higher intensity of water currents and prolonged duration to tidal exposure and inundation. Zonation patterns in mangrove ecosystems often represent a continuing adjustment rather than fixed linear successional development (Lugo & Snedaker, 1974).

The progressive succession of vegetation on Sundarbans mudflats, covering around 70% of the forest area, is: *Porteresia coarctata, Phagmites karka, Sesuvium portulacastrum, Avicennia officinalis* → *Acanthus illicifolius, Aegiceras corniculatum, Avicennia alba, A. marina, Sonneratia apetala* → *Bruguiera gymnorhiza, Salicornia brachiata, Ceriops dacandra* → *Heritiera fomes, Rhizophora apiculata, Xylocarpus mekongenesis* → *Ceriops dacandra, Excoecaria agallocha, Phoenix pelludosa,* and *Suaeda maritima* (Chaudhuri & Choudhury, 1994; Anon, 2003).

4.12.3.5 Ecology of Floral Components in the Indian Sundarbans

The proper functioning and sustainability of the mangrove ecosystem is maintained through intricate interactions among its structural components like plants, microbes, animals, soil and water. The primary productivity of mangrove ecosystems arises from the contribution of three different autotrophic organisms: phytoplankton, benthic algae, and mangrove plants As primary producers, they support the trophic structure by serving as the first level of direct food sources for herbivores (insects, intertidal benthic fauna, forest-inhabiting deer, wild boars, birds, monkeys, etc.). Secondary and tertiary consumers, and other trophic levels, are mainly carnivores and omnivores—finfish, turtles, water monitors, fishing cats, dolphins, and various birds, which channel the flow of energy upward to apex predators, including the Royal Bengal tiger (*Panthera tigris*), and the saltwater crocodile (*Crocodylus porosus*).

Environmental gradients across the Ganga deltaic mangrove dominated islands, from south to north, are considerably characterized by decreasing salinity. The southern zones are dominated by high salinity, while northward areas are influenced by brackish water where saline water is profusely mixed with freshwater at the middle of the delta. The brackish dominated mangrove forests at the Matla–Hariabhanga estuarine complex support freshwater mangrove species like *Heritiera fomes* and *Nypa fruticans*. Further south of this zone, higher salinity influences environmental conditions that favor the settlement of species under the genera *Brugiera*, *Ceriops*, and *Rhizophora*. Mangrove species distribution is not only controlled by salinity, but also by oxygen potential, pH, transparency, turbidity, influx of sediments with a range of cations and anions, hydrodynamic stresses, and interspecific competition patterns (Thom, 1984). The influence of physiography, coupled with ongoing tectonic subsidence and the eastward tilting of Sundarbans delta, play a critical role in restricting mangrove distribution. In addition, sedimentation patterns, the variable intensities of cyclonic storms, and seasonal freshets exert significant influence on the dynamics of the mangrove–estuarine complex of the Indian Sundarbans.

4.12.3.6 Diversity of Bacteria and Fungi

Higher microbial counts, especially various strains of bacteria, were recorded in the upper littoral zones of undisturbed mangrove soils. These microbial components play significant roles in the microbial decomposition of mangrove litter through the exoenzymatic production of several enzymes like cellulase, hemicellulase, cellubiase, pectinance, lignance, chitinase, lipase, and protease (Biswas et al., 1986). A number of bacteria and fungi have been isolated from a wide variety of substrates within the Sundarbans mangrove ecosystem, including 48 bacterial strains, isolated from decomposed litter (Bhaumik et al., 1985). Species of bacteria include *Brevibacterium lypolyticum, B. marinopiscosum, B. minutiferula, B. sociovivum, B. stationis, Listiria monocytogenes, Clostridium carnis, C. pectinivorum, Kurthia bessonii, Lactobacillus brevis, L. delbrueckii, L. fermenti, Micrococcus agilis and M. varians* (Bhaumik et al., 1985). Of these, the genera *Brevibacterium* and *Micrococcus* were found to be abundant. *Bacreia,* identified in decomposed litter, is present within the gut tissues of detritivorus benthic fauna that rely on the substrate as their food sources. Bacteria have been identified from the haemocoelomic fluid of a benthic mud-dwelling globiid fish species, *Boleophthalmus boddarti* and the guts of *Mystus gulio*, and *Uca* spp. Most bacteria belong to ammonifying groups, while others belong to nitrifying groups. A few strains of bacteria, such as chitinolytic bacteria, isolated from a variety of marine and estuarine arthropod sources, help in the breakdown of chitin.

4.12.3.7 Diversity and Distribution of Faunal Components of Mangrove Ecosystem

The vibrant and dynamic Sundarbans mangrove ecosystem sustains a rich biodiversity of aquatic, benthic, and terrestrial organisms through complex interactions among different trophic groups exchanging materials and energy within all those subsystems, largely because of tidal action that interlinks various habitats and promotes ecosystem productivity (Meadows & Tait, 1989; Smith et al., 1992; Weisberg et al., 1993; Wiens et al., 2009; Datta et al., 2018; Ferreira & Faria, 2021). The faunal components inhabiting the intertidal belts function as bioturbators and bring about substantial ecological changes in topography, geomorphology, and biodiversity. In addition to their primary role in the nutrient cycles of a functioning of mangrove ecosystem, mangroves also trap sediments in the intertidal flats, and thereby promote stabilization of the substratum (Marinelli, 1992; Marinelli et al., 2002; Macintosh et al., 2002; Marinelli & Williams, 2003; Manna et al., 2008; Manna et al., 2010; Hansson & Âkesson, 2014).

The fluctuating ecological conditions of the mangrove ecosystem in spatial and temporal contexts promote the settlement, growth, and survivability of a huge number of flora, fauna, and microbial populations in the heterogeneous ecological conditions. The mangrove ecosystem is considered to be the most productive environment, supporting the lives of diversified organisms. A number of research studies have recorded 215 species of fish, 7 species of amphibians, 59 species of reptiles, more than 200 species of birds, 39 species of mammals, as well as numerous species of phytoplankton (> 100 species), algae (150 species), zooplankton (> 100 species), ichthyoplankton (29 families), benthos (26 species of brachyuran crabs; 69 species of polychaetes; 110 Mollusca species, etc.), more than 300 species of soil-inhibiting insects dependent on mangrove plants, along with 44 species of microarthropods (Chaudhuri & Choudhury, 1994; Annon, 2003; Chakraborty et al., 2009; Dey et al., 2010, Chakraborty, 2011). Similar studies have documented numerous commonly-occurring zooplankton orders such as Copepoda (25 species), Rotifera (4 species), Chaetognatha (4 species), and Cladocera (6 species). Important ichthyoplanktons found in this ecozone belong to the families Engraulidae, Megalopidae, and Clupeidae. 79 species of macrobenthos were recorded from various parts of mangrove ecosystem of the Indian Sundarbans, among which *Edwardsia jonesii* and *Paracondylactis indica* represented the most dominant species under the phylum Cnidaria. Polychaetes, a class under the phylum Annelida, are the most diversified group of macrobenthic fauna including major species like *Mastobranchas* sp., *Glycera* spp., *Perinereis* sp., and *Lumbriconereis* sp. Brachyuran crabs are the most bioenergetically significant group of macrobenthic fauna, represented by *Scylla serata, Dotilla blanfordi, Uca* spp., and *Ocypoda macrocera*, representing the classes Crustacea, Littorina, and Nerita, within the phylum Mollusca. A total of 118 finfish species include *Liza, Harpodon, Tenualosa, Chanos, Plotosus, Anguilla,* etc. Faunal diversity further includes 10 species of reptiles (families Varanidae, Colubridae, Crocodylidae, Chelonidae, and Emydidae), about 24 shrimp species, 45 avifauna species (e.g., cormorants, herons, egrets, kingfishers, storks), and 20 species of mammals represented by bats (Chiroptera), rhesus macaque, Indian pangolin, and Bengal fox. Species composition, population dynamics, community structure, and distributional pattern of different groups of fauna have experienced a wide range of spatial and temporal changes because of the prevailing fluctuation in environmental parameters like tidal exposure and inundation, salinity, temperature, and pH (Chakraborty, 2011).

4.12.3.8 Diversity, Distribution and Zonation of Faunal Components

In its hundreds of creeks, tributaries, salt marshes, sandflats, and mudflats, the mangrove ecosystem of Sundarbans supports a rich estuarine faunal diversity in the form of plankton, benthos, and nektons. These assemblages represent complicated food chains and webs at the secondary and tertiary trophic levels within that ecosystem. Diversified faunal components in the biologically productive mangrove ecosystem includes 215 species of fish, 7 species of amphibia, 59 species of reptiles, more than 200 species of birds, 39 species of mammals, as well as numerous species of phytoplankton, zooplankton, ichthyoplankton, benthos, soil inhabiting, and insects dependent on mangrove plants (Chaudhuri & Choudhury, 1994; Ong, 1995; Annon, 2003). Species composition, population dynamics, distribution of various groups of faunal components, and interaction patterns within both the aquatic and forest biotic subsystems, experience a wide range of variations related to the dynamically fluctuating environment. Major groups of macrobenthic fauna like brachyuran crabs, molluscs, and polychaetes have defined distinct zonations along the slopping intertidal belts (Chakraborty & Choudhury, 1992a, 1992b). Copepods dominate the zooplankton community in Sundarbans, comprising 67.8%–90% of total abundance and represented by 36 copepod species across 21 genera and 19 families (Baidya & Choudhury, 1986; Sarkar et al., 1986). Other zooplanktonic faunal groups include holoplankters such as Cladocera, Cumacea, Ostracoda Chaetognatha, Mysidacea, Sergestidae, Amphipoda, and Hydromedusea, as well as meroplankters comprising polychaete larvae, nauplii, zoea, megalopa, fish eggs and larvae, and echinoderm larvae (Annon, 2003; Chakraborty et al., 2009).

One hundred species of icthyoplankton, the juvenile stages of finfish within 29 families constitute an important planktonic group. The major icthyoplankton families are represented by Clupeidae and Megolopidae (Chakraborty et al., 2009). A galaxy of benthic fauna, both as infauna (burrowing, sessile, and semisessile) and epifauna (creepers and crawlers) are found to occur in higher abundance in different ecological habitats and niches within this ecosystem which are subjected to exposures of various oscillating hydrological parameters (Chakraborty et al., 1992; Chaudhuri & Choudhury, 1994). Different benthic faunal components have evolved their own behavioral and physiological adaptative characteristics in order to cope with the environmental stresses because of the impact of varying ecological conditions. Based on body size, benthic fauna is divided into three categories: macrobenthos (> 0.5 mm), meiobenthos (< 0.5 mm), and microbenthos (microscopic). Distinct zonation of microbenthic fauna across the intertidal slope of Sundarbans is driven by differential species-specific adaptations to terrestriality and prevailing ecological parameters (Chakraborty & Choudhury, 1985, 1992a, 1992b). Out of all intertidal macrobenthic fauna, brachyuran crabs constitute the most dominant faunal group having ecological and economic significance, mainly because of their potential as bioturbators, and also because of their foraging activities. In the Sundarbans mangrove ecosystem, early surveys documented 26 species of brachyuran crabs belonging to 15 genera and five families. These crabs exhibit a wide range of behavioral activities that contribute to sediment turnover and nutrient cycling (Chakraborty et al., 1986). A total of 69 species of polychaetes under 45 genera and 25 families were found to inhabit different zones of the intertidal belt of this ecosystem (Misra, 1995; Subba Rao et al., 1995a, 1995b). A total of 110 species of benthic molluscs within the classes Gastropoda (59 species) and Bivalvia (40 species) occupy different habitats of the sandflats and mudflats of the Indian Sundarbans. Both the mangrove forest and associated salt marshes also provide shelter with diversified forms of benthic insects, many of which also inhabit within the soils. A total of 24 dipteran species belonging to five families and 11 genera (Ray & Chowdhury, 1989) and 14 coleopteran species (Poddar et al., 1990) have been recorded from different parts of Sundarbans, India. Mangrove vegetation supports a diverse assemblage of phytophagus insects, with 56 species recorded across several orders: one species of Hymenoptera, eight species of Hemiptera, one species of Homoptera, 16 species of Coleoptera, and 30 species of Diptera (Chakraborty & Naskar, 1988). A rich diversity of insect fauna is found to occur in the ecologically important and climate-sensitive biosphere reserve of the Indian Sundarbans mangrove ecosystems. The heterogeneity of diverse forms of habitat tends to support an array of coexisting ecosystems, including pristine mangrove stands, tidal mudflats, estuarine and freshwater creeks, brackish wetlands, reclaimed agricultural land, aquaculture ponds, and adjoining landscapes with active anthropogenic influences. This ecologically complex mangrove spectrum creates diverse niches suitable for the flourishing of various insect groups.

A total of 91 insect species were previously reported through direct observation, 272 by Panda et al. and Chakraborti et al. (2019) within the mangrove ecosystem of eastern India. Among these, *Apis dorsata* (Chakraborti et al., 2019), *Polistes olivaceus*, *Xylocopa pubescens*, *Allobaccha amphithoe* (Chakraborti et al., 2019), *Chrysomya megacephala* (Chakraborti et al., 2019, *Muscadomestica* (Chakraborti et al., 2019), *Micraspis discolor* (Chakraborti et al., 2019; *Danaus genutia*, *Euploea core* (Chakraborti et al., 2019; and *Melanitis leda* (Chakraborti et al., 2019) are common.

A total of 29 insect species under five orders have been documented in the Sundarbans biosphere reserve of which the highest number of insect species were found under the order Hymenoptera, followed by Diptera, Lepidoptera and Coleoptera. Maximum insect diversity and abundance coincide with the flowering seasons of their respective host plant species. The insect faunal groups were registered in higher abundance during flowering seasons of the mangrove plants like *Avicennia marina*, followed by *Avicennia officinalis, Acanthus ilicifolius* and *Aegiceras corniculatum* where flowers provide important food sources for pollinators, and thereby help maintaining pollinator diversity in the mangrove community.

Insect–plant interaction during flowering season presents a higher possibility of successful pollination and seed-setting, and these relationships play a crucial role in the reproductive success and propagation of mangroves. In the Indian Sundarbans setting, the major mangrove insect visitors belong to order Hymenoptera, Diptera, Lepidoptera and Coleoptera. Among them, hymenopterans, especially bees, are in higher abundance during dry seasons (d'Avila & Marchini, 2008); lepidopterans dominate early dry seasons (Mertens et al., 2021); whereas, coleopterans and dipterans show their peak activity during the wet season (Oliveira et al., 2021; Denlinger, 1980).

In a recent study, a total of 58 dipteran species belonging to 19 families were recorded in association with various mangrove plant species across the Sundarban Biosphere Reserve (SBR), India. Among these, ten species represent new distributional records for the region. Notably, *Eupyrgota (Taeniomastix) pictiventris* was recorded for the first time in the state of West Bengal. Dipteran species belong to 21 families, and are grouped into six ecological guilds—flower visitors, frugivorous, leaf and trunk resting flies, carnivorous, saprophytes and hematophagous—with flower visitors being the most diverse and abundant guild.

Other faunal groups under benthic fauna include actiniarian, sipunculans, echiurans, hemichordates, globiid fish etc., each of which enjoy preferred specific ecological habitats structured by soils having different textural components. A diversified forms of microarthopods were recorded by Dey et al. (2010) belonging to 44 species under six major taxonomic groups viz. Acarina, Collembola, Coleoptera, Diptera, Isoptera, and Hymenoptera. Several meiobenthic fauna including 80 species of nematodes belonging to 26 families were recorded from this ecosystem (Chaudhuri & Choudhury, 1994). A research survey regarding meiofaunal abundance in Sagar Island, Sundarbans was conducted by Rao and Misra which revealed that nematodes constituted the most dominant group followed by Copepoda, Polychaeta, and Ostracoda. 58 species belonging to 31 genera under 21 families and four suborders were documented from an ecological survey on the distribution of intertidal and subtidal foraminiferan faunal components from the mangrove–estuarine belts of coastal West Bengal (Majumder et al., 1996).

Prolific development of coastal fisheries in and around Sundarbans of India are developed in tune with the similar developments following the out-welling hypothesis of Odum (1971), based on the modes of transportation of detritus contributed nutrients to aquatic subsystem. Two major groups of finfish and shellfish constituting nektonic fauna as residents and transients or migrants in mangrove–estuarine system of Sundarbans. In respect of the abilities of the resident fish to tolerate against salinity are grouped into three categories: first, those fish which can tolerate salinity from 6.2% to 32.86% (Mullets, *Liza parsia* and *L. tade*; threadfins—*Polynemus tetradactylus* and *P. indicus*; *Sciaena miles* and *Sciaenoides biauritus;* perches—*Lates calcarifer;* ribbonfish—*Trichiurus savala*; clupeids—*Hilsa sinensis* and *Coilia bornunsis;* catfish—*Tachysurus jella* and *Plotosus canius*). Second, those fish which can tolerate salinity from 4% to 32.86% (*Harpodon nehereus, Setipinna taty*, *Ilisha elongatae*, etc.). Third category is those fish that can tolerate salinity > 32.86% (*Pama pama, Coilia ramcarati, Polynemus paradiscus, Setipinna phasa, Sillaginopsis panijus*, etc.). The migrant species entering the estuarine zone for breeding and feeding can be broadly categorized into three broad categories. First, marine species migrating to upstream in freshwater areas for spawning of the estuary are *Tenulosa ilisha, Polynemus paradiscus*. Second, freshwater species moving to estuarine zones for spawning in saline areas which include the species like *Pangasius pangasius* and the prawn like *Macrobrachium rosenbergi*. Third, marine fish migrating to estuarine zones for spawning in the brackish water of estuary are *Tachysurus jella, Penaeus indicus, Polynemus indicus, P. monodon*. Out of a total of 172 species of fish recorded from the Hooghly–Matla estuarine complex of Sundarbans, 73 species occupied the less saline zone whereas 99 other species preferred to inhabit in the higher salinity zone. A total of 26 species of shrimps and prawns belonging to five families are recorded from the estuarine system of Sundarbans. A total of 250 species of fish were reported from the estuaries of Sundarbans (Chaudhuri & Choudhury, 1994). However, Naskar et al. (2004) reported a total of 137 fish species from the mangrove–estuarine complex of the Indian Sundarbans, of which 13 belong to Chondrichthyes (cartilaginous fishes, including sharks, rays, and skates), and the remainder to Osteichthyes (bony fishes).

4.13 Ecological Perspectives of Biodiversity in the Indian Sundarbans

Temporal and spatial variations in the distribution patterns of different biodiversity components within the Sundarbans mangrove ecosystem are largely governed by key physicochemical parameters of soil and water. These include salinity, cations exchange capacity, water-holding capacity, and textural soil composition. Such gradients drive pronounced shifts in species composition, with a distinct east-to-west transition across the Indian Sundarbans. Additionally, the biological productivity of the mangrove ecosystem is also regulated by different ecological processes like energy flow and decomposition cycle. Fluctuation of ecological parameters and ecological processes are influenced mainly by changing meteorological parameters—wind flow, temperature, and precipitation—together with marine and estuarine physical processes—tides, waves, and currents—which hamper ecological stability compelling the floral and faunal components to restrict their own distributional territory in different parts of Sundarbans (Chakraborty et al., 2009). The structure and species composition of mangrove forests at both regional and local scales are influenced by key ecological factors and mechanisms:

(i) Land building processes that drive autogenic succession of mangroves.
(ii) Species-specific response to variable geomorphologic settings such as sloping gradients of intertidal belts shaped by currents, waves, tidal elevation, and sea-level rise.
(iii) Physiological adaptations to variable ecological stresses such as texture of soil, salinity invasion, duration of tidal exposure and inundation across intertidal zones.
(iv) Dispersal patterns and abilities of seedlings and propagules of mangroves in accordance with their size, shape, buoyancy, and rooting time.
(v) Establishment of propagules overcoming the problems of biotic competition, and predation during post-dispersal phases.

Mangrove species display differential tolerances and requirements for salinity, pH, temperature, and nutrient availability in both water and soil. These differences reflected by variable modes of species recruitment, establishment, and propagation of the dominant species (McKee, 1995). The production abilities, location of plant species, tidal dispersal patterns, and retention abilities of propagules also influence the initial community structure (McKee, 1995). Ecodynamics for the development of floral assemblages are determined by several factors including availability of suitable sites, water quality parameters, textural composition of soil, meteorological parameters, climatic conditions, and different modes of species interactions. In Sundarbans, the types of assemblages of mangrove species and their growth have been changing principally due to steadily declining flows of freshwater into the estuarine water leading to salinity invasion. Continuous exposure to higher salinity is detrimental to the mangroves, causing stunted growth coupled with reduced production of biomass; although several mangrove plant species possess a higher level of tolerance against increased salinity. Additionally, in Sundarbans, composition, survival, growth, distribution, and succession of plant species are controlled by the salinity, which, in turn, is regulated by topography, hydrology, tidal fluctuation, and climatic conditions.

A research survey has shown the occurrence of 24 true mangrove species, 21 mangrove-associated plant species, and nine mesophytic invasive plant species in the northeastern parts of Sundarbans. In southwest parts of Sundarbans, the same survey recorded 17 true mangrove plant species, 19 mangrove-associated plant species, and ten mesophytic invasive plant species; while the central part is represented by only16 true mangrove plant species, and eight mangrove-associated plant species (Chaudhuri & Choudhury, 1994; Chakraborty et al., 2009).

The greatest density and diversity of phytoplanktonic species have been recorded during pre-monsoon, followed by the monsoon and post-monsoon seasons. However, species richness reached its highest level during monsoon, and diminished to its minimum during pre-monsoon (Chaudhuri & Choudhury, 1994).

The tropical evergreen species of the Sundarbans mangrove ecosystem, *Heriteria fomes*, favors distribution to the inland portion of Sundarbans where there is freshwater dominance due to the profuse monsoonal rainfall. However, salt-tolerant mangrove species were found in abundance in the deltaic islands which are closer to the sea where they experience a more marine influence. Others inhabit brackish water patches, receiving freshwater influence mainly from rainfall (Richards, 1993). A marked decline in rainfall has been evident on many seaward deltaic islands over the 40-year period from 1976 to 2015. This reduction has resulted in increased hypersalinity and the expansion of saline blanks (bare, non-vegetated patches) within the interior mangrove forests along the coastal belt of South Bengal (Ray et al., 2013). Mangrove forests in this coastal belt are adversely affected by climatic variability, notably persistent dry winds, elevated salinity in water and soil, and chemical changes in ground water (Blasco, 1984; Wolanski, 2006).

4.13.1 Phytoplankton Diversity of Sundarbans, Northeast Coast of the Bay of Bengal, India

The Indian Sundarban mangrove (21°32′–22°40′N and 88° 05′–89° E) with a network of estuaries has been subject to tremendous stresses because of the input of plant nutrients and toxic residues, released by agrochemical products, industrial waste, and sewage (Cloern et al., 2007). Primarily due to eutrophication, polluted water bodies disrupt seasonal fluctuations of the phytoplankton assemblages. Changes in major ecological parameters impose harmful impacts on the mangrove-dominated estuarine system, especially during the monsoon period of the year. Diatoms were more abundant than dinoflagellates throughout all seasons with 58 species.

Nutrient-rich agricultural runoff triggers explosive algae growth, designated "algal blooms." Although these blooms initially enhance primary productivity in surface waters, the resulting dense biomass impedes light penetration, and forms algal mats on the bottom sediment. The result is the decline of the primary productivity of the deeper waters, due to reduction in the production of oxygen, mainly because of faster consumption of oxygen by the bacteria inhabiting the decomposing dead plant biomass. The newly-developed algal bloom produces toxins, making the ecological condition of the water very much unsuitable for the living organisms. Analysis of the assemblage of indicator phytoplankton with high sensitivity indicates the change in aquatic ecosystems (Valdes–Weaver et al., 2006).

A total of 58 species of phytoplankton is mainly represented by the higher abundance of *Asterionellopsis glacialis* (21.86%) followed by *Skeletonema* cf. *coastatum* (12.03%), *Coscinodiscus eccentricus* (6.34%), *Pleurosigma* sp. (4.55%), *Thalassiothrix* sp. (4.17%), *Chaetoceros decipiens* (3.87), *Nitzschiaseriata* (2.11%), *Thalassiosira subtilis* (1.15%) and *Thalassiosira decipiens* (1.13%). Dinoflagellates represent a higher species diversity (38–58 species), displaying a striking increase in the number of phytoplankton species from 29 in 1990 (De et al., 1991) to 48 species in 2004 (Biswas et al., 2004) and 58 in 2011 according to De et al. This increasing trend in species diversity reflects distinct successional process in the composition and abundance of the phytoplankton community, both spatially and temporally. Average abundance and biodiversity are closely connected to variations in seasonal and annual ecological parameters in tune with the intense seasonal variation for a tropical region. The monsoon season (June–September) brings 1257–2346 mm of rainfall to Sundarbans, accounting for about 80% of the total annual total. This heavy precipitation is driven by the northward annual shift of the Intertropical Convergence Zone (ITCZ) across the equator (Biswas et al., 2009). Conversely, the southward shift of ITCZ during the dry season results in a long duration with negligible rainfall. These sudden seasonal shifts in climate triggers marked seasonal variability in phytoplankton communities, in respect of both diversity and abundance.

4.13.1.1 Seasonal Occurrence and Diversity Indices of Phytoplankton

A recent study of the seasonal occurrence of phytoplankton from the estuarine tracts of Sundarbans has revealed that 58 species of phytoplankton, mainly represented by the species of *A. glacialis* (27.41%), *Skeletonema coastatum* (6.32%), *Coscinodiscus radiatus* (5.31%), *Coscinodiscus lineatus* (5.12), *Thalassiothrix* sp. (5.06%), *Pleurosigma* sp. (5.04%), *Coscinodiscus eccentricus* (5.03), and *C. decipiens* (5.01%) were observed during post-monsoon. The population density of phytoplankton was found to reach its highest level during post-monsoon which is likely due to improved transparency and increased availability of nutrients brought down by river runoff during monsoon. The reduced abundance of the phytoplankton population during pre-monsoon months, is attributed to heightened grazing pressure exerted by zooplankton (Sarkar et al., 1989).

Biotic indices based on species diversity effectively characterize the phytoplankton community. Average values pertaining to monthly variation of different indices for each month showed that the highest values of the dominance index were found during pre-monsoon and the lower values during post-monsoon. A species community with higher values of dominance indices indicate that one or a few species exert powerful control over other species as the degree of dominance concentrates in any one species or, in other words. it can be said that species group can be expressed by an appropriate index of dominance (Odum, 1971; Krebs, 1985). However, under extreme environmental conditions, the controlling influence on community is divided among fewer numbers of species (Odum, 1971). Higher values in the dominance index during pre-monsoon indicates the phytoplankton community was dominated by a fewer number of species out of total number of individuals, and lower values of dominance was attributed by sharing of dominance index by large number of species during post-monsoon. Higher water temperature, increased salinity, and reduced nutrient concentrations tend to impart considerable impact to determine a high dominance index during pre-monsoon. The species richness or variety index refers to the simple ratio between the total number of different species and the total number of individuals. This points out an important message regarding the diversity of organisms. In the Indian Sundarbans, the highest values of the species richness index for phytoplankton were recorded during the monsoon season. This was thought to be due to the decrease in the total number of phytoplankton species. The Shannon–Weaver index (H^-) refers to the ratio between the number of species and the importance values of individual species. The values of diversity index can be used to compare species diversity among different ecozones under the influence of different ecological parameters. A recent study showed an average annual value of the diversity index as 0.87 ± 0.25, indicating that phytoplankton is under a high frequency of disturbance. In the Indian Sundarbans, the Shannon–Weaver index ($H^{-)}$ values of $<$ 2 indicated a high frequency of disturbance through an increased supply of nutrients as hypothesized by Flooder and Sommer. The monthly estimation of the Shannon–Weaver index (H^-) values show its highest values during pre-monsoon.

4.13.2 Faunal Diversity in Sundarbans, India

The Sundarbans mangrove ecosystem, recognized as one of the world's most ecologically and economically important biodiversity hotspots, supports an exceptionally rich faunal diversity that ranges from insects to mammals. This outstanding biological richness has attracted researchers globally, as evidenced by Stoliczka's (1869) pioneering work on the malacofauna of the brackish-water systems of lower Bengal. However, information about the faunal diversity of this Sundarbans deltaic mangrove ecosystem seems to be scattered over several inaccessible periodicals. All animal phyla under the animal kingdom are represented in this biologically productive and ecologically potent mangrove ecosystem. Faunal components in Sundarbans can be broadly categorized into several categories: zooplankton, benthos, nekton (mostly as fish), periphyton, and wild fauna, encompassing reptiles (snakes, turtles, crocodiles), birds, and mammals (Rog et al., 2017).

4.13.3 Ecological Uniqueness of Faunal Components of West Bengal, Coast, India

Faunal diversity of the estuarine–mangrove complexes has been studied, mainly for the purposes of determining their diversity trophic status, their functional roles as bioturbators, and their direct and indirect connections with fishery production (Paterson et al., 1998). Extensive research on the ecology of two major habitat-based faunal categories—plankton and benthos—has been conducted over the past five decades by numerous investigators in the Sundarbans mangrove ecosystem. Sarkar et al. (1986) highlighted a bimodal pattern in population fluctuation of copepods—the dominant zooplanktonic group—with peak population densities observed during the monsoon season and again in the last phase of the post-monsoon period. In contrast, other zooplanktonic groups like Chaetognatha and Cladocera exhibited their highest population densities during the pre-monsoon period, coinciding with elevated saline levels (Chaudhuri & Choudhury, 1994).

The major macrobenthic faunal groups in the Sundarban mangrove ecosystem comprise brachyuran crabs, molluscs, polychaetes, brachiopods, and nemerteans. These communities displayed varying trends of seasonal fluctuations in relation to their population density and community structure. The maximum population density of fiddler crabs was observed during pre-monsoon, followed by post-monsoon, whereas the highest population densities of many species of polychaetes were found to occur during monsoon. The species diversity index, calculated based on the modes of occurrence of total benthic faunal groups, was highest during pre-monsoon, and at the lowest levels during monsoon. However, the indices that were computed on the bases of polychaete and molluscan population densities which registered maximum values during monsoon (Chakraborty et al., 1992, 2009). The spatial and temporal variations of population densities and community indices of macrobenthic species were attributed to the influence of a variety of ecological parameters. The osmoregulatory abilities of macrobenthos play an important role with regard to their distribution patterns in time and space. Temperature plays crucial roles in regulating coastal faunal distribution in in considerable way (Teal, 1958; Kinne, 1963). The texture and nutrient contents of substrates also play important roles in determining the species composition of the various macrobenthic fauna in different habitats (Abele, 1974). The substrates of a habitat is used by one species as a shelter, whereas another species uses it as a feeding site, and yet another utilizes it as a source of nutrition, thus reducing competitive interactions. These multiple modes of sharing resources result in a positive community existence. The intertidal zone, owing to its high habitat heterogeneity, can accommodate a higher number of species in its relatively narrow spatial extent, which in turn enhances species diversity. Sanders (1968) hypothesized that community interactions and species diversity in the intertidal macrobenthic communities are strongly influenced by stable soil environments. Mangrove ecosystems of the tropics also play vital roles in governing the settlements, growth, population density, and community structure of macrobenthic fauna in the littoral zones with higher nutrient enrichment (Heald & Odum, 1970). The highly buffered nature of sea water counterbalances the limiting effect on marine organisms (Rossi & Chapman, 2003).

4.13.3.1 Diversity of Benthic Invertebrates in the Ocean

Benthic fauna are traditionally classified into three size-based categories: *Microbenthos* (microscopic like protists); *Meiobenthos* or interstitial fauna (pass through a sieve with the mesh size of 0.5 mm and examples are nematodes, copepod, kinorhynchs, gastrotrichs, etc.); and *Macrobenthos* (retained by a sieve having mesh size of 0.5 mm—some examples are crabs, polychaetes, sea anemone, bivalves, gastropods, brachiopods).

Based on trophic structure and foraging strategies, benthic fauna are divided into several categories such as herbivores, carnivores, omnivores, detritus feeders, and suspension feeders. Several water quality parameters such as temperature, dissolved oxygen, transparency, pH, nutrient contents of water, soil parameters like texture and nutrients of sediments and different physical processes like tidal amplitudes, upwelling, currents, current, waves, stratification of salinity, temperature and density determine the abundance and distribution of benthic fauna. The animal kingdom comprises approximately 36 phyla, of which about 30 are represented in the Sundarbans mangrove ecosystem. Many species which share common characteristics but differ with respect to some salient characteristics of each phylum. The phylum Porifera, representing simplest form of metazoan body organization, is endowed with several pores (ostia) over the body surface, a hollow body cavity (osculum) bordered by ciliated choanocytes cells, very unique supportive hard structures (spicules) constituted by calcium carbonate or silicon compounds, and a specific type of collagen fibers. Another major phyla is Cnidaria (sea anemones, jellyfish, corals, etc.), characterized by a diploblastic body wall with a layer of dead connective tissue, mesenchyme in between ectoderm and endoderm, and two lifeforms (polyp and medusa) in their life cycle. The phylum, Cnideria, is unique by possessing a radially symmetrical body with stinging cells, cnidoblast. The phylum Nematoda is also known as meiobenthic or interstitial fauna, living within the interstitial spaces of different structural components of sediments. The faunal composition also includes the phylum Platyhelminthes (mostly parasitic with some free-living members, Nemertea (ribbon worms), the phylum Annelida (segmented worms), and the phylum Chaetognatha (arrow worms). The largest phylum, Arthropoda, has more than 50,000 marine–estuarine faunal forms belonging to the class Crustacea. The phylum Mollusca has three major classes: Gastropoda, Bivalvia, and Cephalopoda accommodating thousands of intertidal and subtidal species. Another diverse invertebrate phylum is Echinodermata, with diverse groups of marine–estuarine fauna such as star fish, sea urchins, and sea cucumbers, all characterized by a hemal and perihemal system with rows of water-filled tube feet, and diversified larval forms, among other features.

4.13.3.2 Pollination and Pollinators with Special Emphasis to Diversity of Insects

The flowers of mangrove plants acting as main anatomical structures for facilitating sexual reproduction for producing large numbers of pollen grains, the output of which per flower is 7, 00, 000 in *Rhizophora. Mucronata*. 5, 08, 000 in *R. annamalayana* and 4, 68, 000 in *R.apiculata*. The anatomical configuration of stigma determine the efficiency of flowers as receptacle of flowers for pollens at the time of flower opening and govern the success of pollination. However, such receptivity of flowers is lost within 8 hours in different species of Rhizophora (Kathiresan, 2011). The flowers of mangrove vegetations possessing osmophores, a specialized secretory gland located on different parts of flowers, produce and release volatile aromatic compounds with different flavors to attract pollinators. The production of nectors enhances with sizes of flowers and variable organic substances present in nectors like amino acids (histidine) and sugars (Fructose, Sucrose and Glucose) which determine attraction potential for pollinators (Kavitha and Kathiresan, 2009).

Although, different species of mangrove plants are pollinated by a diverse group of animals like birds and bats but insects acts as prime pollinators. Pollens after being attached and deposited on all those animals during their search for nectors in the flowers, they consequently transfer the pollen grains to the stigma of some other flowers. The preference of nectors of different pollinators make them dependance on different mangrove plant species. For example, *Lumnitzera racemosa* and Bruguiera species are pollinated by insects whereas *L. littorea*, is pollinated primarily by birds. Sunbirds visit and pollinate large and bright colored flowers of *Acanthus ilicifolius* (Aluri, 1990) and large but pale white colored flowers of *Brugiera hainesti* (Noske, 1993, 1995). Mangrove plants mostly having large sized flowers undertake pollination by birds during the dry season. Bats also play their roles as effective pollinators for the large sized mangrove plants like Xylocarpus, Sonneratia etc., the flowers of those plants open up exposing the powdery stamens in the late night/early morning hours. In the absence of nocturnal bat species, different species of hawk moths also sometimes serveb as primary night-time pollinators (Hockey and de Baar, 1991). Two lycaenid butterflies have been reported to act as pollinators of mangroves in Brisbane, whereas there abundance is directly correlated with the abundance of mangrove flowers (Hill, 1992). Different species of Bees regularly visit and pollinate species of Avicennia, Excoecaria, Acanthus, Rhizophora, Scyphiphora and Xylocarpus. Some wasps and flies are highly dependent on mangroves for nesting and are particularly important pollinators of *Lumnitzera racemosa. Ceriops decandra, and Kandelia candel* (Tomilson, 1986). Different species under the genus Rhizophora without having any structural modification of stigma for capturing the wind-blown pollen are supposed to experience wind mediated pollination. Several reports are there about the roles of abundant spiders, flies and carpenter ants, as pollinators in different species under the genus Rhizophora where their bodies are found to wrapped by pollens of flowers (Kavitha, 2009).

A recent study from the Sundarbans Biosphere Reserve, India on plant–pollinator relationship among four true mangrove plants, viz. *Avicennia marina, Avicennia officinalis, Aegialitis rotundifolia* and *Aegiceras corniculatum* have revealed the functional roles of insect visitors with their foraging activities as efficient as pollinators (Chakraborti et al., 2019). Among the common five visitors, viz. *Apis dorsata, Apis mellifera, Danaus chrysippus, Chrysomya megacephala* and *Micraspis discolor, A. dorsata* exhibited the highest visitation rate for all studied plant species whereas M. *discolor* demonstrated the lowest visitation rate both *A. dorsata* and *A. mellifera exhibited* maximum pollen carrying capacities in comparison to other such kind of insect species.

Different kinds of behavioral manifestations of different flower visitors in Sundarbans mangrove ecosystem occur because of the availabilities of multiple sources of pollination resources and varied forms of plant morphology. The nectars in the flowers of different mangrove plants, some amino acids like lysine, valine, phenylalanine, tryptophan, histidine and threonine are present which are required by honey bees. Therefore, these plants possessing all of these chemical constituents can experience more pollination by bees or they exhibit better melittophilous pollination (DeGroot, 1953; Solomon Raju, and Karyamsetty, 2018). Pollen carrying capacity of different insects as being flower visitors tend to determine the plant reproductive success, especially for cross-pollinating plants. It has been found that among the five common flower visitors.

4.13.3.3 Diversity of Insects in the Sundarbans Mangrove Ecosystem, India

Diversified groups of insect fauna are recorded from different corners of Sundarbans, India, include Coleoptera, Diptera, Lepidoptera, Hymenoptera, and Hemiptera. Through an array of plant –insect interactions these insects ensure pollination of mangrove plants, produce some forest-based products like honey, and act as food sources for higher groups of wild fauna such as birds, reptiles (Fig. 4.19). A rich diversity of insect fauna is found to occur in the ecologically important and climate-sensitive mangrove ecosystems, of the Sundarbans Biosphere Reserve, India. The diverse forms of habitat heterogeneity, support an array of coexisting ecosystems, including pristine mangrove stands, tidal mudflats, estuarine and freshwater creeks, brackish wetlands, reclaimed agricultural land, aquaculture ponds, and adjoining landscapes with significant anthropogenic activities. This complex ecological spectrum supports insect faunal components in different niches that are suitable for the flourishing of different groups of insects.

A total of 91 insect species were previously reported through direct observation (Chakraborti et al., 2019) of the mangrove ecosystem of eastern, India. Among these, the following species were found to be common to those studies: *Apis dorsata* (Chakraborti et al., 2019), *Polistes olivaceus*, *Xylocopa pubescens*, *Allobaccha amphithoe* (Chakraborti et al., 2019), *Chrysomya megacephala* (Chakraborti et al., 2019; *Musca domestica* (Chakraborti et al., 2019), *Micraspis discolor* (Chakraborti et al., 2019; *Danaus genutia*, *Euploea core* (Chakraborti et al., 2019; and *Melanitis leda* (Chakraborti et al., 2019).

A total of 29 insect species under five orders have been documented in the Sundarbans biosphere reserve, where the number of insect species under the order Hymenoptera was found to be highest, followed by Diptera, Lepidoptera and Coleoptera (Roy et al., 2019). The maximum abundance of all these species coincides with the flowering seasons of their respective plant species. The insect faunal groups were registered in higher abundance during the flowering seasons of mangrove plants like *Avicennia marina*, followed by *Avicennia officinalis*, *Acanthus ilicifolius,* and *Aegicera scorniculatum.* where the flowers provide important food sources for pollinators and thereby help maintain pollinator diversity in the mangrove community.

Insect–plant interaction during flowering season depicts a higher possibility of successful pollination and seed-setting; thus, such relationships play a crucial role in the reproductive success and propagation of mangroves. In the Indian scenario, the major mangrove insect visitors belong to the orders Hymenoptera, Diptera, Lepidoptera and Coleoptera. Among these, hymenopterans—especially bees—have higher abundance during dry seasons (d'Avila & Marchini, 2008). Lepidopterans are more abundant during the early dry season (Mertens et al., 2021), whereas coleopterans and dipterans show their peak activity during the wet season (Oliveira et al., 2021; Denlinger, 1980).

In a recent study, a total of 58 dipteran species belonging to 19 families were recorded in association with various mangrove plant species across the SBR, India. Among these, ten species represent new distributional records for the region. Notably, *Eupyrgota (Taeniomastix) pictiventris* is recorded for the first time from the state of West Bengal (Roy et al., 2019). These recorded dipteran species belong to 21 families and are categorized in several ecological guilds based on species diversity, richness, and relative abundance, such as flower visitors, frugivorous, leaf- and trunk-resting flies, carnivorous, saprophytes, and hematophagous flies. Of these, flower visitors were found to be the most dominant.

4.13.3.4 Types and Evolutionary Background of Marine Fishes

During the Late Cambrian era, approximately 500 million years ago, an unknown, armor-plated, fish-like creature appeared and began inhabiting the shallow oceanic waters along the seafloor. This creature possessed two eyes and the ability to swim but lacked any kind of fins. This kind of fish-like creature was named a heterostracan, which was considered to be an ancestor of the ostracoderms—a primitive form of fish with bony coverings—and also was considered to be the forerunner of all modern fish. Out of approximately 24 500 species of fish, characterized by a backbone made of repeating components called vertebrae, and also possessing paired and unpaired fins and gills as respiratory organs. These species are divided into two categories: jawless and jawed.

4.13.3.5 Jawless Fish and Jawed Fish

Fish with an ancestral origin like the heterostracans, along with other earlier forms under ostracoderms, do not possess jaws. After the appearance of jawless fish during the Cambrian period, two successful lines of descent evolved in the marine–oceanic environment: cartilaginous fish under the class Chondrichthyes and bony fish under the class Osteichthyes. These groups, characterized by jaws and a skeletal framework attaching to gills, exhibited the ability to move quickly in search of prey and in flight from predators. Other major structural developments include paired, unpaired, and lateral fins, which were meant to achieve balance with maneuverability. All these paired fins are considered the forerunners for the evolution of limbs in higher vertebrates. Cartilaginous marine fish include sharks, chimeras, skates, and rays, all of which have skeletons composed entirely of cartilage and possess osmoregulatory abilities in order to control the salt balance of the body. The fast-moving fish under the class Chondrichthyes are endowed with placoid scales and upwardly directed dorsal lobes in the tail, designated as heterocercal tails, which provide the ability to counteract sinking.

Bony fish under the class Osteichthyes and subclass Actinopterygii have versatile anatomical and physiological features, such as fins with bony fin rays and different types of scales. These features have enabled them to diversify into different ecological niche and habitats within an aquatic ecosystem, extending from deepest trenches of the ocean to shallower parts of coastal water bodies, overcoming harsh ecological conditions caused by wide ranges of fluctuations within numerous ecological parameters.

4.13.3.6 Species Composition and Diversity of Fish in India

India, being one of the biodiversity hotspots of the world and enjoying tropical climatic conditions accounts for 11.72% of all fish species globally, and also shares 27.8% of native fish fauna among Asian countries. As per the National Bureau of

Fish Genetic Resources (NBFGR) India, out of the 2246 finfish species recorded from India, 765 species are freshwater fish species, of which 450 species are identified as SIFS (Small Indigenous Fish Species).

As for fish habitats, 1652 fish species have been reported as marine, 906 species as freshwater, and 201 species as brackish water. Additionally, 332 species inhabit both brackish and marine environments, while 234 fish species have a wide range of distribution across freshwater, brackish, and marine forms. In a recent report from National Bureau of Fish Genetic Resources (NBFGR), Lucknow, has shown that 3145 native fish species—comprising 248 families and 49 orders—occur in different aquatic ecosystems in India. A total of 44 species belonging to 39 genera, 30 families and 16 orders were observed, with Cyprinidae (six species) as the prime contributor.

Fish form an integral part of the faunal assemblage and diversity within the Sundarbans mangrove ecosystem. Some of these fish species spend their entire lifespan within this ecosystem, while other species spend only a part of their life cycle here. The most comprehensive work on the fish diversity of the Hooghly–Matlah estuarine system under Sundarbans was undertaken by Hamilton–Buchanan; subsequently such studies were laid on a strong foundation by the work of researchers like Jayaram (1981), Talwar and Jhingran (1991), and Khan (1994). The influence of freshwater influx and the alteration of salinity profiles of the ecosystem—thereby determining the structure of the fish community and drawing the relationship with planktonic communities—have been undertaken by several other researchers, including Schlacher and Wooldridge (1996) and Gogoi et al. (2019).

Seasonal variation of estuarine fish diversity in different ecozones having their specific ecological variables have become major points of concern in several recent studies. The taxonomic studies projecting the functional roles of estuarine fish assemblages, along with identifying and integrating the major ecological forces structuring population dynamics, have appeared to be instrumental for the conservation and management of estuarine fish resources. Seasonal changes in fish diversity were attributed to sudden changes in salinity caused by freshwater influx from riverine inflows, coupled with heavy rainfall during the monsoon season in the Sundarbans mangrove ecosystem (James et al., 2007; Mukharjee et al., 2013). The mangrove–estuarine complex of Sundarbans supports the rich diversity of fish fauna, mainly because of the availability of all kinds of essential nutrients in the water and soil, which are derived from the decomposition of mangrove plant litter (Mandal et al., 2013). This dynamic ecosystem, providing heterogeneous ecological habitats, helps accommodating both shellfish and finfish fauna of diverse taxonomic components (Mukharjee et al., 2013; (Figs. 4.20, 4.21, 4.22, 4.23, 4.24, 4.25, 4.26, 4.27 and 4.28). A sizable portion of the open land area in the reclaimed deltaic islands has been used for undertaking extensive and semi-intensive forms of brackishwater aquacultural activities in large ponds, bheries, and depressed land accumulating flood water though tidal inundation (Figs. 4.26 and 4.27).

SIFS, or *Small Indigenous Species* (SIS), are those fish species that attain a size of 25 cm (nine inches) in the adult stages of their life cycle (Felts et al., 1996;

Hossain et al., 1999). SIF are endowed with high-quality animal protein, vitamins A and D, and different forms of micronutrients, including, iron, calcium, zinc, and fatty acids. Most of the indigenous fish species which occur in the mangrove ecosystem of Sundarbans possess high nutrient contents and also high ornamental value (Basu et al., 2012). The diversity and density of brackishwater fish in mangrove areas are largely determined by several eco-biological factors such as food availability (mostly in the form of plankton), water flow, conducive temperature, alkaline pH, and the suitable breeding grounds free from anthropogenic activities (Blaber, 2008). Lesser diversity of fish species inhabiting in the canal ecosystem is because of the restricted connectivity between canal and tidal water due to the construction of sluice gates, which disrupt the free movement of brackishwater fish species between canal and river.

Pillay recorded over 120 fish species from the estuarine area of the Indian Sundarbans during this decade-long study. Through a comprehensive survey on the different zones of Sundarbans, 74 fish species belonging to 40 families were recorded by Bose et al. Another long-term study on the fish fauna of the Indian Sundarbans has reported 267 fish species belonging to 81 families and 17 orders (Mandal et al., 2013). A recent study has reported 312 marine and estuarine fish species under 71 families and 18 orders from the Hooghly–Matla estuarine system of the Indian Sundarbans. In that study, the diversity of fish species under the Order Perciformes (164 species) was found to be maximum, followed by the fish species under the Families Clupeiformes (36 species) and Siluriformes (24 species; Bhattacharya et al., 2018).

Mainly based on the modes of transportation of detritus-contributed nutrients from the forest subsystem to the aquatic subsystem, and also following the Odum's (1971) out-welling hypothesis, the prolific development of coastal fisheries in and around the Indian Sundarbans has occured. Two major groups of finfish and shellfish constitute the nektonic fauna, as well as residents, transients, or migrants that inhabit mangrove–estuarine system of Sundarbans.

Resident fish are grouped into three categories based on their abilities to tolerate salinity: first, those fish which can tolerate salinity from 6.2% to 32.86% (Mullets—*Liza parsia* and *L. tade*; thread fins—*Polynemus tetradactylus* and *P. indicus*; *Sciaena miles* and *Sciaenoides biauritus;* Perches—*Lates calcarifer;* Ribbonfish—*Trichiurus savala*; Clupeids—*Hilsa sinensis* and *Coilia bornunsis;* catfish—*Tachysurus jella* and *Plotosus canius*). Second, those fish which can tolerate salinity from 4% to 32.86% (*Harpodon nehereus, Setipinna taty, Ilisha elongatae*, etc.). Third category is those fish which can tolerate salinity > 32.86% (*Pama pama, Coilia ramcarati, Polynemus paradiscus, Setipinna phasa, Sillaginopsis panijus*, etc.). The migrant species entering the estuarine zone for breeding and feeding can be broadly categorized into three broad categories. First, marine species migrating to upstream in freshwater areas for spawning of the estuary are *Tenulosa ilisha, Polynemus paradiscus*. Second, freshwater species move to estuarine zones for spawning in saline areas which include the species like *Pangasius pangasius* and the prawn like *Macrobrachium rosenbergi*. Third, marine fish migrating to estuarine zones for spawning in the brackish water of estuary are *Tachysurus jella, Penaeus indicus, Polynemus indicus,* and *P. monodon*. Out of a total of 172 species of fish recorded from the Hooghly–Matla estuarine complex of Sundarbans, 73 species occupied the less saline zone, whereas 99 other species preferred to inhabit in the higher salinity zone. A total of 26 species of shrimps and prawns belonging to five families are recorded from the estuarine system of Sundarbans. A total of 250 species of fish were reported from the estuaries of Sundarbans (Chaudhuri & Choudhury, 1994). However, Naskar et al. (2004) reported a total of 137 fish species from the mangrove–estuarine complex of Sundarbans, India of which 13 belong to chondrichthys and rest to osteichthys.

4.13.4 Amphibian Diversity

Although amphibian fauna are found to occur in large diversities and densities in freshwater habitats, the saline conditions in Sundarbans restrict the distribution of amphibian species. This is because amphibians are physiologically hypoosmotic, which compels them to lose water and gain ions in the saline water of Sundarbans. Furthermore, all amphibian species lack salt glands, which would otherwise enable them to tolerate salinity. Moreover, they are susceptible to pollution threats, as their highly permeable skin can rapidly absorb toxic substances. In the Sundarbans mangrove ecosystem, ten species of anuran amphibians (toads and frogs) have been recorded (Deulti, 2017a, b).

4.13.5 Reptilian Diversity

The Sundarbans mangrove ecosystem harbors some incredibly unique reptilian fauna, which include snakes, lizards, crocodiles, and turtles (Fig. 4.31).

Specialized groups of reptilian fauna have been recorded in the Indian Sundarbans, including 37 species of snakes, 11 species of freshwater turtles, and one species of estuarine crocodile (Deulti, 2017a, b).

4.13.5.1 Turtles in Sundarbans

Eleven species of turtles have been reported to occur in the Indian Sundarbans. Of these, two species—the Olive Ridley turtle (*Lepidochelys olivacea)*, and the hawksbill sea turtle (*Eretmochelys imbricata*)—nest on sea beaches, and are included in the IUCN-threatened category. Another critically endangered species is the river terrapin (*Batagur baska*), which prefers to nest on the intertidal belts of Indian Sundarban. The zoological description of some of those turtles is as follows:

1. Olive Ridley sea turtle, *Lepidochelys olivacea* (Eschscholtz, 1829): This medium-sized species of sea turtle inhabits tropical warm waters, mostly in the

Pacific and Indian Oceans. Every year, they visit the Gahirmata sea beach of Bhitarkanika National Park in the state of Odisha, India, for nesting, breeding, laying eggs, and nurturing of juveniles. This species also nests in the Indian Sundarbans, especially in the islands protected under the Sundarbans Tiger Project area in the southern parts of Sundarban along the Bay of Bengal.

2. River terrapin, *Batagur baska* (Gray, 1830): This moderately large-sized, critically endangered turtle species is web-footed with four claws on the forelimbs. Normally, they prefer to inhabit the downstream of estuaries with freshwater influence and supported by extensive mangrove vegetation. They usually feed during high tide, scooping up vegetation from low-hanging branches of mangrove plants. Their population has been dwindling over the last several decades because of the absence of suitable sandy but protected beaches in the Indian Sundarbans. Attempts are being made to restore the lost population by culturing the species in the captive breeding center of Sajnekhali, of Sundarbans, India. Sixteen turtles were reported from several village ponds in 11 villages of Sundarbans, specifically: Amtali, Hingalgunj, Kumirmari, Jhingakhali, Bhanderkhali, Mangalchandi, Mollakhali Govindakati, Duldulia, Menaka, and Puinjali (Deulti, 2017a, b).
3. Spotted pond turtle, *Geoclemys hamiltonii* (Gray, 1831): This colorful turtle inhabits shallow-water bodiesm mostly in reclaimed deltaic islands of wetlands such as oxbow lakes, beels, marshes, reservoirs, and ponds. This species is characterized in having an elevated and serrated carapace, a large head, and a short snout.

4.13.5.2 Estuarine Crocodile, *Crocodylus porosus*

The extensively large-sized estuarine crocodile, *Crocodylus porosus*, is found to occur throughout the length and breadth of the core areas of the Sundarbans Biosphere Reserve and in higher density in Bhittarkanika National Park. This largest reptile in the world is declared an endangered species as per IUCN Red Data List due to habitat loss and the skin trade.

4.13.5.2.1 Conservation Measure of Estuarine Crocodile Through Captive Breeding at the Bhagabatpur Crocodile Project, Sundarbans, India (Figs. 4.29 and 4.30)

The Bhagabatpur Crocodile Project was started in the year 1976, when 32 eggs of estuarine crocodiles were collected from the nearby estuaries. Egg collections were first started from the artificial brackishwater impounds within the Bhagabatpur Crocodile Project in the year 1982. Afterward, such a collection of eggs was made from both natural and constructed water bodies under the influence of tides.

The cultural practices include the following:

- Adult crocodiles meet during the month of March, in the early phases of the pre-monsoon season. After 2 months, female crocodiles construct nests at the supra-littoral zones and lay eggs. The eggs are collected from these nests.
- In the artificial cultural process, collected eggs are kept in artificially constructed nests, maintaining natural ecological condition.
- Temperatures are maintained between 29 °C and 32 °C.
- Saline water with a salinity between 15 ppt and 25 ppt is sprinkled over the nests containing the collected eggs.
- Juvenile crocodiles hatch after 80–90 days emitting a mild sound. Unhatched eggs are carefully protected to ensure future hatching success.
- The juveniles start feeding on insects that assemble over the artificial camp placed near the cultural ponds.
- The growing crocodiles are fed fish, crabs, and shrimp.

4.13.5.3 Diversity of Lizards in Sundarbans, India

A total of 13 species of lizards are found to occur in Sundarbans, India, which include three species of Monitor Lizards which are included in vulnerable category by IUCN and Schedule I of the Indian Wildlife (Protection) Act, 1972 (Deulti, 2017a, b; Table 4.14).

4.13.5.4 Diversity of Snakes in Sundarbans, India

A total of 36 species of snakes are known to inhabit different preferred habitats of the Indian Sundarbans. Of these, the poisonous snakes include the king cobra (vulnerable according to IUCN), the monocellate cobra, the Russell's viper, along with the common and banded kraits. Other nonpoisonous species are also observed in large numbers near different marshy lands and water bodies in villages, including typical species like the white-bellied mangrove snake and glossy marsh snake (Table 4.15).

4.13.5.5 Diversity of Turtles

Eleven species of turtles have been recorded from nearby maritime zones, estuaries, forests, and pisciculture ponds in and around Sundarbans, India (Table 4.16).

4.13.6 Avian Diversity of Indian Sundarban

The Sundarbans mangrove ecosystem is rich with its avian diversity, residing in an array of habitats from forested areas to villages, from mangrove swamps to estuaries, and from tall canopies to mudflats within this ecosystem, encompassing both terrestrial and aquatic zones. A total of 428 species of birds have been recorded in Sundarban (Chandra et al., 2020; Table 4.17; Figs. 4.33, 4.34 and 4.35). All these avifauna utilize vast stretches of the Sundarbans mangrove–estuarine complex, which is endowed with multitudinous habitats in the form of extensive mudflats and sandflats, green coverage in the forest areas with the luxuriant growth of mangroves, and their associated plant species (trees, herbs, shrubs, climbers, and grasses) as their preferred habitats and ecological niches. Different tiers of mangrove plants and exposed intertidal zones during low tides are used by different aquatic and terrestrial birds for foraging, where they select their preferred foods (Mukherjee, 1969, 1976).

Different tiers of the mangrove forest are utilized by varied species of birds in tune with their sizes, flying ability, and power of predation.

The highest canopies of the forest are occupied by various large-sized birds that possess superior flying abilities due to their robust body structure. They are considered to be birds of prey like osprey, Brahminy kite, brown fish owl, crested serpent eagle, raptors, white-bellied sea eagle, barn owl, scops owl, and spotted owlet. Some are top predators, found flying high in search of prey like fish and small reptiles like lizards and snakes.

The mangrove canopy is inhabited by passerine birds—flycatchers, warblers, small minivets, spotted doves, green bee-eaters, golden orioles, barbets, and rose-ringed parakeets—which utilize mangrove canopies and branches close to the watermark during high tides for the purpose of perching, catching small fish from the tidal water, and building up nests.

In the lower tier, a predominance of small predatory birds like kingfishers is noted. Sundarban harbors eight species of kingfishers; the most common of which are small blue, white-breasted, brown-winged, pied, black-capped, collard, blue-eared, and ruddy kingfishers. The daily activities of kingfishers—cruising down the rivers and creeks, waiting patiently on a submerged tree stump for a fish, or drifting from branch to branch to hide themselves—are quite common to sight throughout the length and breadth of Sundarbans. However, such activities of different kingfishers vary from season to season, mostly synchronizing with their breeding seasons.

The vast intertidal mudflats of Sundarbans act as foraging grounds for different aquatic and semiaquatic birds, which belong to groups like cormorants, egrets, herons, storks, waders, darters, white-breasted waterhens, sandpipers, little stints, snipes, cinnamon bittern, black and yellow bitterns, redshanks, and red-wattled lapwings. The most common birds of the muddy riverbanks are the curlew and whimbrel, which remain busy in search of crabs and fish, using their elongated and curved beaks to pull their prey from the pneumatophores and slit roots of the intertidal mangrove plants.

Lesser adjutant storks represent the largest of all storks wandering in various parts of Sundarbans and are especially active during low tides. Asian openbill storks are noticed at the outskirts of the Sundarbans Biosphere Reserve, near the sites of cultivation in the reclaimed deltaic islands. All the three species of egrets—cattle, large, and medium—add beauty to the picturesque landscape of Sundarbans. Several species of herons are observed in the Indian Sundarbans; of these, two herons—pond and gray—remain active in and around shallow-water bodies throughout the coastal belts of West Bengal, including Sundarbans. The purple heron is very rarely encountered, and the rarest and largest of all herons is the goliath heron (characterized by a huge wingspan), which originates from East Africa and migrates only to Sundarbans, India, during the nonbreeding season.

The avian fauna of the Sundarbans mangrove ecosystem includes both residential and migratory bird species. The migrant bird species, which include several ducks like the lesser whistling teal, Pallas's gulls, brown-headed gulls, black-headed gulls, gadwalls, wigeons, common pochards, pintails, shovelers, river terns, whiskered terns, and little terns, migrate down to Sundarbans from distant places in good numbers. All these bird species, along with sea gulls, are seen hovering over fishing trawlers in search of fish. At the onset of the post-monsoon season (winter), during the months of November and December in each year, many migratory birds visit various parts of Sundarbans for the purpose of both breeding and feeding from the temperate regions of Northern Hemisphere. All of these birds are seen to congregate both in the mangrove forests and intertidal belts throughout the winter and leave the mangrove zone at the onset of the summer months (Figs. 4.33, 4.34 and 4.35).

4.13.7 Mammals of Sundarbans

Mammals of India represents 9% of the total number of mammalian fauna in the world (De, 2017; Debata et al., 2019). Mammals in India are distributed in an admixture of landscapes within the geographical boundaries of India. Mammals in Sundarbans include both terrestrial and aquatic inhabitants. Among the terrestrial mammals, the most numerous species are deer, (*Axis axis*), wild boar, (*Sus scrofa*) rhesus macaques, (*Macaca mullata*), leopard cats (*Felis bengalnsis*), fishing cats, (*F. viverrina*); gray mongooses, (*Herpestes edwardsi*); small Indian mongooses (*H. auropunctatus*), Chinese pangolins, (*Manis pentadactyla*), crested porcupines, (*Hystrix indica*), large bandicoot rats (*Bandicoota indica*), and lesser bandicoot rats (*B. bengalensis*).

Aquatic mammals found in estuarine water include Irrawaddy dolphins (*Orcaella brevivostris*), Gangetic dolphins (*Platanista gangetica*), and black finless porpoises (*Neophocoena phocaenoides*). Some extinct mammals of Sundarbans are Javan rhinoceros, (*Rhinoceros sondaicus*) wild buffalo (*Bubalus bubalis*), hog deer (*Axis porcinus*), swamp deer (*Cervus devaucelli*), and barking deer (*Muntiacus muntjak*).

Sundarban mammals are endowed with powers of tolerance against variable saline conditions, along with exceptional swimming ability, as they reside in the difficult mangrove terrains with periodic tidal inundation and chase prey from estuarine water sources. Due to escalating anthropogenic pressure, environmental pollution, and shrinkage of the mangrove swamp, the population of several mammalian species has been declining alarmingly. Ongoing ecological perturbations have imposed adverse effect on the mangrove ecosystem of Sundarbans in a profound manner, disrupting the ecological balance of this sensitive and fragile mangrove ecosystem, the ultimate outcome of which is reflected by the colossal loss of biodiversity, including the mammalian fauna. Chiropterans—small arboreal mammals—are represented by species like *Pteropus giganteus* (Indian flying fox), *Cynopterus sphinx* (short-nosed fruit bats), and *Scotophilus kulhi* (the lesser yellow bat). In addition, the greater false vampire (*Megaderma lyra*) is also notable in the Indian Sundarbans.

Some of the charismatic mammalian species occurring in the mangrove–estuarine complex of Sundarbans, India (Fig. 4.36 and Table 4.18) are as follows:

1. Common palm civet, *Paradoxurus hermaphrodites* (Pallas, 1777): Locally known as *bham*, looks like a cat, is nocturnal, and inhabits fringe areas of mangrove forest close to human settlements. This omnivorous mammal is an efficient climber and hunt preys in trees, roofs, and on the ground. Its Sundarbans, diet includes crabs, frogs, squirrels, reptiles, birds and their eggs, and rats.
2. Small Indian civet, *Viverricula indica* (E. Geoffroy Saint–Hilaire, 1803): This tawny gray or grayish-brown mammal, locally known as *gandha gakul*, is nocturnal in habit and prefers scrub habitats close to human habitation in the mangrove forest.
3. Asian small-clawed otter, *Aonyx cinereus* (Illeger, 1815): Locally known as *bhondar*, is active both in the day and the night by the side of tidal creeks. They feed on shrimps, crabs, and small fish, and they have rudimentary claws. The body is dark brown above and paler below. This species is now observed rarely in Sundarbans, and their population is declining due to prey shortage.
4. Smooth-coated otter, *Lutrogale perspicillata* (Geoffroy Saint–Hilaire, 1826): The only extant representative of the genus *Lutrogale* other than the Asian small-clawed otter, which is also found in the Sajnekhali Range of the Sundarbans Tiger Reserve and western parts of Sundarbans. This mammal is characterized as having a smooth, sleek, and almost velvety coat.
5. Rhesus monkey, *Macaca mulatta* (Zimmermann, 1780): Commonly known as *bandar* is frequently noticed in Sundarbans, although they are truly terrestrial species; but in Sundarbans this species has adapted to spend their life in saline tidal water dominated forest ranges. They enjoy human association, especially the tourists of Sundarbans. They are herbivores feeding on plant products like leaves, shoots, fruits, and grains.
6. Spotted deer, *Axis axis* (Erxleben, 1777): Locally known as cheetal, this animal has both diurnal and nocturnal habits and lives in herds of 10–30 individuals, with two or three stags. The presence of profuse white spots on its bright rufous fawn coat compliments the beautiful scenery of green forest and shaded tidal creeks. They demonstrate a good association with other wild animals like monkeys, providing mutual benefit to one another.
7. Gray mongoose, *Herpestes edwardsii* (E. Geoffroy Saint–Hilaire, 1818): Popularly known as *bezi*, it inhabits mostly in scrub lands within the mangrove forest and in cultivated agricultural fields of the reclaimed deltaic islands of Sundarbans. They prey on insects, crabs, frogs, snakes, other small reptiles, and birds.
8. Bengal marsh mongoose, *Herpestes palustris*: A medium-sized mongoose that remains active throughout the day preying on fish, crabs, snakes, lizards.

Characteristically, this marsh mongoose has the habit of standing up on its hind legs for better vision. However, presently, they are rarely sighted in Sundarbans.

9. Fishing cat, *Prionailurus viverrinus* (Bennett, 1833): Declared the state animal of West Bengal, India, is locally known as *mecho beral*. This very charismatic small-sized mammal lives in the mangrove forest tracts associated with shallow-water bodies like tidal creeks, brackish water wetlands, and swampy lands. This nocturnal, wild cat marks its territory using chin rubbing, cheek-rubbing, neck rubbing, head rubbing, and urine spraying to leave scent marks. Their identifying characters include the olive-gray color of their coarse fur, with dark spots arranged in horizontal streaks running along the length of the body. Presently, this animal has been encountering a lot of natural and anthropogenic threats because of the poor availability of prey items and loss of habitats in Sundarbans, India.
10. Gangetic dolphin, *Platanista gangetica* (Roxburgh, 1801): Locally known as *susuk*, frequently noticed at the confluence of freshwater rivers with estuaries of Sundarbans, India. The main food items of dolphins include a variety of shrimp and fish, including carp and catfish. Their body color varies from slaty to sooty black dorsally. They are now under the threat of extinction due to nonjudicious catch, various forms of chemical and biological pollution, habitat destruction, and overfishing in Sundarbans, India.
11. Indian flying fox, *Pteropus giganteus* (Brunnich, 1782): Locally known as *badur*, it can be frugivorous or nectarivorous. This common inhabitant of Sundarbans is rarely noticed in other mangrove ecosystems.
12. Jungle cat, *Felis chaus kutas* (Pearson): Locally known as *junglee biral*, is a solitary wild mammal, characterized by its reddish-brown color on the back of both the ears. Territories are maintained by this wild carnivore by urine spraying and scent marking, where some males are often observed rubbing their cheeks on objects to mark them. The preferred foods of this jungle cat include insects, crabs, fish, frogs, birds, turtles, and small mammals. The population of this mammal has been depleting at a considerable pace due to habitat loss, hunting, and poaching for the fur trade in Sundarban, which is reflected by their rare sighting.
13. Leopard cat, *Prionailurus bengalensis* (Kerr, 1792): Leopard cat, a medium-sized carnivore, prefers to live in the holes of trees within mangrove forests and near human habitation for shelter. Although, this wild cat looks like miniature of leopard in color and pattern, its bold spots without forming rosettes on the flank give it separate look. They feed on birds and other small animals in their nocturnal mode of life. The population of this endangered mammalian species has been dwindling in Sundarbans at alarming pace due to scarcity of prey species, hunting and poaching for its skins, claws, bones which are of very demand and sold with a high price in international market.
14. Wild boar, *Sus scrofa* (Linnaeus, 1758): The omnivorous wild boar, locally called *junglee suyor*, feeds on roots, tubers, crops, insects, snakes, carrion demonstrating both diurnal and nocturnal habits residing in mangrove forests. The Sundarbans' population of pigs, one of the important prey for the Royal Bengal tiger, has been slowly declining due to habitat loss and hunting.

4.13.8 The Tiger and the Sundarbans Mangrove Ecosystem, India

The tiger of Sundarbans is recognized as the most mysterious among all the wild animals of India (Fig. 4.37). The Sundarbans tiger, the Royal Bengal tiger, is considered to be the most intelligent and strongest animal found in this region, as this wild fauna has to overcome difficult living conditions in the tidal mangrove forest by adopting a strenuous process for catching its prey. These tigers also possess extra adaptive agility compared to tigers inhabiting other Indian forests. The main food items of the tigers of Sundarbans are spotted deer, wild pigs, rhesus monkeys, monitors, crabs, and large-sized fish. The name Royal Bengal tiger is thought to have originated from a past incident when the Duke of Edinburgh, as a representive of the Royal family, shot a tiger in the mangrove forest of Sundarbans. Initially, Carolus Linnaeus, the celebrated Swedish taxonomist, scientifically identified and described this wild beast as *Felis tigris* in 1758, based on his original description on a specimen obtained from Bengal. The tiger of Sundarbans (*Panthera tigris tigris*) has adapted—beyond the abilities of other tigers of the Indian forests—with extra power for the strenuous process of hunting, and with improved agility to survive the rigorous living conditions of the tidal mangrove forest. The tiger is a top predator, consuming many animals belonging to lower trophic levels, including crabs, fish, turtles, deer, rhesus monkeys, and monitors. Therefore, the tiger in Sundarbans occupies a unique position in the ecosystem pyramid of mangrove ecosystem, at the pinnacle of both aquatic and terrestrial food webs (Sunquist et al., 1999; Rog et al., 2017).

The tiger, as a consumer, occupies a unique position in this ecosystem as the top predator at the highest position of the ecological pyramids of both aquatic and forest-based terrestrial food chains. The breeding period of the tiger in the mangrove forest is at the very end of the monsoon (September to October). Females tigers usually breed once in a year, with a gestation period of about 16 weeks. Two to five offspring are born at a time, but mortality during childhood is quite high because of the notorious habit of cannibalism among male tigers, and the unavailability of drinking water and prey. The father tiger's behavior (infanticide) is an **evolutionary reproductive strategy**, not the result of the human emotion of jealousy. The biological goal is to eliminate rival genes and bring the female into estrus (ready to mate) sooner in order to pass on his own genes. Territorial marking by urine, as applied in other forest areas in India, is not effective in the tide-dominated mangrove forest of Sundarbans (Sanyal, 1987, 1990; Chakraborty, 2009).

The tiger prefers to live alone reigning over a certain specified area as evident by recent radio-collaring study, although during mating season a male and female live together for some time. Territorial marking by urine as practiced by tigers of other areas cannot be made applicable in Sundarbans area because of the impact of tidal flushing which removes all such indication.

Sundarbans tigers (*Panthera tigris tigris*) inhabit mangrove forests, a unique habitat that gets inundated twice a day by the tidal water of the adjoining estuaries. They are morpho-anatomically distinct from the other recognized tiger subspecies in terms of body size, biomass, muscular strength, and skull morphometrics. The distinct attributes of Sundarbans tigers have necessitated the study of their behavior, demographic parameters, and genetic distinctiveness for understanding the historic and contemporary connectivity among tiger populations between Sundarbans and mainland India. The composite of study information, along with assessment of their ecological roles within the mangrove ecosystem, helps ascertain effective conservation strategies. Demographic analyses suggest that Sundarbans tigers have diverged recently from the peninsular tiger population because of continuous fragmentation and habitat alteration over the past few centuries. This has made them a divergent group—recognized as Royal Bengal tigers—which are non-exchangeable ecologically with other tiger populations. This finding has prompted the management of this tiger as a separate *evolutionarily significant unit (ESU)*, following the adaptive evolutionary conservation (AEC) concept.

4.13.8.1 Man–Tiger Conflict

The Royal Bengal tiger *(Panthera tigris tigris)* represents the most unique biodiversity component of this sensitive, fragile, highly productive tidal forest—the mangrove swamp of Sundarbans wetlands of the world—by virtue of its multidimensional behavioral manifestations, including the typical characteristics of man-eating habits. This gigantic wild carnivorous mammal prefers to feed on all kinds of moving animals as prey, including another unique inhabitant of this ecosystem: crocodiles. However, old and infirm tigers depend more on capturing domestic bovids as food items from the adjoining locality based on physical as well as biological factors. This category of elder, disabled fail to capture moving wild prey in the forest environment.

4.13.8.2 Problem of Tiger Straying: Contributory Factors Compelling Tigers to Go Inside Villages of Sundarban

The following attributes play their roles in determining tiger straying:

(i) Scarcity of prey animals within the mangrove forests.
(ii) Entry of tigers into human habitation adjoining some areas of Sundarbans by crossing narrow creeks, especially during low-tide periods.
(iii) Tigers hardly attack human beings as their prey, but attack only when men are found along the way.
(iv) In the villages the prime target zones for the tigers are cowsheds and shelter sites for other domesticated animals.
(v) The embankment protective mangrove strips near the villages sometimes confuse people by tigers using the area as their own habitat.
(vi) Female tigers during littering season are often seen to stray into the paddy fields around the nearby villages to protect their cubs from the aberrant behavior of the males.
(vii) During the early post-monsoon season, the ripe paddy looks like *Porteresia coarctata*, locally known as *shali ghas*. The tiger camouflage naturally confuses people as its color matches with the ripened paddy.
(viii) Maximum casualties are found to occur during the pre-monsoon time when the salinity is at a maximum in the estuarine ecosystem, which prompts tigers

to come out from their own territories in search of freshwater, and thereby resulting in the increase of tiger straying and man-eating.

(ix) The territories of tigers are marked by pheromone sprayed by the tigers on the tree trunks but due to the flushing by the tidal waves every day these biomarkers get washed away, forcing tigers to stray inside the human habitation.

4.13.8.3 Different Protective Measures Taken by the Forest Department to Combat Such Entry of Tigers

Fences made of nylon netting and branches of mangrove trees covering islands opposite the villages sometimes restrict tigers from cross the rivers and creeks into human habitation zones, even though they are specially adapted for swimming across Sundarbans tidal rivers. Additional strategies and other information pertinent to this topic are as follows:

(i) The Sundarbans mangrove ecosystem is one of the 58 *Project Tiger Reserves* established in 1973, along with the nine other initial Project Tiger Reserves in India.

(ii) Excavation of sweet water ponds inside the forest is another way to stop the straying of tigers in search of freshwaters to the localities adjacent to the tiger reserve.

(iii) Electrified dummies of fishermen, honey collectors, and wood cutters are introduced.

(iv) The development of consciousness among local peoples about the importance of the conservation of tigers, and their steady interactions with Project Tiger authorities, help the formation of ecodevelopment committees through which local people benefit from alternative income-generating sources.

(v) Along the vulnerable areas of the forest–village interface of Sundarbans, nylon rope nets, measuring 4 × 4 inches, are erected. These have been proven to effectively reduce tiger straying incidents in recent times.

(vi) As tigers prefer to avoid illuminated areas, riverbanks of the villages are provided with battery-operated solar lighting, which renders two-fold purposes: helping the local people as well as forest department officials to keep vigilance over the movements of tigers; and to keep tigers away.

(vii) The Project Tiger authorities have intensified regular patrolling by motorboats and by setting up floating and permanent camps at strategic points, which have also proven to be efficient tools.

4.13.8.4 Conservation Challenges of Tigers Based on Understanding of Tigers of Sundarbans

The phenomena of *human killing* inside the mangrove forest and *tiger straying* into nearby human settlements has emerged as the most serious eco-management challenge to ensure the sustainable conservation of biodiversity in the mangrove ecosystem of Sundarbans, India. Human–wildlife conflict mitigation has emerged as one of the major challenges for long-term wildlife conservation. Century-old human–tiger conflict has been magnified because of entry of a considerable number of people into Sundarbans to settle down on reclaimed deltaic islands and develop their livelihoods by harvesting natural resources. The success in the mitigation of human–tiger conflict requires fundamental knowledge of the ecology, population dynamics, reproductive and feeding biology, and behavior of Sundarbans tigers. Several new methodologies have been adopted to monitor the activities of tigers of Sundarbans which include the application of strategies which include: remote sensing and satellite imaging, camera trapping, radio-collaring. The ground truth information generated pertaining to the tiger population, breeding behavior, and prey–predator relationships, with the help of these monitoring methods, provides valuable insight into the prospective tiger management program and helps in human–tiger conflict mitigation, which is essential for the long-term conservation of Sundarbans tigers (Balow, 2009), (Figs. 4.38 and 4.39).

Long-term conservation of tigers (*Panthera tigris*), in the global and national arenas, is a challenge that depends on ecologically sound conservation and management strategies. Based on tiger distribution and the potential for connectivity, there are six tiger landscape complexes in India (Sanderson et al., 2006). The Sundarbans landscape complex is traversed by many tidal channels, separating different deltaic islands, and shared by India and Bangladesh. They represent the last stronghold of Bengal tigers, which have adapted to living in mangrove forests (Sanderson et al., 2006; Jhala et al., 2008, 2011). Tigers of Sundarbans are considered as distinct from other subspecies and are assigned to Bengal tigers (*P. tigris*, *tigris*; Dinerstein et al., 2007). The body weight of a typical Sundarbans tigress varies from 75 to 80 kg, which is actually half of the body weight of the tigress in the mainland, and their skulls are also smaller than the skulls of other tigers (Barlow et al., 2010). Smaller skull size and body weight of Sundarbans tigers, in comparison to mainland tigers, reflects adaptive differences between them and is due to their different ecological niches (Barlow et al., 2010).

Most of the mainland Indian Bengal tiger's inhabit tropical forests and are able to prey upon large size species (> 50 kg, e.g. Sambar, Nilgai and Gaur). They are also able to compete with sympatric species, such as the common leopard (Karanth, 2003; Karanth et al., 2004). However, tigers of the Sundarbans mangrove ecosystem consume small-size prey, without facing competition with other sympatric carnivores (Khan & Monirul, 2008). The small body size of tigers of Sundarbans is energetically advantageous for their survival in the marshy land (Hemmer et al., 2004; Waples & Gaggiotti, 2006). The absence of other large carnivores, like leopards, reduces interspecific as well as intraspecific competition compared to other landscapes. This enables them to adapt to their local environmental conditions and utilize the natural resources, including fish and crabs. Recent deforestation, escalating anthropogenic intervention, and a changing climate over the last couple of centuries have affected Sundarbans tigers considerably, and brought about morphological and behavioral changes, modifying gene flow and local adaptation. All of these dynamics have synergistically shaped the genetic architecture of the tigers of this marshy Sundarbans ecosystem.

4.13.8.5 The Concept of Bonobibi

The local people of Sundarbans accept tiger attacks as natural hazards, but they also live within an inherited culture that believes in the supernatural qualities of these man-eaters. Several traditional rituals co-exist as the forest dwellers call on the blessings of Dakshin Rai, the lord of the forest, and seek blessings from Bonbibi, Narayani Ma, Kalukhan, Sa Jungli, and Gazi Saheb. Moreover, locals depend on spiritual support. To overcome natural hazards, such as storms, cyclones and fear of ghosts, the people are very much dependent on the strength of religious beliefs and supernatural powers. In all Sundarbans villages the goddess Bonobibi is a deity who is worshipped by both Hindus and Muslims in hopes of generating inner strength to fight against all odds. Bonobibi has been considered almost synonymous with the Hindu goddess, *Bonodevi*. However, the former name became more widely accepted with the advent of Muslim rule in lower Bengal. Moreover, another god of the tiger, Dakshin Rai, has been worshipped under a banyan, peepal, or neem tree. Other similar gods include Pir, or Gazi Saheb, who were introduced as tiger gods from the time of the Muslim era. Stories of Dakshin Rai and Gazi have led to popular folk tales like *Raimangal Kabya*.

The tiger reigned as the supreme power throughout the entire Sundarbans region, until the advent of modern weapons. Innumerable legends and myths concerning this magnificent creature have been created from time immemorial in this mysterious part of the world. Considered as both god and demon, the tiger is worshipped with the aim of achieving hypnotic divine power to resist the tiger's man-eating habits. The poor people who enter the mangrove forest for harvesting forest products like timber, honey, and fish have no other option but to believe in the supernatural view of power inaccessibility and helplessness inside the forest because of the ferocity and man-eating habit of the tiger.

In this context, Bonobibi is thought to be the god to protect forest dwellers, whereas Dakshin Rai is the king of tigers. A mythological story tells of a fierce fight between both gods, where Dakshin Rai was defeated. Subsequently, Dakshin Rai was accepted by Bonobibi as a brother. Hence, Bonobibi and Dakshin Rai have been worshipped together (Fig. 4.40).

4.14 Categories of Forests Within Sundarbans

The types of forests within Sundarbans are determined by salinity gradients, the salinity tolerance abilities of different mangrove plant species, and the geomorphological conditions of mangrove habitats. Sir David Prain, in the year 1903, first systematically categorized the mangrove vegetation within the undivided Indian Sundarbans into three types:

(i) Southwest zone and southern coastal strip with well-known dominant mangrove species.
(ii) Central zone habituated by *Heritiera fomes* vegetations.
(iii) Northeastern part of savanna type vegetations with salt marshes and seagrasses.

In the year 1936, Griffith categorized the mangrove forest of Sundarbans into two main categories:

(i) Mangrove swamps.
(ii) Coastal sand dune vegetation.

Champion, in the year 1936, divided the forest of Sundarbans into four types:

(i) Freshwater forest.
(ii) Saltwater mixed forest.
(iii) Mangrove forest of three categories.
(iv) Brackish water mixed with low saline mangrove forests.

After 30 years of Prain's classification, Curtis, in the year 1965, put forward his views on the types of forests of Sundarbans, positing three major types:

(i) Freshwater vegetation.
(ii) Saltwater vegetation with moderate salinity.
(iii) High salinity tolerant vegetation.

Champion and Khattak, in the year 1965, considered the Sundarbans mangrove forest as a mosaic of beach forest and tidal forest. However, a long-term survey of the entire forest tracts within Sundarbans yielded four major categories, based on habitat preferences of different mangrove species.

(i) Namkhana—Sagar Island zone.
(ii) Basanti—Sonakhali zone.
(iii) Hasnabad—Sandeshkhali zone.
(iv) Gosaba—Sundarbans Biosphere Reserve zone.

All of these categories were again subdivided into several types, based on the mangrove vegetation assemblages:

(i) Coastal vegetation, including sand dune creepers.
(ii) Point bar and mid-channel bar vegetation.
(iii) River flood plain and mudflat vegetation, including salt marsh and shrubs.
(iv) Three-tier tidal creek vegetation.

As per the abilities of different mangrove species, higher salinity-tolerant species are *Avicennia officinalis, Rhizophora decandra*, and *Agialitis rotundifolia*, which are stenohaline species experiencing narrow ecological amplitude. In contrast, there are other freshwater-loving stenohaline mangrove plant species, like the very rarely seen, *Heritiera fomes*, and *Nypa fruticansa*, freshwater loving species in the Indian Sundarbans part. Several other mangrove plant species like *Excoecaria agallocha, Ceriops decandra, Xylocarpus* sp., and *Sonneratia apetala*, which is the euryhaline species, possess the ability to tolerate a wide range of salinity with a wide range of ecological amplitude. Therefore, mangroves are classified into three categories based on the degree of tolerance of salinity:

(i) High-salinity tolerant species.
(ii) Medium-salinity tolerant species.
(iii) Low-salinity brackish water species.

The sequential growth of varied species of mangroves prompted researchers to classify the mangrove forest into several stages (I, II, and III; Das & Bhattacharya, 1999b, 1994). Stage I mangroves, also known as mangroves of first generation, include those mangrove plant species that initially occupy the river flood plains and natural levees. However, sediment compaction due to subsidence of banks leads to the burial of stage I mangroves, leaving only the visible portion as upright trunks of stage I mangrove over the water surface in river sections. Several geochemical and biochemical changes of stage I mangroves result in decomposed wood seams which remain sub-horizontally within 30 cm depth. Stage II, the mangroves of second generation, grow along point bars and developed levees on flood plains that subsequently accrete the sediments. The mid-channel point bars intermingle with ebb and flood currents to support stage III, newly-introduced mangrove saplings.

4.15 Conservation Applications Through Developing Biodiversity Maps

Sea level rise is likely to have drastic impacts on mangrove forests worldwide, and particularly in Sundarbans. Under the projected range of sea level rise with range of 30–100 cm by 2100, Sundarbans is likely to lose 10–23% of its present area with an alteration of soil biogeochemistry, and estuarine hydrology. These novel habitat insights can guide forest managers about deciding which mangrove communities or which stressors (siltation, salinity, and pH) to target for future mangrove enhancement (reducing abiotic stresses that caused biodiversity loss), restoration, and reforestation initiatives in Sundarbans. Proper development of biodiversity maps of Sundarbans helps indicate the status of the surviving biodiversity hotspots, major portion of which are located remarkably close to local communities, having a higher vulnerability due to opportunistic tree harvesting (Iftekhar & Islam, 2004). These biodiversity maps can also put forward proper guidelines for formulating sustainable biodiversity conservation initiatives along with contributing successful initiatives and implementation of the REDD+, which, in turn, enhances carbon stock—through biodiversity conservation—as well as economic development.

This development commands more attention for bringing them under protected area management with a serious effort to ensure all-round and immediate protection and also to take eco-management strategies for the long-term conservation of biodiversity, especially the threatened flora and fauna of the Sundarbans mangrove ecosystem.

In the process of building a biodiversity map, coherent and comprehensive quantification is needed. This quantification should be based on ground research data regarding habitatbased alpha, beta, and gamma diversity, combining relevant ecological parameters (such as temperature, salinity, pH, texture of sediments, alkalinity, and biotic interactions), environmental drivers (such as rainfall, climate variability, and deforestation), and historical events in different ecoregions of Sundarbans communities.

4.16 Present Status of Wildlife Conservation and Wildlife–Human Conflicts

Wildlife–man conflicts take into consideration the negative interactions in between wildlife and man, resulting in the injury or even death of human beings, loss of livestock, and damage to human property. All the consequences generated by these conflicts lead to disagreements and negative results for both humans and wild animals. In this contest, ensuring proper wildlife conservation—the practice of protecting wild animals along with their habitats—has become much more challenging, and requires the all-around participation of local people. The conflict between predating animals and their human neighbor is considered to be as old as the history of human civilization. The sudden upsurge in the human population brings about a crisis of land availability. Because of the burgeoning appropriation of land for human domicile and livelihood, the incidence of man–wildlife conflicts have been increasing at an alarming pace across the world. The living space required by predators to lead a normal life is disproportionately higher than the areas required by the growing population of their prey. Therefore, availability of required space for predators brings forth competition with their animal neighbors, especially in developing countries. Indeed, conflict between people and felids has been termed as one of the most urgent wild cat reservation issues of the world (Inskip & Zimmermann, 2009). Sundarbans mangrove wetlands in the lower Ganga Brahmaputra delta, started to be reclaimed beginning in 1770 (Pargiter, 1934). During the next two centuries, approximately 5364 km^2 of former tidal forests were converted to agricultural land in 19 police station areas in the North and South 24 Parganas districts, West Bengal. The present area of the Indian Sundarbans amounts to 4262 km^2. People live in the reclaimed areas of Sundarbans, which were initially under mangrove forests until 1833, and has been colonized very recently. In the process of the expansion of agriculture, the forests have been readily pushed back, and human habitation has extended southward into the interior. The southeastern part comprises a network of tidal waters, covered with dense mangrove forests. The majority of the population in the reclaimed deltaic islands (approximately 95%) depends on agriculture supported by other occupations like fisheries, forestry, and handicrafts. The people of Sundarbans are exposed to a unique set of biotic hazards, ranging from snakebites to tiger attacks, because of their proximity to the mangroves, coupled with underdevelopment, which has influenced their mental makeup and sociocultural framework. Conflicts of interest between authorities protecting the mangrove wildlife and the people utilizing its resources have also become apparent in the last few decades. The common types of vertebrate-induced hazards to humans in these areas are inflicted by snakes, tigers, crocodiles, and sharks. Animal attacks in Sundarbans, especially from tigers and snakes, often prove fatal. The straying of tigers from reserve forests into human habitations poses a major problem for the residents living along the forest boundary. Several conservation activities have been operating with the development of nurseries for mangrove saplings, undertaking mangrove plantation programs involving local people, nongovernmental organization (NGO), environmental researchers, planners, and administrators. These activities, along with the necessary infrastructure development, aim to promoting ecotourism and provide secured habitats for wildlife with the provision available drinking water and food materials (Figs. 4.38 and 4.39).

4.16.1 Understanding Conflicts and Relationships of Plants–Wildlife–Human in the Sundarbans Biosphere Reserve, India: Driving Force for Ecological Sustenance of the Entire Mangrove ecosystem (Flow Chart 4.1)

Under normal conditions, there is a balance of the interaction pathways among the structural components constituting a stable ecosystem. Any kind of qualitative or quantitative alteration of these interaction pathways challenges the sustainability of that stable ecosystem. Degrading this balance jeopardizes the ecosystem's power of self-maintenance and self-regulation, and adversely affects its ecological health. This causes human misery. Therefore, maintaining the sustainability requires minimizing the effects of human intervention in the functional operating processes of the ecosystem.

Different wildlife species, in the various stages of their lifecycle, respond to and tolerate environmental odds for their survivability by virtue of their cognitive capacities, physiological responsiveness, survival strategies, and social organization. A specific wildlife species must demonstrate its ability to locate, capture, and consume as needed to acquire food, shelter, and breeding opportunities. These capacities play a vital role in perpetuating the species. Locating and harvesting food with appropriate foraging strategies involves the development and use of special senses for determining foraging strategies. The postcranial musculoskeletal system ensures locomotory ability and for manifesting postural behavior activities associated with oral movements required for processing and mechanically breaking down of food in the mouth. These crucial facets of behavior supported by morpho-anatomical excellence enable wild animals to cope with the natural or human-induced challenges in the environment.

However, in contrast to all other mammals, primates lack such anatomical specialization to achieve behavioral flexibility. Their success in survivability, especially of haplorrhine primates (monkeys, apes, tarsiers etc.) is due to escaping the reproductive constraints of the estrous cycle. Most orders under the class mammals have been specialized for folivores (foliage specializations—eating leaves or grasses) or faunivory (eating animal matters—vertebrate and/or invertebrate). However, Chiroptera (bats), being nocturnal and arboreal mammals, are adapted for a diversity of diets, and consume insects, fish, amphibians, reptiles, birds, mammals, fruit,

Flow Chart 4.1 This conceptual diagram illustrates the holistic eco-dynamics of the Sundarbans mangrove ecosystem, emphasizing the continuous interactions among water, soil, vegetation, invertebrates, and vertebrate wildlife. Such interactive processes among all the structural components of the mangrove ecosystem of Sundarbans not only ensure the functioning of the entire ecosystem but also provide many ecosystem resources for the benefit of the human population

pollen, and nectar (Wilson, 1975). Adaptive specializations have been acquired by varied species of mammals for chasing, capturing, and eating others, where refined special sense organs help the predators for the detection, chasing, capturing, puncturing, crushing, and sharing of prey, using powerful mandibles supported by teeth. These wild faunae also possess elaborated and elongated small intestines, instead of stomach or colon to digest captured foods because of special categories by virtue of their mobility even after swallowing.

Mangrove forests are unique because of its presence along the interface between land and tidal lotic systems, like estuaries, creeks, and sea. These forests span both aquatic and terrestrial realms, wherein the continuous exchange of materials between the two subsystems plays fundamental roles in the ecological sustenance of both (Beger et al., 2010). The steady decline in the abundance and expanse of mangrove forest across the tropical belts of the world, especially in the South Asian countries like India, Bangladesh, Malaysia, and Philippines, has focused on more negative ecological consequences mainly in coastal–estuarine environment marine environments, largely disregarding their value as terrestrial ecosystems in respect to biological productivity (Buelow & Sheaves, 2014) and suitable habitat for terrestrial fauna (Meades et al., 2002).

Earlier research studies on the diversity and distribution of terrestrial wildlife fauna in mangrove ecosystem were mainly focused on certain charismatic fauna, such as crocodiles, birds, and tigers (Barlow et al., 2011). However, research has given comparatively little attention to the assemblage of other amphibians, reptiles, and smaller mammals, inhabiting the mangrove forests (Nagelkerken et al., 2008; Luther & Greenberg, 2009; Hogarth, 2015). Some terrestrial wildlife species spend some period of their life cycles in mangrove forests, depending on all their critical resource needs, and they are considered to be obligatory and facultative mangrove users (Hansson & Acheson, 2014). The facultative users have been observed to render important ecosystem services, such as pollination of mangrove plants (McKenzie & Rolfe, 1986), and playing roles as macro decomposers helping in the transfer of nutrients from mangrove. Proper understanding of the ecological relationships among wildlife residing in mangroves, terrestrial forests, and intermediary zones helps in planning for effective conservation strategies of the forest ecosystem. It is important also to put emphasis on the roles of the mangrove forest ecosystem in offering habitat and refuge to wild animal species, whether in the nearby forests or in scrub lands. Therefore, a holistic conservation approach of mangrove ecosystem management involves the conservation of adjoining ecosystems along with their biodiversity components.

Research findings have revealed the higher richness of terrestrial wild fauna in those areas within the mangrove forests where there exists a mismatch between floral and faunal diversity. Mangrove plant diversity is found to be in higher magnitudes in the tropical Indo–West Pacific (IWP) regions than it is in the Atlantic, Caribbean, and Eastern Pacific (ACEP) regions (Ellison et al., 1999), and higher species richness of mangrove plants promote species richness of other taxa by determining the variety of food items and habitat structural elements that create niches for other organisms (Hawkins & Porter, 2003). However, low mangrove plant species richness, as observed in Mexico and other parts of South America, promote higher vertebrate species richness. This dynamic is very similar to other parts of Asia, where higher mangrove plant species richness is noted (Qian & Ricklefs, 2008). This finding points out that factors other than low mangrove plant species richness, such as the influence of the resources of adjacent habitats, or the availability of more prey species to support carnivores and omnivores, appears to strengthen richness of facultative wild fauna in some mangrove ecosystems. In the mangroves of Africa and the Middle East, existence of poor correlations between animal and mangrove plant richness is thought to be due to social-political reasons (Reddy & Davalos, 2003).

The inconsistency between observed and predicted species richness was found to be highest in East Africa, which is a biodiversity hotspot, but higher animal richness was observed in West Africa, with higher mangrove richness. Hunting pressure appeared to be one of the major reasons for the disparity between a higher predicted level and the observed lower species richness in Africa (Cawthorn & Hoffman, 2015). The unpredictable relationship between species richness of mangrove plants and animals' richness may be explained by opportunistic occurrences of terrestrial vertebrates in mangrove forests (Reddy & Davalos, 2003). Differences in biogeographic patterns from the predicted floral richness can be explained by developing an understanding of the habitat uses of facultative users of mangroves, which outnumber specialists (Wilson & Yoshimura, 1994). Similar other research studies have focused on obligatory users of mangrove forests, which substantiated the perception that vertebrate species richness is low in these environments (Nagelkerken et al., 2008; Luther & Greenberg, 2009). Similarly, facultative wildlife fauna take shelter in mangrove forests, because of the availability of required resources there, when they flee as a refugee in the face of the anthropocentric disturbance of their primary home due to habitat destruction and subsequent environmental alterations resulting in unavailability of resources (Nowak, 2013). The impacts of the seasonality of climates and variability of tidal phases have been found to impart a considerable influence on the occurrence, abundance, and behavioral patterns of wild fauna in mangrove ecosystems, including lemurs feeding on mangrove flowers in the dry season (Gardner, 2016), and varanids feeding on mangroves during low tide (Kutt, 1997).

Numbers of wild animals considered to be refugees in mangrove forests, although reported to be low, include various groups like felids (Barlow et al. 2011; Nowak, 2013), monkeys (Nowak, 2013), snakes (Nagelkerken et al., 2008), and Wied's marmosets, *Callithrix kuhlii*, (Rodrigues & Martinez, 2014). More information on terrestrial vertebrates—carnivorous and omnivorous—dependent on mangrove forests for their shelter, could strengthen the conservation efforts for mangrove biodiversity as 72 species are classified as *Data Deficient* or *Not Assessed* by the IUCN Red List (Bland et al., 2014). Most of the mangrove forests of the tropical world demonstrate marked seasonality with respect to food sources, such as fish, fruit, nectar, and honey (Chaudhuri & Choudhury, 1994; Fernandes, 1999), and possess leaves with low palatability, coupled with high salt content (Kathiresan & Bingham, 2001). Besides, one important research finding has shown a higher prevalence of carnivores among reptiles and amphibians (Huey, 1982).

However, several charismatic wild fauna, serving as facultative users of mangroves, utilize these areas for certain phases of their life cycles (especially for the purpose of breeding), such as the estuarine crocodile, *Crocodylus porosus* (Hutchings & Recher, 1981). There are also records of some species that take shelter in mangroves to escape heat stress, like chiropteran species, and also kangaroos in Australia (Reef et al., 2014), and by dispersing between primary habitats like the marsh rabbit, *Sylvilagus palustris*, in the United States (Kathiresan & Bingham, 2001). Formulating ecomanagement strategies in the face of anthropogenic disturbances to the functioning of the mangrove ecosystem are enhanced a better understanding of the modes of mutual dependence of both obligate and facultative wildlife in mangrove forests, through the provision of essential ecosystem services, such as pollination of mangroves (Ashraf & Habjoka, 2013), and nutrient transfer (Kristensen, 2008; Reef et al., 2014).

It is really alarming to note that the end of the 21st century is expected to witness almost 75% of Sundarbans mangroves because of the independent or cumulative impacts of different environmental stressors including global sea-level rise. The virgin deltaic islands of the core zones of the Sundarbans Biosphere Reserve, India experience diurnal effects of tidal inundation and exposure. These islands act as protected tiger habitats for tigers (*Panthera tigris tigris*), but the peripheral reclaimed deltaic islands are now currently inhabited by people. The human population density of these reclaimed islands has been increasing steadily during last century, triggering the enhancement of intensities and rising numbers of man–tiger interactions. This imposes several sociopolitical problems, which in turn put development pressure on the remaining islands. Success of long-term strategies for the protection of tigers in the tidal mangrove forests of Sundarbans require the selection of potential new tiger habitats that are currently relatively undeveloped to increase the territory of this large wild mammal, and thereby to reduce ongoing adverse man–tiger interactions.

4.16.2 Causes of Human–Wildlife Conflict in Sundarbans

Large-scale exploitation of forest products, and poaching of wild animals constitutes the root cause of human–wildlife conflict in Sundarbans. It is common for the animals to stray into human habitations and to cause depredations. All of these have led to generating conflict between humans and animals, the root cause of which is mostly socioeconomic. Prevailing conflicts between human and wild animals in Sundarbans have been initiated by a large-scale explosion of human population, excessive loss of natural habitats, and an increase in dependency on forest resources. Conflicts are more acute when a species involved is critically imperiled while its presence in an area poses a significant threat to human welfare. Human–wildlife conflict is potentially instigated in situations where: (1) the behavior of people negatively impacts wildlife; (2) the behavior of wildlife creates negative impact for some stakeholders; and (3) the wildlife-focused behavior of some people creates a negative interaction with other people. Thus, problems from the people–wildlife conflict involve people–wildlife interactions. Sundarbans have a long history of hazards related to man–tiger conflicts. Tigers of Sundarbans are not only famous for their man-eating nature, but are also known for frequent straying into the villages of the fringe areas of Sundarbans. Therefore, man–tiger conflicts arise in two ways: first, through the entrance of people into tiger territory, and second, through straying of tigers into human habitations.

4.16.3 Mental Modeling of the Human–Elephant Conflict Using Fuzzy Cognitive Mapping and Decision Ecology for Conflict Resolution

Human–elephant conflict (HEC) has escalated in recent decades due to habitat fragmentation, agricultural expansion, and human encroachment into elephant habitats. It threatens elephant conservation and severely impacts rural livelihoods, food security, and ecosystem stability. In India, the villages in Bankura district, West Bengal, face frequent and severe HEC incidents (Banerjee et al., 2025). This study explores the socioeconomic and ecological impacts of HEC in the region and endeavors to formulate sustainable mitigation strategies to foster harmonious coexistence. Nine villages in the Barjora block of Bankura district were purposively selected for this research. Key informants were identified, and additional respondents were recruited using snowball sampling. Data was gathered through individual interviews and analyzed using Mental Modeler, a web-based Fuzzy Cognitive Mapping (FCM) tool. Stakeholders' mental models were constructed to identify shared interests and conflicts regarding HEC. These individual models were combined into a shared mental map, enabling a comprehensive understanding of stakeholder perspectives. Scenario analysis was conducted under three hypothetical conditions to inform better decision making. Key findings highlighted critical factors of HEC such as crop damage, elephant attacks, habitat fragmentation, human fatalities or injuries, increased elephant aggression, and reduced food availability within forest areas. Based on these insights, targeted mitigation strategies were proposed, focusing on habitat restoration, reducing grazing activities within the forest area, enhancing food availability for elephants within the forest area, strengthening wildlife corridors, and community-based conflict management approaches including empowerment of forest dwellers. This study underscores the value of participatory tools like FCM in addressing complex human–wildlife conflicts and facilitating sustainable solutions (Banerjee et al., 2025).

The same model can be replicated for human–tiger conflict (HTC) in the Sundarbans Biosphere Reserve by using Mental Modeler, a web-based FCM tool. The community is to be involved, especially those who are sufferers, for *focus group discussion (FGD)*, and for extracting the mental models of the stakeholders. The FCM thus generates needs to be extrapolated for scenario analysis for delineating an intervention plan for an ecofriendly resolution to the problem.

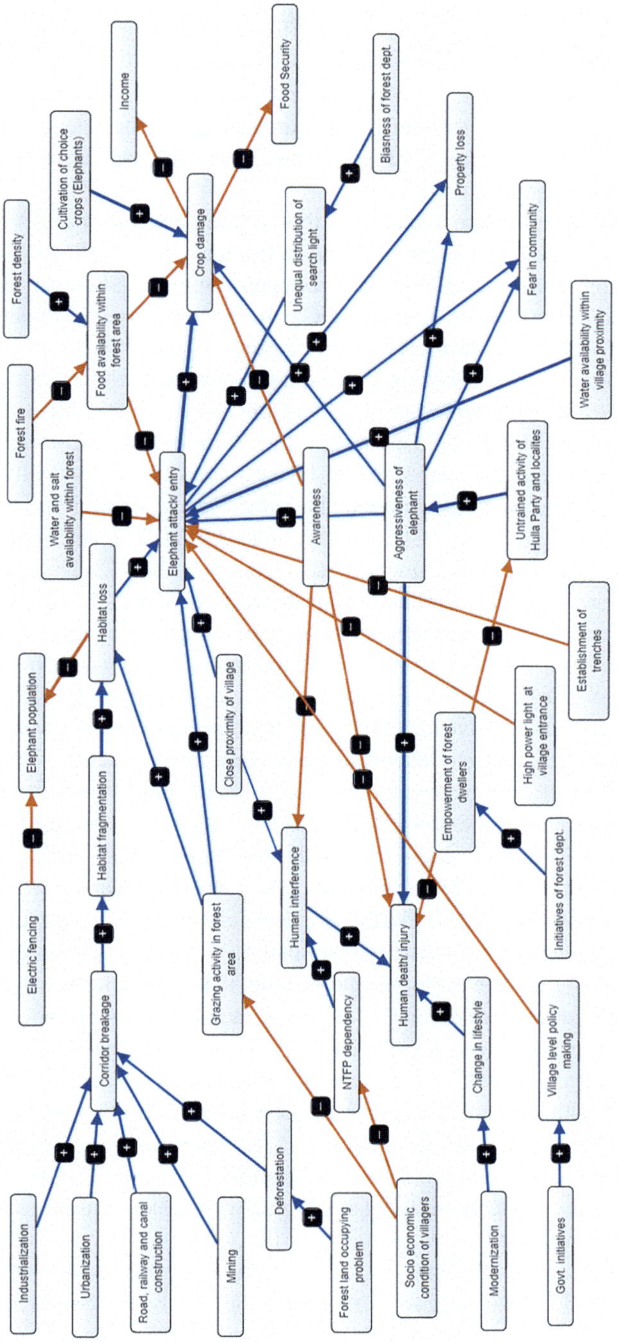
Income
Food Security
Biasness of forest dept.
Cultivation of choice crops (Elephants)
Crop damage
Unequal distribution of search light
Property loss
Forest density
Food availability within forest area
Fear in community
Water availability within village proximity
Forest fire
Water and salt availability within forest
Elephant attack/ entry
Awareness
Aggressiveness of elephant
Untrained activity of Hulla Party and localites
Habitat loss
Elephant population
Close proximity of village
Establishment of trenches
High power light at village entrance
Empowerment of forest dwellers
Habitat fragmentation
Human interference
Initiatives of forest dept.
Electric fencing
Corridor breakage
Grazing activity in forest area
NTFP dependency
Human death/ injury
Change in lifestyle
Village level policy making
Deforestation
Industrialization
Urbanization
Road, railway and canal construction
Mining
Forest land occupying problem
Socio economic condition of villagers
Modernization
Govt. initiatives

4.17 Concluding Remarks Emphasizing New Discoveries of Fauna (Genus and Species) from Sundarbans Mangrove Ecosystem India

The following faunal components have been documented as new to the science (new species or genus) from different corners of Sundarbans, india and its anearby brackish water environment during last one decade.

Crustacea: Suborder–Dendrobranchiata: Family -Panaeidae (*Parapenaeopsis longirostris*, Chandra & Bhattacharya, 2004); 2) Hymenoptera: Apidae (*Braunsapis chandrai* Gupta, Sharma & Kumar, 2015); 3) Nematoda (*Nygellus shamimi*, Sen, 2015); 4) Mollusca (*Salinator fragilis* (Lamarck), Ghosh et al., 2015); 5) Nematoda (*Cobbia bengalensis* Datta, Bhowmik & Choudhury, 2018); 6) Nematoda: Cyatholaimidae (*Cyathoshiva amaleshi* gen. n. sp. n., Datta, et al., 2016); 7) Diptera: Asilidae (*Astochia venkataramani* Roy, Parui et al., 2019); 8) Nematoda (*Trochamus timmi* Dutta & Helal, 2023); 9) Mollusca: Gastropoda, (*Ellobiumin crassatum* H. Adams & A. Adams, 1854, 2023); 10) Arachnida: Acari (*Calvolia summersi* (Mostafa, 1970; 2023); 11) Coleoptera , Lampiridae: (*Triangulara sunderbanensis*, Ghosh, S., Sarkar, S and Chakraborty, S.K 2023); 12) Hymenoptera: Crabronidae (*Brachystegus madhavani* Amal &Girish Kumar, 2024); 13) Nemertea: Eumonostilifera (*Prosadenoporus*, 2025, Majhi et al., 2025); 14) Nemertea: Heteronemertea (*Polydendrorhynchus amaleshii* sp. nov.,Dhara et al., 2025); 15) Crustacea: Brachuran crab (*Petrolisthes tanmayi*, Pahari et al., 2025); 16) Actinopterygii (ray- finned fish): Gobiiformes: (*Butis bargabhimae* sp. Nov, 2026, Maity et al., 2026). Taxonomic survey works are still on process to discover and record more such species which are to be considered as new to the science.

The coastal mangrove belts represent one of the world's most biodiverse enriched zones which are hydrologically linked with numerous ecosystems, including salt marsh, intertidal belts, mangroves, lagoons, corals, seagrasses, and backwaters (Thom, 1967; Hutchings & Richer, 1981; Untawale, 1987; Rossi & Chapman, 2003; Vannucci, 2003; Valdes-Weaver et al., 2006; Hansson & Âkesson, 2014; Chakraborty, 2023). However, this productive and sensitive ecoregion of world has been under tremendous environmental pressure during recent decades. The pressure is expected to be intensified further over next couple of decades as anthropogenic stresses, including population explosion, uncontrolled urbanization, over exploitation of resources, industrialization, drastic changes in the patterns of land use, and different forms of population growth. The situation demands an increased ecological resilience of the coastal ecosystem to cope up with ongoing stresses, a sustainable management strategy integrating different functional inputs of the varied sub-ecosystems, and adhering rigidly to the basic underlying ecological principles within coastal environment in a holistic manner.

Many research studies have established the facts that the mangrove ecosystem on the tropical and subtropical belts of the world represents one of the most productive ecosystems of the world (Alongi, 1988; Robertson & Alongi, 1992). The ecosystem plays an important role in the energy transfer between rivers, estuaries, and inter-estuarine habitats within the mangrove forests. The organic compounds of nitrogen, phosphorus, and sulfur, in the form of detritus, are generated by the decomposition of the falling mangrove leaves, and are either incorporated into the sediment, consumed by detritus-feeding shellfish, or broken up further to become food for bacteria, fungi, and finfish (Tzeng & Wang 1992; Chakraborty, 2018). The mangrove–estuarine complex in Sundarbans is recognized across the world as important nursery grounds for the feeding of the newly hatched juveniles of different commercially important shellfish and finfish, acting as a rich source of food while also ensuring protection from predators (Austin, 1971; Odum & Heald, 1975a, 1975b; Krishnamurthy & Jeyaseelan, 1981; Beckley, 1984; Bell et al., 1984; Blaber et al., 1985, 1989; Janekarn & Boonruang, 1986; Robertson & Duke, 1987; Little et al., 1988; Blaber & Milton, 1990; Tzeng & Wang, 1992; Whitfield, 1999). Most of the populations of fish found in mangrove estuarine complex in the coastal parts of West Bengal, India are short-lived euryhaline, that adapt to a wide range of salinity fluctuations, and some spend a certain phase of their life cycle in these habitats and thereby make the entire eco-region biologically vibrant and productive (Blaber & Milton, 1990; Chakraborty, 2011, 2017).

References

Abele, L. G. (1974). Species diversity of decapod crustaceans in marine habitats. *Ecology, 55*, 156–161.

Ahlstrom, E. H. (1968). An evaluation of the fishery resources available to the California fishermen. In D. W. Gillbert (Ed.), *The future of the fishing industry of the United States* (Vol. 4, pp. 65–80). Uni. Wash.Publ. Fish., New Ser.

Ahlstrom, E. H., & Moser, H. G. (1976). Eggs and larvae of fishes and their role in systemic investigations and in fisheries. *Revue des travaise de L institut des pechesmaritimes, 40*(3&4), 379–398.

Alcock, A. (1985). *Materials for a Carcinological Fauna of India*. IBPSS.

Alfred, J. R. B., & Ramakrishna. (2004). Faunal resources in mangrove ecosystems. *Environmental Forestry Bulletin, 4*, 39–48.

Alfred, J. R. B., & Ramakrishna. (2004). Faunal resources in mangrove ecosystems. *Envis Forestry Bulletin, 4*, 39–48.

Algarte, V. M., Siqueira, T., Landeiro, V. L., Rodrigues, L., Bonecker, C. C., Rodrigues, L. C., Santana, F. N., Thomaz, S. M., & Bini, L. M. (2017). Main predictors of periphyton species richness depend on adherence strategy and cell size. *PLoS One, 12*(7), e0181720.

Alongi, D. M. (1987a). Intertidal zonation and seasonality of meibenthos in tropical mangrove estuaries. *Marine Biology, 95*, 447–458.

Alongi, D. M. (1987b). The influence of mangrove-derived tannins on intertidal meiobenthos in tropical estuaries. *Oecologia, 71*, 537–540.

Alongi, D. M. (1988). Bacterial productivity and microbial biomass in tropical mangrove sediments. *Microbial Ecology, 15*(1), 59–79.

Alongi, D. M. (1989). The role of soft-bottom benthic communities in tropical mangrove and coral reef ecosystems. *CRC Critical Reviews in Aquatic Science, 1*, 243–280.

Alongi, D. M., & Sasekumar, A. (1992). Benthic communities. In A. I. Robertson & D. M. Alongi (Eds.), *Tropical mangrove ecosystem* (Coastal and estuarine studies) (Vol. 41, pp. 137–117). American Geophysical Union.

Aluri, R. J. (1990). Observations on the floral biology of certain mangroves. *Proceedings of the Indian National Science Academy, Part B, Biological Sciences, 56*(4), 367–374.

Amal, S., & Girish Kumar, P. (2024). *Records of the Zoological Survey of India*.

Annon. (2003). *Mangrove ecosystem: Biodiversity and its influence on the natural recruitment of selected commercially important finfish and shellfish species in fisheries*. National Agricultural Technology Project (NATP). Indian Council of Agricultural Research (ICAR). Principal Investigator: George, J. P.; Co–PI: Chakraborty, S.K. and Damory, S. N. pp. 1–514.

Annon. (2005). *Studies on bio-resource assessment and management of degraded mangrove ecosystem of Midnapore Coast, West Bengal*. Research project report, Ministry of Environment and Forests, Govt. of India. Principal Investigator: Susanta Kumar Chakraborty [Sanction No.3/6/2001-CSC (M) Dated 5th Nov.2001] pp. 1–99.

Anonymous. (1987). *A long term multidisciplinary research approach and report on Mangrove ecosystem of Sunderbans*. Department of Marine Science, University of Calcutta.

Anvar Batcha, S. M. (1984). *Studies on bottom Fauna of North Vembanad Lake*. Ph.D. theses, University of Cochin, p. 178.

Anon. (1987). *A long term multidisciplinary research approach and report on mangrove ecosystem of Sundarbans* (p. 92). Department of Marine Science, Calcutta University, Project Investigator: Dr. Amalesh Chowdhury.

Anonymous. (1993). In B. K. Banerjee & H. Singh (Eds.), *The shrimp fry by-catch in West Bengal*. CIBA.

Anvar Batcha, S. M. (1997). Studies on intertidal and benthic macrofauna of Dammam cornichje and half moon bay beaches of the Aranian Gulf. *Journal of the Marine Biological Association of India, 39*(1and 2), 42–43.

Ashraf, M., & Habjoka, N. (2013). Tropical mangroves; biologically most diverse ecosystem: Megachiroptera as key ecological and conservation tool. *Mangrove Action Newsletter, 315*.

Austin, H. M. (1971). A survey of the ichthyofauna of the mangroves of Western Puerto Rico during December 1967-August 1968. *Caribbean Journal of Science, II*(1-2), 27–39.

Bachelet, G. (1990). The choice of a sieving mesh size in the quantitative assessment of marine macrobenthos: A necessary compromise between aims and constraints. *Marine Environmental Research, 30*, 31–35.

Baidya, A. U., & Chaudhury, A. (1984). Ddtibution and abundance of zooplankton ina tidal creek of Sagar Islands, Sunderbanss, West Bengal. *Environmental Ecology, 2*(4).

Balow, D. C. A. (2009). *The Sundarbans tiger adaptation, population status and conflict management*. University of Minnesota PhD. Thesis. pp. 29–54.

Biswas, H., Dey, M., Ganguly, D., De, T. K., Ghosh, S., & Jana, T. K. (2009). Comparative analysis of phytoplankton composition and abundance over a two decade period at the land-ocean boundary of a tropical mangrove ecosystem. *Estuaries and Coasts*. https://doi.org/10.1007/s12237-0099193-5

Bairagi, N. (1998). *Cnidaria: Sea Anemone. State Fauna Series 3: Fauna of West Bengal, Part II* (pp. 29–44). Zoological Survey of India.

Banerjee, S., Ramakrishnan, S. C., & Raghunathan, C. (2023). *Acta BiologicaTurcica, 37*(2), 4–11.

Banerjee, A., Chatterjee, A., & Acharya, S. K. (2025). Mental modeling of human elephant conflict using fuzzy cognitive mapping and decision ecology for conflict resolution. *Environment Systems and Decisions, 45*(3), 33. https://doi.org/10.1007/s10669-025-10032-3

Banerjee, D., Pal, D., Patra, T. K., et al. (1997). Lipids and fatty acids of air breathing fish, *Boleophthalmus boddaerti*. *Food Chemsitry, 60*(3), 303–309.

Banerjee, L. K. (1984). Vegetation of the Bitarkanika sanctuary, Osissa State. *Journal Economic Taxonomy Botany, 5*, 1065–1079.

Banerjee, L. K. (1987/1985). Comparative study on mangroves of the Sundarbans and that of the Mahanadi Delta in Eastern India. In L. J. Bhosale (Ed.), *The Mangroves: Proceedings of National Symposium. Biology. Utilization and Conservation of Mangroves. Kolahpur, India* (pp. 527–530).

Banerjee, L. K., Sastry, A. R. K., & Nayar, M. P. (1989). *Mangroves in India: Identification manual* (p. 113). Botanical Survey of India.

Banerjee, R. K., & Chattopadhyay, G. N. (1980). *Methodology for soil and water analysisin brackishwater culture system*. Central Inland Fisheries Research Institute Miscellaneous Contribution, No. 15.

Barlow, A. C., Smith, J. L., Ahmad, I. U., Hossain, A. N., Rahman, M., & Howlader, A. (2011). Female tiger Panthera tigris home range size in the Bangladesh Sundarbans: The value of this mangrove ecosystem for the species' conservation. *OryoCy, 45*, 125–128.

Barlow, A. C. D., et al. (2010). A preliminary investigation of Sundarbans tiger morphology. *Mammalia, 74*, 329–331.

Basu, S., et al. (2012). Variation of surface water quality in selected tidal creeks of Sagar Island, Indian Sundarban ecoregion: A multivariate approach. *Applied Water Science, 11*, 63.

Beckley, L. E. (1984). The ichthyofauna of the Sundays Estuary, South Africa, with particular reference to the juvenilie marine component. *Estuaries, 7*, 248–259.

Beger, M., Grantham, H. S., Pressey, R. L., Wilson, K. A., Peterson, E. L., Dorfman, D., Mumby, P. J., Lourival, R., Brumbaugh, D. R., & Possingham, H. P. (2010). Conservation planning for connectivity across marine, freshwater, and terrestrial realms. *Biological Conservation, 143*, 565–575.

Bell, J. D., Pollard, D. A., Burchmore, J. J., Pease, B. C., & Middleton, M. J. (1984). Structure of a fish community in a temperate tidal mangrove creek in botany bay, New South Wales. *Australian Journal of Marine and Freshwater Research, 35*, 33–46.

Bhattacharya, M., Kar, A., Chini, D. S., Malick, R. C., Patra, B. C., & Das, B. K. (2018). Multi-cluster analysis of crabs and ichthyofaunal diversity in relation to habitat distribution at tropical mangrove ecosystem of the Indian Sundarbans. *Regional Studies in Marine Science, 24*, 203–211. https://doi.org/10.1016/j.rsma.2018.08.009

Bhowmik, S. K., Rpy, G. N., & Barman, S. (1985). A list of 24 strains of bacteria from the decomposed litter of mangrove areas of Sundarbans complex. In *The mangroves: Proceedings of National Symposium on biology, utililization and conservation of mangroves* (pp. 389–392).

Biggs, B. J. F., & Thomsen, H. A. (1995). Disturbance of stream periphyton by perturbations in shear stress: Time to structural failure and differences in community resistance. *Journal of Phycology, 31*(2), 233–241.

Biggs, J. F., Stevenson, R. J., & Lowe, R. L. (1998). A habitat matrix conceptual model for stream periphyton. *Archiv für Hydrobiologie, 143*, 21–56.

Biswas, A., Bhaumik, S. K., & Roy, G. N. (1985). Chitinolytic bacteria of mangrove swamps of Sundarbans. In *Proceedings of national symposium on biology, utililization and conservation of mangroves* (pp. 382–384).

Biswas, H., Dey, M., Ganguly, D., De, T. K., Ghosh, S., & Jana, T. (2010). Comparative Analysis of Phytoplankton Composition and Abundance over a Two-Decade Period at the Land–Ocean Boundary of a Tropical Mangrove Ecosystem. *Estuaries and Coasts, 33*, 384–394.

Biswas, H., Mukhopadhyay, S. K., De, T. K., Sen, S., & Jana, T. K. (2004). Biogenic controls on the air water carbon dioxide exchange in the Sundarban mangrove environment, of Bay of Bengal, India. *Limnology and Oceanography, 49*(1), 95–101.

Blaber, S. J. M. (1986). Feeding selectivity of a guild of piscivorous fish in mangrove areas of north-west Australia. *Australian Journal of Marine and Freshwater Research, 37*, 329–336.

Blaber, S. J. M. (2008). *Tropical estuarine fishes: Ecology, exploration and conservation*. John Wiley & Sons.

Blaber, S. J. M., & Blaber, T. G. (1980). Factors affecting the distribution of juvenile estuarine and inshore fish. *Journal of Fish Biology, 17*, 143–162.

Blaber, S. J. M., & Milton, D. A. (1990). Species composition, community structure and zoogeography of fishes of mangrove estuaries in the Solomon Islands. *Marine Biology, 105*, 259–267. https://doi.org/10.1007/BF01344295

Blaber, S. J. M., Brewer, D. T., & Salini, J. P. (1989). Species composition and biomasses of fishes in deferent habitats of a tropical northern Australian estuary: Their occurrence in the adjoining sea and estuarine dependence. *Estuarine, Coastal and Shelf Science, 29*, 509–531.

Blaber, S. J. M., Cyrus, D. P., Albaret, J. J., Chong, V. C., Day, J. W., Elliott, M., Fonseca, M. S., Hoss, D. E., Oresanz, J., Potter, I. C., & Silver, W. (2000). Effects of fishing on the structure and functioning of estuarine and near shore ecosystem. *ICES Journal of Marine Science, 57*, 590–602.

Blaber, S. J. M., Young, J. W., & Dunning, M. C. (1985). Community structure and zoogeographic affinities of the coastal fishes of the Dampier region of North-Western Australia. *Australian Journal of Marine and Freshwater Research., 36*, 247–266.

Bland, L. M., Collen, B. E. N., Orme, C. D. L., & Bielby, J. O. N. (2014). Predicting the conservation status of data-deficient species. *Conservation Biology, 29*, 250–259.

Blasco, F. (1977). Outlines of ecology, botany and forestry of the mangals of the Indian subcontinent. In V. Chapman (Ed.), *Wet coastal ecosystems. Ecosystems of the world* (Vol. 1, pp. 241–257). Elsevier Scientific Publishing Co. Ecological water quality—Water treatment and reuse.

Blasco, F. (1977). Outlines of the ecology, botany and forestry of the Mangals of the Indian subcontinent. In V. J. Chapman (Ed.), *Wet ecosystem of the world* (pp. 241–262). Elsevier.

Blasco, F. (1984). Climate factors and the biology of mangrove plants. In S. C. Snedaker & J. G. Snedaker (Eds.), *The mangrove ecosystem: Research methods* (pp. 18–35). UNESCO/SCOR.

Blasco, F., & Aizpuru, M. (2002). Mangroves along the coastalstretch of the Bay of Bengal: Present status. *Indian Journal ofMarine Sciences, 31*, 9–20.

Blasco, F., & Aizpuru, M. (2002). Mangroves along the coastal stretch of the bay of Bengal: Present status. *Indian Journal of Marine Sciences, 31*, 9–20.

Boesch, D. F. (1977). A new look at the zonation of the benthos along the estuarine gradient. In B. C. Coull (Ed.), *Ecology of marine benthos*. University of South Carolina Press.

Boesch, D. F., & Turner, R. E. (1984). Dependency of fishery species on salt marshes: The role of food and refuge. *Estuaries and Coasts, 7*, 460–468.

Bosire, J. O., Dahdouh Guebas, F., Kairo, J. G., Cannicci, S., & Koedam, N. (2004). Spatial variations in macrobenthic fauna recolonisation in a tropical mangrove bay. *Biodiversity and Conservation, 13*(6), 1059–1074.

Boyd, C. E. (1978). *Water quality in warm water fish ponds* (p. 359). Agril. Expt. Stn. Auburn Univ.

Bryars, S. R., & Havenhand, J. N. (2004). Temporal and spatial distribution and abundance of blue swimmer crab (*Portunus pelagicus*) larvae in a temperate gulf. *Marine and Freshwater Research, 55*(8), 809–818.

Buelow, C., & Sheaves, M. (2014). Mangrove forests: A birds-eye view of connectivity in coastal ecosystem mosaics. *Estuarine Coastal and Shelf Science, 152*, 33–43.

Camilleri, J. C. (1992). Leaf-litter processing by invertebrates in a mangrove forest in Queensland. *Marine Biology, 114*, 139–145.

Caspers, H. (1967). Estuaries: Analysis of definitions and biological considerations. In G. H. lauff (Ed.), *Estuaries Publication of American Association of Advance Science* (Vol. 83, pp. 6–8).

Cawthorn, D. M., & Hoffman, L. C. (2015). The bushmeat and food security nexus: A global account of the contributions, conundrums and ethical collisions. *Food Research Internationally, 76*(Part 4), 906–925.

Chakrabarti, K. (1987). Sundarbans honey and the mangrove swamps. *Journal of the Bombay Natural History Society, 1*, 133–137.

Chakraborty, S. K. (1995). Aquaculture potential of mangrove ecosystem of Sundarbans, India. In *Proceedings of seminar on fisheries—A multimillion dollar industry* (pp. 72–83).

Chakraborty, S. K. (1996). Aquaculture potential of mangrove ecosystem of Sundarbans, India. In *Proceedings of Seminar on Fisheries—A Multimillion Dollar Industry* (pp. 72–83). AFI.

Chakraborty, S. K. (1998). Coastal aquaculture and environment in the context of West Bengal, India. *Vidyasagar University Journal of Biological Sciences, 4*, 13–20.

Chakraborty, S. K. (2003). Biodiversity and conservation. In A. D. Mukhopadhyay (Ed.), *Perspectivesand issues in environmental studies* (pp. 267–286). Vidyasagar University.

Chakraborty, S. K. (2009). Biological diversity of Mindapore Coast, West Bengal. *Environmental Forensics, 2*(2), 6.

Chakraborty, S. K. (2010). Coastal environment of Midnapore, West Bengal: Potential threats and management. *Journal of Coastal Conservation, 1*(1), 28–40.

Chakraborty, S. K. (2011). Mangrove ecosystem of Sundarbans, India: Biodiversity, ecology, threats and conservation. In N. M. James (Ed.), *Mangroves: Ecology, biology and taxonomy* (pp. 83–112). NOVA Publishers.

Chakraborty, S. K. (2013). Interactions of environmental variables determining the biodiversity of coastal mangrove ecosystem of West Bengal, India. *Ecoscan Special Issue, 3*, 251–265.

Chakraborty, S. K. (2017). Ecological services on intertidal benthic fauna and the sustenance of coastal wetlands along the Midnapore(East) coast, West Bengal, India. In *Coastal wetlands: Alteration and remediation* (pp. 777–886). Springer.

Chakraborty, S. K. (2018). Bioinvasion and environmental perturbation: Synergistic impact on coastal—mangrove ecosystems of West Bengal, India. In C. Makowski & C. Finkl (Eds.), *Impacts of invasive species on coastal environments: Coast in crisis* (pp. 171–245). Springer.

Chakraborty, S. K. (2021). *Riverine ecology (volume 2): Biodiversity conservation, conflicts and resolution*. Springer Nature. 917 pages.

Chakraborty, S. K., & Choudhury, A. (1985). Distribution of fiddler crabs in Sunderbans mangrove estuarine complex, India. In *The mangrove: Proceedings of national symposium on biology utilization and conservation of mangroves* (pp. 467–472). L. J. Bhosale.

Chakraborty, S. K., Choudhury, A., & Deb, M. (1986). Decapod Brachyura from Sundarbans mangrove estuarine complex, India. *Journal of Bengal Natural History Society, 5*(1), 55–68.

Chakraborty, S. K., Poddar, T. K., & Choudhury, A. (1992). Species diversity of Macrozoobenthos of Sagar Island, Sundarban, India. *Proceedings of Zoological Society, Calcutta, 45*(Suppl. A), 435–444.

Chakraborty, S. K., & Choudhury, A. (1992a). Ecological studies on the zonation of brachyuran crabs in a virgin mangrove Island of Sundarbans, India. *Journal of Marine Biological Association of India, 34*(1 and 2), 189–194.

Chakraborty, S. K., & Choudhury, A. (1992b). Population ecology of fiddler crabs (*Uca.* spp.) of the mangrove estuarine complex of Sundarbans, India. *Tropical Ecology, 33*(1), 78–88.

Chakraborty, S. K., & Choudhury, A. (1993). Community organization of macrobenthic molluscs of Sagar Island, Sundarbans, India. *Indian Biologist, xxv*(1), 33–36.

Chakraborty, S. K., & Choudhury, A. (1994). Community structure of macrobenthic polychaetes of intertidal region of Sagar Island, Hooghly estuary, Sundarbans, India. *Tropical Ecology, 35*(1), 97–104.

Chakraborty, S. K., & Choudhury, A. (1997). Occurrence and abundance of benthic polychaetes in Hoogly Estuary, Sagar Island, India. *Journal of the Marine Biological Association of India, 39*(1& 2), 140–147.

Chakravarty, G., Giri, S., & Chakraborty, S. K. (2004). Distribution of coliform bacteria in different estuaries of West Bengal Coast, India. In A. Kumar, J. R. Dhanze, A. K. Chopra, & D. S. Malik (Eds.), *Protected habitats and biodiversity* (pp. 395–402). Natcon Publication.

Chakraborty, S. K., Giri, S., Chakravarty, G., & Bhattacharya, N. (2009). Impact of eco-restoration on the biodiversity of Sundarbans mangrove ecosystem, India. *Water Air Soil Pollution: Focus, 9*, 303–320.

Chakraborty, S. K., Chatterjee, S., Dey, M. K., et al. (2012). Ecological changes of an eco–rehabilitated delta of hooghly estuary. *Journal of Coastal Environment, 3*(1), 9–27.

Chakraborty, S. K., Maiti Dutta, S., Ghosh, P. B., Ray, R., & Paul, A. K. (2014). Impact of global warming on Sundarbans mangrove ecosystem, India: Role of different assessment tools from ecosystem monitoring to molecular markers. In *Proceedings of the International conference on Green India: Strategic knowledge for combating climate change: Prospects and challenges* (pp. 181–200).

Chakraborty, S. K., Chakravarty, G., Giri, S., Dam Roy, S., & George, J. P. (2005). Crustacea (crab). In J. P. George (Ed.), *Mangrove ecosystem: A manual for the assessment of biodiversity* (pp. 146–149). Central Marine Fisheries Research Institute, India.

Chakravarty, G., Chakravarty, S. K., Kumaraswamy Achary, G. P., & Dam Roy, S. (2005). Sea-anemones. In J. P. George (Ed.), *Mangrove ecosystem: A manual for the assessment of biodiversity* (pp. 146–149). Central Marine Fisheries Research Institute, India.

Chakraborty, U., Mitra, B., & Bhadra, K. (2019). Diversity and ecological role of insect flower visitors in the pollination of mangroves from the Indian Sundarbans. *Current Science, 117*.
Chakrabarti, K. (1992). *Man-eating tigers*. Darbari Print-Pro (P) Ltd.
Chakraborty, P. K., & Naskar, K. R. (1988). Role of mangroves in estuarine fisheries development. In A. G. Jhingran & V. V. Sugunan (Eds.), *Conservation and source management of Inland capture fisheries resources of India* (pp. 229–233). I.F.S.I., India.
Chakraborty, S. K., Paul, A. K., Bandyopadhyay, J., Banerjee, D., Pakhira, H., Middya, S., Jana, S., Sahoo, P., Paria, K., & Bera, K. (2013). Sustainable water management in the drought prone riverine tracts of South West Bengal. *Journal of Ground Water Studies, 1*, 11–34.
Chakraborti, U., Mitra, B., & Bhadra, K. (2019). Diversity and ecological role of insect flower visitors in the pollination of mangroves from the Indian Sundarbans. *Current Science, 117*, 1060–1070.
Chandra, A. K., Chakravarty, G., & Chakraborty, S. K. (2003). Polychaete diversity of Midnapore Coastal Belt, West Bengal, India. *Journal for Nature Conservation, 15*(2), 95–108.
Chandra, A., & Bhattacharya, T. (2004). A new species of the genus Parapenaeopsis, Alcock, 1901 (Penaeoidea:Panaeidae) from Orissa. *Proceedings of Zoological Society, Calcutta, 57*(1), 23–27.
Chandra, K., Majumder, A., & Maheswaran, G. (2020). *Birds of the Sundarban biosphere reserve* (pp. 1–245). Zoological Survey of India, Government of India.
Chaudhuri, A. B., & Choudhury, A. (1994). *Mangroves of the Sundarbans, India* (Vol. 1, pp. 1–247). IUCN.
Chapin, F. S., Zavaleta, E. S., Eviners, V. T., Naylor, R. L., Vitousek, P. M., et al. (2000). Consequences of changing biodiversity. *Nature, 405*, 234–242.
Chong, V. C., Sasekumar, A., Leh, M. U. C., & D'Cruz, R. (1990). The fish and prawn communities of a Malaysian coastal mangrove system, with comparisons to adjacent mud flats and inshore waters. *Estuarine Coastal and Shelf Science, 31*, 703–722.
Clarke, P. J., & Myerscough, P. J. (1991). Floral biology and reproductive phenology of Avicennia marina in south-eastern Australia. *Australian Journal of Botany, 39*, 283–293.
Cloern, J. E. (1996). Phytoplankton bloom dynamics in coastal ecosystem: A review with some general lessons from sustained investigation of San Francisco Bay, California. *Reviews of Geophysics, 34*(2), 127–168.
Cloern, J. E., Alpine, A. E., Cole, B. E., Wong, R. L., Arthur, J. F., & Ball, M. D. (1983). River discharge controls phytoplankton dynamics in the northern San Francisco Bay estuary. *Estuarine Coastal and Shelf Science, 16*, 415–429.
Chakrabarty, S., Chatterjee, T., Gaine, T., & Chaudhuri, P. (2023). Vegetative and reproductive phenology of dominant Mangrove species: A multi-parametric approach from Indian sundarbans. *Journal of Insect Conservation, 27*, 493–503.
Chakravarti, G. (2008). *Temporal and Spatial variation of macrozoobenthic community in Sundarbans mangrove Ecosystem, India*.
Chatterjee, S. (2013). *Ecological studies of Brachyuran crabs with special emphasis on their bioturbatory structures and their role in nutrient recycling of the coastal belt of Midnapore District, West Bengal, India*.
Chatterjee, S., & Chakraborty, S. K. (2014). Feeding behaviour and functional role of some selected species of Brachyuran crabs in nutrient cycle of coastal belt of Midnapore (East), West Bengal, India. *Journal of Biology and Life Science, 5*(1), 106–129.
Chatterjee, S., Das, T. K., & Chakraborty, S. K. (2014). Reproductive biology and bioturbatory activities of two sympatric species of fiddler crabs *Uca lactea annulipes* and *Uca triangularis bengali* (Decopoda:Ocypodidae) at the East Midnapore Coastal belt of West Bengal, India. *Journal of Biology and Life Science, 5*(2), 106–126.
Chapman, V. J. (1976). *Mangrove vegetation*. J Cramer Verlag.
Chapin, F. S., Walker, B. H., Hobbs, R. J., Hopper, D. U., Lawton, J. H., Sala, O. E., & Tilman, D. (1997). Biotic control over the functioning of ecosystems. *Science, 277*, 277–500.

Chatterjee, S., Bhunia, G., & Chakraborty, S. K. (2008). *Bioturbation of Brachyuran crabs and its impact on coastal ecosystem of Midnapore district.* West Bengal. In Proceedings of Fourth International.

Cloern, J. E., Krantz, T., Hogan, C. M., & Duffy, J. E. (2007). Eutrophication. In C. J. Cleveland (Ed.), *Encyclopedia of earth.* Environmental Information Coalition, National Council for Science and the Environment.

Conely, D. J., Carstensen, J., Aertebjerg, G., Christensen, P. B., Dalsgaard, T., Flansen, J. L. S., & Josefson, A. B. (2007). Long-term changes and impacts of hypoxia in Danish coastal waters. *Ecological Applications, 17*, S165–S184.

Costa, C. S. B., & Davy, A. J. (1992). Coastal salt marsh communities of Latin America. In U. Seeliger (Ed.), *Coastal plant communities of Latin America* (pp. 179–199). AcademicPress.

Coupland, G. T., Paling, E. I., & McGuinness, K. A. (2005). Vegetative and reproductive phenologies of four Mangrove species from Northern Australia. *Australian Journal of Botany, 53*, 109–117.

Datta, T. K., Miljutin, D. M., & Chakraborty, S. K., et al. (2016). *Cyathoshiva amaleshi* gen. n. sp. n. (Nematoda: Cyatholaimidae) from the coast of India. *Zootaxa, 4126*(4), 577–586.

Datta, T. K., Bowmick, M., & Choudhury, A. (2018) *Cobbia bengalensios*, sp.nov (Xyalidae: Monhysteridae) from an eroding island of Sundarban, *India Zootaxa 4444*(2): 179–188.

Datta, T. K.,& Al- Helal, M. A. (2023) Review of Euchromadorinae (Nematoda: Chromadorida) with description of a new species of Trochomus, Boucher & De Bovee, 1971 from Sundarban, India. *Zootaxa 5270*(3): 493–510.

Das, G. K., & Bhattacharya, A. (1999a). Destruction of certain encountered shelfish and finish from the coastal waters of Sunderbans and its impact on the coastal ecosystem. *Journal of the Indian Society of Coastal Agricultural Research, 17*(1&2), 216–218.

Das, G. K., & Bhattacharya, A. (1999b). Rampant exploitation of post-larval penaeid prawn (Penaeus mondon) from the salt-water course of the deltaic Sunderbans, West Bengal—a threat to the balance of nature. *Journal of the Indian Society of Coastal Agricultural Research, 17*(1&2), 219–222.

Das, G. K., & Bhattacharya, A. (1994). A piecemeal mechanism of bank erosion following subsidence: A case study from Thakauran river of deltaic Sundarban, West Bengal. *Journal of the Indian Society of Coastal Agricultural Research, 12*, 231–234.

Das Sarkar, S., Naskar, M., Gogoi, P., Raman, R. K., Manna, R. K., Samanta, S., Mohanty, B. P., & Das, B. K. (2019). Impact assessment of barge trafficking on phytoplankton abundance and Chl *a* concentration, in River Ganga, India. *PLoS One, 14*(9), 221–451.

Datta, T. K., & Chakraborty, S. K. (2015). Seasonal zonation pattern of meiobenthic fauna of an intertidal belt in the coastal area of Midnapore (East), West Bengal, India. *International Research Journal of Environmental Sciences, 4*(9), 45–52.

Datta, T. K., Miljutin, D. M., & Chakraborty, S. K. (2016). *Cyathoshiva amaleshi* gen. n. sp. n. (Nematoda: Cyatholaimidae) from the coast of India. *Zootaxa, 4126*(4), 577–586.

Datta, T. K., & Chakraborty, S. K. (2017). Relevance of meiobenthic research: Indian perspectives. *Examines in Marine Biology and Oceanography (Crimson Publishers), 4*(1), 1–5. https://crimsonpublishers.com/eimbo/fulltext/EIMBO.000523.php

Datta, T. K., et al. (2017). *First report of Megodontolaimus coxbazariTimm, 1969 (Nematoda: Chromadorida) from the Indian coast.* SenckenbergGesellschaft fur Naturforschung and Springer—Verlag Berlin Heidelberg (Springer) https://www.researchgate.net/publication/294288826_First_report_of_Megodontolaimus_coxbazari_Timm_1969_Nematoda_Chromadorida_from_the_Indian_coast.

Datta, T. K., Ganguli, A., & Chakraborty, S. K. (2018). Pseudochromad orabenepapillata (Timm 1961) comb. n. (Desmodoridae: Nematoda): Revision of its taxonomic status and distribution. *Zootaxa, 4425*(1), 165–174. (Springer) https://www.mapress.com/zt/article/view/zootaxa.4425.1.10

Dauvin, J. C., Gomez Gesteira, J. L., & Fraga, M. S. (2003). Taxonomic sufficiency: An overview of its use in the monitoring of sublittoral benthic communities after oil spills. *Marine Pollution Bulletin, 46*, 552–555.

Davis, R. A. (1983). *Depositional systems*. Prentice-Hall, Inc.

De, J. K. (2017). Diversity and conservation of mammalian fauna of Indian Sundarbans. In K. Chandra, J. R. B. Alfred, B. Mitra, & B. Roy Chowdhury (Eds.), *Fauna of Sundarban biosphere reserve* (pp. 37–48).

De Alvarenga, A. M. S. B., Botosso, P. C., & Soffiatti, P. (2017). Stem growth and phenology of three subtropical Mangrove tree species. *Brazilian Journal of Botany, 40*, 907–914.

D'Avila, M., & Marchini, L. C. (2008). Faunistic analysis of flower-visiting hymenoptera in cerradao fragment in Itirapina, SP. *Ciencia Florestal, 18*, 271–279.

De, T. K., Ghosh, S. K., Jana, T. K., & Choudhury, A. (1991). Phytoplankton bloom in the Hooghly estuary. *Indian Journal of Marine Sciences, 20*, 134–137.

Debata, S., Panda, R. M., & Palita, S. K. (2019). Chiropteran diversity and key determinants of their distribution in Eastern Ghats, India. *Biodiversity and Conservation, 28*(8–9), 2385–2404.

de Graaf, G. J., & Xuan, T. T. (1998). Extensive shrimp farming, mangrove clearance and marine fisheries in the southern provinces of Vietnam. *Mangroves and Salt Marshes, 2*, 159–166.

DeGroot, A. P. (1953). Protein and amino acid requirements of the honey bee (Apis mellifica L.). *Physiologia Comparata et Oecologia, 3*, 197–285.

Denlinger, D. L. (1980). *Seasonal and annual variation of insect abundance in the Nairobi National park.*

Demaree, G. R., & Rutishauser, T. (2011). From periodical observations to anthochronology and phenology - The scientific debate between Adolphe Quetelet and Charles Morren on the origin of the word phenology. *International Journal of Biometeorology, 55*, 753–761.

Denlinger, D. L. (1980). *Seasonal and annual variation of insect abundance in the Nairobi National.*

Deulti, K. (2017a). Frogs and toads diversity of Indian Sundarbans and threats. In K. Chandra, J. R. B. Alfred, B. Mitra, & B. Roy Chowdhury (Eds.), *Fauna of Sundarban biosphere reserve* (pp. 101–106).

Deulti, K. (2017b). Reptilian diversity in the mangrove of Indian Sundarban. In K. Chandra, J. R. B. Alfred, B. Mitra, & B. Roy Chowdhury (Eds.), *Fauna of Sundarban biosphere reserve* (pp. 83–99).

Dey, M. K. (2009). *Ecological study on the distribution and role of soil micro arthropods in the decomposition of mangrove litter of east Midnapore coast, West Bengal, India.*

Dey, M. K., Hazra, A. K., & Chakraborty, S. K. (2007). Soil microarthropods populations in alluvial and coastal soil in Midnapore District with special reference to relative abundance. *Records of the Zoological Survey of India, 107*(4), 93–99.

Dey, M. K., Hazra, A. K., & Chakraborty, S. K. (2008). Diversity of macroarthropods and their role in the plant litter decomposition in the coastal tract of east Midnapore District, West Bengal, India. *Zoological Research in Human Welfare, 20*, 207–226.

Dey, M. K., Hazra, A. K., & Chakraborty, S. K. (2010). Functional role of microarthropods in nutrient cycling of mangrove–estuarine ecosystem of Midnapore coast of West Bengal, India. *International Journal of Environmental Technology and Management, 10*(10), 1–18.

Dhara, S., Majhi, J., Mandal, B., Bhadury, P., Bhattacharya, S., & Sudipta Ghorai, S. (2025). *Polydendrorhynchus amaleshii* sp. nov., a new species with branched proboscis (Nemertea: Heteronemertea) from the North-East Coastal Bay of Bengal. *Zoological Science, 42*(2), 1–9.

Dittman, S. (1995). Benthos structure on tropical tidal flats of Australia. *Helgoländer Meeresuntersuchungen, 49*, 539–551.

Dittman, S. (2000). Zonation of benthic communities in a tropical tidal flat of north-east Australia. *Journal of Sea Research, 43*, 33–51.

Dinerstein, E., Loucks, C., Wikramnayake, E., Ginsberg, J., Sanderson, E., Seidensticker, J., et al. (2007). The fate of wild tigers. *Bioscience, 57*, 508–514.

Duke, N. C., Bunt, J. S., & Williams, W. T. (1984). Observations on the floral and vegetative phenologies of North-Eastern Australian mangroves. *Australian Journal of Botany, 32*, 87–99.

Ellison, A. M. (2002). Macroecology of mangroves: Large-scale patterns and processes in tropical coastal forests. *Trees, 16*, 181–194.

Ellison, A. M., Mukherjee, B. B., & Karim, A. (2000). Testing patterns of zonation in mangroves: Scale dependence and environmental correlates in the Sundarbans of Bangladesh. *Journal of Ecology, 88*, 813–824. https://doi.org/10.1046/j.1365-2745.2000.00500.x

Ellison, A. M., Farnsworth, E. J., & Merkt, R. E. (1999). Origins of mangrove ecosystems and the mangrove biodiversity anomaly. *Global Ecology and Biogeography, 8*, 95–115.

Felts, R. A., Rajts, F., Akhteruzzaman, M. (1996). *Small indigenous fish species culture in Bangladesh*. Development of Inland Fisheries Technical Brief. IFADP. Sub project 2, p 41.

Fernando, O. J., Khan, S. A., & Rasinathan, R. (1983). Observation on the distribution of benthic fauna in Veller estuary. *Journal of Marine Biological Association of India, 29*, 86–103.

Ferreira, A. S., & Faria, R. G. (2021). Predation risk is a function of seasonality rather than habitat complexity in a tropical semiarid forest. *Scientific Reports, 11*, 16670.

Fernandes, M. E. (1999). Phenological patterns of Rhizophora L., Avicennia L. arid Laguncularia Gaertn. f. in Amazonian mangrove swamps. In *Diversity and function in mangrove ecosystems* (pp. 53–62). Springer.

Fransworth, E. J., & Ellison, A. M. (1991). Patterns of herbivory in Beliean mangrove swamps. *Biotropica, 23*, 555–567.

Fransworth, E. J., & Ellison, A. M. (1997). Global patterns of pre-dispersal propagule predation in mangrove forests. *Biotropica, 29*(3), 318–330.

Frith, D., Tantanasiriwong, W., & Bhatia, O. (1976). Zonation of macrofauna on a macrofauna on a mangrove shore, Phuket Island. *Phuket Marine Biological Centre, Research Bulletin, 10*, 1–37.

Fischer, R., & Sheaves, M. (2003). Community structure and spatial variability of marine nematodes. *Hydrobiologia*. https://doi.org/10.1023/A:1025406624390

Gardner, C. J. (2016). Use of mangroves by lemurs. *International Journal of Primatology, 37*, 317–337.

Ghosh, A., Tripathy, B., & Mukhopadhyay, A. (2015). *Records of the Zoological Survey of India, 115*(3), 277–280.

Ghosh, S., Sarkar, S., & Chakraborty, S. K. (2023a). *Triangulara sunderbanensis* sp. nov. (Coleoptera: Lampyridae)—A new firefly species from the Sunderban Biosphere Reserve in India. *Oriental Insects*. https://doi.org/10.1080/00305316.2023.2289968

Ghosh, D., Acharya, S., & Subramanian, K. A. (2023b). *Sociobiology, 70*(2), e8544.

Giere, O. (1993). *Meiobenthology, the microscopic fauna in aquatic sediments* (Vol. 328). Springer.

Giri, S., Chakravarty, G., Bhakat, R. K., & Chakraborty, S. K. (2004). Ecological status of Jamboo Island, a virgin mangrove area of South West Sundarbans, India, with special reference to its fishery potential. *Nature Conservators, 8*, 219–239.

Giri, S. (2009). *Ecological studies of Ichthyoplankton in Sundarbans Mangrove ecosystem, West Bengal, India*. Ph.D. thesis.

Gogoi, P., Sinha, A., Das, S. S., Chanu, T. N., Yadav, A. K., Koushlesh, S. K., Borah, S., Das, S. K., & Das, B. K. (2019). Seasonal influence of physico-chemical parameters on phytoplankton diversity and assemblage pattern in Kailash Khal, a tropical wetland, Sundarbans, India. *Applied Water Science, 9*, 156.

Gopal, B., & Chauhan, M. (2006). Biodiversity and its conservation in the Sundarban mangrove ecosystem. *Aquatic Sciences, 68*, 338–354.

Goudha, R., & Panigrahy, R. C. (1996). The mangrove of Orissa. *Journal of Indian Ocean Studies, 3*(3), 228–237.

Gray and Elliot. (2009). *Gray JS, and Elliot M (2009) ecology of marine sediments–from science to management* (pp. 1–225). Oxford University Press.

Gupta, S. K. (1987). Some mangrove soils of the Sundarbans ecosystem. *UNESCO Reg Introductory Train Course Estuar Res, Calcutta, 2*, 105–112.

Hall, J. L. W., & Killen, W. D. (2005). Temporal and spatial assessment of water quality, physical habitat, and benthic communities in an impaired agricultural stream in California's San Joaquin Valley. *Journal of Environmental Science and Health, Part A: Environmental Science and Engineering, 40*(5), 959–989.

Hansson, L. A., & Âkesson, S. (2014). Chapter 4: Beyond dispersal: The role of animal movement in modern agricultural landscapes. In K. Birkhofer (Ed.), *Animal movement across scales*.

Hawkins, B. A., & Porter, E. E. (2003). Does herbivore diversity depend on plant diversity? The case of California butterflies. *The American Naturalist, 161*, 40–49.

Hartman, O. (1974). Polychaetous annelids of the Indian Ocean including an account of species collected by members of the international Indian Ocean Expeditions, 1963–64 and a catalogue and bibliography of the species from India. Pt. 1. *Journal of the Marine Biological Association of India, 16*(1), 191–252.

Hatcher, B. G., Johannes, R. E., & Robertson, A. I. (1989). Review of research relevant to the conservation of shallow tropical marine ecosystems. *Oceanography and Marine Biology: An Annual Review, 27*, 337–414.

Heald, E. J., & Odum, W. E. (1970). The contribution of mangrove swamps to Florida fisheries. *Proceedings of the Gulf and Caribbean Fisheries Institute, 22*, 130–135.

Herman, P. M. J., Middelburg, J. J., Van de Koppel, J., & Heip, C. H. R. (1999). Ecology of estuarine macrobenthos. *Advances in Ecological Research, 29*, 195–240.

Hemmer, H., Kahlke, R. D., & Vekua, A. K. (2004). The Old World puma—*Puma pardoides* (Owen, 1846) (Carnivora: Felidae)—In the Lower Villafranchian (Upper Pliocene) of Kvabebi (East Georgia, Transcaucasia) and its evolutionary and biogeographical significance. *Neues Jahrbuch fur Geologie und Palaontologie—Abhandlungen, 233*, 97–231.

Higgins, R. P., & Thiel, H. (1988). *Introduction to the study of meiofauna* (p. 486). Smithsonian Institution Press.

Hill, C. J. (1992). Temporal changes in abundance of two lycaenid butterflies (Lycaenidae) in relation to adult food resources. *Journal of the Lepidopterists' Society, 46*(3), 173–181.

Hockey, M., & de Baar, M. (1991). Some records of moths (Lepidoptera) from mangroves in southern Queensland. *Australian Entomological Magazine, 18*(2), 57–60.

Hogarth, P. J. (2015). *The biology of mangroves and seagrasses*. Oxford University Press.

Hossain, M. A., Afsana, K., & Azad, A. K. M. (1999). Nutritional value of some small indigenous fish species (SIS) of Bangladesh. *Bangladesh Journal of Fisheries, 3*(1), 16–20.

Hutchings, P., & Richer, H. F. (1981). The fauna of Australian mangroves. *Proceedings of the Linnean Society of New South Wales, 106*, 83–121.

Huey, R. B. (1982). Temperature, physiology and the ecology of reptiles. *Biology of Reptilia, 12*, 25–91.

Iftekhar, M. S., & Islam, M. R. (2004). Managing mangroves inBangladesh: A strategy analysis. *Journal of Coastal Conservation, 10*, 139.

Iftekhar, M. S., & Saenger, P. (2008). Vegetation dynamics in the Bangladesh Sundarbans mangroves: A review of forest invento-ries. *Wetlands Ecology and Management, 16*, 291–312. https://doi.org/10.1007/s11273-007-9063-5

Ingole, S. B., et al. (2010). Study of phytoplankton of freshwater reservoir at Majalgaon on Sindphana river district Beed (MS). *International Research Journal, 1*(13), 87–88.

Inskip, C., & Zimmermann, A. (2009). Human-felid conflict: A review of patterns and priorities worldwide. *Oryx43*, 18–34.

Iwaniec, D. M., Childers, D. L., Rondeau, D., Madden, C. J., & Saunders, C. (2006). Effects of hydrologicand water quality drivers on periphyton dynamics in the southern Everglades. In J. C. Trexler, E. E. Gaiser, & D. L. Childers (Eds.), *Interaction of hydrology and nutrients in controlling ecosystem function in oligotrophic coastal environments of South Florida, Hydrobiologia* (Vol. 569, pp. 223–235).

Jhala, Y. V., Gopal, R., & Qureshi, Q. (2008). *Status of the tigers, co-predators, and prey in India* (p. 151). National Tiger Conservation Authority, Government of India, New Delhi, and Wildlife Institute of India.

Jhala, Y. V., Qureshi, Q., Gopal, R., Sinha, P. R. (2011). *Status of the tigers, co-predators, and prey in India, 2010*. National Tiger Conservation Authority, Govt. of India, New Delhi, and Wildlife Institute of India, Dehradun. 302 p.

Janekarn, V., & Boonruang, P. (1986). Composition and occurrence of fish larvae inmangrove areas along the east coast of Phuket Island, western peninsular, Thailand. *Phuket Marine Biological Research Bulletin, 44*, 1–22.

Jennerjahn, T. C., Nasir, B., & Pohlenga, I. (2009). Spatio–temporal variation of dissolved inorganic nutrients related to hydrodynamics and land use in the mangrove fringed Segara Anakan Lagoon, Java, Indonesia. *Regional Environmental Change, 9*(4), 259–274.

Jayaram, K. C. (1999). *The freshwater fishes of the Indian region* (p. 551). NarendraPublishing House.

Jayaram, K. C. (1981). *The freshwater fishes of India, Pakistan, Bangladesh, Burma, Sri Lanka: A hand book* (pp. 47–75). Zoological Survey of India.

James, N. C., Cowley, P. D., Whitfield, A. K., & Lamberth, S. J. (2007). Fish communities in temporarily open/closed estuaries from warm and cool-temperate regions of South Africa: A review. *Reviews in Fish Biology and Fisheries, 17*, 565–580.

Kamruzzaman, M., Osawa, A., Mouctar, K., & Sharma, S. (2017). Comparative reproductive phenology of subtropical mangrove communities at Manko wetland, Okinawa Island, Japan. *Journal of Forest Research, 22*, 118–125.

Khan, H., & Monirul, M. (2008). Prey selection by tigers (*Panthera tigris*) (Linnaeus 1758) in the Sundarbans east wildlife sanctuary of Bangladesh. *Journal of the Bombay Natural History Society, 105*(3).

Karanth, K. U. (2003). Tiger ecology and conservation in the Indian subcontinent. *Journal of the Bombay Natural History Society, 100*, 169–189.

Karanth, K. U., Nichols, J. D., Kumar, N. S., Link, W. A., & Hanes, J. E. (2004). Tiger and their prey: Predicting carnivore densities from prey abundance. *Proceedings of the National Academy of Sciences of the United States of America, 101*, 4854–4858.

Kanavillil, N., & Kurissery, S. (2013). Temporal variation of periphyton communities: A 3-year study from Northwest Lake Simcoe, Ontario Canada. *Inland Waters, 3*(4), 473–486.

Kathiresan, K. (2004). Biodiversity in mangrove ecosystems of India: Status, challenges and strategies. *Envis Forestry Bulletin, 4*.

Kathiresan, K. (2011). Eco-biology of mangroves. In J. N. Metras (Ed.), *Mangrove ecology, biology and taxonomy* (pp. 1–50). Nova Science Publications.

Khan, R. A. (1995). Ecology of the Hugli-Matla estuarine system. In *Estuarine ecosystem series, part 2: Hugli Matla Estuary* (pp. 417–464). Zoological Survey of India.

Kathiresan, K., & Bingham, B. L. (2001). Biology of mangroves and mangrove ecosystems. *Advances in Marine Biology, 40*, 81–251.

Kathiresan, K., & Qasim, S. Z. (2005). *Biodiversity of mangrove ecosystem* (p. 251). Hindustan Publishing Corporation (India).

Karlson, K., Bonsdorff, E., & Rosenberg, R. (2007). The impact of benthic macrofauna for nutrient fluxes from Baltic Sea sediments. *Ambio, 36*, 161–167.

Kavitha, S. (2009). Studies on *Rhizophora annamalayana,* Kathir. And its present species Ph.D. thesis, Annamalai University (pp 1–195).

Kavitha, S., & Kathiresan, K. (2009). Phenological studies in Rhizophora species. In *Conservation and management of mangroves in India* (pp. 245–251). Zool.Surv. India.

Koch, F. (2000). Brown, J. H. & Lomolino, M. V. (1998): Biogeography. 2nd Ed. Sunderland, Massachusetts (Sinauer Associates, Inc. Publishers). - 691 S. ISBN 0-87893-073-6 (hardcover),. £ 35,95. *Biogeography. Zoosystematics and Evolution*. https://doi.org/10.1002/mmnz.20000760118

Kristensen, E. (2008). Mangrove crabs as ecosystem engineers; with emphasis on sediment processes. *Journal of Sea Researchy, 59*, 30–43.

Kutt, A. (1997). Terrestrial vertebrate survey in the coastal wetlands surrounding Cairns International Airport, North Queensland. *Australian Zoologist, 30*, 300–309.

Kennedy, A. D., & Jacoby, C. A. (1999). Biological indicators of marine environmental health: Meiofauna – A neglected benthic component. *Environmental Monitoring and Assessment, 54*, 47–68.

Kinne, O. (1967). *Physiology of estuarine organisms with special reference to salinity and temperature*. General aspect: In Estuaries, A.A.A.S.P. Publ. No. 83. Washington, DC, pp. 525–540.

Krishnamurthy, K., & Jeyaseelan, M. J. (1981). Early life history of fishes from Pichavaram mangrove ecosystem of India. *Rapports Et Procès-Verbaux Des Réunions, Conseil Permanent International Pour L'Exploration De La Mer, 178*, 416–423.

Krishnamurthy, K. A., Choudhury, A., & Untawale, G. (1987). *Status report: Mangroves in India* (p. 150). Ministry of Environment and Forests, Govt. of India.

Kitamori, R. (1963). Studies on the benthos communities of Littoral sea in the Seto-Island Sea and adjacent waers. *Bulletin of Naikai Region Fish Research Laboratory, 21*, 1–90.

Khalua, R. K., Chakravorty, G., & Chakraborty, S. K. (2003). Annotated list of molluscs from the coastal tract of Midnapore District, West Bengal, India. *Journal of the Marine Biological Association of India, 45*(2), 245–249.

Khalua, R. K., Chakravorty, G., & Chakraborty, S. K. (2008). Community structure of macro-benthic molluscs of three contrasting intertidal belts of Midnapore coastal belt, West Bengal. *Zoological Research in Human Welfare, 6*, 47–55.

Khan, M.G. (1994). *Present status and future plan for sustainable marine resources development*. Paper presented at the seminar on sustainable development of marine fisheries resources in Bangladesh, held during Aug 29–30, 1994 at Fisheries Research Institute, FRI/UNDP, Bangladesh, p. 21.

Krebs, C. J. (1985). *Ecology: The experimental analysis of distribution and abundance*. Harper and Row.

Lawton, J. H. (1994). What do species do in ecosystems? *Oikos, 1994*(71), 367–374.

Lee, S. Y. (1995). Mangrove outwelling: A review. *Hydrobiologia, 295*, 203–212.

Lee, O. H. K., & Williams, G. A. (2002). Spatial distribution patterns of *Littoraria* species in Hong Kong mangroves. *Hydrobiologia, 481*, 137–145.

Levin, S. A. (1992). The problem of pattern and scale in ecology. *Ecology, 73*, 1943–1967.

Lie, U., & Kisker, D. S. (1970). Species composition and structure of benthic infauna communities of the coast of Washigton. *Journal of Fishery Research Board, Canada, 27*, 2273–2285.

Lovelock, C. E., Ball, M. C., Feller, I. C., Engelbrecht, B. M. J., & Ling, E. M. (2006). Variation in hydraulic conductivity of mangroves: Influence of species, salinity, and nitrogen and phospho-Rus availability. *Physiologia Plantarum, 127*, 457–464. https://doi.org/10.1111/j.1399-3054.2006.00723.x

Lugo, A. E. (1980). Mangrove ecosystems successional or steady state? *Biotropica, 12*(2), 65–72.

Lugo, A. E., & Snedaker, S. C. (1974). The ecology of mangroves. *Annual Review of Ecology and Systematics, 5*, 39–64.

Luther, D. A., & Greenberg, R. (2009). Mangroves: A global perspective on the evolution and conservation of their terrestrial vertebrates. *Bioscience, 59*, 602–612.

Macintosh, D. J., Ashton, E. C., & Havanon, S. (2002). Mangrove rehabilitation and intertidal biodiversity: A study in the Ranong mangrove ecosystem, Thailand. *Estuarine, Coastal and Shelf Science, 55*, 331–345.

Macnae, W. (1967). Zonation within mangroves aassociated with estuaries in North Queenland. *Estuaries, 83*, 432–441.

Macnae, W. (1968). A general account of the fauna and flora of mangrove swamps and forests in the Indo-West-Pacific region. *Advances in Marine Biology, 6*, 73–720.

Macnae, W. (1974). *Mangrove forest and fisheries*. Indian Ocean Programme Publication Number–34. Indian Ocean Fishery Commission, Rome, p. 351 OFC/DEV/74/34.

Macnae, W., & Kalk, M. (1962). The fauna and flora of sand flats at Inhaca Island, Macambique. *Journal of Animal Ecology, 31*, 93–128.

Maiti, S., Mandal, S., Das, M., Pahari, P. R., & Bhattcharya. (2026). *Butis bargabhimae* sp. nov., a new species of butid fish (Teleostei: Gobiiformes: Butidae) from the Rupnarayan River, Purba Medinipur, West Bengal, India. *Zootaxa, 5763*(4), 588–596.

Maiti Dutta, S., Banerjee Mustafi, S., Raha, S., & Chakraborty, S. K. (2014). Assessment of thermal stress adaptation by monitoring Hsp70 and MnSOD in the freshwater gastropod, *Bellamya bengalensis* (Lamark, 1882). *Environmental Monitoring and Assessment, 186*(12), 8961–8967.

Majhi, J., Dhara, S., & Ghorai, S. (2025). A new species of *Prosadenoporus* (Nemertea: Eumonostilifera) from the North-East Coast of the Bay of Bengal, India. *Zoological Science, 42*(2), 1–8. (Publsd by Zoological Society of Japan).

Manna, B., Banerjee, S., Aditya, G., & Chattopadhyay, J. (2008). Study of marine planktonic diversity in Digha coast, Bay of Bengal. *Zoological Research in Human Welfare, 3*, 49–54.

Manna, S., Chaudhury, K., Bhattacharyya, S., & Bhattacharya, M. (2010). Dynamics of Sundarban estuarine ecosystem: Eutrophication induced threat to mangroves. *Saline Systems, 6*(1), 8.

Mandal, B., Mukherjee, A., & Banerjee, S. (2013). A review on the ichthyofaunal diversity in mangrove-based estuary of Sundarbans. *Reviews in Fish Biology and Fisheries, 23*, 365–374. https://doi.org/10.1007/s11160-012-9300-8

Mare, M. F. (1942). A study of a marine benthic community with special reference to the micro-organisms. *Journal of Marine Biological Association, U.K., 25*, 517–554.

Martosubroto, P., & Naamin, N. (1977). Relationship between tidal forests (mangroves) and commercial shrimp production in Indonesia. *Marine Research Indonesia, 18*, 81–86.

Marshall, N. (1994). Mangrove conservation in relation to overall environmental considerations. *Hydrobiologia, 285*, 303–309.

Masse, H. (1972). Contribution a l′ etude de la macrofauna de peuplements des sables fins infralittoraux des cotes de provence. VI. Donnees sur la biologie des especes. *Tethys, 4*, 63–84.

Marinelli, R. L. (1992). Effects of polychaetes on silicate dynamics and fluxes in sediments: Importance of species, animal activity and polychaete effects on benthic diatoms. *Journal of Marine Research, 50*, 745–779.

Marinelli, R. L., & Williams, T. J. (2003). Evidence for density-dependent effects of infauna on sediment biogeochemistry and benthic-pelagic coupling in nearshore systems. *Estuarine, Coastal and Shelf Science, 57*, 179–192.

Marinelli, R. L., Lovell, C. R., Wakeham, S. G., Ringelberg, D. B., & White, D. C. (2002). Experimental investigation of the control of bacterial community composition in macrofaunal burrows. *Marine Ecology Progress Series, 235*, 1–13.

McHugh, J. L. (1985). The estuarine ecosystem integrated. Foreword. In *Fish community ecology in estuaries and coastal lagoons: Towards an ecosystem integration* (pp. 9–16). Universidad Nacional Autonoma de Mexico-PUAL-ICML, Editorial Universitaria.

McIntyre, A. D. (1969). Ecololgy of marine meiobenthos. *Biological Review, 44*, 245–290.

McKee, K. L. (1995). Soil physico-chemical patterns and mangrove species distribution: Reciprocal effects? *Journal of Ecology, 81*, 477–487.

McKee, K. L., Cahoon, D. R., & Feller, I. C. (2007). Caribbean mangroves adjust to rising sea level through biotic controls on change in soil elevation. *Global Ecology and Biogeography, 16*, 545–556.

McKie, B. E., Woodward, G., Hladyz, S., Nistorescu, M., Preda, E., Popescu, C., et al. (2008). Ecosystem functioning in stream assemblages from different regions: Contrasting responses to variation in detritivore richness, evenness and density. *Journal ofAnimal Ecology, 77*, 495–504.

McKenzie, N., & Rolfe, J. (1986). Structure of bat guilds in the Kimberley mangroves, Australia. *The Journal of Animal Ecology, 55*, 401–420.

Meades, L., Rodgerson, L., York, A., & French, K. (2002). Assessment of the diversity and abundance of terrestrial mangrove arthropods in southern New South Wales, Australia. *Austral Ecology, 27*, 451–458.

Meadows, P., & Tait, J. (1989). Modification of sediment permeability and shear strength by two burrowing invertebrates. *Marine Biology, 101*, 75–82.

Mertens, J. E. J., et al. (2021). Elevational and seasonal patterns of butterflies and hawkmoths in plant-pollinator networks in tropical rainforests of Mount Cameroon. *Scientific Reports, 11*, 9710. https://doi.org/10.1038/s41598-021-89012-x

Mertens, J. E. J., Brisson, L., Janecek, S., Klomberg, Y., Maichar, V., Safian, S., et al. (2021). Elevational and seasonal patterns of butterflies and hawkmoths in plant-pollinator networks in tropical rainforests of Mount Cameroon. *Scientific Reports, 11*, 9710.

Micheli, F. (1993). Feeding ecology of mangrove crabs in north eastern Australia: Mangrove litter consumption by *Sesarma mesa* and *Sesarma smithii*. *Journal of Experimental Marine Biology and Ecology, 171*, 165–186.

Misra, A. (1995). Polychaetes. In *Estuarine ecosystem series- part-2 Hoogly- Matla Estuary West Bengal* (pp. 93–155). Zoological Survey of India.

Mitra, S. C., & Dey, A. (1992). *Land and Freshwater Molluscs. State Fauna Series 3: Fauna of West Bengal. Part 9* (pp. 1–51). Zoological Survey of India.

Mishra, S. S., Acharjee, S. K., & Chakraborty, S. K. (2009). Developmentof tools for assessing conservation categories of siluroid fishes of freshwater and brackish water wetlands of South West Bengal, India. *Journal of Environmental Biology of Fishes, 84*, 395–407.

Misra, S., Choudhury, A., Ghosh, A., & Dutta, J. (1984). The role of hydrophobic substances in leaves in adaptation of plants to periodic submersion by tidal water in a mangrove ecosystem. *Journal of Ecology, 72*, 621–625.

Mukherjee, A. K. (1969). Food habits of water birdsof the Sundarbans, 24-Parganas District, West Bengal, India I. *Journal of the Bombay Natural History Society, 66*(3), 345–360.

Mukherjee, A. K. (1976). Food habits of water birds of the Sundarbans, 24 Parganas District, West Bengal, India. *Journal of the Bombay Natural History Society, 73*(2), 472–486.

Mukherjee, S., Chaudhury, A., Kundu, N., Mitra, S., & Homechaudhury, S. (2013). Comprehensive analysis of fish assemblages in relation to seasonal environmental variables in an Estuarine River of Indian Sundarbans. *Estuaries and Coasts, 36*, 192–202. https://doi.org/10.1007/s12237-012-9558-z

Mitra, A., Pramanik, A., Sarkar, S., Meyyappan, A., Mitra, S., Senthil, S., & Mukherjee, J. (2011). Applications of the microbial biodiversity of the mangroves. In J. N. Metras (Ed.), *Mangrove ecology, biology and taxonomy* (pp. 225–240). Nova Science Publications.

Nagelkerken, I., Blaber, S. J. M., Bouillon, S., Green, P., Haywood, M., Kirton, L. G., Meynecke, J. O., Pawlik, J., Penrose, H. M., Sasekumar, A., & Somerfield, P. J. (2008). The habitat function of mangroves for terrestrial and marine fauna: A review. *Aquatic Botany, 89*, 155–185.

Nadia, T. D., Morellato, L. P. C., & Machado, I. C. (2012). Reproductive phenology of a northeast Brazilian mangrove community: Environmental and biotic constraints. *Flora, 207*, 682–692.

Naskar, K., Sen-Sarkar, N., Ghosh, A., Dasgupta, M., & Sengupta, B. (2004). Status of the mangroves and mangrove ecosystem of Sundarbans in West Bengal: Its impact on estuarine wetlands fisheries. *CIFRI Bulletin, 134*, 8.

Naskar, K., & Guha Bakshi, D. N. (1987). *Mangrove swamps of the Sundarbans. Naya Prakash, Calcutta*, p 282. National Cyclone Risk Mitigation Project (NCRMP) 2016 Implemented through World Bank Assistance (Map showing coastal areas affected by cyclones). Based on IMD data on cyclone hazards.

Nixon, S. W., & Pilson, M. G. (1983). Nitrogen in estuarine and coastal marine ecosystems. In E. J. Carpenter & D. G. Capone (Eds.), *Nitrogen in the marine environment* (pp. 565–648). Academic.

Neira, F. J., Potter, I. C., & Bradley, J. S. (1992). Seasonal and spatial changes in the larval fish fauna within a large temperate Australian estuary. *Marine Biology, 112*, 1–6.

Neira, F. J., & Potter, I. C. (1994). The larval fish assemblage of the Normalup-Walpole estuary, a permanently open estuary on the southern coast of western Australia. *Australian Journal of Marine and Freshwater Research., 45*, 1193–1207.

Nowak, K. (2013). Mangrove and peat swamp forests: Refuge habitats for primates and felids. *Folia Primatologicay, 83*, 361–376.

Noske, R. A. (1993). *Bruguiera hainesii*: Another bird-polinated mangrove ? *Biotropica, 25*(4), 481–483.

Noske, R. A. (1995). The ecology of mangrove birds in Peninsular Malayasia. *Biotropica, 137*(2), 250–263.

Nsor, C. A., & Obodai, E. A. Environmental determinants influencing fish community structure and diversity in two distinct seasons among wetlands of northern region (Ghana). *International Journal of Ecology, 2016, 2016*, 1598701. https://doi.org/10.1155/2016/1598701

Odum, E. P., & Heald, E. J. (1975a). Mangrove forests and aquatic productivity. In A. D. Haster (Ed.), *Coupling of land, water systems. Ecological studies* (Vol. 10, pp. 129–136). Springer.

Odum, W. E., & Heald, E. J. (1975b). The detritus food web of an estuarine mangrove community. In L. Cronin (Ed.), *Estuarine research* (pp. 265–286). Academic Press.

Odum, E. P. (1971). *Fundamentals of ecology*. W.B. Saunders.

Odum, E. P. (1977). *Fundamentals of ecology* (3rd ed.). W. B. Saunders Company.

Odum, W. E., McIvor, C. C., & Smith, T. G., III. (1982). *The ecology of the mangroves of South Florida: A community profile*. U. S. Department of the Interior, Fish and Wildlife Service, Office of Biological Services. FWS/OBS-81/24.

Odum, E. P., & Heald, E. J. (1975c). Mangrove forests and aquatic productivity. In A. D. Haster (Ed.), *Coupling of land and water systems. Ecological studies* (Vol. 10, pp. 129–136). Springer.

Oliveira, C. P. D., et al. (2021). Seasonality and distribution of Coleoptera families (Arthropoda, Insecta) in the Cerrado of Central Brazil. *Revista Brasileira De Entomologia, 65*. https://doi.org/10.1590/1806-9665-rbent-2021-0025

Oliveira, C. P. D., Oliveira, C. M. D., Specht, A., & Frizzas, M. R. (2021). Seasonality and distribution of Coleoptera families (Arthropoda, Insecta) in the Cerrado of Central Brazil. *Revista Brasileira de Entomologia, 65*.

Pahari, P. R., Das, S., Das, M., & Maiti, S. (2025). *Petrolisthes tanmayi* sp. nov., a new species of the porcelain crab (Decapoda: Anomura: Porcellanidae) from G-Plot Island, Sundarbans, India. *Zootaxa, 5631*(2), 369–379.

Parulekar, A. H., Dhargalkar, V. K., & Singbal, S. Y. S. (1980). Benthic studies in Goa Estuaries: Part III–Annual cycle of macrofaunal distribution, production and trophic relations. *International Journal of Molecular Sciences, 9*, 189–200.

Patra, M. K., Acharjee, S. K., & Chakraborty, S. K. (2005). Conservation categories of siluroid fishes in North–East Sundarbans, India. *Biodiversity and Conservation, 14*, 1863–1876.

Paul, A. K. (2002). *Coastal geomorphology and environment* (pp. 1–582). ACB Publication.

Perrings, C., Naeem, S., Farshid, S., et al. (2011). Ecosystem services, targets and indicator for the conservation and sustainable use of biodiversity. *Frontiers in Ecology and the Environment, 9*(9), 512–520.

Panikkar, N. K. (1951). Physiological aspects of adaptation to estuarine conditions. *Proceedings of Indo-Pacific Fisheries Council, 1950*(Sec. 3), 168–175.

Pargiter, F. E. (1934). *A revenue history of the Sundarbans from 1765-1870 analyzing of human wildlife conflicts in Sundarban* (pp. 1–148). Bengal Government Press.

Paterson, D. M., et al. (1998). Microbial mediation of spectral reflectance from intertidal cohesive sediments. *Limnology and Oceanography, 43*(6), 1207–1221.

Pradhan, P., & Chakraborty, S. K. (2008). Ecological study of rotifera and its application for biomonitoring freshwater riverine environment. In *Zoological research in human welfare. Paper-31* (pp. 289–306).

Qian, H., & Ricklefs, R. E. (2008). Global concordance in diversity patterns of vascular plants and terrestrial vertebrates. *Ecology Letters, 11*, 547–453.

Ravichandran, S., Soundarapandian, P., & Kannupandi, T. (2001). Zonation and distribution of crabs in Pichavaram mangrove swamp, southeast coast of India. *Indian Journal Of Fisheries, 48*(2), 221–226.

Reddy, S., & Davalos, L. M. (2003). Geographical sampling bias and its implications for conservation priorities in Africa. *Journal of Biogeography, 30*, 1719–1727.

Reef, R., Feller, I. C., & Lovelock, C. E. (2014). Mammalian herbivores in Australia transport nutrients from terrestrial to marine ecosystems via mangroves. *Journal of Tropical Ecology, 30*, 179–188.

Rees, C. B. (1940). A preliminary study of the ecology of a mud-flat. *Journal of Marine Biological Association of U.K., 24*, 185–199.

Richards, J. F. (1993). Agricultural impacts in tropical wetlands: Rice paddies for mangroves in South and Southeast Asia. In M. Williams (Ed.), *Wetlands: A threatened landscape* (pp. 217–233). Blackwell.

Robertson, A. I. (1986). Leaf-burying crabs: Their influence on energy flow and export from mixed mangrove forests (*Rhizophora* spp.) in northeastern Australia. *Journal of Experimental Marine Biology and Ecology, 102*, 237–248.

Robertson, A. I., & Blaber, S. J. M. (1992). Plankton, epibenthos and fish communities. In A. I. Robertson & D. M. Alongi (Eds.), *Tropical mangrove ecosystems* (pp. 173–224). American Geophysical Union.

Robertson, A. I., & Duke, N. C. (1987). Mangroves as nursery sites: Comparisons of the abundance and species composition of fish and crustaceans in mangroves and other nearshore habitats in tropical Australia. *Marine Biology, 96*, 193–205.

Robertson, A. I., & Alongi, D. M. (Eds.). (1992). *Tropical mangrove ecosystems*. American Geophysical Union.

Rodrigues, N. N., & Martinez, R. A. (2014). Wildlife in our backyard: Interactions between Wied's marmoset *Callithrix kuhlii* (primates: Cailithrichidae) and residents of Ilhéus, Bahia, Brazil. *Wildlife Biology, 20*, 91–96.

Rog, S. M., Clarke, R. H., & Cook, C. N. (2017). More than marine: Revealing the critical importance of mangrove ecosystems for terrestrial vertebrates. *Diversity and Distributions, 23*(1/2), 221–230.

Rossi, F., & Chapman, M. G. (2003). Influence of sediment on burrowing by the soldier crab *Mictyris longicarpus* Latreille. *Journal of Experimental Marine Biology and Ecology, 289*, 181–195.

Roy, S., Parui, P., Mitra, B., & Chakraborty, S. K. (2019). *New species from the genus Astochia Becker, 1913 (Diptera: Asilidae) from the Sundarban, India with an updated checklist of Indian species Oriental Insects*. 54(2), 162–174.

Roy, S., Chakraborty, S. K., Parui, P., & Mitra, B. (2016). Taxonomy of Soldier Flies (Diptera: Stratiomyidae) of Sundarban Biosphere Reserve, India. *Proceedings of Zoological Society*. https://doi.org/10.1007/s12595

Saberwal, V. K., Gibbs, J. P., Chellam, R., & Johnsingh, A. J. T. (1994). Lion –human conflict in the Gir Forest. *India. Conservation Biology, 8*(2), 501–507.

Sanderson, E., Forrest, J., Loucks, C., Ginsberg, J., Dinerstein, E., Seidensticker, J., et al. (2006). *Setting priorities for the conservation and recovery of wild tigers 2005–2015. A technical report*. Wildlife Conservation Society, World Wildlife Fund, Smithsonian, and National Fish and Wildlife Foundation–Save the Tiger Fund. 206 p.

Sarkar, S. K., Singh, B. N., & Choudhury, A. (1989). Seasonal distribution of copepods in the Hooghly estuary, West Bengal. *Indian Journal of Marine Sciences, 15*, 177–180.

Schlacher, T. A., & Woolridge, T. H. (1996). How sieve mesh size affects samploe estimates of estuarine benthic macrofauna. *Journal of Experimental Marine Biology and Ecology, 201*, 159–171.

Sanders, H. L. (1968). Benthic studies in Buzzards Bay. 1. Animal sediment relationship. *Limnology and Oceanography, 3*, 245–258.

Sanyal, P. (1987). Managing the man-eaters in the Sundarbans tiger reserve in India – A case study. In *Tigers of the World* (pp. 427–434). Noyes Publication.

Sanyal, P. (1990). Tiger prey in the Sundarban. *Cat News, 13*, 17.

Schwinghamer, P. (1981). Characteristic size distribution of integral benthic communities. *Canadian Journal of Fishery and Aquatic Science, 38*, 1255–1263.

Schwinghamer, P. (1983). Generating ecological hypotheses from biomass spectra using causal analysis: A benthic example. *Marine Ecological Progress Series, 13*, 151–166.

Sen, D. (2015). *Records of the Zoological Survey of India, 115*(3), 223–229.

Sinha, S., Basak, B., Mandal, B., & Biswas, B. B. (2013). Seasonal variation of epiphytic algae on Pneumatophore of Avicennia alba in Sundarban mangrove forest, India. *International Journal of Applied Biology and Pharmaceutical Technology, 4*(3), 153–159.

Smith, Y., Clough, J., Ekroos, O., Olsson, & Rundlöf, M. (1992). *Book title* (pp. 51–63). Oxford University Press.

Snedaker, J. G. (Ed.). (1984). *The mangrove ecosystem: Research methods* (pp. 3–17). UNESCO/SCOR.

Solomon Raju, A. J., & Karyamsetty, H. J. (2018). Reproductive ecology of Aegialitis rotundifolia Roxb., a crypto-viviparous mangrove plant species in Krishna Mangrove Forest, Andra Pradesh. *Transylvanian Review of Systematical and Ecological Research, 20*(1), 17–30.

Thom, B. G. (1967). Mangrove ecology and deltaic geomorphology: Tabasco, Mexico. *Journal of Ecology, 55*, 301–343.

Stark, J. D. (1998). SQMCI: A biotic index for freshwater macroinvertebrate coded abundance data. *New Zealand Journal of Marine and Freshwater Research, 32*, 55–66.

Spalding, M. (1997). The global distribution and status of mangrove ecosystems. *International Newsletter of Coastal Management Intercoast Network, 1*, 20–21.

Spalding, M., Blasco, F., & Field, C. (1997). *World mangrove atlas*. International Society for Mangrove Ecosystems.

Subba Rao, N. V., Surya Rao, K. V., & Maitra, S. C. (1987). Malacological notes on Sagar Island. *Bulletin of the Zoological Survey of India, 8*(1–3), 149–158.

Subba Rao, N. V., Dey, A., & Barua, S. (1992). *Estuarine and marine molluscs* (State Fauna Series 3: Fauna of West Bengal, Part 9) (pp. 129–268). Zoological Survey of India.

Subba Rao, N. V., Surya Rao, K. V., & Manna, R. N. (1995a). *Mollusca wetland ecosystem series, 1: Fauna of Chilka Lake* (pp. 391–468). Zoological Survey of India.

Subba Rao, N. V., Dey, A., & Barua, S. (1995b). Molluscs. In *Estuarine ecosystem series, Part 2: Hugli Matla Estuary* (pp. 41–91). Zoological Survey of India.

Sasekumar, A. (1974). Distribution of macrofauna on a Malayan mangrove shore. *Journal of Animal Ecology, 43*, 51–69.

Sunquist, M., Karanth, K. U., & Sunquist, F. (1999). Ecology, behaviour and resilience of the tiger and its conservation needs. In J. Seidensticker, S. Christie, & P. Jackson (Eds.), *Riding the tiger: Tiger conservation in human –domunated Landscape* (pp. 5–18). Cambridge University Press.

Talwar, P. K., & Jhingran, A. G. (1991). *Inland fishes of India and adjacent countries. Vol-1 and Vol-2* (p. 1063). Oxford and IBH Publishing Co. Pvt. Ltd.

Teal, J. M. (1958). Distribution of fiddler crabs in Georgia salt marshes. *Ecology, 39*, 185–193.

Tenore, K. R., Tietjen, J. H., & Lee, J. J. (1977). Effect of Meiofauna incorporation of aged eel-grass, *Zostera marina*, detritus by the Polychaete *Nephthys incise*. *Journal of the Fisheries Research Board of Canada, 34*, 563–567.

Thrush, S. F., Hewitt, J. E., Gibbs, M., Lundquist, C., & Norkko, A. (2006). Functional role of large organisms in intertidal communities: Community effects and ecosystem function. *Ecosystems, 9*, 1029–1040.

Thom, B. G. (1984). Coastal landforms and geomorphic processes. In: Snedaker SC, Turner, R. E. 1992. Coastal wetlands and penaeid shrimp habitat. p. 97–104. In R. E. Stroud (Ed.), *Stemming the tide of coastal fish habitat loss* (Marine Recreational Fisheries Publication 14). National Coalition for Marine Conservation, Inc.

Tomlinson, P. B. (1986). *The botany of mangroves* (pp. 1–413). Cambridge University Press.

Tzeng, W. N., & Wang, Y. T. (1992). Structure, composition and seasonal dynamics of the larval and juvenile fish community in the mangrove estuary of Tanshui River, Taiwan. *Marine Biology, 113*, 481–490.

Tzeng, W. N., & Wang, Y. T. (1993). Hydrography and distribution dynamics of larval and juvenile fishes in the coastal waters of the Tanshui River estuary, Taiwan, with reference to estuarine larval transport. *Marine Biology, 116*, 205–217.

Untawale, A. G. (1987). Conservation in Indian mangroves a national perspectives. In T. S. S. Rao, R. Natarajan, B. N. Desai, G. Narayanaswamy, & S. R. Bhat (Eds.), *A special collection of papers to Feliciate Dr. S.Z. Qasim on his sixtieth birthday* (pp. 85–104).

Valdes-Weaver, L. M., et al. (2006). Long-term temporal and spatial trends in phytoplankton biomass and class-level taxonomic composition in the hydro-logically variable Neuse-Pamlico estuarine continuum North Carolina USA. *Limnology and Oceanography, 51*(3), 1410–1420.

Vannucci, M. (1989). *The mangroves and us; a synthesis of insights*. Indian Association for the Advancement of Science.

Vannucci, M. (2003). *Mangrove management and conservation: Present and future* (p. 220). United Nations University Press.

van Schaik, C. P., Terborgh, J. W., & Wright, S. J. (1993). The phenology of tropical forests: Adaptive significance and consequences for primary consumers. *Annual Review of Ecology and Systematics, 24*, 353–377.

Walsh, J. J. (1991). Importance of the continental margins in the marine biogeochemical cycling of carbon and nitrogen. *Nature, 350*, 53–55.

Walsh, J. J., Rowe, G. T., Iverson, R. E., & Mcroy, C. P. (1981). Biological export of shelf carbon is a sink of the global CO_2 cycle. *Nature, 291*, 196–201.

Warwick, R. M., Pearson, T. H., & Ruswahyuni. (1987). Detection of pollution effects on marine macrobenthos further evaluation of the species abundance biomass method. *Marine Biology, 95*(2), 193–200.

Weinstein, M. P. (1979). Shallow marsh habitats as primary nurseries for fishes and shellfish, Cape Fear River, North Carolina. *Fisheries Bulletin, 77*, 339–357.

Whitfield, A. K. (1999). Ichthyofaunal assemblages in estuaries: A South African case study. *Reviews in Fish Biology and Fisheries, 9*, 151–186.

Weisberg, S. B., Frithsen, J. B., Holland, A. F., Paul, J. F., Scott, K. J., Summers, K. J., Wilson, H. T., Valente, R., Heimbuch, D. G., Gerritsen, J., Schimmel, S. C., & Latimer, R. W. (1993). *Virginian province demonstration report, EMAP–Estuaries: 1990. EPA 620/R–93/006*. U.S. Environmental Protection Agency.

Waples, R. S., & Gaggiotti, O. E. (2006). What is a population? An empirical evaluation of some genetic methods for identifying the number of gene pools and their degree of connectivity. *Molecular Ecology, 15*, 1419–1439.

Washington, H. G. (1984). Diversity, biotic and similarity indices: A review with special reference to aquatic ecosystems. *Water Research, 18*, 653–694.

Westheide, W. (1991). The meiofauna of Galápagos. A review. In M. J. James (Ed.), *Galápagos marine invertebrates, taxonomy, biogeography and evolution in Darwin's Islands* (pp. 37–73). Plenum Press.

Wiens, J. A. (1989). Spatial scaling in ecology. *Functional Ecology, 3*, 385–397.

Wiens, J. J., Sukumaran, J., Pyron, R. A., & Brown, R. M. (2009). Evolutionary and biogeographic origins of high tropical diversity in Old World frogs (Ranidae). *Evolution, 63*, 1217–1231.

Wilson, E. O. (1975). Some central problems of sociobiology. *Social Science Information, 14*(6), 5–18.

Wilson, D. S., & Yoshimura, J. (1994). On the coexistence of specialists and generalists. *The American Naturalist, 144*, 692–707.

Wolanski, E. (2006). *The application of ecohydrology for sustainable development and management of mangrove-dominated estuaries*. The ICEMAN 2006 Mangrove Conference in Kuala Lumpur.

Ysebaert, T., & Herman, P. M. J. (2002). Spatial and temporal variation in benthic macrofauna and relationships with environmental variables in an estuarine, intertidal soft-sediment environment. *Marine Ecology Progress Series, 244*, 105–124.

Ysebaert, T., Maeire, P., Coosen, J., & Essink, K. (1998). Zonation of intertidal macrobenthos on estuaries of Schelde and ems. *Aquatic Ecology, 32*, 53–71.

Chapter 5
Ecosystem Services of Sundarbans Mangrove Ecosystem, India

Abstract Ecosystem services denote the contribution of nature to a variety of "goods and services" to mankind in respect of economics and ecology. The products of biodiversity such as food, medicines, clothing materials, shelter, and aesthetics contribute profusely to the very existence of human beings. Coastal biodiversity renders valuable ecosystem services to all compartments (estuaries, mangroves, dunes, deltas, lagoons, intertidal zones, etc.) of the coastal zone, the interface of the land and sea representing an eco-potential ecosystem. Coastal ecosystems encompassing a good number of habitats such as coral reefs; seagrass beds, meadows; deltaic islands; rocky coasts, cliffs; intertidal mud- and sandflats, sandy, pebble, or rocky beaches; dunes, lagoons, estuaries, and coastal river floodplains; salt marshes and mangrove forests act as repositories of biological diversity and provide a wide range of goods and services.

Undervaluation of the goods and services rendered by coastal marine–estuarine ecosystems has led to unsustainable ways of resource exploitation considering those ecosystems as grossly eco-degraded ecosystems and thereby the very least efforts are being made toward proper conservation of the ecosystem in general and biodiversity in particular. Local and regional communities of human beings who are directly dependent on different goods and services from these ecosystems face maximal economic, ecological, and social sacrifice owing to the decline and loss of natural resources. However, the major habitats within the coastal zone have so far been highly modified enormously over millennia by human activities.

The mangrove ecosystem of Sundarbans, India, a reservoir of an array of biodiversity components (phytoplankton, benthic algae, zooplankton (copepods, Cladocera, Chaetognatha, ostracods, etc.), macrobenthic fauna (brachyuran crabs, mollusks, polychaetes, actiniarians, etc.), and meiobenthic fauna (nematodes, foraminifera, copepods, polychates etc. encompass diversified ecotopes within the mangrove forest, transitional intertidal zone, and estuarine flows. All of those biota exhibit different forms of ecological services by offering them as foods for plenty of other faunal components and thereby becoming an integral part of food chains and food-web systems, releasing millions of benthic larvae (meroplankton) into the aquatic system as food of demersal fishes, acting as ecosystem engineers as biotur-

S. K. Chakraborty, S. K. Acharya, *Eco-dynamics and Environmental Perturbations in Mangrove Ecosystems*,
https://doi.org/10.1007/978-3-032-08710-2_5

bators, resulting production of huge amount of nutrients within different strata of sediment, ploughing sediments to maintain textural composition, converting mangrove leaves into detritus to release different kinds of nutrients, bioaccumulating pollutants, serving as bioindicators, and providing aesthetics. Different bioturbatory structures formed by the benthic fauna alongwith several sediments based geomorphological features such as burrows, ridges, tubes, mounds, sand balls, and mud balls help a proper understanding of the geo-morphological processes of shorelines in the deltaic Sundarbans, India.

Keywords Ecosystem goods and services · Mangrove ecosystem · Biodiversity components · Ecosystem engineers · Bioturbation · Geo-morphological processes

5.1 Ecological Services: Concept and Types

Ecosystem Services is defined as The Millenium Ecosystem Assessment where "benefits of people are obtained from ecosystems." All those ecosystem-derived benefits can directly or indirectly cover a wide variety of services such as regulatory or supporting (Costanza & Farber, 2002; Costanza et al., 1997; Beaumont et al., 2007). Ecosystems often frequently lose the services if they are managed only for a single service—such as the production of food, fiber, fuel, or for the control of particular pests or pathogens. One noteworthy advance in the 2020 targets compared with the 2010 targets is the recent development of interpreting the failure or success depending upon the interrelationships and interdependencies of a particular ecosystem having unique social origins (Barrett, 1994, 2003). Accurate quantification of ecosystem services often poses difficulties, partly because they often provide indirect benefits, or difficulty in understanding actual facts about their importance.

Ecosystem services take into consideration different **"goods and services,"** rendered by any ecosystem for human benefit. These services from the point of view of economics can normally be classified under three different categories (Barbier, 2007): (1) "**goods**" in the form of purified water, clean air, and bioresources, along with their genetic constituents; (2) "**services**" catering to the needs of aesthetics along with providing recreational and tourism benefits or supporting habitat functions with ecological regulatory mechanisms such as erosion control, water purification, climate regulation, and habitat maintenance; and (3) **cultural benefits** encompassing spiritual thoughts, religious beliefs, and heritage values (Flow Chart 5.1).

5.2 Ecosystem Services, Types, and Significance

The fruitful existence of human beings has always depended on the natural resources of clean water, fresh food materials, ingredients for medicines, materials for building shelters, biogeochemical and nutrient cycling with several bio-chemical constituents, and through the formation of soil. However, since the start of human civilization, and the domination of human beings through multiple activities, the biosphere has been rapidly altering the structural composition and functions of ecosystems (Vitousek et al., 1997),

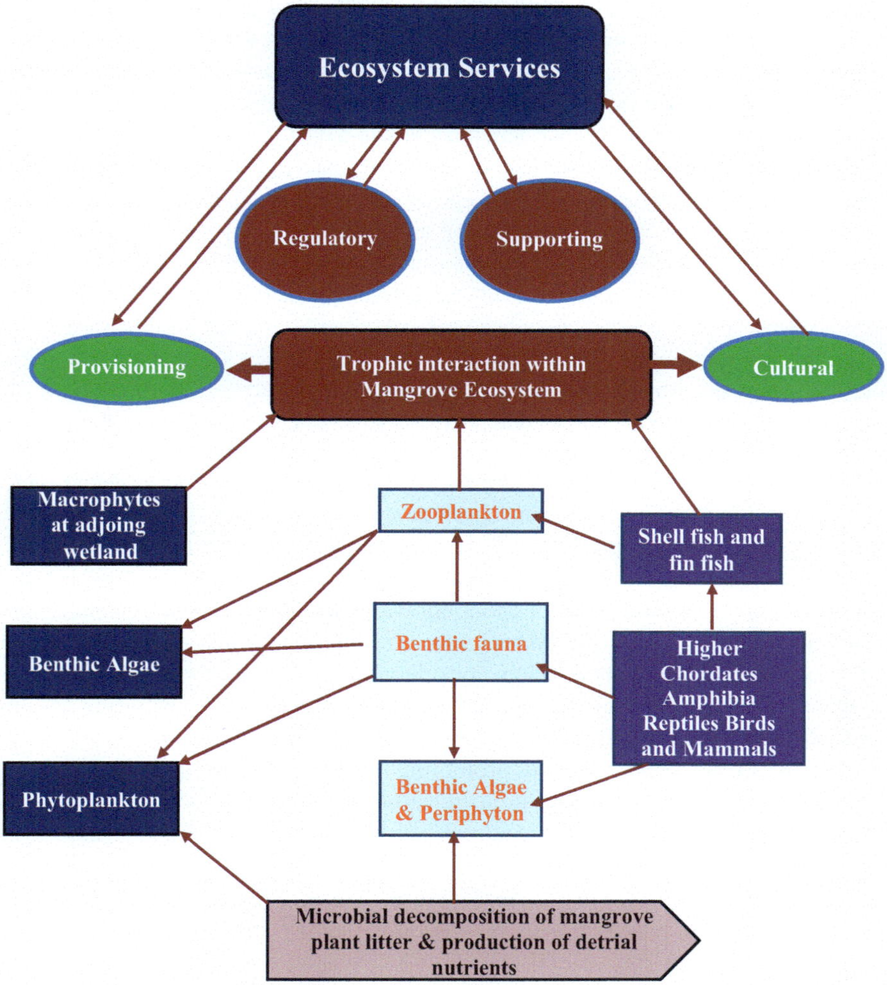

Flow Chart 5.1 Ecosystem services rendered by the mangrove ecosystem though trophic interactions of all of its constituent biotic components, which are very relevant for the Sundarbans mangrove ecosystem

even eroding their abilities to render several crucial services required for human survival (Palmer et al., 2004). Although there has been no consensus on applying only useful taxonomy, several attempts have been made to classify and exhaustively list ecosystem services. The Millenium Ecosystem Assessment emphasized the groupings of ecosystem functions based on goods and services. Four different categories of ecosystem services have recently been identified: (1) the provision of services (food, fiber, timber, fuel, etc.); (2) the regulation of services (flood control, groundwater recharging, climate etc.); (3) the support of services (pollination, soil formation, biodiversity enrichment, and other ecological properties); and (4) cultural services (recreational, spiritual, and aesthetic values). Ecological understanding appears to be a prerequisite for undertaking ecological planning and chalking out strategies for conservation and management

(Balmford et al., 2002). However, half of the world's ecosystem services have been degraded because of the unsustainable utilization of natural resources. Economists and ecologists have been in the process of credibly assigning economic values to several ecosystem processes in order to transform these into tradable, marketable services, emphasizing the potential of ecosystem processes and other forms of natural capital in terms of physical, chemical, biological, social, cultural, and intellectual components, the hidden values of which have as yet been egregiously overlooked. Economic valuations need not only cover all values of ecosystems and routinely use cost–benefit analyses to judge between alternative pathways in the process of such exercises, which facilitate taking difficult decisions on the mode of future developmental works, maintaining parity with ever-expanding human populations (Chakraborty et al., 2014).

In this present world, with the escalating dependence of humans on high technology-based modern lifestyles and management practices, the paradigm of ecosystem services is still not accepted by everyone, even among those who champion this new way of thinking about nature, by inculcating differences of opinion among themselves. For example, although some scientists suggest a virtual one-to-one mapping between ecosystem services and biodiversity, others tend to point out many cases where considerable ecosystem services are rendered by low-diversity systems and modest losses of biodiversity may not substantially undermine ecosystem services. The point of economic valuation is not only to estimate the importance of changes in ecosystem services to human welfare, as many of these services have already been taken into consideration for valuation studies in a holistic manner, taking into consideration all structural components constituting nature components of nature such as commercial fisheries with finfish and non-finfish, utilization of timber and firewood, promoting tourism, generating medicinal remedies, offering nature-based education, constructing physical structures for coastal protection, and maintaining carbon storage (Flow Charts 5.1, 5.2, 5.3, 5.4, and 5.5).

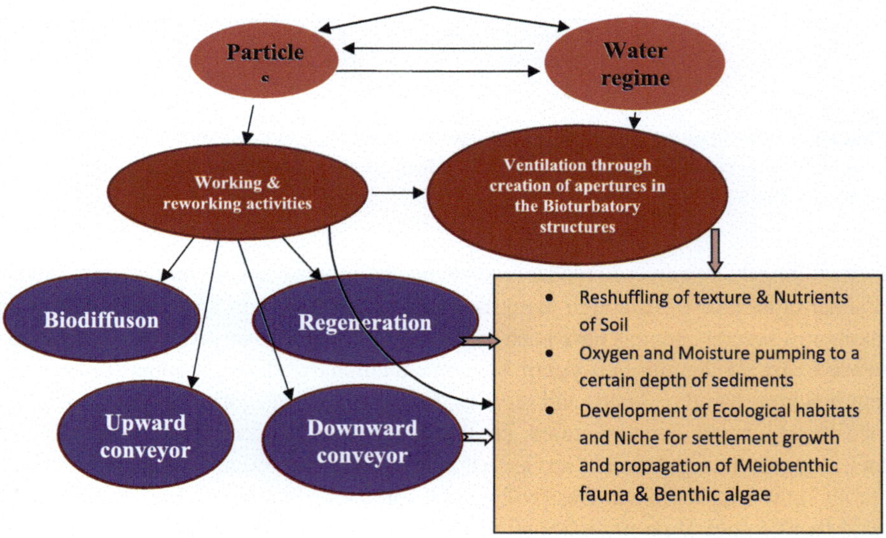

Flow Chart 5.2 Ecosystem services by intertidal macrobenthic fauna through their interaction with the surrounding habitats

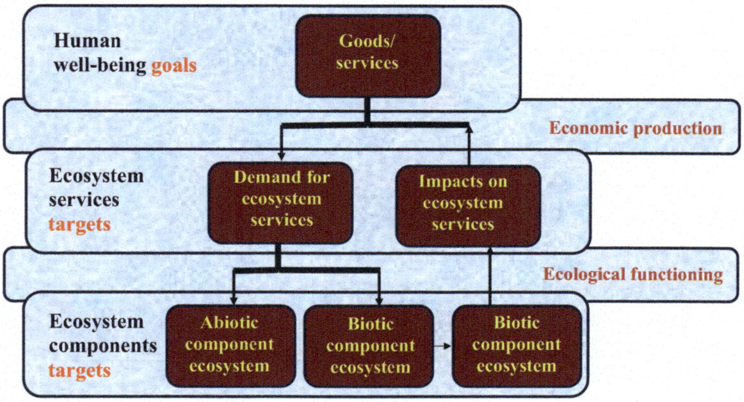

Flow Chart 5.3 Ecosystem services for rendering economic and ecological benefits through abiotic and biotic interaction processes in different tiers of ecosystem functioning

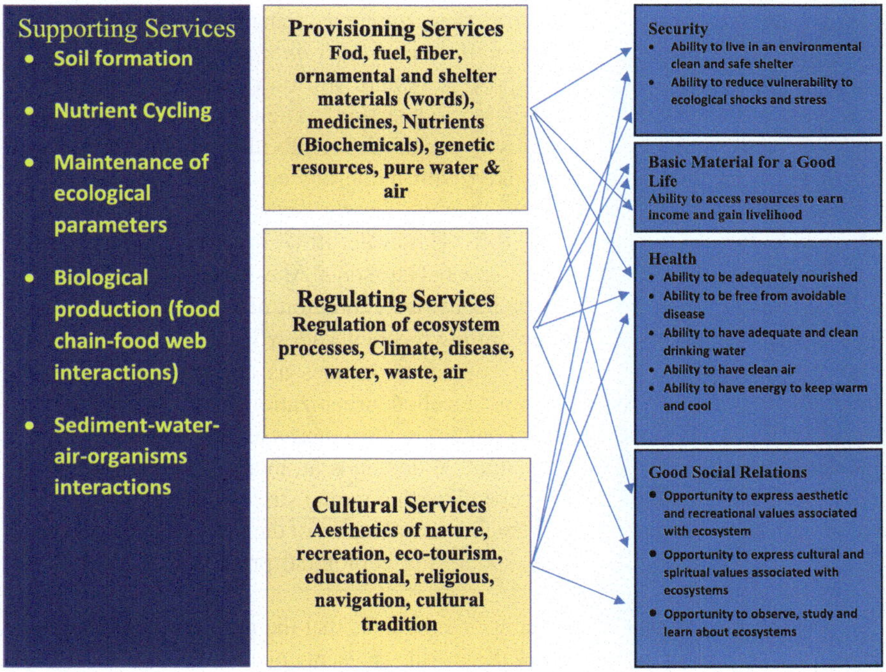

Flow Chart 5.4 A comprehensive representation of different types of ecosystem services of the Sundarbans mangrove ecosystem to ensure a steady supply of basic requirements for maintaining good life with sound health, social relationships, and security

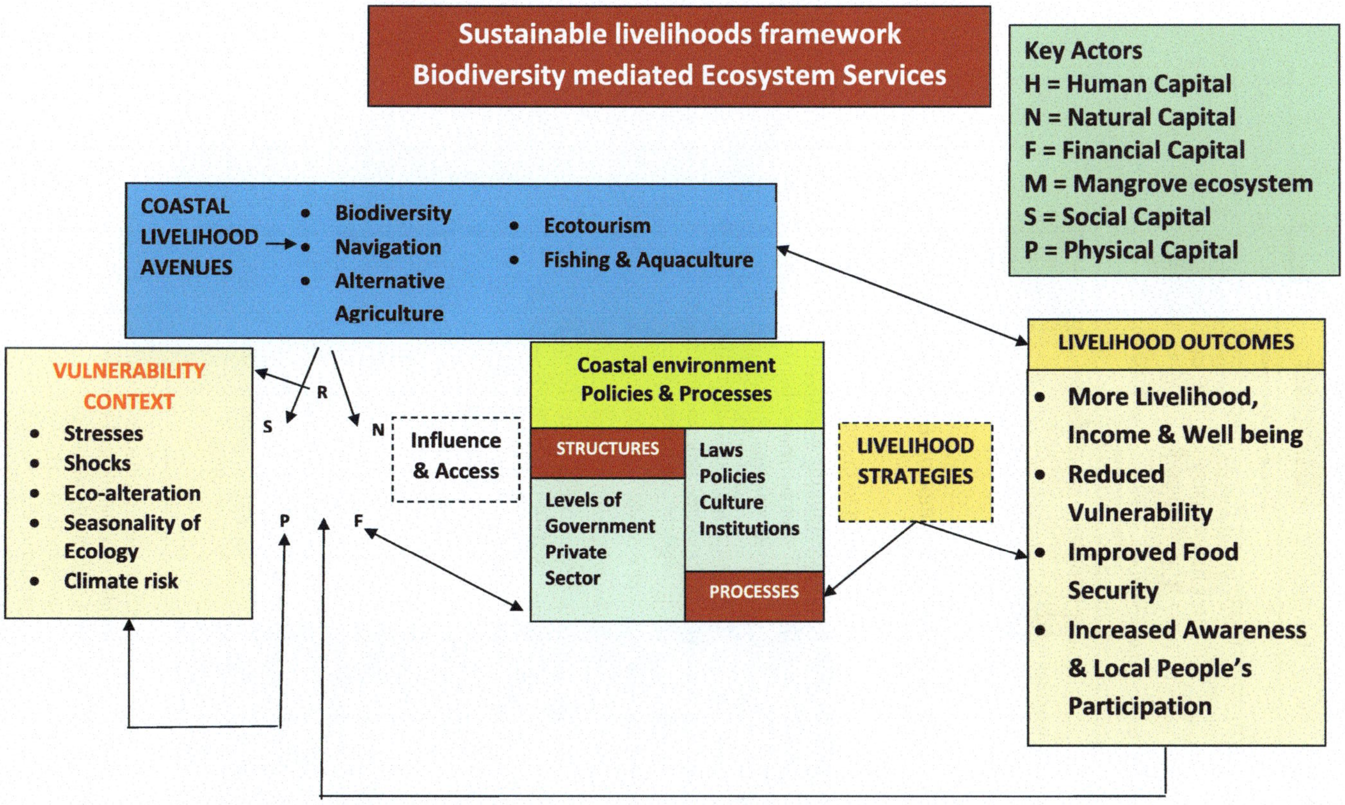

Flow Chart 5.5 Sustainable livelihoods framework through biodiversity-mediated ecosystem services

5.3 Determinants of the Three Ecosystem Services by Several Natural, Social, Political, and Economic Attributes

1. Terrestrial influences mostly enunciated from anthropogenic activities within terrestrial landscapes from changes in land-use patterns, which in turn trigger alteration of hydrological regimes, and nutrient supply though sedimentation transport through land runoff.
2. Marine influences representing natural phenomena, which include weather events such as cyclonic storms, specific wave patterns, and currents of water that in turn influence the transfer processes of heat, material, and nutrients and changes in geomorphological settings.

Eco-degradation and even loss of ecosystems are affected by combined impact of interacting factors, which are considered drivers of the process directly (direct drivers) and indirectly (indirect drivers).

The direct drivers, also referred to as proximate causes, result in eco-degradation and even loss of coastal zone ecosystems, which include:

- Overexploitation of natural resources for promoting livelihood generation and for other commercial reasons.
- Fragmentation and degradation of habitats, by altering land-use patterns such as conversion to aquaculture and agriculture.
- Development of pollution, mostly by nutrient enrichment by drainage of sewage and chemical fertilizers, and the accumulation of toxic pesticides and other harmful chemicals.
- Promotion of the introduction of bio-invasive species and their uncontrolled proliferation, often termed spreading biological pollution.
- Global climate change, reinforcing the effects of other harmful attributes.

The indirect drivers are referred to as root causes or ultimate causes and result in the following:

- Population expansion with the steady increase in populations, which in turn increases demand for natural resources.
- Policy failure necessitating the adoption of policies that take into account resisting unsustainable exploitation in order to maintain the inherent characteristics of the ecosystems.
- Marketing options need to be explored, taking into consideration the proper valuation of different ecosystem goods and services and to find out their marketing opportunities in view of the present state of globalized trade and market liberalization equating prevailing consumption rates, growth, and advancement.

5.4 Interdependencies Between Ecosystem Services in Respect of Biodiversity Targets

Biodiversity, frequently considered as being synonymous with taxonomic diversity, emphasizes tabulation of the recorded number of species from an ecosystem, where individual species contribute in their own way to a particular ecosystem service and represent functional diversity instead of species richness (Kremen & Ostfeld, 2005).

The effectiveness of ecosystem services generally rely on focusing more on functional diversity rather than on taxonomic diversity because functional traits can be used to understand the relationship between biodiversity and ecosystem functioning (Barrett, 1994; Solan et al., 2004; Butchart et al., 2010; Kattge et al., 2011). Prevailing environmental conditions determine the quantity of biodiversity components required to render proper ecosystem services. The greater the expected variation of those conditions, the greater will be the required diversity within the functional groups (Elmqvist et al., 2003).

5.5 Coastal Environment and Biodiversity-Mediated Ecosystem Services

The coast line across the globe is about 440,000 km long (Pethic, 1984) and 66% of the total population of the world reside within very close proximity of the coastal zones, which encompass an array of various geological and geomorphological components such as continental shelf, foreshore and backshore, intertidal belts, littoral zones, ripples, dunes, deltaic islands with mangroves, backwater and brackish water zones, corals, lagoons, and estuaries. The coastal zone, by virtue of continuous interaction processes with lands, atmosphere, and maritime areas, determine the regional climatic conditions, energy flows, food production, and harvesting of bio-resources, human settlements, industrialization, etc.

The three basic tiers of levels of biodiversity are genes, species, and ecosystems/communities/habitats that render benefits as ecosystem services to sustain the environment in general and benefits to humans in particular in the form of goods and services such as food, fuel, and materials, which are essential for human survival. Coastal–estuarine–mangrove ecosystem services encompass all three forms of ecosystem services such as supporting and regulating services, provisioning services, and cultural services, as indicated by the Millennium Ecosystem Assessment. The coastal zone, being the border of the terrestrial landscape with the maritime realm, supports the lives of a galaxy of flora, fauna, and microbes, in the form of mangroves and their associated biodiversity components such as insects, birds, wild mammals and reptiles, benthos (macro-, meio-, and microbenthos), nekton, and plankton in different ecological habitats (estuaries, creeks, intertidal and sub-tidal zones, mangroves, delta, etc.) and ecological niches. All of these floral and faunal inhabitants have been found to display eco-dynamics in tune with various patterns of succession, distribution, changing ecological gradients, and also by enjoying a specific ecological niche (Giere & Pfannkuche, 1982; Costa & Davy, 1992; Branchley, 1982; Chakraborty & Choudhury, 1994; Misra, 1995; Chakravarty et al., 2004; Khalua et al., 2008; Chowdhury et al., 2011).

'Biodiversity,' defined by UNCED 1992 as "variability among living organisms from all sources including inter phyla terrestrial, marine and other aquatic ecosystems and ecological complexes of which they are part: that includes diversity within species, between species and of ecosystems." Self-sustaining structural and functional attributes of biodiversity in time and space are maintained by an intrinsic mechanism of interactions within, between, and among various levels of biodiversity (Chakraborty, 1995, 2003; Chakraborty et al., 2013). The biodiversity of the coastal environment and the hydrologically interconnected and interlinked sub-systems such as salt marsh, swamps, mangroves, dunes, creeks, estuaries, creeks, and lagoons have been continuously encountering natural and human-mediated environmental pressures during last couple of centuries in the Sundarbans mangrove ecosystem, India. Continuous threats of ongoing pollution and overexploitation of resources are being combated by the resilience of the coastal ecosystem. In this context, the management strategy emphasizing the sustainable utilization of coastal resources should focus on prevailing ecosystem integrity by safeguarding the structural components of the ecosystem, interactions among them, and the resultant behavior or dynamic of the system (Chakraborty, 2021a, b, Chakraborty et al., 2023).

5.6 Marine–Coastal–Estuarine–Mangrove Ecosystem and Its Ecosystem Services

The marine–coastal–estuarine–mangrove ecosystem, representing a sensitive, vulnerable, biologically productive, and dynamic ecosystem of the globe, supports the lives of innumerable flora and fauna in its diversified habitats and ecological niches (Carter, 1988; Ysebaert & Herman, 2002; Kennish, 2002; Chakraborty, 2018). However, estuaries along with mangroves located between the sea and freshwater rivers are characterized by highly oscillating ecological variables representing unstable hydrobiological conditions that undergo the rapid pace of eco-degradation because of the impacts of so many ongoing human-mediated developmental activities such as urbanization and industrialization. The critical ecosystem services rendered include yielding a good amount of human foods, mainly fish, the regulation of climate through marine and atmospheric linkage with an estimated 92.2 Gt of carbon per year entering the ocean from the atmosphere and 90.6 Gt of carbon per year entering the atmosphere from the ocean. Another critical service is the production of oxygen. It is estimated that 50% of oxygen used by the global human population is produced directly by marine phytoplankton (Behrenfeld et al., 2006).

Microorganisms of the ecosystem take active roles in decomposing the dumped waste materials to generate biologically required nutrients alongside purifying the coastal environment, but the marine–estuarine biodiversity is affected to a greater extent by global warming, an increase in temperature, sea-level rise, acidification, altered upwelling, and the stratification of density in the water column (NRC, 1993; Koch et al., 2005).

The primary production in the mangrove ecosystem on tropical world mostly because of huge amount of leaf litter based detritus production leads to cause higher biodiversity by facilitating carbon availability to consumers in both mangrove forest and estuarine subsystems, which exceed the production by benthic algae, other macrophytes, and especially phytoplankton (Robertson & Blaber, 1992). Besides, mangrove litter acts as an important trophic link in the food-webs of both estuarine and mangrove subsystems (Nagelkerken et al., 2008). The mangroves on tropical estuaries under the dominance of tidal interplays promote the tidal export of mangrove litter-contributed organic matter (OM) to estuarine areas and thereby trigger higher productivity in the nearshore areas (Lee, 1995; Bouillon et al., 2008). The faunal components of the mangrove ecosystem, especially the nonchordates, appear to utilize locally available food sources (Robertson, 1991) and thereby significantly alter the properties of these sources in the process of export, mineralization, or consumption (Kristensen et al., 2008).

Ecosystem services rendered by mangroves have been estimated to be worth around $57,000 per hectare per year, sustaining more than 100 million people and destroying the world's mangroves, especially by deforestation, resulting in annual economic loss of around $42 billion without adhering to neither ecological nor economic sense as per the United Nations Environment Programme report (UNEP, 2014). Mangroves have the ability to absorb a considerable amount of the ocean's wave energy and help to block the effects of severe storms and sea-level rise. The mangrove ecosystem, being one of the most carbon-rich ecosystems of the world by storing a substantial amount of "blue carbon" within mangrove forests, which have been captured by global coastal and ocean ecosystems. In addition, mangroves act as important carbon sequesters by capturing carbon dioxide (CO_2) from the atmosphere and storing it in their biomass (trunks, canopies, leaves, and roots) and in the soil. By removing greenhouse gases from the atmosphere, mangroves help to mitigate climate change.

The marine–coastal–estuarine–mangrove ecosystems of West Bengal, India, provide a large number of goods and services in the form of resisting soil erosion, driving of nutrient recycling, preservation of biodiversity, developing provision for natural environmental amenities, ensuring climate regulation and carbon sequestration, maintaining cultural heritage, and rendering spiritual benefits, and all of these tend to boost the economic welfare and healthy environmental conditions of the local and global communities (Chakraborty, 2018). However, some estimates have categorically pointed out the rapid declining of coastal wetlands, which amounted to an approximately 1% loss of the total area of global coastal wetland each year in the late phase of the twentieth century.

5.7 Coastal Environment in West Bengal, India

Coastal zones representing the interphase of land and sea margins include that part of the land that is influenced by the adjoining sea and that part of the sea that is influenced by the adjoining land discharges. This coastal zone, by virtue of providing so many human benefits such as ports, tourism, fisheries, and industries trigger economic development alongside opening new avenues for livelihood generation. This important landscape acts as a repository of biodiversity, regulating atmospheric dynamics, promoting nutrient cycling, determining hydrobiological conditions, and helping waste removal (Jennerjahn & Ittekkot, 1999). Coastal landform helps energy inputs and production of inorganic and biological materials after being regulated by several geomorphological and oceanographic variables such as tidal cycles, longshore currents, the presence of ridges, channels, offshore bars, sand banks, and the islands over the shallow seas are the major topographic features in the coastal zone (Hughes et al., 1998; Paul, 2002).

The Indian coast, having a land frontier of 15,200 km and a coastline stretching for about 5700 km on the main land and about 7500 km with the two island territories, possesses most of the known geomorphological features of coastal zones. However, substantial damage to the Indian coastline with different ecosystems, such as mangroves, salt marsh, lagoons, corals, dunes, and estuaries, has resulted in increasing human pressures of multitudinal dimensions (Qasim et al., 1988). The coastal area of West Bengal extends over 0.28 million ha and 220 km of coastline. The coastal tract of Midnapore (East) district, West Bengal, India, sharing 27% (60 km) of the total length of coastal zones of West Bengal, is a transmissible part of the Sundarbans mangrove ecosystem—a World Heritage site, extending along the Hooghly estuary from New Digha (at the confluence of the Subarnarekha with the Hooghly) to the extreme south-west point of Midnapore district and then curving around Junput, Rasulpur, Khejuri, and Haldia in the east to further north east up to Tamluk or even Kolaghat on the banks of the Rupnarayan river (Chakraborty, 2010).

A long history of deposition, vertical accretion and seaward advancement of the coastal plain of West Bengal over the Bengal Basin prevails because of the influence of tectonic settings by neotectonics along with concomitant sea-level changes (Paul, 2002). Continuous sedimentation on the inland and the coastal plain streams followed by erosion on the coastline of West Bengal associated with a rise in sea level have started filling the valleys with sandy sediments. The sand cliffs and clipped clay banks constituting beaches and the backshore undergo continuous and steady erosion at an alarming rate throughout the length and breadth of the West Bengal coast (Paul, 2002). The littoral tracts of the coastal plains of the Midnapore (East) along the inner boundaries correspond to the shorelines of the post-Pleistocene high stand of restored sea level (Paul, 2002). The coastal tract of Midnapore (East), West Bengal, is characterized by extensive sand flats with different forms of ripples, longshore currents, high salinity, less turbidity, large sand dunes, backwater, small creeks, and less coverage with halophytic vegetation (Chakraborty, 2010) in comparison with the Sundarbans mangrove ecosystem, mostly in the district of South

24-Pargana, West Bengal, India. The coast of Midnapore (East), West Bengal, is also provided with a high-energy macro-tidal environment with waves produced by a long fetch across the Bay of Bengal and experiences southwesterly monsoon winds blowing onshore and with southerly and easterly winds strengthened by the visiting cyclones over the Bay of Bengal, mostly during the summer periods.

5.8 Ecosystem and Climate from the Indian Perspective

The coastal zone representing the most dynamic natural systems at the land–sea interphase is endowed with three major spheres of the Earth: the lithosphere, and the hydrosphere, and the atmosphere, which interact with one another to form interconnected environmental systems. The coastal belts undergo geomorphological changes after being governed by different geological, ecological, and climatic processes. In this very unique landscape of the world, sea is influenced by the adjoining land and land is influenced by the nearby sea. Coastal zones within the Indian ocean region represent a geomorphologically highly diverse aquatic realm with several littoral landmasses, deltaic islands, lagoons, corals, and mangroves covering an area between the Australian, African, and Asian continents, reaching the Antarctic landmass. The complexities of ecosystems are manifested by the interactions between living beings and their physical environment. Different oceans and their constituent ecosystems such as coastal, mangroves, corals, and lagoons, constitute essential elements of the chemical, biological, and physical processes of life on Earth. Changes in the climate driven by solar energy and in accordance with the variability of meteorological parameters (e.g., air temperature, rainfall, and humidity) warm the Earth, which in turn alters the circulation of the atmosphere and oceans. The eco-geo-hydrobiological properties of most of the coastal regions are region specific, despite the similar kinds of stresses exerted on coastal regions in a global context, but distinct features in the form of microclimates should be taken into consideration alongside all those effects of climate systems over very large scales.

Different forms of vegetated coastal ecosystems such as mangroves, saltmarshes, seagrass, and meadows render many goods and services, playing fundamental roles in limiting the effects of sea-level rise by dissipating wave energy, acting as biofilters, cycling and redistributing the excess land-derived nutrients, and thereby forming special habitats for the survival of many important species with economic and ecological significance (Beck et al., 2001). However, despite considering the significance of all these benefits, eco-degradation and even loss of these ecosystems are continuously occurring at alarming speeds. In addition, all of these green plants dominating the coastal ecosystems serve as efficient carbon binders, where they can store carbon for millennia (Chmura, 2009, 2013; Chmura et al. 2003). In such a context proper conservation and restoration of these coastal ecosystems constitute one of the best strategies for climate-change mitigation strategies by storing carbon in sediments and live biomass through active and efficient uptake processes of atmospheric CO_2 and preventing the emission of greenhouse gases, minimizing the degradation to a large extent. This observation has helped the term **"blue carbon"** to emerge, by storing carbon in coastal marine environments, and is considered an ecosystem service provider by saltmarshes (Chmura, 2013). The higher productivity of the mangrove ecosystem is attributed to the export of energy in the form of plant litter-contributed detritus to adjacent aquatic systems and a major proportion of the metabolism occurs within the water column rather than binding as OM (Valiela et al., 2000). Sequestration and remobilization of carbon in the mangrove ecosystem depend on factors such as local topography, prevailing geomorphological settings, temperature, precipitation, tidal range, hydroperiod, salinity, nutrient availability, influx of suspended sediment, plant competition, bioturbation, trophic cascades, etc.

5.9 Mangroves and Ecosystem Services

The mangrove ecosystem, being one of the most productive natural ecosystems, provides substantial ecosystem services for human society (Mitsch & Gosselink, 1993; Costanza et al., 1997). The decomposition of mangrove leaf litter to detritus and transformation of the same into valuable nutrients for the growth of halophytic vegetation as well as other primary producers such as benthic algae and phytoplankton on exporting to adjacent ecosystems, triggering the sediment biogeochemical process and promoting metabolism and nutrient cycling (Webb & Eyre, 2004) (Flow Charts 5.1, 5.2, 5.3, 5.4, and 5.5).

Mangroves, woody halophytes growing on the tide-dominated saline areas along estuaries, sheltered bays, and inlets in the tropics and subtropics around the globe, form a coastal forest, which around 50 years ago covered around 75% of the world's coastlines extending 200,000 km^2 (Spalding et al., 1997; Ellison & Farnsworth, 2001). However, 35% of the global mangrove area has been lost since the 1970s with an alarming rate of disappearance of 1–2% annually (Valiela et al., 2001; Alongi, 2002). The removal and disappearance of mangroves worldwide is of great concern because of their contribution of a good number of highly valued goods and ecosystem services for the benefit of humans in general and the coastal environment in particular by way of providing food, supplying materials for shelter, and medicines, ensuring coastal protection, controlling coastal erosion, carrying out water purification, maintaining fishery stocks, facilitating carbon sequestration, promoting tourism, etc. In addition, proper functioning of the mangrove ecosystem and thereby the maintenance of the ecological health of the entire coastal zone determine the following benefits to humans:

1. Supplying of a variety of useful products, such as fishes and shellfishes, fuel wood, timber, raw materials for medicines, resins, and honey.
2. Mangrove ecosystems have a huge amount of nutrients, which makes them nursery and breeding grounds for offshore finfish and shellfish.

3. The propensity of mangrove forests as natural "coastal storm barriers" against periodic wind and storm surge events, such as tropical storms and cyclones, typhoons, and tsunamis. Coastal floods help them to serve protective roles for the entire region (Das & Vincent, 2009).
4. The mangrove ecosystem in its natural state of ecological functioning is closely tied to local culture, religion, mythology, heritage, and traditional knowledge.
5. The power of mangroves by virtue of their intricate root systems (pneumatophores and stilt roots) resists shoreline erosion and thereby stabilizes sediment by retaining soil (Wolanski, 2006).
6. The bioaccumulation ability of mangrove plants and filter-feeding mechanisms of other macrobenthic faunal forms remove a good amount of suspended particles from the water column and thereby reduce the turbidity of the water, thus facilitating water purification processes that in turn help to grow coral reefs, seagrass, and other sensitive organisms and protect them from siltation and pollution pressures (Wolanski, 2006; Wolanski et al., 2009).

5.10 Functional Relationships Among Intertidal Macrobenthic Fauna as Prerequisites for Ecosystem Services

Benthic fauna representing the most important invertebrate faunal components of the marine and estuarine ecosystem with greater diversity, abundance, and biomass of innumerable species accommodating the representatives of almost all invertebrate phyla of the animal kingdom (Chakraborty, 2009, 2011; Chakraborty et al., 2012). The intertidal benthic assemblage forming the structure of the benthic community is determined by an array of ecological parameters, both living and nonliving ones that operate in an intricate manner, resulting in multidimensional relationships.

The benthic faunal contribution to aquatic biological production is converting organic nutrients into biomass, which helps the growth and reproduction of the faunal components, with the development of eggs, sperm, and larvae. The abundant species are usually used by the humans for direct consumption, whereas the innumerable small species contribute to developing the complexity of the respective food-webs and functioning of tropical ecosystems (Hendrickx, 1995). The highly dramatic activities of different benthic fauna in the marine and estuarine coastal zones result in changed contours by modifying the texture of the sediments, nutrient and sediment transport, the development of specific geomorphological settings, and also enhancing primary and secondary productivity.

The intertidal belts of the Midnapore (East) coast-neighboring areas of Sundarbans, India, at the confluence of the Hooghly–Subarnarekha estuaries with the sea, the Bay of Bengal, support diversified forms of macrobenthic fauna, a major part of which is shared by brachyuran crabs, a bioenergetically significant macrobenthic faunal group (Chakraborty & Choudhury, 1985; Chakraborty et al., 1986; Chandra et al., 2003; Khalua et al., 2003; Chakraborty, 2009, 2010). The macrobenthic fauna in the coastal environment render so many ecosystem services directly or indirectly by virtue of their abundance, biomass, life-cycle patterns, and trophic interactions. The abundance and diversity of benthic fauna exhibit positive relationships with demersal fisheries and aquaculture (Chakraborty, 1996). Almost 50% of the world's commercial fish catch takes place from the coastal areas in the form of a large number of demersal fishes that mainly live on benthic animals for harvesting their food materials. The millions of benthic fauna produce crores of larvae from their fertilized ovum in the form of meroplanktons that act as major sources of novel food for both offshore and onshore fish populations, thereby maintaining the ecological equilibrium. Several major macrobenthic fauna, such as brachyuran crabs, play important roles in the degradation of leaf litter in the mangrove ecosystem (Robertson, 1986; Alongi, 2002, 2008; Chatterjee & Chakraborty, 2014; Chatterjee et al., 2014). An innumerable number of micro-arthropods and meiobenthic fauna also play vital roles in the decomposition process in the mangrove ecosystem (Dey et al., 2008; Chakraborty, 2017). Moreover, the intertidal benthic macrofauna and meiofauna can be used as bioindicator species to understand ecological changes in the water and soil quality after being impacted by environmental stresses because of their sensitivity to environmental perturbation (Washington, 1984; Stark, 1998).

The diversity of benthic fauna through their burrowing and feeding activities governs the biological, physical, and chemical environments of the estuary by linking the sediment to the aquatic food-web (Camilleri, 1992). Filter feeders in the benthic community filter a large volume of overlying water by pumping due to the concerted action of their feeding structures and grasping the food, thereby cleaning the water by removing sediments and OM. Unused sediments loaded with OM within the water column are deposited on the bottom of the estuary and the deposit feeders then remineralize those materials into nutrients, which reenter the water column. This process of remineralization of OM forms an important source of nutrients for the aquatic subsystem and acts as a critical contributor for driving primary production, with higher rates in estuaries (Giere, 1993). Established facts of the existing relationship between the benthic diversity, density, and biomass and the higher rates of fishery production due to the higher abundance of demersal fishes and crustaceans have pointed out that the pelagic larvae released from the benthic organisms, being a major biotic component of the planktonic community, influence to a greater extent the planktonic food-web and thereby fishery production (Heald & Odum, 1970; Parulekar et al., 1980; Zimmerman et al., 2000). Environmental pollution mediated ongoing environmental perturbations in the estuarine–coastal ecosystem and their prospective ecological impacts on the ecosystem appear to be very diverse and complex. Only chemical and physical modes of assessment methods hardly provide proper assessment of the intensity of such impacts and thereby necessitate the adoption of biological methods in order to generate actual results (Maiti Dutta et al., 2014; Sanyal et al., 2015).

Different biotic components within the benthic community act as important indicators of the changing ecological health and this kind of bioassessment appears to be very advantageous and cost-effective (Dauvin et al., 2003). Biomonitoring refers to the assessment of environmental changes caused by anthropogenic activities by way of observing biological responses, and the benthic fauna act as the most amenable and suitable faunal group to be involved for this purpose (Warwick, 1988).

The metabolic processes of different biotic components and transformations of energy and matter driven by biotic assemblages within natural ecosystems are termed "ecosystem functions," but functional outcomes of the ecosystems rendering benefits to humans are termed **"ecosystem services"** (Chapin et al., 1997). The diversity and density of intertidal benthic fauna are found to have been determined by a number of ecological parameters such as salinity, pH, dissolved oxygen, sediment texture, and nutrients, which act either in isolation or in concerted ways to determine the vegetation assemblages and eco-biological attributes of faunal components (Chakraborty & Choudhury, 1989, 1992, 1993, 1994; Chakraborty, 2011, 2013; Chakraborty et al., 2009, 2012). Many of the abiotic components of the habitats act as the framework and template on which different species of fauna settle, grow, undergo interspecies competition, and survive successfully, justifying Clement's monoclimax theory of successional changes (Townsend et al., 2008). In a biotic community, each species avoids competitive interactions by using one substrate as a feeding site, another site for shelter, and other as a source of nutrition, but each function in coordination with each other ensures the existence of a community.

The distribution patterns of benthic fauna are also determined by biotic factors, such as different forms of community interactions, for example, competition, amensalism, amelioration, predation, symbiosis, commensalism, and mutualism, which also tend to vary at spatial and temporal scales (Alongi, 1987; Thrush et al., 2012; Fischer & Sheaves, 2003; Dey et al., Das et al , 2014a , 2014b; Chakraborty, 2018). Benthos–sediment interactive processes have been found to modify, stabilize, or destabilize the sediment, mostly by virtue of their feeding activities such as deposit feeding, suspension feeding, filter feeding, herbivorous, and carnivorous (Chakraborty, 2018). Different ecological parameters act on the biotic community either independently or in a combined manner to influence their survival strategies, which in other ways determine nonliving parts of their habitats.

Two prime ecological variables in the marine coastal environment, viz. salinity and temperature, determine the distribution of benthic faunal components (Teal, 1958; Kinne, 1963). Different attributes of substrates are found to be important in determining the species composition of the biotic community of various habitats within the marine–estuarine ecosystem, where a higher number of species are observed to share a different ecological niche within intertidal habitats, resulting in greater species diversity (Abele, 1974. Most diverse biotic communities occur in the tropics and also in the deep sea because of the constancy of their soil environments (Sanders, 1968). The littoral zones of the mangrove ecosystem with greater nutrient enrichment under the tidal influence determine the settlement, growth, proliferation of population, and community interactions of various kinds of faunal components (Heald & Odum, 1970). Bioturbation is an eco-geological process in aquatic depositional systems, where benthic faunal components construct burrows and restructure them in tune with the survival strategies of the concerned fauna and thereby render valuable and critical services, more precisely designated as ecosystem services (Chapin et al., 1997) (Flow Charts 5.1, 5.2, 5.3, 5.4, and 5.5).

5.11 Benthic Fauna as Ecosystem Providers and Ecosystem Engineers: Bioturbation, Nutrient Cycling, and Bioinvasion

An emerging field in the realm of ecology is ecosystem engineering, which denotes an organism's direct or indirect control over resource availability by causing a physical state of changes in structural elements (abiotic or biotic) in an ecosystem. Ecosystem engineering takes into consideration the creation, destruction, or modification of habitats by altering the essential ecological parameters such as temperature, heat, oxygen, and water availability, light, humidity, and sediment characteristics within living space by modulating physical processes such as direction of water flows, currents, nutrients, and sediment loads of water that in turn affect resource availability and distribution patterns.

Bioturbation, referring to the disturbance of habitats by way of stirring or mixing of different layers of sediments within the habitats through different kinds of biological activities, viz. feeding, movement, burrowing, etc., of benthic animals, in turn modifies the geochemistry of sediments and their interstitial water (Woodin, 1981). Different intertidal macrobenthic fauna such as polychaetes, brachyuran crabs, gastropods, brachiopods, actinarians, and gobiid fishes act as the major biotic contributors, causing bioturbation in the Sundarbans mangrove ecosystem, and different bioturbatory structures such as burrows, pseudo-pellets, mud balls, sand balls, sand pyramids, chimneys, semidomes, and hoods have been found to show seasonal variations in their structures and also appearance and thereby facilitate the process of sediment destabilization, soil excavation, pumping of water and oxygen within soil, nutrient reshuffling, etc. Adaptive behavioral strategies associated with all bioturbatory activities are considered for their survival because they extract their requirements for survival by digging the soil and forming pellets, domes, chimneys, etc., for their shelter, forming sediment structures, procuring food, maintaining territories, demonstrating energy efficiencies as a part of sexual selection, and facilitating microbial decomposition of sediment organic materials (Anon, 2003; Chatterjee et al., 2014). In the mangrove ecosystem of Sundarbans, it was observed that macrobenthos along with several microarthropods play vital roles to expedite the process of mangrove litter decomposition by initially breaking the leaves into pieces and thereby increasing the surface area for better activities of the microbes and ready conversion of mangrove litter into detritus (Dey et al., 2010).

Deposit feeders are considered the prime faunal groups undertaking bioturbation by constantly working and reworking over the sediment to harvest food particles, altering the textural composition of the sediment by driving horizontal and vertical movement of the soil particles. Such biogenic alterations of the sediments within the habitats of the bioturbators promote the erodibility of the sediments (Luckenbach, 1986; Kihslinger & Woodin, 2000), the modification of grain sizes (Rhoads & Stanley, 1965; Cadee, 2001; Lim, 2006), and enhancement of the concentrations of pore water by increasing the amount of constituents (Aller, 1982, Aller, 1988, 2001), which in turn bring changes in the ability of individuals within the macrofaunal community to burrow through the sediment. Continuous burrowing and releasing of defecated materials influence recruitment patterns of infaunal components, altering of the chemical constituents of the sediment through infaunal secretions (King, 1986; Woodin, 1981). However, defecation by large infauna impose negative impacts on other infauna, causing a retardation in growth rates, reduced recruitment of new faunal components, and increasing emigration. Bioturbatory activities of intertidal macrobenthic fauna also promote habitat heterogeneities, which in turn lead to the higher diversity of organisms, alteration in the abundance of species, modification of physical conditions, and development of biogenic structures (Lee, 1998, 1999, 2008).

5.12 Bioturbation by Macrobenthic Fauna and Their Roles in the Sustenance of Ecosystem Functioning

Bioturbation, referring to a disturbance of sediment layers by the biological activities of numerous faunal components, mostly benthic fauna in the process of consuming OM trapped between sediment grains, is a significant eco-biological process in the marine–estuarine environment. Macro invertebrates directly release nutrients involving themselves in the processes in the biogeochemical cycle of sediment and OM consuming oxygen. They indirectly trigger the processes of redistributing particulate and dissolved substances in the sediment and mediating key processes for oxygen pumping followed by utilization by an array of burrow-inhabiting organisms. The term bioturbation refers solely to the redistribution of particles and the formation of biogenic structures by burrowing animals resulting in physical disturbances caused by animals on the substratum, in both terrestrial and aquatic environments including particle (reworking) and water (ventilation) movements by way of construction of sediment structures such as burrows, tubes, and mounds, particle ingestion and egestion during foraging, food caching, prey excavation, wallowing, etc. (Fager, 1964; Aller, 1988; Aller & Aller, 1998; Cadee, 2001). The lateral **'ploughing'** of the surface (by moles or heart urchins), particle ingestion and egestion during foraging by faunal components (deposit feeding, geophagy, ornithophagy), and the infilling of abandoned burrow structure.

The geochemistry of sediments along with their interstitial water is influenced by the bioturbation process, which refers to the disturbance of the sediments by biological activities through the stirring or mixing of sediment layers, viz. mobility, feeding, burrowing, etc., of benthic animals for the remove through (Woodin et al., 1998). Out of many types of intertidal fauna, brachyuran crabs, polychaetes, brachiopods, and gobiid fishes represent the major agents of bioturbation of Sundarbans and its adjoining coastal environments. Different bioturbatory structures such as burrows with different shapes and sizes, pseudo-pellets, sand balls, mud balls, sand pyramids, semidomes, chimneys, and hoods have been documented from this coastal environment, along with seasonal variations in the rate of the intensity of soil excavation, sediment destabilization, pumping of water and oxygen within soil, etc. All kinds of biological activities have been hypothesized to be adaptive strategies for predation and foraging, aggression, sexual selection, courtship and mating, territoriality, etc., as manifestations of the strength and energy of the involved bioturbators (Chatterjee et al., 2008).

5.12.1 Bioturbation and Ecosystem Engineering

The process of bioturbation in aquatic environments is now considered **'ecosystem engineering,'** which operates as a major modulator of microbial activities and biogeochemical processes and creates new habitats by changing the availability of nutrients affecting organisms other than their direct competitors, predators, or prey (Kristensen, 2008).

'Bioturbation', as the collective manifestation of biogenic processes, helps the cycling of trace metals, nutrients, and pollutants within sediments and the water column of marine–estuarine ecosystems (Zeitzschel, 1980; Kremling, 1983; Wilkinson et al., 2001).

Biogenic alteration of the sediments by different biological activities of benthic fauna, including defecation, burrowing, tube building, and feeding, in turn affect the distribution and abundance patterns of infauna in ecologically modified habitats that are continually structured by the prevailing physical conditions. The sedimentary status reflected by variable percentages of fine sands, silts, and clay mud is maintained by biogenic activities by bioturbators such as cnidarians, Priapulida, polychaete worms, brachiopods, sipunculans, crustaceans, burrowing crabs and shrimps), mollusks, echinoderms, and other meiobenthic faunal components, as reflected by their burrowing, mound formation, tube building, defecation, and secretion (Woodin, 1974, 1978, 1981, 1999; Rhoads & Boyer, 1982).

Different bioturbatory structures lead to the development of habitat heterogeneity on the sediment surface as a positive ecological developmental process, as these newly developed habitats are being used by other faunal components based on the sediment amelioration process and the ability of individual benthos to burrow through the sediment. A negative impact on other benthic fauna has caused reduced growth rates, decreased recruitment, and increased emigration because of the defecated substances from larger faunal components. Deposit feeders being the major group of bioturbators, constantly take care of processing the sediment in search of food, which results in the horizontal and vertical movement of particles in the sediment. This kind of biogenic alteration expedites the erodibility of the sediments by destabilizing the compactness of the soil habitats (Luckenbach, 1986) and causing several other ecological consequences such as the redistribution of grain sizes

(Rhoads & Stanley, 1965; Cadee, 2001; Lim, 2006), changes in the concentrations of porewater constituents (Aller, 1982), modifying infaunal recruitment patterns, and altering of the chemical constituents of the sediment by infaunal secretions (King, 1986; Woodin, 1981).

Different benthic forms such as brachyuran crabs, brachiopods, polychaetes, gastropods, bivalves, and actinarians in the mangrove–estuarine complex of Sundarbans, India, take part directly and indirectly in the bioturbation process depending on the sediment properties (Rossi & Chapman, 2003; Chatterjee & Chakraborty, 2014; Chatterjee et al., 2014) and some of them can perform those bioturbatory functions even under anoxic conditions or in the high concentrations of hydrogen sulfide. Different bio-geochemical and ecological activities performed by intertidal benthic fauna redistribute textural composition of the sediments and restructure the sedimentary environments, the fundamental building blocks of the habitats by the local physical regime and by the biotic forces facilitating settlement, growth, and development of biotic assemblages that take active roles in burrowing, tube building, defecation, and secretion (Rhoads & Boyer, 1982; Woodin, 1999; Gilbert et al., 2003).

Different benthic fauna undertaking various forms of feeding strategies such as deposit feeding, suspension feeding, filter feeding, herbivorous, omnivorous, and carnivorous modes of feeding determine the fauna–sediment interaction processes.

Deposit feeders as one of the prime groups of bioturbators constantly work and rework on the sediment for harvesting their food and thereby result in horizontal and vertical movement of particles in the sediment. Biogenic alterations in some cases on uncompact sediments expedite the erodibility of the sediments (Kihslinger & Woodin, 2000; Wilkinson et al., 2009), alter the distribution of grain sizes (Lim, 2006), and often modify the concentrations of porewater constituents (Aller, 2001). The chemical constituents in sedimentary habitats are influenced by different burrowing activities, defecation processes and infaunal secretions that affect infaunal recruitment patterns and structuring of the biotic assemblages (Woodin, 1981).

5.12.2 Brachyuran Crabs as a Major Group of Macrobenthic Fauna: Their Roles as Ecosystem Engineers

Intertidal brachyuran crabs with their sheer diversity, abundance, and biomass with an active life on the surface sediments lead a very active life during the exposure period at low tide, constructing their burrows and returning to their shelter during the inundation period of high tide. Their functional roles in a very positive manner by virtue of eco-biological potentiality help to maintain the steady-state balance of the ecosystem. This group of faunal components constitute an important constituent in the food-web of the coastal belt and thereby accelerate the decomposition cycle as macro-decomposers, driving ecosystem functioning of the mangrove–estuarine complex to a large extent. They are morphologically, physiologically and behaviorally well adapted to their respective habitats and render valuable ecosystem services (Kristensen, 2008). Surface topography, with the redistribution of textural constituents of the sediments, enhancement of aeration and moisture levels within the burrows, and a decrease in the concentration of accumulated phytotoxins in the substratum, is found to have been changed owing to bioturbation by crabs. All these changes not only expedite the growth of mangroves but also promote the flourishing of associated faunal components, triggering their reproductive potential and generating the provision of food. The semi-terrestrial and air-breathing grapsid crabs, with their herbivorous feeding habits and tolerance against deoxygenation caused by organic enrichment, exert a strong top–down control on saltmarsh plants (Iribarne et al., 1997; Costa et al., 2003; Alberti et al., 2007). Generation and transformation processes of nutrients, especially carbon, nitrogen, and phosphorus, and altered water movement result in the landward shifting of mangroves, strongly affecting the survival of the crabs.

5.12.3 Excavation of Burrows by Crabs Mimicking the Deposition of Soil

Burrowing by crabs is one of the most common bioturbation activities in mangrove–estuarine complexes, leading to dislodgement of soils and redistribution patterns of associated carbon and nitrogen by crabs and thereby to understanding the influence of biotic factors on material cycling and energy flow. However, the burrowing of soils by crabs does not work alone in coastal mangrove ecosystems but is impacted by physical processes such as water flows, tidal cycles, biological factors such as the body size and biomass of crabs, the root-mat system of mangroves, and other meteorological factors such as rainfall and wind flows, promoting the vertical and horizontal transport of sediments and associated carbon and nitrogen cycling (Woodin, 1981, 1999; Chatterjee et al., 2008).

Three different attributes – excavation, deposition, and net transport – are in use to characterize the soil excavating and depositing processes by the concerted activities of crabs. Net transport, meaning the difference between the amount of soil excavated by crabs and that deposited into a burrow, is related to deposition efficiency, referring to the amount of deposited soil per unit volume of burrow. All of these eco-biological parameters are also influenced by the duration of high tide, the types of burrows, and the diameters of burrows. The construction of burrows alongside excavation of soils for this purpose significantly affects the physical and chemical properties and texture of soils.

The soils deposited into the lower proportion of burrows show the presence of higher percentage of fine grains and higher salinity in comparison with the excavated soil, which is characterized by a lower bulk density and higher water content. Excavated and deposited soils have a lower bulk density and higher nutrient concentrations in comparison with the surrounding background soils. Soil sources have total nitrogen (TN), total organic carbon (TOC), total carbon (TC), and ratios of TOC to TC vary among deposited, excavated soil, background surface soil, and subsurface soil, showing higher to lower ranks in a descending order.

5.12.4 Ecological Services Through the Excavation of Burrows

The burrows, with different configurations constructed by different species of brachyuran crabs, serve as important habitats of different crabs on the one hand and renders protection from aquatic predators during inundation at high tide and from semi-terrestrial predators during aerial exposure at low tide on the other (Crane, 1975; Zeil & Layne, 2002; Chatterjee & Chakraborty, 2014). Burrows of crabs help molting animals to take safe refuge and also provide a safe place for females carrying incubating eggs. The burrow protects the crabs and can save them from desiccation during their activities on the surface of the exposed intertidal belts as they can take shelter within the burrows as well as utilizing moisture and getting access to water in order to undertake normal respiration and feeding. In addition, moisture- and oxygen-enriched burrow walls act as suitable sites for the nutrient cycles in the form of nitrification and de-nitrification processes. In these processes, brachyuran crabs excavate a good amount of soil from the subsurface layers of the intertidal belts to the surface layers and in that process forage in the displaced soils, reshuffle the bottom soils with the surface ones, and create mounds around burrows (Iribarne et al., 1997; Botto & Iribarne, 2000). The burrowing activities of the crabs helps to pump oxygen from the surface to the subsurface layers and oxidizes inorganic compounds such as sulfides that build up, owing to the high rates of organic decomposition.

Construction of burrows and their subsequent maintenance by the continuous activities of the respective crab species facilitate the reshuffling of the textural components of the sediments, pumping oxygen directly from the sediment–atmosphere interphase, enabling movement of water from the surface to the subsurface layers of sediments, and importing of OM produced from the breakdown of litter. Even the entire biomass of some macrophytes within intertidal and shallow subtidal habitats play a great role in the energy flow and cycling of nutrients in coastal systems (Alongi, 1990; Chatterjee & Chakraborty, 2014; Chakraborty, 2018). Of the many families of brachyuran crabs, individuals belonging to the families of Ocypodidae (ghost crabs and fiddler crabs) and Grapsidae (*Sesarma* spp., *Metaplax* spp., and *Metopograpsus* spp.) play their part with their burrowing activities, transporting sediments, and thereby reducing the compactness of sediment, increasing the porosity and percolation efficiencies, breaking the mangrove leaves and other parts of the mangrove plants into smaller pieces to expedite the process of decomposition (Emmerson, 1994; Kristensen & Kostka, 2005; Botto et al., 2005), and also modifying the microtopography by increasing the density of coarse particles on the soil surface (Warren & Underwood, 1986). The burrowing process of brachyuran crabs also influences the alteration of microbial processes, which in turn modify soil chemistry, enhance soil oxygenation, and alter the salinity of interstitial water (Fanjul et al., 2007).

Brachyuran crabs in the process of constructing burrows considerably affect below-ground processes, which in turn can impact marsh plants in at least eight ways.

First, the burrowing processes of crabs expedite the passage of water and gases such as oxygen between the soil and overlying water, which in turn enhances the aerobic decomposition process of OMs (Fanjul et al., 2007) and increases soil oxidation.

Second, the burrowing processes of crabs help to trap finer grain-sized sediments with low density loaded with higher contents of OM through the interactions of the burrows with tidal water, which in this process facilitate OM decompositions and such processes and thereby increase nutrient availability to the growing populations of benthic organisms (Iribarne et al., 1997; Botto et al., 2006).

Third, excavation of soils by crabs for the formation of burrows brings soil and nutrients from the lower strata of the sediments to the upper layers of sediment in the intertidal belt and thereby accelerate the turnover of soil and nutrients (Robertson & Daniel, 1989; Fanjul et al., 2008).

Fourth, the presence of crab burrows influences the hydraulic parameters by almost twofold of hydraulic conductivity by putting up resistance to movement of the overlying tidal water (Hughes et al., 1998).

Fifth, the crab burrows enhance infiltration rates and thereby increase overall surface infiltration (Hughes et al., 1998).

Sixth, burrows of crabs serve as toxin outlet tubes, which reduces the accumulation of metabolic products, such as sulfide.

Seventh, characteristics of plant assemblage in tune with the prevailing soil properties are influenced by excavation by crabs to form their burrows, affecting burrowing processes (Neira et al., 2006).

Eighth, by undertaking distinct horizontal and vertical distributional patterns for avoiding interspecies and intraspecies competitions, the brachyuran crabs exhibit adaptation to different degrees of terrestriality with respect to inundation and exposure, texture of sediments, salinity, temperature, etc. (Lugo, 1980; Chakraborty & Choudhury, 1989).

5.12.5 Roles of Fiddler Crabs in the Maintenance of Its Habitats

Fiddler crabs (genus *Uca*) representing a conspicuous macrobenthic faunal group of the intertidal belts of the mangrove ecosystem in tropical and subtropical regions, are also highly diverse, with about 100 described species (Crane, 1975; Thurman, 1984; Beinlich & von Hagen, 2006; Ng et al., 2008). Fiddler crabs as very common detritivorous macrofauna and their burrowing process brings buried OM to the sediment surface, generating new food sources within burrows and enabling several mangrove shrubs to grow more profusely, receiving more available nutrients (Bertness, 1985; Thurman, 1984; Genoni, 1991).

Out of so many macrobenthic fauna inhabiting the different parts of the intertidal belts, four different species of fiddler crabs, viz. *Uca annulipes annulipes*, *Uca acuta acuta*, *Uca triangularis bengali*, and *Uca dussumieri* have been found to actively play greater roles as bioturbators by virtue of their sheer abundance in comparison with other macrobenthic fauna, except for a ghost crab, *Ocypode macrocera*, in the Sundarbans mangrove ecosystem, India. All four of those species of fiddler crabs avoid interspecies competition by restricting their distribution along the horizontal and vertical tracts of the coastal zones of West Bengal, India, mainly by the influence of the characteristics of the sediments (Chakraborty, 1996). The bioturbatory activities of fiddler crabs, mostly by forming burrows as well as different forms of mounds on their burrow openings, considerably impact the sediment chemistry, the redistribution of the textural components of the sediments, and also development of other ecological niches for the settlement of many other meiobenthic fauna, algae, and microarthropods (Bertness, 1985; Genoni, 1991; Nomann & Pennings, 1998; Chatterjee & Chakraborty, 2014). The formation process of burrows not only alters the textural elements of the sediment but also pumps oxygen from the overlying atmosphere through inlet tubes, increasing moisture content in the deeper layers of the sediment by driving water from the surface to the subsurface zones (Bertness, 1985; Chatterjee & Chakraborty, 2014). Different behavioral attributes of fiddler crabs, such as foraging, morphometrics, allometric growth, claw asymmetry, limb regeneration, sexual selection, reproductive isolation, visual and acoustic display, circadian rhythms, color-changing frequency, osmoregulation, heat tolerance, and visual neurology, are connected directly and indirectly to bioturbatory activities, and by virtue of such eco-ethological characteristics the *Uca* species serve a broad range of ecosystem functions such as environmental monitoring and sustenance of coastal food chain–food-web dynamics (How et al., 2008, 2009; Crane, 1975; Zucker, 1981; Kim et al., 2004a, 2004b, 2006; Chatterjee & Chakraborty, 2014; Chatterjee et al., 2008, 2014).

5.12.6 Functional Roles of Ghost Crabs

The sandy intertidal regions of the West Bengal coast, India, are inhabited by lakhs of red colored ghost crabs, *Ocypode macrocera* (phylum Arthropoda, family Ocypodidae) and serve both beneficial and harmful ecological roles in the marine–estuarine ecosystems of the region. A huge number of these crab-made burrows destabilize the sandy shores as the excavated soils are stored at the openings of the burrows, which are washed away by the receding tidal flows during low tides. Although the sizes and shapes of the burrows of red crabs tend to vary from one zone of an intertidal belt to another, burrows are found with depths ranging from

20 cm to 30 cm in length and with the configurations cylindrical in shape, often with vertically or diagonally inclined branches. Owing to excessive excavation of soils during the burrowing of overpopulated red crabs, erosive processes take place at tidal channels with flood current dominance and after a few months, the sloping gradients of the intertidal zones experience drastic changes owing to the loss of sediments leading to the conversion of an elevated zone into a plain or a depressed zone. The beneficial roles played by the red crabs include the release of millions of newly hatched larvae as meroplanktons to be used as a novel source of food for demersal fishes, pumping oxygen to the deeper parts of the sediments and recharging the moisture content in different strata of the sediments of intertidal zones, paving the way for the fruitful survival of innumerable meiobenthic fauna and benthic algae.

5.12.7 Roles of Grapsid Crabs Towards the Ecological Sustenance of the Mangrove Ecosystem

The invertebrate faunal components of the mangrove ecosystem forage for locally available food sources and with that process considerably alter the quality of these food materials to export, consumption, or mineralization (Kristensen et al., 2008). Lower densities and species richness in tropical mangrove ecosystems in comparison with other intertidal sedimentary environments are thought to be due to the poor nutritive quality of mangrove detritus (Alongi & Sasekumar, 1992; Lee, 2008). In the tropical Indo-west-Pacific mangrove zones, grapsid crabs consume large quantities of mangrove leaf litter (Lee, 1998) and process the mangrove carbon even before microbial colonization (Lee, 1998). Grapsid crabs as "**ecosystem engineers**" exert profound effects on the structure and functioning of mangroves owing to their roles as **'shredders'** consuming mainly fresh but nutritionally poor leaf litter, which reduces tidal export (Robertson, 1986; Robertson & Daniel, 1989; Lee, 2005), and through bioturbation activities transform and egest mangrove-based OMs as fecal materials (Smith III, 1987; Smith III et al., 1991; Lee, 1997, 1998; Kristensen, 2008) in tropical mangrove forests. Fragmented mangrove leaves help to colonize benthic bacteria more easily than intact leaves (Alongi, 1988; Mfilinge et al., 2002). In the process of the degradation of OM as macrodegradators, grapsid crabs provide food for other organisms in mangroves by facilitating microbial colonization (Camilleri, 1992; Lee, 1997), which in turn results in much faster changes in litter quality and the chemical composition of food to make those processed food materials readily available to estuarine macrofauna and zooplankton (Lee, 1997; Werry and Lee, 2005). Unconsumed leaf fragments through **'sloppy feeding'** of grapsid crabs accelerate the mineralization of mangrove OM into sediment, which in turn facilitates bacterial, fungal, and microalgal growth in sediments (Kwok & Lee, 1995). The grapsid crabs in the mangrove ecosystem, by virtue of their physiological and behavioral adaptation to consuming fresh mangrove leaf litter as their food, is probably related to the occurrence of higher cellulases, which help to digest vascular plant materials, but this mode of feeding is occasionally supplemented by opportunistic consumption of animal food (Thongtham & Kristensen, 2005). The mangrove litter derived detritus supplemented by other food sources with lower carbon/nitrogen ratios is obtained from predation, cannibalism, or a scavenging diet supplemented by (Lee, 1997, 2005; Buck et al., 2003). In the natural environment, detritus derived from vascular plants supports the higher biomass of microorganisms, which are used by deposit feeders feeding on their bacterial colonies and also ingest diatoms (Mfilinge et al., 2002). Sesarmid crabs, by virtue of their arboreal habits and dependence on plant materials as food, play a prime role in linking primary and secondary production as well as acting as macro-degradators by initially breaking mangrove leaves of their choice into smaller pieces. Leaves of *Avicennia marina* are reported to contain rich nutrients with greater palatability in comparison with other mangroves leaves (Ravichandran et al., 2001).

5.12.8 Bioturbatory Structures of Crabs and Their Functional Contribution to the Coastal Ecosystem

High activities of brachyuran crabs on the exposed intertidal zones during the period of low tide depict complex behavioral patterns, communicating with other individuals of the same species by waving or drumming their pincers, aggressively interacting with one another in search of food, space, and mates (Zucker, 1974; Chatterjee et al., 2008). Brachyuran crabs and other crustaceans such as *Thalassina* build up various sediment structures (pillars, hoods, chimneys, semidomes, mud balls, sand balls, etc.) from moist mud sand at the entrance of their burrows and nearby places (Chaudhuri & Choudhury, 1994; Fanjul et al., 2007, 2008). Different bioturbatory structures constructed by brachyuran crabs serve different kinds of biological functions such as pillars constructed by *Uca* spp. for sexual attraction (Christy, 1988); similar hoods by *Uca* spp. for mating purposes (Zucker, 1981), creation of mud balls by *Uca* spp. near the burrow entrance to depict the strength of the males to attract females (Oliveira et al., 1998), and also the construction of pyramids by species under the genus *Ocypode* also for courtship and mating between males and females. Some of the bioturbatory structures constructed to project signals of aggressive behavior of the species concerned, such as **hoods** (*Uca*: Zucker, 1974; *Cleistostoma*), some are for invasion or fighting for other burrows (chimneys of *Uca* spp.), and some structures such as the tubular outgrowth of muddy sediments act as barricades (*Ilyoplax* spp.: Wada, 1983). In addition, several other bioturbatory structures such as hoods (*Uca*), chimneys (*Uca*: Crane, 1975; Thurman, 1984), and sand pellets (*Dotilla*) are related to environmental regulation. Development of several biogenic structures that have different configurations and locations in the coastal belts render valuable functional roles in the ecosystem, ensuring their survival strategies vis-a-vis feeding, territory protection, aggression, female attraction and courtship, mating, and breeding on the one hand

and through accelerating different bio-geochemical cycles due to the impact of biotic components on the other (Flores & Paula, 2002; Chatterjee et al., 2008; Chatterjee et al., 2014; Chakraborty, 2018). The creation, appearance, and existence of different types of these bioturbatory structures such as **sand balls** (accumulation of excavated sand mixed with saliva and mucous secretions by *Ocypode macrocera*), **pellets** (residual parts of sediment after being scooped up by the mandible–maxilla of respective for extracting their food materials, *Dotilla blanfordi*), **mud balls** (scooped up mud at the entrance of the burrows of *Dotillopsis brevitarsis, Sesarma taeniolatum, Uca acuta acuta, Uca triangularis bengali, Metopograpsus maculatus*, and *Metaplax intermedia*), **hoods** (like a soft covering for the head, usually worn by women and this structure is formed by soft sediment formed by *Sesarma (Chiromantes) bidens* around their burrow opening), **chimneys** (a vertical shaft that provides a path through which smoke from a fire can pass composed of large wet soil pellets; Uca *acuta acuta*), **semidome** as a roof covering a semicircular space, i.e., a. half a dome, constructed by only *Uca lactea annulipes* by scraping up bits of sediment from the surface with their first four walking legs and depositing the sediment materials around the burrow entrance, forming triangular semidomes) represent the manifestations of different benthic brachyuran crabs about their ecological needs, adaptability, power with energy, means for sexual selection and potential for the nutrient resufflings (Chakraborty, 2017). **Mud balls** undergo seasonal variations in respect of their sizes, shapes, and configurations (Chatterjee et al., 2008, Chatterjee et al., 2014; Chakraborty, 2018) (Figs. 5.1, 5.2, and 5.3).

5.12.9 Harmful Roles of Crabs in the Mangrove Afforestation Program

Burrows of crabs acting as passive traps increase detrital entrapment in the sediment, which is reflected by the higher deposition of carbon content in the burrows as a result of the tidal sediment deposition process than carbon content in the exposed sediment excavated by crabs which result in a net decrease in the amount of export of carbon and particulate OM from the forests to the estuaries (Botto et al., 2006).

It has been reported that the crab community in mangrove forests with the majority of mangrove plants species being *Avicennia marina*, was dominated by microphagous crab species feeding on detritus and belonging to the family Ocypodidae, rather than the leaf-eating sesarmid crabs found in other forest types (Robertson & Daniel, 1989). Decomposition within the mangrove forest is found to have been expedited by tidal inundation, where 78% of the buried plant litters are consumed by crabs within 6 h, but often crabs are referred to as **'sloppy feeders'** as 20% of leaf materials is lost from the mandibles of the respective crabs onto the floor of the burrow (Camilleri, 1989). Ocypodids as deposit feeders primarily feed on

Fig. 5.1 Bioturbation by benthic fauna (1). (**A**) Sand pellets prepared by very small crabs, *Dotilla blanfordi*, under the family Ocypodidae, which create this very small sediment globules around their burrow openings through the continuous process of foraging for food particles from the sedi-

microalgae (France, 1998) or on bacteria (Dye & Lasiak, 1987), whereas sesarmid crabs consuming leaf litter has a distinct effect on litter dynamics in mangrove systems (Twilley et al., 1997; Lee, 1998). However, omnivorous crabs, as well as feeding primarily on algae (Woods, 1993), also take up other foods such as mollusks, polychaete worms, crustaceans, bacteria, fungi, and detritus, and this mixed diet of plant and animal matter promotes a more active life, greatest fitness, and fastest growth (Buck et al., 2003).

5.12.10 Bioturbatory Scores of Different Brachyuran Crabs

Bioturbatory scores are calculated to create a benchmark for assessing the bioturbation potential of different microbenthic fauna and also to interpret the outcomes of the bioturbatory activities of several macrobenthic fauna. With the help of these deducted scores, comparison of functional contributions among different benthic groups involving ten different species of intertidal brachyuran crabs belonging to two major families, **viz.** Ocypodidae (six species) and Grapsidae (four species), has shown maximum scores by *Ocypoda macrocera* (8) followed by *Dotilla blanfordi* (7), *Uca lactea annulipes* (7), *Sesarma (Chiromantes) bidens* (7), *Sesarma taeniolatum* (6), *Metopograpsus maculatus*(6), *Uca acuta acuta* (6), *Uca triangularis bengali* (4), *Dotillopsis brevitarsis* (3), and *Metaplax intermedia* (3).

Bioturbatory scores by 18 different polychaete species belonging to six major families have revealed almost equal bioturbatory potentialities ranging from 3 to 6, where only the species *Eteone ornata* displayed the maximum score, i.e., 6. Only one species, *Lingula anatina*, under the phylum Brachiopoda displayed moderate bioturbatory activities showing maximum bioturbatory scores of 6. All of these findings have clearly pointed out that different major groups of macrobenthic fauna displayed bioturbatory activities with varied intensities, which ultimately help to maintain the stability of the coastal ecosystem in West Bengal, India (Chatterjee, 2014; Chatterjee et al., 2014; Chakraborty, 2018).

ment with the help of mandibles and maxilla and also shaping the sediments in the form of small pellets mixed with their saliva. (**B**) Small mounds mainly composed of silt and clay are constructed by the tiniest crabs, *Ilyoplax gangetica*. (**C**, **D**, and **E**) Habitats of the robust ghost crab, *Ocypoda macrocera*, of the family Ocypodidae, each being an important bioturbator, digs burrows of about 91 cm in depth and thereby reshuffle the sediments, bringing lower ones to the top in the form of unconsolidated soils as sand mounds, and thereby pumps water and oxygen to the lowermost parts of the burrows, making the entire stratum of sediment to a depth of a few meters, suitable as the habitats of many meiobenthic fauna and benthic algae, which together offer a novel source of food for demersal fish. (**A**, **G** and **H**) Habitats of fiddler crab, *Uca annulipes annulipes,* where a huge amount of soil is withdrawn from the lower strata of the soil for the construction of burrows. (**I** and **J**) Bioturbation with the formation of innumerable burrows on the mudflats by the benthic amphipods and brachiopods respectively

Fig. 5.2 Different external sediment structures and markings exhibiting different levels of energy possessed by different macrobenthic fauna inhabiting the open intertidal zones as well as mangrove forest floors and the processes involved in such creative activities reshuffle textural compo-

5.12.11 Meiobenthic Fauna: Diversity and Distribution Toward Ecological Services

The taxa belonging to meiobenthic fauna include Sarcomastigophora, Ciliophora, Turbellaria, Cnidaria, Rotifera, Nemartina, Nematoda, Cladocera, Isopoda, Mystacocarida, Copepoda, Ostracoda, Amphipoda, Cumacea, Gastrotricha, Loricifera, Priapulida, Kinorhyncha, Polychaeta, Oligochaeta, Sipuncula, Tardigrada, Syncarida, Thermosbaenacea, Tanaidacea, Halacarida, Pycnogonida, Palpigradida, Bryozoa, Entoprocta, Brachiopoda, Aplacophora, Gastropoda, Bivalvia, Holothuroidea, and Tunicata. Meiobenthic fauna, distinguished from macrobenthos by their small sizes and by virtue of their sheer density and diversity, play important roles in maintaining ecological balance by way of a specific prey–predator relationship within their community. These faunal groups are exclusively important within any estuarine and marine system as they facilitate biomineralization processes, offer themselves as food for the fauna belonging to higher trophic levels, as excellent biotic components for pollution bio-monitoring. In addition, their large abundance attracts a good number of finfishes and shellfishes to visit the coastal belts in order to forage in the benthic habitats inhabited by meiobenthic fauna, especially in the intertidal and subtidal zones (Bertness & Ellison, 1987; Robertson, 1991; Twilley et al., 1992; Wolanski, 1992; Wolanski, 1995; Twilley & Chen, 1998; Caroline et al., 2009; Coull, 1999).

A recent taxonomic survey has documented about 60 species of meiobenthic fauna, mostly comprising foraminifera, nematodes, polychaetes, harpacticoids, copepods, etc., from the Midnapore coastal tracts, where all those faunal components displayed distinct seasonal variations in respect of their diversity and density, which also synchronized with fishery production vis-à-vis fish landings of the studied zones (Datta & Chakraborty, 2015). Thirty seven species of meiobenthic foraminifera (Ghosh et al., 2016) and more than 20 species of nematodes (Datta & Chakraborty, 2015) along with a number of polychaetes, oligochaetes, harpacticoids copepods, and Kinorhyncha, have been found to be present in this coastal belt (Datta & Chakraborty, 2015; Datta et al., 2016). It has been hypothesized that all

nents of sediments, redistribute sediments, pump atmospheric oxygen, and recharge different strata of the sediment with moisture to a certain depth. (**A**) Sand pellets around the burrow openings of a tiny brachyuran crab, *Dotilla blanfordi*. (**B**) Subsurface presence of small bivalves, *Donax cuneatus* over the ripples formed by sandy loamy clay soils. (**C**) Unconsolidated sandy sediment owing to the foraging activities of the ghost crab, *Ocypoda macrocera* (**D**). (**E**, **F**, and **G**) habitats of fiddler crabs. (**E**) *Uca annulipes annulipes*; (**F**) burrow openings of *Uca triangularis bengali*; (**G**) burrow opening of *Uca annulipes annulipes*, semi hardy moisture-enriched sediment at the interphase of intertidal and subtidal zones serves as an ideal habitat for the colonization of small polychaete species, their high abundance in turn restructures the sediment profiles of their habitats; (**J**): soft muddy sediment in the lower part of the intertidal zones harbored and ploughed by the tunnel-shaped burrows formed by a brachiopod species, *Lingula anatina*

Fig. 5.3 Documentation of bioturbatory structures along with structural configuration of different bioturbatory structures and burrows (depth, amount of excavated sediments, etc.) produced by different brachyuran crab species in the intertidal zones during low tides with the help of plaster of

those meiobenthic fauna demonstrated mutualistic dependence with fishery components advocating the prospect of utilizing these underlying scientific principles of such relationships for developing tools for better fisheries management (Heald, 1969; Wolanski et al., 2009). The presence of a considerable number of free-living marine nematodes in the gut of prawns reflects the food preference of such commercially important shellfishes (Datta & Chakraborty, 2015). Meiofaunal organisms prefer to graze on bacteria and microalgae (Alongi, 1988) or prey on juvenile macrofauna and other meiofauna (Watzin, 1985). They are used as novel sources of food for other macrobenthos and demersal fishes, but they hardly experience top–down control of their populations. Meiobenthic fauna help to make detritus available to macroconsumers (Tenore et al., 1977) and take dissolved OM, diatoms, bacteria, detritus, and protozoans as their own foods (Higgins & Thiel, 1988). Similarly, a total of 51 soil microarthropods belonging to different insect orders, viz. Collembola, Isoptera, Hymenoptera, Diptera, etc., have been recorded from the different parts of the Midnapore (East) coast where they play considerable functional roles in driving nutrient cycles in the estuarine–mangrove ecosystem (Dey et al., 2007, 2008, 2010).

Mangrove leaves are grazed upon predominantly by insects, but they are also a food source for various mangrove crabs. The intertidal belt of the tropical mangrove ecosystem coast promotes higher rates of decomposition by benthic fauna than the subtidal zone, as revealed by more accumulation of organic carbon (Aller, 1998; Alongi, 1988). The intertidal belts in the mangrove estuarine complex function like a sponge with the presence of a complex network of biogenic structures that trigger to-and-fro flows of interstitial water, vertical and lateral movement of nutrients and gases, coupled with tidal advection and drainage. The intensity of the breaking down of leaves can be measured by weight loss of the total amount of generated leaves due to animal feeding, physical fragmentation, microbial activity, and leaching. The biochemical constituents of mangrove detritus undergo profound changes during decomposition process, which involves three physical processes: fragmentation, leaching, and decay (Robertson & Blaber, 1992). Decomposition of mangrove litter to produce detritus and simultaneous remineralization of mangrove

Paris in the open air during low tide. (**G**) Vertical section of mudflat depicting the bioturbatory activities of the benthic brachiopod, *Lingula anatina*, at the confluence of Subarnarekha estuary with the Bay of Bengal at the junction of Midnapore (East) District of West Bengal with the border region of the state of Odisha, India (**A–G**). (**H**) Sand pyramid constructed by male *Ocypoda macrocera* at the entrance of their burrows, which serves for sexual signaling to attract females. This structure also reduces the entry of sunlight in order to retain moisture within the burrow for a longer time. (**I**) Semidomes constructed by fiddler crab, male *Uca lacteal annulipes* with mud balls at the burrow entrance acting as sexual signaling to attract females for mating, but these structures do not trigger aggression. (**J**) Bioturbatory structure as a chimney loosely constructed by the fiddler crab, *Uca acuta acuta*, with the help of mud balls where large gaps remained in the walls of the chimney through which wind flows in the burrows to ventilate the inner space and function for microhabitat regulation and also for burrow protection against wandering crabs litter-derived detritus govern nutrient dynamics within the mangrove forest as well as in offshore systems (Bertness & Ellison, 1987; Aller, 1994; Barbier & Cox, 2003; Awal, 2014; Twilley et al., 1997).

In mangrove ecosystems, macrobenthos along with meiobenthic fauna and several microarthropods through bioturbatory activities accelerate the microbial degradation rate of sediment OMs by primarily acting upon mangrove litter for their ready conversion into detritus (Dey et al., 2010). Viviparous propagules produced by different species of mangrove plants before their settlement within the soft-bottomed floors of mangrove forests and dispersal to invade into new eco-zones, a behavior that has been hypothesized to be an adaptation to their intertidal habits. These large energy-rich propagules attract grazers at the initial stage of their development by insects and after the abscission of grapsid crabs (Smith III, 1987).

5.12.12 Diversity and Succession of Microarthropods: Determinants of Decomposition from a Case Study

On Nayachar Island, an eco-restored deltaic island located at the south-eastern part of Midnapore (East) coast and just opposite the Haldia Industrial complex in the district of West Bengal, India, a long term study to assess the roles of microarthropods in nutrient cycling of an mangrove ecosystem has highlighted their gradual occurrence and succession in the process of decomposition of the organic plant litter. The ecological succession process in the experimental setups had shown the occurrence of one group of microarthropods in the decomposing plant litter of certain mangrove plant species and subsequent replacement of the same groups of microarthropods by another group in the course of onward stages of decomposition. The decomposition was started with the occurrence of a group of microarthropods, Collembola, which was subsequently substantiated by the gradual appearance and involvement as decomposers by Acarina, Coleoptera, Diptera, Isopoda, Hymenoptera, and others, resulting in the generation of the highest amount of nutrients after 6 months of decomposition of mangrove litter (Dey et al., 2010).

5.12.13 Ecological Services by Microarthropods by Way of Mangrove Litter Decomposition

The ecological succession in the occurrence and involvement of varied types of micro-arthropodal faunal components for the onward drive for the process of decomposition of a selected mangrove plant species, *Avicennia officinalis* (L), had revealed 45.45% of relative abundance (RA) of Collembola after 3 months of decomposition followed by RA of Acarina (36.66%) and RA of coleoptera (9.09%). After 6 months of decomposition, prior to the final formation of litter-derived nutrients, the population of Acarina constituted 37.5% of the total microarthropod population followed by Collembola (33.33%), Coleoptera (8.33%), Diptera (8.3%), and Isopoda (4.16%). After 9 months, the Collembola population was recorded as 19.09% followed by Isopoda (17.47%), Acarina (14.28%), Coleoptera (14.28%), and others (9.22%). After 12 months, the population of isopoda was 19.50% followed Hymenoptera (18.29%), Coleoptera (18.29%), Acarina (14.63%), Collembola (10.97%), and other microarthropods (7.31%).

The rate of decomposition of *Avicennia officinalis* (L) after 3 months was 34%, whereas after 6 months, it had reached 57%. After 9 and 12 months, these rates of decomposition were 62% and 70% respectively. After 3 months of decomposition, different physicochemical parameters were recorded, which were pH 7.3, organic carbon 4.3%, total nitrogen 1220 ppm, and available potassium 120 ppm. After 6 months, the recorded parameters were pH 7.1, organic carbon 5.8%, total nitrogen 1230 ppm, total phosphorus 96 ppm, and available potassium 120 ppm. After 9 months, the estimated parameters were pH 6.5, organic carbon 6%, total nitrogen 1230 ppm, total phosphorus 84 ppm, and available potassium 160 ppm. After 12 months, the parameters were pH 6.6, organic carbon 6%, total nitrogen 1220 ppm, total phosphorus 84 ppm, and with pH available potassium 120 ppm (Dey et al., 2010; Chakraborty, 2018). This case study conducted by the research team of first author has clearly highlighted that soil-inhabiting micropods render their roles as ecosystem service providers by virtue of their diversity, density, and ecological functionalities (Dey et al., 2008; Chakraborty, 2018).

Changing ecological conditions in the coastal environment in the general and mangrove estuarine complex of West Bengal, in particular, can be assessed with the diversity, seasonal and spatial distribution patterns, and biomass of macro-, meio- and microbenthic fauna, considering them as bioindicators, even for indicating global climate change (Pradhan & Chakraborty, 2008; Datta & Chakraborty, 2015). Not only in pollution monitoring, all of these benthic faunal groups, especially meiobenthic fauna, also play as key actors for rendering ecological services to maintain the ecological health of their habitats through bioturbation (organic decomposition, redistribution of organic material, nutrient cycling, oxygenation of the sediment), and an effective link in the food web. Benthic faunal groups stimulate bacterial growth by mechanically breaking down the detrital particles by way of producing slime trails and with the secretion of mucus, they excrete nutrients. Microbial mineralization of polycyclic aromatic hydrocarbons such as naphthalene through the top–down control of meiofauna, depending on their feeding and behavioral ecology, has already been proved using molecular tools such as restriction fragment length polymorphism. Therefore, smaller marine meiofaunal organisms such as free-living Nematoda, Gastrotricha, Ostracoda, Kinorhyncha, Foraminifera, Oligochaeta, and Nemertea, can be effectively utilized in translational and regenerative biological research (Datta & Chakraborty, 2015; Chakraborty, 2018).

5.12.14 Ecosystem Services in the Form of Carbon Sequestration in the Mangrove Ecosystem

Out of the many ecosystem services rendered by the mangrove ecosystem in respect of food and fuel resources, nursery grounds for fish, mammals, and other semi-terrestrial and aquatic fauna, depocenters for sediment, protection from coastal erosion, etc., their roles in carbon sequestration need a special mention (Alongi, 2009). However, mangrove forests have been disappearing at an alarming rate (50% of the world's mangrove forests over the past 50 years (Giri et al., 2010). Major causes of such a drastic reduction in mangrove coverage are urban development, aquaculture, mining, overexploitation of timber, fish, crustaceans, and shellfish. Coincidentally, poverty and higher human population density coupled with poverty in the socio-economic conditions around the mangrove ecosystem can partly explain the high annual deforestation rates of these tidal forests (1–3%) (Alongi, 2009).

Plant communities sequester carbon mostly from the atmosphere during the photosynthesis process and store it on converting it into organic forms as biomass. The storing process of carbon occurs in five pools, within the body of plants in the forest ecosystem, namely below ground biomass, above-ground biomass, leaf litter, dead wood, and in the forest floor within soil sediments. Forests act as both a sink and a source of CO_2 at the time of its conservation within biomass and destruction through different physiological processes respectively. The mangrove forest ecosystem is considered one of the leading carbon sinks in the tropics and they help profusely in the mitigation processes required for combating climate change. It has been estimated that the mangrove forest can sequester far more carbon per hectare than any other tropical forest with its ability to store a large amount of carbon (1000 tons per hectare) over a period of thousands of years (UNEP, 2014). Mangroves have been able to sequester four times the carbon per unit area in comparison with any other terrestrial forests in the tropics and this has necessitated the undertaking of all-round efforts to conserve mangroves on a global scale, which will potentially represent a meaningful strategy for mitigating atmospheric greenhouse-gas emissions (Twilley et al., 2018).

The average monetary value of mangroves along with adjoining estuaries with sea-grass belts has been estimated to be first in the world, even greater than the economic value of coral reefs, continental shelves, and the open sea (Berger et al., 2006).

5.12.14.1 Carbon Sequestration in Mangrove Forests

Carbon sequestration, the practice of removing CO_2 from the atmosphere and storing it in the plant biomass or in any other landscape in the earth holds great promise as a solution in combating and stopping unprecedented rates of climate change. Mangrove forests composed of only woody halophytes grow and flourish along the brackish water-dominated coastal environment in the world's subtropical and tropical regions. Although the mangrove forests tend to maintain a continuum of types in relation to ecological gradients mostly structured by temperature, precipitation, humidity, water circulation, and mixing of coastal water on local and regional scales, upstream–downstream variations in sediment load, salinity, precipitation, tides, waves, and influx of fresh water through river flow considerably determine their expanse and biomass across the upstream–downstream of a single estuary (Ellison, 2009). Mangroves play valuable roles in capturing and storing carbon to help to mitigate the impact of climate change (Alongi, 2009; Mills, 2011).

5.12.14.1.1 Production and Sequestration of Carbon in Mangrove Forests

The perennial mangrove vegetation, being an integral component of highly productive coastal forests, contributes a significant fraction of soil carbon to mangrove estuarine complexes (Kristensen et al., 2008) and such plant-derived carbon can play a crucial role in the evaluation process of the rates of net primary productivity of mangroves along with their associated plants, especially benthic microalgae, in terms of carbon production (Duarte et al., 2005).

Like other forests, mangrove plants vary in size and biomass with increasing age and the rates of biological production striking a balance between carbon production and respiration. Studies on the growth rates of mangrove trees over time reveal growth dynamics depicting different phases of growth, like plants of other forests, with an early rapid growth at the time of early establishment followed by colonization, but such growth patterns slow down with the decline in growth rate into maturity and senescence, representing an alternate succession state on the way to reaching the climactic stage (Clough et al., 1997; Fromard et al., 1998; Alongi, 2009).

5.12.14.1.2 Carbon Allocation for Storage and Functioning of the Ecosystem

Although it is difficult to have an accurate understanding on the actual allocation of carbon fixing within mangrove plants, which continuously produce new foliage, stem, branches, reproductive organs, and roots alongside creating storage reserves as a part of providing chemical defense, both the above-ground and below-ground parts of mangrove plants return almost 50% of the total CO_2 assimilated by mangroves through respiration (Clough et al., 1997). However, several ecological factors such as temperature, light intensity, water availability, nutrients, tidal amplitude, intensity of water currents, height of waves, and species composition of the mangrove forest tend to determine the proportional allocation of fixed carbon within trees, which vary from one forest tract to another (Clough et al., 1997). A number of research studies have pointed out that below-ground carbon assimilation is more through pools of roots, with the higher ability of nutrient-conserving mechanisms and soil carbon (Saenger, 2002; Lovelock, 2008).

An estimate has focused on the high potential of mangroves among the most carbon-rich biomes of the world, containing an average of 937 tonnes of carbon/

hectare, facilitating the rapid rates of accumulation, accretion, and burial of carbons within sediment. Mangroves account for 14% of carbon sequestration out of all coastal habitats of the global ocean (Alongi, 2009).

After being accumulated within mangroves, carbon increases the rates by mass sediment accumulation and also follows other pathways such as consumption by living organisms, mostly microbes, remineralization of consumed carbons, and emission back to the atmosphere as CO_2 or removed by dissolved inorganic carbon.

A highly variable proportion of this carbon is mangrove-derived OM brought in by the tides from adjacent seagrass meadows, coral reefs, macroalgae, rivers, and from land-based sources, as well as other marine environments (Kristensen et al., 2008).

Mangroves accumulate carbon in tree biomass, but much of this carbon is eventually lost in the short and medium term by way of clear-cutting and human use, decomposition, and export to adjacent ecosystems. Over the long term, carbon stored primarily below ground as soil carbon is released into the adjoining estuarine ecosystem by soil accretion at the fringing mangrove stands at the sea–forest interface and it also accumulates at intertidal mudflats (Sanders et al., 2010).

5.13 Biochemical Ecology: Ecological Services by Way of the Biotransformation of Nutrients

Different physical processes operating in the mangrove–estuarine–coastal ecosystem of West Bengal, India, such as tides with seasonal variation in amplitudes, run off, longshore currents, sea-level rise with increasing temperature, precipitation, and biological factors such as biotic community interactions, trophic levels, litter fall, decomposition and mineralization, drive nutrient transports and exchange of materials in between forest and aquatic subsystems. The decomposed products, as detritus formed from the decomposition of vascular plant biomass and leaves by both macro- and micro-decomposers (bacteria, fungi, mangrove dwelling faunal components), provide a primary food supply for aquatic and intertidal faunal components (Chakraborty, 2011).

The pathways of conversion of different biochemical constituents through the ecosystem functioning in the mangrove–estuarine system follow specific ecological principles. Research studies conducted in the coastal Midnapore (East) district, India, to understand the mode of bioconversion and biotransformation of lipids, have shown the following pathway:

$$\begin{aligned}&\text{Mangroves} \rightarrow \text{Soil} \rightarrow \text{Water} \rightarrow \text{Macro}\\&\rightarrow \text{decomposers} \rightarrow \text{Micro} - \text{decomposers} \rightarrow \text{Detritus}\\&\rightarrow \text{Detritivores}(\text{Benthos}) \rightarrow \text{Detritus Fauna}\end{aligned}$$

Lipids from both plants and animals are classified into three major types as neutral-, glycol-, and phospholipids (Gurr & James, 1980). The various main classes of the neutral lipids are subcategorized as triacylglycerols or triglycerides, hydrocarbons, wax esters, stearyl esters, and sterols. In addition, some vitamins and terpenoids, after being incorporated, remain present in naturally occurring lipids. Lipids belonging to a heterogeneous class of predominantly nonpolar compounds, characterized by their properties of nonsolubility in water, become soluble in nonpolar organic solvents such as benzene, chloroform, etc. As esters, mostly with fatty acids, and different alcohols, also contain other constituents in many cases. They are compounds constituted mainly by carbon, hydrogen, and oxygen, but also in some cases possess phosphorus, nitrogen, and sulfur. Although the term lipid is sometimes used as a synonym for fat, sub-groups of lipids, called triglycerides also occur. The major biochemical constituents of lipids at molecular levels are fatty acids and their derivatives (including tri-, di-, monoglycerides, and phospholipids), along with other sterol-containing metabolites such as cholesterol. Several biosynthetic pathways exist to break down and synthesize lipids in humans and other mammals but some essential lipids do not follow this process.

Alongside playing vital physiological roles, lipids, being a major biochemical constituent in the physiological functioning of living organisms, also contributing profusely to the flow of organic materials through food chains. Based on some research studies undertaken so far in the Sundarbans mangrove forest, mention can be made of the fractional components of fatty acids and sterols of mangrove leaves (Misra et al., 1983), triterpenoids and sterols of mangrove plant leaves (Ghosh et al., 1985), fatty acids of the detritivores *Boleophthalmus boddarti* (Banerjee et al., 1997), and the biotransformation process of oleanonic acid of mangrove leaves.

Studies on three different macrobenthic fauna (fiddler crab, brachiopod, and nemertean) have revealed the mode of occurrence and biochemical properties of different fractional components of lipids and fatty acids in different body organs of the faunal components studied and the mode of bioconversion of such biochemical entities in the coastal ecosystem. The ultimate products from such bioconversion and their steady biotransformation process ultimately enrich the ecosystem with a large amount of decomposition products through such bio-geo-chemical processes, which render biodiversity components and sustain the prevailing trophic relationships (Chakraborty, 2013).

5.13.1 Bioconversion and Biotransformation of Fatty Acids by an Intertidal Macrobenthos (Brachiopod), **Lingula anatina**

Fatty acids as qualitative markers have been in use to trace or assess predator–prey relationships in the marine estuarine ecosystem for more than 30 years. Moreover, fatty acids are being used as a tool to study trophic ecology and to have an understanding of food-web connections, in contrast to more dependence on traditional gut content analysis methods, which provide information on dietary intake, food constituents, and sequestration processes of lipid reserves over a longer period of time (Auel et al., 2002).

Fatty-acid composition of the organisms representing different trophic levels in marine ecosystems show generally qualitative similarities where the first trophic level in the food chain is occupied by phytoplankton, which are able to synthesize all kinds of fatty acids de novo and the composition of those fatty acids tends to change significantly in the decomposing leaves of mangroves (Bligh & Dyer, 1959; Moreno et al., 1979; Alikunhi et al., 2010). A high concentration of ω3 fatty acid, a unique marine life-based fatty acid, was contributed by phytoplanktonic species such as dinoflagellates, diatoms, etc., to the marine ecosystem. Examination on various organs of *Lingula anatina*, an intertidal detritivorous macrobenthic species, revealed the presence of appreciably high amounts of 20:5ω3 fatty acids, which remain associated with other polyunsaturated fatty acids of the –ω3 series.

This observation has pointed out the intake of a higher amount of the precursor fatty acid (18:3ω3), through primary food sources and their subsequent chain elongations, coupled with desaturation processes, lead de novo to the formation of –ω3 unsaturated fatty acids at higher levels. Intake of this kind of diet enriched with –ω3 acids, explains the mode of accumulation of these polyunsaturates in considerable amounts in the different organs of the studied *Lingula* brachiopod species. α-linolenic acid (ALA 18: 3ω3), the primary precursor molecule for the synthesis of the –ω3 family of fatty acids in animal tissues, is supposed to derive from the diet.

The gut-content analysis to assess the food contents of *L. anatina* have revealed the presence of fragmented mangrove leaves, detritus, and planktonic body parts, which corroborates the previous findings of Emig (1997). The occurrence of eicosapentaenoic acid (EPA) and docosahexaenoic acid (DHA) within different body organs of *L. anatina* has substantiated the very observation that these pharmaceutically significant fatty acids are supposed to have been derived from ALA, which are present in dietary food sources of *L. anatina*, where these fatty acids are included within the foods through biotransformation processes within the body of this rare benthic brachiopod faunal component. The presence of moderate to high levels of EPA and DHA, derived through bioconversion of ALA from food sources within different body parts of the studied species, has pointed out their contribution in the formation of EPA and DHA, which are being considered as the precursors of several metabolites that act as potential lipid mediators. All of those fractions of fatty acids are considered beneficial ingredients for humans, which helps in the prevention or treatment of several diseases (Serhen et al., 2008; Samanta et al., 2014a, 2014b, 2014c, 2015b). Different fractions of fatty acids (EPA and DHA) in the studied brachiopod species *L. anatina* are derived from the fatty acids of the planktons, which are used a food source. In addition, another food source, semi-decomposed products of mangrove leaves, also possess moderate to high amounts of ALA (18:3ω3), which is considered the precursor of long-chain polyunsaturated fatty acids (PUFAs), viz. EPA (20:5ω3) and DHA (22:6ω3) (Samanta et al., 2014a, 2015a, 2015b).

5.13.2 *Bioconversion and Biotransformation of Fatty Acids by an Intertidal Macrobenthos (Fiddler Crab),* **Uca acuta acuta** *(Tables 23.18 and 23.19)*

A recent study on the bioconversion and biotransformation of fatty acids by an intertidal macrobenthos species (fiddler crab: *Uca acuta acuta*) have revealed the occurrence of different fractional fatty-acid components, which are mostly derived from the decomposition product of mangrove leaves on which this brachyuran crab species grazes and thereby consumes a considerable amount of ALA (18:3ω3) (Heald, 1969; Heald & Odum, 1970; Chatterjee et al., 2014). Analysing the lipid and fatty-acid compositions of some organs, viz. body flesh, big chela flesh, *Hepatopancreas*, and gut contents of the studied crab species, *Uca acuta acuta*, feeding mostly on the detritus of the Midnapore (East) coastal tract, West Bengal, India, the following inferences are made:

1. Occurrence of high levels of ALA (18:3ω3) in mangrove plant leaves of the study area, which are recognized as precursors of long-chain PUFAs, viz., EPA (20:5ω3) and DHA (22:6ω3).
2. As a bio-energetically significant macrobenthic faunal component feeding on the leaf detritus, *Uca acuta acuta* consumes a considerable amount of ALA (18:3ω3).
3. The ALA thus consumed by this abundant grazing benthic detritivore inhabiting the estuarine mud flats, is converted in vivo into long-chain PUFAs of the –ω3 series, viz. EPA and DHA.
4. On biochemical analysis, considerably higher amounts of EPA and DHA have been found to occur in the various body organs of *Uca acuta acuta*, particularly in the body flesh of that species.
5. It has been established that the fiddler crab species, *Uca acuta acuta*, has the ability to biosynthesize long-chain PUFAs efficiently (Das et al., 2014a, 2014b).

5.13.3 *Bioconversion and Biotransformation of Total Lipids in the Muscles of an Intertidal Benthic Nemertean,* **Cerebratulus** *sp. Along with Its Primary Food Sources*

Studies on the fractional components of lipids and fatty acids in the intertidal macrobenthic nemertean, *Cerebratulus* sp., along with its primary food sources, have revealed the occurrence of an appreciable amount of ALA in its food components. Nine different types of MUFAs and 12 different types of PUFAs have been found to occur in the muscles of *Cerebratulus* sp. Among the PUFAs, arachidonic acid (AA, 20:4ω6) was estimated to constitute the highest amount (5.00%) followed by eicosatrienoic acid (ETA, 20:3ω3;4.48%), docosahexaenoic acid (DHA, 22:6ω3: 2.89%), etc. However, no such occurrences of AA, ETA, and DHA were noted in the mangrove leaves. The quantity of linoleic acid (LA, 18:2ω6) ranged from moderate to higher amounts (1.84–49.50%) in different studied samples.

An appreciably higher amount of 20:5ω3 fatty acids along with other polyunsaturated fatty acids of the –ω3series have been estimated in muscles of *Cerebratulus* sp. Intake of a higher amount of the precursor acid (18:3ω3) through primary food sources and their subsequent chain elongations and desaturation processes tends to facilitate the formation of –ω3 unsaturated fatty acids at higher trophic levels. This kind of diet enriched with –ω3 acids accounts for the accumulation of these PUFAs in variable amounts in the body organs, especially in the muscles of the studied nemertean species. Lower values of the PUFA/SAFA ratio, as found from the present research investigation, is supposed to be due to the presence of higher levels of palmitic acid (16:0), pointing out a probable contribution of vegetal detritus in the diet of *Cerebratulus* sp. PUFAs enriched with Polyunsaturated fatty acids (PUFA).

In the present study, the occurrence of higher amounts of 18:2ω6 and 18:3ω3 is also thought to be due to the consumption of a considerable amount of green algae (phytoplankton) occurring over the surface of the soil, which predominantly comprise 18:2ω6 and 18:3ω3, very similar to those of terrestrial vascular plants, and also by getting a supply of those ingredients from neighboring mangrove vegetations. In the present study, recording of odd branched fatty acids can be treated as an indicator of bacterial derivative, which highlights a source of food supply for *Cerebratulus* from decaying OM. The presence of the highest amount of AA (20:4ω6) in the muscles of the studied detritivorous species further established the roles of detritus in providing *Cerebratulus* with the precursor biochemical compounds. The presence of the moderate to higher amounts of estimated EPA and DHA within the muscles of the studied species are thought to have been derived from the bioconversion of ALA, which remains present in the food of the nemertean species and serve as the good sources of EPA and DHA, which in turn act as precursors of several metabolites that are potent lipid mediators. All of these fractional components of lipids and fatty acids are recognized to be beneficial biochemical ingredients for humans and can help in the prevention or treatment of several diseases (Serhan et al., 2008).

5.14 Ecosystem Services of the Mangrove–Estuarine Complex in View of Ongoing Global Environmental Perturbations with Special Reference to West Bengal, India

Estuarine and coastal ecosystems (ECEs) represent one of the most heavily used and also threatened natural ecosystem systems of the world (Worm et al., 2006; Halpern et al., 2008). Deterioration of the ecological health of this sensitive ecosystem has been magnified during the last few decades owing to intense anthropogenic activities and the resulting degradation and loss of 50% of salt marshes, 35% of mangroves, 30% of coral reefs, and 29% of seagrasses worldwide (Valiela et al., 2001). At least three major critical ecosystem services have been affected by the global decrease in ECEs, which are (1) decline of a number of viable fisheries by up to 33%; (2) disruption of the provision of nursery habitats of several coastal biotic communities, such as oyster reefs, seagrass beds, and wetlands, with a decline in production of up to 69%, and (3) a decline in the capabilities of submerged vegetation and suspension feeders to provide filtering and detoxification services by up to 63% (Worm et al., 2006). Eco-degradation in ECEs in respect of ecosystem functions, loss of biodiversity, and reduced growth of coastal vegetation may have contributed to paving the way for biological invasions, deterioration of water quality, and disruption of coastal protection from flooding and other natural disasters. Evaluation of some ecosystem services in ECEs for habitat-fishery linkages, coastal protection, control of erosion, and others, requires a proper understanding of the underlying ecological principles for the ecological functions underlying these services, which vary spatially and temporally. In such a context, enhancing the immediate and longer-term eco-management plans for protecting coastal belts to ensure ecosystem services placing emphasis on the connectivity of ECEs across land–sea gradients, which tend to influence and open up provision for other ecosystem services through synergistic effects.

Sustainable ecosystem management can be possible by adhering rigidly to basic operating principles for ecosystem functioning and also taking into consideration understanding of the mode of human activities with regard to ecosystems. Such eco-management encompasses both economic and ecological perspectives for the evaluation of ecosystem goods and services and also requires identification of the values for undertaking the protection measures and assessing the economic costs or gains of preservation. People's perceptions spanning several decades are now also being utilized, not only to understand the trend of change of biodiversity but also to ensure proper conservation strategies (Patra et al., 2005; Mishra et al., 2009).

Intertidal benthic faunal diversity of coastal and estuarine environments, render valuable ecosystem services but their habitats are subject to a wide range of stresses associated with changing patterns of resource use and direct exploitation (Chakraborty, 1998; Qasim et al., 1988). Gaining a better understanding of the relationships between the diversity of mangrove biota and ecosystem processes in shallow coastal habitats in terms of functional relationships between organisms and fluxes of energy or matter, spatial and temporal variability in the distribution of species, and transport and transformation of chemicals within the sediment and across the sediment–water interface can help proper natural resource management strategies (Dye & Lasiak, 1986; Solan et al., 2004). A strong feedback loop exists among food, shelter having specific physical and chemical entities, nonlinear forms of interactions between bioturbation, and the mineralization processes in sediments help benthic fauna to play the role of sustaining the balance in the ecosystem functioning but also to ensure livelihood generation by way of ecological services (Huhta, 2007; Gray & Elliott, 2009).

The intertidal benthic fauna on the Midnapore (East) coast, West Bengal, India, are seen to render valuable ecological services by virtue of their trophic position and functional roles in the food chains and food-webs of the mangrove estuarine coastal wetlands. These services enrich biodiversity by stabilization of the sediments alongside facilitating the development of conducive ecological conditions for the settlement and flourishing of biota such as algae, microbes (bacteria and fungi), meiobenthos, etc., through the process of pumping of sufficient air and water, forming innumerable pores, burrows, tubes, etc., reshuffling sediment texture by bioturbation activities, promoting nutrient cycling, releasing millions of larvae in the form of meroplanktons into the aquatic subsystem and thereby providing the fish with live food. In addition, biotransformation of organic chemicals into inorganic and vice versa, also through the feedback systems, help the sustenance of the coastal ecosystem by the combined activities of millions of individual living organisms of the coastal environment, which in turn tend to contribute in a holistic manner to the growth and propagation of mangroves, resisting bioinvasion, and bioaccumulation of toxic substances also as ecosystem services.

The mangrove ecosystem, alongside providing clean air and water, also renders ecosystem services by maintaining biodiversity, enhancing carbon sequestration, arresting soil erosion, supplying ground freshwater, filtering pollutants, protecting the coast from floods, cyclones, and other natural disasters, reducing saltwater intrusion onto agricultural land, recycling nutrients, and regulating water flows and supplies.

References

Abele, L. G. (1974). Species diversity of decapod crustaceans in marine habitats. *Ecology, 55*, 156–161.

Alberti, J., Daleo, P., Iribarne, O., Silliman, B. R., & Bertness, M. (2007). Local and geographic variation in grazing intensity by herbivorous crabs in SW Atlantic salt marshes. *Marine Ecology Progress Series, 349*, 235–243.

Alikunhi, N. M., Narayanamsamy, R., & Kandasamy, K. (2010). Fatty acids in an estuarine mangrove ecosystem. *Revista de Biología Tropical, 58*(2), 577–587.

Aller, R. C. (1982). The effects of macrobenthos on chemical properties of marine sediment andoverlying water. In P. L. McCall & M. J. S. Tevez (Eds.), *Animal–sediment relations*. Plenum Press.

Aller, R. C. (1988). Benthic fauna and biogeochemical processes in marine sediments: the role of burrow structures. In T. H. Blackburn & J. Sorensen (Eds.), *Nitrogen cycling in coastal marine environments* (pp. 301–338). John Wiley & Sons.

Aller, R. C. (1994). Bioturbation and remineralization of sedimentary organic matter: effects of redox oscillation. *Chemical Geology, 114*, 331–345.

Aller, R. C. (2001). Transport and reactions in the bioirrigated zone. In B. P. Boudreau & B. B. Jørgensen (Eds.), *The benthic boundary layer: Transport processes and biogeochemistry*. Oxford University Press.

Aller, R. C., & Aller, J. Y. (1998). The effect of biogenic irrigation intensity and solute exchange on diagenetic reaction rates in marine sediments. *Journal of Marine Research, 56*, 905–936.

Alongi, D. M. (1987). Intertidal zonation and seasonality of meiobenthos in tropical mangrove estuaries. *Marine Biology, 95*, 447–458.

Alongi, D. M. (1988). Bacterial productivity and microbial biomass in tropical mangrove sediments. *Microbial Ecology, 15*, 59–79.

Alongi, D. M. (1990). The ecology of tropical soft–bottom benthic ecosystems. *Oceanography and Marine Biology: An Annual Review, 28*, 381–496.

Alongi, D. M. (2002). Present state and future of the world's mangrove forests. *Environmental Conservation, 29*, 331–349.

Alongi, D. M. (2008). Mangrove forests: Resilience, protection from tsunamis, and responses to global climate change. *Estuarine, Coastal and Shelf Science, 76*, 1–13.

Alongi, D. M. (2009). *The energetics of mangrove forests*. Springer.

Alongi, D. M., & Sasekumar, A. (1992). Benthic communities. In A. I. Robertson & D. M. Alongi (Eds.), *Tropical mangrove ecosystems* (pp. 137–171). American Geophysical Union.

Anon. (2003). Fisheries in irrigation systems of arid Asia. In *FAO fisheries technical paper No. 430* (p. 150). FAO.

Auel, H., Harjes, M., da Rocha, R., et al. (2002). Lipid biomarkers indicate different ecological niches and trophic relationships of the Arctic hyperiid amphipods Themisto abyssorum and T. libellula. *Polar Biology, 25*, 374–383.

Awal, M. A. (2014). Analysis of causes of disease in Sundarbans natural mangrove. *American Journal of Bioscience and Bioengineering, 2*, 18–32.

Balmford, A., Bruner, A., & Cooper, P. (2002). Ecology–economic reasons for conserving wild nature. *Science, 297*, 950–953.

Banerjee, D., Pal, D., Patra, T. K., Misra, S., & Ghosh, A. (1997). Lipids and fattyacidsof air breathing fish, *Boleophthalmus boddaerti. Food Chemistry, 60*(3), 303–309.

Barbier, E. B. (2007). Valuing ecosystem services as productive inputs. *Economic Policy, 22*, 177–229.

Barbier, E. B., & Cox, M. (2003). Does economic development lead to mangrove loss? A cross–country analysis. *Contemporary Economic Policy, 21*, 418–432.

Barrett, S. (1994). The biodiversity super game. *Environmental and Resource Economics, 4*, 111–122.

Barrett, S. (2003). *Environment and statecraft: The strategy of environmental treaty–making*. Oxford University Press.

Beaumont, N. J., Austen, M. C., Atkins, J. P., Burdon, D., Degraer, S., Dentinho, T. P., Derous, S., Holm, P., Horton, T., van Ierland, E., Marboe, A. H., Starkey, D. J., Townsend, M., & Zarzycki, T. (2007). Identification, definition and quantification of goods and services provided by marine biodiversity:implications for the ecosystem approach. *Marine Pollution Bulletin, 54*, 253–265.

Beck, M. W., Heck, K. L., Jr., Able, K. W., Childers, D. L., Eggleston, D. B., Gillanders, B. M., et al. (2001). The identification, conservation and management of estuarine and marine nurseries for fish and invertebrates. *Bioscience, 51*, 633–641. https://doi.org/10.1641/0006-3568

Behrenfeld, M. J., O'Malley, R. T., & Siegel, D. A. (2006). Climate–driven trends in contemporary ocean productivity. *Nature, 444*, 752–755.

Beinlich, B., & Von Hagen, H. O. (2006). Materials for a more stable subdivision of the genus *Uca* leach. *Zoologische Mededelingen, 80*, 9–32.

Berger, U., Adams, M., Grimm, V., et al. (2006). Modelling secondary succession of subtropical mangroves: causes and consequences of growth reduction in pioneer species. *Perspectives in Plant Ecology, Evolution and Systematics, 7*, 243–252.

Bertness, M. D. (1985). Fiddler crab regulation of *Spartina alterniflora* production on a New England salt marsh. *Ecology, 66*, 1042–1055.

Bertness, M. D., & Ellison, A. M. (1987). Determinants of pattern in a New England salt marsh plant community. *Ecological Monographs, 57*, 129–147.

Bligh, E. G., & Dyer, W. J. (1959). A rapid method for total lipid extraction and purification. *Canadian Journal of Biochemistry and Physiology, 37*, 911–917.

Botto, F., & Iribarne, O. (2000). Contrasting effects of two burrowing crabs (*Chasmagnathus granulata* and *Uca uruguayensis*) on sediment composition and transport in estuarine environments. *Estuarine, Coastal and Shelf Science, 51*, 141–151.

Botto, F., Valiela, I., Iribarne, O., Martinetto, P., & Alberti, J. (2005). Impact of burrowing crabs on N and N sources, control, and transformations in sediments and food webs of SW Atlantic estuaries. *Marine Ecology Progress Series, 293*, 155–164.

Botto, F., Iribarne, G. J., Bava, J., Gagliardini, A., & Valiela, I. (2006). Ecological importance of passive deposition of organic matter into burrows of the SW Atlantic crab *Chasmagnathus granulatus*. *Marine Ecology Progress Series, 312*, 201–210.

Butchart, S., Walpole, M., Collen, B., et al. (2010). Global biodiversity: Indicators of recent declines. *Science, 328*, 1164–1168.

Bouillon, S., Borges, A. V., Castañeda-Moya, E., Diele, K., Dittmar, T., Duke, N. C., Kristensen, E., Lee, S. Y., Marchand, C., Middelburg, J. J., Rivera-Monroy, V. H., Smith, T. J., III, & Twilley, R. R. (2008). Mangrove production and carbon sinks: A revision of global budget and estimates. *Global Biogeochemical Cycles, 22*, GB2013. https://doi.org/10.1029/2007GB003052

Branchley, G. A. (1982). Mechanisms of spatial competition in marine soft-bottom communities. *Journal of Experimental Marine Biology and Ecology, 60*, 17–33.

Buck, T. L., Breed, G. A., Pennings, S. C., Chase, M. E., Zimmer, M., & Carefoot, T. H. (2003). Diet choice in an omnivorous salt–marsh crab: different food types, body size, and habitat complexity. *Journal of Experimental Marine Biology and Ecology, 292*(1), 103–116.

Cadee, G. C. (2001). Sedimentary dynamics by bioturbating organisms. In K. Reise (Ed.), *Ecological comparisons of sedimentary shores* (pp. 127–148). Springer.

Camilleri, J. C. (1989). Leaf choice by crustaceans in a mangrove forest in Queensland. *Marine Biology, 102*, 453–459.

Camilleri, J. C. (1992). Leaf–litter processing by invertebrates in a mangrove forest in Queensland. *Marine Biology, 114*, 139–145.

Caroline, N., Tarik, M., Govrich, P., Hervé Lee, A., & Yip, S. (2009). Role of grapsid crabs, *Parasesarma erythrodactyla*, in entry of mangrove leaves into an estuarine food web: A mesocosm study. *Marine Biology, 156*, 2343–2352.

Carter, R. W. G. (1988). *Coastal environments: an introduction to the physical, ecological and cultural systems of coastlines*. Academic Press.

Chakraborty, S. K. (1996). Aquaculture potential of mangrove ecosystem of Sundarbans, India. In *Proceedings of seminar on fisheries—A multimillion dollar industry* (pp. 72–83).

Chakraborty, S. K. (1998). Coastal aquaculture and environment in the context of West Bengal, India. *Vidyasagar University Journal of Biological Science, 4*, 13–20.

Chakraborty, S. K. (2003). Biodiversity and conservation. In A. D. Mukhopadhyay (Ed.), *Perspectives and issues in environmental studies* (pp. 267–286). Vidyasagar University.

Chakraborty, S. K. (2009). Biological diversity of Midnapore coast, West Bengal. *Environmental Forensics, 2*(2), 6.

Chakraborty, S. K. (2010). Coastal environment of Midnapore, West Bengal: potential threats and management. *Journal of Coastal Environment, 1*(1), 28–40.

Chakraborty, S. K. (2011). Mangrove ecosystem of Sundarbans, India: Biodiversity, ecology, threats and conservation. In J. N. Metras (Ed.), *Mangroves: Ecology, biology and taxonomy* (pp. 83–112). NOVA Publishers.

Chakraborty, S. K. (2013). Interactions of environmental variables determining the biodiversity of coastal mangrove ecosystem of West Bengal, India. *The Ecoscan Special Issue, 3*, 251–265.

Chakraborty, S. K. (2017). Ecological services on intertidal benthic fauna and the sustenance of coastal wetlands along the Midnapore (East) coast, West Bengal, India. In *Coastal wetlands: Alteration and remediation* (pp. 777–886). Springer, Nature.

Chakraborty, S. K. (2018). Bioinvasion and environmental perturbation: Synergistic impact on coastal—Mangrove ecosystems of West Bengal, India. In C. Makowski & C. Finkl (Eds.), *Impacts of invasive species on coastal environments: Coast in crisis* (pp. 171–245). Springer, Nature.

Chakraborty, S. K. (2021a). *Riverine ecology (Vol. 1): Ecofunctionality of the physical environment of rivers resolution*. Springer Nature. 917 pages.

Chakraborty, S. K. (2021b). *Riverine ecology (Vol. 2): Biodiversity conservation, conflicts and resolution*. Springer Nature. 566 pages.

Chakraborty, S. K., & Choudhury, A. (1985). Distribution of fiddler crabs in Sundarbans mangrove estuarine complex, India. In *The mangroves. Proceedings of National Symposium Biology utilization conservation of mangroves* (pp. 467–472).

Chakraborty, S. K., & Choudhury, A. (1989). Ecological studies on the zonation of brachyuran crabs in a virgin mangrove Island of Sundarbans, India. *Journal of Marine Biological Association, India, 34*(1 & 2), 189–194.

Chakraborty, S. K., & Choudhury, A. (1992). Population ecology of fiddler crabs (*Uca.* spp.) of the mangrove estuarine complex of Sundarbans, India. *Tropical Ecology, 33*(1), 78–88.

Chakraborty, S. K., & Choudhury, A. (1993). Community organization of macrobenthic molluscs of Sagar Island, Sundarbans, India. *Indian Biologist, 25*(1), 33–36.

Chakraborty, S. K., & Choudhury, A. (1994). Community structure of macrobenthic polychaetes of intertidal region of Sagar Island, Hooghly estuary, Sundarbans, India. *Tropical Ecology, 35*(1), 97–104.

Chakraborty, S. K., & Choudhury, A. (1994). Community structure of macrobenthic polychaetes of intertidal region of Sagar Island, Hooghly Estuary, Sundarbans, India. *Tropical Ecology, 35*(1), 97–104.

Chakraborty, S. K., Chatterjee, S., Dey, M. K., et al. (2012). Ecological changes of an eco–rehabilitated delta of Hooghly estuary. *Journal of Coastal Environment, 3*(1), 9–27.

Chakraborty, S. K., Choudhury, A., & Deb, M. (1986). Decapod Brachyura from Sundarbans mangrove estuarine complex, India. *Journal of the Bengal Natural History Society, 5*(1), 55–68.

Chakraborty, S. K., Giri, S., Chakravarty, G., & Bhattacharya, N. (2009). Impact of eco-restoration on the biodiversity of Sundarbans mangrove ecosystem, India. *Water Air Soil Pollution: Focus, 9*, 303–320.

Chakraborty, S. K., Dutta, S. M., Ghosh, P. B., Ray, R., & Paul, A. K. (2014). Impact of global warming on Sundarbans mangrove ecosystem, India: Role of different assessment tools from ecosystem monitoring to molecular markers. In *Proceedings of the International conference on Green India: Strategic knowledge for combating climate change: prospects and challenges* (pp. 181–200).

Chakraborty, S. K., Paul, A. K., Bandyopadhyay, J., Banerjee, D., Pakhira, H., Middya, S., Jana, S., Sahoo, P., Paria, K., & Bera, K. (2013). Sustainable water management in the drought prone riverine tracts of South West Bengal. *Journal of Ground Water Studies, 1*, 11–34.

Chakraborty, S. K., Sanyal, P., & Ray, R. (2023). *Wetlands ecology: Eco-biological uniqueness of a Ramsar site (East Kolkata Wetlands, India)*. Springer Nature. 709 pages.

Chakravarty, G., Giri, S., Giri, S., & Chakraborty, S. K. (2004). Distribution of coliform bacteria in different estuaries of West Bengal coast, India. In A. Kumar, J. R. Dhanze, A. K. Chopra, & D. S. Malik (Eds.), *Protected habitats and biodiversity* (pp. 395–402). Natcon Publication.

Chandra, A. K., Chakravarty, G., & Chakraborty, S. K. (2003). Polychaete diversity of Midnapore Coastal Belt, West Bengal, India. *Journal of Nature Conservators, 15*(2), 95–108.

Chapin, F. S., Walker, B. H., Hobbs, R. J., Hopper, D. U., Lawton, J. H., Sala, O. E., & Tilman, D. (1997). Biotic control over the functioning of ecosystems. *Science, 277*, 277–500.

Chatterjee, S., & Chakraborty, S. K. (2014). Feeding behaviour and functional role of some selected species of Brachyuran crabs in nutrient cycle of coastal belt of Midnapore (East), West Bengal, India. *Journal of Biology and Life Science, 5*(1), 106–129.

Chatterjee, S., & Chakraborty, S. K. (2014). Feeding behaviour and functional role of some selected species of brachyuran crabs in nutrient cycle of coastal belt of Midnapore (east), West Bengal, India. *Journal of Biology and Life Science, 5*(1), 106–129.

Chatterjee, S., Bhunia, G., & Chakraborty, S. K. (2008). Bioturbation of brachyuran crabs and its impact on coastal ecosystem of Midnapore district, West Bengal, India. In *Proceedings of Fourth International Conference on Environmental Science and Technology*.

Chatterjee, S., Das, T. K., & Chakraborty, S. K. (2014). Reproductive biology and bioturbatory activities of two sympatric species of fiddler crabs *Uca lactea annulipes* and *Uca triangularis bengali* (Decopoda:Ocypodidae) at the East Midnapore Coastal belt of West Bengal, India. *Journal of Biology and Life Science, 5*(2), 106–126.

Chatterjee, S., Das, T. K., & Chakraborty, S. K. (2014). Reproductive biology and bioturbatory activities of two sympatric species of fiddler crabs *Uca lactea annulipes* and *Uca triangula-*

ris bengali (Decopoda:Ocypodidae) at the east Midnapore coastal belt of West Bengal, India. *Journal of Biology and Life Science, 5*(2), 106–126.

Chaudhuri, A. B., & Choudhury, A. (1994). *Mangroves of the Sundarbans, India* (Vol. 1, pp. 1–247). IUCN.

Chmura, G. L. (2009). What do we need to assess the sustainability of the tidal salt marsh carbon sink? *Ocean and Coastal Management, 83*, 25–31. https://doi.org/10.1016/j.ocecoaman.2011.09.006

Chmura, G. L. (2013). What do we need to assess the sustainability of the tidal salt marsh carbon sink? *Ocean & Coastal Management, 83*, 25–31. https://doi.org/10.1016/j.ocecoaman.2011.09.006

Chmura, G. L., Anisfeld, S. C., Cahoon, D. R., & Lynch, J. C. (2003). Global carbon sequestration in tidal, saline wetland soils. *Global Biogeochemical Cycles, 17*, 1111.

Chowdhury, M. S. N., Hossain, M. S., Das, N. G., & Barua, P. (2011). Environmental variables and fisheries diversity of the Naaf River estuary, Bangladesh. *Journal of Coastal Conservation, 15*, 163–180.

Christy, J. H. (1988). Pillar function in the fiddler crab *Uca beebei* (I): Effects on male spacing and aggression. *Ethology, 78*, 53–71.

Clough, B. F., Ong, J. E., & Gong, W. K. (1997). Estimating leaf area index and photosynthetic production in canopies of the mangrove Rhizophora apiculata. *Marine Ecology Progress Series, 159*, 285–292.

Costa, C. S. B., & Davy, A. J. (1992). Coastal salt marsh communities of Latin America. In U. Seeliger (Ed.), *Coastal plant communities of Latin America* (pp. 179–199). Academic Press.

Costa, C. S. B., Marangoni, J. C., & Azevedo, A. M. G. (2003). Plant zonation in irregularly flooded salt marshes: Relative importance of stress tolerance and biological interactions. *Journal of Ecology, 91*, 951–965. https://doi.org/10.1046/j.1365-2745.2003.00821.x

Costanza, R., & Farber, S. (2002). Complex systems and valuation. *Ecological Economics, 41*, 409–420.

Costanza, R., d'Arge, R., de Groot, R., Sutton, P., Anderson, S. J., & Mulder, K. (1997). The value of the world's ecosystem services and natural capital. *Nature, 387*, 253–260.

Coull, B. C. (1999). Role of meiofauna in estuarine soft-bottom habitats. *Australian Journal of Ecology, 24*, 327–343.

Crane, J. (1975). *Fiddler crabs of the world (Ocypodidae: genus Uca)*. Princeton University Press.

Das, S., & Vincent, J. R. (2009). Mangroves protected villages and reduced death toll during Indian super cyclone. *Proceedings of the National Academy of Sciences, 106*(18), 7357–7360. https://doi.org/10.1073/pnas.0810440106. Published online 2009 Apr 20. PMCID: PMC2678660.

Das, T. K., Ghosh, A., & Chakraborty, S. K. (2014). Phramacologically active fatty acids of fiddler crab *Uca acuta acuta* (Simpson). *International Journal of Life Science and Pharma Research, 4*(2), 21–27.

Das, T., Pal, A. K., Chakraborty, S. K., Manush, S. M., Chatterjee, N., & Mukherjee, S. C. (2004). Thermal tolerance and oxygen consumption of Indian Major Carps acclimated to four temperatures. *Journal of Thermal Biology, 23*, 157–163.

Das, T., Sahu, N. P., Chakraborty, S. K., Chatterjee, N., Mohammed, M. S., Dalvi, R. S., Baruah, K., & Pal, A. K. (2014). Ultrastructural alterations in the gills of Labeo rohita and Cirrhinus mrigala in relation to acclimatation temperatures. *Journal of Fish Biology, 74*, 1487–1498.

Datta, T. K., & Chakraborty, S. K. (2015). Seasonal zonation pattern of meiobenthic fauna of an intertidal belt in the coastal area of Midnapore (east), West Bengal, India. *International Research Journal of Environmental Sciences, 4*(9), 45–52.

Datta, T. K., Miljutin, D. M., Chakraborty, S. K., et al. (2016). *Cyathoshiva amaleshi* gen. n. Sp. n. (Nematoda: Cyatholaimidae) from the coast of India. *Zootaxa, 4126*(4), 577–586.

Dauvin, J. C., Gomez Gesteira, J. L., & Fraga, M. S. (2003). Taxonomic sufficiency: an overview of its use in the monitoring of sublittoral benthic communities after oil spills. *Marine Pollution Bulletin, 46*, 552–555.

Dey, M. K., Hazra, A. K., & Chakraborty, S. K. (2007). Soil microarthropods populations in alluvial and coastal soil in Midnapore District with special reference to relative abundance. *Records of the Zoological Survey of India, 107*(4), 93–99.

Dey, M. K., Hazra, A. K., & Chakraborty, S. K. (2008). Diversity of macroarthropods and their role in the plant litter decomposition in the coastal tract of east Midnapore District, West Bengal, India. *Zoological Research in Human Welfare, 20*, 207–226.

Dey, M. K., Hazra, A. K., & Chakraborty, S. K. (2010). Functional role of microarthropods in nutrientcycling of mangrove-estuarine ecosystem of Midnapore coast of West Bengal, India. *International Journal of Environmental Technology and Management, 12*(1), 67–68.

Duarte, C. M., Middelburg, J. J., & Caraco, N. (2005). Major role of marine vegetation on the oceanic carbon cycle. *Biogeoscience, 2*, 1–8.

Dye, A. H., & Lasiak, T. A. (1987). Assimilation efficiencies of fiddler crabs and deposit–feeding gastro pods from tropical mangrove sediments. *Comparative Biochemistry & Physiology, 87*, 341–344.

Ellison, J. C. (2009). Geomorphology and sedimentology of mangroves. In G. M. E. Perillo, E. Wolanski, D. R. Cahoon, & M. M. Brinson (Eds.), *Coastal wetlands: An integrated ecosystem approach* (pp. 565–591). Elsevier.

Ellison, A. M., & Farnsworth, E. J. (2001). Mangrove communities. In M. E. Hay (Ed.), *Marine community ecology* (pp. 423–442). Sinauer Associates.

Elmqvist, T., Folke, C., & Nystrom, M. (2003). Response diversity, ecosystem change, and resilience. *Frontiers in Ecology and the Environment, 1*, 488–494.

Emig, C. C. (1997). Biogeography of the inarticulated brachiopods. In R. L. Kaesler (Ed.), *Treatise on invertebrate paleontology. Part H. Brachiopoda* (Vol. 1, pp. 497–502). Geological Society of America and University of Kansas.

Emmerson, W. D. (1994). Seasonal breeding cycles and sex–ratios of 8 species of crabs from Mgazana, a mangrove estuary in Transkei. *South African Journal of Crustacean Biology, 14*, 568–578.

Fager, E. W. (1964). Marine sediments: effects of a tube–building polychaete. *Science, 143*, 356–359.

Fanjul, E., Grela, M. A., & Iribarne, O. (2007). Effects of the dominant SW Atlantic intertidal burrowing crab *Chasmagnathus granulates* on sediment chemistry and nutrient distribution. *Marine Ecology Progress Series, 341*, 177–190.

Fanjul, E., Grela, M. A., Canepuccia, A., & Iribarne, O. (2008). The Southwest Atlantic intertidal burrowing crab *Neohelice granulate* modifies nutrient loads of phreatic waters entering coastal area. *Estuarine, Coastal and Shelf Science, 79*, 300–306.

Fischer, R., & Sheaves, M. (2003). Community structure and spatial variability of marine nematodes in tropical Australian pioneer seagrass meadows. *Hydrobiology, 495*, 143–158.

Flores, A. A. V., & Paula, J. (2002). Sexual maturity, larval release and reproductive output of two brachyuran crabs from a rocky intertidal area in Central Portugal. *Invertebrate Reproduction & Development, 42*, 21–34.

France, R. L. (1998). Estimation of the assimilation of mangrove detritus by fiddler crabs in Laguna Joyuda, Puerto Rico, using dual stable isotopes. *Journal of Tropical Ecology, 14*, 413–425.

Fromard, F., Puig, C., Mougin, E., et al. (1998). Structure, above-ground biomass and dynamics of mangrove ecosystems: new data from French Guiana. *Oecologia, 115*, 39–53.

Genoni, G. P. (1991). Increase burrowing by fiddler crabs *Uca rapax* (Smith) [Decapoda]: Ocypodidae in response to low food supply. *Journal of Experimental Marine Biology and Ecology, 147*(2), 267–285.

Ghosh, A., Misra, S., Dutta, A. K., & Choudhury, A. (1985). Pentacyclic triterpenoids and sterols from seven species of mangrove. *Phytochemistry, 24*, 1725–1727.

Ghosh, M., Kumar, L., & Roy, C. (2016). Mapping long-term changes in mangrove species composition and distribution in the Sundarbans. *Forests, 7*, 305. https://doi.org/10.3390/f7120305

Giere, O. (1993). *Meiobenthology, the microscopic fauna in aquatic sediments* (Vol. 328). Springer–Verlag.

Giere, O., & Pfannkuche, O. (1982). Biology and ecology of marine Oligochaeta, a review. *Oceanography and Marine Biology: An Annual Review, 20*, 173–308.

Gilbert, F., Stora, G., & Bonin, P. (1998). Influence of bioturbation on denitrification activity in Mediterranean coastal sediments: An in situ experimental approach. *Marine Ecology Progress Series, 163*, 99–107.

Gilbert, F., Aller, R. C., & Hulth, S. (2003). The influence of macrofaunal burrow spacing and diffusive scaling on sedimentary nitrification and denitrification: An experimental simulation and model approach. *Journal of Marine Research, 61*, 101–125.

Giri, C., Ochieng, E., Tiszen, L. L., et al. (2010). Status and distribution of mangrove forests of the world using earth observation satellite data. *Global Ecology and Biogeography, 20*, 154–159.

Gurr, M. I., Harwood, J. L., & Frayn, K. N. (2002). *Lipid biochemistry: an introduction* (5th ed.). Blackwell Science. ISBN: 0632054093.

Halpern, B. S., et al. (2008). A global map of human impacts on marine ecosystems. *Science, 319*, 948–952.

Heald, E. J. (1969). *The production of organic detritus in a South Florida estuary* (p. 110). Dissertation, University of Miami.

Heald, E. J., & Odum, W. E. (1970). The contribution of mangrove swamps to Florida fisheries. *Proceedings of the Gulf and Caribbean Fisheries Institute, 22*, 130–135.

Hendrickx, M. E. (1995). Checklist of brachyuran crabs (Crustacea: Decapoda) from the eastern tropical Pacific. *Bulletin de l'Institut royal des Sciences naturelles de Belgique, 65*, 125–150.

Higgins, R. P., & Thiel, H. (1988). *Introduction to the study of meiofauna* (p. 486). Smithsonian Institution Press.

How, M. J., Hemmi, J. M., Zeil, J., & Peters, R. (2008). Claw waving display changes with receiver distance in fiddler crabs, *Uca perplexa*. *Animal Behaviour, 75*(3), 1015–1022.

How, M. J., Zeil, J., & Hemmi, J. M. (2009). Variability of a dynamic visual signal: the fiddler crab claw–waving display. *Journal of Comparative Physiology. A, 195*, 55–67.

Hughes, C. E., Binning, P., & Willgoose, G. R. (1998). Characterization of the hydrology of an estuarine wetland. *Journal of Hydrology, 211*, 34–49.

Iribarne, O., Bortolus, A., & Botto, F. (1997). Between–habitat differences in burrow characteristics and trophic modes in the southwestern Atlantic burrowing crab *Chasmagnathus granulata*. *Marine Ecology Progress Series, 155*, 137–145.

Jennerjahn, T. C., & Ittekkot, V. (1997). Organic matter in sediments in the mangrove areas and adjacent continental margins of Brazil: I. Amino acids and hexosamines. *Oceanologica Acta, 20*, 359–369.

Jennerjahn, T. C., & Ittekkot, V. (1999). Changes in organic matter from surface waters to continental slope sediments off the Sao Francisco River, eastern Brazil. *Marine Geology, 161*, 129.

Kattge, J., Diaz, S., Lavorel, S., et al. (2011). TRY—a global database of plant traits. *Global Change Biology, 17*, 2905–2935.

Kennish, M. J. (2002). Environmental threats and environmental future of estuaries. *Environmental Conservation, 29*, 78–107.

Khalua, R. K., Chakravorty, G., & Chakraborty, S. K. (2003). Annotated list of molluscs from the coastal tract of Midnapore District, West Bengal, India. *Journal of Marine Biological Association of India, 45*(2), 245–249.

Khalua, R. K., Chakravorty, G., & Chakraborty, S. K. (2008). Community structure of macrobenthic molluscs of three contrasting intertidal belts of Midnapore coastal belt, West Bengal. *Zoological Research in Human Welfare, 6*, 47–55.

Kihslinger, R. L., & Woodin, S. A. (2000). Food patches and a surface deposit feeding spionid polychaete. *Marine Ecology Progress Series, 201*, 233–239.

Kim, T. W., Christy, J. H., & Choe, J. C. (2004a). Semidome building as sexual signaling in the fiddler crab *Uca lactea* (De Haan) (Brachyura: Ocypodidae). *Journal of Crustacean Biology, 24*, 673–679.

Kim, T. W., Kim, K. W., Srygley, R. B., et al. (2004b). Semilunar courtship rhythm of the fiddler crab *Uca lactea* in a habitat with great tidal variation. *Journal of Ethology, 22*, 63–68.

Kim, T. W., Tae, K. K., Hong, S. K., & Choe, J. C. (2006). Surface mating as an alternative mating strategy in the fiddler crab *Uca lactea*. *Journal of Ecology and Field Biology, 29*(1), 49–53.
King, G. M. (1986). Inhibition of microbial activity in marine sediments by a bromophenol from a hemichordate. *Nature, 323*, 257–259.
Kinne, O. (1963). The effect of temperature and salinity on marine and brackish water animals. *Oceanography and Marine Biology: An Annual Review, 1*, 301–340.
Koch, V., Wolff, M., & Diele, K. (2005). Comparative population dynamics of four fiddler crabs (Ocypodidae), genus (Uca) from a north Brazilian mangrove ecosystem. *Marine Ecology Progress Series, 291*, 177–188.
Kremen, C., & Ostfeld, R. S. (2005). Measuring, analyzing and managing ecosystem services. *Frontiers in Ecology and the Environment, 3*(10), 540–548.
Kremling, K. (1983). Trace metal fronts in European shelf waters. *Nature, 303*, 225–227.
Kristensen, E. (2008). Mangrove crabs as ecosystem engineers; with emphasis on sediment processes. *Journal of Sea Research, 59*, 30–43.
Kristensen, E., & Kostka, J. E. (2005). Macrofaunal burrows and irrigation in marine sediment: microbiological and biogeochemical interactions. In E. Kristensen, R. R. Haese, & J. E. Kostka (Eds.), *Coastal and estuarine studies* (pp. 125–158). America Geophysical Union, Geophysical Union.
Kristensen, E., Bouillon, S., Dittmar, T., et al. (2008). Organic carbon dynamics in mangrove ecosystems: a review. *Aquatic Botany, 89*, 201–219.
Kwok, P. W., & Lee, S. Y. (1995). The growth performances of two mangrove crabs, *Chiromanthes bidens* and *Parasesarma plicata* under different leaf-litter diets. *Hydrobiologia, 295*, 141–148.
Lee, S. Y. (1995). Mangrove outwelling: A review. *Hydrobiologia, 295*, 203–212.
Lee, S. Y. (1997). Potential trophic importance of the faecal material of the mangrove sesarmine crab Sesarma messa. *Marine Ecology Progress Series, 159*, 275–284.
Lee, S. Y. (1998). Ecological role of grapsid crabs in mangrove ecosystems: a review. *Marine and Freshwater Research, 49*, 335–343.
Lee, S. Y. (1999). The effect of mangrove leaf litter enrichment on macrobenthic colonization of defaunated sandy substrates. *Estuarine, Coastal and Shelf Science, 49*(5), 703–712.
Lee, S. Y. (2005). Exchange of organic matter and nutrients between mangroves and estuaries: myths, methodological issues and missing links. *International Journal of Ecology and Environmental Sciences, 31*, 163–175.
Lee, S. Y. (2008). Mangrove macrobenthos: assemblages, services and linkages. *Journal of Sea Research, 59*, 16–29.
Lim, S. L. S. (2006). Fiddler crab burrow morphology: How do burrow dimensions and bioturbative activities compare in sympatric populations of *Uca vocans* (Linnaeus, 1758) and *U. annulipes* (H. Milne Edwards, 1837)? *Crustaceana, 79*(5), 525–540.
Lovelock, C. E. (2008). Soil respiration and belowground carbon allocation in mangrove forests. *Ecosystems, 11*, 342–354.
Luckenbach, M. W. (1986). Sediment stability around animal tubes: the roles of hydrodynamic processes and biotic activity. *Limnology and Oceanography, 31*, 779–787.
Lugo, A. E. (1980). Mangrove ecosystems successional or steady state? *Biotropica, 12*(2), 65–72.
Maiti Dutta, S., Banerjee Mustafi, S., Raha, S., & Chakraborty, S. K. (2014). assessment of thermal stress adaptation by monitoring Hsp70 and MnSOD in the freshwater gastropod, Bellamya bengalensis (Lamark, 1882). *Environmental Monitoring and Assessment, 186*(12), 8961–8967.
Mfilinge, P. L., Atta, N., & Tsuchiya, M. (2002). Nutrient dynamics and leaf litter decomposition in a subtropical mangrove forest at Oura Bay, Okinawa, Japan. *Trees, 16*, 172–180.
Mills RM. (2011) Capturing carbon: The new weapons in the war against climate change. Columbia University Press, 465.
Mishra, S. S., Acharjee, S. K., & Chakraborty, S. K. (2009). Developmentof tools for assessing conservation categories of siluroid fishes of freshwater and brackish water wetlands of South West Bengal, India. *Environmental Biology of Fishes, 84*, 395–407.

Misra, A. (1995). Polychaetes. In *estuarine ecosystem series-part-2 Hooghly-Matla Estuary West Bengal* (pp. 93–155). Zoological Survey of India.
Misra, S., Dutta, A. K., Dhar, T., Ghosh, A., Choudhury, A., & Dutta, J. (1983). Fattyacids of the mudskipper, *Boleophthalmus boddaerti*. *Journal of the Science of food and Agriculture, 34*(12), 1413–1418.
Mitsch, W. J., & Gosselink, J. G. (1993). *Wetlands* (pp. 1–722). Van Nostrand Reinhold.
Mukherjee, B. B., & Mukherjee, J. (1978). Mangroves of Sundarbans. India. *Phytomorphologia, 28*(2), 177–192.
Nagelkerken, I., Blaber, S. J. M., Bouillon, S., Green, P., Haywood, M., Kirton, L. G., Meynecke, J. O., Pawlik, J., Penrose, H. M., Sasekumar, A., & Somerfield, P. J. (2008). The habitat function of mangroves for terrestrial and marine fauna: A review. *Aquatic Botany, 89*(2), 155–185.
Neira, C., Grosholz, E. D., Levin, L. A., et al. (2006). Mechanisms generating modification of benthos following tidal flat invasion by a Spartina hybrid. *Ecological Applications, 16*, 1391–1404.
Ng, P. K. L., Guinot, D., & Davie, P. J. F. (2008). Systema brachyurorum: Part I. An annotated checklist of extant brachyuran crabs of the world. *Raffles Bulletin of Zoology Supplement, 17*, 1–296.
Nomann, B. E., & Pennings, S. C. (1998). Fiddler crab-vegetation interactions in hypersaline habitats. *Journal of Experimental Marine biology and Ecology, 225*, 53–68.
NRC (National Research Council). (1993). *Managing wastewater in coastal urban areas*. National Academies Press.
Nybakken, J. W., & Bertness, M. D. (2005). *Marine biology: An ecological approach* (pp. 1–579). Benjamin Cummings Publishers.
Oliveira, R. F., McGregor, P. K., Burford, F. R. L., et al. (1998). Functions of mudballing behaviour in the European fiddler crab Uca tangeri. *Animal Behaviour, 55*, 1299–1309.
Palmer, M. E., Bernhardt, E., & Chornesky, S. (2004). Ecology for a crowded planet. *Science, 304*, 1251–1252.
Parulekar, A. H., Dhargalkar, V. K., & SYS, S. (1980). Benthic studies in Goa estuaries: part III – annual cycle of macrofaunal distribution, production and trophic relations. *International Journal of Molecular Sciences, 9*, 189–200.
Patra, M. K., Acharjee, S. K., & Chakraborty, S. K. (2005). Conservation categories of siluroid fishes in North-East Sundarbans, India. *Biodiversity and Conservation., 14*, 1863–1876.
Paul, A. K. (2002). *Coastal geomorphology and environment* (pp. 1–575). Ajoy Bhattacharya of ACB publications.
Pethic, J. (1984). *An introduction to coastal geomoephology* (pp. 1–260). Arnold Heiremarn.
Pradhan, P., Giri, S., & Chakraborty, S. K. (2006). Ecological gradients determining the density and diversity of rotifer in a fresh water river systems of South West Bengal, India. *Journal of Aquatic Biology, 21*(1), 19–28.
Pradhan, P., Giri, S., & Chakraborty, S. K. (2008). Zooplanktonic rotifers in river Kansai, West Bengal, India. *Journal of Indian Fishery Society of India, 39*(2), 43–47.
Qasim, S. Z., Sengupta, R., & Kureishy, T. W. (1988). Pollution of the seas around India. *Proceedings of Indian Academy of Science (Animal Science).*
Ravichandran, S., Soundarapandian, P., & Kannupandi, T. (2001). Zonation and distribution of crabs in Pichavaram mangrove swamp, southeast coast of India. *Indian Journal of Fisheries, 48*(2), 221–226.
Robertson, A. I. (1986). Leaf-burying crabs: their influence on energy flow and export from mixed mangrove forests (Rhizophora spp.) in northeastern Australia. *Journal of Experimental Marine Biology and Ecology, 102*, 237–248.
Robertson, A. I. (1991). Plant-animal interactions and the structure and function of mangrove forest ecosystems. *Australian Journal of Ecology, 16*, 433–443.
Robertson, A. I., & Blaber, S. J. M. (1992). Plankton, epibenthos and fish communities. In A. I. Robertson & D. M. Alongi (Eds.), *Tropical mangrove ecosystems* (pp. 173–224). American Geophysical Union.

Rossi, F., & Chapman, M. G. (2003). Influence of sediment on burrowing by the soldier crab *Mictyris longicarpus* Latreille. *Journal of Experimental Marine Biology and Ecology, 289*, 181–195.

Saenger, P. (2002). *Mangrove ecology, silviculture and conservation* (p. 342). Kluwer.

Sahu, S. C., Kumar, M., & Ravindranath, N. H. (2016). Carbon stocks in natural and planted mangrove forests of Mahanadi mangrove wetland, East Coast of India. *Current Science, 110*(12), 2253–2260.

Sanders, L. H. (1968). Benthic studies in Buzzards Bay. I. Animal– sediment relationships. *Limnology and Oceanography, 3*(3), 245–258.

Sanders, C. J., Smoak, J. M., Naidu, A. S., et al. (2010). Organic carbon burial in a mangrove forest, margin and intertidal mud flat. *Estuarine, Coastal and Shelf Science, 90*, 168–172.

Sanyal, P., Bhattacharya, N., & Chakraborty, S. K. (2015). Biomonitoring of four contrasting wetlands of Kolkata, West Bengal based on zooplankton ecodynamics and biotic indices. *Journal of Environmental Protection, 6*(7), 683–699.

Smith, T. J., III. (1987). Seed predation in relation to tree dominance and distribution in mangrove forests. *Ecology, 68*, 266–273.

Smith, T. J., III, Boto, K. G., Frusher, S. D., & Giddens, R. L. (1991). Keystone species and mangrove forest dynamics: the influence of burrowing by crabs on soil nutrient status and forest productivity. *Estuarine, Coastal and Shelf Science, 33*, 419–432.

Solan, M., Cardinale, B. J., Downing, A. L., et al. (2004). Extinction and ecosystem function in the marine benthos. *Science, 306*, 1177–1180.

Spalding, M., Blasco, F., & Field, C. (1997). *World mangrove atlas*. International Society for Mangrove Ecosystems.

Teal, J. M. (1958). Distribution of fiddler crabs in Georgia salt marshes. *Ecology, 39*, 185–193.

Tenore, K. R., Tietjen, J. H., & Lee, J. J. (1977). Effect of Meiofauna incorporation of aged eelgrass, *Zostera marina*, detritus by the Polychaete *Nephthys incise*. *Journal of the Fisheries Research Board of Canada, 34*, 563–567.

Thrush, S. F., Hewitt, J. E., & Lohrer, A. M. (2012). Interaction networks in coastal soft-sediments highlight the potential for change in ecological resilience for change in ecological resilience. *Ecological Applications, 22*(4), 1213–1223.

Thurman, C. L. (1984). Ecological notes on fiddler crabs of South Texas, with special reference to Uca subcylindrica. *Journal of Crustacean Biology, 4*, 665–681.

Townsend, C. R., Begon, M., & Harper, J. L. (2008). *Essentials of ecology* (pp. 1–532). Blackwell.

Twilley, R. R., & Chen, R. (1998). A water budget and hydrology model of a basin mangrove forest in Rookery Bay, Florida. *Marine and Freshwater Research, 49*, 309–323.

Twilley, R. R., Chen, R. H., & Hargis, T. (1992). Carbon sinks in mangroves and their implications to carbon budget of tropical coastal ecosystems. *Water, Air, & Soil Pollution, 64*, 265–288.

Twilley, R. R., Pozo, M., Garcia, V. H., Rivera-Monroy, V. H., Zambrano, R., & Bodero, A. (1997). Litter dynamics in riverine mangrove forests in the Guayas River estuary, Ecuador. *Oecologia, 111*, 109–122.

Twilley, R. R., Rovai, A. S., & Riul, P. (2018). Coastal morphology explains global blue carbon distributions. *Frontiers in Ecology and the Environment, 16*(9), 503–508.

UNCED (United Nations Conference on Environment and Development). (1992). The earth summit (The United Nations Convention on Biological Diversity). Rio de Janeiro—3rd to 14th June.

UNEP (United Nations Environment Programme). (2014). In S. J. Bochove & E. Nakamura (Eds.), *The importance of mangroves to people: A call to action*. United Nations Environment Programme World Conservation Monitoring Centre.

Valiela, I., Cole, M. L., McClelland, J., Hauxwell, J., Cebrian, J., & Joye, S. B. (2000). Role of salt marshes as part of coastal landscapes. In M. P. Weinstein and & D. A. Kreeger (Eds.), *Conceptsand controversies in Tidal Marsh ecology* (pp. 23–36). Kluwer Academic Publisher.

Valiela, I., Bowen, J. L., & York, J. K. (2001). Mangrove forests: one of the world's threatened major tropical environments. *Bioscience, 51*, 807–818.

Vitousek, P. M., Mooney, H. A., Lubchenco, J., et al. (1997). Human domination of earth's ecosystems. *Science, 277*, 494–499.
Warren, J. H., & Underwood, A. J. (1986). Effects of burrowing crabs on the topography of mangrove swamps in New South Wales. *Journal of Experimental Marine Biology and Ecology, 102*, 223–235.
Warwick, R. M. (1988). The level of taxonomic discrimination required to detect pollution effects on marine benthic communities. *Marine Biology, 92*, 557–562.
Washington, H. G. (1984). Diversity, biotic and similarity indices: a review with special reference to aquatic ecosystems. *Water Research, 18*, 653–694.
Watzin, M. C. (1985). Interactions among temporary and permanent meiofauna: observations on the feeding and behaviour of selected taxa. *The Biological Bulletin, 169*, 397–416.
Webb, A. P., & Eyre, B. D. (2004). The effect of natural populations of the burrowing and grazing soldier crab (Mictyris longicarpus) on sediment irrigation, benthic metabolism and nitrogen fluxes. *Journal of Experimental Marine Biology and Ecology, 309*, 1–19.
Wilkinson, D. L., Moore, C., Lopez, M., & Figueroa, M. (2001). *Natural resource damage assessment for a JP-5 fuel spill at Naval Station Roosevelt roads, Puerto Rico. Pre-final report.* Atlantic Division, Naval Facilities Engineering Command.
Wilkinson, M. T., Richards, P. J., & Humphreys, G. S. (2009). Breaking ground: pedology, geological, and ecological implications of soil bioturbation. *Earth-Science Reviews, 97*, 257–272.
Wolanski, E. (1992). Hydrodynamics of tropical coastal marine systems. In D. Connell & D. Hawker (Eds.), *Pollution in tropical aquatic systems* (pp. 3–27). CRC Press.
Wolanski, E. (1995). Transport of sediment in mangrove swamps. *Hydrobiologia, 295*, 31–42.
Wolanski, E. (2006). The application of ecohydrology for sustainable development and management of mangrove-dominated estuaries. In *The ICEMAN 2006 mangrove conference in Kuala Lumpur*.
Wolanski, E., Brinson, M. M., Cahoon, D. R., & Gerado, M. E. (2009). Coastal wetlands: a synthesis. In *Coastal wetlands: an integrated ecosystem approach* (pp. 1–62). Elsevier.
Woodin, S. A. (1974). Polychaete abundance patterns in a marine soft-sediment environment: the importance of biological interactions. *Ecological Monographs, 44*, 171–187.
Woodin, S. A. (1978). Refuges, disturbance, and community structure: a marine soft–bottom example. *Ecology, 59*, 274–284.
Woodin, S. A. (1981). Disturbance and community structure in a shallow water sand flat. Ecology in SA (1999) shallow water benthic ecology: a north American perspective of sedimentary habitats. *Australian Journal of Ecology, 24*, 291–301.
Ysebaert, T., & Herman, P. M. J. (2002). Spatial and temporal variation in benthic macrofauna and relationships with environmental variables in an estuarine, intertidal soft-sediment environment. *Marine Ecology Progress Series, 244*, 105–124.
Zeitzschel, B. (1980). Sediment–water interactions in nutrient dynamics. In K. R. Tenore & B. C. Coull (Eds.), *Marine benthic dynamics* (p. 195). University of South Carolina Press.
Zimmerman, R. J., Minnillo, T. J., & Rozas, L. P. (2000). Salt marsh linkages to productivity of Penaeid shrimps and blue crabs in the northern Gulf of Mexico. In M. P. Weinstein & D. A. Kreeger (Eds.), *Concepts and controversies in marsh ecology* (pp. 293–314). Kluwer Academic.
Zucker, N. (1974). Shelter building as a means of reducing territory size in the fiddler crab, Uca terpsichores (Crustacea: Ocypodidae). *The American Midland Naturalist, 91*, 224–236.
Zucker, N. (1981). The role of hood–building in defining territories and limiting combat in fiddler crabs. *Animal Behaviour, 29*, 387–395.

Chapter 6
Seasonal Dynamics of Physical Processes, Physicochemical Parameters and Biogeochemistry in Mangrove Ecosystems

Abstract An anastomosing network of different estuaries, creeks, tidal inlets, brackish water canals, submerged marshes and swamps constitute the low-saline waterbodies of the Sundarbans mangrove-estuarine complex, which not only supports the world's diversified mangroves but also acts as a breeding ground for a variety of both commercially and ecologically important shellfish and finfish. The freshwater supply of this mangrove ecosystem in Sundarbans, India, has been considerably reduced over the past five decades because of the construction of the Farakha Barrage at Maldah, West Bengal, diverting a huge amount of freshwater flows to Bangladesh, a neighbouring country of India, and the fault movements of neotectonic plates, altering the course of the freshwater flows of the river Hooghly, and thereby reducing freshwater flow towards the Sundarbans. All these changes, in conjunction with changing regional and global environmental issues like climatic change, have caused salinity invasion, leading to a loss of biodiversity, the ecodegradation of soil and the alteration of water-quality parameters that are not conducive to this very sensitive and vulnerable ecosystem.

The physicochemical properties of both water and soil, two major structural components of mangrove ecosystems, play important roles in the normal functioning and sustenance of aquatic life in the mangrove ecosystem of Sundarbans, India, by influencing the growth and propagation of the biodiversity components of this eco-region and thereby the biological productivity of this ecosystem overall. The impacts by an array of ecological parameters, either independently or together, govern the functioning of mangrove and estuarine ecosystems, specifically trophic interactions and biogeochemistry over space and time. Changing hydrology resulting in nonpermissible levels of temperature, salinity, pH, transparency, turbidity, dissolved oxygen, temperature, etc. poses threats to every structural component of this fragile ecosystem, determining their distribution. All those parameters, either independently or combined, operate to maintain the feedback processes required to bring about dynamic properties in the functioning of the ecosystem, through cause and effect relationships among the species or environmental characteristics impos-

S. K. Chakraborty, S. K. Acharya, *Eco-dynamics and Environmental Perturbations in Mangrove Ecosystems*,
https://doi.org/10.1007/978-3-032-08710-2_6

ing intensive implications on ecosystem resilience. The continuous monitoring of different physicochemical parameters helps prediction of seasonal ecological conditions through the proper identification of suitable periods for increased biological productivity in environmentally conducive conditions.

Keywords Sundarbans mangrove ecosystem · Estuaries · Physico-chemical parameters · Bio-geochemistry · Geo-tectonic plates · Ecosystem resilience

6.1 Implications of Physicochemical Parameters on Mangrove Ecosystems

Estuaries, semi-enclosed coastal bodies of water between seas and freshwater rivers with daily tides, serve many human purposes: directly as sources of water for industrial development, both for capture and culture fisheries; water transportation; waste disposal; etc. and indirectly as sites for carbon sequestration, the production of huge amount of oxygen, combating atmospheric pollution, etc. However, an array of human-mediated interventions drastically affects the estuarine ecosystem by disrupting the interlinkages, interrelationships and interactions among biological, geological, hydrological and climatic factors and thereby the ecological health of the estuaries (Giri et al., 2007; Chakraborty, 2022). The water quality of estuarine water exhibits a wide range of variation through different months, seasons and years because of the influence of so many ecological parameters on hydrological cycles in general and estuarine ecosystems in particular, such as meteorological (precipitation, temperature, humidity, wind flow, etc.) and physical parameters such as heat-storing capacity, evaporation and transpiration, tidal amplitudes and flows, currents and waves of water, the depth and volume of waterbodies, etc., chemical factors such as supplies, sinks and the withdrawal of nutrients, geological factors such as the topography and geomorphology of the shores and the bottoms of the estuaries, intensities, the duration and intervals of the occurrences of different natural hazards coupled with geological incidents such as earthquakes, storms and cyclonic incidences releasing magma (Chakraborty, 2022). In addition, the usage of estuarine water for the purpose of hydroelectric generation, aquaculture development, the construction of fishing jetties, the drainage of surface water and the pumping of water from the estuarine beds negatively affects water quality (Wolanski et al., 1992; Rajendran & Mansiya, 2015).

To gain a clear idea about the seasonal dynamics of water-quality parameters among different ecozones within the stretch of an estuary, proper monitoring and assessment through the programmed recording of water-quality parameters (biological, physical and chemical) are required to gauge the proper framework of the hydrochemistry of several estuaries. A full understanding of the physicochemical characteristics of estuarine waters should contribute to the effective ecomanagement of the very fragile and vulnerable mangrove ecosystem.

Out of so many ecological parameters, the seasonal fluctuation of salinity patterns is unique for every estuary and its connecting networks of tidal creeks crisscrossing the mangrove forest. In addition, the geomorphological settings of each estuarine zone structured via different interaction processes, such as rainfall patterns, prevailing winds, fluvial discharge, tidal flows and pressure gradients, play important roles in determining the physicochemical parameters (Officer, 1983; Miranda et al., 2002; Suguio, 2003). All these processes also act as determinants of the ecodynamics of the estuarine environment, through food chain–food web linkages and also sediment mixing and transportation from fluvial origin or drifting from coastal and adjacent marine systems within the different geomorphological units of an estuary (Miranda et al., 2002). The estuarine networks of Sundarbans, along with the tidal creeks in that area, act as the media for transporting and distributing mangrove plant–derived nutrients to adjoining estuaries, triggering an increase in primary productivity and in the diversity among living organisms. The acid sulfate soils within saline micaceous alluvium in the deltaic mangrove ecosystems of Sundarbans are loaded with mangrove litter that contains humic acids, which provide conducive substrates for the growth of a luxuriant mangrove forest. Because substrate soils have a higher concentration of nutrients in low-saline conditions, under such conditions, they promote the rapid growth of mangroves in Sundarbans.

Moreover, the spatial variations in some of the physicochemical parameters of the coastal waters of the Bay of Bengal are mixed up with the river waters of the Hooghly estuary flowing through the industrial belt of Haldia, loaded with the pollutants released from industries, pollutants such as petrochemicals, plastic chemical detergents, batteries, etc., which are in turn the primary causes of nutrient loss, sediment load, eutrophication, hypoxia, toxic contamination, etc. The discharge from the industries in and around the East Kolkata wetlands in the form of urban and industrial sewage contains pollutants from leather industries and fly ash from brick fields, which reach the estuarine networks of the Sundarbans through the Ichamati-Bidyadhari–Raimangal estuaries.

6.2 Physicochemical Parameters: Determinants and Processes

The water quality of an estuary is influenced by the drainage patterns of the macrotidal regime (Dittmar & Lara, 2001). Seasonal meteorological conditions directly influence the periodicity of flow patterns in both rivers and estuaries, triggering chemical cycling called nutrient spiralling, which in turn determines biotic variability even over short distances (El-wood et al., 1981). Moreover, the geomorphology of the estuarine flows also acts as a determinant of different physicochemical parameters and seasonal oscillation, which regulate the abundance of biological organisms, providing specialized ecological niches. However, physicochemical parameters, spatial patterns of drift, currents, waves and detritus determine the diversity and density of living organisms.

6.3 The Properties of Physical Processes

6.3.1 Light as Solar Radiation from the Sun and Hydrodynamics

6.3.1.1 Sunlight

The biological productivity and internal metabolism of estuaries are controlled by solar energy, the prime energy source triggering the process of photosynthesis by green plants where biochemical conversions utilizing both solar energy and available water take place directly by aquatic flora with the formation of organic matter. Therefore, solar radiation, as the major source of energy, drives the biological productivity of aquatic ecosystems, after being incorporated as potential energy through biochemical conversions directly by green aquatic plants (phytoplankton and macro-algae) that grow in estuaries and by terrestrial green plants that grow in water–land interphases, which together ensure the transport of organic matter into waterbodies. The thermal stratification and uneven distribution of density in the different stratified zones of aquatic systems are considerably regulated by the absorption of solar energy in the form of photons or pulses of electromagnetic energy and subsequent dissipation as heat. This in turn controls all biogeochemical cycles, the aquatic ecosystem metabolism and population dynamics. The intensity of sunlight and its power of penetration within a waterbody determines the transparency of water. Therefore, the results of transparency fluctuate and become erratic during dawn and dusk, but such transparency remains consistent with the depth of water, without having much influence on surface light intensity (Wolanski et al., 1992; Sarkar et al., 1989; Sarkar & Bhattacharya, 2010; Sarker et al., 2020).

6.3.1.2 Hydrodynamic Effects

Very complex hydrodynamic processes significantly regulate water-quality parameters by controlling inflows, water temperature, wind and sunlight, all of which modulate algal growth by determining sedimentation, vertical mixing, horizontal transport and primary production. Hydrodynamic transport processes govern the movement of different chemical or biological materials from one part of an estuary to another one. In addition, many other ecological factors, such as temperature, turbidity, salinity and sunlight, influence these kinetic processes. The stratification of temperature and water density influences the growth of algae and other biota, regulating the dissolved oxygen (DO) in that water because of the increase in biological metabolic activities. The stratification of prime water parameters also reduces the vertical mixing between the surface's oxygen-rich water with the bottom's oxygen-deficient water.

6.3.2 Water Movement and its Connection to Ecology

Currents within estuaries carry food, remove waste, renew oxygen and strongly affect the size, shape and behaviour of living organisms. Moreover, the amount of water carried by estuaries differs from one topographic zone with its ecological setup from another part because each receives different amounts of river discharge. The estuaries of Sundarbans, located in tropical climatic conditions, experience variable amounts of river discharge because of the seasonal climatic impacts and exhibit distinct seasonal fluctuations in the currents and volume of water during different seasons.

The chemistry and biology of estuaries are profoundly controlled by the motion of water, which is regulated mostly via wind flow and tidal amplitudes. The combination of temperature, tidal flows and wind make the water move where the water receives the kinetic energy and Coriolis force generated because of the movement of the Earth. Water movements act as prime factors determining the distribution of all forms of nutrients, dissolved gases, energy, momentum, algae, zooplankton and sedimentary materials. Aquatic organisms adapt themselves after being exposed to an array of ecological conditions in the changing state of aquatic environments. Solar and wind energies, after being distributed at the horizontal and vertical profiles of the estuaries because of the water mixing of saline with fresh water, play crucial roles in determining biological productivity. Different estuaries and their associated creeks in the Sundarbans mangrove ecosystems are characterized not only by their geomorphology and green surroundings but also by their operating physicochemical parameters, such as pH, temperature, salinity, dissolved oxygen, turbidity, alkalinity, etc., which are in turn governed by several other macroclimatic processes, such as atmospheric temperature, the intensity of sunlight, precipitation, wind flow, waves, currents, tidal amplitudes, etc., resulting over months, seasons and years. This chapter highlights the characteristics and significance of the major ecological parameters of estuarine systems, citing case studies from the mangrove-estuarine complex of Sundarbans, India, against a backdrop of national and international perspectives (Bhattacharya, 2000; Annon, 2005).

6.3.3 Tides, Levels, Volume and Duration

Tides represent the prime physical process bonding the aquatic subsystem with the mangrove forest subsystem. In the mangrove-estuarine complex of Sundarbans, the tide is semidiurnal with little diurnal inequality and tends to play synchronizing roles along the river courses of the ecological region. Tidal currents impose the strongest hydrodynamic influence on the subaqueous part of the deltaic Sundarbans and on the subaerial delta front (De Boer et al., 1989). Tidal amplitudes often exceed 3.4 m, which is associated with tidal currents of up to 290 cm/s on entering the river mouths and channels of the Hooghly–Matla estuarine complex in this estuarine delta from the sea. In the eastern tidal channels, flood dominance is found to be higher, whereas ebb dominance is observed in the channels of the western part of Sundarbans because of differences in the proportion of water depth: the width in different channels coupled with tidal asymmetry. Tidal length in different estuaries of Sundarbans usually ranges from 60 to 80, whereas it becomes 290 km at the Hooghly estuary, where tidal influence is observed up to that limit. The inland limit of the saline influence of the Hooghly–Matla estuarine complex depends on the size and separation of the distributary channels that dissect the Ganga–Brahmaputra delta front.

The highly indented nature of the entire coastline of West Bengal is supposed to have been developed because of the differential action of erosion and accretion caused by fluvio-tidal processes in the framework of an estuarine delta. The perpendicular elongated shape of most of the shores of the coastal deltaic islands has been formed by the impacts of flooding and ebbing, along with the macrotidal currents of the coastal environment (Davies, 1954). The intertidal landforms, sandflats and mudflats are formed by the deposition of sediments because of longshore currents with high tidal fluctuations from the impacts of unequal inundations and exposures. The highest level of tide is observed from August to September, whereas the lowest tidal amplitudes are noted from February to March. The mean ebb velocity is found to be 58 cm/s in contrast to the mean flood velocity of 48 cm/s (Das, 2009).

The intertidal zone within the continental shelf between two tidal marks—with the highest high-tide level and the lowest low-tide level—is structured by the differential impacts of various tidal amplitudes during different phases of tidal cycles, which are determined in tune with the seasonal and annual variation of the location of the Sun in relation to the Moon and the Earth. The three prime subzones of the intertidal belt are the high-tide level (HTL), mid-tide level (MTL) and low-tide level (LTL). The high-tide level (HTL), being the uppermost part of the intertidal zone, enjoys an expansive area that is determined by the difference between the two tidal marks—the highest tidal level during the maximum spring tide and the lowest tide level during the maximum neap tide. The extent of the low-tide level (LTL) is also determined by the difference between two tidal marks—the lowest tidal level during the maximum spring tide and the lowest tidal level during the maximum neap tide at the lowermost part of the intertidal zone. The mid-tide level (MTL) therefore occupies the maximum area of any intertidal zone, ranging from the lowermost border of the HTL to the uppermost border of the LTL. The flood and ebb tide occurring twice a day exhibits a semidiurnal nature with an interval of 12 h (Chakraborty et al., 2009).

The Bengal Basin, along with deltaic Sundarbans mangrove-estuarine complex, has been gradually tilting eastwards (Morgan & McIntire, 1959). This kind geological and geomorphological change affects the main freshwater discharge, which has also gradually been shifting eastwards (through Bangladesh), severely negatively affecting the freshwater budget for the Hooghly–Matla estuary. In Sundarbans, the tide-dominated estuarine networks demonstrate flow separation with flows of fresh water downstream along the right bank, whereas the saltwater tide characteristics upstream flow along the left bank (eastern) of the channel. The comparatively well-mixed Hooghly–Matla estuaries receive higher rates of freshwater discharge than the estuaries of the eastern part and experience less salinity (Hazra et al., 2002).

6.3.4 *Waves of the Estuarine Environment*

The northern region of the Bay of Bengal nearer to the Sundarbans mangrove ecosystem possesses moderate wave patterns featuring waves with heights ranging from 0.5 to 4 m during high-wave periods. The heights of waves are found to be minimal during the post-monsoon period, but they reach to heights up to 5 m during tropical cyclones occurring in both the pre-monsoon (March–June) and post-monsoon (September–December) periods, endowed with devastating waves of greater heights. Owing to the low elevation of the resulting waves, storm surges can penetrate as far as 100 km inland in the coastal plain of the deltaic Sundarbans because of the lower elevation of the intertidal zones and estuarine basins. Such flushing of saline water waves results in the displacement and death of millions of local people, making them environmental refugees. Seasonal fluctuations in different parameters associated with waves also influence longshore currents and micro- and macrotidal cycles. All these tropical cyclone–induced coastal processes, being very dynamic in nature, alter the geomorphological settings of the coastal region, converting them into areas for reclamation due to the abnormal cliff erosion.

6.4 The Indian Sundarbans Within the Indian Coastal Environment

As the junction between the land and sea and representing the area of land influenced by adjoining the sea and that part of the estuary, the coastal zone is subjected to the impact of adjoining land. This sensitive, vulnerable and productive zone plays a vital role in boosting the Indian economy because it acts as a reservoir of natural resources, harbours tremendous potential for promoting ecotourism and developing fisheries, governs meteorological phenomena, ensures smooth navigation and supports the water transport system. India now has a coastline of 11,098.81 km when including the Andaman and Nicobar Islands (3083.50 km), Gujrat (2340 km), Tamil Nadu (1068.69 km), Andhra Pradesh (1053.07 km), Maharastra (877.97 km), West Bengal (721.02 km), Kerala (600.15 km), Odisha (574.71 km), Karnataka (343.30 km), Goa (193.95 km), Lakshadweep Islands (144.80 km), Daman and Diu (54.38 km) and Pondichery (42.65 km) (Maritime Gateway, 2025). Coastal habitats with a lot of ecological diversity across the world are facing lot of both natural and anthropogenic threats over this past century because of population explosion, industrialization, unplanned urbanization, modern technology-based agricultural practices and drastic changes in lifestyle patterns. Mangrove ecosystems, being unique, sensitive, fragile and biologically productive ecosystems in the intertidal belt of estuarine networks under the influence of tropical climatic condition, are the conglomerations of diversified forms of flora, fauna and microorganisms in fluctuating ecological conditions. Mangrove forests, covering a wide range of tropical and subtropical climatic conditions within the coastal environment, thus support the lives of a galaxy of biodiversity components, encourage sustainable livelihoods and help to protect coastal zones (Robertson & Alongi, 1992; Wolanski, 2006).

The coastal zones of West Bengal, covering over one million hectors of area and extending over 721.02 km of coastal line, includes mostly three coastal districts: South-24 Parganas, North-24 Parganas and Midnapore (east) (Bagchi, K. 1944). The intertidal belt of most of the length of the coast is muddy, 90% of which supports coastal vegetation as mangroves and their associated plant and animal species and the remaining 10% is mostly restricted to the district of Midnapore (east), which is considered to feature sandy belts with a number of dunes and their supporting vegetation. The South-24 Parganas district is supported by the Sundarbans mangrove ecosystem; the North-24 Parganas district has several freshwater-dominated tidal estuaries, creeks, nallahs and brackish water wetlands; and the Midnapore (East) coast has sandflats with dunes and some patches of degraded mangrove. Together, they protect the ecological settings of coastal West Bengal. The coast of the Midnapore (east) district, representing 27% of West Bengal's coastal tract, extends along the west bank of the Hooghly estuary from New Digha and then curves around Junput, Dadanpatrabarh, Soula, Khejuri and Haldia on the east to the further northeast up to Tamluk or even the bank of Rupnarayan.

The Indian Sundarbans (21°32′–22°40′N and 88°05′–89°00′E)—representing the most dynamic, complex and vulnerable bioclimatic zones on the largest Gangetic delta, having an area of 9630 km^2—is encircled by a network of estuaries in the coastal region of the Bay of Bengal. The land–ocean boundary at Sundarbans, on the northeast coast of the Bay of Bengal, is highly irregular and crisscrossed by numerous estuaries, creeks, rivers and waterways, forming of a swampy terrace characterized by extensive fluvial marine deposits. Most of the estuaries within Sundarbans are well-mixed ones under intense semidiurnal tide and wave action, experiencing a meso-macrotidal setting (2–7 m in amplitude). The mean velocities of water current fluctuate between 117 and 108 cm s^{-1} during the periods of low tide and high tide, respectively. The climate of the region is determined by the southwest monsoon (June–September), post-monsoon (October–January) and pre-monsoon (February–May), receiving 70–80% of annual rainfall, which coincides with the high river discharge (2952 and 11,897 m^3 s^{-1}) during the southwest pre-monsoon season, with a gradual and diminishing impact from 900 to 1500 m^3 s^{-1} during the non-monsoonal months (Mukhopadhyay et al., 2006). The entire Sundarbans area, with its large expanse of mangroves along its estuarine networks, suffers from multifaceted forms of ecodegradation because of several human-caused activities, such as illegal human settlements, unplanned tourist activities, intensive boating, deforestation, modern technology-based agricultural activities in the reclaimed lands, monocultural aquacultural practices, the mass collection of shrimp and the collection of honey without valid permission. A significant ecological change has been pronounced in the Hooghly estuarine environment. In addition, a huge amount of domestic, industrial and agricultural waste is discharged from "point sources," whereas the estuary receives the raw sewage coming out of "non-point sources," which originate from the metropolitan cities of Kolkata and Howrah.

6.5 The Roles of Physicochemical Parameters in Coastal Environments in General and Mangrove Ecosystems in Particular

The physicochemical parameters of both soil and water, demonstrating a wide range of seasonal variations, influence the temporal and spatial variation in different biodiversity components and determine their distribution, community interactions and ecological success. Specific ecological parameters, acting as limiting factors, also govern the modes of biological production by sustaining the energy flow and decomposition cycle. Regional seasonal climate-induced changes in the meteorological parameters, such as temperature, precipitation and wind flow, coupled with other physical processes, such as tides, waves and the currents of marine-estuarine ecosystems, result in fluctuations in all ecological parameters, which in turn make the prevailing ecological condition very much unstable and compel the biological components to restrict their own territory in different parts of Sundarbans. Water-quality parameters are also considerably influenced by natural (geological, hydrological and climatic) and anthropogenic factors (Bartram & Balancen, 1996). The Sundarbans mangrove ecosystem, as one of the ecologically fragile ecosystems of the world, on deltaic islands, always remains under threat from extreme climatic anomalies such as tropical cyclones and storm surges. Seawater inundation could be the result of land-use changes in addition to climate change. Global climatic changes coupled with sea-level rise have compounded ecological problems by dramatically increasing high-risk factors due to geomorphological variability on this sensitive and vulnerable ecosystem (Sahana et al., 2019).

6.6 The Relevance and Significance of Physicochemical Parameters in Mangrove Ecosystems

Analyses on the physicochemical parameters of the water and soil of aquatic ecosystems throughout different seasons require in-depth knowledge of morphometric and morphoedaphic attributes such as volume, depth, flows, erosion pressures, rates of nutrient fluxes, heat-storing capacity, thermal stability, diversity, growth, the density and distribution of biota, biological productivity, the configuration of shores, water retention capacity through textural components, porosity, permeability and the sloping gradients of the bottom. Such information from research plays an important role in understanding the ecological, hydrobiological and physicochemical properties of the flowing waterbodies as water movement carries dissolved and suspended substances and determines the horizontal and vertical distribution patterns and the mode of utilization of biological organisms, which in in turn result in the distinct biogeochemistry of an aquatic ecosystem. This kind of ecological process embodies the flux and exchange of materials between the abiotic and biotic components of the ecosystem, and such interactions, ranging from the smallest scales to global ecosystems, influence the ecological characteristics of adjoining landscapes and the composition of the biosphere. The biogeochemistry of mangrove-estuarine ecosystems encompasses the shore and near-shore structural components of ecosystems through the continuous interaction processes between woody halophytic mangrove forest subsystems and adjoining estuarine subsystems and governs nutrient availability, nutrient cycles, the growth of biota and microbial transformation, the atmospheric exchange of gases, sediment transport and the removal of pollutants.

Mangrove ecosystems have met the demands of nutrient-enriched brackish water for the development agriculture, both capture and culture fisheries, recreational avenues and transportation since the inception of human civilization. However, an array of developmental activities—such as the reclamation of virgin forest lands for the development of shelters for human populations, the establishment of industries to meet the escalating needs of people, the conversation of swampy land for aquaculture, the construction of fishing jetties and harbours to facilitate capture fisheries with the help of mechanized boats and trawlers—have created the necessary infrastructure for promoting ecotourism and have caused major alterations to the physicochemical parameters and biogeochemistry of the affected ecosystems (Chakraborty, 1995, 2003, 2009, 2010, 2017).

6.7 Seasonal Changes Influencing the Ecodynamics of Mangrove Ecosystems in Terms of Their Physicochemical Parameters

Seasonal changes in the structural components (soil, water, flora and fauna) of mangrove ecosystems are controlled by seasonal alterations in different climatic factors, such as rainfall, humidity, and temperature, which together cause sea-level rise. Mangrove vegetation—by experiencing daily inundation and exposure twice a day and by possessing prop roots, leaves, pneumatophores, faunal bioturbatory structures and other epiphytic forms—acts as a guard wall and interrupts otherwise continuous water flow throughout the tidal cycle of a certain period. However, to ascertain the actual modes of such interactions and the nature of water flow within mangrove areas, depending on the timescale, different flow models need be developed on the basis of the information on tidal flows, atmospheric processes and groundwater. The prop roots and pneumatophores of mangrove plants remain densely intertwined above the bottom substrate, and by resisting tidal flows, water turbulence and the nature of eddies in the swampy region near the boundary of the open coast are made to mix and diffuse water and materials, altering the distribution of material within the swamp (Wolanski, 1995; Furukawa & Wolanski, 1996; Furukawa et al., 1997). Moreover, sedimentation appears to be enhanced by the prevailing turbulence around the vegetation and results in the formation of new land (Wolanski et al., 1998).

The rising of sea levels in different seasons has been due to a number of factors: (i) seasonal changes in the speed of wind flows and the wind direction on the coastal areas of the ocean (Ridd et al., 1988); (ii) seasonal alteration in water temperature, leading to an expansion in water volume; (iii) modification in atmospheric pressure and (iv) fluctuation in river runoff due to rainfall (Kjerfve, 1990). All these changes in meteorological parameters are directly connected with global climate change and are certainly having profound impacts on the biodiversity of the mangrove-estuarine complexes of Sundarbans, India (Hazra et al., 2002; Mitra et al., 2024; Jafar-Sidik, et al., 2017).

6.8 Water and Soil Quality of the Sundarbans Mangrove Ecosystem: Properties and Significance

6.8.1 The Concept of Water Quality

To judicially utilize the vast array of waterbodies in the mangrove-estuarine complex of Sundarbans, India, a proper understanding of the ecological factors and processes determining the quality of water is a prerequisite because these waterbodies, in the form of estuarine networks, are different kinds of natural and human-mediated environmental threats. The term *water quality* indicates the level of acceptability for water for different human purposes according to its suitability for direct and indirect human consumption. The hydrological cycle, after being influenced by specific agro-climatic conditions, especially in terms of precipitation, transmits water through terrestrial surfaces and the groundwater system to the different strata of various ecosystems. In addition, proper water quality also serves the requirements of both the floral and faunal components of the ecosystem and thereby help to maintain ecosystem health. All such water-quality parameters (pH, temperature, turbidity, salinity, dissolved oxygen (DO), biochemical oxygen demand (BOD) and chemical oxygen demand (COD)) of a functioning ecosystem in general and of aquatic biodiversity in particular have proved their significance in ecological studies.

Aquatic pelagic biodiversity components such as phytoplankton, zooplankton and fish demonstrate water–biota interactions, whereas benthic organisms exhibit sediment–biota interactions, forming the architecture of trophic interaction networks, provide necessary clues for undertaking site-specific but context-dependent ecological research studies on the influence of broadscale factors on the interrelationships, interdependence, and interlinkages among almost all biotic assemblages of aquatic ecosystems (van de Koppel et al., 2001; Widdows et al., 2004; Thrush et al., 2000, 2006; Van Colen et al., 2008; Weerman et al., 2010; Lawton, 1999; Norkko et al., 2006).

All those interactions become complicated by the various responses exhibited by different organisms in the process of harvesting their own requirements from their surrounding environment in the form of nutrients and water loaded with necessary life-supporting ingredients and through the positive feedback system for nutrient regeneration (Lohrer et al., 2004; Thrush et al., 2006). Any shift in these normal pathways of interactions could foreshadow major alterations in the functional roles of aquatic organisms towards nutrient cycling and determining the levels of other ecological parameters (Chakraborty, 2021; Conely et al., 2007). In such a context, any kind of environmental change must control the self-organizing interactions of the ecosystem functions and shifts towards reducing the resilience of communities, leading to a regime shift orchestrated by extreme environmental forces, which in turn result in emergent context-dependent processes along ecological gradients (Daleo & Iribarne, 2009' Needham et al., 2010; Thrush et al., 2009; Chakraborty, 2021a).

The term *water quality* refers to the properties of water (pH, temperature, turbidity, dissolved oxygen (DO), salinity, turbidity, nitrate (N), phosphate (P), total dissolved and suspended solids, biochemical oxygen demand (BOD) and chemical oxygen demand (COD)) that are considered as suitable for the purpose of different human needs according to meeting the international and national standards of permissible levels in terms of several water-quality parameters (Karr, 1981).

Water quality is maintained by several physical, chemical and biological conditions of waterbodies after sustaining natural hazards and human interventions, including complex variations in aquatic ecosystems at temporal and spatial scales (Peters et al., 1997; Vighi et al., 2006; Kernan et al., 2010). However, the application of physicochemical parameters only as the main components off ecological assessments of aquatic ecosystems comes with some limitations because of their partial contributions in the absence of biological parameters that are now being used to develop several biotic indices, because biological parameters, by virtue of their direct or indirect responses against abiotic ones, directly reflect the ecological status of aquatic ecosystems, which are affected by diversity, abundance and the density and distribution of aquatic organisms and therefore play complementary roles in affecting physicochemical parameters (Karr, 1981; Jennings, 2004).

In the mangrove ecosystem, ecological studies depend on the quality of water and are being used for the assessment of the ecodynamics of the entire ecosystem. This conceptual framework on water quality warrants taking a holistic approach to making a proper ecological assessment of waterbodies like mangrove-estuarine costal ecosystems, including all phases of the hydrological cycle: precipitation, percolation within terrestrial land masses, the development of groundwater systems, the recharging of the bottom of waterbodies with water and connecting saline-dominated marine-estuarine ecosystems with freshwater ones (Chakraborty, 2021a, 2022).

6.8.2 Variations in the Physical and Chemical Properties of Water

6.8.2.1 The Structural Excellence of Water Molecules

Basic scientific knowledge of the water molecule is needed to have an understanding about the chemistry of water. An atom of oxygen (O) and two atoms of hydrogen (H) combine to form an electrically neutral water molecule (H_2O) because the slightly negatively charged oxygen atom is electrically balanced by two slightly positively charged hydrogen atoms. The slightly negative parts of one water molecule attract the electrically slightly positive parts of another water molecule, resulting the formation of hydrogen bonding. The water in marine environment boils at 100 °C (212 °F) and freezes at 0 °C (32 °F).

6.8.2.2 The Water-Quality Parameters of Estuaries

6.8.2.2.1 The Properties of Surface Water

The surface water systems of estuaries and coastal waters connect with the atmosphere mainly through hydrological cycles and other ways that support the lives of their many unique categories of aquatic organisms, such phytoplankton, zooplankton, ichthyoplankton and nekton. Therefore, the major common services of surface waters are life-support systems for aquatic organisms, sources of water for human beings and avenues for swimming, boating, fishing and transportation. The water in the surface and subsurface zones of an estuary acts as a receptacle for different forms of waste materials, contaminants and pollutants, which undergo various changes in the realm of surface waters. The surface water is the site for an array of interactions among the water, air, landmass and bottoms of waterbodies through the involvement of many ecological processes, such as hydrological cycles, upwelling, outwelling and air–water intermixing, featuring physicochemical attributes such as water depth, volume, velocity, turbidity, transparency, dissolved oxygen, and nutrients and featuring various biotic components within the water column and bottom-dwelling organisms that make up the ecosystem with profuse quantities of both nutrients and contaminants.

6.9 The Seasonal Dynamics of the Water-Quality Parameters of the Coastal Areas of Sundarbans, India

Freshwater flows from two major rivers, namely the Ganga and the Brahmaputra, along with their associated tributaries, displaying seasonal variation in the quantum of their discharge, which governs the hydrology of the estuarine networks of Sundarbans. The diurnal and seasonal tidal rhythm, ranging from a height of 2 m to that of 5 m, also helps to develop other mixing and distribution patterns. In addition, the tidal ingress and the intensity of freshwater flows from the rivers help to develop the gradients of salinity and trigger other physicochemical parameters to undergo variation both spatially and temporally. However, during the past few decades, all rivers in the discharging fresh water in the western part of the Indian Sundarbans have progressively silted up because of a poor supply of fresh water from upstream, causing a disconnection of the inflow of fresh water with saline water in the Sundarbans mangrove ecosystem. The reduced inflow of freshwater flow has not only increased salinity in the river waters in the western parts of the Sundarbans but also made the rivers shallower over the years. The tidal interplays also play important roles in sustaining the ecosystem functioning of the mangrove's aquatic subsystems by ensuring the supply of essential inputs (water, nutrients, gases, food particles, etc.) to both mangrove subsystems and aquatic subsystems, but simultaneously, the receding water during ebb tides also causes the scouring of topsoil and thereby creates many small creeks that remain crisscrossed throughout the length and breadth of the Sundarbans mangrove ecosystem. The receding tidal water during low tides carries a large amount of dislodged surface sediments which get deposited along the banks of rivers and creeks after being carried by the high tides. This results in increasing the height of the banks and extending eroded channels further inwards, into the islands. Owing to the continuous impact of such forms of interactions on the mangrove ecosystem, different physicochemical parameters like water temperature, pH, conductivity, salinity and biochemical oxygen demand (BOD) temporally and spatially displayed wide variations, which were found to be higher during pre-monsoon, whereas phosphate (PO_4), silicate, nitrite (NO_2), ammoniacal nitrogen (NH_3) and nitrate (NO_3) were found to be at maximum during monsoon. The post-monsoon season, characterized by the lowest temperature of the year, features moderate salinity, turbidity and dissolved oxygen, but salinity came down considerably, almost to freshwater levels, in some parts of Sundarbans because of excessive rainfall and dissolved oxygen (DO) and because turbidity increased (Chakraborty et al., 2009).

6.9.1 Coastal Processes and Corresponding Ecological Development

The dynamic coastal processes prevailing in and around Sundarbans—endowed with the important morphotypes of beaches, such as the mudflats, sandflats, coastal dunes, runnels and inlets, sea grass belts and mangrove forests of Sundarbans, India—are induced by a number natural disasters, like the tidal flooding, storm surge, tropical cyclones of the area. Abnormal cliff erosion expedites the process of coastline changes by exposing the upper and middle foreshores of the sandy coast. The coastal belt, representing the interface between the sea and the landward shoreline, is being restructured by the action of waves, micro- and macrotidal cycles, longshore currents and the intensity of wind flows, resulting in geomorphological changes mostly in the intertidal and subtidal zones. In addition, different bioturbatory structures like burrows and mounds constructed by several macrobenthic fauna like brachyuran crabs, polychaetes, gastropods, brachiopods, and actiniarians, along with mud balls, fragments of wood and pebbles, obstruct the flow during the backwash, disabling the sediments due to scouring.

6.10 Meteorological Parameters and Their Relevance in the Sundarbans Mangrove Ecosystem

6.10.1 Climate and its Seasonal Aspects

Three seasons are well pronounced in the tropical coastal environment of the eastern region, especially the humid tropical conditions of Sundarbans. Each season is 4 months in duration, and they are pre-monsoon (March to June), with the highest temperature (42 °C) and almost no rainfall; monsoon (July to October), with the modest temperature ranging from 24 °centrigade to 32 °centrigade; and post-monsoon, with the lowest temperature (10 °C) and occasional rainfall. Maximum wind speed occurs from May to September from the southeast to the southwest in direction. During the early pre-monsoon period (February to April), winds blow from the south and southwest directions, whereas the calm post-monsoon period (late October to January) has winds flowing from the northwest (Annon, 2003).

The climatic conditions of the Sundarbans estuarine complex are influenced by seasonality in the development of low-pressure areas over the Persian Gulf, which triggers the initiation of the southwest monsoon during the hot and humid state of the late pre-monsoon to early monsoon period (May–September), characterized by stormy winds coupled with more precipitation. Scant rainfall often occurs in the calm and dry climatic conditions of this region because of the northeast monsoon in winter (December–February), developed by the seasonal high-pressure areas over Tibet. In such a context, three principal seasons, namely pre-monsoon, monsoon and post-monsoon, registered mean wind velocities of 11.5, 11.1 and 6.65 km/h, respectively. The wind direction of north-northeast to northeast during the post-monsoon season sharply undergoes changes from the wind direction of the south-southwest to southwest of the pre-monsoon and monsoon seasons.

The flows of water over the surface of estuaries are strongly influenced by changing monsoonal winds (Das, 2010). Strong tidal amplitudes influence the upstream sediment movement by controlling the channel regime, especially during summer, when headwater discharges occur in the mouth of the Hooghly estuary. However, in the monsoon season, headwater discharge and tides combine to influence the channel regime by creating a seaward drift in the sediments.

6.10.2 The Quantum of Freshwater Discharge into the Sundarbans Mangrove-Estuarine Complex of India: Its Impact

The freshwater supply within the coastal belts in and around Sundarbans has been considerably reduced over the past couple of centuries because of the overuse of upstream river waters and the excessive deposition of eroded sediments in the river bed due mainly to large-scale deforestation in the river drainage catchment areas in the river basins of several rivers along the east coast of India. A major portion of fresh water reaches Sundarbans, India, through major rivers such as the Ganga, the Padma and the Bhagirathi-Hugli, and the distributary systems of the Ganga River have sharply declined. A sizeable volume of fresh water carried by the Ganga River is directed towards Bangladesh, a neighbouring country of India that has two-thirds of the portion of mangroves of Sundarbans after the construction of the Farakka Barrage during the mid-1970s. The other freshwater river systems along the east coast of India also release a small amount of fresh water into the coastal areas of West Bengal, except during the months of the rainy season. Such a gradual but steady reduction in the supply of fresh water has resulted in an increase in salinity both in the water and in the soil of Sundarbans, hampering the growth of freshwater-loving floral components. Similar conditions of salinity enhancement occur along the coast of Gujarat, but for different reasons, like the rapidity of evaporation rates from the surface water of estuaries, which increases the concentration of salts in the inner parts of mangrove forests, causing the large-scale degradation of mangroves (Blasco, 1984).

6.10.3 The Influence of the Climate and Local Environmental Settings on Water Quality

The four major ecological attributes of an area, namely topography, sediment properties, the climate and vegetation cover, play important roles in determining the characteristics of the water in the waterbodies surrounding mangrove forests. The topography of the river basin and the foreshore and backshore of estuaries determines the flow patterns of water and thereby controls the stability of a particular area by regulating erosion pressure on the soil. The gradual erosion followed by accretion because of the differential intensities of diurnal and seasonal tidal water cycles and the process of the weathering and decay of the stable sediment also determine the turbidity of water by contributing dislodged soil particles to the runoff water. Different meteorological parameters, such as temperature, humidity, rainfall, and the wind flows constituting the regional climate, directly speed up the altering textural conditions of soil, the rate of the destabilization of sediment, erosion and the weathering and dissolution of minerals. The assemblage of mangrove vegetation resists soil erosion thanks to their root mat system and generates different plant detritus, which contribute biochemical entities through leaf litter decomposition. In addition, the atmospheric deposition of several chemical elements, like particulate matter, chlorine, bromine, iodine, fluorine and pollutants, in the form of aerosols in wet precipitation influence the quality of water for an aquatic ecosystem.

The ecological region of Sundarbans is characterized by the southwest monsoon-influenced climate. Seasonal low-pressure areas that develop in Persian Gulf initialize the southwest monsoon in summer (May–September) with high precipitation rates. Strong tidal action has a considerable impact on the upstream sediment movement by governing the channel regime without having any headwater discharge during the summer, especially in the mouth of the Hooghly estuary. However, during the monsoon season, the channel regime, after being governed by the combined effects of both headwater discharge and tides, tends to determine the seaward drift of sediments. Three seasons in Sundarbans are well pronounced—namely, pre-monsoon, monsoon and post-monsoon, each lasting 4 months because of the influence of rainfall and temperature, two important meteorological factors (Chakraborty et al., 2010).

6.10.4 Tides, the Most Influential Physical Process in the Estuarine Environment: Tidal Amplitudes, Volume and Duration

The tides, the rhythmic rises and falls of water surface with variable tidal amplitudes and drainage intensities within a mangrove ecosystem, act as principal determining physical processes that maintain the flow of water loaded with nutrients and organic materials between a mangrove forest subsystem and an estuarine subsystem in the mangrove ecosystem of Sundarbans. During the monsoon period, the renovation of the estuarine waters with the recharging of more input of organic materials from mangrove forests takes place, as reflected in the intensity of the currents of relatively turbid water through creeks loaded with high concentrations of nitrites and phosphates (Dittmar et al., 2001; Dittmar & Lara, 2001). In a research study on the Hooghly estuary of Sundarbans, India, mangroves represented sources of nutrients primarily during the monsoon period, when around 76% of the reactive dissolved phosphates and 71% of the dissolved inorganic nitrogen were found to have been deposited into the estuary during the summer monsoons (Mukhopadhyay et al., 2006). A similar finding was observed in the Gautami–Godavari estuarine mangrove ecosystem in Andhra Pradesh, India (Tripathy, 2005).

The tides, the rhythmic rises and falls of water levels in the Bay of Bengal, are primarily semidiurnal, which have maximum velocities of 3.8 m s^{-1} and are supported by moderate wave activities. The heights of waves sometimes exceed 2 m during cyclonic storms. The funnel-type configuration of the Hooghly–Matla estuary enables the tides to reach to their highest level (6 m) along the northeastern coast near river mouths. The heights of the water levels during low tides vary from 0.74 to 4.17 m, whereas during high tides, they vary from 1.67 to 5.46 m along this coast (Das, 2009b, 2010).

The intertidal zone, being the most vibrant and ecologically important part of the continental shelf, remains restricted between two tidal marks, where the highest high-tide level and the lowest low tide level in tidal cycles have different tidal amplitudes (spring and neap tides experiencing diurnal, monthly and seasonal variation) according to the specific location of the Sun in relation to the Moon and the Earth. The intertidal zone offers heterogeneity to ecological habitats and niches because they withstand variable exposures and inundations during different tidal cycles and become divided into three subzones—the high-tide level (HTL), the mid-tide level (MTL) and the low-tide level (LTL). HTL is the uppermost part of the intertidal zone; the respective extents of all those tidal levels are determined by the difference between the two tidal marks—the lowest tide level during the maximum neap tide of the year and the highest tidal level during the maximum spring tide of the year. The extent of the LTL, representing the lowermost part of the intertidal zone, is also determined by the difference between two tidal marks—the lowest tidal level during the maximum neap tide of the year and the lowest tidal level during the maximum spring tide of the year. The MTL occupies the maximum part of any intertidal zone and ranges from the lowermost border of the HTL to the uppermost border of the LTL. Similarly, the HTL represents the upper zone of intertidal belt between the maximum tidal mark during the neap tide of the year and the highest tidal mark during the spring tide of the year. The duration of intervals between semidiurnal flood and ebb tides, which occurs twice a day, is 12 h (Chakraborty et al., 2009). Because the dissipation of energy takes place because of the rise and fall of the water level during the high and low tides, it experiences a wide range of both diurnal and seasonal fluctuations, resulting in the migration of energy zones, which are reflected in the different forms of structuring the sediments on the surface of intertidal and subtidal zones, like beach slope, beach morphology, the depth of water, the direction and velocity of flows and the ripples marks of different natures (megaripples marks, crested ripple marks, double-crested ripple marks, swash marks, lingoid ripple marks, rill marks, backwash ripple marks, rhomboid marks and capped-off ripple marks).

6.10.4.1 Function in the Tidal Interplays of Mangrove Ecosystems

The fluctuation in tides in the form of the rise and fall of water levels varies diurnally, monthly and seasonally because of the changing locations of the Sun, the Moon, and the Earth in relation to one another. Such fluctuating trends of water levels in time and space dissipate energy, which imposes imprints on surface sediment, structuring different geomorphological attributes within the coastal belt like the slopping of the beach, local morphology in the form of ripples, the depth of water, waves patterns, the direction and velocity of flow, the textural components of sediments and the stability of the intertidal belts. The diversity in the forms of all those geomorphological attributes, like sandflats, mudflats, beach profiles, different forms of ripple structures (megaripples, small crested ripple marks, sinuous ripple marks, rhomboid marks, swash marks, lingoid ripple marks and rhomboid marks) and the formation of runnels, ridges, creeks and sedimentary structures, is the result of the fluctuation of tides in different intertidal zones. All these indicate hydrodynamic conditions, ranging from the mesotidal coast with a tidal amplitude of less than 4 m to the macrotidal coast with a tidal amplitude of more than 4 m. Proper settlements in and the growth, proliferation and productivity of mangroves depend on topography, tidal inundation, climate and hydrology. The decreasing rate of respiration during a prolonged inundation period is supposed to cause the destruction of mangrove vegetation, but normal inundation twice a day in a semidiurnal situation through rhythmic inundation is required by the mangroves of a tropical region like Sundarbans.

6.10.5 The Pattern of Sedimentation and Formation of Tidal Shoals

The Ganges–Brahmaputra–Meghna (GBM) river system carries a huge amount of sediment load (sand, silt and clay) totalling 2–4 billion tonnes while draining through the deltaic alluvial plain to the sea at the Bay of Bengal (Coleman, 1969). Sands, by virtue of their higher specific density, settle down at the confluence of rivers with the Bay of Bengal and thereby create a sandbar at that point. On settling down at a freshwater-dominated estuarine zone, silts in suspension due to tidal forces cause the formation of mudflats. Depositional processes bring back sediments from the seaside to the riverine tracts, and the continuation of this process forms deltaic islands. In this process, tidal shoals in the form of crescent or linear point bars are formed, the geomorphological alterations of which are made by the settled mangrove vegetation.

In the southern region of Sundarbans, a good amount of nontidal sediment transport takes place during monsoon in the northeast, which tends to raise the river beds of the Bidyadhari, the Raimangal, the Matla and the Herobhanga, prompting the development of several brick kilns and the construction of embankments by utilizing the deposited silts in these rivers and estuarine beds through silt dredging (Barua et al., 1994). The fine silt components of the soil here are deposited more at the river channels than the mouth of river (Bhattacharya, 1999; Rajendran & Mansiya, 2015).

Large sandy shoals, built up predominantly by tides, are frequently visible at low tide after the receding of water from the surface of the sandflat. The major constituents of tidal shoals are sands mixed with a thin layer of silty mud. The developed shoal is separated by a depressed zone from the beach zone by the formation of a shallow channel, designated as an ebb channel. Flood velocity increases on the elevated portions of the shoals, whereas ebb velocity increases in the depressed zone used as an ebb channel.

6.10.6 Rainfall and its Effects on Mangrove Ecosystems

Rainfall is considered the most important determining climatic parameter for the water quality of the estuarine system of Sundarbans. Two peaks of wind velocity were found to synchronize with the northeast and southwest monsoon in the study area. Water flowing across a coastal landscape percolates, facilitated by rainfall within the deeper strata of the sediment with dissolved soluble materials, the concentrations of which vary from one part of the coastal belt to others, determining the salinity of a region. Out of a total of 1800 mm of annual rainfall, the maximum (1600 mm) occurs during the monsoon period, followed by scant rainfall during post-monsoon months and a modest amount during the pre-monsoon period, which is associated with the southwest monsoon and due to the occurrence of cyclonic depression. Of all the physicochemical parameters of marine-estuarine water, the following ones warrant special consideration.

6.10.6.1 Wind Flow

Wind flow is an important meteorological parameter because it plays a role in determining the nature of waves, the direction and intensity of water movement and the mixing of surface water and subsurface water facilitated by the Coriolis force. Seasonal variation in wind velocity was recorded in the Sundarbans region in India, of which the maximum velocity was 16.7–20 km/h during the pre-monsoon season and the minimum was 10.7–11.8 km/h during the post-monsoon season. This important mangrove forest–dominated ecoregion encounters cyclones three to four times in a year, when this extensive mangrove forest acts as buffer against those cyclonic storms with speeds ranging from 80 to 140 km/h. Large-scale littoral drift is due to cyclones which cause devastating modifications in the hydrological variations in coastal zones.

6.10.6.2 Humidity

Humidity in the atmosphere, reflecting the status of moisture within the atmosphere and wind density, influences the hydrological cycle and thereby has some impacts on the water quality of the estuarine networks of Sundarbans. The humidity of the region exhibits variation ranging from a maximum of 95.16 ± 5.51 to a minimum of 33.71 ± 7.48.

6.11 Seasonal Variation in the Physicochemical Factors of Water and Soil, Highlighting Their Interrelationships

6.11.1 The Seasonal Dynamics of Physicochemical Factors in the Sundarbans Mangrove Ecosystem

The impact of huge rainfall (around 1500 mm) during the 4 months of the monsoon period on the Hooghly–Matla estuarine complex system drastically alters both the physicochemical and the biological parameters of the entire mangrove-estuarine ecosystem. Therefore, monsoonal precipitation carried and discharged by feeder rivers into estuaries during this period is considered the single most important factor regulating the occurrence, abundance and dynamics of most of the area's biodiversity components. Monsoonal impacts flood the estuary with the consequent dilution of saline water and a steady deposition of silt and detritus brought to the estuary by feeder rivers, increasing turbidity (Khan, 1995).

Distinct seasonal variations in major physicochemical parameters in the Sundarbans mangrove ecosystem were observed temporally and spatially during the study period. Higher water temperature, pH, turbidity, conductivity, salinity, dissolved oxygen (DO) and biochemical oxygen demand (BOD) were recorded during pre-monsoon, whereas maximum phosphate, silicate, nitrite, ammoniacal nitrogen, and nitrate were found during monsoon. The post-monsoon season was characterized by the lowest temperature, available phosphorous, available potassium and available nitrogen but moderate salinity and moderate amounts of other ecological parameters. During pre-monsoon, higher sand content, temperature, salinity and organic carbon in soil were found, whereas available nitrogen, available phosphorous and available potassium were observed to be at maximum during this period. A canonical-correlation analysis (CCA)—using different macrozoobenthic species and according to 12 physicochemical parameters of water, namely temperature, turbidity, pH, salinity, conductivity, dissolved oxygen, biochemical oxygen demand, silicate and water phosphate (PO_4), nitrate nitrogen (NO_3), ammoniacal nitrogen (NH_3), nitrite nitrogen (NO_2), and ten other ecological parameters of soil, namely organic carbon, temperature, salinity, pH, available phosphorous, available nitrogen, available potassium and textural components (sand, silt and clay)—revealed interrelationships among different parameters with macrozoobenthic species on one hand and a cumulative influence of a group of ecological parameters on the abundance of macrozoobenthic population on another (Chakraborty et al., 2010). The impact of the southwest monsoon was found to have pronounced impacts on the distribution of the inorganic nutrients in the spatially heterogenous mangrove-estuarine complex of Sundarbans, exhibiting the values such as nitrate nitrite (2.42–37.19 μg atm L^{-1}), phosphate (0.41–1.52 μg atm L^{-1}) and silicate (38.5–187.75 μg atm L^{-1}) (Das, 2015).

Different physicochemical parameters at the Midnapore coast exhibited seasonal variations in physicochemical parameters such as water temperature (20.8–32.8 °C), soil temperature (20.1–34.8 °C), the salinity of water in parts per thousands ($‰_0$)

(8.6–26‰), the salinity of soil in parts per thousands (10–33.4‰), dissolved oxygen (DO) from 3.24 to 5.47 mg/L, the pH of water (7.15–8.17) and the pH of soil (7.68–8.72). In addition, the texture of sediment demonstrated variations in the percentage of the occurrence of sand, silt and clay among the tidal levels of the intertidal belt and the ecozones of this coastal environment (Khalua et al., 2008; Chakraborty, 2018).

The biological productivity, biodiversity and ecological sustainability of the estuaries depend on specific water-quality parameters (Jha et al., 2015). Ongoing environmental perturbation, overexploiting resources, sewage from both domestic and industrial activities and uncontrolled urbanization have led to the deterioration of water quality (Ribera & Kjerfve, 2002; Davis & Koop, 2006; Ham et al., 2009; Whitall et al., 2010; Hadibarata et al., 2012; Soo et al., 2016; Saravanane et al., 1998). However, mixing fresh water with sea water within estuaries experiencing the churning action of a water-mixing process hamper modifications to water quality (Ramasamy et al., 2015).

6.11.2 Trending Changes in the Physicochemical Parameters of Water

The freshwater flows from major rivers like the Ganga and the Brahmaputra, along with their associated tributaries, govern the hydrology of the estuarine networks of Sundarbans. The striking seasonal variations in physicochemical parameters are due to seasonal changes in atmospheric temperature, rainfall and tidal cycles in the freshwater rivers in the western part of Sundarbans, which have been steadily and progressively silted up, disrupting the freshwater flows into the mangrove deltaic islands of the Sundarbans mangrove ecosystem. Such geohydrological changes have not only reduced the freshwater flow in the western parts of the Sundarbans, causing an increase in salinity, but also resulted in shallower rivers over the years. Simultaneously, the receding water level during ebb tides has scouring effects on the topsoil of the intertidal belts and thereby creates innumerable small creeks. The receding water, carrying a large volume of eroded and dislodged soil particles, facilitates the deposition of all those sediments along the banks of estuaries and creeks. Such continuous physical processes slowly but steadily increase the height of the banks in comparison to the interiors of the islands. Over time, such newly created channels caused by the scouring effects of ebb tides over the surface of intertidal belts progress further into the islands alongside forming the muddy banks which resist the influx of tidal water during high tides into the interior of the deltaic islands. With regard to seasonal fluctuation in physicochemical parameters in different parts of Sundarbans, water temperature, pH, conductivity, turbidity, salinity, dissolved oxygen (DO) and biochemical oxygen demand (BOD) were found to be higher during pre-monsoon, and the maximum amounts of nitrite (NO_2), ammoniacal nitrogen (NH_3), nitrate (NO_3), silicate, and phosphate (PO_4) were recorded during monsoon. The post-monsoon season was characterized by having the lowest temperature, scant rainfall and moderate salinity and turbidity (Chakraborty et al., 2009).

The various estuaries of the Sundarbans mangrove ecosystem, flowing from the north–south direction on the Indo Gangetic plain and ending in the Bay of Bengal in the extreme south, have been suffering from the impacts of several toxic substances (Sarkar et al., 2002, 2008; Mitra et al., 2024) and have received every year about 616 × 10^6 tonnes of suspended solids (Qasim, 2003). Most of the estuaries with meso-macrotidal influences with an average tidal range between 1.8 m (during neap tide) and 5.5 m (during spring tide) get their influxes of fresh water from rivers such as the Haldi, the Damodar, the Rupnarayan, the Rasulpul, the Champa, the Kelegai and the Ichamati, and saline water from the Bay of Bengal comes from the river delta estuary (Pritchard, 1952; Guha, 2015). Most of these shallow well-mixed estuaries, having an average depth of 6 m, represent the fluvial system influenced by a tropical to subtropical climate with southwest monsoonal rainfall, experience three seasons: pre-monsoon (March–June), monsoon (July–October) and post-monsoon (November–February) (Sadhuram et al., 2005; Mitra et al., 2018a).

Ever increasing human settlements and the exploitation of valuable resources close to estuaries throughout past decades have made this sensitive ecosystem vulnerable to ecodegradation (Lotze et al., 2006). This globally significant deltaic mangrove-estuarine complex of Sundarbans receives a plentiful supply of fresh water from the plains of West Bengal, which maintains ecological balance and which humans used for various activities, such as irrigation, boating, the navigation of small ships and using fishing trawlers (Chakraborty, 2011). However, significant ecological changes have taken place in this region because of the discharge of large amounts of untreated or semi-treated domestic and municipal sewage from the adjoining Kolkata metropolis, discharge that is carried by several rivers, rivulets, canals, nullahs, etc.

6.11.2.1 Variations in the Temperature of Surface Water

Temperature is one of the prime ecological factors influencing the distribution, growth and survivability of living organisms because it regulates the rate of different metabolic activities of living organisms. Increasing ambient temperatures increase the risks associated with extreme weather events, such as storms with lightning activity, cyclones, and floods, posing greater impacts on flowering, the pollination process, the germination of seeds and growth for the species and communities that make up mangrove vegetation. Water temperature is also a vital ecological parameter affecting almost all kinds of physical, chemical, and biological processes (Lotfinasabasl et al., 2018).

Next to rainfall, temperature is considered to be the most important abiotic factor driving the estuarine mangrove ecosystem of Sundarbans, especially because it restricts the seasonal distribution of both pelagic and benthic biodiversity components. The roles of temperature in controlling the distribution of faunal components in a particular area or ecological niche have been emphasized by ecologists, who have warned about the exclusion of species in areas under unstable temperature conditions. A tolerable range of temperatures has been determined to be crucial for maintaining the active lives of benthos in brackish water environments, which are called eurytherm (Kinne, 1963; Khan, 2003; Chakraborty, 2011).

The metabolic activities of faunal components are found to be faster at warm temperatures and slowed down at cold temperatures. The solubility of oxygen in water experiences an inverse relationship with temperature, but the concentration of solutes exhibits a direct relationship with water temperature. The rate of decomposition for organic matter in bottom soil is also regulated by temperature to a great extent. In tropical tidal estuarine environments like that of the Sundarbans, temperature demonstrates a wide range of diurnal and seasonal variation. Several research investigations have shown a wide range of seasonal variations in temperature, fluctuating from 18.0 °C (during post-monsoon) to 34.0 °C (during pre-monsoon), with a mean value of 28.16 ± 3.95 °C, which bears a reciprocal influence on a set of ecological attributes, such as air–water interactions, exposure time to sunlight, inflow from adjacent waterbodies, the evaporation–precipitation rate, the depth of water and vertical and horizontal mixing in water columns (Mitra et al., 2018a; Pillay & Olaniran, 2016; Achara et al., 2014).

A progressive increase in temperature is observed from December to May/June. Minimum air temperatures were noted in the month of December, when the minimum and maximum air temperatures occurred between 15.31 ± 2.44 and 27.01 ± 1.78 °C, and in April, the maximum temperature was found to be 37.83 ± 2.11 °C during a pre-monsoon month. Water temperature triggers the growth of autotrophic algae, which strongly regulate nutrient cycling, speed up the biological decomposition process and sometimes cause eutrophication; additionally, higher temperatures increase salinity but expedite the solubility of dissolved oxygen (DO); and all of these appear to be instrumental in affecting the stratification of the density of water across the water column. Mangroves cannot tolerate temperatures colder than 20 °C for a prolonged duration, and this is why mangroves grow only in the tropical and subtropical parts of the globe, including Sundarbans. The shallow estuarine water usually features a temperature a few degrees higher than other regions of the coastal environment that have a greater depth. Increased water temperature within a locality in the Sundarbans region is thought to be a result of the generation of heat because of the heavy mixing of the silt particles along with different chemicals releasing kinetic energy, which increases heat, and this usually occurs in the vortex of sediment-laden river waters during flood tide, especially because this takes advantage of a symmetric bottom topography situation, where waters vigorously enter from upstream of the river into the Bay of Bengal (Das, 2012; Das & Dutta, 2004).

6.11.2.2 Chlorophyll Contents

In a tropical estuary, increased temperature and higher light intensity in favourable physicochemical conditions, specifically an alkaline pH and the availability of nutrients, promote chlorophyll a (Chl a) synthesis (Wetz et al., 2016). Estimating Chl a helps in assessing the biomass of phytoplankton, acts as an indicator of ecological changes that reflect the integrated effects of water characteristics on biodiversity and helps to identify restoration processes (Boyer et al., 2009). A marine ecosystem can be divided into three categories on the basis of trophic status, which aligns with Chl a concentration: oligotrophic (<1 mg m^{-3}), mesotrophic (1–3 mg m^{-3}), eutrophic (3–5 mg m^{-3}) where Chl content in and hypertrophic (>5 mg m^{-3}) (Smith et al., 1999; Soo et al., 2016).

The Hooghly estuary, one of the largest estuaries in the Sundarbans mangrove ecosystem in India, is a mesotrophic estuary according to this categorization, with a phytopigment concentration ranging from 0.76 to 2.71 mg/m^3. This variation in phytopigment concentration has been found to be related to turbidity, light intensity, temperature, the freshwater discharge from upstream (Godhantaraman, 2002; Soo et al., 2016) and the high grazing rate of zooplankton on phytoplankton (La Rosa et al., 2002).

6.11.2.3 Salinity and its Impact on Biodiversity and Biogeochemical Cycles

The salinity of marine-estuarine water is represented by the total quantity of dissolved salts in a given unit of volume of water and is expressed in parts per thousands (%$_0$). The salinity of marine-estuarine water is determined by mixing of the naturally derived major ions get after being weathered and as eroded products from rocks and leaching out from soils on land. All the essential chemical components for the resulting salinity reach the saline water of estuaries and seas through the flows of rivers loaded also with gases or particulate matter.

Although anthropogenic activities do not contribute major ions to the maritime ecosystem, they play important locally specific roles in releasing sulfate ($SO_4^{=}$) and bicarbonate (HCO_3^-) and affecting the composition of sea water with the introduction of harmful exotic chemical substances. Moreover, some chemical ions from oceans are removed by different biological processes, technically named biogenesis, where biological organisms extract chemicals for biochemical reactions. For example, different planktonic organisms like phytoplankton and zooplankton (crustacean fauna) utilize calcium (Ca^{2+}), silicate ($SiO_3^{=}$) and bicarbonate ($HCO_3^{=}$) ions from seawater for the formation of their exoskeletons. After the death of all such fauna, most of those structural components of the bodies of faunal organisms, such as calcium carbonate ($CaCO_3$) and calcium silicate ($CaSiO_3$), along with other biochemical components, are degraded and washed away, at which point they are deposited on the floor of the saline waterbodies in their elemental forms as nitrogen (N), carbon (C), phosphorus (P), sulfur (S), oxygen (O), etc.

In the process of composing the salinity of water and bottom soils, numerous chemical elements such as carbon (C), sulfur (S), phosphorus (P) and nitrogen (N) move from one compartment of the marine ecosystem to others and thereby influence the chemical and biological budgets of the same ecosystem because all those essential components of nutrients tend to fuel biological productivity. In addition, the faecal matter, after being released by marine animals, gets deposited on the bottom of the marine-estuarine ecosystem, and during this process of the movement of faecal matter through the water column, all this decomposing organic matter is taken up by the deposit and suspension feeders as food. After digestion, egested materials are eaten by several other faunal components that inhabit in the water column and benthic zone.

Seasonal and spatial variations in salinity levels are crucial and unique to each estuary; such variations are determined by climate, hydrology, topography, the geomorphological limits of each ecoregion and the tidal flooding of an area because of the continuous interaction between different ecological processes, like tidal interplays, fluvial discharge, rainfall patterns, pressure gradients and prevailing wind parameters (Officer, 1983; Miranda et al., 2002). Annual and seasonal variations in salinity in the spatial and temporal scales of a particular salinity within a mangrove forest area play vital roles in restricting the species distribution, biological productivity and growth of mangrove-based biodiversity components (Twilley & Chen, 1998). The diversity, density and distribution of estuarine flora and fauna depend on salinity for their growth and reproduction and thereby their biological production. In such a context, salinity appears to be one of the key environmental factors regulating the growth and survival of mangrove species. The distribution, succession, population and diversity of mangrove species in Sundarbans mangrove ecosystem, along with other associated flora and fauna, have been found to vary according to the level of salinity (Chakraborty et al., 2009, Chakraborty, & Choudhury, 1994a; Chakraborty, & Choudhury, 1994b; Ball MC, 1996; Das & Mitra, 2005; Banerjee et al., 2005; Banerjee et al., 2017).

The fluctuation in the salinity of estuarine networks in and around Sundarbans is determined by the seasonal changes of two major meteorological factors, namely temperature and precipitation. Maximum salinity, nearing seawater salinity (33‰), occurs during the pre-monsoon period because the lowest rainfall is associated with highest temperature, and minimum salinity is observed during the monsoon period, when the salinity drops below 10‰ upon receiving heavy rainfall. Moreover, a tidal influx with a lower volume of water also contributes to an increase in salinity in some of the creeks and canals in Sundarbans.

Salinity, being the most important ecological parameter for both water and soil in an estuarine ecosystem and undergoing a continuous mixing of saline water with fresh water, plays a vital role in the production, settlement, management and mortality of the faunal and floral communities of the Sundarbans mangrove ecosystem (Bhunia & Choudhury, 1981; Choudhury et al., 1984a, 1984b). The high rainfall during monsoon, heavy freshwater discharge and the saline marine water of the flood tides are completely nullified. However, these reduced salinity profiles during monsoon are subsequently increased due to reduced discharge and increased surface evaporation during hot pre-monsoon months, a condition common to other tropical estuaries (Fernando, 1987; Khan, 1995). The salinity of the interstitial water of soil and that of open water fluctuate sharply but gradually over the seasons, ranging from 4.62‰ to 31.5‰ and 3.07‰ to 30.45‰, respectively (Chakravarti, 2004; Giri, 2004; Chakraborty, 2011).

Most estuarine faunal components are euryhaline according to their osmoregulatory abilities: they can tolerate a wide range of salinity levels, fluctuating from 10‰$_0$ during monsoon to 33‰$_0$ during pre-monsoon. The zonation and the succession of mangrove plant species are also determined by the drastic changes in salinity in both soil and water from downstream to upstream tidal water courses over different seasons. Increased salinity results in an increase in the osmotic potential of the interstitial water in soil, causing the root system of mangroves to be unable to uptake water. To overcome this difficult condition, mangroves adopt several physiological adaptabilities, such as the excretion, exclusion or storage of excess salts (Bandyopadhyay et al., 2001). According to salinity preferences, mangrove plant species are categorized as species with a higher tolerance for salinity, whose distribution is in the extreme downstream of the estuary, near the confluence with seawater (*Avicennia marina, A. Officinalis, Excoecaria agalocha, Xylocarpus, Rhizophora gymnorhyza, Ceriops decandra* and *Phoenix pelludosa*). Moreover, some estuarine mangrove species (*Heritiera fomeces, Nypa frutigans*, etc.) inhabiting mostly the upstream at the mixing zone with rivers prefer low salinity. Finally, some species, like *Aegialitis, Bruguiera*sp, etc., tolerating wide fluctuations in salinity, show a ubiquitous distribution.

6.11.2.3.1 Salinity and its Changing Trend

Among different parts of the Indian Sundarbans, the western sector showed distinct tidal variations in salinity, within a range of 1.4‰$_0$ to 2.0‰$_0$ between high and low tides, whereas such results were less pronounced in other parts, which had a negligible salinity difference. Also, marked seasonal variations in salinity have been observed throughout the length and breadth of the Sundarbans mangrove ecosystem, ranging from 9.34‰$_0$ to 30.83‰$_0$. The range of salinity in monsoon in most parts of Sundarbans fluctuated from 12‰$_0$ to 14.0‰$_0$, and the range in summer months was from 29‰$_0$ to 30‰$_0$, indicating a lower degree of spatial variations. Significant differences in salinity between upstream and downstream as recorded in winter and pre-summer months showed relatively higher values in larger areas of the estuarine networks, indicating more freshwater influx in this region (Sarkar et al., 2013; Chakraborty et al., 2013). High temperature is considered the prime factor for causing higher salinity during the pre-monsoon season, and it significantly favours a rapid evaporation process in the absence of rainfall. The salinity in an estuarine environment strongly influences the binding property of the elements to particulate matter (Chandrasekaran et al., 2015).

Marked seasonal fluctuations in salinity were noticed, ranging from 9.34‰$_0$ to 30.83‰$_0$ throughout the length and breadth of the entire Sundarbans ecosystem. In the western sector of the Sundarbans mangrove ecosystem, sharp variation in salinity was observed because of the influence of tidal variations, whereas at other parts of this ecoregion, such variation was less pronounced (1.4‰$_0$–2.0‰$_0$) between low and high tides. During the flood period with low discharge conditions, a slight stratification of salinity was noticed, but well-mixed conditions developed during ebb periods. A significant difference in the salinity profiles was found between the upstream and downstream regions of the estuaries, with higher salinities upstream of the estuarine networks due to the higher level of the mixing of saline water from the sea and lower salinities downstream of the estuaries because of the influx of fresh water in this region (Sarkar et al., 2013; Chakraborty et al., 2014). A research study on the variability of salinity from the downstream to upstream gradients has shown variation in salinity from salinities limnetic zone ($0.00\%_0$) to salinities high saline zone ($24.5\%_0$), indicating an increasing trend in the salt concentration downstream, reaching its maximum level at the mouth of the estuary—especially during the pre-monsoon season because of the combined effects of the higher rate of evaporation than that of precipitation (Rajsekar, 2003).

Similar findings with regard to the variation in salinity across the seasons were recorded by other researchers (Arumugum et al., 2016; Perumal et al., 2009). Electrical conductivity was found to be high during pre-monsoon and post-monsoon periods because of the higher ionic concentrations within water, releasing household organic waste and industrial effluents (Fokmare & Mussaddiq, 2000).

6.11.2.4 The Level of pH and its Changing Trend

In the soil and water in the estuarine tracts of the mangrove ecosystem of Sundarbans, pH, being the negative logarithm of the hydrogen-ion concentration level, acts as an important ecological factor regulating many biological activities because enzymatic actions are entirely dependent on pH.

The water pH at subalkaline levels (7.70–8.44) favours trace elements that concentrate in their particulate phase while increasing the likelihood of the sorption of the dissolved elements onto suspended matter (Mitra et al., 2018b). As an important water-quality parameter, pH severely affects the aquatic ecosystem by causing changes in alkalinity, the solubility of metals and the hardness of the water (Osibanjo et al., 2011; Şener et al., 2017). According to research studies, the range of recorded pH in estuarine system of Sundarbans remains in a neutral to subalkaline condition, varying from 7.20 to 8.99, which is within the prescribed the World Health Organization (WHO) and Bureau of Indian Standard (BIS) guidelines of 6.5–8.5 (WHO, 2004; BIS, 1991). Further, pH is controlled by other factors, such as dissolved materials, aerosols, dust particles, human-made waste and byproducts from the photosynthesis of plants (Omer, 2010).

The microbial growth rate is determined by pH, the main actor expediting the biological decomposition of mangrove litters. The most favourable pH condition for the metabolic activities of living organisms varies from 6.0 to 8.0. However, the estimated pH in the bottom sediment and water in the tidal courses of Sundarbans remain in the alkaline range (between 7.2 and 7.8). The pH of estuarine water in the alkaline range (7.60–8.35) corresponds with the optimal range of dissolved oxygen ($5.7–6.23\ mg\ L^{-1}$) in the canals/creeks of Sundarbans, and together, they act as critical limiting factors for the normal growth and survivability of fish (Manna et al., 2010; Chaudhuri et al., 2012). The concentration of trace elements and the sorption of dissolved elements onto suspended matter are enhanced in subalkaline water whose pH is within the range of 7.70–8.44, with an average of 7.94 ± 0.2 (Mitra et al., 2018b).

The pH of coastal water ranges from 7.5 to 8.5 (Chakraborty, 2011). The concentration of carbon dioxide has been increasing in the atmosphere because of human influences. The gradual and steady increase in CO_2 concentration in the Earth's atmosphere over the past three centuries has been due to human interventions in the environment, where a large fraction of that carbon dioxide has dissolved into the saline water of marine–oceanic realms. After being converted to the total amount of dissolved inorganic carbon, such human interventions have contributed to a shifting of seawater chemistry towards a lower pH condition. The result of such changes is reflected in the form of the rising acidification of coastal waters and a decrease in the carbonate ion (CO_3), which is expected impair the ability of many marine-estuarine faunal components to build up their own shells (Mitra et al., 2009). The trending changes in pH over the previous three centuries are due to the increasing concentration of carbon dioxide in the Earth's atmosphere because of human influences. This shows the rise in acidification in coastal waters and the decrease in carbonate ions [CO_3^{2-}], and this ecological state is expected to impair the calcareous shell forming process of marine faunal organisms like molluscs, crustaceans, etc. (Mitra et al., 2009). The estimates put forward by the Intergovernmental Panei on Climate Change (IPCC) in its fourth Assessment Report indicated that the pH of a marine ecosystem will come down gradually from 8.1 to 7.8 within a few decades due to the increasing concentration of CO_2.

6.11.2.5 Alkalinity and its Changing Trend

The carbonate and bicarbonate constituents of water, forming the alkalinity of water, demonstrate their acid-neutralizing capacity in the estuarine environments of Sundarbans. Alkalinity refers to the potential proportionate index of the quantum of the carbon dioxide conversion to organic matter via photosynthesis and the availability of carbon dioxide after being released from biotic respiration and organic matter decomposition. The estuarine water of Sundarbans, having an alkalinity of 50 mg/L or more, is considered productive, and the total alkalinity of this ecozone varies from 50 to 160 ppm (Anonymous, 1987).

6.11.2.6 Turbidity and its Changing Trend

Turbidity refers to a reduction in clarity in water, as revealed from the manifestation of the visual properties of water, which is caused by the occurrence of more total dissolved solids (TDSs) and total suspended solids (TSSs) within a unit of volume of water. The substances responsible for causing turbidity are eroded sediments coming from the nearby terrestrial landscape of the estuarine basins, stirring and mixing bottom sediments in the form of resuspended sediment particles within the water column and by the bloom of planktonic organisms (Nolen et al., 1985).

The turbidity of the estuarine waters is due to the carrying of suspended solids—mostly clay particles, along with microscopic biotic components like plankton—suspended in water. The transparency of water, as reflected by the depth of the penetration of sunlight within water, is determined by turbidity. Transparency within water is vital for the penetration of sunlight and thereby its role in promoting primary production through the photochemical process energized by light. In the process of synthesizing carbohydrate, a long duration of high turbidity caused by suspended clay and silt particles makes the waterbody unproductive, affecting the aquatic organisms that have gills as respiratory organs, which leads to mortality due to the clogging of gills (Anonymous, 1987). Secchi disc transparency in the estuarine waters of Sundarbans varies from 37 to 73 cm.

The estuarine font near the sea will have the least turbidity because of the settling of soil particles downstream and the least influence from the anthropogenic discharge of industrial and municipal waste materials. Helicoidal tidal flows during flood tide, coupled with irregular bottom topography in almost all river estuaries that feature meandering bends in the Sundarbans region, determine the level of turbidity. In addition, seasonal impacts play roles in the resulting fluctuation in turbidity levels, which are the lowest during monsoon (28.8 cm) and the highest during pre-monsoon season (45.2 cm) in several canals and creeks.

The scouring of surface sediment from the wide mudflats along the river banks through the receding flows during low tide and resuspension also influences the increased concentration of TDSs in water, leading to the high turbidity value (Batabyal et al., 2014). The vigorous mixing of eroded particles along the estuarine gradients through flushing activities plays a crucial role in ensuring the wide range of variation in turbidity in the estuary. Turbidity levels have been observed to remain low at the upper sections and extreme southern portions of the estuaries (50 mg/L) and high at the mouth of the estuaries in excess of 150 mg/L because of the high concentrations of suspended solids in the lower estuary (in excess of 150 mg/L), whereas the middle estuary zone has been characterized by lower concentrations (less than 136 mg/L) (Ray et al., 2013; Chakraborty et al., 2014).

An analysis of Landsat TM imagery through the application of remote-sensing tools have shown that turbidity levels were high at the mouth of the estuaries, whereas turbidity levels at the upper sections at the extreme southern portions of the estuary were lower (Ray et al., 2013). High concentrations of suspended solids in excess of 150 mg/L were noted in the lower estuary zone, whereas the middle estuary zone exhibited lower concentrations, lower than 136 mg/L, and the upper estuary zone demonstrated the presence of high suspended-solid concentrations, in excess of 150 mg/L (Chakraborty et al., 2013).

6.11.2.7 Dissolved Oxygen and its Changing Trend: The Effects of Photosynthesis and Respiration

Dissolved oxygen (DO), being a byproduct of aquatic plant photosynthesis, is essential for the growth and survival of aquatic organisms and for various oxidation processes during biochemical interactions. The production and the utilization of dissolved oxygen (DO) as an indicator of the ecological health of an aquatic ecosystem are directly connected with photosynthesis by autotrophic organisms and respiration by all aquatic organisms (Lacerda et al., 1986; Chang, 2002; Jennings & Dulvy, 2005; Sahu et al., 2012). The levels of DO are regulated mainly via photosynthesis and the respiration of the phytoplankton, periphyton and rooted aquatic plants (macrophytes) of a waterbody, which render profound effects on the diurnal and seasonal variability of DO. An increase in DO in a waterbody is made by aquatic plants through photosynthesis, but the DO drops at night due to the respiration of aquatic organisms, threatening the survivability of aquatic organisms. Solar energy triggers the photosynthesis process during daytime, whereas respiration and decomposition processes occur throughout the day and night because they do not depend on solar energy. The photosynthesis process involves the interaction of carbon dioxide and water, which produces glucose and oxygen, leading to a net gain of DO in a waterbody. Conversely, through the process of respiration, aquatic organisms convert glucose and oxygen into carbon dioxide and water with the depletion of the levels of DO in that waterbody. Aquatic plants carry out an oxygen-producing photosynthesis process and side-by-side respiration activities, resulting in the production of more organic matter and more oxygen than they use.

The levels of DO exhibit both seasonal and spatial variations across the length and breadth of the Sundarbans, demonstrating a maximum value of 7.50 ± 0.67 during post-monsoon and a minimum value of 6.53 ± 0.60 during monsoon. Such seasonal variation is due mainly to heavy rainfall and land runoff. The western parts of the Indian Sundarbans have shown an increasing trend in the concentration of DO at a rate of 1 mg/L/year—which is in contrast to the eastern part of this ecoregion over the past few decades, mainly because of a global warming–mediated increase in temperature (Mitra et al., 2009; Chakraborty et al., 2014). A very high level of DO (more than 9 mg/L) in surface water often leads to eutrophication because it increases the growth and proliferation of phytoplankton, microbial populations and zooplankton, which tend to consume oxygen from the water at higher rates causing abrupt depletion of dissolved oxygen from the affected water bodies (Dodds, 2009). The developed eutrophication in turn leads to the drastic depletion of the DO (1.8–7.2 mg/L) concentration, which has detrimental impacts on aquatic life by disrupting both photosynthesis and respiration processes. Most aquatic organisms can survive by overcoming stresses imposed by a wide range of carbon dioxide concentrations because of the presence of a sufficient level of dissolved oxygen in water. Carbon dioxide also plays significant roles in the estuarine ecosystem of the Sundarbans by producing bicarbonate from calcium carbonate and thereby controlling the pH of water by developing a buffer system with the help of CO_2, $CaHCO_3$ and $CaCO_3$. However, a lower concentration of dissolved oxygen results because of the effect of the uptake of oxygen by an appreciable amount of free carbon dioxide (above 15 mg/L), which is considered detrimental to aquatic organisms.

The dissolved oxygen (DO) concentration in the western sector of the Indian Sundarbans in proximity to the sea has exhibited an increasing trend in contrast to the eastern part over the past 30 years. The observed increase in the concentration of dissolved oxygen levels was found to be around 1 mg/L over this period. The fluctuating trend in the concentration of dissolved oxygen is due to other physicochemical factors, such as turbidity, the intensity of light, temperature, air circulation and the mixing patterns of fresh water with sea water (Mitra et al., 2009; Chakraborty et al., 2013).

6.11.2.8 Biochemical Oxygen Demand (BOD)

Biochemical oxygen demand (BOD) refers to the total amount of oxygen biologically or chemically removed from water in a specific time and space and at a specific temperature. It also takes into account the total amount of DO which is needed for the decomposition of organic matter by microbes and the oxidation of several types of inorganic matter. All those chemical compounds, which are known as oxygen-demanding waste, deplete oxygen from an aquatic ecosystem by undergoing chemical reactions. BOD also acts as an indicator of organic pollution rather than a true physicochemical parameter, and it appears at high levels in wastewater-loaded aquatic systems.

6.11.2.9 Chemical Oxygen Demand (COD)

Chemical oxygen demand (COD), referring to the total amount of oxygen required for the decomposition of both inorganic matter and organic matter within a waterbody, indicates the total concentration of nonbiodegradable substances. Different chemical reactions involved in all those chemical reactions considerably reduce DO, disrupting the ecological health of the aquatic ecosystem. COD also determines the levels of organic contamination in a wastewater-loaded aquatic ecosystem. The higher the COD value, the more oxygen utilization is needed to reduce substances (sulfide in salt water or methane in fresh water) within wastewater (Cerco & Cole, 1995; Park et al., 1993). Algal respiration needs oxygen to convert organic carbon to carbon dioxide and water. Chemical and biological processes in sediment beds often uptake oxygen from the water column, which can be a significant fraction of the total oxygen demand in a waterbody.

Sediment oxygen demand refers to oxygen depletion due to benthic reactions, where oxygen is consumed by sediment organisms for respiration and the benthic decomposition of organic material. It highlights oxygen consumption exerted by the organisms in the bottom sediment on the overlying water where methane and sulfide provide additional oxygen demands (Chakravarti et al., 2004; Giri et al., 2007; Dey et al., 2008; Chakraborty, 2013).

6.12 The Adaptability of Mangroves and Associated Fauna in Response to Ecological Changes Caused by Changing Environmental Variables

The species richness of mangrove habitats remains at relatively low levels in comparison with other tropical habitats with high biomass tropical deciduous forests, rain forests and coral reefs (Ricklefs & Latham, 1993). Despite the relatively low biodiversity, mangrove plants exhibit a wide range of structural and functional attributes which help them to survive and propagate in relatively harsh ecological conditions within the intertidal zone (Banerjee L. K., 1985; Ball, 1996; Chowdhury, et al., 2016a, 2016b).

In such a context, the diversity of mangrove-based biota plants should not be assessed only according to the number of species but also according to the ability of each species to adapt within the wide range of environmental conditions that require the application of all of those specialized adaptive features (Duke et al., 1998). The importance of prevailing ecological factors (physical, chemical and climatic factors) can be viewed on the characteristics of mangrove species assemblages that form a distinct forest community, where each species enjoys a distinct range of distributional limits at different geographic scales (Duke, 1992). All those ecological factors sometimes act on each species together, both flora and fauna, within a mangrove ecosystem in a cumulative manner or in isolation, acting independently on a species, species assemblages or any stage of a species' life cycle, such as from the development, dropping and settlement of propagules to dispersal, seedling growth and the attainment of reproductive potential and maturation in plant species, whereas settlement, growth, foraging and reproductive activities take place in animal species.

Each ecological factor is in turn regulated by a wide range of biotic, physical, chemical, geomorphological and climatic factor in tune with global, regional, estuarine and intertidal scales, which synergistically determines the growth and distributional patterns of each species at intertidal belts. Therefore, the settlements, growth, distribution and diversity of mangroves require a careful analysis and a reconsideration of interpretation in view of the relatively wider interplay of influencing factors.

Mangroves, being unique biotic assemblages inhabiting the interspace between land and sea, exhibit a number of morphological and ecological-physiological adaptations, including sunken stomata, viviparous germination, stilt roots, aerial roots (pneumatophores) and different physiological mechanisms to tolerate salinity, inundation and exposure pressures to maintain their growth (Chakraborty et al., 1986, 2012; Chakraborty, & Choudhury, 1994a; Chakraborty & Choudhury, 1994b; Chakravarty et al., 2004; Das & Bhattacharya, 1999a, 1999b; Chatterjee and Chakraborty, 2014a; Chakraborty, 2017, 2018; Chakraborty et al., 2023).

6.12.1 Floral Adaptations

6.12.1.1 Adaptations to Low Oxygen

The pneumatophores of mangrove plants help taking as much atmospheric oxygen as possible during the periods of tidal exposure into the biomass of mangroves, helping them to overcome the oxygen deficiency in the inundated mangrove forest and the moisture-enriched muddy sediments in an anoxic condition. The presence of a wide range of aerenchyma tissue within the roots also helps facilitate oxygen transport within a plant. Different species of the genera *Rhizophora* (red mangroves), supported by stilt roots, prop themselves up to stand, properly erecting their tips above the water level within the intertidal belt, and they can then take in air through pores in their bark (lenticels). Three species of the genus *Avicennia*, known as black mangroves (*A. officinalis*, *A.alba* and *A. marina*), which live at higher tidal levels, are endowed with cultures of pneumatophores with a height of about few metres and are covered in lenticels. Similar types of pneumatophores but with different morphological configurations have been observed in several other species of mangroves, like *Sonneratia* spp. (slender and anastomosing), *Bruguiera* sp. (blunt, fleshy and hard), *Lumnitzera* sp. (upright with modest heights) and *Heritiera fomes*, locally named Sundari, which have numerous woody peg-like pneumatophores with blind root suckers (Tomlinson, 1996).

6.12.1.2 Adaptations for Supporting Mangroves Against Fluctuating Ecological Conditions

Some species of mangrove plants, like herbs such as *Porterasia coarcata*, possesses adventitious roots; others have shrubs with short roots, like *Acanthus ilicifolius*; and others have climbers with a few feeble roots coming out from the nodes of stem, like *Derris* spp. and *Ipomea* sp.—and these all grow in their respective habitats, such as the lowest tidal level, mid-tidal levels, sandy dunes, the edges of rivers, saline waterbodies and marshes, with weak anchorage. In some cases, these short roots obliquely move downwards from near the lower portions of stems, providing additional support for anchorage by acting as stilts (Tomlinson, 1996).

6.12.1.3 Adaptations to High Salinity

Mangrove species, possessing a wide range of salinity tolerance levels, can grow and survive under the inundation pressures within intertidal saline coastal zones. The physicochemical parameters of the water and soil in the mangrove-estuarine complex of Sundarbans tend to interact with the biodiversity components of this ecosystem, mainly the mangrove plants, in three ways: by facilitating the supply of nutrients, by maintaining osmotic inhibition due to the absorption of salt water and by imposing toxicity.

Salinity, as the most important ecological parameter in the mangrove-estuarine complex, determines the extent of the different zones of estuarine tracts through salinity gradients, which are determined by topography, climate, hydrology, the duration of tidal inundation and exposure. This differential pattern of salinity distribution also determines the growth and quantum of productivity in mangrove forests (Sylla et al., 1996; Twilley & Chen, 1998). Salinity also influences the competitive interactions among mangrove species (Ukpong, 1995). Mangrove vegetation has been found to be more luxuriant and highly productive in freshwater-dominated estuarine tracts (Kathiresan et al., 2001). However, some abundant and widely distributed mangrove plant species, namely *Avicennia* spp. *Rhizophora* spp., *Ceriops* and *Brugiera*, can tolerate higher salinity levels, but the level of tolerance varies among species of true mangroves.

All mangroves can exclude most salts in the surrounding water and soil. In such a context, mangroves possess a unique system of regulating ion influx–efflux by regulating the ionic contents of their cells, and according to such capabilities, mangroves are classified into three major categories, namely salt excluding, salt excreting and salt accumulating (Walter, 1961). Mangrove species possessing the ability to exclude salts through an ultrafiltration mechanism of the root system include *Ceriops decandra*, *Rhizophora mucronata* and *Bruguiera gymnorhiza*, and by virtue of this power, these mangrove species can settle, grow and propagate following the ecological principles of ecological succession within the mangrove ecosystem of Sundarbans (Joshi, 1975). Another group of mangrove plant species, including *Avicennia marina*, *Avicennia officinalis*, *Avicennia alba*, *Acanthus ilicifolius* and *Aegiceros corniculatum*, exhibit a salt removal mechanism by regulating their internal salt levels through foliar glands. Several other mangrove plant species, including *Sueda maritima*, *Exoecaria agallocha*, *Sonneratia apetala*, *Lumnitzera racemosa* and *Sesuvium portulacastrum*, can accumulate higher concentration of salts within their biomass. Therefore, ecological succession, zonation and the distribution of mangrove plant species depend on the salt-removing efficiencies of different mangrove species. All these major mangrove plant species of the Sundarbans mangrove ecosystem have shown the ability of sodium-ion accumulation in their leaf tissue at levels ten times higher than those of potassium-ion accumulation. However, mangrove plant species such as *Heritiera fomes* and *Nypa fruticans*, because of their freshwater-philic nature, grow best in freshwater-dominated estuarine tracts and have demonstrated more potassium-ion accumulation power than sodium-ion accumulation power (Karmarkar, 1985; Naskar et al., 2004).

Prolonged higher salinity around the year in some estuaries of Sundarbans has been found to be detrimental to the mangrove community, especially because the hypersalinity retards the growth of mangrove trees like the *Avicennia marina* community (Selvam et al., 1991), reduces the biomass of *Bruguiera gymnorrhiza* (Naido, 1990), decreases the leaf area of mangroves, reduces total NPK and accelerates leaf sap. Osmotic pressure within the leaves of mangroves is retarded because of the hypersalinity in the interstitial water of the bottom soil within mangrove forests (Medina et al., 1995).

A detailed physiological study that analysed sap from mangrove plants has established the fact that more than 90% of salts are excluded due to the ultrafiltration mechanisms of different types of roots of mangrove organisms. Salts accumulating within mesophyll tissue ultimately reach the leaves, and on reaching a certain level of concentration, they are removed by the shedding of the old leaves or are sometimes stored safely in cell vacuoles. White (or grey) mangroves can secrete salts directly; they have two salt glands at each leaf base (their name derives from the fact that they are covered in white salt crystals). The development of salt glands in some species of mangroves has allowed them to secrete ions such as Na^+ and Cl^- (Fahn, 1979).

6.12.1.4 Adaptation for Limiting Water Loss

The nonavailability of fresh water within a mangrove forest has led to salty intertidal soils, and mangrove plants have developed ways and means to compensate for the loss of water through simultaneous and continuous transpiration and evaporation processes in their leaves and in soil, respectively. Moreover, the specific orientation of their leaves and their shrunken stomata (stomatal openings lying below the surface of the leaves) help them to reduce evaporation from their leaves by combating the impact of highly intense sunlight during the midday period.

Mangrove leaves lack vein sheaths surrounding their veins. Sometimes, thick but smooth cuticles, having small hairs, not only make the plants' leaves glossy, thick, fleshy, leathery, coriaceous and translucent with obscure leaf veins but also help the plants to ward off attacks from predators, to reduce the loss of water through transpiration and to increase their conservation of biochemical metabolites (Mitra et al., 2004, Kathiresan, 2000, 2004).

6.12.1.5 Adaptation for Nutrient Uptake

Although mangrove plants grow in nutrient-enriched soil, they encounter a significant problem in the process of nutrient uptake and face difficulty in surviving in waterlogged soil. Among different soil nutrients, nitrogen, phosphorus and potassium (NPK) play important roles in limiting mangrove forest structures in coastal areas in Brazil, South Africa and Florida (Da Cruz et al., 2013; Lovelock et al., 2006; Naidoo, 2009). The higher osmotic potential of the cellular components of mangrove leaf cells helps them to absorb highly dense saline water, which has a higher negative water potential.

However, the release of gaseous nitrogen, inorganic phosphates, sulfides, soluble iron and methane by anaerobic bacteria reduces the nutritious state of mangrove soil and produces a pungent odour. Atmospheric oxygen and different forms of macronutrients like iron, calcium, magnesium and silica are taken up within the bodies of mangrove plants by prop root systems from so-called nonconducive soil (Tomlinson, 1986).

6.12.1.6 Adaptations for Increasing the Survivability of Mangrove Offspring

Mangroves have evolved a special adaptive strategy by possessing a viviparous mode of germination to ensure proper settlements of propagules in the soft bottom soil that resists the dislodging of the growing saplings by tidal water and have developed a mechanism for helping their offspring to survive in very harsh environmental conditions. Upon gaining more buoyancy, mangrove seeds are suitable for water dispersal for a prolonged period (Tomlinson, 1986; Sarkar et al., 1989; Chakraborty et al., 1986, 2012; Chakravarty et al., 2004; Chatterjee & Chakraborty, 2014; Chatterjee et al., 2014; Sarker et al., 2020).

6.13 Faunal Adaptations

6.13.1 Adaptations to Wave Action

Several species of soft-bodied macrobenthic faunal components (actiniarians, bivalves, polychaetes, nemerteans, etc.) found in different ecological habitats and niches at the intertidal zone can resist the fierce action of waves and tidal flows by possessing suctorial feet to anchor themselves to the bottom sediment. The flat calcareous shells of brachyuran crabs are generally round in shape, and they can thereby increase their resistance to wave action. Different species of brachyuran and hermit crabs have strong legs for gripping the surface sediment of the intertidal belts.

6.13.2 Adaptations to Salinity Stress

Bivalves and gastropods close up their shells and opercula to protect their internal organs from high saline water, and they thereby prevent the entry of excessive salty water.

6.13.3 Adaptations to Desiccation or Alternating Wetting and Drying

Intertidal macrobenthic fauna like bivalves, gastropods and brachiopods enclose themselves within their shells and trap a small amount of water, through which they try to overcome desiccation. Some infauna, such as crabs, polychaetes, sipunculans, nemerteans and brachiopods, hide themselves within burrows and escape desiccation.

6.13.4 Adaptations for Gaseous Exchange

The respiratory organs of most intertidal macrobenthic fauna are kept in a protective cavity to prevent desiccation. Mollusks such as gastropods and bivalves have gills as respiratory organs which remain moist by allowing some drops of water within the bronchial chamber of the mantle cavity, protected by their shells. Gaseous exchange occurs in these water drops trapped in the mantle cavity. In gastropods, gills are reduced and the mantle cavity undergoes some modification with reduced gills, which then act as lungs for aerial respiration.

6.13.5 Adaptations for Feeding

Different intertidal macrobenthic faunal components such as bivalves and actiniarians are filter feeders. Mud snails, with the help of a hard ribbon-like structure called radula, graze on the algae on rock surfaces. Brachyuran crabs, by virtue of their strong feeding structures, act as both grazers and deposit feeders that consume detritus. Mudskippers forage on exposed mudflats by hopping and using their limbs like fins. Finally, polychaetes and brachiopods scoop out plankton from the overlying water during low tides.

6.13.6 Responses to Light and Temperature

Exposure to excessive sunlight with high-intensity UV-B radiation results in a drop in the rates of photosynthesis by mangrove plants because it alters their biochemistry, decreasing the production of amino acids and unsaturated fatty acids (Kathiresan, et al., 2001). Prolonged shading also negatively affects mangrove plants by lowering shoot biomass in comparison to mangroves that achieve proper exposure to sunlight.

6.13.7 Responses to Gases

Exchange of gases between mangrove-based biodiversity, water, soil and atmosphere is appeared to be very vital in respect of maintaining ecological health of the mangrove ecosystem by overcoming the problems with anoxic condition, methane loaded acid phosphate soils with pungent odours, elevated concentration of CO_2, etc. and also facilitate process of litter decomposition because of active roles of both macro and micro-decomposers.

The total biomass of *Avicennia germinans* seedlings has been found to diminish by 20–40% (Mckee, 1993). Prolonged inundation and an anoxic condition lead to the incomplete decomposition and higher production of methane, which is counterbalanced by the functional efficiencies of pneumatophores, extracting more and more oxygen, and by the release of more methane through the leaves of mangroves.

6.13.8 Responses to Coastal Changes

Coastal changes brought about by anthropogenic activities, such as deforestation expediting soil erosion, unwanted accretion, the modification of slopping in intertidal belts, increased evapotranspiration, poor freshwater flushing coupled with low rainfall (causing hypersalinity), sea-level rises, enhanced longshore currents and prolonged tidal inundation, stunt the growth of mangroves, cause the replacement of mangroves by salt marsh and cause the development of barren areas devoid of green vegetation cover (Ellison & Stoddart, 1991; Paul, 1999; Raha & Raju, 2016).

6.13.9 Responses to Soil Conditions

Mangroves growing on sediments have their own physical and chemical uniqueness because of continuous inundation and exposure during tidal cycles, harbouring different living organisms (microbes, algae, mangrove plants and micro-, meio- and macrobenthic fauna) acting as sites for different types of biochemical and ecological processes, like litter decomposition. Therefore, all the properties of soil will have major impacts on the nutrition, growth and propagation of mangroves and their associated biota by acting as important limiting factors. Some of the prime properties of soil are texture, pH, electrical conductivity and cation exchange capacity. The cycling of nutrients with the influxes and effluxes of nutrients in the mangrove ecosystem is closely linked to microbial decomposition, mineralization, plant assimilation and food chain–food web dynamics (Alongi, 1996). Nutrient availability and cyclical nutrient movements limit the growth, production, distribution and community interactions of many living organisms (Twilley & Chen, 1998).

In Sundarbans, the texture of mangrove soil has demonstrated spatial variations in respect of textural constituents like sand (18.1–50.5), silt (26.6–60.6) and clay (25.2–39.9) (Naskar et al., 2004) and a fluctuation in sediment constituents among different tidal levels like sand (1.3–87.8), silt (9.2–59.3) and clay (1–47.8) (Chakraborty et al., 2010) and eco-zones was noted.

6.13.10 Responses to Sedimentation

Uncontrolled and large-scale land reclamation, land-use changes and continuous deforestation over the past few centuries have caused the delinking of freshwater flows (creeks and channels of Bidyadhari and Matla estuaries) from the main estuarine networks of Sundarbans and have thereby increased salinity and accelerated the deposition of silts in different deltaic networks of Sundarbans. In addition, continuous deforestation followed by erosion processes have made waterbodies turbid and promoted unwanted accretion, resulting in inconveniences in the water transport system (Chakraborty, 2011).

6.13.11 Responses to Metal Pollution

Mangroves can very efficiently trap trace metals in nonbioavailable forms because of a very unique ecological interaction. Any kind of disturbance in the natural ecological process of a mangrove ecosystem can impair the ability the mangrove soil, causing them to lose their metal-binding capacity and thereby lose their ability mobilize metals. Mangrove soil under such conditions acts as a reservoir and source of heavy metals, shifting it away from its role as a heavy metal sink (Lacerda, 1998).

6.13.12 Responses to Organic Pollution and Inorganic Pollution

Owing to the prevalence of anaerobic and reduced conditions after being periodically flooded by tides, mangrove ecosystems, endowed with high clay and organic matter concentrations, possesses a high capacity to retain pollutants. In view of this, upon receiving wastewater, mangrove ecosystems can tackle such harmful impacts and improve water quality by utilizing the ecological potential and efficiency of wetlands. The productivity of mangroves is thereby connected to the use of anthropogenic waste, rendering beneficial impacts, particularly in areas under nutrient-deficient conditions (Wong et al., 1995).

6.13.13 Responses to Global Changes

Mangrove forests at land–sea interfaces are susceptible to many global changes. Mangrove habitats experience the following environmental stresses: higher temperatures, altered hydrologic regimes, changing patterns of rainfall, changing evapotranspiration, increased runoff and salinity, rising sea levels and a higher likelihood of tropical storms with higher magnitudes and frequencies (Michener et al., 1997). At an increase in temperature exceeding 35 °C, the modification of root structures, the settlement of seedlings on the muddy bottom of mangrove forest floors and rates of photosynthesis were very negatively affected. Moreover, increased salinity and the higher sulfate concentrations in sea water caused by reduced rainfall and freshwater runoff increase the mortality of mangroves and reduce mangrove production.

6.14 The Physicochemical Properties of Soil in Sundarbans

Out of so many ecological variables, the physicochemical properties of sediment and water in a given environment is fully correlated with the occurrence and abundance of any organism in search of food, shelter, spawning and breeding grounds (Yeragi, 2003). The soil quality of mangrove ecosystems plays significant roles in not only ensuring primary productivity but also opening up new avenues for more aquacultural production while maintaining bottom stability.

The characteristics of sediment have often been found to have influential roles in determining the distribution of macrofaunal communities (Sanders, 1968; Kunitzeret al., 1992; Seiderer & Newell, 1999; van Dalfsen et al., 2000). Sediment properties, varying with tidal amplitudes, currents of water, the sloping gradients of intertidal belts and vegetation composition, affect the distribution of both meiobenthic and macrobenthic fauna in the mangroves in Sundarbans (Chakraborty et al., 1992). The seasonal spectrum of ecological variables, especially the supply and deposition of sediments depending on the wave and current of the estuaries, also shapes the temporal and spatial variation in the macrobenthic community among different months and seasons at different tidal levels, which work and rework the deposits in different seasons (Sanders, 1958).

Sanders (1958) also stressed the relationships among benthic fauna as deposit feeders, the percentage of clay content in soil and the determinants for the distribution of filter feeders by medium-size sand particles in sediment with a specific sorting coefficient. This hypothesis was corroborated by the observations of Tenore (1972), based on the research findings on the species distribution in the Pamlico River, where the anoxic state of the overlying water over the mud community impaired the successful establishment of a benthic community. In addition to soil-quality parameters, the seasonal oscillation of other factors in an estuarine environment, such as salinity, pH and dissolved oxygen (DO), play more important roles than sediment composition in directing species distribution (Muus, 1967).

Mostly, the soil texture of the Sundarbans mangrove ecosystem is sandy loam and silty clay. The soil pH is alkaline in nature. Most of the nutrients, like total available nitrogen, phosphate, silicates and organic carbon, with an amount of more than 1% act as major sources of energy for microbes that, through a decomposition process, generate and release nutrients that will undergo further biochemical processes.

6.14.1 *Sediment Fluxes over Geological Time Scales*

Sediment carried by rivers carry a huge amount of sediment particles mostly in the form of silt and clay as a major percentage (>90%) of sand particles get deposited at the upper stretches (upstream) of rivers. Upon mixing sea water at an estuarine stretch, this sediment-loaded freshwater flow affects water by developing turbidity and contributing different forms of chemicals, such as nutrients and discharged pollutants from human activities, and the ultimate result is the deterioration of water quality. After being deposited on the surface of sediment particulate organic matter is decomposed and mineralized through the process of digenesis. The dissolved fraction of inorganic nutrients from the surface of the sediment bed is transported to the water column by way of sediment fluxes. The sediment processes are linked with water-quality processes by two physical processes, namely the sorption and desorption of nutrients such as phosphorus, sulfur and silica, which get suspended over the sediment (Park et al., 1995). Particulate organic matter derived from plant materials (mangroves and algae) settles down to the bottom of a water column and gets decomposed by either an aerobic process or an anaerobic process. During the process of the decomposition of organic matter, oxygen concentrations are depleted, but they are counterbalanced by the growth of green autotrophs producing a good amount of oxygen, which maintains the ecological health of the ecosystem.

All these processes, dominated by the combined action of tidal currents and the flow patterns of the estuary, determine the ecodynamics of the estuarine environment through sediment mixing and the transportation of sediments across different parts of estuarine networks (Miranda et al., 2002; Suguio, 2003). The sediments of the estuarine beds and mangrove forest floors are mostly fluvial in origin, but portions of the sediment are derived from coastal drift and from adjacent marine systems. The composition of sediments is determined to some extent by the flocculation of suspended clayey material, which develops mainly because of variation in pH between the fluvial and estuarine environments, and this in turn regulates the physical and chemical condition of the estuary (Barbosa & Suguio, 1999). Tidal creeks, having very complex geomorphological, chemical and physical properties, transport nutrient-enriched organic material from mangroves to estuaries and thereby trigger an increase in the primary productivity of these ecosystems. The productivity of an ecosystem in turn promotes more fisheries and aquacultural development. The prevailing complexity in the geomorphological processes in estuarine networks in view of the distinct seasonality of other ecological parameters and the tidal cycle determines the quality of the water in these environments. Gaining a proper understanding of all those intricate relationships between fluvial processes and the seasonal dynamics of the physicochemical composition of the waters may contribute to the successful ecomanagement of the ecosystem and thereby maximize the economic benefits that can be derived.

An analysis of nutrient concentrations in the form of ammonia and phosphate in the surface, subsurface, bottom and interstitial waters demonstrated the differences and the higher amount that is found in water within soils. Nitrate becomes denitrified in an anoxic water column and sediment bed, whereas toxic conditions promote the release of ammonia from sediments, which is directly transformed into nitrate via the process of nitrification. Nutrients in the form of phosphate from the sediment bed of an estuarine system is released to the overlying water, and anoxic conditions expedite this process. The sediment digenesis process, enhancing sediment fluxes from the sediment bed, is very slow in a sandy bed, having a negligible amount of organic matter that is deficient in nutrient contents as well as having unstable textural constituents.

The Sundarbans, located at the confluence of several gigantic estuaries that meet the sea at the Bay of Bengal in mesotidal to macrotidal environments, enjoys semi-diurnal tides and is very prone to hazards from frequently occurring tropical storms and tidal surges. During the past couple of centuries, the deltaic Sundarbans has stood firmly against appreciable progradation, exhibiting strong bidirectional currents. Different geo-ecological processes in the entire landscape of Sundarbans promote the formation of numerous intertidal and subtidal shoals. The muddy sediments, after being dislodged from the seaward face of the delta plain and from tidally scrubbed inlets, are carried to further inland areas through the drainage network and are deposited across interfluves. Textural analyses of the tidal flat sediments across the entire Sundarbans have revealed the presence of medium sand and clay. Generally, the horizontal surfaces of tidal channel floors and intertidal flans are areas of sloping depositional surfaces, such as riverbanks and point bars, which are areas of rapid lateral accretion. Five geomorphic zones of sedimentation have been determined according to the mode of alternate processes of erosion and accretion: (i) depositional sand bodies at the mouths of an estuary or a tidal channel are in accordance with the strength of ebb- or flood-tidal currents; (ii) the unidirectional patterns of sedimentation occurs at the margins of lower estuary or tidal channel margins, producing intertidal flats with dominant unidirectional bedforms; (iii) upper intertidal flats and tidal point bars are characterized by accretionary point bar deposits through relatively rapid sedimentation; (iv) higher landward portions of intertidal flats or marshes are endowed with intermittent subaerial exposures; and (v) supratidal flats containing interlaminated sand and clay incorporate scraps or bluffs derived from erosive undercutting by wave actions (Selvam, 2003; WHO, 2011).

6.14.2 Soil Formation and the Determinants of Soil Quality in Sundarbans

The mangrove vegetation of Sundarbans can physiologically tolerate high concentrations of salt by developing its own mechanism to obtain fresh water, counteracting the strong osmotic potential of sediments (Ball, 1996). Many mangrove plant species can exclude salts through an ultrafiltration process carried out by their root systems during the absorption of water from soil. Specialized salt glands are developed in the leaves of some mangrove plant species, which store salts and excrete them during seasonal shedding. Some other mangrove plant species store salt in leaf vacuoles, and over time, succulent configuration develops. Often, the barks of some mangrove plants accumulate salt and transfer salts from leaves to wood.

In Sundarbans, India, a considerable amount of the sediment reaches the sea through two marginal estuaries: the Hooghly in the west and the Padma–Meghna in the east. However, a portion of this sediment load within water is pushed back by the tidal influence of estuaries and creeks. Finer sediments are deposited along the floods plains of the rivers and intertidal belts of estuaries when the silt-laden water spills over those ecozones during high tide and recedes during low tide and the resulting elevation of the water–land interface, in the form of an active delta. The Ganga–Brahmaputra system carries a suspended sediment load amounting to more than one billion tonnes per year (Rudra, 2010).

The clay loam of sediment predominates Sundarbans, whereas silty and sandy loams also occur in many areas because most of the soils are derived from azonal alluvial deposits (transported alluviums) (Chaudhuri & Choudhury, 1994; Sarkar et al., 1999). The formation of numerous tidal flats in the form of tidal deposits on the margins of such estuaries is due to the silting from the headwater flow through the deltaic distributary channels of the Ganga River (Annon, 2003).

6.14.3 The Soils of Sundarbans: The Uniqueness of the Physicochemical Properties of Soil

In the Sundarbans mangrove ecosystem, soils are usually deposited in the convex margin of the meandering creeks of the estuaries and tidal inlets, which form the ideal situation for the luxuriant growth of tidal forests and mangroves in the coastal region (Blasco, 1977). The alluvial mangrove soil, rich in suspended matter in the low-lying coastal plains, helps the newly settled mangrove saplings in the muddy shallower parts of the estuaries, which gradually undergo progradation until they face the Bay of Bengal (Vannucci, 1989).

The biological production in an aquatic ecosystem is directly connected to the physicochemical properties of the soil, which play roles in providing a large number of both macroelements and microelements to living organisms so that they can maintain their metabolic activities and counteract the problems in the saline micaceous deltaic alluvium within humid regions. Higher saline and acid sulfate soils in the coastal tidal forest of Sundarbans, dominated by sulfate and chloride salts, exhibit acidic pH levels, high electrical conductivity (EC), high levels of organic material with humic acids and a lower water-holding capacity and have few avenues for aeration.

6.14.4 The Textural Composition of Mangrove Sediments

Texturally, more than 90% of mangrove soils are represented by silt, whereas the remaining 10% of the bulk is composed of fine sand and clay. The sediments within mangrove ecosystems contain a short range of grain size–sorting values which reflect the high sensitivity of their depositional environment (Das & Bhattacharya, 1991; Bhattacharya, 1999). Granulometric analyses of sediments from the floors of mangrove forests have revealed the percentage of silt content to range from 49.40% to 99.57%, whereas that of clay content ranges from 0.14% to 34.41%. Mineralogically, the silts of mangrove soil are composed of quartz (85.90%), feldspar (5.6%) and mica. Intertidal belts containing mica occur more towards the upper and landward stretch of the river estuaries, which have low-tidal-energy conditions. Fine mangrove debris, the fragmentary parts of the shells of crustaceans, gastropods, bivalves, brachiopods and foraminifers, constitutes the biogenic components of mud deposits, the total amount of which rarely exceeds 10–15%.

A texture analysis of the soil of the Sundarbans mangrove ecosystem showed mostly silty clay soils with the percentage of sand particles ranging from 3.81% on Prentice Island to 42.62% on Sagar Island, whereas the percentage of silt clay ranged from 57.65% on Sagar Island to 94.96% on Prentice Island. In support of variabilities of soil parameters, it was noted that the water-holding capacity of silt clay soil is higher (59.03%) on Lothian Island than the comparatively higher sandy soil (43.16%) on Sagar Island. The soil nutrient statuses of the percentage of humus carbon (0.04–0.24%) and that of organic carbon (0.45–1.86%) in the soil of the mangrove-dominated areas of Sundarbans were found to be higher in the surface layer than in subsurface profiles, mainly because of the unequal distribution of organic residues in the different strata of sediments. The sediments in the mudflats of different islands of Sundarbans generally contain different percentages of textural components like sand (12%), silt (66%) and clay (22%), whereas mudflats restricted in the interfluves of the Hooghly–Matla–Roymangal estuarine complexes show the presence of sand (15%), silt (80%) and clay (5%) (Das, 2011).

Appreciable amounts of soil nutrients were present in the form of available P (10–42 ppm), available N (85–116.4 ppm) and available K (331–630 ppm) (Sahoo et al., 1985; Gupta, 1987). The pH of the mangrove soil varied from 6.5 to 8.1 (Gupta, 1987). Other pronounced parameters of soil, like soil temperature, salinity, organic carbon and the textural components of soil (sand, silt and clay), were found to be higher during pre-monsoon, whereas available nitrogen, available phosphorous and available potassium were found to be at maximum during monsoon. The post-monsoon season in the Sundarbans mangrove ecosystem demonstrated the lowest temperature, available nitrogen, available phosphorous and available potassium but moderate levels of other parameters (Annon, 2003; Bell, 1940; Barbosa and Suguio, 1999; Bhattacharya A, 1999, 2000; Das GK, 2009).

6.14.5 Soil Reaction (pH)

Soil pH, being one of the most important factors for maintaining the productivity of any waterbody by virtue of its roles in regulating the metabolic activities of living biota, exhibited a range from 7.05 and 7.96 in the different soil samples collected from different parts of Sundarbans, India, which establishes that the soil in this environment is alkaline in reaction. Sometimes pH goes down to 3.8 in an anoxic reduced state, making the environment very much nonconducive to the living of organisms. The pH of the soil also controls the availability of the nutrient contents of the floors of mangrove forests through mixing, the rate of decomposition and the mineralization of organic matter, the fixation of N and P and thereby the growth and survival of different species of flora and fauna. According to the range of soil pH, soils are classified into acidic ($pH < 7.0$), neutral ($pH = 7.0$) or alkaline ($pH > 7.0$).

6.14.6 Electrical Conductivity (EC)

Electrical conductivity (EC), referring to the reciprocity of electrical resistance, indicates the levels of salt concentration and is used for the classification of soils on the basis of the exchangeable sodium percentage (ESP) of soil (Bandyopadhyay et al., 2001). EC is expressed as electrical resistivity and acts as the point of separation between saline soil and nonsaline soil (Bandyopadhyay et al., 2001). The electrical conductivity of soil also indicates its osmotic potential (‰), which is connected to the total concentration (ppm) of salts.

6.14.7 The Salinity of Soils and its Roles in Soil Classification

Salinity, being a prime limiting factor in a mangrove-estuarine complex, determines the occurrence, distribution, species diversity and richness of that ecoregion and is associated with several other factors, like the availability of nutrients (NPK), organic carbon, soil texture and the water-holding capacity of soil. It also controls mangrove zonation patterns. Decreasing the osmotic potential of the fluidity of soil has the net effect of reducing the availability of water to plants. However, some tolerant mangrove species, through the production of organic acids, the uptake of salts, etc., can adjust their internal osmotic potential (Bandyopadhyay et al., 2001). Mangroves that grow in salinity-stressed conditions show stunted growth and smaller and thicker leaves with dark green or bluish coloration. The salinity of soil develops due to the capillary rise in saline groundwater; depends on soil texture, elevation, the density of mangrove coverage, rainfall, temperature and drainage in mangrove ecosystems; and plays decisive roles in determining mangrove species diversity and richness and in the zonation of mangroves (Bandyopadhyay et al., 2001).

The soils of Sundarbans are classified into saline soil, saline alkali soil, nonsaline alkali soil and degraded saline alkali soil depending on their salt concentration. The majority of mangrove species in Sundarbans prefer soil with a modest water-holding capacity, a low salinity substratum, scopes for aeration, an alkaline pH, geomorphic units rich in semi-decomposed organic matter and calcium or magnesium in the subsurface sediment (Annon, 2003). The following subsections elaborate on these four types of soils.

6.14.7.1 Saline Soils

Saline soils are characterized by the visible occurrence of white salt encrusted on the soil surface and by having a higher concentration of electrolytes than normal soils do. These soils contain chlorides, nitrates, sulfates of sodium, calcium and magnesium but are devoid of soluble carbonates. The pH of saline soil is less than 8.5. Saline soil with higher concentrations of electrolytes negatively interfere with the availability of nutrients in Sundarbans soils. Small amounts of bicarbonates may be present but are usually absent.

6.14.7.2 Saline Alkali Soils

Depending on the salinity of the soil, ranging between 3 and 12 ppt, the soils of Sundarbans are categorized as saline alkali soils when they have a pH higher than 8.5 (ESP_15). Saline alkaline soils are produced due to leaching by seawater (NaCl) in the presence of magnesite or calcite. However, the leaching away of salts from a saline alkaline soil leads to the development of nonsaline alkaline soils when sodium ions enter the exchange complex (Das, 2011).

6.14.7.3 Nonsaline Alkali Soils

This category of soils, which forms due to the leaching of water with the lowest salt content or mainly due to rain water, has a pH of 7.1–8.2. The salinity (percentage of salts) of this soil ranges between 0.1% and 0.9%. In comparison to other types of soils in Sundarbans, this category of soils contains an average-to-high clay percentage and immobile sesquioxide compounds. This soil is devoid of any carbonate.

6.14.7.4 Acid Sulfate Soils

This type of soil, characterized by extreme acidity (pH < 3.5–4.0), is found in several deltaic islands of Sundarbans. Acidic sulfate soils are formed mainly due to the oxidation of different forms of sulfidic materials and pyrites which accumulate in soil due to a reduction in sulfate salts. These soils are formed due to the prolonged submergence of mangrove forest zones by brackish water and become very rich in clay, soluble sulfur and organic matter, with high humus content and exchangeable Fe, AL and Mn ions but very poor available P content. These soils having higher oxidized products of pyrites (FeS_2), whereas with exceed the neutralizing capacity of soils at pH levels below 4.0 and other byproducts like jarosite [$KFe_3(SO_4)_2.(OH)_6$] or natrojarosite [$NaFe_3(SO_4)_2.(OH)_6$] do not possess that property (Bandyopadhyay et al., 2001).

6.14.8 Soil Distribution in Different Parts of Sundarbans

Different parts of Sundarbans, especially Gosaba, Hasnabad, Hingalganj, Sandeshkhali II, Jaynagar II, Mathurapur II and Kultali, are endowed with saline soils, whereas saline alkaline soils are found in Sagar, Kakdwip and Kulpiblocks. Nonsaline alkali soils have been recorded in the Mathurapur block, and degraded saline alkali soil has been found in the Sandeshkhali, Tushkhali, Ghusighata and Damakhali regions (Anonymous, 1987). The saline soils of the northern parts of Joynagar II, Mathurapur II and Kultalilie south of the Bidya River and the southern portion of Gosaba, Hingalganj and Sandeshkhali II also lie below the Bidya River.

Saline alkaline soils are found in some other parts of Sundarbans, like Mathurapur II's Chuprijhara River, Chatu River, Raidighi River, Mani River, Thakuran River, JalaberiaKhal, Saraldakhal and Kaloakhal on the southern parts and small streams on the northern parts. The same is true of several other areas of Sundarbans, like Sagar. Surface soils with moderate salinity are scattered over different blocks of Sundarbans, like Sandeshkhali I and II, Basanti, Gosaba and Hingalganje. An analysis and interpretation of the Normalized Difference Snow Index (NDSI) map of 2020 has revealed the presence of high to very high soil salinity on land cover, mostly in the southern part of the study area, indicating that those areas are under threat of salinization. Changes in the salinity of soils have been trending towards increasing concentration in the coastal islands of Sundarban, West Bengal, since 1975 and until today, which is alarming for the ecological health of soil. After Aila in 2009, several parts of Sundarbans (Basanti, Sandeshkhali I, Gosaba, and Haroa) have been affected by the intrusion of saline water and the consequent deposition of salts over the surface of soils (Das et al., 2016).

An increase in the salinity of both soil and water has taken place from the eastern to the western parts of Sundarbans, specifically in the Bidyadhari River estuary, which has been slowly dying due to a continuous siltation process since the sixteenth century. This Bidyadhari River estuary, acting as prime supplier of freshwater to the central and eastern parts of Sundarbans, is best known for carrying sewage from Kolkata through several connecting channels for the past couple of centuries (Chakraborty, 1998). The rate of enhancement for salinity (1.3 psu) is found more in the central part of Sundarbans, resulting in a loss of *Hereteria fomes*, locally named the Sundari tree, the pride of the Sundarbans mangrove forest (Banerjee et al., 2012a). The fertility status of the sediments of most parts of Sundarbans is experiencing higher plant mortality from salinity incursion, when salt water percolates through subsoils, leaving the residue salt on the top of most layers during the dry period. However, after receiving fresh water from both river drainage and precipitation during the monsoon season, the level of the salinity comes down to a certain extent (Bagchi & Mukherjee 1978).

6.14.9 Soil Nutrients

The soil nutrients, comprising mostly primary nutrient elements like nitrogen, phosphorus and potassium, trigger the growth and propagation of the mangroves of the Sundarbans in India. Secondary nutrient elements like calcium, magnesium and sulfur and micronutrients such as boron, copper, manganese, zinc, molybdenum, iron and chlorine also render support for the healthy growth and propagation of mangroves.

Among the different forms of soil nutrients, organic carbon acts as a source of energy for the microbes participating in various biochemical processes, resulting in the release of different nutrients. This soil nutrient is very slowly generated within the mangrove swamp because of the influence of oscillating levels salinity during different seasons. Humus, a kind of organic matter in soil, controls the physical, chemical and biological activities in soil while improving the textural properties of soil, contributing to aeration, increasing water-holding capacity, facilitating the buffering and ion-exchange capacity of soil, increasing the solubility of soil minerals and acting as a storehouse of various nutrients required for biological production. Most of the humus is produced within the floor of the mangrove forests by the litter decomposition process and supplied by riverine flows. Organic carbon and other organic substances not only determine the ecological health of soil but also contribute to determining variations in different physicochemical parameters and exhibit significant positive relationships with several trace metals and their oxides, like iron and copper. The rate of organic matter mineralization within mangrove ecosystem has been found to be quite slow in saline soils with a high pH. In contrast, the rate of nitrification is too slow in this soil because of high salinity and alkalinity.

6.14.9.1 Nitrogen

Different forms of nitrogen either in organic combinations or in inorganic nitrogenous compounds (NH_4, NO_3 and NO_2) are released mostly through the bacterial decomposition of organic matter. The easily decomposable form of organic nitrogen is known as available nitrogen, which promotes aquatic productivity. In saline soils, the roots of mangrove plants face difficulty in absorbing nutrients because of reduced root volume and for other physiological reasons.

6.14.9.2 Phosphorus

Phosphorus, being a major nutrient for the biological production, aids in the assimilation of nitrogen into cellular matter, in addition to rendering support for the cell division, metabolism, respiration, growth and synthesis of protein and very often becomes a limiting factor in fauna food organisms' production. Deficiency in the generation of phosphorus is observed in the highly acidic and very low pH sulfate soils of Sundarbans, and this deficiency acts as a limiting factor in the growth of mangrove species and associated plants. The uptake of phosphorus (P) by mangroves is also inhibited under high soil salinity due to the competitive inhibition caused by chloride ions ($Cl^{=}$) given that both chlorides and phosphates, being anions, are absorbed by essentially the same absorption sites.

6.14.9.3 Potassium

Potassium, another nutrient element, aids in the synthesis of protein and chlorophyll and in triggering the growth of aquatic plants. This important nutrient element has high availability in the Sundarbans mangrove ecosystem because of the predominance of micaceous minerals in the saline soils of Sundarbans.

6.15 The Topography of Sundarbans: Unique in Having Mudflats and Sandflats

6.15.1 Mudflats and Their Ecological Relevance

Mudflats represent an important transitional geomorphological unit because it has the very fine silty clay textural components of sediments and specific nutrients within salt marshes or mangrove swamps that extend to the landward side and the shelving shores of the continental shelf that extend to the seaward side (Viles & Spencer, 1995). An expanse of mudflats develops through the continuous deposition of fine clay or silts through a flocculation process in estuaries and sheltered coastal areas (behind sand bars and sand spits). In low-energy areas, mudflats are found to predominate and to experience suspensional depositions, mostly in the middle to the upper stretches of the estuaries in mesotidal to macrotidal settings. Mud deposition occurs mainly in the margins of tidal creeks, often intersecting the intertidal flats that possess mid-channel bars and point bars. The extent of areas under the mudflats ranges from a few square metres to a few hundred square metres, and the thickness of mud blankets have fine rhythmic laminations of silt and clay that vary from a few centimetres to less than a metre.

The sides of the creeks and channels within mangrove swamps or salt marshes are characterized by having mudflats in the lower intertidal zone, producing some significant geomorphological features under the horizontal variation in the tidal velocity regime. The mudflats of the Sundarbans, exhibiting a pronounced break in slope at or near the mid-tide level, are significantly cliffed and notched up by wave actions, below which coarser grain sizes are deposited and extended as sand-dominated flats up to a low watermark. Mudflats in the shallow water estuaries of places like Thakuran, Bidyadhari, Suptamukhi and Matla are terraced and dominated by muddy sediments. Steeper mudflats found along the banks of Hooghly and Ichhamati estuaries are dissected by meandering tidal and rill channels.

Expanded mudflats like sea beaches also dissipate energy from the intensity of waves and tidal current and thereby lend protective support against currents for the channel-side encroachment of mangrove swamps. The formation and expansion of mudflats involves a geomorphologically dynamic process when sediments are transferred over tidal cycles and the seasonal fluctuation in climatic conditions, including storm cycles. Prentice Island and Lothian Island, with extensive mudflats at channel wards at the mouth of the Saptamukhi estuary, promote marked ecological succession with the regeneration, growth and colonization of mangrove forest layers. Therefore, an active process of accumulation represents a significant process triggering the remarkable succession of mangroves. Productive mangrove forests with extensive intertidal mudflats support the lives of a galaxy of benthic fauna, many of which find an ideal condition to work and rework in their favourable habitats with their bioturbatory activities on soft sediment and suspension-feeding organisms, and they not only reduce the turbidity of the overlying water but also produce different surface and subsurface structures on the mudflats in the form of holes, burrows, mounds, pits, pellets and pillars.

6.15.1.1 Mud—Mounds and Mud Balls as Indicators of Coastal Processes

The mud mounds occur in the subaerial environment of the coastal region over a space of tens of metres on many occasions and have elliptical to subelliptical shapes. These mounds with horny projections are generated in the zones of severe erosion; mud mounds, having different configurations (elliptical, subelliptical or with horny projections, scours or grooves at the central parts, supported by rippled sandflats), occur in the subaerial environment of the coastal region over a space of tens of metres, which indicates that they originate from the corrosion of backwash flows along the coastal tract in an estuarine environment (Allen, 1984).

Mud balls, with different shapes and sizes (cylindrical, sphere, spindle, ellipsoid, etc.), are formed from the mud chunks derived from collapsed mud banks due to high tidal intensities and the vigour of wave actions spreading over the stretch of the transitional zone of the coastal region (Bell, 1940). Broken parts of animal remains (the shells of gastropods, brachiopods and pelecypods) are found to become impregnated either within the mud lumps early or inserted later during the formation of mud balls through the rolling of the tidal and wave actions. These mud balls are composed mostly of silty clay textural components with some few fine sand grains and biotic materials added over the surface during rolling. The presence of mud balls in the coastal environment can shed some light on the modes of erosion and also help in determining the hydrodynamic status of estuaries.

6.15.2 Sandflats and Sandy Beaches with Ecological Significances

The foreshore sandy beaches of Sundarbans, India, on different deltaic islands, like Sagar, Kalas, Bakkhali and Fraserganj, possess sea beaches or sandflats with widths ranging from 300 to 600 m. Very fine to coarse sand particles with a good assortment of sediments are made up of siliciclastic materials, which are constituted predominantly by quartzo-feldspathic (95%) materials and otherwise by accessory muscovite, biotite, hornblende and ilmenite materials (Bhattacharya, 1993). A negative correlation exists between grain size, wave energy and beach slope. These sandflats or foreshores include a number of surficial physical structures, such as backwash marks, swash marks, backwash ripples, rhomboid ripples, linguoid ripples and lunate ripples, current crescents and flat-topped ripples. Interference ripples formed by waves and tidal currents, which look like ladder-back chairs, are observed in the lower foreshore zone (Komar, 1976).

6.16 The Alternation of Physicochemical Properties and Their Effects on Sundarbans, India

The current environmental conditions Sundarbans are quite different from those of the past century. The physicochemical characteristics of the brackish waters of the Sundarbans mangrove ecosystem of the Indian subcontinent, especially the prime one, salinity, have been changing along with changes in the river course due to a gradual easterly tilt, mainly due to neotectonic movement (Das, 2009a, 2009b), and this is considered as the most probable cause of reduced freshwater supply to the estuarine rivers of Sundarbans. Additional causes include the reduced supply of fresh water causing salinity invasion, urban sewage discharge, industrial pollutants and the accidental spillage of toxic substances and other inputs from anthropogenic interferences. All these physicochemical parameters, undergoing decadal changes in the estuarine waters of the Sundarbans, have had great impacts on the growth and distribution of mangroves and their associated flora and fauna. The expansion of mangrove forests through transformation into salt marshes has been observed in Sundarbans because of enhanced tidal amplitudes and changes in the water quality of estuarine rivers. The present environmental status of the tidal wetlands of mangrove-dominated Sundarbans requires the regular monitoring of both water quality and soil quality to maintain the proper ecological health of the region and to ensure the proper growth and propagation of halophytic mangrove plant species, along with their associated biodiversity components. Each species in this ecological region has its own environmental requirements from the macrotidal delta. Each species of mangrove has a specific level of tolerance for making physiological adjustments in the absence of the optimum levels of salinity and inundation duration. The disappearance of two prime mangrove plant species—*Heritiera fomes*, locally named Sundari, and *Nypa fruticans*, locally named Goalpata—has caused a gradual increase in salinity level because of a reduction in the freshwater supply and the continuous deposition of alluvium in Sundarbans (Blasco, 1977). The western part of Sundarbans in the Ganga delta of West Bengal, India, has not been receiving fresh water where mangrove species are usually found, even around 5000 years ago, as revealed by the presence the arbore scent vegetation near the Calcutta port region (Mukherjee, 1972).

6.16.1 Sediment Starvation Under Coastal Morphodynamics

Reduced freshwater flows loaded with terrestrial sediments from the riverine networks of Sundarbans result in a shortage in the supply of sediment within the deltaic Sundarbans. Such sediment starvation alters the geo-morphodynamics of coastal processes. Shortages in sediment supply along the stretch of wave attacks have shown that a portion of the delta generates the local source of sediment supply in the system. High rates of erosion, land subsidence, sea-level rise and high water submergence occur frequently at the sea face between the Saptamukhi River mouth to the Gosaba River mouth in the West Bengal portion of Sundarbans. Considerable parts of the sea-facing deltas of Sundarbans are rapidly getting abraded by waves due to disruptions to the longshore sediment transport supply. After being attacked by such wave attacks, the tidal flat platform with mangrove vegetation pushes the sand components of the sediment into the inner parts of the mangrove swamps, thereby reducing the buffer areas of the low-tide shore platform along the base of cliffed mud banks. Often, mangrove vegetation at the partially degraded shore experiences stunted growth because of active saltwater sprays, which fail to act as wave breakers, particularly during high spring tides, as evidenced in the sea face of Lothian Island, Tat Island, Dhanchi Island, Bhangaduni Island, etc. in Sundarbans.

6.17 Littoral Microorganisms and Physicochemical Parameters in the Littoral Zones of Coastal Belts

Different microorganisms inhabiting the intertidal zones of coastal wetlands play crucial roles in nutrient circulation, biodiversity and carbon retention (Bai et al., 2013). The synergetic potential of those microorganisms can have bioconversion potential for sewage-derived components within the littoral (intertidal and subtidal zones) parts of the coastal environment, which simultaneously reduce both pollution and carbon (Averill et al., 2019; Wang et al., 2024). The coupling effect of microorganisms exhibits their ability to interact with several other structural components of the ecosystem, influencing the processes of the bioconversion of organic matter into methane under anaerobic conditions via coupling between anaerobic digestive microorganisms and methane-producing microorganisms (McIlroy et al., 2017). Spatial and temporal variations in several ecological parameters, such as temperature, pH, the transparency of water, salinity, rainfall and the intensity of sunlight, influence microbial processes in the intertidal zone (Ahn et al., 2012; Averill et al., 2019; Kanamori et al., 2017). Seasonal oscillations tend to affect the coupling and composition of the microbial community by influencing the amino acids secreted by plant roots and other carbon sources (Carfora et al., 2021). Oxygen content and soil properties at the different strata of soil affect not only the interactions between microorganisms but also the distribution of aerobic and anaerobic bacteria by regulating moisture content and carbon availability (Robert et al., 2003; D'Hondt et al., 2015). Riverine nutrient inputs to the coastal zone help to drive organic matter cycling and microbe-driven denitrification (Li et al., 2016). Metabolisms of nitrogen and sulfur are being triggered by some interactions between anaerobic ammonia-oxidizing bacteria and sulfur-oxidizing autotrophic denitrifying bacteria (Deng et al., 2022). Coupling among microorganisms can also contribute to the cycling of nutrient elements and metals. Anaerobic methane-oxidizing bacteria coupling with iron-reducing bacteria has been shown to promote the oxidation of methane in sediments (Beal et al., 2009).

References

Achara, M. S., Panigrahi, S. N., Satpathy, K. K., Sahu, G., Mohanty, A. K., & Panigrahy, R. C. (2014). Nutrient dynamics and seasonal variation of phytoplankton assemblages in the coastal waters of southwest Bay of Bengal. *Environmental Monitoring and Assessment, 186*(9), 5681–5695.

Allen, J. R. L. (1984). Sedimentary structures, their character and physical bases. In *Development in sedimentology*. Elsevier.

Alongi, D. M. (1996). The dynamics of benthic nutrient pools and fluxes in tropical mangrove forests. *Journal of Marine Research, 54*(123), 148.

Annon. (2003). Mangrove ecosystem: Biodiversity and its influence on the natural recruitment of selected commercially important finfish and shellfish species in fisheries. National Agricultural Technology Project (NATP). Indian Council of Agricultural Research (ICAR). Principal Investigator: George, J. P.; Co–PI: Chakraborty, S.K., & Damory, S. N. (pp. 1–514).

Annon. (2005). *Studies on bio-resource assessment and management of degraded mangrove ecosystem of Midnapore Coast, West Bengal. Research project report* (pp. 1–99). Ministry of Environment and Forests, Govt. of India. Principal Investigator: Susanta Kumar Chakraborty [Sanction No.3/6/2001-CSC (M) Dated 5th Nov 2001].

Anonymous. (1987). *A long term multidisciplinary research approach and report on mangrove ecosystem of Sundarbans*. Department of Marine Science, University of Calcutta.

Bagchi, K. (1944). *The Ganges Delta*. University of Calcutta.

Ball, M. C. (1996). Comparative eco-physiology of mangrove forest and tropical low land moist forest. In S. S. Mulkey, R. L. Chazdon, & A. O. Smith (Eds.), *Tropical forest plant ecophysiology* (pp. 461–469). Chapman & Hall.

Bandyopadhyay, B. K., Sen, H. S., Maji, B., & Yadav, J. S. P. (2001). Saline and alkali soils and their management. ISCAR monograph series-1. Port Canning.

Batabyal, P., Einsporn, M. H., Mookerjee, S., Palit, A., Neogi, S. B., Nair, G. B., & Lara, R. J. (2014). Influence of hydrologic and anthropogenic factors on the abundance variability of enteropathogens in the Ganges estuary, a cholera endemic region. *Science of the Total Environment, 472*, 154–161.

Bell, H. S. (1940). Armoured mud balls-their origin, properties and role in sedimentation. *Journal of Geology, 48*, 1–31.

Bhattacharya, A. (1999). Embankments and their ecological impacts: A case study from the tropical low-lying coastal plains of the deltaic Sundarbans, India. In H. Grann & M. Vollmer (Eds.), *Largescale constructions in coastal environments* (pp. 171–180). Springer.

Bhattacharya, A. (2000). Some geomorphic observations indicating shoreline variation in the coastal tract of West Bengal (India)—A case study around Bakkhali in Sundarbans. In *Proceedings of the International Quaternary Seminar on INQUA Shoreline* (pp. 32–37). Indian Ocean Sub-Commission, New Academic Publishers.

Bhunia, A., & Choudhury, A. (1981). Observation on the hydrology and quantitative studies on benthic macrofauna on a tidal creek of Sagar Island, Sundarbans, West Bengal. *Proceeding of National Academy of Science India, B-47*(3), 398–407.

BIS. (1991). *Drinking water specification IS:10500:1991*. Bureau of Indian Standard.

Blasco, F. (1977). Outline of the ecology, botany and forestry of the Mangals of the Indian subcontinent. In V. J. Chapman (Ed.), *Ecosystem of the world* (pp. 241–262). Elsevier.

Blasco, F. (1984). Climate factors and the biology of mangrove plants. In S. C. Snedaker & J. G. Snedaker (Eds.), *The mangrove ecosystem: Research methods* (pp. 18–35). UNESCO/SCOR.

Banerjee, L. K. (1985). Comparative study on mangroves of the Sundarbans and that of the Mahanadi Delta in Eastern India. In L. J. Bhosale (Ed.), *The Mangroves. Proceedings of the National Symposium Biological Utilization. Conservation of Mangroves, Kolahpur, India* (pp. 527–530).

Banerjee, K., Das, S., & Mitra, A. (2005). Loss of ichthyoplankton due to wild harvest of shrimp seeds in Indian Sundarbans. *Seshaiyana, 13*(2), 1–2.

Banerjee, K., Gatti, R. C., & Mitra, A. (2017). Climate change-induced salinity variation impacts on a stenoecious mangrove species in the Indian Sundarbans. *Ambio, 46*, 492–499. https://doi.org/10.1007/s13280-016-0839-9

Barbosa, C. F., & Suguio, K. (1999). Bio-sedimentary facies of subtropical microtidal estuary—An example from Southern Brasil. *Journal of Sedimentary Research, 69*, 576–587.

Barton, K. (2015). *MuMIn: Multi-model inference*. R package version 1.10.5.

Boyer, J. N., Kelble, C. R., Ortner, P. B., & Rudnick, D. T. (2009). Phytoplankton bloom status: Chlorophyll a biomass as an indicator of water quality condition in the southern estuaries of Florida, USA. *Ecological Indicators, 9*, 65–67.

Bradshaw, A. L., & Brewer, P. G. (1988). High precision measurements of alkalinity and total carbon in seawater by potentiometric titration: 2. Measurements of standard solution. *Marine Chemistry., 34*, 155–162.

Cerco, C. F., & Cole, T. M. (1995). *User's guide to the CE-QUAL-ICM three dimensional eutrophication model*. Technical report EL-95-15, U.S. Army Engineer Waterways Experiment Station, Vicksburg, MS.

Chakraborty, S. K. (1995). Aquaculture potential of mangrove ecosystem of Sundarbans, India. In *Proceedings of Seminar on Fisheries—A Multimillion Dollar Industry* (pp. 72–83).

Chakraborty, S. K. (1998). Coastal aquaculture and environment in the context of West Bengal, India. *Vidyasagar University Journal of Biological Sciences, 4*, 13–20.

Chakraborty, S. K. (2003). Biodiversity and conservation. In A. D. Mukhopadhyay (Ed.), *Perspectives and issues in environmental studies* (pp. 267–286). Vidyasagar University.

Chakraborty, S. K. (2009). Biological diversity of Mindapore Coast, West Bengal. *Environmental Forensics, 2*(2), 6.

Chakraborty, S. K. (2010). Coastal environment of Midnapore, West Bengal: Potential threats and management. *Journal of Coastal Environment, 1*(1), 28–40.

Chakraborty, S. K. (2011). Mangrove ecosystem of Sundarbans, India: Biodiversity, ecology, threats and conservation. In J. N. Metras (Ed.), *Mangroves: Ecology, biology and taxonomy* (pp. 83–112). NOVA Publishers.

Chakraborty, S. K. (2013). Interactions of environmental variables determining the biodiversity of coastal mangrove ecosystem of West Bengal, India. *The Ecoscan Special Issue, 3*, 251–265.

Chakraborty, S. K. (2017). Ecological services on intertidal benthic fauna and the sustenance of coastal wetlands along the Midnapore (East) coast, West Bengal, India. In *Coastal wetlands: Alteration and remediation* (pp. 777–886). Springer International Publishing AG.

Chakraborty, S. K. (2018). Bioinvasion and environmental perturbation: Synergistic impact on coastal—Mangrove ecosystems of West Bengal, India. In C. Makowski & C. Finkl (Eds.), *Impacts of invasive species on coastal environments: Coast in crisis* (pp. 171–245). Springer.

Chakraborty, S. K. (2021a). *Riverine ecology (Vol. 1): Ecofunctionality of the physical environment of riversresolution*. Springer Nature. 917 pages.

Chakraborty, S. K. (2021b). *Riverine ecology (Vol. 2): Biodiversity conservation, conflicts and resolution*. Springer Nature. 566 pages.

Chakraborty, S. K., & Choudhury, A. (1989). Ecological studies on the zonation of brachyuran crabs in a virgin mangrove island of Sundarbans, India. *Journal of the Marine Biological Association of India, 34*(1&2), 189–194.

Chakraborty, S. K., & Choudhury, A. (1992a). Ecological studies on the zonation of brachyuran crabs in a virgin mangrove Island of Sundarbans, India. *Journal of Marine Biological Association of India, 34*(1&2), 189–194.

Chakraborty, S. K., & Choudhury, A. (1992b). Population ecology of fiddler crabs (*Uca spp.*) of the mangrove estuarine complex of Sundarbans, India. *Tropical Ecology, 33*(1), 78–88.

Chakraborty, S. K., & Choudury, A. (1985). Distribution of fiddler crabs in Sundarbans mangrove Estuarine complex, India. The mangroves. In *Proceedings of National Symposium Biology Utilization Conservation of Mangroves* (pp. 467–472).

Chakraborty, S. K., & Chowdhury, A. (1993). Community organization of macrobenthic molluses of Sagar Island, Sundarbans, India. *Indian Biologist, 25*(1), 33–36.

Chakraborty, S. K., & Chowdhury, A. (1994). Community structure of macrobenthic polychaetes of intertidal region of Sagar Island, Hooghly Estuary, Sundarbans, India. *Tropical Ecology, 35*(1), 97–104.

Chakraborty, S. K., Choudhury, A., & Deb, M. (1986). Decapod Brachyura from Sundarbans mangrove estuarine complex, India. *Journal of the Bengal Natural History Society, 5*(1), 55–68.

Chakraborty, S. K., Giri, S., Chakravarty, G., & Bhattacharya, N. (2009). Impact of eco-restoration on the biodiversity of Sundarbans Mangrove Ecosystem, India. *Water, Air, and Soil Pollution: Focus, 9*, 303–320.

Chakraborty, S. K., Chatterjee, S., Dey, M. K., et al. (2012). Ecological changes of an eco–rehabilitated delta of Hooghly estuary. *Journal of Coastal Environment, 3*(1), 9–27.

Chakraborty, S. K., Paul, A. K., Bandyopadhyay, J., Banerjee, D., Pakhira, H., Middya, S., Jana, S., Sahoo, P., Paria, K., & Bera, K. (2013). Sustainable water management in the drought prone riverine tracts of South West Bengal. *Journal of Groundwater Studies, 1*, 11–34.

Chakraborty, S. K, Maiti Dutta, S, Ghosh, P. B, Ray, R., & Paul, A. K. (2014). Impact of global warming on Sundarbans mangrove ecosystem, India: Role of different assessment tools from ecosystem monitoring to molecular markers. In *Proceedings of the International Conference on Green India: Strategic Knowledge for Combating Climate Change: Prospects and Challenges* (pp. 181–200).

Chatterjee S, anmd Chakraborty SK (2014a) Feeding behaviour and functional role of some selected species of Brachyuran crabs in nutrient cycle of coastal belt of Midnapore (East), West Bengal, India. *Journal of Biology and Life Science, 5*(1):106–129

Chakraborty, S. K., & Choudhury, A. (1994a). Community structure of macrobenthicpolychaetes of intertidal region of Sagar Island, Hooghly Estuary, Sundarbans, India. *Tropical Ecology, 35*(1), 97–104.

Chakraborty, S. K., & Choudhury, A. (1994b). Community structure of macrobenthic polychaetes of intertidal region of Sagar Island, Hooghly Estuary, Sundarbans, India. *Tropical Ecology., 35*(1), 97–104.

Chakraborty, S. K., Sanyal, P., & Ray, R. (2023). *Wetlands ecology: Eco-biological uniqueness of a Ramsar site (East Kolkata Wetlands, India)*. Springer Nature. 709 pages.

Chakravarty, G., Giri, S., Giri, S., & Chakraborty, S. K. (2004). Distribution of coliform bacteria in different estuaries of West Bengal Coast, India. In A. Kumar, J. R. Dhanze, A. K. Chopra, & D. S. Malik (Eds.), *Protected habitats and biodiversity* (pp. 395–402). Natcon Publication.

Chandrasekaran, A., Ravisankar, R., Harikrishnan, N., Satapathy, K. K., Prasad, M. V. R., & Kanagasabapathy, K. V. (2015). Multivariate statistical analysis of heavy metal concentration in soils of Yelagiri Hills, Tamilnadu India—Spectroscopical approach. *Spectrochimica Acta. Part A, Molecular and Biomolecular Spectroscopy, 137*, 589–600.

Chang, H. (2002). Spatial and temporal variations of water quality in the river and its tributaries, Seoul, Korea, 1993–2002. *Water, Air, & Soil Pollution, 161*, 267–284.

Chatterjee, S., & Chakraborty, S. K. (2014). Feeding behaviour and functional role of some selected species of brachyuran crabs in nutrient cycle of coastal belt of Midnapore (East), West Bengal, India. *Journal of Biology and Life Science, 5*(1), 106–129.

Chatterjee, S., Das, T. K., & Chakraborty, S. K. (2014). Reproductive biology and bioturbatory activities of two sympatric species of fiddler crabs *Uca lactea annulipes* and *Uca triangularis bengali* (Decopoda: Ocypodidae) at the East Midnapore Coastal belt of West Bengal, India. *Journal of Biology and Life Science, 5*(2), 106–126.

Chaudhuri, A. B., & Choudhury, A. (1994). *Mangroves of the Sundarbans, India* (Vol. 1, pp. 1–247). IUCN.

Choudhury, A., Das, A., Bhattacharya, S., & Bhunia, A. (1984a). A quantitative assessment of benthic macro-fauna in the intertidal mudflats of Sagar Island, Sundarbans, India. In *Proceedings of Asian Symposium on Mangroves, Environmental Research Management* (pp. 298–310).

Choudhury, A., Bhunia, A. B., & Nandi, S. (1984b). Preliminary survey of macro-benthos of Prentice Island, Sundarbans, West Bengal. *Records of Zoological Survey of India, 81*(3&4), 81–92.

Chowdhury, M. Q., De Ridder, M., & Beeckman, H. (2016a). Climatic signals in tree rings of Heritiera fomes Buch.—Ham. in the Sundarbans, Bangladesh. *PLoS One, 11*, e0149788.

Chowdhury, M. Q., Kitin, P., De Ridder, M., Delvaux, C., & Beeckman, H. (2016b). Cambial dormancy induced growth rings in HeritierafomesBuch.—Ham.: A proxy for exploring the dynamics of Sundarbans, Bangladesh. *Trees, 30*, 227–239.

Coleman. (1969).

Conely, D. J., Carstensen, J., Aertebjerg, G., Christensen, P. B., Dalsgaard, T., Flansen, J. L. S., & Josefson, A. B. (2007). Long-term changes and impacts of hypoxia in Danish coastal waters. *Ecological Applications, 17*, S165–S184.

Daleo, P., & Iribarne, O. (2009). The burrowing crab *Neohelice granulate* affects the root strategies of the cordgrass *Spartina densiflora* in SW Atlantic salt marshes. *Journal of Experimental Marine Biology and Ecology, 373*, 66–71.

Das, G. K. (2009a). Beach processes of Tidal Islands of Hugli Estuary, West Bengal. *Geographical Review of India, 71*(3), 240–248.

Das, G. K. (2009b). Grain size analysis of some beach sands from the Indian coasts. *Geographical Review of India, 71*(1), 10–18.

Das, G. K. (2010). Biodiversity in Sundarbans. In A. K. Dwivedi & S. C. Tripathi (Eds.), *Environment and biodiversity, environmental pollution and risk to biodiversity* (pp. 1–34). Lambert Academic Publishing AG & Saarbrucken.

Das, G. K. (2011). Impact of salinity and nutrients on the changing mangrove floristic—A case study from the River Flood Plains of Sundarbans, India. In *Biotic potential and the abiotic stress* (pp. 119–129). Lambert Academic Publishing AG & Saarbrucken.

Das, G. K. (2012). *Impact of water quality on the changing environmental scenario of Sundarbans. Reason* (Vol. XI, pp. 57–66). Kalyani Government Engineering College.

Das, G. K. (2015). *Estuarine morphodynamics of Sundarbans* (Coastal Research Library 11: Series Editor: C.W. Finkl) (pp. 1–11). Springer.

Das, G. K., & Bhattacharya, A. (1991). Physiographic set up and sensitiveness of textural parameters of tidal river related mangrove sediments. In *Proceedings of the National Symposium on Conservation and Management of Mangroves* (pp. 46–52). Bidhan Chandra Krishi Viswavidyalaya.

Das, G. K., & Bhattacharya, A. (1999a). Destruction of certain encountered shellfish and finish from the coastal waters of Sundarbans and its impact on the coastal ecosystem. *Journal of the Indian Society of Coastal Agricultural Research, 17*(1&2), 216–218.

Das, G. K., & Bhattacharya, A. (1999b). Rampant exploitation of post-larval penaeid prawn (Penaeus mondon) from the salt-water course of the deltaic Sundarbans, West Bengal—A threat to the balance of nature. *Journal of the Indian Society of Coastal Agricultural Research, 17*(1&2), 219–222.

Das, G. K., & Dutta, S. (2004). Studies on the impact of water quality on the adjoining wetland ecosystem of Bidyadhari River, West Bengal. *Indian Science Cruiser, 18*(4), 16–21.

Das, G. K. (2009). Grain size analysis of some beach sands from the Indian Coasts. *Geographical Review of India, 71*(1), 10–18.

Davies, D. H. (1954). The South African pilchard (*Sardinops ocellata*). Development, occurrence and distribution of eggs and larvae, 1950-51. *Investigational Report of the Division of Fisheries ofthe Union of South Africa*, 15, 1–28.

Davis, J. R., & Koop, K. (2006). Eutrophication in Australian rivers, reservoirs and estuaries—A southern hemisphere perspective on the science and its implications. *Hydrobiologia, 559*(1), 23–76.

Deng, Y., et al. (2022). Coupling sulfur-based denitrification with anammox for effective and stable nitrogen removal: A review. *Water Research, 224*, 119051.

Dey, M. K., Hazra, A. K., & Chakraborty, S. K. (2008). Diversity of microarthropods and their role in the plant litter decomposition in the coastal tract of East Midnapore District West Bengal, India. *Zoological Research in Human Welfare, 20*, 207–226.

Dittmar, T., & Lara R. J. (2001). Do mangroves rather than rivers provide nutrients to coastal environments south of the Amazon River? Evidence from long-term flux measurements. *Marine Ecology Progress Series,* 213: 67–77 .

Duke, N. C. (1992). Mangrove floristics and biogeography. In A. I. Robertson & D. M. Alongi (Eds.), *Tropical mangrove ecosystems* (pp. 63–100). American Geophysical Union.

Duke, N. C., Ball, M. C., & Ellison, J. C. (1998). Factors influencing biodiversity and distributional gradients in mangroves. *Global Ecology and Biogeography Letters, 7*, 27. https://doi.org/10.2307/2997695

Ellison, & Stoddart. (1991). Mangrove ecosystem collapse during predicted sea-level rise: Holocene analogues and implications. *Journal of Coastal Research, 7*, 151–165.

Fernando, O. J. (1987). Studies on intertidal fauna of Vellar estuary. *Journal of the Marine Biological Association of India, 29*, 86–103.

Furukawa, K., & Wolanski, E. (1996). Sedimentation in mangrove forests. *Mangroves and Salt Marshes, 1*, 3–10.

Furukawa, K., Wolanski, E., & Mueller, H. (1997). Currents and sediment transport in mangrove forests. *Estuarine, Coastal and Shelf Science, 44*, 301–310.

Giri, C., Pengra, B., Zhu, Z., Singh, A., & Tieszen, L. L. (2007). Monitoring mangrove forest dynamics of the Sundarbans in Bangladesh and India using multi-temporal satellite data from1973 to 2000. *Estuarine, Coastal and Shelf Science, 73*, 91–100.

Godhantaraman, N. (2002). Seasonal variations in species composition, abundance, biomass and estimated production rates of tintinnids at tropical estuarine and mangrove waters, Parangipettai, southeast coast of India. *Indian Journal of Marine Science, 36*, 161–171.

Guha, S. (2015). Coastal dynamics and sediment transport in parts of Hooghly estuary. *International Journal of Trend in Research and Development, 2*(6), 12670. ISSN 2394-9333 Environ Sci Pollut Res (2020) 27:12658–12672.

Hadibarata, T., Abdullah, F., Yusoff, A. R. M., Ismail, R., Azman, S., & Adnan, N. (2012). Correlation study between land use, water quality, and heavy metals (Cd, Pb, and Zn) content in water and green lipped mussels Perna viridis (Linnaeus.) at the Johor Strait. *Water, Air, and Soil Pollution, 223*(6), 3125–3136.

Ham, Y. S., Kobori, H., & Takasago, M. (2009). Effects of combined sewer overflow and storm water on indicator bacteria concentrations in the Tama River due to the high population density of Tokyo metropolitan area. *Environmental Monitoring and Assessment, 152*(1–4), 459–468.

Hazra, S., Ghosh, T., Dasgupta, R., & Sen, G. (2002). Sea level and associated changes in the Sundarbans. *Science and Culture, 68*(9–12), 309–321.

Jafar-Sidik, M., Gohin, F., Bowers, D., Howarth, J., & Hull, T. (2017). The relationship between suspended particulate matter and turbidity at a mooring station in a coastal environment: Consequences for satellite-derived products. *Oceanologia, 59*(3), 365–378.

Jennings, S. (2004). The ecosystem approach to fishery management: A significant step towards sustainable use of the marine environment? *Marine Ecology Progress Series, 274*, 279–282.

Jennings, S., & Dulvy, N. K. (2005). Reference points and reference directions for size-based indicators of community structure. *ICES Journal of Marine Science, 62*, 397–404.

Jha, D. K., Devi, M. P., Vidyalakshmi, R., Brindha, B., Vinithkumar, N. V., & Kirubagaran, R. (2015). Water-quality assessment using water-quality index and geographical information system methods in the coastal waters of Andaman Sea, India. *Marine Pollution Bulletin, 100*(1), 555–561.

Karmarkar, S. M. (1985) Mangroves—A review. In *Proceedings of the National Symposium on Biology, Utilization, and Conservation of Mangroves* (pp. 60–79)

Karr, J. R. (1981). Assessment of biotic integrity using fish communities. *Fisheries, 6*, 21–27.

Kathiresan, K. (2000). A review of studies on Pichavaram mangrove, southeast India. *Hydrobiologia, 430*, 185–205.

Kathiresan, K. (2004). *Biodiversity in mangrove ecosystems of India: Status, challenges and strategies* (Envis Forestry Bulletin No. 4).

Khalua, R. K., Chakravorty, G., & Chakraborty, S. K. (2008). Community structure of macrobenthic molluscs of three contrasting intertidal belts of Midnapore Coastal Belt, West Bengal. *Zoological Research in Human Welfare, 6*, 47–55.

Khan, R. A. (1995). Ecology of the Hugli-Matla estuarine system. In *Estuarine ecosystem series, part 2: Hugli Matla Estuary* (pp. 417–464). Zoological Survey of India.

Khan, R. A. (2003). Population dynamics and biology of aquatic gastropod, *Gyraulus convexiusculus* (Hutton) in a small pond. *Records of Zoological Survey of India, 101*(Part 3–4), 25–37.

Kinne, O. (1963). The effect of temperature and salinity on marine and brackish water animals. *Oceanography and Marine Biology Annual Revenue, 1*, 301–340.

La Rosa, T., Mirto, S., Favaloro, E., Savona, B., Sarà, G., Danovaro, R., & Mazzola, A. (2002). Impact on the water column biogeochemistry of a Mediterranean mussel and fish farm. *Water Research, 36*(3), 713–721.

Lacerda, L. D., Jose, D. V., de Rezende, C. E., Francisco, M. C. F., Wasserman, J. C., & Martins, J. C. (1986). Leaf chemical characteristics affecting herbivory in a New World mangrove forest. *Biotropica, 18*, 350–355.

Lawton, J. (1999). Are there general laws in ecology? *Oikos, 84*, 177–192.

Li, Y., et al. (2016). Sources and fate of organic carbon and nitrogen from land to ocean: Identified by coupling stable isotopes with C/N ratio. *Estuarine, Coastal and Shelf Science, 181*, 114–122.

Lohrer, A. M., Thrush, S. F., & Gibbs, M. M. (2004). Bioturbators enhance ecosystem function through complex biogeochemical inter-ecosystem function through complex biogeochemical inter-actions. *Nature, 431*, 1092–1095.

Lotfinasabasl, S., Gunale, V. R., & Khosroshahi, M. (2018). Applying geographic information systems and remote sensing for water quality assessment of mangrove forest. *Acta Ecologica Sinica, 38*(2), 135–143. https://doi.org/10.1016/j.chnaes.2017.06.017

Lotze, H. K., Lenihan, H. S., Bourque, B. J., Bradbury, R. H., Cooke, R. G., Kay, M. C., Kidwell, S. M., Kirby, M. X., Peterson, C. H., & Jackson, J. B. C. (2006). Depletion, degradation, and recovery potential of estuaries and coastal seas. *Science, 312*, 1806. https://doi.org/10.1126/science.1128035

Lovelock, C. E., Ball, M. C., Feller, I. C., Engelbrecht, B. M. J., & Ling, E. M. (2006). Variation in hydraulic conductivity of mangroves: Influence of species, salinity, and nitrogen and phosphorus availability. *Physiologia Plantarum, 127*, 457–464. https://doi.org/10.1111/j.1399-3054.2006.00723.x

Manna, S., Chaudhury, K., Bhattacharyya, S., & Bhattacharya, M. (2010). Dynamics of Sundarbans estuarine ecosystem: Eutrophication induced threat to mangroves. *Saline Systems, 6*(1), 8.

McKee, K. L. (1993). Soil physic-chemical patterns and mangrove species distribution: Reciprocal effects? *Journal of Ecology, 81*, 477–487.

Mitra, S., Ghosh, S., Satpathy, K. K., Bhattacharya, B. D., Sarkar, S. K., Mishra, P., & Raja, P. (2018a). Water quality assessment of the ecologically stressed Hooghly River estuary, India: A multivariate approach. *Marine Pollution Bulletin, 126*, 592–599.

Mitra, S., Sarkar, S. K., Raja, P., Biswas, J. K., & Murugan, K. (2018b). Dissolved trace elements in Hooghly (Ganges) River Estuary, India: Risk assessment and implications for management. *Marine Pollution Bulletin, 133*, 402–414.

Mitra, S., Ghosh, S., Satpathy, K. K., Deb Bhattacharyaa, B., Sarkar, S. K., Pravakar Mishrad, P., & Rajae, P. (2024). Water quality assessment of the ecologically stressed Hooghly River Estuary, India: A multivariate approach. *Marine Pollution Bulletin*.

Mukhopadhyay, S. K., Biswas, H., De, T. K., & Jana, T. K. (2006). Fluxes of nutrients from the tropical. River Hooghly at the land–ocean boundary of Sundarbans, NE Coast of Bay of Bengal, India. *Journal of Marine Systems, 62*, 9–21.

Muus, B. J. (1967). The fauna of Danish estuaries and lagoons. Distributions and ecology of dominating species in the shallow reaches of the mesohaline zone. *Medd. Dan.Fisk. Havunders, 5*, 1–316.

Naido, G. (1990). Effects of nitrate, ammonium and salinity on growth of the mangrove *Brugeira gymnorrhiza* (L) Lam. *Aquatic Botany, 38*(2–3), 209–219.

Naidoo, G. (2009). Differential effects of nitrogen and phosphorus enrichment on growth of dwarf Avicennia marina mangroves. *Aquatic Botany, 90*, 184–190. https://doi.org/10.1016/j.aquabot.2008.10.001

Naskar, K., Sen-Sarkar, N., Ghosh, A., Dasgupta, M., & Sengupta, B. (2004). Status of the mangroves and mangrove ecosystem of Sundarbans in West Bengal: Its impact on estuarine wetlands fisheries. *CIFRI Bulletin, 134*, 8.

Needham, H. R., Pilditch, C. A., Lohrer, A. M., & Thrush, S. F. (2010). Habitat dependence in the functional traits of Austrohelice crassa, a key bioturbating species. *Marine Ecology Progress Series, 414*, 179–193.

Norkko, A., Hewitt, J. E., Thrush, S. F., & Funnell, G. A. (2006). Conditional outcomes of facilitation by a habitat modifying subtidal bivalve. *Ecology, 87*, 226–234.

Omer, W. M. M. (2010). *Ocean acidification in the Arabian Sea and the Red Sea*. Faculty of Mathematics and Natural Sciences, Institute Chemical Oceanography.

Osibanjo, O., Daso, A. P., & Gbadebo, A. M. (2011). The impact of industries on surface water quality of River Ona and River Alaro in Oluyole Industrial Estate, Ibadan, Nigeria. *African Journal of Biotechnology, 10*(4), 696–702.

Park, R. A., Lee, J. K., & Canning, D. (1993). Potential effects of sea level rise on Puget Sound Wetlands. *Geocarto International, 8*, 99–110.

Paul, A. K. (2009). Man-environment interactions in the single largest mangrove forest along the shoreline of the Bay of Bengal: A case study of Sundarbans. In H. S. Sharma & V. S. Kole (Eds.), *Geomorphology in India* (pp. 263–286).

Perumal, N. V., Rajkumar, M., Perumal, P., & Rajasegar, K. T. (2009). Seasonal variations of plankton diversity in the Kaduviar estuary, Nagapattinam, southeast coast of India. *Journal of Environmental Biology, 30*(6), 1035–1046.

Pillay, L., & Olaniran, A. O. (2016). Assessment sof physicochemical parameters and prevalence of virulent and multiple-antibiotic-resistant Escherichia coli in treated effluent of two wastewater treatment plants and receiving aquatic milieu in Durban, South Africa. *Environmental Monitoring and Assessment, 188*(5), 260.

Pritchard, D. W. (1952). Estuarine hydrography. *Advances in Geophysics, 1*, 243–280.

Qasim, S. Z. (2003). *Indian estuaries* (p. 27). Allied Publishers.

Rajendran, A., & Mansiya, C. (2015). Physico-chemical analysis of groundwater samples of coastal areas of South Chennai in the posttsunami scenario. *Ecotoxicology and Environmental Safety, 121*, 218–222.

Rajsekar, M. (2003). Physico-chemical characteristics of the Vellar estuary in relation to shrimp farming. *Journal of Environmental Biology, 24*, 95–101.

Ramasamy, S. M., Muthukumar, M., & Subagunasekar, M. (2015). Water quality index of estuarine environment. *Current Science, 108*, 1430–1438.

Ray, R., Paul, A. K., & Basu, B. (2013). Application of supervised enhancement technique in monitoring the mangrove forest cover dynamics—A case study on Ajmalmari Reserve Forest, Sundarbans, West Bengal. *International Journal of Remote Sensing and Geosciences, 3*(1), 16–21.

Ribera, C., & Kjerfve, B. (2002). Anthropogenic influence on the water quality in Guanabara Bay, Rio de Janeiro, Brazil. *Regional Environmental Change, 3*(1–3), 13–19.

Ricklefs, R. E., & Latham, R. E. (1993). Global patterns of diversity in mangrove floras. In *Species diversity in ecological communities*. University of Chicago Press.

Robertson, A. I., & Alongi, D. M. (Eds.). (1992). *Tropical mangrove ecosystems* (Coastal and estuarine studies) (Vol. 41, p. 329). American Geophysical Union.

Sadhuram, Y., Sarma, V. V., Ramana Murthy, T. V., & Rao, B. P. (2005). Seasonal variability of physico-chemical characteristics of the Haldia channel of Hooghly Estuary, India. *Journal of Earth System Science, 114*, 37–49.

Sahana, M., Hong, H., Ahmed, R., Patel, P. P., Bhakat, P., & Sajjad, H. (2019). Assessing coastal island vulnerability in the Sundarbans Biosphere Reserve, India, using geospatial technology. *Environmental Earth Sciences, 78*, 304. https://doi.org/10.1007/s12665-019-8293-1

Sahu, G., Satpathy, K. K., Mohanty, A. K., & Sarkar, S. K. (2012). Variations in community structures of phytoplankton in relation to physicochemical properties of coastal waters, southeast coast of India. *Indian Journal of Geo-Marine Sciences, 41*(3), 223–241.

Sanders, H. L. (1958). Marine Benthic diversity a comparative study. *American Naturalist, 102*, 243–282.

Sanders, H. L. (1968). Benthic studies in Buzzards Bay. 1. Animal sediment relationship. *Limnology and Oceanography, 3*, 245–258.

Saravanane, N., Satpathy, K. K., Nair, K. V. K., & Durairaj, G. (1998). Preliminary observations on the recovery of tropical phytoplankton after entrainment. *Journal of Thermal Biology, 23*(2), 91–97.

Sarkar, S. K., & Bhattacharya, B. D. (2010). Water quality analysis of the coastal region of Sundarbans mangrove wetland, India, using multivariate statistical techniques. In S. K. Sarkar (Ed.), *Environmental management* (p. 258) ISBN: 978-953-307-133-6.

Sarkar, S. K., Singh, B. N., & Choudhury, A. (1989). Seasonal distribution of copepods in the Hooghly estuary, West Bengal. *Indian Journal of Marine Sciences, 15*, 177–180.

Sarker, S., Yadav, A. K., Hossain, M. S., Rahman, C. S., Kabird, M. A., & Sharifuzzamanc, S. M. (2020). The drivers of diatom in subtropical coastal waters: A Bayesian modeling approach. *Journal of Sea Research, 163*, 101915. https://doi.org/10.1016/j.seares.2020.101915

Seiderer, L. J., & Newell, R. C. (1999). Analysis of the relationship between sediment composition and biological community structure in coastal deposits: Implications for marine aggregate dredging. *ICES Journal of Marine Sciences, 56*, 757–765.

Selvam, V. (2003). Environmental classification of mangrove wetlands of India. *Current Science, 84*(6), 757–765.

Selvam, V., Mohan, R., Ramasubramaniam, R., & Azariah, K. (1991). Plant communities and soil properties of three mangrove stands of Madras coast. *Indian Journal of Marine Sciences, 20*, 67–69.

Şener, Ş., Şener, E., & Davraz, A. (2017). Evaluation of water quality using water quality index (WQI) method and GIS in Aksu River (SW-Turkey). *Science of the Total Environment, 584–585*, 131–144.

Smith, V. H., Tilman, G. D., & Nekola, J. C. (1999). Eutrophication: Impacts of excess nutrient inputs on freshwater, marine, and terrestrial ecosystems. *Environmental Pollution, 100*(1–3), 179–196.

Soo, C. L., Ling, T. Y., & Lee, N. A. K. (2016). Assessment of the characteristic of nutrients, total metals, and fecal coliform in Sibu Laut River, Sarawak, Malaysia. *Applied Water Science, 6*, 77–96.

Thrush, S. F., Hewitt, J. E., Cummings, V. J., Green, M. O., Funnell, G. A., & Wilkinson, M. R. (2000). The generality of field experiments: Interactions between local and broad-scale processes. *Ecology, 81*, 399–415.

Thrush, S. F., Hewitt, J. E., Gibbs, M., Lundquist, C., & Norkko, A. (2006). Functional role of large organisms in intertidal communities: Community effects and ecosystem function. *Ecosystems, 9*, 1029–1040.

Thrush, S. F., Hewitt, J. E., Dayton, P. K., Coco, G., Lohrer, A. M., Norkko, A., Norkko, J., & Chiantore, M. (2009). Forecasting the limits of resilience: Integrating empirical research with theory. *Proceedings of the Royal Society B, 276*, 3209–3217.

Tomlinson, P. B. (1986). *The botany of mangroves*. Cambridge University Press. 433 pp.

Twilley, R. R., & Chen, R. (1998). A water budget and hydrology model of a basin mangrove forest in Rookery Bay, Florida. *Marine and Freshwater Research, 49*, 309–323.

Van Colen, C., Montserrat, F., Vincx, M., Herman, P. M. J., Ysebaert, T., & Degraer, S. (2008). Macrobenthic recovery from hypoxia in an estuarine tidal mudflat. *Marine Ecology Progress Series, 372*, 31–42.

Van Dalfsen, J. A., Essink, K., Toxvig Madsen, H., Birklund, J., Romero, J., & Manzanera, M. (2000). Differential response of macrozoobenthos to marine sand extraction in the North Sea and the Western Mediterranean. *ICES Journal of Marine Science, 57*, 1439–1445.

van de Koppel, J., Herman, P. M. J., Thoolen, P., & Heip, C. H. R. (2001). Do alternate stable states occur in natural ecosystems? Evidence from a tidal flat. *Ecology, 82*, 3449–3461.

Vannucci, M. (1989). *The mangroves and us; a synthesis of insights*. Indian Association for the Advancement of Science.
Vighi, M., Finizio, A., & Villa, S. (2006). The evidence of the environmental quality concept: From the US EPA Red Book to the European Water Framework Directive. *Environmental Science and Pollution Research, 13*, 9–14.
Weerman, E. J., Herman, P. M. J., & Van de Koppel, J. (2010). Top-down control inhibits spatial self-organization of a patterned landscape. *Ecology, 92*, 487–495.
Wetz, M. S., Hayes, K. C., Fisher, V. B., Price, L., & Sterba-Boatwright, B. (2016). Water quality dynamics in an urbanizing subtropical estuary (Oso Bay, Texas). *Marine Pollution Bulletin, 104*, 44–53.
Whitall, D., Hively, W. D., Leight, A. K., Hapeman, C. J., McConnell, L. L., Fisher, T., Rice, C. P., et al. (2010). Pollutant fate and spatio-temporal variability in the Choptank River Estuary: Factors influencing water quality. *Science of the Total Environment, 408*(9), 2096–2108.
WHO. (2004). *Guidelines for drinking water quality* (Recommendations) (Vol. 1, 3rd ed.). Word Health Organization.
WHO. (2011). *Guidelines for drinking water quality*. Word Health Organization.
Widdows, J., Blauw, A., Heip, C. H. R., Herman, P. M. J., Lucas, C. H., Middelburg, J. J., Schmidt, S., Brinsley, M. D., Twisk, F., & Verbeek, H. (2004). Role of physical and biological processes in sediment dynamics of a tidal flat in Westerschelde Estuary, S.W. Netherlands. *Marine Ecology Progress Series, 274*, 41–56.
Wolanski, E. (2006). The application of ecohydrology for sustainable development and management of mangrove-dominated estuaries. In *The ICEMAN 2006 Mangrove Conference in Kuala Lumpur*.
Wolanski, E., Mazda, Y., & Ridd, P. (1992). Mangrove hydrodyna mics. In A. I. Robertson & D. M. Alongi (Eds.), *Tropical mangrove ecosystems, coastal and estuarine studies* (pp. 436–462). AGU.
Wolanski, E., Spagnol, S., & Ayukai, T. (1998). Field and model studies of the fate of particulate carbon in mangrove-fringed Hinchinbrook Channel, Australia. *Mangroves and Salt Marshes, 2*, 205–221.
Wong, Y. S., Lam, C. Y., Che, S. H., Li, X. R., & Tam, N. F. Y. (1995). Effect of wastewater discharge on nutrient contamination of mangrove soil and plants. *Hydrobiologia, 295*, 243–254.
Yeragi, S. G. (2003). Preliminary survey of marine crabs of Mithbav creek. *Journal of Comparative Toxicology and Physiology, 1*(1), 35–42.

Chapter 7
Environmental Pollution and Ecological Perturbations on the Coastal-Mangrove-Estuarine Complexes of Sundarbans, India (A World Heritage Site)

Abstract Different ecozones within the Sundarbans mangrove ecosystem have been under different environmental threats from various point and nonpoint sources of pollution, deforestation, land reclamation, sea-level rise, temperature elevation, sediment deposition, bioinvasion etc., which have been found to have negative impacts with variable intensities depending on the morphology, meteorology, tidal influence and location of multifarious industries. The mangrove ecosystem, endowed with rich biodiversity resources in its estuarine and tidal forest subsystems, is under serious threats of destruction because of tremendous pressure from the dumping and mixing of a number of pollutants in addition to several socioeconomic impacts from extensive overfishing and indiscriminate capture of post-larval penaeid prawns, from the saltwater courses of the estuarine-deltaic Sundarbans resulting a dwindling a dwindling trend in ecological parameters and biodiversity.

Further hazards in the form of bank erosion not only destabilize the ecology of the mangrove ecosystem but also increase turbidity and disruptions to nutrient recycling. The Sundarbans mangrove ecosystem with its estuarine network and vast expanse of mangrove forests, harbouring an array of biodiversity components, has been under considerable environmental perturbation along with gradually escalating pollution mainly because of multidimensional anthropogenic activities centred on reclamation after encroachment on forest lands, the nonjudicious harvesting of natural resources and alterations in land-use patterns for agricultural, aquacultural and unplanned urbanization coupled with industrialization in and around the coastal areas. Human-mediated pollution includes the dumping of solid waste, the release of industrial sewage loaded with heavy metals, pathogenic microbes, microplastics and petrochemical products, the mixing of agrochemical waste like chemical fertilizers, the collection of prawn seeds at the cost of the lives of many commercially important finfish and shellfish, the spillage of toxic fossil fuels released from water vessels, mechanized boats and fishing trawlers (leading to the suffocation of both plants and animals), tide drops, the deposition of contaminants on sediments during low tide and the introduction of bioinvasive species.

The deterioration of ecological quality and the shrinkage of mangrove forest zones have destroyed the nursery grounds and breeding sites of a large number of

S. K. Chakraborty, S. K. Acharya, *Eco-dynamics and Environmental Perturbations in Mangrove Ecosystems*,
https://doi.org/10.1007/978-3-032-08710-2_7

mangrove-dwelling fauna species, destroyed the grazing sites for herbivores and caused scarcity in the food availability of large carnivorous wildlife because of direct and indirect human interference.

Such problems of pollution and environmental perturbation, posing considerable threats to the mangrove ecosystem in general and the biodiversity of coastal mangrove-estuarine complexes in particular, have become more aggravated because the fragile mangrove ecosystem and other associated biotic components are sensitive due to their higher susceptibility to the negative ecological changes caused by these different types of pollutants. The mangrove ecosystem, as an adaptable ecosystem, can tolerate a wide range of physical changes, but they become vulnerable to oil toxicity and the subsequent impacts of cleanup activities.

Keywords Sundarbans mangrove ecosystem · Anthropogenic activities · Environmental pollution · Ecological perturbation · Pollutants · Global climate change · Vulnerability

7.1 Pollution in the Coastal Environment of Sundarbans, India (Mangroves, Estuaries, Dunes and Intertidal Belts)

The Indian coastal environment, including mangrove-estuarine deltaic systems in Sundarbans, is highly susceptible to a multitude of chemical pollutants, such as sewage, heavy metals, organochlorine pesticides, microplastics, polychlorinated biphenyls and polycyclic aromatic hydrocarbons, which can significantly alter the geochemistry, nutrient cycles and biodiversity of the affected coastal regions. Different types of pollutants reach the affected ecozones within the estuaries through wastewater and sewage discharge and agricultural runoff with agricultural waste, including chemical fertilizers and organochlorine pesticide residue.

7.2 The Indian Coastal Environment and its Extent in Different Parts of India

Peninsular India is surrounded on three sites by three seas, the Indian Ocean at the south, the Arabian Sea in the west and the Bay of Bengal in the east. India now has a coastline of 11,098.81 km after the breakup of the Andaman and Nicobar islands: the Andaman and Nicobar islands 3083.50 km, Gujrat 2340 km, Tamil Nadu 1068.69 km, Andhra Pradesh 1053.07 km, Maharashtra 877.97 km, West Bengal 721.02 km, Kerala 600.15 km, Odisha 574.71 km, Karnataka 343.30 km, Goa 193.95 km, the Lakshadweep Islands 144.80 km, Daman and Diu 54.38 km and Pondicherry 42.65 km. The east coast has deltaic islands with mangrove vegetation and is the site for culture fisheries, whereas the west coast has nonturbid sandy shores ideal for capture fisheries and coral islands. The industrial developments are more common on the west coast in comparison to its eastern counterparts. The development of some kind of disturbance because of hazardous chemicals (the burning of fossil fuels and the use of Chloroflurocarbons (CFCs)), unsafe machinery and risky occupations, having higher intensity and magnitude in the surrounding environment, has had disastrous outcomes. Such unwanted changes in the health of plants and animals, including humans, have affected the entire world. After being directly and indirectly impacted by human actions (drainage for agriculture and forestry; dredging and stream channelization for navigation; flood protection; conversion for aquaculture/mariculture; the construction of schemes for water supply, irrigation and storm protection; the discharge of pesticides, herbicides and nutrients; solid waste disposal; sediment diversion by deep channels and other structures; the mining of wetland soil; groundwater abstraction; hydrological alteration by canals; roads and other structures; etc.), coastal wetlands are now likely to undergo location-specific socioeconomic changes and patterns of land use that in turn result in sea-level rise, alterations in waves, erosion, subsidence, storms, changes in biotic assemblages and landward shifting or the migration of species. All these alterations are difficult to assess on the basis of simplistic assumptions used to predict the future (Chakraborty & Choudhury, 1994; Campos et al., 2000; Chakraborty 2021; Chakraborty et al., 2023; Ganguly and Chakraborty, 2026).

7.3 Pollution Threats to Sundarbans

A multitude of human-made contributions are considered as the major causes of the large-scale ecodegradation and reclamation of mangroves. Several threat factors cause the deterioration of ecological conditions that enable a biologically productive and ecologically healthy Sundarbans, which is endowed with rare endemic flora and fauna, such as birds, crocodiles and the famous royal Bengal tiger; these factors stem from the anthropogenic activity of overexploiting natural resources and releasing toxic substances at the forest–human society interface. All of these have been compounded as result of a growing population that is reclaiming forest land for settlement and aquacultural development and creating salt pans that cut off the mangrove forest—all for wood as fuel, the indiscriminate collection of prawn seeds, the discharging of sewerage and toxic pollutants from urban areas, urbanization and modern technology-based agricultural activities (the cultivation of hybrid varieties of paddies, the overuse of chemical fertilizers and pesticides, the dumping of waste material and garbage, the release of industrial and municipal effluents, the mining of sand and mud and reducing or increasing the quantum of freshwater via dam construction or by diverting water flows, leading to a striking imbalance between exploitation and replenishment, thereby altering the normal ecosystem functioning of this fragile ecosystem and ultimately lead to the environmental degradation of the area).

Mangrove plants can bioaccumulate and retain pollutants in their biomass to survive, grow and propagate in muddy soil enriched with organic matter, even in reduced and anaerobic ecological conditions, where they experience diurnal tidal inundation and exposure by tidal water (Chakraborty, 2011). However, a steadily increasing pollution load and other modes of disruption to normal ecological conditions have become matters of great concern in ensuring the sustainability of the Sundarbans mangrove ecosystem (Figs. 7.1, 7.2, 7.3, 7.4, 7.5, 7.6, 7.7, 7.8, 7.9, 7.10, 7.11, 7.12, 7.13, 7.14, 7.15, 7.16, and 7.17; Flowcharts 7.1, 7.2, 7.3 and 7.4).

Fig. 7.1 Soil erosion, a potential threat to the stabilization of mangrove ecosystem functioning (**A**, **B**, **C**, **D**, **E** and **F**). Different parts of Sundarbans, India, especially the buffer areas, under the threat of massive soil erosion mainly because of the destruction and removal of different indigenous mangrove species, which through their characteristic roots resist soil erosion, but the local people struggle to combat such problems with some locally available materials, as shown in the image with the serial number (**F**)

Fig. 7.2 Mangrove plants' roots as biological antierosion devices (**A**, **B**, **C**, **D**, **E**, **F** and **G**) The manifestation of different natural architectural plans for the roots of different mangrove plant species, which not only help the plants to rigidly hold their positions on the soft sediments on the floor of mangrove forests but also ensure the ecological stability of the mangrove ecosystem of Sundarbans, India, by resisting soil erosion even in the face of the negative ecological impacts of climate change

Fig. 7.3 The collection of seeds of commercially important species, mainly Peneaus monodon, locally named as tiger shrimp by local people (**A**, **B**, **C**, **D**, **E** and **F**) The collection of seeds of shrimp from the shallow water of estuaries with the help of small nylon mesh nets, mostly during spring tides in the lunar cycles by local people, the majority of whom are women

Fig. 7.4 The collection of seeds of species, mainly Penaeus monodon, locaaly named as tiger shrimp by local people: The low-income people of Sundarbans from different age groups and sexes involve themselves in collecting the seeds of commercially important shrimp from the shallow water zones of estuarine networks for a period of 8 days in the spring tide of lunar cycles (3 days earlier and 4 days after the spring tide) during high tides, risking their lives from the attacks of wildlife, the possibility of drowning and developing chronic health-related problems, mainly in girls and women through infections in their reproductive organs (**A**, **D**, **E** and **F**) Blue mosquito nets with a fine mesh size are used for the collection of the seeds of prawns and shrimp by local people. Out of the total collected samples, only the commercially important seeds of shrimp are scooped up and almost all the other seeds and juveniles are thrown out, causing colossal damage to aquatic biodiversity (**B** and **C**)

Fig. 7.5 Deforestation processes by natural and anthropogenic activities, causing damage to the habitats of mangrove-based biodiversity components in Sundarbans. Continuous deforestation has been going on in different parts of Sundarbans, India, especially in the buffer zones and reclaimed areas for the purpose of collecting wood for fuel, shelter, fodder, etc. and for constructing roads, hotels, resorts, etc. (**A**, **B**, **C**, **D**, **E** and **H**)

Fig. 7.6 Afforestation programs of mangroves: (**A** and **B**) which involve the preparation of nursery beds as the first step in afforestation programmes for mangrove plants; (**C** and **E**) show different research organizations, academic institutes and nongovernmental organizations (NGOs) involving local people and getting support from government in such plantation processes; and (**G** and **H**) exhibit continuous monitoring is required to ensure the growth of newly planted saplings of mangrove plants

Fig. 7.7 Climate risk frameworks on the Sundarbans mangrove ecosystem, India, depicting the impact of climate change in conjunction with other natural disasters. The images (**A**, **B** and **C**) show the excessive swelling of sea and estuarine water, especially during high tides, triggers the inundation of mangrove zones, causing a lot of harmful environmental consequences on a chain-wise basis that breaches embankments, inundating vast crop fields and causing salinity intrusion, excessive soil erosion, etc. (**D**) focuses the the straggle of local people struggle to earn a living and face lot of adversity, even in the collection of shrimp seeds. (**E**) shows flooding in the adjoining human settlements in the reclaimed deltaic islands destroys not only paddy fields but also the huts of low-income people. Local people attempt to protect the embankment breaching due to high climate risks with Indigenous methods, with the little investment, by using locally available ingredients with limited financial capacity at the estuary fringes and in sea-facing areas (**F** and **C**). Climate-mediated impacts are reflected in the large-scale ecodegradation in different parts of Sundarbans. (**G**) The large-scale mortality of mangroves in several parts of coastal belts of West Bengal, India, have been cause by climate change–mediated ecological changes, especially the shifting of sands by the higher intensity of wind flows during catastrophic storms, Aila, etc. and the deposition of sandy sediments on the floor of mangrove forests, leading to mass mortality (**I** and **J**)

Fig. 7.8 Bioinvasion on the West Bengal coastal belts, India (1): Waterbodies for Aqaculture -vast area of crop field goes under saline water inundation to be used for aquaculture, especially monoculture of shrimp with chemical inputs (**A** and **B**). Eutrophication of water nearer to coastal zones due to large-scale application of chemical fertilizers in the nearby crop fields andalso receiving hugeamount of sewagewater from hotels and municipality areas (**C** and **D**); Promoting tourism in the coastal areas with the building of built structures and also using traditional boating systems helf introduction of exotic invasive species (**E**). (**F**) reflects the settlement of bioinvasive faunal species due to deterioratinf environmental quality

Fig. 7.9 Bioinvasion on the West Bengal coastal belts, India (2), featuring some of the major exotic species of finfish and shellfish: (**A**) rampant and uncontrolled fishing serves as a major route of entry for exotic finfish and shellfish species; (**B**) grass carp, *Ctenopharyngodo nidella*; (**C**) Pangas, *Pangasius pangasius*; (**D**) African cat fish, *Clarias gariepinus*; (**E**) white Vannamei shrimp, *Litopenaeus vannamei*; (**F**) Tilapia, *Oreochromis mossambicus*

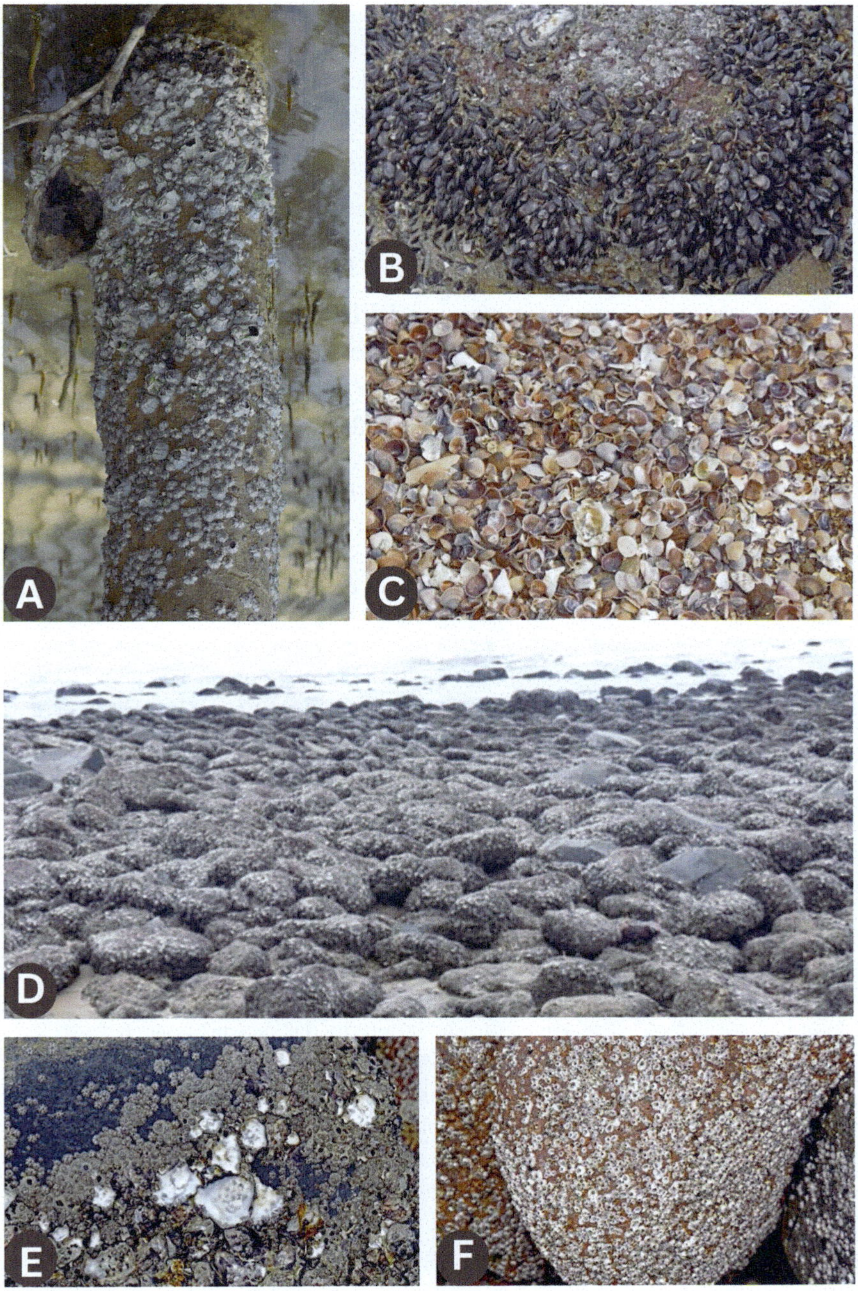

Fig. 7.10 Bioinvasion on the West Bengal coastal belts, India, featuring the settlement, growth and proliferation of different biofouling faunal components: (**A**) the common acorn barnacle, *Balanus nubilus*; (**B** and **C**) a biofouling bivalve, *Enigmonea aenigmata*; (**C**) a heap of calcareous shells of molluscs, especially bivalves; (**D**) a small acorn barnacle, *Balanus aphrodite*, on the granite stones used to protect the coast from erosion; (**E**) bivalves, *Crassostrea madranensis* and *Saccostrea cuculita*, mixed up with barnacles; (**F**) large granites acting as the habitats of barnacles at water–land interfaces

Fig. 7.11 Bioinvasion on the West Bengal coastal belts, India (4): Different bioinvasive plant species one the West Bengal coast, India—(**A**) Intermixing of mangroves, like *Avicennia marina* with *Casuarina equisetifolia* and *Azadirachta indica* (locally named as Neem tree); (**B**) Mesophytic plant species (*Hibiscus tiliaceus*) (**C**) Xerophytic plant like Acacia (*Acacia auriculiformes*) and another exotic plant Eucalyptus (*Eucaliptus gunnii*), two major floral components used for social forestry; (**E**) Climber—*Micania scandens*; (**C**) Tree-*Thespesia populnea*; (**D**) Casuarina and Acacia; (**E**) Climber-*Coccinia grandia*; (**F**) Mixed Mesophytic plants which include *Azadira chta-indica*, *Casuarina equisetifolia* and *Thespesia populnea*

Fig. 7.12 Ecoturism in the Sundarbans mangrove ecosystem, India paving the way for revenue generation in one hand and causing environmental problems on the other: (**A**) the temple of Kapil Muni, a famous religious site located at the highest high tide level at the confluence of the Hooghly estuary at the sea, the Bay of Bengal where millions of people used to visit to offer prayers; (**B**, **C**, **D** and **E**) Hundreds of water vessels in the Indian Sundarbans, which provide communication facilities to local people and water transport facilities to tourists; and (**F** and **G**) different infrastructure facilities, which have developed to cater to the needs of lakhs of tourists visiting this International Heritage Site

Fig. 7.13 The development of different infrastructural facilities for promoting coastal tourism (**A**, **C**, **F** and **G**) and wildlife tourism (**B** and **D**) in the Indian Sundarbans

Fig. 7.14 Tourism that has developed on the basis of cultural and religious beliefs, in the Sundarbans mangrove-estuarine complex in India: (**A**) the temple of Kapil Muni, where a congregation of millions of people takes place each year to experience the bath at the confluence of the Hooghly estuary with the Bay of Bengal on Sagar Island, the largest Gangetic delta in the world, in the extreme western part of Sundarbans, India; (**B**) rituals for the death of a person in the remotest human habitation in Sundarbans, India, by the side of an estuary; (**C**, **D**, **E**, **F**, **G** and **H**) the idols of Banbibi, which are worshipped in parts of Sundarbans, India

Fig. 7.15 Pollution threats to the Sundarbans mangrove ecosystem due to the dumping of biological debris on intertidal belts: (**A**) the dumping of bivalves in the anoxic conditions of mudflats; (**B**)

7.4 Types of Natural Threats on the Coastal Fringes of Forests Dominated by Halophytic Mangrove Plants

The natural death of the Indian coastal mangrove forest has several causes. Mangroves in the tropical coastal environment of India are lost mostly because of the exploitation of natural resources and land-use conversion by human beings for agricultural and aquacultural farming (Blasco, 1984, 1975, 1977; Banerjee et al., 1989; Naskar & Guha Bakshi, 1987; Paul, 2002a, 2002b; Dasgupta et al., 2009, 2014; DasGupta & Shaw, 2013). The restoration of the mangrove ecosystem in India can be realized to a certain extent via afforestation programmes, alongside improving tidal drainage systems by extending artificial channels or ditches, which in turn feed the inner parts of mangrove swamps through high tide water. However, restoring the ecodegraded mangrove forest belts is difficult once they have been naturally degraded by marine-estuarine factors and attributes like catastrophic storms, overwash sand fan lobe encroachment, land subsidence and uplift from faulty tectonic movements, the generation of hydrological stress from altered drainage systems in the catchment areas of river deltas, sediment starvation and hydrological stress–induced coastal morphodynamics, eutrophication, hypersalinity, shifting river banks in the deltaic areas, erosion at the sea face with shoreline retreat, an active barrier rollover process, very low precipitation, a shortage of rainfall and a reduction in freshwater discharge into the coastal zones (determining the ecosystem functioning towards achieving sound ecological health alongside higher biological productivity (Chakraborty & Choudhury, 1985, 1989, 1992; Chakraborty & Chowdhury, 1993, 1994a, b; Chakraborty et al., 1986; Chakraborty, 1995, 1998, 2010; Chatterjee & Chakraborty, 2014; Chatterjee et al. 2014).

7.4.1 Erosion and Accretion: Estuarine and Coastal Processes

Erosion, or the destruction or diminution of sedimentary structures, represents an action of surface processes, mainly governed by some climatic physical processes like wind flows, waves, currents and tidal interplays of water, dredging, deforestation, etc. that dislodge, remove and transport soils or rocky materials from one location of an ecosystem to another area. Accretion, on the other hand, refers to the gradual accumulation of additional newly deposited sediments through fluid flow

Fig. 7.15 (continued) the formation of polychaete tubes in response to the accumulation of chemical pollutants in intertidal belts; (**C** and **D**) thrown-out trash fish in intertidal belts, which invite the growth of pathogenic microbes and cause the mortality of benthic fauna; (**E** and **F**) semi-intensive aquacultural practices, which cause profuse eutrophication; (**G**) plastic- and microplastic-based pollution, causing profuse damage to mangroves; (**H**) the mass mortality of healthy mangrove plants due to climate change–mediated natural disasters, which have shifted a huge amount of sand from eroded dunes to the floors of mangrove forests and have also triggered prolonged inundation under hypersaline tidal water

Fig. 7.16 Communication of people and transport of human commodities mainly by mechanized trwalers operated by oil based engines and construction of infrastructures (roads, hotels, etc.) to tacle overcrowding of tourists cause undesirable deviation from the normal ecological set ups (**A**, **B**, **C**, **D**, **E** and **F**)

Fig. 7.17 The construction of hotels violating laws pertaining to the coastal regulatory zones (CRZs) (**A**, **B** and **C**) impact of torism on West Bengal coast with accummulation of so many organic wastes and synthetic entertainment commodities; (**D**, **E** and **H**) construction of Hotels violating prevailing laws of Coastal Regulatory zones (CRZs) and destroying Dunes on intertidal belts; the destruction of natural sand dunes and the dumping of organic debris and plastics, which contribute to causing coastal pollution in the coastal belts around Sundarbans, India (**G** and **H**) mushroomimg of so many tourist resorts within Sundarbans Biosphere Reserve, violating the laws of land

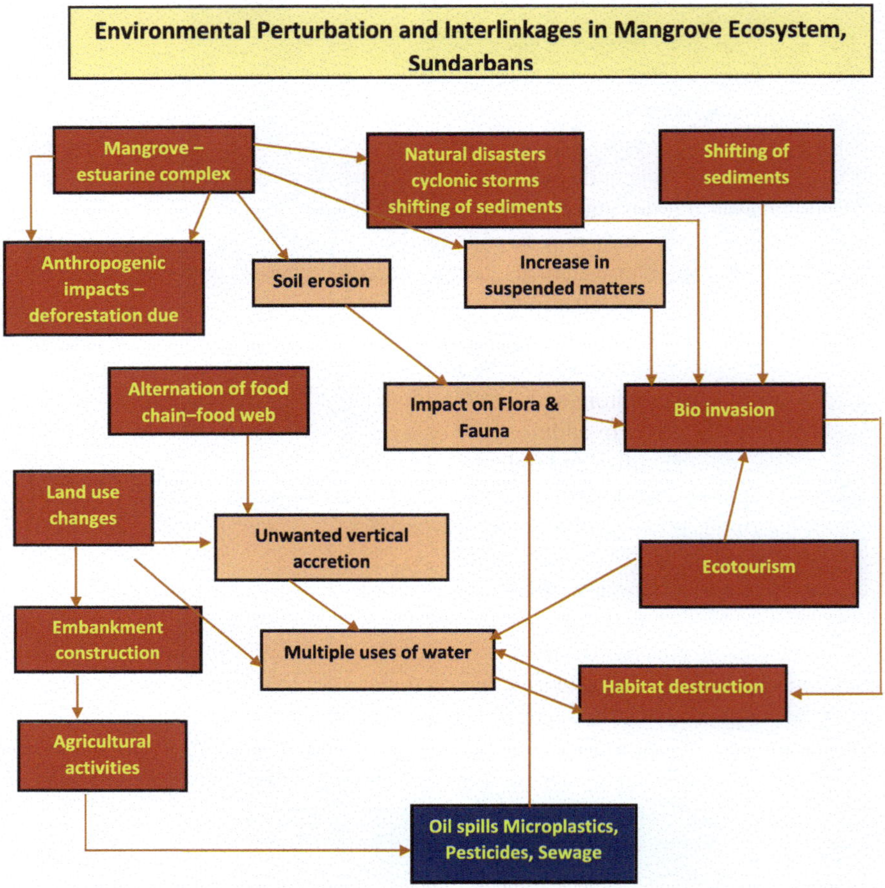

Flowchart 7.1 Environmental perturbation and interlinkages in the mangrove ecosystem of Sundarbans

processes. In a mangrove ecosystem, both erosion and accretion occur simultaneously as the removal of a mangrove results in massive erosion followed by undesirable accretion on the estuarine networks of Sundarbans. The estuarine-deltaic mangrove settings of the West Bengal coast in India experience different erosion and accretion processes with different intensities because of the increased velocity of flood inflow balanced by the frictional dissipation of the hyper-synchronous Hooghly estuary (Paul, 2002a, 2002b; Mazda & Wolanski, 2009; Mazda et al., 2007). In addition, the macrotidal setting in the intertidal belts, the hydrodynamics in the estuarine flows, the funnel-shaped topographical design of different deltas, the movement of large volumes of riverine sediments upstream, coupled with specific sediments with depositional characteristics, have cumulatively contributed to the development of depositional landforms (Paul, 2002a, 2002b; Chakraborty et al., 2012). The geomorphological settings of sediment within the swamp and the fate of waterborne sediment materials have different fractions in mangrove areas, which

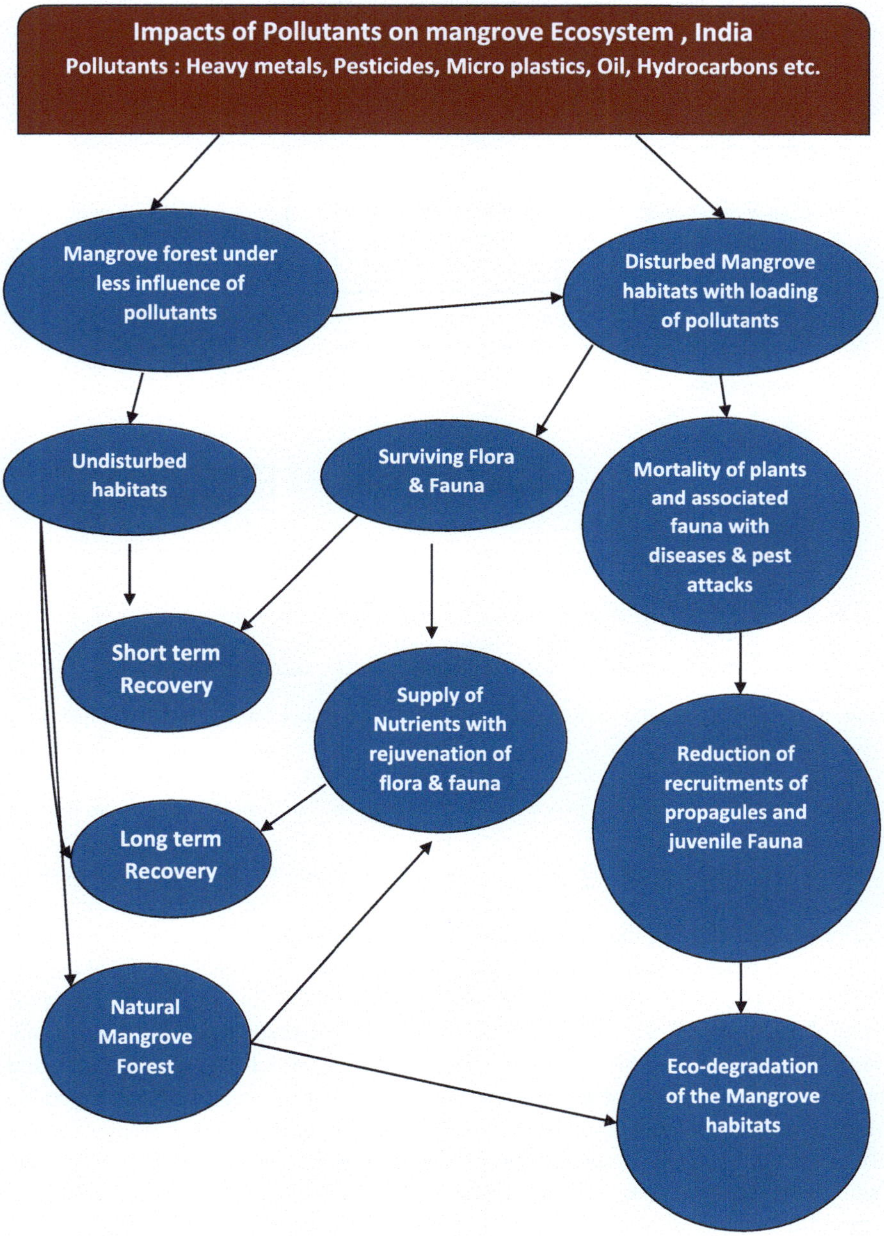

Flowchart 7.2 The modes of action and environmental consequences of several pollutants on the mangrove ecosystem of Sundarbans, India

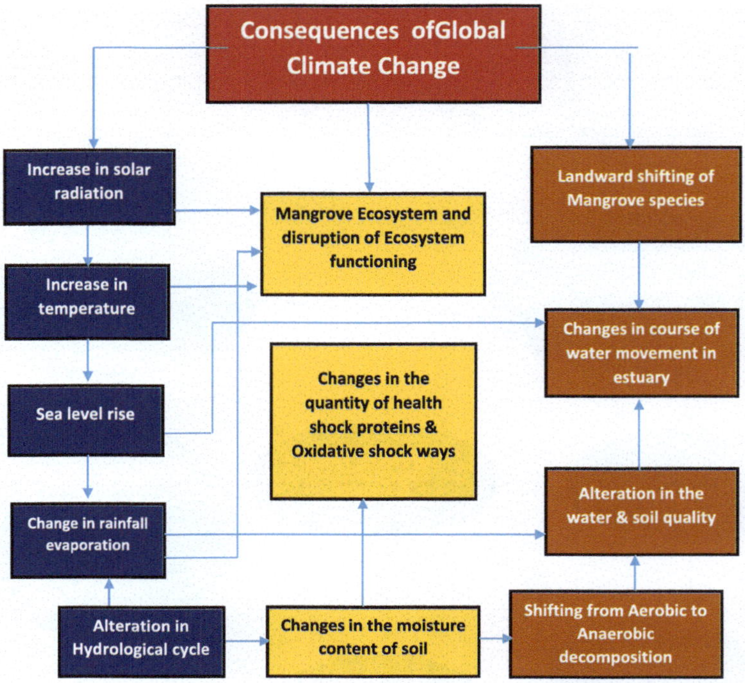

Flowchart 7.3 Ecological consequences of global climate change on the Sundarbans mangrove ecosystem in India

are controlled by dispersion processes, modifying the bathymetry and expansion of the forest area.

On the various deltaic islands of Sundarbans, erosion and accretion processes occur through climate-induced extreme natural events like flooding, tidal surge, cyclones, etc., causing the disappearance of major portions of 12 sea-facing southern deltaic islands. A number of small islands, such as Lohachara, Kabasgadi, Suparibhanga and Bedford, have already vanished from the map of Sundarbans. Several other islands which are vulnerable to erosion pressures include Ghoramara, Sagar, Mousuni, Namkhana, Dalhousie, Dhanchi, Bulchery, Bhangaduni and Jambudwip. The islands on the western banks of estuaries become more vulnerable to erosion than the islands on the eastern banks. Marginal accretion along the eastern and northern margins of islands is localized in the inner estuaries along the coast facing the sea in an east–west orientation. One research study has revealed the reduction in the total land area of Indian Sundarbans from 6402.09 km^2 in 2001 to 6358.05 km^2 in 2009, with a net loss of 44.04 km^2 of land, which includes the erosion of 64.16 km^2 and the accretion of 20.12 km^2. The luxuriant growth and strong establishment of mangroves in the matured islands with less anthropogenic activity compared to the eastern parts of Sundarbans make those islands comparatively stabler. Sagar Island, the largest Gangetic delta in the world, has suffered a major thrust of erosion, with a loss of 30 km^2 of total area, experiencing only marginal accretion (Hazra et al., 2002, Chakraborty et al., 2014).

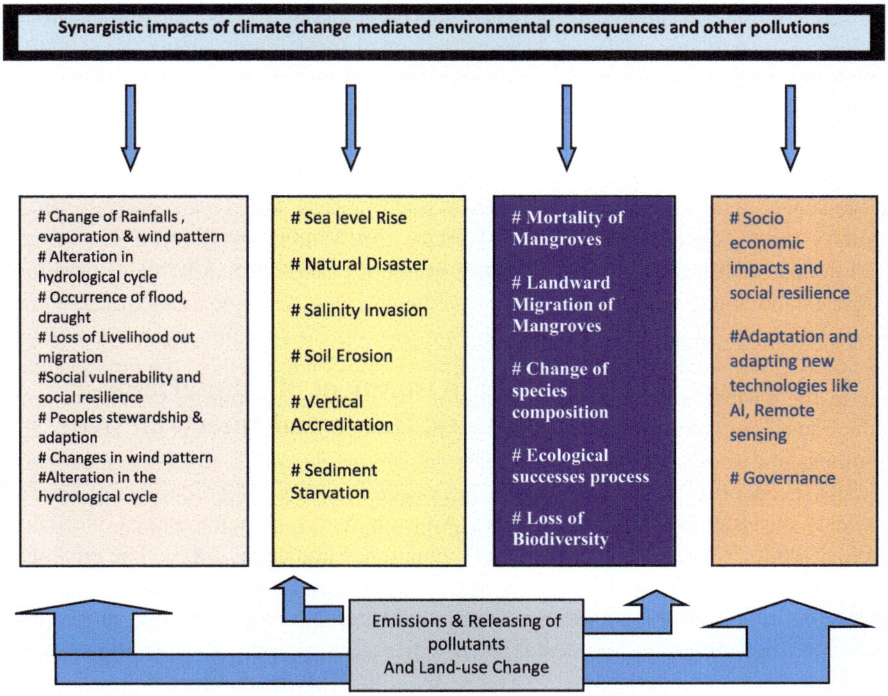

Flowchart 7.4 The synergistic impacts of climate change and other pollutants, including land-use changes on the ecodynamics of the mangrove ecosystem in Sundarbans, India

Generally, erosion occurs on the rivers and estuaries under the influence of ocean-based physical processes like tectonically controlled responses, waves, currents and tides despite the large sediment supply provided by these water-flowing bodies. The observed losses in the Ganges–Brahmaputra river system because of erosion are thought to have been caused by different anthropogenic activities, like deforestation, mining, damming, withdrawals of water and artificial construction for river rerouting, and all of these developments increase eustatic sea-level rise.

In Indian Sundarbans, the process of erosion is controlled primarily by tidal flows and the landward and seaward movements of flood and ebb currents acting along the estuaries. Long-term trends in erosion, as observed on estuarine mouth regions, the banks of estuaries, mid-channel bars, point bars (as marginal and crescentic types) and intertidal swash platforms, are governed mainly by tidal asymmetry, which also facilitates freshwater discharge from upland and drives sea water to coastal brackish zones. The estuarine networks of the drainage system within Sundarbans enjoys dynamic flow patterns that develop a process of erosion, making the geomorphic setup of the estuary complex. Funnel-shaped openings at the mouths of different estuaries, having an average width up to 10 km, result in a higher tidal amplitude, which in turn alters estuarine hydrodynamics. However, during dry seasons, the estuarine plains are completely dried up, and the manifestation of desiccation is evident with salt encrustation, but during the rainy season, much of the estuarine plains get inundated by floodwater (Das, 2008, 2009b). Several erosional sedimentary structures formed as outcomes of erosion, like furrows, ridges, tidal shoals and mud mounds in intertidal zones, render obstacles and scours and help in resisting erosion.

A time-series analysis of alterations in the shape, size and geomorphological properties of different deltaic islands over 32 years (1969–2001) has revealed the large-scale degradation of mangrove zones and mudflats, an increase in the levels of salinity and the development of saline banks within Sundarbans in India. This observation has established that 12 sea-facing southern islands in Sundarbans, including Sagar Island, the world's largest delta, have been under the tremendous threat of erosion. This study quantified that total erosion over the 30-year time span is estimated as 162.879 km^2—in contrast to the accretion of lands on the beds of the inner estuary along the eastern and northern margins of the islands, amounting to 82.505 km^2, which has been taking place mostly over the same period (Hazra et al., 2002). The southeastern, southwestern and northeastern sides of Sagar Island have been experiencing severe erosion pressures due to the concerted actions of a multitude of anthropogenic activities and natural processes. An estimate has shown the rates of erosion from 1996 to 1999 to be 5.47 m/year (Gopinath & Seralathan, 2005). The erosional characteristics induced by waves and tidal currents show a special sequence of different geomorphological formations, such as mud balls, mud mounds, tidal shoals, furrow and ridge structures and destabilized dunes advancing inland from the seaward side. All these erosional characteristics on the morphodynamic processes lead to the landward displacement of the coastal Sundarbans, which emerged with its unique ecological characteristics around 5000 years BP (Das, 2009c). The continued erosion pressures on the deltaic islands of Sundarbans, India, have reduced the mangrove forests to a considerable extent in respect to the density of vegetation and the extent of mangrove coverage (Figs. 7.1 and 7.2). The effects of shoreline retreat because of such ongoing pollution stresses on the mangroves and their fringe forests are revealed from the present position of eroded shore fringe mud banks with exposures of stiff or compact clay beds and vertical stands of mangrove tree stumps, which can be found on deltaic islands like Lothian Island, Prentice Island, Susnir Char, Henry's Island, Dhanchi Island, Kalas Island and Bhanaduni Island.

7.4.1.1 Driving Forces Causing Erosion and Accretion in the Estuarine Networks of Sundarbans, India

The tidal plains of a large number of estuaries (~31) and the tidal inlets of the Ganges–Brahmaputra tidal plain are the most important along the coastline of West Bengal, India. The entire mangrove-estuarine complex is influenced by dominant physical processes like tidal flooding, river fluxes, waves and sea-level change, and the occurrence of several episodic and nonepisodic events like cyclones and storms at certain intervals continually modify this coast (Das, 2009a, 2009b). These river estuaries and tidal inlets carry out two purposes, namely freshwater discharge and the to-and-fro movement of saline water arising from sea faces triggering and structuring the biodiversity (Aller, 1982, 1988, 2001; Allen & Duffy, 1998; Ashton, 2002).

The ebb and flow of floods occurring at different sides of the tide-dominated river-estuarine networks of Sundarbans, determine the hydrodynamics major channels like the Mooriganga, Saptamukhi, Bidyadhari, Thakuran, Raimangal, Matlah estuaries and ecological status of those estuaries are reflected by the diversity of aquatic flora and fauna (Naskar et al., 2004; Sarkar et al., 1986; Sarkar et al., 2013). The asymmetry in the direction of flood tide through the ebb-dominated side and also flood-dominated side within the same tidal channel result in erosion along the tidal channel in the estuarine networks of the region and the erosion trends in these channel systems are governed by the duration of asymmetry (Ginsberg & Perillo, 2004). The tidal cycles within the estuaries, having neap–spring cycles and ebb–flow cycles in the tidal regime, determine the configuration of the river mouth bars, accretionary point bars, mid-channel bars and marginal levees. The accretion process in the estuarine networks of Sundarbans is determined mostly by a higher rate of sediment trapping from the geomorphically elevated areas covered with mangrove vegetation. The subsidence of the banks of estuaries with the simultaneous filling of sediments all along the tidal courses is quite common on newly emerged islands and salt marshes on the banks across the estuaries.

7.4.1.2 Estuarine Banks and the Erosion Process

Estuarine banks, being major sites for erosion, stand on the landward direction of the intertidal belt occupied by marsh vegetation, and they are reinforced and fortified by the construction of embankments in the intertidal area (Bhattacharya, 1999). The top surface of the banks of the estuaries is marked by desiccation cracks and patchy salt-encrusted surfaces after long exposures, and these banks, characterized by fine and obscured laminae of silt and clay profusely impregnated by roots of marsh vegetation and the remains of plants' roots, constitute a substantial component of bank deposits. The bank deposits of Sundarbans are heavily disturbed by the still roots and pneumatophores of mangrove plants; the burrowing activities of macrobenthic fauna, forming curly mud cracks; and bank deposits featuring the remains of plants and plant's roots. In vertical sections, the three strata are clearly visible in bank deposits on the basis of colour from the top to the mean low-water line. The top of the bank is dominated by mangrove species such as *Phoenix paludosa*, *Excoecaria agallocha*, *Avicennia alba*, *Ceriops decandra*, *Acanthus illicifolius* and *Suaeda maritima*. Sediments deposited on the top stratum of the bank with a thickness of up to 2 m are distressed by the penetration of mangrove roots more than 1 m deep. The second tier of sediment strata immediately underlying the top one, where the second tier has a thickness of around 1.5–2 m, is amply bioturbated by the burrows of macrobenthic fauna, especially brachyuran crabs and penetrating mangrove rootlets. The walls of burrows within the muddy substratum and mud cliffs, bearing the markings of roots penetration, feature a yellowish brown colour in the sediments as evidence of an oxidation process (Bhattacharya, 1999). The prevalence of so many furrows and ridges testifies to the ebbing tidal origin of estuarine erosion, but all these structures undergo modification in respect to their designs through wave erosion in several cases (Bhattacharya, 2000). Finally, levee surfaces with cracks, sometimes curly mud cracks, sometimes exhibit salt-encrusted patches due to the evaporation of saline water pools.

Bank erosion takes place in a piecemeal manner, involving different mechanisms: (i) The thrust derived from ebb-and-flood cycles enjoy the seasonal fluctuation of the highest tidal amplitude; (ii) erosion from the surface and subsurface soils of the banks and intertidal zones results from receiving the impact of longitudinal and helicoidal flow patterns in a meandering stream; (iii) the cohesion properties of the textural components of soils, such as sand particles, which have the lowest cohesion properties, are eroded more than those of silts and clay materials; and (iv) differential modes of sediment subsidence cause piecemeal bank erosion, which is later modified by waves near the respective mouths of the rivers.

7.4.1.3 Coastal Processes and Sea-Level Rise on Erosion at Sundarbans, India

The Sundarbans mangrove-estuarine complex has been under grave threats of soil erosion over the past couple of decades for both natural and anthropogenic reasons. Erosion, being an active process all along the shorelines of the Sundarbans and at the interface between land and sea, makes the coastal zone unstable as the waves, currents and wind continuously shift the sands.

The intensity of such erosion processes also varies from one ecozone to another throughout the length and breadth of Sundarbans, mostly in tune with the variabilities of operating factors and attributes causing erosion, like deforestation, changes in the course of water movements, land-use modifications, the mass collection of seeds of shrimp and prawns, tourism and construction activities. In the mouth of estuaries, erosion generally takes place in the tectonically uplifted and well-developed cliff areas, which withstand the beating of waves with different seasonal intensities against the shore. A multitude of geological and tectonic factors, along with marine-oceanographic variables like tidal range, duration and amplitude, water currents, waves and flow volume, the speed and direction of winds, the intensity of storm surges and seasonally fluctuating levels of sea-level rise, continuously dislodge sediments and shift unconsolidated sediments between the estuaries and seas in the coastal belt of West Bengal, India. Erosions on those elevated portions will remain ongoing as long as the landmass becomes parallel to that of the sea surface.

The coastal region of Sundarbans under the continuous impact of erosion leads to the landward displacement of the shore and is generally ornamented with some unique erosional sedimentary structures like erosional furrows and ridges, mud mounds and mud balls. The upper and middle foreshores of the sandy coast possess some erosional structures as obstacles and scours and feature fragments of wood, pebbles, gravel, clay balls, etc. All these nonliving clasts create resistance stream flows during the backwash of waves and form scour troughs. The prevailing biogenic activities on this nonliving process tend to have both short term and long-term effects on the changing configuration of shoreline of the Sundarbans coast of India (Gopinath & Seralathan, 2005). Erosion in the intertidal belts of the Sundarbans coast of India, experiencing meso-macrotidal currents, get inundated twice a day, with a total inundation duration of 12 h. Add to this the gradual denudation of forest belts, along with several catastrophic processes: (i) currents derived from flood-and-ebb tidal cycles; (ii) high tidal amplitudes; (iii) the prevalence of strong littoral drift; (iv) impacts from storm-induced waves; (v) the arrival of cyclonic storms, hurricanes, etc. at certain intervals; and (vi) faulty tectonic movements leading to earthquakes, tsunamis, etc. (Das & Bhattacharya, 1999).

7.4.1.4 Mechanisms of Coastal Erosion: Physical vs. Biological Activities

The process of coastal erosion in deltaic Sundarbans in India occurs through three mechanisms:

1. The erosion of mud via scouring and the deposition of eroded mud after being carried away to the subtidal region by suspension transport.
2. Worn-out sand after erosion, which is deposited because of resuspension and is transferred in a seaward direction by water currents, waves and tides.
3. Coastal structures like dead shells, large sand particles and exotic boulders, in association with biological components, which act as tools that augment and accelerate the process of erosion.

Biological factors also play significant roles in the Sundarbans coastal environment because Sundarbans is considered as a biogenous coast. A galaxy of fauna inhabiting the mangrove ecosystem, especially in the intertidal belts, plays important roles through its bioturbatory activities in developing microstructures and morphological structures as marks of erosional features that significantly contribute to the trapping of sediments and that enable mangrove seeds to settle on the bottom of forest floors. This galaxy of fauna also plays important roles in compensating for the loss of sediment due to erosion.

However, continuous conflicts between different biogenic and physical processes in an intertidal belt exist, where the physical processes become responsible for creating geomorphological structures on a beach, and benthic faunal assemblages act independently or in association with one another to modify original sedimentary structures through their general biogenic and specific bioturbatory activities. All these biological components play important roles in resisting coastal erosion, altering the sedimentation pattern, designing beach morphodynamics and regulating the process of erosion at the intertidal belts of the mangrove ecosystem.

Excessive sedimentation damages mangroves by resisting settlements of seeds and the subsequent colonization of mangrove deposits. Moderate sedimentation at moderate levels helps mangroves to flourish by supplying nutrients and keeping pace with the predicted rise in sea level. Excessive, sudden sedimentation not only causes the mortality of mangroves by fostering the burial of lower portions of mangrove plants and their root structures (stilt roots and pneumatophores), thus inhibiting gas exchange and killing the tissue of roots and trees. Excessive sedimentation also brings about vertical accretion, facilitates the formation of natural disasters such as floods and hurricanes and thereby constrains the normal functioning of the ecosystem.

7.5 Salinity Invasion–Induced Hypersalinity Within the Mangrove Ecosystem of Sundarbans

The salinity of soil and water, being the most pronounced ecological parameter, is considered a prime regulator of biodiversity components in terms of their abundance, distribution and community interaction processes but also becomes a prime threat in the coastal mangrove ecosystem. This ecological variable is in turn determined via the oscillation of seasonal precipitation, tidal cycles, sea-level rise, global warming–mediated temperature increases and an inconsistent supply of fresh water through seasonal flooding or disrupted riverine flows to estuaries.

The fringe mangrove forests in the river deltas within the inner parts of islands within the coastal belt of Sundarbans, far away from the major tidal estuaries, are affected by tidal influxes with high amplitudes, high intensities and high elevations of saline water, which ultimately form tidal drainage with platforms covered by halophytic vegetation. Because tidal drainage cannot reach the upper areas during all daily high tides. Several areas remain uncovered by tides for prolonged periods and thereby become sensitive to evaporation pressure in tropical hot and humid coastal belts with alternating dry and wet periods. Moreover, those areas get deprived of experiencing a further siltation process because of the absence of low tidal drainage on wide flats with concave upward basinal depressions in swampy forest. In addition, all those areas are observed to have been flooded by astronomic tides along with storm-generated saltwater floods across the island platforms. Over time, the entire area of the island platform becomes covered by saltwater at low tide after the end of different phases of astronomic tides alongside storm events in and around the coastal belt. However, the depressed saltwater pockets, at lower depths, look like saltwater pools of tide water within the mangrove island.

After the evaporation of surface water from scorching sunlight at more than 30 °C, the silty clay surface, endowed with high soil moisture, gets desiccated, and with the onward encrustation of salts, this leads to the mass mortality of the mature halophytic vegetation by salt-affected soils. These salt-loaded deposits form hypersaline patches that are designated as saline blanks or natural saltpans within the degraded and partially degraded mangrove ecosystem of Sundarbans, India.

All these saline blanks expand into neighbouring areas thanks to a decrease in the total amount of rainfall due to the increase in the length of the dry phase, lengthening the recurrence interval of major floods in the mangrove-estuarine-coastal belt of Sundarbans, which have been found to develop in the saline blanks of Henrey's Island, Lothian Island, Kalas Island, Dhanchi Island and similar other inner parts of the Sundarbans islands (Indian part). This reveals the presence of many denuded mangrove stands, exposing barren soils without almost any vegetation but developing algal mats on the surface of soils upon getting a profuse supply of water during the rainy season.

A reduction in freshwater flow not only increases the salinity levels of estuarine water but also significantly affects the nutrient cycles in mangrove ecosystems. The outcomes of these changes are manifested in the landward expansion of the mangrove forest, the releasing of ammonium into the overlying tidal water and soils that can take up more dissolved organic carbon, soluble reactive phosphorus and nitrogen in the form of nitrite and nitrate. As phosphorus enrichment does not promote dissolved inorganic nitrogen or dissolved organic carbon flux, this suggests that rapid phosphorus uptake is geochemically mediated by the potential for mangroves to sequester phosphorus.

Recently, the supply of sediments as the products of the erosion of the sea face of an abandoned delta (Indian Sundarbans) is brought into the inner parts of the estuaries and major tidal channels of the Sundarbans, where it settles. The channel banks experience oscillation with the frequent occurrence of tidal scour alongside the deposition of sediments in the process of tidal accretion in response to fluctuating tidal influences, especially during extreme higher and lower tidal interplay with the deltaic intertidal plain areas. In this process, point bars are developed as a result of accretion on one bank in which mangrove seedlings tend to colonize on the accretionary sediment surface, but the mangrove plants on the opposite bank will remain undermined by an advancing channel-induced bank collapse process. The simultaneous growth and mortality of both fully grown and newly settled mangroves take place along the bank of the estuarine networks of Sundarbans and on the Subarnarekha delta at the confluence of the Subarnarekha estuary with the Bay of Bengal at the junction of the Odisha and West Bengal coasts in India.

7.6 Rainfall Shortages and Low Supplies of Fresh Water in the Tropical Hot and Humid Coast: Potent Contributors to Hypersalinity

Although the name Sundarbans is derived from the presence of a unique tropical evergreen mangrove plant species, *Heriteria fomes* were supposed to have been widely distributed in the major portion of the Indian Sundarbans, especially those areas which were under freshwater influence, and the inland environment was developed by occurrences of both profuse monsoonal precipitation and large-scale drainage and by an influx of fresh water through riverine flows. All that fresh water remains stored within the smaller basinal depressions throughout much of the year on the delta plain surface. In contrast, the parts of Sundarbans with large mangrove trees facing the seaside were dominated by salt-tolerant species, whereas a large area between the hypersaline and inland zones included mostly brackish-water and vegetational assemblages. Higher amounts of rainfall during the monsoon period tends to decrease the soil pore runoff process and thereby helps maintain the balance between the productivity of mangrove wetlands and the expansion of the forests. Keeping up with the current pace of the increasing rate of hypersalinity with the development of saline blanks within the core areas of mangrove forests is directly related to very low precipitation and either the diversion or the blockage of freshwater influxes in the coastal belt of South Bengal.

The growth and the propagation of mangrove forests are very much influenced by the combined effects of a number of pronounced ecological parameters (water, soil and atmosphere) that undergo a distinct oscillation through different months, seasons and years (Blasco, 1984; Chakraborty, 2013). One of the major causes of the shortage of freshwater flows is the overuse of river water in drainage catchment areas, mostly by creating a number of large dams on the east coast of India. In Gujarat, with the influence of arid environmental setups, the influence of fresh water is almost nonexistent because it enormously increases the rapidity of the evaporation rates of water between two high tides, leading to an increase in the concentration of salts within the surface sediments of mangrove forests, causing the large-scale mortality of mangrove plants (Blasco, 1984).

The construction of Farakka Barrage in 1975 has resulted in the blockage of the steady flows of fresh water through Padma (Ganga) and the Bhagirathi–Hugli distributary systems of the Ganga River to every nook and corner of Sundarbans, India. The freshwater supply to the Sundarbans by other rivers of eastern India is too little to maintain the brackish-water condition of the mangrove forests. Such a drastic reduction in freshwater flows has resulted in the development of hypersaline conditions for both water and soil, tremendously constraining the normal growth of mangroves, especially freshwater species such as *Heriteria fomes* and *Nypa fruticans*. Similarly, at present, freshwater supplies have been considerably reduced into the Corringa Reserve Forest through the Goutami–Godavari river system because of the construction of a number of cross dams along the Godavari River basin in the states of Telangana and Andhra Pradesh.

The continuous rise in sea level with water of higher salinity than normal is expected to inundate the low-lying fringe areas of estuaries and delta plains endowed with mangrove coverage, and this is in line with the estimation of the Intergovernmental Panel on Climate Change (IPCC) for both coasts of India, specifically the east and west coasts. In addition, a wide range of fluctuation in salinity, which reached 33 ppt during the pre-monsoon period and came down to about 12 ppt during the monsoon period, physiologically stresses mangrove plants and determines the distribution patterns of mangrove vegetation (Coulombier et al., 2012; Chakraborty, 2013). The recession or progradation of the coastal mangrove-estuarine system is determined by the interplay between factors such as erosion, accretion, land-surface elevation, sea-level rise, changes in riverine courses, fluvial sediment deposition and subsidence, the withdrawal of surface water and the lifting of groundwater.

7.7 Land-Use and Land-Cover Change (LULCC) in the Mangrove Habitats of the Sundarbans Region in India

Considerable changes in land-use patterns are due to the development of agriculture, aquaculture, fisheries and the construction of bundhs after removing large chunks of mangrove vegetation. Halophytic vegetation, the dominant structural components of the mangrove ecosystem in tropical countries, has been destructed since 1980s because of land-use and land-cover change (LULCC), which is evidenced by high rates of deforestation coupled with other associated environmental consequences.

Changes in land-use patterns, impacting the species composition and extent of vegetation cover, tend to have negative ecological consequences on the quantum of freshwater inputs, the transport and deposition of sediment, hydrological alterations and the nutrient dynamics of coastal ecosystems (Walsh, 1991; Paul, 1999; Ray et al, 2013). Major changes in land use in tidal wetlands with mangrove vegetation stem from agriculture (rice and vegetable crops), silviculture (timber estates, plants social forestry) and aquaculture (shrimp and fish culture impoundments), largely converting mangroves.

Significant anthropogenic activities have been found to impact the coastal environment surrounding the Bay of Bengal in eastern India and heavily contribute to introducing excess nutrients into many coastal watersheds, which have drastically increased over the past three to four decades (Walsh et al., 1981; Eppley & Peterson, 1979), ultimately resulting in changes in ecosystem structure and function (Cloern, 1996). The deltaic developments in the Ganges–Brahmaputra estuarine networks have been considerably impacted and been made vulnerable because of anthropogenic perturbations, especially with the increased nutrient load from river discharge (Mukhopadhyay et al., 2006). The economic growth within the reclaimed areas of mangrove forests and the releasing of so many toxic elements have caused global warming and associated environmental perturbations, disrupting freshwater flux, nutrient cycling and biological production.

Large-scale land-use alteration in and around Sundarbans over the past couple of centuries has increased mainly because of the accelerated pace of reclaiming deltaic islands to provide shelter and the ingredients for human livelihood, mostly through modern agriculture and aquaculture practices. Increasing land-use changes in coastal watersheds mainly because of deforestation will result in a rise in sediment inputs through erosion within river beds, the outcomes of which are manifested by a reduction in the depth of rivers, the loss of carrying capacity for large volumes of water, an increase in turbidity and a decrease in primary productivity. Besides, a higher amount of nutrients from agricultural lands is washed out into adjoining rivers and ultimately reaches the waterbodies in and around mangrove forests. The supply of sediments and nutrient enrichment within the mangrove ecosystem can increase not only mangrove growth through higher rates of photosynthesis but also mangrove mortality through eutrophication via overenrichment with nutrients and excessive sediment deposition, which also tend to promote the additional accumulation of organic matter. In an impacted mangrove forest in Brazil, the increased presence of organic matter originating from autochthonous sources with either phytoplankton, benthic algae or other allochthonous sources through rapidly accumulated sediments results in an increased accumulation of organic matter and thereby has undesirable ecological consequences (Sanders et al., 2009).

Historically, local people who depend on the resources of the mangrove ecosystem have used it to harvest fuel, food, shelter materials, tannin or charcoal for domestic purposes. Paddy rice cultivation and aquacultural activities for shrimp and fish have caused a considerable amount of mangrove to be lost, with a series of negative impacts on the adjoining environment (Chakraborty, 1996). At each stage of land conversion, considerable losses in belowground and aboveground carbon stocks have taken place, and the emission of a large amount of carbon into the atmosphere as CO_2 occurs from the oxidation of organic compounds. Additional losses of carbon in the process of dissolved and particulate organic carbon export are also assumed to be high.

These various forms of land use bring forth multiple types of land cover on coastal estuarine belts because of soil conversion for human settlements and infrastructural development. Selective and intensive forms of land conversion result in changes in forest cover, species composition and species loss, leading to various phases of ecodegradation:

1. Intact mature forests have been selectively logged.
2. Access to forests needs to be developed by excavating drainage canals through which the harvested timber is carried for commercial use, reducing disturbance to all other structural components of the mangrove ecosystem.
3. Newly excavated and upgraded drainage canals carry saplings to selected intertidal belts for the afforestation programme.
4. The administrative classification of the forest area is to be considered prior to changes, permitting forest removal.
5. The maintenance of the drainage system and a reduction in canopy cover have resulted in an increase in surface temperatures and a reduction in moisture, paving the way for an aerobic decomposition process.
6. The selection of suitable species of mangrove plants for the afforestation programme is required to keep an eye on the natural ecological succession process and to ensure the tolerability of the species concerned against environmental challenges.

7.8 Eutrophication Within Coastal Wetlands: A Contributor of Many Ecological Problems

Land-use change resulting in eutrophication refers to the overenrichment of water with nutrients derived from both allochthonous and autochthonous sources may negatively impact mangrove ecosystem by altering nutrient cycling. The outcomes from the impacts of eutrophication on mangroves are found to be complex ones after being influenced by so many ecological attributes, like tidal inundation frequency and duration, soil texture, salinity, species composition and species position in intertidal zones.

The synergistic effects of litter decomposition and eutrophication exert a strong influence on these processes. Long-term experimental studies with nutrient fertilization have pointed out that decomposers were always nitrogen limited regardless of the effects on net production, which leads to inferring that differential nutrient limitation between decomposers and mangroves exists where the forests are phosphorus limited and where litter quality has changed and resulted in higher rates of decomposition for phosphorus fertilization.

In mangrove ecosystems, higher rates of nitrogen transformation cause eutrophication. An addition of ammonium and nitrate results in faster rates of denitrification and ammonification within mangrove soils. Also, waterbodies in the pristine and undisturbed mangrove zones act as a sink of nitrous oxide (N_2O), and mangrove waters in impacted areas serve as sources of nitrous oxide. Therefore, eutrophication brings about changes within the waterbodies of a mangrove ecosystem, from a sink to a source of nitrous oxide, which is emitted to the atmosphere and thereby represents a positive feedback loop to climate change.

Nutrient pollution, caused mainly by sewage disposal, enriches water and sediment with nutrients, especially with nitrogen and phosphorus, and improves biological productivity such that phytoplankton and the growth of mangrove trees flourish (Twilley & Chen, 1998). But an increase in the uptake rate compared to the disposal rate results in the excessive accumulation of nutrients, where high concentrations lead to eutrophication. The excessive growth of phytoplankton, benthic algae and macrophytes within a mangrove ecosystem obstructs the healthy development of pneumatophores and reduces oxygen exchange through enhanced microbial activity. The profuse growth of benthic algae, forming algal mats, also hinders the growth of mangrove seedlings.

Experimental studies undertaken by Feller et al. noted the regulatory roles of the variation in latitudinal gradients and tidal cycles on nutrient limitation patterns and primary productivity. Nutrient conservation within a mangrove ecosystem forms a direct but complex relationship with nutrient enrichment, which has dramatic effects on growth and nutrient recycling. However, this process is very much dependent on ecological conditions, location and the state of nutrient limitation. In pristine mangroves and impacted mangroves, the rates of nutrient transformation depend on several ecogeological attributes, like the degree of tidal inundation frequency, the duration of tidal inundation and the input of nutrients in soil and water (Reis et al.). Therefore, nutrient transformation processes in a mangrove ecosystem is governed by land-use-change-mediated eutrophication, and all those ecological consequences determine the rate of net primary productivity and the survival of specific mangrove species.

Because of the poor supply of fresh water through overspill waters from the associated riverine networks, nearby estuarine systems convert all those waterbodies into backwater sites, particularly during pre-monsoon summer months, leading to a drastic alteration in food chain–food web dynamics and changes in species composition. The accumulation of hypersaline water and the gradual enhancement of the amount of deposited untreated nutrients from aquacultural activities and the steady release from modern high-tech agricultural practices from the encroached rice paddy fields towards the fringe areas of mangroves have resulted in a series of ecological changes: an abrupt increase in green biomass with the spread of algal blooms covering the surfaces of waterbodies while resisting the penetration of sunlight to deeper parts of the water column; the loss of biodiversity and changes in species composition; the drastic depletion of dissolved oxygen; an increase in acidification, turbidity, etc.; and a reduction in the depth of waterbodies and the volume of water. All such ecological alterations inhibit the regeneration of mangroves by arresting the dispersal and settlement of drifted seeds. All these ecological incidents have been found to occur in Godavari, Mahanadi and the mangroves of Sundarbans in the eastern coastal belt of India.

The domestic sewage of Kolkata and the industrial belts along the bank of the estuaries of West Bengal (Kolkata, Haldia, Diamond Harbour, etc.) are released on a regular basis, especially to estuaries in the Bidyadhari River, through different networks of canals and creeks, which ultimately flow to the Sundarbans mangrove zones of India.

However, the gradual increase in sewage is counteracted by the mangrove systems, mainly because of the following factors: (i) The sewage on its way through the mangrove habitat gets dispersed over vast areas; (ii) the thick blanket of vegetation naturally filters the nutrients needed for its growth (primarily nitrate and phosphate); (iii) the mangrove soil, algae and microbes in conjunction with the prevailing physical processes within the estuaries absorb large amount of pollutants; and (iv) the potential for mangroves to sequester pollutants increases over time. Magnified organic pollution contributes to the development of different waterborne diseases, the mortality of fish and other aquatic fauna and alterations to the species composition within the mangrove ecosystem (Tattar et al., 1994).

A report indicates that an annual amount of 88,600 tonnes of nitrogen and 17,200 tonnes of phosphorus have entered the Ganga Basin (Anonymous, 1982a, 1982b). The escalating rise in population size, growing industrialization and large-scale usage of fertilizer in Southern Asia and Eastern Asia are major providers of nutrients. In addition, ongoing human-caused climate changes have introduced environmental stresses to the coastal environment, affecting the biological properties of phytoplankton and biogeochemical controls over them in the ecological feedback processes due to inadvertent fertilization. Therefore, an understanding of the roles of coastal biota is needed to respond to the anthropogenic ecological changes which will affect natural climatic variability in the coming decades (Cloern et al., 1983; Cloern, 1996).

The mangrove-estuarine complex in the coastal zone of the Bay of Bengal is considerably impacted by cultural eutrophication, which impairs the normal functioning of the complex biogeochemical cycling of this coastal environment. A host of anthropogenic stress factors—such as an escalating population, exploration, deforestation, chemically intensive agricultural practices, semi-intensive aquacultural activities, unorganized tourism, other tourist activities, boating and ship transportation (with the release of petroleum products and oil spills) and rapid economic growth surrounding the Bay of Bengal—make this coastal environment very vulnerable (Millennium Ecosystem Assessment, 2005; Ganguly et al., 2008).

7.8.1 Eutrophication and its Contribution to Bioinvasion

Eutrophication, an ecological process through which an oligotrophic (nutrient poor) aquatic environment is transformed into a eutrophic one (nutrient rich), has resulted in a number of ecological changes in species composition, biodiversity loss, the depletion of dissolved oxygen, the alteration of biogeochemical cycles, etc. The environmental perturbation associated with this unique human-induced environmental phenomenon makes the resultant environmental odds challenging and nonconducive for many indigenous species, and ultimately, invasive species take the opportunity not only to intrude but also to outnumber the native species because of their increased tolerability and resistance to adverse changing ecological consequences. Cultural eutrophication, speeding up the natural eutrophication process with the overenrichment of nutrients, which stimulates the growth and proliferation of algae undergoing population explosions, results in adverse changes in the affected aquatic environment (Cloern et al., 2007a, 2007b). The composition of the biotic community of the affected waterbody changes, which in turn brings about changes in the species composition of fish, deviating from normal conditions and producing toxins from the algal blooms that make the water environment unsuitable for proper ecosystem functioning.

Excessive inputs of nutrients to many coastal watersheds because of anthropogenic activities have dramatically increased over the past three to four decades (Walsh et al., 1981; Eppley & Peterson, 1979), which tend to disrupt ecosystem functioning through eutrophication (Cloern, 1996). Many human activities, like deforestation, ongoing modern technology-based agricultural and aquacultural activities, unplanned urbanization, high tourist activities, intensive boating and the aims of economic growth in Sundarbans near the Bay of Bengal, have made the coastal environment vulnerable and susceptible to the settlement of invasive biota (Millennium ecosystem Assessment, 2005; Ganguly et al., 2008). Several waterbodies, once used for the monoculture of shrimp or polyculture for other commercially important finfish, are presently in an eutrophicated condition because they are used for culturing several exotic species and are infested with a number of macrophytes, such as *Eichhornia crassipes*, *Wolffia microscopic*, *Alternanthera philoxeroides*, *Pistia stratiotes*, *Lemna minor*, *Vallisneria spiralis*, *Ipomoea aquatic*, *Jussiaea repens*, *Nymphaea rubra*, *Nymphaea stellata*, *Nymphaea pubescens* and *Cyperus compressus*.

The disturbance-mediated susceptibility of plants invasions is usually found in the aquatic bodies that suffer the fragmentation of successionally advanced biotic communities, frequent disturbances and a slow recovery rate (Schiffman, 1997; Fensham & Cowie, 1998). Eutrophication is increasing across the globe due to modern anthropogenic contributions and atmospheric nitrogen fertilization and has developed into a frightening prospect because aggressive and highly competitive invaders are preadapted to nutrient-enriched environments, and the Sundarbans mangrove ecosystem is not an exception (Ginsberg & Lowenstam, 1958; Boyd, 1978; Vitousek et al., 1997; Biswas et al., 2009).

The extreme events may provide an ecological advantage to many invasive species that can take advantage of environmental inconsistency because of extreme weather events such as floods, droughts and storms for their steady introduction within a well-established biotic community. This consequently threatens the survival of some native species by affecting population dynamics, morphology, reproduction, species boundaries, behaviour, community structures, composition and ecosystem processes. In a mangrove-estuarine complex, storm surges and sea-level rise have threatening impacts on the continued existence of natural communities.

7.8.2 *Biotechnological Potential of Microbial Load of Mangrove-Estuarine Complex*

A study on biotechnological potential of mangrove-derived microbes inhabiting within the only tide dominated tiger land, mangrove forests of Sundarbans has shown the occurrence of phosphate solubilising, sulfate reducing, nitrogen fixing, biosurfactant producing and methanogenic, bacteria. In addition, several (eight) other obigately halophilic eurihaline cyanobacteria with antimicrobial activities have been isolated from different parts of intertidal zones of this gigantic ecosystem which include genera like Phormidium., Oscillatoria., Synechocystis and Leptolynghbya (Mitra et al., 2011). Diversity of all of these microbial populations settle, grow and colonize into eight different microhabitats which are: a) surface sediments of intertidal belts; b) soil under water at subtidal zones, c) anoxic subsurface layers, d) oxic subsurface layers within burrows, crevises etc formed by benthic fauna; e) rocky surface used for anti-erosion measure at the supralittoral zonesat backshore region; f) as periphytonic organisms on the exposed mangrove root surface or any other immersed hard structures, g) within water columas nanoplankton, h) mangrove plant litterand i) within body especially within gut of nekton and benthos. Twenty six (26) bacterial strains having wide spectrum of antifungal activities with the help of extracellular enzymes like lipase, protease, amylase, etc. have been isolated from the mangrove floor of Bhitarkanika mangrove ecosystem of Odissa, India (Gupta et al., 2007). Antibiotic producing actinomycetes acting against several pathogenic bacteria like *Bacillus subtilis, Shigella flexneri, Vibrio cholera, Pseudomonus vulgaris, Klebsiella pneumonia, Staphylococcus aureus* have been documented from the mangrove sediments of Pichavaram in the state of Tamilnadu, India (Kumar et al., 2005).

7.8.3 *Diversity of Microbes and Spread of Diseases at Amazon and Sundarbans Mangrove Eco-regions*

Comprehensive research study linking between annual hydrological and biogeochemical cycles with the seasonal diversity and abundance of pathogenic bacteria, *Vibrio cholerae*, the causative agent for the gastro-enteric disease, cholera is required as this disease in Sundarbans is endemic one in contrast to Amazon where the disease prevails but not as an endemic ones. Several attributes like organic matter rich brackish water environment coupled with high density human population in and around Sundarbans Biosphere Reserve are responsible for the proliferation of *V. cholerae* (Lara, 2011). In addition, higher preference of Vibrio spp. in brackish water enriched with chitin mainly because of the abundance of crustacean plankta, benthos, detritus etc. with whom these bacterial species remain attached and such association is enhanced by several environmental events such as nutrient enrichments, blooms of crustacean species and anthropogenic inputs (Midya et al., 2019). These pathogenic bacterial species also play important roles in the functioning of aquatic ecosystem due to the strong bondage of *Vibrio* species with crustacean faunal components and also this bacterial species contribute in the cycling of refractory organic matter (Lara, 2011; Midya et al., 2019). A comparison of brackish water dominated mangrove wetlands in Sundarbans of Asia with its South American counterpart in Amazon enjoying almost similar environmental settings (large expanse of mangroves, higher influx of riverine water with greater sediment load), the diversity of flora and fauna is much greater in Asian mangroves. Moreover, higher seasonality of cholera epidemics in Asian Mangrove is due to the natural and climatic cycles, demographic factors like higher human population density, migration trends and settlement patterns reclaiming the mangrove forests (Lara, 2011).

7.9 Sewage Flows and Microbial Loads in Water

Sewage refers to the release and supply of used water from domestic and industrial sources. It contains inorganic materials and some undesirable organic substances, both of which contribute to the growth and propagation of microorganisms, especially some harmful bacteria, which sometimes leads to epidemics among humans. Wastewater contains unwanted substances released from the bodies of humans that are carried away from domestic domains or from industries through sewer systems. Both municipal sewage and industrial sewage contain certain types of organic compounds which after recycling can be used as life-supporting ingredients, such as water, proteins, fats and carbohydrates. Such sewage water is released from both domestic and industrial (paper, cotton, silk, resins etc.) activities because all human activities are associated with those substances, which contain cellulose as a major constituent.

All these organic substances undergo decomposition upon reaching distributary waterbodies, including estuaries, with the help of microorganisms utilizing the required amount of dissolved oxygen from water. However, on the bottom substrates, microbiological activities in the absence of sufficient oxygen and sunlight often create anoxic ecological conditions, and the decomposition of the remaining nondecomposed organic matter takes place in the absence of sufficient oxygen through the activities of anaerobic bacteria that utilize combined oxygen instead of using free oxygen. The outcomes of this anaerobic decomposition process produce some obnoxious end products (organic sulfur compounds, sulfuretted hydrogen, organic amines, phosphine, etc.) characterized by very unpleasant odours, namely methane (NH_3) and hydrogen sulfide (H_2S). Thus, the production of some unpleasant gaseous compounds through anaerobic decomposition drives sensitive species from their own habitats and enables opportunistic or tolerant species to take their place (Chakraborty et al., 2023). Kolkata, the state capital of West Bengal, India, started growing along the levee of the Hooghly River, flowing along the west of the city, and the Kulti–Bidyadhari estuary, which after flowing 30 km eastwards meets the Bay of Bengal. All the while, estuarine networks intermingling with a network of sewage-loaded waterbodies of the east Kolkata wetlands initially processed the sewage of Kolkata and released a considerable amount of the sewage to the Bidyadhari–Thakuran estuaries of Sundarbans, India.

In the estuarine networks of Sundarbans, having a higher amount of total dissolved solids and total suspended solids, have been found to play unique roles in determining the distributional pattern of microbial components. The abundance of coliform bacteria in different waterbodies, as an important indicator of pathogenic microbial groups in water, indicates the pollution status of an aquatic body. Such coliform bacteria include a number of species subsumed under the following genera: *Escherichia*, *Enterobacter*, *Streptococcus*, *Citrobacter*, *Enterobacter*, *Klebsiella*, *Clostridium*, *Salmonella* and *Shigella*. The coliform bacterial group includes all the facultative and aerobic Gram-negative, non-spore-forming rod-shaped bacteria. The faecal coliform group, which is also being considered as an indicator microbial group, originates in the intestinal tract of humans and other animals. Large-scale human settlements in the reclaimed deltaic islands and the banks of the estuarine networks of Sundarbans, India, are the causes of the releasing of large amounts of effluents into the estuarine water flow. A study on the seasonal occurrence of total coliform and faecal coliform bacteria from three major estuaries (Hooghly, Saptamukhi and Matlah) have revealed the maximum microbial load during monsoon followed by that during the pre-monsoon and monsoon seasons. Both of these microbial populations have been found to occur in their highest quantity in the Hooghly estuary, followed by the Saptamukhi estuary and the Matlah estuary. In the same study, out of several physicochemical parameters, salinity, total dissolved solids (TDSs) and total suspended solids (TSSs) might have played decisive roles in determining the distribution of these microbial organisms. The low microbial load was found in the Matlah estuary which always experiences a low salinity level in comparison to the two other studied estuaries (Chakravarty et al., 2004). Similar studies have revealed that along with salinity, temperature also plays pivotal roles in determining the population densities of these pathogenic bacteria.

A low microbial load was recorded in an area that always experienced a low salinity level, namely the Matlah estuary harbour, whose microbial load was lower that that of the two other estuaries, those of Hooghly and Saptamukhi. Statistical analysis has established that temperature exhibited a significant positive correlation with microbial groups in the Hooghly estuary and the Saptamukhi estuary, whereas a significant negative correlation was found in the Matlah estuary (Chakravarty et al., 2004). Salinity demonstrated direct relationships with temperature, promoting the flourishing of the pathogenic micropopulations (Chakravarty et al., 2004). This finding may be interpreted as follows: Salinity alone could not account for their role in determining the density and distribution of microbes, because the excessive dumping of effluents, especially sewage from adjoining municipalities and tourist complexes influence the ecological condition (Bokkhali, Sagar Island, Diamond Harbour, etc.), and enabled the microbial components to flourish in the estuaries, taking advantage of the higher temperature. This research also points out that the fluctuation in microbial populations was not affected by seasonality, but it simultaneously points out that the growth of microbial populations was connected to the location of the estuaries, specifically their proximity to industries and municipalities (Chakravarty et al., 2004).

Cholera, being a devastating disease resulting in epidemics, poses one of the major public health challenges in several tropic areas during certain periods of the year. The causative agent of this disease, *Vibrio cholerae*, instead of being restricted in brackish-water areas, disperses into the adjoining freshwater ponds of coastal areas through coastal flooding (Colwell, 1996). These virulent bacteria remain viable throughout the year in association with other aquatic organisms, such as zooplanktons and phytoplanktons. The facultative human pathogenic *Vibrio cholerae*, the causative agent of the human intestinal disease cholera, has been reported to occur in estuarine ecosystems. Poor sanitation practices in highly populated areas foster sources of epidemics caused by this bacterium, which are due to occasional outbreaks in aquatic systems, and contamination occurs through drinking water loaded with free-living *Vibrio cholerae* cells. Contaminated food, especially seafood products prepared from shellfish such as oysters and crabs, cause the disease to a lesser extent (DePaola, 1981; Kaysner & Hill, 1994).

The excessive discharge of wastewater in the estuarine networks of Sundarbans has some negative impacts on the natural health of the ecosystem, which decrease biopurification efficiency. However, the release of sewage into the mangrove ecosystems of Sundarbans, India, has been an age-old process for several reasons: (i) the proximity (around 40 km) of the east Kolkata wetlands to the Ichamaty–Bidyadhari riverine estuaries; (ii) sewage that flows to the mangrove-estuarine habitats and reaches the target estuarine zones gets dispersed over vast areas; (iii) mangrove vegetation filters, bioaccumulates and bioremediates the sewage loaded with both nutrients needed for its growth and the toxic substances polluting the system; (iv) the soil, algae and microbes in the mangrove ecosystem absorb large amounts of toxic pollutants, and different physical processes, like tidal cycles and an influx of both saline water and fresh water, help to reduce the effects of pollution through a dilution process (Wong et al., 1995); (v) over time, the capacity of sequestering pollutants increases. Nutrients (primarily nitrates and phosphates), being the major constituents of sewage, provide necessary ingredients for the growth of mangroves; mangrove soils cannot remove nitrogen but can trap phosphorus from wastewater.

7.9.1 Pathogens as Contaminants in Water

The physical, chemical or biological properties of a waterbody undergo negative changes in the maintenance of the ecological health of an aquatic ecosystem after being contaminated by the resurgence of pathogens like bacteria and fungi. The prevalence and spread of disease-causing pathogens are associated with two environmental processes: sedimentation and hydrobiological dynamics. Microorganisms not only act as pathogens but also constitute important components in the trophic interactions in the functioning of an estuarine ecosystem by serving as an accelerator of a biochemical decomposition process, and they thereby serve as contributors of biogeochemical cycles. Several microorganisms inhabiting the digestive tract of humans not only help the digestion process but also are removed from the human body as faecal matter loaded with a large number of pathogenic organisms, like disease-causing faecal coliform bacteria, which in turn cause varying degrees of gastrointestinal diseases in humans. Ater being released by human beings and other warm-blooded animals, these pathogenic microorganisms enter estuarine waters through a number of routes, including malfunctioning septic tanks, sewage plants, storm drains and sanitary sewer overflows that bypass treatment during storms or end up in agriculture and urban runoff. Pathogens, being very small organisms, once released into the environment, infect humans through skin contact or the ingestion of contaminated water or food after being transported by water. Examples of pathogens include (1) the bacteria responsible for cholera and typhoid fever; (2) viruses responsible for hepatitis and respiratory disease; and (3) protozoa responsible for giardiasis. Human exposure can occur not only from eating contaminated shellfish but also from swimming, bathing or engaging in water contact sports in contaminated water. Pathogen contamination also occurs in conjunction with other inorganic pollutant contamination, concentrating in the tissues of fish and shellfish, which after being consumed cause illness in the consumer.

7.9.2 Diversity of Microbes and Spread of Diseases at Amazon and Sundarbans Mangrove Eco-Regions

Comprehensive research study linking between annual hydrological and biogeochemical cycles with the seasonal diversity and abundance of pathogenic bacteria, Vibrio cholerae, the causative agent for the gastro-enteric disease, cholera is required as this disease in Sundarbans is endemic one in contrast to Amazon where the disease prevails but not as an endemic ones. Several attributes like organic matter rich brackish water environment coupled with high density human population in and around Sundarbans Biosphere Reserve are responsible for the proliferation of V. cholerae. (Lara, 2011). In addition, higher preference of Vibrio spp. in brackish water enriched with chitin mainly because of the abundance of crustacean plankta, benthos, detritus etc. with whom these bacterial species remain attached and such association is enhanced by several environmental events such as nutrient enrichments, blooms of crustacean species and anthropogenic inputs (Midya et al, 2018. These pathogenic bacterial species also play important roles in the functioning of aquatic ecosystem due to the strong bondage of Vibrio species with crustacean faunal components and also this bacterial species contribute in the cycling of refractory organic matter. (Lara, 2011; Midya et al, 2018). A comparison of brackish water dominated mangrove wetlands in Sundarbans of Asia with its South American counterpart in Amazon enjoying almost similar environmental settings (large expanse of mangroves, higher influx of riverine water with greater sediment load), the diversity of flora and fauna is much greater in Asian mangroves. Moreover, higher seasonality of cholera epidemics in Asian Mangrove is due to the natural and climatic cycles, demographic factors like higher human population density, migration trends and settlement patterns reclaiming the mangrove forests (Lara, 2011)

7.10 Viral and Bacterial Diseases in the Aquacultural Farming System

The viral disease causing white spot syndrome (WSS) and parasitic disease have been reported in the shrimp (*Penaeus monodon* and *Litopenaeus vanammei*) farming systems in the brackish-water impoundments in the reclaimed deltaic islands of Sundarbans, India. *Enterocytozoon hepatopenaei* has been found in *Litopenaeus vannamei*. Parasitic diseases caused by Argulus, Myxobolus, Dactylogyrus, Larnaea and Diplostomum have been reported from finfish aquacultural farms.

Other diseases include fungal infestation, bacterial septicaemia and Ich, caused by several bacterial species, like *Pseudomonas aeruginosa*, *P. putida*, *P. citronelllolis* and *Acinetobacter pitti*, which can degrade phenol and chlorophenol compounds, whereas other bacterial species, like *Chryseobacterium teanense*, *C. geocarposphaerae*, *Bacillus cereus*, *B. flexus* and *B. megaterium*, can effectively degrade protein, lipid and starch.

7.11 The Nonjudicious Harvesting of Natural Resources

The bioresources in the form of timber and fuel wood, golpata leaves (*Nypa fruticans*) as roof thatching material, the fruits and flowers of mangrove plants, finfish and shellfish, grasses as fodder and honey have been collected in a nonjudicious manner by the ever-increasing populations of local people, who use these resources for their own sustenance but at the cost of the ecological health of the mangrove ecosystem.

Different fishery-related activities in the permanent and temporary fisher camps for catching, drying, processing and even local marketing often lead to disturbances in the functioning of the coastal ecosystem and estuarine and mangrove ecosystems by releasing a variety of waste materials and strangulating many marine-estuarine turtles, migratory birds and threatened fish species due to the operations of the increased number of nylon nets with small mesh sizes (Chakravarty et al., 2004).

7.11.1 The Nonjudicious Collection of Prawn Seeds

A huge number of juveniles of more than a hundred species of finfish and shellfish and commercially important shrimp species, including *Penaeus monodon*, are sold to aquacultural farms operating in and around the place of collection for the monoculture of *P. monodon*. An estimate shows that for the collection of one juvenile of *Peneaus monodan*, juveniles of other aquatic fauna, ranging from 27.41 to 31.77 in number, were destroyed at different sites in Sundarbans (Annon, 2003). This kind of problem, being one of the major issues of biodiversity loss within mangrove ecosystems, has been connected to aquacultural practices, especially for the culture of both prawns and shrimp as saline water within brackish water varies about the different ecozones of Sundarbans, which have a high potential for the capture and culture of shellfish. But during the previous three decades, there has been a dwindling trend in the diversity and density of the juveniles of different finfish and shellfish because of the indiscriminate capture of the larval forms of commercially important penaeid shrimp, especially tiger shrimp (*Penaeus monodon*, belonging to the family Penaeidae and the order Decapoda under the class Crustacea of the phylum Arthropoda), along the shallow and marginal water inundating the intertidal zones of different estuaries and the tidal creeks of Sundarbans, coinciding with the spring tides at full moon (locally named as Purnima) and new moon (locally called Amavasya) with the help of nylon nets with mesh like that mosquito curtains. These kinds of activities, dragging nets with a length of 2–4 m fixed in bamboo curtains, further accelerate the soil erosion from the destabilized muddy banks of waterbodies, destroy the natural feeding grounds of many fauna organisms (both benthic and pelagic), impede natural nutrient cycling, increase turbidity, hamper the normal photosynthetic process, reduce biological production and disrupt the prevailing trophic interactions within the ecosystem.

The problem is aggravated to a considerable extent because of the rampant harvesting of huge quantities of estuarine fishery resources, both pelagic and benthic, along with the nonjudicious exploitation of the larval forms of target prawn seeds and nontarget finfish (Figs. 7.3 and 7.4). A steady decline in the availability of prawn seeds leads to a sharp increase in the cost of collecting the required amount of prawn seeds per day to maintain the supply chain for the already-developed aquacultural farms in and around collection sites. However, the earlier potential for the supply of seeds to different kinds of fishery activities has been profusely reduced with the passage of time and since the establishment and blooming of a good number of aquacultural farms since the late 1990s. The proper development, growth and proliferation of prawn seeds require conducive environmental conditions: the physicochemical parameters of water (pH, temperature, dissolved oxygen, turbidity, total dissolved solids, hardness, alkalinity, conductivity and nutrient availability) and soil (texture, detritus and organic matter–based nutrients); the variability of tidal amplitudes; the patterns of water waves; the intensity of currents of water; geomorphological profiles of the bottom and intertidal belts; density; the species composition of mangrove plants; etc. In addition, the colossal loss of the diversity of the larval forms of species of shellfish and finfish alter coastal water biotic communities while disrupting natural networks in the food webs of the water-dwelling biota and thereby causing enormous stress on the mangrove ecosystem and destabilizing its natural ecological health.

Geohydrology, a complex process of hydrodynamics and the sediment movements of the estuarine networks and drainage systems within the Sundarbans mangrove ecosystem, governs deposition towards the formation of point bars on the inner convex of shores and estuarine banks, undercutting the outer concave sides of banks. Being planktonic, the seeds are also spread out by the dominating tidal currents and dispersed into nearby wetlands not.

Such practices developed around five decades back in the estuarine networks of Sundarbans, India, to supply the live spawn of commercially important shrimp species, mainly *Penaeus monodon*, to several commercially operated aquacultural farms in nearby areas—practices that have resulted in the colossal loss of other aquatic biodiversity components suitable for the proper growth of the ichthyoplankton. The expanse, density and species composition of adjoining mangrove forests in natural levees and the upper parts of intertidal point bars of estuaries or tidal creeks help in determining and stabilizing the flows of water. The diurnal inundation of the mangrove forest with variable rhythms of tides facilitates the import of mangrove litter–derived detritus after being determined by lunar cycles, recharging the vast sheet of water with nutrients, which triggers the food chain–food web dynamics of the ecosystem, ultimate resulting in a greater availability of the seeds of prawns and shrimp.

7.11.2 The Species Composition and Availability of Seeds

One research study recorded 19 families of finfish larvae, known as ichthyoplankton, from different corners of the Sundarbans mangrove-estuarine complex that have been experiencing loss in terms of their abundance and diversity over the past almost five decades due to the nonjudicious collection of prawn seeds. These families of finfish with ichthyoplanktonic faunal components were Anguillidae, Mugilidae, Ophichthidae, Muraenidae, Gobiidae, Elopidae, Sciaenidae, Ambassidae, Clupeidae, Lutjanidae, Scatophagidae, Centropoidae, Engraulidae, Carangidae, Teraponidae, Leiognithidae, Sillaginidae, Scombridae and Polynemidae. Icthyoplanktons recorded from study site–I on the Fresergunj and Bokkhali islands harboured the maximum number of families (19); this was followed by study site–II at the Saptamukhi estuary on deltaic islands, namely Prentice Island and Lothian Island, with 16 families; study site–III on Jharkhali Island at the Matlah estuary had 15 families; and study site–IV at the Canning port at the Matlah estuary had 10 families.

Finfish are represented by *Tenulisa ilisha*, *Lates calcarifer*, *Liza tade*, *Mystus. gulio*, *Anguilla bengalensia*, *Herpedon neherius*, *Setipinna phasa*, *Leiognathus daura*, *Corica soborna*, *Trypanchenicthys suratensis* and *Stolephorus* sp., *Stolephorus bucaneeri*, *Stylitera*, *Leiognathus daura*, *Alphaeus* sp., *Pellonaditchela Acetes* sp., *Ophisthopterus tardoore*, *Coilia dussumiery*, *Neenchelys buitendijki*, *Strongylura strongylura* and *Trypanchenichthys suratensis*, while the shellfish species are represented by *Metapeneus monoceros*, *Corica soborna*, *Metapeneus lysianassa*, *Metapeneus monoceros*, *Metapeneus mogiensis*, *Palaemon* (*Exopalaemon*), *Reconda rusheliana*, *Parapeneopsis stylifera* and *Cynoglossus arel*. About 72% of shellfish species, 10% of mullets and 7% of thread fins are the principal constituents of the seed composition.

An estimate from a research study noted that the annual average seed catch of finfish and shellfish is 10,077 nos./net/day, of which 65–75% is composed of shellfish, 3–5% tiger shrimp and 20–30% finfish. The peak period of seed availability, particularly for the period from June to September, featured an average total catch between 9750 nos./net/day and 19,110 nos./net/day (Annon, 2003).

7.11.3 Eco-Ethological Attributes Mediated by Drifting Patterns in Line with Prevailing Hydrodynamics

The swamps of post-larval forms of prawns and shrimp within a time period of about 2 weeks after hatching from fertilized eggs in the estuarine water within conducive ecological conditions tend to move shoreward, especially the shallow zones fringed by mangrove vegetation. The suspended detritus particles of the shallow brackish water constitute the major share of their food, and occasionally, they also harvest food in the form of minute plants' remains and fractional parts of animal bodies from the muddy bottom. Helical flow, with its own distinctness within the meandering channel systems of estuarine networks, drives the drifting of the prawn seeds from the zones of accretion towards the opposite erosion-prone meandering bends of the estuary (Das & Bhattacharya, 1999).

7.11.4 The Life Cycle of Penaeus monodon, *the Most Commercially Important Species*

The life cycle of this species, with four major phases (fertilized ovum or embryo, larva, juvenile and mature), starts from the cross-fertilization of the ovum of females by the sperm of males through the process of spawning, which occurs after moulting through the shedding of old cuticles. The laying of the eggs by mature females coincides with the simultaneous releasing and storing of spermatophores in the seminal vesicles, where the eggs get fertilized and develop into the earliest phase of larva, to be released into the surrounding water. The larval period is characterized by four stages, namely nauplius, metanauplius, zoea and mysis; in the post-larval stage, they are megalopa. After developing within about 13–14 h after spawning, the nauplius stage in the lifecycle of *Penaeus monodon* also includes five more substages (nauplius, pseudometanauplius, metanauplius, calyptopsis and furcilia), followed by other stages, like zoea, which occurs 36–37 h after spawning. Although the nauplius stage depends mostly on stored yolk materials as food, larval forms at the zoea stage start feeding on planktonic organisms. Through three consecutive moulting processes (zoea 1, zoea 2 and zoea 3), zoea-stage larvae transform into another phase of the lifecycle of *Penaeus monodon*, the mysis, and at this stage, larvae subsequently transform into their post-larval stage, from which those in the mature megalopa stage develop by moulting 10–12 times.

7.11.5 Seasonality, Methods of Collection and Transportation

Local people, 70% of whom are girls and women, use drag nets in the lower reaches of Sundarbans for the collection of the fry of *P. monodon*. These rectangular shaped nets are made up of nylon cloth, with a length of 3.0 m and a width of 1.5 m and are firmly wrapped and stitched around a rectangular split bamboo frame (1.25 m × 0.75 m), both ends of which are attached to a long nylon rope by which the net is dragged from behind by one person holding both nylon ropes together during spring tides in shallow water areas. Often, seeds are collected on boats with the help of stationary gears.

Just after collection, all the collected seeds are placed in an open container from where the seeds of target species (*Penaeus monodon*) were first located, and they are categorized on the basis of their colour, size, morphological features and patterns of movement and are thereafter scooped usually by the help of a dried-up shell of a large bivalve (*Lamellidens marginalis*) into a separate small container. The sorted seeds of target species are then sold to traders in nearby marketplaces and then transported to distant aquacultural farms. The prices of the collected seeds are negotiated at all stages in the chain of marketing on the basis of their quality in terms of vitality and growth. As the inhabitants of reclaimed areas of Sundarbans along large estuarine networks, lakhs of people usually under below the poverty level earn their daily income by collecting prawn seeds. A report based on a survey in 1991 from the Directorate of Fisheries in the government of West Bengal reported that collectors of prawns seeds came from 3025 families in the seven blocks of the South-24 Parganas district of West Bengal, India: Sagar, Namkhana, Kakdwip, Diamond Harbour, Kulpi, Kultali and Canning were engaged in this annual collection of 429.1 million prawn fry, of which 78.3 million seeds are collected only during the lean season. Most of these families are below the poverty line, having had a per capita annual income ranging between INR 401 and INR 1200. On average, they used to collect about 2500 prawn seeds a day after 8 h of netting operations during the monsoon period (Annon 2003). In this random operational activity for prawn seed collection, a large number of other species of finfish and shellfish are also killed (Annon, 2003; Das & Bhattacharya, 1999).

7.11.6 The Seasonal Abundance of Ichthyoplankton, Including the Seeds of Prawns

The timing of the collection of prawn seeds and their abundance depend on the tidal cycle and the different seasons of the year. The monsoon is the ideal season for the collection of prawn seeds, followed by the pre-monsoon and post-monsoon periods. The spring tides during the pre- and post-monsoon periods coincide with the maximum availability and seasonal yield of prawn seeds. The seasonal abundance is also correlated with the number of prawn seeds that are transported to the nearby aquacultural farms or rearing ponds of the individual farmers who use the seeds for commercial exploitation.

7.11.7 The Ecological Impact of Seed Capture

The ecology and biodiversity of the Sundarbans mangrove ecosystem face different forms of human-caused threats, both local and regional, from the indiscriminate collection of mainly ichthyoplankton while aiming to collect the seeds of *Penaeus monodon*. This form of human activity for collecting prawn seeds started in the late 1980s with the establishments of many brackish-water aquacultural farms along the bank of estuarine networks with the direct or indirect support of multinational companies. This practice has increased over time at escalating rates to meet the growing needs of the aquacultural farms, with similar trends in demands from both national markets and international markets. Many of the negative impacts of those activities are as follows:

1. Not only the tiger prawn population but also a large number of the fingerlings and seeds of other shellfish and finfish species have diminished.
2. A sudden decline in the populations of different species of prawns, shrimp, crabs, mollusks, fish, etc. from the natural aquatic environment disrupts the food chains of the mangrove-estuarine ecosystems, which ultimately impairs the normal growth of large fish and even the largest aquatic reptiles, such as estuarine crocodiles (*Crocodylus porosus*), which depend on all those small creatures as food.
3. The continuation of such trends in unhindered prawn seed collection, altering natural food chain–food web dynamics, is expected to bring more-deleterious effects in the future.
4. Both the density and the biomass of mangrove vegetation, along with its pneumatophores, change the habitat structures within the mudflats and sandflats, posing perennial problems for the settlement and growth of benthic fauna, turtles and crocodiles.

5. Immature mangrove plants and newly settled saplings are destroyed mainly because of the destabilization of sediments due to the erosion of the banks of estuaries and creeks in line with the vanishing of mangrove plants.
6. Along with the deteriorating ecological health of the affected mangrove areas, several bioinvasive and biofouling species such as Balanus (arthropods; Cirripedia) and oysters (mollusks; Bivalvia) settle on the lower portions of trunks and on pneumatophores, attacking and thus weakening mangrove plants.
7. After settling on mangrove stems and roots, wood borers start feeding op soft conducting tissues like the xylem and phloem of roots and stems, the mesophyll tissue of leaves and newly settled seedlings.
8. Weak mangrove plants are more prone to insect invasions, which are also aggravated in high humidity, especially during late monsoon.

All these deleterious operations during recent decades in most areas of Sundarbans have been bringing about drastic changes in the overall ecology, geomorphology, hydrodynamics, mangrove survivability and growth, ecological succession and zonation, nutrient recycling, biodiversity, etc. Moreover, a declining trend in the collection of the prawn seeds and the rate of mortality for prawn seeds and broods during transportation to rearing tanks requires proper management plans and safeguarding the natural environmental conditions of these highly ecologically sensitive but biologically productive ecoregions of the world. The local people need to be trained to acquaint themselves with basic knowledge about the geomorphology of saltwater courses, primary information on hydrodynamics in terms of the possibility of the availability of prawn seeds with specific knowledge about commercial exploitation, the continuous and systematic monitoring of the ecological conditions of identified areas for the maintenance of the assemblages of mangrove vegetation, textural characteristics (sand, silt and clay) and the nutrient status of the soils of both the intertidal and subtidal zones of the mangrove ecosystem. They need to know about all those attributes to understand the physicochemical properties of the mangrove ecosystem.

7.12 Afforestation to Counteract Deforestation

7.12.1 The Deforestation of Mangrove Forests, the Root Cause of Several Ecological Disasters

This unique ecosystem formed by the amalgamation of soils, water, flora and fauna through their continuous interactions coupled with a steady exchange of materials between forest and aquatic subsystems makes the mangroves of Sundarbans one of the most biologically productive, ecologically sensitive and aesthetically celebrated landscapes in the world. However, this very ecologically viable and economically potential ecosystem of tropical belts has harmed to a considerable extent by the hampering of nutrient cycling, disruptions to normal food chain–food web dynamics, the loss of biodiversity with changes to natural species composition, a reduction in the carbon-sequestration process, threats to wildlife, lowering biological productivity (lowering catches of commercially important products like finfish, shellfish, honey and timber) and promoting more attacks by insect pests and other pathogens.

Being salt-tolerant woody halophytic plants grown in the oscillating ecological conditions of sheltered intertidal belts, mangroves are variously described as the tidal forests and coastal woodlands of Sundarbans. The species richness and species diversity of mangroves are determined by many ecological attributes, like zonation patterns after being controlled by ecological succession patterns, which are dependent on the texture and nutritional condition (N, P, K and organic carbon) of the soil, porosity, permeability, transmissivity, the water-holding capacity of soil, the gradients of the slopping of intertidal belt, the duration of inundation and exposure in line with the intensity of currents and the amplitudes of tides, salinity profiles and the preferred salinity levels of different mangrove species. The mangrove species that experience total inundation twice a day prefer soft sediment with hypersalinity, which is in contrast to the species occurring in the supratidal areas, which have adapted to hard textural zones with low salinity.

The mangrove ecosystem, along the coasts in the tropics, featuring a combination of both forest (salt-tolerant woody halophytic shrubs and trees) and aquatic (with the brackish-water estuaries with tidal influence) subsystems, is a valued ecosystem both economically and ecologically because of its high potential for the production of fish and shellfish, fodder for livestock, wood for fuel and building materials, local medicine, honey and beeswax and plant metabolites like lignin and tannin. Alternative land uses for farming and fishery production have replaced many mangrove areas and continue to pose threats to forests. Moreover, in many coastal states, mangroves are under severe pressure. The demand for shrimp has provided an economic incentive to clear coastal mangroves and develop shallow, open-water shrimp ponds that are developed after the illegal clearing of mangroves in interior tidal areas. These altered land-use changes lead to management and enforcement problems for resource managers. Although natural mangrove regeneration appears to be relatively rapid because of the adaptability of mangrove species along specific hydrologic and salinity gradients coupled with their unique interactive processes to competitive interactions, the replacement of an original mangrove species by introduced species poses a real difficulty because this complex process takes a long time in the changing hydrology. Crucial ecosystem-level functions such as buffer zones, escape cover for commercially important fish, sustainable levels of minor forest product extraction and sediment-trapping ability were lost. The construction of shrimp ponds also results in the ecodegradation of and disruptions to the normal functioning pathways of the mangrove ecosystem because of the excessive withdrawal of freshwater; a reduction in the buffering activities against storm surges; declining protection for nursery grounds; an oversiltation process; more pollution due to the increased accumulation of lime, organic waste, pesticides and other chemicals; the proliferation of disease-causing organisms; etc.

Despite the ecological and economic importance of mangroves, widespread deforestation (the unsustainable practice of cutting of wood) continues and is related mostly to the harvesting of firewood and timber, the reclamation of land for emerging human production needs in agriculture, aquaculture and mariculture. The unregulated exploitation of prime natural resources through deforestation continues despite the many established laws for the protection of mangroves in many countries. The outcomes of such deforestation are reflected in the removal of 20–35% of mangrove areas worldwide since about 1980, and mangrove areas have been disappearing at the alarming rate of approximately 1% per year. The deforestation and overexploitation of mangrove forests in turn cause the ecodegradation of the habitats of an array of faunal components, damage fisheries and expedite the process of shoreline erosion, and ultimately, all ecosystem services rendered by this ecosystem are hampered.

Out of so many contributory factors involved in the process of the removal of mangroves through deforestation, aquacultural expansion, especially the farming of shrimp, ranks first globally in coastal areas (Barbier & Cox, 2003). An estimate of the percentage of the contribution of several mangrove-destroying factors shows that aquaculture accounts for 52% of mangrove loss globally—of which shrimp farming alone accounts for 38%, whereas the harvesting of forest products causes a 26% loss of mangroves globally. Other causative agents leading to the shrinkage in mangrove cover include freshwater diversion (11%), the reclamation of land for other uses (5%), impacts from pesticides, chemical fertilizers as agricultural waste, the construction of salt ponds and other coastal infrastructural developments, like embankments, fish harbours, etc.

The loss of mangroves across the globe at an alarming speed over the past century has increased the significance of estimating the values of such ecological services so that proper public awareness can be built up and an integrated approach towards undertaking coastal management and planning can be adopted. A study conducted in Thailand attempted to estimate the value of ecosystem services rendered by mangrove ecosystems, where a comparison of net income earned by shrimp farming with values for services rendered by mangroves was made in the context of costs incurred for mangrove rehabilitation, and it indicated that higher returns can be derived from a nondisturbed mangrove area than from a disturbed area (Barbier, 2007).

The effects of mangrove deforestation are connected to the seasonal freshwater inflows through river systems, increasing the frequency and intensity of flood flows and causing a drastic decline in a significant proportion of fishery resources in the region because the estuarine fish and shellfish that are used to exploit allochthonous foods in inundated forests increase turbidity, which in turn limits primary production and thereby results in lower fish populations. The removal of forest area along coastal wetlands disrupts the ecological linkages between land alterations, such as deforestation, and the downstream effects of erosion through silt deposition, accelerating the deposition of more sediments, which tends to clog the tidal exchange channels, resulting in severe flooding during the wet seasons and lower-than-normal water levels in the dry season.

The deforestation of Sundarbans was initiated during the thirteenth century with the invasion of Turkic Sultanate rulers and later during the sixteenth to eighteenth centuries under the rule of the Mughal emperors, which continued till the British regime, when the East India Company first took over collecting revenue in Bihar and Bengal in 1765. In that period, the British regime started salt production in different unpopulated areas within Sundarbans as a profitable business venture and established the Salt Department under the company with its headquarters in Khulna, a small port city at the northern boundary of the Sundarbans forest, which is presently a semi-industrial hub. Such an intervention by the then rulers transformed mangrove wetlands through reclamation into revenue-generating agricultural lands and developed wood-processing and salt-producing industries.

Deforestation is the process of losing trees through cutting and disturbing soil, which in turn expedite the rates of soil loss and modify the microbial processes within soils. Concentrations of ammonium in porewater in deforested mangrove soils are found to remain high due to nonuse through the uptake process by the roots of mangrove trees, and the rates of nitrogen fixation are also found to remain low under these conditions, suggesting surface soil disturbance. The removal of mangroves triggers an increase in dissolved inorganic nitrogen concentrations in tidal creeks and the removal of an equivalent amount of reactive nitrogen. The concentrations of soluble reactive phosphate are also found to be lower in deforested areas because of changes in soil geochemistry. The transformation processes in the nitrogen denitrification process also changes in impounded and cleared forests due to disturbances to the surface of the soil.

The mangrove belts during the British regime called for the most rapid destruction of the Sundarbans in that period, which was in sharp contrast to the slower-paced conversion of mangrove forest into arable crop land and the development of human settlements in the low-lying areas during the rule of the Turkic Sultanate invasion and during Mughal rule. Around 70% (2000 km^2) of the land of the mangrove forest tracts of Sundarbans was voided between 1830 and 1873. Protecting the people and their new land-use policies from the attack of the salt tides of the Bay of Bengal in continuation of the rapid pace of deforestation under the British colonization of Bengal (1765–1947) necessitated the construction of embankments, locally called bunds, adjacent to estuarine networks, which, as small earthen dykes, remained projected above the land to protect agricultural rice fields from the intrusion of saline tidewater during the dry season. Initially, most of the bunds of Sundarbans were intentionally breached by the local farmers and cultivators to enable the monsoon to cause floods, locally called *borsha*, which are floods for irrigation and for facilitating the fertility of the arable land, where a *borsha* flood is considered a "blessing of fertility".

7.12.2 The Construction of Impoundments and Their Impacts on Mangrove Belts

The construction of impoundments involves the enclosure of mangroves for the purpose of limiting the movement of water to control problems such as insect infestations, such as those from mosquitoes. After the construction of impoundments, the mangrove ecosystem undergoes alterations in and disturbances to soil horizons and subsequent changes in soil biogeochemistry and geochemistry. Nitrification and denitrification processes meant for transforming nitrogen from one of its chemical states to another through processes like denitrification decline in impounded and cleared forests due to surface disturbances to the soil. The fertilization of plots results in six times more emissions of nitrous oxide while significantly increasing denitrification rates. However, the nitrogen fixation process is retarded in fertilized plots, indicating long-term sinks in nitrogen-limited impounded mangroves. Impoundments cause alterations in hydrology by increasing seasonal flooding, causing higher amounts of soil ammonium and lower amounts of nitrate in water-logged soil and lower rates of nitrification and denitrification. These are manifested in the stunted growth of plants, but the uptake of nitrogen is increased, especially in dwarf and sparse forest types. The construction of impoundments within mangrove belts intensifies soil disturbances by increasing summer flooding, increasing the accumulation of soil ammonium, decreasing soil nitrate and decreasing nitrification and denitrification. Therefore, disturbed mangroves result in the stunted growth of mangrove vegetation, including large trees, which hardly resist the harmful impacts of more storms and cyclones with higher frequency and intensity.

7.12.3 The Afforestation Programme Counteracting the Deforestation of Mangroves

Forest management activities are directly connected to biodiversity development with the propagation of native species and are meant to combat the problems associated with climate change. Massive afforestation programmes of mangroves need to involve local people and use fundings from several national and international schemes.

Over the past couple of centuries, especially during the past several decades after the independence of India, several parts of Sundarbans have been converted into bare open land with sparse vegetation due to the rapid pace of deforestation, followed by the reclamation of large tracts of land through uncontrolled encroachment by local people. Aiming to combat such a trend of damaging mangrove tracts, continuous efforts have been undertaken by the forest departments of the state government of West Bengal, India, for developing an intertidal mangrove nursery, which will play crucial roles in facilitating resource generation through afforestation programmes, which in turn reduce forest coverage and increase the survivability of a number of planted saplings. The following criteria need to be adopted for the development of a mangrove nursery:

1. An intertidal mudflat with gentle sloping needs to be selected, and proposed sites near the upper parts of the selected area need to be marked.
2. Selected areas must be safeguarded from the high intensity of tidal currents and estuarine waves.
3. Selected nursery sites need to be inundated on a regular basis.

The matured seeds, fruits or propagules of different mangrove species, after being collected and sorted from the natural setup of mangrove forests, are kept upright in the collected mud from the intertidal belts. A number of polybags with sizes ranging from 10 to 20 cm keep tapering ends within the mud serially close to each other in a mangrove nursery. Precaution is required to decrease the mortality rate alongside increasing the healthy growth of the saplings into mangrove trees in the future. More than 90% of mangrove seeds get germinated after a certain dormancy period, which varies for different mangrove species. The seeds of *Avicennia* and *Rhizophora* grow in a comparatively faster than those of *Bruguiera*, *Sonneratia*, *Xylocarpus*, *Ceriops*, *Heritiera* and *Nypa*, which take a long time for germination. The initial stage of germination is much better in fresh water than in brackish water. To ensure the success of the afforestation programme, all mangrove saplings must be transplanted within a short span of time during the low tide of lean period of the lunar cycle to the already-identified sites, which have ecological attributes like a brackish-water environment with a steady source of fresh water and a muddy substratum.

7.12.4 Afforestation Versus Deforestation and Their Prospective Outcomes

Continuous sedimentation and accretion processes within a riverbed after deforestation cause the inundation and overflow of creeks, tidal cannels and estuaries, which have greatly declined the mangrove ecosystem of Sundarbans. The process of a well-planned afforestation programme, mainly under the control of both government and nongovernment agencies, involves a large number of local people, mainly women, who engage in the afforestation process and in fishing activities that tend to improve the area.

1. Mangrove plants have been used by local people for the purpose of fuel wood, fodder, building materials or other purposes.
2. The nonjudicious collection of prawn seeds by local people, mostly during ebb tide and mostly in shallow subtidal zones with the help of small mesh nets, in the creeks and inlets results in not only a loss of the biodiversity of aquatic fauna but also the destruction of saplings.
3. The initiation and execution of so many conservative measures, such as declaring the Sundarbans mangrove ecosystem as a biosphere reserve, making it a national park, making it a part of the Tiger Project, protecting it as a breeding centre for crocodiles, and turning it into several sanctuaries, have in one way combated deforestation but also exposed this fragile and vulnerable ecosystem to anthropogenic disturbances. Ecotourism, indiscriminate human settlement,

the illegal settlement of immigrants, populist development activities in and around mangrove areas by political parties have resulted in the fast disappearance of mangrove forests. The newly planted saplings of the mangrove afforestation programme are frequently uprooted during prawn seed collection and capturing operations, navigation, the grazing of bovines, etc.

4. Proper eco-planning and legal administrative steps must be adopted to arrest the rapid pace of mangrove deforestation by providing local people with newer avenues for livelihood generation by diverting their focus from their dependence on mangroves.
5. Newer research developments in the fields of agroforestry, grassland management and agricultural practices must be disseminated to render both biodiversity and climate benefits. These can be made possible by creating conservation villages where more emphasis is placed on recycling waste, continuous efforts are made to cover the soil with different forms of vegetation to reduce soil loss and land degradation via soil erosion, dependence on groundwater is reduced and intercropping is promoted. All those steps tend to check the problems associated with climate change–induced droughts, floods and winds, thereby increasing soil carbon stocks, agricultural productivity and biodiversity.
6. Afforestation programmes are being undertaken by the Department of Forests in the government of West Bengal each year, involving local people, nongovernmental organizations (NGOs), the corporate sector and researchers from different academic institutions (Fig. 7.7).

7.13 The Impact of Sediment Starvation on Coastal Morphodynamics: Impairing Coastal Ecodynamics

The river networks of the lower Gangetic plain that supply fresh water to the estuarine networks of Sundarbans mangrove-estuarine complex also carry large amounts of sediment that get deposited at the estuarine bottom near the Bay of Bengal, and over time, such deposited masses of sediments get elevated and are then reworked and redistributed by waves and tides to form newly accreted lands that are locally called chars (Iftekhar, 1999). These chars are then stabilized by the settled grasses and other saplings of mangrove plant species and stabilize to facilitate deltaic formation. The monsoon period experiences more erosion and forms new, extensive lands with the deposition of eroded sediments within the riverine-estuarine networks of the ecoregion. The increasing rates of erosion of land pose major threats to coastal areas because of the displacement of human settlements and the gradual loss of livelihoods, and the resulting unwanted accretion process represents another problem by causing difficulties in the otherwise-easy movements of water vessels through rivers and estuaries.

Shortages in the supply of sediment trigger morphological responses because of the change in environmental variables along the abandoned portion of the deltaic areas exposed directly to wave attacks, which generate the local source of sediment supply in the system. The loss of wetlands mainly due to wave erosion and high-water submergence is frequently observed at the sea face of a number of estuaries of Sundarbans: Saptamukhi, Matlah, etc. Long shore sediment transport from the western parts of Sundarbans fails to reach the area of the remote eastern parts of Sundarbans, thus changing the region's morphology, which rapidly gets abraded by wave attacks.

The vegetated intertidal zone into the inner parts of mangrove swamps accommodates such wave attacks, which are reflected by the presence of larger sizes of sand particles from the bottom sediment and thereby decrease the buffer areas of low-tide shore platforms along the base of cliffed mud banks. These changes are reflected in the stunted growth of the mangrove vegetation at the shore face, which causes more ecodegradation because of the active spraying of saltwater at the time of the landward advancement of wave breakers, particularly during high spring tides. All these are seen in different areas of Sundarbans, including its sea-facing deltaic islands, such as Lothian Island, Dhanchi Island, Kalas Island and Bhangaduni Island.

7.13.1 Coastal Morphodynamics Under Sediment Starvation

Geomorphological changes associated with shortages in sediment supply along the stretch of abandoned areas within the delta promote multitudinous environmental threats, directly exposing the area to wave attacks, and these changes foster the local source of sediment supply in the system. Moreover, a considerable loss of wetlands by the impacts of wave erosion—causing water submergence in several deltaic islands mostly at the sea face of a number of estuaries (Hooghly, Mooriganga, Saptamukhi Gosdaba, etc. of Indian Sundarbans). Larger sand particles of sediment in the muddy intertidal zones (mudflats with a higher percentage of silt and clay) are endowed with newly settled mangrove saplings and saline grasses in the seaward direction and are endowed with larger mangrove plants in the landward direction—has been pushed by wave attacks into the inner parts of mangrove swamps, thereby minimizing the extent of the buffer areas of the intertidal belts at low tide along the base of cliffed mud banks. The stunted growth of mangrove vegetation at the shore face has also been observed in some sea-facing deltas of Sundarbans, like Sagar Island, Lothian Island, Prentice Island, Susnir Char, Fresergunj, Jamboo Island and Dhanchi Island, in Sundarbans, where mangroves also face partial decaying because they receive saltwater sprays, especially during spring tides during the landward advancement of wave breakers.

7.14 Climate Change and its Impact on Coastal Ecosystems in General and the Sundarbans Mangrove-Estuarine Complex in Particular

7.14.1 A History of Climate Change: Increasing Temperature Across the World

Climate change, being a complex and worldwide phenomenon, has been caused by adverse human activities, which vary from country to country and region to region. Scientists from across the world have identified anthropogenic activities as significant accelerators of climate change, the prime consequence of which have already affected islands and coastal regions around the globe.

The climate of the Earth has passed through many periods of ups and down in temperature over time, which has been evident in the fossil record, demonstrating that all these changes have had important ecological impacts, often resulting in a number of mass extinctions. However, the present state of global warming appears to be very unusual, mostly fuelled by different kinds of anthropogenic activities, mainly the emissions of greenhouse gases from burning fossil fuels, the release of Freon gases from air conditioners, modern agriculture's profuse application of chemical fertilizers and pesticides, changes in land-use patterns and other forms of industrial emissions processes, which together have contributed to increasing climate change. According to estimates, over the past 50,000 years, the Earth's carbon dioxide (CO_2) concentration has increased from 280 parts per million at the lower atmosphere during the preindustrial period to 375 parts per million at the highest level in 2003. Such drastic changes in the concentration of greenhouse gases have led to the present state of climatic changes, where the temperature has increased from 1.4 to 5.8 °C globally alongside increased variability in precipitation patterns, converting many wetter ecoregions into dryer ones and thereby increasing "the risk of water shortage".

7.14.2 Global Warming: Increasing Temperatures and CO_2

Water is the most critical environmental parameter for living organisms, in both terrestrial and aquatic ecosystems, especially in the mangrove ecosystem, which is formed because of the continuous interactions of both of those prime ecosystems. The fundamental distributional limit for all species is determined by the synergistic impact of both water and temperature. Water stress stems from extended droughts and floods and develops due to increased atmospheric carbon dioxide and global warming–mediated disruptions to hydrological cycles, altering seasonal rainfall systems also causing water stress for some plants (Fig. 7.8; Flowcharts 7.3 and 7.4).

The photosynthetic rates of autotrophic organisms have increased mainly because of a gradual but steady increase in carbon dioxide (CO_2) levels in the atmosphere and water until the concentrations of carbon dioxide or any other ecological factor, such as light or nutrients, become limited. Enhanced levels of carbon dioxide also increase the water-use efficiency of green plants by reducing stomatal conductance. The structure and biomass of plants undergo changes because carbon dioxide brings about considerable changes in the chemical composition of plant bodies. The increase in both temperature and carbon dioxide also causes a series of changes in the atmosphere by modifying wind velocity and direction, in the moisture content and density of the atmosphere, and in the interaction processes of various types of particulate matter and aerosols. Also, different waterbodies, including oceans, estuaries and freshwater rivers, absorb more carbon dioxide, making those waterbodies more acidic, which in turn increases the solubility rates of calcium carbonate, a major constituent of the skeletal structures of aquatic organisms.

Although the responses of mangroves are species specific to the higher atmospheric CO_2 concentrations, which promote more growth in mangroves, such responses are hampered by variations in water-use efficiency, salinity and nutrient availability (46–49). However, specific locations within coastal belt act as important attributes in determining the responses of mangroves to the interactive process of increased CO_2 with temperature, salinity and nutrient availability. There are different associated ecological variables, such as soil texture, soil accretion, dissolved oxygen, pH, salinity, nutrient availability, the biogeochemical constituents of particulate and dissolved nutrients, sulfides and iron oxide complexes, etc. Although nutrient cycling and nutrient availability within mangrove ecosystem are not directly connected to increases in atmospheric CO_2, they are probably directly connected to the responses of different mangrove species that achieve their growth with the steady utilization of nutrients such as ammonium and phosphate, promoting a more rapid cycling of these essential elements. The input of nutrients to the mangrove ecosystem from autochthonous (sea, estuary and mangrove forests) and allochthonous (terrestrial and river runoff) sources tiggers the microbial transformation process of organic nitrogen through ammonification to ammonium and thereby releases inorganic phosphate along with dissolved organic phosphorus. In contrast, retardation in mangrove growth indicates the decreased utilization of nutrients with the possibility of building up dissolved soil nutrients in the form of soil ammonium when most of the nitrogen and phosphorus are supposed to remain immobilized.

A steady increase in temperature is likely to cause faster nutrient cycling, increasing the rate of soil nitrogen production, improving the phosphorus transformation processes, increasing the growth of plants and microbes, increasing reproductive activities and photosynthetic activities and increasing the rate of respiration, which with water in a synergistic manner bring forth alterations in species diversity, community composition and extended poleward distributional limits. The current global expansion of mangroves into higher latitudes in North America, Australia, New Zealand, Southern China and Southern Africa is supposed to be the result of increases in mean sea surface temperatures. The present modes of expanding the distribution of mangroves across the world at the expense of salt marshes is due to complex ecological interactions that result in more nitrogen, phosphorus and carbon and more biomass storage above ground and below ground in salt marshes, which are being displaced by rejuvenated mangroves, suggesting greater nutrient retention with rising temperatures. The speed of winds and frequency of catastrophic cyclones are increasing considerably in a warmer climate (Nordhaus et al., 2006). Several mangrove regions in Asia are vulnerable to the consequences of climate change, which are reflected in sea-level rise and other associated environmental consequences, especially posing challenges to low-lying coastal cities such as Kolkata, Mumbai Bangkok, Hong Kong, Karachi, Tokyo and Shanghai (7). Densely populated mega-deltas at the mouths of the Ganges–Brahmaputra intersection in India face real challenges from emerging environmental issues. Unfortunately, sea-level rise has serious negative environmental consequences on poverty-stricken regions such as Vietnam, Sri Lanka and Bangladesh, where a 1- to 5-m increase is expected to take place by 2100, which will submerge large tracts of land and thus displace many thousands of people as environmental refugees intensify sociopolitical unrest (Dasgupta et al., 2007).

7.14.2.1 Temperature, Carbon Cycling and Climate Change

Climate change, being a global environmental issue, has emerged for various reasons and had impacts of varying intensities from one region to another, causing harm to humans. Two of the most important physical factors, climate and topography, play important roles in determining the abundance, behaviour, distribution and survival of all living beings by regulating the ecological requirements of different species. Researchers across the world have identified anthropogenic activities as the major accelerators of climate change. Two important attributes need to be considered: first, the time to assess the variability over time and space of an ecosystem functioning and, second, scales for the measurement of any ecological system. Since the 1950s, many of the observed changes, caused mainly by the warming of the atmosphere and oceans at unprecedented rates over decades, have ultimately led to sea-level rise, the accumulation of high concentrations of greenhouse gases, alterations in hydrological cycles, disruptions to carbon and other biogeochemical cycles, a loss of biodiversity, the replacement of the coral reef community by nonreef systems due to shifts in ecosystem functioning, the reduced calcification of corals due to increased CO_2 levels in the atmosphere and the frequent occurrence of extreme natural events like storms, floods and prolonged droughts affecting ecosystem processes by altering population dynamics, species composition, reproductive behaviour and the community structure of biota.

7.14.2.2 Global Warming–Mediated Changes in Sundarbans, India

Ongoing global warming has been having significant impacts on the functioning of the mangrove ecosystem of Sundarbans because an increase in temperature affects every structural component of the ecosystem, mostly during the pre-monsoon and monsoon periods, which has been established from experimental studies (Maiti Dutta et al., 2014). Temperatures have been reported to have increased by 6.14% in the western sector and by 6.12% in the eastern sector of the Indian parts of Sundarbans over almost three decades at a rate of around 0.05 °C/year. This rate appears to be more than the recorded warming trends observed in the tropical Atlantic Ocean (0.01–0.02 °C/year), the tropical Pacific Ocean (0.01–0.015 °C/year) and across the planet (0.006 °C/year). An analysis of the research pertaining to surface air temperature across different corners of Indian Sundarbans and adjacent parts of the Bay of Bengal has revealed an increasing trend in the yearly rise in temperature. This finding corroborates the prevailing trend of global warming phenomena (Hazra et al., 2002). In the Indian Sundarbans, the western part has exhibited a continuous reduction in its salinity profile (1.67 psu/decade), whereas the eastern sector has shown an increase in salt concentration (~6 psu over 30 years) because of the altered freshwater flow regimes coupled with an influx of fresh water within its estuarine flows.

The mangrove forest of Sundarbans is a critical and very fragile ecosystem with complex geomorphological and environmental settings, increasing population density and drastic shrinkages in the deltaic islands under the impacts of rising sea level (DasGupta & Shaw, 2013). The gradual increase in salinity invasion mostly due to the diversion of freshwater flows, global warming and massive erosion, the faulty movement of tectonic plates and the occurrence of several catastrophic events, like cyclonic storms and fluvial, tidal and fluvio-tidal events, have shifted freshwater-loving mangrove plant species like *Heritiera fomes*, locally named Sundari, and *Nypa fruticans*, locally named golpata, from Indian part of Sundarbans towards the Bangladeshi Sundarbans, which is endowed with more freshwater dominance. In addition, the settlement, growth and propagation of several bioinvasive plant species has in many ways increased the ecological instability of this sensitive ecoregion, compelling several other mangrove plant species, such as *Sonneratia apetala*, *Avicennia alba* and *Acanthus ilicifolius*, to experience landward movement, which is evident from their presence along the banks of the Ganges near Kolkata (Garden

Reach, Strand Road, Aheritala, etc.). All those ecological and physiographic attributes, along with the ecological threats from ecotourism, unplanned and unscientific fisheries and aquacultural activities, have resulted in a considerable loss of the biodiversity of mangroves and other associated flora and fauna (Chakraborty, 2009; Bhakat et al., 2004) (Fig. 7.7; Flowcharts 7.3 and 7.4).

7.14.2.3 The Ecological Impacts of Temperature Increase on Aquatic Fauna

Climate change has permeating impacts on the thermal regime of mangrove-estuarine complexes by warming the water by disordering various ecosystem processes pertaining to the living condition of aquatic life and thereby directly hampers the successful survivability of aquatic life. Almost all aquatic faunal components, including macroinvertebrates and fish, being thermal conformers and obligate poikilotherms, are confronted with problems in maintaining normal physiological functioning from egg to adult because of the elevated temperature of the surrounding water. Five prime types of increases in temperature-mediated impacts on fish and other aquatic macroinvertebrates have been observed, which are listed as follows:

1. Regulating different behavioural activities related to feeding, reproduction, movement and distribution.
2. Disrupting the regulatory powers of metabolic and developmental rates.
3. Triggering responses to the orientation of organisms and their life support systems.
4. Blocking the expression of other environmental factors.
5. Having lethal effects leading directly to killing organisms or indirectly to stress effects.

The aquatic fauna show responses in relation to changing water temperature at various levels of their hierarchical organization, starting from the lowest at the molecular level, to the biotic community level and to the individual and population levels. The thermal tolerance of an organism at the molecular level is gained from eco-physiological adaptability, which involves the genetic constitution of the organism. The impact of increased temperature at the molecular level is manifested in modified biological reactions impacting changes in heat shock proteins, the disordering of biomolecules, the patterns of responses against the stresses and alterations in the functioning of enzymes. Moreover, the altered physiological responses of an organism after being influenced by thermal stress cause the development of several biological and ecological attributes, such as size, biomass, sex and the seasonal variability of water-quality parameters.

7.14.3 Global Sea-Level Variation in and Impacts on the Eco-Geomorphology of Sundarbans Coastal Zones

A gradual but steady increase in sea-level rise in response to global climate change has been recorded since the period of the late Holocene (7000 YBP) and continues today. Global rates of sea-level rise have been hovering around 3.2 mm/year since 1992, according to one estimate by the IPCC. On the Indian subcontinent, sea levels have been rising at a rate of about 2.5 mm/year, and on Sagar Island, the largest Gangetic delta in Sundarbans and its adjoining Bay of Bengal, the relative mean sea-level rise has been recorded at 8 mm/year, according to the record of the past 25 years (Hazra et al., 2002). A rise in relative mean sea level during 2002–2009 was found to be at a rate of 12 mm/year on Sagar Island. Since the deltaic islands of Sundarbans were formed because of gradual and continuous siltation, a drastic decline in sediment supply coupled with, freshwater influxes from upstream has resulted considerable subsidence of land have taken place over the past few decades. In such a context, the subsidence of land results in the localized sinking of the delta, which compounds the effective sea-level rise from 3.84 to 6.74 mm/year (3.84 mm/year + 2.90 mm/year = 6.74 mm/year). The effective sea level works in conjunction with an ever-increasing sea surface temperature (SST), changing rainfall, frequent cyclogenesis, storm surge rise, recurrent floods, bank erosion, topographical condition changes after the shrinkage of islands and retrograding coastal territory.

Sea-level rise (SLR) has also been occurring due to local geomorphological changes, with subsidence and uplift affecting all coastal ecosystems by influencing hydrology in relation to increasing the duration and extent of flooding. Under this changed situation, mangrove ecosystems have adapted to this hydrological change either through vertical accretion for elevating mangrove sediment or by migrating in a landward direction. Freshwater availability and finding new nutrients sources in the newly settled elevated habitats formed through vertical accretion tend to improve the growth and productivity of mangroves by facilitating root growth and the deposition of organic matter.

Climate change–mediated sea-level rise has had considerable impacts on the lower deltaic plain of the Ganga–Brahmaputra delta on the geologically soft and low-lying coastal Sundarbans, altering coastal geomorphic structures, like a reduction in beach width, changes in the water movement of estuaries, an increase in periods of inundation, changing the configuration of intertidal celts (both mudflats and sand flats), accelerating erosion and vertical accretion processes and hindering the natural ecological succession process of the biotic community (Paul, 2002a, 2002b; Chakraborty, 2011).

Rising sea levels with increased tidal amplitude lead to coastal erosion, seawater ingress, tidal shifts, and disruptions to the functioning of the coastal ecosystem, which bring about negative changes in the environment due to an average sea-level rise of 3 mm per year. The altered water currents are faster and have higher waves, which will alter the normal patterns of sediment transport, erosion and deposition.

Such changes in the tidal regime, water movements, sea ingress and salinity profiles have deleterious effects on the coastal ecosystem.

Significant variations in the effects of sea-level rise occur through different geological processes operating on coastal belts in the form of subsidence and uplift through vertical erosion or through human-mediated processes like extracting water, shoreline retreat and flooding, causing changes in ocean currents, shifts in the sediments of storms and tidal surges, hurricanes, tropical cyclones, etc. along with releasing oil and absorbing obnoxious gases. All sea level–induced environmental impacts result in coastal flooding, saltwater intrusion, the salinization of soils, erosion and the landward migration of mangroves and associated biota.

A compiling of several national and international reports shows that the annual mean sea-level variation steadily increased between 1985 and 1998, indicating a minimum 4 cm rise in relative sea level over a period of 14 years. The net rate of sea-level rise in the Ganga–Bramhaputra deltaic system was calculated to be 3.14 mm per year when including the high sedimentation load of 0.1 mm per year. This finding for the Sundarbans deltaic system with a relative sea-level rise of 3.14 mm per year was much higher than the prevailing average global sea-level rise at a rate of 2 mm per year, and the compounded sea-level rise was estimated be very close to the 1 m of sea-level rise by 2050 (Hazra et al., 2002; Chakraborty et al., 2013).

The present state of sea-level rice will be 15–90 cm (average 50 cm) higher by 2100 than now, threatening 92 million people each year with floods if the present trend in the emission of greenhouse gases continues at the current rates. Among the developing countries with mangroves, Bangladesh and the Maldives are considered the most vulnerable to global sea-level rise. When preparing a coastal zone management plan for the coast of West Bengal, this alarming situation in neighbouring countries with similar geological criteria should be kept in mind, giving due importance to the protection and conservation of existing geomorphological units like chains of large dunes in Midnapore (east coast), alternate zones with erosion and accretion potentials, drainage systems etc. through the maintenance of the networks of creeks and channels and patches of saline grasses for their pioneering roles in the mangrove succession process.

Variability in the species composition of mangroves takes place in response to sea-level rise and longer durations in tidal flooding, and such responses appear to be very much species specific. *Avicennia marina* can survive the stresses from prolonged water logging, but its responses to salinity, temperature, water depth, the duration of immersion and other environmental factors have been shown to be highly variable. Other biological responses of a large number of species of mangroves to waterlogging and changes caused by global climate change have exhibited differences in morphology and anatomy in relation to leaf anatomy, diameter, grouping and length; vascular vessel densities; and fibre wall thickness. Prolonged flooding also changes bark anatomy with increased aerenchyma development in the bark, causes the formation of hypertrophied lenticels and causes the formation of adventitious roots in *Avicennia marina* and *Laguncularia racemosa* L. Undoubtedly, some mangroves undergo physiological adaptations to prolonged water logging because of tidal inundation coupled with sea-level rise. Seaward mangroves are likely to experience more suffocation underwater, whereas mangroves inhabiting the supra-littoral zones of intertidal belts move in a landward direction because of lower rates of soil vertical accretion alongside sediment uplift by not keeping pace with sea-level rise. This changed scenario, with the death and burial forming peat lands, in turn results in drastic changes in nutrient dynamics, including dissolved and particulate nitrogen, phosphorus and other nutrient elements in coastal seas and on land.

The impacts of sea-level rise are reflected in the developments of the tidal notches of fingering gully heads, sheet erosion on horizontal tidal platforms, multiple stepped banks on tidal flats and other forms of erosional features in and around the sea-face zones of the Gangetic deltas of Sundarbans, the Subarnarekha delta and the Godavari and Krishna delta at the Bhitarkanika mangrove forest on the east coast of India. The deltaic tidal flats with erosional shore platforms, after exposure during extreme low tides at their sea face, exhibit distinct signs of the development of peat layers or deposits of paleo-marsh at the lower parts of intertidal zone, relics of degraded older tree stumps endowed with decaying pneumatophores at the mid-littoral zone along the stiff clay banks of sea-facing islands. All these findings act as clear testimonies to a considerable loss of mangrove wetlands along the coastal zones by the changing wave energy levels and tidal current energy from sea-level rise. However, mangroves from some areas of the coastal belts of Sundarbans and of Bhitarkanika and Odisha are protected from the rising levels of sea water by the construction of embankments. Mangroves on frontal shore platforms fail to move inland and in many cases remain squeezed between a rising sea and the protective embankment structures of sea-facing deltaic islands like Sagar Island, Henry's Island, Tat Island and Bakhkhali in the Sundarbans mangrove ecosystem and Kalidiania Island and Talchua Island in the Bhitarkanika mangrove ecosystem. The mangrove stands have been ecodegraded along the banks of some tidal creeks because of infilling into the tidal creeks by sand-dominated sediment intruding into the newly inundated tidal wetlands, which are dominated by halophytic mangroves and other related coastal dynamics that affect the fringed shore platforms of other islands. The upper portion of inter tidal zone is transformed into a supratidal zone in the areas of sudden uplift with the acquisition of new ecological characteristics, and shifted mangroves hardly face any problems after undergoing further inundations by regular high tides.

7.14.4 The Impact of Climate Change on the Coastal Ecosystem, Specifically the Mangrove Ecosystem

The coastal zone, representing a transitional zone between land and sea, is one of the most dynamic natural systems where the three major constituents of Earth—the hydrosphere, the lithosphere and the atmosphere—merge and interact to exchange materials. Coastlines are formed by morphological changes governed by climatic and geological processes. The geomorphology of coastal ecosystems is structured and governed by climatic and geological processes. The coastal ecosystems render many goods and services, such as food, fuel, the generation of oxygen and the sequestration of carbons, which benefit humans by regulating most of the biological, physical and chemical processes of the Earth. The geomorphologically diverse Indian Ocean consists of different littoral deltas and islands, encompassing the many large islands of the South Asian subcontinent (Bangladesh, India, Pakistan, Java, Madagascar, Sri Lanka and Sumatra) and numerous atolls and archipelagos. The eco-geobiological characteristics of most of the coastal and marine ecosystems display specificity and distinctiveness in relation to the region in spite of experiencing similar kinds of stresses in the coastal regions across the world due to the uniqueness of region-specific microclimates. The climate of the Earth, after being driven by solar energy, becomes warm, which in turn causes the circulation of the atmosphere and oceans, contributing to the variability of meteorological parameters such as air temperature, precipitation patterns, humidity, storms and droughts. All such climate-driven ecological variabilities are directly connected with physiological processes and the survival of living organisms. The sea-level-rise-mediated flooding of coasts, estuaries and deltas and coastal erosion in turn reduce the habitat availability of the biota, causing disruptions in food chain–food web dynamics. Undesirable changes in drainage patterns and modifications to fluvial flows modify sediment transport, nutrient loading and salinity invasion and threaten mangroves and associated biota. Ever-increasing salinity invasion (because of a drastic reduction in freshwater flows) and saltwater intrusion have seriously magnified the threats to the mangrove ecosystem, one of the most productive and sensitive ecosystems in the world. The coastal-estuarine-mangrove complexes of tropical countries, including the mangrove ecosystem of Sundarbans, are likely to respond to climate change and other associated ecological perturbations—such as enhanced salinity, shifts in rainfall patterns, increased levels of CO_2, changes in the frequency of storms with fierce wave action, the acceleration of soil erosion followed by unwanted accretion, sea-level-rise-mediated prolonged inundation pressure, an increase in the landward migration of biota in search of inland aquatic systems and an ever-increasing loss of suitable habitats for different biodiversity components—which ultimately bring about abrupt changes in the distribution pattern of the indigenous biota and thus alter the species composition of the biotic community. Such disruptions in the normal structure and function of mangrove ecosystems will pave the way for the introduction, settlement and proliferation of alien bioinvasive species (Chakraborty, 2018).

The brunt of climate change impacts on the entire coastal ecosystem of Sundarbans, one of the largest coastal biospheres on this planet, is coming within decades. Many ecosystem services in the form of food, fodder, fuel and shelter materials are rendered by this unique ecosystem (Chakraborty, 2017). The sixth round of caveats on global warming and climate change as submitted through several recommendations of the IPCC has recently offered a bleak scenario of climate change impacting the resilience and dynamics of coastal mangroves, their associated biodiversity, agriculture and livelihood at large. The frequency of cyclones will increase by at least 30% over previous observations, and it indicates that more land will be swallowed by salinity intrusion into crop lands. Sea-level rise will overtake that of previous records, worse enough to over more islands under sea water.

The yields of major crops would be dwindled more alarmingly because of the crusting of salts over agricultural lands due to climate-mediated salinity invasion. The same consequences will happen on the lives many sensitive flora and fauna because of elevating salinity invasion. The social ecology of the Sundarbans mangrove ecosystem is getting more and more vulnerable to territorial and occupational migration, escalating ecological and demographic entropy, causing fiercer human–wildlife conflict and having rippling effects on backyard settlements and habitats, including Kolkata, a city that disproportionately sustains almost 15 million people. The shrinkage of mangrove areas with diminishing amounts of vegetation and a reduction in the expanse of the coastal ecosystem with normal ecological integrity in response to mindless human atrocities has added more disruptions to the existing instability.

7.14.4.1 Impacts of Climate Change on the Eco-Functionality of Mangrove Ecosystems

Climate change, being the predominant challenge of this century, is caused by increased concentrations of greenhouse gases, mostly carbon dioxide (CO_2), in the atmosphere from anthropogenic influences, which have led to an increase in global mean surface air temperatures and consequent weather variability, which includes changes in the seasonal cycles of mean surface air temperature, precipitation patterns, humidity, the direction and speed of wind and the frequency and intensity of storms and droughts. All of these changes can be assessed through basic alterations in the physiological processes and changing patterns of the survival strategies of living organisms. Other modes of assessment include studying and monitoring the consequences of climate change in oceans and coastal areas, including rising sea levels, increasing sea surface temperatures, saltwater intrusion into fresh water, accelerating rates of increasing evapotranspiration from oceans and other waterbodies, a considerable increase in the melting of snow and ice cover and a drastic reduction in the thickness of polar ice sheets and glaciers. The functioning of ecosystems involves an array of interactive processes featuring biological components that tend to exhibit nonlinear responses caused by crossing thresholds, which in turn alter their species diversity and species composition and key ecological processes, thus affecting ecosystem stability, resilience and functioning. Research studies on coastal ecosystems, including coral reefs, dunes, brackish-water and backwater impoundments, mangrove forests and oceans, have revealed that pressures from human activities can bring about dramatic changes where functioning is severely impaired.

7.14.4.2 Climate Change and its Impact on Biodiversity: Case Studies on Sundarbans, India

Climate change has considerably affected the global biodiversity of the world, specifically species richness, ecological distribution patterns, ecosystem boundaries, habitat and niche preferences, the synchronization of reproductive cycles with growing seasons and ecobiological attributes pertaining to complex species interactions (predation, pollination, competition and disease). Species at the edge of their distributional ranges can move to favourable locations after being affected by climate change. Moreover, mobile or opportunistic species are likely to benefit at the expense of those which are incapable of finding any way out for their survivability (Hughes, 2004).

Besides, the impacts of climate change on biodiversity tend to vary between regions and ecosystems. Strategies adopted to mitigate or adapt to climate change also alter in line with the changing levels of greenhouse gas emissions affecting the global climate. Linkages between climate and biodiversity stresses need to be established to draft more-integrated policy responses at international, national and regional levels in terms of reducing energy consumption, increasing energy efficiency and promoting renewable energy technologies as priorities. Activities which meet both climate and biodiversity objectives also include adopting new landscape-management approaches by ensuring effective carbon-sequestration methods and promoting more applications for renewable energy.

Upon reaching a critical threshold limit, even relatively minor stress can hamper ecosystem integrity, causing ecosystem instability. Global climate change, pushing some ecosystems over the limit acting as a driver facilitating a multitude of interactions, exacerbates the impacts of existing environmental and human-mediated harms to the natural world.

Climate change has impacts on biodiversity at different hierarchical levels within an ecosystem, affecting genes, species, populations, communities and habitats. Gradual but steady increases in temperature cause changes in rainfall patterns, drought and coastal flooding, sea-level rise, several natural disasters like storms and cyclones, alterations in several physicochemical parameters and salinity invasion, all of which affect habitat suitability. Individual species, populations and biotic communities react and respond in multiple ways to changing environmental conditions either by remaining unchanged in their normal distribution range or by adopting dispersal to new regions to counteract the possibilities of population decline or extinction. Changes in the ecobiology (physiology, survivorship and reproduction) of a species tend to have impacts on populations or biotic communities within an ecosystem and are counteracted by chain reactions within the food chain–food web dynamics of coastal-estuarine-mangrove complexes, which in turn diminish ecosystem resilience and hamper ecosystem functioning.

7.14.4.3 Major Threats Associated with Aquaculture and Fishery in Sundarbans

The impacts of climate change on aquaculture can become indirect as they affect the natural resources required for aquacultural practices in Sundarbans, namely land, water, fish seed, both living and nonliving food materials, organic manure, etc. Because aquaculture needs significant feed and seed inputs, the impacts of climate change on them will also, in turn, affect the productivity and profitability of aquacultural activities, which depend significantly on the inputs of seeds and feed. Also, the future of local people associated with aquaculture-based activities is now more vulnerable to resource loss due to exposure to extreme weather events. Climate change exacerbates vulnerability to diseases while increasing physiological stress on cultured stock, both of which in turn not only affect the productivity of aquaculture but also increase vulnerability to diseases, which together pose higher risks to health and reduce the income of farmers. Climate change, in conjunction with other extreme weather events, synergistically and profusely impacts farmed stock, alters the genetic diversity of wild stock and affects biodiversity, which includes the availability of living food components for the culturable species used in aquaculture in Sundarbans. The mixed impacts of climate change on aquaculture can be explained as follows: The fluctuating temperature caused by climate change–mediated global warming exposes fish to physiological disturbances that disrupt their internal metabolic homeostasis and force them to shift their habitats, compounding physiological stresses. In tropical regions, stresses have been found to be higher than in other regions, including lower pH, dissolved oxygen and growth for autotrophs, leading even to the mortality of fish (Das et al., 2025). Such environmental perturbations due to aquacultural activities occur in the following ways:

- An increase in atmospheric and water temperature increases salinity by impacting hydrological cycles, which synergistically reduces the viability of amny species of fish used for aquaculture.
- The breaching of dykes surrounding the aquacultural waterbodies is due to tidal surge or catastrophic storms, which speed up the ingression of saline water and thereby affect aquacultural production.
- Climate change–mediated sea-level rise, shifts in tidal cycles, pest attacks, outbreaks of microbial disease, etc. in the Sundarbans region have negative effects on aquacultural operations in this region.
- Most aquacultural farms developed on agricultural lands are presently being converted into brick fields without valid permission for those conversions, which has caused more complications in aquacultural production.
- Poor maintenance embankments adjacent to aquacultural farms have resulted in embankment failure and breaching, which also aggravate aquacultural production.

Therefore, ongoing environmental perturbations from ongoing global climate change and from drastic changes in the sociopolitical and socioeconomical conditions of the region have become major constrains in the optimal use of the recourses from the Indian Sundarbans for local people's benefit.

7.15 Steps for Combating the Nuisances Associated with Climate Change

A comprehensive step needs be taken in the present scenario of climate change through visualization, monitoring, simulation and modelling at global and local levels with the primary objective of engaging in eco-restoration for salinity-affected land, with the help of modern technologies supported by community participation. The phase-by-phase eco-restoration of ecodegraded mangrove environments, integrating eco-rehabilitation and eco-reclamation processes following governmental guidelines and initiatives and involving people's participation and community movement, appears to constitute an effective mitigation strategy against global warming. Appropriate eco-monitoring efforts, involving different types of biomonitoring tools from molecules to ecosystem levels with the help of accurately recording the seasonal oscillation of biodiversity and other physicochemical parameters in different ecozones that are characterized by contrasting ecological attributes, tend to generate baseline research information that can be used for developing appropriate mitigation strategies.

Climate change has been found to have considerable impacts on the vulnerable sections of society—especially in developing countries, where low-income and marginalized people are considered as important bottlenecks in adaptations to climate change. This situation needs the following: (i) ground truth verification and data generation; (ii) the creation of decision support systems (DSSs) at community and district levels; (iii) the transformation of farmers into a climate task force; (iv) monitoring and seamless assessment; (v) community documentation and the visualization of Natural Resource Management (NRM) and risk mitigation; (vi) the promotion of gene biodiversity, the creation of a community gene bank and policy management that factors in short- and long-term perspectives; and (vii) the creation of action research modules through the application of people participatory appraisals (PRAs) and people learning actions (PLA) as tools and techniques. Ongoing climate change has posed a challenge to achieving sustainable development in relation to food security, biodiversity conservation, water requirements and ecological integrity (Patra et al., 2005). Adaptive comanagement by evolving collective and integrated resource-management policies must be undertaken through community resilience and ecosystem-based management to cope up with the ongoing environmental challenges.

In view of the above, the following activities must be undertaken in order to combat ongoing climate-related problems, mainly by implementing the Climate Adaptation and Resilience (CLARE) programme in the mangrove-estuarine ecosystem of Sundarbans, India.

7.15.1 Challenges and Opportunities from Global Climate Change

The primary challenge in dealing with the climate threat to the mangrove ecosystem of Sundarbans, India, commands timely adaptation to global warming–mediated climate change in order to combat all serious ecological, socioeconomic and socio-political problems so that they hardly hamper an eco-planning process that emphasizes climate risks, the vulnerabilities of the ecosystem and the subsequent implementation of adaptive strategies aimed at mitigating the climate crisis at regional, national and international levels.

7.15.2 Constraints Associated with Poverty and Interventions to Adopt

1. Sustainable climate-resilient livelihood generation: The poverty of people at large in Sundarbans and adjoining reclaimed areas is an obvious stressor on fragile ecological diversity and resources. This invariably invites the consequences of denudation and the depletion of resources (deforestation followed by human settlements, the habitat destruction of indigenous floral and faunal species, undesirable alterations in land-use patterns, the intensification of aquacultural practices, the introduction of bioinvasive species, etc.). Sustainable climate-resilient livelihood generation is the primary need of this region.
2. Ecological stewardship: Local people, especially women and youth, can be trained as ecological stewards to undertake constant vigils and monitoring for mangrove protection, restrict rice stubble burning in crop fields, collect and upload micro-level weather data, expand more land under an integrated farming system, socialize ecotourism, etc. All these will have substantial impacts on poverty alleviation and mitigate the impacts of climate change.
3. Constraints in implementing policies: The poverty of society is the most significant barrier to adapting to climate change and should be given the most importance in drafting proper conservation strategies against ongoing biodiversity loss, such as by involving local support from government departments such as that of the environment and forests, fisheries, irrigation and land reform. The uncertainties associated with climate change hamper individual countries' abilities to implement policy because the slowly increasing levels of awareness among the general public impede policy efforts.
4. The loss of wetlands and aquatic biodiversity: The combined action of climate change–mediated sea-level rise and sinking landmasses in coastal environments has destroyed sensitive and vulnerable wetland ecosystems like mangroves and corals. Such destruction has been subsequently compounded by natural disasters such as storms and cyclones.

7.15.3 Climate Change, Feedback Mechanisms and Food Security

Climatic vulnerability implicates the degree of sensitivity and adaptive capacity of different structural components of ecosystems (flora, fauna, soil and water), which are ultimately connected to social and ecological systems. Climate variability prompts extreme weather events, an influx of more sea water and thus salinity invasion, coastal geomorphology changes, sea-level rise, etc., which independently or together exacerbate vulnerabilities to ecological and social sectors in coastal areas. Recurrent and intensive extreme events like cyclones, storm surges and tidal flooding act as mirrors of the climate change–induced threats along coastal zones. The implications of negative impacts of climate change on biodiversity ultimately lead to human food security, outbreaks of diseases, the introduction of vectors into new regions, the ecodegradation of productive waterbodies and agricultural lands and disruptions to the synchronization of plant and animal life cycles with seasonal variations and changes in ecosystem functions that together and in a cumulative manner result in the provision of food, water, and other required ecosystem services.

After being affected by climate change, global biodiversity suffers from global climate through many complex feedback mechanisms. For example, some species of ocean algae have increased in higher ocean temperatures due to global warming, which has ultimately led to algal blooms that release dimethylsulfide into the atmosphere, and this chemical compound acts as a contributor to cloud formation. The development of such clouds in turn helps to reduce temperatures and thereby reduces the heat that can reach the Earth's surface. Similarly, increased temperatures stimulate the growth of phytoplankton in the sea, which in turn increases carbon dioxide uptake. Likewise, the growth and expansion of forest due to higher atmospheric carbon dioxide concentrations and warmer temperatures also in turn increases carbon sequestration.

The existence of many linkages between climate and biodiversity prompts the gathering of more-integrated policy formulation at regional, national and international levels, which is needed for promoting and mainstreaming various sectoral areas of policymaking. Reducing energy consumption, increasing energy efficiency and promoting renewable energy technologies are priorities.

7.15.4 Major Consequences of Climate Change and Ocean Acidification on Coastal Zone Ecosystems

The IPCC Fourth Assessment Report states that coastal environments across the world have been facing numerous adversities because of ongoing climate change, which has resulted in sea-level rise, inundation, erosion, a loss of biodiversity and the malfunctioning of different ecosystems, including corals and mangroves. In addition, coastal zones are vulnerable to extreme events such as altered rainfall and runoff patterns, cyclones, extreme waves, storm surges and ocean acidification. As a result of increasing emissions of carbon dioxide (CO_2) because of unplanned urbanization, intensive industrialization, unorganized and unplanned tourism, deforestation, etc., the rates of absorption for carbon dioxide by estuaries and oceans have been significantly increased, leading to an increase in average oceanic acidity over several decades. Oceans act as major accumulators by absorbing a quarter of all the CO_2 released into the atmosphere by human activities, which in turn creates carbonic acid in seawater. Oceanic-marine ecosystems steadily absorb CO_2 from the atmosphere, and a quarter of all the CO_2 is released into the atmosphere by human activities. This process of the absorption of CO_2 creates carbonic acid in seawater, which has corrosive impacts on the shelled faunal components that constitute the main framework of marine and coastal food webs and promotes the intrusion of bioinvasive species.

Acidification in marine-estuarine systems tends to have large-scale corrosive effects on ecosystems such as corals, mangroves, grass belts, dunes, lagoons and salt marshes and on shelled faunal components such as crustaceans, mollusks and brachiopods, which act as the structural foundation of marine-estuarine-coastal food webs. In addition, emerging acidification problems with naturally occurring high CO_2 levels, which are associated with lowering pH levels, enable a large number of invasive species to intrude and invade natural coastal ecosystem, with the consequence of lowering biodiversity.

Research studies have shown that a gradual decrease in surface water pH is due to the direct or indirect effects of climate change and sea water intrusion in the eastern sector of the mangrove-dominated Indian Sundarbans at spatiotemporal scales (Chakraborty et al., 2013). The development of suitable habitats for the growth and variable biomass of mangroves takes place thanks to the contribution of fresh water from the Bangladeshi Sundarbans through several creeks, channels and tributaries of the Padma–Meghna–Brahmaputra system, counteracting seawater intrusion from the Bay of Bengal into the Indian Sundarbans, and through the photosynthetic activities of a huge chunk of mangrove vegetation. According to a recent report of Intergovernmental Panel on climate Change (IPCC), continuing emissions of carbon dioxide due to climate change are expected to make the oceans up to 150% more acidic by the end of this century (2100) than they were at the beginning of the Anthropocene.

This trend in acidification reduces the ability of the marine-oceanic environment to absorb normal forms of atmospheric carbon dioxide and thereby increases the accumulation of carbon dioxide, which increases the rate of global warming. The consequences of acidification on the carbon system of marine ecosystems reduce the ability of the ocean-marine environment to absorb carbon dioxide (CO_2), which in turn increases the time required for recovery, but coastal vegetation, commonly referred to as the blue carbon bank, plays important roles in counterbalancing the changes in ecological conditions. Ocean acidification, having regulatory impacts on aquatic pH, affects the normal metabolic activities of many saline-water-inhabiting biota, especially marine calcifying organisms, all of which have been rendering valuable goods and services to humans in terms of water, energy and fishing. The average pH of the surface waters of the global ocean has already fallen by about

0.1 units from an average value of about 8.21–8.10 since the onset of the Industrial Revolution. This observation is also reflected in the estuarine networks of the Indian Sundarbans through the direct cause of higher carbon dioxide emissions, and this sensitive ecoregion is close to Kolkata, the state capital of West Bengal, India.

The present trend in the decrease in aquatic pH (0.026) in the eastern sector of the Indian Sundarbans is expected to remain at a much higher rate because this ecoregion used to receive freshwater supply from a network of the riverine channels, estuarine creeks and tributaries of the Padma–Meghna–Brahmaputra system, which have been considerably reduced over the past decades. Additionally, the disrupted supply of fresh water by the Bidyadhari River and Ichamati River is due to rates of sedimentation that have been increasing since the late fifteenth century (IPCC, 2007).

The slow pace of acidification in the estuarine water of the eastern part of the Indian Sundarbans (0.026 unit/decade) in comparison to that of the estuaries of the western part of the Indian Sundarbans (0.04 unit/decade) is explained by the occurrence of dense mangrove forests in the eastern part. The dominant halophytic mangrove species, like *Sonneratia apetala*, *Avicennia alba*, *Avicennia officinalis*, *Avicennia marina* and *Excoecaria agallocha*, more steadily absorb carbon dioxide from the atmosphere and from water columns, and they store this carbon in the vegetative and reproductive parts of those plants, mostly thanks to carbon-fixing enzymes such as ribulose 1 and 5-biphosphate carboxylase/oxygenase (rubisco). Therefore, the occurrence, density, growth and biomass of mangrove vegetation exert control over the prevailing pH of estuarine water because the abilities of these halophytic plant species to absorb carbon dioxide (CO_2) for use in photosynthesis help shift the equilibrium towards alkalinity.

7.15.5 Implications of Climate Change for Biodiversity

7.15.5.1 Climate Change on the Natural Environment

Climate change–mediated impacts on natural systems, including changes to ambient temperatures, scarcity in the availability of water (especially for terrestrial organisms), increases in the concentrations of carbon dioxide, directly and indirectly affect all parts of Sundarbans, even the protected areas that exhibit variations in species and ecosystems.

7.15.5.2 Climate Changes and Potentially Vulnerable Taxonomic Groups

Although mammals can disperse from their own narrow-ranged territories, they are susceptible to rapid climate change consequences. Changes in competition between grazing mammals and carnivorous ones take place because of climate alterations, which tend to impact water availability and provide nutritional foliage for herbivores as a result of CO_2 fertilization.

7.15.5.2.1 Birds

Birds are also highly mobile fauna with the power to disperse in response to the ongoing stresses of climate variabilities, which have resulted in phenological changes, including the natural process of migration and egg-laying behaviour. In view of increased interspecific competition for locating breeding grounds, the breeding sites of waterbirds in coastal zones are susceptible to diminishing freshwater flows, resulting in saltwater intrusion compounded by sea-level rise and natural disasters like cyclones and storm surges. Top predators are more vulnerable to changes in food supply and untimely migrations as a result of seasonal shifts.

7.15.5.2.2 Reptiles

Reptiles have a narrow range of mobility and hardly undertake any dispersal, so the shore-inhabiting species remain subjected to higher temperature due to global warming, which alters sex ratios. However, some species have tried to modify their use of microhabitats.

7.15.5.2.3 Amphibians

Amphibians, possessing high habitat specificity and a different range of mobility, can disperse only to some extent and thereby are considered as the most threatened terrestrial taxa.

7.15.5.2.4 Fish

Freshwater fish species are more susceptible to global climate changes with reduced water flows and a reduced abundance of live food like phytoplankton and zooplankton. Marine species, with greater mobility but with higher habitat specificity (mangroves or rocky or coral reef habitats), are more vulnerable to losing their habitat, changes in ocean currents, a reduced supply of nutrients and the disrupted dispersal of young. In addition, most species are susceptible to the harmful effects of increasing acidity on their physiological growth because they are attuned to distinct sea temperature thresholds for reproduction.

7.15.5.2.5 Invertebrates

Invertebrates appear to be more responsive faunal components than vertebrates because they have short life cycles, have higher rates of reproduction and are more sensitive to climatic variables. Aquatic invertebrates appear to be affected by changes in water-quality parameters and the hydrodynamics of aquatic ecosystems.

Acidic water erodes exoskeletons like shells, plates and spicules that are made of calcium carbonate.

7.15.5.2.6 Plants

Perennial plants appear to be more vulnerable to climate change as it affects recruitment and establishment opportunities. Endemic plants with short life cycles demonstrate higher photosynthetic rates in elevated carbon dioxide, where other factors, like water, temperature and nutrients, are not limiting. A combination of increased carbon dioxide and longer growing seasons increases productivity in some regions by enhancing water-use efficiency, where total water use depicts a decreasing trend due to increased evaporation from soil and decreased total leaf area.

7.15.5.3 Climate-Driven Shifts in Plant and Animal Distributions

The distributional patterns of most living organisms, with the exception of that of *Homo sapiens*, are determined by climate parameters that define their niches and ecological habitats. The movement of species as a result of complex dynamic processes, along with the combination of the dispersal and colonization of new habitats, determines species population dynamics. However, temperature acts as the most determining factor in directing species to move along thermal gradients that run from the equator towards the poles and from low elevations towards higher elevations. However, this general rule becomes complicated by the dispersal capacity of some species, which differ in their capacity to disperse in response to stimuli. The requirements of some species, with their narrow requirements for soil type, pollination, food sources, seasonal climate conditions and habitat types, prevent dispersal even for the most mobile species, demonstrating the requirements for greater connectivity. Such findings are complicated by the uncertainties of changing climate conditions and by drought, heatwaves and fire.

7.15.5.4 The Impacts of Climate Change on Mangroves and Coastal Wetlands

Several subregions in and around Bangladesh, Sri Lanka, China, India and Pakistan within the Indian Ocean coastal region constitute a key area of vulnerability to climate change. Harmful effects have been exacerbated by climate change in recent decades, causing ecodegragation from droughts and declining precipitation, itself leading to a dereliction of wetlands, which are drying up and feature severely degraded ecosystems.

About one-third of the world's mangrove coverage has reportedly been lost over the past 75 years, mostly because of global climate change, a reduction in freshwater flows and saltwater intrusion. Another 25% of mangroves are expected to be lost, mostly in developing countries such as Bangladesh, Indonesia, Sri Lanka, the Philippines, Pakistan and India, by 2025. The ecosystem integrity of mangrove-dominated wetlands plays critical roles in buffering against the flooding and erosion arising from climate change. Mangroves, by sharply responding against climate change, develop important adaptation strategies against sea-level rise, vertical accretion, habitat loss and the migration of biota towards more-inland zones. Climate change also acts as a driver in shifting rainfall patterns, increasing CO_2 levels, increasing the number of storms and exacerbating wave actions and soil erosion, which affect mangroves in different ways, and the constituent species adopt different adaptive mechanisms determining the distribution and composition of the mangrove community.

7.15.6 Climate Change–Mediated Threats to Mangrove Biodiversity

In the past half century, biodiversity enriched ecozones have been exploited both for day-to-day human requirements and for national or international trade and to sustain the harvest and export of natural products from forests, wetlands, grasslands, etc. Some other attributes connected with human developments, like urbanization, industrialization, bioinvasion and the illegal trade of rare species keeping pace with rapid economic and population growth, have considerably impacted biological resource-management processes. In addition, newly emerging pollution-related problems from tourism and shipping, which release considerable amounts of crude oil and human waste like agricultural runoff, urban sewage and industrial effluents, have adversely affected the marine-estuarine environment and associated aquatic life.

Harmful effects on mangrove damage the spawning grounds of numerous aquatic species, triggering other forms of coastal ecosystem degradation. The Sundarbans area at Chakaria in eastern Bangladesh has recently been almost completely cleared out for aquacultural activities, and a considerable area (more than 1 lakh hectares) of mangrove forests in Thailand has disappeared between 1961 and 1993 because of aquacultural activities. The reclamation of lands for human settlements and other coastal developments has also been possible due to the destruction of habitats for wildlife and has invited human–wildlife conflicts.

7.16 Global Warming and its Contribution to Coastal Bioinvasion: Impacts on Sundarbans, India

The impact of climate change on the vulnerability of coastal zones results in a gradual warming of water, a reduced supply of nutrients, sea-level rise, an increased risk of acidification and outbreaks of different diseases. In addition, coastal zones become more suffocated with the increased stratification and reduced mixing, causing a critical loss of oxygen in a water-hypoxic event (Archer & Rahmstorf, 2010). Eutrophication in estuaries is regulated in different ways by climate change, which also triggers changes in mixing characteristics, the density in the stratification of water columns, freshwater runoff, sea-level rise and exchange patterns between coastal zones and their adjoining seas (Kennedy, 1990).

Global warming–mediated rises in temperature affect the biology of a species at its molecular, physiological and biochemical levels, which not only influence the life cycle attributes of different species but also alter their distribution patterns and community interactions (IPCC, 2000; Das et al., 2005). In addition, elevated temperature in conjunction with environmental attributes like altered land-use patterns and the fragmentation of habitats, changes in the courses of estuarine flows, disruptions to nutrient cycling and eutrophication have compounded the problem and hampered the adaptability of species and ecosystems (IPCC, 2007). Climate-driven changes and the warming of ocean water also put constraints on natural biodiversity components, resulting in the invasion of exotic species from warmer regions; the poleward shifting of different species of algae, plankton and fish; and the considerable inhibition of ecosystem functioning (Archer & Rahmstorf, 2010). A gradual but steady increase in the temperature of water, air and soil stunts growth, reduces biomass, and changes geographic distribution and changes the proportionate distribution of flora and fauna. Changing temperature and salinity profiles have direct effects on suspension feeders such as mussels, clams and oysters, which in turn alter the abundance and distribution of their consumers and the abundance of their food components (Alpine & Cloern, 1992).

An increase in invasion by non-native species to the local biotic community is due to global climate change, as noted by researchers such as Dukes and Mooney (1999), who pointed out the compounding threats to biodiversity and other ecological factors (Vitousek et al., 1997). Moreover, larger latitudinal distributional ranges for invasive species than for native species indicate their ability to withstand a wide range of environmental conditions by virtue of their potential to sustain themselves even at increased temperatures (Dukes & Mooney 1999). Also, several other invasive species can colonize and flourish on human-made structures, including ship hulls, seawater pipelines, maricultural farms and other natural hard substrata (Harris & Tyrrell, 2001; Valentine et al., 2007). The Sundarbans mangrove ecosystem of India is also prone to the ecological threats of bioinvasion, mainly because of the ongoing ecological perturbations to this very sensitive and fragile mangrove ecosystem (Bhakat et al., 2004; Chakraborty, 2018) (Figs. 7.8, 7.9, 7.10 and 7.11). All of those species constituting fouling communities are dominated by non-native species, especially in ports and marinas. The invasion of alien species, as observed in some parts of Europe, has pointed out that out of 1700 species across the globe, 81% exhibited shifts in the range of their distribution because of climate change. On the coast of West Bengal, all fouling faunal components have mutually coexisted on hard structures as habitats, sharing preferred ecological spaces governed by a specific set of ecological parameters, such as temperature, moisture, the duration of inundation and food (Connell, 1961).

7.17 Time-Series Analysis on the Eco-Geomorphology of Sundarbans as a Result of Global Warming

A time-series analysis on the changing geomorphic properties of the island in relation to shape and size, erosion and accretion processes, the movement of dunes, the salination of water and sediments, the development of saline banks, etc. over a period of three decades (1969–2001) has shown some important changes, like ecodegradation within mangrove swamps and intertidal belts (mud flats and sand flats) and the shrinkage of land area despite feeble delta-outbuilding processes (Allen & Duffy, 1998; Biswas et al., 2004; Hazra et al., 2002). Coastal erosion followed by unwanted accretion processes have been constantly reshaping the islands of the Indian Sundarbans by modifying shoreline dynamics. In the Indian Sundarbans, the rate of coastal erosion was estimated to be about 5.70 km^2 per year from 1989 to 2014, and the thrust of this erosion process concentrated more on the southwestern edges of around 12 sea-facing islands, including Jambu Island, Bakhkhali Island and Sagar Island. In the Indian Sundarbans, the total land area was 6673 km^2 (approximately) in 1989, which reduced to 5869 km^2 (approximately) in 2014 (Hazra et al., 2002; Raha et al., 2021). Owing to a severe thrust of erosion, a couple islands within the Indian Sundarbans, namely Lohachara Island and Bedford Island (6.212 km^2), have already been disappeared. A research study that evaluated trends in erosion and accretion patterns through remote-sensing and Geospatial Information System (GIS) methods revealed a quantum of erosion measured at 162.879 km^2 and an accretion of 28,275,300 m^2 over 34 years (1989–2014) across the entire Indian Sundarbans (Hazra et al., 2002).

7.18 Natural Hazards at Sundarbans: Tropical Storms, Cyclones and Hurricanes

The very sensitive and fragile mangrove forests has plants with shallow root systems in the noncohesive nature of the forest soils in the intertidal zone. Several climatological factors, like wind flows, the density of atmosphere, precipitation, hydrological cycles and water levels, influence the frequency and intensity of storms and hurricanes. However, the occurrence of storms with low intensities still fells trees via lightning or wind flows, creating forest gaps that spur an important mechanism for natural forest regeneration by expediting sedimentation within the mangrove forest. High-energy storms like typhoons and hurricanes can devastate a considerable part of mangrove forests, leading to significant long-term effects on the ecosystem (Jiménez & Lugo, 1985). However, through ecological resilience, mangrove forests can recover despite such impacts, but this depends on the severity of mangrove mortality, the degree of sediment disturbance, the species composition of the forests and the availability of propagules.

More catastrophic storms and cyclones with different degrees of intensity over the Sundarbans, near the Bay of Bengal, have appeared over the past couple of decades, according to an analysis of available records. This increase will have significant impacts on the extent of coastal flooding, saline water intrusion, soil erosion, precipitation patterns, the shifting of sediments from one part to another (causing the mass mortality of mangrove plants), etc. because of all the storm surges with different intensities over this part of the Bay of Bengal (Hazra et al., 2002; Chakraborty et al., 2013).

Extreme hydro-meteorological events from the climate change produce storm surges and tropical cyclones that usually hit different corners of Sundarbans, but those catastrophic natural disasters are unable to cause much damage to this sensitive ecosystem because of the vast expanse of the mangrove forest. Other types of natural disaster-induced disturbances include the following: (i) a sudden rise in water level in estuaries; (ii) the shifting of depositional bars, sand materials and shoals; (iii) changes in water courses; (iv) the inundation of inland tracts with saline water; and (v) embankment breaches paving way for sudden flooding in agricultural lands (Paul, 2009).

The Sundarbans mangrove ecosystem of India is confronted with three major climate change–induced long-term environmental consequences, namely sea-level rise, the subsidence of land and a loss of wetlands, mainly because of the impacts of several forms of natural hazards, like catastrophic tropical storms, cyclones, seasonal typhoons and hurricanes, and because of prolonged inundation due to flooding from heavy rainfall—all of which frequently happen now with variable intensities caused by extreme climate events. One report indicates that between 1950 and 2004, Asia experienced 157 windstorms, causing 1380 deaths and misery for 2,496,808 other people. The synchronization of devastating cyclonic storms with high spring tides impacts the entire stretch of Sundarbans, resulting in sea-level rise and coastal erosion, which contribute to disruptions to the ecosystem functioning of the vulnerable mangrove ecosystem in Sundarbans, India.

Over several decades, many natural disasters have associated with each other, like frequent cyclone landfalls, causing erosion and inundation; the seismic activities and fault movements of tectonic plates, resulting in tsunamis; storms drifting from overwash sand fan lobes into swales and across cheniers, increasing hypersalinity due to evaporation and drainage loss; and increased sea-level rise. All of them have resulted in great damage to the mangrove forests of Sundarbans and its deltaic islands. The impacts of natural hazards destroying mangrove forests in the carbonate settings of the Andaman islands of India are because of storm-induced log-drift currents and the simultaneous subsidence and upliftment of carbonate platforms due to seismicity and tsunamis. The mangroves of along the Maharashtra and Gujarat coasts are threatened because of the continuous and steady increase in hypersalinity and aridity in their inner parts, due to shortages in river flows into coastal zones.

Sundarbans is bestowed with the eternal beauty of nature, but this sensitive landscape appears to be very vulnerable to a multitude of climatic events. According to the meteorological records, the Sundarbans mangrove ecosystem appears to have experienced many natural disasters with different intensities over the past couple of centuries, including recently Cyclone Aila in 2009, Cyclone Fani in 2019, Cyclone Bulbul in 2019, Cyclone Amphan in 2020, Cyclone Dana in 2024, etc. (IPCC, 2007). Global warming–mediated sea-level rise has the potential to destroy over 75% of the mangroves in Sundarbans by the end of the twenty-first century. However, being endowed with immense aesthetic flavour, this beautiful Sundarbans mangrove ecosystem carries out an important function in acting as a natural buffer against natural disasters, which can save the lives of more than one crore people, not only in the Sundarbans region but also in greater Kolkata.

7.18.1 Neotectonics at the Coast (Subsidence and Emergence)

The neotectonics at the coast has resulted in submergence and the emergence of coastal fringes, but tsunami-deposited sediments counterbalanced this effect with bays and coves along affected shorelines in 2004 at the Andaman and Nicobar groups of islands (Paul, 2005). A sudden uplift at the upper part of intertidal zone converts it into a supratidal zone, where the resultant fringe forest was relatively shifted towards the inland without further inundations by regular high tides. Mangrove stands are lost as a result of such a shift in the environmental gradient of uplifted intertidal shore platforms, leaving only a few species that can adapt in the changed ecological conditions in which they receive seepage flows of salt water during high tides and freshwater supply during monsoon months at the landward parts. However, subsidence in conjunction with permanent inundations and tsunamigenic sediments have resulted in a loss of mangrove forests in some areas of the Andaman and Nicobar groups of islands (Katchal Island, Trinkart Island, Campbell Bay, Garashara, etc.) (Paul, 2005). Catastrophic floods have developed because of faulty geological events that have impacted the coastal fringe forests in the river deltas (the Subarnarekha River in 1950, 1978 and 2008; the Ganga River in 1978 and 2000; and the Mahanadi River in 2003, 2008, 2011 and 2013), which have manifested as the submergence fringe tidal flats of estuaries, bank erosion, active sand splays on bank margins, the deposition of a large volume of sediment, log drift, a reduction in soil pore water salinity and a supply of pollutants.

7.19 The Synergistic Impact of Bioinvasion and Environmental Perturbation on the Coastal Mangrove Ecosystems of West Bengal, India

A bioinvasive species is defined as a species that has been introduced through human activities from outside of its native habitat and that tends to cause both ecological and economical damage. Bioinvasion refers to the invasion of biological organisms (plants, animals and microbes) after moving from a distant place to a new ecosystem. The introduction, settlement, colonization, proliferation and establishment of new species into an ecozone take place away from former habitats (Coblentz, 1990; Vitousek et al., 1997). Biological invasions are now considered as among the main threats to the world's biodiversity, where the displacement of one organism by another takes place. These bioinvasive species can destabilize any ecosystem, even coastal mangrove-estuarine ecosystems. Several other environmental perturbations caused by global warming—including eutrophication, biomagnification and the biotransformation of persistent pollutants—and the negative impacts of bioinvasive species act on marine-estuarine flora and fauna by outcompeting them for basic life support resources, severely threatening this ecologically sensitive region.

Invasive species are defined by the Convention of Biological Diversity as "those alien species that threaten ecosystems, habitats or species." They are non-native species that cause damage beyond any attendant benefits. The synonyms of invasive species are exotic, nonindigenous or alien species that naturally, accidentally or intentionally exhibit two attributes after being introduced to new ecological setup: (i) they are non-native (alien or exotic) to the ecosystem under consideration and (ii) their introduction is likely to cause economic or environmental harm or harm to humans.

7.19.1 Modes of Introduction for Bioinvasive Species

The invasion of an alien species means the arrival of that species into a new ecozone through a natural process after moving from another ecozone (Vitousek et al., 1997). The prime impact of such an invasion is the alteration of ecosystems, where an ecosystem which is defined as the spatially explicit association of biotic and abiotic elements featuring the cyclical movements of nutrients with the formation of biomass and flows of energy in food chain–food web dynamics. Exotic plant species can invade habitats or ecosystems by producing allelochemicals (phenolic compounds) in their vicinity, which drastically act upon and outcompete native species, bringing down their population density.

Four primary ways are recognized by which an exotic species can effect ecosystems: (i) Exotics can differ from natives in their use of resources (nutrients), thus affecting resource availability for other species (Vitousek, 1990); (ii) exotics can alter the flow of energy or biomass by changing food webs; (iii) disturbance regimes are affected by exotics; and (iv) exotics can change the physical structure of the ecosystem (Bertness, 1985; Crooks & Khim, 1999). In such a context, the fragile and sensitive Sundarbans mangrove ecosystem is very prone to the invasion of exotic species due to modern technology-based aquacultural and agricultural activities. Also, the development of unplanned and unorganized tourism can increase the likelihood of such invasion. However, mangrove ecosystems unlike many other ecosystems in the world are endowed with their own functional dynamics, and any such exotic species should encounter difficulty in breaking such naturally designed self-sustaining ecosystems. In addition, as a precautionary measure, timely intervention through effective management and controlling systems that adhere to basic ecological principles is very much required.

7.19.2 *Aquatic Invasive Species: An Indian Perspective*

Emerging life style changes in human civilization—based on prevailing globalized economies towards supporting recreational activities, augmenting aquaculture, managing environmental health (especially human healthcare systems), controlling vectors and promoting hobbies such as gardening with exotic species and ornamental fish cultures with non-native species—facilitate the introduction of species outside their well-established natural range. Unintentional introductions of exotic species by means of global transportation systems (e.g., the ballast water on ships) have enabled such species to establish themselves, causing irreparable destruction to native ecosystems and thereby the economies of new host countries. Invasive species, after establishing themselves in aquatic ecosystems, pose major ecological threats to estuarine-mangrove systems through the displacement of native species, alterations in hydrological parameters, disruptions to nutrient cycles, changes in food chain–food web dynamics, the triggering of outbreaks of diseases and parasites, the promotion of hybridization with native species, changes in fishery composition and biodiversity loss. However, assemblages of flora and fauna with the intermingling of large trees, shrubs and creepers with their root systems, benthic infauna and epifauna, benthic algae, arboreal organisms, etc., with their high biomass and dense canopies of plants, make mangrove floors impenetrable and undisturbed habitats, providing barely any open space for alien species as invaders to intrude, settle and grow, but in contrast, open coastal waterbodies are accessible for notoriously harmful invasive species to settle and flourish.

7.19.3 *Different Invasive Species in India*

India, as 1 of the 18 megadiversity countries of the world, provides shelter for about 8% of the global recorded species despite having only 2.4% of the land area of the world (Ghosh, 2008), and it accommodates almost all major categories of ecosystems in the world (Chakraborty, 2003; Dev Roy and Das, 2000; Debata et al., 2020). In the present era of globalization, flora, fauna and microbes are moving from one region to another, frequently keeping pace with escalating trade. This has resulted in the translocation of plants and animals of a particular geographical region to other regions. Although the realization of the prospective dangers of invasive species to marine-estuarine biodiversity has increased, pathways for the dispersal of those harmful species remain poorly understood in many parts of the world, especially in a country like India. Some of the major invasive species seen in marine ecosystems include *Kappaphycus alvarezii* (algae), *Monostroma oxyspermum*, *Eugymnanthea* sp. (sea anemone), *Ficopomatus enigmaticus* (Annelid), *Carcinus maenas* (crab), *Mytilopsis sallei* (mussel) and many other species, like wood-boring fauna, barnacles, amphipods, isopods, bryozoans and ascidians. A survey on intertidal biodiversity along the Kerala coast documented a total of 258 species of intertidal organisms, such as 48 species of algae, 12 species of sponges, 7 species of cnidarians, 45 species of arthropods, 5 species of flatworms, 6 species of annelids, 3 species of bryozoans, 109 species of mollusks, 2 species of sipunculids, 14 species of echinoderms and 7 species of ascidians. One seaweed (*Caulerpa taxifolia*), one species of bryozoans (*Bugula neritina*), one species of mollusk (*Thecacera pennigera*) and seven species of ascidians (*Phallusia nigra*, *Diplosoma listerianum*, *Microcosmus exasperatus*, *Ascidia liberata*, *Styela plicata*, *Herdmania momus* and *Symplegma braken hielmi*) are the invasive species recorded as part of the intertidal biodiversity of the Kerala coast. Twenty-five "introduced" crustacean species have been introduced in the Indian aquatic system by ballast water; these include five species of decapods, nine species of amphipods, six species of isopods, five species of copepods and seven species of cirripedes. Only one species of decapod, namely *Litopenaeus vannamei*, has been introduced to Indian waters for aquacultural purposes. However, the invasiveness of some of these species has yet to be determined (Figs. 7.11, 7.12, 7.13 and 7.14).

7.19.4 Biofouling Benthic Fauna: Possibilities for Bioinvasion

Biofouling refers to the undesirable development of bioassemblage by both microorganisms and macroorganisms on water surfaces or on any hard structure, and this kind of biogrowth results in operational efficiencies in various human commodities along the coastal belts. Biofouling faunal components are represented by the following: acorn barnacles (*Balanus aphrodite*), belonging to the order Cirripedia under the class Crustacea; several species of bivalves, like *Pinctada imbricate*, *Mytilus galloprovincialis* and *Enigmonea aenigmatica*; shipworms like Bankia, Lyrodus and Toredo; littorinids like *Littorina littorea*; grazing gastropods like *Cerithedia cingulate* and *Crepidula fecunda*; and predators like *Nerita articulata*. They inhabit hard structures like hard lateritic boulders to resist erosion, the bottoms of fishing trawlers and jetties.

The intertidal zone, between high and low tides, is endowed with a number of forms of habitats loaded with rich nutrients, and intertidal taxa include sea weeds, sponges, sea anemones, crustaceans, mollusks, polychaetes, brachiopods, bryozoans and polychaetes. But human developmental activities (the development of ports, fishing jetties, tourism, etc.) have paved the way for the introduction, settlement and flourishing of exotic species, leading to the transportation of viruses, bacteria, protists, algae, zooplankton, benthonic invertebrates and fish in the newer regions at a faster pace, with harmful effects from over two-thirds of recent non-native species (David & Gollasch, 2015).

7.19.5 The Environmental Consequences of Bioinvasion

More potential exotic species have been recognized to pose threats to ecosystems by modifying ecosystems, which in turn have cascading effects on the resident biota. Ecosystems are affected by invasive species through food chain–food web dynamics, nutrient cycling and the availability of food and required physical resources like water, heat, light, living space, etc. However, such ecological consequences become very complex and variable in terms of space and time because ecological principles and ecological interactions in invaded habitats allow the invaders to intrude and flourish.

By jeopardizing ecosystem functioning, bioinvasion has become one of the biggest threats to biodiversity worldwide, having unprecedented economic and environmental effects, including a colossal loss of biodiversity (causing biotic homogenization), a reduction in genetic diversity and even the extinction of native species—affecting 8% of threatened mammals, 30% of threatened birds and 11% of threatened amphibians. The impact of bioinvasion is revealed in altered structures and altered functioning in ecosystems and disruptions to linkages between invasion and ecosystem changes. The International Union for the Conservation of Nature (IUCN) considers the impacts of bioinvasion as "immense, insidious, and usually irreversible," with the more sparkling effects on aquatic ecosystems. Although the prime impact of bioinvasive species centres on ecological, environmental and economic aspects, the major thrust of it is perceived at present in the region in the reduction in the native regeneration from competitive pressures from exotic species.

The environmental consequences of bioinvasion on the coastal estuarine networks of West Bengal, India, with the presence of more than a hundred deltas endowed with one of the largest mangrove forests and associated flora and fauna are the displacement of native species (even leading to their extinction), changes in species composition, disruptions to thenatural litter decomposition process and nutrient cycling, etc., and in such a context, proper holistic ecomanagement strategies are being suggested, especially maintaining the self-sustaining ecosystem functioning of the coastal-estuarine-mangrove ecosystem.

On many coastal islands in several countries, mangroves did not occur naturally but have been found to grow somewhere after being introduced from outside, and therefore, those species under those mangroves are considered as invasive. An example comes from Hawaii, where the settlement, growth and proliferation of one introduced mangrove species, *Rhizophora mangle*, has caused the deterioration of the environmental qualities of some endemic waterbirds and damaged the sensitive landscape—destruction that lines up with the premature infilling of a unique Hawaiian aquatic ecosystem. However, after the introduction and proliferation of these mangroves in the foreign environment, they tend to negative impact the native flora and fauna despite rendering useful ecosystem services like the production of organic matter via litter decomposition, shoreline protection and the maintenance of water quality (Allen, 1998; Chakraborty, 2018).

7.19.6 Bioinvasion on the West Bengal Coast: Interactions with Driving Factors

Interactions among introduced species, physicochemical parameters and anthropogenic stressors tend to control the intensity of the impact of introduced species in the estuaries, bays and mangroves of coastal regions, which represent mostly invaded habitats. Ecodegraded coastal habitats have been invaded with more intensity than undisturbed areas that have better ecological health with reduced competition or predation pressures. In the coastal regions of West Bengal, assemblages of biofouling animals and mesophytic plants occur in areas closer to fishing harbours and other adjoining places with ecotourism activities. Predicting the harmful impacts on an invaded ecosystem from the activity of a single species alone is difficult due to complex interactions, so the ecological consequences that originated from combined effects of several introduced species help to develop the hypothesis that synergistic effects can accelerate and intensify impacts not only on native species but also on ecosystems, effects that will be magnified by the addition of more such species (Annon, 2003).

In the mangrove ecosystem of Sundarbans, India, which has three seasons, the physicochemical parameters of soil and water determine the temporal and spatial variations in biodiversity components and biological productivity by governing energy flow and the decomposition cycle. Seasonal fluctuations in all ecological parameters—due to the impacts of changes in meteorological parameters like temperature, precipitation, wind flow, etc. and changes in marine and estuarine physical processes like tides, waves and water currents—make this environment very unstable and determine their distribution and ecological succession processes, forcing all biological components to restrict their territory in different parts of Sundarbans (Chakraborty, 2009, 2011, 2018).

7.19.7 The Modes of Action of Invasive Plants Threatening the Sundarbans Mangrove Ecosystem

Mangroves, composed of woody halophytic trees and shrubs, normally grow in the intertidal zones along tropical and subtropical coastlines and remain in oscillating environmental conditions, experiencing a wide range of fluctuations in an array of ecological variables (soil, water and climate), by maintaining balance in the intricate relationships among different structural components. The primary ecological impact of plant invasions is observed by noting alterations in the process of detrital inputs, which impairs the natural decomposition process because detritus represents the semi-decomposed plant litter of mangroves and thereby alters the structure of detritus-based food webs. The outcomes of a long-term research study on the detrital pathways in the food webs of native and introduced species in the mangrove ecosystem, dominated by the mangrove plant species *Rhizophora mangle*, have revealed that the introduction of alien species has brought about a change in the diversity and species composition of detritivorous benthic fauna mainly because of changes in the biochemical constituents of detritus derived from mangrove plants.

Mangroves, in their native environment, host habitats for specific, potentially coevolved fauna with limited species overlap with adjacent sediment ecosystems (Sasekumar, 1974). Recently introduced mangroves to Hawaii without their detritivorous faunal associates, having the adaptive capacity for the consumption of tannin-rich detritus derived from *R. mangle*, have failed to sustain themselves, due to the nonuse of tannin by detritivore digestion because tannins are toxic to many mangrove-based organisms (Alongi, 1987). The introduction of tannin-rich mangrove litter to detrital food webs in Hawaii mangrove belts was found to alter the food web structure of communities in Hawaiian coastal ecosystems.

7.19.8 The Modes of Action of Invasive Faunal Components Threatening the Sundarbans Mangrove Ecosystem

Different invading exotic species pose various kinds of threats to ecosystems, directly modifying the functional pathways of ecosystems and thus generating harmful effects on the resident biota. A reduction in the availability of quality nutrients and food particles for organisms, modified flows and the volume of water, shrinkages in living space and changes in water temperature, pH, turbidity, heat or light are the effects of new biotic developments. All these complex modifications are variable from one site to another in an ecoregion, but a proper assessment of their ecological interactions presents an opportunity to predict the prospective impacts of the invaders on invaded ecosystems.

7.19.9 Invasive Mangrove Species in the Indian Sundarbans

A recent study from the three deltaic islands in the Sundarbans Biosphere Reserve, India (Bokkhali, Sagar Island and Jharkhali Island) has revealed a number of potential invasive species, including trees, shrubs and climbers. Trees such as *Eucalyptus*, *Casuarina* and *Leucaena* that have been introduced through human interventions in some parts of the coastal belts of West Bengal appear to have acclimatized to their new ecological setups, while the rest are treated as accidentally introduced weeds, having invaded from adjoining inland areas. And among these, the five most noxious species are *Eichhornia*, *Mikania*, *Eupatorium*, *Lantana* and *Parthenium* (Bhakat et al., 2004). This small piece of research documentation substantiates the already-developed hypothesis that alien invaders tend to occupy and flourish in human-altered landscapes. All three studied deltaic islands have been subjected to human interference and have thus been infested with weeds, but the undisturbed deltaic islands, like Sudhanyakhali, Sajnekhali and Jharkhali, which are protected mangrove forest reserves, harbour almost no exotic species.

The findings from this research have provided a glimpse into the occurrence, extent of distribution and ecological impact of changing species composition and biotic interaction processes. Earlier studies on baseline information from recording the diversity and distribution of the nonmangrove plant species infesting mangrove belts have revealed species such as *Azadirachta indica*, *Ficus bengalensis*, *Acacia nilotica* and *A. juliflora*, which are supposed to pave the way for the invasion of truly invasive species. Other noninvasive plants, like *Casuarina equisetifolia*, *Eucalyptus globulus*, *Clerodendrum viscosum*, *Thespesia populnea*, *Hibiscus tilliaceus*, *Tamarix indica*, *Eupatorium odoratum*, *Ipomoea fistulosa*, *Mikania scandens*, *Coccinea grandis*, *Luffa aegyptiaca*, *Opuntia dillenii* and *Leucaena* sp., occurring in different parts of West Bengal, India, might be potential invasive species in the days to come (Bhakat et al., 2004). However, in ecologically sensitive area like the Sundarbans mangrove ecosystem, the potential threats of invasive species must be taken seriously for conservation purposes and are therefore to be carefully monitored.

7.19.10 Invasive Mangrove Species from the Bangladeshi Sundarbans

Different types of plant invasive species, their intensity and frequency of occurrence, their rates and patterns of invasion, their habitat preferences and their association with invaders in the Sundarbans of Bangladesh have been studied. Among the 23 recorded species, 6 are considered as moderately invasive; 3 are recognized as highly invasive; and the remaining are potentially invasive. Climbers (6 out of 23) constituted the most abundant invasive species, followed by trees (5 out of 23) and shrubs (4 out of 23). The three highly invasive plants were *Eichhornia crassipes* (aquatic shrub), *Derris trifoliata* (climber), and *Chromolaena odorata* (shrub). All these invasive species spread at a faster pace into natural mangrove belts in ecologically disturbed zones where ecological succession is initiated as a natural process. Natural disturbance takes place through a variety of biotic and abiotic processes that vary in frequency, intensity, magnitude and timing, which in turn alters the existing physical environment and biotic community at a particular site (Trudgill, 1977; Connell & Slatyer, 1977; White, 1979; Holling, 1981; Iftekhar, 1999), which in turn leads to invasions by alien species. The successful invasion requires chronic disturbances in terms of extent and intensity, but through cryptic ecological degradation, referring to introgressive mangrove-associated vegetation, invasion slowly starts and steadily proceeds to cover the entire extent of the once-mangrove-species-dominated environment (Dahadough-Guebas et al., 2005). This is evident as a common feature of Sundarbans. Observations of biological invasion in Sundarbans have revealed that land formation is connected with the developmental process of mangroves, which through their establishment on a new substrate contribute to the vertical accretion of land (Davis, 1938; Lugo, 1980).

7.20 Ecotourism and its Impact on the Coastal Environment of West Bengal

Ecotourism refers to nature-oriented tourism through the management and capitalization of the natural ecological settings and resources of a beautiful region, mainly through the involvement of local people. Coastal West Bengal is prone to the ecological impact of unplanned and unorganized ecotourism, which has caused eutrophication and other related forms of coastal ecodegradation. Tourism in the two coastal districts, Midnapore (east) and South 24 Parganas, renders valuable contributions to uplift the economy of the state of West Bengal and helps develop awareness among the local people and tourists of the value of nature, especially by virtue of its enormous wealth of flora and fauna. The township of Digha, being a major tourist hub of the coastal Midnapore (east) district, has changed from a small village to a very busy tourist destination over a period of five decades. Initially, the carrying capacity of this coastal town by the confluence of estuaries with the Bay of Bengal was kept in proper balance by keeping the influx of the tourists at a low order, mainly because of a lack of communication and transport facilities. However, over the past couple of decades, road connections have been improved to a considerable extent, and a fleet of transport operations has led to a significant influx of tourists, which in turn has prompted the large-scale development of private lodges, hotels, holiday houses, etc. The Geological Survey of India, based on long-term research surveys on the Digha coastal belt, has highlighted environmental problems of this coastal tourist hub mainly emphasizing active processes of erosion and vertical accretion which have been accelerated by several human interventions that have removed sand dunes, triggering a boom in the construction of hotels close to seashore. Natural wave actions and catastrophic storms, the mining of sand for different construction purposes and the removal of different mangroves along with Casuarina trees from the dune tops for the purpose of harvesting fuel wood and building materials have resulted in the destruction of sand dunes, the erosion of beaches and the dumping of pollutants on the seashore (Chakraborty, 2017, 2018).

In the Sundarbans of India, tourists are allowed mainly in the buffer zones of the Sundarbans Biosphere Reserve, where a lot of infrastructure has been developed over the past three decades in the form of tourist lodges and hotels. The first accommodation facility was built at the Sajnakhali tourist lodge close to the northern boundary of the Sundarbans Tiger Reserve. This region has a lot pf potentiality for the development of ecotourism, which commands well-designed planning for promoting such ecotourism with infrastructural facilities that highlight the ambience and landscapes of the region. Although Sundarbans is endowed with aesthetic landscapes with diversified forms of flora and fauna, the main attractions of this naturally gifted ecoregion are its wildlife, where nature enthusiasts find an opportunity to sail in the domain of wilderness, especially to learn about the mystery of tidal rhythms and hear the roars of the Bengal tiger.

Tourism in South 24 Parganas is concentrated on Sagar Island, the largest river delta in the world, mainly for religious purposes, but it takes place on other deltaic islands in the western (Fresergunj, Bakhkhali, Henry's Island and Jambu Island) and eastern parts of Sundarbans (Jharkhali and other virgin mangrove-dominated deltaic islands), mostly within the Sundarbans Biosphere Reserve. The development of better infrastructure in the form of communication facilities, grid electricity and accommodation facilities is required to make this unique mangrove ecosystem more attractive to tourists aiming to experience the purity of nature (Figs. 7.12, 7.13 and 7.14). The diversity of flora and fauna in the Sundarbans mangrove ecosystem, including several rare and endangered species, has made this aesthetically celebrated, ecological and economically productive ecoregion a leading destination for tourism in West Bengal, India. The natural habitat of the Bengal tiger in this tidal wetland—along with the presence of other wildlife, like estuarine crocodiles, turtles, birds and spotted deer—has added more attractions for tourists. However, promoting tourism in a way that violates coastal zone regulations imposed by the government of India, the construction of unauthorized hotels, the dumping of nondegradable plastics, the release of municipal sewage, the massive erosion of sediments, etc. have paved the way for bioinvasion.

In such a context, steps need to be taken to promote sustainable tourism, which implies constructing the infrastructure required for ecotourism, such as facilities for water supply, meeting energy requirement, ensuring transportation, developing proper process of waste disposal, controlling of pollution, strengthening existing telecommunications and creating tourism-specific facilities by constructing hotels and resorts as accommodations for tourists, shopping centres, healthcare centres and safety systems. Also, the integration of Indigenous knowledge and innovative scientific knowledge is required for the rehabilitation and restoration of ecodegraded ecosystems, including taking proper steps for the conservation of biological resources while incorporating biology and cultural heritage in regional tourism-development policies, strategies and action plans.

7.21 Pollution in the Mangrove-Estuarine Complex of Sundarbans with Both Organic (Decomposable) and Persistent Pollutants

The mangrove ecosystem in Sundarbans, India, is considerably affected by human-caused pollution, which includes pollution by biodegradable pollutants like organic waste, nutrient pollution from agricultural waste (chemical fertilizers), sewage, etc. and persistent/conservative pollutants like heavy metals, pesticides, polychlorinated biphenyls (PCBs), oil spills and micropollutants (Figs. 7.15, 7.16 and 7.17). Pollution from global warming, due to increases in water temperature, causes chlorotic leaves to have a reduced leaf area and partial defoliation and leads to dwarfed seedlings with a mortality of 100% after a water temperature rise of 7–9 °C above ambient temperatures (Maheim, 1988). Growing industrialization in and around Kolkata and the industrial complex of Haldia, which are situated on the western side of the Hooghly estuary, significantly contributes to the pollution load in the Sundarbans mangrove-estuarine complex and results in the continuous ecodegradation of the mangrove ecosystem (Chakravorty et al., 2004).

7.21.1 *Heavy Metals, Pathways of Entry and Variability in the Sundarbans Mangrove Ecosystem of India*

The Sundarbans mangrove-estuarine complex of India acts as a sink for heavy metals because of the capacity of this ecoregion to retain a good concentration of heavy metals from freshwater rivers, tidal waters and storm water runoff. Heavy metals enter Sundarbans mangrove forests and their associated estuarine networks through two pathways: (i) natural processes like the erosion of sandflats, ore-bearing rocks, wind-blown sediments, volcanic activity and forest fires and (ii) human activities such as atmospheric deposition and the discharging or dumping of industrial waste through rivers. Different types of anthropogenic inputs are associated with the bioturbation activities of different faunal components, such as erosion and vertical accretion processes, which result in the heterogeneous distribution of metals through the physical mixing processes of sediments, restricting some metals (Cr, As and Co) with higher concentrations in surface–subsurface sediments and some other metals (Zn, Fe, and Cu) with higher concentrations in deeper layers (20–30 cm depths).

7.21.1.1 Heavy Metals and Related Stresses on the Sundarbans Mangrove Ecosystem of India

The toxic impacts of heavy metals on different structural components of an aquatic ecosystem are global concerns because of their potential to bioaccumulate in aquatic ecosystems (David et al., 2012; Hall, 2014). Pollution in an aquatic environment by heavy metals is mostly manifested as multielemental compound pollution, causing structural components to endure long-term exposure. The coastal environment of West Bengal, India, and its estuarine networks with estuaries such as Hooghly, Matlah, Saptamukhi, Bidyadhari, Raimongal and Muriganga on the northeast coast of India (Lat. 21–53°N and Long. 88–15°E) have been considerably affected by toxic metal pollution (Ghosh & Choudhury, 1989; Sarkar et al., 2002; Ganguli and Chakraborty, 2026). An appreciable number of heavy metals in the coastal water and adjacent estuaries due to anthropogenic activities such as urbanization, industrialization and aquaculture, tend to cause stresses at the ecosystem level and the species level (Millennium Ecosystem Assessment, 2005; Ganguly et al., 2008; Chakraborty, 2011). The estuarine complex of Sundarbans has been considered the most polluted ecoregion in the world. The increasing usage of heavy metals in industrial activities has caused the release of a large amount of effluent loaded with harmful heavy metals over the past several decades (Agusa et al., 2007; Hajeb et al., 2009). Maximum residual levels of toxic heavy metals like Pb, Cd, Hg, and Cr have been prescribed for humans (FAO, 1983). All these metals have the least beneficial roles in biological systems, but metals like Zn, Cu, Na, K, Ca, Mn, Se and Fe are considered essential metals at low concentrations, playing effective roles in fish metabolism, but after bioaccumulation, these heavy metals reach toxic levels that have deleterious effects on the different structural components of an aquatic ecosystem (Agusa et al., 2007; Hajeb et al., 2009). Through the bioaccumulation of persistent heavy metals by living organisms, biomagnifications take place in the pathways of food chain–food web dynamics in an aquatic ecosystem, with the involvement of phytoplankton, zooplankton and other faunal components from the bottom to the top of the ecological pyramid, and in this way, biomagnifications drive the bioaccumulated pollutants from one trophic level to another (Ghosh & Choudhury, 1989).

7.21.1.2 Metal Content and Metal Oxides in the Mangrove Ecosystems of India

The Sundarbans mangrove-estuarine complex provides shelter and support to many living and nonliving organisms under the influence of tidal fluxes and undergoes various biogeochemical transformations. All these biodiversity components are subjected to the deleterious effects of several metals and nonmetals, which are carried by river water after being released from several sources. All these heavy metals and their compounds are absorbed by sediment particles and slowly get degraded in anoxic conditions.

A recent study on the Bidyadhari River, with Tehatta at its upper limit upstream, a river that enjoys a circuitous tidal course in the district of North 24 Parganas, West Bengal, has demonstrated the special characteristic of carrying tidal domestic sewage and effluents from different industries through a series of artificially made canals, namely Bagjola Khal, Bhangor Kata Khal, the Central Lake canal and the Krishnapur canal. The Bidyadhari River is tidal for most of its entire length within the Ecchumati–Bidyadhari interfluves region, which is located between latitude 22°N to 23°N and longitude 88°30′E to 89°E, but it has been shifting towards the eastern-most side of the Sundarbans delta, enjoying general slope gradients. Ichamati, being a freshwater river with estuarine connections originates from Mathabhanga and carries fresh water to a certain extent, but the freshwater discharge of the river has been reduced over the past few decades because of a developing disconnection upstream that is meeting many other freshwater stream flows, even the Ganga River, due to continuous heavy siltation. The Ichamati River, the only freshwater-supplying river to the eastern part of Sundarbans, fails to reach the lower fringes of the Sundarbans mainly because it experiences a continuous siltation process (Ganguly and Chakraborty, 2026).

The Bidyadhari River, after starting its journey from Haringhata, in the Nadia district of West Bengal, flows through the North 24 Parganas district of West Bengal and joins at the confluence of the Rainmangal River estuary of Sundarbans, India. The Matlah River estuary, originating at the confluence of Bidyadhari, Karatiya and Rampur Khal near Canning, at South 24 Parganas, has tidal flows but has lost its upland freshwater connection, resulting in hypersalinity in the central part. The entire stretch of the Matlah–Bidyadhari–Rainmangal estuaries associated with a network of freshwater channels and brackish-water creeks exhibit manifold environmental dimensions. The northwestern part of this interfluvial region has been subjected to a natural land buildup process and vertical accretion.

The pH of water influences the concentration of heavy metals in an aquatic ecosystem, whereas the concentration of heavy metals in sediments depends not only on its pH but also on the amount of $CaCO_3$ and organic matter. The increased concentration of heavy metals exhibits a direct relationship with the carbonates, sulfides and metallic hydroxides that constitute a portion of the sediments after settling down physically or through precipitation.

Aquatic animals accumulate heavy metals through two intake processes from water: (i) as free ions and simple compounds dissolved in water, taken up through the epithelium of their skin, gills and alimentary canals and (ii) through food organisms, which accumulate heavy metals within their bodies through nutrition. The heavy metals from aquatic environments can directly enter the human body via the gastrointestinal tract by drinking water and consuming food. The bioavailability and the persistent toxicity of heavy metals convert them into potential pollutants. Heavy metals become toxic and injurious to health because toxicity is imparted by prohibiting enzyme action and interfering with blood synthesis, causing damage to the liver, the kidneys and to some extent the brain. Some heavy metals, like mercury (Hg), lead (Pb), cadmium (Cd) and chromium (Cr), having toxic effects even at minimal concentrations in bioavailable forms, enter the food chain and generate a range of deleterious effects in the metabolic activities of the physiological systems of human beings and other living organisms. Moreover, a few more metals, like Fe and Zn, are required at minimal concentrations for the normal development of living organisms, metals that can become detrimental upon reaching a concentration exceeding their permissible levels. All these metals, after entering aquatic ecosystems, bind with particulate matter and upon gaining higher specific gravity settle down in sediment.

7.21.1.3 Sources of Heavy Metals in the Sundarbans Mangrove Ecosystem of India

The main contributors of heavy metal pollution (especially mercury, lead, cadmium, zinc and copper) are mining activities and industrial discharge. Different human-mediated impacts on the Sundarbans mangrove ecosystem include intensive fishing, the drainage of sewage from nearby industrial and urbanized areas, transportation through ferry services, dredging and drilling, and the construction of resorts to promote ecotourism, and all of these appear to be potential sources for increases in these toxic elements. After reaching a mangrove environment, heavy metals are bound to suspended particulates accumulated from sedimentation and pose ecological threats, causing a decrease in the growth and rate of respiration of mangrove plants and having toxic effects on ass fish larvae (causing different physiological stresses) and macrobennthic faunal components (hampering their reproduction), mostly crustaceans.

Finer grains of sands and the silt components of mangrove sediments exhibit higher accumulations of heavy metals than clayey sediments do. However, mangrove plants generally have low concentrations of heavy metals, and they are poor indicators of trace metal concentrations. Among several metals, Zn has been found to exhibit low values in almost all species of mangrove plants and associated animals, followed by Cu and Pb, because the low level of metals is due to their low bioavailability in mangrove sediments (Ganguly and Chakraborty, 2026). The metal-binding capacity and mobilization of these metals are reduced in mangrove soils because of several environmental disturbances, like changes in the duration and frequency of tidal flooding, prolonged dry periods and fluctuations in salinity. All these disruptive forces are also magnified because of anthropogenic stresses. The unstable ecological conditions associated with seasonal variations in physico-chemical parameters and the adjacent land effects of the estuarine region also cause fluctuations in trace metals and metal oxides. Little information is available on the concentrations of different pollutants discharged in domestic sewage by a network of riverine channels and canals into the Bidyadhari River, and little information is available on the resultant effects on bioaccumulation and biomagnifications.

7.21.1.4 Pollution from Heavy Metals in Sundarbans

In the coastal estuarine environment of Sundarbans, metal pollution is generally caused by industrial inputs from land runoff and mining activities. In the coastal belts of Sundarbans, West Bengal, India, high concentrations of Fe, Pb, Zn, Cd, Cu, Ni, Cr and Hg have been found in the surface waters and sediment of Kakdwip, Jambu Island and Frazergaunge and in fish landing stations and different zones of the intertidal belts. Many heavy metals are discharged from the activities of a large number of fishing trawlers and vessels, which use antifouling paints for their regular maintenance, conditioning and protecting water vessels from the attacks of biofoulers like barnacles and bivalves. A recent research report has shown that a number of mangrove plants, like *Avicennia alba*, *A. marina*, *Excoecaria agallocha* and *Acanthus ilicifolius*, act as potential bioaccumulators of heavy metals like Pb, Zn and Cu, where the trend in the accumulation of metals is in the order of Zn > Cu > Pb > Hg in all vegetative parts (Ganguli and Chakraborty, 2026).

A recent study selected four study sites mostly in the reclaimed part of the Sundarbans mangrove ecosystem:

1. Malancha (S-I) is in the extended part of the eastern Calcutta wetlands at the bank of the Bidyadhari River, North 24 Parganas, West Bengal. This site is marked by semi-urban activities such as agriculture, horticulture, aquaculture and fishing, brick fields and leather industries, and it receives a good amount of waste, mainly sewage, from industries and municipalities through sewage channels, mostly released from Kolkata city. Because this study area is connected to the Ecchumati River through a network of rivulets, a sewage canal and expanded swallow wetland networks, mangrove plants such as shrubs and grasses persist.
2. Dhamakhali (S-II) is downstream of the Ecchumati–Bidyadhari interfluvial region, locally known as Kolagachia, and the Dhansa River is in Sandeshkhali II,

North 24 Parganas, West Bengal, India, which flows into the Raimangal River estuary. This tide-dominated ecoregion is considered a gateway to the eastern sector of the Sundarbans Biosphere Reserve, where flows of water in these estuarine networks have been modified and hindered by a number of human developmental projects, like aquaculture, brickfields and the construction of concretized tourist resorts, roads and embankments.

3. Jharkhali (S-III) is situated in Basanti Block, South 24 Paragana, West Bengal, between two mighty estuaries, namely Matlah (on the west) and Bidyadhari (on the east), and is located just opposite to the Herobhanga Reserve Forest and a tiger rehabilitation centre that was established at this site. This site in the Sundarbans mangrove ecosystem is important to aquaculture, fishing, agriculture and ecotourism, where human activities featuring water vessels and launches, the construction of roads and tourist resorts have cleared out the natural mangrove-vegetated area.
4. Canning port (S-IV) is on the bank of the Matlah River at Basanti Block, South 24 Paragana, and is a crucial site for fishing, tourism and other commercial activities that maintain communication between the people of Sundarbans and those of Kolkata.

7.21.1.5 Heavy Metals at Different Study Sites in the Indian Sundarbans

An estimation of several heavy metals, like Zn, Cu, Pb, Hg, Ni, Cd and Cr, in these four these study sites has exhibited different seasonal trends in heavy metal concentration in soil and in water. The concentrations of most of these heavy metals were found to be low during monsoon, which gradually increased during post-monsoon and pre-monsoon for a period of 3 years of this study. Out of the four study sites, the highest concentrations of all heavy metals were observed at study site-I (Malancha). Concentrations of Pb were found in water and soil at the following sites: S-I 6.65 and 5.1 ppm; S-II 2.24 and 1.04 ppm; S-III 2.12 and 1.01 ppm; and S-IV 2.36 and 2.68 ppm. Concentrations of Hg were estimated at the following sites: S-I 5.43 and 2.33 ppm; S-II 1.92 and 1.03 ppm; S-III 1.72 and 1.02 ppm; and S-IV 1.85 and 1.11 ppm. Ni concentrations in water (4.38 ppm) and soil (2.67 ppm) remained high at S-I during post-monsoon and pre-monsoon, and the values of heavy metals concentrations remained close to one another in other study sites. The concentration of Cd in water was estimated to be at maximum during pre-monsoon (2.33 ppm), and in soil, it was highest during post-monsoon (1.88 ppm). The maximum estimated concentrations of Cu were 6.66 ppm in water and 4.85 ppm in soil, and the highest concentrations of Zn were 4.12 ppm in water and 3.43 ppm in soil across all four study sites. The concentrations of Cr in water were found to be at alarming levels at S-I throughout the year, reaching its maximum levels in water (32.19 ppm) and in soil (22.84 ppm) during post-monsoon, and the lowest values of the same metal were observed at S-III at 6.14 ppm in water and 3.43 ppm in soil.

The concentrations of heavy metals (Fe, Zn, Mn, Ni, Pb and Co) in the recently deposited sediment of the Hooghly estuary exhibited higher values in comparison to those in the Matlah and Saptamukhi estuaries. The magnitude of variation in the concentrations of metals among different estuarine sediments was observed to be in the order of Matlah (26–54%), Saptamukhi (14–52%) and Hooghly (9–27%). The rate of sedimentation and the input of metals to the Matlah and Hooghly estuaries are relatively high, mainly because of the deposition of a huge number of silt particles during monsoon and because of the release of other pollutants from upland.

In line with the different rates of sedimentation, different metals demonstrated different ranges of concentration at different sites in the Sundarbans mangrove ecosystem. The maximum values were found to be in the order of Fe (43,613 μg/g), Mn (714 μg/g), Zn (108.0 μg/g), Cr (87.2 μg/g) and Cu (58.5 μg/g). Metal mobilizations are influenced by changes in water-quality parameters and variations in salinity, pH and the redox conditions of sediments, threatening the biota with chronic exposures and influencing the decomposition process of organic substances.

7.21.2 Suspended Particulate Matter (SPM) in the Tidally Dominated Hooghly River Estuary and its Ecotoxicological Relevance

Some toxic, nonbiodegradable and persistent inorganic contaminants, like trace metals with their high concentrations in an aquatic system after getting released from anthropogenic activities such as rapid industrialization and modern lifestyles, cause the deterioration of water quality (Mitra et al., 2018; Bhattacharya et al., 2015) through natural processes. Estuaries filter different metals from micronutrients depending on the duration of the retention period and the absorption potential of the metals associated with aquatic sediment particles (Salomons & Forstner, 1984; Balkis et al., 2009; Karbassi et al., 2015; Nasehi et al., 2013; Lam et al., 2015). Different physical and chemical processes, like coagulation, adsorption, flocculation and the resuspension of the bottom sediments, which are assisted by precipitation and desorption, control these mechanisms. These persistent pollutants directly affect the quality of aquatic food materials, which has harmful impacts on the ecological health of aquatic ecosystems and the health of humans.

The Hooghly–Matlah estuarine complex of Sundarbans, with several major estuaries, like Mooriganga, Saptamukhi, Bidyadhari, Raimangal, Matlah and Hoogly, meet at the Bay of Bengal on the extreme southern part of the Indo-Gangetic plain. They receive fresh water from different shallow and well-mixed freshwater-flowing (average depth <6 m) river estuaries, like Icchamati, Hooghly, Keleghai, Rupnarayan, Haldi and Rasulpur. They also experience meso- and macrotidal influences, with an average tidal range between 1.8 m (during neap tide) and 5.5 m (during spring tide), from the Bay of Bengal (Mitra et al., 2018), and they demonstrate the spatio-seasonal distribution of different elements in the form of suspended particulate matter (SPM), a high load of which (20–3460 mg/L) might have negative impacts on phytoplankton density. A decreasing trend in relation to the relative abundance of different elements (concentration in μg/g and %) was noted as follows: Si

(26.44 ± 3.75%) > Al (7.94 ± 1.52%) > Fe (6.17 ± 1.9%) > K (3.05 ± 1.5%) > Ca (1.97 ± 1.11%) > Mg (1.57 ± 1.71%) > Na (1.45 ± 8.40%) > Mn (1273 ± 2003%) > Zn (178.43 ± 130.95%) > V (151.54 ± 27.13%) > Cr (147.08 ± 32.21%) > Cu (62.06 ± 14.03%) > Ni (49.64 ± 12.09%) > Pb (21.5 ± 10.45%) (Mitra et al., 2020). The abundance of Si, Al and Fe might be attributed to the weathering of basaltic trappean rocks and laterites (Sarkar et al., 2004). The seasonal oscillation of different physicochemical parameters (temperature, pH, alkalinity, turbidity, salinity, etc.) regulates bioabsorption and adsorption. Seasonally, the overall maximum values were observed in late monsoon, followed by early post-monsoon, which were supposed to have been due to the high flushing of flood water and runoff from adjacent lands along the meso- and macrotidal regime during this period (Allen et al., 1980; Guezennec et al., 1999; Suja et al., 2017). Seasonal variations in Al, Fe, and V between the stations were noted during monsoon; Mg, Na and K had maximum fluctuations during pre-monsoon, auguring the arrival of Al, Fe and V from adjacent lands. The average concentration of most of the major elements showed conservative seasonal behaviour, except for Ca, Mg, K and Fe (Mitra et al., 2018, 2020).

Phytoplankton diversity was supposed to have been impacted not only by the SPM load in the water but by the direct influence of algae aggregation, turbidity, light attenuation, etc. (Achara et al., 2014; He et al., 2017), but such interactive processes might cause the depletion of dissolved nutrients up to a considerable concentration in proportion with the concentration of different types of SPM (Middelburg & Herman, 2007). Maximum spatial variation along with the average peak value of SPM was noticed during monsoon due to the terrestrial input from adjacent lands. Higher concentrations of each element were also found to occur in the water zone of the coastal regions of Sundarbans, India, showing the dominance of meso-zooplankton communities (Bhattacharya et al., 2015).

7.21.3 Pollution from Microplastics in the Sundarbans Mangrove Ecosystem

The beginning of the Plasticene Age in the Earth's history, marking the global usage of plastics, will be noticeable around million years from now because old, fossilized plastics have been discovered by scientists while studying the formations of rocks (Cozar et al., 2014). Out of the 359 million tonnes of plastic produced globally, 60–80% of it finds its way to oceans and thereby causing lot undesirable changes in this ecosystem (Plastics Europe, 2018). Thereafter, in the oceans, heavier plastics, after reaching the oceans, get deposited at the bottom, whereas lighter plastics with less density float on the surface of water, are dispersed by water currents and winds, and are deposited on seashores, where they further degrade into smaller particles through the interactive processes of solar radiation and mechanical forcing and by the foraging activities of marine biota (Lebreton et al., 2018; Jonathan et al., 2021).

Pollution by microplastics (MPs < 5 mm) is emerging as an environmental issue causing tremendous harm to various critical coastal habitats, like mangroves, salt marshes, flood plains, estuaries, seagrass beds, back waterbodies and salt pans (Duan et al., 2021; Helcoski et al., 2020; Huang et al., 2020; Nur et al., 2022; Bilal et al.,—; Hossain et al., 2023a, 2023b). Blue carbon habitats like mangroves and salt marshes play more-effective roles in trapping MPs and thereby potentially limit their broader dispersal (Giri et al., 2008; Nor & Obbard, 2014). Additionally, rapid urbanization, industrialization, coastal development, tourism and aquaculture contribute to the extensive spread of MPs (Barnes et al., 2019). In addition, organic carbon stores are increased in these coastal blue carbon habitats by plastic accumulation, preventing the wide of distribution of pollutants and thus preventing them from reaching the open water of marine and estuarine waterbodies while mitigating the emissions of greenhouse gases. Therefore, blue carbon ecosystems help to overcome the adverse effects of plastic pollution on biological systems, avoiding impairing ingestion and resisting disruptions to the food chain for aquatic life (Peng et al., 2020).

Microplastics, as dangerous pollutants, enter coastal belts and remain within "sinks" in coastal habitats, posing many ecological risks to coastal ecosystems in general and biodiversity in particular. Mangrove vegetation acts as a buffer against the accumulation of MPs, reducing the spread of MPs from land to the ocean and thereby helping to mitigate their negative impacts. The movement of MPs (<0.5 mm), specifically their distribution and accumulation, especially in their most prevalent forms as fibres (85%) films (11%), fragments (2.5%) and foams (1.3%), is controlled by vegetation, showing the efficacy of coastal habitats in trapping smaller MPs.

MPs are characterized by their various morphological and chemical characteristics: sizes, shapes, colours and polymer properties (Zhang, 2017; Jung et al., 2021). Smaller particles of MPs formed from the fragmentation of larger plastics through different natural (wave action, currents, sediment characteristics, biological interactions, bacterial degradation, photochemical radiation, etc.) and anthropogenic (industrial and domestic) activities enter the food web, impacting the growth, energy acquisition and survival of aquatic organisms through bioaccumulation and biomagnification processes (Corcoran et al., 2015; Alimi et al., 2018; Peng et al., 2020). In addition, several toxic chemicals, such as polycyclic aromatic hydrocarbons (PAHs), nonylphenols, bisphenol A, polychlorinated biphenyls (PCBs), polybrominated diphenyl ether (PBDE) and phthalates, are transported by MPs, posing serious risks to both aquatic and terrestrial ecosystems (Banik et al. 2022, 2024). Within aquatic ecosystems, coastal ecosystems exhibit instability and unpredictability because of the influence of both terrestrial and marine ecosystems, developing this sensitive landscape into a critical exchange of materials, including MPs, the transport and distribution of which are governed by the intrinsic characteristics of MPs, such as polymer density, size and shape.

Several research studies' findings have emphasized the habitat-specific significance of microplastic pollution to lend support to adopting effective environmental stewardship and a holistic environmental management strategy. These findings also provide valuable insights into microplastic distribution across diverse coastal habitats and the effective roles of mangrove vegetation in reducing microplastic pollution and thereby protecting coastal biodiversity and the ecological health of coastal ecosystems.

7.21.3.1 The Composition, Distribution and Ecological Impacts of Microplastics (MPs)

Plastics are broadly divided into three major categories according to their size: microplastics, mesoplastics and macroplastics (Li et al., 2016). The micro-size plastic fragments are called microplastics (<5 mm) and are ingested by a wide variety of organisms (Ory et al., 2018). The major constituents of microplastics from the highest to the lowest concentration are polyethylene terephthalate (67.5%), polypropylene (10.5%), polyamide (7.4%), polyethylene (5.9%), acrylic (3.9%), rayon (2.9%), and other types (1.9%), the sizes of which vary from smaller than 1 mm (50%) to 3 mm (Aliabad et al., 2019).

The ability of aquatic organisms to ingest microplastics enables them to play predominant roles as vectors in the food chain for the transport of chemical pollutants. Microplastics (MPs) tend to invade even the most remote regions of the marine environment at varying concentrations (Wright et al., 2013). Unable to differentiate MPs from food, aquatic organisms bioaccumulate and biomagnify microplastics along the food chain, including humans. The presence of MPs in the biota often results in harmful physiological and behavioural effects on the integrity and ecological balance of marine environments (Wright et al., 2013; Abbasi et al., 2018). Recent research on MPs has validated that nearly 800 marine species have been contaminated with MPs via ingestion or entanglement (UN Report, 2016). They pose harm to the marine environment with the entanglement of various aquatic organisms, which are entangled with microplastics and are then ingested by fish, seabirds and aquatic mammals despite the large sizes of these plastics (Li et al., 2016; Rodrigue, 2017; Schnurr et al., 2018). The MPs, different additives incorporated with them during manufacturing processes and several environmental contaminants (dioxins, pesticides, polychlorinated biphenyls (PCBs) and heavy metals) that are absorbed over the course of aquatic life spans pose threats to the gastrointestinal tracts of such aquatic life, obstructing and damaging the digestive tract and causing starvation or malnutrition (Baalkhuyur et al., 2018).

The invasion of microplastics and their associated contaminants in aquatic environments give rise to potential risks to innumerable animals—especially fish and therefrom to human beings, where fish elicit several physiological problems (Tanaka et al., 2013; Foekema et al., 2013; Cannon et al., 2016; Naidoo et al., 2016; Ory et al., 2017, 2018; McGoran et al., 2017). The roles of microplastics have been identified as drug carriers in medicine and as facial cleansers in the cosmetic industry (Li et al., 2016). Plastic-derived substances, acting as marine pollutant vectors, are considerably resistant to heat and microbial degradation due to the development of increased adsorption capacity and environmental persistence (Deudero & Alomar, 2015; Avio et al., 2017).

Because microplastics contain noxious chemicals like methyl mercury, they can act as neurotoxicants, which damage brains in a way that causes Alzheimer disease, impair vision, develop allergies and result in kidney failure, depression and hypospermia (Thompson et al., 2009; Bhandari & Nakamura, 2016). Ater being ingested by a range of animals, microplastics are transferred trophically through food chains by following the processes of the intaking, accumulation and translocation of MPs, which depend on the physical and chemical properties of these plastic particles. Plastic debris has been found in the gut contents of seabirds, and other wildlife species has been found to be entangled with microplastics. The chemicals leached from plastics have toxic effects on the marine biota (Ryan, 2013), and seabirds have been profusely affected by plastic debris (Laist, 1987), and different species of wildlife, many of which are recognized as "red-listed" species, have been infected through their own feeding activities (Deudero & Alomar, 2015; Derraik, 2002). Different species in aquatic ecosystems have shown the effects of MPs: reduced feeding activity, increased oxidative stress and neurotoxicity (retarding growth) and impairments in reproduction, all of which ultimately lead to death (Besseling et al., 2015; Nelms et al., 2016; de Sá et al., 2018). Therefore, MPs have harmful effects on human health, disrupt the ecology of the ecosystems they are in and result in a loss of marine biodiversity, and all of these harmful effects can be tackled by undertaking large-scale awareness programmes among people residing near coastal regions about engaging in proper waste management by following regulatory and policy measures (UNCLOS, 1982; Schroeder, 2010; Van Cauwenberghe et al., 2013; Yao et al., 2019; Zhou et al., 2020; Kaly, 2004; Jonathan et al., 2021).

7.21.3.2 Plastics in the Gangetic Delta Region of the Bay of Bengal

Huge amounts of plastics and plastic-based products are reported to have accumulated in the riverbeds, river basins, flood plains, intertidal belts and open estuaries of the Ganga–Brahmaputra–Meghna deltaic systems of the Bengal region, posing severe threats to wildlife and humans through the food chain (Raha et al., 2021). The Gangetic delta region in lower Bengal, near the Bay of Bengal, is considered a major contributor of plastic waste through the Ganga–Brahmaputra–Meghna river basin network, representing the world's sixth-largest marine-polluting rivers, carrying 3,017,170 tonnes of plastics every year (Bandyopadhyay & Ghosh, 2016; Schmidt et al., 2017). The river basins of these rivers, cover nearly 1.75 million km^2 (Elayaperumal et al., 2019). The Ganga River, flowing through India and Bangladesh, represents the second-most-polluted river in the world, after the Yangtze River in China (Lebreton et al., 2017).

The ocean provides life-supporting commodities to more than three billion people whose land-based activities (urbanization, industrialization and the overexploitation of natural resources) result in the dumping of waste materials into waterbodies in the form of marine litter (Guzzetti et al., 2018). Over the years, the spaces in and around global maritime ecosystems are being used as the free dumping grounds of plastics and related persistent polymer products, causing pollution in 40% of the ocean and resulting in the profound depletion of fisheries and the loss of coastal habitats (UNDP, 2015). A large number of olive ridley sea turtles arrive in large numbers on the coast of Odisha, at the Bay of Bengal, every year, many of which ingest plastic fragments and get entangled with abandoned fishing nets, leading to their death (Paul et al., 2016).

Severe damage from the fishing, shipping and tourism industries has occurred because of the plastic-derived marine litter in the Asia-Pacific region, resulting in a loss of around USD 1.265 billion annually (Bruch et al., 2016). The majority of this marine litter is composed of plastics. The manufacturing, use and disposal of plastics have become major issues because of their direct and indirect connection to people and ecosystems around the globe (Villarrubia-Gómez et al., 2018; Windsor et al., 2019). The pollution of coastal environments by the macro-, meso- and micro-components of plastics have emerged as terrible threats to the marine environment and mainly derive from land-based human activities like the burning of fossil fuels, waste management, different recreational activities, plastic-dependent human lifestyles, domestic sewage, commercial fishing and industrial discharge (Lusher et al., 2013; Eriksen et al., 2014). Most of the biopolymers of plastics are biodegradable, and increasing concentrations of them induce alterations in the species composition of biotic communities in the marine ecosystem, causing the mortality of living biota (Moore, 2008; Cole et al., 2011; Deudero & Alomar, 2015; González Carman et al., 2015).

Coastal belts in the South Asian region experience a significant accumulation of plastics, where the Gangetic delta region pours out enormous amounts of plastic into the Bay of Bengal through the rivers of both India and Bangladesh (Gray, 2018; Jebaraj, 2019). The river basins and catchment areas constitute major sources and fluxes of plastic-based pollutants which are transported mainly via hydrological pathways depending on topography, sea surface temperature, wind speed direction, and salinity surface morphology and drainage patterns (Windsor et al., 2019; Rules, 2016; Yao et al., 2019; Zhou et al., 2020; Paray et al., 2025). The riverine networks of the Indian subcontinent, primarily the Ganges, Brahmaputra, and Meghna rivers in their course of water movement accrue large amounts of plastic that are ultimately released into the adjoining sea at the Bay of Bengal (Lebreton et al., 2017). The transboundary of the Ganges River, flowing through India and Bangladesh, accounts for releasing 0.10–0.17 million tonnes of plastics per year into the ocean, against the backdrop of discharging 1.15 and 2.41 million tonnes of plastic per year (1.15 and 2.41) from the rivers into the ocean (Lebreton et al., 2017). The Bay of Bengal, being the new hotspot of plastic litter, surpasses other seas of the Indian Ocean in terms of the accumulation of plastic debris that contains 40% films and 41% fragments (Sakhuja, 2016; Eriksen et al., 2014).

7.21.4 Pollution from Pesticides in the Sundarbans Mangrove Ecosystem

Pesticides represent a particular category of toxic chemicals that are used to eliminate or control undesirable organisms, mostly fungi, insects and rodents, which reduce crop production and cause health problems for livestock. Pesticides have been in use for many years, but since 1945, the rampant usage of synthetic and organic pesticides started mainly to serve two purposes: to control inset vectors and to ensure safety in crop production. Pesticides are persistent organic pollutants (POPs) and broad-spectrum toxicants such as organochlorine pesticides (OCPs), which accumulate in the food web and pose great risks in remote ecosystems to the ecosystem and human health. Some of the compounds are hazardous environmental hormones that disrupt the reproductive cycles of humans and wildlife. Some of the pesticides, being refractory in nature, bioaccumulate in food chains or build up in sediments, posing potential threats to biodiversity and human health. For example, dichlorodiphenyltrichloroethane (DDT), being a highly toxic poison, is capable of eradicating many living organisms within a short span of time. Furans and dioxins, similar kinds of toxic chemicals, are emitted because of combustion and emission processes, with toxic effects on humans and wildlife.

Pesticides, including organochlorine pesticides (OCPs), find their way into aquatic ecosystems through agricultural runoff, the discharge of industrial and sewerage wastewater, runoff from nonpoint sources, industrial effluents or accidental spills and wet and dry deposition. Several research studies on the distribution of organochlorine pesticidal residues in the Sundarbans mangrove-estuarine complex have revealed an erratic pattern of movement from persistent organic components, either top to bottom or vice versa, reflecting nonhomogenous inputs and having deleterious effects on aquatic life through biomagnification. Among several predominating pesticides, the impacts of dichlorodiphenyltrichloroethane (DDT), hexachlorocyclohexane isomers (HCHs), β-HCH and Dichlorodiphenyldichloroethane (DDE), the fractional residues of DDT and HCHs are much pronounced ecotoxicologically. Although fish are killed through exposure to pesticides in high concentrations, sublethal occurrences are more common. However, subacute exposure of aquatic life to pesticides for a long period tends to damage vital organs. The long-term exposure of the mangrove ecosystem of Sundarbans to the disposal of sewage and sludge will result in an increase in the accumulation of toxic pesticides in soils, water and the biota (Bolt, 2005; Aktar et al., 2009).

7.21.4.1 The Effect of Pesticides on Estuarine Ecosystems

Freshwater ecosystems act as major accumulators of persistent pesticides due to their proximity to agricultural fields and industrial belts and are the main contributors of pesticides to estuarine ecosystems as all those chemicals gradually travel from the streams and rivers to the estuaries. Ideal examples of persistent/

nondegradable pesticides are organochlorine insecticides (DDT and toxaphene), which persist for many years without undergoing any change but get accumulated within the bodies of different faunal components. The prevalence of a high lipid–water coefficient in a pesticide indicates a high potentiality for pesticides to be accumulated within living organisms that contain high percentages of lipids, and therefore, those living organisms are likely to take up lipophilic pesticides (several organochlorine insecticides) from their ambient environment (soil or water).

The harmful effects of pesticides cause changes in the stability of an ecosystem by disturbing community interactions. In addition, the reproduction of aquatic animals is affected directly by the toxic effect of pesticides, which is evident from the nonfunctioning reproduction of raptors and the thinning of the eggshells in several species of birds by severely affecting Ca metabolism through the impact of nondegradable organochlorine pesticides. Carnivorous birds in aquatic ecosystems suffer more than terrestrial predatory birds because the former feed more on fish and become more susceptible to the bioaccumulation of pesticides via aquatic food chains. However, the population densities of aquatic faunal invertebrates can recover their population loss more rapidly because of their higher rates of growth and flourishing than those of terrestrial species of birds and mammals, which experience slow rates of population increase. Moreover, in aquatic systems, the dilution of concentrations of pesticides is higher, and the possibility of the dispersal of pesticides is also higher due to the flow of water. Moreover, the duration of pesticide stress from a single exposure is regulated by two factors: the duration of persistence and the intensity of toxicity. The impacts of pesticides not only are manifested by disrupting ecosystem functioning by modifying pathways of energy through food chains and food webs and changing patterns of energy flow and nutrient cycling, reducing species diversity with alterations in species composition and causing undesirable changes in water, soil and air quality but also bring about changes in the stability and resilience of ecosystems.

7.21.5 The Toxicity of Oils and Petroleum Hydrocarbons and Their Effects on Mangrove Ecosystems

The pollution caused by oils and other hydrocarbon products from oil-exploration processes, petroleum production industries and accidental spills of oils severely destroys mangrove ecosystems (Mastaller, 1996). Mangrove soil and several biotic components (plants, algae, fauna and microbes) absorb large amounts of different pollutants, including oil, which cause changes in species compositions, disease and even the death of the biotic community in a mangrove ecosystem (Tattar et al., 1994). Accumulated oil within sediment causes defoliation in mangrove trees and inhibits the establishment and survivability of mangrove seedlings. Mangroves trap excess oil in the sediment of shorelines where mangroves grow but are adversely impacted by oil spills, affecting both land-based and offshore resources. Mangroves around the world have been affected by individual oil spills and chronic oil pollution from long-term exposure to storage tank and refinery discharges. Well-documented oil spills in mangrove areas illustrate some of the complexities and variability in the impacts and response options. Oil spills from the seaward side cause damage to aquatic fauna and seabirds (Qasim et al., 1988). Also, thousands of mechanized boats for carrying passengers and fishing are major sources of oil pollution (Paul, 2002a, 2002b). Moreover, two broad categories of petroleum hydrocarbon–derived synthetic compounds cause major impacts on mangrove ecosystems: (i) PCBs are a group of synthetic organic chemicals which are manufactured as lubricants and coolants for electrical equipment, but these chemical compounds were banned in the 1970s. However, the illegal usage of these chemicals, which are resistant to biological and chemical degradation, causes several problems, like cancer, fertility problems and nervous system impairment. PAHs, complex mixes of organic compounds, formed from the combustion of fossil fuels and as the byproducts of oil burning, cause cancer and breathing difficulties. However, several PAH compounds can be broken down over a period of weeks or months by microorganisms.

Petroleum hydrocarbons are discharged from sources such as marine operations, land-based discharges and atmospheric and natural inputs. Oils, after being discharged from municipal and industrial waste, oceanic dumping and atmospheric fallout, urban and river runoff, transportation, dry docking, tanker accidents, deblasting, etc., enter marine coastal zones and affect biodiversity through bioaccumulation, biotransformation and biomagnification. Upon entering marine-estuarine zones, a major fraction of petroleum hydrocarbons (PHCs) is removed via evaporation, accumulated in sediment, mixed up in water and transferred to the biota.

Mangroves are very susceptible to oil exposure as oil causes their mortality within a time period ranging from a few weeks to several months. Lighter oils are found to be more acutely toxic to mangrove seedlings and plants than heavier oils, which can result in physical smothering and coating impacts. Mangroves and associated biological components in intertidal environments at risk to oil spills are affected mostly in two principal ways: the physical effects of petroleum and the toxicological effects of petroleum. Mangroves, after being impacted by oils, suffer from the yellow colouration of leaves and defoliation, and their trees can die. More-tenuous responses include germination failure, the branching of pneumatophores, a reduction in canopy cover, an increased rate of mutation and increased sensitivity towards other stress factors. The duration of persistence in mangrove ecosystems acts as the determining attribute in assessing the intensity of harmful effects, where prolonged duration eliminates mangroves and results in permanent habitat loss. Physical suffocation and other physiological impairments caused by the impacts of oil pollution vary with the chemical properties of different oils and the duration of exposure.

Because mangroves are highly susceptible to oil pollution, its acute toxic effects causing mortality occur within a few weeks. Mangroves' complex series of physiological mechanisms enable them to survive in high-salinity and low-oxygen conditions in parts of intertidal belts that have been heavily impacted by stranded oil. To overcome the challenges of salt and oxygen, the leaves and submerged roots of mangroves are often far away from the oiled portions of the intertidal belts. The exacerbation of chronic or acute toxicities from the impact of oil leads to the physical smothering of mangrove plants, mostly *Avicennia* spp., which were oiled by oil spills during the Gulf War in 1991, when the surviving pneumatophores tended to develop branched secondary pneumatophores (Böer, 1993). Some studies have pointed out the impacts of toxicity by dispersed or undispersed oil to both mangroves and the associated invertebrate community by increasing the impacts on exposed benthic fauna while reducing the inherent toxicity of oil to mangroves (Lai, 1986). The large-scale mortality of mangroves results in considerable erosion followed by the subsidence of land because of the decomposition of the mangrove root mass.

To control oil pollution, chemical dispersants can be used with support from the turbulence of sea water, which will speed up this dispersion process. Such a natural process of dispersion tends to reduce the intensity of impacts on mangrove-based faunal components, including wildlife like reptiles, birds and mammals, and reduces the amount of floating oil that may reach sensitive nearshore and shoreline habitats.

Persistent organic pollutants (POPs) as synthetic chemicals demonstrate a high potential for long-range transport via water currents or even air movement and persist in the environment and in organisms through bioaccumulation and biomagnification, which has toxic effects on the biota, including humans (SC-POPs, 2009). Given the ecotoxicological potential of these chemicals, the Stockholm Convention on Persistent Organic Pollutants in 2001 adopted resolutions cautioning again the large-scale usage of such toxicants in a global treaty for protecting human health and other biodiversity components (SC-POPs, 2009).

A group of persistent organic pollutants (POPs such as dichlorodiphenyltrichloroethane (DDT), hexachlorocyclohexanes (HCHs), polychlorobiphenyls (PCBs), polychlorobenzenes (PCBzs), and polyaromatic hydrocarbons (PAHs)) have been found on surface sediment samples (0–5 cm) with grain sizes <63 mm along the ecologically stressed Hooghly River estuary in eastern India with a wide range of concentrations (ng/g dry weight) in the following decreasing order: P16PAHs (3.3–630) > P6DDTs (0.14–18.6) > P7PCBs (0.28–7.7) > P2PCBzs (0.01–1.3) > P5HCH (0.10–0.6). All these pollutants are derived from a combination of both pyrolytic and petrogenic sources of PAHs, inputs from weathered DDT and their degradation products in oxidizing environments mostly as industrial inputs over agricultural waste.

Several synthetic chemical compounds like dichlorodiphenyltrichloroethane (DDT), pentachlorobenzene (PeCB), hexachlorocyclohexanes (HCHs) and hexachlorobenzene (HCB), are included among organochlorinated pesticides (OCPs), which have been used as pesticides across the globe. Restrictions have been imposed on using all these compounds, especially by banning most of them, but due to their illegal uses, their inherent chemical properties, having long half-lives, have still been operating in different spheres of environmental matrices (SC-POPs, 2009). Among the organochlorine pesticides, DDT is still permitted for the eradication of mosquito vectors in malaria-prone countries like India, and DDT breaks down into its metabolites, DDE and DDD, upon reaching the environment (Ntow & Botwe, 2011; Botwe et al., 2017).

The Hooghly river estuary is a 260 km long distributary of the Ganges River, and it flows south through West Bengal, India. The Hooghly estuary (21_310Ne23_300N, 87_450E_88_450E) is a major offshoot of the Ganges River, and it claims to be ecologically very sensitive. The estuarine-mangrove complex of Sundarbans features fishing harbours, several human settlements along the banks of rivers and their tributaries and industries with multifarious production activities, from which a huge amount of discharge reaches estuarine networks as pollutants causing great concern about the sustainability of the sensitive mangrove ecosystem. A recent ecotoxicological research study on pesticides from the sediments in a Hooghly estuary has revealed the occasional adverse biological effects of DDTs exceeding the effect range low (ERL) and threshold effect level (TEL) where carcinogenic PAHs (HPAHs) made up 79% of the total PAHs (Mitra et al., 2018, 2019). Local people, particularly those living along the coastal belts in the vicinity of estuarine tracts, are more susceptible to pesticides and pesticidal residues as they have to spend a good part of a year there to maintain their daily livelihood, engaging in bathing, boating and boat cleaning, fishing and catching benthic organisms like intertidal crabs, gastropods and mollusks, mud skippers, etc.

Some research findings from the coastal belts of West Bengal, India, have shown that much higher levels of PHC were estimated in the offshore coastal water of Digha, a coastal tourist hub with visits from more than 50 lakh tourists each year. The concentrations of oils and their derivatives also varied, from 1.17 to 18.5 μg/L in the Hooghly estuary; from 0.15 to 2.33 μg/L in the Saptamukhi estuary; from 0.11 to 1.30 μg/L in the Matlah estuary and from 1.60 to 20.11 μg/L at Sandheads, a shoal 10 km south of Sagar Island. This level of oils and other hydrocarbon products is attributed to spillages during shipping operations and fishing. PHC values were at their highest concentration at the Haldia Port, but they exhibited a fluctuating trend mainly because of the occurrence of the maximum number of petroleum-degrading bacteria in the water there.

7.22 Pollution Abatement and Conservation Efforts

In view of the ongoing pollution threats to the sensitive, fragile and vulnerable but biologically productive mangrove ecosystem of Sundarbans, proper conservation strategies must be carried out with due emphasis on the restoration of ecodegraded mangrove belts and the simultaneous improvement of the tidal drainage artificially extending tide-fed channels or nullahs to push water into the inner parts of mangrove swamps and to contribute to the recovery of mangroves through afforestation programmes. Mangroves in the tropical coastal environment of India are mostly lost because of the nonjudicious utilization of coastal resources, the drastic conversion of land-use patterns and population escalation along the banks of estuaries and reclaimed islands (Blasco, 1984; Banerjee et al., 1989; Naskar & Guha Bakshi, 1987; Paul, 2002a, 2002b; Raha & Raju, 2016; Raha et al., 2021). However, restoring the fringe mangrove forests is difficult after the loss of mangroves by natural marine factors, sudden land subsidence and uplift due to tectonic-related geological events, catastrophic storms, soil erosion, the ongoing imposition of hydrological stress in the drainage catchment areas of river deltas with the consequent modification of coastal morphodynamics, sediment starvation, hypersalinity, eutrophication, encroachment from overwash sand fan lobes and shortages in freshwater discharges into coastal zones.

Conservation strategies must be more proactive in raising awareness in people about utilizing their abilities to mitigate, control or ameliorate the effects of human activities while adhering to the basic ecological principles that protect affected ecosystems. The timely detection of environmental degradation appears to be a prerequisite for understanding the problematic situation associated with the mortality of biological components and the drastic decline in their populations, both of which require appropriate remedial action to overcome such environmental consequences (Heald & Odum, 1970, 1975; Hossain et al., 2022; Macintosh et al., 2002). To abate the harmful effects of pollution and to understand and safeguard the intricate relationships and interconnectedness among the different structural components of the affected mangrove-estuarine ecosystem, environmental administrators must adhere to the basic operating principles of ecology (Flowcharts 7.1 and 7.2).

References

Abbasi, S., Soltani, N., Keshavarzi, B., Moore, F., Turner, A., & Hassanaghari, M. (2018). Microplastics in different tissues of fish and prawn from the Musa Estuary, Persian Gulf. *Chemosphere, 205*, 80–87.

Achara, M. S., Panigrahi, S. N., Satpathy, K. K., Sahu, G., Mohanty, A. K., & Panigrahy, R. C. (2014). Nutrient dynamics and seasonal variation of phytoplankton assemblages in the coastal waters of southwest Bay of Bengal. *Environmental Monitoring and Assessment, 186*(9), 5681–5695.

Agusa, T., Kunito, A., Sudaryanto, T., Monirith, S. K., Klap, A., & Iwata, H. (2007). Exposure assessment for trace elements from consumption of marine fish in Southeast Asia. *Environmental Pollution, 145*, 266–777.

Aktar, M. W., Paramasivam, M., Sengupta, D., Purkait, S., Ganguly, M., & Banerjee, S. (2009). Impact assessment of pesticide residues in fish of Ganga river around Kolkata in West Bengal. *Environmental Monitoring and Assessment, 157*(1–4), 97–104.

Aliabad, M. K., Nassiri, M., & Kor, K. (2019). Microplastics in the surface seawaters of Chabahar Bay, Gulf of Oman (Makran coasts). *Marine Pollution Bulletin, 143*, 125–133. https://doi.org/10.1016/j.marpolbul.2019.04.037

Alimi, O. S., Budarz, J. F., Hernandez, L. M., & Tufenkji, N. (2018). Microplastics and nanoplastics in aquatic environments: Aggregation, deposition, and enhanced contaminant transport. *Environmental Science & Technology, 52*(4), 1704–1724.

Allen, J. A. (1998). Mangroves as alien species: The case of Hawaii. *Global Ecology and Biogeography Letters, 7*, 61–71.

Allen, J. R. L., & Duffy, M. J. (1998). Medium-term sedimentation on high intertidal mudflats and salt marshes in the Severn Estuary, SW Britain: The role of wind and tide. *Marine Geology, 150*, 1–27.

Allen, G. P., Salomon, J. C., Bassoullet, P., Du Penhoat, Y., & De Grandpre, C. (1980). Effects of tides on mixing and suspended sediment transport in macrotidal estuaries. *Sedimentary Geology, 26*(1), 69–90.

Aller, R. C. (1982). The effects of macrobenthos on chemical properties of marine sediment and overlying water. In P. L. McCall & M. J. S. Tevez (Eds.), *Animal–sediment relations*. Plenum Press.

Aller, R. C. (1988). Benthic fauna and biogeochemical processes in marine sediments: The role of burrow structures. In T. H. Blackburn & J. Sorensen (Eds.), *Nitrogen cycling in coastal marine environments*. Wiley.

Aller, R. C. (2001). Transport and reactions in the bioirrigated zone. In B. P. Boudreau & B. B. Jørgensen (Eds.), *The benthic boundary layer: Transport processes and biogeochemistry*. Oxford University Press.

Alongi, D. M. (1987). Intertidal zonation and seasonality of meibenthos in tropical mangrove estuaries. *Marine Biology, 95*, 447–458.

Alpine, A. E., & Cloern, J. E. (1992). Trophic interactions and direct physical effects control phytoplankton biomass and production in an estuary. *Limnology and Oceanography, 27*, 946–955.

Annon. (2003). *Mangrove ecosystem: Biodiversity and its influence on the natural recruitment of selected commercially important finfish and shellfish species in fisheries*. National Agricultural Technology Project (NATP). Indian Council of Agriculture Research (ICAR). Principal Investigator: George, J.P.; Co-PI: Chakraborty, S.K. and Damroy, S.N. (pp. 1–514).

Anonymous. (1982a). *Basin, sub-basin inventory of water pollution: The Ganga basin, part I, the Youmuna sub-basin* (pp. 45–81). Central Board for the Prevention and Control of Water Oollution.

Anonymous. (1982b). *Comprehensive pollution survey and studies of Ganga river basin in West Bengal by Center for Study of Man and Environment* (pp. 34–42). Central Board for the Prevention and Control of Water Pollution.

Archer, D., & Rahmstorf, S. (2010). *The climate crisis—An introductory guide to climate change* (pp. 1–248). Cambridge University Press.

Ashton, E. C. (2002). Mangrove sesarmid crab feeding experiments in peninsular Malaysia. *Journal of Experimental Marine Biology and Ecology, 273*, 97–119.

Avio, C. G., Gorbi, S., & Regoli, F. (2017). Plastics and microplastics in the oceans: From emerging pollutants to emerged threat. *Marine Environmental Research, 128*, 2–11. https://doi.org/10.1016/j.marenvres.2016.05.012

Baalkhuyur, F. M., Bin Dohaish, E.-J. A., Elhalwagy, M. E. A., Alikunhi, N. M., Alsuwailum, A. M., Rostad, A., Coker, D. J., et al. (2018). Microplastics in the gastrointestinal tract of fishes along the Saudi Arabian Red Sea Coast. *Marine Pollution Bulletin, 131*, 407–415.

Balkis, N., Aksu, A., Okus, E., & Apak, R. (2009). Heavy metal concentrations in water, suspended matter, and sediment from Gökova Bay, Turkey. *Environmental Monitoring and Assessment, 167*, 359–370.

Bandyopadhyay, J., & Ghosh, N. (2016). Hydro-political dynamics and environmental security in the Ganges–Brahmaputra-Meghna basin. *Social Science, 24*, 1–25.

Banerjee, L. K., Sastry, A. R. K., & Nayar, M. P. (1989). *Mangroves in India: Identification manual* (p. 113). Botanical Survey of India.

Banik, P., Hossain, M. B., Nur, A. A. U., Choudhury, T. R., Liba, S. I., Yu, J., et al. (2022). Microplastics in sediment of Kuakata Beach, Bangladesh: Occurrence, spatial distribution, and risk assessment. *Frontiers in Marine Science, 9*.

Banik, P., Anisuzzaman, M., Bhattacharjee, S., Marshall, D. J., Yu, J., Nur, A. A. U., et al. (2024). Quantification, characterization and risk assessment of microplastics from five major estuaries along the northern Bay of Bengal coast. *Environmental Pollution, 342*, 123036.

Barbier, E. B. (2007). Valuing ecosystem services as productive inputs. *Economic Policy, 22*, 177–229.

Barbier, E. B., & Cox, M. (2003). Does economic development lead to mangrove loss? A cross–countryanalysis. *Contemporary Economic Policy, 21*, 418–432.

Barnes, D. K. A., Sands, C. J., Richardson, A., & Smith, N. (2019). Extremes in benthic ecosystem services; Blue carbon natural capital shallower than 1000 m in isolated, small, and young Ascension Island's EEZ. *Frontiers in Marine Science, 6*(663), 1–12. https://doi.org/10.3389/fmars.2019.00663

Bertness, M. D. (1985). Fiddler crab regulation of *Spartina alterniflora* production on a New England salt marsh. *Ecology, 66*, 1042–1055.

Besseling, E., Foekema, E. M., Van Franeker, J. A., Leopold, M. F., Kühn, S., Bravo Rebolledo, E. L., Heße, E., Mielke, L., IJzer, J., Kamminga, P., & Koelmans, A. A. (2015). Microplastic in a macro filter feeder: Humpback whale *Megaptera novaeangliae*. *Marine Pollution Bulletin, 95*, 248. https://doi.org/10.1016/j.marpolbul.2015.04.007

Bhakat, R. K., Chakravarty, G., Giri, S., & Chakraborty, S. K. (2004). Invasive species in Sundarbans Mangrove Ecosystem, India. In S. Ayyappan, D. S. Malik, R. Dhanze, & R. S. Chauhan (Eds.), *Fish diversity in protected habitats* (pp. 219–240). Natcon Publication.

Bhandari, B. B., & Nakamura, R. (2016). *The Bay of Bengal a Forgotten Sea* (pp. 295–310)

Bhattacharya, A. (1999). Embankments and their ecological impacts: A case study from the tropical low-lying coastal plains of the deltaic Sundarbans, India. In H. Grann & M. Vollmer (Eds.), *Large-scale constructions in coastal environments* (pp. 171–180). Springer.

Bhattacharya, A. (2000). Some geomorphic observations indicating shoreline variation in the coastal tract of West Bengal (India)—A case study around Bakkhali in Sunderbans. In *Proceedings of the International Quaternary Seminar on INQUA Shoreline* (pp. 32–37). Indian Ocean Sub-Commission, New Academic Publishers.

Bhattacharya, B. D., Nayak, D. C., Sarkar, S. K., Biswas, S. N., Rakshit, D., & Ahmed, M. K. (2015). Distribution of dissolved trace metals in coastal regions of Indian Sundarban mangrove wetland: A multivariate approach. *Journal of Cleaner Production, 5*(56), 233–243.

Biswas, H., Mukhopadhyay, S. K., De, T. K., Sen, S., & Jana, T. K. (2004). Biogenic control on the air water carbon dioxide exchange in the Sundarban mangrove environment, northeast coast of Bay of Bengal, India. *Limnology and Oceanography, 49*(1), 95–101.

Biswas, H., Dey, M., Ganguly, D., De, T. K., Ghosh, S., & Jana, T. K. (2009). Comparative analysis of phytoplankton composition and abundance over a two decade period at the land-ocean boundary of a tropical mangrove ecosystem. *Special Issue, Estuaries and Coasts*. https://doi.org/10.1007/s12237-009-9193-5

Blasco, F. (1975). *The mangroves of India* (Travaux de la section scientifique et technique n°14) (pp. 1–175). Institutional Francais de Pondichery.

Blasco, F. (1977). Outlines of ecology, botany and forestry of the mangals of the Indian subcontinent. In V. J. Chapman (Ed.), *Wet coastal ecosystems, ecosystems of the world* (Vol. 1, pp. 241–260). Elsevier Scientific.

Blasco, F. (1984). Climate factors and the biology of mangrove plants. In S. C. Snedaker & J. G. Snedaker (Eds.), *The mangrove ecosystem: Research methods* (pp. 18–35). UNESCO/SCOR.

Böer, B. (1993). Anomalous pneumatophores and adventitious roots of *Avicennia marina* (Forssk.) Vierh. Mangroves two years after the 1991 Gulf War oil spill in Saudi Arabia. *Marine Pollution Bulletin, 27*, 207–211.

Bolt, H. M. (2005). Vinyl chloride—A classical industrial toxicant of new interest. *Critical Reviews in Toxicology, 35*, 307–323. https://doi.org/10.1080/10408440490915975

Botwe, B. O., Kelderman, P., Nyarko, E., & Lens, P. N. L. (2017). Assessment of DDT, HCH and PAH contamination and associated ecotoxicological risks in surface sediments of coastal Tema Harbour (Ghana). *Marine Pollution Bulletin, 115*, 480–488.

Boyd, C. E. (1978). *Water quality in warm water fish ponds* (p. 359). Agril. Expt. Stn. Auburn Univ.

Bruch, C., Mengerink, K., Harrison, E., Flanagan, D., Carey, I., Casey, T., Davis, M., Hessami, E., Lombardi, J., Michelen, N., Parts, C., Rhodes, L., West, N., Yazykova, S., Savelli, H., Kreilhuber, A., Malvik, P., Beneke, P., Howe, A., ..., Wehrli, J. (2016). Marine litter legislation: A toolkit for policymakers 109.

Cannon, S. M., Lavers, J. L., & Figueiredo, B. (2016). Plastic ingestion by fish in the Southern Hemisphere: A baseline study and review of methods. *Marine Pollution Bulletin, 107*, 286–291.

Chakraborty, S. K. (1995). Aquaculture potential of mangrove ecosystem of Sundarbans, India. In *Proceedings of Seminar on Fisheries—A Multimillion Dollar Industry* (pp. 72–83).

Chakraborty, S. K. (1996). Aquaculture potential of mangrove ecosystem of Sundarbans, India. In *Proceedings of Seminar on Fisheries—A Multimillion Dollar Industry* (pp. 72–83). Madras: AFI.

Chakraborty, S. K. (1998). Coastal aquaculture and environment in the context of West Bengal, India. *Vidyasagar University Journal of Biological Science, 4*, 13–20.

Chakraborty, S. K. (2003). Biodiversity and conservation. In A. D. Mukhopadhyay (Ed.), *Perspectives and issues in environmental studies* (pp. 267–286). Vidyasagar University.

Chakraborty, S. K. (2009). Biological diversity of Mindapore Coast, West Bengal. *Environmental Forensics, 2*(2), 6.

Chakraborty, S. K. (2010). Coastal environment of Midnapore, West Bengal: Potential threats and management. *Journal of Coastal Environment, 1*(1), 28–40.

Chakraborty, S. K. (2011). Mangrove ecosystem of Sundarbans, India: Biodiversity, ecology, threats and conservation. In J. N. Metras (Ed.), *Mangroves: Ecology, biology and taxonomy* (pp. 83–112). NOVA Publishers.

Chakraborty, S. K. (2013). Interactions of environmental variables determining the biodiversity of coastal mangrove ecosystem of West Bengal, India. *The Ecoscan Special Issue, 3*, 251–265.

Chakraborty, S. K. (2017). Ecological services on intertidal benthic fauna and the sustenance of coastal wetlands along the Midnapore(East) coast, West Bengal, India. In *Coastal wetlands: Alteration and remediation* (pp. 777–886). Springer.

Chakraborty, S. K. (2018). Bioinvasion and environmental perturbation: Synergistic impact on coastal—Mangrove ecosystems of West Bengal, India. In C. Makowski & C. Finkl (Eds.), *Impacts of invasive species on coastal environments: Coast in crisis* (pp. 171–245). Springer.

Chakraborty, S. K. (2021a). *Riverine ecology (Vol. 1): Ecofunctionality of the physical environment of riversresolution*. Springer Nature. 917 pages.

Chakraborty, S. K. (2021b). *Riverine ecology (Vol. 2): Biodiversity conservation, conflicts and resolution*. Springer Nature. 566 pages.

Chakraborty, S. K., & Choudhury, A. (1989). Ecological studies on the zonation of brachyuran crabs in a virgin mangrove island of Sundarbans, India. *Journal of the Marine Biological Association of India, 34*(1&2), 189–194.

Chakraborty, S. K., & Choudhury, A. (1992). Population ecology of fiddler crabs (*Uca* spp.) of the mangrove-estuarine complex of Sundarbans, India. *Tropical Ecology, 33*(1), 78–88.

Chakraborty, S. K., & Choudury, A. (1985). Distribution of fiddler crabs in Sundarbans mangrove esturarine complex, India. The mangroves. In *Proceedings of National Symposium Biology Utilization Conservation of Mangroves* (pp. 467–472).

Chakraborty, S. K., & Chowdhury, A. (1993). Community organization of macrobenthic molluses of Sagar Island, Sundarbans, India. *Indian Biologist, 25*(1), 33–36.

Chakraborty, S. K., & Choudhury, A. (1994a). Community structure of macrobenthic polychaetes of intertidal region of Sagar Island, Hooghly Estuary, Sundarbans, India. *Tropical Ecology, 35*(1), 97–104.

Chakraborty, S. K., & Choudhury, A. (1994b). Population ecology of *Metaplax intermedia* (Brachyura: Grapsidae) of SagarIsland, Sundarbans, India. *Proceedings of Zoological Society, Calcutta, 47*(1), 41–45.

Chakraborty, S. K., Choudhury, A., & Deb, M. (1986). Decapod Brachyura from Sundarbans mangrove-estuarine complex, India. *Journal of Bengal Natural History Society, 5*(1), 55–68.

Chakraborty, S. K., Chatterjee, S., Dey, M. K., Bhakat, R. K., & Paul, A. K. (2012). Ecological changes of an eco-rehabilitated delta of Hooghly Estuary. *Journal of Coastal Environment, 3*(1), 9–27.

Chakraborty, S. K., Paul, A. K., Bandyopadhyay, J., Banerjee, D., Pakhira, H., Middya, S., Jana, S., Sahoo, P., Paria, K., & Bera, K. (2013). Sustainable water management in the drought prone riverine tracts of south West Bengal. *Journal of Ground Water Studies, 1*, 11–34.

Chakraborty, S. K., Maiti Dutta, S., Ghosh, P. B., Ray, R., & Paul, A. K. (2014). Impact of global warming on Sundarbans mangrove ecosystem, India: Role of different assessment tools from ecosystem monitoring to molecular markers. In *Proceedings of the International Conference on Green India: Strategic Knowledge for Combating Climate Change: Prospects and Challenges* (pp. 181–200).

Chakraborty, S. K., Sanyal, P., & Ray, R. (2023). *Wetlands ecology: Eco-biological uniqueness of a Ramsar site*. East Kolkata Wetlands, India. 709 pages.

Chakravarty, G., Giri, S., Giri, S., & Chakraborty, S. K. (2004). Distribution of coliform bacteria in different estuaries of West Bengal Coast, India. In A. Kumar, J. R. Dhanze, A. K. Chopra, & D. S. Malik (Eds.), *Protected habitats and biodiversity* (p. 395). Natcon Publication.

Chatterjee, S., & Chakraborty, S. K. (2014). Feeding behaviour and functional role of some selected species of brachyuran crabs in nutrient cycle of coastal belt of Midnapore (East), West Bengal, India. *Journal of Biology & Life Sciences, 5*(1), 106–129.

Chatterjee, S., Das, T. K., & Chakraborty, S. K. (2014). Reproductive biology and bioturbatory activities of two sympatric species of fiddler crabs Uca lactea annulipes and Uca triangularis Bengali (Decopoda: Ocypodidae) at the East Midnapore Coastal belt of West Bengal, India. *Journal of Biology & Life Sciences, 5*(2), 106–126.

Cloern, J. E. (1996). Phytoplankton bloom dynamics in coastal ecosystem: A review with some general lessons from sustained investigation of San Francisco Bay, California. *Reviews of Geophysics, 34*(2), 127–168.

Cloern, J. E., Alpine, A. E., Cole, B. E., Wong, R. L., Arthur, J. F., & Ball, M. D. (1983). River discharge controls phytoplankton dynamics in the northern San Francisco Bay Estuary. *Estuarine, Coastal and Shelf Science, 16*, 415–429.

Cloern, J. E., Krantz, T., Hogan, C. M., & Duffy, J. E. (2007a). Eutrophication. In C. J. Cleveland (Ed.), *Encyclopedia of Earth*. Environmental Information Coalition, National Council for Science and the Environment.

Cloern, J. E., Jassby, A. D., Thompson, J. K., & Hieb, K. A. (2007b). A cold phase of the East Pacific triggers new phytoplankton blooms in San Francisco Bay. *Proceedings of the National Academy of Sciences, 104*(47), 18561–18565.

Coblentz, B. E. (1990). Exotic organisms: A dilemma for conservation biology. *Conservation Biology, 4*(3), 261–265.

Cole, M., Lindeque, P., Halsband, C., & Galloway, T. S. (2011). Microplastics as contaminants in the marine environment: A review. *Marine Pollution Bulletin, 62*, 2588–2597. https://doi.org/10.1016/j.marpolbul.2011.09.025

Colwell, R. R. (1996). Global climate and infectious disease: The cholera paradigm. *Science, 274*, 2025–2031.

Connell, J. A. (1961). The influence of interspecific competition and other factors on the distribution of the barnacle, Chthalamus stellatus. *Ecology, 42*, 710–723.

Connell, J. H., & Slatyer, R. O. (1977). Mechanisms of succession in natural communities and their role in community stability and organization. *The American Naturalist, 111*, 1119–1144.

Corcoran, P. L., Norris, T., Ceccanese, T., Walzak, M. J., Helm, P. A., & Marvin, C. H. (2015). Hidden plastics of Lake Ontario, Canada and their potential preservation in the sediment record. *Environmental Pollution, 204*, 17–25. https://doi.org/10.1016/j.envpol.2015.04.009

Coulombier, T., Neumeier, U., & Bernatchez, P. (2012). Sediment transport in a cold climate salt marsh (St. Lawrence Estuary, Canada), the importance of vegetation and waves. *Estuarine, Coastal and Shelf Science, 101*, 64–75.

Cozar, A., Echevarria, F., Gonzalez-Gordillo, I., Irigoien, X., et al. (2014). Plastic debris in the open ocean. *Proceedings of the National Academy of Sciences of the United States of America, 111*(28), 10239–10244.

Crooks, J. A., & Khim, H. S. (1999). Architectural vs biological effects of a habitat alluring, exotic mussel, Musculista senhousia. *Journal of Experimental Marine Biology and Ecology, 240*, 53–75.

Dahadough-Guebas, F., Hettiarachchi, S., Leo Seen, D., Batelaan, O., Sooriyarachchi, S., Jayatissa, L. P., & Koedam, N. (2005). Transition in ancient inland freshwater resource management in Sri Lanka affect biota and human populations in and around coastal lagoons. *Current Biology, 15*, 579–586.

Das, G. K. (2008). Evidences of some erosional features along the river bank of Sundarbans, India. *Geographical Review of India, 70*(2), 139–145.

Das, G. K. (2009a). Beach processes of Tidal Islands of Hugli Estuary, West Bengal. *Geographical Review of India, 71*(3), 240–248.

Das, G. K. (2009b). Beach erosion for crab-burrowing. *Indian Science Cruiser, 23*(1), 58–59.

Das, G. K. (2009c). Grain size analysis of some beach sands from the Indian Coasts. *Geographical Review of India, 71*(1), 10–18.

Das, G. K., & Bhattacharya, A. (1999). Rampant exploitation of post-larval penaeid prawn (Penaeus mondon) from the salt-water course of the deltaic Sundarbans, West Bengal—A threat to the balance of nature. *Journal of the Indian Society of Coastal Agricultural Research, 17*(1&2), 219–222.

Das, T., Nayak-Tripathy, S., Midya, S., Chakravorty, P. P., Chakraborty, S. K., Pal, A. K., & Dalvi, R. S. (2025). Histological changes in the vital organs of *Labeo rohita* acclimated to increased temperatures and exposed to critical temperatures. *Environment and Ecology, 43*(1), 122–128.

Das, T., Pal, A. K., Chakraborty, S. K., Manush, S. M., Dalvi, R. S., Sharma, S., & Mukherjee, S. C. (2005). Thermal dependence of embryonic development and hatching rate in Labeo rohita (Hamilton, 1822). *Aquaculture, 255*, 536–541.

DasGupta, R., & Shaw, R. (2013). Changing perspectives of mangrove management in India—An analytical overview. *Ocean and Coastal Management, 80*, 107–118.

Dasgupta, S., Laplante, B., Meisner, C., Wheeler, D., & Yan, J. (2009). The impact of sea level rise on developing countries: Acomparative analysis. *Climatic Change, 93*, 379–388.

Dasgupta, S., Huq, M., & Wheeler, D. (2014). *Facing the hungry tide: Climate change, livelihood threats and household responses incoastal Bangladesh*. Policy Research Working Paper (forthcoming), Development Research Group, World Bank.

David, M., & Gollasch, S. (2015). Introduction. In *Global maritime transport and ballast water management, invading nature* (Springer series in invasion ecology) (Vol. 8, p. 306). Springer.

David, I. G., Matache, M. L., Tudorache, A., Chisamera, G., Rozylowicz, L., & Radu, G. L. (2012). Food chain biomagnification of heavy metals in samples from the Lower Prut Floodplain Natural Park. *Environmental Engineering and Management Journal, 11*, 69–73.

Davis, J. H. (1938). Mangroves makers of land. *Nature, 31*, 551–553.

Debata, S., Panda, R. M., & Palita, S. K. (2020). Chiropteran diversity and the key determinants of their distribution in Eastern Ghats, India. *Biodiversity and Conservation, 28*, 2385–2404.

de Sá, L. C., Oliveira, M., Ribeiro, F., Rocha, T. L., Futter, M. N., et al. (2018). Studies of the effects of microplastics on aquatic organisms: What do we know and where should we focus our efforts in the future? *Science of the Total Environment, 645*, 1029–1039. https://doi.org/10.1016/j.scitotenv.2018.07.207

DePaola, A. (1981). Vibrio cholerae in marine foods and environmental waters: A literature review. *Journal of Food Science, 46*, 66–70.

Derraik, J. G. B. (2002). The pollution of the marine environment by plastic debris: A review file. *Marine Pollution Bulletin, 44*, 842–852. ///C:/Users/Erica Parenti/Desktop/rimandi ARTICOLI/1-s2.0-S0025326X8780019X-main.pdf file:///C:/Users/Erica Parenti/Desktop/rimandi ARTICOLI/1-s2.0-0025326X82900388-main.pdf.

Deudero, S., & Alomar, C. (2015). Mediterranean marine biodiversity under threat: Reviewing influence of marine litter on species. *Marine Pollution Bulletin, 98*, 58–68. https://doi.org/10.1016/j.marpolbul.2015.07.012

Dev Roy, M. K., & Das, A. K. (2000). Taxonomy, ecobiology and distribution pattern of Brachyura

Duan, J., Han, J., Cheung, S. G., Chong, R. K. Y., Lo, C. M., Lee, F. W. F., et al. (2021). How mangrove plants affect microplastic distribution in sediments of coastal wetlands: Case study in Shenzhen Bay, South China. *Science of the Total Environment, 767*, 144695.

Dukes, J. S., & Mooney, H. A. (1999). Does global change increase the success of biological invaders? *Trends in Ecology & Evolution, 14*, 135–139.

Elayaperumal, V., Hermes, R., & Brown, D. (2019). An ecosystem based approach to the assessment and governance of the Bay of Bengal Large Marine Ecosystem. *Deep Sea Research Part II: Topical Studies in Oceanography, 163*, 87–95. https://doi.org/10.1016/j.dsr2.2019.01.001

Eppley, R. W., & Peterson, B. J. (1979). Particulate organic matter flux and planktonic new production in the deep ocean. *Nature, 282*, 677–680.

Eriksen, M., Lebreton, L. C. M., Carson, H. S., Thiel, M., Moore, C. J., Borerro, J. C., Galgani, F., Ryan, P. G., & Reisser, J. (2014). Plastic pollution in the world's oceans: More than 5 trillion plastic pieces weighing over 250,000 tons afloat at sea. *PLoS One, 9*, 1–15. https://doi.org/10.1371/journal.pone.0111913

FAO. (1983). *Compilation of legal limits for hazardous substances in fish and fishery products. FAO Fishery Circular No. 464* (pp. 5–100). FAO.

Fensham, R. J., & Cowie, I. D. (1998). Alien plant invasions on the Tiwi Islands. Extent, implications and priorities for control. *Biological Conservation, 83*, 55–68.

Fleming, M., Lin, G., & Sternberg, L. (1990). Influence of mangrove detritus in an estuarineecosystem. *Bulletine of Marine Science, 47*, 663–669.

Foekema, E. M., De Gruijter, C., Mergia, M. T., van Franeker, J. A., Murk, A. J., & Koelmans, A. A. (2013). Plastic in North sea fish. *Environmental Science & Technology, 47*, 8818–8824.

Ganguly, A., & Chakraborty, S. K. (2026). Synergistic impact of urbanization of Kolkata Metropolitan and East Kolkata Wetlands on the mangrove-estuarine complex of Sundarbans, India. In *Wetlands and urban resilience: A global outlook* (Sustainable Development Goals Series) (pp. 375–398). Springer Nature.

Ganguly, D., Dey, M., Mandal, S. K., De, T. K., & Jana, T. K. (2008). Energy dynamics and its implication to biosphere–atmosphere exchange of CO_2, H_2O and CH_4 in a tropical mangrove forest canopy. *Atmospheric Environment, 42*, 4172–4184.

Ghosh, A. K. (2008). Biodiversity conservation: Issues and concern. *Zoological Research in Human Welfare, 2*, 19–22.

Ghosh, P. B., & Choudhury, A. (1989). Cu, Zn and Pb in the sediments of Hooghly Estuary. *Environment and Ecology, 7*, 427–430.

Ginsberg, R. N., & Lowenstam, H. A. (1958). The influence of marine bottom communities on the depositional environment of sediments. *Journal of Geology, 66*, 310–318.

Ginsberg, S. S., & Perillo, G. M. E. (2004). Characteristics of tidal channels in a mesotidal estuary of Argentina. *Journal of Coastal Research, 20*(2), 489–497.

Giri, C., Zhu, Z., Tieszen, L. L., Singh, A., Gillette, S., & Kelmelis, J. A. (2008). Mangrove forest distributions and dynamics (1975–2005) of the tsunami-affected region of Asia. *Journal of Biogeography, 35*(3), 519–528.

González Carman, V., Machain, N., & Campagna, C. (2015). Legal and institutional tools to mitigate plastic pollution affecting marine species: Argentina as a case study. *Marine Pollution Bulletin, 92*, 125–133. https://doi.org/10.1016/j.marpolbul.2014.12.047

Gopinath, G., & Seralathan, P. (2005). Rapid erosion of the coast of Sagar Island, W.B. India. *Environmental Geology, 48*, 1058–1067.

Gray, A. (2018). *90% of plastic polluting our oceans comes from just 10 rivers* [www.document]. World Economic Forum. https://www.weforum.org/Agenda/2018/06/90-of-plastic-polluting-our-oceans-comes-from-just-10-rivers/

Guezennec, L., Lafite, R., Dupont, J. P., Meyer, R., & Boust, D. (1999). Hydrodynamics of suspended particulate matter in the tidal freshwater zone of a macrotidal estuary (the Seine Estuary, France). *Estuaries, 22*(3), 717–727.

Gupta, N., Das, S., & Basak, U. C. (2007). Useful extracellular acivities of bacteria isolated from Bhitarkanika mangrove ecosystem of Orissa coast. *Malaysian Journal of Microbiology, 3*, 15–18.

Guzzetti, E., Sureda, A., Tejada, S., & Faggio, C. (2018). Microplastic in marine organism: Environmental and toxicological effects. *Environmental Toxicology and Pharmacology, 64*, 164–171. https://doi.org/10.1016/j.etap.2018.10.009

Hajeb, P., Jinap, S., Ismail, A., Fatimah, A. B., Jamilah, J., & Rahim, M. A. (2009). Assessment of mercury level in commonly consumed marine fishes in Malaysia. *Food Control, 20*, 79–84.

Hall, J. E. (2014). Bioconcentration, bioaccumulation, and the biomagnification in Puget sound biota: Assessing the ecological risk of chemical contaminants in Puget sound. *Tahoma West Literary Arts Magazine, 8*, 40–51.

Harris, L., & Tyrrell, M. (2001). Changing community states in the Gulf of Maine: Synergism between invaders, overfishing and climate change. *Biological Invasions, 3*, 9–21.

Hazra, S., Ghosh, T., Dasgupta, R., & Sen, G. (2002). Sea level and associated changes in the Sundarbans. *Science and Culture., 68*(9–12), 309–321.

He, Q., Qiu, Y., Liu, H., Sun, X., Kang, L., Cao, L., Li, H., & Ai, H. (2017). New insights into the impacts of suspended particulate matter on phytoplankton density in a tributary of the Three Gorges Reservoir, China. *Scientific Reports, 7*, 13518. https://doi.org/10.1038/s41598-017-13235-0

Heald, E. J., & Odum, W. E. (1970). The contribution of mangrove swamps to Florida Fisheries. *Proceedings of the Gulf and Caribbean Fisheries Institute, 22*, 130–135.

Heald, E. J., & Odum, W. E. (1975). The contribution of mangrove swamps to Florida Fisheries. *Proceedings of the Gulf and Caribbean Fisheries Institute, 22*, 130–135.

Helcoski, R., Yonkos, L. T., Sanchez, A., & Baldwin, A. H. (2020). Wetland soil microplastics are negatively related to vegetation cover and stem density. *Environmental Pollution, 256*, 113391.

Holling, C. S. (1981). Forest insects, forest fires and resistance. In H. Mooney, T. M. Bonnicksen, N. L. Christensen, J. E. Lotan, & W. A. Reiners (Eds.), *Fire regimes and ecosystem properties. USDA Forest Service, general technical report WO–26* (pp. 445–464). US Department of Agriculture.

Hossain, M. J., AftabUddin, S., Akhter, F., Nusrat, N., Rahaman, A., Sikder, M. N. A., et al. (2022). Surface water, sediment, and biota: The first multi-compartment analysis of microplastics in the Karnafully river, Bangladesh. *Marine Pollution Bulletin, 180*, 113820.

Hossain, M. B., Yu, J., Nur, A. A. U., Banik, P., Jolly, Y. N., Al-Mamun, M., et al. (2023a). Distribution, characterization and contamination risk assessment of microplastics in the sediment from the world's top sediment-laden estuary. *Journal of Environmental Management, 344*, 118472.

Hossain, M. B., Yu, J., Nur, A. A. U., Banik, P., Jolly, Y. N., Al-Mamun, M., et al. (2023b). Microplastics in surface water from a mighty subtropical estuary: First observations on occurrence, characterization, and contamination assessment. *Environmental Research, 226*, 115594.

Huang, Y. Z., Xiao, X., Xu, C. C., Perianen, Y. D., Hu, J., & Holmer, M. (2020). Seagrass beds acting as a trap of microplastics—Emerging hotspot in the coastal region? *Environmental Pollution, 257*, 113450.

Hughes, R. G. (2004). Climate change and loss of saltmarshes: Consequences for birds. *Ibis, 146*(Suppl 1), 21–28.

Iftekhar, M. S. (1999). *Vegetation dynamics in the Sundarbans and the contribution of salinity between 1985–1995* [B.Sc. thesis, Forestry and Wood Technology Discipline, Khulna University, Khulna, Bangladesh] (pp. 1–54.

IPCC. (2000). In N. Nakicenovic & R. Swart (Eds.), *Special report on emissions scenarios: A special report of Working Group III of the fourth assessment report of the Intergovernmental Panel on Climate Change*. Cambridge University Press.

IPCC (Intergovernmental Panel on Climate Change). (2007). *Climate change 2007: The physical science basis. Contribution of Working Group I to the fourth assessment report of the IPCC*. Cambridge University Press.

Jebaraj, P. (2019) *70% towns along Ganga let out garbage directly into the river: Study* [www. document]. https://www.thehindu.com/sci-tech/energyand-environment/70-towns-along-ganga-let-out-garbage-directly-into-theriver/article25981284.ece

Jiménez, & Lugo. (1985).

Jonathan, M. P., Sujitha, S. B., Rodriguez-Gonzalez, F., Elizabeth, L. C. V., Hern'andez-Camacho Claudia, J., & Sarkar, S. K. (2021). Evidences of microplastics in diverse fish species off the Western Coast of Pacific Ocean, Mexico. *Ocean and Coastal Management, 204*, 105544.

Jung, S., Cho, S. H., Kim, K. H., & Kwon, E. E. (2021). Progress in quantitative analysis of micro-plastics in the environment: A review. *Chemical Engineering Journal, 422*, 130154.

Kaly, U. L. (2004). Review of land-based sources of pollution to the coastal and marine environments in the BOBLME Region. Theme Rep. 10, 1. GCP/RAS/179/WBG.

Karbassi, A. R., Tajziehchi, S., & Afshar, S. (2015). An investigation on heavy metals in soils around oil field area. *Global Journal of Environmental Science and Management, 4*(1), 275–282.

Kaysner, C. A., & Hill, W. E. (1994). Toxigenic Vibrio cholerae O1 in food and water. In K. I. Wachsmuth, P. A. Blake, & O. Olsvik (Eds.), *Vibrio cholerae and cholera: Molecular to global perspectives* (pp. 27–39). American Society for Microbiology.

Kennedy, V. S. (1990). Anticipated effects of climate change on estuarine and coastal fisheries. *Fisheries, 15*, 16–25.

Kumar, K. S., Sahu, M. K., & Kathiresan, K. (2005). Isolation and characterization of Streptomycetes producing antibiotic, from a mangrove environment. *Asian Journal of microbiology, Biotechnology and environmental Sciences, 7*, 457–464.

Lai, H. C. (1986). Effects of oil on mangrove organisms. In J. L. Maclean, L. B. Dizon, & L. V. Hosillos (Eds.), *Proceedings of the First Asian Fisheries Forum* (pp. 285–288).

Laist, D. (1987). Overview of the biological effects of lost and discarded plastic debris in the marine environment. *Marine Pollution Bulletin, 18*, 319–326. https://doi.org/10.1016/S0025-326X(87)80019-X

Lam, P. J., Twining, B. S., Jeandel, C., Roychoudhury, A., Resing, J. A., Santschi, P. H., & Anderson, R. F. (2015). Methods for analyzing the concentration and speciation of major and trace elements in marine particles. *Progress in Oceanography, 133*, 32–42.

Lara, R. J. (2011). Mangroves in the change: The need of new synthesis of climate, ecohydrology and biomedical research. In J. N. Metras (Ed.), *Mangrove ecology, biology and taxonomy* (pp. 291–299). Nova Science Publications.

Lebreton, L. C. M., Van Der Zwet, J., Damsteeg, J. W., Slat, B., Andrady, A., & Reisser, J. (2017). River plastic emissions to the world's oceans. *Nature Communications, 8*, 1–10. https://doi.org/10.1038/ncomms15611

Lebreton, L., Slat, B., Ferrari, F., Sainti-Rose, B., Aitken, J., et al. (2018). Evidence that the Great Pacific Garbage Patch is rapidly accumulating plastics. *Scientific Reports, 8*, 4666.

Li, W. C., Tse, H. F., & Fok, L. (2016). Plastic waste in the marine environment: A review of sources, occurrence and effects. *Science of the Total Environment, 566–567*, 333–349. https://doi.org/10.1016/j.scitotenv.2016.05.084

Lugo, A. E. (1980). Mangrove ecosystems successional or steady state? *Biotropica, 12*(2), 65–72.

Lusher, A. L., McHugh, M., & Thompson, R. C. (2013). Occurrence of microplastics in the gastrointestinal tract of pelagic and demersal fish from the English Channel. *Marine Pollution Bulletin, 67*, 94–99. https://doi.org/10.1016/j.marpolbul.2012.11.028

Macintosh, D. J., Ashton, E. C., & Havanon, S. (2002). Mangrove rehabilitation and intertidal biodiversity: a study in the Ranong mangrove ecosystem, Thailand. *Estuarine, Coastal and Shelf Science, 55*, 331–345.

Maheim, B. S. J. (1988). Annex V of the MARPOL convention: Will it stop marine plastic pollution. *Georgetown International Environmental Law Review, 71*, 1.

Maiti Dutta, S., Banerjee Mustafi, S., Raha, S., & Chakraborty, S. K. (2014). Assessment of thermal stress adaptation by monitoring Hsp70 and MnSOD in the freshwater gastropod, Bellamya bengalensis (Lamark, 1882). *Environmental Monitoring and Assessment, 186*(12), 8961–8967.

Massel, S. R. (1999). Tides and waves bin mangrove forests. In *Fluid mechanics for marine ecologists* (pp. 418–425). Springer.

Mastaller, M. (1996). *Destruction of mangrove wetlands—Causes and consequences. A biannual collection titled natural resources and development—Focus: Mangrove forests*. Institute for Scientific Cooperation. 43/44.

Mazda, Y., & Wolanski, E. (2009). Hydrodynamics and modeling of water flow in mangrove areas. In *Coastal wetlands: An integrated ecosystem approach*. Elsevier.

Mazda, Y., Wolanski, E., & Ridd, P. V. (2007). *The role of physical processes inmangrove environments: Manual for the preservation and utilization of mangrove ecosystems* (p. 598). Tokyo.

McGoran, A., Clark, P., & Morritt, D. (2017). Presence of microplastic in the digestive tracts of European flounder, *Platichthys flesus*, and European smelt, *Osmerus eperlanus*, from the River Thames. *Environmental Pollution, 220*, 744–751.

Middelburg, J. J., & Herman, P. M. J. (2007). Organic matter processing in tidal estuaries. *Marine Chemistry, 106*, 127–147.

Midya, S., Ganguly, R. K., Islam, S. S., & Chakraborty, S. K. (2019). Attachment and antimicrobial susceptibility of bacterial associates of zooplanktonic copepod: Lesson for environmental safety. *Ecotoxicology and Environmental Safety, 180*, 749–755.

Millennium Ecosystem Assessment. (2005). *Ecosystem and human well-being, synthesis*. Island Press.

Mitra, A., Pramanik, A., Sarkar, S., Meyyappan, A., Mitra, S., Senthil, S., & Mukherjee, J. (2011). Applications of the microbial biodiversity of the mangroves. In J. N. Metras (Ed.), *Mangrove ecology, biology and taxonomy* (pp. 225–240). Nova Science Publications.

Mitra, S., Ghosh, S., Satpathy, K. K., Bhattacharya, B. D., Sarkar, S. K., Mishra, P., & Raja, P. (2018). Water quality assessment of the ecologically stressed Hooghly River Estuary, India: A multivariate approach. *Marine Pollution Bulletin, 126*, 592–599.

Mitra, S., Corsolini, S., Pozo, K., Audy, O., Sarkar, S. K., & Biswas, J. K. (2019). Characterization, source identification and risk associated with polyaromatic and chlorinated organic contaminants (PAHs, PCBs, PCBzs and OCPs) in the surface sediments of Hooghly Estuary, India. *Chemosphere, 221*, 154–165.

Mitra, S., Sudarshan, M., Jonathan, M. P., Sarkar, S. K., & Thakur, S. (2020). Spatial and seasonal distribution of multi-elements in suspended particulate matter (SPM) in tidally dominated Hooghly River Estuary and their ecotoxicological relevance. *Environmental Science and Pollution Research, 27*, 12658–12672.

Moore, C. J. (2008). Synthetic polymers in the marine environment: A rapidly increasing, long-term threat. *Environmental Research, 108*, 131–139. https://doi.org/10.1016/j.envres.2008.07.025

Mukhopadhyay, S. K., Biswas, H., De, T. K., & Jana, T. K. (2006). Fluxes of nutrients from the tropical River Hooghly at the land–ocean boundary of Sundarban, NE coast of Bay of Bengal, India. *Journal of Marine Systems, 62*, 9–21.

Naidoo, T., Smit, A., & Glassom, D. (2016). Plastic ingestion by estuarine mullet *Mugil cephalus* (*Mugilidae*) in an urban harbour, KwaZulu-Natal, South Africa. *African Journal of Marine Science, 38*, 145–149.

Nasehi, F., Hassani, A. H., Monavari, S. M., Karbassi, A. R., & Khorasani, N. (2013). Evaluating the metallic pollution of riverine water and sediments: A case study of Aras River. *Environmental Monitoring and Assessment, 18*(5), 197–203.

Naskar, K. R., & Guha Bakshi, D. N. (1987). *Mangrove swamps of Sundarbans—An ecological perspectives* (pp. 1–264). Naya Prakash.

Naskar, K., Sen Sarkar, N., Ghosh, A., Dasgupta, M., & Sengupta, B. (2004). *Status of the mangroves and mangrove ecosystem of Sundarbans in West Bengal: Its impact on estuarine wet lands fisheries* (Bulletine No. 134 CIFRI, Barrackpore, Kolkata) (pp. 1–40)

Nelms, S. E., Duncan, E. M., Broderick, A. C., Galloway, T. S., Godfrey, M. H., Hamann, M., Lindeque, P. K., & Godley, B. J. (2016). Plastic and marine turtles: A review and call for research. *ICES Journal of Marine Science, 73*(2), 165–181. https://doi.org/10.1093/icesjms/fsv165

Nor, N. H. M., & Obbard, J. P. (2014). Microplastics in Singapore's coastal mangrove ecosystems. *Marine Pollution Bulletin, 79*(1–2), 278–283.

Nordhaus, I., Wolff, M., & Diele, K. (2006). Litter processing and population food intake of the mangrove crab *Ucides cordatus* in a high intertidal mangrove forest in northern Brazil. *Estuarine, Coastal and Shelf Science, 67*, 239–250.

Ntow, W. J., & Botwe, B. O. (2011). Contamination status of organochlorine pesticides in Ghana. In B. G. Loganathan & P. K. S. Lam (Eds.), *Global contamination trends of persistent organic chemicals* (pp. 393–411). CRC Press.

Nur, A. A. U., Hossain, M. B., Banik, P., Choudhury, T. R., Liba, S. I., Umamaheswari, S., et al. (2022). Microplastic contamination in processed and unprocessed sea salts from a developing country and potential risk assessment. *Chemosphere, 308*, 136395.

Ory, N. C., Sobral, P., Ferreirad, J. L., & Thiel, M. (2017). Amberstripe scad *Decapterus muroadsi* (Carangidae) fish ingest blue microplastics resembling their copepod prey along the coast of Rapa Nui (Easter Island) in the South Pacific subtropical gyre. *Science of the Total Environment, 586*, 430–437.

Ory, N., Chagnon, C., Felix, F., Fern'andez, C., Ferreira, J. L., Gallardo, C., Ordóñez, O. G., et al. (2018). Low prevalence of microplastics contamination in planktivorous fish species from the Southeast Pacific Ocean. *Marine Pollution Bulletin, 127*, 211–216.

Paray, B. A., Riya, K. K., Marshall, D. J., Hossain, M. K., Banik, P., Sultana, S., Jimmy, Y., Norhayati Ngah, M., & Hossain, B. (2025). Microplastic contamination, comparative retention efficiency, and ecological hazard in saltmarsh, mangrove, sandy, and muddy habitats along the northern Bay of Bengal coast. *Ocean & Coastal Management, 266*, 107695.

Patra, M. K., Acharjee, S. K., & Chakraborty, S. K. (2005). Conservation categories of siluroid fishes in North-East Sundarbans, India. *Biodiversity and Conservation, 14*, 1863–1876.

Paul, A. K. (1999). Geomorphic significance of mangrove Swamps—A study in the Sundarban. In K. R. Naskar & P. Sanyal (Eds.), *Mangal of the Sundarban* (pp. 39–52). Naya Prokash.

Paul, A. K. (2002a). *Coastal geomorphology and environment* (p. 582). ACB Publications.

Paul, A. K. (2002b). *Tsunami the disaster in Indian Ocean Nations* (p. 125). ACB Publications.

Paul, A. K. (2005). Oceanic Islands under acute shortage of fresh water resource, a case study in Andaman and Nicobar Archipelago. DELF/FAEA II:41–49. Article 7.

Paul, A. K. (2009). Man-environment interactions in the single largest mangrove forest along the shoreline of the Bay of Bengal: A case study of Sundarbans. In H. S. Sharma & V. S. Kole (Eds.), *Geomorphology in India* (pp. 263–286).

Paul, R., Chatterjee, D. P., & Dutta, K. (2016). Concerns regarding "plastic" pollution: Reasons, effects and needs to generate public awareness. *International Journal of Humanities and Social Science Studies, 3*, 123–148.

Peng, L. C., Fu, D. D., Qi, H. Y., Lan, C. Q., Yu, H. M., & Ge, C. J. (2020). Micro- and nanoplastics in marine environment: Source, distribution and threats—A review. *Science of the Total Environment, 698*, 134254.

Plastics Europe. (2018). *Plastics-The Facts 2018: An analysis of European plastics production, demand and waste data*. Plastics Europe, Association of Plastics Manufacturers.

Qasim, S. Z., Sengupta, R., & Kureishy, T. W. (1988). Pollution of the seas around India. *Proceedings of Indian academy of Science (Animal Science), 92*(1), 73–85.

Raha, U. K., & Raju, K. D. (2016). The Enrica Lexie case at the permanent court of arbitration: An analysis. *Indian Journal of International Law, 56*, 221–238. https://doi.org/10.1007/s40901-017-0043-x

Raha, U. K., Ramesh Kumar, B., & Sarkar, S. K. (2021). Policy framework for mitigating land-based marine plastic pollution in the Gangetic Delta Region of Bay of Bengal—A review. *Journal of Cleaner Production, 278*, 123409.

Ray, R., Paul, A. K., & Basu, B. (2013). Application of supervised enhancement technique in monitoring the mangrove forest cover dynamics—A study on Ajmalmari Reserve Forest, Sundarban, West Bengal. *International Journal of remote Sensing and Geoscience, 2*(1), 16–21.

Rodrigue, B. (2017). The effect of macroplastic debris on marine vertebrates.

Rules. (2016). *Plastic waste management rules*. Governemnt of India, 2016 [www.document]. http://www.moef.gov.in/sites/default/files/EWM%20Rules%202016%20english%20 23.03.2016.pdf

Ryan, P. G. (2013). A simple technique for counting marine debris at sea reveals steep litter gradients between the Straits of Malacca and the Bay of Bengal. *Marine Pollution Bulletin, 69*, 128–136. https://doi.org/10.1016/j.marpolbul.2013.01.016

Sakhuja, V. (2016). *Fisheries and the plastic threat in Bay of Bengal*. [www.document]. Natl. Marit. Found. http://www.maritimeindia.org/View Profile/636074271086249914

Salomons, W., & Forstner, U. (1984). *Metals in the hydrocycle* (pp. 1–349). Springer-Verlag.

Sanders, C. J., Smoak, J. M., & Naidu, S. A. (2009). Mangrove forest sedimentation and its reference to sea level rise, Cananeia, Brazil. *Environmental Earth, 60*(6), 1291–1301.

Sandwell, D. R., Pilditch, C. A., & Lohrer, A. M. (2009). Density dependent effects of an infaunal suspension-feeding bivalve (Austrovenus stutchburyi̇) on sandflat nutrient fluxes and microphytobenthic productivity. *Journal of Experimental Marine Biology and Ecology, 373*, 16–25.

Sarkar, S. K., Singh, B. N., & Choudhury, A. (1986). The ecology of copepods from Hooghly Estuary of West Bengal, India. *Mahasagar-Bulletin of the National Institute of Oceanography, 19*(2), 103–112.

Sarkar, S. K., Bhattacharya, B., & Debnath, S. (2002). Heavy metal concentration from biota of Sundarbans mangrove ecosystem. *Aquatic Ecosystem Health & Management, 5*, 467–472.

Sarkar, S. K., Franciscovic-Bilinski, S., Bhattacharya, A., Saha, M., & Bilinski, H. (2004). Levels of elements in the surficial estuarine sediments of the Hugli river, Northeast India 478 and their environmental implications. *Environment International, 30*, 1089–1098.

Sarkar, S., Ghosh, P. B., Das, T. M., Som Mazumdar, S., & Saha, T. (2013). Environmental assessment in terms of salinity distribution in the tropical mangrove forest of Sundarbans, North East Coast of Bay of Bengal, India. *Archives of Applied Science Research, 5*(6), 109–118.

Schiffman, P. M. (1997). Animal-mediated dispersal and disturbance: Driving forces behind alien plant naturalization. In J. O. Luken & J. W. Thieret (Eds.), *Assessment and management of plant invasions* (pp. 87–94). Springer.

Schmidt, C., Krauth, T., & Wagner, S. (2017). Export of plastic debris by rivers into the sea. *Environmental Science & Technology, 51*, 12246–12253. https://doi.org/10.1021/acs.est.7b02368

Schnurr, R. E. J., Alboiu, V., Chaudhary, M., Corbett, R. A., Quanz, M. E., Sankar, K., Srain, H. S., Thavarajah, V., Xanthos, D., & Walker, T. R. (2018). Reducing marine pollution from single-use plastics (SUPs): A review. *Marine Pollution Bulletin, 137*, 157–171. https://doi.org/10.1016/j.marpolbul.2018.10.001

Schroeder, M. (2010). Forgotten at sea—An international call to combat islands of plastic waste in the Pacific Ocean. *Southwestern Journal of International Law*.

SC-POPs. (2009). *Stockholm convention on persistent organic pollutants*. http://chm.pops.int/TheConvention/Overview/TextoftheConvention/tabid/2232/Default.aspx

Suja, S., Kessarkar, P. M., Fernandes, L. L., Kurian, S., & Tomer, A. (2017). Spatial and temporal distribution of metals in suspended particulate matter of the Kali Estuary, India. *Estuarine, Coastal and Shelf Science, 196*, 10–21.

Tanaka, K., Takada, H., Yamashita, R., Mizukawa, K., Fukuwaka, M.-A., & Watanuki, Y. (2013). Accumulation of plastic-derived chemicals in tissues of seabirds ingesting marine plastics. *Marine Pollution Bulletin, 69*, 219–222.

Tattar, T. A., Klekowski, E. J., & Stern, A. I. (1994). Dieback and mortality in red mangrove, Rhizophora mangle L., in Southwest Puerto Rico. *Arboricultural Journal, 18*, 419–429.

Thompson, R. C., Moore, C. J., Saal, F. S. V., & Swan, S. H. (2009). Plastics, the environment and human health: Current consensus and future trends. *Philosophical Transactions of the Royal Society B: Biological Sciences, 364*, 2153–2166. https://doi.org/10.1098/rstb.2009.0053

Titus, J. G. (1991). Green house effect and coastal wetland policy: How Americans could abandon an area with a size of Massachusetts at minimum cost. *Environmental Management, 15*(1), 39–58.

Trudgill, S. T. (1977). *Soil and vegetation systems*. Clarendon Press.

Twilley, R. R., & Chen, R. (1998). A water budget and hydrology model of a basin mangrove forest in Rookery Bay, Florida. *Marine and Freshwater Research, 49*, 309–323.

UN Report. (2016). Secretariat of the convention on biological diversity marine debris: Understanding, preventing and mitigating the significant adverse impacts on marine and coastal.

UNCLOS. (1982). The United Nations Convention on the Law of the Sea 1982, 1833 UNTS 3

UNDP. (2015). Goal 14: Life below water. Sustainable development goals (pp. 1–12). https://doi.org/10.1017/CBO9781107415324.004

Valentine, P. C., Carman, M. R., Blackwood, D. S., & Heffron, E. J. (2007). Ecological observations on the colonial ascidian Didemnum sp. in a New England tide pool habitat. *Journal of Experimental Marine Biology and Ecology, 342*, 109–121.

Van Cauwenberghe, L., Vanreusel, A., Mees, J., & Janssen, C. R. (2013). Microplastic pollution in deep-sea sediments. *Environmental Pollution, 182*, 495–499. https://doi.org/10.1016/j.envpol.2013.08.013

Vernberg, W. B., & Vernberg, F. J. (1972). *Environmental physiology of marine animals*. Springer.

Villarrubia-Gómez, P., Cornell, S. E., & Fabres, J. (2018). Marine plastic pollution as a planetary boundary threat—The drifting piece in the sustainability puzzle. *Marine Policy, 96*, 213–220. https://doi.org/10.1016/j.marpol.2017.11.035

Vitousek, P. M. (1990). Biological invasions and ecosystem processes: Towards an integration of population biology and ecosystem studies. *Oikos, 57*, 7–13.

Vitousek, P. M., Mooney, H. A., Lubchenco, J., et al. (1997). Human domination of Earth's ecosystems. *Science, 277*, 494–499.

Walsh, J. J. (1991). Importance of the continental margins in the marine biogeochemical cycling of carbon and nitrogen. *Nature, 350*, 53–55.

Walsh, K. (2004). Tropical cyclones and climate change: Unresolved issues. *Climate Research, 27*, 77–83.

Walsh, J. J., Rowe, G. T., Iverson, R. E., & Mcroy, C. P. (1981). Biological export of shelf carbon is a sink of the global CO_2 cycle. *Nature, 291*, 196–201.

White, P. S. (1979). Pattern, process and natural disturbance in vegetation. *The Botanical Review, 45*, 229–299.

Windsor, F. M., Durance, I., Horton, A. A., Thompson, R. C., Tyler, C. R., & Ormerod, S. J. (2019). A catchment-scale perspective of plastic pollution. *Global Change Biology, 25*, 1207–1221. https://doi.org/10.1111/gcb.14572

Wong, Y. S., Lam, C. Y., Che, S. H., Li, X. R., & Tam, N. F. Y. (1995). Effect of wastewater discharge on nutrient contamination of mangrove soil and plants. *Hydrobiologia, 295*, 243–254.

Wright, S. L., Thompson, R. C., & Galloway, T. S. (2013). The physical impacts of microplastics on marine organisms: A review. *Environmental Pollution, 178*, 483–492.

Yao, W., et al. (2019). Micro- and macroplastic accumulation in a newly formed Spartina alternifora colonized estuarine saltmarsh in southeast China. *Marine Pollution Bulletin, 149*, 110636.

Zhang, H. (2017). Transport of microplastics in coastal seas. *Estuarine, Coastal and Shelf Science, 199*, 74–86.

Zhou, Q., Tu, C., Fu, C., Li, Y., Zhang, H., Xiong, K., Zhao, X., Li, L., Waniek, J. J., & Luo, Y. (2020). Characteristics and distribution of microplastics in the coastal mangrove sediments of China. *Science of the Total Environment, 703*, 134807. https://doi.org/10.1016/j.scitotenv.2019.134807

Chapter 8
Community Mobilization, Method-Based Innovation, Adaptation and Ecomanagement Strategies of Sundarbans, Mangrove Ecosystem

Abstract The Sundarbans biosphere is becoming more vulnerable to erosion, extortion and extinction every day, mostly due to anthropogenic intrusion and aggression. Rising sea levels, increases in the salinity, massive erosion, unwanted accretion, sediment starvation, the subsidence of sediment, the loss of biodiversity, simmering human–wildlife conflicts, the steady decline in agricultural productivity and the outmigration of local people have increased concerns about one of the world's most beautiful and diverse biospheres, the Sundarbans of the Indian subcontinent, jointly shared by two neighbouring countries, Bangladesh and India. The problem is huge, yet opportunities are also immense. The completion of national and international research projects with some interesting and trailblazing research findings on the prevailing problems and future prospects of this biologically productive, economically viable and aesthetically celebrated mangrove ecosystem of Sundarbans and execution of enormous research information could have ephemeral impacts. In such a context, understanding and standardizing the process of community mobilization through people's participation with their participatory skills are required to usher in a movement rather than only write a text of execution. Participatory rural appraisal (PRA) tools and techniques can be of great help in generating hard evidence, mapping perception, projecting future strategies and integrating tiny and discrete endeavours into a people's movement. These are extremely effective in space, time, flow and decision analyses aimed at verifying problems, processes, resources and opportunities. People's Indigenous Indicators have natural efficacy in resource monitoring and microplanning.

This chapter has taken into consideration some people-based research tools like PRA tools, participatory learning and action (PLA), decision support systems (DSSs), management information systems (MISs), climate-resilient adaptive tools (CRATs), artificial neural network (ANN), analytical hierarchial process (AHP), fuzzy cognitive map (FCM), time line and trend analysis, matrix ranking, rank based quotient (RBQ) and artificial intelligence (AI) with proper illustrations citing case studies which can successfully be extrapolated to the Sundarbans biosphere for its protection, regeneration and sustenance with the involvement of local people. In addition, the modes of community metabolism have been assessed to understand the

S. K. Chakraborty, S. K. Acharya, *Eco-dynamics and Environmental Perturbations in Mangrove Ecosystems*,
https://doi.org/10.1007/978-3-032-08710-2_8

balance between biological production through ecofriendly biotechnological tools (leading to the accumulation of organic matter) and the easy means of its utilization as insight into potential support for the biological wealth of this unique ecoregion.

Keywords The Sundarbans biosphere · Participatory rural appraisal (PRA) tools · People's Indigenous Indicators · Microplanning · Resource monitoring · Community mobilization

8.1 An Integrated Ecosystem- and Community-Based Approach to Conserving the Mangrove Ecosystem of Sundarbans: Community Mobilization and Method-Based Innovation

Out of so many environmental issues, like global warming–mediated climate change, eutrophication, deforestation, soil erosion, land subsidence, sea-level rise, salinity invasion, the release of industrial waste, breached embankments, crop failures, biodiversity loss and need for proper understanding on the development of different push factors (increased in poverty, unemployment, outmigration, etc.) and pull factors (new avenues of livelihood generation by diversifying agriculture, promoting ecotourism, etc.) and their counterbalancing effects are being considered as an important prerequisite for achieving the sustainable natural resource management options. Out of all of these, global climate change has become an unprecedented threat that will unfold over a long timescale, unevenly, in a nonlinear fashion and unpredictably. In response, focus has increased on programming and policies that promote transformative and incremental adaptation and enhance resilience to climate change. Emerging approaches that build the resilience of socioeconomic and ecological systems, including through the community-based sustainable management of natural resources, are gaining recognition as viable tools for climate change adaptation and mitigation.

These integrated ecosystem- and community-based solutions present new opportunities to plan for an uncertain future, using frameworks that include the most vulnerable people and important ecosystems. For their effective implementation, policymakers need a clearer understanding of how these initiatives function, ways to measure their effectiveness and scale them up and a stronger quantitative evidence base to demonstrate their benefits (Chambers, 1997; Dudley, 2008; Zimmermann & Nicholls, 2015).

The community based ecosystem management process involves People's Indigenous Indicators which as matrics designed to assess, quantify and monitor the benefits which are derived from the nature and natural resources based on the traditional knowledge and perception of the local vis-a-vis indigenous people. The interactive and integrated interactions between ecosystem and community have been evolving to sustain the complex ecological system of Sundarbans with their ecosystem services after being supported by soil, water, biodiversity, sunlight, etc. The orchestration of ecosystem function, resilience and services is extremely important for the sustenance of a coastal ecosystem like the Sundarbans mangrove ecosystem,

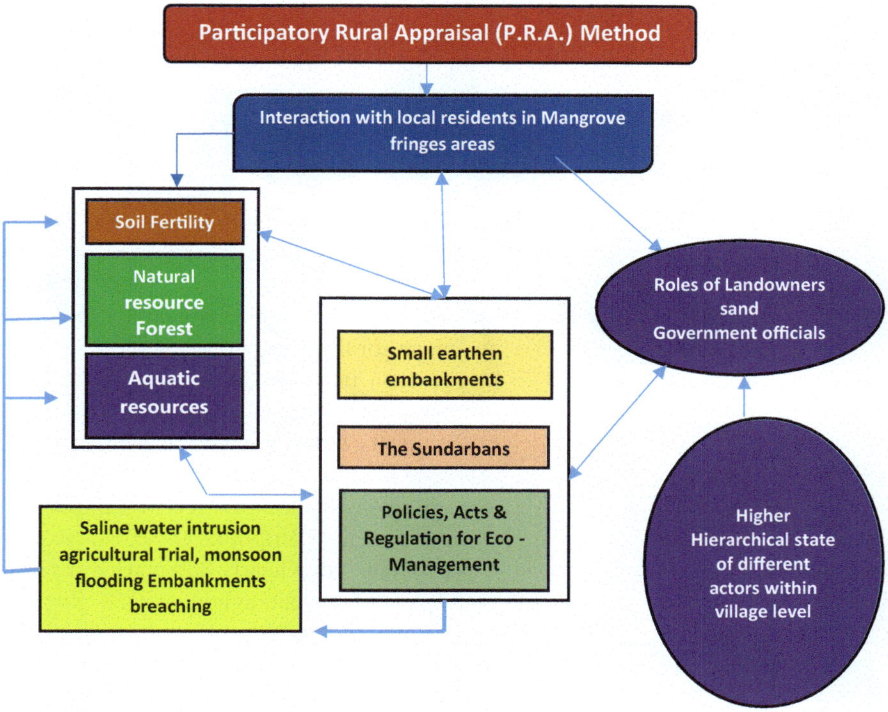

Flowchart 8.1 Different facets of the participatory rural appraisal (PRA) method as applied in the Sundarbans mangrove estuarine complex of India

but this highly productive, ecologically sensitive, economically viable and aesthetically celebrated ecosystem is becoming counterproductive because of coercive anthropogenic activities. The techniques and processes of community mobilization and an ecosystem transfer process of energy and materials are crucial for participatory tools and techniques, methodological interventions, finding new avenues for sustainable livelihood generation, etc. through comprehensive landscape management. The main need of adopting different eco-sociological approaches suitable in the Sundarbans ecoregion for achieving environmental resilient and sustainable livelihood generation includes the development of social entropy, understanding the root causes of chaos and forming appropriate environmental policy, the development of synergy and assessing uncertainty and unpredictability, social resilience and adaptability and means for achieving proper community (Conference on Rapid Rural Appraisal, 1987; Robinson-Pant, 1995; Pretty et al., 1997; Leurs, 1997; Chambers, 1997; Sontheimer et al., 1999; Flower et al., 2000; Ellis & Allison, 2004; Castelloe et al., 2005; Dudley, 2008; Ahammad et al., 2013). Some methods and tools, like participatory rural appraisal (PRA) tools and participatory learning and action (PLA), need to be used for participatory data generation and data collection for decision support systems (DSSs) as important steps towards ecomanagement with management information systems (MISs) and applications of artificial intelligence (AI) for disaster forecasting and for analyses of responses (Flowcharts 8.1, 8.2, 8.3, 8.4, 8.5 and 8.8).

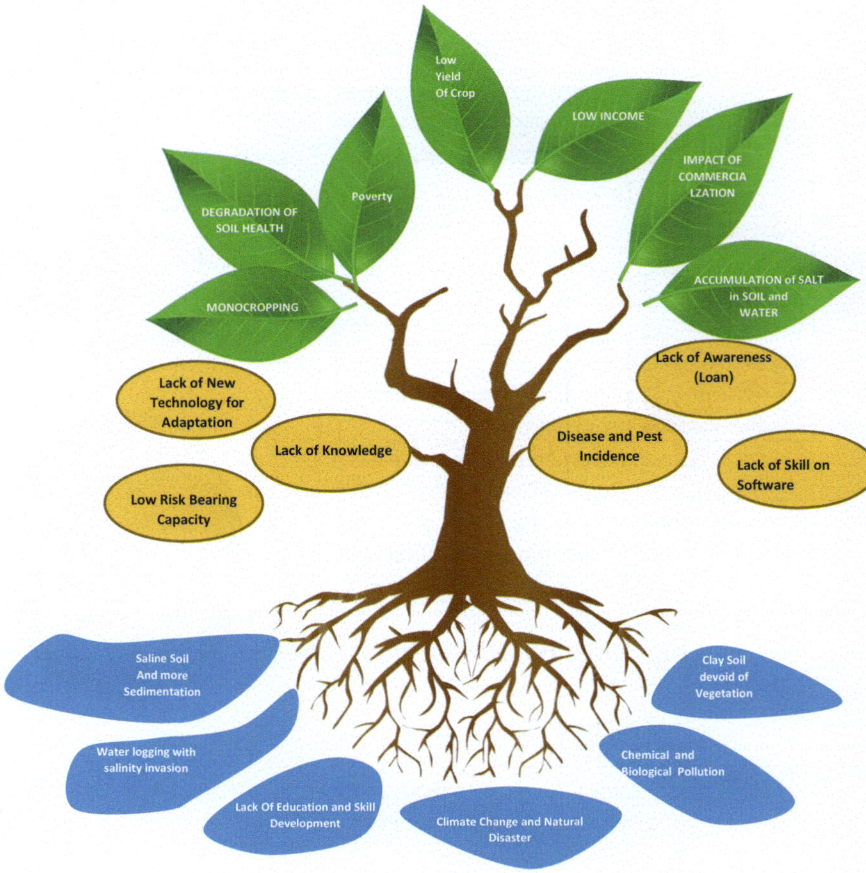

Flowchart 8.2 The presentation and depiction of micro-level information derived from the PRA and PLA methods from the village community in the form of an information tree to help track resource management and new livelihood-generation strategies

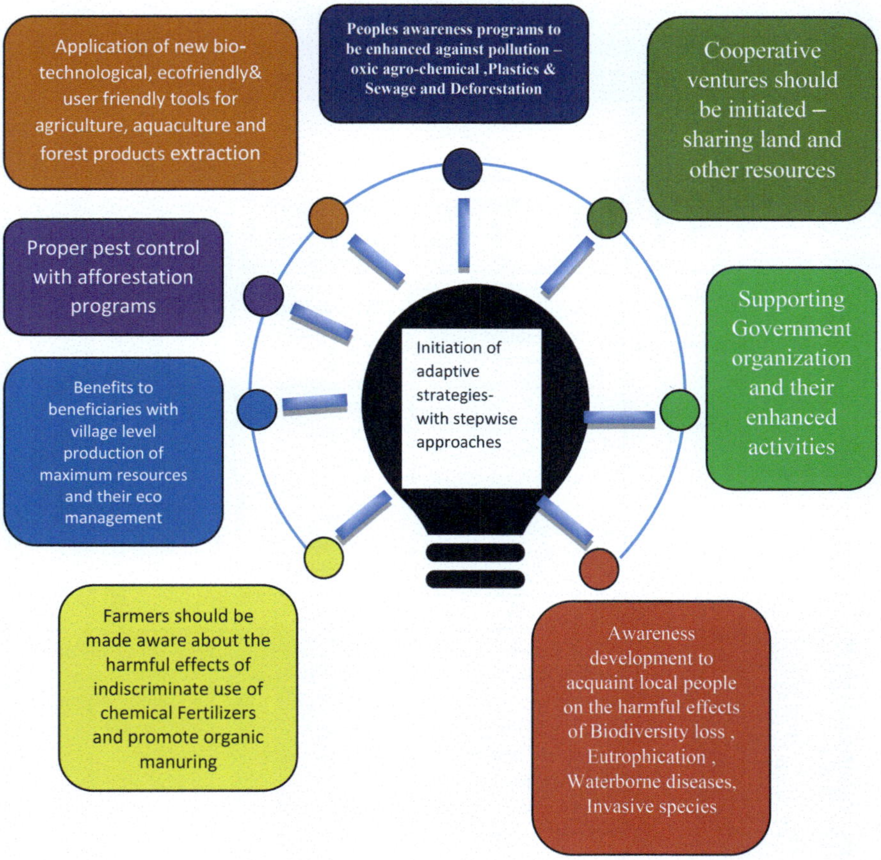

Flowchart 8.3 The development of adaptive strategies based on people's perceptions and with the application of modern technological tools on a stepwise basis against different forms of pollution and the negative environmental consequences of climate change

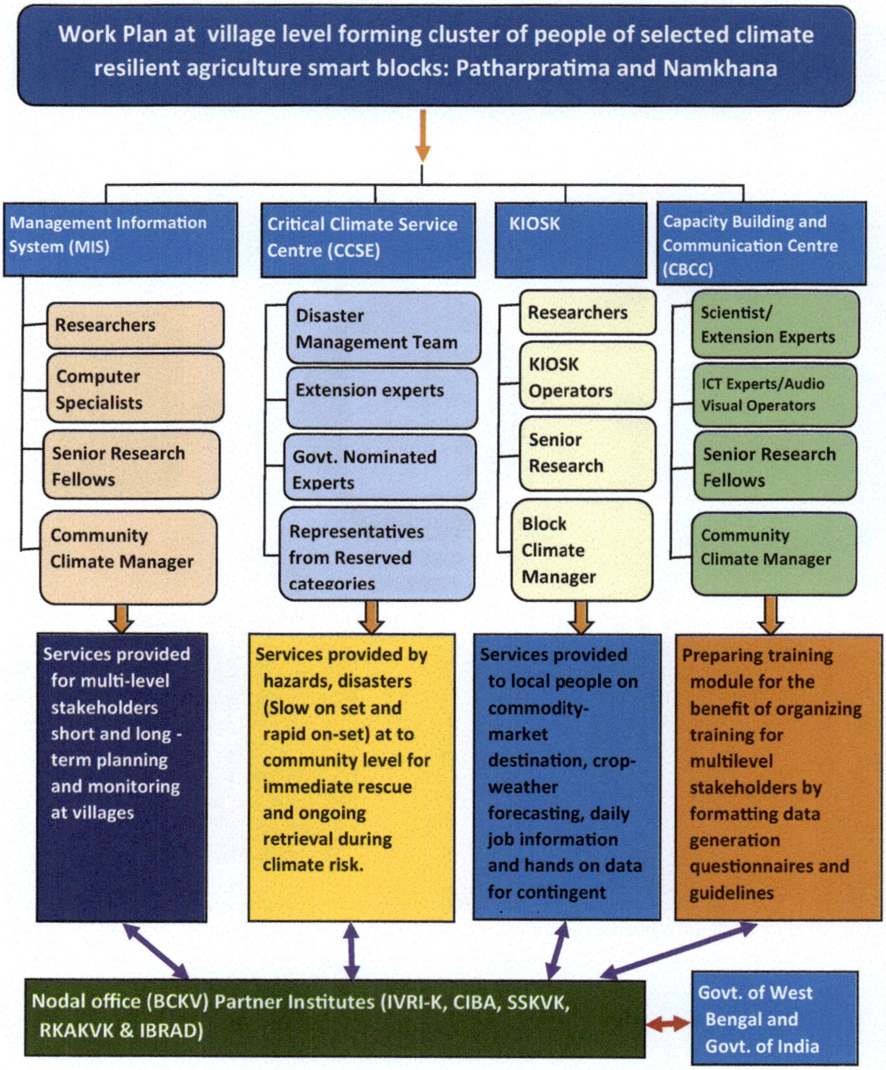

Flowchart 8.4 Micromanagement plans at the village level, which combat the problems of environmental perturbations with the application and usage of management information systems (MISs), critical climate service centres (CCSCs), the Kommunikasjon Integrate Offentlig Service Kontor (KIOSK), capacity-building and communication centres (CBCCs), etc., with help from researchers, of Bidhan Chandra Krishi Viswavidaylay (BCKV) and their partner institutes, a disaster-management team and climate managers

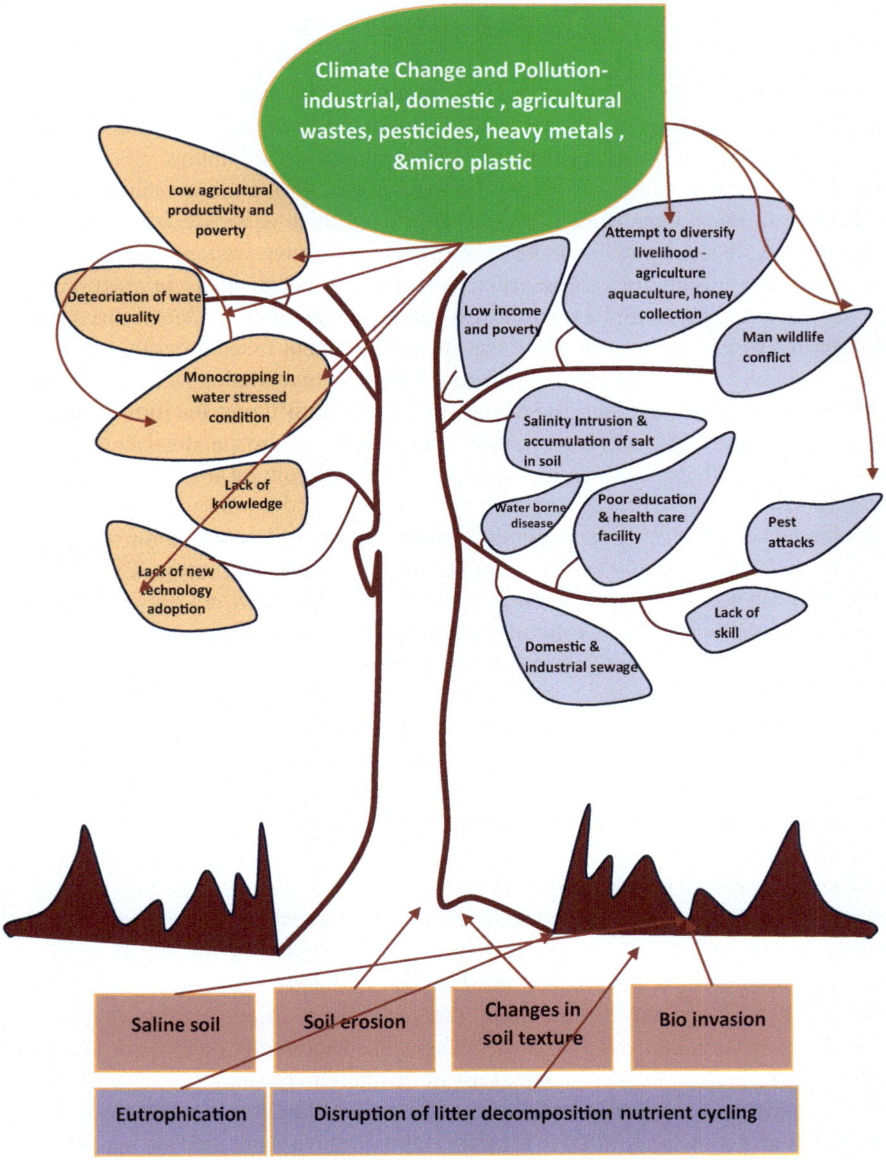

Flowchart 8.5 A diagramatic representation of the multidimensional interlinkages among different facets of the structure, function, ecosystem functioning, societal interactions, pollution, disruptions to the normal pathways of trophic interaction and nutrient cycling and ecodynamics of the mangrove ecosystem through an information tree, the proper understanding of which lends support to proper ecomanagement strategies for the mangrove ecosystem

8.2 The Resilience of Natural Mangrove Ecosystems

Ecological resilience in the context of the mangrove ecosystem is its to withstand ecodegradation and return to its normal functioning, which it can thanks to its unique assemblage of different structural components (flora, fauna, microbes, soil and water), which exhibit persistence in maintaining ecosystem stability by being resistance to most negative ecological changes and thereby withstanding the initial disturbances. The specific attributes contributing to ecosystem resilience in a mangrove ecosystem are their structural complexity, the continuous interaction processes between forest and estuarine subsystems, their unique adaptabilities, their ability to yield ecosystem services and the development of socioecological bonds that promote social resilience. Social resilience refers to the capability of human community to withstand and overcome stresses from different environmental perturbations and to return to a normal state by developing and maintaining positive relationships with environmental and social parameters and by adapting to changing conditions.

Mangroves are deeply intermingled with their nearby coastal community, rendering social benefits to them; providing ecological goods and services in the form of food, fodder, medicine, shelter material and fuel; generating avenues for tourism, and thereby helping to overcome food insecurity and alleviate poverty.

Improving the ecomanagement of coastal habitats and associated mangroves to reduce threats, such as the overharvesting of natural resources, drastic land-use changes with more reclamation, ever-increasing pollution from different sources, the arrival and proliferation of invasive alien species, contributes to building healthier and more-resilient ecosystems. To make coastal environments more ecologically stable, drafting a large-scale mangrove plantation programme with suitable mangrove plant species and the involvement of local people, especially women, appears to be sustainable and cost-effective. Such long-term biodiversity conservation is expected to reinforce the ecosystem services of mangroves in coastal areas by increasing forest coverage and building capacity to adapt to increasing climatic threats. Large coastal forests also need to be protected again increasing climatic threats, such as prolonged inundation, increased salinization and higher-intensity and -frequency cyclonic winds and storm surges. Functionally diverse communities residing within mangrove ecosystems are more likely to be more resilient to climate change–mediated environmental changes, which tend to destroy mangrove-based resources like spawning grounds for fish, and thereby, human settlements that produce livelihood assets can be protected by maintaining and restoring mangroves along with their coastal communities, which are highly dependent on climate-sensitive natural resources for their livelihood and which develop adaptations to overcome different climate risks. The eco-restoration of mangroves is a critical issue in the Indian Sundarbans given the downsizing trend in natural mangroves, with both climatic and nonclimatic pressures threatening valuable biodiversity wealth and prevailing ecosystem services. Thus, well-assessed and appropriate actions need to be taken for sustainable conservation.

8.3 The Ecological Vulnerability of Coastal Environments

The term *ecological vulnerability* refers to the state of being susceptible to the negative impacts from ongoing environmental perturbations. The mangrove ecosystem of Sundarbans is vulnerable to a number of natural and human-mediated activities, like habitat destruction, the nonjudicious exploitation of natural resources, manifold pollution threats and changes in ongoing global climate change. All these factors together or in isolation increase the loss of mangrove-based biodiversity, destroy fisheries and bring about negative ecological conditions on the coasts, with changes in the ecological succession process, species composition and the landward migration of mangrove species. Sea-level rise and a smaller supply of sediments submerge mangroves in eroded bottom soils, causing their mortality. Proper protection of sensitive and vulnerable ecosystems like the mangrove ecosystem in coastal zones requires buffers against climate impacts to thereby sustain coastal livelihoods, which are possible only by adhering to the basic principles of ecology.

8.4 Adaptation Options as a Unified Strategy to Tackle Altered Ecological Conditions for Resource Management

Intrusions of salinity with rises in sea level render ecological conditions that are unsuitable for agricultural activities, especially for traditional rice farming, which leads to the diversification of livelihood options utilizing abandoned areas for the culturing of valued shrimp and other estuarine fish species. These kinds of shifts from one mode of resource generation to another bring about major changes in supply chains, which are then investigated by administrative planners for planning and forecasting. Sea-level rise also causes flooding, where the additional intrusion of saline water causes ecological and habitat changes within a mangrove ecosystem, which acts as nursery grounds for the nurturing, breeding and growth of many euryhaline species.

Adaptation involves crop diversifications, increasing aquacultural productivity, and improvements in fisheries and the maintenance of livestock in an integrated manner, aiming at improving the rural economy, alleviating poverty and reducing the risk of economic loss by diversifying income sources. Improvements in community resilience happen alongside a simultaneous reduction in environmental degradation caused by exacerbation from climate change. A common framework needs to be developed at the village level to highlight ways and means for integrating and amalgamating all the relevant ecomanagement strategies for proper resource management at the micro level, starting from Gram Panchayat (village council) to integrated coastal zone management. In this practice, given that vulnerability is not uniform, most affected areas range from moderately vulnerable to extremely vulnerable to storm and tidal surges. The most vulnerable areas in Sundarbans are Sagar Island, Namkhana, Patharpratima, Kultali, Basanti, Gosaba, Sandeshkhali-II, Minakhan and Hingalganj (Hazra et al., 2010).

The coastal ecosystems of West Bengal, including the mangrove estuarine complex of Sundarbans, provide a wide range of ecosystem services for the benefit of humans (Chakraborty, 2017). However, negative climate change–mediated ecological consequences like salinity invasion, sea-level rise and soil erosion have pushed this sensitive and at-risk ecoregion, subjected to serious disruptions to its ecosystem services, which is caused by impairing its normal ecosystem functioning at different spatial and temporal scales (Nichlls et al., 2007; Nandy & Ahammad, 2012). Adaptation to coastal changes has assumed great importance in combatting different climate change–mediated risks, generating much variability in the ecological parameters of the coast, and help us to understand the responses of natural systems to the degree of vulnerability in the world's coastal regions (Klein et al., 1998). Diverse socioeconomic and cultural facets based on human vulnerability are broadly connected with development and the ecological capacity to adapt to newly developed sea-level rise and climate change (Flowcharts 8.1, 8.2, 8.3 and 8.4; Tables 8.1, 8.2, 8.3 and 8.4).

Table 8.1 Timeline analysis

Year	General events	Breakthrough events
1898	Dwelling started in Akshayanagar village	Chemical fertilizer and pesticides were first used by the villagers in 1968 First (HYV) was introduced in the village in 1988 Electricity comes to the village in 2006 First use of LPG in the village was in 2009 Severe losses have been incurred because of Cyclone Alia in 2009 and Cyclone Amphan in 2020
1953	Villagers got a facility for drinking water	
1961	A pucca (concrete) house was first built in the village	
1965	Water-pumping machine was first used	
1968	Fertilizer shop was first established	
1972	Panchayati raj started	
1975	Power tiller was introduced	
1984	Water supply was introduced in the village by the government	
1986	Television was first introduced	
1994	Health centre was established	
1994	First poultry firm was started	
1996	ICDS was established	
1998	High-yield variety was introduced	
2006	Electricity facility built	
2008	First SSK was established in the village	
2009	First use of LPG in the village	

Full forms of abbreviations: High Yielding of Varieties (HYV) denoting genetically developed improved crop seeds in order to boost up higher crop production, which were started using after Green Revolution in India and within a span of seven decades, all those seeds have shown to require huge amount of chemical fertiliers, pesticides and irrigation supports causing lot of environmental and socio-economic nuisences; Liberalisation, Privatization and Globalisation (LPG): since the starting of the decade ninety, consideration changes in the human perceptions have been occurred due to this which are also connected with the socio economic and ecological conditions of Sundarbans; Integrated Child Development Services (ICDS) and Shishu Shiksha Kendra (SSK) have been in operation as important pro-people village level developmental components straightening adaptability of local people

Table 8.2 Time-trend analysis on trends related to variables over various timespans

Item/Character	1990–2000	2000–2010	2010–2020
Yield of rice	20 md/bigha	18 md/bigha	15 md/bigha
Count of livestock	400–450	200–250	100
Net income from crop	7000	5000	2000
Disease pest problem	30%	50%	70%
Application of organic manure	8–10 cartload/bigha	5	1–2 cartload
Number of Totos	0	02	140
Number of vans	25	15	3

Table 8.3 Seasonality graph of wages: wage analysis (Sundarbans social ecology is deeply impacted by wage variability and accessibility throughout the year. Wages are low, uncertain and variable depending on gender. That is why migration, both territorial and occupational, is so acute and inevitable in Sundarbans)

Months	Male	Female
Boishakk (April to May)	400	350
Jyoistho (May to June)	400	350
Ashar (June to July)	400	350
Shrabon (July to August)	400	350
Bhadro (August to September)	400	350
Ashwin (September to October)	400	350

Table 8.4 Rice cultivation: resistant varieties and their characteristics

Characteristics	Bangla Pan (Bengal Beetleven)	Mitha Pan (Sweat Beetleven)
Cultivated area	8	6
Yield	7	5
Market value	6	9
Resistance to pests and disease	8	6
Profit	7	9
Total	36	35
Mean	7.2	7

8.5 Climatic Vulnerability, Impacts and Coastal Adaptations

Climatic vulnerability implies the degree of the sensitivity and adaptive capacity of the people of a region with distinct socioeconomic and cultural characteristics against climate-induced extreme weather events like sea-level rise, embankment breaching, salinity invasion and changes in coastal geomorphology in coastal areas. The intense and recurrent forms of all these extreme events are associated with catastrophic cyclones, storm surges and tidal flooding, as clear manifestations of climate change–induced threats in different coastal zones (Flowchart 8.5).

The severity of the physical impacts from climate changes is reflected in agriculture, the availability of usable water, changing water regimes (triggering pathogenic activity), disaster-prone conditions and coastal flooding (Hossain et al., 2012). The major sources of livelihood generation in the Sundarbans coastal areas of West Bengal are different forms of agricultural activities, aquacultural practices and fisheries in coastal reclaimed lands, but almost 50% of the coastal arable lands have been affected by salinity because of different natural disasters, especially flooding in the reclaimed lands by saline water, which encrusts those lands with deposited salts even after the floodwater has been removed. The livelihoods of coastal communities are profoundly affected by climate change–induced shocks as local people depend mainly on the monsoon rainfall to make the reclaimed lands suitable for agriculture. But in the dry season, no further crops can be cultivated because of a lack of fresh water, except that the cultivation of some salt-tolerant rice varieties and pulses with the application of limited fresh water can continue. The lower productivity of agriculture indicates a lower adaptive capacity of marginalized farmer communities. Additionally, the cultivation of agricultural crops, capture fisheries and extensive aquacultural practices serve to supplement the loss of livelihood for a large coastal population that has been affected by climatic threats such as sea-level-rise-induced salinity and cyclone-induced flooding. All these natural causes and climatic risk–related impacts on the ecological condition of rivers, estuaries, mangrove swamps, fish habitats and breeding grounds make the eco-restoration process of the ecodegraded mangrove ecosystem, through afforestation and mangrove rehabilitation programmes, very difficult, hampering the resumption of normal ecosystem functioning in Sundarbans. But large-scale plantation programmes involving local people in subsidence can ensure the protection of mangroves and can promote ecological adaptation against climate shocks and mitigation efforts.

8.6 Community Mobilization for Coastal Ecosystem Management in Sundarbans: Tools and Techniques

8.6.1 The Definition and Importance of Community Mobilization

Community mobilization refers to the process of engaging and empowering local communities to take action on a particular issue or cause (WHO, 2017). In the context of coastal ecosystem management in Sundarbans, community mobilization is essential for promoting sustainable livelihoods, conserving biodiversity and enhancing the resilience of local communities to climate change. The interaction and interdependence between ecosystem and community are integral, evolving and organic. Life is supported by soil, water, biodiversity, sunlight and a complex ecological system and its services. The orchestration of ecological services, functions and resilience are crucial for coastal ecosystems, especially for the Sundarbans biosphere, which is becoming fragile, denuded and counterproductive due to coercive human interventions. The techniques and processes of community mobilization are crucial to methodological interventions, participatory tools and techniques and sustainable livelihood generation through comprehensive landscape management (Beer et al., 2000; Ayer & Maginnis, 2005; Blann, 2006; De Haan et al., 2007; Das & Saha, 2008).

8.6.2 Community Mobilization and Its Implications for the Ecomanagement of Mangrove-Based Resources

The Sundarbans coastal ecosystem of the Indian subcontinent, in the eastern coastal belts of India and southwestern Bangladesh, represents a unique and fragile environment which supports an array of diverse forms of flora and fauna. However, in the face of numerous challenges encountered by this ecosystem, including habitat destruction, overfishing, climate change, salinity invasion, erosion, unwanted accretion and sediment starvation, threatening the livelihoods of local communities, proper ecomanagement for the long-term sustainability of the ecosystem is possible through people's stewardship and community mobilization. Community mobilization is a critical component of coastal ecosystem management in Sundarbans because it enables local communities to take ownership of conservation efforts and promotes sustainable livelihoods. Therefore, environmentally sustainable livelihood generation is the primary need for the region under consideration.

8.6.2.1 Tools and Techniques for Community Mobilization

Several tools and techniques can be used for community mobilization in the Sundarbans Biosphere Reserve, which are as follows (Flowcharts 8.1, 8.2, 8.3, 8.4, 8.5, 8.8 and 8.9).

1. Participatory rural appraisals (PRAs): PRA is a method that involves local communities in the planning and decision-making process (Chambers, 1994). It is a useful tool for identifying community needs, priorities and concerns. This is basically an approach for the visualization of resources, erosion, opportunities, threats and interventions. Rural people, as insiders, act as facilitators in stimulating the process of active data generation. This method improves agro-ecosystem analysis through time, space, flow and decision analysis. Participatory timeline and time-trend analyses immensely help to visualize the decline of agricultural and aquacultural production, the status of livestock, scopes for the further utilization of animal resources and the identification of pests, weed and nondescript plants.
2. Participatory learning and action (PLA): PLA is a qualitative research approach with the active involvement and participation of community members in the recording of resource availability and management. This method collects micro-level information from a community to understand local contexts with less bias.
3. Community-based natural resource management (CBNRM): CBNRM is also a useful tool for the management of natural resources that features a local community–based approach that promotes sustainable livelihoods while conserving local biodiversity (Kellert et al., 2000).
4. Management information systems (MISs): A community-driven MIS will provide and seamlessly organize data generation, data stage, data analysis, data retrieval, data disposal to multilayer stakeholders, disaster forecasting and

responses, applications of the global positioning system (GPS) and drones to document resource availability and act as weather intelligence.

5. Ecotourism: Ecotourism is emerging as a new form of tourism that promotes the conservation of natural resources and supports local communities by generating income and promoting sustainable livelihoods (Cater, 1993).
6. Community-led conservation (CLC): CLC is an approach that involves local communities in the conservation of natural resources with the prospect of promoting community ownership and responsibility for conservation efforts (Borrini-Feyerabend et al., 2004).
7. Decision support systems (DSSs): A DSS is a blend of human intelligence and artificial intelligence (AI) that effectively makes action-driven decisions. It would be extremely operative during cyclones, droughts, and sudden salinity aggression due to the breaching of sea dikes or floods. The role of drones is also complementary to predicting, planning and pursuing effective and precise interventions by saving resources and minimizing risk.

   Analytic Hierarchical Process (AHP) is structured method to analyze complex decision and then break them down into goal, criteria b and alternatives. By pairwise ranking, it is possible to estimate the relative weigh of each of the alternatives. It is immensely important to estimate relative weightages of each community preference in dealing with ecological trade offs.
8. Fuzzy Cognitive Map (FCM) is a basically a cognitive map within which the relations between the elements (e.g.concepts, preferences, project resources, events etc.) of a mental landscape can be used to compute the strength of impact of these elements. As an unique analytical tool it helps understanding community response to participatory ecology management.
9. Applications of artificial intelligence (AI): AI can be used to make conclusions on escalating chaos and entropy in an ecosystem, both of which pose threats to its resilience and stability.

Since the beginning of the Industrial Revolution in the eighteenth century and continuing even today, the chronicle of civilizations has depicted more-horrible consequences than we have predicted. Ecological changes have undermined the quiet and placid pathway of evolution as delineated by Charles Darwin, where evolution has become more serendipitous, ephemeral and phenomenal. The changes are becoming nonlinear in ecosystems, far away from simple harmony. The Bay of Bengal has become the global capital of severe cyclonic storms, such that the resilience of ecosystems has been unexpectedly disrupted and made fragile. Continuous perturbation and chaos lead to increased entropy and disorder. An analogy can take a serpentine look at the cyclone-induced destruction of mangrove-flooded crops from the perspectives of in-group migration vs outgroup conflict and chaos. When uncertainty and unpredictability cripple an ecosystem, artificial intelligence (AI) and machine learning can be appealed to for help. Peoples' soft perceptions need to undergo a quantification process to better understand their roles in and contributions to mainstreaming processes, and as a result, tools like Artificial Neural Networks (ANN).

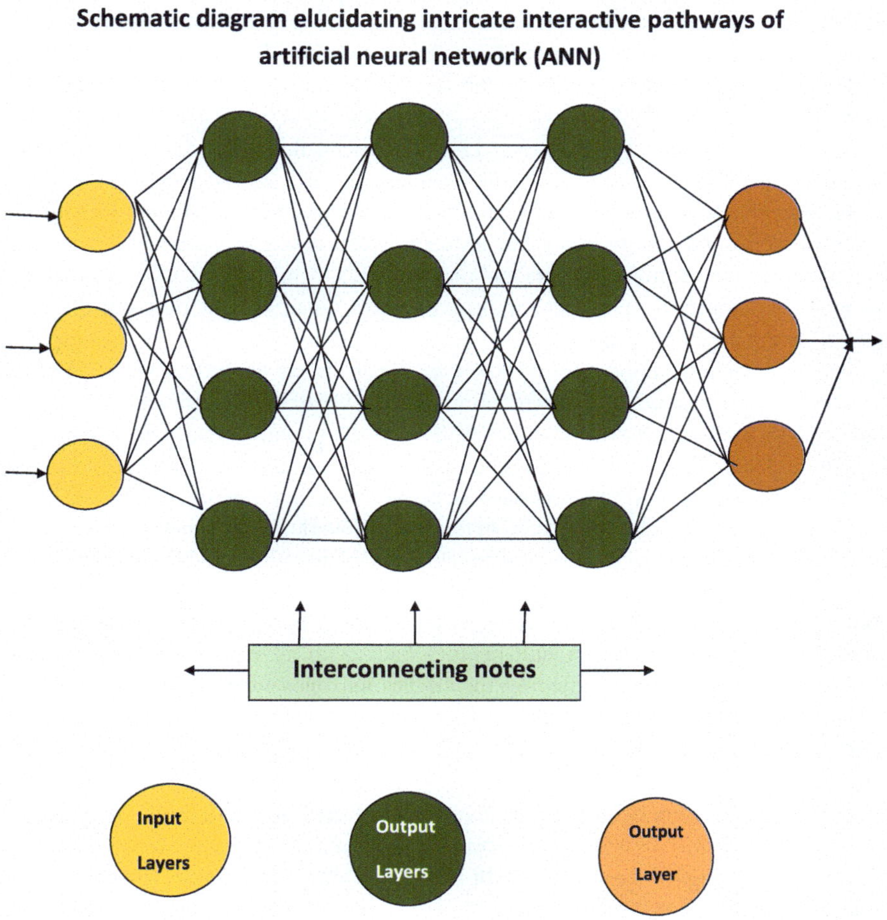

Artificial Neural Network (ANN) mooting to as the computational model mirroring the function of human brain and interconnected nodes or neurons. It operates through input layers, hidden layers and output layers. A non linear ANN is a powerful tool for identifying complex, multi-parametric discriminant function directly derived from data. Since mangrove of ecosystem of Sundarbans is very complex, agile and evolving, application of ANN as a non linear analytical stands highly effective in understanding socio-enonomic and ecological dynamics pertaining to that ecosystem. Owing to the power of ANN to deduct complex and non-linear ecological models to depict ecological relationships among physical, chemical and biological parameters functioning in variable ecosystems, like forest, aquatic, etc.

This is an extremely important mathematical tool for estimating non linear cause-effect pathway in ecological terrains. For Example, in Sundarbans ingracing of saline water engulfs crop lands with the enhancement of salinity within cultiviable land impair the production of crops where introduced varieties replacing the traditional one almost century half century back with advent of green revolution mediated hybrid varieties replacing the traditional ones won't work and lead to change the composition of biodiversity components. Such steady decline of agricultural productivity and crop failures trigger enhanced migration from farm to off farm economy.

Artificial Neural Network (ANN) mooting to as the computational model mirroring the function of human brain and interconnected nodes or neurons. It operates through input layers, hidden layers and output layers. A non linear ANN is a powerful tool for identifying complex, multi-parametric discriminant function directly derived from data. Since mangrove of ecosystem of Sundarbans is very complex, agile and evolving, application of ANN as a non linear analytical stands highly effective in understanding socio-enonomic and ecological dynamics pertaining to that ecosystem. Owing to the power of ANN to deduct complex and non-linear ecological models to depict ecological relationships among physical, chemical and biological parameters functioning in variable ecosystems, like forest, aquatic, etc.

This is an extremely important mathematical tool for estimating non linear cause-effect pathway in ecological terrains. For Example, in Sundarbans ingracing of saline water engulfs crop lands with the enhancement of salinity within cultiviable land impair the production of crops where introduced varieties replacing the traditional one almost century half century back with advent of green revolution mediated hybrid varieties replacing the traditional ones won't work and lead to change the composition of biodiversity components. Such steady decline of agricultural productivity and crop failures trigger enhanced migration from farm to off farm economy.

This is an extremely important mathematical tool for estimating non linear cause-effect pathway in ecological terrains. For Example, in Sundarbans ingracing of saline water engulfs crop lands with the enhancement of salinity within cultiviable land impair the production; Analytic Hierarchical Process (AHP) pointing out a structured method to analyze complex decision and then break them down into goal, criteria and alternatives. By pair wise ranking, the relative weightage of each of the alternaives can be estimated. It is immensely important to estimate relative weightages of local people's preference in dealing with ecological trade offs.and Artificial Neural Network (ANN) mooting to as the computational model mirroring the function of human brain and interconnected nodes or neurons. It operates through input layers, hidden layers and output layers. A non linear ANN is a powerful tool for identifying complex, multi-parametric discriminant function directly derived from data. Since mangrove of ecosystem of Sundarbans is very complex, agile and evolving, application of ANN as a non linear analytical stands highly effective in understanding socio-enonomic and ecological dynamics pertaining to that ecosystem.

Fuzzy Cognitive Map (FCM), Analytical Hierarchical Process (AHP) and Auto Regressive Integrated Moving Average (ARIMA) are used for deep historical data analysis and long term future prediction, as for example what would the ecological status of mangrove ecosystem of Sundarbans sustaining the present pace of environmental perturbation and climate change driven ecological consequences and all of these are supposed to be more effective and inevitable. From historical data to current assessments and from assessed algorithms to practised scenarios, there will be more mathematics and strategies of analysis. In the beginning of ecosystem analysis, only historical data were available to help to draw inferences, but since then, AI has been applied in a futuristic approach to ecosystem analysis to develop more-comprehensive exercises for the preparation of status reports with data clubbing, cataloguing and networking, including intervention plans and scenario analysis. In such a context, the amount of prospective uncertainty for an individual can be quantified so that they can sustain, confront and adapt to changing environmental conditions. Whether the honey collectors in Sundarbans, for example, can adapt to abrupt attacks by tigers remains a million-dollar question. The biggest triumph of humankind depends on how close one's predictions mirror future reality. Hypothesizing consequences is easy, yet post facto revelation is crucial in science, which plays a role in linking certainties and uncertainty with predictability.

8.6.2.2 Case Studies on Community Mobilization in Sundarbans, India

Several case studies from the West Bengal coast, including Sundarbans, India, have demonstrated the effectiveness of community mobilization in coastal ecosystem management.

1. The Sundarbans Biosphere Reserve: The Sundarbans Biosphere Reserve is a UNESCO-designated reserve for mangroves and associated flora and fauna that promotes the conservation of biodiversity and supports local communities (UNESCO, 2019). The reserve has implemented several community-based initiatives, including ecotourism and community-led conservation.
2. The West Bengal Coastal Zone Management Authority: The West Bengal Coastal Zone Management Authority is a government agency that promotes the sustainable management of coastal resources (WBCZMA, 2020). The authority has implemented several community-based initiatives, including participatory rural appraisal and community-based natural resource management.

8.6.2.3 The Challenges and Limitations of Community Metabolism

Although community mobilization is a critical component of coastal ecosystem management in Sundarbans, there are several challenges and limitations to its implementation. Community mobilization is a critical component of coastal ecosystem management in Sundarbans, India. By engaging and empowering local communities, community mobilization initiatives can promote sustainable livelihoods, conserve biodiversity and enhance the resilience of local communities to climate change. Although the implementation of community mobilization initiatives comes with several challenges and limitations, the benefits of community mobilization make it an essential tool for coastal ecosystem management in Sundarbans, India.

1. Limited institutional capacity: The lack of institutional capacity and resources can limit the effectiveness of community mobilization initiatives (Kumar, 2002).
2. Power dynamics: Community mobilization initiatives can be influenced by power dynamics between different stakeholders, which can limit the participation and empowerment of local communities (Ghosh et al., 2017).
3. Scalability: Community mobilization initiatives can be difficult to scale up to larger areas, which can limit their impact on the Sundarbans coastal ecosystem as a whole (Mitra et al., 2019).

8.7 Core–Periphery Interlinkages in the Sundarbans Biosphere Reserve in India: The Severity of Climate Change Impacts and Nonclimate Stressors

The existence, growth and regeneration of any robust ecosystem across the planet has to confront, comply with and reconcile three sub-ecosystems: core, periphery and far-periphery ecological dynamics. Sundarbans is no exception. The existence, growth, regenerative behaviours, dimensions and trajectories of any ecosystem depend on the kinetic interplay between these ecological factors.

The population living in the reclaimed deltaic islands of Sundarbans in India, specifically in the tidally active lower deltaic plain of the Ganges–Brahmaputra–Meghna river basin, has been predominantly dependent on agriculture, although the current prospects of agricultural investments in terms of labour and money have suffered setbacks over the past several decades due to low irrigation support, a loss of land fertility, increases in salinity and lower crop yields in all most all areas under cultivation (Hazra et al., 2015). Only the people engaged in paddy cultivation, who depend on the Kharif season (June to October) to cultivate only the aman variety of paddy seeds, can make use of the water of the rainy season, but none of the areas can support boro cultivation during the period of water scarcity (February to May), when the water deficit is high. For the livelihood of the people residing throughout the length and breadth of the Indian Sundarbans, agriculture is their best option. However, around 25% of current nonmigrant households have a propensity to migrate to other places for another occupation because of the development of multiple impoverishment risks. Climate change–mediated ecological changes have compounded the harmful consequences of other environmental perturbations and aggravated conditions to the extent that local people have been drive to choose other method of livelihood generation elsewhere. Confronted by increasing threats, households on the front lines of climate change impacts on local people become "environmental refugees" by working to seek earnings from outside their homelands, resulting in a hollowed-out economic condition.

Villagers who live on several deltaic islands in the southwestern parts of Sundarbans, like Sagar Island, Namkhana and Patharpratima, are no longer interested in conventional paddy-based agriculture, because of adversity due to salinity invasion and alternative occupations like animal husbandry and fishing. But those people, after being occupationally and physically displaced, are trapped in new adverse socioeconomic conditions due to the absence of required financial and social capital. Under this changed scenario, mainly due to development-forced displacement, impoverishment risks include: (i) landlessness; (ii) a loss of access to common property and services; (iii) joblessness without income; (iv) marginalization; (v) homelessness; (vi) increased morbidity and mortality; (vii) food insecurity; and (8) social disarticulation (Cernea, 2004).

Increased flood risks and the inundation of vast paddy fields in the Indian Sundarbans have necessitated restoring at least some of the intertidal areas along with agricultural fields in the reclaimed zones by constructing embankments so that these areas can once again function as part of the ecosystem. High rates of erosion coupled with salinization in different parts of the Indian Sundarbans expose the landscape of deltaic islands like Sagar Island and Ghoramara Island to excessive coastal flooding, catastrophic storms, cyclones, etc., leading to a considerable loss of landmass, including agricultural fields (Hazra et al., 2010; Ghosh et al., 2014). Ongoing climate change–associated environmental problems have exacerbated these situations, prompting higher rates of outmigration and displacement (Ghosh et al., 2014; Nandi, et al. 2016). The construction of embankments either by governments or by local people constituted the early stages of the development of agricultural activities, resisting the overflowing of brackish water on land during the process of cultivation (Mortreux et al., 2018).

The net value is calculated by estimating the benefits and subtracting the costs of ex situ adaptation, which is considered as the most acceptable economy among the available options. This development creates a case for an organic population movement to activate and internalize the proposed framework in the decision-making process while giving due importance to the comprehensive economic valuation of economic benefits and costs derived from ecosystem services. This kind of understanding helps to achieve a higher measure of well-being in terms of the costs and the benefits, which can be achieved in three ways:

- The first involves the identification of the different socioeconomic and ecological outcomes of ongoing human actions like planned retreat.

- The second involves the identification of the various components of costs and the various components of benefits.
- The third involves the estimation of the costs and the benefits associated with each of the scenarios and comparing them across the scenarios.

In the context of core–periphery interlinkages, the connectivity of Sundarbans with Kolkata has to be taken into consideration. The functioning of Kolkata depends on the existence of mangrove vegetation; otherwise, a cyclonic hit could lash off the life process of Kolkata, especially for those people dwelling on the peripheral and core ecological *holons*. Similarly, this urban settlement is the recipient of any kind of migrant population, both people and animals, from the regions in and around the mangrove-forest-based estuarine ecosystem.

The transformation of forest lands into agricultural lands may be better in the initial phase by creating a backup for sustenance in the form of food, livelihood and economy, but in the long run, serious problems and threats would be posed by heavy metals, pesticides, the destruction of natural flora and fauna and land-related conflicts, generating both ecological chaos and social entropy. The seamless destruction of mangroves, the extraction of shrimp and *hilsa* and the mindless aggression of new settlements that blatant deny the subtle ecological fabric of the region will lead to ecological chaos and disruption. The increasing incidents of human–wildlife conflicts, the intensification of human in- and outmigration, the engulfing of croplands by a mantle of salts (leading to crop failures) and higher numbers of cyclones every year are all adding to the ecological entropy in this area.

Therefore, interlinkages between the core areas of the Sundarbans Biosphere Reserves and the periphery occupied by human populations with their own life-supporting activities, centred mainly on agricultural practices, are important in conserving the entire ecoregion in an integrated and holistic manner.

8.8 The Necessity of People's Participation in the Conservation of Sundarbans, India

People's participation in development programmes designed to meet their basic needs is not controversial. The postmortem of any unsuccessful programme has shown that programmes have invariably failed when they did place adequate importance on people's participation in planning and analysis.

A body of philosophy and a set of assumptions undergird people's participation in rural development.

(i) The programme should be a people-based programme with government participation.
(ii) Communities need to build up their own capacity, and agents play a catalytic role.
(iii) Self-imposed changes need more permanence than changes imposed from outside.

(iv) Holistic approaches are far better than fragmented approaches.
(v) People need help to solve unique problems.
(vi) Barriers that prevent active participation also inhibit personal development.
(vii) Given the opportunity, people can solve their own problems.
(viii) Seemingly helpless communities can use their wisdom to help themselves.

8.9 The Concept and Definition of Participatory Rural Appraisal (PRA) Tools

Participatory rural appraisal (PRA) tools help in interacting with local communities, understanding them and learning from them. PRA has been defined as an "approach and method for learning about rural life and conditions from, with and by rural people" and thus goes beyond that to analysis, planning and action. PRA tools help to involve local communities, such as by engaging in Indigenous knowledge–building exercises. PRA is a way of learning from and with community members to investigate, analyse and evaluate constraints and opportunities and make informed and timely decisions regarding development projects (Flowcharts 8.1, 8.2, 8.3, 8.4, and 8.5).

PRA involves a set of principles, a process of communication and a menu of techniques for seeking villagers' participation in putting forward their points of view about any issue and enabling them to carry out their own analysis and offer contributions in different forms by using such learning. It initiates a participatory process and sustains it. Its principles and its menu of techniques help in organizing participation.

PRA is a means of generating different kinds of data, identifying and mobilizing intended groups, evoking their participation and opening ways for intended groups to participate in decision-making, project design, execution, monitoring and evaluation. It provides an alternative framework for data collection and analysis. Because of its participatory nature, it is a useful method to focus attention on people, their livelihood and their interrelationships with socioeconomic and ecological factors.

8.9.1 Subcategories of the PRA Method

PRA is sometimes known as rapid rural appraisal (RRA), where the emphasis is on both rapidity and participation. The emphasis on rapidity, however, is more in terms of data collection and less in terms of the process of development or even the implementation of plans. Other names for PRA include participatory rapid rural appraisal (PRRA) and participatory learning method (PALM). PRA, as practised in the field, has given rise to different schools of PRA with differences in style and emphasis (Chambers, 1981).

8.10 A Detailed Analysis of the PRA Method

8.10.1 The Participatory Resource Appraisal (PRA) Method: Modes of Exercising as Tools

Participatory rural appraisal (PRA) is frequently used as an alternative and complementary method to conventional survey methods for resource assessment and utilization. Narayanasamy gave a complete elaboration of this innovative approach and tool: “a growing family of approaches and methods to enable local people to share, enhance and analyze of their knowledge of life and conditions to plan and act accordingly” (Narayanasamy, 2009).

Community participation in the sustainable ecosystem management of Sundarbans, with well-designed community mobilization tools and techniques to account for its large expanse (around 10,000 km^2), rich diversity of bioresources and fragile ecosystem functioning, is needed. Humankind and ecology are organically linked and inseparable. Livelihoods should be earned through agricultural cultivation and through the collection of honey, shrimp and crabs while protecting mangroves by combatting sea-level rise and salinity invasion. Centuries-old traditional practices and stewardship are to be interwoven with governmental initiatives as a blend of people and technology for successful, sustainable development. The participatory delineation and modelling for time, space, decision and flow analyses can generate a sui generis dataflow to help long-term planning for ecosystem conservation and stewardship (Gupta et al., 2022).

8.10.2 The Origin of the PRA Method

The philosophical roots of participatory rural appraisal techniques can be traced back to activist adult-education methods such as those of Paulo Freire and the study clubs of the Antigonish Movement. In this view, an actively involved and empowered local population is essential to the successful development of a rural community in relation to maintaining the ecological health of a certain piece of landscape while sustainably using resource bases. Robert Chambers (1983), a key exponent of PRA, argued that the approach owes much to “the Freirian theme, that poor and exploited people can and should be enabled to analyze their own reality.”

By the early 1980s, there was growing dissatisfaction among development experts with both the reductionism of formal surveys and the biases of typical field visits. In 1983, Robert Chambers, a fellow at the Institute of Development Studies (UK), used rapid rural appraisal (RRA) to describe techniques that could bring about a “reversal of learning,” to learn from rural people directly. Then, 2 years later, the first international conference was held in Thailand, and their participants shared their experiences related to RRA. This was followed by a rapid acceptance of its methods, which involved rural people in examining their own problems, setting their own goals and monitoring their own achievements. By the mid-1990s, the term *RRA* had been replaced by a number of other terms, including *participatory rural appraisal* (*PRA*) and *participatory learning and action* (*PLA*).

Robert Chambers (1993a, b) acknowledged the significant breakthroughs and innovations that informed the method, which came from community development practitioners in Africa, India and elsewhere. Chambers (1993a, b) helped PRA gain acceptance among practitioners by explaining the function of participatory research in PRA in the following manner: The central thrust of the new paradigms involves decentralization and empowerment. Decentralization means that resources and discretion are devolved, turning back the inward and upward flows of resources and people. Empowerment means that people, especially poorer people, are enabled to take more control over their lives and secure a better livelihood by having ownership and control of productive assets as one key element. Decentralization and empowerment enable local people to exploit the diverse complexities of their own conditions and to adapt to rapid change.

8.10.3 Basic Attributes Needed to Apply the PRA Technique

- Understanding group dynamics – e.g. through learning contracts, role reversals and feedback sessions
- Surveying and sampling – e.g. transect walks, wealth ranking and social mapping
- Interviewing – e.g. focus-group discussions, semi-structured interviews and triangulation
- Community mapping – e.g. Venn diagrams, matrix scoring, ecograms and timeline analysis

8.11 Relevant Methods for Community Development

8.11.1 Asset-Based Community Development (ABCD)

ABCD is a method for the sustainable development of communities based on their strengths and potentials. It involves assessing the resources, skills and experience available in a community, organizing the community around issues that move its members to action and thereafter making the final decision on taking appropriate action. This method uses a community's own assets and resources as the basis for development by empowering the people of that community and by encouraging them to use what they already possess.

8.11.2 Community-Based Programme Design

Community-based programme design is a social method for designing programmes that enable social service providers, organizers, designers and evaluators to serve specific communities in their own environment. This programme-design method depends on the participatory approach of community development often associated with community-based social work, and it is often employed by community organizations. In this approach, programme designers assess the needs and resources within a community, involve community stakeholders in the process and attempt to create a sustainable and equitable solution to address that community's needs.

Several community-based programme models are available:

(i) The social-ecological model, which provides a framework for programme design
(ii) The logic model, which is a graphical depiction of logical relationships between the resources, activities, outputs and outcomes of a programme
(iii) The social action model, whose objectives are to recognize the change around a community to preserve or improve standards, to understand the social action process/model as a conceptualization of how directed change takes place and to implement social action as a successful community problem-solving tool

Finally, programme evaluation involves the ongoing systematic assessment of community-based programmes.

8.11.3 Public Participation in the Decision-Making Process

Social science refers to different mechanisms for the public to express opinions and ideally exert influence in making political, economic, management or other society-related decisions. Participatory decision-making can take place in any realm of human social activity, including economic (participatory economics), political (participatory democracy or polity), management related, cultural (polyculturalism), familial or gender specific (feminism). For well-informed participation to occur, some version of transparency, such as radical transparency, is arguably necessary but not sufficient. Some have also argued that those who are most affected by a decision should have the most say and those who are least affected should have the least say on a topic.

8.11.4 Participatory Monitoring

The term *participatory monitoring* embraces a broad number of approaches, including the monitoring of harvests by local resource users, censuses run by local rangers and inventories kept by amateur naturalists. The term includes techniques referred to as self-monitoring, ranger-based monitoring, event-based monitoring, participatory assessment, the monitoring and evaluation of biodiversity, community-based observing, community-based monitoring and community-based information systems.

Many of these approaches are directly linked to resource management, but the entities being monitored vary widely, from individual animals and plants to habitats and ecosystem goods and services. However, all the approaches have in common that the monitoring be carried out by individuals who live in the monitored places and rely on local natural resources and have in common that local people or local government staff be directly involved in the formulation of research questions, data collection, data analysis (in most instances) and the implementation of management solutions based on research findings.

Participatory monitoring (also known as collaborative monitoring, community-based monitoring, locally based monitoring or volunteer monitoring) is the regular collection of measurements or other kinds of data (monitoring), usually of natural resources and biodiversity, undertaken by local residents of the monitored area, who rely on local natural resources and thus have more local knowledge of those resources. Those involved usually live in communities with considerable social cohesion, where they regularly cooperate on shared projects.

Participatory monitoring has emerged as an alternative or addition to professional scientist-executed monitoring. Scientist-executed monitoring is often costly and hard to sustain, especially in those regions of the world where financial resources are limited. Moreover, scientist-executed monitoring can be logistically and technically difficult and is often perceived to be irrelevant by resource managers and local communities. Involving local people and their communities in monitoring is often part of the process of sharing the management of land and resources with local communities. It is connected to the devolution of rights and power to locals. Aside from potentially providing high-quality information, participatory monitoring can raise local awareness and build the kind of expertise in communities and local governments that is needed to address the management of natural resources.

Participatory monitoring is included in the term *participatory monitoring and management*, which has been defined as "approaches used by local and Indigenous communities, informed by traditional and local knowledge, and, increasingly, by contemporary science, to assess the status of resources and threats on their land and advance sustainable economic opportunities based on the use of natural resources." The term *participatory monitoring and management* is especially used in tropical, Arctic and developing regions, where communities are most often the custodians of valuable biodiversity and extensive natural ecosystems.

8.12 The Rural Agricultural Experience Programme: Case Studies in Sundarbans, India

Rural people have been living in the agro-ecosystem of Sundarbans for several generations. They have a greater ability to accurately and diagrammatically represent their surroundings whether they are literate or illiterate. If properly facilitated, they can analyse, plan, implement, monitor or evaluate their programme with a high level of precision.

8.12.1 A Field-Based Research Team Carrying Out an Interactive Process with Local People

At the outset of the interactive process, villagers with traditional and Indigenous knowledge were directly contacted to earn their support and earn their help in applying their perceptions and knowledge to learn about problems and prospective solutions. This process had the following features:

(i) Mixed groups were formed, and they entered the village in small groups.
(ii) A group leader was selected, and they acted as the leading facilitator among us.
(iii) Proper time schedules were prepared, coinciding times that were suitable to villagers.
(iv) Necessary cultural protocols were followed before interaction.
(v) The team expressed excitement and enthusiasm in learning from villagers.
(vi) Innovative ideas were shared by the villagers.
(vii) Nothing was done to interrupt the villagers' concentration.
(viii) Six words (who, what, where, when, why and how) were emphasized to smoothly conduct the interactive process and to enable the helpers to provide necessary information, where respect was shown to the opinions of all members of the groups.

8.12.2 Icebreaking for Creating a Conducive Situation

In the Rural Agricultural Work Experience (RAWE) programme, icebreaking is the most indispensable task. Icebreaking means reducing the difficulties in starting a conversation, and it refers to interacting with, making rapports with and introducing people to accelerate access to their activities.

(i) Some amount of energy is required to break the psychological barriers of participants.
(ii) Similarly, rapid interaction is required to remove hesitation, fear and suspicion and feelings of superiority. For example, among two communities, namely the farming community and the student community, the farmers are habituated to their rural life, and the student communities are outsiders. The farmers are a heterogeneous mass, and they naturally feel hesitation in interacting with students because students are seen as outsiders or intruders. There is gap between the farmers' mentality and the students' mentality, and moreover, some may exhibit egotism.
(iii) To overcome this, icebreaking is needed, and it must be casual because an artificial approach may undermine connecting with the farmers. Our approach will be such that they may feel that we are very close to them.
(iv) We espoused out fondness for rustic people and their way of living in the lap of nature far away from crowds and artificial grandeur.

8.12.3 Stepwise Methods for Interacting with Local People

At the core of the procedure lie fundamental concepts such as the system itself, its hierarchical structure, its inherent properties and the agro-ecosystem. The participants initiated the process by outlining the objectives of the analysis, identifying the relevant systems, establishing their boundaries and arranging them in hierarchical order. Subsequently, a pattern analysis ensued, with all participating disciplines examining the systems from various angles, including the space, time, flow and decision aspects. These discerned patterns play a crucial role in determining the key properties of agro-ecosystems: productivity, stability, sustainability and equity. The outcome of these analyses can be either a set of agreed-upon key questions for future research or a set of preliminary development guidelines (Conway, 1985).

8.13 Developing Tools Based on Space Analysis

Space analysis is an analytical technique used in strategic management and planning. The analysis allows us to create an appropriate idea of the agro-ecosystem of the village (Chambers, 1984). A space analysis of the village has been done with the help of a social map, a hydrology map and an enterprise map. All of these attributes are highlighted in the following subsections.

8.13.1 A Social Map of a Village

A social map is delineated by community people who reflect their perceptions of the social dimensions of their reality with a high degree of authenticity. It seeks to explore spatial dimensions by depicting habitation patterns, the nature of housing and social infrastructure, such as roads, drainage systems, schools, drinking-water facilities, etc. (Flowchart 8.6).

A social map of a village can lay out who lives where and with what resources by drawing it with the help of the people in the village. This map shows the village's social structures, such as households, health centres, schools, churches and other major institutions in the area. The map indicates the social and economic characteristics of the households (e.g. high income vs low income, female-headed household vs male-headed household, etc.). Social mapping is best carried out at the beginning of the appraisal, and it can provide what's needed for other appraisals, such as wealth ranking or Venn diagrams (Singh, 2001). The components of social maps are as follows: (i) households; (ii) roads; (ii) religious places; (iv) integrated child development service (ICDS) centres; (v) schools; (vi) cultivable lands; (vii) orchards and vegetable gardens; (viii) the location of cash crops like beetle vines; and (xi) social activities like afforestation with the involvement of nongovernmental organizations (NGOs) and local clubs.

Flowchart 8.6 Social map prepared from grassroots-level interactions with local people

A social map can indicate both the temporal and spatial changes in response to changes in population, livelihood, settlement and ecology. This participatory tool can be highly effective in Sundarbans areas where the spatial and temporal change signatures are fast, dramatic and deeply impactful on life and ecology.

8.13.1.1 The Salient Features of a Social Map

- The social map was used as an interesting and enjoyable tool to engage community members in discussing their village.
- The social map was used to understand the physical structure of the village in relation to cast and gender, cultivated crops, wealth and well-being rankings and identified issues that need further supervision by experts.
- From this map, we learned that this is a mainly agriculture-based village, where crops like gourd, watermelon, brinjal, chilli, onion, mango and lemon are cultivated in upland area.
- In the lowland area, mainly rice is cultivated.
- The village has a health centre, a primary school and several ponds that are used for fish cultivation.
- The fields are mostly irrigated with underground water through irrigation channels.

8.13.2 A Hydrological Map of a Village

A hydrological map of an area denotes various types of features on the basis of land type, water resources, drainage facilities and direction, the water table, etc. Because the main enterprises of a village, like crop production and fishing, are strictly

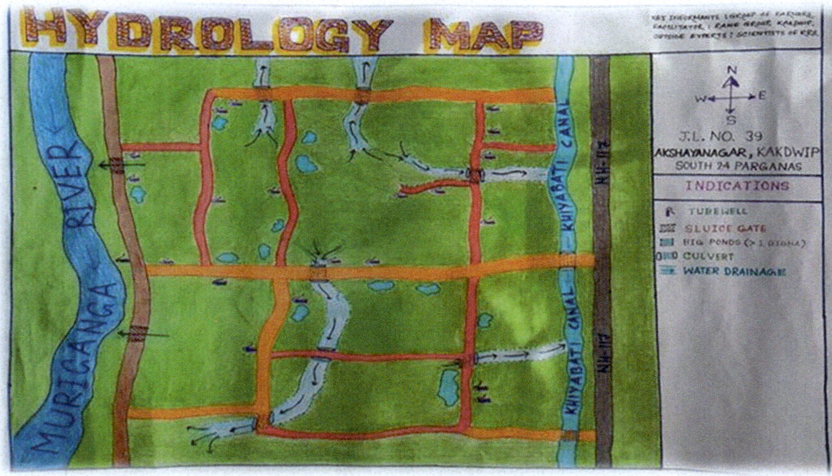

Flowchart 8.7 Hydrological map prepared from grassroots-level interactions with local people

dependent on water availability, a hydrological map shows an overview of water resources and their utilization patterns. A hydrology map of the village was drawn with the help of village residents (Flowchart 8.7).

Water is the most important resource for any village to have a decent and dependable livelihood. Participatory mapping by local villagers includes the distribution of water resources, both surface water and groundwater. It delineates the direction of rainwater runoff and the distribution of ponds and lakes. Swamps and bogs play a crucial role in rural agro-ecosystems. Better stewardship can change a placid aquacultural system into a vibrant economy through fishing enterprises and ecotourism. Fluctuations in ponds and groundwater tables have decisive impacts on the farming economy of a village. Arsenic contamination in groundwater can be correlated with the heavy extraction of groundwater to support summer rice cultivation.

The participatory hydrological mapping can be of great importance when planning for composite fish cultures, rearing duck and poultry and raising livestock.

8.13.2.1 The Salient Features of a Hydrology Map

A hydrology map provides an outline of the positioning of different water sources and the physical locations of waterbodies throughout a village, generating research information. The components of a hydrology map include (i) big ponds with different sizes, shapes and depths (0–10 katha (1 katha = 720 feet^2); 10–20 katha; and greater than 1 bigha); (ii) tidal canals with specific characteristics like connectivity with large nearby creeks and estuaries, the intensity of the water current, tidal amplitude, the depth of water, bottom soil composition, the drainage patterns of fresh water and associations with mangrove vegetation; and (iii) estuarine creeks, where creeks of different lengths and breadths, most of which crisscross the mangrove forests and help with the mixing up of fresh water with saline water, are selected. Most of the freshwater ponds in reclaimed deltas are used for pisciculture and for the monoculture of commercially important prawn species (*Macrobrachiium rosenbergii*) and shrimp species (*Penaeus monodon*). Local irrigation to paddy fields is jointly carried out by pond water fed by rainwater and, in boro season during the summer months, by a submersible pump that extracts subsurface water. Drinking water is supplied by both tube wells and submersible pumps. Occasionally, waterline tap water is used for drinking purposes.

8.13.3 An Enterprise Map of a Village

Enterprise mapping is the abstract representation, description and definition of the structures, processes, information and resources of an identifiable business, government body or other large organization. The enterprise map of Akshyanagar was carried out with the help of local villagers. The components of an enterprise map includes (i) poultry farms; (ii) the location of organic manures like vermicompost and green manure; (iii) dry fish production units; and (iv) fishing-related and aquacultural activities (Flowchart 8.8).

Four patterns are chosen as likely to reveal the key functional relationships that determine system properties. Three of these – space, time and flow – are known as to be important in determining the properties of ecological systems. All three attributes play significant roles in ensuring increased productivity. Variability or heterogeneity in space is also considered an important promoter of stability and sustainability. Variability in time, however, can be destabilizing given that systems with long time lags are often unstable (May, 1975). Flows are either stabilizing or

Flowchart 8.8 Enterprise map prepared through interactions with people at the village level

destabilizing depending on the feedback loops at play, which are either negative or positive (Levins, 1974).

Time-trend and timeline analyses are crucial to delineate the chronological changes and patterns in an evolving ecosystem. Here, the time trend acts as a participant. The generated research data have delineated the pattern of changes in some attributes under study. The drastic decline in livestock count on the deltaic islands of the Sundarbans mangrove ecosystem has certainty contributed to the deteriorating soil health. An occupational shift of rural people from farming to other livelihood options, including driving battery-driven Toto vehicles (electric rickshaws), is noticeable.

8.14 Assessing Temporal Changes Via Timeline Analysis

Graphs provide the most effective means of representing temporal patterns, and three specific patterns hold significance in agro-ecosystems. The first pattern revolves around seasonal variations, which can be analysed by overlaying or concurrently plotting cropping sequences, labour, credit peaks, prices and other relevant factors against various agro-meteorological parameters. This approach is particularly useful for pinpointing crucial periods during the year when the timing of operations and the availability of resources significantly impact productivity and stability. On the other hand, longer-term changes in variables such as prices, production, climate conditions and demographic parameters can be graphically depicted in a conventional manner, with a minimum of 10 years' worth of data required for a meaningful analysis. These include components, timelines, time trends and space trends.

8.14.1 Objectives of Timeline Analysis

1. To draft a historical profile of the agro-ecosystem indicating the major qualitative and quantitative changes
2. To access and understand the development of the village

8.14.2 Details About the Timeline Analysis (Connecting Historical Events to Establish a Link Between Eventualities and Logic)

Timeline analysis is a valuable method that enables individuals to gain a deeper understanding of data by systematically organizing them on the basis of specific criteria. This organization can involve sorting data according to chronological dates or categorizing data according to specific types of events (Table 8.1).

8.14.3 Time-Trend Analysis

Trend analysis is a popular component in the PRA method, which is used to explore the temporal dimension by focusing on the changes that it captures and trends related to certain variables over different spans of time (Tables 8.1 and 8.2).

8.14.4 Seasonality Analysis

- A seasonality analysis is a participatory time analysis with a seasonal or ephemeral focus on resource management, migration, wages, livelihood, variability in groundwater, etc.
- Seasonality analysis helps in identifying the peak period or gluts of different characteristics and possible reasons.

8.14.5 Seasonality Analysis for Upland

The following case study project for the villages of Sundarbans:

- Boro paddy is cultivated from the first week of Poush (January) to the first week of Boishakh (April), where the cultivated variety is Pratikshya.
- Aman paddy is cultivated from the first week of Ashar (the middle of July) to the second or third week of Kartik (end of October).
- Agrohayon is cultivated from November to December, where the cultivated variety is Santashi.
- Mustard is cultivated from the second week of Agrohayon (November) to the third week of Magh (January), where the cultivated variety is Binay.
- Grass pea is cultivated from the second week of Agrohayon (November) to the first or second week of Falgun (March), where the cultivated variety is Ratan.
- Lentil is cultivated from the second week of Agrohayon (November) to the first week of Falgun (March), where the cultivated variety is Maitrayi.
- Sunflower is cultivated from the second week of Poush (December) to the third week of Boishakh (April), where the cultivated variety is Svnbred.
- Mung bean is cultivated from the second or third week of Falgun (March) to the third week of Boishakh (April), where the cultivated variety is Samrat.
- Black gram is cultivated from the second or third week of Falgun (March) to the third week of Boishakh (April), where the cultivated variety is Pant-U-31.

Intervention

- Farmers have a wide scope of crops, which is depicted through a seasonality analysis of crops cultivated in medium land (Flowcharts 8.11 and 8.12).

8.14.6 *Seasonality Analysis for Low Land (Table 8.3)*

- In this season, crops are cultivated in low land with irrigation water. The source of water is rainfall.
- Boro rice is cultivated from January to April. The cultivated local variety of paddy seed is Pratikshya.
- Aman paddy is cultivated from the third week of June to the second week October. The cultivated local variety of rice is Dudheswar.

Intervention
In the summer, aman rice can be utilized for raising short-duration leafy vegetables like ipomea or Amaranthus. Low-lying land can be developed in the form of integrated farming systems, including paddy and fish cultures, to mitigate both the stress of salinity and low yields when farmers have less area for crops.

8.15 Wage Analysis

The social ecology of Sundarbans is deeply impacted by wage variability and accessibility throughout the year. Wages are low, uncertain and variable depending on gender. That is why migration, both territorial and occupational, is so acute and inevitable in Sundarbans (Flowchart 8.9).

There has been a stark difference between male rural labourers and female rural labourers. For certain months of the year, especially from September to November, the main days for earning are plunging to their lowest levels, leaving a serious crisis for wage earners because women fail to undertake this manual hardship. For any ecosystem, when its peripheral sub-ecosystem, which in this case is the agro-ecosystem, fails to provide income backup, the core area of the main ecosystem is subjected to stress and made vulnerable. The loss of such a sub-ecosystem would lead to ecological chaos and entropy (Acharya & Sharma, 2019). The seamless outmigration from different parts of the Indian Sundarbans can be attributed to the failures of supporting the agro-ecosystem in generating necessary returns from the land. Punitive poverty associated with uncertainty in income has made the social ecology of Sundarbans extremely vulnerable to disorder, chaos and entropy. The lands of the small and marginal farmers are forcibly extorted in favour of brackish fishery, which jeopardizes both the local ecology and the economy, worse enough to add to sociopolitical anomie, which is extremely lethal to sustainable ecosystem services. The eco-restoration of the ecologically degrading mangrove ecosystem of Sundarbans is possible by engaging in a large-scale eco-rehabilitation programme with the active participation of local people by planting mangroves.

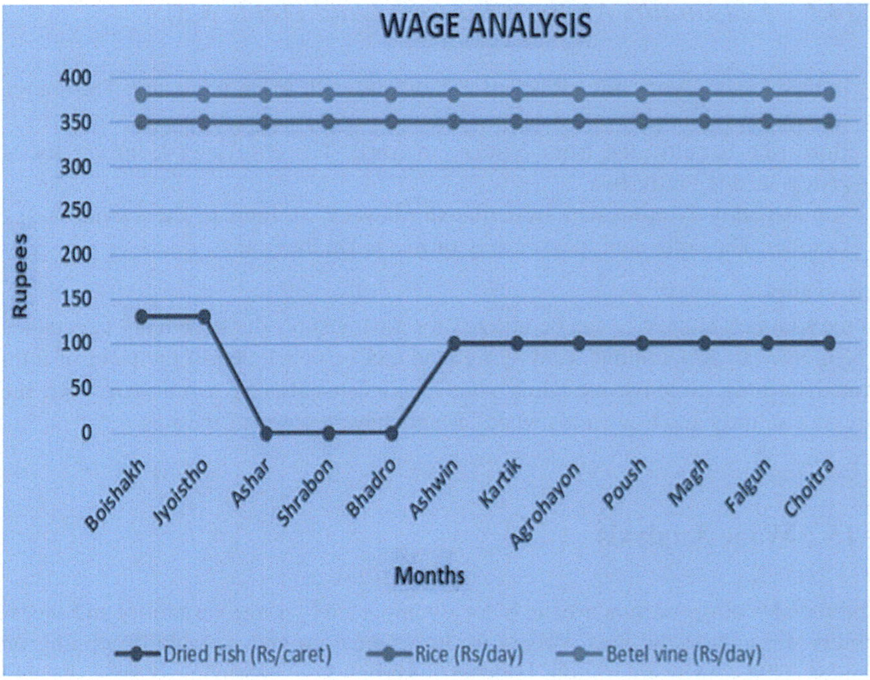

Flowchart 8.9 Wage analysis through different Bengali months of the year in relation to three major crops produced in the reclaimed areas of Sundarbans (Boishakh refers to the first month of the Bengali year (the middle of April to the middle of May), and Choitra refers to the last month of the Bengali year (the middle of March to the middle of April))

8.16 Flow Analysis for Evaluating Complex and Dynamic Social Networks

Complex and dynamic social network in Sundarbans need to be analysed through participatory Venn diagrams. Such a diagram nicely depicts the role of different institutions and their interdependencies. The exercise will be very helpful in identifying different organizations, their impacts and core–peripheral linkages. Within this category, the patterns related to the movement and transformation of various elements need to be examined within agro-ecosystems, such as energy, materials, money and information. Although these patterns can be represented by using traditional flow diagrams, the primary objective should not be an exhaustive depiction of every intricate relationship. The focus should be on analysing flows to identify the primary causes and effects and the presence of stabilizing or destabilizing feedback loops. To maintain clarity and effectiveness, flow diagrams need to be kept as straightforward as possible. Other tools, like tables, matrices, bar histograms and regression graphs, can also prove valuable in highlighting significant relationships within an agro-ecosystem. These alternative representations contribute to a more comprehensive understanding of the system without overcomplicating the analysis with excessive detail (Conway, 1985).

8.16.1 Venn Diagram for Assessing the Community Interaction Process

A Venn diagram is one kind of diagram pictorially representing logical sets as circles or closed curves within an enclosing rectangle (the universal set), common elements of the sets being represented by intersections of the circles. The diagram shows the relationship of various institutions, organizations or individuals with the village as perceived by the villagers (Flowchart 8.12).

Objectives

- To identify groups and institutes operating in the community and show their interactions.
- To show the accessibility of the institutes.
- To discover the importance of decision-making for the community.

Revelation:

Most of the amenities are near the village.

Description of the Diagram

- After being closely and actively acquainted with the village, the assistant directorate of agriculture (ADA) of the Kakdwip block, the Kakdwip Regional Research Station (RRS) of Bidhan Chandra Krishi Viswa-Vidyalaya, the local agro-chemical shop, pan Mistri and the local Gram Panchayat play significant roles in the planning of cultivation practices in their respective villages.
- The local post office and bank, which are in proximity of the village, are also involved.
- The local hospital is quite far from the village.
- The local state poultry farm (SPF) and the Central Institute of Brackish Water Aquaculture (CIBA), which is part of the government of India, do not play any significant role in local cultivation practices (Flowchart 8.5).

Intervention

- The SPF and CIBA should try to encourage the establishment of more enterprises in the village and help in the process.
- Healthcare facilities should be increased.
- More contact with Panchayat is needed.
- Schools have to pay more attention to the development of rural communities and agricultural activities.

Venn diagrams and problem-tree delineation are vital tools for a participatory assessment of institution–community linkages in the fabric of ecosystem functioning. Through community-based mapping exercises, critical problems can be identified. Improvements in transport, storage and marketing facilities can act as a magic wand. The community perception is that unless agricultural and livestock enterprises are improved, the extraction of forest and water resources will continue to hamper ecological health. The terrains and trajectories of ecological resilience bank on core–periphery linkages and dynamics. The quality of rural live needs to be improved such that agriculture, livestock, fishery and tourism can become game changers.

8.17 Using Problem Trees to Depict the Flow of Resources

A problem tree is a schematic diagram of resource inflow into an area and outflow from that area. This diagram shows the flow of resources from the original source to its final output and visualizes the use and control of the resources (Flowcharts 8.2 and 8.5). These schematic diagrams depict problem identification and elucidation through problem–cause diagramming. (Different problems and causes are deeply interwoven in the sociocultural and sociopolitical conditions of Sundarbans. These problem trees have been drawn to depict the present status of the human community of Sundarbans.) Local people need to decide to address problems and need to understand how problems impose barriers to achieving goals. The identification of problems is carried out by a group of farmers. All the problems of that area are identified democratically by villagers. To tackle the selected problems, the time was scheduled in according with priorities. The problems are generally prioritized on the basis of three (3) major attributes: (i) the seriousness of the problem; (ii) the frequency at which problems occur and (iii) the distribution of problems.

Processes Involved: Ranking the Main Problems

1. The area consists mainly of lowlands where water logging occurs.
2. The infestation of several pests and plant pathogens has resurged.
3. Climate change has led to an uneven distribution of rainfall.

Intermediary Problems

- Farmers and villagers have no awareness of the negative effects of excessively using chemical fertilizer.
- Villagers lack education facilities.
- Farmers lack of knowledge about using advanced technology.

Peripheral Problems

- Farmers are far behind in modern-age cultivation skills.
- Most of the local farmers are very poor, so their risk-bearing capacity is low.
- The area lacks proper infrastructure for the transport, storage and marketing of crops.

Flowcharts 8.1, 8.2, 8.3, 8.4 and 8.5 explain community-based decision analysis, pointing out effective microplanning, which is essential to elicit the choices of local people on occupations, enterprises, crop varieties, livestock breeds, selecting brackish-water finfish and shellfish like shrimp and prawns. Also, alternative livelihood features honey collection and the collection of nontimber forest resources from all corners of the forests. People's intention to enter and venture depends on the profitability of honey and other forest products in response to market belligerence. Matrix ranking will be helpful in determining community facts and figures about various choices.

Decision analysis allows farmers to make decisions among suitable alternatives (Table 8.4). Decisions about agro-ecosystems involve a wide spectrum of interests, including national agricultural policies and the day-to-day choices made by individual farmers. These decisions occur at various levels within the hierarchy of agro-ecosystems, and two patterns are particularly noteworthy. The first pattern revolves around the choices made within a specific agro-ecosystem under varying conditions. This pattern is best elaborated through the construction of a decision tree. Developing such a tree serves two purposes: It reveals the objectives of the decision maker, and it sheds light on the constraints that shape decision-making within an agro-ecosystem. For instance, when decision trees were created for agro-ecosystems in northeastern Thailand, they highlighted the significance of constraints from labour availability and land types on farm and village production. These decision trees offer valuable insights into decision-making processes and the factors that influence them within the specific context of an agro-ecosystem (Conway, 1985). This exercise is important for two reasons: It helps decision makers to select the best course of action, and it provides a structured framework to quantify the outcomes and the probabilities of achieving them.

8.18 Matrix Ranking for Adjudging the Quality of Agricultural Products

(i) Using matrix ranking on rice: After using the matrix to study and analyse various rice varieties and their attributes, we found that among these varieties, the Dudheswar rice variety has the maximum attribute value according to matrix ranking due to its having the best taste, a high yield and a high price and being in high demand. After Dudheswar, another rice variety, namely Pratikshya, is also a preferable variety due to its high yield and resistance to pests and diseases. Soil suitable for the cultivation of resistant varieties of rice should be used (Flowcharts 8.10, 8.11, 8.12 and 8.13).

(ii) Using matrix ranking on betel vine, a climber, the leaves of which serve as a cash crop: After using the matrix to study and analyse two betel vine varieties and their attributes, we found that among these varieties, one variety, locally named Bangla Pan, has the maximum attribute value according to matrix ranking due to its having the highest cultivated area, a high yield and low levels of pest attacks and disease. After Bangla Pan, Mitha Pan, another variety, is also preferable due to its high price and high demand. Soil suitable for the cultivation of resistant varieties of betel vine should be used.

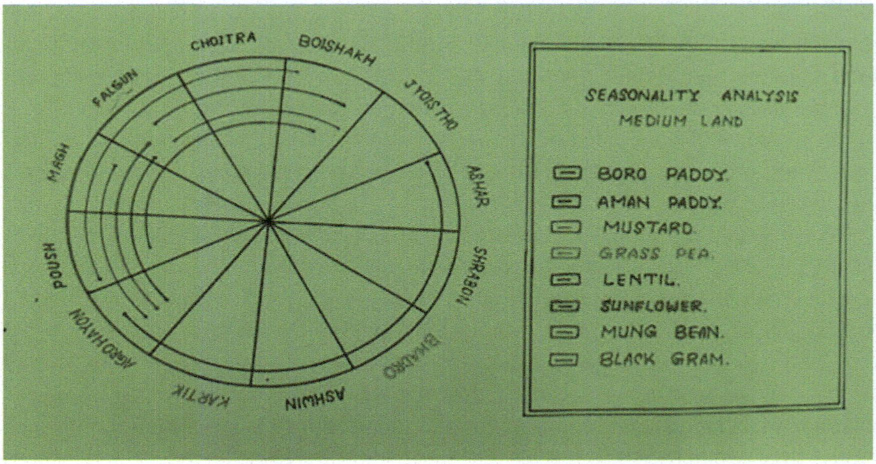

Flowchart 8.10 A seasonality analysis of different agricultural products during different periods of the Bengali year in the peripheral zones of the core area of Sundarbans

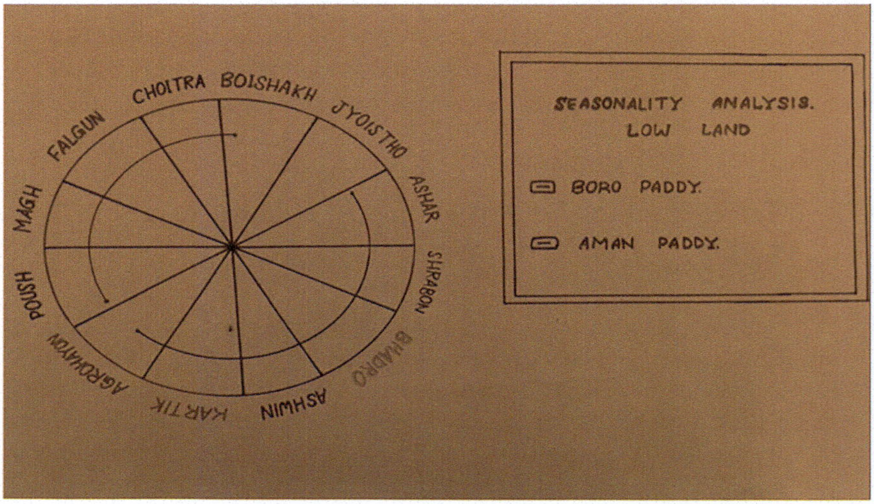

Flowchart 8.11 A seasonality analysis of two major crops of paddy during different periods of the Bengali year in the peripheral zones of the core area of Sundarbans, India

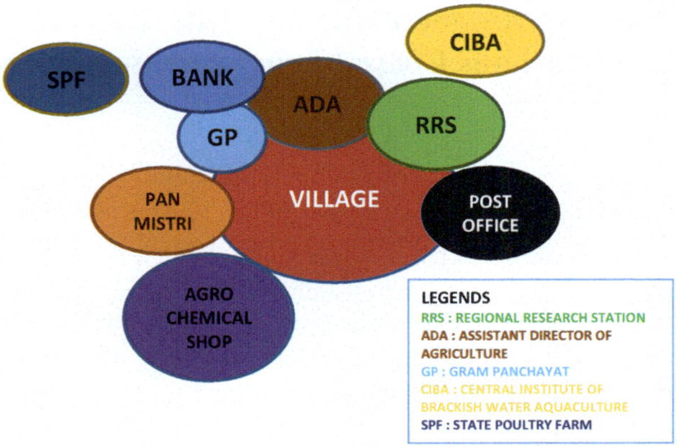

Flowchart 8.12 Venn diagram depicting different institutional organizations in different villages within the reclaimed areas of Sundarbans, India, to assess the community-interaction process

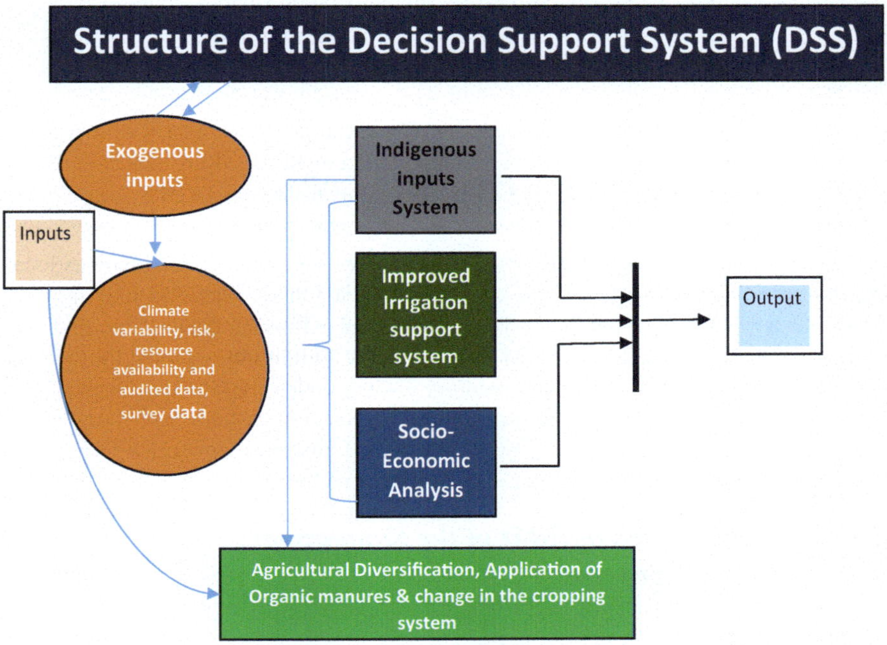

Flowchart 8.13 A diagrametric representation of the different components of a decision support system (DSS), including its inputs and outputs, the combined activities of which help in understanding and mitigating climate variability

8.19 Participatory Resource Management (PRM) in the Sundarbans Coastal Ecosystem: A Review

The Sundarbans coastal ecosystem, located in the eastern part of India and south-western Bangladesh, is a unique and fragile environment that supports a diverse range of flora and fauna. The ecosystem is facing numerous challenges, including climate change, overfishing and habitat destruction, which threaten the livelihoods of local communities and the long-term sustainability of the ecosystem. Participatory resource management (PRM) has been recognized as a promising approach to address these challenges and promote sustainable management in the Sundarbans coastal ecosystem. This very participatory method is devised to assess Maximum Sustainable Yield (MSY) and Optimum Sustainable Yield (OSI). An ecosystem is supposed to operate within a resilient framework set by its intrinsic capacity to move from highest possible limit of productivity to the base level yield. MSY denotes the maximum level which may touch the tipping point after which the ecosystem is expected to suffer serious disruption. OSI is the average level of of yield disposition by allowing both sides buffering space to move up and down to make the system resilient.

8.19.1 The Definition and Principles of PRM

PRM is a collaborative approach to natural resource management that involves local communities, stakeholders and government agencies in the decision-making process. The approach is based on the principles of participation, decentralization and empowerment (Kumar, 2002). PRM recognizes the importance of local knowledge and expertise in managing natural resources and seeks to build capacity and promote ownership among local communities.

8.19.2 Applications of PRM in the Sundarbans Coastal Ecosystem

Several studies have demonstrated the effectiveness of PRM in promoting sustainable management in the Sundarbans coastal ecosystem. For example, a study by Ghosh et al. (2017) found that the participatory management of fisheries in Sundarbans has led to increased fish catches and improved livelihoods for local fishing communities. Another study, by Mitra et al. (2019), found that community-led conservation efforts in Sundarbans resulted in significant reductions in habitat destruction and improved biodiversity.

8.19.3 Benefits of PRM in the Sundarbans Coastal Ecosystem

The application of PRM in the Sundarbans coastal ecosystem has several benefits:

(i) PRM has been shown to improve livelihoods for local communities by providing them with greater control over natural resources and promoting sustainable use practices (Kumar, 2002).
(ii) PRM has been found to promote biodiversity conservation in Sundarbans by involving local communities in conservation efforts and promoting sustainable use practices (Mitra et al., 2019).
(iii) PRM has been shown to increase the resilience of local communities to climate change and other external shocks by promoting adaptive management practices and providing them with greater control over natural resources (Ghosh et al., 2017).

8.19.4 Challenges and Limitations of PRM in the Sundarbans Coastal Ecosystem

Despite the benefits of PRM, its application has several challenges and limitations in the Sundarbans coastal ecosystem:

(i) The lack of institutional capacity and resources can limit the effectiveness of PRM in Sundarbans (Kumar, 2002).
(ii) PRM can be influenced by power dynamics between different stakeholders, which can limit the participation and empowerment of local communities (Ghosh et al., 2017).
(iii) PRM can be difficult to scale up to larger areas, which can limit its impact on the Sundarbans coastal ecosystem as a whole (Mitra et al., 2019).

In view of the above, PRM has been recognized as a promising approach to promoting sustainable management in the Sundarbans coastal ecosystem. The application of PRM in Sundarbans has several benefits, including improved livelihoods, enhanced biodiversity and increased resilience. However, its application also has several challenges and limitations in Sundarbans, including limited institutional capacity, power dynamics and scalability. Addressing these challenges will be critical to ensuring the long-term sustainability of the Sundarbans coastal ecosystem.

8.20 Cases Study on the Application of PRA Tools for Designating Conservation Categories for Fish

India is blessed with having rich fish diversity and germ plasma bioresources in diversified aquatic habitats. A case study in different types of wetlands, including estuaries, tidal creeks, bheries, aquaculture farms, large ponds and inundated paddy fields with brackish water in different parts of northeast Sundarbans, have revealed interesting information on the declining trend in the diversity and density of many finfish, especially those belonging to the family Siluridae, and the identification of the causal factors responsible for such a decline in view of continuous anthropogenic stresses. With the advancement of modern technologies applicable to the fishery sector and the use of a modernization process to meet the ever-growing demand for fish, this rich diversity is being eroded across the world, and some species have been pushed into the "threatened" category. The resolutions adopted at the convention of the Earth Summit held at Rio de Janeiro in 1992 aimed to conserve biological diversity and ensure the sustainable use of biological resources and the equitable sharing of such benefits to all involved in this process. In addition to the ecodegration of different freshwater and brackish-water habitats, the fishery resources of Sundarbans have been gradually but steadily declining, making it a nonproductive sector of earning revenue, owing to varied natural and anthropogenic factors (Jhingran, 1988; Pandit et al., 199; Kothari, 1997; Patra et al., 2004).

In diversified agro-climatic regions and a heterogeneous rural society like India, the problem of such a decline in bioresources is very much needs based and location specific (Meena, 2001). Owing to the culture potentiality of several species of catfish under the family Siluridae in the stock fishery of northeast Sundarbans, most of them are very prone to the harmful effects of environmental degradation because they inhabit different aquatic habitats with different ecological characteristics, which results in a decline in their population. In this research study, an attempt has been made to assess the present conservation status of the freshwater and brackish-water fish of the order Siluriformes in northeast Sundarbans in order to identify and prioritize their needs-based problems, such as their depletion, through the techniques of participatory rural appraisal (PRA), which has emerged as an intensive and systematic learning experience carried out in a local community (Mukherjee, 1995).

8.20.1 Participatory Techniques

The participatory and objective methods involved in this PRA tool are meant for assigning threat categories from several identified fish species and deriving recommendations for conservation actions through interactive group dynamics from a number of stakeholders. Selected PRA techniques were carefully designed and applied to generate useful primary information pertaining to the prevailing conservation status of fish under consideration. The tools and techniques involved several other method-based elements: trend analysis, fish magnitude value (FMV), matrix ranking and rank-based quotient. Trend analysis was assessed to analyse the nature of increasing or declining numbers of the target fish population over decades by taking into consideration the age and experience profiles of 50 stakeholders. FMV was estimated from a participatory approach where both geographical and social attributes were considered in temporal dimensions. The contention was that fish species had declined not only over a geographical area but also over time scales. In such a context, each individual stakeholder was interviewed to assess the availability of a fish species in terms of biomass (kg) for a given decade, and that was multiplied by the visibility range – that is, the area that fish occupy (km^2) – to help determine the abundance status of that species over that decade. This type of calculation helps to delineates the percentage of decadal increase or decline since the benchmark decade. To avoid biases in perceptual assessment, averages were used.

8.20.2 Diversity of Ichthyofauna from Different Brackish-Water Impoundments of Northeastern Sundarbans

Fish were collected from aquatic subsystems like rivers, estuaries, creeks and brackish-water bheries by using conventional fishing technologies. All the collected fish were identified according to the standard taxonomic literature (Jhingran, 1988; Pandit et al., 199; Kothari, 1997; Patra et al., 2005). Relevant information pertaining to different fishing technologies has been gathered through participatory interaction with stakeholders by using a set of preformulated questionnaires in a village-level group who underwent physical verification (Mukherjee, 1995).

8.20.3 Participatory Methods: A Tool for Bio-assessment and Conservation

Participatory methods include the spontaneous involvement of local communities or stakeholders because each person performs their own analysis through such methods with their own ways of explaining the findings and proposing action on that basis. These simple and handy methods are both "verbal" and "visual" and involve the use of signs, symbols, objects, models, measures, etc. (Patra et al., 2005; Mishra et al., 2009).

8.20.3.1 Participatory Rural Appraisal (PRA) Techniques

A methodological tool based on people's perceptions, including a "basket of techniques," that is adopted by local communities to understand their perceptions and to learn from them is designated as a PRA method. This method, as alternative to data collection and analysis, promotes lively interaction with local communities for Indigenous knowledge–building exercises. Manifold research data are collected from locals with their subsequent identification and mobilization by the intended groups, and their active participation opens up various avenues for the intended groups to participate in decision-making, project design, execution and monitoring. PRA, being a useful strategy, helps a person to concentrate on socioeconomic, political and ecological surroundings with a special focus on livelihoods and the interrelationships between people through the participation of a local community.

8.20.3.1.1 Questionnaire Survey

A questionnaire survey refers to a unique means of investigation into the people of an area of specific ecological distinctiveness, where they are treated as passive entities because they are not involved or they are unable to influence the objectives and direction of the ongoing enquiry. In this research study, a picture-based questionnaire survey was conducted to evaluate the knowledge of the villagers in an ecozone of Sundarbans, India, about the trend showing a decline in natural resources and the problem of developing ecological potentialities.

Before conducting the interviews and taking necessary photographs, the verbal consent of willing persons was taken. The questionnaire survey was conducted to explore two dimensions of people's cognizance to gather information of the background socioeconomic status of the respondent, viz. age, gender, occupation and level of education, which can help understand the effect of these variables on the perception of respondents with respect to the possible causes of a reduction in ecological potentialities (Patra et al., 2005, Kasina, 2007; Mishra et al., 2009). After taking the age profiles of the respondents into account, 30 willing participants aged 40 years and above were considered as the "principal informants" to categorize the commonly identified problems of bioresource decline.

Selected participatory rural appraisal techniques, namely (a) trend analysis, (b) fish magnitude value, (c) matrix ranking and (d) rank-based quotient, were carefully designed and applied to extract primary information relating to the conservation status of fish. These techniques are elaborated below.

8.20.3.1.2 Trend Analysis

Trend analysis is used as a methodological component of participatory rural appraisal techniques and is applied to depict the mode of decline in a species over decades. It has been carried out to analyse the nature of the increasing or declining numbers of a target fish population over decades. In this analysis, local people were interviewed to understand the trends of different variables from a historical perspective by using the experience profiles of stakeholders. The method helps by providing a background for any issue through trend analysis (Mukherjee, 1997). In applying this method of trend analysis, 75 respondents were to identify and assess the decadal decline in different fish species (17) in terms of time and the visibility of the target species in nearby surroundings. The scoring by villagers was low amid declining populations of various indigenous fish species in their natural habitats (irrigation canal, tributaries and rivers), which have reduced drastically from 1960 to 2000 (Patra et al., 2005).

8.20.3.1.3 Fish Magnitude Value (FMV)

The FMV was calculated in a participatory approach where both the geographical and temporal dimensions were considered. For the designing, formulation and calculation of the fish magnitude value (FMV), vertical and horizontal logics were considered. To compute findings through this method, both the geographical distributions and temporal dimensions were taken into consideration. The calculation of the fish magnitude value was considered in conformity with the International Union for the Conservation of Nature (IUCN) guidelines (IUCN, 2001; Anon, 2001).

The prime objective was to find out the modes of decline in fish species both over geographical space and over time. This is an innovative technique relying on perceptual assessments by fishers of different age groups (60–70 years old). This assessment is related to the decline in different fish species numbers, not only over time but also over distribution. For example, fishers assessed availability through the biomass (kg) of a fish species in a given decade in terms of scores designed by themselves, and these score values were multiplied by the visibility range – i.e. the extent of their occurrence (km^2) – to help to determine the status of that species over that decade. This type of calculation over decades helps in the delineation of the percentage of the rate of decline since the benchmark decade. To avoid introducing biases in the perceptual assessment made by an individual interviewer, the total score value of a group was divided by the number of members of that group to reach an interpolated value in the form of a mean:

$$\text{Mean of FMV} = \frac{\text{Yield} \times \text{Area of Occupancy}}{\text{Number of Respondents}}$$

8.20.3.1.4 Matrix Ranking

Matrix mixing is a unique method used to derive local perceptions and quantify the perceptions to draw a logical inference on why one ecosystem can become superior to another. By applying this method, one can reach an excellent estimation through a participatory mode of different attributes or values for different sites. Different attributes and criteria which were mentioned in a matrix scoring exercise were ranked according to their relative importance on the basis of fixed scoring (e.g. a score out of ten). The perceptions of rural people were fruitfully revealed through an exercise in matrix scoring. A number of causal factors were identified as instrumental in the declining fish populations in their natural habitats (brackish-water bheries, irrigation canals, tributaries, floodplains, tidal creeks and rivers) in north-eastern Sundarbans, India. Group meetings with the local fisher community of some selected villages were organized to collect information on the factors responsible for declining fishery resources.

Mostly the male members of the community assembled in the meeting place. The set of criteria was discussed by the fishing community. They were also asked to add to the list and to score the set of criteria generated. Habitat destruction, agricultural waste (fertilizers and pesticides), industrial effluents with pollutants and contaminants, overfishing, nonrestricted nylon thread fishing gear, ichthyotoxic materials derived from local plants, fishing during breeding seasons, fish diseases (epizootic ulcerative syndrome and others), the excessive and nonjudicious use of water, human population explosion, the occurrence of flood, the introduction of exotic fish species, siltation and fish bodies without scales (species under the order Siluriformes) were found to be the major causes of the decline in fishery resources. To initiate the process of scoring by individuals, participating people were supplied with the criteria written on a piece of paper in their vernacular (such criteria were read aloud and understood by the participants, although many of them could not read or write) and arranged in a column on the ground. Each participant was given a handful of stone chips (ten stone chips) and asked to score each criterion as much or as little as they wished to. It was free scoring, as explained to the participants, where the relative importance of a criterion would be indicated by assigning more stone chips to more-important ones and fewer stone chips to less-important ones (Patra et al., 2004, 2005; Mishra et al., 2009).

8.20.3.1.5 Rank-Based Quotient (RBQ)

In this type of approach, to calculate the findings from different forms of interactions with the stakeholders, generated data pertaining to different quality responses in both the vertical and horizontal dimensions were interpolated and interpreted. At first, the individual choices for ranking against the relative position in the rank were calculated, then the same were multiplied by the rank values, and then the total values were divided by the total number of members in the participatory groups. This approach is a unique concept used to eradicate the outliers among quality responses (Patra et al., 2005). To calculate the rank-based quotient (RBQ) of each problem, the formula found in Sabarathnam (1988) and Sabarathnam and Sengottaiyan (1996) was used. Ranking was done with the help of rank-based quotient (RBQ). A focus group was formed. They were put to a discussion to identify ten causal factors. The formula was made to re-rank the identified causes according to their vertical and horizontal dimensions:

$$\text{RBQ} = \frac{\Sigma \text{fi}(n+1-i)}{N * n} \times 100$$

where

fi = the frequency of respondents reporting the problem under *i*th rank
I = the rank of problems
N = the total number of respondents
n = the total number of problems

To identify the causal factors responsible for the sharp decline in the ecological potentialities of fish habitats over decades, a meeting involving 60 local people was organized. On identifying some problems through participatory interactions, they were categorized, sorted and screened. In these interaction processes, the number of participants was fixed at 20 for each focus group, and three such groups were formed to engage in the required interactions, which over time generated adequate data.

This type of participatory tool (RBQ) had already been successfully used over decades to assess and document the nature and degree of the decline in scores for selected bioresources (Patra et al., 2005).

8.20.3.1.6 Market Surveys and Assessments of Fish Production

Extensive market surveys carried out on a regular basis over different months and seasons of the year provide readily available information of the diversity of fish species in a particular region. The data on the average monthly landing of 17 species of fish in the order Siluriformes collected from the major fish markets of the study areas from 1997 to 2001 provide such information. In this process, eight focus groups comprising five fish retailers in each group with an age profile of above 50 years were taken into consideration from different landing centres. They were asked to present an account of fish landings at their respective markets in terms of kg/month for the past 5 years (Patra et al., 2004, 2005).

8.20.3.1.7 The Categorization of Fish: A Primary Step for Conservation

The International Union for the Conservation of Nature and Natural Resources (IUCN) categorizes taxa in a given scheme depending on various criteria (IUCN, 2001, Red List Categories and Criteria, Version 3.1). In this research study, the species were categorized according to the percentage of population size reduction and the mean decline in market landing. The percentage of population size reduction was calculated on the basis of FMV and was compared with the percentage given on the list. For critically endangered (CR) species, the decline according to the FMV was 80%, and the mean decline in market landing was 60%. For endangered (EN) species, the decline according to the FMV was 70%, and the mean decline in market landing was 40%. For vulnerable (VU) species, the decline according to the FMV was 50%, and the mean decline in market landing was 30%. For near-threatened (NT) species, the decline according to the FMV was <50%, and the mean decline in market landing was 15%. The IUCN standardizes the percentage of population reduction over decades, which has been taken into consideration with the FMV (Patra et al., 2004, 2005; Mishra et al., 2009).

8.20.3.1.8 Causal Analysis: Understanding Cause and Effect

The perceptions of the outcomes from using the PRA method cannot be assessed as if they were the results of some laboratory-based research. The best way of determining the benefits is to discuss them with the community. Representing a PRA process involves some diagrams, maps, rankings, flowcharts, etc., which elicit and present local knowledge and experiences. The causes of the decline in or even the extinction of bioresources in a given ecosystem can better be represented in a form of flow analysis known as a causal diagram. The relationship between the cause and effect of an event, process or entity can be delineated by using this method. PRA teams consist of people who are involved in the consumption, production and maintenance of fish diversity. Those participating people, having heterogeneous experiential profiles, social and economic backgrounds, etc., were involved in a structured manner to determine the prospective type of analysis. The diversity created in the group composition was meant to identify, assess and enumerate different causes, viz. biotic, abiotic and social, from multiple perspectives.

8.20.3.1.9 The Identification and Assessment of Anthropogenic and Environmental Stresses (Through the RBQ Method)

A number of causal factors were found to be instrumental in the decline in fish populations in their natural habitats (irrigation canals, tributaries and rivers) in southern West Bengal, India (Patra et al., 2005; Mishra et al., 2009). Group meetings with the local fisher communities of some selected villages were organized to collect information on the factors responsible for declining fishery resources. Mostly the male members of the community (30 members) assembled at a meeting place. The set of criteria was discussed by the fisher community, who were also asked to add to the list and to score according to the set of criteria generated. Habitat destruction, multifarious water uses, water pollution from industries and human settlements, the usage of nonrestricted nylon thread fishing gear, the application of ichthyotoxic materials (plants and chemicals), fishing during breeding seasons, siltation, human population pressure, the sluggish behaviour of some fish, fish diseases (epizootic ulcerative syndrome and others) and the introduction of exotic fish species were identified as the major causes of the declining fishery resources. To initiate the process of scoring by individuals, the criteria were written on a piece of paper in the respondents' vernacular (such criteria were read aloud and understood by the participants, although many of them could not read or write) and arranged in a column on the ground.

Each person was given a handful of stone chips (ten stone chips) and asked to score each criterion as much or as little as they wished to. It was seen as free scoring, where participants explained their own views on the relative importance of a criterion, which they indicated by assigning more stone chips to more-important ones and fewer stone chips to less-important ones. Ranking was done with the help of rank-based quotient (RBQ). A focus group was formed.

They were put into a discussion to identify ten causal factors and to extract their relative value in explaining the declining trend in fish populations in general and some selected fish species in particular. Afterwards, they assessed the extent of the impacted area and the intensity of the problems to finally obtain the village's fish magnitude values, which were subsequently ranked to provide a holistic overview of the problem.

8.21 Climate Change and Adaptation

Adaptation is defined through many words, including *adjustment*, *acclimatization*, *resilience*, *steps*, *processes* and *outcomes*, but the applicability of those terms often invites ambivalence, and often, broad and open-ended terms explain a particular subject, especially responses against environmental perturbations. Adaptation, as a context-specific approach, refers to the acclimatization of living organisms, including humans, to changing environmental conditions and the development of attributes that allow them to adjust to natural or human-induced changes in states of development in response to the actual or expected impacts of climate change (Smit et al., 2001) by reducing vulnerability. This is possible to some extent by reducing the period and intensity of exposure to environmental hazards and by evolving a set of actions to reduce vulnerability to environmental perturbations, especially climate change–mediated problems. The set of actions differs from place to place and target to target, occurs at different time scales and depends on the degree of spontaneity, which may be reactive, anticipatory, planned or autonomous. Autonomous adaptation refers to changes in humans in response to sudden alterations in their immediate surroundings without making policy-based decisions and without expecting invention from others, including government organizations. Planned adaptation is the manifestation of actions that stem from raising awareness about the actual states of changed conditions so that desired situations can be achieved. All these adaptive manifestations require outside interventions to reduce suffering, prevent loss and change location to suitably protect themselves.

8.21.1 Conceptual Framework for an Analysis of Adaptive Capacity

Local people along the coastal belts across the globe have been experiencing a multitude of ecological, social and economic changes that have caused a drastic deviation in normal ecological conditions, resulting in environmental degradation, biodiversity loss and a decline in fishery and aquacultural production, which have compounded the harmful consequences of climatic change (Bennett et al., 2014). The present conceptual framework pertaining to adaptive capacity has been rooted in academic research exercises on resilience in ecological, social, political and economic dimensions (Holling, 1973; Berkes & Folke, 1998; Gunderson & Holling, 2002; Berkes et al., 2003). Adaptive capacity reduces the effects of vulnerability to natural hazards (Burton et al., 1993, Mustafa, 1998, Smith, 2013), the deterioration of economic conditions and subsequent famine (Sen, 1982; Swift, 1989; Watts & Bohle, 1993) and climate change (Smithers & Smit, 1997; McCarthy et al., 2001; Turner et al., 2003). The identification of indicators for an assessment of adaptive capacity and social resilience involves different components of society, like diversification in livelihood options and the mobility of people (Ellis, 1998; Adger et al., 2002) in the existing state of natural capital, social, economic and cultural components (Berkes & Folke, 1998; Carney, 1998; Lemos et al., 2013), modes of administrative governance, the availability of facilities in institutions, a culture of learning (Ostrom, 1999; Hagmann & Chuma, 2002; Armitage & Plummer, 2011; Berman et al., 2012), a supply of resources (Scoones, 1998), resource dependence (Bailey & Pomeroy, 1996), redundancy in the functioning of governing organizations (Staber & Sydow, 2002), prevailing public policies (Kalikoski et al., 2010), leadership in resource management (Hill, 2013) and increases in income and social stability (O'Garra, 2007). In such a context, adaptive capacity is linked directly to social capital, national adaptive capacity through relationships based on trust, the capabilities of governance, political rights and the reciprocity of willingness among different members of groups (Adger, 2003; Adger & Vincent, 2005). As adaptive capacity is interlinked directly with economic development and the well-being of the people (McCarthy et al., 2001; Brooks et al., 2005), both ecological and social resilience systems limit the changes while retaining control over the structure and function of the self-sustaining ecosystem, which is capable of self-organization. The same system also increases the capacity for learning and adaptation (Resilience Alliance, 2010). Importantly, therefore, concept of resilience points out the inherent complexity, unpredictability and uncertainty in ecological-social-political systems due to the interdependence between natural systems and human interests, resulting in adaptive responses that maintain long-term ecosystem functioning and thereby higher productivity (Holling, 2001; Berkes et al., 2003; Carpenter et al., 2009).

Social adaptive capacity is defined as "the ability to respond to challenges through learning, developing new knowledge and devising effective approaches for managing risks and harmful impacts." Adaptive capacity in the social sphere acts as a yardstick of social resilience, which tends to reduce the vulnerability of individuals, communities and social institutions to ongoing changes due to environmental, social, political or economic perturbations and shocks (Marshall et al., 2010). Although communities have been experiencing the manifold impacts of several ecological, social, political and economic stressors, previous research applications for adaptive capacity have tended to focus on a particular stressor (O'Brien & Leichenko, 2000; Ommer, 2007; Leichenko & O'Brien, 2008; Bunce et al., 2010). Moreover, adaptive abilities are influenced to by multiple scales of action and by individual, community, national and international perspectives (Adger et al., 2005; Cinner et al., 2009).

The discussion on adaptation against ongoing perturbation centres on two dimensions: ecosystem-based adaptation and community-based adaptation. Ecosystem-based adaptation, being a natural adaptive process, integrates the derived benefits from biodiversity (food, fodder, clothing and shelter materials) and ecosystem services (fresh air, clean water, carbon sequestration, etc.) into the process of a developing strategy to be adopted by people to help them to adapt to the adverse impacts of environmental perturbations, including climate change. The main objective of this mode of adaptation is to reduce vulnerability while enhancing resilience to ongoing environmental problems. A proper understanding of this adaptive approach helps in identifying the prospective applications of pollution-controlling technologies and climate-resilient infrastructure, which need to be cost-effective for coastal communities (Colls et al., 2009).

Community-based adaptation refers to the all-round involvement of a local community to take action against their region's vulnerability to environmental perturbations, including harmful environmental climate change–mediated consequences, while aiming to protect environmental resources and enhance adaptive capacities. Such objectives can be achieved by gaining a proper understanding of integrating the impacts of environmental perturbations into activities for livelihood maintenance and increased resilience to immediate environmental problems, which will thereby improve the adaptive capacity of local communities (Enser & Berger, 2009).

In the Sundarbans mangrove ecosystem, ecological health is measured mostly via biological productivity, mostly in the form of green biomass. Fishery and aquacultural products have been declining because of population explosions in coastal communities, increasing dependence on natural resources, increasing vulnerability to climate-induced changes, etc. All these changes have necessitated adopting proper adaptive measures that promote management interventions and conservation initiatives with livelihood implications. In addition, mangrove-dependent coastal communities on the Andaman and Nicobar islands have also been presented with several economic opportunities associated with their livelihoods, mainly in tourism, fishery, agriculture and coastal bioresource-based products. Proper eco-planning and the effective implementation of ecomanagement strategies are required to increase the adaptive capacity of communities so that they may respond to these complex and ongoing changes. Conservation strategies have experienced a paradigm shift over the past few decades, relying on in-depth prior research on prospective conservation actions involving local people with their power of adaptation in changing environmental setups. Ecosystem valuation through regional eco-planning has emerged as an important prerequisite in such a transformation process for conservation (Colette et al., 2007). The intensity of ecodegradation and the potential prospect of recovery critically depend on the identities, motivations and activities of involved actors within the coastal community to tackle three major kinds of anthropogenic disturbances in mangrove ecosystems: (i) the intensity, extent and duration of ecological perturbations; (ii) the ability of natural or manually operated seedling banks to regenerate; and (iii) the rate of re-establishment for seedlings (Farnsworth & Ellison, 1997). Monitoring ecological changes with the tools required for conservation, like ecologically sound afforestation programmes, long-term demographic studies and avoiding labour- and time-intensive seedling replantation, helps in identifying the possible causes of mangrove decline. Special emphasis is to be placed on the local assessment of altered geomorphology in soil chemistry, erosion, hypersalinity, acidification, the accumulation of soil sulfides, tidal wetlands and bio-invasion by early colonizers, which cumulatively impair the regrowth of mangroves (Gong et al., 1984; Srivastava et al., 1987; FAO, 1994; De, 1990; Gong & Ong, 1990). Environmental administrators and ecological planners need to remember that without providing a proportionate share of economic benefits and incentives to mangrove-dwelling local communities, sustainable resource management might incur setbacks (Smith & Berkes, 1993) (Flowchart 8.14).

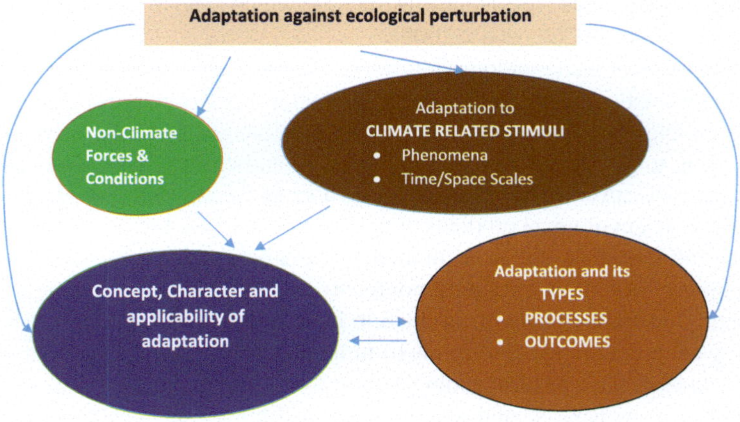

Flowchart 8.14 Different strands of adaptation against climate variability and other forms of environmental perturbations

8.21.2 *Climate Change with Cumulative Effects and Counterbalancing Adaptability in Sundarbans, India*

Sea-level rise will have significant impacts worldwide this century, which may exceed the prevailing estimations across the world (IPCC, 2018). One estimate has pointed out that a significant percentage of coastal populations is at risk of inundation (Neumann et al., 2015). In the Indian Sundarbans, a loss of land is instrumental in causing the ecodegradation of mangrove ecosystems, which in turn adversely affects people's livelihoods and promotes displacement and outmigration (Ghosh et al., 2014). The construction of embankments to save people's habitation and their agriculture lands is an important task to reduce sea-level-rise-mediated flood risks, because these structures restrict sediment movement and deposition, leading to enhanced subsidence, which aggravate the ecological conditions of estuarine ecosystems (Auerbach et al., 2015; Takagi et al., 2015).

The future impacts of climate change on the Sundarbans are expected to be considerably more intense, like many other parts of the world in comparison to earlier negative consequences due to climate changes since the 1880s, which have necessitated planned adaptation. The entire Sundarbans region is currently already about 1 °C warmer than what it was in 1880s and is expected to be 3.7 °C warmer by 2100 (Carbon Brief, 2018). Out of several options from planned adaptation, the well-designed retreat of organisms and human population from vulnerable coastal areas, migrating to safer places in more-landward directions to lower-risk areas, aims to reduce the potential loss of life and property in environmentally perturbed areas. Consequently, in setting aside space for the natural manoeuvring of altered coastal dynamics in response to catastrophic storms, cyclones, storm surges, projected sea-level rise, etc., a natural buffer can be created to protect sensitive habitats and bring about long-term benefits, counterbalancing up-front costs (Cornwall, 2018).

8.21.2.1 Outcomes of Climate-Mediated Changes in the Sundarbans Mangrove Ecosystem in India

The Sundarbans of India, historically an underdeveloped region in the country, is demographically characterized by its high population density and high percentage (around 40%) of poor people, most of whom live below the poverty line with deficits in electricity, healthy drinking water and healthcare infrastructure and experience food shortages with high food insecurity, poor education facilities with high dropout rates, few banking facilities, no prospects for agricultural insurance, an absence of major industries, few disaster warnings, few disaster shelters and negligible marketing infrastructure for resources generated from agricultural, fishery and nontimber forest products. The prime prospects for earning a livelihood are in agriculture and labour.

The following are the observed changes due to the cumulative effects of climatic and nonclimatic factors:

(i) The sea surface temperature (SST) in Sundarbans has been increasing at a rate of 0.5 °C per decade, whereas the global rate of increase is 0.06 °C per decade.
(ii) The rising sea level in Sundarbans also exhibits a higher rate than the global average. An estimate reveals that sea level has risen at a rate of 8 mm/year in Sundarbans over the past three decades, which appears to be almost double the global average.
(iii) Sediment starvation in the Indian Sundarbans, due to the construction of several dams and other built structures upstream of freshwater rivers over the past seven decades, has not only reduced freshwater inflow but also drastically reduced the supply and deposition of sediments within Sundarbans. This deficit of supply causes land subsidence, which together with sea-level rise increases inundation pressure.
(iv) The modes of rainfall have become more erratic, resulting in incidences of unpredictable flood or draught.
(v) The intensity and frequency of the occurrence of cyclones have also increased.
(vi) Cultivable lands in Sundarbans are being lost due to submergence and erosion at a fast pace, which are substantiated by an observed land loss of about 250 km^2 over 80 years. Four islands have so far been completely wiped out due to such an increasing pace of land loss.
(vii) A good amount of reclaimed agricultural land on 54 deltaic reclaimed islands has been lost due to the rising population and due to changes in land-use patterns.

(viii) Soil salinity has increased because of seawater ingress and retention, mostly because of cyclonic activities causing floods to inundate large chunks of agricultural land, thereby reducing the productivity of agricultural land.

(ix) The agricultural production that sustains local people has been considerably reduced due to farmers' adhering to the old traditional system of agriculture because of a lack of electricity and irrigation facilities, increased soil salinity and increased land fragmentation, among others.

(x) Fishing resources and catches of commercially important fish have been dwindling because of the synergistic effects of environmental pollution, habitat destruction, climate change and overexploitation.

(xi) A steady decrease in the yields of nontimber forest products (NTFPs) like honey, fodder, fruits and wax is due to manifold social-economic-political changes in and around Sundarbans, India.

(xii) Owing to all of those abovementioned changes, the people of Sundarbans hardly maintain a modest form of lifestyle and have started migrating to survive.

8.22 Different Adaptation Challenges in Sundarbans

The adaptation of people in Sundarbans is shown in their taking up complex challenges in conserving the mangrove ecosystem to ensure the sustainability of people who depend largely on natural resources. Better resource management and sustainable conservation will help address people's survival struggles from increasing population growth, which conflicts with the task of conserving the ecosystem and ensuring a just benefit-sharing mechanism. Adaptation debates on the strategies, synergies and priorities that benefit local people take precedence. Several adaptive measures can be undertaken.

(i) Disaster management and community radio: Cyclone warning centres as disaster-management systems need to be activated to face problems like cyclones, one of which hit Sundarbans in 2019, and the area received active support from the government. The absence of cyclone warning centres and the absence of efforts such as an All India Radio (AIR) and other community radio stations in the nearby places of the Indian Sundarbans are the main reasons why the vulnerability of the delta has increased. Also, the functioning of a centre of this kind would be effective as a way to make public announcements from its loudspeakers.

(ii) Information and communication: Knowledge about climate change, sea-level rise, soil erosion, rainfalls, disasters, etc. needs to be shared and widely circulated among people so that they can make informed choices. In adverse conditions, information communicated to local people allows them to tackle cross-sectoral challenges by raising awareness in them about the environmental consequences of such conditions.

(iii) Migrating to survive: Monsoon is considered the festival season in Sundarbans because of its heavy precipitation (around 1500 mm in a year) and is celebrated and worshipped as a god.
(iv) Saving the ecosystem and protecting biodiversity: The fragile mangrove ecosystem of Sundarbans can be protected by conserving its unique biodiversity, which forms an integral part of the total ecosystem.
(v) Socioeconomic assessment: The impacts of climate change on the entire ecosystem can be assessed by including the effects of global warming on the people in the region, the state of infrastructural development, social changes and deficits in governmental administration.
(vi) Potential bioresource bases: Sundarbans is a nursery for various kinds of animals and plants found along the entire east coast of the country. This ecosystem and biodiversity must be conserved and its resources managed efficiently so that the people will not be pushed further into ecological poverty.
(vii) The management of land and the maintenance of embankments: Higher productivity from agriculture can be assured both by managing lands in a way that increases fertility and by securing embankments to protect lands from flooding with saline water. Land resources are not only already diminishing in size because of erosion, but the quality is also deteriorating fast because of high salinity.
(viii) Awareness and information: As climatic conditions become more severe, people need to be equipped with necessary information to adapt. At a social level, the traditional knowledge of the people should be given importance in tackling many local problems, provided that such traditional knowledge does not compromise the application of scientific knowledge.

8.23 Recommended Responses to Reduce Vulnerability

According to a joint recommendation from the World Wildlife Fund (WWF) and India, about a million people residing in and around the Indian Sundarbans need to be relocated from vulnerable locations by 2050 (Danda et al., 2019). This joint recommendation also emphasizes not only planned retreat but also ecosystem regeneration in the vacated region, which will yield ecosystem services to the community in the form of ecotourism, fishing, carbon sequestration, etc. Diversifying avenues for new livelihoods and opening up new markets will enable relocated communities to survive, and they will also derive benefits from selling carbon credits by holding rights over the vacated land. Given the various possible benefits from ecosystem services and the various costs incurred to combat vulnerability, two possible steps are required to render benefits to local people: (i) The populations subjected to current and projected risks from coastal flooding and thereby recurrent inundation and erosion need to be identified, and (ii) a mechanism needs to be devised to allow the residents of specific at-risk locations to monetize their land assets at rates above the government-defined "market value," so that voluntary relocation is possible (Danda et al., 2019).

In such a context, relaxing land policies on holding size (land ceiling) and land-use patterns could pave the way for the establishment of aquacultural farms, tourist resorts and other revenue-earning establishments at certain identified places within the Sundarbans Biosphere Reserve. Suitable locations within the territory of the Sundarbans region, but in reclaimed areas, outside the high-inundation-risk zones, need to be identified to improve economic options and to accommodate more physical infrastructure for more market opportunities. Innovative financing options need to be considered to better manage the retreat of local people (the Environment-Forced Displacement and Resettlement (EFDR) programme). This could include private capital, the issuance of bonds and public funding. Using vacated land to regenerate mangrove forests can open up possibilities for cash flows through a governmental revenue-earning process. Those rehabilitated sites with mangrove vegetation can act as climate-mitigation sites in the face of rapid erosion, increased flood risk and the discontinuation of agricultural production where mangroves tend to sequester 36–220 tonnes of carbon per hectare (Brown & Pearce, 1994). These rejuvenated sites, with improved ecological health, could also be attractive for the development of income-earning ecotourism.

8.24 Adaptation: Needs, Ideas, Types and Processes

According to a report published in October 2010, India ranks as the second-most-vulnerable country to climate change, after Bangladesh. A new Climate Change Vulnerability Index (CCVI), prepared by the UK-based global risk advisory firm Maplecroft, has suggested an evaluation of 42 social, economic and environmental factors. National vulnerabilities stem from three core areas: exposure to climate-related natural disasters and increasing sea-level rise; the demography of local populations and the availability and modes of using natural resources; and the dependency of village people on agriculture and conflicts between human activities and natural ecosystems. People's adaptive capacity and infrastructural development by a county's government to combat climate change are also taken into consideration in the preparation of such an index. Adaptation, being an integral component of climate change action plans must be ecosystem based, but not state specific. Otherwise, it will be incongruous and unscientific because the horizontal and vertical extents of an ecosystems are not restricted by any boundary but instead overlap with some of the structural components of neighbouring ecosystems. Moreover, impacts from climate change are not confined within political or administrative boundaries; rather, livelihood practices such as agriculture, fishing, honey collection and the harvesting of forest products are specific to locations and cultures.

Different structural components of the mangrove ecosystem continuously interact with local people, who harvest the necessary ingredients for their survivability and reap the benefits of various ecosystem services. Combatting the ongoing climate change–mediated changes within the mangrove ecosystem demands collective action from governments, local communities, conservationists, nongovernmental organizations (NGOs) and other interested stakeholders. They must plan and empower local action to strengthen ecological and social resilience in the face of a changing climate (Flowchart 8.14).

8.24.1 Adaptation in Sundarbans: One of the Most Threatened Ecosystems in India

Sundarbans, an underdeveloped coastal area, is one of the most visible victims of the ravages of climate change in India. The fragility of the ecosystem and its underdevelopment have made this region particularly vulnerable. The increasing vulnerability of the people living on the deltaic islands of both Bangladesh and India, mainly because of climatic change, has imposed many constraints on agriculture, fishing and other natural resource-based livelihood generation practices. Resident populations need to increase their abilities to cope with changing ecosystems and to adopt proper adaptive measures in order to reduce the increase in their vulnerability.

The socioeconomic profiles of the people of Sundarbans are characterized by poverty and excessive population pressure, which contribute to the development of the precarious ecological conditions of the mangrove ecosystem, and noninclusive developmental processes undertaken mainly by governments to overcome different forms of vulnerabilities have hardly improved people's welfare and well-being. The main challenges in overcoming the problems pertaining to the lives and livelihoods of local people are additionally compounded by climate change. Policymakers and governments have taken two steps to counteract these challenges: protecting an ever-increasing number of people living in these ecosystems and mitigating the impacts of climate change on these ecosystems.

8.24.1.1 Land: The Bone of Contention

Land in the Sundarbans delta has developed into its present form with its specific shape and size over the past 300 years through the continuous deposition of suspended sediment, amounting to more than 1 billion tons per year through the Ganges–Brahmaputra–Meghna system. This low marshy landmass formed as alluvial plains, which are still in the process of forming through siltation and powerful tidal currents, a continuous process of erosion and accretion. New land is being added to the existing mass through an accretion process, followed by some erosion that has reduced the landmass as part of a natural cycle. The deltaic islands of Sundarbans are characterized by sandy beaches, mudflats, sandflats, coastal dunes, estuaries, creeks, mangrove swamps, tidal cannels, etc., which all face the environmental crisis imposed on different parts of Sundarbans. The contours of the land have been changing constantly within this active delta through the natural processes of erosion and accretion. Erosion makes sediment on the floor of mangrove forest unstable, and unwanted accretion through the deposition and dumping of eroded soil material within the estuarine beds disrupts the navigation channel, forming a new, elevated landmass, locally called a char. Construction activities within Sundarbans and in the upstream areas, such as dams, hydroelectric power plants and embankments, block the flow of fresh water with sediments, causing salinity invasion and sediment starvation. In and around Sundarbans, especially the confluence of the estuaries with the sea, an enormous amount of silt is deposited depending on some geomorphological and hydrological characteristics associated with estuarine flows. Depending on the sloping gradient and the depth and velocity of water flows within the estuaries, a large portion of deposited sediment is pushed downstream and into the surrounding deltaic islands by high tides, and this process increases the height of the land. A recent research study has revealed that ten of the sea-facing deltaic islands of southern Sundarbans – namely Sagar Island, Ghoramara Island, Jambudwip, Namkhana, Mousuni Island, Dakhsin Surendranagar Island, Dhanchi Island, Dalhousi Island, Bulchery Island and Bhangaduani Island – have been identified as the most vulnerable (Danda et al., 2019).

8.24.1.2 Embankments: Roles and Crisis Management

To make Sundarbans habitable, earthen embankments were initially constructed to block the ingress of saline water into peripheral human settlements. Around 3500 km of embankments have so far been developed in this region. During the colonial era, the first attempts were made by the British in the eighteenth and early nineteenth centuries to construct concretized embankments in the reclamation region of Sundarbans, which extended up to the fringes of Kolkata. The initiative to start agriculture on the flood plains of the Sundarbans region cleared forests, but these attempts were unsuccessful, mainly because the area was submerged under saline water during high tides. Large parts of the embankments in Sundarbans were constructed between 1873 and 1939 to respond to the high demands for their erection from local people, who after having settled down on the islands secured permission from the then British rulers to start engaging in agriculture. However, these artificial structures have been playing negative ecological roles in unsettling the land, seriously impeding delta-building processes, increasing erosion, triggering the settling of dislodged sediments on river beds, raising the relative water levels and disrupting food chain–food web relationships. Breaching earthen embankments intermittently poses serious problems to local people who want to save their waterbodies and agricultural lands. They call for different solutions, especially immediately repairing the breached embankments in the Indigenous way as a physical adaptation strategy for Sundarbans. The deposition of eroded sediments on the estuarine beds has made the estuaries shallow, reduced their water-holding capacity, increased turbidity, blocked the movement of water vessels and narrowed river channels. A high volume of water flowing through narrow tidal channels with high-intensity water flows can strike banks with a great force and escalate flood risks.

8.24.1.3 Education and Adaptation

To cope up the brunt of climate change in the interaction processes of the Sundarbans mangrove ecosystem, proper adaptive strategies need to be adopted, such as providing basic education. Hundreds of primary, secondary and higher education schools have been established on the reclaimed deltaic islands (54 islands), mostly after the independence of India. Among them, primary education imparts basic skills, which ultimately offer collective social benefits. But primary education cannot render many social benefits in terms of employability. Higher education for the people in the Sundarbans brings about prospects for employability, which in turn improve levels of adaptive capacity, offer choices to people and reduce their dependence on natural resources for survival and sustenance.

8.24.2 Nature-Based Solutions in the Mangrove Ecosystem of Sundarbans

Nature-based solutions use ecomanagement approaches to address ecodegradation-related problems in the mangrove ecosystem, focusing on the mode that capitalizes on the merits of natural processes for coastal protection, climate change mitigation and biodiversity enrichment. The success of this kind of solution rests on the underlying principles of properly understanding and utilizing the ecological functions of mangroves, such as decreasing the intensity of waves, stabilizing sediments, increasing biodiversity and sequestering carbon alongside considering social and economic aspects. The components of nature-based solutions are as follows:

i) The plantation and development of a large extent of mangroves with green biomass, canopies and specialized root systems, which can act as natural barriers against several natural disasters, like catastrophic storms and cyclones, and which thereby prevent coastal flooding and erosion by reducing the energy of wind and waves
ii) The frequent breaching of embankments, which can be prevented by building a bio-shield to resist storm surges
iii) Steps are to be taken for soil trapping from tidal currents by wooden and cemented structures
iv) The excavation of old silted channels as much as possible

v) Improving fresh water inflow by creating drainage channels to drain out monsoonal excess rainwater from the terrestrial upland to the creeks and estuaries of Sundarbans
vi) Expanding salt-tolerant crop cultivation
vii) Constructing cyclone-resistant low-cost housing

In addition, the uniqueness of the mangrove ecosystem acts as a nursery, renders many ecosystem services and improves coastal protection, shoreline stabilization, carbon sequestration and the mitigation of climate-related risks. Nature-based solutions also address ecological issues like ecological mangrove rehabilitation (EMR) and community-based ecological mangrove restoration (CBEMR) by focusing on restoring degraded mangrove areas through community engagement and sustainable management practices with community involvement. In a sustainable ecomanagement process, nature-based solutions are important and acceptable options for reducing impacts on the livelihoods and well-being of people and communities (Schuerch et al., 2018).

In the process of formulating conservation policies, constitutional provisions for Sundarbans should be like those for the rural regions and populations in India, but many other policy changes and legislative actions would be useful:

(i) The preparation of legal frameworks for the relocation of climate change–affected people to new lands
(ii) The initiation of a decision-making process based on assessments by affected local populations
(iii) The establishment of financial mechanisms for both anticipatory planning and capacity building

In view of the above, a rights-based approach that provides opportunities for poverty alleviation, livelihood generation, social justice and strengthening the human rights of populations who have been severely affected by climate change impacts needs to be adopted.

8.25 Diversification in Resource Management Systems as an Adaptive Strategy

Diversification in the livelihoods of the people residing in and around Sundarbans is an important dimension of the adaptation of people against environmental adversity, which mainly originates from climate-induced ecological changes. Fisheries and the aquacultural sector in India, as the providers of nutritious food, have a high potential for rural development through employment, providing nutritional security. Both capture and culture fisheries have grown considerably over the past few decades with the involvement of several modern biotechnological techniques, and such successful growth has had important implications for food security and incomes, which together help to alleviate poverty and combat the direct and indirect effects of climate change (Amritage et al., 2011; Chakraborty, 2021b; Chakraborty et al., 2023).

8.25.1 The Significance of Fisheries and Aquaculture on the Gangetic and Deltaic Islands of West Bengal, India

The land-use patterns of West Bengal, where 63% of the land is used for agriculture and around 14% for other reasons, have recently shown perceptible indications of change due to the ongoing changes in climate. Since the 1980s, the coastal belts of West Bengal have endured natural disasters like severe storms, floods and droughts, which indicate the severity of the impacts of climate change, which are making the coastal belt, including the Sundarbans mangrove-based deltaic islands, more vulnerable.

The state of West Bengal, possessing the largest impounded brackish-water area in the country, mostly within the territory of the Sundarbans mangrove estuarine complex, produces 40% of the country's total shrimps and prawns from aquaculture. The high production of brackish-water fisheries is due to the productive base of Sundarbans, specifically its saline soil and water, biologically productive estuaries, the traditional knowledge systems of local people and favourable agro-climatic conditions. The profuse quantity of mangrove detritus from the mangrove forest subsystem to the adjoining estuarine subsystems has made this ecological habitat a nursing ground for a large number of prawn and estuarine finfish (Jhingram, A. G., 1988).

The ecologically sound management practices of the resources and ecosystems on the Indo-Gangetic plain can transform this important fishing sector into an area of great food security potential. However, fishery and aquaculture activities are also heavily influenced by external factors such as a pollution runoff, land-use transformation and competing aquatic resource uses. In addition, the impact of climate change could have an important compounding effect. The many people who depend on coastal fisheries and aquaculture either as producers, consumers or intermediaries are prone to vulnerability because of the direct and indirect impacts of climatic change.

8.25.2 Climate Change, Natural Disasters and Fishery: Alternative Adaptive Options

The government of India, based on some observed changes in climatic parameters, has made assessments on global warming and climate change in relation to inland fisheries and concluded as follows:

(i) The surface air temperature at the national level has increased by 0.4 °C over the past century.
(ii) Although no significant trend in the monsoonal rainfall at the all-India level has been observed, a significant trend in regional monsoon variations has been recorded.

(iii) Specific trends have been noted in more-frequent droughts, followed by less-severe droughts over many decades. An overall increasing trend has been observed in severe storm incidence along the coast of West Bengal and Gujrat and a rising trend in the frequency of heavy-rain events.
(iv) Sea-level rise ranging from 1.06 mm to 1.07 mm per year has been recorded, where such rates are in line with the global sea-level rise estimates of the IPCC, ranging from 1 mm to 2 mm per year.
(v) Changes in air and water temperature on the Gangetic plains (West Bengal), with tropical climatic conditions, have shown a tendency towards increasing temperature at the end of the winter months (January–February, being spring) and in summer (April–May). During these periods, the increase in temperature has not been linear, but a sudden temperature jump has been noted within a short period of time, where the months from January to April act as transitional months from winter to summer.
(vi) Because rainfall is a key meteorological parameter and an important stage in the hydrological cycle, it plays vital roles not only in the production of green biomass but also in promoting fishing and aquacultural production because the early maturation of brood fish requires influence from rainfall patterns. On the Gangetic deltaic islands, maximum rainfall (around 70% of the total annual rainfall, where the total is 1800 mm/year) occurs during the monsoon months (June–October).

In view of the abovementioned climatic and ecological conditions, the people of Sundarbans engage in both brackish-water and freshwater aquaculture by culturing *Lates calcarifer*, *Penaeus monodon*, etc. in saline waterbodies and culturing Indian major carps (*Catla catla*, *Labeo rohita* and *Cirrhina mrigala*) by applying modern culture techniques like composite fish cultures and hypophysation for captive breeding.

For the breeding of fish in aquacultural farms, several fish seed hatcheries have been established, where the owners of those hatcheries utilize the recorded information from the past few decades on the performance and sustainability of hatcheries in terms of fish species breed, brood management, procedures and techniques in breeding, breeding season duration and the price and marketing of spawn. During the 1980s, the breeding season for aquacultural farm hatcheries on the Gangetic plains of West Bengal used to commence at the onset of monsoon during June and continued till the end of September, lasting 4 months. But because of global climate change, the breeding season was extended by 45–60 days. The duration of previous breeding seasons (1980–1985) was 110–120 days, whereas that of the current breeding season (2000–2005) is 160–170 days (Chand et al.; CIFRI). Interactive responses from fishers and hatchery owners indicated that 90–95% of them attributed the increased breeding period on hatchery farms to water temperature rise. Fluctuating temperatures because of global climatic change often disrupt the physiological homeostasis of the cultured fish species, subjecting them to physiological stress and thereby shifting their ecological niches or habitats. These effects are often synergistic with low pH, less phytoplankton availability, algae growth and oxygen shortages, bringing about more-adverse ecological conditions that ultimately lead to the mortality of fish species. Earlier studies have established that temperature increases cause various stresses on the sensitive blood and tissue parameters of fish such as *Labeo rohita* and *Rita rita*, which are normally acclimatized to 29 °C but which have been subjected to a rapid sublethal rise to 35 °C (Das et al., 2002). Another experimental study has pointed out that exposing the juveniles of *L. rohita* at lower temperatures by gradual lowering the ambient temperature from 28 °C to 13 °C resulted in a significant rise in plasma cortisol, hyperglycaemia and the cessation of feeding and all activity (Das et al., 2008).

Climate change, in conjunction with sea-level rise, brings about many natural disasters like cyclones, which in turn cause a number negative environmental changes in West Bengal, like land loss and erosion, an increase in salinity, the breaching of embankments, flooding and the inundation of agricultural lands, disruptions to fish cultural bheries and freshwater ponds, disruptions to the process of coastal aquaculture and the displacement of local people, making them environmental refugees. All those changes compel local fishers, who were totally dependent on fishing and wild fish seed collection from natural resources as their only source of income, to adopt new, adaptive measures by culturing saline-tolerant fish species, namely *Mugil parsia*, *M. tade* and *Lates calcarifer*, in temporarily developed saline ponds. After being inundated by the ingresses saline flood water, paddy fields have become unfit for agriculture, so people are carrying out paddy-cum-fish culturing instead. Diversification in the application of fish species in aquacultural practices is an adaptive option in response to climate change–mediated draught conditions, where more than 80% of the hatcheries have been diverted from rearing Indian major carps to rearing other species, such as pangasius (*pangasius sutchi*), *Puntius javanicus* and African catfish (*Clarius gariepinus*), which can adapt to water stress and high temperature conditions.

In such a context, adaptation options for the people affected by climate change include the following:

1. Integrating aquaculture into other farming systems, emphasizing multiple-use and reuse processes
2. Using smaller waterbodies, locally called ponds (100–200 m^2), for rearing and culturing suitable species of fish or prawns
3. Intensifying aquacultural practices for the generation of resources by reusing wastewater and degraded ground saline water
4. Increasing infrastructural facilities for hatcheries to increase the production of seeds and the spawns of fish and shrimp
5. Promoting carbon sequestration because fish species used in aquaculture in a tropical country such as India are used to feeding on species low on the food chain
6. Deterring species of carnivorous finfish and cultured shrimp because they feed mainly on fishmeal and fish oil and thus contribute to carbon emissions
7. Raising exotic carp species because they have potential significance in the carbon cycle by fixing CO_2 through phytoplankton and utilize minimal industrial energy

8. Controlling the culturing of fish on the tropical belts of coastal West Bengal as an alternative practice in response to climate change and as an adaptive measure that will reduce greenhouse gas (GHG) emissions after being adopted
9. Triggering phytoplankton and periphyton growth as a major energy-saving measure in aquacultural practices, in which periphyton-based practices are used to catch fish in open waters – a practice known as komor or huri in West Bengal and katha in Bangladesh

References

Acharya, S. K., & Sharma, N. K. (2019). *Entropy, chaos and social ecology: The stress and response of Indian farmers*. Rathore Academic Research Publication. ISBN: 978-81-941991-8-2.

Adger, W. N. (2003). Social capital, collective action, and adaptation to climate change. *Economic Geography, 79*(4), 387–404.

Adger, W. N., & Vincent, K. (2005). Uncertainty in adaptive capacity. *Comptes Rendus Geosciences, 337*(4), 399–410.

Adger, W. N. P., Kelly, M., Winkels, A., Huy, L. Q., & Locke, C. (2002). Migration, remittances, livelihood trajectories, and social resilience. *Ambio, 31*(4), 358–366.

Adger, W. N., Arnell, N. W., & Tompkins, E. L. (2005). Successful adaptation to climate change across scales. *Global Environmental Change, 15*(2), 77–86.

Ahammad, R., Nandy, P., & Husnain, P. (2013). Unlocking ecosystem based adaptation opportunities in coastal Bangladesh. *Journal of Coastal Conservation, 17*, 833–840. https://doi.org/10.1007/s11852-013-0284-x

Amritage, P., Terry, N., & Jason, P. (2016). *Building performamce evaluation programme: Findings from non-domestic projects*. Innovate UK.

Anon. (2001). Fruit Reaeaexh and Development Institute, Science and Education.

Auerbach, L. W., Goodbred, S. L., Jr., Mondal, D. R., Wilson, C. A., Ahmed, K. R., Roy, K., & Ackerly, B. A. (2015). Flood risk of natural and embanked landscapes on the Ganges–Brahmaputra tidal delta plain. *Nature Climate Change, 5*, 153–157.

Bailey, C., & Pomeroy, C. (1996). Resource dependency and development options in coastal Southeast Asia. *Society and Natural Resources, 9*(2), 191–199.

Bennett, N., Dearden, P., & Peredo, A. M. (2014). Vulnerability to multiple stressors in coastal communities: A study of the Andaman coast of Thailand. *Climate and Development, 7*, 124.

Berkes, F., & Folke, C. (1998). *Linking social and ecological systems: Management practices and social mechanisms for building resilience*. Cambridge University Press.

Berkes, F., Colding, J., & Folke, C. (2003). *Navigating social-ecological systems: Building resilience for complexity and change*. Cambridge University Press.

Brooks, N., Adger, W. N., & Kelly, P. W. (2005). The determinants of vulnerability and adaptive capacity at the national level and the implications for adaptation. *Global Environmental Change, 15*(2), 151–163.

Brown, K., & Pearce, D. W. (1994). *The causes of tropical deforestation: The economic and statistical analysis of factors giving rise to the loss of tropical forests*. UBC Press.

Bunce, M., Rosendo, S., & Brown, K. (2010). Perceptions of climate change, multiple stressors and livelihoods on marginal African coasts. *Environment, Development and Sustainability, 12*(3), 407–440.

Burton, I., Kates, R., & White, G. F. (1993). *The environment as hazard* (2nd ed.). Guilford.

Carbon Brief. (2018, September). Retrieved from https://www.carbonbrief.org/mapped-how-everypart-of-the-world-has-warmed-and-could-continue-to-warm

Carpenter, S. R., Mooney, H. A., Agard, J., Capistrano, D., DeFries, R. S., Díaz, S., Dietz, T., Duraiappah, A. K., Oteng-Yeboah, A., Pereira, H. M., Perrings, C., Reid, W. V., Sarukhan, J., Scholes, R. J., & Whyte, A. (2009). Science for managing ecosystem services: Beyond the millennium ecosystem assessment. *Proceedings of the National Academy of Sciences, 106*(5), 1305–1312.

Castelloe, P., Gamble, & Dorothy, N. (2005). Participatory methods in community practice: Popular education and participatory rural appraisal. In M. Weil & M. Reisch (Eds.), *Handbook of community practice* (pp. 261–275). SAGE Publications.

Cernea, M. M. (2004). Impoverishment risks, risk management, and reconstruction: A model of population displacement and resettlement. In *Social aspects of hydropower development. UN symposium on hydropower and sustainable development, Beijing*. Retrieved from https://commdev.org/wp-content/uploads/2015/06/Impoverishment-Risks-Risk-Management-and--Reconstruction.pdf

Chakraborty, S. K. (2017). Ecological services on intertidal benthic fauna and the sustenance of coastal wetlands along the Midnapore (East) coast, West Bengal, India. In *Coastal wetlands: Alteration and remediation* (pp. 777–886). Springer.

Chakraborty, S. K. (2021b). Riverine ecology (volume 2): Biodiversity conservation, conflicts and resolution. Springer Nature. 566 pages.

Chakraborty, S.K. Sanyal, P and Ray, R 2023. Wetlands ecology: Eco-biological uniqueness of a Ramsar site (East Kolkata Wetlands, India), 709 pages).

Chambers, R. (1983). *Rural development: Putting the last first* (Vol. 198). Longman.

Chambers, R. (1981). Rapid rural appraisal: rationale and repertoire. *Public Administration and Development, 1*(2), 95–106. https://doi.org/10.1002/pad.4230010202

Chambers, R. (1984). *[1983]. Rural development: putting the last first*. Longman. ISBN 978-0582644434. OCLC 9196254.

Chambers, R. (1993a). *Participatory rural appraisal (PRA): Challenges, potential and paradigm*. Institute of Development Studies Sussex, July.

Chambers, R. (1993b). *Challenging the professions: Frontiers for rural development*. Intermediate Technology Publications. ISBN 978-1853391941. OCLC 28470414.

Chambers, R. (1994). The origins and practice of participatory rural appraisal. *World Development, 22*(7), 953–969.

Chambers, R. (1997). *Whose reality counts?: Putting the first last* (p. 106). Intermediate Technology Publications.

Cinner, J. E., Fuentes, M. M. P. B., & Randriahmahazo, H. (2009). Exploring social resilience in Madagascar's marine protected areas. *Ecology and Society, 14*(1), 41.

Colette, A., et al. (2007). *Case studies on climate change and world heritage*. UNESCO World Heritage Centre.

Colls, A., Ash, N., & Ikkala, N. (Eds.). (2009). *Ecosystem- Based Adaptation: A natural response to climate change*. IUCN.

Conference on Rapid Rural Appraisal: KhonKaen University. KhonKaen, Thailand: Published by Rural Systems Research Project and Farming Systems Research Project for KhonKaen University. (1987). ISBN 9789745552517. OCLC 19025969.

Cornwall, W. (2018). *As sea levels rise, Bangladeshi islanders must decide between keeping the water out–or letting it in*. Retrieved from https://www.sciencemag.org/news/2018/03/sealevels-rise-bangladeshi-islanders-must-decide-between-keeping-water-out-or-letting

Danda, A. A., Ghosh, N., Bandyopadhyay, & Hazra, S. (2019). Managed retreat: Adaptation to climate change in the Sundarbans ecoregion in Bengal Delta. *Journal of the Indian Ocean Region, 15*(3), 317–335.

Das, L. (2008). *Content analysis: A method of social science research* (pp. 174–193). Rawat Publication.

De, S. (1990). *The Sundarbans*. Oxford University Press.

Dudley, N. (Ed.). (2008). *Guidelines for applying protected area management categories*. IUCN.

Ellis, F., & Allison, E. (2004). *Livelihood diversification and natural resource access*. Food and agriculture Organization of the United Nations (FAO).
Ellis, J. (1998). Household strategies and rural diversification. *Journal of Development Studies, 35*, 1–38.
Enser, J., & Berger, R. (2009). Community-based adaptation and culture in theory and practice. In W. N. Adger, I. Lorenzoni, & K. O'Brien (Eds.), *Adapting to climate change: Thresholds, values, governance*. Cambridge University Press.
Farnsworth, E. J., & Ellison, A. M. (1997). The global conservation status of mangroves. *Ambio, 26*(6), 328–334.
Flower, C., Mincher, P., & Rimkus, S. (2000). Overview – participatory processes in the North. *PLA Notes, 38*, 14–18.
Food and Agriculture Organization (FAO) of the United Nations. (1994). *Mangrove forest management guidelines* (FAO forestry paper 117). FAO.
Ghosh, T., Hajra, R., & Mukhopadhyay, A. (2014). Island erosion and afflicted population: Crisis and policies to handle climate change. In W. L. Filho (Ed.), *International perspectives on climate change* (pp. 217–225). Springer.
Gong, W. K., & Ong, J. E. (1990). Plant biomass and nutrient flux in a managed mangrove forest in Malaysia. *Estuarine, Coastal and Shelf Science, 31*, 519–530.
Gong, W. K., Ong, J. E., Wong, C. H., & Dhanarajan, G. (1984). Productivity of mangrove trees and its significance in a managed mangrove ecosystem in Malaysia. In E. Soepadmo, A. N. Rao, & D. J. Macintosh (Eds.), *Proceedings of the UNESCO Asian symposium on mangrove environments research and management* (pp. 216–225). Universiti Malaya.
Gunderson, L. H., & Holling, C. S. (2002). *Panarchy: Understanding transformations in human and natural systems*. Island.
Gupta, M., Acharya, S. K., & Sharangi, A. B. (2022). *Stewardship and conservation in agriculture: The small farm reality* (p. 191., ISBN 978-93-5461-175-9). Astral International Pvt. Ltd.
Hazra, S., Samanta, K., Mukhopadhyay, A., & Akhand, A. (2010). *Temporal change detection (2001–2008) study of Sundarban*. School of Oceanographic Studies, Jadavpur University.
Hazra, S., Bhadra, T., & Sinha Roy, S. P. (2015). *Proceedings of the international seminar on challenges to ground water management: Vision 2050*. Centre for Ground Water Studies.
Hill, M. 2013. Climate change and water governance. Springer, .
Holling, C. S. (1973). Resilience and stability of ecological systems. *Annual Review of Ecology, Evolution, and Systematics, 4*(1), 1–23.
Holling, C. S. (2001). Understanding the complexity of economic, ecological, and social systems. *Ecosystems, 4*(5), 390–405.
Hossain, S. M., Reza, M. I., Rahman, S., & Kayes, I. (2012). Climate change and its impacts on the livelihoods of the vulnerable people in the South-western coastal zone in Bangladesh. In W. L. Filho (Ed.), *Climate change and the sustainable use of water resources*. Springer. Iftekhar MH (2006) Conservation and management of the Bangladesh.
IPCC. (2018). Summary for policymakers. In V. Masson-Delmotte, P. Zhai, H. O. Pörtner, D. Roberts, J. Skea, P. R. Shukla, et al. (Eds.), *Global warming of 1.5°C. An IPCC special report on the impacts of global warming of 1.5°C above pre-industrial levels and related global greenhouse gas emission pathways, in the context of strengthening the global response to the threat of climate change, sustainable development, and efforts to eradicate poverty*. World Meteorological Organization. Irrigation & Waterways Department, Government of West Bengal. Retrieved from https://wbiwd.gov.in/index.php/applications/aila
Jhingram, A. G. (1988). *Reservoir fisheries management in India*. Bulletin 45, Central Inland Capture Fisheries Research Institute, FAO/CIFRI Publication.
Kalikoski, D. C., Neto, Q., & Pand Almudi, T. (2010). Building adaptive capacity to climate variability: the case of artisanal fisheries in the estuary of the Patos lagoon, Brazil. *Marine Policy, 34*(4), 742–751.
Kasina, J. M. (2007). *Bee pollinators and economic importance of pollination incrop production: Case of Kakamega, Wester Kenya* (p. 152).
Kothari, C. R. (1997). *Research methodology: Methods and techniques*. Wishwa Prakashan.

Leichenko, R., and O'Brien, K. 2008. Environmental change and globalization: Double exposures. Oxford University Press, .

Lemos, M. C., Kirchoff, C. J., & Dessai, S. (2013). Actionable knowledge for environmental decision making: Broadening the usuability of climate science. *Annual Review of Environment and Resources, 38*.

Leurs, R. (1997). Critical reflections on rapid and participatory rural appraisal. *Development in Practice, 7*(3), 290–293. JSTOR 4029070.

Levins, R. (1974). Qualitative analysis of partially specified systems. *Annals of the New York Academy of Sciences, 231*, 123–138.

Marshall, N. A., Marshall, P.A., Tamelander, J., Obura, D., Malleret-King, D and Cinner, J.E 2010. A framework for social adaptation to climate change: sustaining tropical coastal communities and industries. International Union for Conservation of nature (IUCN), .

May, R. M. (1975). *Stability and complexity in model ecosystems* (2nd ed.). Princeton University Press.

McCarthy, J. J., Canziani, O. F., Leary, N. A., Dokken, D. J., & White, K. S. (2001). *IPCC third assessment report: climate change 2001: impacts, adaptation, and vulnerability*. Cambridge University Press.

Mortreux, C., de Campos, R. S., Adger, N. W., Ghosh, T., Das, S., Adams, H., & Hazra, S. (2018). Political economy of planned relocation: A model of action and inaction in government responses. *Global Environmental Change, 50*, 123–132.

Mukherjee, N. (1997). *Participatory appraisal of natural resourses* (p. 186). Concept Publishing Company.

Mustafa, D. (1998). Structural causes of vulnerability to flood hazard in Pakistan. *Economic Geography, 74*(3), 289–305.

Nandi, S., Ghosh, M., Kundu, A., Dutta, D., & Baksi, M. (2016). Shoreline shifting and its prediction using remote sensing and GIS techniques: A case study of Sagar Island, West Bengal (India). *Journal of Coastal Conservation, 20*, 61–80.

Nandy, P., & Ahammad, R. (2012). Navigating mangrove resilience through the ecosystem based adaptation: lessons from Bangladesh. In D. J. Macintosh, R. Mahindapala, & M. Markopoulos (Eds.), *Sharing lessons on mangrove restoration* (p. 243). Mangroves for the Future and Gland, Switzerland, IUCN.

Narayanasamy, N. (2009). *Participatory rural appraisal: principles, methods and application*. SAGE Publications. ISBN 9788178298856. OCLC 265732154.

Neumann, B., Vafeidis, A. T., Nicholls, R. J., Marinova, N., Lowe, J. A., Brown, S., Vellinga, P., Gusmao, D., et al. (2015). Sea level rise and its possible impacts given a 'beyond 4°C world' in the twenty-first century. *Philosophical Transactions of the Royal Society A: Mathematical, Physical and Engineering Sciences, 369*, 161–181.

Nichlls, R. J., Wong, P. P., Burkett, V. R., Codignotto, J. O., Hay, J. E., McLean, R. F., Ragoonaden, S., & Woodroffe, C. D. (2007). Coastal systems and low-lying areas. In M. L. Parry, Canziani OF, J. P. Palutikof, P. J. van der Linden, & C. E. Hanson (Eds.), *Climate change 2007: impacts, adaptation and vulnerability. Contribution of working group II to the fourth assessment report of the intergovernmental panel on climate change*. Cambridge University Press.

O'Brien, K. L., & Leichenko, R. M. (2000). Double exposure: Assessing the impacts of climate change within the context of economic globalization. *Global Environmental Change, 10*(3), 221–232.

O'Garra, T. (2007). *Supplementary livelihood options for Pacific Island communities: a review of experiences*. Foundation of the Peoples of the South Pacific International.

Ommer, R. E. (2007). *Coasts under stress: restructuring and social-ecological health*. McGill-Queens University Press.

Patra, M. K., Acharjee, S. K., & Chakraborty, S. K. (2005). Conservation categories of siluroid fishes in North-East Sundarbans, India. *Biodiversity and Conservation, 14*, 1863–1876.

Pretty, J. N., Vodouhê, A., & Simplice, D. (1997). Using rapid or participatory rural appraisal. In B. E. Swanson, R. P. Bentz, & A. J. Sofranko (Eds.), *Improving agricultural extension: A*

reference manual (2nd ed.). Food and Agriculture Organization of the United Nations. [1984], ISBN 978-9251040072. OCLC 456586976.
Resilience Alliance. (2010). *Resilience.* [online] http://www.resalliance.org/576.php
Robinson-Pant, A. (1995). PRA: A new literacy? *PLA Notes., 24*, 78–82.
Scoones, I. (1998). *Sustainable rural livelihoods: a framework for analysis* (IDS working paper 72). Institute of Development Studies.
Sen, A. (1982). *Poverty and famines: an essay on entitlement and deprivation.* Oxford University Press.
Singh, K. (2001). Handing over the stick: the global spread of participatory approaches to development. In M. Edwards & J. Gaventa (Eds.), *Global citizen action* (pp. 175–187). Lynne Rienner Publishers. ISBN 978-1555879686. OCLC 45879585.
Smit, B., Pilifosova, O., Burton, I., Challenger, B., Huq, S., Klein, R. J. T., & Yohe, G. (2001). Adaptation to climate change in the context of sustainable development and equity. In J. J. McCarthy, O. F. Canziano, & N. Leary (Eds.), *Climate change 2001 – Impacts, adaptation and vulnerability* (Contribution of working group II to the third assessment report of the intergovernmental panel on climate change) (pp. 877–912). Cambridge University Press.
Smith, A. H., & Berkes, F. (1993). Community-based use of mangrove resources in St. Lucia. *International Journal of Environmental Studies, 43*, 123–132.
Smithers, J., & Smit, B. (1997). Human adaptation to climatic variability and change. *Global Environmental Change, 7*(2), 129–146.
Sontheimer, S., Callens, K., & Seiffert, B. (1999). *PRA tool box. Conducting a PRA training and modifying PRA tools to your needs: An example from a participatory household food security and nutrition project in Ethiopia.* Food and Agriculture Organization of the United Nations.
Srivastava, P. B. L., Keong, G. B., & Muktar, A. (1987). Role of Achrostichum species in natural regeneration of Rhizophora species in Malaysia. *Tropical Ecology, 28*, 274–288.
Staber, U.,and J. Sydow.J 2002. Organizational adaptive capacity: A structuration perspective. Journal of Management Inquiry 11 (4):408–424.
Swift, J. (1989). Why are rural people vulnerable to famine? *IDS Bulletin, 20*(2), 41–49.
Takagi, H., Ty, T. V., Thao, N. D., & Esteban, M. (2015). Ocean tides and the influence of sea-level rise on floods in urban areas of the Mekong Delta. *Journal of Flood Risk Management, 8*(4), 292–300.
Turner, B. L., Kasperson, R. E., Matson, P. A., McCarthy, J. J., Corell, R. W., Christensen, L., Eckley, N., Kasperson, J. X., Luers, A., Martello, M. L., Polsky, C., Pulsipher, A., & Schiller, A. (2003). A framework for vulnerability analysis in sustainability science. *Proceedings of the National Academy of Sciences of the United States of America, 100*(14), 8074–8079.
Watts, M. J., & Bohle, H. G. (1993). The space of vulnerability: the causal structure of hunger and famine. *Progress in Human Geography, 17*(1), 43–67.
Zimmermann, J., & Nicholls, R. J. (2015). Future coastal population growth and exposure to sea-level rise and coastal flooding – A global assessment. *PLoS One, 10*(3), e0118571.

Chapter 9
Stewardship, Social Entropy, Adaptive Capacity and Ecological Chaos: Conservation Strategies for the Sundarbans Mangrove Ecosystem in India

Abstract The entire ecology of Sundarbans has been suffering stresses, shocks and vulnerability. Different environmental perturbations from manifold sources of pollution, like global warming–mediated sea-level rise, the massive erosion of soils followed by vertical accretion, a reduction in the drainage of fresh water coupled with a reduced supply of sediments (causing a steady increase in salinity) and transforming the ecoregion by subjecting it to sediment starvation. Salinity invasion not only drives the freshwater-loving estuarine species of plants and animals to adjoining freshwater-dominated areas but also engulfs crop fields in the reclaimed deltas and disrupts the normal ecological states of soil-water-forest systems under the domain of the mangrove ecosystem. Gradual depletion in natural resources and ecosystem service–based goods and products in the Indian Sundarbans has prompted the outmigration of humans and escalated human–wildlife conflicts, demanding immediate interventions. This needs a transformative conservation approach turning aside from– the traditional dependence on government administrators to a stewardship approach involving local people. While conservation aims at maintaining biotic and abiotic counts, stewardship goes further by focusing on ecological services and human needs.

Conservation is a systematic approach, whereas stewardship includes systemic viewpoints. Stewardship practices generate community mobilization in that resource sustenance leads to livelihood generation in commensuration with ecological functions. In the context of a loss of natural resources and ongoing ecodegradation, poverty has increase so much that local people are turning to outmigration and the non-judicious exploitation of existing natural resources in a response to push and pull factors. Stewardship and using proper adaptive strategies can ensure an myriad of socioecological holons in the terrains and trajectories of mangrove ecosystems. Application of Lorentz's Butterfly models on Sundarbans explains the intricate relationships among other similar interconnected sensitive complex ecosystems like Amazon, Everglades etc. emphasizing the contributions of the tiniest parts of one ecosystem (holons) can result long-term ecological consequences on others through cascading effects in a chain wise process. In such a context, this chapter deals with a conceptual framework applied to provide a theoretical base depending generated

S. K. Chakraborty, S. K. Acharya, *Eco-dynamics and Environmental Perturbations in Mangrove Ecosystems*,
https://doi.org/10.1007/978-3-032-08710-2_9

research information for conducting empirical studies and analyses on co-adaptation against environmental problems and addressing them though stewardship, community mobilization mobilization mediated social entropy, gender issues and ecological chaos to sustainably conserve mangrove ecosystems.

Keywords Sundarbans · Community mobilization · Stewardship · Co-adaptation · Sustainable conservation · Push and Pull factors · Ecological chaos

9.1 Conceptual Framework for Ecological Assessment and Conservation

In view of the escalating conflicts between humans and wildlife and manifold environmental perturbations such as global climate change–mediated sea-level rise, more and more-intense outmigration is becoming a significant issue for the entire biosphere. Furthermore, 66% of the total population of humans reside along the world's coastlines, and they reap different kinds of ecosystem services from these regions, including geological and geomorphological components such as continental shelves, the foreshore and backshore, intertidal belts, littoral zones, ripples, dunes, deltaic islands with mangroves, backwater and brackish-water zones, corals, lagoons and estuaries (Pethic, 1984, 1992). Through continuous interaction processes with land, the atmosphere and marine areas, the coastal zones determine the regional energy flows and food production, which humans use for harvesting bioresources, making settlements, industrialization, etc.

The Indian coasts extend across 15,200 km and the Indian coastline stretches across 5700 km on the mainland and about 7500 km when including the two island territories. However, substantial damage across the Indian coastline, which has ecosystems like mangroves, salt marshes, lagoons, corals, dunes and estuaries, has resulted in increasing human pressures (Qasim et al., 1988). The coastal area of West Bengal extends across 0.28 million hectares and 220 km of coastline, mostly in the district of South 24 Parganas, India. The coastal tract of the Midnapore (east) district, West Bengal, India, sharing 27% (60 km) of the total length of the coastal zones in West Bengal, India, is a transmissible part of the Sundarbans mangrove ecosystem, which is a World Heritage Site extending along the Hooghly estuary from New Digha (at the confluence of Subarnarekha with Hooghly) to the extreme southwest point of the Midnapore district, at which point it curves around Junput, Rosulpur, Khejuri and Haldia in the east to further northeast up to Tamluk or even Kolaghat on the banks of Rupnarayan (Chakraborty, 2010). The influence of tectonic settings through the geotectonic process and a long history of deposition, vertical accretion and seaward advancement in the coastal plain of West Bengal over the Bengal Basin have resulted in sea-level rise (Paul, 2002).

Around 300 BCE, the Earth's geomorphology offered the most beautiful and diverse biosphere in the world, namely Sundarbans, subsequently declared as a reserved forest in 1865. In 1987, Sundarbans National Park was declared as a World Heritage Site. In the lap of the Bay of Bengal with the world's largest delta in the background, it is seamlessly flanked and fed by the Ganges–Meghna–Brahmaputra river system. The Indian Sundarbans has ever been evolving with the diverse terrain of the large extent of mangrove forests, varieties of wild animals (including the Bengal tiger (*Panthera tigris*)), countless marine and estuarine invertebrates and fish. The deep, dense and beautiful yet dangerous Sundarbans has been invaded by human settlements and human migration and subjected to mangrove destruction only to welcome agriculture and livestock for the livelihoods of humans. Such endeavours have resulted in the reclamation of 54 of 104 deltaic islands across the length and breadth of the Indian Sundarbans. All those dozens of deltaic islands appear like pearls of hope, fascinating enough to create human settlements within their robust ecosystem. The bold yet blatant denials of the wild and traditional inhabitants of Sundarbans, like tigers, crocodiles, snakes, birds, benthos, honeybees and mangroves, have certainty contributed to ecological disruptions, including human–wildlife conflicts, which are sometimes brutal and lethal.

The existence and ecodynamics of any ecosystem depend on its core–periphery interactions, which have generated an interconnected ecological interaction process with the holistic development, functioning and growth of the Sundarbans ecosystem. Therefore, a proper understanding of the complex territorial interdependences between core–periphery–far periphery interactions is a prerequisite for sustainable conservation (Turner II, 1990; Schaller, 1993; Whitten et al., 2003; Schroth et al., 2004; Tomich et al., 2004; Wojtkowski, 2004; Thompson et al., 2007; Suding et al., 2008; Water Stewardship in Agriculture (WSA), 2015). While Sundarbans is flanked by three sub-ecosystems, viz. far peripheral, peripheral and core sub-ecosystems, with nuance and nomenclature, ecological transmission, transformation and trajectories are quite conspicuous. The grid and spill network between these three echelons is extremely important. Even in human–wildlife conflicts, the issues are less within the forest and more in the social ecology of human cultures and economies. The approach for the ecomanagement of Sundarbans can be scripted by using the socioecological resilience paradigm. First, we need to train local communities within the vicinity of the Sundarbans ecosphere in adaptive coping strategies, especially in response to slow-onset disasters like embankment erosion, tidal overflow and abiotic stresses. Next, we can move on to perennial adaptive capacity by changing agricultural technologies, generating livelihoods and restoring mangroves. Yet adaptation alone cannot totally solve this problem; we need transformative adaptation. Farmers will need to transform into climate managers; fishers will become stewards of aquatic ichthyofaunal components, honey collectors will transform into honeybee protectors and the women of Sundarbans will need to learn a family repository of landraces, the types that are quickly becoming endangered.

9.2 The Concept of Farm Stewardship and Its Applicability in a Mangrove Ecosystem

Farm stewardship and ecology are interconnected concepts that emphasize a sustainable agricultural production process. Farm stewardship refers to the ethical and responsible management of agricultural resources. It includes natural biodiversity, soil and water for resilience and productivity. Ecology, on the other hand, is the scientific study of the interrelationship between biological organisms and their surrounding environment. The conceptual framework of farm stewardship rigidly adheres to the basic operating principles of ecology and integrates ecological principles into ecomanagement practices not only in the sectors of agriculture but also in other natural resources, like forests and fisheries, to promote sustainability and environmental conservation (Pretty Jules, 1995; Power, 1999; Redford & Richter, 1999; Molnar, 2003; Ricketts, 2004; Robertson & Swinton, 2005; Sayer & Maginnis, 2005; Ruark et al., 2006; Blann, 2006; Scherr et al., 2007; Molden et al., 2007). This framework recognizes that ecosystems, being complex and dynamic systems, are impacted by both natural factors and anthropogenic factors.

This conceptual framework has several key components.

1. An ecosystem-based approach: The framework emphasizes the importance of viewing farms as ecosystems and managing them in a way that mimics natural ecological processes. This includes promoting biodiversity, improving soil health and minimizing the use of synthetic inputs.
2. Sustainable agricultural practices: The framework advocates for the adoption of sustainable agricultural practices, including conservation agriculture, crop diversity, agroforestry and ecofriendly pest management. These practices help to reduce the environmental impact of farming while maintaining or improving productivity.
3. Resource conservation: Farm stewardship involves the responsible use and conservation of natural resources. This includes practices such as water conservation, soil conservation and nutrient management to prevent pollution and the degradation of natural resources.
4. Community engagement: The framework emphasizes the importance of engaging with local communities and stakeholders to ensure that agricultural practices are socially acceptable and beneficial. This includes sharing knowledge, building partnerships and supporting local economies.
5. Adaptive management: Given the dynamic nature of agricultural ecosystems, the framework promotes adaptive management approaches that allow farmers to respond to changing environmental conditions and uncertainties. This includes monitoring and evaluating farming practices to assess their effectiveness and make adjustments as needed.

Overall, the conceptual framework of people-based stewardship in agriculture in the peripheral areas of mangrove forests and reclaimed deltaic areas provides a holistic and integrated ecological approach to agricultural management that promotes productivity, sustainability and resilience.

9.3 The Origin and Adoption of the Concept of Stewardship

Stewardship is a concept with its Western origins and has been traditionally practised by professionals to protect water, soil and forests from the present rapid and aggressive depleting trend due to human activities. In such a context, the potential application of functional stewardship needs to be investigated in Indian agriculture and other environmental resource management sectors. Such ventures can be adapted to fulfil stewardship functions by extrapolating conservation approaches.

A landowner's assertion of stewardship over their land for a specific purpose carries significant responsibility. The steward must not regard the land as solely their own but rather must manage it as if acting on behalf of others, as if the land belonged to others. In an ecological context, land stewardship requires the landowner to manage the land with the goal of minimizing their ecological footprint. To claim to be practising sound ecological stewardship, the landowner must demonstrate not only that their land is ecologically significant but also that its management does not jeopardize the ecological integrity of neighbouring lands. This is the true test of land stewardship in ecological terms, and it is the standard against which any claims by farmers to be acting as responsible ecological stewards must be measured.

This declaration embodies the core belief of the "first stewards of the land" because people rely on their land for biological production and to earn their livelihoods by having intimate knowledge of and affection for the natural resources considered as environmentally friendly or even beneficial. This proposition points out that if farmers are allowed to utilize the natural resources that come naturally to them, driven by their love, it will result an understanding of the all-round utility of the land and thereby resolve environmental concerns (Tables 9.1 and 9.2).

The stewardship function is multifaceted, originating from land-based production systems that operate across time, cost and resource continua. In the sphere of agricultural practices in and around Sundarbans, especially in the reclaimed deltaic

Table 9.1 Differences among steady-state resource management, ecosystem management and resilience-based ecosystem stewardship

Key features of steady-state resource management	Ecosystem management considerations	Resilience-based ecosystem stewardship factors
It relies on a reference state, typically a historical condition.	Historical conditions	Considering the trajectory of change
It focuses on managing a single resource or species.	Managing multiple ecosystem services	Managing fundamental social-ecological properties
It aims to maintain a single equilibrium state with sustainable properties.	Recognizing multiple potential ecosystem states	Recognizing multiple potential outcomes
	Accepting the historical range of variability	Promoting variability and diversity
It seeks to reduce variability in the managed system.	Accepting natural disturbances	Encouraging disturbances that sustain social-ecological properties
It aims to prevent natural disturbances.	Acknowledging people as part of the social-ecological system	Acknowledging people's responsibility to sustain future options
It acknowledges that ecosystems are utilized by people.	Collaborating with multiple stakeholders to define goals	Collaborating with multiple stakeholders to define goals
Managers play a crucial role in defining the function of the operating system.	Managing multiple uses, even if efficiency is reduced	Maximizing flexibility for future options
It aims to maximize sustained yields and economic efficiency.		
The management structure is designed to protect current management goals.	Adapting management goals to a changing human ethos	Adapting management strategies to respond to and shape human values

Table 9.2 Ecosystem services as well as the diversity of nature-based components and the mechanism of their operations

Ecosystem services	Diversity among nature-based components and the mechanism of their operation
1. Production by societally important plants	Functional composition: Fast-growing species tend to produce more biomass. Species vary in their timing and spatial patterns of resource use, with complementarity allowing for more-efficient resource utilization. Species number: A larger species pool is more likely to include productive species.
2. Stability of crop production	Genetic diversity: Such diversity protects production from losses caused by pests and environmental fluctuations. Species number: Growing multiple species together in the same area maintains high production across a wider range of conditions. Functional composition: Different species respond differently to environmental changes and disturbances, which helps stabilize production.
3. Maintenance of soil resources	Functional composition: Fast-growing species contribute to soil fertility. Dense root systems help prevent soil erosion.
4. Regulation of water quantity and quality	Landscape diversity: Preserving riparian corridors reduces erosion. Functional composition: Fast-growing plants with high transpiration rates can reduce stream flow.
5. Pollination for food production and species survival	Functional composition: The decline in specialized pollinators can reduce both the fruit set and the diversity of plants that reproduce successfully. Species number: A decrease in pollinator species leads to a reduction in the diversity of plants that can successfully reproduce, resulting in genetic impoverishment. Landscape diversity: Large, well-connected landscape units allow pollinators to promote gene flow among habitat patches.
6. Resistance to invasive species with negative ecological/ cultural effects	Functional composition: Certain competitive species can resist the invasion of exotic species. Landscape structure: Roads can act as corridors for the spread of invasive species, whereas natural habitat patches can resist this spread. Species number: Communities with high species richness are likely to have fewer unused resources and more-competitive species, which can help resist invaders.
7. Pest and disease control	Genetic diversity or species control: Reducing the density of suitable hosts for specialized pests and diseases. Landscape diversity: Offering habitat for the natural enemies of pests.
8. Biophysical climate regulation	Functional composition: This dictates water and energy exchange, thereby influencing local air temperature and circulation patterns. Landscape structure: This affects the convective movement of air masses, consequently influencing local temperature and precipitation.
9. Climate regulation by carbon sequestrations	Landscape structure: Fragmented landscapes have a higher edge-to-edge ratio, resulting in greater carbon loss at the edges. Functional composition: Small, short-lived plants store less carbon. Species number: A high number of species reduces pest outbreaks, which can cause carbon loss.
10. Protection against natural hazards (e.g. floods, hurricanes and fires)	Landscape structure: This structure influences the spread of disturbances and/or provides protection against natural hazards. Functional composition: Extensive root systems help prevent erosion and uprooting. Deciduous species are less flammable than evergreens.

islands, the most severely impacted components of farm ecology are soil, water and biodiversity. The ecological products or byproducts, such as paddy, straw, betelvine, coconuts and fish, lack adequate market prices, indicating that such agricultural activities have become increasingly nonremunerative. For the cultivators and agricultural producers, the costs of production have been increasing, whereas returns are uncertain or declining. The strains from market uncertainties have proven detrimental to both social and farm ecologies.

There is a growing sense of desperation among local farmers, leading them to opt for shorter crop durations and aim for moderate to higher productivity to keep up with market demands. This has resulted in an increased use of agro-chemicals, fertilizers and unsustainable resource management practices. The short-term focus has compromised the prospects of sustainability and resilience in the ecosystem. There is a concerning trend where higher agricultural production is ironically causing more distress to farmers due to market fluctuations and livelihood losses.

Merely focusing on soil and water conservation is not enough for ecological restoration, which thereby necessitates a qualitative shift towards more-proactive stewardship interventions. For instance, cultivating short-duration leafy vegetables by using residual moisture and minimal inputs can significantly contribute to water conservation, making it a form of responsible stewardship. Similarly, excavating a seepage tank to conserved runoff water and then using this water to harvest fast-growing seasonal fish not only adds economic value but also qualifies as a socially beneficial stewardship practice (Bunch, 2000; Chapin III et al., 2002; Buck et al., 2004; Buck et al., 2007; Chapin III et al., 2009).

9.4 Understanding Change: Conservation and Stewardship

9.4.1 The Definition and Concept of Stewardship

Stewardship can be defined as the responsible and sustainable management of land and its resources. Land stewards, which can include individuals, organizations and communities, are tasked with making decisions that prioritize the needs of the environment, society and economies. This may include practices such as sustainable agriculture, reforestation, habitat restoration and soil conservation. Land stewardship recognizes that land is a valuable and finite resource that must be carefully managed to ensure its continued health and productivity for future generations.

Soil stewardship refers to the responsible and sustainable management of soil resources. It involves practices that aim to improve and maintain the health, fertility and productivity of soils. Soil stewards, which can include farmers, land managers and conservationists, implement conservation tillage and nutrient management to protect soil structure, prevent erosion and improve soil health. Soil stewardship recognizes that soil is a vital and finite resource that is essential for food production, biodiversity and ecosystem health and that it must be carefully managed to ensure its long-term sustainability.

Aidan et al. (2016) identified four main areas of stewardship practices:

1. Examining the link between soil stewardship practices and health indicators at both farm and landscape levels.
2. Assessing how soil stewardship practices can reduce exposure to chemical contaminants, such as those found in veterinary pharmaceuticals used in manures, and their transfer through the food chain and to air and water.
3. Studying how soil stewardship practices, including landscape design, influence the dynamics of pathogenic organisms that affect crop, animal and human health.
4. Analysing the connections and trade-offs between ecosystem services and health indicators at the landscape and regional scales. Utilizing large-scale environmental and soil surveys (e.g., the land-use/land-cover-area survey (LUCAS), and the countryside survey) and research networks (e.g., long-term ecological research networks) that include various measurements can help to investigate these relationships. Additionally, incorporating risk assessments for soil security, as suggested, could be expanded to include risks to animal, human and ecosystem health, thereby considering potential cascading effects on those reliant on these soils.

9.4.2 The Development of Stewardship Concepts from Agricultural Sectors in the Reclaimed Deltaic Islands of Sundarbans, India

In the agricultural sector, production serves as a cornerstone for both biological and ecological functions within a broader ecological framework, intertwining with other ecological units to form what can be called ecological networking. This networking aspect facilitates the integration of ecological and sub-ecological functions, fostering a multifaceted ecological functionality. Currently, the world is experiencing unprecedented changes in various factors that define its core characteristics and their impact on society (Flowcharts 9.1, 9.2, 9.3, 9.4 and 9.5; Tables 9.1 and 9.2).

Throughout history, humans have engaged with and moulded ecosystems to drive social and economic progress (Turner II, et al., 1990; Redman, 1999; Jackson, 2001). Over the past 50 years, human actions have significantly and swiftly altered ecosystems more than in any previous era (Foley, et al., 2005; Diamond, 2005; MEA 2005c). Currently, the Earth's climate is warmer than it has ever been in recorded human history, due in part to the buildup of greenhouse gases (GHGs).

Mostly CO_2 but also methane, oxides of nitrogen in the atmosphere from burning fossil fuels and other forms of anthropogenic activities have led to global warming. The global population explosion has starkly encroached upon natural habitats and resources, once considered nature's blessings. With the burgeoning human population, the increasing need for food and natural resources has spurred on the expansion of agriculture, forestry and other anthropogenic activities and interventions.

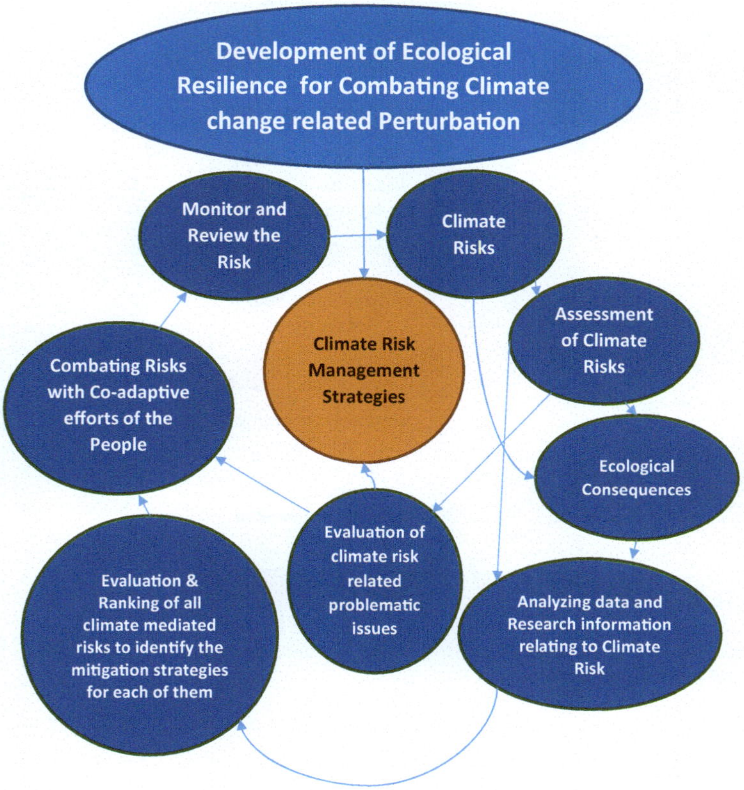

Flowchart 9.1 Climate risk management strategies through ecological resilience and co-adaptation

This expansion has resulted in significant land-cover changes, habitat loss and a decline in biological diversity. Currently, about 50% of the global population resides in urban settlements, relying on links with rural regions worldwide for sustenance, water and waste management. Furthermore, increased human mobility is accelerating the spread of plants, animals, diseases, industrial goods and cultural perspectives at an unprecedented pace. The rise in worldwide mobility, combined with enhanced connectivity through global markets and modern communication methods, interlinks the economies and cultures of the world. As a result, decisions made in one location frequently have international repercussions.

The planet's life-support system is largely modified by the globalization of economies, cultures and urbanization (Odum, 1989), which is its ability to fulfil the needs of all organisms, including humans. The sharp rise in the extinction rate of species implies that global changes have had disastrous effect on many species. These global changes have unleashed both beneficial and harmful effects on human societies. While some parts of the world have had increased food production, higher incomes and living standards, improved treatment for many diseases and longer life

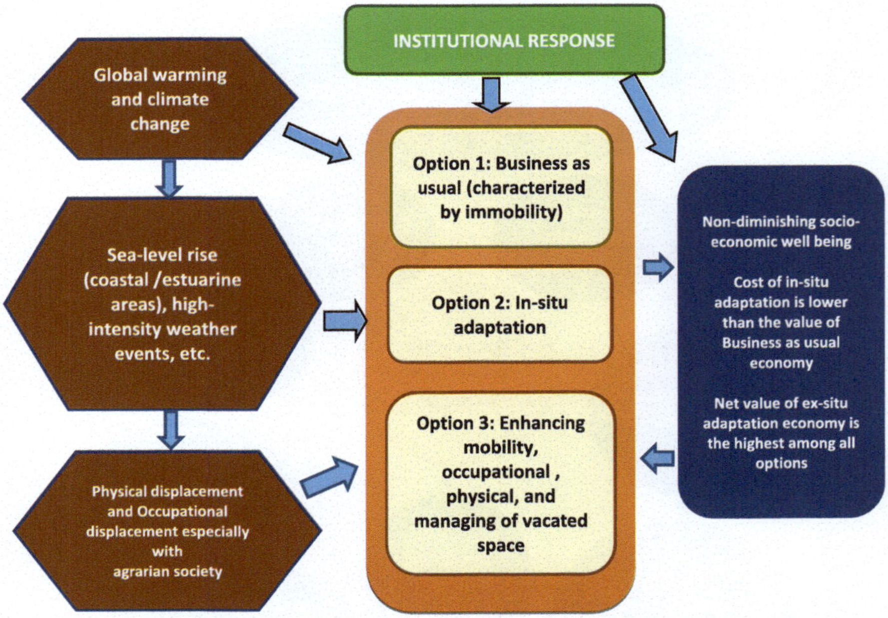

Flowchart 9.2 Mangement strategies for global warming–mediated climate changes through societal and institutional frameworks

expectancies, these have been offset by a decline in ecosystem services—the benefits that society derives from ecosystems.

A high quality of life has been inadvertently degraded as people strive to accomplish their material desires and needs (MEA, 2005). Humans have demonstrated our ability to significantly impact the planet's life-support system. With responsible stewardship, this capacity can be harnessed to not only restore but also improve the Earth's ability to sustain societal development. A key characteristic of these changes is their directionality, indicating a consistent trend over time. Many of these trends have intensified since the mid-twentieth century and are likely to persist and be exacerbated in the coming decades. This dynamic approach planning a planning dilemma emphasises the perception that the future will resemble the past is unreliable, but one can rely on past experiences to guide future planning. This challenge is predominantly sharpened against the background of sustainable natural resource management.

The outlook of conservationists differs significantly from that of stewards. Conservationists resist changes in ecosystems and strive to protect them from being exploited by surrounding stakeholders. Stewards, on the other hand, align anthropogenic needs and functions with ecosystem services, allowing for various functional changes by fostering resilience. Managing resources to enhance resilience—responding to and shaping change to maintain and develop fundamental functions, structures, identities and feedback—appears crucial for humanity and the future of

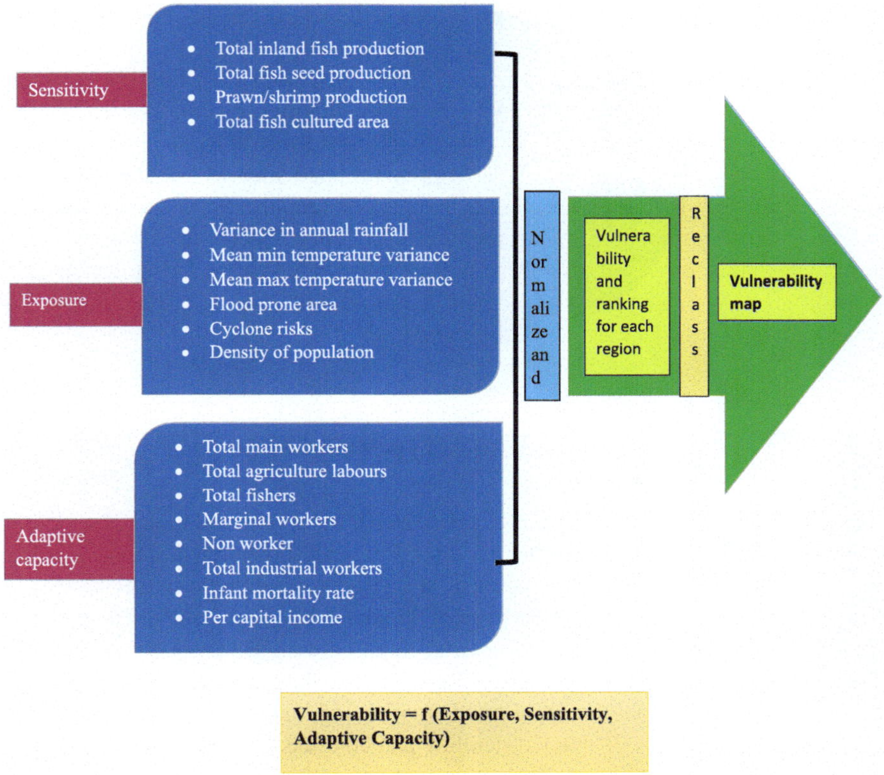

Flowchart 9.3 Vulnerability map depicting the major components of vulnerability and prospective outcomes

the Earth's systems. Resilience-based ecosystem stewardship represents a fundamental shift from the approach of steady-state resource management, which aims to minimize variability and prevent change rather than adapting to and shaping change for the benefit of society (Flowcharts 9.1 and 9.5; Tables 9.1 and 9.2).

Ecological resilience is being prioritized as a concept acknowledging change as a fundamental aspect of the world's functioning and development, making it particularly relevant during periods characterized by significant change. The main focus centres on ecosystems that offer a range of services, rather than single resources like fish or trees. Underscoring the importance of stewardship is required to impart to managers as integral components of the systems that they oversee and entails a sense of responsibility for the system's conditions (Leopold, 1949). The key challenge is to anticipate and shape change in a sustainable manner that preserves future options (Folke, 2003). Ecosystem stewardship recognizes the need for society's resource use to align with ecosystems' capacity to provide services, which is limited by the planet's life-support system.

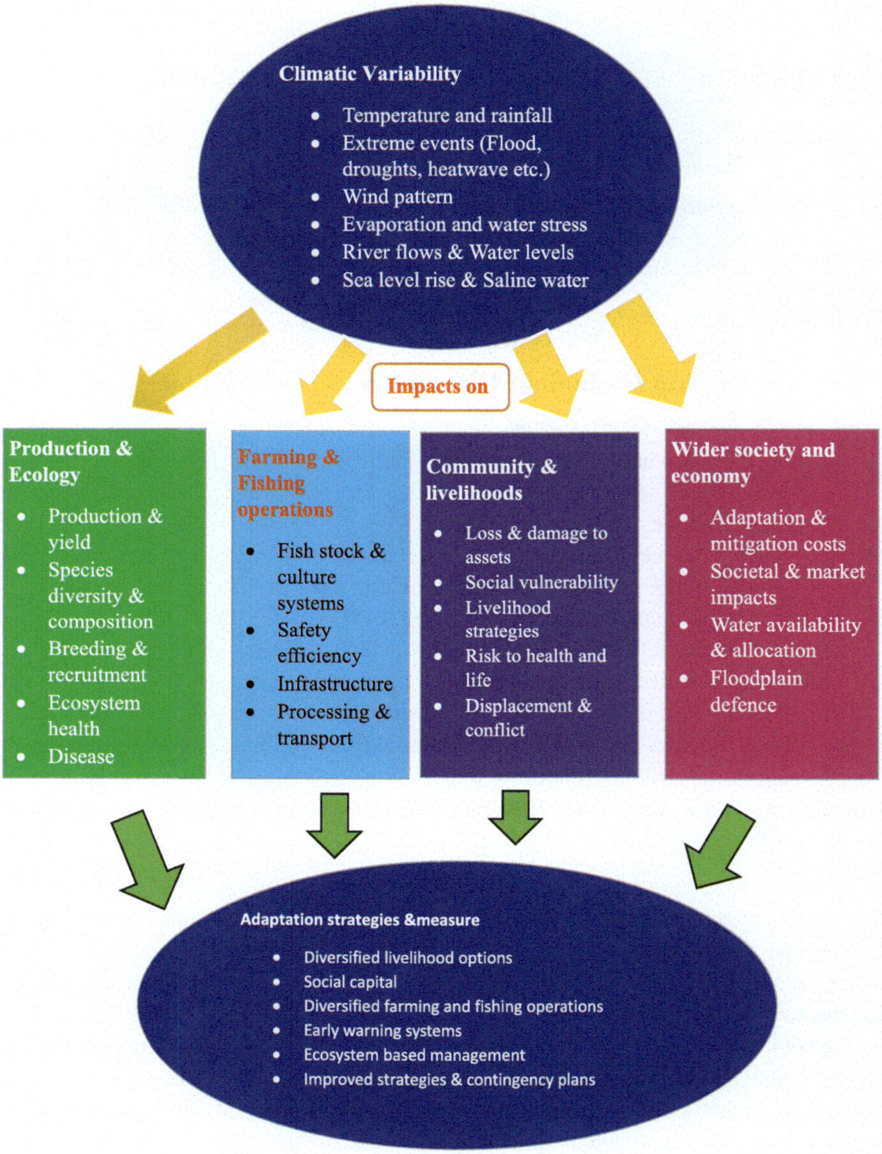

Flowchart 9.4 Impact of climate variability on ecology, biological production system, social vulnerability and regional economies, with a focus on adaptation strategies

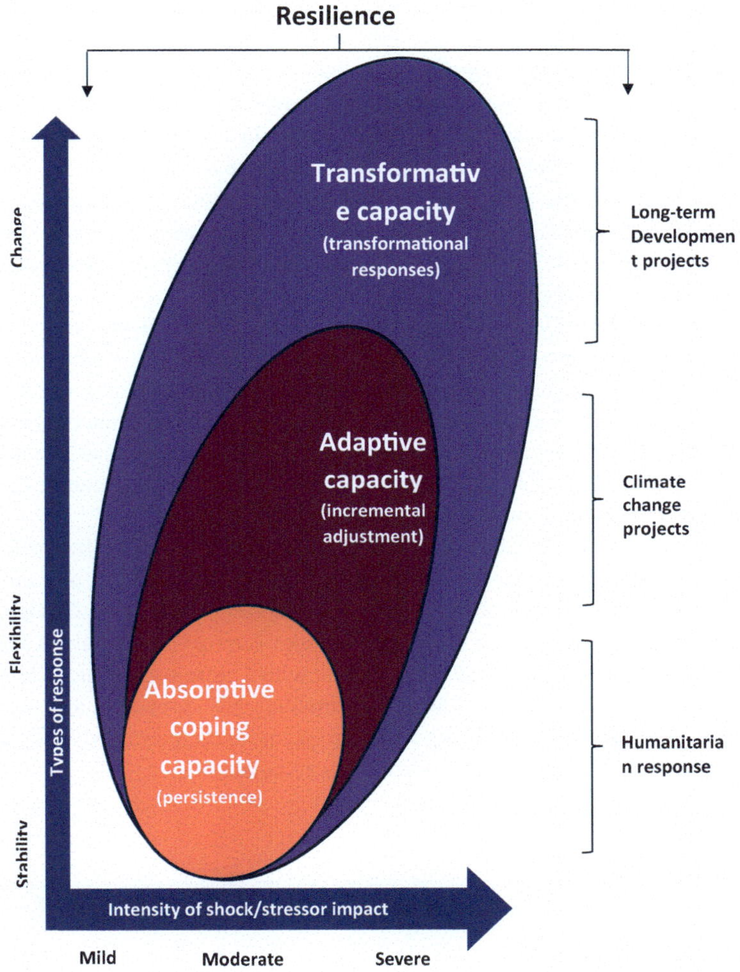

Flowchart 9.5 Resilience (ecological and social) centring on changes, flexibility and stability in relation to transformative, adaptive and absorptive capacities against long-term development and climate change projects, emphasizing humanitarian responses and an adaptable toolkit to tackle the sustainability challenges in a world experiencing directional changes

9.5 Stewardship and Sustainability Issues

Achieving sustainability involves addressing the dynamic nature of complex adaptive systems. Sustainability must be contextualized within the context of the ongoing directional changes in the factors that shape most social-ecological systems. Three main outcomes are supposed to emerge: (i) maintaining the current system's fundamental properties through adaptation; (ii) transforming the system into a substantially different, more-desirable state; and (iii) enduring passive changes from the system's failure to adapt or transform. Intermediate outcomes may also occur, where some components of the system perform, others evolve and others decline (Turner, 2003). Sustainability entails maintaining the system's fundamental properties or actively transforming them by intentionally replacing other forms of capital to meet community needs in renewed ways. Conversely, degradation involves the erosion of systemic wealth and subsequently the ability to attain sustainability.

Four key approaches have been identified to promote sustainability in the face of directional change in the processes of persistence or transformation: (i) reducing vulnerability; (ii) improving adaptive capacity; (iii) increasing resilience; and (iv) fostering transformability. Each approach highlights distinct processes that contribute to fostering sustainability.

Reducing vulnerability involves understanding the nature of stressors that drive change, the system's sensitivity to these stressors and its capacity to adapt to change. Enhancing adaptive capacity focuses on the ability of individuals or groups to adjust, minimizing the negative impacts of change. Resilience encompasses not only adaptive capacity but also additional system-level attributes that provide flexibility for adjustment. Transformability involves taking effective measures to uplift the system to a potentially more-desirable state.

All these approaches initially developed somewhat independently by the interdisciplinary contributions of anthropologists, ecologists and geographers, and a trent has been a growing towards integration (Berkes et al., 2003; Turner et al., 2003; Janssen et al., 2006; Young et al. 2006). This merging of ideas equips policymakers and managers to undertake natural resource management with a more sophisticated approach.

Resilience-based ecosystem stewardship interactively considers all the factors delineated in Flowchart 9.5, ensuring that ecosystem functions can be sustained and providing essential ecological services to meet societal needs.

9.5.1 Vulnerability

Assessing vulnerability is crucial for organizing conservation or stewardship efforts. If an ecosystem is not protected against vulnerability, its components become more fragile to internal and external stresses. Vulnerability refers to the degree to which a system suffers from exposure to a specific hazard or stress (Turner II et al., 2003; Adger, 2006). Although rooted in socioeconomic studies of events (such as floods or wars) or chronic stresses (like food insecurity) on social systems, vulnerability theory has evolved to address responses from entire social-ecological systems. Vulnerability analysis explicitly includes human ethical considerations such as uniformity and quality livelihoods. To reduce vulnerability to a particular stress, one can undertake three approaches: (i) mitigate exposure to the stress; (ii) decrease the system's sensitivity to stress by maintaining natural capital and the components of well-being, especially for disadvantaged groups; and/or (iii) improve adaptive capacity and resilience to cope with stress (Flowchart 9.3).

Downscaling a stress can involve reducing its severity, propensity and span. For instance, preventing pollution or prohibiting the use of toxic pesticides can lower the vulnerability of individuals. Mitigating exposure is particularly difficult when the stress results from cumulative processes occurring at scales larger than the managed system. For example, human activities that contribute to global warming through fossil fuel combustion are globally dispersed, precluding the ability to reverse the effects solely through actions taken in regions experiencing the greatest impacts of climate change (McCarthy et al., 2005).

Reducing sensitivity to a stress can be achieved in at least three ways: (i) by preserving the placid ecological factors that rely on natural capital; (ii) by upholding key components of well-being; and (iii) by focusing on the needs of disadvantaged communities. For instance, low-income people are particularly susceptible to food shortages and economic downturns, whereas people residing in and around rivers or estuarine floodplains or at wild–urban interfaces experience risks from natural calamities like flooding or wildfires. Understanding the reasons behind varying kinds of vulnerability can lead to targeted interventions to decrease the overall vulnerability of the social-ecological system.

9.5.2 Adaptive Capacity

Adaptive capacity, also known as adaptability, refers to the ability of actors, including communities with their responsive behaviour, to adapt to change (Folke, 2003 Walker et al., 2004; Adger, 2006). When considering all living organisms as actors in social-ecological systems, the main focus is primarily on people's roles in adaptive capacity in response to social-ecological change. Human stakeholders act not only on past experiences but also on their ability to cope with future unpredictability. This is in contrast to evolution, which shapes organisms' properties only on the basis of genetic responses to past events, without a forward-looking component. Adaptive capacity relies on four attributes: (i) resource diversity, behavioural economics and societal variations, which provide the foundational elements for adapting to change; (ii) individuals' and groups' ability to learn about how their system functions; (iii) empirical interventions to test their learning; and (iv) effective governance capacity to select, communicate and implement suitable solutions.

9.5.3 Resilience—Both Ecological and Social Perspectives

Resilience refers to a social-ecological system's ability to withstand a range of shocks or reorganize in a new context (Holling, 1973; Douglass, 1984; Walker, et al., 2006; Folke, 2003; Gunderson & Holling, 2002). The significance of understanding resilience, according to ecological and social perspectives, is contingent on anticipating, planning and reducing disaster-mediated risks to effectively protect individuals, communities and countries, along with their livelihoods, health, cultural heritage, socioeconomic assets and ecosystems. Terminology like *spring forward*, *bounce back* and *build back better* are often used in the context of resilience.

What sets resilience theory apart is its acknowledgement and identification of several system properties that support renewal and reorganization following disturbances (Holling, 1973). Resilience relies on five attributes: (i) adaptive capacity (as discussed earlier); (ii) biophysical and social legacies that enhance diversity and offer established pathways for recovery; (iii) individuals' ability to plan for the long term amid uncertainty and chaos; (iv) a balance between customizing feedback mechanisms that shield the system from stress and disruption on one hand and innovation that fosters change on the other; and (v) the ability to adapt governance structures to meet evolving needs (Holling & Gunderson, 2002; Folke, 2003; Walker et al., 2006).

The decline in resilience breaks a system down to its restrictions followed by ecodegradation. When resilience is diminished, a disturbance that previously shook and rejuvenated the system might now push the fragile system beyond a critical threshold into an alternate state (known as a regime shift) with a new trajectory of change. Such alterations drastically disrupt the flow of ecosystem services and affect the livelihoods and well-being of individuals and societies. Clearly, resilience is a crucial aspect of resource stewardship, particularly in the face of uncertainty and change. Therefore, this approach to resource management is more important today than ever before.

Minimizing variability and change contributes to resilience by maintaining components within the system. This reduces the likelihood of catastrophic effects from inevitable disturbances. Conversely, suppressing small-scale disturbances tends to eliminate components that have adapted to disturbances, thereby reducing the system's ability to cope with disruptions. Early-succession species, well adapted to post-disturbance conditions but possibly uncommon in mature forests, illustrate this point. Similarly, the narratives and recollections of elders, along with written histories, often offer insights into how people managed past crises and suggest options for posterity. This often involves drawing on social memory, the collective wisdom of how to navigate various circumstances (Chapin III et al., 2007).

An ecosystem is composed of organisms such as plants, microbes, animals (including humans) and physical components like soil and water, along with other ecological variables that interact with one another, resulting in energy flow, biological productivity and nutrient recycling. Although all ecosystems are influenced, to varying degrees, by social processes, the study of ecosystems typically emphasizes biological interactions.

Studies on social ecology have aimed to understand social resilience and social entropy with the interactions and responses between individuals and their surrounding ecological conditions and have shown how these interactions affect society and the environment as a whole. In the field of social ecology, the word *social* denotes its recognition of the often-overlooked fact that nearly all our present ecological problems have arisen from deep-rooted social problems, whereas the word *perception*, being the sensory experience of the world, tends to highlight an idea, belief or image of developing the power of visualizing or understating something and involves recognizing both environmental changes and actions in response to these changes (Flowcharts 9.1, 9.2 and 9.5).

A considerable proportion of ecosystem services rendered to a society that depends on them for survival and leading a good life have already been disrupted to meet the material desires and needs of humans (MEA, 2005). Consequent changes in ecological resilience leading to ecodegradation and a loss of ecosystem services have posed real challenges for society to overcome these problems. Affected people cannot tackle this trend of adverse ecological and social changes only by engaging in appropriate stewardship so that reinforced human capacity can be mobilized not only to repair but also to amplify the capacity of the Earth's life-support system to support societal development towards necessary persistence. However, a dilemma in determining sustainable conservation strategies has arisen because foreseeing the trend in changes in societal perceptions and modes of using natural resource bases in the future world is difficult (Kremen et al. 2002; Swift et al., 2004; Molnar et al., 2004; Nierenberg, 2005; Patz et al., 2005; Pearce, 2005; STCP (Sustainable Tree Crops Programme, 2005); McBratney et al., 2014).

9.5.4 Factors Driving People to Engage in Outmigration

Outmigration and in-migration processes tend to structure societal fabric of an ecosystem, the maintenance of ecological health of which ensure the well-being of the inhabitants, the concerted activities of whom in turn help to achieve sustainable ecological development.

9.5.4.1 Push and Pull Factors

Push and pull factors are the reasons which compel people to migrate from one place to another. Push factors are negative conditions that encourage people to leave their homes, whereas pull factors are positive conditions that attract people to a new location. When an ecosystem like a mangrove ecosystem fails to accommodate a population and provide livelihoods to local people due to negative changes in their surroundings, like the nonavailability of resources, the development of pollution, soil erosion, the loss of fertility in agricultural fields, disruptions to social normalcy and the origin of chaos, the push factors take effect and eco-dwellers drift out. Similarly, when an ecosystem is resilient, full of resources and capable of unleashing ecological services, people from outside the ecosystem are attracted to move to a new social ecology, so pull factors act as attractants.

9.5.4.2 Using Social Entropy Theory to Explain Entropy in Social Systems

This unique theory, bridging the gap between societies and ecosystems, is rooted in the application of thermodynamics to social systems, which suggests that societal structures tend to be inclined towards disorder or increased entropy. It posits that factors like wealth, population distribution and technology, akin to entropy in physics, can drive societal disorganization. Social entropy, as theorized by Kenneth D. Bailey, applies the concept of entropy from physics to social phenomena. In physics, entropy refers to the measure of disorder or randomness in a system. In social systems, it can be seen in the increasing complexity, unpredictability and potential for conflict within societies. Social entropy theory offers a way to analyse societal changes and challenges through the lens of entropy—a concept borrowed from physics. By examining how population, technology, information, spatial distribution, organization and livelihood interact, the theory helps us to understand the processes that can lead to social order or social disorder (Bailey, 1968, 1982, Berkes et al., 2003; Chapin III et al., 2006).

9.5.4.3 The Ecodynamics of Ecosystems

Every ecosystem is in a seamless process of evolution, and the ecodynamics of the Sundarbans forest is no exception. The ongoing transfer of energy, input and output, can take place in a systemic way, such as in the ecology of a forest, wherein factors from the periphery and isochronous and isoterrestrial situations act on the same location and at the same point in time. This invites mutual changes and metamorphosis in ecological parameters (Berkes et al., 2003; Folke, 2006; Cassman et al., 2003).

9.6 The Roles of Women in Resource Management Through Adaptive Co-management

The approach of co-management in Sundarbans can be scripted by using the socio-ecological resilience paradigm. First, local communities need to be trained within the vicinity of the Sundarbans ecosphere in "adaptive coping strategies," especially for slow-onset disasters like embankment erosion, tidal overflow and abiotic stresses. The second step includes moving on to perennial adaptive capacity by changing agricultural technologies, generating livelihoods and restoring mangroves. Yet adaptation alone cannot totally solve all problems, and it necessitates the adoption of transformative adaptation. Local producers need to transform into climate managers, fishers, honey collectors, etc. and thereby become protectors of the structural components of Sundarbans. Women will experience a transformation into a family repository of landraces, the types that are quickly becoming endangered (Chapin III et al., 2008; Biggs and Rogers, 2003).

9.6.1 Women's Empowerment and Stewardship in Coastal Ecosystems

The role of women in the stewardship of coastal ecosystems is crucial for the sustainability of these ecosystems and for the maintenance of the prevailing opportunities for the livelihoods of coastal communities. Women, representing almost 50% of the local population, traditionally take care of most of the household responsibilities, nurture and provide firsthand education to their children, care for their livestock, earn a living through innovative production processes of manufacturing different food items, stitch clothing and assist in agriculture. Therefore, women's empowerment is essential for effective coastal zone management, including the protection of the mangrove-based biota, so that they can thereby participate in and contribute to decision-making processes as active members of a decision support system (DSS). Their steady access to locally available resources like forest, agricultural and fishery products allows them to extract more economic benefits for local people and maintain the social-economic-ecological health of the region (UNEP, 2016; World Bank, 2012).

9.6.2 The Roles Played by Women in the Stewardship Process for Coastal Zone Management

1. Fisheries management: Women are involved in small-scale fisheries, specifically the processing and marketing of fish products (Williams, 2016).
2. Coastal conservation: Women participate in coastal conservation efforts, such as mangrove restoration and the management of protected marine areas (Kumar, 2018).
3. Disaster risk reduction: Women play a critical role in disaster risk reduction and management, particularly in coastal areas prone to cyclones and tsunamis (UNISDR, 2015).

9.6.3 Barriers to Women's Empowerment in Coastal Ecosystems

Despite their critical roles, women face several barriers to empowerment in coastal ecosystems:

1. Limited access to natural resources: Women often have limited access to resources, such as credit, training and technology, which hinders their ability to participate in coastal management (FAO, 2017).
2. Social and cultural norms: Social and cultural norms often restrict women's participation in decision-making processes and limit their access to education and training (IUCN, 2018).
3. A lack of representation: Women are often underrepresented in coastal management institutions and decision-making processes, which limits their ability to influence policy and practices (UNEP, 2016).

9.6.4 Strategies for Empowering Women in Coastal Ecosystems

Several strategies can be employed to empower women in coastal ecosystems:

1. Capacity building: Providing women with training and capacity-building programmes can enhance their skills and knowledge in coastal management (Kumar, 2018).

2. Institutional support: Strengthening institutions and policies that support women's participation in coastal management can help to address the barriers that they face (FAO, 2017).
3. Community-based initiatives: Community-based initiatives that involve women in decision-making processes and provide them with access to resources and services can help to empower them (IUCN, 2018).

Women's empowerment is critical for the effective stewardship of coastal ecosystems. Addressing the barriers that women face and providing them with opportunities for capacity building, institutional support and community-based initiatives can help to empower them. By empowering women, more-equitable and -sustainable coastal management practices can be promoted for the all-round benefit of both people and the environment.

9.7 The Social Entropy, Ecological Chaos and Geopolitical Anomalies of Sundarbans, India

The existence, lustre, beauty and pristineness of forests across the world have repeatedly been extorted, exploited and robbed by geopolitical mafias. This is true whether it takes place in Sundarbans in India or in the Amazon in Africa. The land sharks here in Sundarbans have been able to grab considerable acres of land from crop fields and mangrove forest zones to convert those areas into shrimp fisheries, called bheries, and to develop infrastructure for ecotourism (Fig. 9.6), which in turn invites ecological chaos. Such anthropogenic invasions are destroying more than a thousand kilometres of muddy embankments by destroying mangroves, and as a result, the high intensity of brackish water enters the crop lands. Although such a tidal wave of brackish water will recede, it leaves behind a mantle of salt that will destroy or disrupt rice or vegetable cultivation, which has been in operational condition for a couple of centuries on the reclaimed deltaic islands. With the advent of muddy embankment erosion, phenomena have been both repetitive and predictable, and the threats have aggressively increased the frequency of cyclonic storms (IPCC report, 2022). Participatory time trend analysis shows a steady and fast decline in livestock populations near peripheral locations, which threatens to degrade soil health and reduce the microbial count in soil. When agriculture fails to generate livelihood, hungry people will enter the forest, deeper and deeper, to eke out a livelihood. At the same time, with the reduction in the population of dear or boars, due to the masking effect of salt mantles, hungry tigers desperately enter nearby human settlements in search of food, whether that is livestock or a human inhabitant. The rampant development of hotels and other infrastructure to promote tourism, not properly adhering to the rules and regulations of governments, both central and state, has invited many environmental problems, such as pollution in the form of oil, hydrocarbons or microplastics, and habitat destruction, which has caused the removal of mangroves, biodiversity loss, erosion, vertical accretion, etc. In such a context, a proper understanding of the ecological functioning of the mangrove ecosystem, which is now confronted with human needs, human greed and conflict, is needed. To improve

Sundarbans, even amid climate change–mediated sea-level rise and salinity invasion, community mobilization and ecological stewardship, propped up by traditional wisdom and appropriate technology, appear to be urgently needed.

9.8 Adaptive Co-management and Ecological Chaos: Understanding the Complex Dynamics of Ecosystems

9.8.1 Adaptive Co-management

The approach for adaptive co-management in Sundarbans can be scripted by using the socioecological resilience paradigm. First, we need to train the local community within the vicinity of Sundarbans ecosphere in *adaptive coping strategies*, especially for slow-onset disasters like embankment erosion, tidal overflow and abiotic stresses. Achieving perennial adaptive capacity by changing agricultural technology, generating livelihoods and restoration of mangroves is needed. Yet adaptation alone cannot totally achieve this and thus demands the adoption of transformative adaptation. Farmers need to transform into climate managers and fishers to become stewards for aquatic ichthyofaunal components; honey collectors will transform themselves into honeybee protectors; and Sundarbans women will experience a transformation into a family repository of landraces, the types that are quickly becoming endangered (Jackson & Jackson, 1999; Kremen et al., 2002; Siebert, 2002; De Haan et al., 2007; Ayer & Maginnis, 2005; Rainforest Alliance, 2006; Rhodes et al., 2013).

The existence, growth and regeneration of any robust ecosystem has to confront and reconcile with three sub-ecosystems: core, periphery and far periphery. The mangrove ecosystem of Sundarbans is no exception. The existence, growth and regenerative behaviours and the dimensions and trajectories of any ecosystem depend on the kinetic interplay between these ecological factors.

The functioning of Kolkata depends on the existence and expanse of mangrove vegetation; otherwise, a cyclonic hit will lash off the life process of all of Kolkata, not only that of the populace dwelling on the peripheral and core ecological holons. Similarly, this urban settlement is the recipient of migrant populations, both human and animals, looking to exploit the forest ecosystem (Beer et al, 2000; Berkes et al, 2003; De haan et al, 2007).

The transformation of forests into agricultural land may initially better serve people by creating a backup for sustenance in the forms of food, livelihood and the rural economy, but in the long run, there would be serious problems and threats contributed by the destruction of natural flora and fauna, heavy metals, pesticides and land-related conflicts, generating both ecological chaos and social entropy.

The destruction of mangroves, extraction of shrimps and hilsa and mindless aggression of new settlements will lead to ecological chaos and disruption. The increased incidents of human–wildlife conflicts, intensification of human trafficking (including in and outmigration) and engulfing of croplands by a mantle of salts, leading to crop failures and more cyclones every year, are exacerbating ecology entropy in this area.

9.8.2 Ecological Chaos and Its Connection to Ecosystems

Human conduct seems "chaotic" or, at the very least, erratic and unpredictable to many people. This is most often the case for actions that occur in intricate circumstances. However, if one limits oneself to describing behavioural probabilities that can be found in a large population (or group) of individuals, stochastic behavioural models may be built, at least for relatively simple scenarios. The gas-kinetic pedestrian model has used this concept (Helbing & Molnar, 1995). Ecological chaos refers to the complex and unpredictable behaviour of ecosystems, which can be influenced by various factors, including climate change, human activities and natural disturbances. Understanding ecological chaos is crucial for managing ecosystems and mitigating the impacts of environmental changes. This article examines the concept of ecological chaos, its causes and its consequences, and it highlights the need for a holistic approach to ecosystem management.

9.8.2.1 Different Aspects of Ecological Chaos in a Mangrove ecosystem

Disruptions in ecosystem functioning lead to ecological chaos. The continuum, connectivity and composition of an ecosystem sustains and releases ecological services that a myriad of flora and fauna rely on to thrive and survive, and they interact to make the system agile, dynamic and resilient. Losing a so-called insignificant species from an ecosystem can disrupt the functioning of the entire ecosystem, which is constituted by thousands of structural components in the form of innumerable fauna and flora living under the governance of many chemical and physical factors and for various ecological processes. A simple burning of dry leaves in a forest can not only reduce microbial count but also invite disastrous consequences by driving out millions of life-forms from the ecosystem. The ripple effects of ecological chaos cannot be limited by geospatial confinement. These effects collect, connect and coalesce all forms of chaos, which ultimately diffuses into every nook and corner of the globe. When all the apparently discrete disruption and chaos interconnects, from Sundarbans to the Amazon, global chaos takes shape as a Lorenz butterfly. Therefore, a comprehensive policy needs to be formulated to address the global chaos and entropy affecting mangrove ecosystems.

9.8.2.2 Ecological Chaos and Butterfly Effects: Chaos Theory and the Butterfly Effect

The functioning of an ecological systems takes place in a nonlinear direction, meaning that the cause-and-effect relationships among different structural components of the same ecosystem do not operate in a straightforward way. Therefore, a small change in one part of the system influences other parts, even leading to a proportional change in one or more other parts.

The butterfly effect centres on the idea that even minute differences in the starting conditions of a system can yield vastly different outcomes over time. Chaos theory holds that small changes that develop in an ecosystem at its initial phase of functioning can lead to drastically different and unpredictable outcomes.

The preliminary idea revolves around four related attributes.

1. Unpredictability: Predicting the long-term behaviour of complex ecological systems is difficult because the inbuilt sensitivity within the ecosystem makes them inherently unpredictable.
2. Interconnectedness: The butterfly effect emphasizes the interconnectedness of ecological systems, where apparently isolated events can have cascading effects throughout the system.
3. Resilience and vulnerability: Understanding that the reasons behind major consequences are due to the potential impacts of small changes will play a crucial role in assessing the resilience and vulnerability of ecosystems to disturbances.
4. Conservation and management: The knowledge from the interconnectedness of different small cases of changes can improve conservation and management strategies for biodiversity by minimizing human impacts and anticipating potential cascading effects from interventions.

According to the butterfly effect, "the flap of a butterfly's wings" is a metaphor for a small change in an ecological system, the cumulative impacts of which, through a complex chain of interactions, potentially cause major events like flooding with high tidal amplitudes, catastrophic storms or a tornado in a different location. Simultaneously, one must realize that the butterfly effect can cause major changes all the time. Therefore, predicting which changes will have significant consequences is difficult.

The complexity of mangrove ecosystems rests on the complex interconnected food webs and dependencies of each species on ecological support from an array of ecological variables through continuous interactions between different species and their environment. However, mangrove ecosystems also experience ecological chaos from various threats, including deforestation for aquaculture, a decline in agricultural productivity, encroachment due to urban development, soil erosion, salinity intrusion and pollution from many sources, and the butterfly effect, stemming from chaos theory, may exacerbate these issues. The butterfly effect points out that small changes in the initial stage in part can lead to large, unpredictable outcomes in the form of large changes, posing challenges to effectively managing mangrove ecosystems, especially when they are already subjected to continuous anthropogenic interventions.

9.8.2.3 The Ecological Ladder and Hierarchical Steps that Explain Ecological Chaos in Different Strata

Ecology is basically a manifestation of a vertical and horizontal continuum that maintains function and harmony in a complex function. The lowest to highest levels on the ecological hierarchical ladder include five major components:

1. Holon: The smallest ecological entity which can become a unit of an ecosystem
2. Echelon: A stratum of diverse ecology comprising diverse floral and faunal components

3. Ecosystem: An identifiable unit comprising both holons and echelons
4. Biosphere: All the areas where floral and faunal components around the world can live
5. Ecosphere: An area on and around the Earth where life exists, which differs from the biosphere in that it includes nonliving components too

The influence of the butterfly effect in the stability and resilience of the complex ecodynamics of a mangrove ecosystem can be understood better only after conducting more research studies. Out of so many contributions of butterfly models to explain ecological changes across the world, the butterfly effect can emphasize the concerning impact of climate change, such as rising temperatures, sea-level rise and intensifying storm surges, through complex and unpredictable interactions with mangrove ecosystems. For the protection, conservation and even eco-restoration of the mangrove ecosystem of Sundarbans, developing flexible adaptive management strategies by taking into account all the small changes within the ecosystem and anticipating unpredictable outcomes as a conservation effort against changing ecological conditions is required.

9.8.2.4 The Butterfly Model by Lorenz

The term *butterfly effect* was coined by internationally acclaimed meteorological scientist Edward Lorenz, who from one of his lectures at the annual meeting of the American Association for the Advancement of Science in 1972, which was titled "Predictability: Does the Flap of a Butterfly's Wings in Brazil Set Off a Tornado in Texas?" It emphasized a fascinating idea: the flapping of wings of butterflies in Brazil could potentially set off a tornado in Texas. The butterfly effect is a metaphor aiming to explain how many small changes in different parts of a global environment can have large effects in the complex systems of an ecosystem, such as the mangrove ecosystem in Sundarbans on the Indian subcontinent, in the Everglades in Florida (USA) or in the Amazon forest tracts in Brazil. This concept in other way substantiating chaos theory, as being an extension of chaos theory, emphasizes that even a minor change in the initiation of ecosystem functioning can lead to significant but unpredictable changes in the future via the amplification of small differences in a system's dynamics.

Lorenz's seminal research (1963, 1969 and 1972) introduced Sensitive Dependency on Initial Conditions (SDIC), which is well known as the butterfly effect, and has revolutionized the perception and understanding of humans on the weather and climate predictability for more than five decades. This idea was first proposed by Maxwell and Poincare, according to Ghys (2015). Lorenz made a distinction between two types: Butterfly Effect1 (BE1), which stands for SDIC in chaotic systems, and BE2, which is a metaphor that implies that minor disruptions lead to significant atmospheric changes, such the creation of tornadoes and other catastrophic storms which in turn affect other compartments of the environment. Even though Lorenz rediscovered the butterfly effect, his main contribution to chaos theory was the models and techniques which were developed by him for advancing research on climate predictability and highlighting the inherent limitations of making forecasts about large dynamic systems (Lorenz, 1963; Scheffer et al., 2001; Shen, 2022).

9.8.2.5 Causes of Ecological Chaos

Several factors contribute to the development of ecological chaos: (i) climate change is a major driver of ecological chaos, altering the delicate balance of ecosystems and leading to changes in species distributions, behaviour and extinction risk; (ii) human activities, such as deforestation, overfishing, and pollution, can disrupt ecosystem processes and lead to ecological chaos; and (iii) natural disturbances, such as hurricanes, wildfires and droughts, can also contribute to ecological chaos by altering ecosystem structure and function.

9.8.2.6 The Consequences of Ecological Chaos

1. Biodiversity loss: Ecological chaos can lead to biodiversity loss, including the extinction of plant and animal species.
2. Changes in ecosystem processes: Ecological chaos can alter ecosystem processes, including nutrient cycling, decomposition and primary production.
3. Decreased ecosystem services: Ecological chaos can decrease ecosystem services, including carbon sequestration, water regulation and soil conservation.
4. Impacts on human well-being: Ecological chaos can have significant impacts on human well-being, including livelihoods, health and cultural values.
5. Managing ecological chaos: Ecosystem-based management approaches can help mitigate the impacts of ecological chaos by considering the complex interactions within ecosystems.
6. Adaptive management: Adaptive management strategies can help managers respond to the unpredictable nature of ecological chaos.
7. Conservation and restoration: Conservation and restoration efforts can help to maintain ecosystem resilience and promote biodiversity.

9.8.2.7 Butterfly Models on Mangroves

The butterfly effect reveals the intricate relationships and sensitive interconnected complex systems of nature that maintain a delicate ecological balance. In the complex functioning of an ecosystem, many small changes play important roles, emphasizing that the contributions of the tiniest parts of an ecosystem can result in long-term and unpredictable consequences. This can be evaluated by following nonlinear dynamics and by determining the future course of a chain of ecological events. The cascading effects from a small change in one part of a system trigger a chain reaction of ecological interactive events over time and the simultaneous development of feedback loops magnifying the initial perturbation.

According to the butterfly model designed by Edward Lorentz, known as the butterfly effect, any initial change in the micro-environmental level can bring about, through a chain-wise process, a series of other changes in an unpredictable manner in larger environmental sectors in different parts of the world. In the context of global mangrove ecosystems, similar chain-wise impacts on the sensitive coastal habitats play vital roles in ensuring interconnectedness among the different components of mangrove ecosystems, whether in Sundarbans in India, in the Amazon of Brazil or in the Everglades of the United States. This unique model mimics a butterfly's wing flap, which can, theoretically, influence environmental changes far away. Even small ecological changes in mangrove ecosystems can have cascading effects throughout all such ecosystems across the world. This kind of hypothetical interpretation of global research on mangroves crucial for undertaking sustainable conservation strategies to safeguard the valuable ecosystem services rendered by mangroves, like biodiversity enrichment, carbon sequestration, coastal protection, ecotourism and fishing, and therefore, any decline in the coverage of mangroves can trigger a series of negative consequences. In such a context, interconnected with different hierarchical levels from the tiniest levels to the largest levels in an ecosystem, no part is isolated from the others; instead, all of them remain linked to both terrestrial and marine environments. Disturbances in the ecological health of mangrove ecosystems, such as biodiversity loss; increasing soil erosion, water turbidity and salinity; and reduced water-quality parameters, sediment stability and food chain–food web dynamics tend ripple outwards, affecting nearby coastal communities, fisheries and even the wider ocean. The butterfly model signifies interconnectedness and the potential for far-reaching consequences within mangrove ecosystems by recognizing and emphasizing that even small actions can degrade these vital habitats, which are restricting not only mangrove ecosystems but also human environments.

9.8.2.7.1 The Everglades Mangroves in Florida, USA

The mangrove forests of the Everglades ecosystem in the United States, covering about 900 km^2, play crucial roles in reducing erosion, increasing sediment stability, protecting coastlines, slowing down tidal flows, promoting fisheries, storing carbon and providing critical habitats for diverse wildlife. This mangrove forest ecosystem, constituting an integral part of the coastline of Florida is dominated by mangrove plant species such as *Avicennia gaminans*, *Laguncularia racemosa* and *Rhizophora mangle*. The mangroves of the Everglades are especially vulnerable to sea-level rise. Overall resilience by adapting to salinity fluctuations in response to sea-level rise results in the landward migration of mangroves. Sea-level rise also results in coastal flooding. This mangrove ecosystem has been experiencing eco-degradation over the past few centuries due to human activities, which has necessitated re-establishing the predevelopment biota and landscape patterns to the greatest extent possible so that alongside the proliferation of newly settled mangroves, higher volumes of freshwater influx support seasonal cycles. However, the rapidly accelerating rate of sea-level rise makes restoring the coastal Everglades a difficult task.

A recent monitoring study has shown that the increasing trend has been halted by natural disaster like recent hurricanes. The mangroves of the Everglades can be divided into four categories: healthy (no change), disturbed (affected by storms), recovering (rebounding to a healthy state) and declining (not recoverable naturally after a disturbance).

9.8.2.7.2 Amazon Mangroves in Brazil

The world's largest continuous mangrove forest tract, spanning around 11,200 km^2, is in the Amazon River delta forest along the northern coast of Brazil, mostly in three states of Maranhão, Pará and Amapá, representing 80% of Brazil's mangrove area. Amazon coastal tracts in Brazil support both freshwater floodplain forests and abundant brackish-saline mangroves. The mangroves in the Amazon River delta of Brazil are characterized by their occurrence in freshwater tidal environments, experiencing driving forces from large tides with amplitudes of 4–8 m while receiving significant freshwater discharge from the Amazon River.

These mangroves, like other parts of the world, play crucial roles in biodiversity enrichment, carbon storage, erecting natural barriers and rendering ecosystem benefits to coastal communities through fishing, shellfish aquaculture, the harvesting of forest products, etc., despite ecological and anthropogenic saltwater intrusion, deforestation, soil erosion, sea-level rise and pollution (Bernardino et al., 2022). Brazilian mangroves are dominated by salt-tolerant mangrove species like *Rhizophora mangle* and *Avicennia germinans*. In addition to mangrove species, woody freshwater species like açaí berries (*Euterpe oleracea*), Buritis palm (*Mauritia flexuosa*), cork oak (*Pterocarpus* sp.) and ferns (*Acrostichum aureum*) are also found.

9.8.2.7.3 The Mangroves of Sundarbans in Southeast Asia

Sundarbans constitutes the world's largest delta with tidal halophytic mangrove forests, which is encircled by numerous large, medium and small estuaries; which covers an area of 26,000 km^2; and which is jointly shared by two neighbouring countries, Bangladesh and India. In the global context, this chunk of mangrove forest in tropical climatic conditions has been recognized as one of the most productive natural ecosystems in the world and has earned the status of being a Worlds Heritage Site and a Ramsar Site, a wetland of international importance.

Mangrove forests are globally recognized as highly carbon-rich tropical ecosystems that provide a range of critical economic and ecological services to surrounding coastal populations (Barbier et al., 2011; Donato et al., 2011). However, mangroves have been heavily impacted by degradation and deforestation, with 20%–35% of global mangrove extent lost over the past 50 years (Polidoro et al., 2010). Although the large-scale conversion of mangroves to shrimp and rice aquaculture—exacerbating large-scale losses from land-use changes (Thomas et al., 2017), which are supposed to be the primary causes of loss during the twentieth century due to global aquaculture demand (Friess et al., 2016)—climate change–driven ecological consequences, like the warming of surface water of estuaries, salinity invasion, sea-level rise, soil erosion and the subsidence of mangrove floors, are expected to increase global sea levels, wave energy (Reguero, et al., 2019) and the intensity and frequency of extreme weather events (EWEs) such as droughts and tropical cyclones (Bhatia, et al. 2018; Murakami et al., 2020).

Anthropogenic activities across the global hotspots of mangroves of Southeast Asia, including Sundarbans and around the world, have been identified as the main causative factors of erosion and can play significant roles in forest vulnerability. Random forest-based analysis at the global scale using of over one million Landsat images, have provided an understanding of the modes of decline and the extent of human and natural threats as the drivers of mangrove loss (Goldberg et al., 2020). Additionally, 62% of global mangrove losses between 2000 and 2016 were the result of drastic land-use changes, primarily through the conversion of mangrove forest tracts to agriculture and aquaculture. Around 80% of these human-driven losses of mangroves occurred in six Southeast Asian nations, reflecting the regional emphasis on enhancing protection measures. Although necessary efforts are being made to restore and protect mangroves over decadal timescales, the need for natural drivers to emerge to compensate for such of losses becomes an immediate challenge, especially by undertaking coastal adaptation.

9.8.3 *Explosive Shock Wave Theory and Its Relevance in the Functioning of the Sundarbans Mangrove Ecosystem*

The gifted internationally acclaimed Sundarbans mangrove ecosystem on the Indian subcontinent is going to be pounded into the debris of geomorphological remnants because it is unable to absorb ecological shock waves. An explosion is a process of physical and chemical transformations. A powerful shock wave is created in the air at the instant of the explosion when high-temperature, high-pressure gas is created in the core region and quickly spreads out. One of the most significant effects of shock waves is overpressure, which is the temporary pressure that surpasses the ambient atmospheric pressure. With distance, the shock wave overpressure decreases. Before entertaining this interesting theory, one should think about the following undermentioned statement on the present case of Sundarbans on the Indian subcontinent: “Is it that beautiful Sundarbans, the gift of nature, if it fails to absorb ecological shock waves, is going to be pounded into the debris of geomorphologic remnants?”

The explosion’s damage and peak overpressure decrease with increasing distance from the explosion’s centre (Zhang et al., 2021). The term *shock wave* is often used metaphorically by social scientists to characterize swift and profound social transformations brought on by calamities like pandemics, wars or terrorist acts. For instance, the 11 September 2001 attack on the twin towers of the World Trade Center in the United States generated a “shock wave” in the security strategy of the international community, which ultimately resulted in a financial crisis in the form of “economic shock waves” (Klein, 2007).

9.8.4 *Ecological Chaos: Understanding the Complex Dynamics of Ecosystems*

Chaos theory has far-reaching implications for environmental science, highlighting the potential for small changes to have significant and unpredictable consequences in complex systems. By understanding the principles of chaos theory and its applications in ecological and environmental science, researchers can better predict and manage environmental phenomena.

Lorenz's (1963) deterministic nonperiodic flow has been modified for ecodynamic studies in cross-territorial ecological entanglement to determine a pattern from chaos. Given the analogy of Lorenz's butterfly effect, the deterministic nonperiodic flow can be extrapolated to the ecodynamics of the Sundarbans mangrove ecosystem in Southeast Asia. If tiny ecological perturbations and ripples are not addressed, the small amount of ensuing chaos generated in furthest area is expected to entangle continental forest systems like the Amazon River or the Congo River.

Ecological chaos refers to the complex and unpredictable behaviour of ecosystems, which can be influenced by various factors, including climate change, human activities and natural disturbances. Understanding ecological chaos is crucial for managing ecosystems and mitigating the impacts of environmental changes. This article examines the concept of ecological chaos, its causes and its consequences, and it highlights the need for a holistic approach to ecosystem management.

9.8.5 *Managing Ecological Chaos*

- Ecosystem-based management approaches can help to mitigate the impacts of ecological chaos by considering the complex interactions within an ecosystem.
- Adaptive management strategies can help managers to respond to the unpredictable nature of ecological chaos.
- Conservation and restoration efforts can help to maintain ecosystem resilience and promote biodiversity (10).

9.8.6 *Social Entropy Theory and the Sundarbans Mangrove Ecosystem*

Sundarbans has been experiencing huge social entropy due to land transformations, population and cultural changes, sea-level rise and the submergence of islands into deep, saline water. The physical laws governing energy, entropy and equilibrium can metaphorically—and occasionally mathematically—inform our understanding of social entropy. A key idea in information theory, statistical mechanics and thermodynamics, entropy is the degree of disorder or unpredictability in a system.

Ecological chaos is a complex and pressing issue that requires a holistic approach to ecosystem management. By understanding the causes and consequences of ecological chaos, we can develop effective strategies to mitigate its impacts and promote ecosystem resilience. For Sundarbans and its survival, biophysical resilience alone is insufficient; we need to do more through social resilience.

Rudolf Clausius first used entropy in thermodynamics in the nineteenth century as a way to quantify the amount of energy that was unavailable for work in a thermodynamic process. The mathematical expression of Clausius's definition of entropy change (ΔS) is the heat added to a system divided by the temperature at which it was added: $\Delta S = Q$ tev/T, where Q tev is the reversible heat transfer and T is the absolute temperature (Clausius, 1865).

Behaviour, collective dynamics and human systems form the foundation of the link between thermodynamics and social physics. The evolution and stability of complex systems, including societies, have long been compared to thermodynamics, particularly the idea of entropy. Entropy is a measure of disorder or the number of microstates.

9.9 Supporting Services: Sustaining Ecosystem Functioning

Foundational processes known as supporting services enable all other ecosystem services to function. These services are vital for the overall health and productivity of ecosystems. One key supporting service is the sustenance of ecosystem functioning, which entails maintaining ecological processes that drive ecosystem dynamics and support the provision of other ecosystem services. Ecosystem functioning relies on various processes (Costanza et al., 1997; Daily, 1997; Convention on Biological Diversity (CBD), 2006).

1. Nutrient cycling: Essential nutrients like nitrogen, phosphorus and carbon are cycled through ecosystems, supporting soil fertility and plant growth.
2. Soil formation: Soil, crucial for plant growth and habitats, forms through rock weathering, organic matter decomposition and soil particle accumulation.
3. Primary production: Plants and photosynthetic organisms convert sunlight into energy, forming the basis of the food chain and supporting ecosystem productivity.
4. Water cycling: Water moves through the atmosphere, soil and vegetation, maintaining soil moisture, supporting plant growth and regulating the climate.
5. Habitat provision: Ecosystems offer habitats for diverse plants and animals, supporting biodiversity and ecosystem functioning.

Sustaining ecosystem functioning is crucial for maintaining ecosystem health, productivity and resilience. This foundational service is essential for supporting human well-being and ecosystem sustainability because it underpins the provision of other ecosystem services.

Supporting services are fundamental ecological processes that govern ecosystem structure and functioning. All ecosystem services fundamentally depend on supporting services that are more directly valued by society.

9.10 Significance of Maintaining Soil Resources

Soil and sediments play crucial roles in regulating ecosystem processes by providing resources necessary for organisms. Managing the formation, degradation and resource-supplying potential of soils and sediments is key to sound ecosystem management. The volume of soil in an ecosystem is governed largely by the balance between inputs from weathering, deposition and erosion. Global ecosystems are being disrupted by accelerating soil erosion. Soil erosion from land leads to sediment input into lakes and estuaries. Globally, the augmented sediment input into oceans from faster erosion is somewhat thwarted by better sediment capture by reservoirs. Lakes, including reservoirs, and estuaries are thus the aquatic ecosystems most strongly affected by terrestrial erosion.

9.11 Maintaining Balance Between Water and Soil Subsystems of Sundarbans Mangrove Ecosystem Cycling

Water is a critical soil resource, often the most significant limiting factor in terrestrial ecosystem productivity. It enters ecosystems primarily as precipitation. Two major pathways of water loss are transpiration and runoff. These sources are frequently utilized by humans for domestic and industrial purposes, irrigation and hydropower. Consequently, ecosystem services experience inherent trade-offs. The most significant challenges in water management arise from managing these trade-offs. The subtle ecological tradeoffs of Sundarbans are entangled with the dynamics of hydrological cycle. The balance between fresh and saline water in this eco-region is the main driver for agricultural sustainability, fisheries potentiality wild life diversity ,ecological interaction within mangrove forests and embankment stability. Rainwater harvesting in this entire eco-region helps tradeoffs for resource generation (agriculture, fisheries, apiculture and plants 'biomass) by reclaiming and managing salinity both in water and soil into a cultivable landscapes to support crops, fishery resources, animal livestocks managent, honey production etc.

Climatic factors ultimately dictate the amount of water that enters ecosystems through precipitation, setting a fundamental limit on the water cycled within them. Within this limit, the division of water between transpiration and runoff is influenced by three processes: (1) the soil surface's level of compaction, affecting water infiltration; (2) the soil's water-holding capacity, determined by its quantity and particle-size distribution; and (3) the vegetation's ability to transpire water (Rockström, et al., 1999, 33).

Vegetation plays a crucial role in promoting water seepage and subsequent storage. The canopy of the plant and residue help prevent soil compaction caused by raindrops, which can otherwise seal soil pores. Additionally, root and soil organisms associated with vegetation create channels for the movement of water. By improving water infiltration (through reduced compaction), increasing water-holding capacity (by promoting the production of organic matter in soil) and reducing soil erosion (by decreasing overland flow), vegetation creates stabilizing feedback loops that maintain soil productivity and decrease ecosystem susceptibility to drought. However, human interventions often disrupt ecosystem resilience. For instance, ploughing can significantly reduce the organic content of soil, often by 50% within a few years, thereby decreasing the water-holding capacity of soil and the ability of soil to support crop growth solely with natural rainfall.

9.11.1 Agricultural Water Stewardship for Agricultural and Ecological Resilience

Agricultural water stewardship involves efficiently using water in agriculture while considering its benefits to food production, the environment and human health. It recognizes that water-management decisions should consider ecological, social and economic factors. This approach considers local factors like soil type, ecology, topography, agronomic practices and ecosystem health. It emphasizes holistic farm practices that recognize agriculture's role in the local watershed and the local community. Effective stewardship results in various ecosystem services that benefit humans, including those in agricultural systems. The best water-management practices for a farm depend on its local and regional context, aiming to maximize water efficiency, minimize environmental impact and balance societal needs. Agricultural water stewardship considers many solutions, including advanced technology and cultural and ecological practices that improve water quality and supply management by integrating farming with the environment.

9.11.2 Beneficial Management Practices

Numerous practices are already known to enhance performance across a host of water stewardship indicators, and their adoption should be encouraged. Some examples of beneficial on-farm stewardship practices include the following:

- Using irrigation systems designed for optimal "distribution uniformity," coupled with appropriate management to maximize efficiency for irrigation and each location
- Improving the timing of water applications and chemical and nutrient inputs to minimize aquifer pollution
- Implementing tail water ponds and return systems
- Establishing vegetated filter strips
- Restoring natural flows and hydrograph cycles
- Rehabilitating riparian corridors and other waterways to provide multiple ecosystem services (e.g. infiltration, soil and water quality and wildlife habitats)
- Employing soil-management techniques to enhance water infiltration and retention, such as cover crops, soil amendments and minimum tillage
- Monitoring soil moisture and crop water status
- Utilizing wind turbines for frost control
- Implementing dry farming techniques
- Constructing off-stream storage ponds for irrigation and frost protection
- Implementing managed groundwater recharge by using surplus irrigation water when chemical or nutrient runoff is minimal
- Establishing wetlands to enhance water quality and provide wildlife habitats

- Recycling municipal water for agricultural use, ensuring proper attention to water quality
- Adopting measures to match water quality with intended use, including drinking water for rural communities
- Modifying conveyance capabilities to redirect flows to recharge areas

9.11.3 *Freshwater Facts*

1. Water security threats affect nearly 80% of the world's population.
2. Over 64% of the world's wetlands were lost in the twentieth century.
3. The Freshwater Living Planet Index, tracking trends in vertebrate species populations, indicates a 76% decline from 1970 to 2010.
4. Hydropower generates more than one-fifth of the world's electricity.

9.11.4 *The Water Stewardship Approach in Lotic Waterbodies, Including Estuaries*

Estuaries are supported by luxuriant mangroves, and their many rivers and basins are at risk of closure due to excessive water resource allocation. We have an opportunity to advocate for basin-level strategies and gain a deeper understanding of water-management decisions and their consequences. For businesses, the key question is how they can implement interventions that can drive the necessary changes in catchment water management to address their water-related challenges. Although improving local hydrology to meet operational water needs and mitigate the risk of water scarcity is feasible in some case, in other cases, multiple water users rely on the same water sources. As a result, any water savings from individual facilities or interventions may not exclude other users from accessing this water.

Moreover, this fails to consider the potential reputational or regulatory risks that may arise when interventions unintentionally increase water consumption and decrease downstream flows. In most cases, isolated field-level interventions by individual businesses are unlikely to be implemented at the necessary scale or with sufficient continuity to significantly impact water flows. Addressing supply chain risks requires more than just farm-level water management. It necessitates collective action with other water users and engagement with water-governance institutions to achieve optimized, sustainable water allocation and the preservation of environmental flows and ecosystem services. To effectively mitigate fundamental water risks, direct business interventions in agricultural water management must be part of a broader strategy to enhance water governance and improve river basin planning. These interventions must explicitly acknowledge other stakeholders within the catchments and actively address shared water challenges that contribute to corporate water risks. This involves engaging with and learning from the water-management agencies and regulatory bodies responsible for managing water at various scales. It also requires a re-evaluation of crop choices and their purposes while considering alternative water uses, such as for energy, ecosystem services or drinking water security. Collective action should prioritize collaborative decision-making among water users to ensure optimal water resource utilization rather than letting influential commercial entities dictate water-management practices. Bottom-up collective efforts can help consolidate various small-scale interventions and align them with larger-scale infrastructure and systems when applicable. While individual field-level initiatives may not be adequate to fully address all water-related business risks, more-comprehensive and -rigorous monitoring must be implemented to assess their impact on water savings and catchment hydrology. If these interventions prove to be effective, they can be expanded by entities mandated by water-management governance.

As the field of water stewardship progresses and transitions from theoretical discussions to practical implementation in business operations occur, a simultaneous mobilization of practitioners and academics who scrutinize both the intended and unintended outcomes also occurs. Therefore, businesses involved in water stewardship must fully comprehend the hydrological and sociopolitical implications of their water-management interventions. Failing to grasp these complexities may result in simplistic, supply-side, technology-focused solutions that do not effectively mitigate risks or achieve sustainable water management at the basin scale.

Businesses engaging in agricultural water management need to factor in some key considerations.

1. Enhancing the transparency of water information at the site level: The lack of transparency and data hinders the verification of actual water savings.
2. Improving the availability of basin-level water stocks and flows: Information on surface and groundwater levels and fluxes is essential for developing effective solutions.
3. Addressing shared water challenges, not just efficiency solutions: Companies should ensure that interventions effectively tackle water challenges and reduce water risk rather than merely transferring these challenges and risks downstream.
4. Understanding hydrological scales, from local to catchment or basin: Failure to understand how water flows through a system can undermine the effectiveness of interventions.
5. Recognizing water's role in the economy: Evaluating trade-offs between using water for food, energy, ecosystems and greater water security.
6. Clarifying definitions: Paying particular attention to terms such as *water efficiency*, *water productivity* and *saved water*.
7. Understanding water allocations: Relating water-management activities to the basin and considering how they support water management in policy and practice.
8. Simultaneously pursuing solutions at the field and policy levels: Undertaking field-scale interventions in water management should be accompanied by policy engagement to establish an effective framework of water allocation.

9. Acknowledging intense competition for "saved" resources: Businesses should understand that saved resources are subject to competition and should initiate dialogue with interested parties (IUCN, 2015).

9.12 Preserving Biological Diversity

The impact of biodiversity on ecosystem functioning is primarily linked to traits that influence how species affect ecosystem processes and how they respond to environmental changes. To promote these effects of biodiversity change—whether due to the loss of key species or the introduction of species with significant impacts—species that cover a wide range of effect-and-response diversity within the system need to be maintained (Rockström, et al. 1999; Elmqvist, 2003).

Species diversity's functional redundancy acts as insurance against severe environmental changes, like those resulting from human mismanagement of ecosystems or climate change. In diverse ecosystems, radical environmental shifts are unlikely to eradicate all species of a particular functional type, enabling surviving species to thrive and uphold crucial functions that might otherwise be severely impacted. (Douglass, 1984; Farnsworth & Ellison, 1997; Elmqvist et al., 2003; Convention on Biological Diversity (CBD) 2006; Díaz et al., 2006).

9.13 The Operational Concept of Farm Stewardship and Its Applicability in the Sundarbans Ecosystem

Human civilization embarked on a new journey, filled with enchantment, joy, and adventure, when it first embraced agriculture in its primitive form. Although agricultural production was largely guided by nature and executed through human innovation, the "modernization of agriculture" in the 1950s and 1960s was primarily a response to the political economy of hunger, poverty and stagnation. This era ushered in the green revolution and a host of lingering challenges.

These residual challenges, both biophysical and socioecological, have left a lasting impact. The remnants of the green revolution manifest in contaminated groundwater, soil salinization and a stark erosion of biodiversity. "While water contamination and soil erosion have complicated the agricultural production process, the loss of biodiversity has narrowed India's agricultural gene pool, leaving it fragile (quoted from *The Silent Spring*" (Carlson, 1965)).

What is the role of stewardship in farming or agricultural production processes on a broader scale? Farm stewardship is both a process and an approach aimed at enhancing the resilience, adaptability and productivity of farming operations within a specific ecological context. It involves creating conditions that support ecological benefits while providing for the needs of local communities and other living organisms.

Therefore, this can be considered as the operational definition of farm stewardship, representing a conceptual and empirical advancement over traditional conservation strategies. The fundamental components of stewardship for Indian farms can be outlined as follows:

1. Gathering baseline information on input application, management practices and production status to understand their interrelationship with environmental factors
2. Shifting towards a production process driven by organic inputs to reduce reliance on costly chemical inputs
3. Studying the sub-ecological interactions and relationships of farms to promote integrated and resilient farm-production processes
4. Ensuring that social compatibility, economic profitability and ecological suitability align with a shared goal of sustainable outcomes, considering both design and impact

The present empirical research proposes to apply both inductive and deductive approaches. The approaches are outlined as follows:

1. Studying the connectivity between the conservation of soil, water and biodiversity in terms of their interdependence and interactions, which will involve the application of structured schedules and a participatory approach
2. Examining the connectivity between the stewardship of soil, water and biodiversity in terms of their interdependence and interactions, which will also involve the application of structured schedules and a participatory approach
3. Exploring the feasibility of applying farm stewardship approaches through the transformation, upgrading and acculturation of conservation strategies currently practised by farmers
4. Identifying conceptual, motivational and institutional constraints that act as barriers to this desired transformation
5. Assessing the readiness of people through participatory design and analysis to adopt farm stewardship in terms of its economic viability, social compatibility and ecological resilience

9.14 Stewardship: Ecological Goods and Services

Stewardship work is all amplifying ecological capacities to produce environmental products and services individually. Whereas the expressed acknowledgement of biological system administrations may be a modern wonder, the idea that common environments offer assistance to society can be traced back to ancient times. Although the term *ecosystem services* started to be used in the late 1960s, only within the most recent decade has the concept taken hold within broader investigations, approaches and the asset administration community (Whitten et al., 1999). A brief history of the biological products and administrations concept is given by Mooney and Ehrlich (1997).

Ecological services are the conditions and forms through which characteristic environments, and the species that make them up, support and sustain human life. Biophysical frameworks with biological astuteness keep up biodiversity and the generation of environmental products such as fish, timber, biomass fills, common fibres, numerous pharmaceuticals, mechanical items and their components. In addition to the generation of ecosystem services and products, biological system administration enables life-support capacities such as cleansing, reusing and recharging, and they bestow numerous social benefits (Daily, Gretchen, 1997). Costanza et al. noted that biological system capacities refer to habitats, natural or framework properties or forms of environments. Biological system products (such as nourishment) and administration (such as digestion) mark the benefits that humans reap, specifically or indirectly, from ecosystem capacities (Daily, Gretchen, 1997).

9.15 Defining Sustainable Agriculture as a Foundational Concept to Be Applied in Other Sectors of the Environment

According to Schaller and Neill (1993), sustainable farming, just like the concept of sustainability itself, requires not only the central characteristics of a goal for horticulture but also specific cultivating hones that seem to move horticulture towards that goal. He focuses on characteristics that constitute the definition. "As a destination," he writes, "sustainability isn't easily captured in brief definitions. On the other hand, particular cultivating hones have to be custom fitted to the specific area and circumstance, nonstop adjustment and adaptation." Because of these difficulties, he notes, numerous individuals lean towards the centre on the specialized side of maintainable horticulture, around the distinctive sorts of hones that will moderate soil and water, protect the environment and give farmers decent benefits. Researchers in specific, Schaller argues, tend to think of the causes of maintainability as "givens" and the part of science as determining how best to attain them. For that reason and because issuing a top-down definition of maintainable cultivating hones would be troublesome and hasty, Schaller proposes that people's convictions and values play a part in deciding what maintainability is and how it can and ought to be accomplished (Costanza, 1947).

Schaller notes that current convictions and values markedly contrast. Some people accept that traditional agribusiness is essentially sound, needing on fine-tuning or alteration. This view is still broadly held by numerous individuals and organizations inside the conventional agrarian community, such as the US Division of Farming, agribusiness colleges, product cultivation organizations and trade firms. On the other hand, many others argue that a on a very basic level diverse ways of considering horticulture are required to make it maintainable (Schaller & Neill, 1993).

Another valuable system for understanding the diverse conceptualizations of economical farming is provided by Douglass (Douglass; Gordon, 1984). He noted that agricultural sustainability can be characterized in different ways and looked for distinct examples. He characterized it as having three primary approaches. The primary approach characterizes agricultural sustainability in financial or generation terms, focusing on nourishment adequacy. The second approach characterizes it in biophysical terms and stewardship. Finally, the third approach, advocated by "alternative" agriculturalists, is concerned with maintaining the lasting carrying capacity of re-establishing agrarian assets, but it also focuses on the impacts of diverse agrarian frameworks on the social organization and culture of rural life. This third approach includes a concern for the values of stewardship, self-reliance and holism. Advocates of the third approach also accept that the agricultural information in the social practices of the conventional social order is as imperative to the creation of economical agrarian frameworks as scientific information (Kloppenburg, et al., 2000).

The rise of this "alternative" approach can be traced back to Rachel Carson's 1962 work *Quiet Spring*, a commentary on pesticides, which served as a catalyst for the development and growth during the 1970s and 1980s of a rural maintainability that centred on the social and natural externalities related to present-day agrarian innovations (Carson, 2015). Kloppenburg et al. (2000) noted that the victory that agrarian activists delighted in by deconstructing the authenticity of industrial agriculture was reflected in the 1989 National Investigate Board report "Elective Agribusiness," which called for making elective hones broader.

Kloppenburg et al. (2000) noted that although the development of "alternative" rural supportability found its roots in agricultural-environmental issues, globalization in the 1990s incited a broader approach to the social investigation into nourishment. The disappointment of numerous advocates of economical farming in satisfactorily treating social treachery (livelihood, sex, and especially hunger) encouraged Kloppenburg et al. (2000) to point out significant feedback. Allen (1993) for illustration, has argued that, generally, maintainable horticulture has centred on the natural, but human needs and social relations need to be incorporated too. Working to fulfil fundamental human needs, she contends, "requires reframing our concept of supportability to incorporate a social dimension and a concomitant extension of our approach to sustainable agricultural research" (Allen, 1993). Allen called for "new epistemologies and [inquiries into] approaches that integrate the normal forms of environment combined with the social relations compatible with supportability. This will include, for occurrence, looking at not as it were strategies for diminishing soil disintegration and water exhaustion, but too unused shapes of social organization that alternative rural hones presuppose." Conceptions of and activism on nourishment issues have by and large currently come to encompass the bigger concerns of social equity and natural interface in conventional rural problems. Those working towards change in the nourishment sector now commonly outline their aspirations not in terms of maintainable horticulture per se but rather as the realization of a feasible nourishment framework (Kloppenburg et al., 2000).

9.15.1 Sustainable Agriculture as Alternative Agriculture

The understanding of feasible farming featured in this proposition takes after Douglas's third approach, advocated by "alternative" agriculturalists. It may be an all-encompassing approach to maintainable agribusiness which includes the objectives of nourishment adequacy and natural assurance but which also addresses human needs and social relations. The definition of *maintainable agribusiness* in this proposal extends beyond our current generation to include a complete nourishment framework. (Dorfman & Dorfman, 1972; Douglass, 1984; Daily, 1997; Cairns & Garrity, 1999; Cassman et al., 2003; Elmqvist et al., 2003; Fischer et al., 2007).

Feasible agribusiness coordinates three fundamental objectives—natural well-being, financial benefits and social and financial value. Supportability rests on the rule that we must meet current demands without compromising the capacity of future generations to meet their needs. It recognizes that characteristic assets are limited, recognizes limits on financial development and builds value in asset allotment. Hence, stewardship for both nonhuman and human needs is paramount. Stewardship of human assets incorporates consumer well-being and security, social responsibilities, the social ecology of workers and the needs of rural communities both today and in the long run. Stewardship of assets includes maintaining or improving this crucial asset base in the long run.

Taking into account the ideas of equity and value, Douglass characterizes as unsustainable any structure of agrarian generation that specifically or indirectly worsens the disperion of opportunities for livelihood generation within communities, and he judges as unsustainable all structures that fall flat in making strides towards the dispersion of openings or incomes by a particular degree within a given period of time. Agribusiness will be found to be economical, he suggests, "when ways are found to meet future requests for foodstuffs without forcing on society genuine increments within the social costs of generation and without causing the dispersion of openings or incomes to worsen" (Allen, 1993).

According to Lovely, economical farming is any framework of nourishment generation that methodically pursues objectives: a more intensive consolidation of natural processes; a reduction in the use of off-farm, outside and nonrenewable inputs; a more even-handed distribution of beneficial assets and openings; an advancement towards more socially just shapes of agriculture; a more beneficial use of neighbourhood information and hones; and an increase in self-reliance among farmers and other rural people (Swaminathan, 1994).

Ikerd has contended that maintainable horticulture must utilize cultivating frameworks that preserve assets, protect the environment, deliver proficiently, compete commercially and improve the quality of life of farmers and society at large (Ikerd, 1993).

Sustainable farming depends on the administration of the internal resources of cultivation (i.e. administration aptitudes, information and work); maximizes dependence on normal, renewable and on-farm inputs; and reduces dependency on acquired commercial inputs in endeavouring to decrease negative environmental impacts while financially maintaining ranches. The goal is to create efficient, biological frameworks that don't require high levels of fabric inputs. Maintainability implies greater dependence on human assets in terms of the quality and amount of work and administration and implies relatively less dependence on capital. In this way, feasible cultivating frameworks may require more cultivation administrators, more cultivation workers and more cultivation families than do customary cultivating frameworks (Ikerd, 1993).

9.16 Ecosystem Stewardship

This is an action-oriented system planning to cultivate the social-ecological supportability of a quickly changing planet. Later advancements distinguish three techniques that make ideal use of our current understanding of an environment of unavoidable instability and unexpected change: lessening the impact and introduction of and the sensitivity to known stresses; centring on proactive approaches that shape change; and dodging or avoiding unsustainable social-ecological traps. All social-ecological frameworks are powerless to later and anticipated changes but have sources of versatile capacity that can uphold ecosystem services and human welfare through dynamic biological system stewardship.

9.16.1 A Call for Ecosystem Stewardship

Human activities are having huge and quickening impacts on the climate, the environment and the biological systems of soil, subsequently undermining biological system administration. This unsustainable direction requires a dramatic change in human relationships with the environment and the life-support framework of the planet. Here, we address recent developments in considering the maintainable use of biological systems and assets by society within the setting of fast and frequent environmental changes. Western asset administration models have advanced from exploitation, where maintainability isn't an imperative thought, to steady-state asset management aimed at ideal feasible administration (MSY or OSY, separately); to the proficient generation of a single asset, such as trees; and to environment administration to maintain a broader suite of ecosystem services. In spite of its maintainability objective, administration for MSY or OSY tends to overexploit resources because of excessively hopeful presumptions around the capacity to support efficiency, dodge unsettling influences, control harvesters' conduct and expect extraordinary financial or natural occasions. Biological system administration aims to support different types of biological system administration but frequently employ, as a reference point, notable conditions that are not achievable in a quickly changing world. Given the challenges of the economical use of environments amid quick changes, we advocate a move to environmental stewardship. Its central objective is to maintain the capacity to supply biological system administration that bolsters human well-being under conditions of vulnerability and change. Instability has always characterized social-ecological frameworks and ought to subsequently not be an obstruction to activity. Such a paradigm move involves vital trade-offs, especially between proficiency and adaptability and between short-term benefits and long-term benefits. Ecosystem stewardship coordinates three broad maintainability approaches: (i) diminishing helplessness in the face of anticipated changes; (ii) fostering resilience to support beneficial conditions in the face of annoyances and instability; and (iii) changing from undesirable directions when openings rise. Versatile capacity contributes to all three supportability approaches. By building on past inquiries into defencelessness, adjustments, flexibility and change, environment stewardship helps to prepare society to oversee a range of challenges that benefit or harm society. The point is to distinguish practical techniques that increase the probability of socially advantageous results and reduce the hazard of unwanted results. This approach is explicitly focused on human standards, values and well-being and must therefore ceaselessly be talked about and reassessed by partners (Chapin, et al., 2009).

9.17 The Vision of Natural Resource Stewardship

Although copying the characteristic frameworks that were described earlier for settlements is in fact unachievable, through honest intercession, one can sensibly anticipate that a more robust environment can be re-established over the city for future generations to appreciate. This is the basic introduction of the characteristic assets of stewardship vision.

9.17.1 The Philosophy of Natural Resource Stewardship

Stewardship vision advances an ecosystem-based approach to reclamation and administration. A biological system is essentially where things live and includes associations between various physical components (soil, water, fauna, flora, etc.). All these components need to be considered in setting objectives and plans for administration. Ecosystem-based management sees individuals as a portion of the community, where maintaining a healthy environment is the perfect way to meet human needs and the needs of other life-forms in the community. Common objectives of this logic are as follows:

- To improve the well-being of the biological systems of the city
- To upgrade the organic characteristics of local territories
- To make suitable adjustments that factor in asset conservation, recreational use and community development

- To keep up the common judgement of Lino Lakes
- To set up associations and partner associations with an assortment of agencies and citizens in the community to sustain maintainable assets in Lino Lakes

The stewardship plan centres on achieving asset quality, which means that Lino Lakes can keep up a certain level of asset quality within the setting of realistic limits—which relies on two essential variables:

1. An open understanding of and commitment to normal asset conservation and stewardship programmes
2. Undertaking logically sound biological restoration and administration programmes

Through a well-defined stewardship programme and a concerted, continuous effort by Lino Lakes, a level of certainty can be gained such that the current ecological conditions and patterns can be turned around and a more sustainable and higher-quality landscape achieved. Note, however, that stewardship programmes moreover need to be adaptable due to the changing nature of the living frameworks affected by the programme. For these reasons, the programme ought to be treated as immutable or optimal. It may be a beginning point in a continuous effort that relies on observing and inquiring into programme viability.

9.18 Eco-agriculture: Integrating Production and Conservation at the Landscape Scale

The original green transformation, maintainable horticulture and natural resource administration (with its primary centre on sustaining the resources supporting generation), or even the ecotechnology approach of Swaminathan (Wittenberg et al., 2001) with its focus on farmers' fields, have major components to contribute. Approaches to biodiversity conservation also have to move past wild biodiversity to focus also on secure zones and the unassuming objectives of integrated preservation and improvement projects. Eco-agriculture—a completely coordinated approach to farming, preservation and provincial livelihoods, within an environment context—is required in many districts. This approach of eco-agriculture can well be incorporated to improve the outcomes of environmental stewardship (McNeely & Scherr, 2003).

9.18.1 Eco-agricultural Landscapes

Eco-agriculture recognizes the financial and biological connections and common interdependency among agribusiness, biodiversity and environment administration. Eco-agriculture includes zones in the natural environment and their agricultural generation. Eco-agriculture frameworks depend on maximizing biological, economic and community integration. The term *landscape* is defined as a territory or region managed by stakeholders working together to accomplish biodiversity, livelihood generation and vocation goals. Eco-agriculture landscapes can be one of the following:

1. "natural" regions, which are overseen to benefit agricultural jobs through positive synergies with production or other livelihood benefits
2. agricultural generation regions, which are designed and overseen to provide a "matrix" with generous or positive ecological qualities for wild biodiversity and biological system services
3. institutional components, which arrange activities to achieve production, preservation and livelihood objectives at the landscape, cultivate and community scales, by relying on synergies and overseeing trade-offs among them

The concept of eco-agriculture further recognizes that agriculture-dependent rural communities are critical stewards of biodiversity and ecosystem services. While protected natural zones are essential in eco-agriculture landscapes to guarantee the basic environment required for at-risk species, maintain water sources and provide cultural resources, these assets are often overseen by local communities and farmers.

9.18.2 Biodiversity and Ecosystem Services in Eco-agriculture Landscapes Applicable to the Reclaimed Deltaic Islands of Sundarbans

The preservation of biodiversity in eco-agriculture landscapes embraces all three components of agricultural biodiversity characterized by the Convention on Biological Diversity: hereditary differences among tamed crops, animals and trees; differences among wild species on which agricultural generation depends; and diversity among wild species and biological communities that utilize rural areas as their habitats (CBD, 2006). Although wild biodiversity and ecosystem services are closely connected, they are not synonymous. A landscape with moderate wild biodiversity is likely to supply comprehensive ecosystem services. However, ecosystem administration can also be carried out via a combination of local and non-native species in heavily managed settings. The implication is that indeed where wild biodiversity has been essentially diminished to make way for agricultural generation, high levels of ecosystem services can regularly still be extracted through intentional land-management strategies. On the other hand, managing an eco-agricultural area for biological system services does not automatically guarantee that wild biodiversity will be satisfactorily ensured. Hence, wild biodiversity and biological system administration both require unequivocal consideration in eco-agriculture systems (Jackson & Jackson, 2002; Jackson et al., 2005; Harvey et al., 2005; Foley et al., 2006; International Institute of Tropical Agriculture (IITA), 2007).

9.18.3 Eco-agriculture Approaches to the Sundarbans Ecosystem

Broadly, eco-agricultural regions depend on six basic strategies of asset administration, three of which centred on the rural portion of the region and other three on the surrounding network.

Agriculturists increase agricultural yields and decrease costs in ways that enhance habitat quality and environment administration:

1. By minimizing agricultural losses and contamination
2. By managing assets in ways that preserve water, soil, wild vegetation and fauna
3. By using trim, grass and tree combinations to imitate the ecological structure and work of characteristic habitats

Farmers, local people and other biodiversity other preservation supervisors secure and expand normal zones in ways that moreover provide benefits to adjoining agriculturists and communities:

1. By minimizing or transforming characteristic zones having its own set up of biodiversity, fertility status of the land and productive potential of water bodies
2. By protecting and extending patches of high-quality natural habitats
3. By developing viable biological systems cultivation coupled with harvesting systems of indigenous natural resources paving the way for their onward passage for commercialization passages (McNeely, and Scherr, 2003)

The relative zone and spatial setup of agrarian and normal segments are key planning issues (Forman, 1995). The preservation of wild species that are exceedingly delicate to territory changes—and those most imperilled or uncommon globally—requires huge well-connected patches of natural habitats. But numerous wild species, including those that are undermined and imperilled, can contradictorily coexist in overseen rural areas, evening high-yielding systems. Numerous approaches to agribusiness, conservation and rural improvement contribute components, management practices and arranging systems that can be applied in eco-agricultural areas. The results of planning and arrangements among the different stakeholders in any specific region depend on the context of nearby societies and the philosophies of agricultural administration.

Genuine costs are related to cross-functional planning, integration and specialized innovations needed for any region. The top priorities are as follows:

1. Agrarian areas in or around critical environment zones for wild species of local, national or universal significance
2. Corrupted rural areas where restored ecosystem administration will be fundamental to achieving both rural and biodiversity benefits
3. Agrarian areas that enable basic environment administration
4. Peri-urban agrarian frameworks, where careful management is required to ensure ecological, wildlife and human health

No evaluation of the geological scale and area of such need ranges for eco-agriculture development methods has been conducted, but undertaking such an analysis may be a basic step to direct arrangement activity.

9.19 The State of Eco-agricultural Knowledge and Practices

Minor efforts have been committed to pursuing agricultural advancement and biodiversity preservation targets together at a regional scale, so encounters are poorly reported, and the science is novel and has been poorly synthesized over disciplines (Leakey, 1999; Leakey & Tchoundjeu, 2001; Lefroy et al., 1999; MEA (Millineum Ecosystem Assessment), 2005b; Millard, 2007). Evaluating the major barriers to the far-reaching advancement of eco-agriculture landscapes requires answering questions like the following:

- How can rural generation frameworks contribute to moderating biodiversity while maintaining or increasing efficiency?
- How can rural and normal zones be jointly managed to deliver satisfactory ecosystem services, including from natural environments, at a regional scale?
- How can eco-agricultural approaches be more financially practical for farmers, ranchers and other stakeholders?
- How can communities be educated about and governments mobilize education and the policies required for eco-agricultural areas?

The current state of information on these four questions is considered below.

9.19.1 Production Systems that Support Biologically Diverse Eco-agriculture Landscapes

Since the 1960s, the improved seed–fertilizer–pesticide (irrigation) worldview has characterized both industrial agriculture and the green revolution in agriculture. This production model included short-term, plot-level production of a small number of crops, for the most part as monocultures. Wild flora and fauna were considered coordinate competitors for resources or collected services. More biologically generous generation frameworks were retained in numerous conventional frameworks, which for environmental, social or financial reasons were not successfully added to mechanical strategies. Such frameworks aimed to construct, rather than supplant, common biological systems. Many advanced approaches have centred on diverse angles of environmental synergy, arising from changes in topographical conditions. Agro-ecology, permaculture, conservation farming, agroforestry, natural agriculture and economical farming have centred on keeping up the asset base for resource generation, by managing nutrient cycles, protecting pollinators and other beneficial micro-organisms, maintaining healthy soils and conserving water. They aimed to diminish the ecological footprint of cultivated zones and the harm to wild species from toxins, soil degradation and water pollution, but most centred on farm-scale action rather than the planning efforts from farmers, ranchers and others to reap biodiversity benefits at a regional scale (Redford & Richter, 1999; Power, 1999; Molnar, 2003; Robertson & Swinton, 2005).

1. Reducing pollution in natural ecosystems

Reducing agrochemical use and animal waste in high-input-generation frameworks can enormously benefit wildlife. For example, high-nutrient or harmful runoff into waterbodies can significantly reduce aquatic biodiversity. Major progress has been made in methods that reduce the use of fertilizers or increase the productivity of fertilizer use through better timing and better strategies. Agricultural pesticides may also kill nontarget insects and plants. Integrated pest management systems have viably utilized many crop blends through administration practices aimed at reducing the amount of required pesticides (Kogan,1998). Cellular science and atomic science have been utilized to tailor pesticides to target specific pests. Environmental and biochemical strategies have discovered host–pest relations that might revolutionize agrarian pest control in the future (Wittenberg & Cock, 1998). A meta-review of the agriculturist involvement found increases in both efficiency and biodiversity from reduced chemical use in developing districts. For example, the framework of rice mobilizes biological interactions in plant–soil frameworks in a way that increases yields while reducing costs (Uphoff et al., 2006). In the interim, modern whole-farm approaches minimize the runoff of agrochemicals and animal waste into waterbodies by increasing capacity frameworks, by overseeing areas to increase agrochemical penetration and decrease runoff and by building buffer zones to channel runoff away from streams (Coombe, 1996; IPCC (Intergovermental Panek on Climate Change) 2007; Chakraborty et al. 2014).

2. Managing production systems to improve habitat quality

Agriculturists have converted areas into valuable habitats for species or improved their value as "corridors" interfacing with the normal habitat ranges of an area (Forman, 1995; Clay, 2004). Buck et al. (2006, 2007) evaluated 79 biodiversity factors (more often than not species richness) related to 18 particular rural hones. The procedure most often connected with the preservation of wild biodiversity was the support of adjoining hedgerows, windbreaks or woodlots; 18 factors reported positive relationships with eight taxa. Natural horticulture was related to an increase in seven taxa in eight factors (Peach, et al., 2001; Kleijn & Sutherland, 2003; Kleijin et al., 2006). Shaded tropical crop production (particularly coffee and cacao) had higher species richness for three higher taxa thanks to eight factors (Buck et al., 2007). Many of these hones give extra benefits to farmers and ranchers, such as useful byproducts, a diminished chance of dry spells, various nourishment and livelihood resources and decreased vulnerability to natural dangers. For example, after Tropical Storm Mitch in October 1998 (one of the worst catastrophe to strike Central America in 200 years), analysts found that farmers who used agro-ecological hones endured 58% less harm in Honduras, 70% less in Nicaragua and 99% less in Guatemala than those who used customary cultivating strategies (Bunch, 2000).

The impacts of preservation hones may be species specific. For example, cotton is a habitat for numerous warblers, especially due to its exceptionally high levels of pesticide use in customary frameworks. But approaches such as preservation culturing and strip cover trimming decrease the environmental effect of cotton areas. Amid relocation and breeding periods, bird densities in strip cover areas were 2–6 and 7–20 times more prominent than in preservation and routine areas, respectively. Although the clover treatment pulled in the most elevated avian and arthropod densities, preservation areas still provide more natural life and agronomic benefits than customary administration. The diminishment of inputs in the clover framework permits farmers to reduce their costs related to routine cotton generation. Transgenic cotton was created to reduce the required levels of pesticides, aiming to increase biodiversity (Cattaneo et al., 2006).

The agricultural industry has lately started to create measures that address the preservation of wild biodiversity. Wild taxa, including birds and invertebrates, and arable greenery can benefit from natural administration through increases in species richness (Hole et al., 2005) Management hones are especially useful for farmland. Although yields from natural systems are still regularly lower than those from customary frameworks, the gap is narrowing and studies are showing that agricultural generation frameworks that fundamentally or solely rely on natural inputs can increase agronomic and financial benefits (Uphoff et al., 2006).

Carefully focusing on administration hones that are connected to trimmed or noncropped environments within ordinary farming may also bring about biodiversity benefits. Weibull et al. found that wild species abundance for the most part expanded with regional heterogeneity on a cultivation scale, and territory type had a major impact on species abundance for most species found in grasslands and leys. Having farmers preserve biodiversity on their farmstead enriched biodiversity more than using particular hones.

9.20 Achieving Globally Significant Farm Stewardship and Its Applicability in Mangrove Ecosystems

The present pool of knowledge has been insufficient to meet these objectives across diverse ecosystems and production systems. Scaling up farm stewardship approaches to a globally significant level will require three key elements: new knowledge, enhanced institutional capacity and a supportive policy and market environment.

9.20.1 Generating and Disseminating Knowledge for Farm Stewardship

Key priorities for shaping agricultural landscapes include understanding the interactions and dynamics between conservation and production areas; developing production systems that explicitly promote soil stewardship, water stewardship, pesticide stewardship and biodiversity stewardship objectives while mimicking natural ecosystems; and ensuring that more aspects of farming systems are ecologically sustainable, including industrial processing, packaging and transportation. Successful eco-agricultural systems need to be thoroughly documented, both for their landscape-scale outcomes and for specific interventions. Mapping the spatial overlays between important agricultural areas and important biodiversity will be crucial.

9.20.2 Enhancing Capacity for Farm Stewardship

Knowledge innovation systems must be reconfigured to offer services to rural resource stewards and facilitate the rapid intake of empirical knowledge across sectors. Rural communities should be recognized as vital stewards of biodiversity conservation, leading professional conservation organizations, public agencies and others to accordingly realign their efforts to support community-based organizations and other stakeholder groups. Conservation organizations should actively engage with farming partners, cultivate agricultural expertise and advocate for sustainable agriculture investment in alignment with conservation strategies (Coombe, 1995; Cairns & Garrity 1999; Buck et al., 2004).

9.20.3 Promoting Markets and Policies that Bolster Farm Stewardship

The success of technical and local organizational efforts for farm stewardship hinges on the removal of major policy barriers and the development of supportive policies. Farm stewardship presents cost-effective methods for national investment strategies to meet the United Nations' (UN) Millennium Development Goals. Strategic shifts in the food industry, institutional procurement practices, the eco-certification of agricultural products and the financial oversight of agricultural investments by investors can be leveraged to redirect financial incentives towards farm stewardship. Internationally, people have opportunities to incorporate stewardship strategies into the work programmes of global environmental conventions to ensure that World Trade Organization rules align with eco-agriculture landscapes.

9.20.4 *Managing Food Production Systems for Stewardship*

Managing food production systems sustainably is a paramount challenge for humanity given the fundamental importance of food for survival. Ecosystem health is crucial both as a means to achieve and as an outcome of resilient crop and animal production. These systems fundamentally rely on the world's ecosystem and atmosphere, generating some of the most vital ecosystem services on the planet. Nevertheless, they are important contributors to air and water pollution in major locations (Matson et al., 1997; Naylor, 2001, Smil, 2000; Vitousek et al., 1997; Conway, 1997). The inherent connection between human well-being, food production and the status of natural resources undermines the necessity of managing these systems for resilience, anticipating and shaping change to promote long-term environmental quality.

The challenge of the twenty-first century is to upgrade food production systems that support rural incomes, increase yields, efficiently use inputs (especially water and nutrients), minimize environmental impacts and provide a healthy diet for the human population (Conway, 1997; Power, 1999; Robertson et al., 2005; Tilman et al., 2002). Although increasing productivity remains essential for sustainability, the goal should not always be to maintain or increase productivity in existing systems. Instead, the focus should be on redesigning and realigning the said systems. Achieving this goal requires efforts in small-scale farming systems that primarily serve local and regional demands that cater to national and global needs (Vitousek et al., 1997).

9.21 A View on Farm Stewardship

The variety of agricultural systems worldwide should not be viewed as representative of any specific region's agriculture. The methods used to maintain resilience and sustainability in one location may not be applicable to other similar niches, and they may not be transferable across production scales.

The global food production system is inherently complex. Despite its complexity, transitioning towards resilience involves three main steps (Lambin, 2007). The first step requires recognizing that the biophysical and socioeconomic conditions of the twenty-first century are markedly different from those of the past, necessitating fundamental redesigns of many crop and animal production systems. Some principles could help guide this redesign process. The second step involves operationalizing government policies that influence the direction of food production systems. The third step towards resilience transformation involves promoting a scientific and informational toolkit. Integrated management practices, particularly water and nutrients, will be equally critical for increasing crop productivity, incomes and environmental services worldwide (109). Additionally, the use of new information technologies significantly improves farmers' capabilities in both affluent nations and impoverished nations over time (Cassman, 1999; Lobell & Ortiz-Monasterio, 2008).

The key point is the necessity of developing a comprehensive toolkit as an integrated package worldwide (Solow, 1991). The generation of this toolkit needs the institutional commitment of international agriculture (Falcon & Naylor, 2005). Many farmers are already reconfiguring their agricultural systems to align with community needs and barriers.

Farm stewardship, the ethical and responsible management of agricultural resources, is undergoing a paradigm shift from conventional conservation agriculture to building ecological resilience. This shift is driven by the recognition that conventional conservation agriculture, while beneficial in many respects, may not be able to respond to the complex challenges in agriculture. Building ecological resilience involves enhancing the capacity of agricultural ecosystems to cope up and recover from disruptions while maintaining productivity and ecosystem services. Ecosystem stewardship, the responsible management of ecosystems for the benefit of current and future generations, faces several challenges that must be addressed to ensure the long-term sustainability of natural resources. The following list features some of the key challenges.

1. Climate change: One of the biggest challenges facing ecosystem stewardship is climate change. Rising temperatures, changing precipitation patterns and more-frequent extreme weather events are already having significant impacts on ecosystems around the world. Stewardship efforts must adapt to these changes and develop strategies to mitigate their effects.
2. Biodiversity loss: Ecosystems are facing unprecedented rates of biodiversity loss due to factors such as habitat destruction, pollution and climate change. Stewardship efforts must focus on protecting and restoring biodiversity to ensure the health and resilience of ecosystems.
3. Land degradation: Soil erosion, desertification and deforestation are examples of land degradation that can have serious consequences on ecosystems. Stewardship efforts must address these issues through sustainable land-management practices that promote soil health and prevent further degradation.
4. Water scarcity: Many ecosystems are facing water scarcity due to factors such as overuse, pollution and climate change. Stewardship efforts must focus on sustainable water-management practices that ensure ecosystems have access to an adequate supply of clean water.
5. Invasive species: Invasive species can have devastating effects on ecosystems by outcompeting native species and disrupting ecosystem processes. Stewardship efforts must focus on preventing the intrusion and diffusion of aggressive species and managing existing populations.
6. Human activities: Human activities such as agriculture, urbanization and resource extraction can have significant impacts on ecosystems. Stewardship efforts must address these impacts through adaptive practices that minimize disruptions to ecosystems.

7. A lack of awareness and education: A lack of awareness about and education on the importance of ecosystem stewardship can hinder conservation efforts. Stewardship efforts must focus on raising awareness about and educating the public on the value of ecosystems and the need to protect them.
8. Limited natural resources: Limited financial and human resources can also be a challenge for ecosystem stewardship efforts. Stewardship efforts must find ways to maximize the effectiveness of available resources and secure additional support where needed.

Addressing these challenges will require a coordinated and collaborative effort by ensuring the participation of local groups, communities, businesses professionals and individual actors to ensure the long-term health and sustainability of ecosystems for generations to come.

9.22 Interactions Between Humans and Nature: Pathways for the Sustainable Conservation of Mangrove Ecosystem

The people-centric conservation approach ensures sustainable conservation by effectively protecting ecosystems and biodiversity as benefits of such conservation initiatives, which in turn rewards local communities, encouraging a sense ownership and responsibility for their immediate surroundings. Such an ecosystem-based approach for the conservation of mangrove ecosystems encompasses pragmatic, holistic initiatives, adhering to the basic operating principles of ecology, such as the interconnectedness and interdependence of all the structural components of the concerned ecosystem. The perceptions and behavioural attitudes of coastal communities are rigidly linked to different coastal processes, like tidal interplays and the seasonal oscillation of ecological variables after being influenced by changing climatic conditions. The survival strategies of different biodiversity components inhabiting mangrove-estuarine complexes have also been changing because of ongoing environmental perturbation like sea-level rise, a reduction in freshwater flows, sediment starvation, land subsidence, massive soil erosion, vertical accretion, hypersalinity, eutrophication, acidification and an increase in the nonfertility of soil for agriculture (Hazra, 2003; Hazra et al., 2002). In such a context, an ecosystem approach for conservation must include systematic, strategic and effective planning to address ongoing ecological, social and political problems and to find remedies to save the mangrove ecosystem of Sundarbans in general and the precious biodiversity wealth of this unique ecosystem in particular. The relationship between humans and nature is determined by the ecological changes due to the continuous interactions between three elements, namely ecological systems, structural dynamics and changes in biotic assemblages, in which any two elements affect the third. The science of conservation biology has emerged in integrating the complex issues of environmental policy and ecosystem management with other disciplines, namely the social, physical and life sciences—together developing conceptual frameworks in line with research information on biotic interactions, trophic relationships, ecological succession patterns and other such eco-biological attributes (Ayer & Maginnis, 2005).

Ecosystem services provided by the structural components of mangrove ecosystems help to open up new opportunities for livelihood generation for coastal people (Upadhyay et al., 2002; Naylor et al., 2000; Sayer & Maginnis, 2005; Verhulst et al., 2006; Chakraborty, 2017). The ongoing loss of mangroves across the world, according to the estimates of Spalding et al. (1997), was substantiated by a case study on the rate of disappearance for mangrove areas: 41% in Southern Thailand between 1961 and 1996 (Plathong, 1998). This has necessitated introducing alternative sources of food, fodder, fuel and shelter materials by the government and nongovernment organizations, aiming to promote ecofriendly biotechnologies (Agoramoorthy & Hsu, 2008). The ecological health of mangrove ecosystems need to be continuously monitored with the active involvement and participation of local communities (Macintosh & Ashton, 2002). which in turn help harvesting natural resources in the face of climate driven harmful consequences (Das and Saha, 2008; Chand et al, 2014).

Mangrove ecosystems throughout the globe have been suffering from ecodegradation because of logging, coastal development, land-use changes and pollutants and contaminants like pesticides, oil spills, chemical fertilizers, heavy metals and microplastics. An effective eco-rehabilitation programme followed by eco-restoration strategies should take into account adequate tidal exchange to re-establish natural hydrologic and tidal regimes; planting mangrove propagules or mangrove saplings according to research information on their ecological requirements; and facilitating the colonization of mangrove plants and their associated fauna. Ecological niches for mangrove recruitment within mangrove ecosystems are synergistically determined by sediment flow and deposition, erosion and accretion, the levels of freshwater inflow and its mixing with saline water, impacts from catastrophic storms and tidal flooding, seasonal hydrology, invasion by exotic species, prey–predator relationships, etc. (Lewis & Streever, 2000). Prior knowledge on the modes of production for organic matter, on abiotic ecological states (temperature, moisture, pH, soil texture, etc.) and on biotic activities (microbial and faunal) and the development of functional food webs, nutrient cycling, energy movement, etc. are important functional aspects in the eco-restoration of mangrove ecosystems.

Extensive survey-based research over the past few decades on the different parts of Sundarbans has been undertaken, applying proper methods for the recording of the occurrence, diversity, density and distribution of different biodiversity components and for the evaluation of soil and water, which have revealed useful information on the ecodynamics of the mangrove ecosystem of Sundarbans, India, and on the possible developmental prospects of brackish-water aquaculture (Figs. 9.1, 9.2, 9.3, 9.4, 9.5, 9.6 and 9.7). All this generated local research information can be used to draft proper ecomanagement plans to achieve sustainable conservation and gain proper support from both government and nongovernment agencies in the form of building necessary infrastructure facilities for nursery development, mangrove plantation and wildlife protection (Chakraborty et al., 2009; Chakraborty, 2013, 2014, 2021).

Fig. 9.1 The developmental activities of the Government of West Bengal aiming at ensuring all-round conservation for wildlife (**A**, **B**, **C**, **D**, **E** and **F**) by providing necessary requirements in terms of food, water and habitats for their successful survivability, facilitating the mangrove afforestation programme even within the core area of the Sundarbans Biosphere Reserve through the creation of nursery beds (**G** and **H**), generating awarness for nature conservation with nature interpretation centres (**A**, **B** and **C**) and promoting tourism with the necessary expansion of required facilities (**A**, **E** and **F**)

Fig. 9.2 The creation of necessary infrastructure in and around the coastal belts of Sundarbans, India, for the conservation of wildlife, providing them with foraging grounds (**A**, **B** and **C**), promoting awarness in tourists through nature interpretation centres (**D**) and protecting fragile coastal geomorphological structures like dunes, natural vegetation and other biota (**E** and **F**)

Fig. 9.3 Interacting with local people in Sundarban, India (**A**, **B**, **C**, **D** and **E**) , for surveys and market survery (**F**) and to collect information on ecological changes, natural resource availability and people's perceptions of both; visiting tourist lodges to interact with vising tourists (**G** and **H**) and to undertake discussion with fisher communities (**I**, **J** and **K**); and endeavours for sustainable conservation strategies that rely on the stewardship of local people

Fig. 9.4 The continuous ecomonitoring of ecological changes in different corners of the Sundarbans mangrove ecosystem by visiting sampling sites; collecting water, soil and biological materials; and estimating seveval collected samples—where all such field-based research activities (**A**, **B**, **C**, **D**, **E**, **F** and **G**) are instrumental for generating local information on the structure and function of the mangrove ecosystem of Sundarbans, which can be effectively used for drafting a proper conservation strategy

Fig. 9.5 The rigorous environment of Sundarbans poses problems to collecting local information on the structural components of this mangrove ecosystem, and images (**A**, **B**, **C**, **D**, **E** and **F**) highlight the committed efforts by a team of researchers, including the first author of this book, in undertaking field studies

Fig. 9.6 Field studies by the researchers, who made observations about local issues related to the diversity of ecological habitats (**A**, **B**, **D**, and **H**), the potential of mangrove biodiversity (**A**, **B**, **C** and **E**) and the availability of facilities such as government-sponsored tourist lodges (**F** and **G**)

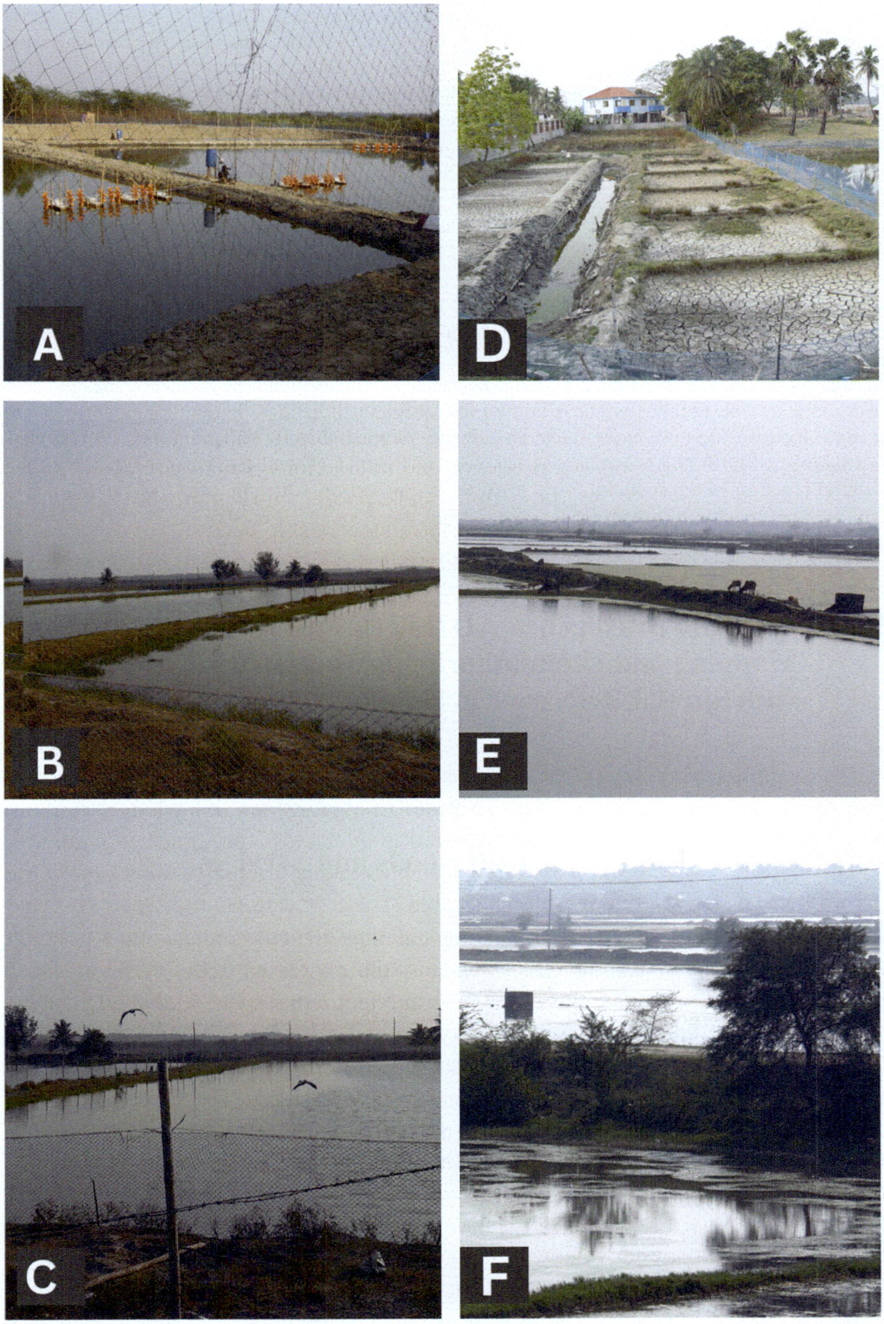

Fig. 9.7 As a part of sustainable environment conservation in the Sundarbans mangrove ecosystem in India, the diversification of livelihoods is an important adaptive proposition in view of the climate change–mediated vulnerability of the ecosystem, and the development of brackish-water a

Fig. 9.7 (continued) as depicted in where quaculture has emerged as one of the prominent livelihood development options thoughout the length and breadth of the Indian Sundarbans, where the tidal water of the nearby estuaries or creeks are utilized with the application of modern aquacultural tools like aerators, the dredging of deposited mud and the occasional application artificial feed and medicine (**A**, **B**, **C**, **D**, **E** and **F**). In such aquaculture activities (**A**–**D**), semi-intensive aquaculture on Henry Island in the vicinity of the mangrove-estuarine complex of Sundarbans, where a good number of aquacultural farms, each with an area of 1 acre, was developed as a joint initiative of the government of West Bengal and the World Bank around four decades ago to promote livelihood generation and revenue for the local government. In aquacultural practices, water impoundments are developed by digging up soil to develop shallow, depressed land and construct the banks of waterbodies with the excavated soil, thereby separating one aquacultural site from another. Brackish water from the main estuarine flows enters waterbodies through a pumping system and goes out through exit channels. The to-and-fro movement of natural water provides necessary nutrients, salinity, dissolved oxygen, pH, etc., which improve the growth of the spawns of both finfish and shellfish, and maintaining the necessary ecological conditions of waterbodies requires using aerators time to time to remove bottom soil, providing necessary nutrients in the form of fish feed from outside. In most cases, the monoculturing of commercially important shellfish (*Paeneus monodon*) or finfish (*Lates calcarifer*) has been undertaken (**E** and **F**): Mismanagement in the cultural practices of aquacultural farms leads to eutrophication, which in turn results in failures in aquaculture production (Levin, 1992, 1998, 2005; Chakraborty, 2011, 2017)

9.23 Artificial Intelligence(AI) Driven Fuzzy Cognitive Mapping and Ecomonitoring in Sundarban Biosphere with Special Reference to Human-Tiger Conflicts

Fuzzy Cognitive Mapping (FCM) emphasizes to build a model of a system with the help of boxes and connecting links in order to capture new knowledge without empirical data. The boxes, normally referred to as ‘factors’ or ‘concepts’, can represent anything which is expressed as a variable making a sense to think of it going up or down, instead of making it quantifiable. The connections, mostly represented by arrows are often referred to as ‘edges’, and they represent causal links between factors and are meant to show direct causal influence. The exploration of how these causal influences propagate through a system when it is subject to change or intervention is at the heart of what an FCM is intended to do. Two main approaches centre around this in common use within the community and it is often not explicitly stated which approach is being used.

The first approach of FCM takes into account the strength of links between factors as ‘causal’ representing how certain the roles of one factor causing, or suppressing another (Helfgott et al., 2015), Values of factors produced in analysis represent how strongly caused (or ‘activated’) they are by changes in other factors based on level of uncertainty about whether causal links in a map exist or not. It can be possible to how do the changes in some factors would cause changes in others, instead of making it quantifiable as how large those changes might be. The output of analyzing a causal FCM is a list of numerical values of factors under different scenarios with each number representing how relatively strongly cause a factor to be relevant ones. It aims to answer the question: if anything is changed in the ecosystem, what would its implications on other structural components of the ecosystem.

The second approach, referring to as 'dynamical' (Helfgott et al., 2015), takes into consideration the propagation of effects of one factor on another, producing a simple dynamical representation of the relative magnitude of changes in factor values, that is, a plot of all factor values at each iteration. Here, magnitude of factor value does indicate how relatively large an effect is. It is used to help understand which factors and connections are most important or influenced in a system, with larger factor values interpreted as being more important or influenced/influential and how changing map structure changes that. It can also be used to gauge how relative changes in factor values might play out dynamically, for example, whether change accelerates, stabilizes, or dies away.

Modern Artificial Intelligence (AI), especially Machine Learning (ML) and Deep Learning (DL) , has become an established and dominant discipline in many activity sectorsembracing new technologies. The feature development of human society lies in ML and DL to solve intricate problems and offer reliable solutions. It is often discussed that the potential of ML and DL may transform human-oriented processes into automatic every day tasks, wherein human intervention is no longer required.

The best possible eco monitoring in Sunrbans biosphere, the robust and complex, is possible through community mobilization and participatory mapping-modelling and action. Ai driven FCM can simulate the complexity, vulnerability and actionable points by driving community intelligence and action mapping. AI is expected to develop the algorithm through Machine Learning and Deep learning for problem identification, vulnerability analyses and scenario analysis. Subsequently, Focus Group Discussion (FGD) can usher the process of alternative solutions. The edge and impact of Fuzzy Maps will be effective in designing participatory eco monitoring in Sundarban biosphere. The empirical study on man –elephant conflict and it Fuzzy solution can also be extrapolated to man-tiger conflict in Sundarban.

Steps to be followed:

Organize Focus Group Discussion to elicit highlighted problems and nature. ii) Programme AI algorithm for training data and Neural Network. iii) Execute FCM by delineating out-degree and in-degree or edge-impact binary.

Project FCM results for scenario analysis.

Build up community based micro plan for regeneration-recycling-renewable and sustainable eco-monitoring for SundarbanBiosprhere.

The microplanning output can be integrated with policy framing.

When open-ended interactions are on, their options are recorded, and extrapolated to the Fuzzy Mental Modelling. Both the out degree and in degree points are noted. Out degrees are nodes where possible options are generating from and, In degrees are nodes where different solutions are converging. The blue lines indicates additive impact and red li

The colored lines in the network indicate additive impact and coercive impact. Here elephant depredations add on crop damages, wherein poverty enhances grazing of cattle on grasslands. The FCM depicts that village level policy making has reduces elephant attack or intrusion. The same exercises in Sundarbans mangrove ecosystem can be replicated on human-tiger conflict. FCM increase community empowerment, strategy delineation and orchestrated intervention to save both people and wild life of Sundarbans of Indian sub-continent (Banerjee et al., 2025).

References

Adger, W. N. (2006). Vulnerability. *Global Environmental Change, 16*(3), 268–281.

Agoramoorthy, G., & Hsu, M. J. (2008). Biogas plants ease ecological stress in India's remote villages. *Human Ecology, 36*, 435–441.

Allen, P. (1993). Connecting the social and the ecological in sustainable agriculture. In P. Allen (Ed.), *Food for the future*. Wiley.

Ayer, J. A., & Maginnis, S. (2005). *Forest in Landscapes:Ecosystem Approaches to Sustainability* (p. 71). Earthscan Publications.

Bailey, R. W. (1968). *An analytical bibliography of statistical stylish*. Sage.

Bailey, O. K. (1982). *Post functional system analysis*. Willey.

Banerjee, A., Chatterjee, A., & Acharya, S. K. (2025). Mental modeling of human elephant conflict using fuzzy cognitive mapping and decision ecology for conflict resolution. *Environment Systems and Decisions, 45*, 1–17.

Beer, J., Ibrahim, M., & Schlonvoigt, A. (2000). *Timber production in tropical agroforestry systems of Latin America*. XXI IUFRO World Congress.

Berkes, F., Colding, J., & Folke, C. (2003). *Navigating social-ecological systems: Building resilience for complexity and change*. Cambridge University Press.

Birkeland, P. W. (1999). *Soils and geomorphology* (3rd ed.). Oxford University Press.

Blann, K. (2006). *Habitat in agricultural landscapes: How much is enough? A state of the science literature review*. Defenders of Wildlife.

Buck, L. E., Gavin, T. A., Lee, D. R., & Uphoff, N. T. (2004). *Eco agriculture: A review and assessment of its scientific foundations*. Cornell University.

Buck, L. E., Milder, J. A., Gavin, T. A., & Mukherjee, I. (2006). *Understanding Ecoagriculture: A framework for measuring landscape performance* (Eco Agriculture Discussion Paper Number 2). Cornell University.

Buck, L. E., Gavin, T. A., Uphoff, N. T., & Lee, D. R. (2007). Scientific assessment of ecoagriculture systems. In *Farming nature, science and practice of eco-agriculture*. Island Press.

Bunch, R. (2000). *Reasons for resiliency: Toward a sustainable recovery after Hurricane Mitch*. World Neighbours.

Cairns, M., & Garrity, D. P. (1999). Improving shifting cultivation in Southeast Asia by building on indigenous fallow management strategies. *Agroforestry System, 47*(1/3), 37–48.

Carson, R. (2015). Silent spring. In *Thinking about the environment* (pp. 150–155). Routledge.

Cassman, K. G. (1999). Ecological intensification of cereal production Systems: Yield potential, soil quality, and precision agriculture. *Proceedings of the National Academy of Sciences of the United States of America, 96*(11), 5952–5959.

Cassman, K. G., Dobermann, A., Walters, D. T., & Yang, H. (2003). Meeting cereal demand while protecting natural resources and improving environmental quality. *Annual Review of Environment and Resources, 2*, 316–351.

Cattaneo, M. G., Yafuso, C., Schmidt, C., Huang, C. Y., Rahman, M., Olson, C., Ellers-Kirk, C., Orr, B. J., Marsh, S. E., Antilla, L., Dutilleul, P., & Carrière, Y. (2006). Farm scale evaluation of the impacts of transgenic cotton on biodiversity, pesticide use and yield. *Proceedings of the National Academy of Sciences of the United States of America, 103*(20), 7571–7576.

Chakraborty, S. K. (2013). Interactions of environmental variables determining the biodiversity of coastal mangrove ecosystem of West Bengal, India. *The Ecoscan Special Issue, 3*, 251–265.

Chakraborty, S. K. (2017). Ecological services on intertidal benthic fauna and the sustenance of coastal wetlands along the Midnapore(east) coast, West Bengal, India. In *Coastal wetlands: Alteration and remediation* (pp. 777–886). Springer.

Chakraborty, S. K. (2021). *Riverine ecology (Volume 2): Biodiversity conservation, conflicts and resolution*. Springer. 917 pp.

Chakraborty, S. K. (2021a). *Riverine ecology (Vol. 1): Ecofunctionality of the physical environment of rivers resolution*. Springer Nature. 917 pages.

Chakraborty, S. K. (2021b). *Riverine ecology (Vol. 2): Biodiversity conservation, conflicts and resolution*. Springer Nature. 566 pages.

Chakraborty, S. K., Poddar, T. K., & Choudhury, A. (1992). Species diversity of Macrozoobenthos of Sagar Island, Sundarban, India. *Proceedings of Zoological Society, Calcutta, 45*(Suppl. A), 435–444.

Chakraborty, S. K., Chatterjee, S., Dey, M. K., et al. (2012). Ecological changes of an eco–rehabilitated delta of hooghly estuary. *Journal of Coastal Conservation, 3*(1), 9–27.

Chakraborty, S. K., Paul, A. K., Bandyopadhyay, J., Banerjee, D., Pakhira, H., Middya, S., Jana, S., Sahoo, P., Paria, K., & Bera, K. (2013). Sustainable water management in the drought prone riverinetracts of South West Bengal. *Journal of Ground Water Studies, 1*, 11–34.

Chakraborty, S. K., Giri, S., Chakravarty, G., & Bhattacharya, N. (2009). Impact of eco-restoration on the biodiversity of Sundarbans Mangrove Ecosystem, India. *Water Air Soil Pollution: Focus, 9*, 303–320.

Chakraborty, S. K., Maiti Dutta, S., Ghosh, P. B., Ray, R., & Paul, A. K. (2014). Impact of global warming on Sundarbans mangrove ecosystem, India: Role of different assessment tools from ecosystem monitoring to molecular markers. In *Proceedings of the International Conference on Green India: Strategic Knowledge for Combating Climate Change: Prospects and Challenges* (pp. 181–200).

Chakraborty, S. K., Sanyal, P., & Ray, R. (2023). *Wetlands ecology: Eco-biological uniqueness of a Ramsar site (East Kolkata Wetlands, India)*. Springer Nature. 709 pages.

Chakraborty, S. K., Giri, S., Chakravarty, G., & Bhattacharya, N. (2009). Impact of eco-restoration on the biodiversity of Sundarbans mangrove ecosystem, India. *Water Air Soil Pollution: Focus, 9*, 303–320.

Chakraborty, S. K., Dutta, S. M., Ghosh, P. B., Ray, R., & Paul, A. K. (2014). Impact of global warming on Sundarbans mangrove ecosystem, India: role of different assessment tools from ecosystem monitoring to molecular markers. In *Proceedings of the International conferenceon Green India: Strategic knowledge for combating climate change: prospects and challenges* (pp. 181–200).

Chand, B. K., Trivedi, R. K., & Dubey, S. K. (2014). *Climate change in Sundarban and adaptation strategy for resilient aquaculture*. CIFRI.

Chapin, F. S., III, Matson, P. A., & Mooney, H. A. (2002). *Principles of terrestrial ecosystem ecology*. Springer.

Chapin, F. S., III, Hoel, M., Carpenter, S. R., Lubchenco, J., Walker, B., Callaghan, T. V., Folke, C., Levin, S. A., Mäler, K. G., Nilsson, C., Barrett, S., Berkes, F., Crépin, A. S., Danell, K., Rosswall, T., Starrett, D., Xepapadeas, A., & Zimov, S. A. (2006). Building resilience and adaptation to manage Arctic change. *Ambio, 35*(4), 198–202.

Chapin, F. S., III, Carpenter, R. S., Kofinas, G. P., Folke, C., Abel, N., Marketer, W. C., Olsson, P., Smith, D. S., et al. (2009). *Ecosystem stewardship: Sustainability strategies for a rapidly changing planet*. Elsevier.

Chapin, F. S., Danell, K., Elmqvist, T., Folke, C., & Fresco, N. L. (2007). Managing climate change impacts to enhance the resilience and sustainability of Fennoscandian forests. *Ambio, 36*, 528–533.

Chapin, F. S., Randerson, J. T., McGuire, A. D., Fole, J. A., & Field, C. B. (2008). Changing feed backs in the earth-climate system. *Frontiers in Ecology and the Environment, 6*, 313–320.

Clausius, R. (1865). Concerning several conveniently applicable forms for the main equation of the mechanical heat theory. *Annalen der Physik und der Chemie, 125*, 353–400.

Clay, J. (2004). *World agriculture and the environment: A commodity-by-commodity guide to impacts and practices*. Island Press.

Convention on biological diversity 2002 text of theconvention on biological diversity. See http://www.biodiv.org. Convention on biological diversity. Decision VI/9: Global Strategy for Plant Conservation, 2006.

Conway, G. (1997). *The doubly green revolution: Food for all in the 21st century*. Cornell University Press.

Coombe, R. I. (1995). Watershed protection: A better way. In *Environmental enhancement through agriculture. Proceedings conference held in Boston*, 15–17 November, Lockeretz, W., Ed., Organized by the Tufts University School of Nutrition Science, The American Farmland Trust, and the Henry A; Wallace Institute of Alternative Agriculture; Center for Agriculture, food, andEnvironment; Tufts University, 1996, pp 25–34.

Costanza, R., d'Arge, R., de Groot, R., Farber, S., Grasso, M., Hannon, B., Limburg, K., Naeem, S., O'Neill, R. V., Paruelo, J., Raskin, R. G., Sutton, P., & van den Belt, M. (1997). The value of the world's ecosystems services and natural capital. *Nature, 387*(6630), 253–260.

Daily, G. (1997). *Nature's services: Societal dependence on natural ecosystems*. Island Press.

Das, M. K., & Saha, P. K. (2008). *Impact and adaptation of inland fisheries to climate change in India*. CIFRI. Bulletin 151.

De Haan, L. R., Cox, T. S., Van Tassal, D. L., & Glover, J. D. (2007). Perennial grains. In S. J. Scherr & J. A. McNeely (Eds.), *Farming with nature: The science and practice of eco-agriculture*. Island Press.

Diamond, J. (2005). *Collapse. How socities choose or fail to succeed*. Viking Press.

Díaz, S. J., Fargione, J., Chapin, F. S., III, & Tilman, D. (2006). Biodiversity loss threatens human Well-being. *PLoS Biology; Plant Library of Science (PLoS), 4*(8), e277.

Dorfman, R., & Dorfman, N. S. (Eds.). (1972). *Economics of the environment: Selected readings* (pp. 19–187). W. W. Norton.

Douglass, G. K. (1984). *Agricultural sustainability in a changing world order*. Westview Press.

Ellis, J. (1998). Household strategies and rural diversification. *Journal of Development Studies, 35*, 1–38.

Elmqvist, T., Folke, C., Nyström, M., Peterson, G., Bengtsson, J., Walker, B., & Norberg, J. (2003). Response diversity, ecosystem change, and resilience. *Frontiers in Ecology and the Environment, 1*(9), 488–494.

Falcon, W. P., & Naylor, R. L. (2005). Rethinking food security for the 21st century. *American Journal of Agricultural Economics, 87*(5), 1113–1127.

Farnsworth, E. J., & Ellison, A. M. (1997). The global conservationstatus of mangroves. *Ambio, 26*(6), 328–334.

Fischer, J., Manning, A. D., Steffen, W., Rose, D. B., Daniell, K., Felton, A., Garnett, S., Gilna, B., Heinsohn, R., Lindenmayer, D. B., Macdonald, B., Mills, F., Newell, B., Reid, J., Robin, L., Sherren, K., & Wade, A. (2007). Mind the sustainability gap. *Trends in Ecology & Evolution, 22*(12), 621–624.

Food and Agricultural Organization (FAO). (2017). *The future of food agriculture: Trends and challenges*. FAO.

Foley, J. A., DeFries, R., Asner, G. P., Barford, C., Bonan, G., Carpenter, S. R., Chapin, F. S., Coe, M. T., Daily, G. C., Gibbs, H. K., Helkowski, J. H., Holloway, T., Howard, E. A., Kucharik, C. J., Monfreda, C., Patz, J. A., Prentice, I. C., Ramankutty, N., & Snyder, P. K. (2005). Global consequences of land use. *Science, 309*(5734), 570–574.

Foley, J. A., DeFries, R., Asner, G. P., Barford, C., Bonan, G., Carpenter, S. R., Chapin, F. S., Coe, M. T., Daily, G. C., Gibbs, H. K., Helkowski, J. H., Holloway, T., Howard, E. A., Kucharik, C. J., Monfreda, C., & Folke, C. (2006). Resilience: The emergence of a perspective for social—Ecological system analysis. *Global Environmental Change, 16*(3), 253–267.

Folke, C. (2003). Freshwater for resilience: A shift in thinking. *Philosophical Transactions of the Royal Society of London. Series B: Biological Sciences, 358*(1440), 2027–2036.

Folke, C. (2006). Resilience: The emergence of a perspective for social—Ecological system analysis. *Global Environmental Change, 16*(3), 253–267.

Forman, R. T. (1995). *Land mosaic: The ecology of landscapes and regions*. Cambridge University Press.

Gunderson, L. H., & Holling, C. S. (2002). *Panarchy: Understanding transformations in human and natural systems*. Island Press.

Harvey, C. A., Sanez, J. C., & Montero, J. (2005). Conservación de la Biodiversidaden Paisajes Fragmentados y Rurales de Mesoamerica? Que HemosAprendido y Que todavı'aNecesitamosConocer? ConfederaciónUnitaria de Trabajadores: CATIE. Draft. 70.

Hazra, S. (2003). *Vulnerability assessment in a climate change scenario: A pilot study on ecologically sensitive Sundarban Island system*. Technical Report, MoEF, Govt. of India.

Hazra, S., Ghosh, T., DasGupta, R., & Sen, G. (2002). Sea level and associated changes in the Sundarbans. *Science and Culture, 68*(9/12), 309–321.

Helfgott, A., Lord, S., Bean, N., Wildenberg, M., Gray, S., Gray, S., Vervoort, J., Kok, K., & Ingram, J. (2015). Clarifying fuzziness: Fuzzy Cognitive Maps, neural networks and system dynamics models in participatory social and environmental decision-aiding processes. *Processes EUFP7 TRANSMANGO Working Paper*)

Hole, D. G., Perkins, A. J., Wilson, J. D., Alexander, I. H., Grice, P. V., & Evans, A. D. (2005). Does organic farming benefit biodiversity? *Biological Conservation, 122*(1), 113–130.

Holling, C. S. (1973). Resilience and stability of ecological systems. *Annual Review of Ecology and Systematics, 4*(1), 1–23.

Holling, C. S. Gunderson, L. H. Resilience and adaptive cycles. In Panarchy: Understanding transformations in human and natural systems, Gunderson, L. H., Holling, C. S., Eds.; Island Press, 2002, pp. 25–62.

IITA (International Institute of Tropical Agriculture). (2007). *Delivering the African agricultural development potential: The IITA vision (strategic plan 2001–2010)*. IITA.

Ikerd, J. E. (1993). The need for a systems approach to sustainable agriculture. *Agriculture, Ecosystems and Environment, 46*(1–4), 147–160.

IPCC (Intergovermental Panek on Climate Change). (2007). *Climate change: The physical science basis, contribution of Working Group I to the Fourth Assessment Report of the Intergovernmental Panel on Climate Change*. Cambridge University Press.

Jackson, J. B. C. (2001). What was natural in the coastal oceans? *Proceedings of the National Academy of Sciences, 9*, 5411–5418.

Jackson, W., & Jackson, L. L. (1999). Developing high Seedyielding perennial polycultures as a mimic of mid-grass prairie. In E. C. LeFroy, R. J. Hobbs, M. H. O'Connor, & J. S. Pate (Eds.), *Agriculture as a mimic of natural ecosystems* (pp. 1–37). Kluwer Academic Publishers.

Jackson, D. L., & Jackson, L. L. (2002). *The farm Ashabitat: Reconnecting food systems with ecosystems*. Island Press.

Jackson, L., Bawa, K., Pascual, U., & Perrings, C. C. (2005). *Agro biodiversity: A new science agenda for biodiversity insupport of sustainable agroecosystems*. DIVERSITAS report no. 4.

Janssen, M. A., Schoon, M. L., Ke, W., & Börner, K. (2006). Schlorly networks on resilience, vulnerability and adaptation within the human dimensions of global environmental change. *Global Environmental Change, 16*(3), 240–252.

Kleijn, D. J., & Sutherland, W. J. (2003). How effective are European Agri-environment schemes in conserving and promoting biodiversity? *Journal of Applied Ecology, 40*(6), 947–969.

Kleijn, D., Baquero, R. A., Clough, Y., Díaz, M., De Esteban, J., Fernández, F., Gabriel, D., Herzog, F., Holzschuh, A., Jöhl, R., Knop, E., Kruess, A., Marshall, E. J., Steffan-Dewenter, I., Tscharntke, T., Peach, W. J., Lovett, L. J., Wotton, S. R., & Jeffs, C. (2001). Countryside stewardship delivers Cirl buntings (Emberizacirlus) in Devon, UK. *Biological Conservation, 101*(3), 361–373.

Kloppenburg, J., Lezberg, S., De Master, K., Stevenson, G., & Hendrickson, J. (2000). Tasting food, tasting sustainability: Defining the attributes of an alternative food system with competent, ordinary people. *Human Organization, 59*(2), 177–186.

Kogan, M. (1998). Integrated pest management: Historicalperspectives and contemporary developments. *Annual Review of Entomology, 43*, 243–270.

Kremen, C., Williams, N. M., & Thorp, R. W. (2002). Crop pollination from native bees at risk from agricultural intensification. *Proceedings of the National Academy of Sciences of the United States of America, 99*(26), 16812–16816.

Lambin, E. F. (2007). *The middle path: Avoiding environmental catastrophe*. University of Chicago Press.

Leakey, R. (1999). Agroforestry for biodiversity in farming systems. In W. W. Collins & C. O. Qualset (Eds.), *Biodiversity in agroecosystems* (pp. 127–145). CRC Press.

Leakey, R. R. B., & Tchoundjeu, Z. (2001). Diversification of tree crops: Domestication of companion crops for poverty reduction and environmental services. *Experimental Agriculture, 37*(3), 279–296.

Lefroy, R., et al. (1999). *Agriculture as a mimic of natural ecosystems* (Current Plant Science in Biotechnology and Agriculture) (Vol. 37). Kluwer Academic.

Leopold, A. (1949). *A Sand County almanac*. Oxford University Press.

Levin, S. A. (1992). The problem of pattern and scale in ecology. *Ecology, 73*, 1943–1967.

Levin, S. A. (1998). Ecosystems and the biosphere as complex adaptive systems. *Ecosystems, 1*, 431–436.

Levin, S. A. (2005). Self-organization and the emergence of complexity in ecological systems. *Biocontrol Science, 55*(12), 1075–1079.

Lewis, R. R., & Streever, B. (2000). *Restoration of mangrove habitat*. WRP Technical Notes Collection (ERDC TN-WRP-VN-RSW-3.2).

Lobell, D. B.'., & Ortiz-Monasterio, J. I. (2008). Satellite monitoring of yield responses to irrigation practices across thousands of fields. *Agronomy Journal, 100*(4), 1005–1012.

Macintosh, D. J., & Ashton, E. C. (2002). *A review of mangrove biodiversity conservation and management final report 10/06/2002*. Centre for Tropical Ecosystems Research.

Matson, P. A., Parton, W. J., Power, A. G., & Swift, M. J. (1997). Agricultural intensification and ecosystem properties. *Science, 277*(5325), 504–509.

Marshall, N. (1994). Mangrove conservation in relation to overall environmental considerations. *Hydrobiologia, 285*(1–3), 303–309

McBratney, A., Field, D. J., & Koch, A. (2014). The dimensions of soil security. *Geoderma, 213*, 203–213.

McCarthy, J. J., Martello, M. L., Corell, R., Selin, N. E., Fox, S., et al. (2005). Climate change in the context of multiple stressors and resilience. In ACIA (Ed.), *Arctic climate impact assessment* (pp. 945–988). Cambridge University Press.

McNeely, J. A., & Scherr, S. J. (2003). *Ecoagriculture: Strategiesfor feeding the world and conserving wild biodiversity*. Island Press.

MEA (Millineum ecosystem assessment). (2005). *Ecosystem and human being: Current status and trends*. Island Press.

MEA (Millineum Ecosystem Assessment). (2005b). *Ecosystem and human being: Scenarios*. Island Press.

MEA (Millineum Ecosystem Assessment). (2005c). *Ecosystem and human being: Synthesis*. Island Press.

Millard, E. (2007). Restructuring the supply chain. In S. J. Scherr & J. A. McNeely (Eds.), *Farming with nature: The science and practice of eco-agriculture*. Island Press.

Molden, D., Tharme, R., Abdullaev, I., & Puskar, R. (2007). Irrigation. In S. J. Scherr & J. A. McNeely (Eds.), *Farming with nature: The science and practice of ecoagriculture* (p. 75). Island Press.

Molnar, A. (2003). *Forest certification and communities: Looking forward to the next decade*. Forest Trends.

Molnar, A., Scherr, S. J., & Khare, A. (2004). *Who conserves the world's forests? A new assessment of conservation andinvestment trends*. Forest Trends and Ecoagriculture Partners.

Mooney, H., & Ehrlich, P. R. (1997). Ecosystem services: A fragmented history. In G. C. Daily (Ed.), *Nature's services: Societal dependence on natural ecosystems* (pp. 11–19). Island Press.

Naylor, R. L., Goldburg, R. J., Primavera, J. H., Kautsky, N., Beveridge, M. C. M., Clay, J., Folke, C., Lubchenco, J., Mooney, H., & Troell, M. (2000). Effect of aquaculture on world fish supplies. *Nature, 405*(6790), 1017–1024.

Nierenberg, D. (2005). *Happier meals: Rethinking the global Meatindustry*. Worldwatch Institute.

Odum, E. P. (1989). *Ecology and our endangered life-support systems*. Sinauer Associates.

Ong, J. E. (1995). The ecology of mangrove conservation and management. *Hydrobiologia, 295*, 343–351.

Patz, J. A., Prentice, I. C., Ramankutty, N., & Snyder, P. K. (2005). Global consequences of land use. *Science, 309*(5734), 570–574.

Paul, A. K. (2002). *Coastal geomorphology and environment* (pp. 1–575). Ajoy Bhattacharya of acb publications.

Peach, W. J., Lovett, L. J., Wotton, S. R., & Jeffs, C. (2001). Countryside stewardship delivers Cirl buntings (*Emberiza cirlus*) in Devon, UK. *Biological Conservation, 101*(3), 361–373.

Pearce, D. (2005). *Managing environmental wealth for poverty reduction*. Poverty and Environment Partnership MDG7Initiative-Economics.

Pethic, J. (1984). *An introduction to coastal geomoephology* (pp. 1–260). Arnold Heiremarn.

Pethic, J. (1992). *Natural change in coastal zone and management* (pp. 49–63). Thomas Telford.

Pethick, J., & Orford, J. D. (2013). Rapid rise in effective sea-level in southwest Bangladesh: Its causes and contemporary rates. *Global and Planetary Change, 111*, 237–245.

Plathong, J. (1998). *Status of mangrove forests in southern Thailand* (p. 128). Wetlands International-Thailand programme/PSU, Publication No5.

Power, A. G. (1999). Linking ecological sustainability and world food need. *Environment, Development and Sustainability, 1*(3/4), 185–196.

Pretty Jules, N. (1995). *Regenerating agriculture: Policies and practices for sustainability and self-reliance*. Joseph Henry Press.

Qasim, S. Z., Sengupta, R., & Kureishy, T. W. (1988). Pollution of the seas around India. *Proceedings of Indian Academy of Science (Animal Science), 97*(2), 117–131.

Rainforest Alliance. (2006). *The Certified Sustainable Products Alliance*, 79.

Redford, K. H., & Richter, B. D. (1999). Conservation Ofbiodiversity in a world of use. *Conservation Biology, 13*(6), 1246–1256.

Redman, C. L. (1999). *Human impacts on ancient environments*. University of Arizona Press.

Rényi, A. (1961). On measures of entropy and information. *Entropy, 547*, 547–561.

Rhodes, G., Henrys, P., Thomson, B. C., & Pickup, R. W. (2013). Mycobacterium avium subspecies paratuberculosis is widely distributed in British soils and waters: Implications for animal and Human health. *Environmental Microbiology, 15*(10), 2761–2774.

Ricketts, T. H. (2004). Tropical Forest fragments enhance pollinator activity in nearby coffee crops. *Conservation Biology, 18*(5), 1262–1271.

Robertson, G. P., & Swinton, S. M. (2005). Recounciling agricultural productivity and environmental integrity: A grand challenge for agriculture. *Frontiers in Ecology and the Environment, 3*(1), 38–46.

Rockström, J., Gordon, L., Folke, C., Falkenmark, M., & Engwall, M. (1999). Linkages among water vapor flows, food production, and terrestrial ecosystem services. *Conservation Ecology, 3*(2), 5.

Ruark, G., Scott, J., Riemenschneider, D. & Volk, T. (2006). Perennial crops for bio-fuels and conservation. In: *Presented at 2006 United States of America agricultural outlook forum—Prospering in rural America*.

Sayer, J. A., & Maginnis, S. (2005). *Forest in Landscapes:Ecosystem Approaches to Sustainability*. Earthscan Publications.

Schaller, N. (1993). Sustainable agriculture and the environment. *Agriculture, Ecosystems and Environment, 46*, 89–97.

Scherr, S. J., Milder, J. C., & Inbar, M. (2007). Paying farmers for Stewardship. In S. J. Scherr & J. A. McNeely (Eds.), *Farming with nature: The science and practice of ecoagriculture*. Island Press.

Scheffer, M., Carpenter, S. R., Folke, C., & Walker, B. (2001). Catastrophic shifts in ecosystem. *Nature, 413*, 591–596.
Schroth, G., da Fonseca, G., Harvey, C., Gascon, C., Vasconcelos, H., & Izac, M. (2004). *Agroforestry and biodiversity conservation in tropical landscapes*. Island Press.
Siebert, S. F. (2002). From shade- to sun-grown perennial Cropsin Sulawesi, Indonesia: Implication for biodiversity conservation and soil fertility. *Biodiversity and Conservation, 11*(11), 1889–1902.
Smil, V. (2000). *Feeding the world: A challenge for the 21st century*. MIT Press.
Solow, R. M. (1991). Sustainability: An economist's perspective. In *Eighteenth J. Lecture, S. J.* Marine Policy Center.
Spalding, M. D., Blasco, F., & Field, C. D. (1997). *World mangrove atlas* (p. 178). International Society for Mangrove Ecosystems.
STCP (Sustainable Tree Crops Programme). (2005). *Sustainable tree crops programme: A public-private partnership*.
Suding, K. N., Lavorel, S., Chapin, F. S., III, Cornelissen, J. H. C., Díaz, S., Garnier, E., Goldberg, D., Hooper, D. U., Jackson, S. T., & Navas, M. (2008). Scaling environmental change through the community level: A trait based response and effect framework for plants. *Global Change Biology, 14*(5), 1125–1140.
Swaminathan, M. S. (1994). *Ecotechnology and rural employment: A dialogue*. Macmillan India.
Swift, M. J., Izac, A.-M. N., & van Noordwijk, M. (2004). Biodiversity and ecosystem services in agricultural landscapes—Are we Asking the right questions? *Agriculture, Ecosystems & Environment, 104*(1), 113–134.
Thomas, B. F., Caineta, J., & Nanteza, J. (2017). Global awareness of groundwater sustainability on storage anomalies. *Geophysical Research Letters, 44*(22), 11445–11455.
Thompson, J., Hodgkin, T., Atta-Krah, K., Jarvis, D., Hoogendoorn, C., & Padulosi, S. (2007). Biodiversity in agroecosystems. In S. J. Scherr & J. A. McNeely (Eds.), *Farming with nature: The science and practice of ecoagriculture*. Island Press.
Tilman, D., Cassman, K. G., Matson, P. A., Naylor, R., & Polasky, S. (2002). Agricultural sustainability and intensive production practices. *Nature, 418*(6898), 671–677.
Tomich, T. P., van Noordwijk, M. V., & Thomas, D. E. (2004). Onbridging gaps. *Agriculture, Ecosystems and Environment, 104*(1), 1–3.
Trewavas, A. J. (2001). The population/biodiversity paradox: Agriculture efficiency to save wilderness. *Plant Physiology, 125*(1), 174–179.
Turner, B. L., II, Clark, W. C., Kates, R. W., Richards, J. S., Mathews, J. T., et al. (Eds.). (1990). *The earth as transformed by human action: Global and years* (p. 1). Cambridge University Press.
Turner, B. L., II, Kasperson, R. E., Matson, P. A., McCarthy, J. J., Corell, R. W., Christensen, L., Eckley, N., Kasperson, J. X., Luers, A., Martello, M. L., Polsky, C., Pulsipher, A., & Schiller, A. (2003). A framework for vulnerability analysis in sustainability science. *Proceedings of the National Academy of Sciences of the United States of America, 100*(14), 8074–8079.
Turner II, B. L.; Clark, W. C.; Kates, R. W.; Richards, J. S.; Mathews, J. T., et al., Eds. The earth as transformed by human action: Global and years; Cambridge University Press, 1990.
Upadhyay, V. P., Ranjan, R., & Singh, S. (2002). Human mangrove conflicts: The way out. *CurrSei, 83*(11), 1328–1334.
Uphoff, N., et al. (Eds.). (2006). *Biological approaches to sustainable soil systems*. CRC Press.
Verhulst, J., West, T. M., & Yela, J. L. (2006). Mixed biodiversity benefits of agri-environment schemes in five European countries. *Ecology Letters, 9*(3), 243–254.
Vitousek, P. M., Mooney, H. A., Lubchenco, J., & Melillo, J. M. (1997). Human domination of earth's ecosystems. *Science, 277*(5325), 494–499.
Walker, B. H., Holling, C. S., Carpenter, S. R., & Kinzig, A. P. (2004). Resilience, adaptability and transformability in social-ecological Systems. *Ecology and Society, 9*(2), 5.
Walker, B. H., Gunderson, L. H., Kinzig, A. P., Folke, C., Carpenter, S. R., & Schultz, L. A. (2006). Handful of heuristics and some propositions for understanding resilience in socioecological systems. *Ecology and Society, 11*(1), 13.

Water Stewardship in Agriculture, WWF Brief. (2015). *Together in partnership with IUCN*.

Weibull, A. C., Ostman, O., & Granqvist, A. (2003). Species richness in agroecosystems: The effect of landscape, habitat and farm management. *Biodiversity and Conservation, 12*(7), 1335–1355.

Whitten, S., Salzman, James, Shelton, D., & Proctor, W. (2003). *Markets for ecosystem services: Applying the concepts*. Commonwealth Scientific and Industrial Research Organization.

Wittenberg, R., & Cock, M. J. W. (2001). *Invasive alien species: A toolkit of best prevention and management practices*. CAB International.

Wojtkowski, P. (2004). *A landscape agroecology*. Food Products Press.

Young, O. R., Berkhout, F., Gallopin, G. C., Janssen, M. A., Ostrom, E., & van der Leeuw, S. (2006). The globalization of socioecological systems: An agenda for scientific research. *Global Environmental Change, 16*(3), 304–316.

Chapter 10
Conclusion

Abstract The Sundarbans mangrove ecosystem, an integral part of the dynamic Ganges–Brahmaputra deltaic tidal plain of the coastal environment of West Bengal, earned the prestigious status of being a World's Heritage Site, by the International Union for the Conservation of Nature (IUCN), in 1997. This area includes a tiger reserve, with an area of 2585 km^2 (1970), three wildlife sanctuaries (Lothian Island, with an area of 3800 ha (1948); Sojnekhali Island, with an area of 35,240 ha (1960); and Haliday Island, with an area of 595 ha (1976)) and one national park, with an area of 1330 km^2 (1972). So many ecosystem services are rendered by the mangrove ecosystem of Sundarbans, which also acts as a buffer against natural disasters; an arrester of soil erosion; a provider of food, fodder and ingredients for medicines; facilitator of biodiversity enrichment; a bioremediator of pollutants; and a potential ecosystem structure carbon sequesters. Mangroves, being a predominant floral assemblage in tropical coastal habitats, are constantly threatened by various anthropogenic pressures, because of which a large extent of mangrove coverage has been considerably deteriorated. The integrity of the mangrove ecosystem is compromised for economic concerns, which need to be counterbalanced by carrying out appropriate ecomanagement strategies. Physicochemical characteristics like salinity, pH, alkalinity, total dissolved solids, turbidity and transparency are considered as key regulators limiting species richness in coastal ecosystems. The construction of embankments and seasonal cyclonic storms on deltaic islands result in ecological imbalance, causing significant changes in salinity, temperature and other ecological variables that deplete mangrove vegetation and its diverse floral and faunal habitats.

In addition, mangroves and their associated flora and fauna are lost because of natural causes like the movement of tectonic plates, leading to result earthquakes, cyclonic storms, etc., and other human-induced environmental changes, like deforestation, changes in land-use patterns, increasing temperatures, an increasing rate of sedimentation on river beds, the formation of mid-channel bars, the construction of concretized embankments, the nonjudicious collection of prawn seeds, changes in the course of water movements, changes in the topography of deltaic islands, disruptions to ecological balance, increasing salinity within estuarine networks and the

S. K. Chakraborty, S. K. Acharya, *Eco-dynamics and Environmental Perturbations in Mangrove Ecosystems*,
https://doi.org/10.1007/978-3-032-08710-2_10

invasion of salinity into inland waterbodies. All such environmental changes have ultimately resulted in the present ecodegradation of the Indian Sundarbans. Many negative ecological consequences from sea-level rise have changed the coastal areas, result harmfull ecological consequences that include increases in erosion, submergence, flooding, salinization, biodiversity loss, the landward migration of mangrove species, etc., which necessitate undertaking proactive adaptive strategies through a series of measures, such as reducing possible impacts through planned retreat and pre-emptive measures aimed at saving sensitive coastal areas and enabling coastal dynamics–related ecological processes to occur naturally providing more space in order to minimize inhibition.

The changing ecodynamics of various geomorphological attributes, in conjunction with several types of anthropogenic interference, have brought about several ecological alterations, like sea-level rise and a high rate of erosion followed by accretion in the deltaic system, which substantially impair the normal ecosystem functioning of the Sundarbans mangrove ecosystem. In view of these facts, comprehensive multidisciplinary ecomanagement strategies are required to protect the Sundarbans mangrove ecosystem, such as generating people's awareness about the importance of mangroves, undertaking appropriate measures for the adaptation of people living in Sundarbans against environmental odds and for the conservation of the biodiversity of this region.

Keywords Sundarbans mangrove ecosystem · Hydrobiology · Geo-morphological attributes · Environmental perturbations · Biodiversity loss · Salinity invasion · Global climate change · Comprehensive eco-management

10.1 The Indian Sundarbans and Its Physiography

The Indian Sundarbans has been labelled as World Heritage Site by the International Union for the Conservation of Nature (1987) and declared a biosphere reserve under the Man and the Biosphere (MAB) programme of the United Nations Educational, Scientific and Cultural Organization (UNESCO, 2001). The Hooghly–Matlah estuarine complex, classified as a positive estuary in the mixohaline range with the mixing of fresh water and saline water from regular tidal influxes, creates a steady gradient from saltwater to freshwater conditions (Khan, 1995). This hydrological gradient is governed mainly by the total amount of freshwater discharge during different seasons of the year, and the seasonal changes in turn cause the Hooghly–Matlah estuarine system to change with the changes in tidal interactions. The Indian Sundarbans, having 104 deltaic islands, both reclaimed (50) and virgin (54), are surrounded by a network of estuaries and creeks. The Indian Sundarbans, as the second-largest area of mangroves in the world, next to that of Bangladesh, represents a fragile ecosystem inhabited by a considerable number of threatened flora and fauna. An area of 4200 km^2 of reserve forest in India is a tide-dominated low-lying coastal wetland crisscrossed by anastomosing channels, meandering creeks, tidal distributaries with larger tidal mouths, bifurcated channels, link channels and smaller tidal passes along the shoreline, which are major drainage features that play determining roles in land-building activities on the tidal flats of the Sundarbans mangrove-estuarine complex. Its vegetation covers the underwater environment. The extent and development of mangrove vegetation provides sediment stability and acts as a cohesive force for sediment accretion under depositional surfaces, varying in size from 1.7 m to 2.6 m above the mean sea level under tidal inundation, over the extensive coastal plain surface of Sundarbans (Paul, 2002).

Three seasons are well pronounced in this region, each with a 4-month duration, namely pre-monsoon (March to June), monsoon (July to October) and post-monsoon (November to February), which are determined by mainly two factors, namely temperature, ranging from a minimum of 10°C to a maximum of 38°C, and rainfall, varying from 1600 mm to 1800 mm. Three variations in terms of changes in tidal regime have been observed in this estuarine system: (i) During monsoon (July to October) with heavy rainfall, the flow tides get nullified by freshets with strong ebb tides; (ii) during post-monsoon (November to February), the second tidal rhythm is due to the banking of fresh water from monsoonal effects when the strength of the flood tide is minimum over the ebbing tide; and (iii) the third tidal rhythm occurs during pre-monsoon (March to June), when the effects of flood tide are considerably accentuated over the ebbing tide, bringing about a marked change in the salinity gradient of the estuary.

Many severe cyclonic storms occur mostly from the late-pre-monsoon to early-monsoon periods. Out of the 104 deltaic islands under the influence of tidal river water, 54 remain virgin with mangroves biodiversity components, whereas 50 have been reclaimed by human settlements, converting the saline forest belts into agricultural land. The only tiger tidal mangrove wetland on Earth was declared a World Heritage Site by the IUCN in 1997, considering this landscape as a highly valued ecosystem in terms of economy, ecology and aesthetics. It includes three wildlife sanctuaries, one tiger reserve (2585 km^2, 1970) and one national park (1330 km^2; 1972). However, the ecological settings of the Indian Sundarbans have been severely changing over the past century because of natural and anthropological activities like land-use changes, the reclamation of virgin islands for human settlement, agriculture, embankment construction, drainage control, sedimentation, sediment starvation, protective measures against tidal flooding, neo-tectonic movement, ecotourism, aquaculture and fishery development and international treaties on sharing fresh water. The physicochemical properties of water and soil in the mangrove ecosystem have been undergoing many changes because of salinity invasion, massive shoreline erosion coupled with unwanted accretion, the shifting of mudflats and sand dunes, increasing turbidity, increasing temperature, increasing tidal amplitude, decreasing transparency, reduced nutrients, lower pH levels, etc. The synergistic impacts of such changes have resulted in the modification of the species composition of biotic communities, leading to a decline in and a loss of mangrove-dependent detritivores, herbivores and carnivores.

10.2 The Network of Deltaic Islands

In total, 104 deltaic islands are in the Indian Sundarbans, 54 of which are virgin with luxuriant mangrove forest areas and the remaining 50 of which are reclaimed with anthropogenic activities like agriculture, aquaculture, poultry farms and brick production. The deltaic islands have been subjected to modifications to their topography by tidal river water courses in the estuarine network system of Sundarbans, India. All these islands of Sundarbans are characterized by mangrove forests but with different intensities of growth and biomass, from scrubby wood jungles to luxuriant mangrove assemblages. The main estuarine islands at the mouths of the rivers from west to east are as follows: (i) the Nayachar, Jambu, Lohachura and Ghoramara islands; the largest estuarine delta in the word on Sagar Island, between Muriganga or Bartala and the Hooghly estuary; (ii) Mohisani Island, between the Chenair (Pitt's creek) and Muriganga estuaries; (iii) Bakkhali, Fraser Ganj and Namkhana between the Saptamukhi and Chenair (Pitt's creek) estuaries; (iv) Frederick Island and Henry Island between the Saptamukhi and Bakkhali estuaries; (v) Dalhousie Island between the Guasuba and Matlah estuaries; (vi) Bangaduni Island in the Bangaduni River; (vii) the Prentice and Lothian islands in the Saptamukhi; (viii) Dhanchi Island between the Thakuran estuary and Jagadal River; (ix) Bulcherry Island between the Thakuran and Matlah estuaries; and (x) Haliday Island in the Matlah estuary. Three major mechanisms are involved in the formation of the deltaic islands of the Indian Sundarbans: (i) Depositional islands developed from the settling down of sediment particles on the river bed to build up the Channel Islands and mid-channel bars. These flood-tidal bars, formed primarily under the influence of flood currents, tend to move upstream, whereas the river mouth island (bar) formed by the influence of ebb currents tends to move seaward; (ii) the mid-channel bars developed through the disconnection of the bank-ward edge of the bar, which remains connected by active or abandoned spill channels; and (iii) other deltaic islands formed through the shallowing of river beds after cutting off the mainland at the bifurcation of channels. Intermittent sedimentation and the subsidence of the Bengal delta are associated with all these geomorphological and topographical changes.

10.3 The Uniqueness of the Mangrove Ecosystem in Sundarbans, India

Mangroves, representing woody angiospermic halophytes, grow in the intertidal belts of tropical and subtropical coastal environments, especially between 25°N and 25°S, and experience tidal inundation and exposure. A mangrove ecosystem, as an ecotone, is composed of the structural components (flora, fauna, microbes, soil, water, etc.) of both estuarine and halophytic forest ecosystems, and all those components possess their own structural and adaptive abilities, housing viviparous embryos and tolerating inundation and exposure, saline water and anoxic soils in waterlogged states (Alongi, 2009). Fluctuating ecological variables like salinity, pH, turbidity, soil texture, nutrient availability, etc. make the habitats of flora and fauna in mangrove communities ecologically very unstable, but all those biotic components take on the challenge of tolerating the oscillating seasonal salinities and water scarcity in oxygen-poor and physiologically stressed soil by developing several adaptive features, like trap roots (stilt roots), pneumatophores, sunken stomata, thick lignin-coated leaves, viviparous germination and physiological mechanisms for excluding salts.

The circumtropical global distribution of mangroves shows the highest concentration in the Indo-Pacific region. These mangroves occupy around one-fourth of the tropical coastlines in the world, amounting to about 75% of the tropical coasts (Chapman, 1976; Farnsworth & Ellison, 1997). The total mangrove area coverage, reaching about 1,000,000 km^2, is distributed among 30 countries, including various island nations (Annon, 2003). A considerable number of ecosystem services have been rendered by mangrove ecosystems across the world—which include food, fodder, medicine, shelter materials, resistance against natural disasters, protection for coastal tracts by arresting soil erosion, trapping sediments in clean water, acting as a natural supplier of nutrients, carbon sequestration, nursery grounds for finfish and shellfish, canopies for animal species (insects, reptiles, birds and mammals) from which they can obtain their life-support materials. The mangrove forest accommodates the habitats and niches of many benthic fauna species (actiniarians, crabs, polychaetes, gastropods, bivalves, hemichordates, brachiopods, horseshoe crabs, globoid fish, etc.) and a hoard of meiobenthic fauna species (nematodes, copepods, foraminifera, kinorhynch, Gastrotricha, etc.). The mangroves of Sundarbans are furnished with the sheltered areas of tidal creeks, channels, shallow bays, strait fringes and the embayed shores of deltaic islands. The open-shore platforms, experiencing direct thrusts of tidal waves, are not preferred by mangroves for growth or for dense colonization because of the non-availability of favourable environmental gradients. The mass mortality of mangroves can be seen in different parts of Sundarbans mostly because sand has been shifting over the mudflats at high wave energy levels and because of salinity intrusion from natural disasters like storms, cyclones, earthquakes and tsunamis. The abrupt cementation of carbonate sediments affects the mangrove regeneration process on the shore platforms in many places. Different forms of anthropogenic activities, especially changes in land-use patterns for shrimp aquaculture and mariculture, for the collection of fuel wood and shelter materials and for urban or industrial development, have caused the accumulation of pollutants like oil, microplastics, pesticides and heavy metals (Chakraborty, 2003, 2009, 2017; Chakraborty & Choudhury, 1985, 1992; Chakraborty & Chowdhury, 1993, 1994; Chakraborty et al., 1986, 2004, 2009, 2012, 2013, and 2014; Chandra et al., 2003).

10.4 A Holistic Approach to the Functioning of Mangrove Ecosystems

The mangrove ecosystem is to be considered a holistic and unique ecosystem because of the presence of one important physical process, the tides which during high tide and low tide inundate and expose mangrove forests, facilitating the exchange of materials in the form of water, nutrients and energy between the two subsystems constituting the mangrove ecosystem, namely the forest and the estuarine (brackish-water) subsystem. Innumerable structural components like soil, water, flora, fauna, microbes and detritus in a mangrove ecosystem not only interact with one another but remain interlinked in a way that renders two important functions: a very high level biological production, initiated by three autotrophs (mangrove plants, benthic algae and phytoplankton), and a transfer of energy through intricate food chain–food web linkages at the different ecozones of the entire mangrove ecosystem and at different timescales. The holistic approach to understanding the mangrove ecosystem requires a proper understanding of ecological hydrology; an ecological interpretation of the associated riverine networks and avenues for the freshwater drainage system of the river basin; and the mode of interconnectivities (such as among brackish-water canals, creeks and estuaries and other coastal waterbodies, like backwater zones and lagoons) through which the exchange and transformation processes of water-loaded soil particles and dissolved materials and biotic interactive processes take place (Odum & Heald, 1975b; Blasco, 1977; Blasco & Aizpuru, 2002; Wolanski, 2006a). Modern ecological approaches emphasizing both qualitative and quantitative aspects in analysing the ecosystem functioning process involve two simple components: documentation through a series of collections and afterwards the classification of species on the basis of quantitative criteria. In contrast, the primary objective of a holistic approach is to maintain the community structure, emphasizing more the important values of different indices (Ellison & Farnsworth, 1997; Chakraborty, 2011, 2017).

The growth and zonation patterns of mangrove species depend strongly on the ecological hydrology of coastal estuarine environments, such as the tidal interplays between the duration and depth of inundation, the period of exposure, slopping gradients with the elevation of intertidal zones, the textural composition of sediment, salinity, the water saturation of soils, etc. (Bunt et al., 1995), which in turn help to explain the simplified classification model for mangroves (Watson, 1928). According to such an understanding, four major factors play important roles in determining the functioning of mangrove ecosystems individually and in conjunction with others: (i) biodiversity components like mangrove trees, pelagic components (phytoplankton, zooplankton and nektons) and benthic faunal groups (actiniarians, crabs, mollusks, polychaetes; (ii) the profiles of sediment topography; (iii) ecological processes within aquatic bodies (the intensity of water flow and the amplitudes of tides and waves) and water-quality parameters; (iii) atmospheric processes (the intensity and direction of wind, humidity, rainfall, etc.); and (iv) nutrient cycling. Every factor interacts with one of other factors, and the results of their interactions ultimately shape the structure and function of mangrove ecosystems (Mazda et al. 2007). The duration and intensity of impacts from the inundation of mangrove swamps depends on not only the volume of the water but also the density of the vegetational assemblage, which hinders water inundation (Mazda et al., 1999). Similarly, the impacts of tidal flows also develop resistance thanks to mangrove vegetation and are modified accordingly in offshore areas (Mazda & Kamiyama, 2007). The physicochemical properties also determine the division of mangrove assemblages into several ecotonal zones, characterized by a specific category of mangrove plant species that has adapted to this set of physicochemical variables. (Chaudhuri and Chowdhury, 1994). According to the zonation patterns, after being determined by different ecological, eco-edaphic, geomorphological and hydrological variables, the Sundarbans mangrove forest is divided into four major tidal levels: (i) mangroves on supratidal levels, (ii) mangroves on mid-tidal levels, (iii) mangroves that experience occasional inundation and (iv) frequently inundated mangroves at low tide levels.

10.5 The Sundarbans Mangrove Ecosystem: Its Uniqueness

The Sundarbans of India and its Bangladesh counterpart represent the single largest mangrove forest in the world. The largest area of tidal mangrove forest, enjoying fluctuations in the very high amplitude of tides, is the only habitat of the largest tiger population in the country. In addition, the mangrove vegetation of the Sundarbans mangrove forest bioaccumulates a huge number of pollutants discharged from the nearby city of Kolkata. This forest stabilizes coastal tracts by resisting soil erosion and acts as a buffer against several natural disasters. This mangrove ecosystem supports the lives of one of the most species-diverse places in the world, including many threatened, endemic and critically endangered species, like river terrapin, which was supposed to have gone extinct.

10.6 Why Are Mangrove Ecosystems one of the Most Productive Ecosystems in the World?

Mangrove ecosystems, which are dominated by intertidal salt-tolerant halophytic vegetation and enjoy the influences of two high and two low tides a day, offer a unique ecological setup for biodiversity development, shoreline stabilization, resistance against natural disasters and livelihood generation (Chakraborty, 1996). During high tide, major parts of the forest subsystem get inundated and receive major inputs from estuarine water in the form of the moisture-recharging textural constituents of the bottom sediment and a supply of anionic chemical constituents, which provides nutrients in the form of macroelements, microelements and trace elements to soil and water, expediting the process of biological production. The receding water during low tide moves a huge amount of mangrove litter to an adjoining aquatic subsystem, which is utilized by phytoplankton along with sunlight and plenty of water available in the open water system for the generation of green biomass in the forest. A group of perennial evergreen plants produce a huge amount of leaf litter within the mangrove forest diurnally, monthly and seasonally throughout the year. The mangrove leaves, after falling on the moisture-rich surface of the silty clay bottom soil, are fragmented by the foraging activities of a galaxy of benthic fauna (crabs, gastropods, microarthropods and meiobenthic fauna) into smaller pieces, thereby creating more opportunities for microbial communities (bacteria, fungi and protozoa) to act on the exposed mesophyll tissues of the decomposing plant litter through an aerobic litter decomposition process, releasing a considerable quantity of nutrients. Deposit feeders (crabs, mollusks, polychaetes, nematodes, etc.), through their feeding activities, turn over the subsurface sediment layer, exposing new litter surfaces to microbial actions for the production and reshuffling of nutrients (Chakraborty, 2011). The holistic functioning of the detritus-based mangrove ecosystem, integrating the forest and aquatic subsystems, becomes highly biologically productive, with a productivity level about 20 times higher than the average production of oceanic ecosystems (Goudha & Panigrahy, 1996). The biological productivity of mangrove ecosystems is due to the synergistic impact of four attributes: (i) three biotic types of primary production components (mangrove plant species, benthic algae and phytoplankton); (ii) the ebb and flow of water movements during tidal action, inundating and exposing mangrove forests; (iii) the abundant production and exchange of nutrients; and (iv) the rapid regeneration and bio-transformation of nutrients due to the activity of microorganisms and filter feeders (Schelake & Odum, 1962).

10.7 Physicochemical Parameters and Seasonal Dynamics

Both the water-quality parameters (salinity, dissolved oxygen, pH, turbidity, alkalinity, transparency, dissolved solids, nutrients, etc.) and soil-quality parameters (the texture of soil, pH, salinity, nutrients as organic carbon, nitrate, phosphate, sulfur, potassium, calcium, etc.) in the tropical climate of mangrove ecosystems demonstrate distinct seasonal fluctuations, determined mainly by major meteorological parameters, temperature and rainfall. However, many parts of coastal West Bengal, India, with salt marshes, mangroves, floodplains, estuaries, back water, etc., over the past century have been radically modified by the draining and constructing of many tidal restriction structures, such as water drains, culverts, floodgates and weirs (Williams and Watford, 1997). All these have led to adverse environmental consequences, including radical alterations in salt marshes and mangrove wetlands, which adversely affect normal hydrology, disrupt species abundance, bring about biodiversity loss, reduce biological productivity and constrain litter decomposition within mangrove forests (Hutchings and Saenger, 1987; Hughes et al., 1998). Research findings have pointed out the substantial decline in freshwater supply from adjoining rivers into the estuarine networks of Sundarbans—from 3700 m^3/s in 1974 to 364 m^3/s in 1997—because of the construction of the Farakka Dam in India (Mirza, 1998). Another estimate has revealed an average increase in soil salinity by 60% since 1980 (Aziz & Paul, 2015). Moreover, the most burning problem is associated with the opportunistic harvesting of trees, leading to soil erosion and heavy siltation in the estuarine networks of Sundarbans, in both Bangladesh and India (Rahaman et al., 2015). A considerable reduction in headwater discharge has not only caused salinity invasion but also altered the dynamics of ecological succession and the vegetation and modified both the landscape and estuarine courses, having negative impacts on the ecological, hydrological and geomorphological setups in Sundarbans. Changes in the intensity of water current and the duration of the inundation of forest subsystems impair nutrient cycling, causing a shortage in the supply of surface organic matter and a gradual reduction in soil fertility, as reflected in the stunted growth of mangrove plants. Moreover, mangroves enjoy luxuriant growth in low-saline conditions, and therefore, hypersalinity and high nutrient concentrations encourage the rapid growth of mangrove saplings, especially during monsoon. Certain physicochemical parameters, especially temperature, pH, oxygen and the moisture of soil, promote microbial activities in a mangrove forest, causing the decomposition of plant litter, the production of nutrients and the subsequent export of said nutrients to adjoining estuaries during low tide (Gupta, 1987;Chakraborty, 2011, 2013, 2018). However, excessive precipitation (1500 mL) during monsoon and the subsequent water runoff create waterlogged conditions, preventing the penetration of atmospheric oxygen into soil and thereby lowering the activities of decomposers (microbes and soil-inhabiting detritivores). During pre-monsoon, higher air temperature enhances the diffusion of oxygen in soil, which in turn increases its redox potential—a measure of electron pressure or availability in a solution—and steadily increases the rate of decomposition (Mandal et al., 2009). Oxidation occurs both through the uptake of oxygen and through the removal of hydrogen, as a chemical giving up an electron. In contrast, reduction represents an opposite process where oxygen is removed with the gaining of hydrogen or an electron (Mitsch & Gosselink, 1993). Seasonal fluctuations in physicochemical parameters are heavily influenced by meteorological parameters, mainly temperature and precipitation, which lead to many changes in ecological processes like nutrient cycling and energy flows and in the ecodynamics of biodiversity components.

10.8 The Zonation of Mangroves: Flora and Fauna

The zonation of different flora and fauna, in the horizontal and vertical profiles of the coastal tracts of the Sundarbans mangrove ecosystem, can be cited as a unique structural excellence of mangrove forests (Woodroffe, 1992). Factors that contribute to the development of such zonation patterns include plant succession, responses to tidal levels, physiological adaptations, the size of propagules, geomorphology, varying levels of tolerance against physiological stress, seed predation and interspecific interactions (Schwamborn & Saint-Paul, 1996).

Mangrove species demonstrate distinct zonation patterns in relation to their responses and adaptability to different environmental stimuli and levels of resource availability. The species composition of different biotic assemblages within a mangrove ecosystem is determined by its responses to various environmental stimuli. Therefore, the species assemblage of one ecozone will differ from that of another ecozone, and even the different tidal levels within an intertidal belt (high tide level, mid tide level and low tide level) reflect the level of the tolerability of different species of mangroves to different ecological-hydrological factors like tidal inundation and exposure, salinity and erosion. Their associated faunal components are also influenced by ecological-ethological attributes like predation, competition and herbivory. The zonation and distribution patterns of benthic organisms like algae, mollusks, polychaetes, crabs and bryozoans are determined not only by the changing textural properties of the bottom sediment but also by the variability of the intricate root systems of mangrove plants, which provide a hard substratum for anchoring those organisms. The behavioural manifestations of benthic faunal components, by virtue of their movement, namely feeding and reproductive activities, help to impede water flow alongside enhancing the deposition of sediment in their habitats. In addition, the accumulation of different toxic substances, like microplastics, heavy metals fine and anoxic sediments, within sediment and the overlying seawater of mangrove forests acts as a sink that is scavenged by biotic components. Of the 110 mangrove species distributed across the globe, only about 54 belonging to 20 genera under 16 families are recognized as "true mangroves," species that occur almost exclusively under the influence of marine-estuarine habitats. A mangrove ecosystem, supporting the lives of a galaxy of the pelagic and benthic fauna of ecological and economic importance, plays a crucial role in coastal food webs by exporting a huge amount of carbon fixed in mangroves. Convergent evolution has enabled many species to overcome the problems from tidal interplays, anaerobic soils, variable salinity and the availability of intense sunlight, all characteristics of tropical belts (Chakraborty & Choudhury, 1989; Hazra & Choudhuri, 1990; Hazra & Sanyal, 1996).

According to the zonation patterns of mangroves, the Sundarbans Biosphere Reserve, covering an area of 9630 km^2, has been categorized into three ecozones of conservation significance: (i) the core zone, which is in an absolutely undisturbed condition and which has been designated as a national park; (ii) the buffer zone under the Sundarbans Tiger Reserve, which serves as an area for tourists, fishers, wood cutters and honey collectors; and (iii) the transitional zone, which features both reclaimed agricultural lands and a small area of forest.

10.9 The Ecology and Distribution of Mangroves and Associated Organisms

Different kinds of marine, estuarine and even freshwater organisms demonstrate their own survival strategies and select preferred habitats and ecological niches, including shallow water littoral zones and the highest high-tide levels. Many ecological processes, like tides and water currents, act on these habitats and stir both the overlying water and the bottom sediments in the process of the generation, distribution and transformation of nutrients throughout the ecosystem. Additionally, many nutrients are derived from allochthonous sources, from the runoff of adjoining landmasses. All those organisms get life-supporting ingredients not only from the intertidal zone but also from the adjoining subtidal zone because of the close interactions and interconnections between these two important habitats of various biodiversity components (Ellison & Farnsworth, 1997; Perrings et al., 2011).

Distinct bio-geographic zones with characteristic ecological setups develop in tropical belts, supporting a number of trophic levels on the ecological pyramid of the mangrove ecosystem of Sundarbans because they use a sufficient amount of sunlight and plenty of available water loaded with sufficient nutrients, which tends to support a higher diversity of species in high abundance but with a proportionately low biomass. The sizes of faunal components inhabiting such ecozones, including zooplankton, benthos and nekton, are small, but food chains are long, accommodating six trophic levels. This is in contrast to temperate regions, which have less available sunlight and a low temperature; they harbour large zooplankton with less diversity but with shorter-length food chains, which can accommodate only three or four trophic levels.

10.10 The Ecological Characteristics of Intertidal and Subtidal Zones on Different Deltaic Islands

In the Sundarbans mangrove ecosystem, the main estuarine islands, formed from west to east at the mouths of the rivers, are as follows: (i) the Nayachar, Lohachara, Ghoramara, Jambu and Sagar islands between the Muriganga or Bartala and Hugli estuaries; (ii) Susnir Char and the Namkhana, Fraser Ganj and Bakkhali islands between Saptamukhi and Chenair (Pitt's creek); (iii) Mohisani Island between the Muriganga estuary and Chenair (Pitt's Creek); (iv) Henry Island and Frederick Island between the Saptamukhi and Bakkhali estuaries; (v) the Lothian and Prentice islands encircled by the Saptamukhi estuary; (vi) Dhanchi Island between the Thakuran and Jagadal estuaries; (vii) Kalas-Bulcherry Island between the Matlah and Thakuran estuaries; (viii) Haliday Island on the Matlah estuary; (ix) Dalhousie Island between the Matlah and Guasuba estuaries; and (x) Bangaduni Island on the Bangaduni estuary. Deltaic island formation takes place through three major ecological processes within the Sundarbans mangrove-estuarine complex: (i) Depositional islands developed by the settling down of particles under the influence of flood currents in conjunction with tidal actions, forming the Channel Islands and the mid-channel bars; (ii) a mid-channel bar evolves through the disconnection of the bank-ward edge of the bar, which is occupied by either active or abandoned spill channels; and (iii) deltaic islands are formed by cutting off the mainland at the time of the bifurcation of estuarine channels, resulting from the shallowing of water depth, intermittent subsidence and sedimentation.

The intertidal zone, as a part of the littoral zone, receives sunlight down to its bottom, and this shallower zone gets inundated and exposed by the rising and falling tides. This zone is the starting point of the continental shelf of the marine ecosystem where land meets with the sea. The fluctuating tidal cycles over this zone cause the development of very harsh ecological conditions, which experience changing ecological parameters in the overlying water, variable currents and waves of water, seasonal variation in the influx and outflux of water from the rivers and seas in line with changing meteorological conditions, like rainfall and temperature. All these have made the lives of intertidal biota quite challenging. To overcome these adverse ecological conditions, all intertidal organisms adopt mechanisms to withstand desiccation followed by inundation pressures, including different feeding mechanisms (filter feeding, suspension feeding, deposit feeding, etc.), to avoid competition among themselves, to tackle the problems of having an uncertain supply of food and to develop morphological and eco-physiological adaptabilities to resist fierce water currents. The upper parts of intertidal zones intermittently experience submergence and are quite frequently drenched by spray from crashing waves. The middle intertidal zone always gets submerged during the tide and offers more-stable ecological conditions for the biota to settle, grow and flourish. The lower parts of the intertidal belt, being a part of the littoral zone, merge with the subtidal zone but always remain under shallow seawater.

10.11 The Ecological Characteristics of the Aquatic Subsystem of the Mangrove Forest Ecosystem

The mangrove ecosystem remains surrounded by a network of tide-dominated brackish-water waterbodies in the form of numerous estuaries, channels, creeks, rivulets, etc. The pelagic habitats within this ecosystem are represented by water columns supporting the lives of numerous aquatic faunal components, which are broadly divided into two categories: plankton, which lack a locomotory ability in their waterbody habitats, and nekton, which are endowed with strong locomotory abilities that let them change their direction of movement. An estuary is a semi-enclosed coastal body of water that remains connected with the open sea and freshwater rivers; seawater here is measurably diluted by the discharge of fresh water from rivers with a terrestrial origin; and the entire stretch of an estuary always experiences tidal impacts. The water column of a vast sheet of estuarine water is stratified into several zones on the basis of ecological variables like light, transparency, temperature, density and water pressure. Those zones include the epipelagic zone, which receives plenty of sunlight for the process of photosynthesis; the mesopelagic zone, which receives a modest amount of sunlight that is insufficient to sustain photosynthesis; and the bathypelagic zone, at a depth of a few hundred metres, lacks autotrophic food chains. Different strata within the water column support the lives of different groups of pelagic fauna, which display different modes of adaptations for their survivability and distribution. The bottom-doweling benthic fauna are divided into macrobenthic fauna and meiobenthic fauna, which adopt different modes of feeding strategies (like deposit feeding for detritovores and suspension feeding for filter feeders) and thereby find avenues for their coexistence in these nutrient-loaded habitats. The benthic fauna are also differentiated into infauna (crabs, bivalves, polychaetes, actiniarians, etc.), residing within the bottom sediments, and epifauna (hermit crabs, gastropods, globoid fish, etc.), which restrict their movement and presence only to the surface layers of sediments and thereby avoid competition for resources. Also, another group of deposit-feeding benthic fauna, known as scavengers, feed only on dead or semi-decomposed organic material. Suspension feeders flourish in sandy environments with low turbidity in water, where they can harvest and consume tiny organic material from drifting detritus along with planktonic organisms in the water column overlying them. Infaunal filter feeders (sea pens, sea anemones, crustaceans, polychaetes, brachiopods, Ectoprocta, phoronids, etc.) filter the overlying brackish water loaded with organic detritus and microscopic planktonic organisms by using well-coordinated feeding mechanisms with the help of well-designed feeding appendages like tentacles and cilia. Several suspension and deposit feeders also simultaneously act as active predators (echinoderms, mollusks, demersal fish, etc.), which graze on sediments and harvest both meiobenthic fauna and benthic algae.

10.12 The Biodiversity of Mangrove Ecosystems: Global Perspectives and the Interactions of Biodiversity Components with a Changing Ecosystem

Ecozones with mangrove habitats in the tropical belt, extending from 30°S to 30°N, have tremendous ecological potential for generating resources towards achieving sustainability in socioeconomic development. However, this gift of nature across the globe in general and on the Indian subcontinent in particular has been under tremendous natural and anthropogenic stresses, which are reflected in the considerable shrinkage in mangrove coverage (3–6).

The ecological and structural properties of any community can be assessed by making spatial predictions about biodiversity and by monitoring the temporal dynamics and spatial distributions of biodiversity, which are required for immediate and long-term protection and conservation efforts (Socolar, et.al, 2015). Understanding and assessing the three aspects of biodiversity, namely alpha biodiversity, beta biodiversity and gamma biodiversity, reveals the fundamental attributes of natural communities that are used as prerequisites for undertaking specific conservation strategies (Whittaker, 1960). In mangrove-estuarine complexes, both alpha biodiversity and gamma biodiversity have been observed to increase along estuarine flows to the interior of the forest and along water flow gradients from downstream to upstream. In addition, decreasing salinity and downstream-to-upstream gradients promotes heterogeneity in species composition (Ellison, 2001; Osland et al., 2017; Record et al., 2013). Spatially mapping alpha biodiversity can indicate the most species-rich habitats, whereas mapping for beta diversity can focus on the occurrence of heterogeneous biotic communities with more biodiversity components. Similarly, from an assessment of gamma biodiversity, areas with the highest biodiversity components can be identified. Although, in a regional context, mangrove plant communities within mangrove ecosystem exhibit the least diversity among species in a spatially homogeneous ecological setup in comparison with upland tropical forests, more heterogeneity in vegetation structure is noticed at finer scales, promoting more diversity among the biota (Farnsworth & Illius, 1998; Hutchinson 1958). In such a context, ascertaining the response of biodiversity components to ecological parameters such as biotic and abiotic ones appears to be crucial for constructing biodiversity maps with more information on not only the presence or absence of species but also their spatiotemporal variations in relation to their habitats, and all that baseline information helps when engaging in spatial conservation planning (Devictor et al., 2010).

Mangrove ecosystems must have a recognizable boundary after being structured by the interacting processes of an array of living organisms, like the microbes, plants and animals inhabiting the mangrove forest. The estuarine-mangrove complex, composed of a vast sheet of water with distinct ecological features and a large expanse of green perennial halophytic forest, accommodates an array of living organisms, the continuous interactions among them sustaining the functioning of the mangrove ecosystem, which include subsystems with smaller dimensions, such as the stretch of shore on intertidal belts, mangrove forests and dunes.

10.12.1 Competition in the Biotic Community's Interaction Process

Faunal components, in the process of selecting their own ecological niche within a habitat, exhibit different kinds of behavioural activities for the overall functioning of feeding, reproduction, etc. A particular ecological niche is occupied by individuals of the same species within an ecosystem. Two species engage in intense competition for the similar ecological niche within the same community when both species require similar resources. In the process of competition, one species survives after causing the mortality of less-successful species, and this elimination of one species by another within the same ecological niche is called the competitive exclusion principle. However, in some cases, two or more species share the same resources by developing specializations within their morphoanatomical structures, and this kind of sharing of resources among two competitive species is called resource partitioning, which ensures their coexistence.

10.12.2 Feeding Relationships for the Functioning of the Mangrove Ecosystem

Mangrove ecosystems use solar energy, the driving force behind ecosystem functioning, thanks to autotrophs (phytoplankton, benthic algae and mangrove plants), which through the process of photosynthesis act as primary producers of the organic matter on which other organisms, mostly heterotrophs in the same biotic community (herbivores, omnivores and carnivores), depend for their food supplies. Unidirectional energy flows in the form of organic matter within food are exchanged between ecosystems and their surroundings by food chains and food webs. Representative organisms in each tier of a food chain represent a trophic (feeding) level, like primary producers (phytoplankton, algae and mangrove plants), herbivores (zooplankton and some benthos) as secondary producers, omnivores and carnivores as tertiary producers, etc. Faunal components within estuarine-mangrove complexes are food for carnivores, which feed on green autotrophs (primary producers) and other heterotrophic faunal components (herbivores and omnivores) to derive nutrients. The microbial organisms (bacteria, fungi, and animals such as protists) within the mangrove ecosystem, acting as decomposers, facilitate the cyclical movement of matter by breaking down several kinds of organism-derived organic substances and forming detritus and other decomposed products as nutrients. Many faunal components living mostly on detritus are known as detritivores, which develop another pathway for circulating organic matter back into the ecosystem. In an open water system, alongside food chains and food webs, other kinds of exchange pathways, known as the microbial loop for materials and energy, have been developed by the decomposers and microscopic detritivores that feed on them. In the sphere of ecosystem functioning, only a few species play a disproportionate role in maintaining the integrity of an ecosystem because complexity appears to be equal to resilience given that most complex food webs, encountering the loss of only few such species, influence the structure and function of the rest of the community.

Deposit feeders, feeding on the surface and subsurface soils at the bottom of forests and waterbodies, preferentially colonize during the post-monsoon season, coinciding with the increasing availability of a higher amount of detritus in sediments. The population densities of all those benthic faunal components also vary depending on the season (Chakraborty and Chowdhury, 1996; Chakraborty et al., 1998). The abundance and distribution of different faunal components depend not only on physicochemical parameters like the pH of water and soil, dissolved oxygen (DO) content, salinity, temperature, turbidity, textural components and the level of nutrients in soil but on the quantity of detritus within habitats. All living organisms, organizing themselves as a vital biotic complex, flourish when they can derive all their requirements from their ecosystem. Faunal diversity is higher in ecologically heterogeneous habitats than in ecological homogeneous habitats.

Biodiversity is determined by the species diversity and genetic diversity within a given area of mangrove ecosystems. An analysis of the biodiversity within mangrove ecosystems revealed much greater species richness among fungi, bacteria, protists, viruses and other invertebrate phyla in comparison to that of macrofaunal diversity as species diversity declines, where faunal groups have been reduced to fewer than 50 species within a given forest area (Alongi & Sasekumar, 1992) and where the highest faunal diversity was found among crustaceans (Kathiresan & Bingham, 2001). Highly diverse species groups are represented by only temporary visitors like insects and birds, with the number of species often exceeding 300 within a single mangrove forest per estuary. Fish represent the most diverse among vertebrate phyla, with 100 to 250 species per estuary (Robertson & Blaber, 1992). In total, 260 fish species were recorded in a Southeast Asian mangrove estuary, reflecting the potential of that ecosystem to harbour such a good number of fish fauna in an ecologically healthy ecosystem constituted by the interactions of a variable number of environmental parameters (Hong & San, 1993).

This observation is substantiated by a comparative account of the macrofaunal diversity of a major species of benthic fauna like brachyuran crabs, which exhibit lower species richness in east Africa (about 35 species) in comparison with that in Southeast Asia (> 100 species) (Gillikin & Schubart, 2004), reflecting differences in the diversity of faunal components between two ecoregions of mangrove-based forests in the world. However, at the local scale, average metazoan diversity was found to be more on the tree epibiont assemblages encrusting on tree trunks or on the surface of the forest floor in comparison with the faunal diversity in the tidal waters of adjoining estuaries (Alongi, 1990). The diversity of fauna and flora inhabiting the mangrove canopy is also important to the structure of species composition and to the functioning of food webs throughout the ecosystem, constituted by forest and aquatic subsystems like any other forest ecosystem in the world (Ellison & Farnsworth, 2001; Kathiresan & Bingham, 2001). According to estimates, insects ordinarily consume 5% of the net primary production of mangrove biomass (Robertson, 1991) but render valuation ecosystem services by affecting the establishment and growth of seedlings (Minchinton and Dalby-Ball, 2001; Burrows, 2003; Sousa et al., 2003) as pollinators (Ellison & Farnsworth, 2001). Birds and mammals, being temporary or permanent inhabitants of mangrove forests, use the bioresources of mangrove forests as shelter, breeding grounds or food sources. Research has shown the importance of mangroves as a home for many species of birds; 315 species of birds prefer to reside within mangrove forests enjoying the plentiful supply of food in the form of insects and other invertebrates living in the habitats of the forest floor, on tree branches and in tidal water (Lefebvre et al., 1992, 1994; Lefebvre & Poulin, 1996, 1997).

10.13 Biodiversity in Two Coastal Districts of West Bengal: Sundarbans in South-24-Parganas and the Coastal Belts of Midnapore (East)

Mangroves develop, flourish and are distributed in their relatively sheltered coastal tidal flats of the Ganges–Brahmaputra deltas of the east coast of India. However, on the west coast of India, mangroves are concentrated in smaller pockets in areas of estuary fringes, in the barrier areas and in the sheltered tidal flats of gulf areas. Alterations in several ecological processes and factors because of natural and anthropogenic influences have resulted in increased vulnerability to cyclones, sea-level rise, climate variability and a reduction in freshwater discharge by river systems, causing hypersalinity in the mangrove forests on the eastern coast of India. Major formations of mangroves occur in the regions of the Indus riverine plains, Sundarbans, the Mahanadi deltas, Bittorkonika, the Gulf of Kutch, Andaman, the mangrove wetlands in Kerala, the Nicobar group of islands and part of the west coast. The floral diversity in the mangrove ecosystem of the Indian subcontinent is composed of 72 mangrove species and 165 species of angiosperms as associate species. However, a number of species, such as *Heretiera macrophylla*, *Xylocarpus obovata*, *Bruguiera rheedii*, *Sonneratia persica*, *Rhizophora annamalayana*, and *Xylocarpus obovat*, need some taxonomic confirmation in relation to both morphotaxonomy and molecular taxonomic assessment. Out all those species, the maximum number of species occur in the Ganga–Brahmaputra deltaic regions of Sundarbans and the deltaic islands of the Bittarkanika mangrove forest at the Mahanadi delta on the east coast of India. A good number of species with high abundance also occur in Andaman and the Nicobar group of islands. Furthermore, 22 species of mangroves are found in Sundarbans, in both Bangladesh and India, enjoying similar kinds of topographical, geological, geomorphological, hydrological and ecological setups with the dominance of species under the genera Avicennia, Sonneratia, Rhizophora, Acanthus, Sonneratia, Bruguiera, Agecerius, Excoecaria, etc. Restricted distributions of some abundant mangrove species located in a few ecozones, having specific ecological variables, clearly indicate the historical background of the mangrove-based delta-formation process (Aziz & Paul, 2015; Ghosh et al., 2016; Gopal & Chauhan, 2006). Although the Sundarbans mangrove-estuarine complex harbours a large number of species of finfish and shellfish, the random exploitation of them and damage caused to me through the collection of prawn seeds result in a loss of a number of commercially and economically important species. Continuous prawn seed collection damages the colony and biomass of both benthic algae and epiphytic algae grown on pneumatophores, which affects the habitat structure of mangrove plants and that of aquatic organisms. This causes deviations in the natural balance of the coastal ecosystem.

The mangroves ecosystem of Sundarbans, being a very specialized, vulnerable, sensitive and fragile ecosystem, supports the lives of a large number of faunal components, some of which spend their entire lives within the mangrove swamp and adjoining waterbodies for their shelter, feeding and breeding. Any alteration in the natural ecosystem fabric of the mangrove-estuarine environment will result in vulnerable effects on both flora and fauna, which is revealed by the changing trends in the mangrove vegetation complex, which is turning into salt marshes due to the expansion of mangrove swamps. Increases in the amplitude of tidal cycles along with changes in nutrient supply and the sedimentation process to the mangrove ecosystem facilitate the transformation of mangroves into salt marsh areas.

The Indian Sundarbans has immense potential thanks to its faunal diversity, which includes microarthropods, meiobenthic fauna, plankton, nekton, periphyton, insects, fish, birds, reptiles and mammals and is therefore considered as a highly important biodiversity zone. Mangrove-dominated deltaic islands, having unique faunal diversity, have drawn the attention of many researchers, naturalists and scientists since the first publication of Stoliczka (1869), on the brackish-water malacofauna of lower Bengal. Mangrove communities in highly silted areas of the Sundarbans of the Indian subcontinent represent very poor alpha and gamma diversities but exhibit almost similar patterns in species composition. These observations corroborate earlier findings, where the limited growth and regeneration of many mangroves occurred due to the sediment burial of aerial roots in Sundarbans (Mitra and Zaman, 2016). The sediment burial of aerial roots, prohibiting root aeration, acts as a major factor in the mortality of mangroves (De Deurwaerder et al., 2016).

However, the sensitivity of individual species to sediment burial at the species level substantially varies, according to experimental observations that 100% mortality in *Avicennia officinalis*, 70% in *Rhizophora mucronata* and 40% in *Sonneratia caseolaris* were caused under an extreme sediment accretion level down to a depth of 32 cm (Thampanya et al., 2002). The mangrove ecosystem at the Sundarbans on the Indian subcontinent, being an active deltaic coastal environment with a network of estuaries, annually transports around 2.4 billion tons of sediments (Mitra and Zaman, 2016). This observation prompted the undertaking of research into species-specific sensitivities and adaptations against the unique architectural plans of mangrove root systems, oscillating seasonal salinity, sedimentation patterns and the duration of exposure and inundation from tidal cycles.

10.13.1 Biodiversity in the Indian Sundarbans

The Sundarbans mangrove ecosystem supports 34 true mangrove species and around 50 mangrove-associated plant species, 150 species of algae, 163 species of fungi, 32 species of lichens and thousands of true aquatic, semiaquatic and terrestrial species (including tigers, crocodiles, turtles, snakes, birds and insects). Among true mangrove plant species, special mention may be made of *Avicennia marina*, *Rhizophora apiculata*, *Sonneratia apetala*, *Excoecaria agallocha*, *Bruguiera cylindrica* and *Acanthus ilicifolius*. The mangrove-associated plants are represented by species such as *Sarcolobus carinatus*, *Suaeda maritima* and *Pandanus tectorius*. Some examples of the mesophytic bioinvasive plants in the three study sites are *Casuarina equisetifolia* and *Alternanthera* sp. Important phytoplankton species include Nitzschia sp., Peridinium sp. and Ceratium sp. (Chakraborty & Choudhury, 1994; Chakraborty, 2010).

The faunal biodiversity of the Sundarbans mangrove ecosystem includes 215 species of fish, seven species of amphibians, 59 species of reptiles, more than 100 species of birds, 39 species of mammals and numerous species of phytoplankton, zooplankton, ichthyoplankton, benthos and soil-inhabiting and mangrove plant–dependent insects. The biodiversity of Sundarbans includes numerous species of phytoplankton, zooplankton, microorganisms, benthic invertebrates, mollusks, amphibians and mammals (Gopal & Chauhan, 2006). The faunistic composition of zooplankton in this system includes copepods as principal components of the total zooplankton population (67.8% to 90%). In total, 36 copepod species belonging to 19 families and 21 genera have been recorded. Other zooplanktonic forms are cladocera, ostracoda, cumacea, chaetognatha, mysidacea, sergestidae, amphipoda, hydromedusea, etc. (holoplankters) and polychaete larvae, nauplius, zoea, megalopa, fish eggs and larvae, echinoderm larvae, etc. (meroplankters) (Annon, 2003; Chakraborty, 2010).

Nektonic faunae include hundreds of species of fish under 29 families. Important ichthyoplanktons belong mainly to families of fish like Clupeidae, Engraulidae, Megalopidae, etc. A hoard of benthic fauna, both infauna (sessile, semi-sessile and burrowing) and epifauna (crawlers and creepers), are the residents of these habitats.

Benthic faunae are divisible into three broad groups according to their body size: macrobenthos, meiobenthos and microbenthos. Among the intertidal macrobenthic fauna, 26 species of brachyuran crabs belonging to 15 genera and five families have been recorded in the deltaic Sundarbans estuarine complex. In total, 69 species of polychaetes under 45 genera and 25 families have been documented in this ecosystem. Also, 110 species of benthic Mollusca classified under the classes Gastropoda (59 species) and Bivalvia (40 species) and a rich abundance of benthic insects have also been identified. In this region, 44 species of microarthropods belonging to six major taxonomic groups, viz. Acarina, Collembola, Coleoptera, Diptera, Isoptera and Hymenoptera, have been identified. Actiniarian, sipunculans, echiurans, hemichordates, globoid fish, etc. are other important benthic faunal groups. Finally, 80 species of nematodes belonging 26 families have been documented, which are the major meiofaunal groups (Chakraborty et al., 2011).

10.13.2 The Biodiversity of the Midnapore Coastal Tracts, a Contagious Part of the Indian Sundarbans

The Midnapore coast in the Midnapore district (east) is characterized by macrotidal intertidal zones, longshore currents, a good number of large dunes, both sandy and muddy intertidal flats, creeks, etc. The floral species, belonging 32 families of mangroves, 28 species of benthic algae under 4 families and 8 phytoplankton species under 3 families, have been recorded on the Midnapore coastal belt. A good number of faunal components in the form of both pelagic and benthic forms are found in this ecozone. Here, 17 species of zooplankton, mainly comprising rotifer, Cladocera, copepod and chaetognath and a considerable number of nauplius larvae, have been recorded. In total, 48 intertidal mollusk species, belonging to three classes, 15 orders and 36 families, have been documented as inhabiting several ecological habitats. In total, 22 polychaete species belonging to 10 families have been documented within intertidal belts. Also, 12 actiniarians belonging to 2 classes, 3 orders and 6 families have been observed in different parts, mostly at the junction of intertidal and subtidal zones. A number of very interesting benthic fauna, like sea pens (Cnidaria), sea cucumbers (Holothuroidea) and *Lingula anatine* (Brachiopoda), were found in the mudflats of the Talsari, Shankarpur, Junput and Nayachar islands of this district. Species of horseshoe crabs, viz. Carcinocorpius rotundicauda and Taphypleus gigas, have also been observed in Digha–Talsary intertidal flats. Out of the 68 arthropod species recorded on this coastal belt, 13 are represented by brachyuran crabs, whereas the others are crustaceans, prawns and shrimp (13 species). Additionally, 21 species of insects belonging to 33 families represent another major group of fauna. In total, 51 soil microarthropods belonging to four major insect orders, viz. Collembola, Hymenoptera, Diptera and Isoptera, have been recorded in different parts of this coast. Finally, 51 fish species under 2 classes, 9 orders and 25 families have been documented in different fish markets and landing centres (Chakraborty, 2010).

10.14 The Geomorphic Environments of Coastal West Bengal, Including Sundarbans

The Indian Sundarbans, representing an estuarine-mangrove complex dominated by a large expanse of a network of tidal channels, macrotidal intertidal flats, mangrove swamps and marshes and deltaic islands, experiences a continuous process of erosion and deposition, which results in geomorphologic variability and exhibits sediment transport dynamics (Ginsberg & Perillo, 2004).

The meso-macrotidal influence on the coastline of West Bengal, India, promotes the development of all those geomorphological settings. A complex network of different geomorphical morphotypes in the Indian Sundarbans include estuaries and creeks, beaches and inlets, natural levees, mid-channel bars, point bars, mangroves, marshes, swamps, mudflats, sandflats and coastal dunes. Both sandy and muddy intertidal flats have developed at the sea face after being determined by high- and low-energy zones. Tidal rivers or inlets represent mostly muddy systems, except for some sandy flats occurring in the mid-channel bars and some sandy swash bars occurring at the mouths of rivers or inlets. The creeks have mostly developed due to the scrapping effects of tidal water on muddy surface sediments at the bottom. Mangrove swamps have developed on the intertidal mudflats of salinity-dominated estuaries, inlets and creeks and in freshwater-dominated rivers and channels.

The occurrence of different natural hazards, including episodic and nonepisodic events like storms and cyclones, cause the variability in different physical processes, like tidal flooding, longshore currents, river fluxes, waves and ultimately sea-level changes that in turn cause the structural modification of the coast and thereby underline the morphodynamic changes in the coastline. Long-term tectonic and geological processes occur as well (Biswas 1963; Sengupta 1972; Allison 1998; Kuehl et al. 1997; Bhattacharya, 1999). At the mouths of all the estuaries and inlets at the confluence with the sea, the Bay of Bengal has widened into funnel-shaped configurations. The geomorphological features of the mangrove-estuarine complex with seasonal tidal interplays on the coast of West Bengal, India, play dominant roles in determining the morphological evolution of this coast, which can be designated as an estuarine delta coast (Galloway, 1975). Most of the beaches on the Sundarbans coasts of India exist in a dissipative dynamic state with rhythmic beach profiles (Das and Bhattacharya 2002). Sandy beaches—constituted predominantly by quartzofeldspathic elements (95%) along with other accessory biotic, muscovite, hornblende and ilmenite components (5%) of fine to very fine sand grains of sediments at the foreshores of the coastal belts within Sundarbans, India—can be found on deltaic islands like Fraser Ganj on Namkhana Island, Gangasagar on Sagar Island, Kalas on Bulchery Island, etc., where the width of those beaches ranges from 250 to 500 m (Bhattacharya 1993). The foreshore sandy beaches possess a number of surficial physical structures like different forms of ripples: rhomboid ripples, flat-topped ripples, backwash ripples, backwash marks, swash marks and current crescents. The creeks on the foreshore contain unidirectional current ripples of both linguoid and lunate types, whereas the lower foreshore zone includes distinct current- and wave-induced interference ripples (of the ladderback type) (Komar 1976).

10.15 Ecosystem Services of the Mangrove Ecosystem

The mangrove ecosystem is considered an ecologically healthy, economically productive and aesthetically celebrated area for rendering many ecosystem services, like providing food, fodder, materials for medicines and shelter; transporting and exporting living resources to various parts of the world, creating avenues for livelihood generation for local people, acting as a natural coastal defence against cyclonic storm surges, protecting the land from episodic erosion, facilitating carbon sequestration and promoting ecotourism. Coastal wetlands, boundaries between land and ocean like the mangrove-estuarine complex of Sundarbans with different geomorphological settings under the influence of brackish water, render many ecological services that maintain the health of the Earth's environment in general and that provide benefits to coastal communities in particular. Tidal salt marshes and mangroves resist flood waters from the land runoff from upland and filter out sediments and nutrients, improving water quality. The habitats and niches of mangrove ecosystems are critical for the juveniles and adults of many economically important marine shellfish and finfish species that spend part of their lives in the oceans.

Although habitats for mangroves and their associated organisms are among the most rapidly vanishing landscapes on Earth, with a loss of 50% of mangrove coverage since the 1950s (Feller et al., 2010), and although it is among the most threatened landscapes in the world, tropical and subtropical mangrove ecosystems (between 30°N and 30°S) provide numerous ecosystem services and generate livelihoods for coastal people worldwide (Lee et al., 2014; Polidoro et al., 2010; Richards and Friess, 2016). Local people have developed their own Indigenous knowledge for developing aquaculture with the collection of finfish, shellfish, prawn seeds and crab potting, reducing the large-scale exploitation of mangrove flora and fauna, reducing the loss of mangrove vegetation cover, expediting natural succession, regenerating mangrove swamps and conserving the overall biological diversity of the ecosystem.

10.15.1 Mangrove Ecosystems: Ecosystem Services, Goods and Value

The ecosystem services provided by mangrove ecosystems in the form of goods and services to human beings are well recognized and appreciated at the national and global levels (Hamilton & Snedaker, 1984; Stafford-Deitsch, 1996; Dahdouh-Guebas et al., 2000; Kairo & Kivyatu, 2000). Many of these services include resistance against soil erosion; buffering against natural disasters like seasonal storms, cyclones, etc.; and local people collecting food (honey, fish, etc.), fodder, building materials, fuel wood and medicine from bioactive substances from plants and animals, which also act as bioremediators by providing accumulation sites for sediment, nutrients and other toxic elements. Luxuriant bioresources in the Sundarbans mangrove ecosystem render potential benefits to forestry, fisheries, aquaculture, agriculture, pollinators and honey-producing insects, reptiles, birds and other wildlife. The ecological potential of the mangrove ecosystem for biological production stems from (1) its large quantities of energy in the form of mangrove detritus, which are exported from the mangrove forest to open waterbodies (Odum & Heald, 1975a) and from (ii) the positive relationship between the extent of mangroves and total bioresources from adjacent waterbodies (Macnae, 1974).

In addition, profitable regional and international markets for high-quality biological products can use the products derived from the mangrove ecosystem. The ecozones of mangrove ecosystems are used as important nursery grounds and breeding sites for innumerable faunal components, like finfish, shellfish, turtles, crocodiles, birds and mammals. Honey from Sundarbans is world famous because of its taste, purity and nutritional contents, the maximum natural production of which is reached during late pre-monsoon with the synchronization of the flowering seasons during late post-monsoon and early pre-monsoon season, when all of Sundarbans is in full blossom. The collection of honey starts in March and April and continues for about the next 2 to 3 months. During this period, the honey collectors, locally known as mahals, enter the deeper parts of the mangrove forests of Sundarbans, getting necessary permission from the Department of Forests for a restricted period. The potential parts of forests have vegetation assemblages with flowers, and after tracing, the targeted beehives move in rows to track the pathways of flying honeybees. Two species of honey bees, *Apis dorsata* (rock bees) and *Apis sarana* (small bees), are the main producers of honey in Sundarbans, hovering over the flowers of some suitable host plant species, like "Khalsi" (*Aegiceras corniculatum*), "Bani" (*Avicennia alba*) "Horkocha" (*Acanthas illicifoloius*) and "Keora" (*Sonneratia apetala*) trees.

Fisheries and aquaculture form important sources of resource and livelihood generation in the Sundarbans ecosystem in India thanks to the many species of fish and shellfish within estuaries, creeks, rivers and channels, which make this ecosystem inviting to lively interactions between nature and humans, who harvest natural bioresources for their livelihoods.

10.16 The Effects of Natural Disasters and Environmental Pollution on the Mangrove Ecosystem of Sundarbans

Mangrove ecosystems, being among the most threatened ecosystems in the world, are still considered poorly understood ecosystems in relation to their functioning, their potential for ecosystem services and the causes of their loss. Mangroves and their associated flora and fauna have been under threat from catastrophic storms, cyclones, tsunamis, floods, invasive species, diseases, etc. and have become very susceptible due to human stressors like habitat alterations, varied forms of pollutants and uncontrolled urbanization. In view of all those ongoing stress factors, mangroves respond in commensuration with the evolution of their inhabited environmental settings (Giri et al., 2010).

Different environmental stresses from natural processes and anthropogenic activities need to be assessed before undertaking proper preventive measures and controlling mechanisms. In this respect, two overriding trends characterize the recent environmental management system: (i) The global diversity of mangrove ecosystems is threatened by imbalances in biological productivity, affecting the distribution process of ecosystem goods and services, and (ii) the entire world has been in the process of all-round changes at an accelerating pace despite internationally coordinated environmental stewardship effects. The unsustainable advancement of society with extremes of wealth and poverty threatens the stability of the normal conditions of the global environment, and such deviations have culminated in the emerging problems of pollution. Alongside mangrove ecosystems, other types of biotic communities, such as forests, grasslands and wetlands, are facing ecodegradation, resulting in a conversion or transformation of natural ecosystems into a fragmented state, causing further biodiversity loss (Dearden, P. 1996).

The problem of environmental pollution dates back to the era of the Industrial Revolution in Europe during the eighteenth century, after which the overdependence of human civilization on modern technology prompted an increase in the additional usage of industrial commodities to keep pace with modern lifestyles, resulting in a drastic alteration in human environments. Mangrove ecosystems and their structural components (flora, fauna, water and soil) have faced severe global environmental problems, like the release of toxic substances from both industrial and municipal activities, eutrophication, acid rain, habitat destruction and deforestation, increasing emissions of greenhouse gases, the destruction of ozone layer and biodiversity loss, and all these have severely damaged marine ecosystems.

Many of the environment-related problems in the mangrove-estuarine complex of Sundarbans, India, originate directly from land-based human activities such as sewage outfalls, runoff loaded with agricultural waste, the combustion of gases, discharge from power station and, emissions from automobiles, which tend to enter this ecosystem from the atmosphere. Eutrophication caused by the excessive enrichment of nutrients due to the excessive influx of nitrogenous and phosphorus compounds in estuarine water leads to excessive blooms of harmful algae, which in turn bring about a chain-reaction process with many harmful ecological consequences, like oxygen depletion, lower pH values, changes in species composition, increases in turbidity and reductions in biological productivity. Such undesirable ecological changes in mangrove ecosystems, in combination with an influx of other nondegradable chemical pollutants (plastics and microplastics, high-density metals, organic pollutants, polychlorinated biphenyls, oil-related products, chlorinated hydrocarbons, etc.), have aggravated coastal waterbodies. Most of these persistent pollutants and nonbiodegradable pollutants, through the process of biological magnification (the magnification of the concentration of toxic substances through mangrove-based food chain–food web dynamics), accumulate in the bodies of faunal components at the highest trophic levels.

10.16.1 Increasing Greenhouse Gas Emissions: Impacts on the Marine Environment

Since the Industrial Revolution, the level of greenhouse gases (GHGs), such as CO_2, CH_4, H_2O, N_2O_5, and $CFCl_3$, in the atmosphere has considerably increased, mainly because of dramatic changes in the lifestyle patterns of humans, who increasingly use air conditioners, burn fossil fuels and overuse chemical fertilizers and pesticides in the agricultural sector (IPCC, 2001). This increase in the level of GHGs over the past couple of centuries has resulted in an increase in the global mean temperature, the impacts of which are profound on mangrove ecosystems across the globe, mostly in the form of increasing atmospheric temperature; increasing sea-level rise; changing the currents, waves and tides of estuaries; increasing coastal erosion; changing the species composition of biodiversity assemblages within fragile and sensitive ecosystems like mangrove ecosystems; expediting bioinvasion; threatening wildlife; increasing storm surges and the inundation of low-lying coastal areas via coastal flooding and the landward intrusion of seawater into estuaries and aquifers (Alley et al. 2005; Wong 2003; IPCC, 2001); and generating uncertainty and unpredictability in global meteorological and climatic conditions (UNFCCC 2005).

10.16.2 Acidification and Its Impacts on Mangrove-Estuarine Organisms

Marine-estuarine ecosystems act as sinks for the absorption of a large amount of carbon dioxide (CO_2) from the atmosphere, which after mixing with water results in the formation of a weak acid, called carbonic acid. Therefore, increasing the rate of CO_2 emissions in the atmosphere reduces the average pH of marine-estuarine water to an alarming level, which in turn poses real threats to several sensitive species in mangrove ecosystems, especially faunal components with calcium carbonate in their outer shells (calcified organisms) (IPCC, 2007).

10.17 Socioeconomic Changes and Their Consequent Impacts on Mangrove Biodiversity in Sundarbans

The major threats to the Sundarbans mangrove ecosystem are increased salinity, sewage pollution, cyclonic destruction, eutrophication through nutrient enrichment, semi-intensive aquaculture and human–wildlife conflicts. Globally, the coastal zones with mangroves ecosystems are under different kinds of stresses, mostly through water pollution, and have become vulnerable because of multidimensional anthropogenic activities, including unplanned urbanization and industrialization, deforestation, massive soil erosion, trade and tourism (Mondal and Bandyopadhyay, 2014; Lakshmi and Edward, 2010). Over the past couple of centuries, both the banks of estuarine networks and mangrove-based deltaic islands have experienced large-scale human settlements, which are prominent in different parts of the Indian Sundarbans. This situation was aggravated after the independence of India, which took on the responsibility of all the consequences of partitioning India and creating a new country (i.e. Pakistan). Bangladesh is endowed with two-thirds of the mangrove coverage of Sundarbans across the Indian subcontinent. Unplanned urbanization coupled with consequent industrialization in and around Sundarbans has brought about environmental degradation, which stresses the coastal ecosystems of India by adversely impacting the biota, water- and soil-quality parameters and productivity (Stanley 2004). Most of the tide-fed canals in and around the Indian Sundarbans once enabled fish to enter the main estuarine flows of Sundarbans because those channels were connected to estuarine networks through channels. But over time, all those connecting channels lost their connections with the major estuaries of Sundarbans and lost their water-carrying capacity and thus became redundant.

The alpha biodiversity and gamma biodiversity in the biotic communities within mangrove ecosystems experienced stresses because of large-scale forest exploitation, the spread of disease outbreaks, siltation and decreases in water pH and soil alkalinity. The occurrence of tropical storms—cyclones like Aila (2009), Bulbul (2019), Amphan (2020), Yash (2021) and Dana (2024)—over Sundarbans has resulted in the colossal destruction of coastal engineering structures in human settlements on reclaimed deltaic islands, has caused changes in the species composition of the mangrove biota, has enabled bioinvasive species to settle and flourish, has expedited salinity invasion within estuaries and has enhanced soil salinization, leading to disruptions to agricultural productivity. And this process has brought about associated socioeconomic unrest, producing environmental refugees irrespective of age and sex (Banerjee, 2013; Bera, 2013). Enhanced salinity and the shifting of sands over the mangrove mudflats have resulted in the mass mortality of mangrove plants with simultaneous increases in bank erosion, silt deposition and a loss of compaction in sediments due to more tidal influx upstream, which have ultimately made the Sundarbans estuarine ecosystem more vulnerable in the face of sea-level rise.

Remote-sensing data analyses have revealed that healthy mangrove vegetation has decreased over the past few decades (20.545 ha) at Gosaba, Basanti, Canning 2, part of Bangar of South 24 Parganas, Sandeshkhali 1 and 2, Hasnabad and a major part of Hingalganj at North 24 Parganas from 2010 to 2020 due to increasing human settlements, modern technology–based agricultural and aquacultural activities. Massive changes in healthy vegetated area have been observed along the river banks of Bidyadhari, Dansa, Gosaba, Bidya and Matlah in North 24 Parganas and South 24 Parganas, West Bengal. The degradation of forest cover in different parts of Sundarbans, India, is found only because of the elimination of shrubland (open forest or less-dense vegetation cover) and a decrease in healthy vegetation cover over the study period, which are influenced by changing edaphic factors like reduced freshwater discharge from the upstream region with nutrients, enhanced soil salinity, the increased anthropogenic release of surface runoff from the agro-farming sector and wastewater containing heavy metals. A significant integrated relationship was established between vegetation density, a decline in forest cover and vegetation type on one hand and a rapid increase in human settlements on Jharkhali Island on the other from 1986 to 2009 with the help of Geospatial Information System (GIS) (Samanta and Hazra, 2012).

10.18 Assessment of Land-use Changes and Their Impacts on Mangrove Ecosystems

Remote-sensing data on the trends in land-use/land-cover changes have shown the submergence of low-lying land as the interconnecting drainage network that is supposed to have been chocked up due to siltation or sediment deposition. A large-scale decrease in wetland area (1064.58 ha) and an increase in aquacultural area (3728 ha) have been observed over a 10-year period (2009 to 2019), where the probable causes for such changes are the conversion of wetlands into aquacultural land and intensive fish farming. Humans have changed the forest land (mostly open forest area) to cultivable land, agricultural land into settlements or industrial land (brick kiln) for economic purposes (Vink, 1975). A decline in the extent of agricultural land (2905.9 ha) also indicates the general conversion of agricultural land into aquacultural development for short-term economic benefits for settlements, which have increased due to population pressure (Hazra et al., 2010).

10.19 Different Environmental Stresses from River Drainage in the Sundarbans Mangrove Ecosystem

The dynamic deltaic islands in Sundarbans face different threats, both natural (the movement of tectonic plates, catastrophic storms and cyclones, changes in the course of water flows, etc.) and human caused (habitat destruction, the reclaiming of islands, deforestation, the construction of embankments, salinity invasion, pollution by eutrophication, the discharge of effluents from industries and aquacultural farms and human–wildlife conflicts). Due to interconnections between different freshwater rivers and estuaries, the saline water received through tidal influx from the sea is measurably diluted with floodwater or freshwater flows from rivers. The riverine flow not only ensures freshwater drainage into the estuarine system but also releases waste material loaded with sediments and toxic substances carried from the uplands of urban areas, agricultural land, surface runoff and industrial effluents to estuarine networks. The release of toxic substances and the sedimentation of solid materials through water columns influence water- and soil-quality parameters. Natural impacts on the deltaic islands of Sundarbans also include large-scale sedimentation in active deltas, where about 2.4 billion tons of sediments are deposited annually by river network transports (Mitra and Zaman, 2016). This requires the mandatory monitoring of all the ecological variables to determine the influence of human perturbations on physicochemical parameters at temporal and spatial scales. Different human activities in and around Sundarbans, India, include fishing, tourism, brick manufacturing, the tannery industry and modern agricultural activities, which overuses chemical fertilizers and pesticides, degrading the ecological conditions of the drainage basin. The transportation of sediments by water increases the rate of siltation on the river bed due to land reclamation, cultivation, sewage dumping, waste from sludge and fishery waste and thereby further worsens ecological conditions.

10.20 Impacts from the Construction of Embankments

The large-scale construction of concrete embankments to protect the 50 reclaimed islands of the Sundarbans, accommodating about eight million people, must have major impacts on the fragile ecology of the mangrove ecosystem. These concrete embankments prohibit the free movement of the bottom-dwelling fauna inhabiting river floodplains and the natural levees of mangrove ecosystems, which has detrimental effects on the mangrove ecosystem. The loss of any species because of such human-built structures leads to a gap in the food chain and food web systems, which ultimately impairs the functioning of the entire mangrove ecosystem. The intertidal belts, the floodplains of adjoining rivers and the riverbanks within Sundarbans act not only as habitats for thousands of macrobenthic and meiobenthic fauna but also as breeding grounds for the birds such as kingfishers and seagulls. The embankments, made up of mud, temporarily act as protection for human settlements but are very prone to erosion by tidal actions, as shown on several reclaimed islands, like Sagar Island, Mahishani Island and Ghoramara Island. Sometimes, polypropylene sheets are used instead of the concretization of mud-built embankments, but they are harmful to the estuarine environment because polypropylene is not a biodegradable ingredient.

10.21 Climate Change and Its Impacts

Deltaic islands, after emerging in estuarine rivers mouths and converging with the sea across the coastal tracts of Sundarbans, have been under the process of gradual disappearance because of global warming–mediated climatic changes and consequent sea-level rise. The environmental consequences are reflected in the vanishing of the Ghorama and Lohachura islands in the Indian Sundarbans, but this differ from the ecological consequences on Jambudwip Island, a deltaic island near the confluence of the Hooghly estuary with the sea at the Bay of Bengal, where the environment has been degraded because of overstress from human activities. Such activities include fish drying and the seasonal settlements of fishers for around 4 months of the year, when these fishers exploit mangrove resources, which are crucial causes of the deterioration of the overall environmental quality of this island.

The anthropogenic activities on this island not only cause the clearance of mangroves for fish-drying operations but also accelerate erosion processes, especially in the southern part of Jambudwip. In addition, all such activities further trigger the modification of estuarine courses, reduce freshwater flow, promote climate variability and bring about sea-level rise. The cumulative outcomes of all these continuing deleterious operations over recent decades have been bringing about serious changes in geomorphology, the hydrodynamics of watercourses and nutrient recycling in Sundarbans, India.

Alongside these fishing activities, an increase in ecotourism due to climate change–mediated environmental consequences has compounded problems (Coombes et al. 2009). Nonconducive environmental conditions, with higher temperatures and higher humidity because of the ongoing impact of global warming, for mangrove forests restrict tourist entry to only four winter months of the year, especially when tourists prefer to visit the only tiger habitat in a mangrove forest in the world, the Sundarbans.

An alternate approach to dealing with climate change is being practised to compensate for the ineffectiveness of the international framework for fighting the negative impacts of climate change and to fulfil the targets set by this framework by incorporating, among so many other things, lifestyles and behavioural changes to reduce the problems associated with ongoing environmental perturbations. In such a context, adaptation should be prioritized as a mitigation measure. This is supposed to be achieved by increasing incomes through the effective and sustainable management of natural resources by devising innovative, ecofriendly technologies for aquaculture, honey production, dairy management, etc. and by training local people through workshops on computer data and forecasting natural disasters. Because the harmful consequences of climate change are unstoppable, people need to adapt by increasing the resilience among communities that are prone to greater risks. All such measures are included as integral components of climate action.

The dwindling abundance of mangrove vegetation, barring certain pollution-tolerant species, is represented by mangrove species like *Sonneratia* sp., *Avicennia* sp. and Ceriops, Rhyzophora and macrobenthic faunal species like brachyuran crabs, Metapopgrasus, *Scylla searata*, polychaetes worms, *Mastrobranchus indicus*, *Diopatra cuprea cuprea* and *Dendronereisheteropoda*, which have become quite conspicuous over the past two decades. The soft bottom polychaetes in particular act as opportunistic species in this area, eliminating the other polychaete species.

10.22 The Root Causes of Biodiversity Loss in the Mangrove Ecosystem of Sundarbans

The loss of biodiversity within mangrove ecosystems across the globe over the past century has become a pressing environmental issue because the elimination of a couple of species, either flora and fauna, will directly cause imbalances in the natural fabric of the food chain–food web dynamics of mangrove ecosystems. In addition, different ecological habitats and niches, accommodating numerous living organisms, generate life-supporting ingredients through continuous interactions between themselves and substances that encourage the growth and flourishing of other living organisms, most of which are very sensitive to ecological changes in their surrounding environment (Odum & Heald, 1975b).

Owing to the ongoing removal of mangroves across the tropical belts of the world because of a multitude of anthropogenic activities, several other ecological problems have emerged, like the loss of flora and fauna, erosion and unwanted accretion, changes in the course of water movement in estuaries and creeks, salinity invasion and sea-level rise. All those ecological changes ultimately alter the structure and function of the remaining endangered mangrove forests in particular those in Sundarbans, a UNESCO World Heritage Site and the world's largest contiguous mangrove ecosystem (Duke et al., 2007). The IUCN estimated that 40% of the listed mangrove tree species have become threatened (Polidoro et al., 2010). Ever-escalating anthropogenic pressures on mangroves, such as the reclamation of land for aquaculture and salt pond construction (Chakraborty, 1998a, b; Primavera, 1995), overharvesting for timber and fuel wood (Semesi, 1998), mining, pollution and the damming of rivers, have increased salinity levels in water (Qasim et al., 1988; Lewis, 1990, Wolanski, 1992) and oil spills (Sengupta and Qasim, 1988; Ellison & Farnsworth, 1996; Burns et al., 1994). The global range of mangrove forests has been reduced to less than 50% of their original total cover (Saenger et al., 1983; Spalding et al., 1997). After the development of intensive shrimp-farming techniques in Taiwan in the 1970s, there was a sudden rush towards modern shrimp farming in Southeast Asia (Chakraborty, 1998a, b) and in the Indo-Pacific region (Primavera, 1995).

The destruction and overexploitation of mangroves and their associated bioresources have hampered the spawning grounds of numerous aquatic species, which in turn has triggered other forms of coastal ecosystem degradation. For example, Sundarbans at Chakaria in eastern Bangladesh experienced a total clearance of mangroves for aquaculture, and more than half of the total area (208,220 ha) has disappeared because of the large-scale felling of mangroves between 1961 and 1993 in Thailand. The International Union for the Conservation of Nature (IUCN) evaluated the current conservation status of different species of mangroves as threatened, endangered or critically endangered, including Xylocarpus spp., *Kandelia candel*, *Bruguiera gymnorhiza*, *Sonneratia apetala*, *Ceriops tagal*, *Heretiera fomes* and *Nypa fruiticans*, which have been reported to be disappearing from the mangrove belts of the Indian subcontinent.

10.23 The Changing Trend in Mangrove Coverage and the Restoration of the Mangrove Ecosystem

After getting damaged, mangrove forests within deltaic systems can recover rapidly, undergoing a recolonization process through mangrove establishment (Gilman et al., 2008, Ellison, and Coleman, 2007). Before undertaking eco-restoration through eco-rehabilitation processes, a proper understanding of the extent of the distribution of mangroves is required. The seaward extent of deltaic systems is governed by the interplay between three principal factors that ultimately affect mangrove distribution:

1. The input of sediment, keeping pace with the fluvially controlled sediment supply, sediment starvation and subsidence, which increase the risk of coastal flooding
2. A continuous supply of sediment, which encourages deltas and their associated mangroves to either extend seaward into deeper waters or keep pace with the ongoing subsidence process, sediment starvation and erosion pressures, transgressing the original shoreline
3. The rate of sediment supply, land-use practices (Meade, 1996) and the construction of dams and embankments upstream at high mountain elevations (Milliman and Syvitski, 1992; Syvitski et al., 2005, 2009; Vorosmarty et al., 2009)

The most striking causes of the rate of subsidence are sediment compaction and geologically controlled crustal movement through a tectonic process in deltaic systems (Ericson et al., 2006; Syvitski et al., 2009, Becker and Sultan, 2009). Several estuarine processes, like degrading the local sea level and thereby causing storms and storm surges, have considerable effects on the morphology of coastal landforms (Suursaar and Kullas, 2009; Vespremeanu-Stroe et al., 2007). Observations of the consistent patterns of change in mangrove assemblages can help us to understand the interplay between these factors, which determine spatial extent, recession or progradation in the deltaic coastal systems supported by mangroves (Gilman et al., 2008).

Any eco-restoration programme in the Sundarbans mangrove ecosystem can be successful only if proper attention is paid to the basic operating principles of the functioning of mangrove ecosystems and the simultaneous involvement of two constituent subsystems—estuarine and mangrove forests and their structural components, like the aquatic and forest-based biota (plants, animals and microbes), water and soil. Afforestation programmes, planting suitable plant species, represent the first stage in mangrove restoration efforts, which are to be followed by improving tidal drainage systems through the excavation and extension of artificial channels or ditches fed by high-tide water and through the upgradation of creeks so that tidal water can reach the innermost parts of the mangrove forests and swamps for at least half of the day. The selection of plant species for the afforestation programme requires a basic understanding of the ecological succession process and zonation patterns of mangroves for a particular ecozone with specific ecological parameters. Background knowledge of the trends in mangrove coverage in a particular area of a mangrove forest is required to combat that process of destruction either naturally or artificially. Mangroves in the tropical coastal environment of India are lost mostly due to the exploitation of bioresources in the form of food, fodder, shelter material, medicines, etc. within a mangrove forest for human uses and the land-use conversion of mangrove belts for the purpose of agriculture, aquaculture and tourism (Richards 1993; Blasco, 1984; Banerjee et al., 1989; Naskar & Guha Bakshi, 1987;

Paul, 2002). Moreover, all-round restoration strategies require incorporating the history of natural disasters, storm catastrophes, marine-oceanographic variables, soil erosion, tectonics-based land subsidence and uplift, coastal geomorphology, hydrological stress factors like hypersalinity and eutrophication, drainage systems in the catchment areas of river deltas, sediment starvation, the shifting of riverbanks in deltaic areas, overwash, sand fan lobe encroachment, shoreline retreat and shortages in rainfall, causing a reduction in freshwater discharge into coastal zones.

10.24 Conservation Strategies: Conservation and Adaptation in the Indian Sundarbans

The geomorphology of the Sundarbans mangrove ecosystem, which is one of the world's richest biodiversity hotspots, has been determined by the deposition of sediment, erosion pressure, vertical accretion, the development of geomorphological structures and influxes of both saline water and fresh water (Ginsberg & Perillo, 2004). This is how islands are formed. Replenishing the floodplains and the low-lying coastal areas is carried out by the deposition of organic and detrital substances, which maintains the surface equilibrium of the deltaic islands, which have been formed by the deposition of sediments and vertical accretion, resulting in the bifurcation of rivers into new channels (Alam et al., 2003; Vorosmarty et al., 2009). The landscape of the Indian Sundarbans has been modified mainly because of the construction of polders for human settlements, which have brought about significant alterations in channel hydrodynamics, the siltation process, the amplification of tidal range and several other human-induced stress factors (Pethic and Orford, 2013; Wilson et al., 2017; Higgins et al., 2018).

The mangrove habitats of Sundarbans, India, render many ecosystem services (food, fodder, shelter material, etc.) to the people of that region, where a sizeable portion of commercially important living resources are transported and exported to various parts of the world, thereby providing livelihood support for many people in Indian. They also protect against coastal erosion and recurrent tropical cyclones like Aila (2009), Bulbul (2019), Amphan (2020) and Yash (2021) over Sundarbans, which led to the massive destruction of built structures on reclaimed human-dominated islands; hamper mangrove vegetation composition; and cause crop failures and the overtopping of saline water inland, resulting in soil salinization and therefore bringing about associated socioeconomic problems such as creating environmental refugees and promoting outmigration (Banerjee, 2013; Bera, 2013).

The mangrove ecosystem of Sundarbans has been under considerable stresses from both natural and human-mediated threats like the overexploitation of forest resources; changes in coastal land use; climate change; sea-level rise; natural disasters; pollution by heavy metals, microplastics and agricultural waste (pesticides and chemical fertilizers); oil spills; upstream water flow reduction; increased erosion followed by vertical accretion; salinity invasion; disease outbreaks; a lack of knowledge about forest conservation; uncontrolled tourism; a lapse in upholding conservation rules; a lack of adequate planning and management; and an increase in human–wildlife conflicts.

Any kind of qualitative or quantitative change in structural components or in functional efficiency disrupt the stability of the ecosystem, which in turn undermines sustainability. Disruptions to the self-maintenance and self-regulation of the ecosystem jeopardize the normal functioning of the concerned ecosystem and adversely affect its ecological health. Therefore, reducing human interventions helps to maintain the sustainability of the functioning of the ecosystem, which is of great concern to ecologists, conservationists, environmental planners and naturalists. An integrated ecomanagement approach for the protection of this unique ecosystem through various pro-environmental steps. This can be achieved through the all-round involvement of a large number of mangrove forest fringe villagers, by enabling them to undertake proper adaptation through training on fishing and the collection of fuel, fodder, honey etc. as means for their livelihood. Any governmental effort towards the sustainable conservation of natural resources will not be successful unless the concerted effort and active involvement of concerned local people are incorporated into the planning and management of resources. Alternate livelihood options need to be created for the local inhabitants of Sundarbans, a World Heritage Site, in the fields of mangrove afforestation, honey collection, fisheries and natural disaster management for their existence and this area's sustainable development. In view of the ongoing imposition of anthropogenic stresses on mangrove regions like changes in land-use patterns (e.g. the conversion of mangrove wetlands into agricultural land and for aquacultural activities), urbanization and industrialization, deforestation followed by soil erosion, increasing turbidity and salinity invasion followed by disruptions to the natural flows of fresh water from the river drainage system. In addition, regional climate change projections suggest that the mangrove ecosystem of the Sundarbans will experience more warm days with temperatures > 30°C, an 8% increase in precipitation with more-intense rainfall, prolonged monsoons, longer dry seasons and an annual rise in sea level of 1–2 mm (Unnikrishnan and Shankar 2007, START 2010). To achieve sustainable and equitable ecomanagement strategies and to help restore and protect biodiversity in particular and the mangrove ecosystem in general, effective conservation measures for bioresources are required. Currently, the mangrove ecosystem of Sundarbans, India, like other coastal communities around the globe, has been experiencing a broad array of biophysical and socioeconomic changes, including the shifting of economic status and political settings, environmental pollution, declines in fisheries and climatic changes (Bennett et al. 2014a). In such a context, appropriate protection measures and conservation strategies need to be adopted for the mangrove ecosystems in India, with the prime objective of further inhibiting the deterioration of healthy ecological conditions and genetic depletion. This can be made possible by undertaking large-scale afforestation with modern afforestation techniques, ecology-based coastal wetland development, and improvements to nursery grounds; rigidly adhering to the natural fabric of ecosystem functioning; and imparting education and increasing awareness on the ecological and economic importance of mangroves at different strata and at various levels of society. However, economic growth by adopting new livelihood opportunities for local communities, particularly in tourism, agriculture, harvesting bioresources like fish and honey and even engaging educated youths in disaster-management efforts, which could diversify livelihoods and reduce pressure on marine resources (Ellis et al., 2004). Coastal human communities are also subjected to environmental vulnerability because of global climate-related changes, rising temperatures and sea-level rise, extreme weather events, increasing ocean acidification and the seasonal occurrence of natural disasters. All these environmental problems have direct impacts on local communities through infrastructure damage, through erosion and by impairing the ecosystems services on which they depend due to coral bleaching and mangrove retreat and have indirect impacts on resource-dependent livelihoods (Marshall et al., 2010). The complexity in the changing processes of natural resource distribution patterns affects ecological stability and socioeconomic conditions, which necessitates adopting mixed-method approaches to explore possibilities for enabling coastal communities to undertake proper adaptive measures to counteract those changes in a such a manner that helps to bring about positive social-ecological outcomes. A multiscale analysis can help draft recommendations for affected communities to allow them to make the following adaptations: (i) the development of innovative local ecomanagement or conservation strategies against environmental degradation or declining fishery production; (ii) the development of acceptable economic or livelihood opportunities for nonfishing sectors, to reduce stresses on marine-estuarine-mangrove systems; (iii) tackling climate-related changes; and (iv) the generation of a strong knowledge base to assess their opportunities from the availability of freshwater inputs, the textural properties of sediments and the salinity profiles of soil, and the judicious use of groundwater and surface water (Belperio, 1993; Semeniuk, 1994; Blasco et al., 1996; Alongi, 2008; Alongi, 2008; Erwin, 2009; Gilman et al., 2008).

Against this background, an integrated approach is proposed, one that refines the existing management strategy and combines updated information from supplementary scientific studies and expeditions into mangrove forests. Properly implementing the proposed resilient strategies, namely the incorporation of all stakeholders in protecting mangroves developing, awareness programmes, reducing the dependency of local people on mangrove forests, increasing monitoring undertaking, ecological restoration adopting, the implementation of legally binding rules on disaster management and endeavoring adequate research and planning and all of which can promote the sustainable management of Sundarbans and similar mangrove forests around the world.

The timely adoption of proper adaptation by local people appears to be the main challenge in addressing the climate threat to coastal zones due to global warming. Too much delay in this process impairs the applicability of such adaptation for human benefits. A multidimensional long-term approach is needed to take care of the imposed stresses on mangrove ecosystems by continuously monitoring the impacts of climate change on such a complex and sensitive ecosystem. Larger-scale remote-sensing data appear to be reliably effective in the evaluation process of the total area covered by mangroves. The protected mangrove regions under the conservation process need to be constantly monitored for changing patterns in hydrological, ecological and biological characteristics. Global and national biodiversity regulation policies need to be rigidly implemented to mitigate the emission of GHGs and conserve marine habitats.

Adaptation activities should emphasize the capacity-building activities of local people in response to early-warning systems for natural disasters, where such responses must strictly adhere to national disaster-management policies, and community-level programmes need to be adopted to reduce prospective risks. These programmes are expected to reduce the overall loss of life from cyclonic storms, flooding and other types of climate-related events affecting coastal zones. Special emphasis should be placed on counterbalancing the prevailing ecological, political, socioeconomic and technical problems hampering the planning and implementation of adaptive strategies.

References

Alongi, D. M. (1990). The ecology of tropical soft–bottom benthic ecosystems. *Oceanography and Marine Biology: An Annual Review, 28*, 381–496.

Alongi, D. M. (2008). Mangrove forests: Resilience, protection from tsunamis, and responses to global climate change. *Estuarine, Coastal and Shelf Science, 76*, 1–13.

Alongi, D. M. (2009). *The energetics of mangrove forests*. Springer.

Alongi, D. M., & Sasekumar, A. (1992). Benthic communities. In A. I. Robertson & D. M. Alongi (Eds.), *Tropical mangrove ecosystems* (pp. 137–171). American Geophysical Union.

Annon. (2003). *Mangrove ecosystem: Biodiversity and its influence on the natural recruitment of selected commercially important finfish and shellfish species in fisheries*. National Agricultural Technology Project (NATP). Indian Council of Agricultural Research (ICAR). Principal Investigator: George, J. P.; Co–PI: Chakraborty, S.K. and Damory, S. N.: 1–514.

Aziz, A., & Paul, A. R. (2015). Bangladesh Sundarbans: Present status of the environment and biota. *Diversity, 7*(3), 242–269. https://doi.org/10.3390/d7030242

Banerjee, L. K., ARK, S., & Nayar, M. P. (1989). *Mangroves in India: Identification manual* (p. 113). Botanical Survey of India.

Bhattacharya, A. (1999). Embankments and their ecological impacts: A case study from the tropicallow-lying coastal plains of the deltaic Sunderbans, India. In H. Grann & M. Vollmer (Eds.), *Largescaleconstructions in coastal environments* (pp. 171–180). Springer.

Blasco, F. (1977). Outlines of the ecology, botany and forestry of the Mangals of the Indian subcontinent. In V. J. Chapman (Ed.), *Wet ecosystem of the world* (pp. 241–262). Elsevier.

Blasco, F. (1984). Climate factors and the biology of mangrove plants. In S. C. Snedaker & J. G. Snedaker (Eds.), *The mangrove ecosystem: Research methods* (pp. 18–35). UNESCO/SCOR.

Blasco, F., & Aizpuru, M. (2002). Mangroves along the coastal stretch of the bay of Bengal: Present status. *Indian Journal of Marine Sciences, 31*, 9–20.

Blasco, et al. (1996). Estimating the extent of floods in Bangladesh using SPOT data. *Remote Sensing of the Environment, 39*, 167–178.

Bunt, J. S. (1995). Continental scale patterns in mangrove litterfall. *Hydrobiologia, 295*(1–3), 135–140.

Burrows, D. W. (2003) *The Role of Insect Leaf Herbivory on the Mangroves Avicennia marina and Rhizophora stylosa*. PhD Thesis, James Cook University, Townsville. p 286.

Chakraborty, S. K., & Choudhury, A. (1994). Community structure of macrobenthicpolychaetes of intertidal region of Sagar Island, Hooghly Estuary, Sundarbans, India. *Tropical Ecology, 35*(1), 97–104.

Chaudhuri, A. B., & Chowdhury, A. (1994). *Mangroves of the Sundarbans, India* (Vol. I, pp. 1–247). IUCN.
Chakraborty, S. K. (1996). Aquaculture potential of mangrove ecosystem of Sundarbans, India. In Proceedings of Seminar on Fisheries—A Multimillion Dollar Industry (pp. 72–83). AFI.
Chakraborty, S. K. (1998a). Coastal aquaculture and environment in the context of West Bengal, India. *Vidyasagar University Journal of Biological Science, 4*, 13–20.
Chakraborty, S. K. (1998b). Coastal aquaculture and environment in the context of West Bengal, India. *Vidyasagar University Journal Biological Sciences, 4*, 13–20.
Chakraborty, S. K. (2003). Biodiversity and conservation. In A. D. Mukhopadhyay (Ed.), *Perspectivesand issues in environmental studies* (pp. 267–286). Vidyasagar University.
Chakraborty, S. K. (2009). Biological diversity of Mindapore coast, West Bengal. *Environmental Forensics, 2*(2), 6.
Chakraborty, S. K. (2010). Coastal environment of Midnapore, West Bengal: Potential threats and management. *Journal of Coastal Environment, 1*(1), 28–40.
Chakraborty, S. K. (2011). Mangrove ecosystem of Sundarbans, India: Biodiversity, ecology, threats and conservation. In J. N. Metras (Ed.), *Mangroves: Ecology, biology and taxonomy* (pp. 83–112). NOVA Publishers.
Chakraborty, S. K. (2013). Interactions of environmental variables determining the biodiversity of coastal mangrove ecosystem of West Bengal, India. *The Ecoscan Special Issue, 3*, 251–265.
Chakraborty, S. K. (2017). *Ecological services on intertidal benthic fauna and the sustenance of coastal wetlands along the Midnapore(east) coast, West Bengal, India (published in the book, coastal wetlands: Alteration and remediation)* (pp. 777–886). Springer International Publishing AG part of Springer Nature.
Chakraborty, S. K., & Choudhury, A. (1989). Ecological studies on the zonation of brachyuran crabs in a virgin mangrove Island of Sundarbans, India. *Journal of the Marine Biological Association of India, 34*(1 & 2), 189–194.
Chakraborty, S. K., & Choudhury, A. (1992). Population ecology of fiddler crabs (*Uca. spp.*) of the mangrove estuarine complex of Sundarbans, India. *Tropical Ecology, 33*(1), 78–88.
Chakraborty, S. K., & Choudhury, A. (1994). Community structure of macrobenthic polychaetes of intertidal region of Sagar Island, Hooghly estuary, Sundarbans, India. *Tropical Ecology, 35*(1), 97–104.
Chakraborty, S. K., & Choudhury, A. (1994). Population ecology of *Metaplax intermedia* (Brachyura: Grapsidae) of SagarIsland, Sundarbans, India. *Proceedings of Zoological Society, Calcutta, 47*(1), 41–45.
Chakraborty, S. K., & Choudury, A. (1985). Distribution of fiddler crabs in Sundarbans mangrove esturarine complex, India. The mangroves. In *Proceedings of National Symposium Biology Utilization conservation of mangroves* (pp. 467–472).
Chakraborty, S. K., & Chowdhury, A. (1993). Community organization of macrobenthic molluses of Sagar Island, Sundarbans, India. *Indian Biology, 25*(1), 33–36.
Chakraborty, S. K., & Chowdhury, A. (1994). Community structure of macrobenthic polychaetes of intertidal region of Sagar Island, Hooghly Estuary, Sundarbans, India. *Tropical Ecology, 35*(1), 97–104.
Chakraborty, S. K., Choudhury, A., & Deb, M. (1986). Decapod Brachyura from Sundarbans mangrove estuarine complex, India. *Journal of Bengal Natural History Society, 5*(1), 55–68.
Chakraborty, S. K., Giri, S., Chakravarty, G., & Bhattacharya, N. (2009). Impact of eco-restoration on the biodiversity of Sundarbans mangrove ecosystem, India. *Water Air Soil Pollution: Focus, 9*, 303–320.
Chakraborty, S. K., Chatterjee, S., Dey, M. K., et al. (2012). Ecological changes of an eco-rehabilitated delta of hooghly estuary. *Journal of Coastal Conservation, 3*(1), 9–27.
Chakraborty, S. K., Paul, A. K., Bandyopadhyay, J., Banerjee, D., Pakhira, H., Middya, S., Jana, S., Sahoo, P., Paria, K., & Bera, K. (2013). Sustainable water management in the drought prone riverinetracts of South West Bengal. *Journal of Ground Water Studies, 1*, 11–34.
Chakraborty, S. K., Maiti Dutta, S., Ghosh, P. B., Ray, R., & Paul, A. K. (2014). Impact of global warming on Sundarbans mangrove ecosystem, India: Role of different assessment tools from ecosystem monitoring to molecular markers. In *Proceedings of the international conference*

on green India: Strategic knowledge for combating climate change: Prospects and challenges (pp. 181–200).
Chakravarty, G., Giri, S., Giri, S., & Chakraborty, S. K. (2004). Distribution of coliform bacteria in different estuaries of West Bengal coast, India. In A. Kumar, J. R. Dhanze, A. K. Chopra, & D. S. Malik (Eds.), *Protected habitats and biodiversity* (pp. 395–402). Natcon Publication.
Chandra, A. K., Chakravarty, G., & Chakraborty, S. K. (2003). Polychaete diversity of Midnapore Coastal Belt, West Bengal, India. *Journal of Nature Conservators, 15*(2), 95–108.
Chapman, V. J. (1976). *Mangrove vegetation.* J. Cramer, Germany Chen, R., and R. R. Twilley. 1998. A gap dynamic model of mangrove forest development along gradients of soil salinity and nutrient resources. Journal of Ecology 86: 37–52.
Dahdouh-Guebas, F., Mathenge, C., Kairo, J. G., & Koedam, N. (2000). Utilization of mangrove wood products around Mida Creek (Kenya) among subsistence and commercial users. *Economic Botany, 54*, 513–527.
De Deurwaerder, H., Okello, J. A., Koedam, N., Schmitz, N., & Steppe, K. (2016). How are anatomical and hydraulic features of the mangroves Avicennia marina and Rhizophora mucronata influenced by siltation? *Trees, 30*, 35–45. https://doi.org/10.1007/s00468-016-1357-x
Duke, N. C., Meynecke, J.-O., Dittmann, S., Ellison, A. M., Anger, K., Berger, U., & Dahdouh-Guebas, F. (2007). A world without mangroves? *Science, 317*, 41b–42b. https://doi.org/10.1126/science.317.5834
Ellis, J., Nicholls, P., Craggs, R., Hofstra, D., & Hewitt, J. (2004). Effects of terrigenous sedimentation on mangrove physiology and associated macrobenthic communities. *Marine Ecology Progress Series, 270*, 71–82.
Ellison, A. M. (2001). Macroecology of mangroves: Large-scale patterns and processes in tropical coastal forests. *Trees, 16*, 181–194. https://doi.org/10.1007/s00468-001-0133-7
Ellison, A. M., & Farnsworth, E. J. (1996). Anthropogenic disturbance to Caribbean mangrove ecosystems: Past impacts, current trends, and future impacts. *Biotropica, 28*, 549–565.
Ellison, A. M., & Farnsworth, E. J. (1997). Simulated sea level change alters anatomy, physiology, growth, and reproduction of red mangrove (Rhizophora mangle L.). *Oecologia, 112*, 435–446.
Ellison, A. M., & Farnsworth, E. J. (2001). Mangrove communities. In M. D. Bertness, S. D. Gaines, & M. E. Hay (Eds.), *Marine community ecology* (pp. 423–442). Sinauer.
Erwin, K. L. (2009). Wetlands and global climate change: The role ofwetland restoration in a changing world. *Wetlands Ecology and Management, 17*, 71–84.
Farnsworth, K. D., & Illius, A. W. (1998). Optimal diet choice for herbivores: An extended contingency model. *Functional Ecology, 12*(1), 74–81.
Feller, I. C., Lovelock, C. E., Berger, U., McKee, K. L., Joye, S. B., & Ball, M. C. (2010). Biocomplexity in mangrove ecosystems. *Annual Review of Marine Science, 2*, 395–417. https://doi.org/10.1146/annurev.marine.010908.163809
Galloway, W. E. (1975). *Process framework for describing the morphologic and stratigraphic evolution of deltaic depositional system.*
Ghosh, M., Kumar, L., & Roy, C. (2016). Mapping long-term changes in mangrove species composition and distribution in the Sundarbans. *Forests, 7*, 305. https://doi.org/10.3390/f7120305
Gillikin, D. P., & Schubart, C. D. (2004). Ecology and systematics of mangrove crabs of the genus Perisesarma (Crustacea: Brachyura: Sesarmidae) from East Africa. *Zoological Journal of the Linnean Society, 141*, 435–445.
Gilman, E., Ellison, J., Duke, N. C., & Field, C. (2008). Threats to mangroves from climate change and adaptation options: A review. *Aquatic Botany, 89*, 237–250.
Ginsberg, S. S., & Perillo, G. M. E. (2004). Characteristics of tidal channels in a mesotidal estuary of Agentina. *Journal of Coastal Research, 20*(2), 489–497.
Giri, C., Ochieng, E., Tiszen, L. L., et al. (2010). Status and distribution of mangrove forests of the world using earth observation satellite data. *Global Ecology and Biogeography, 20*, 154–159.
Gopal, B., & Chauhan, M. (2006). Biodiversity and its conservation in the Sundarban mangrove ecosystem. *Aquatic Sciences, 68*, 338–354. https://doi.org/10.1007/s00027-006-0868-8
Goudha, R., & Panigrahy, R. C. (1996). The mangrove of Orissa. *Journal of Indian Ocean Studies, 3*(3), 228–237.

Gupta, S. K. (1987). Some mangrove soils of the Sundarbans ecosystem. *UNESCO Reg Introductory Train Course Estuaries Research, Calcutta, 2*, 105–112.

Hamilton, L. S., & Snedaker, S. C. (Eds.). (1984). *Handbook for mangrove area management* (p. 123). Environment and Policy Institute, East-West Center.

Hazra, A. K., & Choudhuri, D. K. (1990). Ecology of subterranean macro and micro arthropods fauna in different degraded and polluted soil environment of West Bengal. India. *Record of The Zoological Survey Of India Occassional Paper, 120*, 1–280.

Hazra, A. K., & Sanyal, A. K. (1996). Ecology of collembola in a periodically inundated newly emerged alluvial island in the river Hooghly, West Bengal. *Proceeding of Zoological Society. Calcutta, 49*(2), 157–169.

Hong, P. N., & San, H. T. (1993). *Mangroves of Vietnam* (p. 173). IUCN.

Hughes, C. E., Binning, P., & Willgoose, G. R. (1998). Characterization of the hydrology of an estuarine wetland. *Journal of Hydrology, 211*, 34–49.

Hutchinson, G. E. (1958). Concluding remarks. *Cold Spring Harbor Symposia on Quantitative Biology, 22*, 415–427.

IPCC. (2007). Climate change 2007: The physical science basis. In S. Solomon, D. Qin, M. Manning, Z. Chen, et al. (Eds.), *Contribution of working group I to the fourth assessment report of the intergovernmental panel on climate change*. Cambridge University Press.

IPCC (International Panel on Climate Change). (2001). Climate change 2001: Impacts, adaptation, and vulnerability. In *Technical summary, and summary for policymakers. Third assessment report of working group I of the intergovernmental panel on climatic change*. http://www.ipcc.ch

Kairo, J. G., & Kivyatu, B. (2000). *Mangrove management survey within and adjacent to Kiunga marine National Reserve. WWF/KE: 0089/01, final technical project* (p. 113). World Worldlife Fund for Nature East Africa Regional Programme Office (WWF–EARPO).

Kathiresan, K., & Bingham, B. L. (2001). Biology of mangroves and mangrove ecosystems. *Advances in Marine Biology, 40*, 81–251.

Khan, R. A. (1995). Ecology of the Hugli-Matla estuarine system. In *Estuarine Ecosystem Series, Part 2: Hugli Matla Estuary, Zoological Survey of India* (pp. 417–464).

Lee, S. Y., Primavera, J. H., Dahdouh-Guebas, F., McKee, K., Bosire, J. O., Cannicci, S., & Record, S. (2014). Ecological role and services of tropical mangrove ecosystems: A reassessment. *Global Ecology and Biogeography, 23*, 726–743. https://doi.org/10.1111/geb.12155

Lefebvre, G., & Poulin, B. (1996). Seasonal abundance of migrant birds and food resources in Panamanian mangrove forests. *Wilson Bull, 108*, 748–759.

Lefebvre, G., & Poulin, B. (1997). Bird communities in Panamanian black mangroves: Potential effects of physical and biotic factors. *Journal of Tropical Ecology, 13*, 405–415.

Lewis, R. R. (1990). Creation and restoration of coastal wetlands in Puerto Rico and the US Virgin Islands. In J. A. Kusler & M. E. Kentula (Eds.), *Wetland creation and restoration: The status of science, Vol. I: Regional reviews*. Island Press.

Macnae, W. (1974). *Mangrove forest and fisheries. Indian Ocean Programme publication number–34* (p. 351 OFC/DEV/74/34). Indian Ocean Fishery Commission.

Mandal, B., Mukherjee, A., & Banerjee, S. (2009). A review on the ichthyofaunal diversity in mangrove-based estuary of Sundarbans. *Reviews in Fish Biology and Fisheries, 23*, 365–374. https://doi.org/10.1007/s11160-012-9300-8

Marshall, N. A., Marshall, P. A., Tamelander, J., Obura, D., Malleret-King, D., & Cinner, J. E. (2010). *A framework for social adaptation to climate change: Sustaining tropical coastal communities and industries*. International Union for Conservation of nature (IUCN). http://www.iucn.org/about/work/programmes/marine/marine_resources/marine_publications/?4943/A-Framework-for-Social-Adaptation-to-Climate-Change

Mazda, Y., & Kamiyama, K. (2007). Tidal deformation and inundation characteristics within mangrove swamps. *Mangrove Science, 4*(5), 21–29.

Mazda, Y., Kanazawa, N., & Kurokawa, T. (1999). Dependence of dispersion on vegetation density in a tidal creek-mangrove swamp system. *Mangroves and Salt Marshes, 3*, 59–66.

Mazda, Y., & Kamiyama, K. (2007). Tidal deformation and inundation characteristics within mangrove swamps. *Mangrove Science, 4/5*, 21–29.

Mazda, Y., & Wolanski, E. (2009). Hydrodynamics and modeling of water flow in mangrove areas. In *Coastal wetlands: An integrated ecosystem approach*. Elsevier.

Mazda, Y., Wolanski, E., & Ridd, P. V. (2007). *The role of physical processes in mangrove environments: Manual for the preservation and utilization of mangrove ecosystems* (p. 598). TERRAPUB.

Mitsch, J. W., & Gosselink, J. G. (1993). *Wetlands*. Van Nostrand Reinhold.

Naskar, K., & Guha Bakshi, D. N. (1987). *Mangrove swamps of the Sundarbans* (p. 282). National Cyclone Risk Mitigation Project (NCRMP) 2016 Implemented through World Bank Assistance (Map showing coastal areas affected by cyclones). Based on IMD data on cyclone hazards.

Odum, W. E., & Heald, F. J. (1975a). The detritus based food web of an estuarine mangrove community. In L. E. Cronin (Ed.), *Estuarine research* (Vol. 1, pp. 265–286). Academic Press.

Odum, E. P., & Heald, E. J. (1975b). Mangrove forests and aquatic productivity. In A. D. Haster (Ed.), *Coupling of land, water systems. Ecological studies* (Vol. 10, pp. 129–136). Springer-Vesllog.

Paul, A. K. (2002). *Coastal geomorphology and environment* (pp. 1–582). ACB Publication.

Paul, A. K. (2002). *Coastal geomorphology and Environment* (pp. 1–575). Ajoy Bhattacharya of acb publications.

Perrings, C., Naeem, S., Farshid, S., et al. (2011). Ecosystem services, targets and indicator for the conservation and sustainable use of biodiversity. *Frontiers in Ecology and the Environment, 9*(9), 512–520.

Polidoro, B. A., Carpenter, K. E., Collins, L., et al. (2010). The loss of species: Mangrove extinction risk and geographic areas of global concern. *PLoS One, 5*, el0095.

Primavera, J. H. (1995). Mangroves and brackish water pond culture in The Philippines. *Hydrobiologia, 295*, 303–309.

Qasim, S. Z., Sengupta, R., & Kureishy, T. W. (1988). Pollution of the seas around India. *Proceedings of Indian Academy of Science (Animal Science), 97*(2), 117–131.

Rahaman, S. M. B., Rahaman, M. S., Ghosh, A. K., Gain, D., Biswas, S. K., Sarder, L., & Sayeed, A. B. (2015). A spatial and seasonal pattern of water quality in the Sundarbans river systems of Bangladesh. *Journal of Coastal Research, 300*, 390–397. https://doi.org/10.2112/JCOASTRES-D-13-00115.1

Robertson, A. I., & Blaber, S. J. M. (1992). Plankton, epibenthos and fish communities. In A. I. Robertson & D. M. Alongi (Eds.), *Tropical mangrove ecosystems* (pp. 173–224). American Geophysical Union.

Saenger, P., Hegerl, E. J., & Davie, J. D. S. (1983). Global status of mangrove ecosystems. *The Environmentalist, 3*, 1–88.

Schelake, C. L., & Odum, E. P. (1962). Mechanisms for maintaininghigh productivity in Georgia estuaries. *Proceedings of Gulf Caribb Fishery Institute, 14*, 75–80.

Schwamborn, R., & Saint-Paul, U. (1996). Mangroves—forgotten forests? *Nature Resource Development, 43*(44), 13–36.

Spalding, M., Blasco, F., & Field, C. (1997). *World mangrove atlas*. International Society for Mangrove Ecosystems.

Thampanya, U., Vermaat, J. E., & Terrados, J. (2002). The effect of increasing sediment accretion on the seedlings of three common Thai mangrove species. *Aquatic Botany, 74*, 315–325. https://doi.org/10.1016/S0304-3770(02)00146-8

Watson, J. G. (1928). *Mangrove forests of the Malay peninsula. Malayan Forest records no. 6* (p. 275). Forest Department, Federated Malay States, Kuala Lumpur.

Wafar, S., Untawale, A. G., & Wafar, M. (1997). Litter fall and energy flux in a mangrove ecosystem. *Estuarine, Coastal and Shelf Science, 44*(11), 1–124.

Williams, R. J., & Watford, F. A. (1997). Identification of structures restricting tidal flow in New South Wales. *Wetlands Ecology and Management*.

Wolanski, E., Mazda, Y., & Ridd, P. (1992). Mangrove hydrodyna mics. In A. I. Robertson & D. M. Alongi (Eds.), *Tropical mangrove ecosystems, coastal and estuarine studies* (pp. 436–462). AGU.

Index

S. K. Chakraborty, S. K. Acharya, *Eco-dynamics and Environmental Perturbations in Mangrove Ecosystems*,
https://doi.org/10.1007/978-3-032-08710-2

The manufacturer's authorised representative in the EU is Springer
Nature Customer Service Centre GmbH, Europaplatz 3, 69115 Heidelberg,
Germany. If you have any concerns regarding our products, please
contact ProductSafety@springernature.com

Printed and bound by CPI Group (UK) Ltd, Croydon, CR0 4YY
07/10/2026
02229350-0018